Biology Data Book
Second Edition
VOLUME II

COMPILED AND EDITED BY

Philip L. Altman and Dorothy S. Dittmer

Federation of American Societies for Experimental Biology

BETHESDA, MARYLAND

PRINTED IN THE UNITED STATES OF AMERICA

Library of Congress Catalog Card Number: 72-87738

International Standard Book Number: 0-913822-07-8

FOREWORD

Volume II of the *Biology Data Book* includes a wealth of new information. The section on BIOLOGICAL REGULATORS AND TOXINS has completely new tables on hypothalamic releasing and inhibiting hormones, a four-part table on chemicals produced by arthropods, a three-page table on mycotoxins, and an important table on neurotoxins in man. The ENVIRONMENT AND SURVIVAL section, in addition to revision of tables on temperature, radiation, and pollutants, has a new table on the effects of low concentrations of sulfur dioxide on vascular plants. PARASITISM—of plants on plants, animals on animals, and each on the other—has new tables on external and internal parasites of laboratory animals. The section on SENSORY AND NEURO- BIOLOGY, contains 29 tables that have not appeared in any of the previous handbooks.

The availability of new, critically evaluated information—that can be presented in tabular or graphic form—resulted in approximately 250 more pages than the originally announced estimate. The increase in size has delayed publication but provides a bargain for the purchaser. The 96 tables are based on contributions made by 260 eminent scientists who have given generously of their time and advice. FASEB acknowledges with appreciation their intellectual contributions.

Each volume is independently indexed and can be purchased separately by those wishing to have data limited to particular fields of interest. Volume I (published in 1972) covers genetics, cytology, reproduction, development, and growth. It includes tables giving the properties of biological substances and information about some of the many widely used materials and methods. It has tables on diet, culture media, and chemical indicators. In the appendixes there will be found keys to the scientific and corresponding common names of animals and plants. There are tables giving the taxonomic classification of all living things, and an estimation of the number of species in phyla and classes of the plant and animal kingdoms. Past history of living things is illustrated by a table on geologic distribution. In addition to chemical, physiological, and mathematical constants, there is also a bibliography on sources of organisms and equipment.

Volume III (to be published in 1974) will contain sections on nutrition, digestion, and excretion; metabolism; respiration and circulation; and blood and other body fluids. Those familiar with the Data Book Series will recognize that the last three sections have appeared as complete specialized handbooks: *Blood and Other Body Fluids,* 1961—2nd printing, 1966, and 3rd printing, 1971; *Metabolism,* 1968; and *Respiration and Circulation,* 1971. Other specialized handbooks published by FASEB were *Environmental Biology,* 1966, and *Growth, including Reproduction and Morphological Development,* 1962.

The FASEB Publications Committee listed on the following page has the responsibility for general guidance of the data book program and the selection of fields to be covered. A special Biology Data Book Advisory Committee also listed on the following page was chosen to determine what should be included and what should be excluded from the three volumes described briefly above. On the basis of their extensive experience in research and teaching, Advisory Committee members have also made suggestions as to authorities in particular fields who should be asked to contribute their services in the preparation of a table or a part of a table. Tables or portions sent in by more than one contributor are integrated by the handbook staff and then sent to two or more reviewers for critical evaluation. With the aid of Committee members, the staff has obtained remarkable cooperation in securing data for these volumes. The tables are organized to conform to established standards and are subject to critical evaluation and another review. Because of the intricate nature of the compilation, it has been found more efficient to have composition, editing, indexing, and the preparation of camera-ready copy done entirely within the Office of Biological Handbooks.

This volume is the first in a long series of biology data books that has been produced without financial support from any federal government agency. The Federation of American Societies for Experimental Biology has advanced the funds needed for its production. Let us hope that biologists everywhere in educational, industrial, and governmental activities will do their part in promoting wide distribution of this valuable compilation of quantitative data of the life sciences.

15 August 1973
Bethesda, Maryland

Raymund L. Zwemer, Chairman
Biology Data Book Advisory Committee

The Handbooks Project lost a good friend when Thurlo B. Thomas, professor of zoology and senior faculty member of Carleton College, died in his sleep of an apparent heart attack on 14 April 1973. Dr. Thomas gave generously of his time and many talents as a member of the 1964 Biology Data Book Advisory Committee, and also as a member of the Advisory Committee for the current revision of that volume. He will also be remembered as chief contributor of Appendix VIII in the 1964 edition of the *Biology Data Book,* and of Table 71 and Appendix IX in Volume I of this second edition.

FASEB PUBLICATIONS COMMITTEE

FRANK G. STANDAERT, *Chairman*
AMERICAN SOCIETY FOR PHARMACOLOGY AND EXPERIMENTAL THERAPEUTICS
Georgetown University School of Medicine and Dentistry
Washington, D.C. 20007

ELIJAH ADAMS
AMERICAN SOCIETY OF BIOLOGICAL CHEMISTS
University of Maryland School
of Medicine
Baltimore, Maryland 21201

EUGENE L. HESS*
Federation of American Societies
for Experimental Biology
9650 Rockville Pike
Bethesda, Maryland 20014

EDWIN M. LERNER, II
AMERICAN ASSOCIATION OF IMMUNOLOGISTS
Leonard Wood
Memorial
Washington, D.C. 20037

DONALD B. HACKEL
AMERICAN SOCIETY FOR EXPERIMENTAL
PATHOLOGY
Duke University Medical Center
Durham, North Carolina 27706

WILLIAM G. HOEKSTRA
AMERICAN INSTITUTE OF NUTRITION
University of Wisconsin
Madison, Wisconsin 53706

CHARLES S. TIDBALL
AMERICAN PHYSIOLOGICAL SOCIETY
George Washington University
Medical Center
Washington, D.C. 20005

BIOLOGY DATA BOOK ADVISORY COMMITTEE

RAYMUND L. ZWEMER, *Chairman*
Federation of American Societies for Experimental Biology
Bethesda, Maryland 20014

ROBERT H. BURRIS
University of Wisconsin
Madison, Wisconsin 53706

RAY G. DAGGS
American Physiological Society
Bethesda, Maryland 20014

NOBLE O. FOWLER
University of Cincinnati College
of Medicine
Cincinnati, Ohio 45229

KARL F. HEUMANN*
Federation of American Societies
for Experimental Biology
Bethesda, Maryland 20014

ROSS A. McFARLAND
Harvard University School of
Public Health
Boston, Massachusetts 02115

ARTHUR B. OTIS
University of Florida College
of Medicine
Gainesville, Florida 32601

NATHAN W. SHOCK
Baltimore City Hospitals
Baltimore, Maryland 21224

WALTER SHROPSHIRE, JR.
Radiation Biology Laboratory of the
Smithsonian Institution
Rockville, Maryland 20852

THURLO B. THOMAS†
Carleton College
Northfield, Minnesota 55057

BETTY M. TWAROG
Tufts University
Medford, Massachusetts 02155

HAROLD L. WILCKE
Ralston Purina Company
St. Louis, Missouri 63199

HANDBOOK STAFF

PHILIP L. ALTMAN, *Director*   DOROTHY S. DITTMER, *Editor*

PATRICIA BRYANT
ELLEN R. BUTCHART
ELSIE COMSTOCK

CLAIRE L. DOYLE
JEAN M. GIEGOLD
SAKI HIMEL

GERALDINE M. JOHNSON
ETHEL E. KETCHUM
MARCIA I. LISSAUER

* *ex officio*
† deceased

iv

CONTRIBUTORS AND REVIEWERS

ADOLPH, EDWARD F.
University of Rochester School of
Medicine and Dentistry
Rochester, New York 14642

ALLEN, MARY BELLE
University of Alaska
College, Alaska 99701

ALLEN, WILLIAM W.
University of California
Berkeley, California 94720

AMARAL, AFRANIO do
Butantan Institute
São Paulo, Brazil

ANDERSEN, AXEL L.
Michigan State University
East Lansing, Michigan 48823

ANDREWS, RICHARD V.
Creighton University
Omaha, Nebraska 68131

ANGEVINE, JAY B., JR.
University of Arizona College of
Medicine
Tucson, Arizona 85721

ARIMURA, AKIRA
Tulane University School of Medicine
New Orleans, Louisiana 70112

ARNOLD, ELSIE P.
Arthur D. Little, Inc.
Cambridge, Massachusetts 02140

BAILEY, LOWELL F.
University of Arkansas
Fayetteville, Arkansas 72701

BAIRD, M. B.
Masonic Medical Research Laboratory
Utica, New York 13501

BAKER, EDWARD W.
USDA, Systematic Entomology
Laboratory
Beltsville, Maryland 20705

BARDACH, JOHN E.
University of Hawaii
Kaneohe, Hawaii 96744

BARDOS, THOMAS J.
State University of New York
Buffalo, New York 14214

BARLOW, JOHN S.
Massachusetts General Hospital
Boston, Massachusetts 02114

BARNES, J. M.
Medical Research Council Laboratories
Carshalton, Surrey, England

BARRINGTON, E. J. W.
University of Nottingham
Nottingham, NG7 2RD, England

BARTELS, PAUL G.
University of Arizona
Tucson, Arizona 85721

BENEKE, E. S.
Michigan State University
East Lansing, Michigan 48823

BENNETT, ALBERT F.
University of Michigan
Ann Arbor, Michigan 48104

BERN, HOWARD A.
University of California
Berkeley, California 94720

BERNSTEIN, LEON
U.S. Salinity Laboratory
Riverside, California 92502

BICKFORD, REGINALD G.
University of California School of
Medicine, San Diego
La Jolla, California 92037

BIEBL, RICHARD
University of Vienna
Vienna, Austria

BLOCKLEY, VINCENT W.
Physiometrics, Inc.
Malibu, California 90265

BLUM, MURRAY S.
University of Georgia
Athens, Georgia 30601

BOWMAN, THOMAS E.
National Museum of Natural History
Washington, D.C. 20560

BRANDT, C. STAFFORD
Maryland Environmental Health
Administration
Baltimore, Maryland 21201

BRAWER, JAMES R.
Tufts University School of Medicine
Boston, Massachusetts 02111

BRAZIER, MARY A. B.
University of California Medical
Center
Los Angeles, California 90024

BRETT, J. R.
Fisheries Research Board of Canada
Nanaimo, British Columbia, Canada

BRINLEY, F. J., JR.
Johns Hopkins University School of
Medicine
Baltimore, Maryland 21205

BROOKHART, JOHN M.
University of Oregon Medical School
Portland, Oregon 97201

BROOKS, CHANDLER McC.
State University of New York
Brooklyn, New York 11203

BROWN, JAMES W.
468 Barrello Lane
Cocoa Beach, Florida 32931

BURG, ALBERT
University of California
Los Angeles, California 90024

BURGESS, P. R.
University of Utah School of Medicine
Salt Lake City, Utah 84112

BYAS, WALTER J.
National Museum of Natural History
Washington, D.C. 20560

CALDWELL, PETER C.
University of Bristol
Bristol, BS8 1UG, England

CARLSON, RICHARD W.
University of Southern California
School of Medicine
Los Angeles, California 90033

CASKEY, JAMES E., JR.
Environmental Science Information
Center
Rockville, Maryland 20852

CASTRO, ANTHONY J.
Louisiana State University Medical
Center
New Orleans, Louisiana 70112

CHEN, K. K.
7975 Hillcrest Road
Indianapolis, Indiana 46240

CHRISTIE, JESSE R.
202 South Summerlin Avenue
Orlando, Florida 32801

CLARK, ARNOLD M.
University of Delaware
Newark, Delaware 19711

CLAYTON, R. K.
Cornell University
Ithaca, New York 14850

CLIVER, DEAN O.
University of Wisconsin
Madison, Wisconsin 53706

CLOSE, R.
Australian National University
Canberra, Australia

COLLINS, J. C.
University of Liverpool
Liverpool, L69 3BX, England

CRABILL, RALPH E., JR.
National Museum of Natural History
Washington, D. C. 20560

CUMMING, BRUCE G.
University of New Brunswick
Fredricton, New Brunswick, Canada

DAINTY, JACK
  University of Toronto
  Toronto 181, Ontario, Canada
DALY, JAMES J.
  University of Arkansas Medical Center
  Little Rock, Arkansas 72201
DARDEN, T. R.
  University of Washington
  Seattle, Washington 98105
DAWSON, WILLIAM R.
  University of Michigan
  Ann Arbor, Michigan 48104
DEXTER, RALPH W.
  University of South Carolina
  Columbia, South Carolina 29208
DICKEY, ROBERT S.
  Cornell University
  Ithaca, New York 14850
DICKSON, DONALD W.
  University of Florida
  Gainesville, Florida 32601
DOWBEN, ROBERT M.
  Brown University
  Providence, Rhode Island 02912
DOWNS, R. J.
  North Carolina State University
  Raleigh, North Carolina 27607
DUNHAM, PHILIP B.
  Syracuse University
  Syracuse, New York 13210
DUPRÉ, MARGARET V.
  14 Granger Place
  Buffalo, New York 14222

ECCLES, JOHN C.
  State University of New York
  Buffalo, New York 14214
EDERSTROM, H. E.
  University of North Dakota School
    of Medicine
  Grand Forks, North Dakota 58201
EDGAR, S. ALLEN
  Auburn University
  Auburn, Alabama 36830
EHRLER, W. L.
  U.S. Water Conservation Laboratory
  Phoenix, Arizona 85040
EICHBAUM, FRANCISCO W.
  University of São Paulo
  São Paulo, Brazil

FARNER, DONALD S.
  University of Washington
  Seattle, Washington 98105
FAUST, ERNEST CARROLL
  Tulane University
  New Orleans, Louisiana 70112

FISHER, KENNETH D.
  Federation of American Societies
    for Experimental Biology
  Bethesda, Maryland 20014
FITZHUGH, O. G.
  Environmental Protection Agency
  Washington, D.C. 20460
FLYNN, ROBERT J.
  Argonne National Laboratory
  Argonne, Illinois 60439
FOLK, G. EDGAR, JR.
  University of Iowa
  Iowa City, Iowa 52240
FOX, CLEMENT A.
  Wayne State University School
    of Medicine
  Detroit, Michigan 48207
FRANZINI-ARMSTRONG, CLARA
  University of Rochester School of
    Medicine and Dentistry
  Rochester, New York 14620
FREYGANG, WALTER H.
  National Institute of Mental Health
  Bethesda, Maryland 20014
FRIEDMAN, LORRAINE
  Tulane University School of Medicine
  New Orleans, Louisiana 70112
FROBISHER, MARTIN
  101 Benjamin Franklin Drive
  Sarasota, Florida 33577
FURMAN, DEANE P.
  University of California
  Berkeley, California 94720

GALE, GEORGE O.
  American Cyanamid Company
  Princeton, New Jersey 08540
GEORG, LUCILLE K.
  Center for Disease Control
  Atlanta, Georgia 30333
GERGELY, J.
  Boston Biomedical Research Institute
  Boston, Massachusetts 02114
GIBBS, ERNA L.
  University of Illinois College of
    Medicine
  Chicago, Illinois 60612
GIBBS, FREDERIC A.
  University of Illinois College of
    Medicine
  Chicago, Illinois 60612
GÓMEZ-LEIVA, MARCO A.
  University of Costa Rica School
    of Medicine
  San José, Costa Rica
GOODMAN, LOUIS S.
  University of Utah College of
    Medicine
  Salt Lake City, Utah 84112

GORBMAN, AUBREY
  University of Washington
  Seattle, Washington 98105
GORDON, MORRIS A.
  State of New York Department of
    Health
  Albany, New York 12201
GRAYDON, JOHN J.
  Commonwealth Serum Laboratories
  Parkville, Victoria 3052, Australia
GRENELL, ROBERT G.
  University of Maryland School of
    Medicine
  Baltimore, Maryland 21201
GRILL, RENATE
  University of Erlangen-Nürnberg
  Schloszgarten 4, Germany

HAAGEN-SMIT, A. J.
  416 South Berkeley Avenue
  Pasadena, California 91107
HALDE, CARLYN
  University of California School of
    Medicine
  San Francisco, California 94122
HARRINGTON, WILLIAM F.
  Johns Hopkins University
  Baltimore, Maryland 21218
HARRIS, J. DONALD
  U.S. Naval Medical Research Labora-
    tory
  Groton, Connecticut 06342
HART, J. SANFORD
  National Research Council of Canada
  Ottawa 7, Ontario, Canada
HAUPT, WOLFGANG
  University of Erlangen-Nürnberg
  Schloszgarten 4, Germany
HAYES, D. K.
  USDA, Northeastern Region Agricul-
    tural Research Center
  Beltsville, Maryland 20705
HAYS, KIRBY L.
  Auburn University
  Auburn, Alabama 36830
HEGGESTAD, H. E.
  USDA, Northeastern Region Agricul-
    tural Research Center
  Beltsville, Maryland 20705
HENDRIX, F. F., JR.
  University of Georgia
  Athens, Georgia 30601
HENNEMAN, ELWOOD
  Harvard Medical School
  Boston, Massachusetts 02115
HEPTING, GEORGE H.
  11 Maplewood Road
  Asheville, North Carolina 28804

HEYNEMAN, DONALD
  University of California
  San Francisco, California 94122
HITCHCOCK, FRED A.
  133 Amazon Place
  Columbus, Ohio 43214
HODGKIN, ALAN L.
  Cambridge University
  Cambridge, CB2 3EG, England
HOLTZER, ALFRED M.
  Washington University
  St. Louis, Missouri 63130
HOROWICZ, PAUL
  University of Rochester School of
    Medicine and Dentistry
  Rochester, New York 14642
HOWE, R. W.
  Pest Infestation Control Laboratory
  Slough, Bucks, England
HUBBARD, JOHN I.
  University of Otago Medical School
  Dunedin, New Zealand
HULL, HERBERT M.
  USDA, Plant Science Research
    Division
  Tucson, Arizona 85719
HUNT, CARLTON C.
  Washington University School of
    Medicine
  St. Louis, Missouri 63110

JACOBSON, STANLEY
  Tufts University School of Medicine
  Boston, Massachusetts 02111
JAFFE, LIONEL F.
  Purdue University
  West Lafayette, Indiana 47907
JANES, BYRON E.
  University of Connecticut
  Storrs, Connecticut 06268
JAQUES, LOUIS B.
  University of Saskatchewan
  Saskatoon, Saskatchewan, SN7 0W0,
    Canada
JIMÉNEZ-PORRAS, JESÚS M.
  University of Costa Rica School of
    Medicine
  San José, Costa Rica
JOHNS, RICHARD J.
  Johns Hopkins University
  Baltimore, Maryland 21205
JOHNSON, ROBERT J.
  University of Pennsylvania School
    of Medicine
  Philadelphia, Pennsylvania 19104
JOHNSON, THOMAS N.
  George Washington University Medi-
    cal School
  Washington, D.C. 20005

JONES, ALAN L.
  Michigan State University
  East Lansing, Michigan 48823
JUKES, THOMAS H.
  University of California
  Berkeley, California 94720

KANAZAWA, ROBERT
  National Museum of Natural History
  Washington, D.C. 20560
KANDEL, ERIC R.
  New York University Medical Center
  New York, New York 10016
KASTIN, ABBA J.
  Tulane University School of Medicine
  New Orleans, Louisiana 70112
KETELLAPPER, H. J.
  University of California
  Davis, California 95616
*KIESSELBACH, T. A.
KIM, KE CHUNG
  Pennsylvania State University
  University Park, Pennsylvania 16802
KLEMM, W. R.
  Texas A & M University
  College Station, Texas 77843
KNIGGE, KARL M.
  University of Rochester School of
    Medicine and Dentistry
  Rochester, New York 14620
KOHN, ALAN J.
  University of Washington
  Seattle, Washington 98195
KOWALSKI, JOSEPH J.
  Michigan State University
  East Lansing, Michigan 48823
KRAUSS, BEATRICE
  University of Hawaii
  Honolulu, Hawaii 96822
KRAVITZ, E. A.
  Harvard Medical School
  Boston, Massachusetts 02115
KUFFLER, STEPHEN W.
  Harvard Medical School
  Boston, Massachusetts 02115

LAGIOS, MICHAEL D.
  Children's Hospital of San Francisco
  San Francisco, California 94119
LAPORTE, Y.
  College of France
  75231 Paris, France
LARSEN, SIGURD
  Royal Veterinary and Agricultural
    College
  DK-1871 Copenhagen, Denmark
LARSON, EDWARD
  University of Miami
  Coral Gables, Florida 33124

LAUER, EDWARD W.
  University of Michigan Medical School
  Ann Arbor, Michigan 48105
LAVERACK, M. S.
  Gatty Marine Laboratory
  St. Andrews, KY16 8LB, Scotland
LeBARON, FRANCIS N.
  University of New Mexico School of
    Medicine
  Albuquerque, New Mexico 87106
LEES, MARJORIE B.
  McLean Hospital
  Belmont, Massachusetts 02178
LELE, P. P.
  Massachusetts Institute of Technology
  Cambridge, Massachusetts 02139
LELLINGER, DAVID B.
  National Museum of Natural History
  Washington, D.C. 20560
Le QUESNE, PAMELA M.
  Middlesex Hospital Medical School
  London, W1N 8AA, England
LESSEL, ERWIN
  American Type Culture Collection
  Rockville, Maryland 20852
LEVINE, NORMAN D.
  University of Illinois
  Urbana, Illinois 61801
LEWIS, R. A.
  University of Washington
  Seattle, Washington 98105
LINTS, F. A.
  Catholic University of Louvain
  3030 Heverlee, Belgium
LLINÁS, R.
  University of Iowa
  Iowa City, Iowa 52242
LOWEY, SUSAN
  Children's Cancer Research Founda-
    tion, Inc.
  Boston, Massachusetts 02115

MacROBBIE, E. A. C.
  University of Cambridge
  Cambridge, CB2 3EA, England
McDONALD, W. IAN
  National Hospital
  London, WC1N 3BG, England
McFARLAND, ROSS A.
  Harvard School of Public Health
  Boston, Massachusetts 02115
McINTYRE, A. K.
  Monash University
  Clayton, Victoria 3168, Australia

MACHLIS, LEONARD
  University of California
  Berkeley, California 94720

---

* Deceased

MACKIE, G. O.
 University of Victoria
 Victoria, British Columbia, Canada
MAINS, R. E.
 Harvard Medical School
 Boston, Massachusetts 02115
MARTINI, LUCIANO
 University of Milan
 20129 Milan, Italy
MEINWALD, JERROLD
 University of California, San Diego
 La Jolla, California 92037
METCALF, ROBERT L.
 University of Illinois
 Urbana, Illinois 61801
MEYERS, EDWARD
 Squibb Institute for Medical Research
 Princeton, New Jersey 08540
MINTON, SHERMAN A., JR.
 Indiana University School of
  Medicine
 Indianapolis, Indiana 46202
MITCHELL, J. E.
 University of Wisconsin
 Madison, Wisconsin 53706
MODLIBOWSKA, IRENA
 East Malling Research Station
 Maidstone, Kent, England
MOKRASCH, LEWIS C.
 Louisiana State University Medical
  Center
 New Orleans, Louisiana 70112
MORGAN, KARL Z.
 Georgia Institute of Technology
 Atlanta, Georgia 30332
MORTON, JULIA F.
 University of Miami
 Coral Gables, Florida 33124
MUIR, ROBERT M.
 University of Iowa
 Iowa City, Iowa 52240
MULLINS, L. J.
 University of Maryland School of
  Medicine
 Baltimore, Maryland 21201
MUTCHMOR, JOHN A.
 Iowa State University
 Ames, Iowa 50010

NACHMIAS, JACOB
 University of Pennsylvania
 Philadelphia, Pennsylvania 19104
NARAHASHI, TOSHIO
 Duke University Medical Center
 Durham, North Carolina 27710
NICHOL, CHARLES A.
 Wellcome Research Laboratories
 Research Triangle Park, North
  Carolina 27709

NICK, M. SUSAN
 Arthur D. Little, Inc.
 Cambridge, Massachusetts 02140
NISHI, SYOGORO
 Loyola University Stritch School
  of Medicine
 Maywood, Illinois 60153

PAGE, SALLY
 University College London
 London, WC1E 6BT, England
PALKA, JOHN
 University of Washington
 Seattle, Washington 98105
PALM, PAUL E.
 Arthur D. Little, Inc.
 Cambridge, Massachusetts 02140
PAPENFUSS, HERBERT D.
 Boise State College
 Boise, Idaho 83707
PAULY, JOHN E.
 University of Arkansas Medical Center
 Little Rock, Arkansas 72201
PAWSON, DAVID L.
 National Museum of Natural History
 Washington, D.C. 20560
PEACHY, LEE D.
 University of Pennsylvania
 Philadelphia, Pennsylvania 19104
PERL, E. R.
 University of North Carolina School
  of Medicine
 Chapel Hill, North Carolina 27514
PHILIP, CORNELIUS B.
 California Academy of Sciences
 San Francisco, California 94118
PINE, RONALD H.
 National Museum of Natural History
 Washington, D.C. 20560
PINKSTON, J. O.
 State University of New York
 Brooklyn, New York 11203
PISEK, A.
 University of Innsbruck
 A-6020 Innsbruck, Austria
PLATZ, BARBARA B.
 Arthur D. Little, Inc.
 Cambridge, Massachusetts 02140
PLESS, CHARLES D.
 University of Tennessee
 Knoxville, Tennessee 37901
PRICE, ROGER D.
 University of Minnesota
 St. Paul, Minnesota 55101
PRITHAM, GORDON H.
 Kansas City College of Osteopathic
  Medicine
 Kansas City, Missouri 64124

PYSH, JOSEPH J.
 Northwestern University Medical
  School
 Chicago, Illinois 60611

RAPER, JOHN R.
 Harvard University
 Cambridge, Massachusetts 02138
RAVEN, J. A.
 The University
 Dundee, DD1 4HN, Scotland
RAWLINS, STEPHEN L.
 U.S. Salinity Laboratory
 Riverside, California 92502
REICHLIN, SEYMOUR
 University of Connecticut School of
  Medicine
 Hartford, Connecticut 06105
REINERT, RICHARD A.
 North Carolina State University
 Raleigh, North Carolina 27607
ROBINSON, HAROLD E.
 National Museum of Natural History
 Washington, D.C. 20560
RODRÍGUEZ-BARQUERO, JORGE A.
 University of Costa Rica School of
  Medicine
 San José, Costa Rica
RUSSELL, FINDLAY E.
 University of Southern California
  School of Medicine
 Los Angeles, California 90033

SALO, RICHARD J.
 University of Wisconsin
 Madison, Wisconsin 53706
SAMIS, H. V.
 Masonic Medical Research Laboratory
 Utica, New York 13501
SARGENT, P. B.
 Harvard Medical School
 Boston, Massachusetts 02115
SASSER, J. N.
 North Carolina State University
 Raleigh, North Carolina 27607
SCHALLY, ANDREW V.
 Tulane University School of Medicine
 New Orleans, Louisiana 70112
SCHEVING, LAWRENCE E.
 University of Arkansas Medical Center
 Little Rock, Arkansas 72201
SCHMITT, JOHN A.
 Ohio State University
 Columbus, Ohio 43210
SCHÖNBOHM, E.
 University of Marburg
 Pilgrimstein 4, Germany
SCHWARTZ, JAMES H.
 New York University Medical Center
 New York, New York 10016

SCOTT, HOWARD A.
  University of Arkansas
  Fayetteville, Arkansas 72701
SEALANDER, JOHN A., JR.
  University of Arkansas
  Fayetteville, Arkansas 72701
SENAY, LEO C., JR.
  Saint Louis University School of
  Medicine
  St. Louis, Missouri 63104
*SILBERSCHMIDT, KARL M.
SIMMONDS, RICHARD C.
  NASA, Lyndon B. Johnson Space
  Center
  Houston, Texas 77058
SJODIN, RAYMOND A.
  University of Maryland School of
  Medicine
  Baltimore, Maryland 21201
SMART, GROVER C., JR.
  University of Florida
  Gainesville, Florida 32601
SMITH, CHARLES W.
  Ohio State University
  Columbus, Ohio 43210
SNIDER, RAY S.
  University of Rochester School of
  Medicine and Dentistry
  Rochester, New York 14620
SOROKIN, CONSTANTINE
  University of Maryland
  College Park, Maryland 20742
SOUTHERLAND, JAMES H.
  U.S. Environmental Protection
  Agency
  Research Triangle Park, North
  Carolina 27711
SPANSWICK, ROGER M.
  Cornell University
  Ithaca, New York 14850
SPENCER, WM. ALDEN
  New York University Medical Center
  New York, New York 10016
STEINBACH, H. BURR
  Woods Hole Oceanographic Institution
  Woods Hole, Massachusetts 02543
STROTHER, G. K.
  Pennsylvania State University
  University Park, Pennsylvania 16802
SWEENEY, BEATRICE M.
  University of California
  Santa Barbara, California 93106
SZENT-GYORGYI, ANDREW G.
  Brandeis University
  Waltham, Massachusetts 02154

TASAKI, ICHIJI
  National Institute of Mental Health
  Bethesda, Maryland 20014

TAYLOR, ROBERT E.
  NIH, Laboratory of Biophysics
  Bethesda, Maryland 20014
TERRELL, EDWARD E.
  USDA, Plant Taxonomy Laboratory
  Beltsville, Maryland 20705
THORNBERRY, H. H.
  University of Illinois
  Urbana, Illinois 61801
TYREE, M. T.
  University of Toronto
  Toronto 181, Ontario, Canada

UECKER, FRANCIS A.
  USDA, National Fungus Collection
  Beltsville, Maryland 20705

VERNBERG, F. J.
  University of South Carolina
  Columbia, South Carolina 29208
VON EULER, U. S.
  Physiology Institute
  Stockholm 60, Sweden
VON MAYERSBACH, H.
  Medizinische Hochschule Hannover
  3 Hannover-Kleefeld, Germany

WALD, GEORGE
  Harvard University
  Cambridge, Massachusetts 02138
WALKER, HOMER W.
  Iowa State University
  Ames, Iowa 50010
WALLACE, B. G.
  Harvard Medical School
  Boston, Massachusetts 02115
WATERMAN, TALBOT H.
  Yale University
  New Haven, Connecticut 06520
WEATHERSTON, JOHN
  Insect Pathology Research Institute
  Sault Sainte Marie, Ontario, Canada
WEBB, RAYMON E.
  USDA, Northeastern Region Agricul-
  tural Research Center
  Beltsville, Maryland 20705
WEINTRAUB, ROBERT L.
  George Washington University
  Washington, D.C. 20005
WEISS, MITCHELL J.
  University of Washington
  Seattle, Washington 98105
WELCH, C. D.
  Texas A & M University
  College Station, Texas 77843
WEST, JOHN A.
  University of California
  Berkeley, California 94720

WHITEHEAD, MARVIN D.
  Georgia State University
  Atlanta, Georgia 30303
WHITTOW, G. CAUSEY
  University of Hawaii School of
  Medicine
  Honolulu, Hawaii 96822
WILLIS, JOHN S.
  University of Illinois
  Urbana, Illinois 61801
WILLIS, W. D., JR.
  Marine Biomedical Institute
  Galveston, Texas 77550
WILSON, VICTOR J.
  Rockefeller University
  New York, New York 10021
WOGAN, G. N.
  Massachusetts Institute of Technology
  Cambridge, Massachusetts 02139
WOLEDGE, R. C.
  University College London
  London, WC1E 6BT, England
WOLFENBARGER, D. O.
  University of Florida
  Homestead, Florida 33030
WOLKEN, JEROME J.
  Carnegie-Mellon University
  Pittsburgh, Pennsylvania 15213
WOODBURY, J. WALTER
  University of Washington
  Seattle, Washington 98105
WRIGHT, IRVING S.
  Cornell University Medical School
  New York, New York 10021

YUHAS, JOHN M.
  Oak Ridge National Laboratory
  Oak Ridge, Tennessee 37831
YUNKER, CONRAD E.
  U.S. Naval Medical Research Unit
  No. 3
  Fleet Post Office, New York 09527

ZAUMEYER, WILLIAM J.
  USDA, Northeastern Region Agricul-
  tural Research Center
  Beltsville, Maryland 20705
ZUG, GEORGE R.
  National Museum of Natural History
  Washington, D.C. 20560

---

* Deceased

CONTENTS

VI. BIOLOGICAL REGULATORS AND TOXINS

VII. ENVIRONMENT AND SURVIVAL

VIII. PARASITISM

IX. SENSORY AND NEURO- BIOLOGY

APPENDIXES

INTRODUCTION

The first edition of the *Biology Data Book,* published in 1964, was a 630-page compendium of "broad scope and limited coverage designed to serve as a basic reference in the field of biology." The scope of the second edition of the *Biology Data Book* is broader, and the coverage is not so limited. This newer edition should therefore be even more useful, than was the original publication, in providing information in subject areas outside the user's own field of competence.

Since it was impractical, as well as impossible, to include data for all species, contributors were instructed to restrict coverage to man and the more important laboratory, domestic, commercial, and field organisms. Despite this restriction, data for many more species—than the 400 covered in the 1964 volume—can now be found in the second edition.

As a result of the broadened scope and coverage, and the inclusion of data for additional species, the revised *Biology Data Book* will appear as three volumes totaling more than 1900 pages. Publication dates and a brief description of the contents of Volumes I and III are given in the Foreword to this volume.

Contents and Review

Volume II of the *Biology Data Book* is arranged in four sections, with the data organized in the form of 96 tables (quantitative and descriptive) and graphs plus two appendixes. Contents of this volume were verified by 260 outstanding authorities in the fields of biology and medicine. The review process to which the data were subjected was designed to eliminate, insofar as possible, material of questionable validity and errors of transcription.

Headnote

An explanatory headnote, serving as an introduction to the subject matter, may precede a table. More frequently, tables are prefaced by a short headnote containing such important information as units of measurement, abbreviations, definitions, and estimate of the range of variation. To interpret the data, it is essential to read the related headnote.

Exceptions

Occasionally, differences in values for the same specifications, certain inconsistencies in nomenclature, and some overlapping of coverage may occur among tables. These result, not from oversight or failure to choose between alternatives, but from a deliberate intent to respect the judgment and preferences of the individual contributors.

Conventions and Terminology

The main conventions used throughout this volume were adapted from the third edition of the *CBE Style Manual,* published in 1972 for the Council of Biology Editors by the American Institute of Biological Sciences. Terminology was checked against *Webster's Third New International Dictionary,* published in 1961 by G. & C. Merriam Company.

Contributors and References

Appended to the tables are the names of the contributors, and a list of the literature citations arranged in alphabetical sequence. The reference abbreviations conform to those in *ACCESS: Key to the Source Literature of the Chemical Sciences,* published by the American Chemical Society in 1969.

Animal and Plant Classification

Animal and plant taxa are arranged according to the classification outlines designated Appendixes III and IV at the back of Volume I of this *Biology Data Book.* The outlines were compiled from information provided by specialists at the Smithsonian Institution's National Museum of Natural History, the U.S. Department of Agriculture, and the Americal Type Culture Collection. The classifications reflect some of the recent agreements reached by the International Commissions on Nomenclature in the biological sciences.

Scientific Names

In the tables, a synonym following the scientific name of an organism indicates that the synonym, although cited in the reference, is no longer the preferred name. No other attempt was made to provide taxonomic synonymy. All scientific names were either verified in standard taxonomic checklists and classification lists, or submitted for authentication to the appropriate authorities at the institutions listed above.

Upon the advice of these experts, some subspecies of plants appearing in Volume I have been changed to varieties in Volume II.

To aid the user in identifying an organism, the index includes the taxonomic orders for animals, and the families for plants. Two appendixes provide cross-reference to scientific and equivalent common names occurring in this volume.

Range of Variation

Values are generally presented as either the mean, plus and

continued

minus the standard deviation, or the mean and the lower and upper limit of the range of individual values about the mean. The several methods used to estimate the range—depending on the information available—are designated by the letters "a, b, c, or d" to identify the type of range in descending order of accuracy.

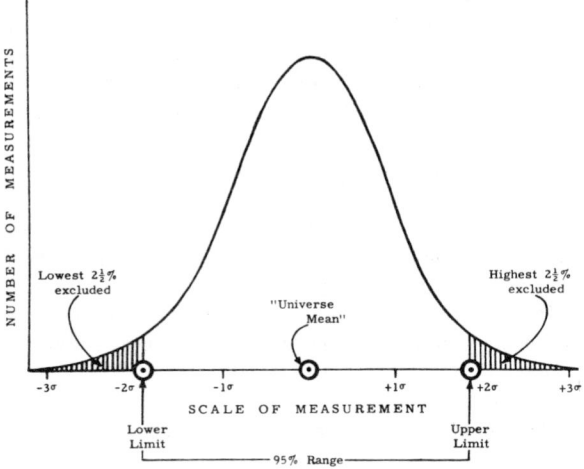

"a"—When the group of values is relatively large, a 95% range is derived by curve fitting. A recognized type of normal frequency curve is fitted to a group of measured values, and the extreme 2.5% of the area under the curve at each end is excluded (*see* illustration).

"b"—When the group of values is too small for curve fitting, as is usually the case, a 95% range is estimated by a simple statistical calculation. Assuming a normal symmetrical distribution, the standard deviation is multiplied by a factor of 2, then subtracted from and added to the mean to give the lower and upper range limits.

"c"—A less dependable, but commonly applied, procedure takes as range limits the lowest value and the highest value of the reported sample group of measurements. It underestimates the 95% range for small samples and overestimates for larger sample sizes, but where there is marked asymmetry in the position of the mean within the sample range, this method may be used in preference to the preceding one.

"d"—Another estimate of the lower and upper limits of the range of variation is based on the judgment of an individual experienced in measuring the quantity in question. The trustworthiness of such limits should not be underestimated.

ABBREVIATIONS AND SYMBOLS

Only those abbreviations and symbols not generally defined in the
headnote, body, or footnotes of a table are included in this list.

Measurements			
yr	= year
mo	= month
wk	= week
da	= day
hr	= hour
min	= minute; minimum
s	= second
ms	= millisecond
ps	= picosecond
ht	= height
m	= meter
km	= kilometer
cm	= centimeter
mm	= millimeter
μ	= micron
nm	= nanometer
Å	= Angstrom
mi	= mile
yd	= yard
ft	= foot
in.	= inch
wt	= weight
g	= gram
kg	= kilogram
mg	= milligram
μg	= microgram
ng	= nanogram
pg	= picogram
lb	= pound
vol	= volume
dl	= deciliter
ml	= milliliter
μl	= microliter
%	= parts per hundred
ppm	= parts per million
pphm	= parts per hundred million
RH	= relative humidity
temp	= temperature
°C	= degrees Celsius
°F	= degrees Fahrenheit
°K	= degrees Kelvin
cal	= calorie
kcal	= kilocalorie
atm	= atmosphere
mm Hg	= millimeters of mercury
μbar	= microbar
ft·lb	= footpound
cpm	= cycles per minute
rpm	= revolutions per minute
dB	= decibel
cd	= candela
rem	= roentgen equivalent man

mA	= milliampere
F	= farad
μF	= microfarad
pF	= picofarad
Hz	= hertz
MHz	= megahertz
MΩ	= megohm
kΩ	= kilohm
Ω·cm	= ohm centimeter
mV	= millivolt
μV	= microvolt
no.	= number
max	= maximum
approx.	= approximate; approximately
$\sim$	= approximately
$\approx$	= approximately equal to
$\cong$	= congruent to
$\pm$	= plus or minus
$<$	= less than
$>$	= more than
$\not<$	= not less than
$\not>$	= not more than
°	= degree (angular)
′	= arc minute

Biological and Chemical Specifications	
♂	= male
♀	= female
♂♀	= male and female
sp.	= species (singular)
spp.	= species (plural)
subg.	= subgroup
var.	= variety
f. sp.	= forma specialis
RBC	= red blood cells
WBC	= white blood cells
CNS	= central nervous system
ECG	= electrocardiogram
VCG	= vectorcardiogram
DNA	= deoxyribonucleic acid
RNA	= ribonucleic acid
IU	= international unit
LD	= lethal dose
log	= logarithm
pH	= hydrogen ion concentration (negative log)
pI	= isoelectric point
U.S.P.	= United States Pharmacopeia
DL or *dl*	= racemic mixture
D	= dextro (configuration)
L	= levo (configuration)
d	= dextro (rotation)
-f	= furanose ring form
m	= *meta*

o	= *ortho*
p	= *para*
n or N	= normal
N	= normal; *nitro*
M or M	= molar
m*M*	= millimolar
μ*M*	= micromolar
O	= *oxy*
S	= *sulf; sulfo*
ad lib.	= *ad libitum* (as desired)
NAD$^+$	= nicotinamide adenine dinucleotide, oxidized form
PABA	= *p*-aminobenzoic acid
Ala	= alanine; alanyl
Arg	= arginine; arginyl
Asn	= asparagine; asparaginyl
Asp	= aspartic acid; aspartyl
Cys	= cysteine; cysteinyl
CyS	= cystine; cystinyl
Gln	= glutamine; glutaminyl
Gly	= glycine; glycinyl
His	= histidine; histidyl
Ile	= isoleucine; isoleucyl
Leu	= leucine; leucyl
Lys	= lysine; lysyl
Met	= methionine; methionyl
Phe	= phenylalanine; phenylalanyl
Pro	= proline; prolyl
Ser	= serine; seryl
Thr	= threonine; threonyl
Trp	= tryptophan; tryptophyl
Tyr	= tyrosine; tyrosyl
Val	= valine; valyl

Miscellaneous	
Fn	= footnote
A.M.	= *ante meridiem* (before noon)
P.M.	= *post meridiem* (after noon)
°N	= degrees north (latitude)
°S	= degrees south (latitude)
vs.	= versus
e.g.	= *exempli gratia* (for example)
i.e.	= *id est* (that is)
etc.	= *et cetera* (and so forth)
Jan	= January
Feb	= February
Mar	= March
Aug	= August
Sept	= September
Nov	= November
Dec	= December

Biology Data Book
Second Edition

VOLUME II

VI. BIOLOGICAL REGULATORS AND TOXINS

72. ENDOCRINE ORGANS AND HORMONES: INVERTEBRATES

Chemistry: mol wt = molecular weight.

	Phylum & Class[1] (Synonym)	Possible Endocrine Organs or Areas	Possible Endocrine Factors	Chemistry	Effects
1	Chordata Tunicata[2] (Urochordata)	Neurosecretion in brain ⟨?⟩	....	....	Related to reproduction ⟨?⟩
2		Subneural gland	....	....	Gonadotropic in some species
3		Endostyle	Minimal thyroxinogenesis (greater in tunic)	....	....
4	Hemichordata	Neurosecretion in brain ⟨?⟩	....	....	....
5		Endostyle	Minimal thyroxinogenesis ⟨?⟩	....	....
6	Echinodermata	Neurosecretion in circumoral nerve ring ⟨?⟩	....	....	Water balance ⟨?⟩
7		Pyloric caeca ⟨?⟩ (stainable cells not insulin-secreting)	Insulin-like factor[3]	....	Hypoglycemic
8	Asteroidea	Secretion by supporting cells of radial nerve ⟨?⟩	Gonad-stimulating substance (GSS)	Polypeptide; mol wt = 2100 (22 amino acids)	Gonadotropic
9		Ovarian follicle cells and testis	1-Methyladenine (MIS; maturation-inducing substance)	....	Spawning
10	Arthropoda Arachnida	Brain - and - neurohemal organ neurosecretory system	....	....	Water balance
11		Retrocerebral glands	....	....	Molting in ticks
12		Corpus luteum-like structure ⟨?⟩	....	....	Reproduction in scorpions
13	Crustacea	X-organ - sinus gland neurosecretory system (eyestalk)	Somatic chromatophorotropins	....	Somatic pigmentation
14			Erythrophore-concentrating hormone (ECH)	Octapeptide	Concentration of red pigments
15			"Melanophore"-dispersing hormone (MDH)	Peptide	Dispersion of ommochromes
16			Retinal chromatophorotropin[4] (DRPH)	Peptide	Distal retinal pigment dispersion
17			Molt-inhibiting hormone (MIH)	Peptide	Tropic (inhibitory) on Y-organ (ecdysone production)

[1] Unless otherwise indicated. [2] Subphylum. [3] Also claimed for tunicates, gastropods, and crustaceans. [4] May be MDH (*see* entry 15).

continued

	Phylum & Class (Synonym)	Possible Endocrine Organs or Areas	Possible Endocrine Factors	Chemistry	Effects
18			Gonad-inhibiting hormone[5] (OIH; TIH)	Peptide	Inhibitory of ovary development in ♀, of testis development in ♂ crabs, and of sex accessory development
19			Hyperglycemic hormone	Protein	Blood sugar regulation ("diabetogenic")
20			Other metabolic factors ⟨?⟩	....	O_2 consumption, Ca^{2+} metabolism, water metabolism (probably related to molting); lipid metabolism; protein metabolism
21		Brain - and - postcommissure organ neurosecretory system	Somatic chromatophorotropins ⟨?⟩	....	Somatic pigmentation ⟨?⟩
22		Brain	Hydromineral factor(s)	....	Favors increased ion levels in freshwater species, decreased ion levels in marine species
23		Thoracic ganglion - and - sinus gland neurosecretory system	....	....	Water movement (storage in foregut) at ecdysis
24		Thoracic ganglion - and - pericardial organ neurosecretory system	....	Peptides	Frequency and amplitude of heartbeat
25		Neurosecretion in brain and ventral ganglia ⟨?⟩	....	....	Brain: limb regeneration
26		Y-organ[6] (ecdysial gland)	β-Ecdysone (20-Hydroxy-ecdysone)	Steroid	Rate of molting in adults
27		Androgenic gland[7]	....	....	Development of gonad, gonoduct, and secondary sex characteristics in ♂
28		Ovary	....	....	Secondary sex characteristics in ♀
29	Insecta	Brain - and - corpus cardiacum neurosecretory system	Ecdysiotropin ("Brain hormone")	Protein	Stimulates ecdysial gland
30			Bursicon[8]	Protein; mol wt approx. 40,000	Tanning of cuticle
31			Diuretic hormone	....	Water excretion by malpighian tubules
32			Egg development hormone	....	Yolk deposition in Diptera
33			....	....	Gonad and accessory sex development in *Lampyris*

[5] May be MIH (*see* entry 17). [6] Controlled by MIH (*see* entry 17; there is no evidence that the Y-organ actually secretes β-ecdysone (*see* entry 45). [7] Controlled by TIH (*see* entry 18) in crabs. [8] Not released from corpus cardiacum.

continued

	Phylum & Class (Synonym)	Possible Endocrine Organs or Areas	Possible Endocrine Factors	Chemistry	Effects
34			....	....	Protein and carbohydrate metabolism
35		Corpus cardiacum (probable intrinsic factors)	Myotropic factors	....	Motility of oviduct, gut, and malpighian tubules
36			Cardioaccelerator (produced by pericardial cells under corpus cardiacum stimulation ⟨?⟩)	....	Heartbeat
37			Hyperglycemic factor	....	Blood sugar regulation ("diabetogenic")
38			Antidiuretic factor	....	Rectal water absorption
39			Metabolic factors ⟨?⟩	....	Lipid metabolism
40			....	....	Depression of spontaneous activity in nerve cord
41		Stomatogastric (sympathetic) system - and - perisympathetic organs neurosecretory systems (median nerve; ventral ganglia neurosecretion)	....	....	Functions similar to intrinsic corpus cardiacum factors (see entries 35-40), especially in cardioacceleration and water metabolism ⟨?⟩
42		Tritocerebral neurosecretion ⟨?⟩	....	....	Chromatophorotropic in phasmids and Chaoborus (Corethra) larvae
43		Subesophageal ganglion neurosecretion ⟨?⟩	....	....	Egg diapause in Bombyx
44		Ventral ganglia neurosecretion ⟨?⟩	....	....	Color change
45		Ecdysial (molting) gland: prothoracic, ventral, etc.⁹/	β-Ecdysone (20-Hydroxyecdysone), 20,26-dihydroxyecdysone, and 3-deoxy-β-ecdysone	Steroids	Larval and pupal molting; differentiation of adult structures; termination of diapause
46		Corpus allatum	Juvenile hormone (Neotenin)	Methyl 10, 11-epoxy-7-ethyl-3, 11-dimethyl-2,6-tridecadienoate (also 7-methyl derivative)	Larval and nymphal development; gonadotropic in many species; vitellogenesis; sex pheromone secretion (♀) and sensitivity (♂); accessory sex development (♂♀)
47		Testis—apical tissue	....	....	In Lampyris (like crustacean androgenic gland, see entry 27), stimulates sex differentiation in ♂ —testis, gonoduct, sex characteristics
48	Symphyla, Chilopoda, Diplopoda, & Pauropoda	Brain - and - cerebral gland neurosecretory system	....	....	Molt inhibition; larval molting in chilopods

⁹/ The ecdysial gland is necessary for the formation of β-ecdysone elsewhere in the insect body, but apparently does not secrete it.

continued

	Phylum & Class (Synonym)	Possible Endocrine Organs or Areas	Possible Endocrine Factors	Chemistry	Effects
49	Onychophora	Neurosecretion in brain, ventral cord, pedal nerves ⟨?⟩; neurohemal infracerebral organs ⟨?⟩	....	....	....
50	Annelida Hirudinea	Neurosecretion in brain and ventral nerve cord; neurohemal area at base or posterior surface of brain	....	....	Gonad stimulation; pigmentation ⟨?⟩
51	Oligochaeta	Neurosecretion as in Hirudinea (*see* entry 50)	....	....	Gonad stimulation; secondary sex characteristics; regeneration; pigmentation; diapause; hydromineral metabolism
52	Polychaeta	Neurosecretion as in Hirudinea (*see* entry 50) and infracerebral "gland" in Nereidae and some other groups[10]	....	....	Inhibition of gonad maturation and of heteronereid transformation; oogenesis and vitellogenesis; growth and regeneration
53		Neurosecretion in brain and neurohemal area at dorsal surface of brain in *Arenicola*	....	....	Gonad stimulation
54	Echiura (Echiuroidea)	Neurosecretion in ventral nerve cord ⟨?⟩	....	....	....
55	Sipuncula (Sipunculoidea)	Neurosecretion in brain ⟨?⟩; anterior neurohemal area ("finger organs") ⟨?⟩	....	....	Reproduction ⟨?⟩; myotropic factor ⟨?⟩
56	Mollusca Cephalopoda	Neurosecretion in brain ⟨?⟩	....	....	....
57		Optic glands[11]	....	....	Gonadotropic
58		Neurosecretory system of the vena cava, including neurons at posterior end of brain	....	Not 5-hydroxytryptamine or catecholamines	Cardio-accelerator ⟨?⟩
59		Branchial glands	....	Ethanol-soluble; heat-stable	Survival
60	Bivalvia	Neurosecretion in most ganglia ⟨?⟩	....	....	....
61		Cerebral ganglia	....	....	Inhibition of spawning
62		Visceral ganglion ⟨?⟩	....	....	Osmoregulation
63	Scaphopoda	Neurosecretion in various ganglia ⟨?⟩	....	....	....

[10] In syllids, the reproduction-inhibiting factor is in the proventriculus. [11] Inhibited by light; under nervous control from the subpedunculate lobe.

continued

	Phylum & Class (Synonym)	Possible Endocrine Organs or Areas	Possible Endocrine Factors	Chemistry	Effects
64	Gastropoda	Neurosecretion in all ganglia; numerous subcapsular neurohemal areas in ganglia, connectives, and nerves	....	....	Carbohydrate metabolism
65			....	....	Reproduction
66		Cerebral ganglia	....	....	Spermatogenesis
67		Abdominal ganglion	....	....	Ovulation in *Aplysia*
68		Pleural, parietal, and visceral ganglia	Diuretic and ion-uptake factors	....	Water and ion balance in *Lymnaea*
69		Dorsal bodies—mediodorsal and laterodorsal bodies in pulmonates, juxtaganglionic tissue in prosobranchs	....	....	Reproduction—vitellogenesis; growth and differentiation of reproductive tract
70		Optic tentacles in slugs ⟨?⟩	....	....	Regulation of ovotestis development and steroid metabolism
71		Gonads—♂-phase ovotestis	C_{19}-Steroids of the vertebrate type	Androstenedione, androsterone, androstanediols	Reproductive tract development; acts on tentacles (reciprocal relation)
72	Nematoda	Neurosecretion in cephalic nerve ring ganglia ⟨?⟩	....	....	Production of molting-fluid enzymes for larval ecdysis
73	Rhynchocoela (Nemertina)	Neurosecretion in brain ⟨?⟩	....	....	Gonad inhibition (♀); electrolyte metabolism
74		Cerebral organ	....	....	Spawning ⟨?⟩
75	Platyhelminthes Cestoda	Neurosecretion in rostellar neurons ⟨?⟩	....	....	Proglottid shedding ⟨?⟩
76	Turbellaria	Neurosecretion in brain ⟨?⟩	....	....	Regeneration ⟨?⟩; asexual reproduction
77	Coelenterata (Cnidaria)	Subhypostomal neurosecretion in hydra	....	....	Growth stimulation; reproduction inhibition

Contributor: Bern, Howard A.

General References

[1] Bern, H. A., and I. R. Hagadorn. 1965. In T. H. Bullock and G. A. Horridge. Structure and Function in the Nervous System of Invertebrates. W. H. Freeman, San Francisco. pp. 353-429.

[2] Durchon, M. 1967. L'Endocrinologie des vers et des mollusques. G. Masson, Paris.

[3] Engelmann, F. 1970. Physiology of Insect Reproduction. Pergamon Press, Oxford.

[4] Fingerman, M. 1963. The Control of Chromatophores. Macmillan, New York.

[5] Gabe, M. 1966. Neurosecretion. Pergamon Press, Oxford.

[6] Gersch, M. 1964. Vergleichende Endokrinologie der Wirbellosentiere. Geest and Portig, Leipzig.

[7] Gottfried, H., and R. I. Dorfman. 1969. Excerpta Med. Found. Int. Congr. Ser. n184:368

[8] Highnam, K. C., and L. Hill. 1969. The Comparative Endocrinology of the Invertebrates. E. Arnold, London.

[9] Hoar, W. S., and H. A. Bern, ed. 1972. Gen. Comp. Endocrinol., Suppl. 3.

[10] Jenkin, P. M. 1962 & 1970. Animal Hormones. Pergamon Press, Oxford. pt. I & II.

[11] Novak, V. J. A. 1966. Insect Hormones. Methuen, London.

[12] Scharrer, B., and M. Weitzman. 1970. In W. Bargmann and B. Scharrer, ed. Aspects of Neuroendocrinology. J. Springer, Berlin. pt. 1.

[13] Stutinsky, F., ed. 1967. Neurosecretion. J. Springer, Berlin.

[14] Tombes, A. S. 1970. An Introduction to Invertebrate Endocrinology. Academic Press, New York.

The following Subclasses are in addition to the taxonomic classification scheme selected for the *Biology Data Book* (consult Appendix III): **Chondrichthyes**—Elasmobranchii (includes the Orders Chlamydoselachiformes, Hexanchiformes, Heterodontiformes, Squaliformes, and Rajiiformes); Holocephali (includes the Order Chimaeriformes). **Acti-**

	Gland or Tissue	Agnatha	Chondrichthyes	Actinopterygii [1]	Crossopterygii [2]
	Hypophysis [102,278] **Adenohypophysis [110]**				
1	Pars distalis	Organized as two or more masses, often sharply demarcated from one another, with associated tendency towards marked compartmentalization of functional cell types and specialized regional vascular patterns (tendency most elaborately demonstrated in Teleostei): (i) a rostral pars distalis (proadenohypophysis), and (ii) a proximal pars distalis (mesoadenohypophysis) which contains gonadotropes, somatotropes, & thyrotropes			
		Separated from brain by layer of connective tissue [132,214] Myxiniformes: Not divided into regions; may be penetrated by small diverticula from nasopharyngeal duct [132,214] Petromyzontiformes: Rostral zone (irregular cell cords) histologically differentiated from proximal zone (broad cell cords) [132,214] **Gen. Ref.:** 1,6,62,78, 79,216,228	A portion occurs as a variably, but often completely, separated & independently vascularized mass without direct hypothalamic innervation or neurovascular contact with infundibulum; called ventral lobe [3] in Elasmobranchii, rachendachhypophyse in Holocephali [63, 228] **Gen. Ref.:** 6,56,62, 130,184,187,204, 216,249	Chondrostei, Holostei, Elopiformes, Clupeiformes, Salmonidae: Rostral pars distalis has follicular structure Polypteriformes, some Elopiformes, Clupeiformes: Rostral pars distalis associated with persistent buccohypophyseal canal [221,228] Chondrostei, Holostei: Regional segregation of cell types poorly developed [6,9,62,134,137,149, 151,152,216,228] Teleostei: Rostral pars distalis characteristically contains prolactin & adrenocorticotropic hormone (corticotropin) activities [6,7,62, 70,71,122,129,131,159,180,182, 212,213,216,228,256]	A greatly attenuated & focally intermittent "cord" of tissue lying between internal carotid arteries along their entire intracranial course; probably bipartite as in Actinopterygii; attenuated rostral portion has follicular structure [153, 195] **Gen. Ref.:** 196
2	Pars tuberalis	Absent	Absent [290]	Absent	Probably absent

[1] According to Romer [245]. [2] Only surviving species, *Latimeria chalumnae.* [3] The ventral lobe in Elasmobranchii and the lateral lobes of the adenohypophyseal anlage in amniotes have similar embryology, and a homology has been suggested [278]. In all other respects, however, they differ. The ventral lobe of Elasmobranchii has defined hormonal activities (thyrotropic and gonadotropic), has no association with a median eminence or portal vessels, and is constantly present in Elasmobranchii. The lateral lobes of amniotes develop into the inconstantly present pars tuber-

nopterygii—Chondrostei (includes the Orders Acipenseriformes and Polypteriformes); Holostei (includes the Orders Semionotiformes and Amiiformes); Teleostei (includes all other Orders listed under Osteichthyes). **Gen. Ref.** = general references.

Dipnoi	Tetrapoda			
	Amphibia	Reptilia	Aves	Mammalia
Organized as a single mass of well-vascularized cell cords, with variable, but generally limited, regional segregation of cell types [1]				
A prominent segregation of erythrosinophils (prolactin cells) in "cephalic lobe" seen in Aves, less clearly in Dipnoi, Amphibia, & Reptilia				Situated anteriorly & ventrally ("anterior lobe" in older terminology) [105,114]
Organized as a single mass, closely resembling that of Amphibia, in contrast to other fish-like vertebrates *Protopterus:* Situated ventrally; separated from pars intermedia by hypophyseal cleft [276] **Gen. Ref.:** 6, 62,138,216, 228	Situated at posterior end of gland

Gen. Ref.: 133,135,136, 158,215-217 | Well-developed; situated ventrally & often posteriorly [275]

Gen. Ref.: 165,246, 247 | Large; situated ventrally or anteroventrally to neurohypophysis; histologically distinguishable into cephalic & caudal regions [275]

Gen. Ref.: 24,277 | **Gen. Ref.:** 10,19,111,252 |
| Probably absent [138, 276] | Derived from lateral lobes of adenohypophyseal anlage[3]; not always present; becomes secondarily associated with developing median eminence [2] | | | |
| | Gymnophiona, Caudata: Vestigial lateral processes Caudata: Represented by lateral lobes connected with pars distalis [221] Salientia: Developed, but usually reduced (sometimes absent); forms two small plates on ventral surface of tuber cinereum [221]

Gen. Ref.: 65 | Conservative (primitive) orders—Chelonia, Rhynchocephalia, Crocodylia: Always developed Sauria: Not always present; distinguishable only as a few cells in floor of brain [275] Serpentes: Absent, except in Boidae & Colubridae

Gen. Ref.: 246,247 | Consists of (i) pars tuberalis proper, a layer of cells on ventral surface of diencephalon and within pia mater; (ii) a portal zone of cell cords & blood vessels connecting pars tuberalis proper with (iii) pars tuberalis interna which is fused with pars distalis [275] **Gen. Ref.:** 277 | Surrounds infundibular process, forming base for primary plexus of hypophyseal portal system which vascularizes pars distalis [105, 114] Two cell types: (i) ciliated "chromophobes" associated with follicular (stellate) cells of pars distalis, and (ii) small numbers of "basophils" for which gonadotropic functional association has sometimes been demonstrated **Gen. Ref.:** 19,26,111,162, 163,219,285 |

alis which lacks hormonal activity and is characteristically closely associated with the median eminence or its portal vessels. The ventral lobe of Elasmobranchii may be functionally homologous with the ventral portion of the proximal pars distalis of Teleostei, while the rostral pars distalis of Elasmobranchii contains prolactin-like and corticotropic activities as in Teleostei.

continued

	Gland or Tissue	Agnatha	Chondrichthyes	Actinopterygii[1]	Crossopterygii[2]
3	Pars intermedia	Myxiniformes: Histologically distinct zone of cells in posterior adenohypophysis, in contact with neurohypophysis in adults, described in *Myxine;* relationship to a pars intermedia unclear at present Petromyzontiformes: Represented by most posterior part of adenohypophysis; thin, with entire length in close contact along neurohypophysis which it tends to surround [132, 214]	Interdigitates with infundibular process to form neurointermediate lobe [63,228]	Elaborately developed; characteristic interdig or neurohypophysis with which it shares vas Represented by meta-adenohypophysis, the tissue most intimately associated with the neurohypophysis [105,221, 228]	Thick trabeculae contain numerous large follicles; two secretory cell types identified within pars intermedia [154]
		Gen. Ref.: 1,6,62,78,79	Gen. Ref.: 6,62,187, 204,249	Gen. Ref.: 6,62,129,134,137, 149,151,152	Gen. Ref.: 196
4	Neurohypophysis Anterior[4] (Median eminence)	Myxiniformes: A caudally directed, sac-like projection of the hypothalamus represents infundibular process; thin-walled & unpaired, with many terminations of neurosecretory fibers; no contact with adenohypophysis; well-vascularized with a folded surface [132,214]; neurosynaptic endings described abutting on poorly vascularized fibrous tissue between ventral surface of infundibulum & adenohypophysis; although no organized portal vessels as such, capillaries supplying adenohypophysis pass through the fibrous tissue and could function as more proper portal vessels; neurohumors, if present, possibly transported more through convection & diffusion than in conventional neurohumoral fashion [1,62,78,79,103,121,203,226, 228] Petromyzontiformes: Formed by thickened floor of diencephalon, particularly posterior part [132, 214]	Organized as a median eminence or homologous neurohemal contact mologous tuberal nuclei in Tetrapoda; characteristic feature is a pri ohypophysis or a part thereof (pars distalis); portal vessels may oc 137,151,152,232,233]		
			Median eminence-like organization on ventral surface of hypothalamus, with portal vessels vascularizing pars distalis, but no interdigitation or more direct contact between the two [186] Elasmobranchii [183, 185,187] Holocephali [130,187, 249]	*Amia:* Limited neurohypophyseal extension into pars distalis Teleostei: No distinguishable portal vascular system[5]; characteristic interdigitation with adenohypophysis; direct synaptic (secretomotor) contact with pars distalis [80,82,122, 129,131,141,266,286]	Median eminence region identified; well-defined portal vessels
			Gen. Ref.: 6,62,226, 228	Gen. Ref.: 62,226,228	Gen. Ref.: 153,196

[1] According to Romer [245]. [2] Only surviving species, *Latimeria chalumnae*. [4] Tuberoinfundibular tract, with monoaminergic, 800- to 1000-Å-diameter, dense-colored axonal (secretory, or transmitter) vesicles. [5] A portal-like

Dipnoi	Tetrapoda			
	Amphibia	Reptilia	Aves	Mammalia
itation with posteri- cularization	Reduced and not always present; no interdigitation, although occasional synaptic contact with pos- terior			3
Protopterus: Dorsal to hypophyseal cleft; tube-shaped lobules with wide cavities; closely attached to infun- dibular process ly- ing rostral to pars intermedia [276]	Situated pos- terodorsal to pars distalis [221]	Variable development; usually separated from pars distalis by hypophyseal cleft [275] Chelonia, *Sphenodon,* some Sauria: Thin- walled sac [275] Sauria: No synaptic contacts occur; com- plete separation by perivascular space [244] Serpentes: Compact organ [275]	Absent	Located between pars distalis & neurohypophysis; often sepa- rated from pars distalis by re- mains of hypophyseal cleft; occasionally absent [105, 114]; may occur as cellular rests in posterior part of pars distalis, and infiltrates pars nervosa Prototheria, Metatheria: Sepa- rated from pars distalis by persistent hypophyseal cavity Cetacea: Absent
Gen. Ref.: 6,62, 138,287	**Gen. Ref.:** 124, 135,136,199	**Gen. Ref.:** 189,246, 247,270	**Gen. Ref.:** 275	**Gen. Ref.:** 111

area on ventral surface of diencephalon (hypothalamus); neurons originate in nucleus lateralis tuberis in fishes & in ho- 4
mary capillary bed on which monoaminergic & other axons abut; primary capillary bed then vascularizes most of aden-
cur, or vascularization may be direct; no direct contact with pars distalis (except Teleostei, *see* below) [110,117,134,

Dipnoi	Amphibia	Reptilia	Aves	Mammalia
Monoaminergic nerves extend in- to pars distalis along portal ves- sels [287] *Protopterus:* Medi- an eminence be- lieved to be repre- sented by area in floor of third ven- tricle just rostral to anterior tip of adenohypophysis [276]	*Rana:* Median eminence with portal vessels [105]	Median eminence is differentiated out of infundibular floor; ranges from simple form, differing little from pars nervosa, to thickened structure with capillaries of hypophyseal portal system buried in it [105,275]	Median eminence develops from floor of third ven- tricle or infundibular stem anterior to pars ner- vosa, from which it is not sharply delimited; covered by capillary net of hypophyseal portal system [275]; partial di- vision of median emi- nence into anterior & posterior zones, with rel- atively discrete portal distribution in pars dis- talis	Median eminence is poorly de- fined region of infundibular stem on floor of the dienceph- alon, usually anterior to pars nervosa and always connected to pars distalis by portal blood vessels [105,275]
Gen. Ref.: 62,138, 226,228	**Gen. Ref.:** 60, 64,66,75,104, 218,288	**Gen. Ref.:** 246,247	**Gen. Ref.:** 191,272	**Gen. Ref.:** 2-4,111,140,142, 280

vascular system has been claimed for some Teleostei, but for these species [119,250].
characteristic neurohemal contacts have not been described

continued

	Gland or Tissue	Agnatha	Chondrichthyes	Actinopterygii[1]	Crossopterygii[2]
5	Posterior[6] (Pars nervosa)	Largely consists of terminations of preoptico-hypophyseal tract *Myxine:* Limited contact with posterior adenohypophysis [1, 79] *Petromyzon:* Well-developed meta-adenohypophysis	Extensive interdigitation & contact with pars intermedia (so-called neurointermediral portion represents termination of hypothalamo-hypophyseal tract, with its origin cleus (pars magnocellularis in Teleostei), or homologous supraoptic & paraventricu niotes)		
			Infundibular process interdigitates with the pars intermedia; a diffuse structure formed by endings of neurosecretory fibers originating in preoptic area [63,228]; separate, independent vascular supply	Vascularization shared to a variable extent with anterior ysis & pars distalis	
				Chondrostei, Holostei [62, 117,134,137,149,151,152, 226,232,233] Teleostei [62,80,82,122,129, 131,141,160,161,226,228, 260,286]	Tubular posterior neurohypophyseal processes are closely associated with pars intermedia trabeculae to form a neurointermediate lobe; axonal endings largely peptidergic (classic neurosecretory type [154]
			(Saccus vasculosus, an associated well-vascularized evagination of posterior infundibular wall, with specialized [Krönchenzellen] ependyma, apparently a sensory epithelium)		
		Gen. Ref.: 62,78,103, 121,203,226,228	**Gen. Ref.:** 62,130,183, 185,187,188,226,249		**Gen. Ref.:** 153,196
6	Thyroid[7]	Myxiniformes: Separate, large follicles scattered in adipose & connective tissue along most of ventral surface of pharynx, and between gill pouches [264] Petromyzontiformes: Elongated follicles tend to occur from second to fifth pharyngeal pouches, dorsal to medioventral cartilage of branchial basket [83] **Gen. Ref.:** 88,100-102, 118,268	Compact & encapsulated, lying between basihyoid & bifurcation of the ventral aorta, & between coracomandibular & coracohyoid muscles **Gen. Ref.:** 55,100,102, 281	Chondrostei, Holostei: Diffuse follicles around ventral aorta [181] Teleostei: Characteristically, no organized compact gland; follicles scattered in small groups along course of ventral aorta & afferent branchial arteries Some exceptional Teleostei (e.g., *Thunnus, Scarus*): Encapsulated, compact, organized glands [181] Many experimental Teleostei (e.g., Poeciliidae & others): Follicles can occur throughout anterior body & head region, including eye & pronephros, especially under stimulation of thyroid-stimulating hormone (thyrotropin) **Gen. Ref.:** 100,102,282	Compact and encapsulated; does not contain ultimobranchial cells [289]
7	Parathyroid[8]	Not observed	Not observed	Not observed	Not observed

Dipnoi	Tetrapoda				
	Amphibia	Reptilia	Aves	Mammalia	
ate lobe); neu- in preoptic nu- lar nuclei (am- ⎯⎯⎯⎯ neurohypoph-	A true pars nervosa (neural lobe), with an independent blood supply, is present as posterodorsal thick- ening of infundibular process				5
	Relatively large in terrestrial species, smaller in aquatic spe- cies [105]	Well-defined neural lobe [275]	Distinct pars ner- vosa (neural lobe) lying poste- riorly and carried by an infundibu- lar process [275]	Pars nervosa usually globular; formed from a distal extremity of infundibulum; may or may not contain an extension of third ventricle [105,275]	
Gen. Ref.: 62, 138,226,228, 276,287	Gen. Ref.: 64, 66,104,218,288	Gen. Ref.: 189,246,247,270	Gen. Ref.: 272	Gen. Ref.: 111,140,142	
....	Paired structures, widely sepa- rated, & closely associated with carotid bifurca- tion & hyoid bone Caudata: Small, scattered groups of accessory follicles lie scat- tered on super- ficial muscle fasciae of throat	Single lobulate organ lying close to trachae [40] Chelonia, Rhynchocephalia, certain Squamata: Single mass ventral to trachae Chrysemys: Flattened mass ven- tral to truncus arteriosus [40] Many Sauria: Bilobed, paired, & compact glands may occur (of- ten within a single genus) Crocodylia: Well-defined lateral lobes extending on either side of trachae, and connected by isthmus, as in Aves	One pair of glands, situated at boundary of neck & thoracic cavity, and asso- ciated with tra- chae	Two lateral lobes & median isthmus; lateral lobes at sides of trachae, with cranial ends at lev- el of caudal edge of cricoid car- tilage Echidna (Monotremata): Un- paired; forms single intrathorac- ic mass	6
	Gen. Ref.: 102, 258	Gen. Ref.: 102,176	Gen. Ref.: 18,102	Gen. Ref.: 100,102	
Not observed	Parathyroid tissue derives from paired embryonic pharyngeal pouches (Roman numerals refer to pouches numbered from anterior to posterior)				7
	Caudata: One pair in convexi- ty of aortic arch [208] Salamandra: Spherical bod- ies derived from ventral ends of pharyngeal	Chelonia: Anterior pair (III) embedded in thymus (requires microscopic identification); posterior pair (IV) embedded in fibrous tissue at aortic arch in association with left ultimo- branchial tissue Sauria: Pair III at carotid bifur- cation at angle of jaw, often	Develop from pha- ryngeal pouches III & IV [18]; tendency for close approxima- tion of pairs III & IV, with fusion in some species; characteristically	Usually two pairs closely associ- ated with dorsolateral border of thyroid gland (usual position in Homo sapiens); four glands nor- mally persist, although supernu- merary pairs are common in Ho- mo sapiens & Rattus; histologic variation striking, particularly in hyperfunctional states & in aged	

chemistry, can be traced to prevertebrate chordates. 8/ De- rived from the endoderm of pharyngeal pouches III and IV; distributed between the levels of the hyoid bone and the diaphragm, about the carotid sheaths and aortic arch, or in the anterior mediastinum; often associated with vessels de- rived from the aortic arch, usually the carotid arteries; de- spite considerable cytologic variation, apparently consists of only one functional cell type.

continued

Gland or Tissue	Agnatha	Chondrichthyes	Actinopterygii [1]	Crossopterygii [2]
8 Ultimo-branchial tissue	Not observed [52,269]	Vesicles with colloid contents on left side only; dorsal to pericardium (suprapericardial body) [31,269] *Squalus:* Ventral to pharyngo-esophageal junction, just anterior to pericardium between cerato & basibranchial cartilages [54] **Gen. Ref.:** 52,261	Vesicular, gland-like bodies, often on left side only, or fused to median body above pericardial wall; source of peptide hormone thyrocalcitonin (calcitonin) [52,69,102,269] Chondrostei, Holostei [52,235] *Lepisosteus:* Developed from terminal pharyngeal pouches [52,235] Teleostei: In peritoneopericardial septum, just ventral to esophagus [54] and in close association with sinus venosus [52,59,97,98,102, 235,241,269]	Probably discrete, separate from thyroid [289]
9 Endocrine pancreas [3] (Islets of Langerhans)	Intimate association with major exocrine ducts & intercalation of islet elements in duct epithelium Myxinoidei: Represented by cluster of cell cords in intestinal wall adjacent to common (major) bile duct [12, 15,274] Petromyzontiformes: In ammocoete larva, groups of cells in intestinal wall adjacent	*Carcharhinus, Dasyatis, Mustelus:* Islet cells form second epithelial layer around duct cells of pancreas [262] *Raja:* Solid cords of islet cells contiguous to pancreatic ducts [262] *Squalus acanthias:* Small islands of cell cords, fre-	Tendency for islet tissue to be independent of exocrine pancreas as encapsulated, macroscopic "Brockmann's" bodies [73,90] Chondrostei, Holostei: Diffuse, scattered islet tissue in *Polypterus & Polyodon* [154], *Amia & Lepisosteus* [73] Teleostei: Associated as separate bodies with portal vein and/or common (major) bile duct [41,72,73,102,123,144,231,267]	On dorsal surface of spiral intestine, in intimate association with major exocrine ducts and in-

[1] According to Romer [245]. [2] Only surviving species, *Latimeria chalumnae.* [3] Vertebrates in most classes contain a few islet cells that do not stain like A or B cells, are often argyrophylic, and have characteristic large, uncondensed granule profiles by electron microscopy; they are termed D or Al cells [89]. There has been an effort to

Dipnoi	Tetrapoda			
	Amphibia	Reptilia	Aves	Mammalia
	pouches III & IV widely separated from thyroids, lateral to aortic arches; absent in some Perennibranchiata (e.g., *Ambystoma, Necturus*) [85] Salientia: Two pairs in tadpoles [279]; recent study suggests one pair situated near lingual artery & external jugular vein on either side [237] *Rana:* Two pairs, lying against jugular veins [85] **Gen. Ref.:** 102	embedded in adventitia of carotid artery; pair IV said to be frequently suppressed Serpentes: Pair III at carotid bifurcation at angle of jaw; pair IV between thymic lobes at level of pericardium Rhynchocephalia: Pair III at carotid bifurcation; presumptive pair on posterior wall of aortic arch Crocodylia: Apparent fusion of pairs III & IV at carotid-subclavian bifurcation, as in Aves; often in close association with thymus; suppression of pair IV also described **Gen. Ref.:** 47-49,61,102,209	located at carotid-subclavian bifurcation [201] **Gen. Ref.:** 102	animals [36,106] *Homo sapiens:* One to three pairs; variations in location most thoroughly studied; common anomalous locations are intrathyroid & intrathymic sites, the latter in anterior mediastinum [36,106] *Homo sapiens, Canis familiaris, Felis catus:* Pair III may be embedded in anterior thyroid [36,106] *Rattus:* Only pair IV present, usually visible on surface of thyroids [106]; pair III said to be suppressed; a recent study cites 30% incidence of intrathyroid & intrathymic parathyroid tissue [37] **Gen. Ref.:** 102
No available description	Derived from ventral wings of pharyngeal pouches III & IV; separate from thyroid in hyoid area [102] *Salamandra:* Small single body near heart [85] Gymnophiona, Salientia: Bilateral; vesicular; at level of truncus arteriosus [85] **Gen. Ref.:** 238-240,242,269	Usually paired, with a tendency to asymmetric development of a larger left gland; variable position [269], usually lying near carotid-subclavian bifurcation Chelonia: Adjacent to thyroid Sauria: Only on left side at level of vagus ganglion, adjacent to esophagus or tracheal bifurcation Serpentes: Against thyroid [269] **Gen. Ref.:** 61,198,253	Closely associated with, and may surround, parathyroid glands; cell cords with some colloid-filled vesicles [269] **Gen. Ref.:** 38,53, 102,198,224,254	Derived from the most posterior pharyngeal pouches; becomes incorporated with thyroid primordium where it is recognized as C (para-follicular) cells in adults, or as a differentiated type of follicle; may also be found in parathyroids [57], lateral neck tissues [102], & thymus Source of peptide hormone thyrocalcitonin (calcitonin) **Gen. Ref.:** 27,33,34,99,223,224, 257,269
Tendency for islet tissue to be independent of exocrine pancreas as encapsulated, macroscopic "Brockmann's" bodies [73,90] *Protopterus:* On	Islet tissue independent of intestinal wall & major exocrine ducts; tendency for segregation of islet cell types within exocrine pancreas			
	Within single islets, central B-cell & peripheral A-cell orientation Caudata: B cells predominate [102]	Islets of Langerhans in pancreas; more A than B cells [102] Chelonia, Rhynchocephalia, Crocodylia (conservative orders): Within single islets, central B-cell & peripheral A-cell orientation [91,192,194] Sauria: Not segregated by cell types; large, confluent islets [72,89,192-194,236,251,267]	Segregation of islet tissue into specific single or major cell class [72,89,102,145]	Within single islets, central B-cell & peripheral A-cell orientation; principal cell types are A & B

equate this cell type with intra-islet gastrin secretion, or to suggest it as a source of a third islet hormone [22,32,89, 146,174]. Despite recent evidence suggesting that D cells can represent A- or B-cell intergrades [84,166,194], gastrin has been identified in mammalian islets with labeled immunochemical methods [107].

continued

	Gland or Tissue	Agnatha	Chondrichthyes	Actinopterygii [1]	Crossopterygii [2]
		to bile duct; in adults, groups of cell cords within intestinal wall [11,15,77] Gen. Ref.: 72,73,102, 197,273	quently separated from duct system of pancreas [262] Gen. Ref.: 72,73,89,102,220, 222		tercalation of islet elements in duct epithelium [109]
10	Suprarenal cortical-interrenal tissue & associated chromaffin tissue [102, 139]	Suprarenal cortical tissue present as small clusters of cells against ventral & lateral walls of posterior cardinal veins in trunk & tail; also as groups of cells in pronephros & around kidney vessels [96,112,128] Chromaffin tissue probably represented by groups of cells distributed along main arteries & veins in trunk [95]	Elasmobranchii: Interrenal tissue tends to aggregate in discreet masses or "glands" associated with posterior cardinal veins & posterior portion of mesonephros; occasionally occurs as single left-sided "gland"; no association with segmentally distributed chromaffin tissue [42,126,127] *Raja:* Suprarenal cortical tissue forms horseshoe-shaped body, usually asymmetrical, with two limbs extending anteriorly against inner sides of kidneys [42,115] *Squalus:* Suprarenal cortical tissue present as elongated body between kidneys and close to cardinal vein, with additional small groups of cells anteriorly [42,115] *Torpedo:* Suprarenal cortical tissue present as oval body lying posteriorly on left kidney [42,115]	Suprarenal cortical tissue present as areas of cells in the usually lymphoid pronephros; may be scattered (e.g., *Salmo*), or arranged as layers of cells closely associated with cardinal veins [5,42] Chromaffin tissue mainly in pronephros as scattered cells or islets near or in walls of cardinal veins & their branches [5,42] *Acipenser, Polypterus, Lepisosteus, Amia:* Interrenal tissue scattered along posterior cardinal veins & within mesonephros & its capsule; more numerous anteriorly in *Acipenser* [10] [45,58,94, 125] Teleostei: Interrenal tissue associated with posterior cardinal veins, but largely limited to adult hematopoietic pronephros (head kidney); association with chromaffin tissue in some species is apparently incidental [7,30,43,45,76,94, 200]	No available description
11	Juxtaglomerular granular epithelioid cells	Absent [207,254]	Absent [23,35,207,254]	Granular epithelial cells are modified smooth muscle Chondrostei, Holostei: Not observed [147,154,206], but angiotensin is demonstrable upon incubation in rat bioassay [202] Teleostei: Renin acts on an uncharacterized, heat-labile, dialyzable blood pep-	Present in renal arteries, similar morphologically

[1] According to Romer [245]. [2] Only surviving species, *Latimeria chalumnae.* [10] Stannius (1839) originally described interrenal tissue in *Acipenser* [58], but the term "corpuscles of Stannius" now refers by convention to sep-

Dipnoi	Tetrapoda			
	Amphibia	Reptilia	Aves	Mammalia
dorsal wall of intestine, just anterior to spiral region [164]		Serpentes: Large aggregations of islets macroscopically visible at splenic end of pancreas [14]		
	Gen. Ref.: 13,72,89,108, 143,157,251			Gen. Ref.: 72,74,89,102,145, 166,251

Dipnoi, Amphibia, & embryos of amniotes: A specialized, steroid-synthesizing mesodermal tissue derived from mesial portion of urogenital ridge, and diffusely distributed in association with mesonephros and its proper venous drainage, the posterior cardinal veins & their branches — 10

Dipnoi	Amphibia	Reptilia	Aves	Mammalia
Suprarenal cortical tissue distributed along ventromedial border of mesonephros in association with posterior cardinal veins [45, 128]		In adults, suprarenal cortical tissue is compact & encapsulated as a pair of "glands" usually situated at anterior pole of metanephric kidney and associated with its proper venous drainage from embryonic subcardinal system; developmental association with portion of extensive chromaffin tissue, a part of the sympathetic nervous system, neural elements of which are specialized for neurovascular secretion		
Chromaffin tissue present as groups of cells distributed along the posterior cardinal veins; also associated with the intercostal arteries [42]	Chromaffin cells interspersed with cords of suprarenal cortical cells [85,115] Caudata, Salientia: Suprarenal cortical tissue associated with a portion of chromaffin tissue, and may occur in part within mesonephros Caudata: Suprarenal cortical tissue discontinuous Salamandra: Series of orange patches along ventromesial border of kidneys; may extend forward in association with sympathetic ganglia to subclavian artery; brighter & more conspicuous in breeding season [85,115] Salientia: Suprarenal cortical tissue aggregated into pair of elongated masses paralleling mesonephros	Chromaffin tissue irregularly dispersed throughout suprarenal cortical "glands" One pair of "glands," elongated in shape; suprarenal cortical tissue forms anastomosing cords; chromaffin tissue intermingles and forms peripheral layer [42,275] Chelonia: On ventral surfaces of kidneys; may fuse [42,275] Sauria, Serpentes: Anterior to kidneys [42,275] Crocodylia: Anterior to & between kidneys [42,275]	One pair of "glands" (occasionally fused) situated at anterior ends of kidneys; wholly or partly covered by gonads; chromaffin tissue irregularly dispersed in meshes of anastomosing cords formed by suprarenal cortical tissue [115]	One pair of ovoid-shaped glands; "cortex" of suprarenal cortical tissue surrounds compact gland formed by that portion of chromaffin tissue associated with suprarenal cortical tissue; in many species, well-developed zonation reflects different enzymatic capabilities of different layers Prototheria: Chromaffin tissue interdigitates with suprarenal cortical tissue [42] Echidna: Cortical zonation lacking Theria: Cortex formed by suprarenal cortical tissue, medulla by chromaffin tissue [42] Eutheria: Situated immediately anterior to kidneys, with which suprarenal glands may be in contact
	Gen. Ref.: 225,229,230	Gen. Ref.: 29,92,173		Gen. Ref.: 259

elements of renal arteries, and secrete a peptide, renin — 11

Dipnoi	Amphibia	Reptilia	Aves	Mammalia
Protopterus: Present [206]			Macula densa, associated with vascular pole of glomerulus, occurs in some species in which renal portal	Confined to efferent arterioles & extraglomerular mesangium; associated with specialized segment of distal tubule, the macula densa, which, along with an architecture

arate endocrine tissue (*see* entry 12) of pronephric origin which does not occur in Chondrostei, and hence not in *Aci-*

penser, and has no association with chromaffin tissue.

continued

	Gland or Tissue	Agnatha	Chondrichthyes	Actinopterygii [1]	Crossopterygii [2]
				tide similar to angiotensin of Mammalia; present in glomerular & aglomerular marine forms; generally less well-developed in freshwater forms [23,25,35,50,51,147,150,190,254]	to those in Teleostei [154]
12	Corpuscles of Stannius	Absent [35,45,52]	Absent [35,45,52]	Chondrostei: Absent [58,147] *Amia:* Multiple corpuscles; derived from pronephric duct & distal tubular segments of mesonephros (similar in *Lepisosteus*) Teleostei: Derived from pronephric duct; occur as paired structures in posterior mesonephros; several pairs in Salmonidae [44,45,52,58,147, 175,205,234,255,263]; recent evidence suggests role in water/electrolyte adaptation similar to granular epithelioid cells (renin) [175, 205], but role in calcium regulation complementary to that of ultimobranchial tissue is also suggested [39]	No available description

| 13 | Gonadal tissue
Testis | *Myxine:* Interstitial cells absent; boundary cells probably present
Petromyzontiformes: Interstitial cells present | Two distinct tissues: (i) interstitial (Leydig) cells which occur as truly interstitial, or tions, have been recently shown to have steroidogenic capacity in Mammalia, or an leostei and in Amphibia, Reptilia, & Aves | | Interstitial cells in testis [168] |

Interstitial cells of usual sort [46,102,168]
Squalus acanthias: Interstitial cells perhaps absent [168]
Scyliorhinus, Chimaera: Probably interstitial cells between tubules of testis [178]

Teleostei: Interstitial cell elements occur in two forms, true interstitial cells &, in some species, lobule boundary cells (steroidogenic interstitial cells embedded or intercalated in walls of seminiferous tubules); both forms may occur within a species, but may be extremely difficult to demonstrate during periods of inactivity, even using hydroxysteroid dehydrogenase techniques; Sertoli cells may also show signs of steroidogenesis [81,102,168-171,210,271]
Salmoidae: Great variation in cell types present within individual species [168,178,179,243]

Gen. Ref.: 168

| 14 | Ovary | Steroidogenesis, as minimally substantiated by positive 3-β-hydroxysteroid dehydrogenase activity has been theca interna | | | |

Corpus luteum-like structures develop both in postovulatory follicles and as corpora atretica; efforts to demonstrate role for these structures (analogous to their role in Eutheria) in Elasmobranchii & Teleostei have not been entirely convincing, despite evidence they may contain 3-β-hydroxysteroid dehydrogenase activity

[1] According to Romer [245]. [2] Only surviving species, *Latimeria chalumnae.*

Dipnoi	Tetrapoda			
	Amphibia	Reptilia	Aves	Mammalia
	Gen. Ref.: 17,51,67,68,116,211, 254	Gen. Ref.: 254	system is reduced or absent Gen. Ref.: 68, 254	permitting countercurrent distribution of solutes, is characteristic; renin acts on angiotensin, a circulating enzyme precursor Gen. Ref.: 8,68,254	
No available description	Absent	Absent	Absent	Absent	12

as (in some Teleostei) lobule boundary cells, & (ii) Sertoli cells which, in addition to sustentacular & phagocytic func- 13
ultrastructural organization and/or histologic & enzymatic reactions characteristic of steroidogenic tissue in certain Te-

Dipnoi	Amphibia	Reptilia	Aves	Mammalia	
Interstitial cells in testis [168]	Caudata: Boundary cells in testis [168] Salientia: Interstitial cells in testis [168]	*Mauremys caspica leprosa (Emys leprosa):* Endocrine cells lie in seminiferous tubules, and are called boundary cells [168]			
	Gen. Ref.: 102,167,169,227	Gen. Ref.: 102, 169,172	Gen. Ref.: 18, 109,168,169	Gen. Ref.: 16,102,168,177,221	
	demonstrated in granulosa & variably in	Steroidogenesis associated with granulosa & theca interna		Four histologically differentiated tissues capable of endocrine steroidogenesis: (i) granulosa (epithelial layer of follicle) intimately associated with ovum; steroidogenesis during late phases of oocyte maturation as part of follicle, and in postovulatory state as part of corpus luteum; (ii) theca interna—a luteinized portion of ovarian stroma which condenses around developing oocyte and forms part of corpus luteum in postovulatory period (as	14
	Corpus luteum-like structures develop both in postovulatory follicle and as corpora atretica; efforts to demonstrate role for these structures (analogous to their role in Eutheria) in viviparous toads has not been entirely convincing, despite evidence they may contain 3-β-hydroxysteroid dehydrogenase activity; e.g., in the microlecithal *Nectophrynoides,* gestation is associated with development & persistence of corpora lutea which exhibit dehydrogenase	Corpora lutea formed, but not clear that they have functional significance, even in ovoviviparous forms latory follicles retained, probably steroidogenic, but functional significance not clearly defined	Usually, only left ovary develops to functional state; testicular elements in undeveloped right ovary; postovu-		

continued

Gland or Tissue	Agnatha	Chondrichthyes	Actinopterygii [1/]	Crossopterygii [2/]
	Gen. Ref.: 28,113	Gen. Ref.: 46,102,156, 210	Gen. Ref.: 102,155,265	Gen. Ref.: 196
15 Urophysis (Urohypophysis; caudal neurosecretory system)	Absence indicated in preliminary studies	Variably developed neurohemal organ, reminiscent of neurohypophysis in organization, on ventral surface of caudal-most spinal cord (urophysis as such found only in Teleostei); axons of neurosecretory neurons, located in caudal spinal cord, abut on perivascular space (associated with glial elements); small, clear vesicles & large, dense-cored vesicles observed in nerve terminals; suggested [148] that pharmacology of at least one of neurosecretory products extracted is indistinguishable from arginine vasotocin (at least two other urotensins have been demonstrated); venous drainage is through renal portal system via caudal vein [20,21,86,87,102,120,248]		....
		Elasmobranchii: Present as Dahlgren cells [102]	*Polypterus, Lepisosteus:* Has been described	
		Holocephali: Absence indicated in preliminary studies	Teleostei: Forms discrete bulbous mass ventral to spinal cord just rostral to filum terminale	

[1/] According to Romer [245]. [2/] Only surviving species, *Latimeria chalumnae.*

Contributor: Lagios, Michael D.

References

[1] Adam, H. 1963. In H. Brodal and R. Fange, ed. The Biology of Myxine. Universitetsforlaget, Oslo, Norway. pp. 459-476.

[2] Adams, J. H., et al. 1964. Endocrinology 75:120.

[3] Adams, J. H., et al. 1965. Neuroendocrinology 1: 193.

[4] Akmayev, I. G. 1971. Z. Zellforsch. Mikrosk. Anat. 116:178, 195.

[5] Baecker, R. 1928. Z. Mikrosk. Anat. Forsch. 15:204.

[6] Ball, J. N., and B. I. Baker. 1969. In W. S. Hoar and D. J. Randall, ed. Fish Physiology. Academic Press, New York. v. 2, pp. 1-110.

[7] Ball, J. N., and M. Olivereau. 1966. Gen. Comp. Endocrinol. 6:5.

[8] Barajas, L. 1970. J. Ultrastruct. Res. 33:116.

[9] Barannikova, I. A. 1949. Dokl. Akad. Nauk SSSR 69:117.

[10] Barnes, B. G. 1963. Colloq. Int. Cent. Nat. Rech. Sci. 128:91.

[11] Barrington, E. J. W. 1942. J. Exp. Biol. 19:45.

[12] Barrington, E. J. W. 1945. Quart. J. Microsc. Sci. 85:391.

[13] Barrington, E. J. W. 1951. Ibid. 92:205.

[14] Barrington, E. J. W. 1953. Ibid. 94:281.

[15] Barrington, E. J. W. 1963. Introduction to General and Comparative Endocrinology. Clarendon Press, Oxford.

[16] Bell, J. B. G., et al. 1971. Proc. Roy. Soc. B176:433.

[17] Bellocci, M., et al. 1971. Z. Zellforsch. Mikrosk. Anat. 114:203.

[18] Benoit, J. 1950. Traite Zool. 15:290.

[19] Bergland, R. M., and R. M. Torack. 1969. Amer. J. Pathol. 57:273.

[20] Bern, H. A. 1969. In W. S. Hoar and D. J. Randall, ed. Fish Physiology. Academic Press, New York. v. 2, pp. 399-418.

[21] Bern, H. A., and N. Takasugi. 1962. Gen. Comp. Endocrinol. 2:96.

[22] Blair, E. L., et al. 1969. Acta Pathol. Microbiol. Scand. 75:583.

| Dipnoi | Tetrapoda | | | |
	Amphibia	Reptilia	Aves	Mammalia
	activity specific for conversion of pregnenolone to progesterone and become atretic at time of parturition; in apparent contradiction to this association, gestation is not interrupted by castration **Gen. Ref.:** 93,102,283,284	**Gen. Ref.:** 102	theca-lutein cells); (iii) ovarian stroma apparently undifferentiated; large capacity for steroidogenesis; may be induced to functional activity, either normally as part of theca, or in certain abnormal states; (iv) hilar cells—considered homologous to interstitial cells of testis; regarded as rests in human ovarian tissue **Gen. Ref.:** 102 **Gen. Ref.:** 102,168,169,221	
Absence indicated in preliminary studies [21,86]	....	....	....	 15

[23] Bohle, A., and F. Walvig. 1964. Klin. Wochenschr. 42:415.

[24] Brasch, M., and T. W. Betz. 1971. Gen. Comp. Endocrinol. 16:241.

[25] Bulger, R. E., and B. F. Trump. 1969. Amer. J. Anat. 124:77.

[26] Burlet, A., and H. Legait. 1967. C. R. Ass. Anat. 137:286.

[27] Bussolati, G., and A. G. E. Pearse. 1967. J. Endocrinol. 37:205.

[28] Busson-Mabillot, S. 1967. J. Microsc. (Paris) 6:807.

[29] Butler, D. G., and W. H. Knox. 1970. Gen. Comp. Endocrinol. 14:551.

[30] Butler, D. G., et al. 1969. Ibid. 12:503.

[31] Camp, W. E. 1917. J. Morphol. 28:369.

[32] Capella, C., et al. 1969. Nippon Soshikigaku Kiroku 30:479.

[33] Capen, C. C., and D. M. Young. 1967. Lab. Invest. 17:717.

[34] Capen, C. C., and D. M. Young. 1969. Amer. J. Pathol. 57:365.

[35] Capreol, S. V., and L. E. Sutherland. 1968. Can. J. Zool. 46:249.

[36] Castleman, B. 1952. Armed Forces Inst. Pathol. Fasc. 15.

[37] Caswell, M. W. 1970. Brit. J. Exp. Pathol. 51:197.

[38] Chan, A. S., et al. 1969. Rev. Can. Biol. 28:19.

[39] Chan, D. K. O. 1972. Gen. Comp. Endocrinol., Suppl. 3:411.

[40] Charipper, H. A. 1929. Anat. Rec. 44:117.

[41] Chavin, W., and J. E. Young. 1970. Gen. Comp. Endocrinol. 14:438.

[42] Chester Jones, I. 1957. The Adrenal Cortex. Cambridge Univ. Press, London.

[43] Chester Jones, I., et al. 1964. J. Endocrinol. 29:155.

[44] Chester Jones, I., et al. 1966. Ibid. 34:393.

[45] Chester Jones, I., et al. 1969. In W. S. Hoar and D. J. Randall, ed. Fish Physiology. Academic Press, New York. v. 2, pp. 321-376.

continued

[46] Chieffi, G. 1962. Gen. Comp. Endocrinol., Suppl. 1:275.

[47] Clark, N. B. 1965. Ibid. 5:297.

[48] Clark, N. B. 1968. Ibid. 10:99.

[49] Clark, N. B. 1970. In C. Gans and T. S. Parsons, ed. The Biology of the Reptilia. Academic Press, New York. v. 3, pp. 235-262.

[50] Colombo, L., et al. 1971. Gen. Comp. Endocrinol. 16:74.

[51] Connei, G. M., and G. Kaley. 1964. Biol. Bull. 127:366.

[52] Copp, D. H. 1969. In W. S. Hoar and D. J. Randall, ed. Fish Physiology. Academic Press, New York. v. 2, pp. 377-398.

[53] Copp, D. H., et al. 1967. Can. J. Physiol. Pharmacol. 45:1095.

[54] Copp, D. H., et al. 1968. Calcif. Tissue Res. 2 (Suppl.):29.

[55] Daniel, J. F. 1934. The Elasmobranch Fishes. Univ. California Press, Berkeley.

[56] De Roos, R., and C. De Roos. 1967. Gen. Comp. Endocrinol. 9:267.

[57] Des Marchais, J., and P. Jean. 1970. Union Med. Can. 99:290.

[58] De Smet, W. 1962. Acta Zool. (Stockholm) 43:201.

[59] Deville, J., and E. Lopez. 1970. Arch. Anat. Microsc. Morphol. Exp. 59:393.

[60] Dierickx, K. 1966. Z. Zellforsch. Mikrosk. Anat. 74:53.

[61] Dix, M. W., et al. 1970. Gen. Comp. Endocrinol. 14:243.

[62] Dodd, J. M., and T. Kerr. 1963. Symp. Zool. Soc. London 9:5.

[63] Dodd, J. M., et al. 1960. Ibid. 1:77.

[64] Doerr-Schott, J. 1970. Z. Zellforsch. Mikrosk. Anat. 111:413.

[65] Doerr-Schott, J. 1971. Gen. Comp. Endocrinol. 16:516.

[66] Doerr-Schott, J., and E. Follenius. 1970. Z. Zellforsch. Mikrosk. Anat. 106:99, 111:427.

[67] Dongen, W. J. van, and C. A. van der Heijden. 1969. Ibid. 94:40.

[68] Edwards, J. G. 1940. Anat. Rec. 76:381.

[69] Eggert, B. 1938. Endokrinologie 20:1.

[70] Emmart, E. W. 1969. Gen. Comp. Endocrinol. 12:519.

[71] Emmart, E. W., et al. 1966. Ibid. 7:571.

[72] Epple, A. 1968. Endocrinol. Jap. 15:107.

[73] Epple, A. 1969. In W. S. Hoar and D. J. Randall, ed. Fish Physiology. Academic Press, New York. v. 2, pp. 275-319.

[74] Esterhuizen, A. C., et al. 1968. Diabetes 17:33.

[75] Etkin, W. 1965. J. Morphol. 116:371.

[76] Fagerlund, U. H. M., and J. R. McBride. 1969. Gen. Comp. Endocrinol. 12:651.

[77] Falkmer, S., and L. Winbladh. 1964. Struct. Metab. Pancreatic Islets Proc. Int. Wenner-Gren Symp., 3rd, 1963, p. 381.

[78] Fernholm, B. 1969. Acta Zool. (Stockholm) 50:169.

[79] Fernholm, B., and R. Olsson. 1969. Gen. Comp. Endocrinol. 13:336.

[80] Follenius, E. 1965. Arch. Anat. Microsc. Morphol. Exp. 54:195.

[81] Follenius, E. 1968. Gen. Comp. Endocrinol. 11:198.

[82] Follenius, E. 1970. C. R. Acad. Sci. 271:1034.

[83] Fontaine, M., et al. 1952. Ann. Endocrinol. 13:55.

[84] Forsmann, W. G., and L. Orci. 1969. Z. Zellforsch. Mikrosk. Anat. 101:419.

[85] Francis, E. T. B. 1934. The Anatomy of the Salamander. Clarendon Press, Oxford.

[86] Fridberg, G. 1962. Gen. Comp. Endocrinol. 2:249.

[87] Fridberg, G., and H. A. Bern. 1968. Biol. Rev. Cambridge Phil. Soc. 43:175.

[88] Fujita, H., and Y. Honma. 1966. Z. Zellforsch. Mikrosk. Anat. 73:559.

[89] Fujita, T. 1968. Nippon Soshikigaku Kiroku 29:1.

[90] Gabe, M. 1969. Arch. Anat. Microsc. Morphol. Exp. 58:21.

[91] Gabe, M. 1969. Arch. Biol. 80:71.

[92] Gabe, M. 1970. In C. Gans and T. S. Parsons, ed. The Biology of the Reptilia. Academic Press, New York. v. 3, pp. 263-318.

[93] Gallien, L. 1959. Comp. Endocrinol. Proc. Columbia Univ. Symp. Cold Spring Harbor, N.Y., 1958, p. 479.

[94] Garrett, F. D. 1942. J. Morphol. 70:41.

[95] Gaskell, J. F. 1912. J. Physiol. (London) 44:59.

[96] Giacomini, E. 1902. Monit. Zool. Ital. 13:143.

[97] Giacomini, E. 1912. Rend. Accad. Sci. Ist. Bologna 16:77.

[98] Giacomini, E. 1936. Boll. Soc. Ital. Biol. Sper. 11:1012.

[99] Godwin, M. C. 1937. Amer. J. Anat. 60:299.

[100] Gorbman, A. 1959. Comp. Endocrinol. Proc. Columbia Univ. Symp. Cold Spring Harbor, N.Y., 1958, p. 266.

[101] Gorbman, A. 1963. In A. Brodal and R. Fange, ed. The Biology of Myxine. Universitetsforlaget, Oslo, Norway. pp. 477-480.

[102] Gorbman, A., and H. A. Bern. 1962. A Textbook of Comparative Endocrinology. J. Wiley, New York.

[103] Gorbman, A., et al. 1963. Gen. Comp. Endocrinol. 3:505.

[104] Green, J. D. 1947. Anat. Rec. 99:21.

[105] Green, J. C. 1951. Amer. J. Anat. 88:225.

[106] Greep, R. O. 1948. Hormones 1:255.

[107] Greider, M. H., and J. E. McGuigan. 1971. Diabetes 20:389.

[108] Grossner, D. 1967. Z. Zellforsch. Mikrosk. Anat. 82:82.

[109] Grossner, D. 1968. Ibid. 84:417.

[110] Hansen, G. N. 1971. Kgl. Dan. Vidensk. Selsk. Biol. Skr. 18(1):1.

[111] Hänstrom, B. 1966. In G. W. Harris and B. T. Donovan, ed. The Pituitary Gland. Univ. California Press, Berkeley. v. 1, pp. 1-57.

[112] Hardisty, M. W., and M. E. Baines. 1971. Experientia 27:1072.

[113] Hardisty, M. W., and K. Barnes. 1968. Nature (London) 218:880.

[114] Harris, G. W. 1955. Neural Control of the Pituitary Gland. E. Arnold, London.

[115] Hartman, F. A., and K. A. Brownell. 1949. The Adrenal Gland. Lea and Febiger, Philadelphia.

[116] Hartroft, P. M. 1966. Fed. Proc. Fed. Amer. Soc. Exp. Biol. 25:238.

[117] Hayashida, T., and M. D. Lagios. 1969. Gen. Comp. Endocrinol. 13:403.

[118] Henderson, N. E., and A. Gorbman. 1971. Ibid. 16:409.

[119] Hill, J. J., and N. E. Henderson. 1968. Amer. J. Anat. 122:301.

[120] Holmgren, U. 1959. Anat. Rec. 135:51.

[121] Honma, S., and Y. Honma. 1970. Nippon Suisan Gakkaishi 36:125.

[122] Honma, Y., and E. Tamura. 1967. Ibid. 33:303.

[123] Honma, Y., and E. Tamura. 1968. Ibid. 34:555.

[124] Hopkins, C. R. 1971. Gen. Comp. Endocrinol. 16:112.

[125] Idler, D. R., and M. J. O'Halloran. 1970. J. Endocrinol. 48:621.

[126] Idler, D. R., and B. Szeplaki. 1968. J. Fish. Res. Bd. Can. 25:2549.

[127] Idler, D. R., et al. 1969. Gen. Comp. Endocrinol. 13:303.

[128] Janssens, P. A., et al. 1965. J. Endocrinol. 32:373.

[129] Jasinski, A. 1961. Acta Biol. Cracov. Ser. Zool. 4:79.

[130] Jasinski, A., and A. Gorbman. 1966. Gen. Comp. Endocrinol. 6:476.

[131] Jensen, J., and N. E. Henderson. 1968. Amer. J. Anat. 122:301.

[132] Kamer, J. C. van de, and A. F. Schreurs. 1959. Z. Zellforsch. Mikrosk. Anat. 49:605.

[133] Kemenade, J. A. M. van. 1969. Ibid. 96:466.

[134] Kerr, T. 1949. Proc. Zool. Soc. London 118:973.

[135] Kerr, T. 1965. Gen. Comp. Endocrinol. 5:232.

[136] Kerr, T. 1966. Ibid. 6:303.

[137] Kerr, T. 1968. J. Morphol. 124:23.

[138] Kerr, T., and P. G. W. J. van Oordt. 1966. Gen. Comp. Endocrinol. 7:549.

[139] Kjaerheim, Å. 1968. Z. Zellforsch. Mikrosk. Anat. 91:426, 456.

[140] Knigge, K. M., and D. E. Scott. 1970. Amer. J. Anat. 129:223.

[141] Knowles, F., and L. Vollrath. 1966. Phil. Trans. Roy. Soc. London B250:311, 329.

[142] Kobayashi, H., et al. 1966. Z. Zellforsch. Mikrosk. Anat. 71:387.

[143] Kobayashi, K. 1969. Gunma J. Med. Sci. 17:60.

[144] Kobayashi, K., and Y. Takahashi. 1970. Nippon Soshikigaku Kiroku 31:433.

[145] Kobayashi, S., and T. Fujita. 1969. Z. Zellforsch. Mikrosk. Anat. 100:340.

[146] Kobayashi, S., et al. 1970. Nippon Soshikigaku Kiroku 31:477.

[147] Krishnamurthy, V. G., and H. A. Bern. 1969. Gen. Comp. Endocrinol. 13:313.

[148] Lacanilao, F. 1972. Ibid. 19:413.

[149] Lagios, M. D. 1968. Copeia, p. 401.

[150] Lagios, M. D. 1968. Gen. Comp. Endocrinol. 11:248.

[151] Lagios, M. D. 1968. Ibid. 11:300.

[152] Lagios, M. D. 1970. Ibid. 15:453.

[153] Lagios, M. D. 1972. Ibid. 18:73.

[154] Lagios, M. D. Unpublished. Children's Hospital of San Francisco, Calif., 1972.

[155] Lambert, J. G. D. 1970. Gen. Comp. Endocrinol. 15:464.

[156] Lance, V., and I. P. Callard. 1969. Ibid. 13:255.

[157] Lange, R. 1967. Z. Zellforsch. Mikrosk. Anat. 82:156.

[158] Larsen, L. O., et al. 1971. Gen. Comp. Endocrinol. 15:165.

[159] Leatherland, J. F. 1969. Z. Zellforsch. Mikrosk. Anat. 98:122.

[160] Leatherland, J. F., and J. M. Dodd. 1969. Gen. Comp. Endocrinol. 13:45.

[161] Leatherland, J. F., et al. 1966. Ibid. 7:234.

[162] Legait, H., and J. L. Contet. 1969. C. R. Soc. Biol. 163:489.

[163] Legait, H., and E. Legait. 1969. C. R. Ass. Anat. 142:1153.

[164] Leiner, M., and D. Schmidt. 1956. Photogr. Forsch. 7:129.

[165] Licht, P., and C. S. Nicoll. 1969. Gen. Comp. Endocrinol. 12:526.

[166] Like, A. A. 1967. Lab. Invest. 16:937.

[167] Lofts, B. 1964. Gen. Comp. Endocrinol. 4:550.

continued

[168] Lofts, B. 1968. In E. J. W. Barrington and C. B. Jørgensen, ed. Perspectives in Endocrinology. Academic Press, New York. pp. 239-304.

[169] Lofts, B., and L. Y. L. Choy. 1971. Gen. Comp. Endocrinol. 17(3):588.

[170] Lofts, B., and A. J. Marshall. 1957. Quart. J. Microsc. Sci. 98:79.

[171] Lofts, B., et al. 1966. Gen. Comp. Endocrinol. 6: 74.

[172] Lofts, B., et al. 1966. Ibid. 6:466.

[173] Lofts, B., et al. 1971. Ibid. 16:121.

[174] Lomsky, R., et al. 1969. Nature (London) 223: 618.

[175] Lopez, E. 1969. Gen. Comp. Endocrinol. 12:339.

[176] Lynn, G. 1970. In C. Gans and T. S. Parsons, ed. The Biology of the Reptilia. Academic Press, New York. v. 3, pp. 201-234.

[177] Markwald, R. R., et al. 1971. Gen. Comp. Endocrinol. 16:268.

[178] Marshall, A. J., and B. Lofts. 1956. Nature (London) 177:704.

[179] Marshall, A. J., and B. Lofts. 1957. Quart. J. Microsc. Sci. 98:79.

[180] Mattheij, J. A. M. 1970. Z. Zellforsch. Mikrosk. Anat. 105:91.

[181] Matthews, S. A. 1948. Anat. Rec. 101:251.

[182] McKeown, B. A., and A. P. van Overbeeke. 1971. Z. Zellforsch. Mikrosk. Anat. 112:350.

[183] Mellinger, J. C. A. 1960. Bull. Soc. Zool. Fr. 85: 123.

[184] Mellinger, J. C. A. 1962. C. R. Acad. Sci. 255:2294.

[185] Mellinger, J. C. A., et al. 1962. Ibid. 254:1158.

[186] Meurling, P. 1960. Nature (London) 187:336.

[187] Meurling, P. 1967. Sarsia 28:1, 30:83.

[188] Meurling, P., and A. Björklund. 1970. Z. Zellforsch. Mikrosk. Anat. 108:81.

[189] Meurling, P., and A. Willstedt. 1970. Acta Zool. (Stockholm) 51:211.

[190] Meyer, D., et al. 1967. Z. Zellforsch. Mikrosk. Anat. 83:508.

[191] Mikami, S.-I., et al. 1970. Ibid. 106:155.

[192] Miller, M. R. 1962. Gen. Comp. Endocrinol. 2:407.

[193] Miller, M. R. 1963. Ibid. 3:579.

[194] Miller, M. R., and M. D. Lagios. 1970. In C. Gans and T. S. Parsons, ed. The Biology of the Reptilia. Academic Press, New York. v. 3, pp. 319-349.

[195] Millot, J., and J. Anthony. 1955. C. R. Acad. Sci. 241:114.

[196] Millot, J., and J. Anthony. 1965. Anatomie de *Latimeria chalumnae*. Centre Nationale de la Recherche Scientifique, Paris. v. 1-2.

[197] Morris, R., and D. S. Islam. 1969. Gen. Comp. Endocrinol. 12:72.

[198] Mosely, J. M., et al. 1968. Lancet 1:108.

[199] Nakai, Y., and A. Gorbman. 1969. Gen. Comp. Endocrinol. 13:108.

[200] Nandi, J. 1962. Univ. Calif. Berkeley Publ. Zool. 65:129.

[201] Nevalainen, T. 1969. Gen. Comp. Endocrinol. 12: 561.

[202] Ogawa, M., et al. 1972. Ibid., Suppl. 3:374.

[203] Nishioka, R. S., and H. A. Bern. 1966. Ibid. 7:457.

[204] Norris, H. W. 1941. The Plagiostome Hypophysis. Science Press, Grinnell, Iowa.

[205] Ogawa, M. 1967. Z. Zellforsch. Mkirosk. Anat. 81: 174.

[206] Ogawa, M., et al. 1972. Gen. Comp. Endocrinol., Suppl. 3:374.

[207] Oguri, M., et al. 1970. Nippon Suisan Gakkaishi 36:881.

[208] Oguro, C. 1969. Annot. Zool. Jap. 42:21.

[209] Oguro, C. 1970. Gen. Comp. Endocrinol. 15:313.

[210] O'Halloran, M. J., and D. R. Idler. 1970. Ibid. 15: 361.

[211] Okkels, H. 1929. C. R. Acad. Sci. 188:193.

[212] Olivereau, M. 1963. Colloq. Int. Cent. Nat. Rech. Sci. 128:315.

[213] Olivereau, M., and J. N. Ball. 1964. Gen. Comp. Endocrinol. 4:523.

[214] Olsson, R. 1959. Z. Zellforsch. Mikrosk. Anat. 51: 97.

[215] Oordt, P. G. W. J. van. 1963. Colloq. Int. Cent. Nat. Rech. Sci. 128:301.

[216] Oordt, P. G. W. J. van. 1968. In E. J. W. Barrington and C. B. Jørgensen, ed. Perspectives in Endocrinology. Academic Press, New York. pp. 405-467.

[217] Oordt, P. G. W. J. van, and E. J. M. de Kort. 1969. Colloq. Int. Cent. Nat. Rech. Sci. 177:345.

[218] Oota, Y., and H. Kobayashi. 1963. Z. Zellforsch. Mikrosk. Anat. 60:667.

[219] Oota, Y., and K. Kurosumi. 1966. Nippon Soshikigaku Kiroku 27:501.

[220] Ostberg, H., et al. 1966. Gen. Comp. Endocrinol. 7:475.

[221] Parkes, A. S., ed. 1956-66. Marshall's Physiology of Reproduction. Ed. 3. Little, Brown; Boston. v. 1-3.

[222] Patent, G. J., and A. Epple. 1967. Gen. Comp. Endocrinol. 9:325.

[223] Pearse, A. G. E. 1966. Proc. Roy. Soc. B164:475.

[224] Pearse, A. G. E. 1970. Calcitonin Proc. 2nd Int. Symp., 1969, p. 125.

[225] Pehlemann, F.-W., and W. Hanke. 1968. Z. Zellforsch. Mikrosk. Anat. 89:201.

[226] Perks, A. M. 1969. In W. S. Hoar and D. J. Randall, ed. Fish Physiology. Academic Press, New York. v. 2, pp. 111-205.

[227] Picheral, B. 1968. J. Microsc. (Paris) 7:115.

[228] Pickford, G. E., and J. W. Atz. 1957. The Physiology of the Pituitary Gland of Fishes. New York Zoological Society, New York.

[229] Piezzi, R. S. 1967. Gen. Comp. Endocrinol. 9:143.

[230] Piezzi, R. S., and M. H. Burgos. 1968. Ibid. 10:344.

[231] Planas, J., et al. 1968. Acta Anat. 69:520.

[232] Polenov, A. L. 1966. Dokl. Akad. Nauk SSSR 169:1467.

[233] Polenov, A. L. 1968. Arch. Anat. Histol. Embryol. 51:553.

[234] Rasquin, P. 1956. Biol. Bull. 111:399.

[235] Rasquin, P., and L. Rosenbloom. 1954. Bull. Amer. Mus. Natur. Hist. 104:359.

[236] Rhoten, W. B. 1970. Anat. Rec. 167:401.

[237] Rizkalla, W. 1969. Acta Vet. (Budapest) 19:1.

[238] Robertson, D. R. 1965. Z. Zellforsch. Mikrosk. Anat. 67:584.

[239] Robertson, D. R. 1968. Ibid. 85:453, 90:273.

[240] Robertson, D. R. 1969. Gen. Comp. Endocrinol. 12:479.

[241] Robertson, D. R. 1969. J. Anat. 105:115.

[242] Robertson, D. R. 1971. Gen. Comp. Endocrinol. 16:329.

[243] Robertson, O. H. 1958. U.S. Fish Wildl. Serv. Fish. Bull. 58:9.

[244] Rodriguez, E. M., and J. La Pointe. 1969. Z. Zellforsch. Mikrosk. Anat. 95:37.

[245] Romer, A. S. 1966. Vertebrate Paleontology. Ed. 3. Univ. Chicago Press, Chicago.

[246] St. Girons, H. 1963. Colloq. Int. Cent. Nat. Rech. Sci. 128:275.

[247] St. Girons, H. 1970. In C. Gans and T. S. Parsons, ed. The Biology of the Reptilia. Academic Press, New York. v. 3, pp. 73-91.

[248] Sano, Y. 1961. Ergeb. Biol. 24:191.

[249] Sathyanesan, A. G. 1965. J. Morphol. 116:413.

[250] Sathyanesan, A. G., and S. Haider. 1971. Gen. Comp. Endocrinol. 17:360.

[251] Sato, T., et al. 1966. Ibid. 7:132.

[252] Schechter, J. 1971. Ibid. 16:1.

[253] Sehe, C. T. 1965. Ibid. 5:45.

[254] Sokabe, H., et al. 1969. Tex. Rep. Biol. Med. 27:867.

[255] Sokabe, H., et al. 1970. Gen. Comp. Endocrinol. 13:313.

[256] Stahl, A. 1963. Colloq. Int. Cent. Nat. Rech. Sci. 128:331.

[257] Stoeckel, M. E., and A. Porte. 1970. Z. Zellforsch. Mikrosk. Anat. 106:251.

[258] Stone, L. S., and H. Steinitz. 1953. J. Exp. Zool. 124:469.

[259] Sucheston, M. E., and M. S. Cannon. 1968. Gen. Comp. Endocrinol. 11:603.

[260] Sundararaj, B. I., and N. Viswanathan. 1971. J. Comp. Neurol. 141:95.

[261] Tan-Tue, V. 1969. Arch. Anat. Microsc. Morphol. Exp. 59:21.

[262] Thomas, T. B. 1940. Anat. Rec. 76:1.

[263] Tomasulo, J. A., et al. 1970. Amer. J. Anat. 129:307.

[264] Tong, W., et al. 1961. Biochim. Biophys. Acta 52:299.

[265] Ulrich, E. 1969. J. Microsc. (Paris) 8:447.

[266] Vollrath, L. 1967. Z. Zellforsch. Mikrosk. Anat. 78:234.

[267] Watari, N., et al. 1970. Nippon Soshikigaku Kiroku 31:371.

[268] Waterman, A. J., and A. Gorbman. 1963. Gen. Comp. Endocrinol. 3:58.

[269] Watzka, M. 1933. Z. Mikrosk. Anat. Forsch. 34:485.

[270] Weatherhead, B. 1971. Z. Zellforsch. Mikrosk. Anat. 119:21.

[271] Weibe, J. P. 1969. Gen. Comp. Endocrinol. 12:256.

[272] Wilson, F. E. 1967. Z. Zellforsch. Mikrosk. Anat. 82:1.

[273] Winbladh, L. 1966. Gen. Comp. Encodrinol. 6:534.

[274] Winbladh, L. 1966. Ibid. 9:505.

[275] Wingstrand, K. G. 1951. The Structure and Development of the Avian Pituitary. C. W. K. Gleerup, Lund, Sweden.

[276] Wingstrand, K. G. 1956. Dan. Naturhist. Foren. Videnskab. Medd. 118:193.

[277] Wingstrand, K. G. 1963. Colloq. Int. Cent. Rech. Sci. 128:243.

[278] Wingstrand, K. G. 1966. In G. W. Harris and B. T. Donovan, ed. The Pituitary Gland. Univ. California Press, Berkeley. v. 1, pp. 58-126.

[279] Witschi, E. 1949. Z. Naturforsch. 4b:230.

[280] Wittkowski, W. 1967. Z. Zellforsch. Mikrosk. Anat. 82:434.

[281] Woodhead, A. D. 1966. J. Zool. 148:238.

[282] Woodhead, A. D., and S. Ellett. 1966. Exp. Gerontol. 1:315.

[283] Xavier, F. 1970. C. R. Acad. Sci. 270(D):2018.

[284] Xavier, F., and R. Ozon. 1971. Gen. Comp. Endocrinol. 13:255.

[285] Young, B. A., et al. 1965. J. Endocrinol. 31:279, 35:101.

[286] Zambrano, D. 1970. Z. Zellforsch. Mikrosk. Anat. 110:2, 496.

[287] Zambrano, D. 1972. Gen. Comp. Endocrinol., Suppl. 3:22.

[288] Zambrano, D. and E. de Robertis. 1968. Z. Zellforsch. Mikrosk. Anat. 90:230.

[289] Chavin, W. 1972. Nature (London) 239:340.

[290] Mellinger, J. 1972. Gen. Comp. Endocrinol. 18:608.

Abbreviations: AcSer = Acetylseryl; (pyro)Glu = pyroglutamyl; 3-iodoTyr = 3-iodotyrosyl or 3-iodotyrosine; diiodoTyr = diiodotyrosyl or diiodotyrosine; mol wt = molecular weight; mp = melting point; pI = isoelectric point; $s_{20,w}$ = sedimentation coefficient corrected to water at 20°C; D_{20} = diffusion coefficient at 20°C; $[\alpha]_D$ ($[\alpha]_D^{20}$, $[\alpha]_D^{25}$, $[\alpha]_D^{29.50}$) = specific optical rotation based on a sodium light wavelength (D) of 589 nanometers (at 20°C, 25°C, 29.50°C); absorption max = wavelength(s) of maximal absorbancy; insol. = insoluble; sol. = soluble; sl. = slightly; BMR = basal meta-

	Hormone [Synonym]	Structure & Properties	Sources	Assay Methods	Metabolites
			Hypothalamus		
1	Corticotropin-releasing hormone [Adrenocorticotropin-releasing hormone; CRH or factor: CRF; at least 2 structures, α & β]	Basic peptide, with NH_2-terminal structure like that of α-MSH (entry 16), and COOH-terminal structure like that of lysine vasopressin; tentative structure of β-CRH—AcSer·Tyr·Cys·Phe·His·Asn·Gln·Pro·Val·Lys·Gly·NH_2 Inactivated by oxidation, reduction, peptic & tryptic proteolysis, disulfide linkages	Hypothalamus & neurohypophysis of man, cattle, dog, rabbit, rat, sheep, & swine	In vivo release of ACTH (entry 9) (i) into peripheral circulation by injection of sample into rats pretreated with cortisone monoacetate, dexamethasone (entry 38), or chlorpromazine, morphine, & pentobarbital Na; (ii) measured by ascorbic acid depletion from suprarenal cortex in rats with stereotaxic hypothalmic lesions In vitro release of ACTH into medium from one-half of adenohypophysis, as compared to release from other untreated half	Unknown
2	Follicle-stimulating hormone—releasing hormone [Follicle-stimulating hormone—releasing factor; FRH; FSH-RH; FSH-RF]	Amino acid sequence: (pyro)Glu·His·Trp·Ser·Tyr·Gly·Leu·Arg·Pro·Gly·NH_2 sol. in 0.1 N HCl; stable to heat	Ventral hypothalamus of man, cattle, rat, sheep, & swine	Increase in plasma FSH (entry 10) in oophorectomized rats pretreated with estrogen & progesterone (entry 46); depletion of hypophyseal FSH in ♂ rats, normal or castrated In vitro release of FSH from rat adenohypophysis in vivo or in vitro	Unknown
3	Growth hormone—releasing hormone [Growth hormone—releasing factor; somatotropin-releasing hormone; GRH; GRF; SRH; SRF]	Probably a polypeptide; acidic Mol wt = 2500; stable in boiling 0.1 N HCl; inactivated by proteolysis; not inactivated by thioglycollate	Hypothalamus of man, cat, cattle, dog, guinea pig, rat, & swine	Elevation of plasma GH (entry 11) in monkey In vivo or in vitro depletion of GH from rat hypophysis	Unknown
4	Luteinizing hormone—releasing hormone [Luteinizing hormone—releasing factor; LH-RH; LRH; LH-RF; LRF]	Amino acid sequence same as for entry 2 Stable in boiling 0.1 N HCl	Hypothalamus of man, cattle, dog, monkey, rabbit, rat, sheep, & swine	Increase in plasma ICSH (entry 12) in castrated ♂ rats pretreated with testosterone (entry 41), or oophorectomized ♀ rats pretreated with estradiol (entry 43), & progesterone (entry 46); depletion of ovarian ascorbic acid in pseudopregnant rats In vitro release of LH (entry 12) from adenohypophysis from oophorectomized ♀ rats pretreated with estrogen & progesterone	Unknown
5	Thyrotropin-releasing hormone [Thyrotropin-releasing factor; thyroid-stimulating hormone—releasing hormone; TRH; TRF]	Bovine, ovine, porcine sources: L-(pyro)Glu·His·Pro·NH_2 (synthetic tripeptide has activity) sol. in dilute acetic acid; destroyed by boiling in 6 N HCl; not inactivated by proteolysis	Hypothalamus & neurohypophysis of man, cattle, goat, rabbit, rat, sheep, & swine	Release of ^{131}I in iodine-deficient mice pretreated with codeine & T_4 (entry 22) In vitro release of TSH (entry 14) from incubated rat adenohypophysis (0.01 ng of TRH releases 200-2000 ng TSH)	Unknown

bolic rate; CNS = central nervous system; DNA = deoxyribonucleic acid; RNA = ribonucleic acid; mRNA = messenger ribonucleic acid; NADH = nicotinamide adenine dinucleotide (reduced form); NADPH = nicotinamide adenine dinucleotide phosphate (reduced form); ATP = adenosine 5'-triphosphate; cAMP = cyclic adenosine 3',5'-monophosphate; P_i = inorganic phosphate.

Targets	Principal Functions & Effects	Effects of Deficiency (−) & Excess (+)	Inhibitors (I) & Stimulators (S) of Secretion	Reference	
Hypothalamus					
ACTH-secreting cells of adeno-hypophysis	Stimulates: Release of ACTH from basophils of adenohypophysis into blood; also increases rate of synthesis of ACTH by basophilic cells	(−): Insufficient release of ACTH into circulation (+): Not definitely known; presumably excessive secretion of ACTH and therefore of glucocorticoids	(I): High blood levels of corticosteroids (long-term) & of ACTH (short-term)—negative feedback from suprarenal cortex normally controls secretion of CRH; final target organ probably more important (S): Neural impulses from CNS; neurosecretory cells function as transducers; low blood levels of corticosteroids & ACTH; cyclic nyctohemeral control through hypothalamus; stress	4,10,12, 18,24, 30,31, 42,49, 52,57, 60	1
FSH-LH-secreting basophils (delta cells) of adenohypophysis	Stimulates: Release of FSH & LH into peripheral blood; ovulation	(−): Excessive storage & decreased release of FSH in cells of hypophysis; absence of ovulation (+): Stimulation of ovulation (more frequent or multiple) through increased release of FSH into blood	(I): High levels of FSH or of estrogens in blood (negative feedback) (S): Neural stimulation from CNS; constant exposure to light; low levels of FSH or of estrogens; dopamine	4,10,12, 18,24, 30,31, 42,49, 52,53, 57,60	2
GH-synthesizing acidophils (alpha cells) of adenohypophysis	Stimulates: Release of GH from adenohypophysis into blood	(−): Decreased release of GH; consequent stunting of growth (+): Excessive release of GH (self-limiting)	(I): High blood levels of GH (negative feedback) (S): CNS stimulation; decreased blood levels of GH (depletion of GH in the hypophysis)	4,10,12, 18,24, 30,31, 42,49, 52,53, 57,60	3
ICSH- (LH-) secreting basophils of adenohypophysis	Stimulates: Release of ICSH (LH) into blood Decreases: Ovarian ascorbic acid	(−): Presumably, decreased release of LH from adenohypophysis, with consequent decreased levels of estrogen & progesterone (+): Excessive release of LH (to a maximum); premature ovulation	(I): High levels of LH & of estrogens or progesterone in blood (negative feedback); suckling stimulus (S): Neural stimulation from CNS; proestrus, puberty; exposure to light	4,10,12, 18,20, 24,30, 31,42, 49,52, 53,57, 60	4
TSH-secreting cells of adeno-hypophysis	Stimulates: Release of TSH from basophils of adenohypophysis; secretory phase of thyroid gland	(−): Not definitely known; presumably inadequate release of TSH (+): Excessive release of TSH into peripheral circulation	(I): High blood levels of TSH & thyroid hormones (negative feedback) (S): Neural impulses from CNS; low blood levels of thyroid hormones & TSH	4,10,12, 18,20, 24,30, 31,42, 49,52, 53,57, 60	5

continued

	Hormone [Synonym]	Structure & Properties	Sources	Assay Methods	Metabolites
6	MSH release-inhibiting hormone [1] [MSH release—inhibiting factor; MRIH; MIF]	Structure unknown; not a polypeptide Mol wt <1000; stable to heat	Hypothalamus of man, cat, rabbit, rat, sheep, swine, & frog	Inhibition of release of MSH (entry 15) from hypothalamus of rat; increase of hypothalamic and decrease in plasma concentrations of MSH	Unknown
7	Prolactin release-inhibiting hormone [PRIH; PRH; PIF]	Polypeptide of low mol wt; different from epinephrine (entry 39), oxytocin (entry 18), & ADH (entry 19), or any kinin sol. in 0.1 N HCl; stable to boiling	Same part of hypothalamus that secretes ICSH (entry 12) of cattle, rat, sheep, & swine	Inhibition of depletion of hypophyseal LTH (entry 13) after stimulation of cervix of rat in estrus; inhibition of release of LTH from adenohypophysis of rat in vitro	Unknown
8	Natriuretic hormone	Polypeptide; 1 basic amino acid, 1 free terminal NH_2 group Mol wt <20,000; dialyzable; stable to acid, heat; inactivated by proteolysis by aminopeptidase, chymotrypsin, trypsin	Hypothalamus; posterior nucleus; plasma	Inhibition of Na^+ transport through frog skin or through proximal tubule of rat or dog kidney	Unknown
			Adenohypophysis		
9	Adrenocorticotropic hormone [Adrenocorticotropin; corticotropin; ACTH]	Unbranched polypeptide chain of 39 amino acids; NH_2-terminal Ser, COOH-terminal Phe; residues 1-24 [2] essential for activity in all species Mol wt = 4500 ± 50 (varies among species); pI = 4.7-4.8; $s_{20,w}$ = 2.08; sol. in water, dilute HCl in 80% acetone, acetic acid glacial; stable in hot acids; destroyed by 0.1 N NaOH at 100°C	Adenohypophysis—chromophobic basophils of pars distalis	Bioassay: Rate of decrease of suprarenal ascorbic acid in hypophysectomized rat; rate of increase of cholesterol & other lipids in atrophic suprarenal cortex; maintenance of normal cortical wt in hypophysectomized rat; urinary excretion of corticoids after injection of sample into guinea pig; increased corticosterone (entry 35) concentration in suprarenal venous blood of hypophysectomized dog or rat Radioimmunoassay: Inhibition of antibody-binding by radioactive labeled hormone [3] (usually higher results)	No specific metabolites definitely known; probably polypeptides of lower mol wt; half-life of ACTH in blood = 3-5 min
10	Follicle-stimulating hormone [FSH]	Glycoprotein containing sialic acid & small amounts of hexoses & hexosamines (up to 20% carbohydrate in man) Mol wt = 36,000 for man (dimer), 67,000 for sheep, 29,000 for swine; pI = 4.5 for	Basophils of adenohypophysis of man (♂♀), sheep (♂♀), & swine (♂♀); postmenopausal serum & urine	Bioassay: Wt increase in ovaries of intact immature rats, in intact immature ♀ rats or mice treated with excess HCG (entry 47), or in uterus of intact immature mice; increase in testicular wt in hypophysectomized immature rats treated with excess HCG; follicular growth & repair of interstitial tissue, or increase in ovarian wt in hypophysectomized immature rats	Not definitely known; probably a polypeptide

[1] There is a possibility of an MSH-releasing hormone (MSH-releasing factor, MRH, MRF). [2] Amino acid sequence of active portion: Ser·Tyr·Ser·Met·Glu·His·Phe·Arg·Trp·Gly· Lys·Pro·Val·Gly·Lys·Lys·Arg·Arg·Pro·Val·Lys·Val·Tyr·Pro (the synthetic tricosapeptide has full activity). A heptapeptide "core" is common to ACTH, LPH, and MSH (see

Targets	Principal Functions & Effects	Effects of Deficiency (−) & Excess (+)	Inhibitors (I) & Stimulators (S) of Secretion	Reference	
MSH-secreting cells of pars intermedia of hypothalamus	Inhibits: Release of MSH from hypophysis into circulation	(−): Not specifically known; possibly allows excessive pigmentation (+): Probably blanching of skin	(I): High blood levels of MSH (negative feedback) (S): Exposure to dark background; neural stimulation from CNS	4,10,12, 18,24, 30,31, 42,49, 52,57,60	6
LTH-secreting basophils of adenohypophysis	Inhibits: Release of LTH from adenohypophysis of ♂ & ♀	(−): Unknown; probably permits release of LTH in greater than normal amounts (+): Unknown; presumably prevents release of LTH, thereby inhibiting milk release or reaccumulation of milk after suckling-induced depletion	(I): Lactation; high blood levels of PRIH (negative feedback); increased blood levels of gonadal steroids (S): Unknown; presumably excessive lactation	4,10,12, 18,24, 30,31, 42,49, 52,57, 60	7
Epithelial cells in proximal tubule of kidney	Inhibits (regulates?): Na^+ transport (reabsorption) from lumen of proximal tubule of kidney into capillary—much more effective than oxytocin (entry 18) or ADH (entry 19)	(−): Not definitely known; presumably decreased excretion of Na^+ in urine (+): Increased loss of Na^+ in urine	(I): Unknown (S): Released from posterior hypothalamus by hemodynamic stimuli (e.g., carotid occlusion)	5,27,29	8

Adenohypophysis

Targets	Principal Functions & Effects	Effects of Deficiency (−) & Excess (+)	Inhibitors (I) & Stimulators (S) of Secretion	Reference	
Zonae fasciculata & reticularis of suprarenal cortex; mitochondria & enzyme systems in cytoplasm of many tissues; chromatophores	Decreases: Ascorbic acid & cholesterol in suprarenal cortex; renal transport of P_i & urate; urea production from exogenous amino acids; liver degeneration of corticosteroids Increases: Secretion of cortical hormones; oxidative phosphorylation; lipolysis in adipose tissue; fatty acid transport & oxidation; protein synthesis; I uptake by thyroid, synthesis of cAMP, activating α-glucan phosphorylase and increasing NADPH in suprarenal cortex; stimulates DNA synthesis by activating DNA nucleotidyltransferase	(−): Atrophy & hyposecretion of suprarenal cortex, especially zona fasciculata; decreased 17-hydroxy- & 17-ketosteroids in blood & urine; fasting hypoglycemia & increased insulin (entry 27) sensitivity; decreased skin pigmentation (+): Hypertrophy & hypersecretion of zonae fasciculata & reticularis; increased output of 17-hydroxy- & 17-ketosteroids; darkening of skin, especially scars & pressure areas (Cushing's disease)	(I): Increased blood levels of glucocorticoids (negative feedback, or servomechanism)—mediates neural regulation of circadian rhythm of corticoid secretion; blocked by dexamethasone (entry 38) (S): CRH (entry 1) of hypothalamus or neurohypophysis; stimulation of median eminence; psychic trauma, acting through hypothalamus; decreased level of circulating glucocorticoids, mainly hydrocortisone (entry 37)	4,10,12, 16,18, 24,27, 28,36, 41,48, 50,52, 53,57, 60,62	9
Seminiferous tubules (♂); Graafian follicles of ovary (♀)	Stimulates: Growth & maturation of ovarian follicles Increases: Spermatogenesis; growth of seminiferous tubules; testosterone (entry 41) secretion; supporting (Sertoli) cell hormone secretion in testis; incorporation of α-amino-	(−): Atrophy or immaturity of gonads; no maturation of ova or spermatozoa; obesity; decreased libido, potency, & hair growth; decreased blood levels of estrogen (+): Hypertrophy of secondary sex organs; increased growth & maturation of numerous follicles; increased	(I): Increased levels of circulating estrogens (S): Low blood levels of estrogens or possibly androgens; castration; menopause; hypothalamic-neurohypophyseal stimulation; external factors (e.g., light) through hypothalamus; FRH (entry 2) of hypothalamus	4,10,12, 16,18, 24,27, 28,30, 41,47, 49,50, 53,57, 60,62	10

entries 9, 13, and 15, respectively): Met·Glu·His·Phe·Arg· Trp·Gly. 3/ Depends on competition between radioactively labeled and unlabeled hormones for specific binding sites of the antibody; requires COOH-terminal Phe residue.

continued

	Hormone [Synonym]	Structure & Properties	Sources	Assay Methods	Metabolites
		sheep, 5.1-5.2 for swine; $s_{20,w}$ = 2.97 for man, 4.7 for sheep, 2.49 for swine; sol. in water, 50% acetone, 50% dioxane, 70% ethanol, half-saturated $(NH_4)_2SO_4$; stable at pH 7-8 for 30 min at 75°C; destroyed by reducing disulfide bonds		after treatment with diethylstilbestrol Immunoassay: Microcomplement fixation (hemolysis), hemagglutination, & radioimmunoassay with human ^{131}I-FSH as a reference	
11	Growth hormone [Somatotropin; GH]	Polypeptide [4/] (size varies with species); NH_2-terminal & COOH-terminal amino acid is Phe for all species, but also Ala for NH_2-end in cattle & sheep; no. of disulfide bonds in GH in man = 2, in swine & whale = 3, in cattle & monkey = 4, and in sheep = 5	Acidophils of adenohypophysis; plasma (in man, GH levels decrease with age from 2-16 yr)	Bioassay: Increase in tail length & tibial epiphyseal cartilage of hypophysectomized rat; body growth, wt gain in hypophysectomized ♀ rats Radioimmunoassay: Radioactivity of ^{131}I-GH released from anti-GH-antigen complex by GH in sample; chromatoelectrophoresis coupled with radioimmunoassay	Unknown; probably similar to metabolites of other polypeptides
		Mol wt = 21,700 in man, 45,000 in cattle (dimer?), 23,000 in monkey, 46,000 in rabbit (dimer?), 24,000 in rat, 47,800 in sheep (dimer?), 41,600 in swine (dimer?), 39,900 in whale (dimer?)			
		pI = 5.12 in man, 6.8 in cattle & sheep, 5.5 in monkey, 6.3 in swine, 6.2 in whale; $s_{20,w}$ = 2.18 in man, 3.19 in cattle, 1.88 in monkey, 3.21 in rabbit, 2.2 in rat, 2.76 in sheep, 3.02 in swine, 2.84 in whale; $D_{20} \times 10^7$ = 8.88 in man, 7.23 in cattle, 7.20 in monkey, 5.25 in sheep, 6.54 in swine, 6.56 in whale; $[\alpha]_D^{25}$ in 0.1 M acetic acid = −39° in man, −36° in cattle, −55° in monkey, −49° in sheep, −47° in swine, −52° in whale; sol. in dilute solutions of neutral salts & in ethanol; sl. sol. in water; inactivated by HNO_2, acetylation; after removal of fractions by chymotryptic proteolysis, 10% of activity retained in man (40% by peptic proteolysis), 25% in cattle, sheep, swine, & whale, 20% in monkey			
12	Interstitial cell-stimulating hormone [ICSH, for ♂; luteinizing hormone, LH, for ♀]	Glycoprotein containing hexose, sialic acid, galactosamine & glucosamine; adsorbed on cation-exchange resins Mol wt = 33,150 in man (dimer), 28,000-30,000 in sheep (dimer), 100,000 in swine; pI = 5.4 for man, 7.3 for sheep, 7.45 for swine; $s_{20,w}$ = 2.79 for man (dimer), 2.47 for sheep, 5.39 for swine; $D \times 10^7$ = 7.1 for man, 7.54 for sheep, 5.9 for swine; sol. in water, dilute solutions of acidic or neutral salts, 40% ethanol	Basophils of adenohypophysis; postmenopausal plasma or urine	Bioassay in immature, intact rats or mice: Increase of hyperemia in ovary; ascorbic acid depletion in rat ovary; cholesterol depletion in ovary of rat pretreated with HCG (entry 47) or PMSG [5/] Bioassay in immature hypophysectomized rats: Ovarian interstitial cell repair; wt of ventral prostate or of seminal vesicles; increase in prostatic alkaline phosphatase Radioimmunoassay: With pure [125]I- or ^{131}I-ICSH	Unknown; probably similar to other glycoproteins
13	Lactogenic hormone [Luteotropin; prolactin;	1 peptide chain; 205 amino acid residues in sheep, slightly more in cattle & swine; no. of disulfide bridges = 6 in cattle & sheep, 14 in swine; in man, lactogenic hormone	Basophils of adenohypophysis of cattle, rat, sheep, & swine; in man, may be	Bioassay: Increase in secretory activity of breast tissue of pregnant mice (best for human assay), or in wt of pigeon crop (other species); in rat & mouse, increase in luteal cell size in hypophy-	....

[4/] Amino acid sequence for human GH: Phe·Pro·Thr·Ile·Pro·Leu·Ser·Arg·Leu·Phe·Asp·Asn·Ala·Met·Leu·Arg·Ile·Ser·Leu·Leu·Leu·Ile·Gln·Ser·Trp·Leu·Glu·Pro·Val·Glu·Phe·Ala·His·Arg·Leu·His·Gln·Leu·Ala·Phe·Asp·Thr·Tyr·Glu·Glu·Phe·Glu·Glu·Ala·Tyr·Ile·Pro·Lys·Glu·Gln·Lys·Tyr·Ser·Phe·Leu·Gln·Asp·Pro·Glu·Thr·Ser·Leu·CyS·Phe·Ser·Glu·Ser·Ile·Pro·Thr·Pro·Ser·Asn·Arg·Glu·Glu·Thr·Gln·Lys·Ser·Asn·Leu·Gln·Leu·Leu·Arg·Ser·Val·Phe·Ala·Asn·Ser·Leu·Val·Tyr·Gly·Ala·Ser·Asn·Ser·Asp·Val·Tyr·Asp·Leu·Leu·Lys·Asp·Leu·Glu·Glu·Gly·Ile·Glu·Thr·Leu·Met·Gly·Arg·Leu·Glu·Asp·Pro·Ser·Gly·

Targets	Principal Functions & Effects	Effects of Deficiency (−) & Excess (+)	Inhibitors (I) & Stimulators (S) of Secretion	Reference	
	isobutyric acid into proteins of ovaries (rat); growth & maturation of ovarian follicles—small amounts of LH (entry 12) required; transport of glucose & amino acids across cell membrane in ovary	estrogen secretion (FSH + LH required); sexual precocity			
Granular endoplasmic reticulum of liver & most tissues; epiphyseal cartilage; fibroblasts	Decreases: Blood amino acid & glucose levels after short-time administration; urinary excretion of inorganic ions Increases: Skeletal & soft-tissue growth; protein anabolism; fibroblastic activity; swelling of liver mitochondria (rat); activity or synthesis of RNA nucleotidyltransferase (RNA polymerase); transport of neutral amino acids, glucose, & other hexoses; transport of fatty acids from fat depots to liver, with lowering of the respiratory gas exchange ratio; produces positive balances of Na⁺, K⁺, N, & P into tissues; erythropoiesis; renal function; tubular reabsorption of SO_4^{2-} (dog); blood glucose levels after prolonged administration	(−): Dwarfism and/or infantilism; delayed closure of epiphyses; increased sensitivity to insulin (entry 27); marked decrease in nuclear replication (thymidine uptake) and in turnover of acid mucopolysaccharides in skin; hypoglycemia (+): Gigantism before puberty or acromegaly after puberty; hypertrophy of viscera; diabetes mellitus; hyperglycemia; increased liver glycogenolysis; lipolysis & fatty acid oxidation; increased synthesis of nuclear RNA in liver; increased renal blood flow & extracellular volume; osteoporosis; muscle asthenia; negative Ca^{2+} balance	(I): Hyperglycemia or high glucose-glycogen metaboiism; maintenance of positive N balance; increased plasma corticosteroid levels (Cushing's syndrome); hyperlipemia; increase in plasma levels of GH; not species specific (S): Hypoglycemia, either fasting- or insulin-induced; exercise while fasting; prolonged fasting; emotional or traumatic stress; rapid growth (i.e., protein synthesis in infants); pregnancy (?); anorexia nervosa; intravenous infusion of Arg; hypolipemia; GRH (entry 3) of hypothalamus; L-DOPA causes rise in human GH persisting <2 hr; increased secretion of GRH by neural stimulation	4,10,12, 16,18, 21,14, 27,28, 30,41, 45,48, 49,51- 53,57, 60,62	11
In ♂, interstitial (Leydig) cells of testes; in ♀, Graafian follicles "primed" to maturity by FSH (entry 10), & corpora lutea of ovaries	Increases (♂): Spermiogenesis by seminiferous tubules; biosynthesis of testosterone (entry 41) by interstitial (Leydig) cells—synergistic with FSH Increases (♀): Follicle maturation, with FSH; production of corpora lutea—requires FSH & LTH (entry 13) for maintenance; estrogen & progesterone (entry 46) biosynthesis by corpora lutea; uptake of glucose by ovaries	(−): Decrease or loss of estrogen, progesterone, or testosterone biosynthesis; atrophy of interstitial & seminiferous cells in testis; slow or no maturation of ovarian follicles; no corpora lutea (+): Increased estrogen (with FSH) or androgen secretion; hypertrophy of interstitial cells of testis; precocious ovulation & luteinization of prepared follicles	(I): High blood concentrations of progesterone or testosterone, fluctuating with menstrual cycle in ♀; plasma levels much higher in postmenopausal ♀ (S): Low or moderate blood levels of gonadal hormones; LH-RH (entry 4) of hypothalamus	10,12, 15,18, 20,27, 28,39, 41,47, 49,50, 52,53, 57,60	12
Corpora lutea, matured by LH (entry 12); secretory cells of mammary	Increases: Milk secretion & ejection by mammary glands prepared by estrogen & progesterone (entry 46); nidation of zygote;	(−): Failure of lactation; progesterone deficiency; possible abortion (+): Precocious lactation of functional mammary glands	(I): Inhibition of oxytocin (entry 18) release; nervous inhibition; PRIH (entry 7) from hypothalamus (S): Oxytocin, in turn stimulated	10,11, 12,15, 18,24, 26-28, 41,42,	13

Arg·Thr·Gly·Gln·Ile·Phe·Lys·Gln·Thr·Tyr·Ser·Lys·Phe·Asp· Thr·Asn·Ser·His·Asn·Asp·Asp·Ala·Leu·Leu·Lys·Asn·Tyr·Gly· Leu·Leu·Tyr·CyS·Phe·Arg·Lys·Asp·Met·Asp·Lys·Val·Glu· Thr·Phe·Leu·Arg·Ile·Val·Gln·CyS·Arg·Ser·Val·Glu·Gly·Ser· CyS·Gly·Phe. Purified human GH is similar to LTH (*see* entry 13). A polypeptide with properties of human GH has been synthesized. 5/ PMSG = pregnant mare serum gonadotropin.

continued

Hormone [Synonym]	Structure & Properties	Sources	Assay Methods	Metabolites
mammotropin; LTH; MH]	similar to but distinct from human growth hormone Mol wt = 25,000 (?) in cattle, 23,000 in sheep; pI = 5.73 in cattle & sheep; $D_{20,w} \times 10^7 = 9$; viscosity coefficient = 6.65; $[\alpha]_D^{25} = -40.5°$; sol. in acid ethanol, methanol, dilute acids; sl. sol. in water, dilute salt solution; not inactivated by partial hydrolysis	identical with GH (entry 11)	sectomized immature rats pretreated with PMSG & HCG Radioimmunoassay: In species for which specific antisera are available	
14 Thyrotropic hormone [Thyrotropin; TSH]	Glycoprotein (single peptide chain) containing covalently bound fucose, mannose, galactosamine, & glucosamine (species differences, but immunologic cross-reaction among TSH from man, cattle, & swine); COOH-terminal amino acid sequence in man & cattle = His·Tyr·Lys·Ser·Tyr Mol wt = ∿28,000 in man & cattle, 25,000 in swine (2 subunits); inactivated by boiling, Cys, ketene; similar in structure & action to LH	Basophils of adenohypophysis; blood	Bioassay: Increase in colloid droplets in follicular cells; increase in thyroid wt in cattle; increase in plasma iodine in guinea pig & mouse; depletion of ^{131}I from guinea pig thyroid slices; McKenzie mouse assay; release of ^{125}I from thyroids of mice on low-iodine diet and injected with ^{125}I & T_4 (entry 22); iodine depletion in thyroids of 1-da-old chicks; growth of hindlimb of tadpole when metamorphosis has been arrested by starvation (stasis) Radioimmunoassay: Inhibition of binding of ^{125}I- or ^{131}I-TSH by thyroid in hypophysectomized rat or mouse Immunofluorescence: Of follicular epithelium using rabbit antibovine TSH	....
		Pars Intermedia		
15 Melanocyte-stimulating hormone [Melanotropin; MSH; melanophore-stimulating hormone]	Dialyzable; smallest sequence retaining activity is Met·Glu·His·Phe·Arg·Trp·Gly; adsorbed on oxycellulose Sol. in water, acidified acetone; moderately heat-stable; destroyed by tryptic proteolysis	Pars intermedia of hypophysis of many species including mammals; anterior lobe in birds & porpoises; synthesized in vitro	Bioassay: In vivo, intensity of reflected light from microscopic observation of frog or toad skin after injection of sample into dorsal lymph sac, compared with response produced by a standard extract of hypophyseal lobe; in vitro, intensity of darkening at the end of 60 min of isolated sections of skin from light-adapted, hypophysectomized frogs or toads (changes in reflection correspond to amount of MSH in solution) Radioimmunoassay: Of ^{131}I-α-MSH or -β-MSH (half-life in plasma much shorter by bioassay than by radioimmunoassay) (All above methods measure total effects of α-MSH, β-MSH, & ACTH)	Unknown; probably similar to other polypeptides
16 α-MSH[7]	Tridecapeptide; same residues as those for ACTH (entry 9) pI = 10.5-11	Basophils of hypophysis of all mammals; synthesized in vitro		
17 β-MSH[8] [Intermedin]	18 amino acid residues in cattle, horse, monkey, sheep, & swine, 22 in man; core sequence essential for activity in all species is Met·Glu·His·Phe·Arg·Trp·Gly pI = 4.1	Basophils of hypophysis; synthesized in vitro		

[6] Central "core" of Met·Glu·His·Phe·Arg·Trp·Gly is common to MSH, ACTH, & β-LPH (beta-lipotropin); β-LPH stimulates lipolysis in adipose tissue [12,60]. [7] Amino acid sequence: AcSer·Tyr·Ser·Met·Glu·His·Phe·Arg·Trp·Gly·

Targets	Principal Functions & Effects	Effects of Deficiency (−) & Excess (+)	Inhibitors (I) & Stimulators (S) of Secretion	Reference	
glands; pigeon crop	protein anabolism (GH-like in most species); in mammals, lipogenesis & fat deposition (also in birds & teleosts), sebaceous gland size & activity, and growth of ♂ accessory sex organs (synergistic with androgens); in pigeon, growth & secretion of crop gland; in reptiles & amphibians, regeneration of tail & legs		by suckling	49,50, 57,60	
Secretory epithelial cells of thyroid gland; adipose tissue cells (?)	Decreases: Iodine & colloid content of thyroid Increases: Synthesis & mobilization of thyroid hormones; serum protein-bound I; thyroid RNA & protein; proteolytic activity; oxidase granules; O_2 consumption by thyroid cells; entry of glucose into cells; entry of I^- into thyroid cells; rate of glucose oxidation by hexose monophosphate, glycolytic, & tricarboxylic acid pathways; lipolysis in adipose tissue; activities of enzymes mediated by cAMP, e.g., iodine-trapping & "organification," coupling of mono- & diiodoTyr, NADPH oxidation to pentose shunt, uptake of glucose & amino acids, phospholipid turnover	(−): Decreased synthesis of thyroid hormones; low serum protein-bound I; decreased I^- uptake by thyroid; normal or increased thyroid iodine (+): Increased synthesis of thyroid hormones; serum protein-bound I, I^- uptake by thyroid, & BMR; decreased thyroid iodine, blood cholesterol, & goiter	(I): Increased circulating thyroid hormones; high ambient temp; inhibition of hypothalamus (S): Decreased circulating thyroid hormones; low ambient temp; TRH (entry 5) of hypothalamus	4,10,12, 18,20, 24,27, 28,30, 39,41, 42,44, 49-53, 57,58, 60,62	14
Pars Intermedia					
Melanocytes of skin—most noticeable in species with chromatophores, but active in all species	Expands chromatophores rapidly (within a few minutes), causing pigment granules to disperse and color skin in vivo or in vitro (not species-specific—MSH from animals expands chromatophores in human skin); increases during most of pregnancy; expands chromatophores, contracts guanophores in anuran larvae Increases: Synthesis of melanin by chromatophores in human skin—most apparent in areas under pressure or trauma (scars)	(−): Chromatophore contraction & guanophore expansion; complete lack causes permanent blanching of skin (+): Hyperglycemia; darkening of skin (potentiated or supplemented by ACTH)	(I): High blood level of corticosteroids(?); MRIH (entry 6) from hypothalamus (S): Low level of corticosteroids[6/] (e.g., Addison's disease); MRH of hypothalamus; sympathetic postganglionic potentials from superior cervical ganglion	4,10,12, 18,24, 27,31, 41,49- 53,57, 60	15

16

17 |

Lys·Pro·Val·NH$_2$.   [8/] Amino acid sequence for human β-MSH: Ala·Glu·Lys·Lys·Asp·Glu·Gly·Pro·Tyr·Arg·Met·Glu·   His·Phe·Arg·Trp·Gly·Ser·Pro·Pro·Lys·Asp.

continued

Hormone [Synonym]	Structure & Properties	Sources	Assay Methods	Metabolites
Neurohypophysis				
18 Oxytocin [Oxytocic hormone]	Octapeptide—Cys·Tyr·Ile·Gln·Asn·Cys·Pro·Leu·Gly·NH₂ Inactivated by reduction of disulfide by alkali; hydrolyzed by oxytocinase in serum of pregnant women, pepsin, chymotrypsin	Neurohypophysis; blood of man, cattle, horse, rat, & sheep; synthesized in vitro	Stimulation of uterine contraction of pregnant women in vivo, or of rat uterus in vitro; increased milk ejection in guinea pig, rabbit	Unknown; probably similar to other polypeptides; half-life in plasma 1-4 min; rapidly metabolized in kidney
19 Vasopressin [Antidiuretic hormone; ADH]	Octapeptide—Cys·Tyr·Phe·Gln·Asn·Cys·Pro·Arg·Gly·NH₂ [10]/; disulfide bond essential for activity Mol wt = 1228; pI = 10.8; sol. in acid ethanol, phenol; stable in acid, not in alkali; inactivated by vasopressinase in serum of ♂ & nonpregnant ♀, by oxytocinase in the serum of pregnant ♀, and by trypsin & chymotrypsin, but not by pepsin or pancreatic carboxypeptidase	Neurohypophysis & blood [11]/ of man, cattle, dog, horse, rat, & sheep	Bioassay: Antidiuresis in conscious dog, rabbit, rat, or hydrated toad; antidiuresis or increased blood pressure in anesthetized rats Immunoassay: Bovine serum albumin as antigen and radioiodinated lysine vasopressin as reference	Unknown; probably similar to other polypeptides
Pineal Body				
20 Melatonin [5-Methoxy-*N*-acetyl-tryptamine]	C₁₃H₁₆O₂N₂ Mol wt = 222; sol. in aqueous organic solvents; blue color with *p*-dimethylaminobenzaldehyde	Pineal body of mammals (isolated from man, cattle, monkey, & rat) & birds; small amounts stored in peripheral nerves; in reptiles, amphibians, & fishes also occurs in brain & eye Synthesized in pineal body by oxidation & decarboxylation of Trp to serotonin, which is acetylated & then methylated to melatonin by the action of hydroxyindole-*O*-methyltransferase (HIOMT); also synthesized in vitro	Degree of blanching of isolated sections of skin of amphibians (e.g., *Rana pipiens*)	Hydroxylated to 6-hydroxymelatonin by liver enzyme; excreted in urine as glucuronide or sulfate conjugate of hydroxy group In pineal body, 5-methoxytryptamine (deacetylation) & 5-methoxyindoleacetic acid (oxidative deamination) also present

[9]/ Not essential for parturition; sensitivity of uterus to oxytocin greatly increased by estrogens, decreased by progesterone (*see* entry 46). [10]/ In hogs Lys replaces Arg. Amino acid sequence for vasotocin present in birds, reptiles, and some amphibians: Cys·Tyr·Ile·Gln·Asn·Cys·Pro·Arg·Gly·

Targets	Principal Functions & Effects	Effects of Deficiency (−) & Excess (+)	Inhibitors (I) & Stimulators (S) of Secretion	Reference	
Neurohypophysis					
Uterine & other smooth muscles; myoepithelial cells of mammary alveoli	Facilitates: Sperm movement up uterine (fallopian) tubes Stimulates: Release of LTH (entry 13) Increases: Uterine muscle contraction [9/]—may initiate labor; permeability of myometrial cell membrane to K^+, thus decreasing membrane potential & excitability threshold	(−): Delayed, weak pre- or post-partum uterine contractions; decreased milk secretion & ejection (+): Rapid, forceful uterine contractions; danger of abortion; excessive milk secretion	(I): No specific inhibitors known (may be a feedback mechanism) (S): Stimuli from birth canal; in animals with multiple fetuses, suckling of firstborn stimulates birth of more young (neural through hypothalamus); oxytocin level rises sharply halfway through labor	4,12,18, 24,29-31,42, 46,52, 60,62	18
Capillaries; arterioles; coronary vessels; vascular bed & tubules of kidney; smooth muscle	Increases: Water reabsorption in renal tubules by increased membrane permeability in distal tubules; excretion of Na^+, Cl^-, & urea; blood pressure by arteriole constriction; activity of adenyl cyclase, and thereby level of cAMP in blood & sizes of water "pores" in distal & collecting tubules (possibly forms a disulfide bond with cell membrane)	(−): Extreme polyuria (diabetes insipidus)—idiopathic & familial diabetes insipidus responsive to exogenous ADH, nephrogenic type (in ♂) not responsive to ADH; polydipsia; decreased renal excretion of Na^+ & urea (+): Oliguria; water retention (intoxication); Na^+ loss resulting from increased Na^+ excretion in urine; hyponatremia; increased blood pressure; smooth muscle contraction	(I): Decreased blood plasma osmotic pressure; increased extracellular fluid volume (S): Increased blood osmotic pressure, stimulating osmoreceptors in diencephalon; decreased volume of extracellular fluid (e.g., hemorrhage), stimulating volume ("stretch") receptors; nephrogenic stimuli through hypothalamus; drugs, e.g., morphine, nicotine, anesthetics	4,12,18, 24,29-31,52, 53,60, 62	19
Pineal Body					
CNS & sympathetic nervous system; hypothalamus; adenohypophysis; pars intermedia; gonads; chromatophores in skin	Decreases: Pigment exposure in skin by contraction of chromatophores—antagonistic to MSH (entries 15-17); synthesis & release of ICSH (entry 12) & LH-RH (entry 4) in hypothalamus & adenohypophysis; wt of testes; ovulation & onset of puberty; ACTH (entry 9) synthesis (?) Increases: Brain levels of serotonin & γ-aminoisobutyric acid—may function as neuroendocrine transducer and regulate circadian rhythm	(−): Darkening of skin of animals with expandable chromatophores; increased estrus, ovarian wt, & gametogenesis; precocious puberty (?) (+): Blanching of skin; decreased or delayed estrus & gametogenesis; decreased ovarian wt	(I): Exposure to light (through eyes) (S): Absence of light; blinding (feedback mechanism doubtful)	1,12,30, 56,57, 60,62, 63,65	20

NH$_2$. Amino acid sequence for ichthyotocin, present in some teleosts: Cys·Tyr·Ile·Ser·Asn·Cys·Pro·Ile·Gly·NH$_2$.

[11/] Levels in circulating blood: normal, 1-5 μunits/ml; after water deprivation, 6-21 μunits/ml; in adrenal insufficiency, 20-100 μunits/ml.

continued

	Hormone [Synonym]	Structure & Properties	Sources	Assay Methods	Metabolites
			Thyroid Gland		
21	Thyroglobulin	Glycoprotein containing Ala, Arg, Asp, Cys, Glu, Gly, His, Ile, Leu, Lys, Met, Phe, Pro, Ser, Thr, Trp, Tyr, Val, 3-iodo-Tyr, diiodoTyr, T_4 (entry 22) & iodo-T_4 Mol wt = 660,000; pI = 4.6; $s_{20,w}$ = 19.3	Follicular colloid of thyroid gland (stored form of thyroid hormones); hydrolyzed by proteases in the colloid, diffusion of T_3 & T_4 out of follicular cells facilitated by hyaluronidase in colloid	Secretory activity (iodine turnover) of thyroid gland—uptake of ^{131}I by thyroid in vivo, either by time lapse or scan of neck	Not specifically known; globulin part metabolized in same way as any globulin; proteases & hyaluronidases present in thyroid follicles For fate of iodinated residues, *see* T_4 & T_3 (entries 22 & 23)
22	Thyroxine [L-3,5,3′, 5′-Tetra-iodothyronine; β-/(3′,5′-diiodo-4-hydroxy-phenoxy)-3,5-diiodophenyl/alanine; T_4]	$C_{15}H_{11}O_4NI_4$; crystalline form—needles Mol wt = 777; mp = 232-233°C; $[\alpha]_D$ in NaOH = −3.2°; sol. in alkaline water, acid or alkaline ethanol; insol. in water, ethanol, volatile solvents	Follicles & epithelial cells of thyroid; in blood plasma, 99.9% bound to plasma proteins, mainly globulins Synthesis in vivo involves oxidation of I^- to I_2, iodination of Tyr, coupling of 2 moles of diiodoTyr to form T_4; in vitro, from p-methoxyphenol & 3,4,5-triiodo-dinitrobenzene (peroxidases needed)	Measurement of serum T_4 by displacement of T_4 from pool of T_4-binding globulin; displacement of ^{125}I-T_4 from protein-bound I by ^{125}I-T_4 extracted from serum and comparison with curve prepared by repeating with standard T_4; free T_4 in plasma by determining % dialyzable ^{131}I-T_4 added to serum; butanol-extractable iodine; adsorption on an anionic resin, elution, & digestion; paper, gel, or thin-layer chromatography (For additional methods, *see* T_3, entry 23)	Degraded through di- & monoiodothyronine, Diiodo-Tyr, & 3-Iodo-Tyr derivatives by deiodinases [12]/ in salivary glands, liver, kidney, thyroid glands, & plasma (deiodinase for T_4 found in muscle)
23	3,5,3′-Tri-iodothyronine [T_3]	$C_{15}H_{12}O_4NI_3$ Mol wt = 651; mp = 233°C; $[\alpha]_D^{29.50}$ of a 4.75% solution = +21.5°; sol. in alkaline solution, HCl	Follicles & epithelial cells of thyroid; in circulating blood plasma at concentrations of 4.4-11% of normal T_4 (entry 22) range; mean conc in human serum = 0.2 μg/100 ml Synthesized in vivo, by coupling 1 mole of 3-iodoTyr with 1 mole of diiodoTyr or -T_4 dialyzable from serum; circulating T_3 & T_4 in blood by protein-bound I by digestion & redox analysis; protein-bound ^{131}I in serum after oral dose of ^{131}I, by precipitation, digestion, distillation, & iodimetry; chromatographic separation of iodinated thyronines & Tyr in serum after oral dose of ^{125}I; chromatographic separation of T_3 from serum, measured by displacement of ^{131}I-T_3 from T_4-binding protein; uptake of ^{131}I-T_3 by erythrocytes after addition of T_3 to oxalated blood (also uptake of ^{131}I-T_3 in a resin)	Physiological effects: Increase in BMR and/or of O_2 consumption of hypothyroid mammals; prevention or regression of goiter in mammals fed goitrogens or deficient in I^-; calcaneus (Achilles') tendon reflex time in hypothyroid humans; increase in growth of thyroidectomized rats; acceleration of metamorphosis of amphibian larva Chemical assay: "Free" T_3 and/or T_4 by measurement of percent ^{131}I-T_3	May be excreted in bile, mainly as conjugates of glucuronic acid May be deaminated to form iodinated derivatives of pyruvic acid In rat: Deiodination of β (phenolic) ring more rapid than deiodination of α ring (converse true in dog)

[12]/ Deiodinases are important for conserving I by recycling it to the thyroid secretory epithelium.

Targets	Principal Functions & Effects	Effects of Deficiency (−) & Excess (+)	Inhibitors (I) & Stimulators (S) of Secretion	Reference	
		Thyroid Gland			
....	....	....	Same as for T_3 & T_4	10,12, 18,27, 28,39, 41,42, 49-53, 55,57, 58,60, 62	21
Most cells of the body, especially bones, striated muscle, heart, liver, & kidney— synergistic with GH (entry 11)	In large amounts, uncouples phosphorylation from oxidation in mitochondria Stimulates: Growth; maturation; neuromuscular function; skin development; hematopoiesis; gametogenesis; lactation; O_2 consumption (T_3 5 times as active as T_4); absorption through intestinal wall Decreases: TSH (entry 14) secretion (T_3 is 4 times as active as T_4) Increases: Rate of O_2 uptake without increased P uptake; rate of protein synthesis in cells by increasing RNA synthesis; rate of lipid, protein, carbohydrate, mineral, & water metabolism; rate of tadpole metamorphosis (T_3 is 7 times as active as T_4) (No qualitative differences between activities of T_3 & T_4)	(−): In children—cretinism, dwarfism, delayed maturation (calcification, epiphyseal closure) & growth, incomplete differentiation or metamorphosis, mental retardation, increased serum cholesterol; in adults—myxedema with decreased mental processes, dry & brittle hair, facial edema, thick tongue & speech, goiter, and decreased circulation, respiration, cardiac output, & intestinal absorption (+): Excessive acceleration of maturation, growth, BMR, & metamorphosis; thyrotoxicosis (toxic goiter, or Graves' disease); exophthalmos; mitochondrial swelling; lymphocytosis; decreased serum cholesterol; increased protein-bound I & butanol-extractable I	(I): Severe stress, such as fear, rage, hemorrhage, inflammatory processes, & severe muscular exercise inhibits secretion for first 24-48 hr; prolonged stress stimulates increased uptake of I^- or of antithyroid drugs, e.g., 2-thiouracil, thiourea, thiocyanate; increased levels of thyroid hormones; high ambient temp (S): Low ambient temp; pregnancy; prolonged stress; TSH (entry 14); prolonged dietary I^- deficiency; long-acting thyroid stimulator (LATS) in plasma of thyrotoxic patients (an A 7-S γ-globulin, possibly an immunoglobulin)		22 23

continued

	Hormone [Synonym]	Structure & Properties	Sources	Assay Methods	Metabolites
24	Thyrocalci-tonin [Cal-citonin; TCT]	Polypeptide[13]; 32 amino acid residues; a 1,7-disulfide ring at NH_2-terminal end Crude preparations stable at neutral pH & room temp (purified preparations labile); crude & purified preparations stable at pH 3; no loss of activity by alkylation or oxidation of Met; inactivated by reduction of disulfide bond, proteolysis by trypsin, chymotrypsin, or pepsin, and oxidation by o-diphenol oxidase (pyrocatechol oxidase), H_2O_2 or light	Tissue of ultimobranchial origin (C cells, parafollicular cells, "light" cells) in thyroid; in man, in parathyroids & thymus; distinct ultimobranchial bodies are main source in birds, reptiles, & fish; synthesized in vitro	Bioassay: Decrease in blood Ca^{2+}—sensitivity increased by administration of P_i concurrently to 5-wk-old rats pretreated 1 da on low Ca diet; inhibition of release of ^{45}Ca from labeled rat bone tissue culture pretreated with PTH (entry 25) Radioimmunoassay: With guinea-pig serum (as source of antibodies against porcine thyrocalcitonin) & purified ^{131}I-thyrocalcitonin (as reference standard)—by this method, the mean concentration in normal rabbit blood = 0.14 ng/ml, in human blood = 30-85 ng/ml	Unknown; probably similar to other polypeptides
			Parathyroid Glands		
25	Parathyroid hormone [Parathormone; PTH]	Polypeptide; 84 amino acid residues in man, cattle, & swine; first 29 residues have been synthesized & are required for activity; 20 amino acids at COOH-terminal end are active portion of molecule[14] Mol wt = ~8500 (varies with species); sol. in water, saline, aqueous ethanol, 94% acetic acid, concentrated phenol, 50% glycerol; insol. in volatile organic solvents; stable in dilute HCl; inactivated by oxidation of Met, Trp, or Tyr, or by acetylation or esterification	Parathyroid glands; malignant tumors, especially carcinomas of lung & kidney; urine, especially in hyperparathyroid humans	Bioassay: Hypercalcemia & phosphaturia (decrease in serum P_i) in dogs; increase in serum Ca^{2+} in parathyroidectomized rats; prevention of fall in, or elevation of, serum Ca^{2+} in thyroparathyrocauterized mice or thyroparathyroidectomized rats maintained on normal diet plus T_4 (entry 22), as compared with standard controls; stimulation of uptake of P_i by isolated rat liver mitochondria; release of ^{45}Ca from fetal rat bones Radioimmunoassay: With pure bovine ^{125}I-PTH as antigen and plasma of hyperparathyroid humans as standards (nonuniform standards)	Unknown; presumably similar to other polypeptides
			Pancreas		
26	Glucagon	Single polypeptide chain of 29 amino acids[15] Mol wt = 3485; pI = 7.5-8.5; sol. in dilute acid & alkali; insol. in water; stable in alkali; forms fibrils when heated to 40°C in acidic solution and then cooled	α cells of islets of Langerhans in pancreas	Bioassay: Hyperglycemia at definite intervals in fasted anesthetized cats; release of glucose into medium from rabbit liver slices; reactivation of α-glucan phosphorylase in liver slices & homogenates Radioimmunoassay: With ^{125}I- or ^{131}I-glucagon & antibodies from guinea pig	Similar to other polypeptides

[13] Amino acid sequence of porcine thyrocalcitonin: Cys·Ser·Asn·Leu·Ser·Thr·Cys·Val·Leu·Ser·Ala·Tyr·Trp·Arg·Asn·Leu·Asn·Asn·Phe·His·Arg·Phe·Ser·Gly·Met·Gly·Phe·Gly·Pro·Glu·Thr·Pro·NH₂. [14] Amino acid sequence of bovine PTH: Ala·Val·Ser·Glu·Ile·Gln·Phe·Met·His·Asn·Leu·Gly·Lys·His·Leu·Ser·Ser·Met·Glu·Arg·Val·Glu·Trp·Leu·Arg·Lys·Lys·Leu·Gln·Asp·Val·His·Asn·Phe·Val·Ala·Leu·Gly·Ala·Leu·Gly·Ala·Ser·Ile·Ala·Tyr·Arg·Asp·Gly·Ser·Ser·Gln·Arg·Pro·Arg·Lys·Lys·Glu·Asp·Asn·Val·Leu·Val·Glu·Ser·His·Gln·Lys·Ser·Leu·

Targets	Principal Functions & Effects	Effects of Deficiency (−) & Excess (+)	Inhibitors (I) & Stimulators (S) of Secretion	Reference	
Blood-bone membrane of osteoclasts, osteoblasts, & osteocytes	Causes hypophosphatemia Regulates: Ca^{2+} homokinesis; skeletal development & maintenance by altering Ca exchange Inhibits: Bone resorption, by decreasing permeability of bone membrane to Ca^{2+}, thereby producing hypocalciuria & possible hypocalcemia, depending on dietary intake of Ca & vitamin D; requires vitamin D for action	(−): Increased Ca^{2+} in serum; symptoms of hyperparathyroidism (+): Decreased Ca^{2+} in serum; symptoms of hypoparathyroidism (may be used therapeutically to prevent bone resorption & osteoporosis in certain diseases and to lower blood Ca^{2+} in pathological hypercalcemias)	(I): Decreased serum Ca^{2+} in pathological hypocalcemias (S): Increase of Ca^{2+} above 10 mg/100 ml	2,6,9, 10,12, 19,27, 37,42, 50-53, 57,60, 62	24

Parathyroid Glands

Targets	Principal Functions & Effects	Effects of Deficiency (−) & Excess (+)	Inhibitors (I) & Stimulators (S) of Secretion	Reference	
Bone-blood membrane of osteoblasts, osteoclasts, & osteocytes; cells of renal proximal tubules & of intestinal mucosa (?)	Regulates: Ca^{2+} & P_i levels in serum by mineral exchange between blood & bone, and by Ca^{2+} reabsorption & P_i excretion by kidney; renal tubule excretion of K^+, HPO_4^{2-}, & $H_2PO_4^-$ (actions on both bone & renal cells mediated through increased activity of cAMP; increase of PTH causes increased cAMP in both tissues; vitamin D required Stimulates: DNA & RNA synthesis & binding of enzymes, cAMP, & Ca^{2+} to membranes of bone cell	(−): Hypocalcemia; tetany; convulsive seizures; hypersensitivity of autonomic nervous system; chronic mental sluggishness; spotty alopecia (+): Hypercalcemia; renal calculi; decreased serum P_i (increased renal HPO_4^{2-} excretion); increased serum alkaline phosphatase activity; depolymerization of skeletal mucopolysaccharides	(I): Increased concentration of free Ca^{2+} (from diet or bone); decreased concentration of P_i (?) (S): Decrease of serum Ca below 10 mg/100 ml; increased serum P_i	2,6,9, 10,12, 19,27, 37,42, 50-53, 57,60, 62	25

Pancreas

Targets	Principal Functions & Effects	Effects of Deficiency (−) & Excess (+)	Inhibitors (I) & Stimulators (S) of Secretion	Reference	
Similar to insulin (entry 27)	Antagonistic to insulin Inhibits: Synthesis of fatty acids from precursors Promotes: Glycogenesis in liver but not in muscle; glycogenolysis in liver by stimulation of cAMP synthesis which mediates (i) conversion of phosphorylase b to α-glucan phosphorylase (phosphorylase a), (ii) release of nonesterified fatty acids from adipose tissue by glucagon & epinephrine (entry 39), (iii) increased protein metabolism (and increased urine levels of creatinine & urea), (iv) stimulatory effect	(−): Hypoglycemia; symptoms of hyperinsulinism (+): Increased N excretion; negative N balance; decreased plasma amino acids; decreased volume & acidity of gastric juice; decreased gastrointestinal motility & pancreatic secretion; permanent diabetes mellitus by destruction of β cells of the islets of Langerhans	(I): Hyperglycemia; glucagon & insulin regulate blood glucose levels by opposite feedback mechanisms, involving liver glycogen as storage compound; increased conc of plasma fatty acids (S): Hypoglycemia; pancreozymin (entry 31)—potency in stimulating glucagon secretion $^1/_6$ that of stimulation of insulin secretion; increased conc of plasma amino acids	10,12, 14,18, 27,28, 40,49- 53,57, 60,61	26

Gly·Glu·Ala·Asp·Lys·Ala·Asp·Val·Asp·Val·Leu·Ile·Lys·Ala· Lys·Pro·Gln. 15/ Amino acid sequence: His·Ser·Gln·Gly·

Thr·Phe·Thr·Ser·Asp·Tyr·Ser·Lys·Tyr·Leu·Asp·Ser·Arg·Arg· Ala·Gln·Asp·Phe·Val·Gln·Trp·Leu·Met·Asn·Thr.

continued

	Hormone [Synonym]	Structure & Properties	Sources	Assay Methods	Metabolites
27	Insulin	Polypeptide consisting of an A chain (21 amino acids) & a B chain (30 amino acids) connected by disulfide bridges *16*/; species differences in residues 8, 9, & 10 of disulfide ring in A chain and in COOH-terminal residue of B chain (no effect on activity); more than one type may be present in a single species; crystallized with Zn (3 atoms/mol wt of 35,000) Mol wt = 5734 at pH 2.5 in cattle (minimum), possibly 12,000-48,000 at pH 7-8; unknown in circulating blood; pI = 5.3; stable in acid solution; resistant to denaturing agents; inactivated in alkaline solution; completely inactivated by cleavage of disulfide bonds between A & B chains by oxidation or reduction, & by proteolysis	β cells in islets of Langerhans in pancreas; proinsulin (an inert peptide chain of 33 residues connecting A & B chains), synthesized in granular endoplasmic reticulum, transferred to Golgi bodies, and packaged in granules; hydrolyzed to insulin (sol. insulin also present); released into circulation by appropriate stimulation Synthesized in vitro	Bioassay: Uptake of glucose by isolated mouse or rat diaphragm; liberation of $^{14}CO_2$ from ^{14}C-glucose in rat epididymal fat pads Chemical assay: Precipitation by salts, ethanol, or antibodies; paper chromatoelectrophoresis; adsorption on anion exchange resins or molecular gels Radioimmunoassay: Rate of displacement of ^{125}I- or ^{131}I-insulin (bovine) from complexes with antibodies (guinea pig) to insulin type being assayed	Similar to other polypeptides
			Gastrointestinal Tract		
28	Cholecysto-kinin-Pancreozymin [CCK-PZ]	Polypeptide consisting of 33 amino acid residues *17*/ (low mol wt, dialyzable); identical to pancreozymin (entry 31) pI = 5.0-5.5; sol. in water	Mucosa of upper intestine	Rise in pressure within gallbladder (in situ or isolated) of anesthetized cat, dog, or guinea pig	Similar to other polypeptides

16/ Amino acid sequence of porcine insulin: A chain: Gly·Ile·Val·Glu·Gln·Cys·Cys·Thr·Ser·Ile·Cys·Ser·Leu·Tyr·Gln·Leu·Glu·

B chain: Phe·Val·Asn·Gln·His·Leu·Cys·Gly·Ser·His·Leu·Val·Glu·Ala·Leu·Tyr·

17/ COOH-terminal octapeptide is active portion: Lys·Ala·Gly·Pro·Ser·Arg·Val·Ile·Met·Ser·Lys·Asn·Asn·Gln·His·Leu·Leu·Pro·

Targets	Principal Functions & Effects	Effects of Deficiency (−) & Excess (+)	Inhibitors (I) & Stimulators (S) of Secretion	Reference	
	of epinephrine on cerebral cortex, and (v) increased force of heartbeat (inotropic effect) by glucagon & epinephrine Decreases: Protein synthesis in liver Increases: Blood glucose level; O_2 consumption by tissues; gluconeogenesis from amino acids by inducing & stimulating aminotransferases (synergistic with glucocorticoids); conversion of acetate to ketone bodies				
All tissues & cells involved in metabolism of carbohydrates, lipids, & proteins, as well as chiefly enzymatic systems in cell membranes, mitochondria, & other cytoplasmic organelles of muscle, adipose tissue, & liver	Antagonistic to glucocorticoids Regulates: Carbohydrate & fatty acid metabolism Activates: Ribosomes; glycogen synthetase Stimulates: Glycolytic enzymes; phosphorylases; lipogenesis; transport & oxidation of lipids; oxidation of carbohydrates; glucose & amino acid uptake by cells; synthesis of protein & mucopolysaccharides Decreases: Gluconeogenesis; ketogenesis; lipolysis in adipose tissue; free sulfhydryl content of muscle; activity of adenyl cyclase, and thereby the concentration of cAMP Increases: Rates of transport of glucose into striated muscle & adipose tissue, and of K^+ into muscle; synthesis of mRNA (30 min) & DNA (24-36 hr) in liver, stimulating protein synthesis upon increased uptake of amino acids by striated muscle, coupled with transport of Na^+; activities of lipogenic enzyme systems, and thereby the deposition of triglycerides in adipose tissue; incorporation of SO_4^{2-} into chondroitin sulfates	(−): Diabetes mellitus—hyperglycemia, glycosuria, ketonuria, decreased wt & blood volume, negative N balance; wound healing delayed, gangrene; polyphagia, dehydration, polydipsia, & polyuria; dwarfism, increased response to GH (entry 11); decreased HPO_4^{2-} uptake by cells; decreased ATP, glycogenesis, lipogenesis, & proteogenesis; increased lipolysis, free fatty acids in plasma, & H^+ secretion by renal tubules (+): Hyperinsulinism—hypoglycemia, convulsions, nausea, muscular weakness, anxiety, confusion; increased food intake & fat & protein deposition; hypermobility & hypersecretion of stomach	(I): Hypoglycemia during fasting or prolonged exercise; hyperinsulinemia (negative feedback?); epinephrine (entry 39); norepinephrine (entry 40)—α adrenergic stimulation (S): Hyperglycemia (glucose must be phosphorylated or further metabolized); high-protein diet (increased amino acids in blood); stimulants of adenyl cyclase synthesis; short-chain fatty acids; sulfonylureas; tolbutamide; cAMP; gastrointestinal hormones—gastrin, pancreozymin, & secretin (entries 30-32); glucagon? (entry 26); increased ACTH (entry 9), GH (entry 11), T_3 (entry 23), or T_4 (entry 22); vagal stimulation; all stimulants probably induce synthesis of proinsulin by β cells and facilitate release of insulin into blood	2,10,12, 14,18, 27,28, 40,49-53,57, 60,64	27

Gastrointestinal Tract

Targets	Principal Functions & Effects	Effects of Deficiency (−) & Excess (+)	Inhibitors (I) & Stimulators (S) of Secretion	Reference	
Muscle of gallbladder (may be identical with pancreozymin)	Increases: Contraction & emptying of gallbladder (Cholagogue effect)	(−): Unknown; presumably sluggish emptying of the gallbladder (+): Unknown	(I): Anticholecystokinin in blood & urine, secreted by wall of gallbladder (S): Fat, fatty acids, polypeptides or acidic chyme in duodenum	12,25, 40,42, 53,59, 60,62	28

Asn·Tyr·Cys·Asn

Leu·Val·Cys·Gly·Glu·Arg·Gly·Phe·Phe·Tyr·Thr·Pro·Lys·Ala.
Ser·Ser·Arg·Ile·Asp·Ser·Arg·Asp·Tyr·Met·Gly·Trp·Met·Asp·Phe·NH$_2$ (last 5 residues have gastrin activity).

SO_3H

continued

	Hormone [Synonym]	Structure & Properties	Sources	Assay Methods	Metabolites
29	"Enterogastrone" [18]	Polypeptide; dialyzable Sol. in water; insol. in organic solvents; destroyed by peptic hydrolysis	Mucosa of duodenum (?)	Decrease in contractions & acid secretion by stomach	Unknown; probably similar to other polypeptides
30	Gastrin	Polypeptide[19]; active portion pentopeptide Gly·Trp·Met·Asp·Phe·NH$_2$; dialyzable pI = 5.5; stable to heat; destroyed by ultraviolet light, alkali, peptic hydrolysis	Gastric mucosa, mainly in pyloric antral area, of man, dog, sheep, & swine Synthesized in vitro	Stimulation of gastric HCl secretion in conscious cats, with histamine as the reference (1 histamine unit = stimulation of HCl by 1 mg/kg body wt = 0.001 mg histamine)	Unknown; probably similar to other polypeptides
31	Pancreozymin-Cholecystokinin [PZ-CCK]	Polypeptide; identical to cholecystokinin (entry 28) Physical & chemical properties very similar to those of cholecystokinin; stable to heat, acid; destroyed by alkali	Mucosa of upper intestine (first meter)	Increase in rate of secretion and concentration of enzymes in pancreatic juice	Similar to other polypeptides
32	Secretin	Polypeptide, with 27 amino acid residues[20]; strongly basic; synthetic polypeptide has full activity Sol. in dilute acids, alcohols; salted out by NaCl, tricholoroacetic acid	Mucosa of upper intestine	Increase in flow of pancreatic juice and concentration of HCO$_3^-$ in anesthetized cat or dog	Similar to other polypeptides

Suprarenal Cortex[21]

	Hormone [Synonym]	Structure & Properties	Sources	Assay Methods	Metabolites
33	Aldosterone [Electrocortin; mineralocorticoid; 11β,21-dihydroxy-18-oxopregn-4-ene-3,20-dione]	C$_{21}$H$_{28}$O$_5$; in equilibrium between 11,18-hemiacetal & 18-aldehyde structures when in solution Mol wt = 360; mp = 164°C; $[\alpha]_D$ in chloroform = +152.2°; absorption max = 240 nm; fluoresces at 560 nm in alkaline solution	Zona glomerulosa (chiefly) of suprarenal cortex of virtually all vertebrates; plasma, 5-15 ng/100 ml (35% free); urine, 1-7 μg/24 hr (5-12% free) Synthesized in vivo from cholesterol through pregnenolone, progesterone (entry 46), deoxycorticosterone (entry 34), & 18-hydroxycorticosterone	Bioassay: In vitro perfusion of suprarenals from intact or hypophysectomized rats with Krebs-Ringer HCO$_3^-$ solution; measurement of aldosterone by one of the methods described below Plasma assay: Double isotope dilution using aldosterone with ^{14}C- or ^{3}H-labeled nucleus (as recovery indicator) & ^{35}S-tosan, ^{35}S- or ^{131}I-pipsyl, or ^{3}H-acetic anhydride (as esterifying agent) to determine mass after repeated extractions, chromatographic separations, & preparation of necessary derivatives Urine assay: Extraction & purification of either acid-labile aldosterone-18-glucuronide or tetrahydroaldosterone-3-glucuronide, followed by fluorimetric or photometric measurement of color produced by (i) NaOH—soda fluorescence,	Most important urinary metabolite is 3α,5β-tetrahydroaldosterone, the product of stepwise reduction; bicyclic acetals also produced in small amounts by reduction at C-20

[18] Existence of a distinct hormone very doubtful; most actions produced by secretin (entry 32). [19] Amino acid sequence in human gastrin: Glu·Gly·Pro·Trp·Leu·Glu·Glu·Glu·Glu·Glu·Ala·Tyr·Gly·Trp·Met·Asp·Phe·NH$_2$. Porcine:

$\overset{\uparrow}{SO_3H}$

residue 5 = Met; other species vary in residues 5-10. [20] Amino acid sequence: His·Ser·Asp·Gly·Thr·Phe·Thr·Ser·Glu·Leu·Ser·Arg·Leu·Arg·Asp·Ser·Ala·Arg·Leu·Gln·Arg·Leu·Leu·Gln·Gly·Leu·Val·NH$_2$. [21] All corticosteroids have some effect on electrolyte and water metabolism (mineralocorticoid effect), and some effect on carbohydrate metabolism

Targets	Principal Functions & Effects	Effects of Deficiency (−) & Excess (+)	Inhibitors (I) & Stimulators (S) of Secretion	Reference	
Mucosa & muscle of stomach (?)	Decreases: Acid secretion & motility of stomach—may prevent peptic ulcers (?)	(−): Unknown; perhaps excess HCl secretion (+): Unknown	(I): Not definitely known (S): Fat (triglycerides) or acid in, or distention of, duodenum	12,40, 42,48, 60	29
Parietal & chief cells of gastric mucosa; biliary system of liver	Stimulates: Secretion of gastric HCl, pepsinogen, & Castle's intrinsic factor; also liver bile & pancreatic enzymes	(−): Decreased acid secretion (+): Unknown; probably predisposition to gastric ulcer	(I): High H^+ concentration in stomach (S): Polypeptides in contact with pyloric mucosa	12,40, 42,48, 60	30
Enzyme-secreting cells of pancreas (may be identical with cholecystokinin)	Inhibits: Gastric motility & secretion Stimulates: Secretion of enzymes by pancreas, possibly by promoting transport across cell membranes (no effect on volume)	(−): Unknown; possibly inadequate secretion of pancreatic enzymes (+): Unknown	(I): Absence of duodenal chyme (S): Fatty acids; fats; polypeptides; certain amino acids	12,40, 42,48, 60	31
Acinar or exocrine cells of pancreas; biliary system of liver	Inhibits: Gastric motility & secretion Stimulates: Volume flow rate of pancreatic juice, including ions but not enzymes; bile flow from liver; intestinal juices	(−): Unknown; presumably sluggish flow of pancreatic juice and/or bile (+): Unknown	(I): Absence of food in duodenum (S): Acidic chyme, water, ethanol, fatty acids, polypeptides, some amino acids in duodenum	12,40, 42,48, 60	32

Suprarenal Cortex [21]/

Targets	Principal Functions & Effects	Effects of Deficiency (−) & Excess (+)	Inhibitors (I) & Stimulators (S) of Secretion	Reference	
Proximal & distal convoluted tubules of kidney	Helps regulate blood pressure Increases: K^+ excretion & Na^+ reabsorption in renal tubules, & sweat & salivary glands; reabsorption of Na^+ by intestinal mucosa [22]/; reabsorption of Na^+ in distal tubule accompanied by reabsorption of water (salt-retaining potency of aldosterone 20-100 times that of deoxycorticosterone, 40-1000 times that of the glucocorticoids)	(−): Excessive loss of Na^+ in urine; dehydration & decreased plasma Na^+; increased plasma K^+, resulting in cardiac arrhythmia & heart block; acidosis resulting from H^+ retention & Na^+ loss (+): Increased plasma Na^+ & extracellular fluid volume; decreased plasma K^+ & H^+, resulting in muscle spasms, polyuria, polydipsia, alkalosis, hypertension, & congestive heart failure	(I): Increased plasma Na^+, blood flow (negative feedback through renal juxtaglomerular apparatus); decreased plasma K^+ (S): Renin-angiotensin II system from juxtaglomerular apparatus (except in rat); ACTH (entry 9), by steroidogenesis; increased plasma K^+; decreased plasma Na^+ (e.g., low-salt diet); hemorrhage	3,7,9, 10,12, 17,19, 22,27, 28,49-53,57, 60,62	33

(glucocorticoid effect). Aldosterone (*see* entry 33) and DOC (*see* entry 34) have much stronger mineralocorticoid and inflammatory actions in all species, while cortisone (*see* entry 36) and hydrocortisone (*see* entry 37) have much stronger glucocorticoid and anti-inflammatory actions; corticosterone (*see* entry 35) is intermediate. Corticosterone is the principal glucocorticoid in rodents & lagomorphs; cortisone and hydrocortisone are the principal glucocorticoids in man. [22]/ Probable mechanism of action is preferential binding of aldosterone to nucleoproteins in tubule cells and stimulation of transcription from DNA to a unique mRNA which regulates synthesis of an enzyme that is rate-limiting for synthesis of ATP by mitochondria. Increase in ATP at site of sodium pump promotes transport of Na^+ through epithelial cells.

continued

	Hormone [Synonym]	Structure & Properties	Sources	Assay Methods	Metabolites
				(ii) K *tert*-butoxide, (iii) H_2SO_4, (iv) blue tetrazolium, (v) modified Porter-Silber (phenylhydrazineHCl at pH 4.1) Secretion rate: Measurement of aldosterone in suprarenal vein blood of conscious, pretreated, trained Merino sheep after infusion of [3]H-, ^{14}C-aldosterone into suprarenal artery of left suprarenal gland transplanted into a carotid artery-jugular vein skin loop	
34	Deoxycorticosterone [Cortexone; Δ^4-pregnen-21-ol-3,20-dione; DOC; DOCA]	$C_{21}H_{30}O_3$ Mol wt = 330; mp = 141-142°C; $[\alpha]_D^{20}$ in ethanol = +178°; sol. in acetone, benzene, chloroform, other volatile solvents, vegetable oils; insol. in water	Suprarenal cortex Synthesized commercially, from cholesterol, diosgenin	Inhibits aldosterone (entry 33) production in essential hypertension; decrease in Na^+/K^+ ratio in suprarenalectomized rats	Corticosterone; aldosterone; 11-dehydrocorticosterone; 5β-pregnane-3α,20α-diol; 17-ketosteroids
35	Corticosterone [11β,21-Dihydroxy-pregn-4-ene-3,20-dione; Kendall's compound B]	$C_{21}H_{30}O_4$ Mol wt = 346; mp = 180-182°C; $[\alpha]_D^{20}$ in ethanol = +262°; sol. in organic solvents; sl. sol. in vegetable oils; very sl. sol. in water	Zona fasciculata of suprarenal cortex; blood; urine Synthesized in vitro from deoxycholic acid; pregnenolone; progesterone (entry 46), cholesterol	Chemical assay of unconjugated steroids: Extraction with organic solvents; purification by paper, column, & gas-liquid chromatography; measurement by acid or alkali fluorescence; Porter-Silber reaction (phenylhydrazine); reduction of tetrazolium salts; double isotope dilution (*see* Aldosterone, entry 33) Chemical assay of conjugated steroids: Preliminary removal of unconjugated steroids by solvent extraction, followed by hydrolysis of conjugated steroids in aqueous phase by incubation with β-glucuronidase; determination by one of above methods	Dihydro- & tetrahydro-derivatives, by reduction (*see also* Aldosterone, entry 33)
36	Cortisone [Kendall's compound E; 17-hydroxy-11-dehydrocorticosterone; 17α,21-dihydroxy-pregn-4-ene-3,11,20-trione]	Mol wt = 360; mp = 220-224°C; $[\alpha]_D^{20}$ in ethanol = +209°; sol. in acetone, benzene, chloroform, ethyl ether, vegetable oils; sl. sol. in water	Zona fasciculata of suprarenal cortex; also, in placenta, blood, & urine Synthesized in vivo from cholesterol through progesterone (entry 46) & hydrocortisone (entry 37)—in equilibrium with hydrocortisone; in vitro from squalene, cholesterol, pregnenolone, progesterone	Same as for aldosterone (entry 33) Bioassay (used for comparison of actions, especially for the newly synthesized analogues): Life maintenance of suprarenalectomized rats on limited salt & water diets; deposition of liver glycogen 1 hr after injection of sample into suprarenalectomized, fasting mice or rats; excretion of Na^+ & K^+ by suprarenalectomized rats after controlled intake of granuloma (produced artificially by small cotton ball in wound of suprarenalectomized rat pretreated with corticosteroids) Chemical assay: Porter-Silber reaction; *m*-dinitrobenzene reaction; fluorimetric	Reduction to dihydro- & tetrahydro-reduction products (cortolone); also same as for hydrocortisone

Targets	Principal Functions & Effects	Effects of Deficiency (−) & Excess (+)	Inhibitors (I) & Stimulators (S) of Secretion	Reference	
					34
Action & regulation intermediate between those of the mineralocorticoids, aldosterone & deoxycorticosterone (entries 33 & 34), and those of the glucocorticoids, cortisone & hydrocortisone (entries 36 & 37)				3,9,10, 12,19, 22,27, 28,42, 49-54, 57,60, 62	35
Most cells, notably liver, striated muscles, kidneys, blood vessels, integument, bone marrow, lymphoid & adipose tissue, reticuloendothelial system, pancreas, & gastrointestinal mucosa	Antagonistic to insulin (entry 27) Decreases: Nucleotide & protein synthesis in extrahepatic tissues; synthesis of triglycerides & other lipids (lipogenesis); extrahepatic synthesis of sulfated mucopolysaccharides; synthesis & activities of enzymes involved in oxidative catabolism of glucose Increases: Deamination of amino acids to glucose; activities of liver & kidney glucose-6-phosphatase, fructose diphosphatase, pyruvate decarboxylase, glycogen synthetase & arginase, liver pyruvate kinase & alkaline phosphatase, & muscle aminopeptidase;	(−): Asthenia; hemoconcentration; chronic Addison's disease; decreased blood glucose, liver glycogen, body wt, blood pressure, & stress resistance; delayed or minimal secondary sex characteristics; hyperpigmentation; increased K^+/Na^+ ratio in serum; emaciation; vomiting; diarrhea. In acute deficiency, hypotension, circulatory collapse, adrenal apoplexy (Waterhouse-Friderichsen syndrome). (+): Cushing's syndrome; abnormal distribution of fat; "buffalo hump;" moon face; truncal obesity with thin extremities; negative N balance; hyperglycemia, unless	(I): Increased plasma levels of glucocorticoids—negative feedback through ACTH (entry 9) & CRH (entry 1) (S): Decreased plasma levels of glucocorticoids; stressful conditions of any kind (traumatic or emotional through cerebro-hypothalamo-hypophyseal-adrenal pathways)	3,9,10, 12,19, 22,24, 27,28, 32,33, 49-53, 57,60, 62	36

continued

	Hormone [Synonym]	Structure & Properties	Sources	Assay Methods	Metabolites
37	Hydrocortisone [Cortisol; 17-hydroxy-corticosterone; Kendall's compound F; 11β,17α,21-trihy-droxypregn-4-ene-3,20-dione]	$C_{21}H_{30}O_5$ Mol wt = 362; mp = 217-220°C; $[\alpha]_D^{20}$ in ethanol = +167°; sol. in chloroform, ethyl ether, vegetable oils; sl. sol. in water	Zona fasciculata of suprarenal cortex; also in blood & urine Synthesized in vivo from cholesterol through pregnenolone, 17α-hy-droxyprogesterone, & 11-deoxycortisol	Same as for aldosterone (entry 33) & cortisone (entry 36); double isotope method very precise; competitive protein binding (globulin) is accurate & practical	Reduction to tet-rahydrocortisol & cortol (hexa-hydrocortisol); in equilibrium with cortisone
38	Dexamethasone [9α-Fluoro-16α-methyl-prednisolone]	$C_{22}H_{29}FO_5$ Mol wt = 392.4; mp = 262-264°C; $[\alpha]_D^{25}$ = +78°	Synthetic compound	Anti-inflammatory (see also aldosterone, entry 33, & hydro-cortisone, entry 37)	Not definitely known; similar to other corti-costeroids
			Suprarenal Medulla		
39	Epinephrine [Adrenalin; suprarenine; adrenine]	Mol wt = 183.2; mp of l-form = 211-212°C; $[\alpha]_D^{20}$ in dilute HCl = −50.72°; sol. in acid, alkali (with decomposition); sl. sol. in water; insol. in chloroform, ethanol, ethyl ether	Suprarenal medulla of mammals Synthesized in chro-maffin cells or sym-pathetic neurons (l-form) from l-Tyr through DOPA & norepinephrine (entry 40); commer-cially from pyrocat-echol	Bioassay: Rise in blood pressure of specially prepared rats; inhi-bition of contractions in isolated rat uterus; infusion or perfusion of isolated organs, such as guinea-pig heart or strips of rat stomach Chemical assay: Extraction & purification by adsorption on aluminum oxide & ion-exchange resins, and by paper chromatog-raphy, followed by determina-tion by trihydroxyindole fluo-rescence, Na ascorbate, Na_2SO_3, K ferricyanide, or ethylene di-amine	Oxidative deami-nation & meth-ylation to 4-hy-droxy-3-meth-oxymandelic acid (vanillyl-mandelic acid, VMA) at rate of 2-8 mg/24 hr; some meta-nephrine (3-O-methylepineph-rine), free & conjugated, ex-creted in urine at rate of 1 mg/ 24 hr
40	Norepinephrine [Noradrenalin, L-arterenol]	Mol wt = 169.2; mp of l-form = 216.5-218°C (with decomposi-tion), of HCl salt of l-form = 146-147°C;	Suprarenal medulla; various nerves, espe-cially splenic; spleen; heart; blood vessels	Same as for epinephrine in all tests except pressor & contrac-tion of gravid uterus	Principal metab-olite is 4-hy-droxy-3-meth-oxymandelic acid; normeta-

Targets	Principal Functions & Effects	Effects of Deficiency (−) & Excess (+)	Inhibitors (I) & Stimulators (S) of Secretion	Reference
	activities of liver tryptophan oxygenase, tyrosine aminotransferase, & other amino acid oxidases & aminotransferases, thereby increasing liver synthesis of protein from amino acids; lipolysis in adipose tissue; fatty acids in plasma; secretion of HCl & pepsinogen by gastric mucosa, and enzymes by pancreas	insulin also high; osteoporosis; wasting of muscle; alkalosis; increased gastric acidity; adrenogenital syndrome; hypertension; inhibition of inflammatory repair of wounds & antibody formation		
Same as for cortisone	....			37
....	....	....	....	38
Suprarenal Medulla				
Sympathetic nervous system; striated, cardiac, & smooth muscle, especially in arterioles; cells producing enzymes involved in glycogenolysis & lipolysis; β cells of islets of Langerhans	Inhibits: Intestinal motility Stimulates: CNS; synthesis & activity of adenyl cyclase, thereby increasing level of cAMP which in turn increases concentrations of α-glucan phosphorylase, thus increasing phosphorylation & hydrolysis of liver glycogen to glucose-1-phosphate; release of ACTH (entry 9) Decreases: Insulin (entry 27) secretion; eosinophils in peripheral circulation (indirectly by ACTH-induced secretion of glucocorticoids); activity of glycogen synthetase; peripheral resistance; gastrointestinal movement Increases: Cardiac output; BMR; dilation of bronchi, blood vessels of liver & of striated & cardiac muscle; neutrophilia, eosinopenia, & marked hyperglycemia following administration; cAMP (when insulin levels low) which activates lipases in adipose tissue, increasing free fatty acids in plasma	(−): No known clinical disease (+): Pheochromocytoma; marked hypertension, pigmentation, tachycardia, headache, nausea, blanching of skin, sweating; increased blood levels of nonesterified fatty acids	(I): No known inhibitors or feedback mechanism (S): Splanchnic impulses from sympathetic nervous system; stressful stimuli, e.g., trauma, exposure to low temp, hemorrhage, acute hypoxia, severe & prolonged exercise, emotional states such as fear or anger	10,12, 19,27, 28,48, 50-53, 57,60, 62 — 39
Same as for epinephrine	Does not stimulate CNS, increase O$_2$ consumption, or cause eosinopenia Decreases: Cardiac output Increases: Blood pressure	Same as for epinephrine	(I): Unknown (S): Splanchnic impulses from sympathetic nervous system; neurohumor (see also epinephrine)	10,12,19 27,28, 48,50- 53,57, 60,62 — 40

continued

	Hormone [Synonym]	Structure & Properties	Sources	Assay Methods	Metabolites
		$[\alpha]_D^{25}$ in dilute HCl = $-37.3°$; solubility same as for epinephrine (entry 39)			nephrine (3-O-methylnorepinephrine) also excreted
				Testis	
41	Testosterone [Δ^4-Androsten-17β-ol-3-one]	$C_{19}H_{28}O_2$ Mol wt = 288; mp = 155-156°C; $[\alpha]_D$ in ethanol = +109°; absorption max = 238 nm; sol. in ethanol, ethyl ether, volatile solvents; sl. sol. in vegetable oils; insol. in water	Interstitial (Leydig) cells of testis; plasma Main pathway for peripheral conversion—from acetate through cholesterol, pregnenolone, progesterone (entry 46), & androstenedione—from suprarenal cortex (dehydroepiandrosterone is main cortical C_{19}-steroid)	Bioassay: Increase in comb size of castrated cock (capon); blackening of bill of English sparrow (Pfeiffer's test); increase in wt in seminal vesicles of castrated ♂ rat (Mathieson & Hays test); rate & amount of axillary hair growth in men Chemical assay (plasma): m-Dinitrobenzene & KOH (Zimmerman); ^{35}S-thiosemicarbazide; thin-layer chromatographic purification of doubly labeled (^{14}C-ring & ^{3}H-acetate ester) testosterone; measurement of ^{3}H-testosterone chloroacetate by gas-liquid chromatography & electron capture Chemical assay (urine): Extraction with ethyl ether and measurement by gas-liquid chromatography; scintillation counting of [3]H-, ^{14}C-testosterone & -epitestosterone; testosterone glucuronide by column chromatography	Reduction of ring A & of 17-keto group to androsterone (entry 42), etiocholanone, epiandrosterone; some conversion to estrogens
42	Androsterone [Androstan-3α-ol-17-one]	$C_{19}H_{30}O_2$ Mol wt = 290; mp = 185.5°C; $[\alpha]_D^{20}$ in dioxane = +87.8°; sol. in volatile solvents; sl. sol. in vegetable oils; insol. in water; not precipitated by digitonin	Urine of man, bull, & pregnant cow Synthesized from cholesterol or sitosterol	Same as for testosterone (entry 41)	Dehydroepiandrosterone; 4-androstene-3,17-dione; 17α-hydroxyprogesterone. Since only found in urine, probably a metabolite of testosterone (entry 41)
				Ovary	
43	Estradiol [Estradiol-17β; dihydrofollicullin; dihydroestrone; 1,3,5-estratriene-3,17β-diol]	$C_{18}H_{24}O_2$ Mol wt = 272; mp = 178°C; $[\alpha]_D$ in dioxane = +81°; sol. in alkali, volatile solvents; sl. sol. in vegetable oils; insol. in water	Ovaries; placenta; suprarenal cortex (♂♀); blood & urine (♂♀); testes Synthesized in vivo from acetate through cholesterol, pregnenolone, testosterone (entry 41), & intermediates; in vitro from cholesterol	Bioassay (mainly for comparative studies): Cornified cells and/or increased mitotic figures in vaginal smear from castrated ♀ mice; increase in uterine wt in intact immature rats or mice Chemical assay: Splitting protein bond by boiling with dilute HCl, hydrolysis of conjugates, extraction in ethyl ether, separation by paper, column, & thin-layer chromatography, and measurement by colorimetric (quinol-H_2SO_4), fluorometric (H_2SO_4),	Excreted in urine as conjugate (acetate or glucuronate); small amounts of α-estradiol; some dehydrogenated to estrone (entry 45) or oxidized to various hydroxy & keto derivatives; several methoxy derivatives also

Targets	Principal Functions & Effects	Effects of Deficiency (−) & Excess (+)	Inhibitors (I) & Stimulators (S) of Secretion	Reference	
	by vasoconstriction; BMR; blood glucose—moderate hyperglycemia following administration (*see also* epinephrine, entry 39)				
		Testis			
All ♂ sex organs; adenohypophysis; muscle; hair follicles; epiphyses of long bones; vocal cords	Androgenic; little effect on wt of heart, urinary bladder; produces positive balances of Ca, N, P, K, & S Inhibits: Growth of thymus, suprarenals Decreases: Folliculoid & luteoid activity in immature ♀; amino acid catabolism; creatinuria; Na$^+$ & Cl$^-$ excretion; kidney alkaline phosphatase Increases: Development of ♂ secondary sex organs & sex characteristics; libido; BMR; protein anabolism by DNA-based stimulation of mRNA synthesis; amino acid transport in striated muscle; amino acid incorporation & activities of D-amino acid oxidase, arginase, acid phosphatase, & β-glucuronidase in kidney cortex; renal transport of Na$^+$ & K$^+$; rate of synthesis of fatty acids, citrate, & fructose in seminal vesicles; respiration rate in seminal vesicles & prostate	(−): Immaturity or atrophy of accessory sex organs; lack of secondary sex characteristics & ♂ behavior patterns; poor muscle development & function; delayed closure of epiphyses; decreased excretion of 17-ketosteroids in urine; functional decrease with age, inducing ♂ climacteric characterized by "hot flashes," vasomotor & neurologic disturbances similar to those in menopause; deficiency effects more pronounced before puberty than after (+): Precocious sex development; hypertrophy of ♂ accessory sex organs; increased muscle mass; increased skeletal growth until epiphyseal closure; increased excretion of 17-ketosteroids in urine; decreased scalp hair (?); increased hirsutism	(I): High levels of testosterone—negative feedback through ICSH (entry 12) & FSH (entry 10) (S): FSH + ICSH mediated through cAMP (second messenger)	3,8,10, 12,19, 22,27, 28,33, 34,42, 48,50-53,56, 57,60, 62	41 42
		Ovary			
All ♀ sex organs; mammary glands; mucous membranes; adenohypophysis; osteoblasts	Estrogenic; antagonizes androgen effects; potentially carcinogenic; mildly vasodilator; slows growth of the skeleton Affects: Endometrial proliferation & development and maintenance of vaginal mucosa (cornification of superficial layer) Promotes: Epiphyseal closure Stimulates: Transhydrogenases in human placenta;	(−): Immaturity or atrophy of accessory sex organs; lack of secondary ♀ sex characteristics & behavior patterns; decreased mammary gland development; delayed epiphyseal closure and continued growth of long bones; osteoporosis (+): Precocious maturity; hypertrophy of secondary ♀ sex organs & mammary glands; estrus changes; cystic hyperplasia of endometrium;	(I): Increased level of estradiol, which inhibits FSH (entry 10) production (S): Combined FSH & LH (entry 12); cyclic changes in blood levels with estrus or menstrual periods	3,8,10, 12,16, 19,22, 27,28, 33-35, 42,48, 50-53, 57,60, 62	43

continued

	Hormone [Synonym]	Structure & Properties	Sources	Assay Methods	Metabolites
				or enzymatic (hydroxysteroid dehydrogenases) assay Isotope dilution: Using ^{3}H- or ^{14}C-estrogen standards or formation of radioactive derivatives with ^{14}C- or ^{3}H-acetic anhydride Conjugates: Separated by extraction with water, hydrolysis, and measurement by one of above assay methods	isolated from urine; in horses, equilenin ($\Delta^{1,3,5:10,6,8}$-estrapentaen-3-ol-17-one) also found in urine (*see also* reactions for estriol, entry 44)
44	Estriol [Trihydroxyestrin; $\Delta^{1,3,5:10}$-estratriene-3,16α,17β-triol]	$C_{18}H_{24}O_3$ Mol wt = 288; mp = 282°C; $[\alpha]_D^{20}$ in dioxane = +53° to +63°; sol. in alkali, volatile solvents; sl. sol. in vegetable oils; insol. in water	Urine of pregnant women; human placenta Synthesized from estrone (entry 45) & estradiol (entry 43)	Same as for estradiol	Estradiol ⇌ estrone ⇌ α-estradiol; also, estrone → estriol (principal metabolite in urine; excreted as water-soluble, inactive glucuronide)
45	Estrone [Folliculin; ketohydroxyestrin; $\Delta^{1,3,5:10}$-estratrien-3-ol-17-one]	$C_{18}H_{24}O_2$ Mol wt = 270.3; mp = 260°C; $[\alpha]_D$ in chloroform or dioxane = +160°; sol. in alkali, volatile solvents; sl. sol. in vegetable oils; very sl. sol. in water	Suprarenal cortex; placenta; urine of man & of ♂ cattle (bull & steer); testes of horse Synthesized in vivo from cholesterol (*see* estradiol, entry 43)		Also, 16-oxo-(keto)-estrone, 16-oxo-(keto)-(17β)-estradiol, & various isomers & derivatives of pregnenolones
46	Progesterone [Progestin; Δ^4-pregnene-3,20-dione]	$C_{21}H_{30}O_2$ Mol wt = 314; mp of α-form = 128°C, of β-form = 121°C; $[\alpha]_D$ in dioxane = +172° to +182°; absorption max = 240 nm; sol. in volatile solvents; sl. sol. in vegetable oils; insol. in water	Corpora lutea; placenta; suprarenal cortex; testes Synthesized in vivo from cholesterol through pregnenolone; in vitro from cholesterol or stigmasterol	Bioassay: Proliferation of endometrium after injection into isolated uterine or uterine horn segments from immature ♀ rabbits primed with estrogen; hypertrophy of stromal nuclei in endometrium of oophorectomized mice by injection of sample into uterus; increase in carbonic anhydrase content of uterine endometrium Physicochemical assay: Ultraviolet absorption by progesterone at 240 nm; dinitrophenylhydrazone, isonicotinic acid hydrazide, & thiosemicarbazide derivatives; H_2SO_4-ethanol chromogen; gas-liquid chromatography Double isotope derivates, with ^{3}H & ^{35}S or ^{3}H & ^{14}C	Principal urinary metabolites: Pregnanediol (pregnane-3α, 20α-diol), pregnanolone (3α-hydroxypregnan-20-one), & allopregnanediol (allopregnan-3α,20α-diol), plus 6β-hydroxyprogesterone & smaller amounts of pregnanolones
	Placenta				
47	Human chorionic gonadotropin [HCG]	Glycoprotein containing sialic acid, hexoses, hexosamines, & amino acids (Tyr·Arg·Trp); 135,000 IU/mg Mol wt = 55,000-	Chorionic villi of human placenta; blood & urine of pregnant women—peak at 8th wk of pregnancy (60-75 da after 1st da of last menses),	Bioassay: Increase in ovarian wt or ovarian hyperemia in intact immature rats; expulsion of spermatozoa in amphibians; increase in wt of uterus, prostate, seminal vesicle, or total accessory reproductive organs in rats	Excreted in urine of pregnant women

Targets	Principal Functions & Effects	Effects of Deficiency (−) & Excess (+)	Inhibitors (I) & Stimulators (S) of Secretion		Reference
	calcification of bones (moderately); peripheral circulation Decreases: Erythrocyte P uptake; plasma cholesterol; β-lipoproteins Increases: Mammary gland duct development; uterine motility; growth of uterine (fallopian) tubes, axillary & pubic hair (♀ humans), down (♀ birds), and all secondary ♀ sex organs; tissue growth & cell division; uterine uptake of Na⁺, K⁺, glucose, & water; endometrial alkaline phosphatase & glycogen; vaginal glycogen; protein anabolism in uterus by stimulation of RNA nucleotidyltransferase (polymerase) & serine hydroxymethyltransferase synthesis; retention of Ca^{2+}, Na^+, & water	blocking of ovulation; skeletal growth deceleration; excessive calcification of tissues if parathyroids are normal			44 45
Endometrium of uterus; lobules & alveoli of mammary gland; kidney tubules; adenohypophysis	Luteinizing, affecting preparation of endometrium for implantation of zygote; antagonistic to aldosterone, promoting excretion of Na^+, Cl^-, & water Stimulates: Protein catabolism; galactose oxidation Decreases: Uterine contractions; alkaline phosphatase in uterus Increases: Growth of lining epithelial cells in estrogen-primed endometrium with increased glycogen, mucin, & fat; lobule development in mammary glands; BMR; body temp	(−): Lack of normal cyclic changes and of endometrial development for implantation & gestation; danger of abortion (+): Pregestational changes; pregnancy prolongation; inhibition of uterine growth, especially endometrium; increased Na^+ & K^+ excretion, catabolism	(I): Degeneration of corpora lutea in estrus & menstrual cycles; decreased LH (entry 12) secretion (S): LH; LH + LTH (entry 13) in mouse & rat		3,8,11, 12,16, 19,22, 27,28, 33-35, 42,48, 50-53, 57,60, 62	46
		Placenta			
Ovaries, especially corpora lutea	Actions similar to those of LH (entry 12) & LTH (entry 13); synergistic with hypophyseal gonadotropins; maintains corpora lutea during early pregnancy (high titer in	(−): Degeneration of corpora lutea with decreased progesterone secretion; danger of abortion (+): Produced in excessive amounts by hydatid mole, chorioepithelioma, & some	(I): ♀ sex steroids, notably progesterone (S): Enzyme systems of developing placenta (specific stimulant unknown)		10,12, 15,18, 26-28, 42,48, 50-53, 57,60, 62	47

continued

	Hormone [Synonym]	Structure & Properties	Sources	Assay Methods	Metabolites
		60,000 (dimer?); pI = 2.95; $[\alpha]_D^{20} = -68°$; sol. in water, 50% acetone, 60% ethanol; inactivated by removal of N-acetylneuraminic acid (NANA) end-groups	decreasing to 15% of peak at 16th wk	or ♂ hamsters Immunoassay: Inhibition of agglutination of "stable" HCG-coated sheep erythrocytes by rabbit antiserum to HCG; complement fixation with sensitized sheep erythrocytes & rabbit antiserum to HCG; radioimmunoassay, with ^{131}I-HCG as reference, rabbit antiserum to HCG, & guinea-pig antiserum to rabbit γ-globulin (double antibody)	
48	Human placental lactogen [HPL]	Polypeptide[23] Mol wt = 30,000-38,000 (dimer, or 2 chains?)	Villous trophoblast of human placenta; blood of pregnant women (concentration higher than that of human GH—entry 11)	Using rabbit antibodies to HPL & ^{131}I-lactogen bound to antibody	Unknown; probably similar to other polypeptides
49	Relaxin	Polypeptide Mol wt = 9000; pI = 5.5; sol. in water, 95% ethanol	Corpora lutea of pregnant woman, mouse, swine; placenta; endometrium; blood of pregnant woman, cat, dog, horse, rabbit, swine	Degree of relaxation of guinea-pig pubic ligament	Similar to other polypeptides
			Accessory Genital Glands		
50	Prostaglandins[24]	Cyclic, oxygenated C_{20} fatty acids; derivatives of prostanoic acid; $C_{20}H_{34}O_5$ & $C_{20}H_{36}O_5$ Mol wt of prostaglandin E_1 = 354; absorption max = 275 nm; sol. in alkaline solution, fats, & fat solvents; insol. in water;	Semen in man & sheep; seminal vesicle in sheep; ♂ accessory genital glands in most mammals; lung in man, cattle, guinea pig, monkey, sheep, & swine; menstrual fluid; iris of cat, cattle, dog, rabbit, & sheep; brain & CNS of	Bioassay: Contraction of intestinal smooth muscle strips from mammals; decrease in blood pressure (depressor activity) in rats; inhibition of lipolysis by epinephrine (entry 39) in isolated rat adipose tissue Chemical assay: Ether extraction of homogenized tissue, separation by column chromatography, esterification, thin-layer chromatogra-	7 metabolites produced by reduction in lung or liver of the C-13:14 double bond and/or by oxidation of the C-15 alcohol group (with significant loss of activity); 80% of metabolites excreted in urine, 20% in feces

[23] Amino acid sequence: Val·Gln·Lys·His·Arg·Asp·Thr·Ser·Glu·Pro·Gly·Ala·Cys·Met·Ile·Leu·Tyr·Trp·Phe. HPL is similar to human GH (*see* entry 11) in gel diffusion, immuno-

electrophoresis, antibiological activity, and binding affinity.
[24] The prostaglandins include prostaglandin E_1 (PGE$_1$; 11α,-15α-dihydroxy-9-keto-*trans*-13-prostenoic acid), prostaglan-

Targets	Principal Functions & Effects	Effects of Deficiency (−) & Excess (+)	Inhibitors (I) & Stimulators (S) of Secretion	Reference	
	urine 1 mo after first missed menses is basis for pregnancy test (e.g., Ascheim-Zondek, and agglutination-inhibition reactions) Stimulates: Luteal production of progesterone (entry 46) until placenta attains full function; premature descent of testes in immature primates Increases (by injection): Follicle maturation & ovum release in nonpregnant ♀; androgen secretion by interstitial (Leydig) cells of testes	neoplasms of testes			
Myo-epithelial cells of mammary glands; corpora lutea	Synergistic with progesterone (entry 46), LTH (entry 13), & oxytocin (entry 18) in stimulating & maintaining lactation, and in milk ejection; augments GH	(−): Not definitely known; presumably difficult ejection of milk (+): Unknown	(I): Unknown (S): Not definitely known; probably gonadotropins or GH	13,28, 42,48, 57,60, 62	48
Pubic ligaments	Effects relaxation of pubic ligament and separation of symphysis pubis, chiefly during last stages of pregnancy in most mammals; after pre-sensitization with estrogen, connective tissues of symphysis become more vascular, collagen fibers are dissolved, and mucopolysaccharides are depolymerized, rendering ligament more flexible; effects softening & relaxation of uterine cervix after pretreatment with estrogen; water imbibition by myometrium in rat; synergistic with estrogen & progesterone (entry 46) in mammary development in rat Inhibits: Rhythmic uterine contractions (in rat) & uterine motility	(−): Not definitely known; may prolong labor (+): Not definitely known	(I): Not definitely known; possibly changes in balances of estrogen & progesterone during parturition (levels in blood peak during terminal stages of pregnancy and decline rapidly just prior to, during, and 1 da after delivery) (S): Not definitely known; probably increased levels of progesterone during pregnancy	10,12, 15,19, 48,49, 60	49
Accessory Genital Glands					
Most tissues & cells of mammals	Specific physiological functions not definitely known; block lipolysis by epinephrine; prostaglandin E types cause vasodilation and increase in cardiac output; prostaglandin F types constrict veins, causing hypertension & increased cardiac output Regulate: cAMP formation & action in many tissues, notably effects of LH (entry 12) in stimulating estrogen	(−): Not definitely known (+): Not definitely known	(I): No specific natural inhibitors known (S): Released from cells by variety of stimuli—catecholamines, serotonin, histamine, neural, & mechanical (distention, stirring)—mediated through cAMP	12,23, 27,38, 42,43, 48-51, 53,60	50

din E₂ (PGE₂; 11α,15α-dihydroxy-9-keto-*cis*-5,*trans*-13-prostadienoic acid), prostaglandin E₃ (PGE₃; 11α,15α-dihydroxy-9-keto-*cis*-5,*trans*-13,*cis*-17-prostatrienoic acid), prostaglan-

dins F₁, F₂, and F₃ (PGF₁; PGF₂; PGF₃; the corresponding trihydroxy derivatives), and the prostaglandin A and B series (PGA; PGB; monohydroxy derivatives).

continued

Hormone [Synonym]	Structure & Properties	Sources	Assay Methods	Metabolites
	destroyed by concentrated alkali	cat, cattle, dog, horse, rabbit, fowl, & frog; thymus of cat & young cattle; suprarenal gland of cat; kidney of swine; intestine of frog	phy, gas-liquid chromatography (electron capture detection); measurement of 15-dehydroprostaglandin by spectrometry or of NADH by fluorimetry, cycling, or photokinetic analysis after action of 15-hydroxyprostaglandin dehydrogenase on sample	

Contributor: Pritham, Gordon H.

References

[1] Axelrod, J. 1970. Endeavour 29:144.
[2] Behrens, O. K., and E. L. Grinnan. 1969. Annu. Rev. Biochem. 38:83.
[3] Briggs, M. H., ed. 1970. Advan. Steroid Biochem. Pharmacol. 1.
[4] Burgus, R., and R. Guillemin. 1970. Annu. Rev. Biochem. 39:499.
[5] Cort, J. H., et al. 1968. Lancet 1:230.
[6] DeLuca, H. F. 1971. Recent Progr. Horm. Res. 27:479.
[7] Edelman, I. S., and G. M. Fimognari. 1969. Ibid. 24:1.
[8] Eik-Nes, K. B. 1971. Ibid. 27:517.
[9] Eliel, L. P., et al. 1971. Pediatrics 47:229.
[10] Escamilla, R. F., ed. 1971. Laboratory Tests in Diagnosis and Investigation of Endocrine Function. Ed. 2. F. A. Davis, Philadelphia.
[11] Frantz, A. G., et al. 1972. Recent Progr. Horm. Res. 28:527.
[12] Frieden, E., and H. Lipner. 1971. Biochemical Endocrinology of the Vertebrates. Prentice-Hall, Englewood Cliffs, N.J.
[13] Friesen, H. G., et al. 1969. Recent Progr. Horm. Res. 25:161.
[14] Fritz, I. B., ed. 1972. Insulin Action. Academic Press, New York.
[15] Fuchs, F., and A. Klopper. 1971. Endocrinology of Pregnancy. Harper and Row, New York.

[16] Garren, L. D., et al. 1971. Recent Progr. Horm. Res. 27:433.
[17] Glaz, E., and P. Vecsei. 1971. Aldosterone. Pergamon Press, New York.
[18] Gray, C. H., and A. L. Bacharach, ed. 1967. Horm. Blood, Ed. 2(1).
[19] Gray, C. H., and A. L. Bacharach, ed. 1967. Ibid. (2).
[20] Gual, C., et al. 1972. Recent Progr. Horm. Res. 28:173,201.
[21] Hagen, T. C., et al. 1972. Metab. Clin. Exp. 21(7):603.
[22] Heftmann, E. 1970. Steroid Biochemistry. Academic Press, New York.
[23] Horton, E. W. 1972. Prostaglandins. J. Springer, New York.
[24] James, V. H. T., ed. 1968. Recent Advan. Endocrinol., Ed. 8.
[25] Jorpes, J. E. 1968. Gastroenterology 55:157.
[26] Li, C. H., et al. 1971. Science 173:56.
[27] Litwack, G., ed. 1972. Biochemical Actions of Hormones. Ed. 2. Academic Press, New York. v. 1.
[28] Loraine, J. A., and E. T. Bell. 1971. Hormone Assays and Their Clinical Application. Ed. 3. Williams and Wilkins, Baltimore.
[29] Margoulis, M., ed. 1969. Protein Polypeptide Horm. Int. Symp., 1968.
[30] Martini, L., and W. F. Ganong, ed. 1967. Neuroendocrinology 2.

75. HYPOTHALAMIC RELEASING AND INHIBITING HORMONES: VERTEBRATES

Structure or Chemical Nature: Ala = alanyl; Arg = arginyl; Glu = glutamyl; Gly = glycyl; His = histidyl; Leu = leucyl; Lys = lysyl; Pro = prolyl; (pyro)Glu = pyroglutamyl; Ser = seryl; Trp = tryptophyl; Tyr = tyrosyl; Val = valyl. Activity: Effective Dose—values are dry weight; PPM = partially purified material, HP = highly purified, S = synthetic, P = pure, CRE = crude extract, RIA = radioimmunoassay. Figures in heavy brackets are reference numbers. Data in light brackets refer to the column heading in brackets. For additional information, consult reference 60.

	Hormone (Abbreviation) [Synonym]	Function [Synonym]	Structure or Chemical Nature	Occurrence	Activity		
					Animal	Method	Effective Dose
1 2	Corticotropin-releasing hormone (CRH) [60] [Corticotropin-releasing factor (CRF)]	Stimulation of synthesis & release of ACTH [Corticotropin]	Polypeptide?	Cattle, dog [13, 60]; rat [3,5]; sheep [20,60]; swine [60]	Rat [2,60] Rat[L/] [2,20, 60]	In vitro In vivo, intravenous	1-2 μg PPM 1-4 μg PPM [2]

L/ Pretreated with (i) dexamethasone, (ii) morphine & pentobarbital sodium, or (iii) chlorpromazine, morphine, & pentobarbital sodium.

continued

Targets	Principal Functions & Effects	Effects of Deficiency (−) & Excess (+)	Inhibitors (I) & Stimulators (S) of Secretion	Reference
	& progesterone (entry 46) synthesis in the ovaries; transmission of impulses in sympathetic nervous system Inhibit: HCl secretion by gastic mucosa Stimulate: Contraction of smooth muscle, notably that in uterus			

[31] McCann, S. M., and J. C. Porter. 1969. Physiol. Rev. 49:240.
[32] McKerns, K. W., ed. 1968. Functions of the Adrenal Cortex. Appleton-Century-Crofts, New York. v. 1-2.
[33] McKerns, K. W. 1969. Steroid Hormones and Metabolisms. Appleton-Century-Crofts, New York.
[34] McKerns, K. W. 1971. The Sex Steroids: Molecular Mechanism. Appleton-Century-Crofts, New York.
[35] Mueller, G. C., et al. 1972. Recent Progr. Horm. Res. 28:1.
[36] Munro, H. N., ed. 1970. Mammalian Protein Metab. 4.
[37] Munson, P. L. 1968. Recent Progr. Horm. Res. 24:589.
[38] Pharriss, B. B., et al. 1972. Ibid. 28:51.
[39] Pierce, J. G., et al. 1971. Ibid. 27:165.
[40] Pincus, G., et al., ed. 1964. Hormones 4.
[41] Pincus, G., et al., ed. 1964. Ibid. 5.
[42] Prunty, F. T. G., and H. Gardiner-Hill, ed. 1972. Mod. Trends Endocrinol. 4.
[43] Ramwell, P., and J. E. Shaw, ed. 1971. Ann. N.Y. Acad. Sci. 180.
[44] Reichlin, S., et al. 1972. Recent Progr. Horm. Res. 28:229.
[45] Root, A. W. 1972. Human Pituitary Growth Hormone. C. C. Thomas, Springfield, Ill.
[46] Rudinger, J., et al. 1972. Recent Progr. Horm. Res. 28:131.
[47] Ryan, R. J., et al. 1970. Ibid. 26:105.
[48] Sawin, C. T. 1969. The Hormones: Endocrine Physiology. Little, Brown; Boston.

[49] Schwartz, T. B., ed. 1968. Yearb. Endocrinol. 1967/1968.
[50] Schwartz, T. B., ed. 1969. Ibid. 1969.
[51] Schwartz, T. B., ed. 1970. Ibid. 1970.
[52] Schwartz, T. B., ed. 1971. Ibid. 1971.
[53] Schwartz, T. B., ed. 1972. Ibid. 1972.
[54] Smellie, R. M. S., ed. 1971. The Biochemistry of Steroid Hormone Action. Academic Press, London.
[55] Sterling, K. 1970. Recent Progr. Horm. Res. 26:249.
[56] Stollerman, G. H., ed. 1970. Advan. Intern. Med. 16.
[57] Sunderman, F. W., and F. W. Sunderman, Jr. 1971. Laboratory Diagnosis of Endocrine Diseases. C. V. Mosby, St. Louis.
[58] Taurog, A. 1970. Recent Progr. Horm. Res. 26:189.
[59] Thaysen, E. H., ed. 1971. Gastrointestinal Hormones and Other Subjects. Alfred Benzon Foundation, Aalborg, Denmark.
[60] Turner, C. D. 1971. General Endocrinology. Ed. 5. W. B. Saunders, Philadelphia.
[61] Unger, R. H., and P. J. Lefebre. 1972. Glucagon: Molecular Physiology, Clinical and Therapeutic Implications. Pergamon Press, New York.
[62] Williams, R. H., ed. 1968. Textbook of Endocrinology. Ed. 4. W. B. Saunders, Philadelphia.
[63] Wolstenholme, G. E. W., and J. Knight. 1971. Pineal Gland Ciba Found. Symp., 1970.
[64] Wool, I. G., et al. 1968. Recent Progr. Horm. Res. 24:139.
[65] Wurtman, R. J., and F. Anton-Tay. 1969. Ibid. 25:493.

75. HYPOTHALAMIC RELEASING AND INHIBITING HORMONES: VERTEBRATES

	Hormone (Abbreviation) [Synonym]	Function [Synonym]	Structure or Chemical Nature	Occurrence	Activity		
					Animal	Method	Effective Dose
3	Follicle-stimulating hormone—releasing hormone (FSH-RH) [60] [Follicle-stimulating hormone—releasing factor (FSH-RF)]	Stimulation of synthesis & release of FSH [Follicle-stimulating hormone] [4,39]	(pyro)Glu·His· Trp·Ser·Tyr· Gly·Leu·Arg· Pro·Gly—NH$_2$	Man [58,60,63]; cattle [56]; rat [25,39]; sheep [15]; swine [64]	Man, ♂♀ [31-33]	Intravenous, subcutaneous	300-700 μg HP [31-33]; 10 μg S
4					Rat, ♂♀	In vivo, intracarotid [64]	1.0 ng P [64]
5						In vitro [40]	0.5 ng P [64]
6	Growth hormone—releasing hormone (GH-RH) [60] [Growth hormone—releasing factor (GH-RF)]	Stimulation of synthesis & release of growth hormone	Polypeptide, possibly Val· His·Leu·Ser· Ala·Glu·Glu· Lys·Glu·Ala	Man [58,60,63]; cat, dog, mouse [41]; cattle, sheep [14,41, 60]; rat [12, 46]	Monkey [35]	Intravenous	10-20 mg CRE [35]
7					Rat [61]	In vivo, intracarotid	1 ng [61]
8						In vitro	0.1 pg [61]
9						Radioimmunoassay	Not active by RIA [59]

continued

Hormone (Abbreviation) [Synonym]	Function [Synonym]	Structure or Chemical Nature	Occurrence	Activity		
				Animal	Method	Effective Dose
Luteinizing hormone—releasing hormone (LH-RH) [60] [Luteinizing hormone—releasing factor (LH-RF)]	Stimulation of synthesis & release of LH [Luteinizing hormone]	(pyro)Glu·His·Trp·Ser·Tyr·Gly·Leu·Arg·Pro·Gly—NH$_2$ [4,37]	Man [58,60,63]; cattle [10,52,60]; monkey [10]; rabbit [17]; rat [38]; sheep [22]; swine [57,64]	Man, ♂♀ [31-33]	Intravenous, subcutaneous	10-700 µg HP [31-33]; 2 µg S
				Rabbit [24]	Intrahypophyseal	0.2 µg HP [4]
				Rat	In vivo, intravenous [47]	0.2 ng P, S [64]
					In vitro [51]	0.5 ng P, S [64]
				Sheep [50]	Intracarotid	1 µg HP, S
Melanocyte-stimulating hormone—releasing hormone (MRH) [MSH-releasing factor (MRF)]	Stimulation of release of melanotropin [MSH]	Unknown	Rat	Rat [16,27]	Intravenous	80-800 µg CRE
			Sheep?	....	In vitro	....
Melanocyte-stimulating hormone—release-inhibiting hormone (MRIH) [MSH release-inhibiting factor (MIF)]	Inhibition of release of melanotropin [MSH]	Pro·Leu·Gly—NH$_2$ [11,43]	Man, cat, cattle, rabbit, rat, sheep, swine, frog [27,28]	Rat [26,27,29,30]	In vitro, intravenous	60 ng HP
				Frog [28,34]	Intrahypophyseal	100 ng HP; 10 ng S [34]
Prolactin-releasing hormone (PRH) [Prolactin-releasing factor (PRF)]	Stimulation of synthesis & release of luteotropin [Prolactin]	Unknown	Blackbird, chicken, duck, pigeon, quail, turkey [44]	Pigeon, blackbird	In vitro	CRE
Prolactin release-inhibiting hormone (PRIH) [60] [Prolactin release-inhibiting factor (PIF)]	Inhibition of synthesis & release of luteotropin [Prolactin]	Unknown	Cattle, sheep, swine [55]; rat [45,67]	Rat	In vivo, intravenous [1,18,36]	5-40 mg CRE [36]
					In vitro [45,55,67]	10-20 mg CRE [55]
Thyrotropin-releasing hormone (TRH) [60] [Thyrotropin-releasing factor (TRF)]	Stimulation of synthesis & release of thyrotropin	(pyro)Glu·His·Pro—NH$_2$ [9,42]	Man [63]; cattle [60]; goat [19]; rabbit [65]; rat [66]; sheep [21,60]; swine [60,62]	Man, ♂♀ [7,23,53]	Radioimmunoassay Orally	20 mg S
					Intravenous	100-800 µg P, S
				Goat, sheep [49]	In vitro	500 ng
				Mouse, ♂♀ [48,60,62]	In vivo, intravenous [6]	1 ng P, S
				Rat	In vivo, intravenous [8]	20 ng S
					In vitro [54,60,62]	10 pg P, S

Contributors: Schally, Andrew V., Arimura, Akira, and Kastin, Abba J.

References

[1] Amenomori, Y., and J. Meites. 1970. Proc. Soc. Exp. Biol. Med. 134:492.

[2] Arimura, A., et al. 1967. Endocrinology 81:235.

[3] Arimura, A., et al. 1969. Ibid. 83:300.

[4] Baba, Y., et al. 1971. Biochem. Biophys. Res. Commun. 44:459.

[5] Berthold, K., et al. 1970. Acta Endocrinol. (Copenhagen) 63:431.

[6] Bowers, C. Y., and A. V. Schally. 1970. In J. Meites, ed. Hypophysiotropic Hormones of the Hypothalamus. Williams and Wilkins, Baltimore. p. 74.

[7] Bowers, C. Y., et al. 1971. J. Med. Chem. 14:477.

[8] Burgus, R., and R. Guillemin. 1970. In J. Meites, ed. Hypophysiotropic Hormones of the Hypothalamus. Williams and Wilkins, Baltimore. p. 227.

[9] Burgus, R., et al. 1970. Nature (London) 226:321.

continued

[10] Campbell, H. J., et al. 1964. J. Physiol. (London) 170:474.

[11] Celis, M. E., et al. 1971. Proc. Nat. Acad. Sci. U.S. 68:1428.

[12] Deuben, R. R., and J. Meites. 1964. Endocrinology 74:408.

[13] DeWied, D., et al. 1969. Ibid. 85:561.

[14] Dhariwal, A. P. S., et al. 1965. Ibid. 77:932.

[15] Dhariwal, A. P. S., et al. 1965. Proc. Soc. Exp. Biol. Med. 118:999.

[16] Dhariwal, A. P. S., et al. 1966. Ibid. 121:996.

[17] Endroczi, E., and J. Hilliard. 1965. Endocrinology 77:667.

[18] Grosvenor, C. E., et al. 1967. Ibid. 81:1021.

[19] Guillemin, R. 1965. Excerpta Med. Found. Int. Congr. Ser. 87:284.

[20] Guillemin, R., and A. V. Schally. 1963. Tex. Rep. Biol. Med. 21:541.

[21] Guillemin, R., et al. 1962. C. R. Acad. Sci. 255:1018.

[22] Guillemin, R., et al. 1963. Endocrinology 73:564.

[23] Hershman, J. M., and J. A. Pittman. 1970. J. Clin. Endocrinol. Metab. 31:457.

[24] Hilliard, J., et al. 1971. Endocrinology 88:730.

[25] Igarashi, M., and S. M. McCann. 1964. Ibid. 74:446.

[26] Kastin, A. J., and A. V. Schally. 1966. Ibid. 79:1018.

[27] Kastin, A. J., and A. V. Schally. 1966. Gen. Comp. Endocrinol. 7:452.

[28] Kastin, A. J., and A. V. Schally. 1967. Ibid. 8:344.

[29] Kastin, A. J., et al. 1968. Endocrinology 83:137.

[30] Kastin, A. J., et al. 1969. Ibid. 84:20.

[31] Kastin, A. J., et al. 1969. J. Clin. Endocrinol. Metab. 29:1046.

[32] Kastin, A. J., et al. 1970. Ibid. 31:689.

[33] Kastin, A. J., et al. 1970. Amer. J. Obstet. Gynecol. 108:177.

[34] Kastin, A. J., et al. 1971. Proc. Soc. Exp. Biol. Med. 137:1437.

[35] Knobil, E., et al. 1968. Excerpta Med. Found. Int. Congr. Ser. 158:226.

[36] Kuroshima, A., et al. 1966. Endocrinology 78:216.

[37] Matsuo, H., et al. 1971. Biochem. Biophys. Res. Commun. 43:1334.

[38] McCann, S. M., et al. 1960. Proc. Soc. Exp. Biol. Med. 104:432.

[39] Mittler, J. C., and J. Meites. 1964. Ibid. 117:446.

[40] Mittler, J. C., and J. Meites. 1966. Endocrinology 78:500.

[41] Muller, E. E., et al. 1967. Gen. Comp. Endocrinol. 9:349.

[42] Nair, R. M. G., et al. 1970. Biochemistry 9:1103.

[43] Nair, R. M. G., et al. 1971. Biochem. Biophys. Res. Commun. 43:1376.

[44] Nicoll, C. S. 1965. J. Exp. Zool. 158:203.

[45] Pasteels, J. L. 1961. C. R. Acad. Sci. 254:2664.

[46] Pecile, A., et al. 1965. Endocrinology 77:241.

[47] Ramirez, V. D., and S. M. McCann. 1963. Ibid. 73:193.

[48] Redding, T. W., et al. 1966. Ibid. 79:229.

[49] Redding, T. W., et al. 1970. Gen. Comp. Endocrinol. 14:598.

[50] Reeves, J. J., et al. 1970. J. Anim. Sci. 31:933.

[51] Schally, A. V., and C. Y. Bowers. 1964. Endocrinology 75:312.

[52] Schally, A. V., and C. Y. Bowers. 1964. Ibid. 75:608.

[53] Schally, A. V., and C. Y. Bowers. 1971. Proc. 6th Midwest Conf. Thyroid Endocrinol., p. 25.

[54] Schally, A. V., and T. W. Redding. 1967. Proc. Soc. Exp. Biol. Med. 126:320.

[55] Schally, A. V., et al. 1965. Ibid. 118:350.

[56] Schally, A. V., et al. 1966. Endocrinology 79:1087.

[57] Schally, A. V., et al. 1967. Ibid. 81:77.

[58] Schally, A. V., et al. 1967. J. Clin. Endocrinol. Metab. 27:755.

[59] Schally, A. V., et al. 1968. Ann. N.Y. Acad. Sci. 148:372.

[60] Schally, A. V., et al. 1968. Recent Progr. Horm. Res. 24:497.

[61] Schally, A. V., et al. 1969. Endocrinology 84:1493.

[62] Schally, A. V., et al. 1969. J. Biol. Chem. 244:4077.

[63] Schally, A. V., et al. 1970. J. Clin. Endocrinol. Metab. 31:291.

[64] Schally, A. V., et al. 1971. Biochem. Biophys. Res. Commun. 43:393.

[65] Schreiber, V., et al. 1961. Experientia 17:264.

[66] Sinha, D., and J. Meites. 1965-66. Neuroendocrinology 1:4.

[67] Talwalker, P. K., et al. 1963. Amer. J. Physiol. 205:213.

76. SEX HORMONES: THALLOPHYTES

Symbols: Plus (+) and minus (−) indicate two compatible mating types in the absence of morphological differentiation as male and female.

Part I. Algae

For additional information, consult references 3, 28, 31-34, 42-44, 57, and 58.

	Family & Species	Hormone	Source	Organ Affected	Specific Activity	Reference
	Chrysophyta [1/]					
1	Coscinodiscaceae *Stephanopyxis turris*	Existence postulated but not demonstrated	Eggs	Sperm	Chemotaxis	52

[1/] All species listed are members of the class Bacillariophyceae.

continued

Part I. Algae

	Family & Species	Hormone	Source	Organ Affected	Specific Activity	Reference
2	Fragillariaceae *Rhabdonema adriaticum*	Existence postulated but not demonstrated	Vegetative cells of ♂ & ♀ strains	Vegetative cells of ♂ & ♀ strains	Induction of eggs & sperm	51
3	Eunotiaceae *Eunotia arcus*	Existence postulated but not demonstrated	Gametes	Gametes	Chemotropism of copulation tubes	16
4	Achnanthaceae *Cocconeis placentula*	Existence postulated but not demonstrated	Gametes	Gametes	Chemotropism of copulation tubes	17
	Chlorophyta					
5	Polyblepharidaceae *Dunaliella salina*	Not fully characterized	+ vegetative cells	− vegetative cells	Induction of clumping	30
6		Not fully characterized	− vegetative cells	+ vegetative cells	Induction of clumping	
7	Chlamydomonadaceae *Chlamydomonas eugametos*	Glucoprotein	+ gametes	− gametes	Induction of clumping	10,
8		Glucoprotein	− gametes	+ gametes	Induction of clumping	11, 37
9	*C. moewusii rotunda*	Not fully characterized	− gametes	+ gametes	Chemotaxis	54-56
10	*C. paupera*	Existence postulated but not demonstrated	Sedentary gametes	Motile gametes	Chemotaxis	40
11	*C. reinhardti*	Not fully characterized	+ gametes	− gametes	Induction of clumping	12
12		Not fully characterized	− gametes	+ gametes	Induction of clumping	
13	*C. suboogamum*	Existence postulated but not demonstrated	♀ gametes	♂ gametes	Chemotaxis	53
14	*Chlorogonium oogamum*	Existence postulated but not demonstrated	Eggs	Sperm	Chemotaxis	41
15	Volvocaceae *Eudorina conradii*	Existence postulated but not demonstrated	♀ colonies	Sperm packets	Chemotaxis	18
16	*Gonium pectorale*	Existence postulated but not demonstrated	Vegetative colonies of + & − strains	Vegetative colonies of + & − strains	Differentiation & liberation of gametes	50
17	*Pandorina morum*	Existence postulated but not demonstrated	Vegetative colonies of + & − strains	Vegetative colonies of + & − strains	Differentiation & liberation of gametes	2
18	*Volvox aureus* M5, DS	Not fully characterized (protein or polypeptide)	Sexual ♂ colonies	Vegetative ♀ colonies	Induction of sexual ♂ colonies	7,8
19	*V. carteri* f. *nagariensis*	Not fully characterized (protein or polypeptide)	Sexual ♂ colonies	Vegetative ♂ & ♀ colonies	Induction of sexual ♂ & ♀ colonies	49
20	*V. carteri* f. *weismannia*	Not fully characterized (protein or polypeptide)	Sexual ♂ colonies	Vegetative ♀ colonies	Induction of sexual ♀ colonies	26
21	*V. gigas*	Not fully characterized (protein or polypeptide)	Sexual ♂ colonies	Vegetative ♂ colonies	Induction of sexual ♂ colonies	48
22	*V. rousseletii*	Not fully characterized (protein or polypeptide)	Sexual ♂ colonies	Vegetative ♂ & ♀ colonies	Induction of sexual ♂ & ♀ colonies	36
23	*Volvulina pringsheimii*	Existence postulated but not demonstrated	Vegetative colonies of + & − strains	Vegetative colonies of + & − strains	Differentiation & liberation of gametes	47

continued

Part I. Algae

	Family & Species	Hormone	Source	Organ Affected	Specific Activity	Reference
	Tetrasporaceae					
24	*Tetraspora lubrica*	Not fully characterized	+ gametes	− gametes	Induction of clumping	15
25		Not fully characterized	− gametes	+ gametes	Induction of clumping	
	Hydrodictyaceae					
26	*Hydrodictyon reticulatum*	Existence postulated but not demonstrated	Sedentary gametes	Motile gametes	Chemotaxis	35
	Ulvaceae					
27	*Ulva lactuca*	Existence postulated but not demonstrated	+ gametes	− gametes	Induction of clumping	14
28		Existence postulated but not demonstrated	− gametes	+ gametes	Induction of clumping	
	Sphaeropleaceae					
29	*Sphaeroplea* sp.	Not fully characterized	"Eggs"	"Sperm"	Chemotaxis	40
	Cladophoraceae					
30	*Cladophora suhriana*	Existence postulated but not demonstrated	+ gametes	− gametes	Induction of clumping	13
31		Existence postulated but not demonstrated	− gametes	+ gametes	Induction of clumping	
	Oedogoniaceae					
32	*Bulbochaete hiloensis*	Existence postulated but not demonstrated	Eggs	Sperm	Chemotaxis	4
33	*Oedogonium* sp.	Not fully characterized	Dwarf ♂	Oogonial mother cells	Cell division	33,45
34		Not fully characterized	Oogonial mother cells	Androspores	Chemotaxis	
35		Not fully characterized	Oogonial mother cells	Dwarf ♂	Directional growth	
36		Not fully characterized	Oogonia	Sperm	Chemotaxis	
37	*O. cardiacum*	Not fully characterized	Oogonia	Sperm	Chemotaxis	23,24
	Zygnemataceae					
38	*Zygnema circumcarinatum*	Existence postulated but not demonstrated	Mixed conjugating cells of strains 42 & 43	Vegetative cells of strain 43	Induction of papilla formation in cells of strain 43	25
	Desmidiaceae					
39	*Cosmarium botrytis*	Existence postulated but not demonstrated	+ vegetative cells	− vegetative cells	Chemotaxis	1
	Caulerpaceae					
40	*Caulerpa cupressoides; C. serrulata; C. sertularioides*	Existence postulated but not demonstrated	Gametes	Gametes	Clumping	19
	Derbesiaceae					
41	*Derbesia tenuissima*	Existence postulated but not demonstrated	♀ gametes	♂ gametes	Chemotaxis	59
	Dasycladaceae					
42	*Acetabularia mediterranea; A. wettsteinii*	Existence postulated but not demonstrated	+ gametes	− gametes	Chemotaxis, clumping, & copulation	20
43		Existence postulated but not demonstrated	− gametes	+ gametes	Chemotaxis, clumping, & copulation	
44	*Dasycladus clavaeformis*	Not fully characterized	+ gametes	− gametes	Induction of clumping	27
45		Not fully characterized	− gametes	+ gametes	Induction of clumping	
	Pyrrophyta					
	Glenodiniaceae					
46	*Glenodinium lubiniensiforme*	Not fully characterized	Vegetative cells of each strain	Vegetative cells of each strain	Induction of gametes in opposite strain	9

continued

Part I. Algae

	Family & Species	Hormone	Source	Organ Affected	Specific Activity	Reference
			Phaeophyta			
47	Ectocarpaceae *Ectocarpus siliculosus*	Existence postulated but not demonstrated	♂ gametes	♀ gametes	Induction of sedentary habit	21
48		Ectocarpin:	♀ gametes	♂ gametes	Chemotaxis	39
49	*Giffordia mitchellae*	Existence postulated but not demonstrated	♀ gametes	♂ gametes	Chemotaxis	38
50	Cladostephaceae *Cladostephus spongiosus*	Existence postulated but not demonstrated	♀ gametes	♂ gametes	Chemotaxis	46
51	Cutleriaceae *Cutleria multifida*	Not fully characterized	♀ gametes	♂ gametes	Chemotaxis	22
52	Punctariaceae *Colpomenia sinuosa*	Existence postulated but not demonstrated	♀ gametes	♂ gametes	Chemotaxis	29
53	Fucaceae *Fucus serratus; F. spiralis; F. vesiculosus*	Hydrocarbon	Eggs	Sperm	Chemotaxis	4,5

Contributors: West, John A., and Machlis, Leonard

References
[1] Brandham, P. E. 1967. Can. J. Bot. 45:483.
[2] Coleman, A. W. 1959. J. Protozool. 6:249.
[3] Coleman, A. W. 1962. In R. A. Lewin, ed. Physiology and Biochemistry of Algae. Academic Press, New York. pp. 711-729.
[4] Cook, A. H., and J. A. Elvidge. 1951. Proc. Roy. Soc. B138:97.
[5] Cook, A. H., et al. 1948. Ibid. B135:293.
[6] Cook, P. W. 1962. Trans. Amer. Microsc. Soc. 81: 384.
[7] Darden, W. H., Jr. 1966. J. Protozool. 13:239.
[8] Darden, W. H., Jr. 1966. Ibid. 15:412.
[9] Diwald, K. 1933. Flora (Jena) 132:174.
[10] Forster, H., and L. Wiese. 1954. Z. Naturforsch. 9b: 548.
[11] Forster, H., and L. Wiese. 1955. Ibid. 10b:91.
[12] Forster, H., et al. 1956. Ibid. 11b:315.
[13] Foyn, B. 1934. Arch. Protistenk. 83:1.
[14] Foyn, B. 1934. Ibid. 83:154.
[15] Geitler, L. 1931. Biol. Zentralbl. 51:173.
[16] Geitler, L. 1951. Oesterr. Bot. Z. 98:292.
[17] Geitler, L. 1952. Z. Naturforsch. 7b:411-4.
[18] Goldstein, M. 1964. J. Protozool. 11:317.
[19] Goldstein, M., and S. Morrall. 1970. Ann. N.Y. Acad. Sci. 175:660.
[20] Hammerling, J. 1934. Arch. Protistenk. 83:57.
[21] Hartmann, M. 1934. Ibid. 83:110.
[22] Hartmann, M. 1950. Pubbl. Sta. Zool. Napoli 22: 120.
[23] Hill, G. J. C., and L. Machlis. 1970. Plant Physiol. 46:224.
[24] Hoffman, L. R. 1960. Southwest. Natur. 5:111.
[25] Hoshaw, R. W. 1968. In D. F. Jackson, ed. Algae, Man and the Environment. Syracuse Univ. Press, Syracuse, N.Y. pp. 135-184.
[26] Kochert, G. D. 1968. J. Protozool. 15:438.
[27] Köhler, K. 1957. Arch. Protistenk. 102:209.
[28] Köhler, K. 1966. Handb. Pflanzenphysiol. 18:282.
[29] Kunieda, H., and S. Suto. 1938. Shokubutsugaku Zasshi 52:539.
[30] Lerche, W. 1937. Arch. Protistenk. 88:236.
[31] Lewin, R. A. 1954. In D. H. Weinrich, et al, ed. Sex in Microorganisms. American Association for the Advancement of Science, Washington, D.C. pp. 100-133.
[32] Linskens, H. 1954-60. Fortschr. Bot. 17-23.
[33] Machlis, L. 1962. Proc. Annu. Biol. Colloq. Oreg. State Univ. 22:79.
[34] Machlis, L., and E. Rawitscher-Kunkel. 1967. In C. B. Metz and A. Monroy, ed. Fertilization. Academic Press, New York. v. 1, pp. 117-161.
[35] Mainx, F. 1931. Arch. Protistenk. 75:502.

continued

Part I. Algae

[36] McCracken, M. D., and R. C. Starr. 1970. Ibid. 112: 262.

[37] Moewus, F. 1933. Ibid. 80:469.

[38] Müller, D. 1969. Naturwissenschaften 56:220.

[39] Müller, D., et al. 1971. Science 171:815.

[40] Pascher, A. 1931. Jahrb. Wiss. Bot. 75:551.

[41] Pascher, A. 1931. Beih. Bot. Zentralbl. 48:466.

[42] Raper, J. R. 1952. Bot. Rev. 18:447.

[43] Raper, J. R. 1957. Symp. Soc. Exp. Biol. 11:143.

[44] Raper, J. R. 1966. Handb. Pflanzenphysiol. 18:214.

[45] Rawitscher-Kunkel, E., and L. Machlis. 1962. Amer. J. Bot. 49:117.

[46] Schreiber, L. 1931. Ber. Deut. Bot. Ges. 49:235.

[47] Starr, R. C. 1962. Arch. Mikrobiol. 42:130.

[48] Starr, R. C. 1968. Proc. Nat. Acad. Sci. U.S. 59: 1082.

[49] Starr, R. C. 1969. Arch. Protistenk. 111:204.

[50] Stein, J. R. 1958. Amer. J. Bot. 45:664.

[51] Stosch, H. A. von. 1958. Ber. Deut. Bot. Ges. 71: 241.

[52] Stosch, H. A. von, and G. Drebes. 1964. Helgolaender Wiss. Meeresunters. 11:209.

[53] Tschermak-Woes, E. 1962. Planta 59:68.

[54] Tsubo, Y. 1956. Shokubutsugaku Zasshi 69:1.

[55] Tsubo, Y. 1957. Ibid. 70:327.

[56] Tsubo, Y. 1961. J. Protozool. 8:114.

[57] Wiese, L. 1961. Fortschr. Zool. 13:119.

[58] Wiese, L. 1969. In C. B. Metz and A. Monroy, ed. Fertilization. Academic Press, New York. v. 2, pp. 135-188.

[59] Ziegler, J. R., and J. M. Kingsbury. 1964. Phycologia 4:105.

Part II. Fungi

For general information, consult references 9, 19, 36, 48, 69, and 72.

	Family & Species	Hormone	Source	Organ Affected	Specific Activity	Reference
			Phycomycetes			
	Blastocladiaceae					
1	*Allomyces* sp.	Sirenin [1]	♀ gametes	♂ gametes	Chemotaxis	11,22,23, 40-48,51, 52
	Saprolegniaceae					
2	*Achlya ambisexualis;*	Antheridiol [2]	♀ vegetative mycelia	♂ vegetative mycelia	Induction of antheridial branches & antheridia	1,8,9,26,50
3	*A. bisexualis;* in part, other *Achlya* spp. & *Thraustotheca* spp.			Antheridial branches	Chemotropism to oogonia	
4		Hormone A [3]; not fully characterized	♀ vegetative mycelia	♂ vegetative mycelia	Induction of antheridial hyphae	7,24,60-68, 70,73,74
5		Hormone A1; not fully characterized	♂ vegetative mycelia	♂ vegetative mycelia	Augmentation of hormones A & A2	7-10,19,24, 44,60-68, 70,73,74
6		Hormone A2; not fully characterized	♀ vegetative mycelia	♂ vegetative mycelia	Induction of antheridial hyphae	
7		Hormone A3; not fully characterized	♂ vegetative mycelia	♂ vegetative mycelia	Inhibition of hormones A, A1, & A2	
8		Hormone B; not fully characterized	♂ antheridial hyphae	♀ vegetative mycelia	Induction of oogonial initials	7,24,60-68, 70,73,74
9		Hormone C; not fully characterized	♀ oogonial initials	♂ antheridial hyphae	Chemotropic growth to oogonial initials; induction of antheridial differentiation	7-10,19,24, 44,60-68, 70,73,74

[1] Structure:

[2] Structure:

[3] Probably identical to antheridiol.

continued

Part II.   Fungi

	Family & Species	Hormone	Source	Organ Affected	Specific Activity	Reference
10		Hormone D; existence postulated but not demonstrated	♂ antheridia	♀ oogonial initials	Induction of oogonial differentiation	7,24,60-68, 70,73,74
11	Rhipidiaceae *Sapromyces reinschii*	Not fully characterized	♂ vegetative mycelia	♀ vegetative mycelia	Induction of oogonial initials	12
12		Not fully characterized	♀ vegetative mycelia	♂ vegetative mycelia	Induction of antheridial hyphae	
13		Existence postulated but not demonstrated	♀ oogonial initials	♂ antheridial hyphae	Chemotropic growth to oogonia	
14	Mucoraceae *Mucor hiemalis; M. mucedo; Phycomyces blakesleeanus; Rhizopus nigricans* [4]	+ progamone; not fully characterized	+ vegetative mycelia	− vegetative mycelia	Induction of production of zygogenic hormone	4,18,34,35, 53-59,74, 78
15		− progamone; not fully characterized	− vegetative mycelia	+ vegetative mycelia	Induction of production of zygogenic hormone	
16		Trisporic acid B [5]; trisporic acid C [6]	Mixed + & − mycelia or + mycelium in filtrate of − mycelium	+ & − vegetative mycelia	Induction of formation of zygophores	2,4,20,21, 27-32,53, 60,75
17		+ zygogenic hormone; not fully characterized	+ vegetative mycelia	− vegetative mycelia	Induction of zygophores	4,18,34,35, 53-59,74, 78
18		− zygogenic hormone; not fully characterized	− vegetative mycelia	+ vegetative mycelia	Induction of zygophores	
19		+ zygotropic hormone; not fully characterized	+ zygophores	− zygophores	Chemotropism	
20		− zygotropic hormone; not fully characterized	− zygophores	+ zygophores	Chemotropism	
21	Pilobolaceae *Pilobolus crystallinus*	Existence postulated but not demonstrated	+ vegetative mycelia	− vegetative mycelia	Initiation & control of entire sexual interaction prior to gametantial fusion	37,38
22		Existence postulated but not demonstrated	− zygophores	+ zygophores		
			Ascomycetes			
23	Saccharomycetaceae *Saccharomyces cerevisiae*	a-Hormone; not fully characterized (sterol?)	a vegetative cells	α vegetative cells	Induction of swelling	76,79,80, 81
24		α-Hormone; not fully characterized (sterol?)	α vegetative cells	a vegetative cells	Induction of swelling	

[4] *Blakeslea trispora* (family, Choanephoraceae) also possesses this group of hormones.   [5] Structure:

[6] Structure:

continued

Part II. Fungi

	Family & Species	Hormone	Source	Organ Affected	Specific Activity	Reference
25		Not fully characterized (peptide?)	α vegetative cells	a vegetative cells	Inhibition of cell division; induction of conjugation tubes	25,39,77
26	Sordariaceae *Bombardia lunata*	Not fully characterized	Spermatia	Compatible trichogyne	Chemotropism	82,83
27	*Neurospora sitophila*	Existence postulated but not demonstrated	Conidia, spermatia	Compatible trichogyne	Chemotropism	3
28		Existence postulated but not demonstrated	Vegetative mycelia	Conidia	Inhibition of conidial germination	
29	Diaporthaceae *Glomerella cingulata*	Existence postulated but not demonstrated	Gametangial knot	Compatible vegetative mycelia	Chemotropism	50
30	Pezizaceae *Ascobolus stercorarius*	Not fully characterized	A antheridia or oidia	a vegetative mycelia	Induction of ascogonia	13-17,70
31		Not fully characterized	a antheridia or oidia	A vegetative mycelia	Induction of ascogonia	
32		Not fully characterized	A or a antheridia or oidia	A or a trichogyne	Chemotropism	
33		Not fully characterized	A vegetative mycelia	a oidia or vegetative mycelia	Sexual activation	
34		Not fully characterized	a vegetative mycelia	A oidia or vegetative mycelia	Sexual activation	
35		Not fully characterized	A or a vegetative mycelia	A or a oidia	Suppression of vegetative development	
36		Existence postulated but not demonstrated	Ascogonia	Ensheathing hyphae	Stimulation of growth; chemotropism	
	Basidiomycetes					
37	Tremellaceae *Tremella mesenterica*	Not fully characterized	A vegetative cells	a vegetative cells	Induction of conjugation tubes	5,6
38		Not fully characterized	a vegetative cells	A vegetative cells	Induction of conjugation tubes	

Contributor: Raper, John R.

References

[1] Arsenault, G. P., et al. 1968. J. Amer. Chem. Soc. 90:3635.
[2] Austin, D. J., et al. 1969. Nature (London) 223: 1178.
[3] Backus, M. P. 1939. Bull. Torrey Bot. Club 66:63.
[4] Banbury, G. H. 1955. J. Exp. Bot. 6:235.
[5] Bandoni, R. J. 1963. Can. J. Bot. 41:467.
[6] Bandoni, R. J. 1965. Ibid. 43:627.
[7] Barksdale, A. W. 1960. Amer. J. Bot. 47:14.
[8] Barksdale, A. W. 1963. Mycologia 55:627.
[9] Barksdale, A. W. 1970. Science 166:831.
[10] Barksdale, A. W. 1970. Mycologia 62:411.
[11] Bhalerao, U. T., et al. 1970. J. Amer. Chem. Soc. 92:3429.
[12] Bishop, H. 1940. Mycologia 32:505.
[13] Bistis, G. N. 1956. Amer. J. Bot. 43:389.
[14] Bistis, G. N. 1957. Ibid. 44:436.
[15] Bistis, G. N. 1965. In K. Esser and J. R. Raper, ed. Incompatibility in Fungi. J. Springer, Berlin. pp. 23-31.
[16] Bistis, G. N., and L. S. Olive. 1968. Amer. J. Bot. 55:629.
[17] Bistis, G. N., and J. R. Raper. 1963. Ibid. 50:880.
[18] Burgeff, H. 1920. Ber. Deut. Bot. Ges. 38:318.
[19] Burnett, J. H. 1970. Fundamentals of Mycology. St. Martin, N.Y.
[20] Caglioti, L., et al. 1966. Tetrahedron, Suppl. 7:175.
[21] Cainelli, G., et al. 1967. Chim. Ind. (Milan) 49:629.
[22] Corey, E. J., and K. Achiwa. 1969. Tetrahedron Lett. 23:1837.
[23] Corey, E. J., et al. 1969. J. Amer. Chem. Soc. 91: 4318.
[24] Couch, J. N. 1926. Ann. Bot. (London) 40:848.
[25] Duntze, W., et al. 1970. Science 168:1472.

continued

Part II. Fungi

[26] Edwards, J. A., et al. 1969. J. Amer. Chem. Soc. 91:1248.
[27] Ende, H. van den. 1967. Nature (London) 215:211.
[28] Ende, H. van den. 1968. J. Bacteriol. 96:1928.
[29] Ende, H. van den, and D. Stegwee. 1971. Bot. Rev. 37:22.
[30] Ende, H. van den, et al. 1970. J. Bacteriol. 101:423.
[31] Gooday, G. W. 1968. Phytochemistry 7:2103.
[32] Gooday, G. W. 1968. New Phytol. 67:815.
[33] Grieco, P. A. 1969. J. Amer. Chem. Soc. 91:5660.
[34] Kehl, H. 1937. Arch. Mikrobiol. 8:379.
[35] Kohler, F. 1935. Planta 23:358.
[36] Kohler, K. 1967. Handb. Pflanzenphysiol. 18:121.
[37] Krafczyk, H. 1931. Ber. Deut. Bot. Ges. 49:141.
[38] Krafczyk, H. 1935. Beitr. Biol. Pflanz. 23:349.
[39] Levi, J. D. 1956. Nature (London) 177:753.
[40] Machlis, L. 1958. Ibid. 181:1790.
[41] Machlis, L. 1958. Physiol. Plant. 11:181.
[42] Machlis, L. 1958. Ibid. 11:845.
[43] Machlis, L. 1962. Proc. Annu. Biol. Colloq. Oreg. State Univ. 22:79.
[44] Machlis, L. 1966. In G. C. Ainsworth and A. S. Sussman, ed. The Fungi. Academic Press, New York. v. 2, pp. 415-433.
[45] Machlis, L., and E. Rawitscher-Kunkel. 1963. Int. Rev. Cytol. 15:97.
[46] Machlis, L., and E. Rawitscher-Kunkel. 1967. In C. B. Metz and A. Monroy, ed. Fertilization. Academic Press, New York. v. 1, pp. 117-161.
[47] Machlis, L., et al. 1966. Biochemistry 5:2147.
[48] Machlis, L., et al. 1968. J. Amer. Chem. Soc. 90:1674.
[49] McGahen, J. W., and H. E. Wheeler. 1951. Amer. J. Bot. 38:610.
[50] McMorris, T. C., and A. W. Barksdale. 1967. Nature (London) 215:320.
[51] Plattner, J. J., and H. Rapoport. 1971. J. Amer. Chem. Soc. 93:1758.
[52] Plattner, J. J., et al. 1969. Ibid. 91:4933.
[53] Plempel, M. 1957. Arch. Mikrobiol. 26:151.
[54] Plempel, M. 1960. Naturwissenschaften 47:472.
[55] Plempel, M. 1960. Planta 55:254.
[56] Plempel, M. 1963. Naturwissenschaften 50:226.
[57] Plempel, M. 1963. Planta 59:492.
[58] Plempel, M., and G. Braunitzer. 1958. Z. Naturforsch. 13b:302.
[59] Plempel, M., and W. Dawid. 1961. Planta 56:438.
[60] Raper, J. R. 1939. Amer. J. Bot. 26:639.
[61] Raper, J. R. 1939. Science 89:321.
[62] Raper, J. R. 1940. Amer. J. Bot. 27:162.
[63] Raper, J. R. 1942. Ibid. 29:159.
[64] Raper, J. R. 1942. Proc. Nat. Acad. Sci. U.S. 28:509.
[65] Raper, J. R. 1950. Ibid. 36:524.
[66] Raper, J. R. 1950. Bot. Gaz. 112:124.
[67] Raper, J. R. 1951. Amer. Sci. 39:110.
[68] Raper, J. R. 1952. Bot. Rev. 18:447.
[69] Raper, J. R. 1954. In D. H. Wenrich, et al., ed. Sex in Microorganisms. American Association for the Advancement of Science, Washington, D.C. pp. 42-81.
[70] Raper, J. R. 1957. Symp. Soc. Exp. Biol. 11:143.
[71] Raper, J. R. 1960. Amer. J. Bot. 47:794.
[72] Raper, J. R. 1966. Handb. Pflanzenphysiol. 18:214.
[73] Raper, J. R., and A. J. Haagen-Smit. 1942. J. Biol. Chem. 143:311.
[74] Ronsdorf, J. 1931. Planta 14:482.
[75] Sutter, R. P. 1970. Science 168:1590.
[76] Takao, N., et al. 1970. Develop. Growth Differ. 12:299.
[77] Throm, E., and W. Duntze. 1970. J. Bacteriol. 104:1388.
[78] Verkaik, C. 1930. Proc. Kon. Ned. Akad. Wetensch. 33:656.
[79] Yanagishima, N. 1969. Planta 87:110.
[80] Yanagishima, N. 1971. Physiol. Plant. 24:260.
[81] Yanagishima, N., et al. 1968. Proc. Int. Congr. Genet., 12th, 1:65.
[82] Zickler, H. 1937. Z. Indukt. Abstamm. Vererbungsl. 73:403.
[83] Zickler, H. 1952. Arch. Protistenk. 98:1.

77. RELATIVE ACTIVITY OF GROWTH REGULATORS: PLANTS

Part I. Cell Elongation of Oat Coleoptiles

Promotive or inhibitory effects were determined by floating fifteen 3-mm-long subapical segments of decapitated *Avena* coleoptiles, dark-grown for 90-92 hours, on the surface of 25 ml of a solution of the specified regulator, 2.5 mM potassium maleate at pH 4.5, and 90 mM sucrose in a covered petri dish, at 24°C for 24 hours. Values are concentrations of the regulator giving either promotion or inhibition of growth as compared with controls.

Growth Regulator	Concentration, in μM Giving			Growth Regulator	Concentration, in μM Giving	
	10% Promotion	5% Inhibition			10% Promotion	5% Inhibition
Phenoxyacetic Acids			5	(3-Acetylphenoxy)acetic acid	100	2500
			6	(4-Acetylphenoxy)acetic acid	Inactive	1000
1 Phenoxyacetic acid	300	1000	7	(3-Aminophenoxy)acetic acid	500	....
2 (3-Acetamidophenoxy)acetic acid	Inactive	....	8	(4-Aminophenoxy)acetic acid	Inactive	....
3 (4-Acetamidophenoxy)acetic acid	Inactive	....	9	(2-Bromophenoxy)acetic acid	50	....
4 (2-Acetylphenoxy)acetic acid	10	500	10	(3-Bromophenoxy)acetic acid	0.6	150

continued

Part I. Cell Elongation of Oat Coleoptiles

	Growth Regulator	Concentration, in μM Giving				Growth Regulator	Concentration, in μM Giving	
		10% Promotion	5% Inhibition				10% Promotion	5% Inhibition
11	(4-Bromophenoxy)acetic acid	3	....		50	[3-(Methylsulfonyl)phenoxy]-acetic acid	3000	....
12	[3-(n-Butyl)phenoxy]acetic acid	Inactive	70		51	{3-[(Trifluoromethyl)sulfonyl]-phenoxy}acetic acid	30	....
13	(3-Carboxyphenoxy)acetic acid	Inactive	....					
14	(4-Carboxyphenoxy)acetic acid	Inactive	....			Phenylacetic Acids		
15	(2-Chlorophenoxy)acetic acid	50	....		52	Phenylacetic acid	10	1000
16	(3-Chlorophenoxy)acetic acid	2	250		53	[3-(Acetamido)phenyl]acetic acid	50	500
17	(4-Chlorophenoxy)acetic acid	0.4	200		54	[4-(Acetamido)phenyl]acetic acid	Inactive	....
18	(2,4-Dichlorophenoxy)acetic acid	1	200		55	(3-Acetylphenyl)acetic acid	20	....
19	(3-Cyanophenoxy)acetic acid	30	3300		56	(4-Acetylphenyl)acetic acid	Inactive	200
20	(4-Cyanophenoxy)acetic acid	Inactive	....		57	(3-Aminophenyl)acetic acid	10	4000
21	(2-Ethylphenoxy)acetic acid	Inactive	500		58	(4-Aminophenyl)acetic acid	50	3000
22	(3-Ethylphenoxy)acetic acid	5	320		59	(2-Bromophenyl)acetic acid	0.3	460
23	(2-Fluorophenoxy)acetic acid	500	1000		60	(3-Bromophenyl)acetic acid	0.2	200
24	(3-Fluorophenoxy)acetic acid	300	700		61	(4-Bromophenyl)acetic acid	10	200
25	(4-Fluorophenoxy)acetic acid	0.5	700		62	(2-Carboxyphenyl)acetic acid	Inactive	3000
26	(2-Hydroxyphenoxy)acetic acid	1000	....		63	(3-Carboxyphenyl)acetic acid	500	2000
27	(3-Hydroxyphenoxy)acetic acid	200	10,000		64	(4-Carboxyphenyl)acetic acid	Inactive	2000
28	(4-Hydroxyphenoxy)acetic acid	Inactive	.....:....		65	(2-Chlorophenyl)acetic acid	0.7	650
29	(2-Iodophenoxy)acetic acid	50	300		66	(3-Chlorophenyl)acetic acid	0.3	310
30	(3-Iodophenoxy)acetic acid	0.5	100		67	(4-Chlorophenyl)acetic acid	3	300
31	(4-Iodophenoxy)acetic acid	Inactive	50		68	(2-Cyanophenyl)acetic acid	5	1000
32	(2-Methoxyphenoxy)acetic acid	Inactive	200		69	(3-Cyanophenyl)acetic acid	3	1250
33	(3-Methoxyphenoxy)acetic acid	200	800		70	(4-Cyanophenyl)acetic acid	Inactive	500
34	(4-Methoxyphenoxy)acetic acid	700	....		71	(3-Fluorophenyl)acetic acid	3	700
35	[3-(Trifluoromethoxy)phenoxy]-acetic acid	Inactive	10		72	(4-Fluorophenyl)acetic acid	2	600
36	[4-(Trifluoromethoxy)phenoxy]-acetic acid	Inactive	10		73	(2-Hydroxyphenyl)acetic acid	10	1000
					74	(3-Hydroxyphenyl)acetic acid	20	3000
37	(2-Methylphenoxy)acetic acid	200	....		75	(4-Hydroxyphenyl)acetic acid	40	2200
38	(3-Methylphenoxy)acetic acid	300	500		76	(2-Iodophenyl)acetic acid	0.6	3500
39	(4-Methylphenoxy)acetic acid	Inactive	600		77	(3-Iodophenyl)acetic acid	0.05	170
40	[3-(Trifluoromethyl)phenoxy]-acetic acid	0.3	200		78	(4-Iodophenyl)acetic acid	Inactive	100
41	(2-Nitrophenoxy)acetic acid	300	....		79	(2-Methoxyphenyl)acetic acid	5	....
42	(3-Nitrophenoxy)acetic acid	50	1400		80	(3-Methoxyphenyl)acetic acid	5	1300
43	(4-Nitrophenoxy)acetic acid	Inactive	1000		81	(4-Methoxyphenyl)acetic acid	50	1100
44	(2-Phenylphenoxy)acetic acid	Inactive	50		82	(2-Methylphenyl)acetic acid	1	1000
45	(3-Phenylphenoxy)acetic acid	Inactive	60		83	(3-Methylphenyl)acetic acid	2	500
46	(4-Phenylphenoxy)acetic acid	Inactive	50		84	(4-Methylphenyl)acetic acid	10	500
47	[3-(n-Propyl)phenoxy]acetic acid	300	110		85	[3-(Trifluoromethyl)phenyl]acetic acid	0.2	100
48	[3-(Methylthio)phenoxy]acetic acid	50	....		86	(3-Nitrophenyl)acetic acid	0.7	500
49	[3-(Pentafluorosulfo)phenoxy]-acetic acid	20	....		87	(4-Nitrophenyl)acetic acid	Inactive	100
					88	[3-(n-Propyl)phenyl]acetic acid	10	100
					89	[3-(Methylthio)phenyl]acetic acid	0.4	....

Contributor: Muir, Robert M.

General References

[1] Hansch, C., et al. 1963. J. Amer. Chem. Soc. 85: 2817.

[2] Muir, R. M., et al. 1967. Plant Physiol. 42:1519.

continued

Part II. Stem Curvature of Slit Pea and Leaf Expansion of Bean

Slit Pea Stem Curvature: Values are percent of activity of 3-indoleacetic acid [1,3-5]. **Bean Leaf Expansion:** Values are activity expressed as the reciprocal of the dose, in micromoles, causing 50% repression of leaf expansion [2,6].

	Growth Regulator	Slit Pea Stem Curvature	Bean Leaf Expansion		Growth Regulator	Slit Pea Stem Curvature	Bean Leaf Expansion
	Benzoic acids			33	(2,4-Dichlorophenoxy)ethyl-amine	....	296
1	2-Chlorobenzoic acid	....	<5	34	(2,4-Dichlorophenoxy)thioacetic acid	....	20,300
2	3-Chlorobenzoic acid	....	<15	35	(2,5-Dichlorophenoxy)acetic acid	15	<69
3	4-Chlorobenzoic acid	....	<8	36	(2,6-Dichlorophenoxy)acetic acid	3-4	137
4	2,4-Dichlorobenzoic acid	0	<19	37	(3,5-Dichlorophenoxy)acetic acid	<0.05	<44
5	2,5-Dichlorobenzoic acid	....	204	38	(2,4,5-Trichlorophenoxy)acetic acid	500	<4740
6	3,4-Dichlorobenzoic acid	0	<19	39	(2,4,6-Trichlorophenoxy)acetic acid	0.4	294
7	2,3,5-Trichlorobenzoic acid	....	2130	40	(2,3,4,6-Tetrachlorophenoxy)acetic acid	1	....
8	2,3,6-Trichlorobenzoic acid	200	....	41	(4-Chloro-2-methylphenoxy)acetic acid	500	513
9	3,4,5-Trichlorobenzoic acid	....	<45	42	(2,4-Difluorophenoxy)acetic acid	12	5360
	Indole compounds			43	(2,4-Diiodophenoxy)acetic acid	....	344
10	3-Indoleacetic acid	100	<18	44	(2,4,6-Trimethylphenoxy)acetic acid	0	....
11	β-(3-Indole)propionic acid	....	<19	45	D-α-(2,4-Dichlorophenoxy)propionic acid	1200	....
12	γ-(3-Indole)butyric acid	190	<40	46	DL-α-(2,4-Dichlorophenoxy)propionic acid	600	16,800
	Naphthalene compounds			47	β-(2,4-Dichlorophenoxy)propionic acid	....	<47
13	1-Naphthaleneacetic acid	250-370	<100	48	γ-(2,4-Dichlorophenoxy)butyric acid	....	18,500
14	2-Naphthaleneacetic acid	100	<19		Phenyl compounds		
15	1-Naphthaleneacetamide	10	....	49	Phenylacetic acid	3-10	<3
	Naphthoxy compounds			50	(2,4-Dichlorophenyl)acetic acid	15	....
16	1-Naphthoxyacetic acid	....	<40	51	(2,4-Dinitrophenyl)acetic acid	0.1	<23
17	2-Naphthoxyacetic acid	....	319	52	N-(2,4-Dichlorophenyl)glycine	....	204
18	1-Naphthoxyacetamide	....	25	53	Phenylthioacetic acid	0	....
	Phenoxy compounds			54	(4-Chloro-2-methylphenyl)thioacetic acid	200	....
19	Phenoxyacetic acid	0	<11	55	S-(2,4-Dichlorophenyl)thioglycolic acid	....	<47
20	(2-Bromophenoxy)acetic acid	....	<23	56	γ-Phenylbutyric acid	2	0
21	(4-Bromophenoxy)acetic acid	....	6160	57	(4-Bromophenyl)butyric acid	15	....
22	(2,4-Dibromophenoxy)acetic acid	....	11,500				
23	(2,4,6-Tribromophenoxy)acetic acid	0.1	....				
24	(2-Chlorophenoxy)acetic acid	4	<19				
25	(3-Chlorophenoxy)acetic acid	....	<37				
26	(4-Chlorophenoxy)acetic acid	200	18,700				
27	(2,4-Dichlorophenoxy)acetic acid	200-1200	23,500				
28	Butyl (2,4-dichlorophenoxy)acetate	....	23,100				
29	(2,4-Dichlorophenoxy)acetyl chloride	....	19,900				
30	(2,4-Dichlorophenoxy)acetamide	....	7760				
31	(2,4-Dichlorophenoxy)acetanilide	....	30,800				
32	(2,4-Dichlorophenoxy)ethanol	....	22				

Contributors: Brown, James W., and Weintraub, Robert L.

References
  [1] Bonner, J. 1950. Plant Biochemistry. Academic Press, New York.
  [2] Brown, J. W., and R. L. Weintraub. 1950. Bot. Gaz. 111:448.
  [3] Thimann, K. V. 1951. In F. Skoog, ed. Plant Growth Substances. Univ. Wisconsin Press, Madison.
  [4] Thimann, K. V. 1952. Plant Physiol. 27:392.
  [5] Thimann, K. V. Unpublished. Univ. California, Santa Cruz, 1971.
  [6] Weintraub, R. L., et al. Unpublished. George Washington Univ., Washington, D.C., 1971.

Data are for selected examples of molecular modifications changing metabolites to cytotoxic compounds or to inhibitors of biochemical functions in certain microbial or mammalian cells. In most cases, a competitive relationship or reversal of toxic action by the metabolite has been demonstrated. In some cases, biological antagonism extends to metabolites other than the one most similar in structure. Names in brackets "-( )-" refer to the column heading in brackets.

	Metabolite	Antimetabolite -(Synonym)-	Structural Alteration	Antimetabolite Action
			Vitamins	
1	Thiamine	3-o-Aminobenzyl-4-methyl-thiazolium chloride	2 CH for 2 N, loss of side chains	Inhibits fish thiaminase
2		Butylthiamine	Butyl for CH_3	....
3		Oxythiamine	OH for NH_2	Inhibits cocarboxylase synthesis
4		Pyrithiamine	CH=CH for S	Inhibits cocarboxylase synthesis
5	Riboflavin	Corresponding phenazine	2 CH for 2 N, 2 NH_2 for 2 OH	Inhibitory effects in animals & micro-organisms readily reversed by riboflavin
6		5-Deoxyriboflavin	H for OH	
7		6,7-Dichlororiboflavin	2 Cl for 2 CH_3	
8		Galactoflavin	Dulcityl for ribityl	
9		Hydroxyethyldimethylisoal-loxazine	Replacement of side chain	
10	Nicotinic acid (or amide)	6-Aminonicotinamide	NH_2 for H	Forms cofactor analogue
11		5-Fluoronicotinamide	F for H	Forms cofactor analogue
12		Methyl pyridyl ketone -(3-Acetylpyridine)-	$COCH_3$ for COOH	Induces deficiency in mammals
13		Pyridine-3-sulfonic acid or amide	SO_3H for COOH	Activity in animals & microorganisms
14	B_6	5-Deoxypyridoxal	H for OH	Inhibits pyridoxal kinase
15		4-Deoxypyridoxine	H for OH	Activity in animals & microorganisms
16		5-Homopyridoxine	CH_2 added	Activity in microorganisms
17		Isoniazid -(Pyridine-4-carboxylic acid hydrazide)-	....	Antitubercular drug; inhibits pyridoxal kinase
18		4-Methoxypyridoxine	OCH_3 for OH	Activity in mouse & chick
19		ω-Methylpyridoxine	C_2H_5 for CH_3	Replaces B_6 in some test systems
20		Toxopyrimidine	Pyrimidine portion of thiamine	Convulsant activity
21	Biotin	Biotin sulfone	SO_2 for S	Inhibition of various microorganisms prevented by biotin; in some species analogues can replace biotin
22		Dethiobiotin	2 H for S	
23		Homobiotin	Extra CH_2 group in side chain	
24		Ureylenecyclohexylvaleric acid	2 C for S	
25	Pantothenic acid	Dexpanthenol -(Pantothenyl alcohol)-	CH_2OH for COOH	Inhibits microorganisms, not animals
26		ω-Methylpantothenic acid	CH_3 for H	Induces deficiency in animals
27		Pantoyltaurine & derivatives	SO_3H & derivatives for COOH	Inhibits microorganisms & pantothenate-utilizing enzymes
28		Phenylpantothenone	COC_6H_5 for COOH	Inhibits microorganisms
29		Salicylyl β-alanine	o-Hydroxybenzoyl for pantoyl	Inhibits microorganisms
30	Folic acid	Aminopterin -(4-Aminopteroylglutamic acid)-	NH_2 for OH	Antileukemic drug [1]
31		Diaminopteridines, e.g., 2,4-diamino-6-methylpteridine	....	....
32		Diaminopyrimidines, e.g., pyrimethamine -(2,4-Diamino-5-(p-chlorophenyl)-6-ethylpyrimidine)-	....	Antimalarial drug [1]

[1] Inhibitors of dihydrofolate reductase.

continued

	Metabolite	Antimetabolite (Synonym)	Structural Alteration	Antimetabolite Action
33		Diaminotriazines, e.g., 2,4-Diamino-5-p-chlorophenyl-6-dimethyl-s-dihydrotriazine	....	....
34		Methotrexate (4-Amino-N^{10}-methylpteroylglutamic acid)	NH_2 for OH, CH_3 for H	Antileukemic drug [1]
35		Homofolic acid	Additional CH_2 in side chain	Substrates for dihydrofolate reductase
36		Methopterin (10-Methyl-pteroylglutamic acid)	CH_3 for H	
37		Ninopterin (9-Methylpteroylglutamic acid)	CH_3 for H	
38		Pteroylaspartic acid	Replacement of glutamic acid	
39		Tetrahydrohomofolic acid	Additional CH_2 in tetrahydrofolic acid	Competes with cofactor of thymidylate synthetase
40	p-Aminobenzoic acid	p-Aminoacetophenone & derivatives	COR for COOH	....
41		p-Aminobenzamide	$CONH_2$ for COOH	....
42		p-Aminosalicylic acid	OH for H	Antitubercular drug
43		Arsanilic acid	$AsO(OH)_2$ for COOH	....
44		Heterocyclic acids	N or S for C	....
45		p-Nitrobenzoic acid	NO_2 for NH_2	....
46		Sulfanilamide & derivatives	SO_2NH_2 or derivatives for COOH	Antimicrobial drugs
47	Inositol	Lindane (Hexachlorocyclohexane)	6 Cl for 6 OH	Acts on fungi & plants
48		γ isomer	....	Insecticide
49	B_{12}	5,6-Dichlorobenzimidazole	2 Cl for 2 CH_3	Replaces benzimidole of B_{12} in some bacteria
50	Choline	2-Amino-2-methylpropanol	....	....
51		α,α-Dimethylcholine	....	....
52		Triethylcholine	Ethyl for methyl	Competes with or replaces choline
53	Ascorbic acid	Glucoascorbic acid	Addition of CHOH & optical inversion	Inhibits growth of animals
54	α-Tocopherol	α-Tocopherol quinone	Opening of ring by addition of O	Antagonizes vitamins E & K
55	K	Bishydroxycoumarin & derivatives	O for C, side-chain alterations	Anticoagulant
56		2-Chloro-1,4-naphthoquinone	....	....
57		Dichlone (2,3-Dichloronaphthoquinone)	2 Cl for alkyl side-chains	....
58		Methoxynaphthoquinone	OCH_3 for CH_3	....
			Amino Acids [2]	
59	Alanine	Cycloserine (4-Amino-3-isoxazolidine)	....	Antibiotic; inhibits D-alanine utilization for cell wall formation
60	β-Alanine	β-Aminobutyric acid	CH_3 for H	....
61		Propionic acid	H for NH_2	
62	Arginine	Canavanine	O for CH_2	....
63	Asparagine	β-Aspartylhydrazine	NH_2 for H	Antagonist of asparagine & glutamine
64	Aspartic acid	α-Aspartophenone	C_6H_5 for OH	....
65		Hadacidin (N-Formyl-N-hydroxyglycine)	....	Antibiotic; inhibits purine synthesis in tumors; reversible with aspartate
66		β-Hydroxyaspartic acid	OH for H	....

[1] Inhibitors of dihydrofolate reductase. [2] L-α-Amino acids, unless otherwise specified.

continued

	Metabolite	Antimetabolite (Synonym)	Structural Alteration	Antimetabolite Action
67		β-Methylaspartic acid	CH_3 for H	....
68	Glutamic acid	N-Alkylglutamines	N-Alkyl for OH	....
69		β-Hydroxyglutamic acid	OH for H	....
70		δ-Hydroxylysine	Aminomethylcarbinol group for COOH	Inhibits glutamine synthetase
71		Methionine sulfoxide	$SOCH_3$ for COOH	Convulsant
72	Glutamine	Azaserine (O-Diazoacetyl-L-serine)	$COCHN_2$ for $CONH_2$, O for CH_2	Inhibits glutamine utilization for purine biosynthesis
73		N-Benzylglutamine	N-Benzyl derivative	....
74		6-Diazo-5-oxo-L-norleucine (DON)	$COCHN_2$ for $CONH_2$	Inhibits glutamine utilization for purine biosynthesis
75		N-Ethylglutamine	N-Ethyl derivative	....
76		γ-Glutamylhydrazine	NH_2 for H	....
77		Methionine sulfoximine	....	Convulsant; inhibits glutamine synthesis
78	Histidine	2-Thiazolealanine	Replacement of imidazole	Many analogues have antihistamine activity
79		1,2,4-Triazole-3-alanine	Replacement of imidazole	tivity
80	Isoleucine	α-Amino-3-chlorobutyric acid	Replacement of side chain for Cl	Valine acts as antagonist in microbial & mammalian systems; some analogues
81		2- & 3-Cyclohexeneglycine	Replacement of branched chain	are antagonists of both valine & iso-
82		Cyclopentaneglycine	Bridging methyl groups with methylene group	leucine
83		ω-Dehydroisoleucine (2-Amino-3-methylpentenoic acid)	....	
84		O-Methylthreonine	O-Methyl for ethyl	
85	Leucine	2-Amino-4-methylhexanoic acid	CH_3 extension of branched chain	Competitive antagonists in microbial metabolism; optical isomer D-leucine
86		Cyclopentanealanine	Replacement of branched chain	acts as an antimetabolite
87		Methallylglycine (2-Amino-4-methyl-4-pentenoic acid)	....	
88	Lysine	S-(2-Aminoethyl)cysteine	S for CH_2	Inhibits lysine incorporation into bone marrow
89		Hexahomoserine (α-Amino-ε-hydroxycaproic acid)	OH for NH_2	....
90	Methionine	Ethionine	C_2H_5 for CH_3	Toxic; interferes with protein synthesis
91		Methionine sulfone	....	....
92		Methoxinine	O for S	Replaces methionine for some organisms
93		Norleucine	CH_2 for S	....
94		Selenomethionine	Se for S	....
95	Phenylalanine	4-Chlorophenylalanine	Cl for H	Many related modifications of phenyl
96		1-Cyclopentenealanine	Cyclopentene for benzene ring	group or side chain; toxic effects of
97		3- & 4-Fluorophenylalanine	F for H	some analogues reversed by tyrosine;
98		2- & 3-Furylalanine	O for CH=CH	studied mainly in microbial & cell cul-
99		β-Hydroxyphenylalanine	OH for H	ture systems; several show antiviral ac-
100		2-Thiophenealanine	S for CH=CH	tivity
101	Proline	3,4-Dehydroproline	Unsaturated bond in ring	Competitive inhibitor in bacteria
102		Hydroxyproline	OH for H	Inhibits some fungi
103	Serine	α-Methylserine	CH_3 for H	Threonine acts as an antagonist of serine
104	Tryptophan	2-Azatryptophan	N for CH	Competitive antagonism[3]

[3] Other methylated, halogenated, & ring modifications studied; several analogues of indole & o-aminobenzoic acid competitive with tryptophan; studied mainly in microbial metabolism.

continued

	Metabolite	Antimetabolite (Synonym)	Structural Alteration	Antimetabolite Action
105		7-Azatryptophan	N for CH	Competitive antagonism[3]
106		2-Benzimidazolealanine	Replacement of indole ring	Fn[3]
107		Benzothienylalanine	S for N	Fn[3]
108		5-Fluorotryptophan	F for H	Inhibits *o*-aminobenzoic acid (anthranilic acid) conversion[3]
109		Indoleacrylic acid	Loss of NH_3	Fn[3]
110		5-Methyltryptophan	CH_3 for H	Noncompetitive in bacteria[3]
111		1-Naphthyleneacrylic acid	Loss of NH_3, C=C for N	Fn[3]
112	Tyrosine	*p*-Aminophenylalanine	NH_2 for OH	Antagonizes
113		3,5-Difluorotyrosine	F for H	Less toxic than monofluorotyrosine
114		3-Fluorotyrosine	F for H	Toxic for mice & rats
115		5-Hydroxy-2-pyridinealanine	Pyridine for benzene ring	Competitive antagonist

Purines[4]

	Metabolite	Antimetabolite (Synonym)	Structural Alteration	Antimetabolite Action
116	Adenine	Adenine arabinoside	....	Arabinosides of adenine, thymine, & uracil isolated from sponges
117		3'-Amino-3-deoxyadenosine	....	Antibiotic
118		2-Azaadenine	Ring N for CH	....
119		Benzimidazole	....	....
120		Cordycepin (3'-Deoxyadenosine)	....	Antibiotic
121		Decoyinine	CH for 6'-CH_2OH of psicofuranine	Antibiotic
122		2,6-Diaminopurine	NH_2 for H	....
123		2,6-Diaminopurine riboside	....	....
124		Formycin	C for 9-N, N for 8-C of adenosine	Antibiotic; antitumor activity
125		6-Methylthiopurine riboside	CH_3S for NH_2 of adenosine	Antileukemic agent
126		Nebularine[5]	....	Antibiotic; also occurs in mushrooms
127		Nucleocidin	F for 4'-H, SO_2NH_2 for PO_3H_2 of adenosine monophosphate	Antitrypanosomal antibiotic
128		Psicofuranine	CH_2OH for 1'-H of adenosine	Antibiotic
129		Purine	H for NH_2	....
130		Pyrazolopyrimidines, e.g., 4-aminopyrazolo[3,4-*d*]pyrimidine	....	....
131		Toyocamycin (4-Amino-5-cyano-7-(D-ribofuranosyl)-pyrrolo[2,3-*d*]pyrimidine)	....	Antibiotic
132		Tubercidin (4-Amino-7-(D-ribofuranosyl)-7*H*-pyrrolo-[2,3-*d*]pyrimidine; 7-Deazaadenosine)	....	Antibiotic
133		9-Xylosyladenine	3'-Epimer of adenosine	Triphosphate is antagonist of adenosine triphosphate in nucleotide synthesis
134	Hypoxanthine	Azathiopurine (6-[(1-Methyl-4-nitroimidazol-5-yl)thio]purine)	Derivative of 6-mercaptopurine	Immunosuppressive agent
135		7-Deazainosine	CH for N	....
136		6-Mercaptopurine	S for O	Antileukemic agent

[3] Other methylated, halogenated, & ring modifications studied; several analogues of indole & *o*-aminobenzoic acid competitive with tryptophan; studied mainly in microbial metabolism. [4] Generally, substrates for anabolic and catabolic enzymes. Diverse metabolic effects on biosynthesis and function of nucleic acids after conversion to nucleotides. Primarily of interest as potential anticancer and antiviral agents. [5] Purine riboside.

continued

	Metabolite	Antimetabolite ~~(Synonym)~~	Structural Alteration	Antimetabolite Action
137		6-Mercaptopurine arabinoside	2′-Epimer, SH for O	....
138		6-Mercaptopurine riboside	SH for O	Substrate for inosine kinase
139	Guanine	Guanazolo ~~(5-Amino-7-hydroxytriazolo[4,5-d]pyrimidine; 8-Azaguanine)~~	....	....
140		6-Thiodeoxyguanosine	....	
141		6-Thioguanine	....	
142		6-Thioguanosine	....	
143	Xanthine	Allopurinol ~~(4-Hydroxypyrazolo[3,4-d]pyrimidine)~~	....	Inhibits xanthine oxidase

Pyrimidines[4]

	Metabolite	Antimetabolite ~~(Synonym)~~	Structural Alteration	Antimetabolite Action
144	Cytosine	5-Azacytidine	Ring N for CH of cytidine	Incorporates into nucleic acids
145		6-Azacytidine & 2′-deoxy analog	Ring N for CH	Inhibits bacteria & tumors
146		6-Azacytosine	Ring N for CH	Inhibits bacteria
147		Cytosine arabinoside	2′-Epimer of cytidine	Antagonist of deoxycytidine; inhibits DNA[6] viruses
148		5-Fluorocytosine	F for H	....
149		5-Iododeoxycytidine	I for H of deoxycytidine	....
150	Orotic acid	5-Azaorotic acid	Ring N for CH	Inhibits orotate metabolism in animals
151		3-Diazocitrazinic acid	COCHN₂ for CONH of pyrimidine ring	Antagonist of orotic acid in microorganisms
152		Fluoroorotic acid	F for H	More potent than Br or Cl analogues
153		Uracil-6-methylsulfone	SO₂CH₃ for COOH	Inhibits formation of orotidylate
154		Uracil-6-sulfonic acid	SO₃H for COOH	
155	Thymine	6-Azathymidine	Ring N for CH	Nucleosides inhibit DNA[6] viruses and act as antagonists of thymidine
156		6-Azathymine	Ring N for CH	
157		5-Bromodeoxyuridine	Br for CH₃	
158		5-Bromouracil	Br for CH₃	
159		5-Iodouracil	I for CH₃	
160		Idoxuridine ~~(5-Iododeoxyuridine)~~	Br for CH₃	
161		5-Trifluoromethyldeoxyuridine	....	
162		5-Trifluoromethyluracil	CF₃ for CH₃	
163		5-Mercaptodeoxyuridine	CF₃ for CH₃	Substrate for thymidine kinase; nucleotide inhibits thymidylate synthetase
164		5-Mercaptouracil	SH for CH₃	Competes with thymine in microorganisms
165		5-Methylmercaptodeoxyuridine	CH₃S for CH₃ of thymidine	Substrate for thymidine kinase inhibits herpes virus
166	Uracil	5-Azauracil	Ring N for CH	Inhibits pyrimidine synthesis
167		6-Azauracil	Ring N for CH	Nucleotide inhibits orotidine-5′-phosphate decarboxylase (orotidylate decarboxylase)
168		6-Azauridine	Ring N for CH of uridine	....
169		3-Deazauridine	CH for N of uridine	Inhibits tumor cell cultures
170		Floxuridine ~~(5-Fluorodeoxyuridine)~~	F for H of 2′-deoxyuridine	Deoxyribonucleotide inhibits thymidylate synthetase; ribonucleotide incorporates into RNA[7]
171		5-Fluorouracil	F for H	
172		5-Fluorouridine	F for H of uridine	

[4] Generally, substrates for anabolic and catabolic enzymes. Diverse metabolic effects on biosynthesis and function of nucleic acids after conversion to nucleotides. Primarily of interest as potential anticancer and antiviral agents. [6] DNA = deoxyribonucleic acid. [7] RNA = ribonucleic acid.

continued

	Metabolite	Antimetabolite -(Synonym)-	Structural Alteration	Antimetabolic Action
173		Showdomycin -(3-β-D-Ribo-furanosylmaleimide)-	....	Antitumor antibiotic; inhibits uridine metabolism
174		2-Thiouracil	S for O	....
175		4-Thiouridine	S for O	Inhibits tumor cell cultures
	Miscellaneous[8/]			
176	Acetate	Fluoroacetate	F for H	Condenses with oxalacetate to form fluorocitrate
177	Glucose	D-Glucosamine	NH_2 for OH	Inhibits anaerobic glycolysis
178	Mevalonate	Fluoromevalonate	F for H	Inhibits cholesterol synthesis

[8/] Hormone analogues comprise large groups of synthetic compounds, some of which have either antagonistic or enhanced hormonal activity. Other classes of drugs provide many examples of biological antagonism by structural analogues; examples (not listed in table) include antihistamines, antagonists of serotonin (5-hydroxytryptamine), and inhibitors of monoamine oxidase (amine oxidase) and cholinesterase. Hexose and pentose antimetabolites, inhibitory analogues of intermediates on pathways of carbohydrate and lipid metabolism, and analogues retaining metabolite activity also do not appear in the table.

Contributors: Nichol, Charles A.; Bardos, Thomas J.

General References

[1] Balis, M. E. 1968. Antagonists and Nucleic Acids. Wiley-Interscience, New York.
[2] Bardos, T. J. 1972. Hematol. Rev. 3:53.
[3] Bloch, A. 1973. In E. A. Ariens, ed. Drug Design. Academic Press, New York. v. 4.
[4] Fox, J. J., et al. 1966. Progr. Nucl. Acid Res. Mol. Biol. 5:251.
[5] Hochster, R. M., and J. H. Quastel, ed. 1963. Metab. Inhibitors 1-2.
[6] Montgomery, J. A. 1965. Progr. Drug Res. 8:431.
[7] National Cancer Institute. 1959-72. Cancer Chemother. Rep. 1-56.
[8] Roy-Burman, P. 1970. Analogues of Nucleic Acid Components. J. Springer, New York.
[9] Schnitzer, R. J., and E. Hawking, ed. 1963-67. Exp. Chemother. 1-5.
[10] Sexton, W. A. 1963. Chemical Constitution and Biological Activity, Van Nostrand, Princeton.
[11] Woolley, D. W. 1952. A Study of Antimetabolites. J. Wiley, New York.

79. ANTICOAGULANTS

Part I. Dosages: Mammals and Fowl

Anticoagulant dosage is usually determined in vitro from the clotting time for heparin and heparinoid compounds, or from the prothrombin time for indirect anticoagulants [4, 18, 19, 26]. In vivo, the dosage required to prevent coagulation may be many times that indicated by the test in vitro [29].

Direct Anticoagulants (lines 1-26): Heparin and heparinoid compounds injected intravenously give a prolonged coagulation time. For some purposes, the high peak blood levels resulting from intravenous administration are essential [25]. Heparin is inactive orally [12], and intramuscular and subcutaneous injections are not as generally effective as those given intravenously [32]. The subcutaneous route, however, is the commonly accepted method of choice in doses of 100-150 units given every 8-12 hours, depending on clotting time [32]. It is desirable that the clotting time be approximately twice that of the normal control before the next dose is administered [32]. The effect on clotting time is increased and prolonged by p-(benzylsulfonamido)benzoic acid (caronamide) [28] and phosphorylated hesperidin [9]. The value for maximal clotting time can be estimated from the effect of clotting time in vitro [20]. One international unit of heparin = $1/130$ mg of the international standard; commercial heparin = 15-170 units/mg [24].

Indirect Anticoagulants (lines 27-72) are usually given orally. Dosage is dependent on the technique used to detect the change in prothrombin time and on individual susceptibility. A number of individuals in various species are refractory to one or more of these drugs, as judged by the effect on prothrombin time [23]. Significantly different results are obtained with tests for coagulation factors. The action of indirect anticoagulants is cumulative, and can be greatly enhanced by following the initial dose with a series of smaller doses. Such administration also avoids toxic side

continued

Part I. Dosages: Mammals and Fowl

effects. Bishydroxycoumarin (Dicumarol) must be dissolved with a small amount of 5 N sodium hydroxide for intravenous use; it can be given intraperitoneally in suspension in propylene glycol (10-100 mg/ml). Ethyl biscoumacetate (Tromexan) and warfarin are more soluble than bishydroxycoumarin, and can be given intraperitoneally in neutral solution.

Route: i.v. = intravenous; i.m. = intramuscular; s.c. = subcutaneous; i.p. = intraperitoneal. Values in parentheses are ranges, estimate "c" unless otherwise indicated (see Introduction).

	Species	Dose & Route	Maximal Clotting Time or Prothrombin Time		Achievement of Therapeutic Effect	Recovery Time	Remarks	Reference
			Control	Experimental				
			Dextran Sulfate					
1	Oryctolagus cu-	7 mg/kg	....	>80 min	....	50 min	....	2
2	niculus	17 mg/kg	....	>80 min	....	100 min		
3		21 mg/kg	....	>80 min	....	3 hr		
			Heparin [1]					
4	Homo sapiens	1000 units, i.v.	....	<4.5 min	....	....	28 hyporeactors	8
5				(4.5-7) min	....	....	50 normal reactors	
6				>7 min	....	....	7 hyperreactors	
7		5000 units, i.v.	15 min	115 min	....	....	....	
8		2500, 5000, & 7500 units, i.m.	12 min	18 min	1 hr	3-4 hr	....	25
9		10,000 units, i.m.	12 min	27 min	....	6 hr	....	
10		15,000 units, i.m.	12 min	30 min	....	10 hr	....	
11		5000 units, i.v.	....	55 min	....	2 hr	....	
12		7500 units, i.v.	....	80 min	....	4 hr	....	
13		10,000 units, i.v.	....	>80 min	....	4 hr	Incoagulable for 0.5 hr	
14		15,000 units, i.v.	....	>80 min	....	4 hr	Incoagulable for 1 hr	
15	Canis familiaris	2-3 units·kg^{-1}· min^{-1} for 2 hr, i.v.	20 s	5 min	....	4 hr	Thrombin time used; heparin appeared in lymph in 10-90 min	31
16		65-290 units/kg, i.v.	20 s	5 min	....	200 min		
17		30 units/kg, i.v.	....	34 min	....	32 min	....	20
18		100 units/kg, i.v.	....	>2 hr	....	....		
19		450 units/kg, s.c.	....	(0.25-3.5) hr	....	2-13 hr	Prolonged by anesthetics	18
20	Mus musculus	50 units every 8 hr, i.p.	(66-125) min	>20 min	....	7-8 hr	60 units for evening dose	30
21	Oryctolagus cu- niculus	2000 units every 8 hr, i.v.	....	>60 min	....	Up to 8 hr	Duration proportional to dose; adjusted to clotting time	30
22	Rattus norvegi-	330 units/kg, i.v.	....	>24 hr	....	2 hr	Blood levels are small percentage of dose	30
23	cus	1100 units/kg, s.c.	....	>24 hr	....	8 hr		
24		10,000 units/kg, i.p.	....	>24 hr	....	16 hr		

[1] In man and dog, intra-arterial injection of 1 mg heparin/100 ml of blood, plus 1.2 mg protamine/100 ml of blood into the venous outflow of the organ (e.g., limb, kidney), gives satisfactory local heparinization without general heparinization [14]. To the neutralized heparin, slowly inject protamine in an amount equal to 1.2-2.5 times the weight of heparin in the blood as determined by in vitro titration; an excess of protamine will be anticoagulant. [6,14,21]

continued

Part I. Dosages: Mammals and Fowl

	Species	Dose & Route	Maximal Clotting Time or Prothrombin Time		Achievement of Therapeutic Effect	Recovery Time	Remarks	Reference
			Control	Experimental				
	Xylan Polysulfuric Acid Ester [2]							
25	*Oryctolagus cuniculus*	15 mg/kg	....	65 min	....	1.5 hr	....	2
26		30 mg/kg	....	>60 min	....	2.5 hr		
	Bishydroxycoumarin [Dicumarol; 3,3'-methylenebis(4-hydroxycoumarin)]							
27	*Homo sapiens*	250 mg, oral	....	30 s	....	....	....	17
28		500 mg, oral	....	63 s	3-4 da	Variable		
29		750 mg, oral	....	106 s	....	....		
30		300 mg, oral [3]	....	....	....	....	Minimal effective dose	1
31	*Canis familiaris*	5 mg/kg, oral	10.7(6.7-14.7)[b] s	37.8(30.4-45.2)[b] s	....	6 da	Quick prothrombin time	23
32	*Mesocricetus auratus*	5 mg/kg, oral	....	No effect	....	....	Effective only with vitamin-K deficient diet	23
33	*Mus musculus*	5 mg/kg	10.6(7.8-13.4)[b] s	14.0(12.4-15.6)[b] s	....	24 hr	Quick prothrombin time	23
34	*Oryctolagus cuniculus*	0.37 mg	28 s	32 s	1 da	3 da	Susceptible animals only; 12.5% plasma prothrombin time	27
35		0.75 mg	28 s	40 s	1.67 da	4.67 da		
36		1.5 mg	28 s	46 s	2 da	8 da		
37		3 mg	28 s	67 s	2.5 da	9 da		
38		6 mg	28 s	85 s	3 da	10 da		
39		2.5 mg	8.0(7.6-8.4)[b] s	10.6(6.2-15.0)[b] s	....	5 da	Quick prothrombin time	23
40		5 mg/kg	8.0(7.6-8.4)[b] s	25.4(8.4 to >60) s	....	10 da		
41		10 mg/kg	15-16 s	27 s	1 da	4 da	Susceptibility decreased in puerperium	17
42		25 mg/kg	15-16 s	80 s	1 da	5 da		
43		50 mg/kg	15-16 s	6.5 min	2.5 da	8 da		
44		100 mg/kg	15-16 s	20 min	3 da	8 da		
45	*Rattus norvegicus*	800 ppm, in food	34 s	4 min	....	24 hr	Quick prothrombin time	30
46	*Gallus gallus* [4]	100 mg/kg [3]	....	....	....	....	Minimal effective dose	26
47		300 mg/kg	....	....	....	....	Effective dose	
	Diphenadione [Dipaxin; 2-diphenylacetyl-1,3-inandione]							
48	*Homo sapiens*	20 mg [3]	....	....	1-2 da	6-10 da	2-4 mg maintenance dose	11
49	*Mus musculus*	5 ppm, in drinking water	17.1 s	65 s	1 da	....	Dose increase for ♀ and after 2nd da	30
	Ethyl biscoumacetate [Tromexan; 3,3'-carboxymethylenebis(4-hydroxycoumarin)ethyl ester]							
50	*Homo sapiens*	1200 mg	14 s	25 s	30 hr	56 hr	....	5
51		1500 mg	14 s	32 s	30 hr	60 hr		
52	*Canis familiaris*	10 mg/kg	15 s	40 s	2 da	5 da	....	15
53		40 mg/kg	15 s	55 s	3 da	5 da		
54	*Oryctolagus cuniculus*	400 mg	12 s	30 s	30 hr	48 hr	....	5
55	*Gallus gallus* [4]	100-500 mg/kg [3]	....	....	....	....	Minimal effective dose	26

[2] As reported in the reference, a xylan polysulfuric acid ester, thrombocid, was used. In present terminology, thrombocid is sodium polyanhydromannuronic acid sulfate. [3] Initial dose. [4] Synonym: *G. domesticus.*

continued

Part I. Dosages: Mammals and Fowl

Species	Dose & Route	Maximal Clotting Time or Prothrombin Time		Achievement of Therapeutic Effect	Recovery Time	Remarks	Reference
		Control	Experimental				
Ethylidenebis(4-hydroxycoumarin) [EDC]							
56 Homo sapiens	0.5 g³/	....	Great individual variation	....	....	0.2 g maintenance dose	10
57 Oryctolagus cuniculus	20 mg, oral	7 s	10 s	....	....	....	10
58	30 mg, oral	7 s	27 s	....	....		
Phenindione [2-Phenyl-1,3-inandione]							
59 Canis familiaris	50 mg/kg	(9-13) s	30 s	26 hr	36 hr	....	22
60 Oryctolagus cuniculus	50 mg/kg	(10-12) s	22 s	25 hr	40 hr	....	22
61 Rattus norvegicus	1-1.5‰, in food³/	21 s	53 s	1 da	....	0.85-1.0‰ in food for maintenance	30
Phenprocoumon [Liquamar; 3-(1-Phenylpropyl)-4-hydroxycoumarin]							
62 Homo sapiens	18 mg	20.9 s	40 s	96 hr	144 hr	....	3
63	21 mg³/	20.9 s	36 s	96 hr	168 hr	Minimal effective dose	3,26
64 Oryctolagus cuniculus	2.5 mg	14 s	22 s	48 hr	96 hr	....	3
65	4 mg	14 s	22 s	72 hr	160 hr		
66 Gallus gallus⁴/	50 mg/kg³/	....	....	....	....	Minimal effective dose	26
Warfarin [3-(α-Acetonylbenzyl)-4-hydroxycoumarin]							
67 Homo sapiens	45-50 mg³/	12 s	28 s	....	....	5-10 mg/da for maintenance	7,32
68	100 mg, rectal	....	40 s	18-24 hr	6 da	....	13
69 Mus musculus	2.5 ppm, in food	22 s	(150-200) s	24 hr	....	Quick prothrombin time	30
70 Oryctolagus cuniculus	5 mg/kg, i.p.	7.9 s	35.2 s	48 hr	....	....	16
71	40 mg/kg, i.p.	9.6 s	36.7 s	48 hr	....		
72 Rattus norvegicus	6.7 ppm, in drinking water	22.6 s	70-200 s	24 hr	24 hr	Quick prothrombin time	30

³/ Initial dose. ⁴/ Synonym: *G. domesticus.*

Contributors: Jaques, Louis B.; Wright, Irving S.

References
[1] Allen, E. V. 1947. J. Amer. Med. Ass. 134:323.
[2] Astrup, T., et al. 1955. Scand. J. Clin. Lab. Invest. 7:204.
[3] Bourgain, R., et al. 1954. Circulation 10:680.
[4] Brown, A., and A. S. Douglas. 1952. Glasgow Med. J. 33:225.
[5] Burke, G. E., and I. S. Wright. 1951. Circulation 3:164.
[6] Chargaff, E., and K. B. Olson. 1937. J. Biol. Chem. 122:153.
[7] Clatanoff, D. V., and O. O. Meyer. 1956. Arch. Intern. Med. 97:753.
[8] De Takats, G. 1943. Surg. Gynecol. Obstet. 77:31.
[9] Evans, J. M., et al. 1955. Amer. Surg. 21:745.
[10] Fantl, P., and M. H. Nance. 1947. Med. J. Aust. 2:133.
[11] Field, J. B., et al. 1952. Proc. Soc. Exp. Biol. Med. 81:678.
[12] Fisher, A., and T. Astrup. 1939. Ibid. 42:81.
[13] Freeman, D. J., and O. O. Meyer. 1956. Ibid. 92:52.
[14] Gordon, L. A., et al. 1956. N. Engl. J. Med. 255:1026.
[15] Gruber, C. M., et al. 1951. Arch. Int. Pharmacodyn. Ther. 87:402.
[16] Heisey, S. T., et al. 1956. J. Agr. Food Chem. 4:144.
[17] Jansen, K. F. 1944. Dikumarin. E. Munksgaard, Copenhagen.
[18] Jaques, L. B. 1950. N. Engl. J. Med. 243:395.
[19] Jaques, L. B. 1955. Rev. Hematol. 10:379.
[20] Jaques, L. B., and A. G. Ricker. 1948. Blood 3:1197.
[21] Jaques, L. B., et al. 1938. Acta Med. Scand., Suppl. 90:190.

continued

79. ANTICOAGULANTS

Part I. Dosages: Mammals and Fowl

[22] Jaques, L. B., et al. 1950. Conf. Blood Clotting Allied Probl. Trans. 3:11.

[23] Jaques, L. B., et al. 1957. Arch. Int. Pharmacodyn. Ther. 111:478.

[24] Jaques, L. B., et al. 1967. Arzneim. Forsch. 17:774.

[25] Jorpes, J. E. 1950. Acta Chir. Scand. 99:476.

[26] Koller, T., and W. R. Merz, ed. 1955. Thromb. Embolism Proc. Int. Conf., 1st, 1954.

[27] Link, K. P. 1943. Harvey Lect. Ser. 39:162.

[28] Sirak, H. D., et al. 1948. Surgery 24:811.

[29] Solandt, D. Y., and C. H. Best. 1940. Lancet 1:1042.

[30] University of Saskatchewan, Department of Physiology. 1954-71. Theses. Saskatoon, Can.

[31] Willis, P. W., et al. 1953. J. Lab. Clin. Med. 42:968.

[32] Wright, I. S. Unpublished. Cornell Univ. Medical School, New York, 1971.

Part II. Interactions of Coumarin Drugs with Other Drugs

In addition to the drugs listed below, it is becoming evident that many others may interact with coumarin or phenylindanedione compounds. Therefore, patients on anticoagulant therapy of any type should be instructed not to take new drugs or discontinue the use of drugs without consulting the prescribing physician. The prothrombin time should be checked more frequently during the first 7-10 days after a change in the medication regimen.

Drug (Synonym)	Interaction with Coumarins	Management	Drug (Synonym)	Interaction with Coumarins	Management
1 Chloramphenicol 2 Clofibrate 3 p-Hydroxyphenyl-butazone (Oxyphenylbutazone) 4 Indomethacin 5 Norethandrolone 6 Phenylbutazone 7 Phenyramidol 8 Quinidine 9 Quinine 10 Salicylates 11 Sulfisoxazole	Potentiate anticoagulant effect	Reduce coumarin dosage while patient is on concurrent medication; increase coumarin dosage when concurrent medication is terminated	12 D-Thyroxine sodium salt (Dextrothyroxine; detrothyronine) 13 Allopurinol 14 Glutethimide 15 Griseofulvin 16 Haloperidol 17 Meprobamate 18 Phenobarbital	Inhibit anticoagulant effect	Increase coumarin dosage while patient is on concurrent medication; reduce coumarin dosage when concurrent medication is terminated
			19 Diphenylhydantoin 20 Tolbutamide	Potentiated by coumarins	Reduce diphenylhydantoin dosage while patient is on coumarins; increase drug dosage after coumarins are withdrawn

Contributor: Wright, Irving S.

Reference: Wright, I. S. 1971. Drug Ther. 1:9.

80. CHEMICALS PRODUCED BY ARTHROPODS

Part I. Exocrine Defensive Substances

Order, Family, & Species (Synonym)	Composition of Secretion (Synonym)	Reference
Arachnida		
1 Phalangida *Heteropachyloidellus robustus*	2,3-Dimethylquinone (2,3-dimethyl-*p*-benzoquinone); 2,5-dimethylquinone (2,5-dimethyl-*p*-benzoquinone); 2,3,5-trimethylquinone (2,3,5-trimethyl-*p*-benzoquinone)	53,54
2 Thelyphonida *Mastigoproctus giganteus*	Acetic acid; caprylic acid	46

continued

Part I. Exocrine Defensive Substances

Order, Family, & Species (Synonym)	Composition of Secretion (Synonym)	Reference
	Insecta	
	Hymenoptera—Formicidae	
3 Acanthomyops claviger	Citral; citronellal; 2,6-dimethyl-5-heptenaldehyde (2,6-dimethyl-5-hepten-1-al); 2,6-dimethyl-5-hepten-1-ol; formic acid; n-hendecane (n-undecane); n-tridecane	36,94
4 Aphaenogaster longiceps	α-Farnesene	32
5 Atta capiguara; A. laevigata	Citronellol	20
6 A. sexdens (A. sexdens rubropilosa)	Citral; geraniol; phenylacetic acid	20-22,64
7 Camponotus sp.; C. maculatus; Cataglyphis bicolor	Formic acid	129
8 Dolichoderus clarki	Dolichodial; 4-methyl-2-hexanone	29
9 D. dentata	Dolichodial	29
10 D. scabridus	Dolichodial; iridodial; isoiridomyrmecin; 6-methyl-5-hepten-2-one	29
11 Dorymyrmex bicolor (Conomyrma pyramicus bicolor)	6-Methyl-5-hepten-2-one	74
12 D. pyramicus (C. pyramicus pyramicus)	Iridodial; 6-methyl-5-hepten-2-one	74
13 D. pyramicus flavopectus (C. pyramicus flavopectus)	2-Heptanone; iridodial	14,74
14 Formica polyctena (F. rufa polyctena)	Formic acid	84
15 F. rufa	Formic acid	129
16 Harpogoxenus sublaevis	Phenylacetic acid	64
17 Iridomyrmex conifer	Iridodial; 6-methyl-5-hepten-2-one	31
18 I. detectus	Iridodial; 6-methyl-5-hepten-2-one	29,31
19 I. humilis	Iridomyrmecin	31
20 I. itinerans (I. nitidiceps)	Iridodial; 6-methyl-5-hepten-2-one	29
21 I. myrmecodiae	Dolichodial; 4-methyl-2-hexanone	29
22 I. nitidus	Isodihydronepatolactone; isoiridomyrmecin	28,31
23 I. pruinosus	2-Heptanone	18
24 I. pruinosus analis	Iridodial; iridomyrmecin	74
25 I. rufoniger	Dolichodial; iridodial; 6-methyl-5-hepten-2-one	29
26 Lasius alienus	Citronellal; citronellol; 2,6-dimethyl-5-heptenaldehyde (2,6-dimethyl-5-hepten-1-al); 2,6-dimethyl-5-hepten-1-ol; n-hendecane; n-undecane; n-tridecane	95
27 L. fuliginosus	Citral; dendrolasin; farnesal; n-hendecane (n-undecane); n-pentadecane; perillen; n-tridecane	8
28 L. spathepus	Citronellal	11
29 L. umbratus	Citronellal; citronellol; n-hendecane (n-undecane)	19,93
30 Manica rubida (Myrmica rubida)	Phenylacetic acid	64
31 Myrmecia gulosa	n-Heptadecane; cis-8-heptadecene; n-pentadecane	30
32 Myrmecina graminicola; Myrmica laevinodis	Phenylacetic acid	64
33 Myrmicaria natalensis	Limonene	57
34 Paltothyreus tarsatus	Dimethyldisulfide; dimethyltrisulfide	27
35 Tapinoma sessile	Iridodial; isoiridomyrmecin	74
36 T. simrothi karavaievi (T. nigerrimum)	Iridodial; 2-methyl-4-heptanone (iso-butyl propyl ketone); 6-methyl-5-hepten-2-one	88,130
37 Tetramorium caespitum	Phenylacetic acid	64
38 Ants[1]	Formic acid	140
Lepidoptera		
Notodontidae		
39 Cerura vinula	Formic acid	92

[1] Many species.

continued

Part I. Exocrine Defensive Substances

	Order, Family, & Species (Synonym)	Composition of Secretion (Synonym)	Reference
40	*Schizura leptinoides*	Formic acid	140
	Papilionidae		
41	*Baronia brevicornis; Graphium marcellus*	Isobutyric acid; 2-methylbutyric acid (α-methylbutyric acid)	51
42	*G. sarpedon choredon; Papilio aegeus aegeus; P. anactus*	Isobutyric acid; 2-methylbutyric acid (α-methylbutyric acid)	41
43	*P. cresphontes*	Isobutyric acid; 2-methylbutyric acid (α-methylbutyric acid)	51
44	*P. demoleus sthenelus*	Isobutyric acid; 2-methylbutyric acid (α-methylbutyric acid)	41
45	*P. demodocus; P. glaucus*	Isobutyric acid; 2-methylbutyric acid (α-methylbutyric acid)	51
46	*P. machaon*	Isobutyric acid; 2-methylbutyric acid (α-methylbutyric acid)	45
47	*P. palamedes; P. polyxenes; P. troilus*	Isobutyric acid; 2-methylbutyric acid (α-methylbutyric acid)	51
	Coleoptera		
	Alleculidae		
48	*Prionychus ater*	2-Ethylquinone (2-ethyl-*p*-benzoquinone); *p*-toluquinone	119
	Cantharidae		
49	*Chauliognathus lecontei*	Dihydromatricaria acid	77
	Carabidae		
50	*Abacomorphus asperulus*	Angelic acid; formic acid; methacrylic acid; tiglic acid	83
51	*Abax ater; A. ovalis; A. parallelus*	Methacrylic acid; tiglic acid	113,126
52	*Acinopus* sp.	Formic acid	119
53	*Agonum assimilis (Platynus assimilis)*	Formic acid	126
54	*A. dorsalis (P. dorsalis)*	*n*-Decane; formic acid; *n*-hendecane (*n*-undecane); methyl salicylate	104,124,126
55	*A. marginatum; A. moestum; A. sexpunctatum; A. viduum*	Formic acid	126
56	*Amara familiaris; A. similata*	*n*-Decane; *n*-hendecane (*n*-undecane); methacrylic acid; tiglic acid; *n*-tridecane	104
57	*Amblytelus curtus*	Formic acid	83
58	*Anisodactylus binotatus*	Formic acid	109,126
59	*Arthropterus* sp.	2-Ethylquinone (2-ethyl-*p*-benzoquinone); *p*-toluquinone	83
60	*Asaphidion flavipes*	Salicylaldehyde; *n*-valeric acid	104
61	*Badister bipustulatus*	Formic acid	126
62	*Bembidion andreae; B. lampros*	Isobutyric acid; isovaleric acid	126
63	*B. quadriguttatum*	Salicylaldehyde; *n*-valeric acid	104,126
64	*Brachinus crepitans*	Quinone (*p*-benzoquinone); *p*-toluquinone	102
65	*B. explodens; B. sclopeta*	Quinone (*p*-benzoquinone); *p*-toluquinone	105
66	*Broscus cephalotes*	Isobutyric acid; isovaleric acid	126
67	*Calathus fuscipes*	Formic acid	126
68	*C. melanocephalus*	Formic acid	119,126
69	*Callistus lunatus*	2-Ethylquinone (2-ethyl-*p*-benzoquinone); quinone (*p*-benzoquinone); *p*-toluquinone	125,126
70	*Calosoma affini; C. alternans sayi*	Salicylaldehyde	65
71	*C. externum*	Salicylaldehyde	73
72	*C. macrum*	Salicylaldehyde	65
73	*C. marginalis*	Methacrylic acid; salicylaldehyde	73
74	*C. oceanicum*	*n*-Caproic acid; methacrylic acid; salicylaldehyde	83
75	*C. parvicollis*	Salicylaldehyde	65
76	*C. perigrinater*	Salicylaldehyde	72
77	*C. prominens*	Salicylaldehyde	47
78	*C. schayeri*	*n*-Caproic acid; methacrylic acid; salicylaldehyde	83
79	*C. scrutator*	Methacrylic acid	73
80	*C. sycophanta*	Methacrylic acid; salicylaldehyde; tiglic acid	26

continued

Part I. Exocrine Defensive Substances

	Order, Family, & Species (Synonym)	Composition of Secretion (Synonym)	Reference
81	*Carabus auratus; C. granulatus; C. problematicus*	Methacrylic acid; tiglic acid	113,126
82	*Carenum bonelli*	Angelic acid; hexenoic acid; isocrotonic acid; methacrylic acid	83
83	*C. interruptum*	Angelic acid; isocrotonic acid; methacrylic acid	83
84	*C. tinctillatum*	Angelic acid; isocrotonic acid; methacrylic acid; tiglic acid	83
85	*Castelnaudia superba*	Acetic acid; methacrylic acid; tiglic acid	83
86	*Chlaenius australis*	*m*-Cresol	83
87	*C. bipunctatus*	*m*-Cresol	104
88	*C. chrysocephalus*	*m*-Cresol	126
89	*C. cordicollis*	*m*-Cresol	48
90	*C. festivus; C. tristus*	*m*-Cresol	126
91	*C. vestitus*	2-Ethylquinone (2-ethyl-*p*-benzoquinone); quinone (*p*-benzoquinone); *p*-toluquinone	125
92	*Clivina basalis*	Quinone (*p*-benzoquinone); *p*-toluquinone	83
93	*C. fossor*	2-Methoxy-3-methylquinone (2-methoxy-3-methyl-*p*-benzoquinone); quinone (*p*-benzoquinone); *p*-toluquinone	126
94	*Craspedophorus* sp.	*m*-Cresol; *n*-tridecane	83
95	*Cratoferonia phylarchus; Cratogaster melas*	Methacrylic acid; tiglic acid	83
96	*Cychrus rostratus*	Tiglic acid	113
97	*Diachromus germanus*	Formic acid	126
98	*Diaphoromenus edwardsi*	Formic acid	83
99	*Dicaelus dilatatus*	Formic acid	67
100	*D. purpuratus*	Formic acid	71
101	*D. splendidus*	Formic acid	67
102	*Dichirotrichus obsoletus*	Formic acid	126
103	*Dicrochile brevicollis; D. goryi*	Formic acid	83
104	*Drypta dentata*	Formic acid	126
105	*Elaphrus ripareus*	Isobutyric acid; isovaleric acid	126
106	*Eudalia macleayi*	Formic acid	83
107	*Eurylychnus blagravei*	Methacrylic acid; tiglic acid	83
108	*E. olliffi*	Isovaleric acid; methacrylic acid; tiglic acid	83
109	*Harpalus atratus; H. azurus (Ophonus azurus)*	Formic acid	126
110	*H. caliginosus*	Formic acid	66
111	*H. dimidiatus; H. distinguendus*	Formic acid	119,126
112	*H. griseus (Pseudophonus griseus)*	Formic acid	109,126
113	*H. luteicornis*	Formic acid	126
114	*H. pubescens (Psuedophonus pubescens)*	Formic acid	109,126
115	*H. tardus*	Formic acid	126
116	*Helluo costatus*	Formic acid; *n*-nonyl acetate; nonyl formate	83
117	*Helluomorphoides ferrugineus; H. latitarsus*	Formic acid; *n*-nonyl acetate	50
118	*Laccopterum foveigerum*	*n*-Caproic acid; crotonic acid; ethacrylic acid; hexenoic acid; isocrotonic acid; methacrylic acid; tiglic acid	83
119	*Lebia chlorocephala*	Formic acid	126
120	*Leistus ferrugineus*	Methacrylic acid; tiglic acid	126
121	*Licinus nitidior*	Formic acid	126
122	*Loricera pilicornis*	Isobutyric acid; isovaleric acid	126
123	*Loxandrus longiformis*	Salicylaldehyde	83

continued

Part I. Exocrine Defensive Substances

	Order, Family, & Species (Synonym)	Composition of Secretion (Synonym)	Reference
124	*Loxodactylus carinulatus*	Methacrylic acid	83
125	*Molops elatus*	Methacrylic acid; tiglic acid	126
126	*Mystropomus regularis*	2-Ethylquinone (2-ethyl-*p*-benzoquinone); quinone (*p*-benzoquinone); *p*-toluquinone	83
127	*Nebria livida*	Methacrylic acid; tiglic acid	126
128	*Notiophilus biguttatus*	Isobutyric acid; isovaleric acid	126
129	*Notonomus angustibasis; N. crenulatus; N. miles; N. muelleri; N. opulentus; N. rainbowi; N. scotti; N. triplogenioides; N. variicollis*	Formic acid	83
130	*Odacantha melanura*	Formic acid	126
131	*Omophron limbatum*	Isobutyric acid; isovaleric acid	126
132	*Pamborus alternans; P. guerini; P. pradieri; P. viridis*	Ethacrylic acid; methacrylic acid	83
133	*Panagaeus bipistulatus*	*m*-Cresol	126
134	*Pasimachus californicus; P. duplicatus*	Methacrylic acid	70
135	*Philophloeus australis*	Formic acid	83
136	*Philoscaphus tuberculatus*	Crotonic acid; isocrotonic acid; methacrylic acid; 4-methyl-valeric acid (isocaproic acid); tiglic acid	83
137	*Polystichus connexus*	Formic acid	126
138	*Promecoderus* spp.	*n*-Butyric acid; *n*-caproic acid; isovaleric acid	83
139	*Prosopognus harpaloides*	Methacrylic acid	83
140	*Pseudoceneus iridescens*	Methacrylic acid; tiglic acid	83
141	*Pterostichus cupreus (Poecilus cupreus); Pterostichus macer; P. melas*	*n*-Decane; *n*-hendecane (*n*-undecane); methacrylic acid; tiglic acid; *n*-tridecane	104,126
142	*P. metallicus; P. niger*	*n*-Decane; *n*-hendecane (*n*-undecane); methacrylic acid; tiglic acid; *n*-tridecane	104,113,126
143	*P. vulgaris*	*n*-Decane; *n*-hendecane (*n*-undecane); methacrylic acid; tiglic acid; *n*-tridecane	104,126
144	*Rhytisternus laevilaterus*	Methacrylic acid; tiglic acid	83
145	*Sarticus cyaneocinctus*	Formic acid	83
146	*Scaphinotus andrewsi gerari*	Methacrylic acid; tiglic acid	144
147	*S. andrewsi montana; S. viduus*	Methacrylic acid	144
148	*S. webbi*	Methacrylic acid; tiglic acid	144
149	*Siagonyx blackburni; Sphallomorpha colymbetoides; Sphodrosomus saisseti*	Formic acid	83
150	*Stenaptinus catoirei (Pheropsophus catoirei)*	Quinone (*p*-benzoquinone); *p*-toluquinone	105
151	*S. verticalis (P. verticalis)*	Quinone (*p*-benzoquinone); *p*-toluquinone	83
152	*Stenolophus mixtus*	Formic acid	126
153	*Trichosternus nudipes*	Methacrylic acid; tiglic acid	83
154	Cerambycidae *Aromia moschata*	Salicylaldehyde	59
155	Chrysomelidae *Melasoma populi*	Salicylaldehyde	59
156	*Phylodecta vittelinae*	Salicylaldehyde	135
157	*Plagiodera* sp.	Salicylaldehyde	59
158	Dytiscidae *Acilius sulcatus*	Benzoic acid; deoxycorticosterone (cortexone); *p*-hydroxybenzaldehyde; methylparaben (methyl *p*-hydroxybenzoate); 4,6-pregnadiene-3,20-dione; 4,6-pregnadien-21-ol-3,20-dione; 4,6-pregnadien-20α-ol-3-one; 4-pregnen-20α-ol-3-one	103,104,122

continued

Part I. Exocrine Defensive Substances

	Order, Family, & Species (Synonym)	Composition of Secretion (Synonym)	Reference
159	*Agabus bipustulatus*	Deoxycorticosterone (cortexone); *p*-hydroxybenzaldehyde; methyl 3,4-dihydroxybenzoate; methylparaben (methyl *p*-hydroxybenzoate)	103,104
160	*A. sturmi*	Benzoic acid; *p*-hydroxybenzaldehyde; methylparaben (methyl *p*-hydroxybenzoate); 4,6-pregnadiene-15α,20β-diol-3-one isobutyrate; 4,6-pregnadien-15α-ol-3,20-dione; 4,6-pregnadien-15α-ol-3,20-dione isobutyrate	103,104
161	*Colymbetes fuscus*	Benzoic acid; hydroquinone; *p*-hydroxybenzaldehyde; *p*-hydroxybenzoic acid; methyl 3,4-dihydroxybenzoate; methylparaben (methyl *p*-hydroxybenzoate)	104
162	*Copelatus ruficollis*	*p*-Hydroxybenzaldehyde; methylparaben (methyl *p*-hydroxybenzoate)	104
163	*Cybister confusus*	Deoxycorticosterone (4-pregnen-21-ol-3,20-dione)	37
164	*C. lateralimarginalis*	Benzoic acid; ethyl 3,4-dihydroxybenzoate; *p*-hydroxybenzaldehyde; methyl 3,4-dihydroxybenzoate; methylparaben (methyl *p*-hydroxybenzoate); 4,6-pregnadiene-12β,20α-diol-3-one; 4,6-pregnadiene-3,20-dione; 4,6-pregnadien-21-ol-3,20-dione; 4,6-pregnadien-20α-ol-3-one	103,104,119, 121,123
165	*C. limbatus*	Deoxycorticosterone (4-pregnen-21-ol-3,20-dione); 4,6-pregnadien-12β-ol-3,20-dione; 4,6-pregnadien-21-ol-3,20-dione; 4,6-pregnadien-20α-ol-3-one; 4-pregnen-12β-ol-3,20-dione; 4-pregnen-20α-ol-3-one	37,127
166	*C. tripunctatus*	Benzoic acid; deoxycorticosterone (4-pregnen-21-ol-3,20-dione); *p*-hydroxybenzaldehyde; methylparaben (methyl *p*-hydroxybenzoate); 4,6-pregnadien-12β-ol-3,20-dione; 4,6-pregnadien-20α-ol-3-one; 4-pregnen-20β-ol-3-one	37,103,104, 107
167	*Dytiscus latissimus*	Benzoic acid; *p*-hydroxybenzaldehyde; methylparaben (methyl *p*-hydroxybenzoate)	119,123
168	*D. marginalis*	Benzoic acid; deoxycorticosterone (cortexone); *p*-hydroxybenzaldehyde; methyl 3,5-dihydroxyphenylacetate; methylparaben (methyl *p*-hydroxybenzoate); 4,6-pregnadien-20α-ol-3-one; 4-pregnen-20α-ol-3-one	103,104,106, 108,119, 123
169	*Graphoderus cinereus*	Benzoic acid; *p*-hydroxybenzaldehyde; *p*-hydroxybenzoic acid; methyl 3,4-dihydroxybenzoate; methylparaben (methyl *p*-hydroxybenzoate)	104
170	*Hydroporus pallustris*	*p*-Hydroxybenzaldehyde	119,123
171	*Ilybius fenestratus*	Benzoic acid; dehydrotestosterone (1,4-androstadien-17β-ol-3-one); estradiol; estrone; hydroquinone; *p*-hydroxybenzaldehyde; *p*-hydroxybenzoic acid; methyl 8-hydroxyquinoline-2-carboxylate; methylparaben (methyl *p*-hydroxybenzoate); 4-pregnen-20β-ol-3-one; testosterone	103,104,120
172	*I. fuliginosus*	Testosterone	120
173	*Rhantus exsoletus*	Benzoic acid; hydroquinone; *p*-hydroxybenzaldehyde; *p*-hydroxybenzoic acid; methylparaben (methyl *p*-hydroxybenzoate)	104
174	Silphidae *Oecoptoma thorica; Phospuga atrata; Silpha obscura*	Ammonia	104
175	Staphylinidae *Stenus bipunctatus*	1,8-Cineole; isopiperitenol; 6-methyl-5-hepten-2-one	104
176	Tenebrionidae *Alphitobius diaperinus*	2-Ethylquinone (2-ethyl-*p*-benzoquinone); *p*-toluquinone	133

continued

Part I. Exocrine Defensive Substances

	Order, Family, & Species (Synonym)	Composition of Secretion (Synonym)	Reference
177	*Blaps gigas*	2-Ethylquinone (2-ethyl-*p*-benzoquinone); quinone (*p*-benzoquinone); *p*-toluquinone	119
178	*B. lethifera*	2-Ethylquinone (2-ethyl-*p*-benzoquinone); quinone (*p*-benzoquinone); *p*-toluquinone	111,119
179	*B. mortisaga; B. mucronata; B. requienii*	2-Ethylquinone (2-ethyl-*p*-benzoquinone); *p*-toluquinone	111,119
180	*B. sulcata; B. wiedemanni*	2-Ethylhydroquinone; 2-ethylquinone (2-ethyl-*p*-benzoquinone); hydroquinone; 2-methylhydroquinone; quinone (*p*-benzoquinone); *p*-toluquinone	61
181	*Diaperis boleti*	2-Ethylquinone (2-ethyl-*p*-benzoquinone); *p*-toluquinone	119
182	*D. maculata*	2-Ethylquinone (2-ethyl-*p*-benzoquinone); *p*-toluquinone	98
183	*Eleodes hispilabris*	2-Ethylquinone (2-ethyl-*p*-benzoquinone); *p*-toluquinone	12
184	*E. longicollis*	Caprylic acid; 2-ethylquinone (2-ethyl-*p*-benzoquinone); quinone (*p*-benzoquinone); *p*-toluquinone; *n*-tridecane	35,60,78
185	*Gnaptor spinimanus*	2-Ethylquinone (2-ethyl-*p*-benzoquinone); *p*-toluquinone	111
186	*Helops aeneus; H. quisquilus*	2-Ethylquinone (2-ethyl-*p*-benzoquinone); *p*-toluquinone	119
187	*Morisisa planta tingitiana*	*p*-Toluquinone	111
188	*Opatroides punctulatus*	2-Ethylquinone (2-ethyl-*p*-benzoquinone); *p*-toluquinone	119
189	*Pimelia confusa*	*p*-Toluquinone	111
190	*Scarus uncinus*	2-Ethylquinone (2-ethyl-*p*-benzoquinone); quinone (*p*-benzoquinone); *p*-toluquinone	119
191	*Tenebrio molitor*	*p*-Toluquinone	102,119
192	*T. obscurus*	Quinone (*p*-benzoquinone)	111
193	*Tribolium castaneum*	2-Ethylquinone (2-ethyl-*p*-benzoquinone); 2-methoxyquinone (2-methoxy-*p*-benzoquinone); *p*-toluquinone	1,63,98
194	*T. confusum*	2-Ethylquinone (2-ethyl-*p*-benzoquinone); *p*-toluquinone	1,52,58
195	*Zophobas rugipes*	*m*-Cresol; 2-ethylquinone (2-ethyl-*p*-benzoquinone); quinone (*p*-benzoquinone); *p*-toluquinone	132
	Hemiptera Cimicidae		
196	*Cimex lectularius*	Acetaldehyde; 2-butanone (methyl ethyl ketone); *trans*-2-hexenaldehyde (*trans*-2-hexenal); *trans*-2-octenaldehyde (*trans*-2-octenal)	39,116
	Coreidae		
197	*Acanthocephala declivis*	*trans*-2-Hexenaldehyde (*trans*-2-hexenal)	68
198	*A. femorata*	*trans*-2-Hexenaldehyde (*trans*-2-hexenal)	17
199	*A. granulosa*	*trans*-2-Hexenaldehyde (*trans*-2-hexenal)	65,68
200	*Acanthocoris sordidus*	Caproic aldehyde (*n*-hexanal); *trans*-2-hexenaldehyde (*trans*-2-hexenal)	134,145
201	*Agriopocoris frogatti; Amorbus alternatus*	Caproic aldehyde (*n*-hexanal); hexyl acetate	137
202	*Amorbus rhombifer*	Acetic acid; butyl butyrate (butyl butanoate); *n*-butyraldehyde (*n*-butanal); caproic aldehyde (*n*-hexanal); hexyl acetate	137
203	*A. rubiginosus*	Caproic aldehyde (*n*-hexanal); hexyl acetate	137,139
204	*Aulacosternum nigrorubrum*	Caproic aldehyde (*n*-hexanal); hexyl acetate	137
205	*Hygia opaca*	Caproic aldehyde (*n*-hexanal)	134,145
206	*Leptoglossus clypeatus*	Caproic aldehyde (*n*-hexanal)	69
207	*Libyaspis anglolensis*	*n*-Butyraldehyde (*n*-butanal); *trans*-2-decenaldehyde (*trans*-2-decenal); *trans*-2-hexenaldehyde (*trans*-2-hexenal); *trans*-4-keto-2-hexenaldehyde (*trans*-4-oxo-hex-2-enal); propionaldehyde (propanal)	38
208	*Mictis caja*	Caproic aldehyde (*n*-hexanal); hexyl acetate	137

continued

Part I. Exocrine Defensive Substances

	Order, Family, & Species (Synonym)	Composition of Secretion (Synonym)	Reference
209	*M. profana*	Caproic aldehyde (*n*-hexanal); hexyl acetate	2,137,139
210	*Pachycolpura manca*	Caproic aldehyde (*n*-hexanal); hexyl acetate	137
211	*Plinachtus bicoloripes*	Caproic aldehyde (*n*-hexanal); caprylic aldehyde (octanal)	134,145
212	*Pternistria bispina*[2/]	*trans*-2-Hexenaldehyde (*trans*-2-hexenal); *trans*-4-keto-2-hexenaldehyde (*trans*-4-oxo-hex-2-enal); *trans*-2-octenaldehyde (*trans*-2-octenal)	3
213	*P. bispina*[3/]	*n*-Butyraldehyde (*n*-butanal); caproic aldehyde (*n*-hexanal); hexyl acetate; hexyl butyrate (hexyl butanoate)	4
214	*Riptortus clavatus*	Butyraldehyde (butanal)	145
215	*Stenocoris apicalis (Leptocoris apicalis)*	Caproic aldehyde (*n*-hexanal); *trans*-2-decenaldehyde (*trans*-2-decenal); *trans*-2-octenaldehyde (*trans*-2-octenal); *n*-octyl acetate	2
	Corixidae		
216	*Corixa dentipes*	*trans*-4-Keto-2-hexenaldehyde (*trans*-4-oxo-hex-2-enal)	91
217	*Sigara falleni*	*trans*-4-Keto-2-hexenaldehyde (*trans*-4-oxo-hex-2-enal)	90,91
	Cydnidae		
218	*Macroscytus* sp.	*trans*-2-Decenyl acetate; *n*-dodecane; *trans*-4-keto-2-hexenaldehyde (*trans*-4-oxo-hex-2-enal); *trans*-2-octenyl acetate; *n*-tridecane	2
219	*Scaptocoris divergens*	*trans*-2-Heptenaldehyde (*trans*-2-heptenal); *trans*-2-hexenaldehyde (*trans*-2-hexenal); *trans*-2-octenaldehyde (*trans*-2-octenal)	97
	Hyocephalidae		
220	*Hyocephalus* sp.	Caproic aldehyde (*n*-hexanal)	137
	Miridae		
221	*Leptoterna dolabrata*	Acetaldehyde; *trans*-2-octenaldehyde (*trans*-2-octenal)	40
	Naucoridae		
222	*Ilyocoris cimicoides*	*p*-Hydroxybenzaldehyde; methylparaben (methyl *p*-hydroxybenzoate)	128
	Notonectidae		
223	*Notonecta glauca*	*p*-Hydroxybenzaldehyde; methylparaben (methyl *p*-hydroxybenzoate)	87
	Pentatomidae		
224	*Aelia fieberi*	*trans*-2-Decenaldehyde (*trans*-2-decenal); *trans*-2-octenaldehyde (*trans*-2-octenal)	134,145
225	*Biprorulus bibax*	*n*-Hendecane (*n*-undecane); *trans*-2-hexenaldehyde (*trans*-2-hexenal); *trans*-2-octenaldehyde (*trans*-2-octenal); *n*-tridecane	86
226	*Brochymena quadripustulata*	*trans*-2-Hexenaldehyde (*trans*-2-hexenal)	9,86
227	*Carpocoris purpureipennis*	*n*-Tridecane	96
228	*Dolycoris baccarum*	*trans*-2-Decenaldehyde (*trans*-2-decenal); *trans*-2-hexenaldehyde (*trans*-2-hexenal); *trans*-2-octenaldehyde (*trans*-2-octenal)	115,116
229	*Eurygaster* sp.	*trans*-2-Hexenaldehyde (*trans*-2-hexenal); *trans*-2-octenaldehyde (*trans*-2-octenal)	116
230	*Euschistus servus*	*trans*-2-Heptenaldehyde (*trans*-2-heptenal); *n*-tridecane	13
231	*Graphosoma rubrolineatum*	Caproic aldehyde (*n*-hexanal); *trans*-2-decenaldehyde (*trans*-2-decenal)	134,145
232	*Menida scotti*	*trans*-2-Decenaldehyde (*trans*-2-decenal)	145
233	*Musgraveia sulciventris (Rhoecocoris sulciventris)*	*trans*-2-Hexenaldehyde (*trans*-2-hexenal); *trans*-2-octenaldehyde (*trans*-2-octenal); *n*-tridecane	86,139
234	*Nezara antennata*	*trans*-2-Decenaldehyde (*trans*-2-decenal)	134,145

[2/] Nymph. [3/] Adult.

continued

Part I. Exocrine Defensive Substances

	Order, Family, & Species (Synonym)	Composition of Secretion (Synonym)	Reference
235	N. viridula	trans-2-Decenaldehyde (trans-2-decenal)	13,134,145
236	N. viridula var. smaragdula	trans-2-Decenaldehyde (trans-2-decenal); trans-2-decenyl acetate; n-dodecane; n-hendecane (n-undecane); trans-2-hexenaldehyde (trans-2-hexenal); 2-hexenyl acetate; trans-4-keto-2-hexenaldehyde (trans-4-oxo-hex-2-enal); trans-2-octenaldehyde (trans-2-octenal); trans-2-octenyl acetate; n-tridecane	55,139
237	Oebalus pugnax	trans-2-Heptenaldehyde (trans-2-heptenal); n-tridecane	13,16
238	Palomena viridissima	trans-2-Decenaldehyde (trans-2-decenal)	116
239	Piezodorus teretipes	trans-2-Hexenaldehyde (trans-2-hexenal)	56
240	Poecilometis strigatus	trans-2-Hexenaldehyde (trans-2-hexenal); trans-2-octenaldehyde (trans-2-octenal)	139
241	Scotinophora lurida	trans-2-Decenaldehyde (trans-2-decenal); trans-2-hexenaldehyde (trans-2-hexenal); trans-2-octenaldehyde (trans-2-octenal)	134,145
242	Tessaratoma aethiops	trans-2-Hexenaldehyde (trans-2-hexenal); trans-4-keto-2-hexenaldehyde (trans-4-oxo-hex-2-enal); trans-2-octenaldehyde (trans-2-octenal); trans-2-octenyl acetate; n-tridecane	2
	Plataspidae		
243	Ceratocoris cephalicus	n-Tridecane	2
	Pyrrhocoridae		
244	Dysdercus intermedius[4/]	Caproic aldehyde (n-hexanal); n-dodecane; trans-2-hexenaldehyde (trans-2-hexenal); trans-4-keto-2-hexenaldehyde (trans-4-oxo-hex-2-enal); 4-keto-2-octenaldehyde (4-oxo-oct-2-en-1-al); trans-2-octenaldehyde (trans-2-octenal); n-pentadecane; n-tridecane	23,24
245	D. intermedius[3/]	Acetaldehyde; caprylic aldehyde (n-octanal); trans-2-hexenaldehyde (trans-2-hexenal); trans-2-octenaldehyde (trans-2-octenal)	23
	Dermaptera—Forficulidae		
246	Forficula auricularia	2-Ethylquinone (2-ethyl-p-benzoquinone); p-toluquinone	110
	Isoptera		
	Mastotermitidae		
247	Mastotermes darwiniensis	Quinone (p-benzoquinone)	82
	Termitidae		
248	Amitermes herbertensis	α-Phellandrene; terpinolene	82
249	A. laurensis	Limonene; α-phellandrene; 2-pinene (α-pinene)	82
250	A. vitiosus	Limonene; myrcene; α-phellandrene; 2-pinene (α-pinene); terpinolene	82
251	Drepanotermes rubriceps	Limonene; α-phellandrene; terpinolene	82
252	Nasutitermes exitiosus; N. graveolus	Limonene; nopinene (β-pinene); 2-pinene (α-pinene)	81,82
253	N. longipennis	Limonene; nopinene (β-pinene); α-phellandrene; 2-pinene (α-pinene)	82
254	N. magnus	Limonene; nopinene (β-pinene); 2-pinene (α-pinene); terpinolene	82
255	N. triodiae	Nopinene (β-pinene); 2-pinene (α-pinene)	82
256	N. walkeri	Limonene; nopinene (β-pinene); 2-pinene (α-pinene)	81,82
257	Tumulitermes pastinator	Limonene; nopinene (β-pinene); 2-pinene (α-pinene)	82
	Orthoptera		
	Acrididae		
258	Romalea microptera	Romallenone[5/]	76,101

[3/] Adult. [4/] Larva. [5/] Name used by contributor for complex allenic structure.

continued

Part I. Exocrine Defensive Substances

	Order, Family, & Species (Synonym)	Composition of Secretion (Synonym)	Reference
	Blaberidae		
259	*Diploptera punctata*	Ethylquinone (2-ethyl-*p*-benzoquinone); quinone (*p*-benzoquinone); *p*-toluquinone	98
	Blattidae		
260	*Desmosteria scripta; Drymaplaneta communis; D. semivitta; D. shelfordi*	*trans*-2-Hexenaldehyde (*trans*-2-hexenal)	136
261	*Eurycotis biolleyi; E. decipiens*	Gluconic acid; *trans*-2-hexenaldehyde (*trans*-2-hexenal)	42,43
262	*E. floridana*	Gluconic acid; *trans*-2-hexenaldehyde (*trans*-2-hexenal)	42,43,100
263	*Euzosteria nobilis*	*trans*-2-Hexenaldehyde (*trans*-2-hexenal)	42,136
264	*Maoriblatta novaeseelandiae (Platyzosteria novaeseelandiae)*	*trans*-2-Hexenaldehyde (*trans*-2-hexenal)	99
265	*Megazosteria patula*	*trans*-2-Hexenaldehyde (*trans*-2-hexenal)	136
266	*Methana convexa*	2-Methylenebutyraldehyde (2-methylene butanal); 2-methylenevaleraldehyde (2-methylene pentanal)	136
267	*Pelmatosilpha coriacea*	*trans*-2-Hexenaldehyde (*trans*-2-hexenal)	10
268	*Platyzosteria* sp.	2-Methylenebutyraldehyde (2-methylene butanal); 2-octenaldehyde (2-octenal)	136
269	*P. armata*	2-Heptanol; 2-heptanone; 2-pentanol	136
270	*P. castenea*	2-Methylenebutanol; 2-methylenebutyraldehyde (2-methylene butanal)	138
271	*P. coolgardiensis*	*trans*-2-Hexenaldehyde (*trans*-2-hexenal)	136
272	*P. jungi; P. morosa*	2-Methylenebutanol; 2-methylenebutyraldehyde (2-methylene butanal)	138
273	*P. nitidella*	*trans*-2-Hexenaldehyde (*trans*-2-hexenal)	136
274	*P. occidentalis*	2-Methylenebutyraldehyde (2-methylene butanal)	136
275	*P. ruficeps*	2-Methylenebutanol; 2-methylenebutyraldehyde (2-methylene butanal)	138
276	*P. scabra; P. scabrella*	*trans*-2-Hexenaldehyde (*trans*-2-hexenal)	136
277	*P. soror (Cutilia soror)*	*trans*-2-Hexenaldehyde (*trans*-2-hexenal)	34
278	*P. stradbrokensis*	*trans*-2-Hexenaldehyde (*trans*-2-hexenal)	136
279	*Polyzosteria cuprea*	Caproic aldehyde (*n*-hexanal); caprylic aldehyde (octanal); *trans*-2-hexenaldehyde (*trans*-2-hexenal); hexenoic acid; hex-2-en-1-ol; *trans*-2-octenaldehyde (*trans*-2-octenal)	136
280	*P. limbata*	Caproic aldehyde (*n*-hexanal); *trans*-2-hexenaldehyde (*trans*-2-hexenal); hexenoic acid; hex-2-en-1-ol	136
281	*P. mitchelli*	*trans*-2-Hexenaldehyde (*trans*-2-hexenal)	136
282	*P. oculata*	Caproic aldehyde (*n*-hexanal); *trans*-2-hexenaldehyde (*trans*-2-hexenal); hexenoic acid; hex-2-en-1-ol	136
283	*P. pulchra*	Caproic aldehyde (*n*-hexanal); caprylic aldehyde (octanal); *trans*-2-hexenaldehyde (*trans*-2-hexenal); hexenoic acid; hex-2-en-1-ol; *trans*-2-octenaldehyde (*trans*-2-octenal)	136
284	*P. viridissima*	Caproic aldehyde (*n*-hexanal); *trans*-2-hexenaldehyde (*trans*-2-hexenal); hexenoic acid; hex-2-en-1-ol	136
285	*Zonioploca bicolor; Z. pallida*	*trans*-2-Hexenaldehyde (*trans*-2-hexenal)	136
	Phasmatidae		
286	*Anisomorpha buprestoides*	Anisomorphal	75
	Diplopoda		
	Cambalida		
287	*Cambala hubrichti*	2-Methoxy-3-methylquinone (2-methoxy-3-methyl-*p*-benzoquinone); *p*-toluquinone	49
	Spirostreptida		
288	*Doratogonus annulipes; Orthoporus conifer; O. flavior; O. punctilliger*	2-Methoxy-3-methylquinone (2-methoxy-3-methyl-*p*-benzoquinone); *p*-toluquinone	49

continued

Part I. Exocrine Defensive Substances

	Order, Family, & Species (Synonym)	Composition of Secretion (Synonym)	Reference
289	*Spirostreptidus castaneus*	Quinone (*p*-benzoquinone)	6
290	*Spirostreptus virgator*	*p*-Toluquinone	6
	Spirobolida		
291	*Chicobolus spinigerus; Floridobolus penneri*	2-Methoxy-3-methylquinone (2-methoxy-3-methyl-*p*-benzo-quinone); *p*-toluquinone	80
292	*Narceus annularis*	2-Methoxy-3-methylquinone (2-methoxy-3-methyl-*p*-benzo-quinone); *p*-toluquinone	80,89
293	*N. gordanus*	2-Methoxy-3-methylquinone (2-methoxy-3-methyl-*p*-benzo-quinone); *p*-toluquinone	80
294	*Pachybolus laminatus*	*p*-Toluqinone	6
295	*Rhinocricus insulatus*	*trans*-2-Dodecenaldehyde (*trans*-2-dodecenal); *p*-toluquinone	143
	Julida		
296	*Archiulus sabulosus*	2-Methoxy-3-methylquinone (2-methoxy-3-methyl-*p*-benzo-quinone); *p*-toluquinone	131
297	*Blaniulus guttulatus*	Cetyl acetate (hexadecan-1-ol acetate); 9-hexadecen-1-ol acetate; 2-methoxy-3-methylquinone (2-methoxy-3-meth-yl-*p*-benzoquinone); 9-octadecen-1-ol acetate; *p*-toluqui-none	142
298	*Brachyiulus ulineatus; Cylindroiulus teutonicus*	2-Methoxy-3-methylquinone (2-methoxy-3-methyl-*p*-benzo-quinone); *p*-toluquinone	112
299	*Julus terrestris*	Quinone (*p*-benzoquinone)	7
300	*Oriulus delus*	*o*-Cresol; 2-methoxy-3-methylquinone (2-methoxy-3-meth-yl-*p*-benzoquinone)	62
301	*Trigonoiulus lumbricinus*	2-Methoxy-3-methylquinone (2-methoxy-3-methyl-*p*-benzo-quinone); *p*-toluquinone	6,80
302	*Uroblaniulus canadensis*	2,3-Dimethoxyquinone (2,3-dimethoxy-*p*-benzoquinone); quinone (*p*-benzoquinone)	141
	Polydesmida		
303	*Apheloria corrugata; Cherokia georgiana*	Benzaldehyde; hydrogen cyanide	44
304	*Gomphodesmus pavani*	Benzaldehyde; hydrogen cyanide	5
305	*Nannaria* sp.; *Oxidus gracilus*	Benzaldehyde; hydrogen cyanide	44
306	*Pachydesmus crassicutus*	Benzaldehyde; hydrogen cyanide	15
307	*Polydesmus collaris collaris*	Benzaldehyde; hydrogen cyanide	25
308	*P. vicinus*	Cuminaldehyde; hydrogen cyanide	85
309	*P. virginiensis; Pseudopolydesmus serratus*	Benzaldehyde; hydrogen cyanide	44
	Chordeumida		
310	*Abacion magnum*	*p*-Cresol	48
	Glomerida		
311	*Glomeris marginata*	1,2-Dimethyl-4(3)-quinazolinone; 1-methyl-2-ethyl-4(3)-qui-nazolinone	79,114,117, 118

Contributor: Weatherston, John

References

[1] Alexander, P., and D. H. R. Barton. 1943. Biochem. J. 137:463.

[2] Baggini, A. R., et al. 1966. Rev. Espan. Entomol. 42:7.

[3] Baker, J. T., and P. A. Jones. 1969. Aust. J. Chem. 22:1793.

[4] Baker, J. T., and P. A. Kemball. 1967. Ibid. 20:395.

[5] Barbetta, M., et al. 1966. Estrat. Mem. Soc. Entomol. Ital. 45:196.

[6] Barbier, M., and E. Lederer. 1957. Biokhimiya 22: 236.

[7] Behal, A., and M. Phisalix. 1900. Bull. Mus. Nat. Hist. Natur. (Paris) 6:338.

[8] Bernardi, R., et al. 1967. Tetrahedron Lett., p. 3893.

continued

80. CHEMICALS PRODUCED BY ARTHROPODS

Part I. Exocrine Defensive Substances

[9] Blum, M. S. 1961. Ann. Entomol. Soc. Amer. 54: 410.

[10] Blum, M. S. 1964. Ibid. 57:600.

[11] Blum, M. S. 1969. Annu. Rev. Entomol. 14:57.

[12] Blum, M. S., and R. D. Crain. 1961. Ann. Entomol. Soc. Amer. 54:471.

[13] Blum, M. S., and J. G. Traynham. 1960. Int. Kongr. Entomol. Verh. 11th, Vienna, 3(Symp.3):48.

[14] Blum, M. S., and S. L. Warter. 1966. Ann. Entomol. Soc. Amer. 59:774.

[15] Blum, M. S., and J. P. Woodring. 1962. Science 138: 512.

[16] Blum, M. S., et al. 1960. Ibid. 132:1480.

[17] Blum, M. S., et al. 1961. Nature (London) 189:245.

[18] Blum, M. S., et al. 1966. J. Insect Physiol. 12:419.

[19] Blum, M. S., et al. 1968. Ann. Entomol. Soc. Amer. 61:1354.

[20] Blum, M. S., et al. 1968. Comp. Biochem. Physiol. 26:291.

[21] Butenandt, A. 1959. Naturwissenschaften 46:461.

[22] Butenandt, A., et al. 1959. Arch. Anat. Microsc. Morphol. Exp. 48:13.

[23] Calam, D. H., and G. C. Scott. 1969. J. Insect Physiol. 15:1695.

[24] Calam, D. H., and A. Youdeowei. 1968. Ibid. 14: 1147.

[25] Casnati, G., et al. 1963. Experientia 19:409.

[26] Casnati, G., et al. 1965. Ann. Soc. Entomol. Fr. 1: 705.

[27] Casnati, G., et al. 1967. Chim. Ind. (Milan) 49:57.

[28] Cavill, G. W. K., and D. V. Clark. 1967. J. Insect Physiol. 13:131.

[29] Cavill, G. W. K., and H. Hinterberger. 1960. Aust. J. Chem. 13:514.

[30] Cavill, G. W. K., and P. J. Williams. 1967. J. Insect Physiol. 13:1097.

[31] Cavill, G. W. K., et al. 1956. Aust. J. Chem. 9:288.

[32] Cavill, G. W. K., et al. 1967. Tetrahedron Lett., p. 2201.

[33] Cavill, G. W. K., et al. 1968. Aust. J. Chem. 21:2819.

[34] Chadha, M. S., et al. 1961. Ann. Entomol. Soc. Amer. 54:642.

[35] Chadha, M. S., et al. 1961. J. Insect Physiol. 7:46.

[36] Chadha, M. S., et al. 1962. Ibid. 8:175.

[37] Chadha, M. S., et al. 1970. Tetrahedron 26:2061.

[38] Cmelik, S. 1969. Hoppe Seylers Z. Physiol. Chem. 350:1076.

[39] Collins, R. P. 1968. Ann. Entomol. Soc. Amer. 61: 1338.

[40] Collins, R. P., and T. H. Drake. 1965. Ibid. 58:764.

[41] Crossley, A. C., and D. F. Waterhouse. 1969. Tissue Cell 1:525.

[42] Dateo, G. P., and L. M. Roth. 1967. Ann. Entomol. Soc. Amer. 60:1023.

[43] Dateo, G. P., and L. M. Roth. 1967. Science 155:88.

[44] Eisner, H. E., et al. 1963. Chem. Ind. (London), p. 124.

[45] Eisner, T., and Y. C. Meinwald. 1965. Ibid. 150: 1733.

[46] Eisner, T., et al. 1961. J. Insect Physiol. 6:272.

[47] Eisner, T., et al. 1963. Ann. Entomol. Soc. Amer. 56:37.

[48] Eisner, T., et al. 1963. Psyche 70:94.

[49] Eisner, T., et al. 1965. Ann. Entomol. Soc. Amer. 58:247.

[50] Eisner, T., et al. 1968. Ibid. 61:610.

[51] Eisner, T., et al. 1970. Ibid. 63:914.

[52] Engelhardt, M., et al. 1965. Science 150:632.

[53] Estable, C., et al. 1955. J. Amer. Chem. Soc. 77: 4942.

[54] Fieser, L. F., and M. Ardao. 1956. Ibid. 78:744.

[55] Gilby, A. R., and D. F. Waterhouse. 1965. Proc. Roy. Soc. B162:105.

[56] Gilchrist, T. L., et al. 1966. Proc. Roy. Entomol. Soc. London A41:55.

[57] Grunanger, P., et al. 1960. Atti Accad. Naz. Lincei Cl. Sci. Fis. Mat. Natur. Rend. 28:293.

[58] Hackman, R. H., et al. 1948. Biochem. J. 43:474.

[59] Hollande, A.-C. 1909. Ann. Univ. Grenoble, Sci. Med. 21:459.

[60] Hurst, J. J., et al. 1964. Ann. Entomol. Soc. Amer. 57:44.

[61] Ikan, R., et al. 1970. J. Insect Physiol. 16:2201.

[62] Kluge, A. F., and T. Eisner. 1971. Ann. Entomol. Soc. Amer. 64:314.

[63] Loconti, J. D., and L. M. Roth. 1953. Ibid. 46:281.

[64] Maschwitz, U., et al. 1970. J. Insect Physiol. 16: 387.

[65] McCullough, T. 1966. Ann. Entomol. Soc. Amer. 59:1018.

[66] McCullough, T. 1966. Ibid. 59:1020.

[67] McCullough, T. 1967. Ibid. 60:861.

[68] McCullough, T. 1967. Ibid. 60:862.

[69] McCullough, T. 1969. Ibid. 62:673.

[70] McCullough, T. 1969. Ibid. 62:1492.

[71] McCullough, T. 1969. Ibid. 62:1493.

[72] McCullough, T. 1969. Ibid. 62:1498.

[73] McCullough, T., and A. J. Weinheimer. 1966. Ibid. 59:410.

[74] McGurk, D. J., et al. 1968. J. Insect Physiol. 14:841.

[75] Meinwald, J., et al. 1962. Tetrahedron Lett., p. 29.

[76] Meinwald, J., et al. 1968. Ibid., p. 2959.

[77] Meinwald, J., et al. 1968. Science 160:890.

[78] Meinwald, Y. C., and T. Eisner. 1964. Ann. Entomol. Soc. Amer. 57:513.

[79] Meinwald, Y. C., et al. 1966. Science 154:390.

[80] Monro, A., et al. 1962. Ann. Entomol. Soc. Amer. 55:261.

[81] Moore, B. P. 1964. J. Insect Physiol. 10:371.

[82] Moore, B. P. 1968. Ibid. 14:33.

continued

Part I. Exocrine Defensive Substances

[83] Moore, B. P., and B. E. Wallbank. 1968. Proc. Roy. Entomol. Soc. London B37:62.

[84] Otto, D. 1960. Zool. Anz. 164:42.

[85] Pallares, E. S. 1946. Arch. Biochem. 9:105.

[86] Park, R. J., and M. S. Sutherland. 1962. Aust. J. Chem. 15:172.

[87] Pattenden, G., and B. W. Staddon. 1968. Experientia 24:1092.

[88] Pavan, M., and R. Trave. 1958. Insectes Soc. 5:299.

[89] Percy, J. E., and J. Weatherston. 1971. Can. J. Zool. 49:278.

[90] Pinder, A. R., and B. W. Staddon. 1965. J. Chem. Soc. London, p. 2955.

[91] Pinder, A. R., and B. W. Staddon. 1965. Nature (London) 205:106.

[92] Poulton, F. B. 1888. Brit. Ass. Advan. Sci. Meet. 57th, p. 765.

[93] Quilico, A., et al. 1957. Rend. Ist. Lomb. Sci. Lett. Cl. Sci. Math. Natur. 91:271.

[94] Regnier, F. E., and E. O. Wilson. 1968. J. Insect Physiol. 14:955.

[95] Regnier, F. E., and E. O. Wilson. 1969. Ibid. 15: 893.

[96] Remold, H. 1963. Nature (London) 198:764.

[97] Roth, L. M. 1961. Ann. Entomol. Soc. Amer. 54: 900.

[98] Roth, L. M., and B. Stay. 1958. J. Insect Physiol. 1:305.

[99] Roth, L. M., and E. R. Willis. 1960. Smithson. Misc. Collect. 141:1.

[100] Roth, L. M., et al. 1956. Science 123:60.

[101] Russell, S. W., and B. C. L. Weedon. 1969. J. Chem. Soc. D(1):85.

[102] Schildknecht, H. 1959. Angew. Chem. 71:524.

[103] Schildknecht, H. 1968. Angew. Chem. Suppl. Nachr. Chem. Tech. 18:311.

[104] Schildknecht, H. 1970. Angew. Chem. Int. Ed. Engl. 9:1.

[105] Schildknecht, H., and K. Holoubek. 1961. Angew. Chem. 73:1.

[106] Schildknecht, H., and D. Hotz. 1967. Ibid. 79:902.

[107] Schildknecht, H., and W. Kornig. 1968. Angew. Chem. Int. Ed. Engl. 7:62.

[108] Schildknecht, H., and U. Maschwitz. 1966. Ibid. 5:421.

[109] Schildknecht, H., and K. H. Weis. 1960. Z. Naturforsch. 15b:361.

[110] Schildknecht, H., and K. H. Weis. 1960. Ibid. 15b: 755.

[111] Schildknecht, H., and K. H. Weis. 1960. Ibid. 15b: 757.

[112] Schildknecht, H., and K. H. Weis. 1961. Ibid. 16b: 810.

[113] Schildknecht, H., and K. H. Weis. 1962. Ibid. 17b: 439.

[114] Schildknecht, H., and W. F. Wenneis. 1966. Ibid. 21b:552.

[115] Schildknecht, H., et al. 1962. Ibid. 17b:350.

[116] Schildknecht, H., et al. 1964. Angew. Chem. Int. Ed. Engl. 3:73.

[117] Schildknecht, H., et al. 1966. Z. Naturforsch. 21b: 121.

[118] Schildknecht, H., et al. 1967. Naturwissenschaften 54:196.

[119] Schildknecht, H., et al. 1964. Angew. Chem. Int. Ed. Engl. 3:73.

[120] Schildknecht, H., et al. 1967. Ibid. 6:558.

[121] Schildknecht, H., et al. 1967. Justus Liebigs Ann. Chem. 703:182.

[122] Schildknecht, H., et al. 1967. Z. Naturforsch. 22b: 938.

[123] Schildknecht, H., et al. 1968. Naturwissenschaften 55:230.

[124] Schildknecht, H., et al. 1968. Z. Naturforsch. 23b: 46.

[125] Schildknecht, H., et al. 1968. Ibid. 23b:637.

[126] Schildknecht, H., et al. 1968. Naturwissenschaften 55:112.

[127] Sipahimalani, A. T., et al. 1970. Ibid. 57:40.

[128] Staddon, B. W., and J. Weatherston. 1967. Tetrahedron Lett., p. 4567.

[129] Stumper, R. 1960. Naturwissenschaften 20:457.

[130] Trave, R., and M. Pavan. 1956. Chim. Ind. (Milan) 38:1015.

[131] Trave, R., et al. 1959. Ibid. 41:19.

[132] Tschinkel, W. R. 1969. J. Insect Physiol. 15:191.

[133] Tseng, Y. L., et al. 1971. Ann. Entomol. Soc. Amer. 64:425.

[134] Tsuyuki, T., et al. 1965. Agr. Biol. Chem. 29: 419.

[135] Wain, R. L. 1943. Long Ashton Agr. Hort. Res. Sta. (Univ. Bristol) Annu. Rep., p. 108.

[136] Wallbank, B. E., and D. F. Waterhouse. 1970. J. Insect Physiol. 16:2081.

[137] Waterhouse, D. F., and A. R. Gilby. 1964. Ibid. 10:977.

[138] Waterhouse, D. F., and B. E. Wallbank. 1967. Ibid. 13:1657.

[139] Waterhouse, D. F., et al. 1961. Ibid. 6:113.

[140] Weatherston, J. 1967. Quart. Rev. 21:287.

[141] Weatherston, J., and J. E. Percy. 1970. Can. J. Zool. 47:1389.

[142] Weatherston, J., et al. 1971. Chem. Phys. Lipids 7:98.

[143] Wheeler, J. W., et al. 1964. Science 144:540.

[144] Wheeler, J. W., et al. 1970. Ann. Entomol. Soc. Amer. 63:469.

[145] Yamamoto, Y., and T. Tsuyuki. 1964. Int. Union Pure Appl. Chem. Int. Symp. Chem. Natur. Prod. Kyoto, p. 133.

continued

Part II. Nonexocrine Defensive Substances

Compounds listed are produced by insects for their defense. The substances, however, are not necessarily biosynthetically elaborated by the insect, as they may be obtained from the plants the insect feeds on.

	Family & Species (Synonym)	Composition of Secretion (Synonym)	Reference
	Lepidoptera		
	Arctiidae		
1	*Arctia caja*	Acetylcholine; β,β-dimethylacrylylcholine; histamine	7,10,17
2	*A. villica*	Acetylcholine; histamine	7,17
3	*Cycnia tenera*	Acetylcholine; histamine	11,17
4	*Diacrisia mendica (Spilosoma lubricipeda)*	Acetylcholine; histamine	2,7,17
5	*Phragmatobia fuliginosa*	Acetylcholine; histamine	7,17
6	*Tyria jacobaeae*	Acetylcholine; histamine; integerrimine; senecionine; seneciphylline	1,2,7,17
7	*Utetheisa bella*	β,β-dimethylacrylylcholine	14,17
	Danaidae		
8	*Amauris lobengula; Danaus chrysippus; D. plexippus*	Calactin; calotropin	12,13
	Papilionidae		
9	*Battus philenor; B. polydamus; Ornithoptera priamus*	Aristolochic acid	17
10	*Papilio antimachus*	Cardenolides	17
11	*P. aristolochiae*	Aristolochic acid	15
12	*Troides helena*	Acetylcholine	17
	Sphingidae		
13	*Laothoe populi*	Histamine	3,17
	Zygaenidae		
14	*Procris geryon*	Hydrogen cyanide	8
15	*Zygaena filipendulae; Z. lonicerae*	Acetylcholine; histamine; hydrogen cyanide	8,17
16	*Z. trifolii*	Hydrogen cyanide	8
	Coleoptera		
	Meloidae		
17	*Epicauta adspersa; E. gorhami; E. pestifera; E. pilma; E. ruficeps; E. waterhousei*	Cantharidin	18
	Staphylinidae		
18	*Paederus fuscipes*	Pederin; pederone	4,5,9
	Homoptera		
	Aphididae		
19	*Aphis nerii*[1/]	Adynerin; odoroside-H; strospeside	16
20	*A. nerii*[2/]	Calotropin; proceroside	16
	Orthoptera		
	Acrididae		
21	*Phymateus baccatus; P. viridipes; Poekilocerus bufonius; P. pictus*	Calactin; calotropin	6,12

[1/] Fed on their natural food. [2/] Reared in laboratory on milkweed, *Asclepias curassavica,* for several generations.

Contributor: Weatherston, John

References

[1] Aplin, R., et al. 1968. Nature (London) 219:747.
[2] Bisset, G. W., et al. 1959. J. Physiol. (London) 146: 38.
[3] Bisset, G. W., et al. 1960. Proc. Roy. Soc. B152:255.
[4] Cardani, C., et al. 1965. Ann. Soc. Entomol. Fr. 1: 813.

continued

80. CHEMICALS PRODUCED BY ARTHROPODS

Part II. Nonexocrine Defensive Substances

[5] Cardani, C., et al. 1967. Tetrahedron Lett., p. 4023.

[6] Euw, J. von, et al. 1967. Nature (London) 214:35.

[7] Frazier, J. F. D., and M. Rothschild. 1960. Int. Kongr. Entomol. Verh. 11th, Vienna, 3(Symp.4):249.

[8] Jones, D. A., et al. 1962. Nature (London) 193:52.

[9] Matsumoto, T., et al. 1968. Tetrahedron Lett., p. 6297.

[10] Morley, J., and M. Schachter. 1963. J. Physiol. (London) 168:706.

[11] Parsons, J. A., and M. Rothschild. Unpublished. Ashton, Peterborough, England.

[12] Reichstein, T. 1967. Naturwissenschaften 20:499.

[13] Reichstein, T., et al. 1968. Science 161:861.

[14] Rothschild, M., and P. T. Haskell. 1966. Proc. Roy. Entomol. Soc. London A41:167.

[15] Rothschild, M., et al. 1968. Entomol. Soc. Convers. 6:3.

[16] Rothschild, M., et al. 1970. J. Insect Physiol. 16: 1141.

[17] Rothschild, M., et al. 1970. Toxicon 8:293.

[18] Walter, W. G., and J. F. Cole. 1967. J. Pharm. Sci. 56:174.

Part III. Alarm Pheromones

Alarm pheromones are restricted to hymenopterous and isopterous insects. They are the least specific of all insect pheromones, and often function as defensive substances as well.

	Family [1] & Species (Synonym)	Composition of Secretion (Synonym)	Reference
		Hymenoptera	
	Apidae		
1	*Apis dorsata; A. florea; A. indica*	Isoamyl acetate	22
2	*A. mellifera*	2-Heptanone; isoamyl acetate	27
3	*Lestrimelitta limao*	Citral	4
4	*Trigona postica; T. tubiba*	2-Heptanone; 2-nonanone	5
	Dolichoderinae [2]		
5	*Azteca* spp.	2-Heptanone	5
6	*Dolichoderus clarki*	4-Methyl-2-hexanone	12
7	*D. scabridus*	6-Methyl-5-hepten-2-one	11
8	*Dorymyrmex pyramicus (Conomyrma pyramicus pyramicus)*	2-Heptanone; 6-methyl-5-hepten-2-one	6,20
9	*Forelius foetidus*	4-Methyl-2-hexanone	5
10	*Iridomyrmex conifer*	6-Methyl-5-hepten-2-one	13
11	*I. detectus*	6-Methyl-5-hepten-2-one	10
12	*I. itinerans (I. nitidiceps)*	6-Methyl-5-hepten-2-one	11
13	*I. pruinosus*	6-Methyl-5-hepten-2-one	7
14	*I. rufoniger*	6-Methyl-5-hepten-2-one	11
15	*Liometopum microcephalum*	6-Methyl-5-hepten-2-one	9
16	*Tapinoma simrothi karavaievi (T. nigerrimum)*	2-Methyl-4-heptanone; 6-methyl-5-hepten-2-one	28
	Formicinae [2]		
17	*Acanthomyops claviger*	Citral; citronellal; citronellol; 2,3-dihydrofarnesal; 2,6-dimethyl-5-heptenaldehyde (2,6-dimethyl-5-hepten-1-al); 2,6-dimethyl-5-hepten-1-ol; *n*-hendecane; 2-tridecanone	2, 17, 25, 26
18	*Formica cinerea; F. fusca; F. polyctena*	Formic acid	18
19	*F. sanguinea*	*n*-Hendecane	1,2
20	*Lasius alienus*	*n*-Hendecane	26

[1] Unless otherwise indicated. [2] Subfamily of Formicidae.

continued

Part III. Alarm Pheromones

	Family [1] & Species (Synonym)	Composition of Secretion (Synonym)	Reference
21	*L. niger*	*n*-Hendecane	2
22	*L. spathepus; L. umbratus*	Citronellal	5
	Myrmicinae [2]		
23	*Atta bisphaerica; A. capiguara; A. colombica; A. laevigata; A. robusta; A. sexdens*	4-Methyl-3-heptanone	8
24	*A. texana*	4-Methyl-3-heptanone	23
25	*Crematogaster africana*	2-Hexenaldehyde (2-hexenal)	3
26	*C. peringueyi*	3-Octanone	16
27	*Myrmica brevinodis brevinodis (M. brevinodus)*	6-Methyl-3-octanone; 3-nonanone; 3-octanol; 3-octanone	14
28	*M. lobicornis fracticornis (M. fracticornis); M. punctiventris punctiventris (M. punctiventris); M. rubra; M. ruginodis; M. sabuleti americana (M. americana); M. sabuleti sabuleti (M. sabuleti); M. scabrinodis*	6-Methyl-3-octanone; 3-nonanone; 3-octanol; 3-octanone	15
29	*Myrmicaria natalensis*	Limonene	24
30	*Pogonomyrmex badius; P. barbatus; P. californicus; P. desertorum; P. occidentalis; P. rugosus*	4-Methyl-3-heptanone	19
	Isoptera		
	Termitidae		
31	*Amitermes herbertensis*	Terpinolene	21
32	*Drepanotermes rubriceps*	Limonene	21

[1] Unless otherwise indicated. [2] Subfamily of Formicidae.

Contributor: Weatherston, John

References

[1] Bergström, G., and J. Löfqvist. 1968. J. Insect Physiol. 14:995.

[2] Bergström, G., and J. Löfqvist. 1970. Ibid. 16:2353.

[3] Bevan, C. W. L., et al. 1961. J. Chem. Soc. London, p. 488.

[4] Blum, M. S. 1966. Ann. Entomol. Soc. Amer. 59:962.

[5] Blum, M. S. 1969. Annu. Rev. Entomol. 14:57.

[6] Blum, M. S., and S. L. Warter. 1966. Ann. Entomol. Soc. Amer. 59:774.

[7] Blum, M. S., et al. 1963. J. Insect Physiol. 9:881.

[8] Blum, M. S., et al. 1968. Comp. Biochem. Physiol. 26:291.

[9] Casnati, G., et al. 1964. Boll. Soc. Entomol. Ital. 94:147.

[10] Cavill, G. W. K., and D. L. Ford. 1953. Chem. Ind. (London), p. 351.

[11] Cavill, G. W. K., and H. Hinterberger. 1960. Aust. J. Chem. 13:514.

[12] Cavill, G. W. K., and H. Hinterberger. 1962. Int. Kongr. Entomol. Verh. 11th, Vienna, 3(Symp.3):53.

[13] Cavill, G. W. K., et al. 1956. Aust. J. Chem. 9:288.

[14] Crewe, R. M., and M. S. Blum. 1970. J. Insect Physiol. 16:141.

[15] Crewe, R. M., and M. S. Blum. 1970. Z. Vergl. Physiol. 70:363.

[16] Crewe, R. M., et al. 1969. Ann. Entomol. Soc. Amer. 62:1212.

[17] Ghent, R. L. 1961. Ph.D. Thesis. Cornell Univ., Ithaca, N.Y.

[18] Maschwitz, U. 1964. Z. Vergl. Physiol. 47:596.

[19] McGurk, D. J., et al. 1966. J. Insect Physiol. 12:1435.

[20] McGurk, D. J., et al. 1968. Ibid. 14:814.

[21] Moore, P. B. 1968. Ibid. 14:33.

[22] Morse, R., et al. 1967. J. Apicult. Res. 6:113.

[23] Moser, J., et al. 1968. J. Insect Physiol. 14:529.

[24] Quilico, A., et al. 1960. Int. Kongr. Entomol. Verh. 11th, Vienna, 3(Symp.3):66.

[25] Regnier, F. E., and E. O. Wilson. 1968. J. Insect Physiol. 14:955.

[26] Regnier, F. E., and E. O. Wilson. 1969. Ibid. 15:893.

[27] Shearer, D. A., and R. Boch. 1965. Nature (London) 206:530.

[28] Trave, R., et al. 1956. Chim. Ind. (Milan) 38:1015.

continued

Part IV. Attractant Pheromones

Species (Synonym)	Sex Producing Pheromone	Composition of Secretion (Synonym)	Reference
		Hymenoptera	
1	*Apis cerana indica; A. dorsata*	♀	Queen substance (9-oxodec-*trans*-2-enoic acid)	26,28
2	*A. florea*	♀	Queen substance (9-oxodec-*trans*-2-enoic acid)	26
3	*A. mellifera*	♀	Queen substance (9-oxodec-*trans*-2-enoic acid)	7,26,28
		Lepidoptera	
4	*Apamea monoglypha*	♂	Pinocarvone	2
5	*Argyroploce leucotreta*	♀	*trans*-7-Dodecen-1-ol acetate	18
6	*Argyrotaenia velutinana*	♀	*cis*-11-Tetradecen-1-ol acetate	20
7	*Bombyx mori*	♀	*trans*-10,*cis*-12-Hexadecadien-1-ol	6
8	*Bryotropha similis*, B$_1$	♀	*cis*-9-Tetradecen-1-ol acetate	22
9	*B. similis*, B$_2$	♀	*trans*-9-Tetradecen-1-ol acetate	22
10	*Cadra cautella (Ephestia cautella)*	♀	*cis*-9-*trans*-12-Tetradecadien-1-ol acetate	5,14
11	*Choristoneura rosaceana*	♀	*cis*-11-Tetradecen-1-ol acetate	24
12	*Danaus gilippus berenice; D. gilippus strigosus*	♂	2,3-Dihydro-7-methyl-1*H*-pyrrolizin-1-one; *trans*-3,-*trans*-7-dimethyl-2,6-decadiene-1,10-diol	17
13	*Galleria mellonella*	♂	*n*-Hendecanaldehyde (*n*-undecanal)	25
14	*Grapholitha molesta*	♀	*cis*-8-Dodecen-1-ol acetate	23
15	*Holomelina aurantiaca; H. ferruginosa; H. fragilis; H. immaculata; H. laeta; H. lamae; H. nigricans; H. rubicundaria; Isia isabella (Pyrrharctia isabella)*	♀	2-Methylheptadecane	21
16	*Leucania conigera; L. impura*	♂	Benzaldehyde; isobutyric acid	1,2
17	*L. pallens*	♂	Benzaldehyde; isobutyric acid	2
18	*Lycorea ceres ceres*	♂	Cetyl acetate (*n*-hexadecan-1-ol acetate); 2,3-dihydro-7-methyl-1*H*-pyrrolizin-1-one; *cis*-11-octadecen-1-ol acetate	15,16
19	*Ostrinia nubilalis*	♀	*cis*-11-tetradecen-1-ol acetate	13
20	*Pectinophora gossypiella*	♀	Propylure (10-propyl-*trans*-5,9-tridecadien-1-ol acetate)	12
21	*Phlogophora meticulosa*	♂	Benzaldehyde; 2-methylbutyric acid (2-methylbutanoic acid); 6-methyl-5-hepten-2-ol; 6-methyl-5-hepten-2-one	1,2
22	*Plodia interpunctella*	♀	*cis*-9-*trans*-12-Tetradecadien-1-ol acetate	5,14
23	*Polia nebulosa; P. persicariae (Mamestra persicariae)*	♂	Benzaldehyde; benzyl alcohol; phenethyl alcohol (2-phenylethanol)	2
24	*Porthetria dispar*	♀	*cis*-7,*cis*-8-Epoxy-2-methyloctadecane	4
25	*Spodoptera eridania (Prodenia eridania)*	♀	*cis*-9-*trans*-12-Tetradecadien-1-ol acetate; *cis*-9-tetradecen-1-ol acetate	11
26	*S. frugiperda*	♀	*cis*-9-Tetradecen-1-ol acetate	27
27	*Trichoplusia ni*	♀	*cis*-7-Dodecen-1-ol acetate	3
		Coleoptera	
28	*Acanthoscelides obtectus*	♂	(−)Methyl-*trans*-2,*trans*-4,*trans*-5-*n*-tetradecatrienoate	9
29	*Anthonomus grandis*	♂	*cis*-3,*cis*-3-Dimethyl-Δ$^{1,\alpha}$-cyclohexaneacetaldehyde; *trans*-3,*trans*-3-dimethyl-Δ$^{1,\alpha}$-cyclohexaneacetaldehyde; *cis*-3,*cis*-3-dimethyl-Δ$^{1,\beta}$-cyclohexaneethanol; *cis*-2-isopropenyl-1-methylcyclobutaneethanol	31
30	*Attagenus megatoma*	♀	*trans*-3,*cis*-5-Tetradecadienoic acid	30
31	*Costelytra zealandica*	♀	Phenol	8

continued

Part IV. Attractant Pheromones

	Species (Synonym)	Sex Producing Pheromone	Composition of Secretion (Synonym)	Reference
32	*Dendroctonus brevicomis*	♂♀	Endo-brevicomin; exo-brevicomin; frontalin; *trans*-verbenol; verbenone	29
33	*D. frontalis*	♂♀	Frontalin; *trans*-verbenol; verbenone	29
34	*D. ponderosae*	♀	*trans*-Verbenol	29
35	*Ips confusus*	♂	(+)2-Methyl-6-methylene-2,7-octadien-4-ol; (−)2-methyl-6-methylene-7-octen-4-ol; (+)-*cis*-verbenol	29
36	*Limonius californicus*	♀	*n*-Valeric acid	10
37	*Trogoderma inclusum*	♀	(−)14-Methyl-*cis*-8-hexadecen-1-ol; (−)methyl-14-methyl-*cis*-8-hexadecenoate	19

Contributor: Weatherston, John

References

[1] Aplin, R. T., and M. C. Birch. 1968. Nature (London)217:1167.

[2] Aplin, R. T., and M. C. Birch. 1970. Experientia 26:1193.

[3] Berger, R. S. 1966. Ann. Entomol. Soc. Amer. 59:767.

[4] Bierl, B. A., et al. 1970. Science 170:87.

[5] Brady, U. E., et al. 1971. Ibid. 171:804.

[6] Butenandt, A., et al. 1961. Hoppe Zeylers Z. Physiol. Chem. 324:84.

[7] Butler, C. A. 1967. Proc. Roy. Entomol. Soc. London A42:71.

[8] Henzell, R. F., and M. D. Lowe. 1970. Science 168:1005.

[9] Horler, D. F. 1970. J. Chem. Soc. C, p. 859.

[10] Jacobson, M., et al. 1968. Science 159:208.

[11] Jacobson, M., et al. 1970. Ibid. 170:542.

[12] Jones, W. A., et al. 1966. Ibid. 152:1516.

[13] Klun, J. A., and T. A. Brindley. 1970. J. Econ. Entomol. 63:779.

[14] Kuwahara, Y., et al. 1971. Science 171:802.

[15] Meinwald, J., and Y. C. Meinwald. 1966. J. Amer. Chem. Soc. 88:1305.

[16] Meinwald, J., et al. 1966. Science 151:583.

[17] Meinwald, J., et al. 1969. Ibid. 164:1174.

[18] Read, J. S., et al. 1968. Chem. Commun., p. 792.

[19] Rodin, J. O., et al. 1969. Science 165:904.

[20] Roelofs, W. L., and H. Arn. 1968. Nature (London) 219:513.

[21] Roelofs, W. L., and R. T. Carde. 1971. Science 171:684.

[22] Roelofs, W. L., and A. Comeau. 1969. Ibid. 165:398.

[23] Roelofs, W. L., and A. Comeau. 1969. Nature (London) 224:723.

[24] Roelofs, W. L., and J. Tette. 1970. Ibid. 226:1172.

[25] Roller, H., et al. 1968. Acta Entomol. Bohemoslov. 65:208.

[26] Sannasi, A., and G. Sundara Rajula. 1971. Life Sci. 10:195.

[27] Sekul, A. A., and A. N. Sparks. 1967. J. Econ. Entomol. 60:1270.

[28] Shearer, D. A., et al. 1970. J. Insect Physiol. 16:1437.

[29] Silverstein, R. M. 1970. In M. Beroza, ed. Chemicals Controlling Insect Behaviour. Academic Press, New York. pp. 21-40.

[30] Silverstein, R. M., and J. O. Rodin. 1967. Science 157:85.

[31] Tumlinson, J. H., et al. 1970. In M. Beroza, ed. Chemicals Controlling Insect Behaviour. Academic Press, New York. pp. 41-59.

81. ANIMAL TOXINS

Part I. Reptiles

Species, Distribution, and **Max Length:** Consult references 30, 31, 73, 105, 139-141. **Venom Yield:** Expressed as weight of dry venom; maximal yields may be several times the average or the upper limit of the average range. **Mouse** LD_{50}: Toxicity expressed as LD_{50} of dry venom when venom is administered intravenously to albino mice, unless

continued

Part I. Reptiles

otherwise indicated. **Venom Components: Nonenzymatic** are lethal proteins or polypeptides, according to reference 252. In addition to the nonenzymatic components listed, there is a nerve growth factor found in all Crotalidae, Viperidae, and Elapidae venoms [6,9,118]. Many of the references quoted deal with specific isolation of homogeneous components. Those components shown in *italics* are known to be absent, or present in very low concentrations. Neu-

tralizing antivenins are available for all snake venoms listed [31,141,179,213]. Figures in heavy brackets are reference numbers. Data in light brackets refer to the column heading in brackets. For recent reviews and special publications on snake venoms, consult General References and references 19, 30, 31, 45, 56, 114, 115, 118, 126, 130, 164, 171, 179, 180, 195, 201, and 242.

VENOM COMPONENT ABBREVIATIONS

Enzymatic

AaOx	=	L-amino-acid oxidase
AChe	=	acetylcholinesterase
Acoag	=	unidentified anticoagulant factors
Adp	=	ADPase (adenosine 5′-diphosphatase)
Afib	=	antiplasmin (antifibrinolysin)
Amp	=	adenosine 5′-monophosphatase (AMPase, 5′-nucleotidase)
Amyl	=	α-amylase
AnChe	=	anticholinesterase
Aproth	=	antiprothrombic and/or antithrombic factors
Arges	=	BAEE-, TAME-, or BAME-hydrolyzing L-arginine esterases
$Arges_{bk}$	=	bradykinin-releasing L-arginine esterase (kininogenase)
$Arges_{cthr}$	=	cothrombic activator
$Arges_{th}$	=	thrombinlike L-arginine esterase
$Arges_x$	=	F X-activating L-arginine esterase
Athr	=	antithromboplastic factors
Atp	=	adenosine 5′-triphosphatase (ATPase)
Cat	=	catalase
Chon	=	chondroitinase (chondroitin sulfate lyase)
Cmt	=	catechol O-methyl transferase
Coag	=	unidentified coagulant factors
Cpep	=	carboxypeptidases A and/or B
Dna	=	deoxyribonuclease (DNase, endonuclease)
Expep	=	exopeptidases—aminopeptidase, dipeptidase, or tripeptidase

Fib	=	plasmin (fibrinolysin or fibrinogenolysin)
HtDd	=	5-hydroxytryptophan decarboxylase and/or DOPA decarboxylase
Hyal	=	hyaluronidase (hyaluronate lyase)
Inv	=	β-fructofuranosidase (invertase)
Lip	=	lipase
Moa	=	monoamine oxidase
Nad	=	nicotinamide-adenine-dinucleotide pyrophosphatase (NADase, diphosphopyridine nucleotidase, DPNase)
Nadp	=	NADP pyrophosphatase (NADPase, triphosphopyridine nucleotidase, TPNase)
Pdes	=	phosphodiesterase (exonuclease)
Phos	=	nonspecific phosphomonoesterases (alkaline and/or acid phosphatases)
PLipA	=	phospholipase A (phosphatidase A, lecithinase, cephalinase)
PLipB	=	phospholipase B (lysophospholipase)
PLipC	=	phospholipase C
PmAc	=	direct or indirect plasminogen activators
Prot	=	proteases—bradykininase, caseinase, gelatinase, hemoglobinase
Pta	=	prothrombinase (a direct prothrombin activator)
Rna	=	ribonuclease (RNase, endonuclease)
TrAm	=	transaminase (aminotransferase)

Nonenzymatic

CT	=	cardiotoxins
DLF	=	direct lytic factor
HR	=	hemorrhagins (vasculotoxins)

NT	=	neurotoxins
SF	=	cardiovascular-shocking factors

continued

Part I. Reptiles

Species (Synonym) [Distribution]	Max Length cm	Venom Yield [Mouse LD$_{50}$]	Venom Components		Symptoms & Signs of Envenomation in Man	Mortality %
			Enzymatic	Nonenzymatic		
Crotalidae[1]						
1 *Agkistrodon contortrix (Ancistrodon contortrix)* [Eastern & southern United States, except peninsular Florida]	125 [141]	45-56 mg [45, 73] [54-236 μg/20 g [45,106,141, 149]; 512 μg/ 20 g[2] [138]]	AaOx [106,133,242] *AChe* [133,243] Amp [133] Arges [31,106,133, 153,220,229] Arges$_{bk}$ [31,133,153] *Arges$_{th}$* [31,133] Athr [31] *Atp* [133] *Cpep* [229] Expep [217,229] Fib [31,109] *Pdes* [106,133] *Phos* [19] PLipA [106,133] *PmAc* [109] Prot [31,106,133,149, 153,229] *Pta* [31]	HR [68,149, 188] *NT* [15,180] SF [180]	Similar to envenomation by *A. piscivorus* (see entry 3), but less severe; far less local tissue reaction, and rarely any tissue loss; local pain, swelling, & serous blisters; lymphangitis & lymphadenitis; sweating, nausea, & vomiting; in severe cases, petechiae, bloody stools, & shock [177,178]	0.01 [155]
2 *Agkistrodon halys (Ancistrodon halys)* [From Caspian Sea eastward to Japan, southward to Yangtze River & Penghu (Pescadores) Islands]	91 [141]	[19 μg/20 g [149]]	AaOx [133,201] *AChe* [133] Amp [133,201,202] Arges, Arges$_{bk}$ [133, 153,201] *Arges$_{th}$* [31,133,153] Athr [31] Atp [133,201] Dna [201] Expep [217] Hyal [201] *Nad* [201,203] Pdes [133,201,202] Phos [202] PLipA [133,201] Prot [133,149,152, 153,201] *Pta* [31] Rna [201]	HR [149,188, 201] SF [201]	High morbidity (2000-3000 bites yearly), mostly mild cases on fingers & toes. Swelling of bitten limb and regional lymphadenopathy; severe, necrotic complications rare. Death from prolonged shock or acute renal failure. [206]	0.1 [206]
3 *Agkistrodon piscivorus Ancistrodon piscivorus)* [Southeastern United States to central Texas]	187 [141]	98-120 mg [45, 73] [41-80 μg/20 g [45,106,149]; 510 μg/20 g[2] [138]]	AaOx [106,133,242] *AChe* [133,243] Amp [133,174] Aproth [62] Arges [31,106,133, 153,220,221,229] Arges$_{bk}$ [31,133,153] *Arges$_{th}$* [31,62,133, 153,221]	*CT* [113] *DLF* [59,113, 180] HR [68,149, 188] *NT* [15] SF [180]	Predominantly vasculotoxic & local, like *Bothrops* (see entries 4-7) or North American *Crotalus* (see entries 9-10, 13-14) envenomation, but no defibrination syndrome, erythrocyte destruction, tissue sloughing, or neurologic symptoms; hemorrhage, hy-	0.1 [156]

[1] Fangs in front, movable, hollow; pit between eye and nostril; approximately 120 species. [2] Subcutaneous administration.

continued

Species (Synonym) [Distribution]	Max Length cm	Venom Yield [Mouse LD$_{50}$]	Venom Components		Symptoms & Signs of Envenomation in Man	Mortality %
			Enzymatic	Nonenzymatic		
			Atp [133,244] Cmt [76] Cpep [229] Dna [174] Expep [217,229] Fib [30,31,62,109] HtDd [76] Hyal [19,195,242] Moa [76] Nadp [44] Pdes [19,106,133, 174] Phos [19,174] PLipA [7,59,60,106, 133,180] PLipB [60] PmAc [109] Prot [31,106,133,149, 153,229,230] Pta [31,62]		potensive shock, other systemic effects, & local tissue reactions also less prominent. [141,156,177,178]	
4 Bothrops alternatus [Southern Brazil; Uruguay; Paraguay; northern Argentina]	156 [141]	66-70 mg [30, 188] [256 µg/20 g 2/ [188]]	AaOx [30,133,242] AChe [133,243] Amp [133] Arges, Arges$_{bk}$ [31, 133] Arges$_{th}$ [31, 133] Atp [133] Chon, Dna [30] Fib [30,31] Hyal [30,242] Pdes [133] PLipA [30,133] PLipC [30] Prot [30,31,133] Rna [30]	HR [188] NT [11,25] SF [11]	High morbidity. Similar to envenomation by B. atrox (see entry 5), but more severe necrosis & sloughing; plastic surgery required. Self-amputation of leg observed. [1,30]	<1 [30]
5 B. atrox [Central & eastern Mexico to northeastern Argentina; Tobago; Trinidad; absent from Uruguay & Chile]	250 [30]	51-230 mg [30, 102,188] [6.5-85 µg/20 g [45,102,106, 141,149,214]; 1256-1984 µg/ 20 g 2/ [190]]	AaOx [30,98,106,133, 242] AChe [133,193,243] Amp [98,133,174,200] Aproth [62] Arges [31,102,106, 133,153,221] Arges$_{bk}$ [31,41,153, 176] Arges$_{th}$ [31,41,53,62, 66,98,100,133,153] Arges$_x$ [53] Atp [30,98,133,244] Cat [97] Cpep [102]	CT [41] DLF [97] HR [149,188, 214] NT [15,25, 41] SF [41,102]	High morbidity. Predominantly hemorrhagic, rapidly developed. Outstanding early features: intense local reactions (immediate, burning pain followed by erythematous, deforming edema extending to trunk & even head), spontaneous bleeding from gums & bite wounds, & blood incoagulability through defibrination syndrome. Lymphadenopathy, sanguineous blisters, ecchymoses on vessel paths; paresthesia, tingling or numb-	<1 [30]

2/ Subcutaneous administration.

continued

Part I. Reptiles

Species (Synonym) [Distribution]	Max Length cm	Venom Yield [Mouse LD$_{50}$]	Venom Components		Symptoms & Signs of Envenomation in Man	Mortality %
			Enzymatic	Nonenzymatic		
			Dna [30,77,98,174] Expep [102,217] Fib [30,31,41,53,98] Hyal [30,98,193] Nad [98] Pdes [98,106,133,174, 200,236] *Phos* [19,97,174] PLipA [30,98,106, 133,193] Prot [30,31,98,106, 133,149,153] *Pta* [53] Rna [30,98]		ness; profuse sweating, dizziness, faintness, nausea, vomiting, thirst, general pain & prostration, intense headache, visual disturbances, tachycardia, hypotension; anemia, thrombocytopenia; sometimes, epistaxis, hematemesis, melena, or hematuria. Severe cases: clouded consciousness, oliguria, & shock. Death from systemic massive hemorrhage, shock, or renal failure. Autopsy: hemorrhage into brain, lungs, heart, & many other organs. Histopathology: renal cortical necrosis; thrombi in glomerular capillaries. Sequelae: extensive necrosis & sloughing; sometimes, denuding of bones or loss of limbs. [1,3,30,51,95,100,136,141,161,163]	
6 *B. jararaca* [Southern Brazil; eastern Paraguay; northeastern Argentina]	187 [141]	25-36 mg [30, 188] [22 µg/20 g [149]; 40-452 µg/20 g[2/] [191,192]]	AaOx [30,133,193, 242] *AChe* [132,133,243] Amp [133] *Aproth* [62] Arges [31,89,133,153, 220] Arges$_{bk}$ [31,41,56,133, 153,176] Arges$_{th}$ [31,41,53,54, 62,89,133,146,153] Arges$_x$ [53,54] Atp [30,133,244] Chon, Dna [30] Expep [217] Fib [30,31,53] Hyal [30,195,242] Pdes [132,133] Phos [193] PLipA [30,133] Prot [30,31,89,133, 149,153] *Pta* [53] Rna [30]	*CT* [41] HR [149, 188] *NT* [11,25, 41] SF [11,41]	Similar to envenomation by *B. atrox* (*see* entry 5), but hemolysis slight. Autopsy: generalized visceral hemorrhages, cerebral hemorrhage, hemoglobinuric nephrosis. [30,138,140]	<1 [30]
7 *B. nummifer* [Central America (southern Mexico to Panama)]	98 [102]	120 mg [102] [48-79 µg/20 g [102,214]]	AaOx [99,133,242] *AChe* [97,133,243] Amp [99,101,133] *Aproth* [62] Arges [101,133,221] Arges$_{bk}$ [133] Arges$_{th}$ [31,62,99,221]	*DLF* [99,101] HR [97,99, 214]	Very few cases recorded (information inadequate). Usual local symptoms of *Bothrops* poisoning. Systemic poisoning rarely severe. [163]	Unknown

2/ Subcutaneous administration.

continued

Part I. Reptiles

Species (Synonym) [Distribution]	Max Length cm	Venom Yield [Mouse LD$_{50}$]	Venom Components Enzymatic	Venom Components Nonenzymatic	Symptoms & Signs of Envenomation in Man	Mortality %
			Atp [99,101,133] Cat [97] Cpep [101] Dna [99,101] Expep [101,217] Fib [99,101] Hyal [97] Nad [99,101] Pdes [99,101,133] Phos [99,101] PLipA [97,99,101,133] Prot [99,101,133] Pta [97] Rna [99,101]			
8 Calloselasma rhodostoma (Agkistrodon rhodostoma) [39] [Malaysia, Cambodia, Thailand, Vietnam, Laos, Burma, Ceylon, Java, Sumatra]	104 [141]	51 mg [45] [94-124 µg/20 g [106,141]]	AaOx [106] Arges [37,69,106, 222] Arges$_{bk}$ [37] Arges$_{th}$ [31,37,53,69, 172] Arges$_x$ [53] Fib [31,53] Hyal [19] Pdes, PLipA [106] Prot [37,106,222] Pta [31]	DLF [37] HR [37,69, 222] NT [15,37, 180]	Outstanding feature & most valuable sign of systemic poisoning: benign defibrination syndrome prolonged for several wk. Most cases: remarkable general well-being apart from variable pain & rapidly developed local swelling; spontaneous, systemic hemorrhage conspicuously absent despite complete incoagulability of blood. Death in severe cases from hemorrhagic shock & anemia within 5 hr to 10 days. Autopsy: tightly swollen limb; subcutaneous tissues packed with blood or hemorrhagic exudate; pronounced subdural & subarachnoid hemorrhage; ecchymotic patches in pericardium; blood from heart unclottable. Histology: mainly negative. [31, 37,45,171,172]	Low
9 Crotalus adamanteus [Southeastern United States to extreme eastern Louisiana, in lowlands]	251 [141]	240-437 mg [45,73] [22-34 µg/20 g [106,149, 181]; 290 µg/ 20 g²/ [104, 138]]	AaOx [50,52,55,106, 133,242] AChe [55,133,243] Amp [55,64,133,174] Aproth [31] Arges [31,55,106,133, 153,220] Arges$_{bk}$ [31,55,133, 153] Arges$_{th}$ [31,62,122, 133,153] Atp [55,133] Dna [55,174] Expep [223] Fib [62] Hyal [195,242] Nad [55]	CT [16,17] CT [113] DLF [113, 180] HR [17,68, 149] NT [15-17] SF [180,181]	Similar to envenomation by Bothrops atrox (see entry 5), but neurologic symptoms significant or dominant. Immediate local pain, progessive edema, ecchymosis, petechiae; hypotension, pallor, rapid pulse; widespread bleeding, defibrination syndrome, hemolytic anemia. Some cases: severe weakness, vertigo, numbness, fasciculation, tingling, cramps or paralysis, disorientation, delirium, pupil contraction & general parasympathetic stimulation,	<0.5 [158]

²/ Subcutaneous administration.

continued

	Species (Synonym) [Distribution]	Max Length cm	Venom Yield [Mouse LD$_{50}$]	Venom Components		Symptoms & Signs of Envenomation in Man	Mortality %
				Enzymatic	Nonenzymatic		
				Pdes [55,64,106,133, 174,181] *Phos* [174] PLipA [55,60,106, 133,180,235] PLipB [60] Prot [31,55,106,133, 149,153] *Pta* [31,62] Rna [55]		phlebothrombosis; convulsions in children. Very severe cases: generalized hemorrhage, anemia, bleb formation, spontaneous skin rupture; marked twitching, paresthesia, difficult speech, yellow vision or blindness from conjunctival hemorrhage; early nausea, vomiting, incontinence of urine & feces; renal, hepatic, & cardiovascular damage; shock, with marked tachycardia & unconsciousness; coma or death shortly after bite, sometimes within 30 min. Death frequently from extensive hemorrhage in major organs & cavities. Sequelae: limb disability or amputation. [5,124-127,141,177,178,233]	
10	*C. atrox* [Southwestern United States through most of Texas and into northern Mexico]	218 [141]	175 mg [73] [53-84 μg/20 g [106,149, 181]; 150 μg/ 20 g[2/] [104, 138,188]]	AaOx [55,96,106, 133] *AChe* [55,96,133, 243] Amp [55,96,133, 174] *Aproth* [62] Arges [31,55,106, 133,153,220,221, 237,245] Arges$_{bk}$ [31,55,133, 153] *Arges$_{th}$* [31,62,96, 133,221] Athr [55] Atp [55,96,133,244] *Cpep* [229] Dna [55,96,174] Expep [217,223,229] Fib [31,62,96] Hyal [96,195] Nad [55,96] Pdes [55,96,106,133, 174,181] *Phos* [55,96,174] PLipA [55,96,106, 133,180,237] Prot [31,55,96,106, 132,149,153,229, 245] *Pta* [31,62] Rna [55,96]	*CT* [113] *DLF* [96,97, 113,180] HR [68,149, 188] *NT* [15] SF [180,181]	Similar to envenomation by *C. adamanteus* (*see* entry 9) but neurologic symptoms & local tissue damage less marked. Immediate, sharp, burning pain; petechiae & bluish discoloration with rapidly progressing edema; weakness, nausea, cold perspiration, protracted vomiting, intense thirst; tingling, numbness, feeling of suffocation, spasms of respiratory muscles; local sloughing of skin & subcutaneous tissues. Severe cases: profound shock. [67,126,177,178]	<1 [126, 178]

2/ Subcutaneous administration.

continued

Part I. Reptiles

#	Species (Synonym) [Distribution]	Max Length cm	Venom Yield [Mouse LD$_{50}$]	Venom Components		Symptoms & Signs of Envenomation in Man	Mortality %
				Enzymatic	Nonenzymatic		
11	*C. durissus durissus* [Central America (southern Mexico to Costa Rica)]	167 [102]	98-140 mg [73, 102] [13-25 µg/20 g [102,106]]	AaOx [97,106,227] *AChe* [97,193] Amp [97] Arges [102,106] Arges$_{th}$ [31,97,193, 227] Atp, *Cat* [97] *Cpep* [102] Dna [97] Expep [102,223] Fib [30,31,97] Hyal [97,193,195] Nad [97] Pdes [97,106] *Phos* [97] PLipA [97,106,227] Prot [30,97,106,193, 227] *Pta*, Rna [97]	*DLF* [97] *NT*, SF [11, 227]	Low morbidity. Clinical picture shares features of envenomation by either North American *Crotalus* spp. or *C. durissus terrificus* (*see* entries 9-10, 12-14). Local, sharp pain & intense hemorrhagic reaction; transient blindness, delirium, slight pulse, marked pallor; broken-neck syndrome. [3,4,163]	Unknown
12	*C. durissus terrificus* [South America to central Argentina, except in dense forest; absent west of the Andes Cordillera]	171 [140]	24-33 mg [73, 188] [2.0-5.3 µg/20 g [26,45,106, 140,149]]	AaOx [10,106,133, 185,227,242] *AChe* [133,193,243] Amp [26,133] *Aproth* [62] Arges [31,106,133, 153] Arges$_{bk}$ [31,79,153, 176] Arges$_{th}$ [31,53,62,86, 133,153,193,227] *Arges$_x$* [53] Atp [133,244] Dna [26] Expep [217,223] *Fib* [30,53,62] Hyal [86,193,242] Nad [26,79] Pdes [64,106,133] Phos [193] PLipA [10,26,86,88, 106,133,180,227] Prot [106,133,149, 153,193,227] *Pta* [53]	*DLF* [180] *HR* [149] NT [10,11, 15,25,26,28, 40,56,79,86, 88,164,180, 185,188, 227] SF [11,27, 164,227]	Paramount features are neurotoxic, hemolytic, & nephrotoxic. Usually no local reaction, except immediate, intense pain; sometimes, slight edema, & swelling of regional lymph nodes. Latent period, 30-60 min, followed by pathognomonic, neurotoxic facies: blepharoptosis & ophthalmoplegia with vision impairment & strabismus. Tingling or numbness, muscle weakness, flaccid paralysis; neck pain & stiffness (broken-neck syndrome); loss of equilibrium; hemolysis followed by methemoglobinuria, oliguria, & anemia. Severe cases: completely paralyzed eyeballs, generalized pain, delayed pupillary reflexes, mydriasis (fatal sign), bilious vomiting, unclottable blood,	12 [30]

& anuria. Death usually from hemoglobinuric nephrosis, but also from shock & respiratory depression within 1-3 days in massive envenomation causing prostration, agitation, & coma within a few hr. Autopsy: kidneys enlarged. Histopathology: intermediate nephron nephrosis. Sequelae: none, except for transient motor difficulty of the eyes. [1,26,30, 51,141]

continued

Part I. Reptiles

Species (Synonym) [Distribution]	Max Length cm	Venom Yield [Mouse LD$_{50}$]	Venom Components		Symptoms & Signs of Envenomation in Man	Mortality %
			Enzymatic	Nonenzymatic		
13 *C. scutulatus* [From the Mojave Desert of California to central Mexico]	125 [141]	50-90 mg [141] [3.6-4.2 µg/20 g [106,141]]	AaOx [106] Arges [106,221] *Arges*$_{th}$ [221] Pdes, PLipA, Prot [106]	*HR* [68] NT [15,68]	Immediate, intense, prolonged pain at bite area; considerable swelling & edema with lymph node involvement, but no hemorrhagic changes. Symptoms apparently more neurological than in envenomation by other North American rattlesnakes; tingling of lips, dryness of mouth, difficulty in swallowing; severe nausea & vomiting; involuntary defecation & other evidences of general parasympathetic stimulation. Victim may lapse into immediate, fatal unconsciousness. [173, 177,178,233]	Low
14 *C. viridis* [Southwestern Canada through central United States and into northern Mexico]	156 [141]	25-100 mg [141] [22-32 µg/20 g [106,141, 149]; 144 µg/ 20 g$^{2/}$ [138]]	AaOx [106,133] *AChe* [133,243] Amp [133] Arges [106,133,153, 220] Arges$_{bk}$, *Arges*$_{th}$ [133, 153] Atp [133] Expep [217,223] Pdes [19,106,133, 181] *Phos* [19,212] PLipA [106,133] Prot [106,133,149, 153]	HR [68,149] NT [14,15, 56] SF [56,181]	Same symptoms as envenomation by *C. adamanteus* (*see* entry 9), but less severe. Immediate, local, burning pain, & swelling; contiguous or patchy skin discoloration, edema, & ecchymosis; tingling of face & scalp; nausea, vomiting, abdominal pain, weakness, headache, dizziness, faintness, profuse cold sweating, diarrhea. Severe cases: excitement, muscular tremor, numbness & paresthesia, paralysis, dyspnea (rapid shallow, or slow stertorous, respiration), cyanosis, semiconsciousness or shock; convulsions in children. [61,91,177,178]	<1 [126, 178]
15 *Lachesis muta* [Southern Nicaragua to tropical Brazil, including Colombia, Venezuela, the Guianas, Trinidad, Ecuador, eastern Peru, & Bolivia [228]]	400 [30]	115-450 mg [45,102] [33-73 µg/20 g [45,102]; 740 µg/20 g$^{2/}$ [139]]	AaOx [97] Arges [31,102] Arges$_{bk}$ [31] Arges$_{th}$ [31] *Cpep,* Expep [102] Fib [30,31] PLipA [31] Prot [30,31,102]	HR [188]	Bites scarce, few recorded; lacerating, as if caused by a dog. Less severe necrotic effects but more intense neurologic action than in envenomation by *Bothrops atrox* (*see* entry 5); local, intense pain, & swelling rapidly extending to whole limb, but no hemorrhage; marked difficulty in neck movements; faintness & profuse sweating, sometimes shock. Sequela: edema in lower limbs. [30,51,163]	6 [30]
16 *Sistrurus catenatus* [Southern Ontario; central to	97 [158]	14-31 mg [73, 158] [104 µg/20 g$^{2/}$ [104,138,140, 188]]	AaOx [133,242] *AChe* [133,243] Amp, Arges, Arges$_{bk}$, *Arges*$_{th}$ [133] Atp [133,244]	HR [188] *NT* [15]	Deleterious effects of same degree as those of envenomation by *Agkistrodon contortrix* (*see* entry 1); pain, edema, ecchymoses, weakness,	<0.5 [158]

$^{2/}$ Subcutaneous administration.

continued

Part I. Reptiles

Species (Synonym) [Distribution]	Max Length cm	Venom Yield [Mouse LD$_{50}$]	Venom Components		Symptoms & Signs of Envenomation in Man	Mortality %
			Enzymatic	Nonenzymatic		
southwestern United States]			Pdes, PLipA, Prot [133]		sweating, vomiting. Severe cases: hemolytic anemia. [140,177]	
17 *Trimeresurus flavoviridis* [Japan—certain of the Ryukyu Retto Islands (Amami & Okinawa Archipelagos)]	234 [141]	300 mg [142] [64-74 mg/20 g [106,149, 151]]	AaOx [106,133,201] *AChe* [133] Amp [133,201,202] Arges [106,133,153, 201,221] *Arges$_{bk}$* [133,153] *Arges$_{th}$* [31,133,153, 221] Athr [31,130] Atp [133,201] Dna [201] Expep [217] Fib [30,31,130] Hyal [142,201] Pdes [106,133,201, 202] Phos [202] PLipA [106,121,133, 201] Prot [106,133,142, 149,153,201,204] *Pta* [31] Rna [201]	*DLF* [142] HR [111,142, 149,151, 188] SF [111]	Outstanding clinical features: local, intense edema, swelling, pain, hemorrhage, & rapidly progressing myonecrosis. Vomiting, hypotension, & shock. Severe cases: the stronger the systemic symptoms the more severe & extensive the myonecrosis, followed by permanent ankylosis or deformity of limb. Vaccination against venom needed and currently practiced. Histopathology of amputated legs: endarteritis from thrombosis-induced ischemia. Death from acute circulatory failure 24 hr after envenomation. [150, 183]	2-3 [206]
18 *T. mucrosquamatus* [Taiwan (Formosa); southeastern China; northern Vietnam & Laos; Burma; northeastern India; Bangladesh (East Pakistan)]	125 [141]	41-75 mg [73, 90] [20 µg/20 g [90]]	AaOx [201,240] *AChe* [38,240] Amp [201,202,240] Aproth [154] Arges [153,201,220] Arges$_{bk}$ [153] *Arges$_{th}$* [153,154] Athr [130,154] Atp [201,240] Dna [201] Expep [217] Fib [130,154] Hyal [201,240] Nad [240] Pdes [201,202,240] Phos [202,240] PLipA [201,240] Prot [153,201,240] Rna [201] TrAm [211]	HR [182, 201] SF [162]	Similar to envenomation by *T. flavoviridis* (*see* entry 17). Little systemic reaction; local pain, ecchymoses, blistering; hemorrhage usually confined to bite area; motor disturbances after wound healing. [140,141,182,184]	1.5 [182]
Viperidae[3/]						
19 *Bitis arietans* (*B. lachesis*) [Africa (most-	156 [141]	66-180 mg [45, 73] [7.8-25 µg/20 g	AaOx [31,42,106, 133,242] *AChe* [31,133,193,	HR [149, 224] SF [180]	Massive & often severe local reactions: pain & tenderness, edema, swelling, extravasa-	5 [31]

[3/] Fangs in front, movable, hollow; approximately 60 species.

continued

Part I. Reptiles

Species (Synonym) [Distribution]	Max Length cm	Venom Yield [Mouse LD$_{50}$]	Venom Components		Symptoms & Signs of Envenomation in Man	Mortality %
			Enzymatic	Nonenzymatic		
...ly central & southeastern); southern Arabia]		[31,45,106, 149]; 156-200 µg/20 g[2/] [139]]	243] Amp [31,42,133] *Aproth* [62] Arges [106,133,153, 220,224] *Arges$_{bk}$* [133,153] *Arges$_{th}$* [31,62,133, 153] Athr [31,70] Atp [31,133,244] Expep [217] Fib [31,62] Hyal [31,193] Pdes [19,31,42,106, 133] Phos [19,42,193] PLipA [31,42,106, 133,193] Prot [42,106,133,149, 153,193,224,226] *Pta* [31,62]		tion, blisters, ecchymosis, bleeding from fang marks, thrombophlebitis, superficial or deep necrosis, limb-denuding sloughing. Prime features—lymphangitis & lymphadenitis; no defibrination syndrome; restlessness, weak pulse, dyspnea, gastrointestinal hemorrhages; vomiting, abdominal pain, dizziness, sweating, drowsiness. Conjunctival hemorrhage & hematuria rare. Severe cases: lower limb can become markedly & solidly fixed at the knee. Death from oligemic shock due to extravasation extending well beyond limb involved; delayed, sudden, & often fatal collapse due to concealed fluid loss. Rarely, rapid death from overwhelming envenomation. [31,140,171]	
20 *B. gabonica* [Southern Sudan; Uganda; Tanganyika; Congo; Gabon; Angola; Zambia; eastern Rhodesia; Mozambique; Kwazulu (Zululand)]	209 [141]	200-500 mg [45,73] [14-36 µg/20 g [31,45,106, 149]]	AaOx [31,106,133, 242] *AChe* [31,133,193, 243] Adp [186] Amp [132,133] Aproth [31,53] Arges [106,133,153, 221,224] Arges$_{bk}$ [133,153] *Arges$_{th}$* [31,53,133, 153] *Arges$_x$* [31,53] Athr [31,72] Atp [31,133,244] Coag [117] Expep [217] Fib [72] Hyal [19,31,193] Pdes [31,106,133] Phos [193] PLipA [31,42,106,133] *PmAc* [72] Prot [106,133,149, 153,193,224] *Pta* [31,53]	HR [149, 224] *NT* [180]	Few cases of attack on man recorded. Profound effects resembling those from bites of both Viperidae (*see* entries 19-26) & Elapidae (*see* entries 30-49); marked, rapid, & very painful hemorrhagic edema followed by tremendous destruction of local tissues; no blood-clotting defect; depressed cardiac action; hematuria, extreme dyspnea. [31,80,171]	Unknown

[2/] Subcutaneous administration.

continued

Part I. Reptiles

Species (Synonym) [Distribution]	Max Length cm	Venom Yield [Mouse LD₅₀]	Venom Components		Symptoms & Signs of Envenomation in Man	Mortality %
			Enzymatic	Nonenzymatic		
21 *Causus rhombeatus* [Africa, south of the Sahara except: Eritrea, rain forests of western & central Africa, South West Africa, western Bechuanaland, northwestern Cape Province]	94 [141]	80 mg [42] [205 µg/20 g [31]]	AaOx, *AChe* [133] Acoag [130] *Amp* [133] Arges [133,153,221] *Arges*bk, *Arges*th [133, 153] *Atp* [133] Coag [130] Expep [217] Pdes [133] PLipA [31,133] Prot [133,149,153]	HR [149]	Large number of bites, often at night, but envenomation mild; considered not to be a serious threat to man. [31]	Almost 0 [31]
22 *Echis carinatus* [Africa, north of the equator; Arabia; Iraq; southern Russian Asia; India; Republic of Pakistan; Ceylon]	83 [141]	19 mg [45] [10-46 µg/20 g [31,45,56, 106,141]; 66 µg/20 g[2]/ [138]]	AaOx [31,133,242] *AChe* [31,133,144, 193,243] Acoag [31] Amp [31,133] *Amyl* [144] Arges [133,153,221] Argesbk [133,153] *Arges*th [133,153,221] Atp [31,133,244] Expep [217] Fib [56] Hyal [19,31,93,193, 242] Pdes [31,133] Phos [193] PLipA [31,133,143, 193] PLipB [143] PmAc [71] Prot [31,133,144,153, 193] Pta [56] Rna [8] TrAm [144]	HR [31,117] NT [21,31]	Similar to envenomation by *Vipera russelii* (*see* entry 25). Poisoning minimal or absent in many Asiatic cases. Early features of poisoning: local pain, swelling, & hemoptysis; later, nonclottability of blood, bleeding from gums, & discoid ecchymoses. Hypovolemic shock: profound apathy; persistent thirst; rapid, thready pulse; increased sweating; hypotension. Death, 24 hr to 12 days. Autopsy: massive intestinal, & retroperitoneal hemorrhages. Sequelae: local subcutaneous necrosis common in finger or toe bites; very rarely, permanent impairment of vision or renal cortical necrosis followed by renal calcification. [31,45,117,140, 171]	<5 [84] 20 [31]
23 *E. coloratus* [Eastern Egypt; Arabian peninsula; Israel; Lebanon]	94 [141]	38 mg [78] [11 µg/20 g [78]]	AaOx [78,133] *AChe* [133,144] Amp [133] *Amyl* [144] Arges, Argesbk [46, 133] *Arges*th [31,53,133] Argesx [53] Atp [133] Fib [31,57,78,167] Hyal [78] Pdes [133]	HR [57,78, 167] NT [57,78]	Very low morbidity (inadequate information). Severe defibrination syndrome, thrombocytopenia, generalized bleeding. [31,65]	Unknown

[2]/ Subcutaneous administration.

continued

Part I. Reptiles

Species (Synonym) [Distribution]	Max Length cm	Venom Yield [Mouse LD$_{50}$]	Venom Components Enzymatic	Nonenzymatic	Symptoms & Signs of Envenomation in Man	Mortality %
			PLipA [57,78,133, 143] PLipB [143] Prot [78,133,144] Pta [31,57,167] TrAm [144]			
24 *Vipera berus* [British Isles; across northern Europe & Asia to Korea]	88 [141]	10 mg [73] [11 μg/20 g [141]; 130 μg/20 g[2/] [189]]	AaOx [31,242] *AChe* [31] Acoag [31,130] Arges$_{th}$ [193] Coag [31,130] Fib [31] Hyal [31,193] Phos [193] PLipA [31,193] Prot [193]	HR [31] *NT* [180]	Local, sharp pain, & massive hemorrhagic edema extending up limb into trunk or even head; hemorrhagic discoloration along lymphatics. Varying systemic reactions: from slight weakness to profound collapse; vomiting prominent, sometimes hematemesis; abdominal pain, drowsiness, cyanosis; oliguria or anuria; swelling of lips, face, & tongue common. Ptosis common after bite by *V. berus bosniensis.* [31,140,232]	<1 [138] 1-5 [232]
25 *V. russellii* [Sinkiang; Republic of Pakistan; Bangladesh; Nepal; India; Ceylon; Burma; Thailand; Laos; Cambodia; Vietnam; southeastern China; Taiwan (Formosa); Indonesia]	169 [141]	38-200 mg [45, 73,90] [1.6-15 μg/20 g [45,56,90, 106,141,149]; 100 μg/20 g[2/] [112,166]]	AaOx [55,106,133, 240,242] *AChe* [38,55,133,240, 243] Amp [55,132,240] Aproth [31,108] Arges [106,133,153, 222,224] Arges$_{bk}$ [133,153] Arges$_{th}$ [31,53,132, 153] Arges$_x$ [31,53,108, 187] Atp [55,133,240,244] Dna [55] Expep [217] Fib [30,31] Hyal [93,240,242] *Inv, Lip* [140] Nad [240] Pdes [19,55,106,133, 181,240] Phos [55,240] PLipA [55,59,60,106, 133,240] PLipB [60] Prot [55,106,132,149, 153,222,224,240] *Pta* [31,53] Rna [55,129] TrAm [211]	*CT* [113] *DLF* [59, 113] HR [149] *HR* [222, 224] *NT* [15,180] SF [180,181]	Immediate pain variable; marked swelling within a few min of bite; blood extravasation & subcutaneous discoloration; necrotic sequelae rare. Outstanding systemic symptoms: generalized hemorrhage due to vascular damage aggravated by defibrination syndrome; hemoptysis as early as 20 min after bite. Continuous oozing from bite or injection site; epistaxis; hematemesis, melena, & hematuria less common. Severe cases: swelling above the knee or elbow within 1-2 hr of bite; serous or sanguinolent blisters extending up limb; hemorrhagic signs besides hemoptysis (e.g., gum bleeding, ecchymosis) develop; unconsciousness. Biopsy of patients showing acute renal failure & bilateral ureteric obstruction by thrombi shows acute tubular necrosis; Death within 15 min to 9 days (usually 22 hr) from bleeding into brain, peritoneum, or other organs, & hypovolemic shock. Autopsy: subcutaneous hemorrhages near bite area, meningeal congestion, blood in lungs. [31,56, 140,171,238]	Almost 0 [182] <5 [84]

[2/] Subcutaneous administration.

continued

Species (Synonym) [Distribution]	Max Length cm	Venom Yield [Mouse LD$_{50}$]	Venom Components		Symptoms & Signs of Envenomation in Man	Mortality %
			Enzymatic	Nonenzymatic		
26 *V. xanthina palestinae (V. palestinae)* [Jordan; Israel; Lebanon; Syria; western Asiatic Turkey]	108 [107]	32 mg [107] [8.3 µg/20 g [149]]	AaOx [31,56] AChe [193] Aproth [31,57] Arges, Arges$_{bk}$ [153] Arges$_{cthr}$ [57,108, 167] *Arges$_{th}$* [31,153] Athr [108,167] Hyal [19,31,193] Phos [193] PLipA [31,56,57,83, 128] Prot [31,57,83,149, 153,193] *Pta* [31]	*DLF* [49] HR [13,31, 57,83,128, 149,167] NT [13,31, 56,57,145] SF [13]	Latent period, 20-30 min, followed by overwhelming weakness, abdominal pain, vomiting, & diarrhea; profuse sweating, moderate or severe hypotension, semiconsciousness, extreme restlessness; slurred speech due to conspicuous edema of the tongue and/or lip; local pain, swelling, ecchymosis, & serous or bloody blisters; lymphangitis & regional lymphadenitis; widespread edema; generalized hemorrhagic tendency (epistaxis, hematemesis, melena), but no defibrination syndrome. Most severe cases: edema extended to opposite side of body, peripheral shock within 1st hr of envenomation, and death within 24-48 hr from intractable shock & internal hemorrhage. Some cases: delayed, sudden, shock-induced death after 1 wk of improvement. Autopsy: severe hemorrhage in subcutaneous tissues, deep muscles, and in brain, heart, lungs, kidneys, & other organs; intensive damage to endothelial cells; necrosis of capillary walls. [31,65]	Almost 0 [65]

Hydrophidae[4]

Species (Synonym) [Distribution]	Max Length cm	Venom Yield [Mouse LD$_{50}$]	Enzymatic	Nonenzymatic	Symptoms & Signs of Envenomation in Man	Mortality %
27 *Enhydrina schistosa* [Persian Gulf to South China Sea and south to Ceylon & northern Australia; mostly coastal waters]	156 [141]	9-15 mg [45, 169,219] [0.2-2.8 µg/20 g [45,56,141, 213,219]; 3.0 µg/20 g[2] [137]]	*AaOx* [31] *AChe* [31,218,243] Acoag [31] Coag, Dna [218] Expep [218,219,221] Hyal [31,218] Pdes, Phos [218] PLipA [31,218,219]	*CT* [40] NT [31,40, 218] *SF* [40]	No local effects. Latent period of 30-60 min remarkably constant; giddiness. Prominent early diagnostic symptoms: generalized myalgia (muscle-movement pains) & stiffness, followed within 3-5 hr by myoglobinuria. Later, muscle weakness, ptosis, trismus, hypertension, & flaccid paresis; marked elevation of serum aminotransferase (transaminase). Severe cases: myoglobinuria as early as 1-2 hr after bite; within a few hr, development of respiratory failure; slow recovery, resembling muscular dystrophy. Death from respiratory paralysis, hyperkalemia-induced cardiac arrest, or acute renal failure. Autopsy: marked, widespread, hyaline myonecrosis even in rapidly fatal cases; renal congestion with distal tubular necrosis. [31,123,168, 169,171]	Low [169] 5-20 [123, 168]
28 *Laticauda semifasciata* [China, Taiwan (Formosa), Ryukyu Retto Islands of Japan, Philippines]	200 [219]	16-17 mg [219] [4.2-9 µg/20 g [219,225]]	*AaOx* [205,225] *AChe* [219,225] *AnChe* [225] *Arges* [219,221] Arges$_{th}$ [221] Fib [219] Pdes, Phos [225] PLipA [115,219,225] *Prot* [219,225]	*DLF, HR* [225] NT [56,114, 115,147, 205,219, 225] *SF* [225]	Species does not bite man. *Enhydrina schistosa (see* entry 27) antivenin expected to be effective. [141,205]	....

[2] Subcutaneous administration. [4] Fangs in front, short, permanently erect; approximately 50 species, strictly marine.

continued

Part I. Reptiles

Species (Synonym) [Distribution]	Max Length cm	Venom Yield [Mouse LD$_{50}$]	Venom Components		Symptoms & Signs of Envenomation in Man	Mortality %
			Enzymatic	Nonenzymatic		
29 *Pelamis platurus* [Tropical regions of Indian & Pacific Oceans: west to coasts of Arabia & northeastern Africa; east to west coast of the Americas from Baja California to Peru; north to East China Sea; & south along west, north & east coasts of Australia as far as Tasmania & New Zealand	114 [141]	2 mg [219] [3.4-3.6 µg/20 g [102,213]]	AChe [102]	*HR*, NT [163]	Bites to man extremely rare; no cases recorded in the Americas. Similar to envenomation by *Enhydrina schistosa* (*see* entry 27): local symptoms negligible; vomiting & nasal regurgitation; muscular aching & stiffness; later, flaccid, sometimes spastic, paralysis, respiratory failure; cyanosis, hypertension, & sweating; myoglobinuria & albuminuria. *E. schistosa* antivenin effective, but clinical data lacking. [3,30, 31,141,163,213]	Unknown
			Elapidae [5]			
30 *Acanthophis antarcticus* [Most of Australia excluding Tasmania & Victoria; New Guinea & nearby islands]	94 [141]	65-85 mg [45, 73,75] [8 µg/20 g [45]]	AaOx [242] AChe [242,243] Acoag [130] Arges [221] *Arges*$_{th}$, *Arges*$_x$ [53] Atp [244] *Cpep*, Expep [216] Hyal [242] PLipA [59,60] PLipB [60] Pta [53]	*DLF* [59] NT [30,32, 34]	Similar to poisoning by *Notechis scutatus* (*see* entry 45), except that peripheral circulatory failure is more common, and there is no blood clotting defect. Little pain or edema; early onset of vomiting, faintness, pain in regional lymph nodes; visual difficulty with ptosis; difficulty in speaking & swallowing; progressive paralysis of voluntary muscles. Death from respiratory obstruction & paralysis. [30,32,35,209]	7 [34]
31 *Bungarus caeruleus* [India, Republic of Pakistan, Ceylon]	156 [141]	8 mg [45] [1.8-4.8 µg/20 g [45,56,106, 141]; 20 µg/ 20 g[2] [94, 166]]	AaOx [106,242] AChe [242,243] *Arges* [106] Hyal [93] *Pdes*, PLipA, Prot [106]	CT, NT [194] SF [180,194]	Usually no pain or other local effects. Neurotoxic symptoms similar to those of envenomation by *Naja* spp. (*see* entries 41-44). Latent period may extend to 8-12 hr, followed by abdominal pain, sometimes severe; staggering gait, dysphagia, dyspnea, ptosis, stiffness of jaws, coma; no primary cardiovascular depression recorded. Death from respiratory paralysis in 3-60 hr, usually 18 hr. [31,85,141,171]	>5 [84]
32 *B. multicinctus* [Burma through southern China to Hainan & Taiwan (Formosa)]	156 [141]	4-6 mg [73,90] [1.4-1.5 µg/20 g [90,106]]	AaOx [106,201,240] AChe [38,240] Amp [201,202,240] *Aproth* [154] *Arges* [106,153,201, 220] *Arges*$_{bk}$ [153] *Arges*$_{th}$ [153,154] Atp [201,240] Dna [201]	NT [87,114, 115]	Similar to envenomation by *B. caeruleus* (*see* entry 31); inadequate information. [184]	4 [182]

[2] Subcutaneous administration. [5] Fangs in front, grooved, though virtually fused for most of length; approximately 180 species.

continued

Part I. Reptiles

Species (Synonym) [Distribution]	Max Length cm	Venom Yield [Mouse LD$_{50}$]	Venom Components		Symptoms & Signs of Envenomation in Man	Mortality %
			Enzymatic	Nonenzymatic		
			Expep [217] Fib [154] Hyal [201,240] Nad [201,203,240] Pdes [201,202,240] Phos [202,240] PLipA, Prot [106,201, 240] Rna [201] TrAm [211]			
33 Demansia tex-tilis[6]/ (Pseu-donaja tex-tilis) [Eastern Australia except Tasmania; New Guinea]	218 [141]	2 mg [75] [5 µg/20 g[2,7] [140]]	AaOx [242] AChe [242,243] Arges$_{th}$, Arges$_x$ [53] Atp [244] PLipA [59] Pta [53]	DLF [59] NT [30,75]	Latent period to 12 hr, followed by abdominal pain, vomiting, headache, dizziness, weakness, rapid pulse, & subnormal temp; hemolysis, hemoglobinuria, peripheral thrombosis & marked defibrination syndrome; neurotoxic effects quite significant. Death from respiratory & cardiovascular failure. [30,35,140,209]	7 [34] 10-40 [82, 138]
34 Dendroaspis angusticeps [Eastern Africa from Ethiopia to Natal]	280 [141]	100 mg [42] [9-58 µg/20 g [31,141,180]; 70 µg/20 g[2]/ [43]]	AaOx [132,133,180] AChe [31,132,133, 180,193] Afib [119,120] Amp [132,133] Arges [133,153,220] Arges$_{bk}$ [132,133] Arges$_{th}$ [132,133,153] Athr [119] Atp [31,133,180, 244] Coag [180] Expep [217] Fib [119,120] Hyal [31,180,193] Pdes [31,42,133,180] Phos [193] PLipA [42,132,133, 193] Prot [42,119,132,133, 153,180,193]	DLF [180] NT [42,180, 199]	Similar to envenomation by D. polylepis (see entry 35); local pain & swelling; severe hypotension; collapse or drowsiness followed by coma; dyspnea & respiratory failure. [31]	>40 [138]
35 D. polylepis [Central Africa from Ethiopia & Somali Republic; & South Africa to Natal; not	436 [141]	100 mg [45] [5.5-11 µg/20 g [31,45,180]]	AaOx [180] AChe [31,180,193] Afib [119,120] Arges$_{th}$ [193] Athr [119] Coag [180] Fib [119,120] Hyal [31,180,193]	DLF [180] NT [180, 199]	Pain, but minimal reaction at site of bite. Essential effects paralyzing & rapidly produced; hypotension & very profuse sweating; abdominal cramps, vomiting; fasciculation in lower limbs; ptosis, pupils widely dilated, diplopia; glossopharyngeal palsy, paralysis of	25 [110, 198]

2/ Subcutaneous administration. 6/ Replaced in western and northern Australia by D. nuchalis nuchalis and in southern Australia by D. nuchalis affinis, both of which probably are similar in toxicity and results of envenomation. 7/ Mouse LD$_{100}$.

continued

Part I. Reptiles

Species (Synonym) [Distribution]	Max Length cm	Venom Yield [Mouse LD$_{50}$]	Venom Components		Symptoms & Signs of Envenomation in Man	Mortality %
			Enzymatic	Nonenzymatic		
in rain forests]			Pdes [31,180] Phos [193] PLipA [31,180,193] Prot [119,180]		vocal cords, salivation, masklike expression; intercostal respiratory paralysis & progressive dyspnea. Death from respiratory failure & obstruction within 24 hr (usually 8 hr), or not at all. Victim fully alert, with sensorium intact, and restless. Autopsy: nothing. [31,110,198]	
36 Denisonia superba [Southeastern Australia; Tasmania]	187 [141]	19-35 mg [45, 73,75] [10 μg/20 g [45]]	AaOx [140,242] AChe [242,243] Acoag [130] Arges [221] Arges$_{bk}$ [176] Arges$_{th}$ [221] Atp [244] Coag [75] Cpep, Expep [216] Hyal [195,242] PLipA [59,60] PLipB [60]	DLF [59] NT [30,75]	Similar to poisoning by Notechis scutatus (see entry 45), but nervous symptoms less evident. Rapid loss of muscle tone & consciousness; peripheral circulatory failure. [30,35,140,209]	7 [34] 10-40 [82, 138]
37 Hemachatus haemachatus [South Africa, from Natal to eastern Cape Province]	156 [141]	84-100 mg [42, 45] [24-31 μg/20 g [31,45]; 52 μg/20 g 2/ [137]]	AaOx [31,42,133, 242] AChe [31,133,193, 242,243] Amp [31,42,115, 133] Aproth [62] Arges, Arges$_{bk}$ [132, 133,153] Arges$_{th}$ [62,132,133, 153] Atp [31,133,244] Fib [62] Hyal [31,193] Pdes [19,31,42,133] Phos [19,42,193] PLipA [31,42,49,56, 115,133,180,193] PLipB [115] Prot [42,133,153] Pta [62]	CT [42,56, 113,115] DLF [2,49, 56,113,115, 180] NT [21,42, 114,115, 199] SF [42,115]	Local pain; dyspnea; weak, thready pulse; extreme respiratory paralysis with cyanosis; collapse. Venom frequently sprayed at eyes, producing acute, intense ophthalmia with violent pain & photophobia, but systemic poisoning does not occur from such contact; symptoms usually resolved within 2-3 days; permanent damage to vision rare. [31,140]	Low [31] 2-10 [138]
38 Micrurus corallinus [Southern Brazil; Paraguay; northeastern Argentina]	125 [140]	15 mg [73]	AaOx [242] AChe [242,243] Aproth, Arges$_{th}$ [62] Atp [244] Coag [130] Fib [62] Hyal [140] PLipA [25]	NT [25]	Low morbidity. Immediate burning pain followed by paresthesia spreading to proximal region; no other local reactions. Drowsiness, muscle hypotonia, headache. Neurotoxic facies (ptosis & ophthalmoplegia) outstanding feature; more rapidly developed & severe than in envenomation by Crotalus durissus terrificus (see entry 12); dysphagia & dysphonia from glossopharyn-	7 [30]

2/ Subcutaneous administration.

continued

Part I. Reptiles

Species (Synonym) [Distribution]	Max Length cm	Venom Yield [Mouse LD$_{50}$]	Venom Components		Symptoms & Signs of Envenomation in Man	Mortality %
			Enzymatic	Nonenzymatic		
				geal & laryngeal palsy; profuse, thick salivation; cranial nerve involvement; generalized pain; dyspnea & widespread, flaccid paralysis. No blood-clotting defect or hemolysis. Some cases: vomiting, visual disturbances, swelling of face & lips, agitation, cramps, & prostration followed by coma. Fatal cases: death within 4-6 hr from asphyxia due to diaphragm paralysis & glottis spasm. Autopsy: pulmonary edema. [1,25, 30,51,140,141]		
39 *M. fulvius* [Southeastern United States to western Texas and into northeastern Mexico]	122 [141]	3 mg [48] [4.8-7.7 µg/20 g [45,47, 106]; 26 µg/ 20 g[2/] [137]]	AaOx [106] Acoag [165] *Aproth* [62] *Arges* [106] *Arges$_{th}$, Fib* [62] Hyal [242] Pdes [106] PLipA [47,106,165] *Prot* [106,165,234]	CT [165, 234] DLF [47] *DLF*, HR [165] *HR* [68] NT [47,165, 234] SF [180,234]	Many bites so superficial that no envenomation ensues. Usually no pain or swelling; sometimes, burning pain, slight swelling, & discoloration limited to bite area. Latent period, several hr. Early symptoms: euphoria & drowsiness; tremors & convulsions in children; followed by numbness & paresthesia in injured limb, muscle fasciculation, dyspnea, giddiness, weakness, nausea & vomiting, headache, sweating. Later: bulbar paralysis with ptosis; fixed, contracted pupils; blurred vision; feeling of thickening of the tongue & throat, dysphonia; alarmingly rapid peripheral, respiratory, & bulbar paralysis, & respiratory failure, aggravated by excessive salivation, invariably present; cardiac failure. Death within 24 hr, if at all. Sensorium remains clear in respirator-treated patients. [124,126, 141,157,177]	<10 [157]
40 *M. nigrocinctus* [Central America (southern Mexico to northwestern Colombia)]	105 [102]	5 mg [102] [14 µg/20 g [102]]	AaOx [102] AChe [102] *Aproth* [62] *Arges* [102] *Arges$_{th}$* [62] *Cat*, Expep [102] *Fib* [62] PLipA [97] *Prot* [102]	DLF, NT [97]	Very low morbidity. Similar to envenomation by *M. corallinus* (*see* entry 38). Immediate sharp, burning pain; swelling, but no other local reactions. Rapid onset of systemic symptoms: initially, agitation, pallor, dizziness, & hypothermia; followed by depression & progressive paralysis. Victims surviving >24 hr generally recover. Autopsy: subdural & subarachnoid congestion. [95,163]	Low
41 *Naja haje* [Natal to central & eastern Africa; Egypt; western & southern Arabian peninsula]	250 [141]	40 mg [73] [21 µg/20 g [31]]	AaOx [31,42,132, 242] AChe [31,132,144, 193,242,243] Acoag [31] Amp [42,132] *Amyl* [144] *Arges* [132,220] *Arges$_{bk}$, Arges$_{th}$* [132] Athr [130]	CT [40,113] DLF [113] NT [21,22, 31,40,42,56, 114,115, 147,199] SF [40,115]	Bites to man not common (inadequate information). Similar to envenomation by *Dendroaspis polylepis* (*see* entry 35). [31]	Unknown

[2/] Subcutaneous administration.

continued

Part I. Reptiles

Species (Synonym) [Distribution]	Max Length cm	Venom Yield [Mouse LD$_{50}$]	Venom Components		Symptoms & Signs of Envenomation in Man	Mortality %
			Enzymatic	Nonenzymatic		
			Atp [22,132] Coag [131] Dna [22] Expep [217,223] Hyal [19,31,42,193, 242] Pdes [19,22,31,132] Phos [19,42,193,212] PLipA [31,42,115, 132,143,193] PLipB [143] *Prot* [42,132,144, 193] Rna [22] TrAm [144]			
42 *N. naja* [Southern Asia to Indonesia; Taiwan (Formosa); Philippines; Ceylon]	203 [141]	38-317 mg [45,73] [4.9-16 μg/20 g [45,56,106, 135,141, 149]; 4.0 μg/ 20 g[2]/ [166]]	AaOx [6,55,106,132, 242] AChe [6,55,115,133, 242,243] Afib [120] Amp [6,24,55,133, 201,202] Aproth [31,62,130] *Arges* [6,106,133, 153,201,220-222] *Arges*$_{bk}$ [55,132, 176] *Arges*$_{th}$ [31,62,133, 153,221] Athr [31,130] Atp [24,55,133,201, 244] Dna [24,55,201] Expep [6,217,223] Fib [120] Hyal [6,19,93,242] *Inv* [140] Lip [6] Nad [55] Pdes [6,19,24,55,106, 133,201,202] Phos [6,19,24,55,201, 202] PLipA [6,13,55,56, 60,106,115,133,180, 196] PLipB [60,115] PLipC [23,24,115] *Prot* [6,55,106,132, 149,201,222]	CT [23,40, 55,56,113, 115,194, 196] DLF [56,113, 115,180, 196] HR [18] *HR* [149, 222] NT [12,13, 15,21,25,40, 56,103,115, 147,180, 194,199] SF [13,40, 115,180, 194,196]	Local features of poisoning: immediate pain radiating from site of bite; swelling, usually starting within 1-3 hr; dusky discoloration, blebs; delayed subcutaneous necrosis which can be very extensive without systemic symptoms. Both neurotoxic & cardiotoxic systemic effects. Latent period, 15 min to 5 hr, followed by drowsiness, numbness, limb weakness, giddiness, salivation, vomiting & thirst; bilateral ptosis, external ophthalmoplegia, blurred or double vision, pupil dilation; glossopharyngeal palsy leading to difficulty in swallowing, difficult, indistinct, nasal speech, and even to quick death from respiratory obstruction; progressive respiratory paralysis. Severe cases: neurotoxic signs within 1 hr or less; facial palsy & drooping of head; flaccid limb paresis; rapid development of respiratory depression & failure; mental confusion; cardiovascular shock (sweating, cyanosis, tachycardia, & hypotension); & terminal convulsions or muscular twitching. Death, 15	>5 [84]

[2]/ Subcutaneous administration.

continued

Part I. Reptiles

	Species (Synonym) [Distribution]	Max Length cm	Venom Yield [Mouse LD$_{50}$]	Venom Components		Symptoms & Signs of Envenomation in Man	Mortality %
				Enzymatic	Nonenzymatic		
				Pta [31,62] Rna [6,24,55,201]		min to 60 hr. [31,170,171]	
43	*N. naja atra* [Taiwan (Formosa); Hainan; southern China; Vietnam; Thailand]	172 [141]	62-88 mg [90] [4.1-9.4 µg/20 g [90,149]]	AaOx [29,133,201, 240] AChe [29,38,115,133, 240] Amp [29,115,133, 201,202,240] AnChe [38] Aproth [154] *Arges* [133,153,201, 220] *Arges*$_{bk}$ [132,133, 153] *Arges*$_{th}$ [133,153,154] Athr [154] Atp [29,133,201,240] Dna [201] Expep [217,223] *Fib* [154] Hyal [29,201,240] Nad [29,240] Pdes [29,133,201, 202,240] Phos [201,202,212, 240] PLipA [29,115,133, 201,240] PLipC [115] *Prot* [133,149,153, 201,240] Rna [29,201] TrAm [211]	CT [56,113, 115,116, 148] DLF [29,56, 113,115] *HR* [149] NT [56,114- 116,147, 148,239, 241] SF [115]	Similar to envenomation by *N. naja* (*see* entry 42), including extensive local necrosis. Sequelae: motor disturbances after wound healing. [184]	Almost 0 [182]
44	*N. nigricollis* [Africa, in savanna area, in southern Egypt, & south of the Sahara to the borders of Cape Province]	228 [141]	85 mg [73] [12-25 µg/20 g [20,31]; 61.0 µg/20 g²/ [137]]	AaOx [31,133,242] *AChe* [31,132,144, 193,242] Adp [20] Afib [120] Amp [20,42,133,174] *Amyl* [144] AnChe [20,207] *Arges* [133,153,221] *Arges*$_{bk}$ [132,133, 153] *Arges*$_{th}$ [133,153,221] Athr [20,120,131] Atp [20,31,133,244] Dna [174] Expep [20,217,223] Fib [120]	CT [40,42,56, 92,113] DLF [20,56, 59,92,113] HR [18] NT [21,40, 42,56,63, 114,115, 147,180, 199,207] SF [40,42, 115,207]	Similar to envenomation by *N. naja* (*see* entry 42). Venom usually spit at eyes of victim; effects resemble those produced by *Hemachatus haemachatus* (*see* entry 37). Dizziness, restlessness, drowsiness, ptosis, strabismus, incoordination of speech, shock (profuse sweating & hypotension), incontinence of urine & feces, drooping head, limb flaccidity. Pain and/or swelling at bite area not features of poisoning. Outstanding symptoms: paralysis of respiration	Low [31] 11-40 [138]

²/ Subcutaneous administration.

continued

Part I. Reptiles

#	Species (Synonym) [Distribution]	Max Length cm	Venom Yield [Mouse LD$_{50}$]	Venom Components — Enzymatic	Venom Components — Nonenzymatic	Symptoms & Signs of Envenomation in Man	Mortality %
				Hyal [19,20,31,93, 193] Nad [20] Pdes [20,42,133,174] Phos [20,174,193] PLipA [20,21,42,59, 115,133,143,193, 231] PLipB [143] *Prot* [20,42,133,153, 193] Pta [20,131] *Rna* [20] *TrAm* [144]		& swallowing (victim can drown in own saliva). Death usually preceded by quiet unconsciousness & shallow breathing. [31]	
45	*Notechis scutatus* [Most of eastern & southern Australia; Tasmania]	250 [141]	35-36 mg [45, 73] [0.4 µg/20 g [45]; 3.50 µg/ 20 g2/ [139]]	AaOx [140,242] AChe [58,242,243] Amp [58] *Aproth* [62] *Arges* [221] *Arges*th [53,62] *Arges*x [53] Atp [58,244] *Cpep,* Expep [216] *Fib* [31,62] Hyal [242] Nad [36] Pdes, *Phos* [19] PLipA [58-60] PLipB [60] Pta [31,53]	*DLF* [59] NT [30,58]	Little or no immediate local reaction. Latent period, 15-60 min, followed by faintness, profuse sweating, severe headache; pain in abdomen & regional lymph nodes; nausea, vomiting; local erythema, edema, & bruising; hemolysis, defibrination syndrome, blood oozing from bite wound; hematemesis, hemoglobinuria, hemorrhage, ecchymoses; dizziness, apathy, drowsiness, generalized muscular weakness. Later, dullness of sensation, staggering, ptosis & aphasia, slurred speech, dysphagia; widely dilated pupils not reacting to light; rapid, weak pulse & respiration; widespread, severe flaccid paralysis with progressive dyspnea & cyanosis. Death from respiratory failure & obstruction, toxic myocarditis, & peripheral circulatory collapse. Autopsy: bite area swollen & discolored; unclottable blood; lungs emphysematous, hemorrhagic, & thrombotic. [30,34,35,171,209]	7 [34] 10-40 [35, 209] >40 [138]
46	*Ophiophagus hannah (Naja hannah)* [Peninsular India to the Himalayas; Burma; southern China; Laos; Cambodia; Vietnam, Thailand; Malaysia; Indonesia; as far east as Luzon, Philippines [74]]	561 [141]	421 mg [74] [18-64 µg/20 g [56,106,149]; 34.0 µg/20 g2/ [137]]	AaOx [106,133,201, 242] AChe [133,242,243] Afib [120] Amp [133,201,202] Arges [106,132,221, 222] *Arges*bk [132,133,153] *Arges*th [133,153,221] Athr [120] Atp [133,201,244] Dna [201] Expep [217,223] Fib [120] Hyal [201]	*CT, DLF* [113] HR [149] *HR* [222] NT [180]	Reported cases not common, as species is rare. Similar to envenomation by *Naja* spp. (*see* entries 41-44). Slight initial bleeding from fang marks. Symptoms develop early: bleb formation at bite area, but minimal local necrosis follows wound healing even in victims overwhelmingly envenomated by massive amounts of venom; early local pain; swelling progressing up limb; followed by drowsiness, drooping of eyelids, difficulty of deglutition & breathing, apnea, & unconsciousness; pupils not	>40 [138]

2/ Subcutaneous administration.

continued

Part I.   Reptiles

#	Species (Synonym) [Distribution]	Max Length cm	Venom Yield [Mouse LD$_{50}$]	Venom Components		Symptoms & Signs of Envenomation in Man	Mortality %
				Enzymatic	Nonenzymatic		
				Nad [203] Pdes [106,133,201, 202] Phos [202,212] PLipA [106,132,201] Prot [106,132,149, 153,222] Rna [201]		reacting to light, no reflexes in extremeties; blood pressure drop. Death often within 20 min to 6 hr. [31,74]	
47	Oxyuranus scutellatus [Northeastern Australia, southeastern New Guinea]	335 [141]	100-120 mg [45,75] [0.46 μg/20 g [45]]	Arges [220] Arges$_{th}$, Arges$_x$ [53] Cpep [216] Expep [216,217] PLipA [59,60] PLipB [60] Pta [53]	DLF [59] NT [30,33, 34,209] SF [33]	Similar to envenomation by Notechis scutatus (see entry 45). Slight local pain; regional lymph nodes painful; sudden loss of consciousness; severe hypotension with absence of carotid & femoral pulses; prolonged bleeding from bite wound; spitting, coughing, or vomiting of blood; incontinence of blood-stained or black urine; bloody feces; muscle fasciculations, numbness of affected area, restlessness, mental confusion; salivation; dilated, fixed pupils; marked drowsiness, muscular weakness, blurred or double vision, myasthenic facies (ptosis & external ophthalmoplegia), glossopharyngeal palsy (slurred speech & dysphagia); respiratory & general paralysis. Often rapidly fatal. [30,33-35,141,208,209]	7 [34] >40 [138]
48	Pseudechis porphyriacus [Eastern & southern Australia excluding Tasmania]	218 [141]	35-40 mg [73, 75] [40.0 μg/20 g[2/] [139]]	AaOx [140,242] AChe [242,243] Arges [221] Atp [140] Coag [32,130] Cpep, Expep [216] Hyal [195] PLipA [59,60] PLipB [60]	DLF [59] NT [30,75]	Local pain & swelling; considerable hemolysis & thrombosis; hematemesis, hemorrhage from nose & mouth, hemoglobinuria; prostration. Neurotoxic & cardiotoxic effects minimal. Adults nearly always survive. [30,35,209]	<1 [30]
49	Tropidechis carinatus [210] [Eastern Australia, mostly New South Wales & Queensland]	106 [75]	[9.0-18.0 μg/ 20 g[2,7/]]	....	....	Similar to envenomation by Notechis scutatus (see entry 45). [35,209]	10
				Colubridae[8/]			
50	Dispholidus typus [Africa south of the Sahara, forested portions]	187 [141]	1.8-15 mg [45, 73,81,175] [1.3 μg/20 g [31]; 250 μg/ 20 g[2/] [139]]	Arges [175] Coag [31,42,81,197] PLipA [31,42,81] Prot [81,175]	HR [42,197] NT [81] SF [31]	Attack on man rare. Early symptoms: local pain & swelling, throbbing headache, drowsiness & confusion, nausea, severe vomiting, collapse. Outstanding features: prolonged defibrination syndrome; pronounced spontane-	25 [31]

2/ Subcutaneous administration.   7/ Mouse LD$_{100}$.   8/ Fangs in rear, immovable, grooved in venomous species; dangerous if handled; numerous species, mostly harmless.

continued

Part I. Reptiles

Species (Synonym) [Distribution]	Max Length cm	Venom Yield [Mouse LD$_{50}$]	Venom Components		Symptoms & Signs of Envenomation in Man	Mortality %
			Enzymatic	Nonenzymatic		
					ous bleeding from nose & mouth, & sometimes from all mucous membranes & skin; melena. Renal failure; jaundice. Recovery slow. Autopsy: hemorrhage in numerous organs including muscle & brain; fibrin thrombi in capillaries & larger vessels of many organs; acute renal tubular necrosis; marked hepatic necrosis. [31,42,45,197]	
Helodermatidae 9/						
51 *Heloderma suspectum* [Southwestern United States, chiefly Arizona; northwestern Mexico, chiefly Sonora]	62 [140]	[16.4-40 µg/20 g2/ [134,135, 215]]	*AaOx, AChe, Amp* [135] *Aproth* [160] Arges [215] Arges$_{bk}$, Arges$_{th}$ [135] *Athr* [160] *Atp, Dna* [135] Expep [215] Hyal, *Pdes, Phos,* PLipA, Prot, *Rna* [135]	*NT* [15] SF [159]	Very low morbidity. Prolonged local pain; high sensitivity to cold, heat, & touch in bite area; swelling, edema, & hyperemia (red, phlebitis-like streaks); weakness, faintness, dyspnea, nausea, blood-spotted vomiting; tinnitus, sensation of blinding lights. Syndrome rapidly developed. Death from respiratory paralysis & cardiovascular failure. No specific antivenin available; terrestrial snake antivenins not effective. [30,134,140]	Unknown

2/ Subcutaneous administration. 9/ Venom glands in lower jaw; teeth in front and often also in rear; 8-10 upper and 6-11 lower teeth, grooved, venom-conducting; 2 species— only truly venomous lizards.

Contributors: Jiménez-Porras, Jesús M., Gómez-Leiva, Marco A., and Rodriguez-Barquero, Jorge A.; Minton, Sherman A., Jr.; Graydon, John J.; Amaral, Afranio do

Specific References

[1] Abalos, J. W., and I. Pirosky. 1963. Venomous Poisonous Anim. Noxious Plants Pac. Reg., p. 363.

[2] Aloof-Hirsch, S., et al. 1968. Biochim. Biophys. Acta 154:53.

[3] Ambrose, M. S. 1956. In E. E. Buckley and N. Porges, ed. Venoms. AAAS, Washington, D.C. v. 44, p. 323.

[4] Amorim, M. 1971. In W. Bücherl and E. E. Buckley, ed. Venomous Animals and Their Venoms. Academic Press, New York. v. 2, p. 319.

[5] Andrews, C. E. 1960. Arch. Surg. (Chicago) 81:699.

[6] Angeletti, R. H. 1970. Proc. Nat. Acad. Sci. U.S. 65(3):668.

[7] Augustyn, J. M., and W. B. Elliott. 1970. Biochim. Biophys. Acta 206:98.

[8] Babkina, G. T., and S. K. Vassilenko. 1964. Biokhimiya 29(2):268.

[9] Banks, B. E. C., et al. 1968. Biochem. J. 108:157.

[10] Barrio, A. 1961. Acta Physiol. Lat. Amer. 11(4):224.

[11] Barrio, A., and O. V. Brazil. 1951. Ibid. 1:291.

[12] Bhargava, V. K., et al. 1970. Brit. J. Pharmacol. 39(2):455.

[13] Bicher, H. I. 1966. Mem. Inst. Butantan (São Paulo) 33(2):523.

[14] Bonilla, C. A., and M. K. Fiero. 1971. J. Chromatogr. 56:253.

[15] Bonilla, C. A., et al. 1971. Ibid. 56:368.

[16] Bonilla, C. A., et al. 1971. Chem. Biol. Interact. 4:1.

[17] Bonilla, C. A., et al. Unpublished. Colorado State Univ., Ft. Collins, 1973.

[18] Bonta, I. L., et al. 1970. Eur. J. Pharmacol. 13:97.

[19] Boquet, P. 1964. Toxicon 2(1):5.

[20] Boquet, P., et al. 1967. Ann. Inst. Pasteur 112:213.

[21] Boquet, P., et al. 1969. Ibid. 116:522.

[22] Botes, D. P., and D. J. Strydom. 1969. J. Biol. Chem. 244(15):4147.

[23] Braganca, B. M., and V. G. Khandeparkar. 1966. Life Sci. 5:1911.

[24] Braganca, B. M., et al. 1967. Biochim. Biophys. Acta 136:508.

[25] Brazil, O. V. 1965. O Hospital (Rio de Janeiro) 68(4):183.

[26] Brazil, O. V., et al. 1966. Mem. Inst. Butantan (São Paulo) 33(3):973.

continued

Part I. Reptiles

[27] Brazil, O. V., et al. 1966. Ibid. 33(3):993.

[28] Brazil, O. V., et al. 1967. Cienc. Cult. (São Paulo) 19(4):658.

[29] Brisbois, L., et al. 1968. J. Chromatogr. 37:463.

[30] Bücherl, W., and E. E. Buckley, ed. 1971. Venomous Animals and Their Venoms. Academic Press, New York. v. 2.

[31] Bücherl, W., et al., ed. 1968. Ibid. v. 1.

[32] Campbell, C. H. 1966. Med. J. Aust. 2:922.

[33] Campbell, C. H. 1967. Ibid. 1(15):735.

[34] Campbell, C. H. 1969. Toxicon 7(1):25.

[35] Campbell, C. H. Unpublished. Univ. Sydney, New South Wales, Australia, 1972.

[36] Chain, E. 1939. Biochem. J. 33:407.

[37] Chan, K. E. 1964. Doctoral Thesis. St. Andrews Univ., Scotland.

[38] Chang, C. C., and C. Y. Lee. 1955. J. Formosan Med. Ass. 54(3):9.

[39] Chernov, S. A. 1957. Zool. Zh. (Moscow) 36(5):790.

[40] Cheymol, J., et al. 1966. Mem. Inst. Butantan (São Paulo) 33(2):541.

[41] Cheymol, J., et al. 1968. Bull. Soc. Pathol. Exot. 61(4):673.

[42] Christensen, P. A. 1955. South African Snake Venoms and Antivenoms. South African Institute for Medical Research, Johannesburg.

[43] Christensen, P. A. Unpublished. South African Institute for Medical Research. Johannesburg, 1971.

[44] Chung, A. E., and R. G. Langdon. 1963. J. Biol. Chem. 238:2317.

[45] Clarke, E. G. C., and M. L. Clarke. 1969. Vet. Annu. 10:27.

[46] Cohen, I., et al. 1970. Biochem. Pharmacol. 19:785.

[47] Cohen, P., and E. B. Seligmann. 1966. Mem. Inst. Butantan (São Paulo) 33(1):339.

[48] Cohen, P., et al. 1968. Amer. J. Trop. Med. Hyg. 17(2):308.

[49] Condrea, E., and P. Rosenberg. 1968. Biochim. Biophys. Acta 150:271.

[50] Curti, B., et al. 1968. J. Biol. Chem. 243(9):2306.

[51] Dao, L. 1966. Dermatol. Int. 5:132.

[52] De Kok, A., and A. B. Rawitch. 1969. Biochemistry 8(4):1405.

[53] Denson, K. W. E. 1969. Toxicon 7(1):5.

[54] Denson, K. W. E., and W. E. Rousseau. 1970. Ibid. 8(1):15.

[55] Devi, A., and N. K. Sarkar. 1966. Mem. Inst. Butantan (São Paulo) 33(2):573.

[56] De Vries, A., and E. Kochva, ed. 1972. Toxins of Animal and Plant Origin. Gordon and Breach, New York.

[57] De Vries, A., et al. 1962. New Istanbul Contrib. Clin. Sci. 5:151.

[58] Doery, H. M. 1958. Biochem. J. 70:535.

[59] Doery, H. M., and J. E. Pearson. 1961. Ibid. 78:820.

[60] Doery, H. M., and J. E. Pearson. 1964. Ibid. 92:599.

[61] Doughty, J. F. 1928. Calif. West. Med. 29(4):237.

[62] Eagle, H. 1937. J. Exp. Med. 65:613.

[63] Eaker, D., and J. Porath. 1967. Proc. Int. Congr. Biochem. 7th, Tokyo, Abstr. 3:499.

[64] Eckstein, F., and H. P. Bär. 1969. Biochim. Biophys. Acta 191:316.

[65] Efrati, P. 1969. Toxicon 7(1):29.

[66] Egberg, N., and S. Nordström. 1970. Acta Physiol. Scand. 79:493.

[67] Ehrlich, S. P. 1928. Bull. Antivenin Inst. Amer. 2(3):65.

[68] Emery, J. A., and F. E. Russell. 1963. Venomous Poisonous Anim. Noxious Plants Pac. Reg., p. 409.

[69] Esnouf, M. P., and G. W. Tunnah. 1967. Brit. J. Haematol. 13:581.

[70] Forbes, C. D., et al. 1965. E. Afr. Med. J. 42(11):565.

[71] Forbes, C. D., et al. 1966. Nature (London) 211:989.

[72] Forbes, C. D., et al. 1969. J. Clin. Pathol. 22:312.

[73] Freyvogel, T. A., and E. Hofmann. 1965. Acta Trop. 22(1):11.

[74] Ganthavorn, S. 1971. Toxicon 9(3):293.

[75] Garnet, J. R. 1968. Venomous Australian Animals Dangerous to Man. Commonwealth Serum Laboratories, Parkville, Victoria, Australia.

[76] Gennaro, J. F., et al. 1968. Comp. Biochem. Physiol. 25:285.

[77] Georgatsos, J. G., and M. Laskowski. 1962. Biochemistry 1(2):288.

[78] Gitter, S., et al. 1960. Amer. J. Trop. Med. Hyg. 9:391.

[79] Gonçalves, J. M. 1956. In E. E. Buckley and N. Porges, ed. Venoms. AAAS, Wash., D.C. v. 44, p. 261.

[80] Grasset, E. 1946. Acta Trop. 3(2):97.

[81] Grasset, E., and A. W. Schaafsma. 1940. S. Afr. Med. J. 14:236.

[82] Graydon, J. J. Unpublished. Commonwealth Serum Laboratories, Parkville, Victoria, Australia, 1971.

[83] Grotto, L., et al. 1967. Biochim. Biophys. Acta 133:356.

[84] Gupta, P. S., et al. 1960. J. Indian Med. Ass. 35(9):387.

[85] Haast, W. E., and M. L. Winer. 1955. Amer. J. Trop. Med. Hyg. 4(6):1135.

[86] Habermann, E. 1957. Biochem. Z. 329:405.

[87] Hamaguchi, K., et al. 1968. J. Biochem. (Tokyo) 64(4):503.

[88] Hendon, R. A., and H. Fraenkel-Conrat. 1971. Proc. Nat. Acad. Sci. U.S. 68(7):1560.

[89] Henriques, O. B., et al. 1966. Mem. Inst. Butantan (São Paulo) 33(2):359.

[90] Huber, G. S. 1961. U.S. Nav. Med. Res. Unit 2, Rep. 61-3:1.

[91] Imriet, R. J. 1968. Appl. Therap. 10(6):394.

[92] Izard, Y., et al. 1969. C. R. Acad. Sci. 269:666.

[93] Jaques, R. 1956. In E. E. Buckley and N. Porges, ed. Venoms. AAAS, Washington, D.C. v. 44, p. 291.

[94] Jaques, R. Unpublished. Ciba, Basel, Switzerland, 1972.

continued

Part I. Reptiles

[95] Jiménez, E., and I. García. 1969. Rev. Méd. Hosp. Nal. Niños (Costa Rica) 4(2):91.

[96] Jiménez-Porras, J. M. 1961. J. Exp. Zool. 148(3): 251.

[97] Jiménez-Porras, J. M. 1963. Doctoral Thesis. Louisiana State Univ., Baton Rouge.

[98] Jiménez-Porras, J. M. 1964. Toxicon 2(3):155.

[99] Jiménez-Porras, J. M. 1964. Ibid. 2(3):187.

[100] Jiménez-Porras, J. M. 1967. Proc. Int. Congr. Biochem. 7th, Tokyo, Abstr. 3:499.

[101] Jiménez-Porras, J. M. 1967. In F. E. Russell and P. R. Saunders, ed. Animal Toxins. Pergamon Press, Oxford and New York. p. 307.

[102] Jiménez-Porras, J. M., et al. 1973. 53rd Annu. Meet. Amer. Soc. Ichthyol. Herpetol. (San José, Costa Rica) Abstr. Vol. (in press).

[103] Karlsson, E., et al. 1972. Biochim. Biophys. Acta 257:235.

[104] Klauber, L. M. 1956. Rattlesnakes. Univ. California Press, Berkeley. v. 2.

[105] Klemmer, K. 1963. In Behringwerk-Mitt. Die Giftschlangen der Erde. Marburg/Lahn. p. 255.

[106] Kocholaty, W. F., et al. 1971. Toxicon 9(2):131.

[107] Kochva, E. 1960. Amer. J. Trop. Med. Hyg. 9(4): 381.

[108] Kochwa, S., et al. 1960. Ibid. 9(4):374.

[109] Kornalík, F. 1966. Mem. Inst. Butantan (São Paulo) 33(1):179.

[110] Krengel, B., and J. Walton. 1967. S. Afr. Med. J. 41:1150.

[111] Kurihara, Y., et al. 1967. Gunma J. Med. Sci. 16:61.

[112] Kuwajima, Y. 1953. Jap. J. Exp. Med. 23:457.

[113] Larsen, P. R., and J. Wolff. 1968. J. Biol. Chem. 243(6):1283.

[114] Lee, C. Y. 1970. Clin. Toxicol. 3(3):457.

[115] Lee, C. Y. 1971. In L. L. Simpson, ed. Neuropoisons: Their Pathophysiological Actions. Plenum Press, New York. v. 1, p. 21.

[116] Lee, C. Y., et al. 1968. Naunyn Schmiedebergs Arch. Exp. Pathol. Pharmakol. 259:360.

[117] Lefrou, G., and J. Martignoles. 1954. Ann. Inst. Pasteur 86:446.

[118] Levi-Montalcini, R., and P. U. Angeletti. 1968. Physiol. Rev. 48(3):534.

[119] MacKay, N., et al. 1968. Brit. J. Haematol. 15: 549.

[120] MacKay, N., et al. 1969. J. Clin. Pathol. 22:304.

[121] Maeno, H., et al. 1962. Jap. J. Exp. Med. 32(1): 55.

[122] Markland, F. S., et al. 1970. Lancet 1:1398.

[123] Marsden, A. T. H., and H. A. Reid. 1961. Brit. Med. J. 1:1290.

[124] McCollough, N. C., and J. F. Gennaro. 1963. J. Fla. Med. Ass. 49:959.

[125] McCollough, N. C., and J. F. Gennaro. 1968. Ibid. 55:327.

[126] McCollough, N. C., and J. F. Gennaro. 1970. Clin. Toxicol. 3:483.

[127] McCreary, T., and H. Wurzel. 1959. J. Amer. Med. Ass. 170(3):268.

[128] McKay, D. G., et al. 1970. Lab. Invest. 22(5): 387.

[129] McLennan, B. D., and B. G. Lane. 1968. Can. J. Biochem. 46:93.

[130] Meaume, J. 1966. Toxicon 4(1):25.

[131] Meaume, J., et al. 1966. Mem. Inst. Butantan (São Paulo) 33(3):929.

[132] Mebs, D. 1968. Hoppe Seylers Z. Physiol. Chem. 349:1115.

[133] Mebs, D. 1970. Int. J. Biochem. 1:335.

[134] Mebs, D. 1970. Salamandra 6(3/4):135.

[135] Mebs, D., and H. W. Raudonat. 1966. Mem. Inst. Butantan (São Paulo) 33(3):907.

[136] Mekbel, S. T., and R. Céspedes. 1963. Acta Méd. Costarric. 6(2):111.

[137] Minton, S. A., Jr. 1967. Toxicon 5:47.

[138] Minton, S. A., Jr. Unpublished. Indiana Univ. Medical Center, Indianapolis, 1972.

[139] Minton, S. A., Jr., and M. R. Minton. 1969. Venomous Reptiles. Scribner's Sons, New York.

[140] Minton, S. A., Jr., et al. 1964. In P. L. Altman and D. S. Dittmer, ed. Biology Data Book. Federation of American Societies for Experimental Biology, Washington, D.C. pp. 328-335.

[141] Minton, S. A., Jr., et al. 1968. Poisonous Snakes of the World. U.S. Government Printing Office, Washington, D.C.

[142] Mitsuhashi, S., et al. 1959. Jap. J. Microbiol. 3(1): 95.

[143] Mohamed, A. H., et al. 1969. Toxicon 6(4):293.

[144] Mohamed, A. H., et al. 1969. Ibid. 7(3):185.

[145] Moroz, C., et al. 1966. Biochim. Biophys. Acta 124:136.

[146] Morse, E. E., et al. 1967. J. Lab. Clin. Med. 70: 106.

[147] Nakai, K., et al. 1970. Naturwissenschaften 57(8): 387.

[148] Narita, K., and C. Y. Lee. 1970. Biochem. Biophys. Res. Commun. 41(2):339.

[149] Ohsaka, A., et al. 1966. Mem. Inst. Butantan (São Paulo) 33(1):193.

[150] Okonogi, T., et al. 1964. Gunma J. Med. Sci. 13(2): 101.

[151] Omori-Satoh, T., and A. Ohsaka. 1970. Biochim. Biophys. Acta 207:432.

[152] Oshima, G., et al. 1968. J. Biochem. (Tokyo) 64(2): 227.

[153] Oshima, G., et al. 1969. Toxicon 7(3):229.

[154] Ouyang, C. 1957. J. Formosan Med. Ass. 56(9): 435.

[155] Parrish, H. M., and C. A. Carr. 1967. J. Amer. Med. Ass. 201(12):927.

[156] Parrish, H. M., and H. D. Donnell. 1967. S. Med. J. 60:429.

[157] Parrish, H. M., and M. S. Khan. 1967. Amer. J. Med. Sci. 253:561.

[158] Parrish, H. M., and G. H. Wiechmann. 1968. S. Med. J. 61(2):118.

[159] Patterson, R. A. 1967. Toxicon 5(1):5.

continued

[160] Patterson, R. A., and I. S. Lee. 1969. Ibid. 7(4): 321.

[161] Peña-Chavarría, A., et al. 1970. Amer. J. Trop. Med. Hyg. 19(2):342.

[162] Peng, M. T. 1951. Mem. Fac. Med. Nat. Taiwan Univ. 1(4):215.

[163] Picado, C. 1931. Serpientes venenosas de Costa Rica. Imprenta Alsina, San José.

[164] Prado-Franceschi, J. 1970. Doctoral Thesis. Campinas State Univ., São Paulo, Brazil.

[165] Ramsey, H. W., et al. 1972. Toxicon 10(1):67.

[166] Rao, S. S., and M. E. Kulkarni. 1952. Haffkine Inst. (Bombay) Rep., p. 49.

[167] Rechnic, J., et al. 1960. Bull. Res. Counc. Isr. 8:81.

[168] Reid, H. A. 1961. Brit. Med. J. 1:1284.

[169] Reid, H. A. 1962. Ibid. 2:576.

[170] Reid, H. A. 1964. Ibid. 2:540.

[171] Reid, H. A. 1970. Clin. Toxicol. 3(3):473.

[172] Reid, H. A., and K. E. Chan. 1968. Lancet 1:485.

[173] Rhoten, W. B., and J. F. Gennaro. 1968. J. Fla. Med. Ass. 55:324.

[174] Richards, G. M., et al. 1965. Biochemistry 4:501.

[175] Robertson, S. S. D., and G. R. Delpierre. 1969. Toxicon 7(3):189.

[176] Rocha e Silva, M. 1949. Arq. Inst. Biol. (São Paulo) 19:1.

[177] Russell, F. E. 1962. In G. M. Piersol, ed. Cyclopedia of Medicine, Surgery and Specialties. F. A. Davis, Philadelphia. v. 2, p. 197.

[178] Russell, F. E. 1969. Toxicon 7(1):33.

[179] Russell, F. E., and L. Lauritzen. 1966. Trans. Roy. Soc. Trop. Med. Hyg. 60(6):797.

[180] Russell, F. E., and P. R. Saunders, ed. 1967. Animal Toxins. Pergamon Press, Oxford.

[181] Russell, F. E., et al. 1963. Toxicon 1(3):99.

[182] Sawai, Y., and C. S. Tseng. 1969. The Snake 1(1):9.

[183] Sawai, Y., et al. 1967. Jap. J. Exp. Med. 37(1):51.

[184] Sawai, Y., et al. 1970. The Snake 2(1):13.

[185] Schenberg, S. 1959. Science 129:1361.

[186] Schenberg, S., and F. A. Pereira Lima. 1961. Acta Physiol. Lat. Amer. 11:233.

[187] Schiffman, S., et al. 1969. Biochemistry 8(4):1397.

[188] Schöttler, W. H. A. 1951. Amer. J. Trop. Med. 31: 489.

[189] Schöttler, W. H. A. 1951. Ibid. 31:836.

[190] Schöttler, W. H. A. 1952. Bull. W. H. O. 5:293.

[191] Schöttler, W. H. A. 1955. Ibid. 12:877.

[192] Schöttler, W. H. A. 1958. Ibid. 19:341.

[193] Schwick, G., and F. Dickgiesser. 1963. In Behringwerk-Mitt. Die Giftschlangen der Erde. Marburg/Lahn. p. 35.

[194] Singh, G. S., and R. K. Sanyal. 1967. Indian J. Med. Res. 55:1092.

[195] Slotta, K. H. 1955. Fortschr. Chem. Org. Naturst. 12:406.

[196] Slotta, K. H., and J. A. Vick. 1969. Toxicon 6(3): 167.

[197] Spies, S. K., et al. 1962. S. Afr. Med. J. 36(40): 834.

[198] Strover, H. M. 1967. Cent. Afr. J. Med. 13(8):185.

[199] Strydom, D. J., and D. P. Botes. 1970. Toxicon 8(3):203.

[200] Sulkowski, E., et al. 1963. J. Biol. Chem. 238(7): 2477.

[201] Suzuki, T. 1966. Mem. Inst. Butantan (São Paulo) 33(2):389.

[202] Suzuki, T., and S. Iwanaga. 1958. J. Pharm. Soc. Jap. 78(4):354.

[203] Suzuki, T., et al. 1960. Ibid. 80(7):868.

[204] Takahashi, T., and A. Ohsaka. 1970. Biochim. Biophys. Acta 198:293.

[205] Tamiya, N., and H. Arai. 1966. Biochem. J. 99: 624.

[206] Tateno, I., et al. 1963. Jap. J. Exp. Med. 33(6):331.

[207] Tazieff-Depierre, F., and J. Pierre. 1966. C. R. Acad. Sci. D263:1785.

[208] Trinca, J. C. 1969. Med. J. Aust. 1:514.

[209] Trinca, J. C., and J. J. Graydon. Unpublished. Commonwealth Serum Laboratories, Parkville, Victoria, Australia, 1973.

[210] Trinca, J. C., et al. 1971. Med. J. Aust. 2:801.

[211] Tsai, F. T. 1961. Fukuoka Acta Med. 52:145.

[212] Tu, A. T., and A. Chua. 1966. Comp. Biochem. Physiol. 17:297.

[213] Tu, A. T., and S. Ganthavorn. 1969. Amer. J. Trop. Med. Hyg. 18:151.

[214] Tu, A. T., and M. Homma. 1970. Toxicol. Appl. Pharmacol. 16:73.

[215] Tu, A. T., and D. S. Murdock. 1967. Comp. Biochem. Physiol. 22(2):389.

[216] Tu, A. T., and P. M. Toom. 1967. Aust. J. Exp. Biol. Med. Sci. 45:561.

[217] Tu, A. T., and P. M. Toom. 1967. Experientia 23: 439.

[218] Tu, A. T., and P. M. Toom. 1971. J. Biol. Chem. 246(4):1012.

[219] Tu, A. T., and T. Tu. 1970. In B. W. Halstead, ed. Poisonous and Venomous Marine Animals of the World. U.S. Government Printing Office, Washington, D.C. v. 3, p. 885.

[220] Tu, A. T., et al. 1965. Toxicon 3(1):5.

[221] Tu, A. T., et al. 1966. Ibid. 4(1):59.

[222] Tu, A. T., et al. 1967. Biochem. Pharmacol. 16(11): 2125.

[223] Tu, A. T., et al. 1967. In F. E. Russell and P. R. Saunders, ed. Animal Toxins. Pergamon Press, Oxford and New York. p. 351.

[224] Tu, A. T., et al. 1969. Toxicon 6(3):175.

[225] Uwatoko-Setoguchi, Y. 1970. Acta Med. Univ. Kagoshima. 12(1):73.

[226] Van der Walt, S. J., and F. J. Joubert. 1971. Toxicon 9(2):153.

[227] Vellard, J. 1938. Rev. Soc. Argent. Biol. 14:409.

[228] Vial, J. L., and J. M. Jiménez-Porras. 1967. Amer. Midl. Natur. 78:182.

[229] Wagner, F. W., and J. M. Prescott. 1966. Comp. Biochem. Physiol. 17:191.

[230] Wagner, F. W., et al. 1968. J. Biol. Chem. 243(17): 4486.

continued

Part I. Reptiles

[231] Wahlström, A. 1971. Toxicon 9(1):45.

[232] Walker, C. W. 1945. Brit. Med. J. 2:13.

[233] Watt, C. H., and J. F. Gennaro. 1966. Trans. S. Surg. Ass. 77:378.

[234] Weis, R., and R. J. McIsaac. 1971. Toxicon 9(3): 219.

[235] Wells, M. A., and D. J. Hanahan. 1969. Biochemistry 8(1):414.

[236] Williams, E. J., et al. 1961. J. Biol. Chem. 236(4): 1130.

[237] Wu, T. W., and D. O. Tinker. 1969. Biochemistry 8(4):1558.

[238] Wyon, P. H. 1945. Brit. Med. J. 2:919.

[239] Yang, C. C. 1967. Biochim. Biophys. Acta 133: 346.

[240] Yang, C. C., et al. 1960. J. Biochem. (Tokyo) 48(5):714.

[241] Yang, C. C., et al. 1968. Biochim. Biophys. Acta 168:373.

[242] Zeller, E. A. 1948. Advan. Enzymol. 8:459.

[243] Zeller, E. A. 1949. Helv. Chim. Acta 32(1):94.

[244] Zeller, E. A. 1950. Ibid. 33(4):821.

[245] Zwilling, R. von, and G. Pfleiderer. 1967. Hoppe Seylers Z. Physiol. Chem. 348:519.

General References

[246] Amaral, A. do. 1960. Proc. Haffkine Inst. (Bombay) Symp. 159:128.

[247] Amaral, A. do. 1966. Mem. Inst. Butantan (São Paulo) 33(1):293.

[248] Boquet, P. 1966. Toxicon 3(4):243.

[249] Condrea, E., and A. De Vries. 1965. Ibid. 2(4): 261.

[250] Gans, C., and W. B. Elliott. 1968. Advan. Oral Biol. 3:45.

[251] Jiménez-Porras, J. M. 1968. Annu. Rev. Pharmacol. 8:299.

[252] Jiménez-Porras, J. M. 1970. Clin. Toxicol. 3(3): 389.

[253] Meldrum, B. S. 1965. Pharmacol. Rev. 17(4):393.

[254] Minton, S. A., Jr., ed. 1971. Snake Venoms and Envenomation. M. Dekker, New York.

[255] Miranda, F., et al. 1972. In A. De Vries and E. Kochva, ed. Toxins of Animal and Plant Origin. Gordon and Breach, New York. p. 251.

[256] Rosenfeld, G., and E. M. A. Kelen, ed. 1969. Bibliography of Animal Venoms, Envenomations, and Treatments. Industria Gráfica Saraiva, São Paulo.

[257] Russell, F. E., and R. S. Scharffenberg, ed. 1964. Bibliography of Snake Venoms and Venomous Snakes. Bibliographic Associates, W. Covina, Calif.

[258] Tu, A. T. 1973. Annu. Rev. Biochem. 42:235.

[259] Vidal, J. C. 1970. Doctoral Thesis. Univ. Buenos Aires, Argentina.

Part II. Toads

Various species of *Bufo* produce a venom in a pair of special glands, so-called "paraotid" glands, behind each eye and in the warts of the dorsal skin. When the venom comes in contact with the face, there is a tickling sensation of the nose followed by intermittent sneezing, the tip of the tongue becomes numb, and the throat is dry for a brief period followed by an increase in salivation. These effects may last for an hour, depending on the amount of venom present. **Toxin Component:** Although the digitalis-like action of toad venom was discovered in the middle of the last century [38], only in the last 60 years have the active principles been isolated in crystalline form [10], the first of which was marinobufagin (bufagin) [1]. In 1970, derivatives of cardenolides were reported [17,18]; previously, the structures of all components reported were of the bufodienolide type (a six-membered lactone ring of the α-pyrone type at carbon-17) [20]. Neither bufodienolides

nor cardenolides of toad origin conjugate with carbohydrates at carbon-3 to form glycosides, but bufodienolides may react with suberylarginine, resulting in the formation of a bufotoxin [26]. The steroid compounds—the bufodienolides, the cardenolides, and their conjugates—mimic the action of digitalis; several are more active than ouabain while a few are inactive at dosage levels tested. Each species of toad is capable of synthesizing several cardiac sterols [2,8,34]. Potentially, the active sterols could cause death if they came into direct contact with the circulatory system, but since they are sparingly absorbed after ingestion, there are few or no fatalities in humans from toad venom. Other pharmacologically active substances—epinephrine [1], norepinephrine [23], and the indolealkylamines [15,30,40,41] including serotonin (5-hydroxytryptamine) [36] and dehydrobufotenine [27]—have been isolated; these bases, as a group, contract the smooth muscle

continued

Part II. Toads

of the uterus, while others have a marked pressor action in pithed cats; their clinical action on the central nervous system is controversial. Cholesterol, ergosterol, and/or γ-sitosterol constitute the noncardiotonic fraction of the toxin. Their discovery prompted the demonstration of the biosynthesis of some bufodienolides (e.g., marinobufagin and marinobufotoxin) from cholesterol [35]; the common occurrence of cholesterol in many toad toxins may indicate the occurrence of this biosynthesis in many species of toad. **Activity**: Cardiotonic potency is measured by the lethal dose for 50% of test cats (LD_{50}); the reciprocal, LD_{50}/mg, gives rise to a direct expression of activity [10]. Figures in brackets are reference numbers.

	Species [Distribution]	Toxin Component	Activity LD_{50}/mg		Species [Distribution]	Toxin Component	Activity LD_{50}/mg
1	*Bufo alvarius*	Alvarobufotoxin [10]	1.32 ± 0.32	43		Gamabufotalin [7]	9.9 ± 0.50
2	[Southwestern U.S.]	Telocinobufagin [34]	9.8 ± 0.64	44		Hellebrigenin [1]	13.0 ± 0.93
				45		Marinobufagin [2]	0.7 ± 0.12
3	*B. america-nus* [Eastern U.S.]	Bufagin present, but not crystallized [9]	Unknown	46		3-Methylsuberyldigitoxigenin	Unavailable
4		Bufotoxin present, but not crystallized	....	47		3-Methylsuberyl-14,15β-epoxydigitoxigenin	Unavailable
5	*B. arenarum*	Arenobufagin [12,32]	13.0 ± 1.14	48		Oleandrigenin	4.6 ± 0.27
6	[Argentina,	Arenobufotoxin	2.46 ± 0.07	49		Periplogenin	1.4 ± 0.16
7	Uruguay,	Argentinogenin	Unavailable	50		Resibufogenin	Inactive
8	southern	Bufalin	7.3 ± 0.55	51		Sarmentogenin	2.2 ± 0.15
9	Brazil]	Bufarenogin	Unavailable	52		Telocinobufagin	9.8 ± 0.64
10		Bufotalinin	1.6 ± 0.31	53	*B. bufo for-*	Argentinogenin	Unavailable
11		Hellebrigenin [1]	13.0 ± 0.93	54	*mosus* [Ja-*	Bufalin	7.3 ± 0.55
12		Hellebrigenol	Unavailable	55	pan]	Bufalone	Unavailable
13		Marinobufagin [2]	0.7 ± 0.12	56		Bufatalone	Unavailable
14		Resibufogenin	Inactive	57		Cinobufagin	5.0 ± 0.44
15		Telocinobufagin	9.8 ± 0.64	58		Cinobufotalin	5.0 ± 0.62
16	*B. asper* [Ja-*	Bufotalin [21]	7.6 ± 0.40	59		Deacetylcinobufagin [6]	Inactive
17	va, Indone-*	Bufotoxin unknown	....	60		Deacetylcinobufotalin [8]	Inactive
18	sia]	Marinobufagin [2]	0.7 ± 0.12	61		3-Epibufalin	Unavailable
19		Resibufogenin	Inactive	62		Gamabufotalin [7] [22,31]	9.9 ± 0.50
20	*B. blombergi* [Colombia]	Bufagin present, but not identified	Unavailable	63		Gamabufotoxin [14]	2.67 ± 0.19
				64		Hellebrigenin [1]	13.0 ± 0.93
21		Bufotoxin unknown	....	65		Marinobufagin [2]	0.7 ± 0.12
22	*B. bufo bu-*	Bufotalin [39]	7.6 ± 0.40	66		Resibufogenin	Inactive
23	*fo [3]* [Eur-*	Bufotalinin	1.6 ± 0.31	67		Telocinobufagin	9.8 ± 0.64
24	rope]	Bufotoxin [26]	3.43 ± 0.21	68	*B. granulosus*	Gamabufotalin [7] [3]	9.9 ± 0.50
25		Hellebrigenin [1]	13.0 ± 0.93		*fernandeze*		
26		Marinobufagin [2]	0.7 ± 0.12		[Panama to Argentina]		
27		Telocinobufagin [37]	9.8 ± 0.64	69	*B. marinus [9]*	Argentinogenin	Unavailable
28	*B. bufo gar-*	16β-Acetoxy-14,15β-epoxyperiplogenin	Unavailable	70	[Circumtrop-*	Bufalin [2,34]	7.3 ± 0.55
	garizans [4]			71	ical]	Gamabufotalin [7]	9.9 ± 0.50
29	[China]	Arenobufagin [16]	13.0 ± 1.14	72		Hellebrigenin [1]	13.0 ± 0.93
30		Artebufogenin	Inactive	73		Hellebrigenol	Unavailable
31		Bufalin [28,29,33]	7.3 ± 0.55	74		Jamaicobufagin	Unavailable
32		ψ-Bufarenogin [19]	Inactive	75		Marinobufagin [2]	0.7 ± 0.12
33		Bufogenin B [5]	3.8 ± 0.28	76		Marinobufotoxin [8]	2.40 ± 0.12
34		Bufotalin	7.6 ± 0.40	77		Resibufogenin	Inactive
35		Cinobufagin [11]	5.0 ± 0.44	78		Telocinobufagin	9.8 ± 0.64
36		Cinobufaginol [25]	Unavailable	79	*B. mauretani-*	Arenobufagin	13.0 ± 1.14
37		Cinobufotalin	5.0 ± 0.62	80	*cus* [South-*	Bufalin	7.3 ± 0.55
38		Cinobufotoxin	2.79 ± 0.19	81	west Africa]	Bufotalinin	1.6 ± 0.31
39		Deacetylcinobufagin [6]	Inactive	82		Hellebrigenin [1] [6,24,34]	13.0 ± 0.93
40		Digitoxigenin [17,18]	2.2 ± 0.17	83		Marinobufagin [2]	0.7 ± 0.12
41		14,15β-Epoxydigitoxigenin	Unavailable	84		Resibufogenin	Inactive
42		14,15β-Epoxyoleandrigenin	Unavailable	85		Telocinobufagin	9.8 ± 0.64

[1] Synonym: bufotalidin. [2] Synonym: bufagin. [3] Synonym: *B. vulgaris*. [4] Venom is source of Ch'an Su, a commercial preparation of Chinese toad venom used in medicine for external application. [5] Synonym: desacetyl-bufotalin.

[6] Synonym: desacetyl-cinobufagin. [7] Synonym: gamabufagin. [8] Synonym: desacetyl-cinobufotalin. [9] Synonym: *B. agua*.

continued

Part II. Toads

#	Species [Distribution]	Toxin Component	Activity LD$_{50}$/mg
86	B. melanos-	Bufalin	7.3 ± 0.55
87	tictus [Java,	Bufotalin	7.6 ± 0.40
88	Indonesia;	Hellebrigenin [1/]	13.0 ± 0.93
89	Canton, Chi-	Marinobufagin [2/] [21]	0.7 ± 0.12
90	na; Assam,	Resibufogenin	Inactive
	India; Bang-		
	kok, Thailand]		
91	B. peltoceph-	Arenobufagin	13.0 ± 1.14
92	alus [Cuba]	Argentinogenin	Unavailable
93		Gamabufotalin [7/]	9.9 ± 0.50
94		Marinobufagin [2/]	0.7 ± 0.12
95		Telocinobufagin [3]	9.8 ± 0.64
96	B. quercicus	Bufotoxin present, but not crystallized	....
97	[Southeastern U.S.]	Quercicobufagin [9]	10.3 ± 0.39
98	B. regularis	Arenobufagin	6.5 ± 0.25 [10/]
99	[South Afri-	Bufotalinin	1.6 ± 0.31
100	ca]	Gamabufotalin [7/]	6.5 ± 0.25 [10/]
101		Hellebrigenin [1/]	13.0 ± 0.93
102		Hellebrigenol [4,5,7,34]	Unavailable
103		Regularobufotoxin [7]	2.10 ± 0.12
104		Resibufogenin	Inactive
105	B. spinulosus	Arenobufagin	13.0 ± 1.14
106	[Chile; Peru]	Gamabufotalin [7/] [3]	9.9 ± 0.50
107		Marinobufagin [2/]	0.7 ± 0.12
108		Resibufogenin	Inactive
109		Telocinobufagin	9.8 ± 0.64
110	B. valliceps	Arenobufagin	13.0 ± 1.14
111	[Texas &	Bufotoxin present, but not crystallized	....
112	Louisiana; eastern	Hellebrigenin [1/]	13.0 ± 0.93
113	Mexico]	Hellebrigenol	Unavailable
114		Marinobufagin [2/]	0.7 ± 0.12
115		Telocinobufagin	9.8 ± 0.64
116		Vallicepobufagin [3,9]	5.0 ± 0.43
117	B. viridis viri-	Viridobufagin [13]	9.0 ± 0.66
118	dis [Hamburg, W. Germany]	Viridobufotoxin	3.70 ± 0.16
119	B. wood-	Arenobufagin	13.0 ± 1.14
120	housei fow-	Fowlerobufagin	4.6 ± 0.24
121	leri [South-	Fowlerobufotoxin	1.26 ± 0.09
122	eastern	Gamabufotalin [7/]	9.9 ± 0.50
123	U.S.]	Marinobufagin [2/] [3]	0.7 ± 0.12

[1/] Synonym: bufotalidin. [2/] Synonym: bufagin. [7/] Synonym: gamabufagin. [10/] Value for mixed crystals of arenobufagin and gamabufagin.

Contributor: Chen, K. K.

References

[1] Abel, J. J., and D. I. Macht. 1912. J. Pharmacol. Exp. Ther. 3:319.
[2] Barbier, M., et al. 1959. Helv. Chim. Acta 42:2486.
[3] Barbier, M., et al. 1961. Ibid. 44:362.
[4] Bharucha, M., et al. 1961. Ibid. 44:651.
[5] Bharucha, M., et al. 1961. Ibid. 44:844.
[6] Bolliger, R., and K. Meyer. 1957. Ibid. 40:1959.
[7] Chen, K. K., and A. L. Chen. 1933. J. Pharmacol. Exp. Ther. 49:503.
[8] Chen, K. K., and A. L. Chen. 1933. Ibid. 49:514.
[9] Chen, K. K., and A. L. Chen. 1933. Ibid. 49:526.
[10] Chen, K. K., and A. Kovaříkova. 1967. J. Pharm. Sci. 56:1535.
[11] Chen, K. K., et al. 1931. J. Pharmacol. Exp. Ther. 43:13.
[12] Chen, K. K., et al. 1933. Ibid. 49:1.
[13] Chen, K. K., et al. 1933. Ibid. 49:14.
[14] Chen, K. K., et al. 1933. Ibid. 49:26.
[15] Heinzelman, R. V., and J. Szmuszkovics. 1963. Progr. Drug Res. 6:98, 142.
[16] Hofer, P., and K. Meyer. 1960. Helv. Chim. Acta 43:1495.
[17] Höriger, N., et al. 1970. Ibid. 53:1503.
[18] Höriger, N., et al. 1970. Ibid. 53:1993, 2051.
[19] Huber, K., et al. 1967. Ibid. 50:1994.
[20] International Union of Pure and Applied Chemistry. 1951. Helv. Chim. Acta 34:1680.
[21] Iseli, E., et al. 1964. Helv. Chim. Acta 47:116.
[22] Iseli, E., et al. 1965. Ibid. 48:1093.
[23] Lasagna, L. 1951. Proc. Soc. Exp. Biol. Med. 78:876.
[24] Linde, H., and K. Meyer. 1958. Pharm. Acta Helv. 33:327.
[25] Linde, H., et al. 1966. Helv. Chim. Acta 49:1243.
[26] Linde-Tempel, H. O. 1970. Ibid. 53:2188.
[27] Märki, F., et al. 1961. J. Amer. Chem. Soc. 83:3341.
[28] Meyer, K. 1949. Pharm. Acta Helv. 24:222.
[29] Meyer, K. 1952. Helv. Chim. Acta 35:2444.
[30] Michl, H., and E. Kaiser. 1963. Toxicon 1:186.
[31] Ohno, S., et al. 1961. J. Jap. Pharm. Ass. 81:1345.
[32] Rees, R., et al. 1959. Helv. Chim. Acta 42:2400.
[33] Ruckstuhl, J.-P., and K. Meyer. 1957. Ibid. 40:1270.
[34] Schröter, H., et al. 1958. Ibid. 41:140.
[35] Siperstein, M. D., et al. 1957. Arch. Biochem. Biophys. 67:154.
[36] Udenfriend, S., et al. 1952. Experientia 8:379.
[37] Urscheler, H. R., et al. 1955. Helv. Chim. Acta 38:883.
[38] Vulpian, E. F. A. 1854. C. R. Soc. Biol., Ser. 2, 1:133.
[39] Wieland, H., and F. J. Weil. 1913. Ber. Deut. Chem. Ges. 46:3315.
[40] Wieland, H., and T. Wieland. 1937. Justus Liebigs Ann. Chem. 528:234.
[41] Wieland, H., et al. 1934. Ibid. 513:1.

continued

Part III. Marine Organisms

Chemistry & Toxicology: mol wt = molecular weight; pK$_a$ = negative logarithm of the dissociation constant of the acid form. Data in brackets refer to the column heading in brackets.

	Type of Poisoning [Cause]	Species (Synonym)	Distribution	Symptoms & Signs	Chemistry & Toxicology	Reference
			Chordata			
1	Boxfish & soapfish poisoning	Lactophrys bicaudalis (Rhinesomus bicaudalis)	West Indies to Florida	....	Toxin from skin secretion of O. lentiginosus (pahutoxin) is a choline chloride ester of 3-acetyoxypalmitic acid [L/] (C$_{23}$H$_{46}$NO$_4$Cl); crude dried product toxic to fish at concentration of 1 ppm; ∿50-100 mg present in 1 adult Lactoria fornasini. Hemolytic activity of toxin said to correlate with lethality. Mouse LD$_{min}$ = 200 mg/kg.	7,8, 37, 53, 73, 97, 101
2	[Skin secretions of certain fishes. Toxin probably released as alarm substance to deter possible predators.]	Lactoria fornasini	Indo-Pacific area; Indian Ocean; eastern Africa			
3		Ostracion lentiginosus	Indo-Pacific area			
4		Rypticus saponaceus	Tropical & subtropical Atlantic Ocean			
5	Ciguatera poisoning	Acanthurus glaucopareius	Indo-Pacific area	Onset within 36 hr; tingling about lips, tongue, & throat, followed by numbness of some parts; dryness of mouth, nausea, vomiting, & abdominal cramps common; headache, dizziness, pallor, restlessness, weakness, blurring of vision, itching, ataxia, & convulsions may occur. Deaths reported.	Ciguatoxin considered to be a lipid containing a quaternary N atom, 1 or more hydroxyl groups, and a cyclopentanone moiety (probable empirical formula, C$_{35}$H$_{65}$NO$_8$). Alters cardiovascular & respiratory activities, but mechanism not known; inhibits action potential of frog sciatic nerve; appears to have a direct effect on neuromuscular junction. LD$_{50}$ (appears to differ with preparation) = 0.1-100 mg/kg i.v.	4,36, 42, 52, 71-73, 78, 90
6	[Ingestion of the flesh, & especially the liver, of any of several reef & semipelagic marine fishes. In most cases, toxicity related to food-chain cycle, probably originating in benthic algae.]	A. triostegus	Hawaiian Islands; Johnson Island of Johnston Atoll			
7		Albula vulpes	Red Sea to Hawaiian Islands			
8		Aluterus scriptus (Alutera scripta)	Warm seas			
9		Aprion virescens	Indo-Pacific area			
10		Balistoides conspicillum	Indo-Pacific area; China; Japan			
11		Caranx hippos	Tropical Atlantic Ocean			
12		Cephalopholis argus	Tropical Indo-Pacific area			
13		Clupanodon thrissa	Tropical Pacific Ocean; Japan; China; Korea			
14		Engraulis japonicus	China; Taiwan (Formosa); Korea; Japan			
15		Epibolus insidiator	Indo-Pacific area			
16		Epinephelus fuscoguttatus	Indo-Pacific area			
17		Euthynnus pelamis	Circumtropical areas			
18		Gnathodentex aureolineatus	Tuamotu Archipelago west to eastern Africa			
19		Gymnothorax flavimarginatus	Hawaiian Islands west to eastern Africa			
20		G. javanicus	Hawaiian Islands west to eastern Africa			
21		G. meleagris	Indo-Pacific area			
22		G. pictus	Polynesia to eastern Africa			

[L/] Synonym: 3-acetoxyhexadecanoic acid.

continued

Part III. Marine Organisms

	Type of Poisoning [Cause]	Species (Symptoms)	Distribution	Symptoms & Signs	Chemistry & Toxicology	Reference
23		*G. undulatus*	Hawaiian Islands to Red Sea & eastern Africa			
24		*Lactophrys trigonus*	Western Atlantic Ocean; tropical coasts of the Americas			
25		*Lactoria cornutus*	Tropical Pacific Ocean; Japan; South Africa			
26		*Lethrinus miniatus*	Polynesia west to eastern Africa			
27		*Lutjanus bohar*	Indo-Pacific area; Red Sea; East Indies			
28		*L. gibbus*	Tropical Indo-Pacific area			
29		*L. monostigmus*	Polynesia west to Red Sea			
30		*L. nematophorus*	Australia			
31		*L. vaigiensis*	Polynesia west to eastern Africa & Australia			
32		*Mycteroperca venenosa*	Western tropical Atlantic Ocean			
33		*Pagellus erythrinus*	Mediterranean & Black Seas; eastern Atlantic Ocean from the British Isles & Scandanavia to the Azores, Canary Islands, & Fernando Poo			
34		*Pagrus pagrus*	Eastern Atlantic Ocean; Mediterranean Sea			
35		*Parupeneus chryserydros*	Polynesia west to eastern Africa			
36		*Plectropomus oligacanthus*	Indo-Pacific area			
37		*Scarus coeruleus*	Florida; West Indies; Panama			
38		*S. microrhinos*	Indo-Pacific area			
39		*Sphyraena barracuda*	All tropical seas except the eastern Pacific Ocean			
40		*Tetragonurus cuvieri*	Temperate waters			
41		*Upeneus arge*	Indo-Pacific area			
42	Scorpionfish sting [Puncture from dorsal, pelvic, or anal fin spines. Venom glands are within integumentary sheath (differ in *Pterois, Scorpaena,* & *Synanceja*). Spines used in defense.]	*Centropogon australis*	Australia	Immediate, severe, local pain, subsequently radiating; pallor & increased skin temp around puncture wound; primary shock in some cases; weakness, nausea, dyspnea, headache, paresthesia in involved extremity. In *Synanceja* sting, symptoms usually more severe, often with	Lethal activity associated with labile proteins (molecular wt >400,000) & sulfhydryl group integrity. Direct effect on pulmonary circulation & heart, producing hypotension & pulmonary congestion; little effect on peripheral nervous system. *Scorpaena guttata* venom releases acetylcholine from muscarinic sites &	3,9, 12, 13, 37, 40, 82- 86
43		*Choridactylus multibarbis*	Indo-Pacific area; India; China			
44		*Inimicus didactylus*	Indo-Pacific area			
45		*I. japonicus*	Japan			
46		*Minous monodactylus*	Indo-Pacific area; China; Japan			
47		*Notesthes robusta*	Australia			
48		*Pterois antennata*	Indo-Pacific area; Indian Ocean; China			
49		*P. radiata*	Indo-Pacific area; Indian Ocean; Red Sea			
50		*P. volitans*	Indo-Pacific area; Australia; Japan; China; Indian Ocean; Red Sea			

continued

Part III. Marine Organisms

	Type of Poisoning [Cause]	Species (Synonym)	Distribution	Symptoms & Signs	Chemistry & Toxicology	Ref-er-ence
51		*Scorpaena guttata*	Central California into Gulf of California	respiratory & cardiovascular changes, including shock & coma. Localized necrosis & deaths reported.	endogenous norepinephrine. Hemolysis observed in vitro. $LD_{50} = 0.2$ mg protein/kg i.v.	
52		*S. plumieri*	Eastern coast of the United States to the West Indies & Brazil			
53		*S. porcus*	English Channel to the Canary Islands; Mediterranean & Black Seas			
54		*S. scrofa*	Western coast of France; northwestern Africa; Mediterranean Sea			
55		*Scorpaenopsis diabolis*	Indo-Pacific area; Australia			
56		*Synanceja horrida*	Indo-Pacific area; Australia; China; Japan			
57		*S. verrucosa*	Indo-Pacific area; Australia; Indian Ocean; Red Sea			
58	Tetraodon poisoning	*Arothron hispidus*	Tropical Pacific to Japan & Red Sea	Onset within 15-20 min, with tingling & numbness about mouth, lips, & tongue; excessive salivation, weakness, nausea, vomiting, difficulty in swallowing; paresthesia & paralysis may occur over different parts of body; convulsions & coma reported. Mortality rate ~50%.	Aminoperhydroquinazoline ($C_{11}H_{17}N_3O_8$) produces nondepolarizing neuromuscular block by acting selectively to prevent or reduce usual increase in Na^+ permeability without affecting K^+ movement; direct effect on cardiovascular & respiratory systems. $LD_{50} = 12$ µg/kg i.p.	6,16, 36, 44, 61, 73, 100, 105
59	[Ingestion of tetraodontiform fishes. Liver & gonads most toxic, intestines & skin less so; in most instances, toxicity related to reproductive cycle.]	*A. meleagris*	Indo-Pacific area; Japan to eastern Africa & Red Sea			
60		*A. nigropunctatus*	Indo-Pacific area; Japan to eastern Africa; Red Sea			
61		*Canthigaster margaritatus*	Indo-Pacific area; eastern Africa; China; Red Sea			
62		*C. rivulatus*	Japan; Indo-Pacific area			
63		*Chilomycterus spinosus*	West Indies; Brazil; South Africa			
64		*Colomesus psittacus*	Rivers of Guiana; northern Brazil; West Indies			
65		*Diodon holocanthus*	Tropical Atlantic, Pacific, & Indian Oceans			
66		*Fugu basilevskianus*	Northern China; northwestern Korea			
67		*F. chrysops*	Pacific coast of Japan			
68		*F. niphobles*	China; Japan; Philippines			
69		*F. ocellatus*	China; Japan; Philippines			
70		*F. pardalis*	China; Japan			
71		*F. pseudommus*	Eastern China; Yellow Sea			
72		*F. rubripes*	China to Korea; Japan; Pacific Ocean			
73		*F. stictonotus*	Southern Korea; East China Sea; Japan			
74		*F. vermicularis*	East China Sea; Japan			
75		*F. xanthopterus*	China; Korea; southern Japan			
76		*Lagocephalus laevigatus inermis*	Eastern Africa; Indian Ocean; Australia; East China Sea; Japan			
77		*L. lunaris*	Red Sea; southern & eastern Africa; India to Australia; China; Japan			

continued

Part III. Marine Organisms

	Type of Poisoning [Cause]	Species (Synonym)	Distribution	Symptoms & Signs	Chemistry & Toxicology	Reference
78		L. sceleratus	East coast of Africa to the Philippines; southern Japan; Australia; Tahiti			
79		Mola mola	Temperate & tropical seas			
80		Sphoeroides annulatus	Baja California to Peru; Galápagos Islands			
81		S. maculatus	Atlantic coast of North America to Guiana			
82		S. spengleri	West Indies; Brazil; Canary Islands; west coast of Africa; Gulf of Mexico; tropical Atlantic			
83		Tetraodon lineatus	Rivers of Africa			
84		Torquigener hamiltoni	Australia; Indo-Pacific area			
85	Weeverfish sting [Puncture from opercular or dorsal spines. Venom contained in grooves of spine within integumentary sheath; spines used in defense.]	Trachinus araneus	Mediterranean Sea	Immediate, intense, local pain, with increased skin temp near wound; some local edema & discoloration; weakness in involved extremity; dyspnea; marked weakness, nausea, & vomiting in severe cases. Coma & deaths reported.	Lethal fraction is non-dialyzable, heat-labile protein of unknown size; crude venom also contains histamine, epineph-rine[2], norepineph-rine[3], an unknown catecholamine, & cholinesterase. Small doses produce vasodilation, large doses vasoconstriction; direct effect on heart, but not on central or peripheral nervous systems. LD$_{50}$ of crude venom = 15 mg/kg.	11, 32, 37, 38, 71, 73, 76, 92, 93
86		T. draco	Norway to northern Africa; Mediterranean & Adriatic Seas			
87		T. radiatus	Mediterranean Sea			
88		T. vipera	Southern North Sea; English Channel; Mediterranean Sea			
89	Stingray sting [Puncture from bilaterally serrated caudal spine or from sting on tail. Venom contained within glandular triangle of ventrolateral grooves of sting's integumentary sheath; venom apparatus used in defense.]	Aetobatus narinari	Tropical & warm, temperate areas of Atlantic & Pacific Oceans, Indo-Pacific area, & Red Sea	Immediate, intense, local pain, subsequently radiating; lacerated wound with ragged, discolored edges; local edema & elevated skin temp; primary shock, weakness, nausea, vomiting, diaphoresis, muscle spasms, pain; diarrhea, local necrosis, & secondary shock reported. Deaths extremely rare.	Lethal activity associated with labile proteins (molecular wt >100,000). Dose-dependent direct effect on heart & cardiovascular system; no significant effect on nerve or neuromuscular transmission. LD$_{50}$ of lethal fraction ∿2.9 mg protein/kg i.v.	14, 34, 37, 39, 65, 75, 77-81
90		Aetomylaeus nichofii	Indo-Pacific area; Australia; Indian Ocean			
91		Dasyatis brevis	Gulf of California; Peru; Hawaiian Islands			
92		D. dipterura	British Columbia to Central America			
93		D. guttata	West Indies; Gulf of Mexico to southern Brazil			
94		D. longus	Gulf of California to Panama & Galápagos Islands			
95		D. pastinaca	Northeastern Atlantic Ocean; Mediterranean Sea; Indian Ocean			
96		D. sabina	Western North Atlantic Ocean from the Chesapeake Bay to Brazil; Gulf of Mexico			

[2] Synonym: adrenaline. [3] Synonym: noradrenaline.

continued

Part III. Marine Organisms

	Type of Poisoning [Cause]	Species (Synonym)	Distribution	Symptoms & Signs	Chemistry & Toxicology	Reference
97		*D. violacea*	Mediterranean Sea			
98		*Gymnura marmorata*	California into Mexico			
99		*Myliobatis californica*	Oregon to Baja California			
100		*Potamotrygon falkneri*	Rivers of the Panama region			
101		*P. hystrix*	Rivers of the Amazon region			
102		*P. labradori*	Rivers of the Panama region			
103		*P. motoro*	Rivers of Paraguay; Amazon River south to Rio de Janeiro			
104		*Urolophus armatus*	Indo-Pacific area			
105		*U. halleri*	Southern Pacific coast of North America			

Echinodermata

	Type of Poisoning [Cause]	Species (Synonym)	Distribution	Symptoms & Signs	Chemistry & Toxicology	Reference
106	Sea cucumber poisoning [Contact or ingestion]	*Actinopyga agassizi*	West Indies	On contact: local pain, pruritus, tenderness, & swelling; conjunctivitis in swimmers. On ingestion: gastric distress, with nausea & vomiting. Deaths extremely rare.	Holothurian A ($C_{50-52}H_{81-85}O_{25-26}$; mol wt = 1155); mixture of several closely related sulfate ester glycosides containing a steroid aglycone of 26-28 C atoms, 4-5 O atoms, & 1 molecule of H_2SO_4 as the Na salt.	1,26-28, 33, 35, 62, 69, 75, 95, 98
107		*Bohadschia argus* (*Holothuria argus*)	Indo-Pacific area; Polynesia to Indian Ocean			
108		*Cucumaria echinata*	Japan			
109		*Holothuria atra*	Indo-Pacific area; Polynesia; Australia			
110		*H. impatiens*	Circumtropical areas			
111		*H. poli*	Mediterranean Sea & adjacent coasts			
112		*Patinapta ooplax* (*Leptosynapta ooplax*)	Japan		A potent saponin surfactant which lyses erythrocytes and has hemolytic activity greater than digitonin; direct effect on mammalian nerve & muscle.	
113		*Stichopus variegatus*	Polynesia westward to the Indian Ocean; Australia; Red Sea			
114	Sea urchin sting [Contact with secondary spines of globiferous pedicellariae of a number of species]	*Asthenosoma varium*	Indonesia; Indian Ocean; Gulf of Suez	Immediate, local pain, with subsequent swelling & redness around wound; often aching sensation in involved part. In more severe cases: primary shock, weakness or partial paralysis of extremity, & respiratory distress.	Pedicellariae toxin has hemolytic activity against erythrocytes of man, cattle, rabbit, & fish; i.v. injection produces marked fall in systemic arterial pressure & direct effect on heart; releases histamine & serotonin, but not acetylcholine, and causes block in indirectly elicited nerve-muscle response of crayfish, and a block to muscle response to intracellular stimulation. LD_{50} = 15.9 μg precipitable protein/kg.	1,2, 23, 24, 58, 78
115		*Diadema setosum*	Indo-Pacific area; China; Japan			
116		*Echinothrix calamaris*	Indo-Pacific area; Red Sea; Australia; Japan			
117		*E. diadema*	Indo-Pacific area; Red Sea; Australia; Japan			
118		*Phormosoma bursarium*	Indo-Pacific area			
119		*Toxopneutes pileolus*	Indo-Pacific area; Japan			
120		*Tripneustes gratilla*	Indo-Pacific area; Polynesia to East Africa; Australia; Japan			

continued

Part III. Marine Organisms

	Type of Poisoning [Cause]	Species (Synonym)	Distribution	Symptoms & Signs	Chemistry & Toxicology	Reference
			Annelida			
121	Worm bites & bristle worm	*Chloeia flava*	Indo-Pacific area; Japan; China Sea; Indian Ocean	Worm bites: immediate pain, local	Nereistoxin (C_5H_9-NS_2), a cyclic disul-	31, 35,
122	stings	*Eunice aphroditois*	Circumtropical areas	redness & swelling, numbness &	fide, produces increase in parasympa-	41,
123	[Contact with	*Eurythoe brasiliensis*	Brazil	pruritus. Setae:	thetic activity; site	63,
124	chitinous fangs (*Glycera* spp.)	*Glycera dibranchiata*	East coast of the United States	stinging sensation.	of action unknown	66
125	or bristlelike	*G. ovigera*	New Zealand			
126	setae (other species shown)]	*Hermodice caruncu-lata*	Florida; West Indies			
			Mollusca			
127	Cone sting	*Conus aulicus*	Polynesia to the Indian Ocean	Immediate, sometimes intense, lo-	Venom protein or bound to protein.	20, 35,
128	[Contact with venom appara-	*C. californicus*	California	cal pain, with subsequent numbness	Duct extracts yield homarine, γ-butyro-	47,
129	tus consisting of radular	*C. catus*	Red Sea to the Hawaiian Islands; throughout the Indo-Pacific area	& ischemia around wound. More se-	betaine, *N*-methyl-pyridinium, amines,	48, 71, 75,
130	teeth, radular sheath, venom	*C. geographus*	Indo-Pacific area; Polynesia to eastern Africa	vere cases: possibly tingling or	5-hydroxytrypt-amine, lipoproteins,	78, 102-
131	duct, & bulb]	*C. imperialis*	Indo-Pacific area	numbness around	& carbohydrates.	104
132		*C. lividus*	Polynesia; Red Sea	mouth; tremor,	Causes decrease in	
133		*C. magus*	Indo-Pacific area	muscle fascicula-	systemic arterial	
134		*C. marmoreus*	Polynesia to the Indian Ocean	tions & incoordination, nausea &	pressure & heart rate, with increase in res-	
135		*C. quercinus*	Red Sea to the Hawaiian Islands; throughout the Indo-Pacific area	vomiting, increased salivation & lacrimation.	piration; larger doses produce paralysis, cardiac failure, & re-	
136		*C. spurius*	Florida; West Indies	Deaths reported.	spiratory arrest. *C.*	
137		*C. striatus*	Indo-Pacific area; Australia to eastern Africa		*geographus* & *C. ma-gus* venom: mouse LD = 0.2-1.3	
138		*C. textile*	Polynesia; Red Sea		mg/kg s.c.; *C. californicus* venom:	
139		*C. tulipa*	Polynesia; Red Sea		LD_{50} = 2.4 mg protein/kg i.v.	
140	Octopus bite	*Eledone aldrovandi*	Europe	Pain or burning	From *Eledone*: eledoi-	29,
141	[Contact with	*E. moschata*	Mediterranean & Red Seas	around puncture	sin, an endecapep-	35,
142	venom appara-tus consisting	*Octopus apollyon*	Pacific coast of North America	site (may involve entire extremity	tide producing marked vasodilation,	43, 71,
143	of beak or	*O. bimaculatus*	Southern California	in severe poison-	leading to hypoten-	91,
144	mandibles, sali-	*O. fitchi*	Gulf of California	ings), local red-	sion in mammals.	96,
145	vary ducts, an-	*O. flindersi*	Australia	ness, pruritus, &	Identified in salivary	99
146	terior & posterior salivary glands, and buccal mass]	*O. macropus*	Europe; Mediterranean & Red Seas; Indian Ocean; Malaysia; coasts of China, Japan, & Australia	swelling reported. Severe cases: respiratory distress, vomiting, & car-	secretions of *Octopus*: tyramine, epinephrine [2], norepinephrine [3], seroto-	
147		*O. vulgaris*	Warm seas	diovascular collapse. Deaths reported.	nin, histidine, histamine, dopamine, tryptophan, polyphenols, indoleamines, & guanidine bases, among others.	

[2] Synonym: adrenaline. [3] Synonym: noradrenaline.

continued

Part III. Marine Organisms

	Type of Poisoning [Cause]	Species (Synonym)	Distribution	Symptoms & Signs	Chemistry & Toxicology	Reference
			Coelenterata (Cnidaria)			
148	Hydroid sting	*Lytocarpus nuttingi*	Coast of southern California	Local burning sensation & itching, with subsequent erythematopapular lesions in some cases; occasionally pustular lesions; rarely serious	*See* Jellyfish sting, entries 153-159	17,
149	[Contact with stinging nematocysts]	*Millepora alcicornis*	Tropical Pacific Ocean; Indian Ocean; Carribean & Red Seas			18, 35, 68,
150		*Pennaria tiarella*	Coast of Maine to Brazil; Bermuda; Bahama Islands; California; Baja California; tropical Pacific Ocean			71
151		*Rhizophysa eysenharti*	Warmer waters of all oceans			
152		*Sarsia tubulosa*	Pacific coasts of North & South America; Hawaiian Islands; western North Atlantic Ocean			
153	Jellyfish sting	*Chironex fleckeri*	Australia	Local pain, pruritus, edema, & wheals, or papular eruptions, which may give way to vesicles & pustulation; weakness, nausea, muscle spasms, tremor, & vertigo. Severe cases, wheals, erythema, edema, severe pain, weakness, vomiting, painful muscle spasms & fasciculations, dyspnea, & cardiovascular collapse. Deaths from *Chironex fleckeri* & *Chiropsalmus quadrigatus* stings, but rare from stings of other species.	Nematocysts contain proteins, amines, hydroxyproline, tyrosine, aspartic acid, a succinate dehydrogenase[4] inhibitor; pyrocatechols[5], mineral salts, acid & alkaline phosphatases, serotonin, & cholinesterase, among other substances. Lethal component apparently a polypeptide. Crude venom has direct effect on the heart & neuromuscular junction, and is antigenic; mechanisms unknown. *Chironex fleckeri*: LD ∿0.005 ml/kg.	5,25,
154	[Contact with nematocysts, usually with the tentacles]	*Chiropsalmus quadrigatus*	Australia; Philippines; Indian Ocean			35, 46,
155		*Chrysaora hysoscella*	Atlantic coast of Europe; Mediterranean Sea; Malayan archipelago; New Zealand; Japan			50, 51, 67,
156		*Cyanea capillata*	North Atlantic & North Pacific Oceans; Baltic Sea; Japan; China; Australia			75, 78
157		*Dactylometra quinquecirrha*	Atlantic & Indian Oceans; western Pacific Ocean from the Malayan archipelago to Japan & the Philippines			
158		*Pelagia noctiluca*	Warmer parts of the Atlantic, Pacific, & Indian Oceans			
159		*Rhizostoma pulmo*	Europe; Mediterranean Sea			
160	Physalia sting	*Physalia physalis*	Tropical Atlantic	Local pain, pruritus, & erythematous wheals which may give rise to papules; weakness; nausea; tremor, pain, & spasms in large muscle masses, with respiratory difficulties	Toxin contains serotonin & several polypeptides. Direct effect on cardiovascular system. ∿55,000,000 nematocysts/g. Supernatant of homogenized nematocysts: mouse LD = 1.7 mg/kg.	18,
161	[Contact with nematocysts, usually with the tentacles]	Related species	Indo-Pacific area; Hawaiian Islands; southern Japan			35, 70, 78
162	Sea anemone sting	*Actinia equina*	Eastern Atlantic Ocean to the Gulf of Guinea; Mediterranean & Black Seas; Sea of Azov	Stinging sensation on contact, sometimes followed by pruritus & urticaria; local redness	*See* Jellyfish sting, entries 153-159	10,
163	[Contact with nematocysts]	*Adamsia palliata*	Norway to Spain; Mediterranean Sea			18, 35, 64, 71

[4] Synonym: succinoxidase. [5] Synonym: *o*-diphenols.

continued

Part III. Marine Organisms

	Type of Poisoning [Cause]	Species (Synonym)	Distribution	Symptoms & Signs	Chemistry & Toxicology	Reference
164		*Anemonia sulcata*	Eastern Atlantic Ocean from Norway & Scotland to the Canary Islands; Mediterranean Sea	& edema reported		
165		*Sagartia elegans*	Iceland to the Atlantic coast of France; Mediterranean Sea; coast of Africa			
			Porifera			
166	Sponge sting [Contact]	*Haliclona viridis*	West Indies	Burning sensation, itching, local redness, swelling in some cases; vesicles, weakness, malaise, & primary shock reported	Nature of toxin unknown, but ∿30 substances identified in crude extract. Water-extractable toxin of *H. viridis:* LD_{50} = 20 μg/kg.	19, 35, 49, 74, 75, 106
167		*Neofibularia nolitangere (Fibulia nolitangere)*	West Indies			
168		*Tedania ignis*	West Indies			
169		*T. toxicalis*	California coast			
			Protozoa			
170	Paralytic shellfish poisoning[6] [Ingestion of certain echinoderms, arthropods, & mollusks which have fed on toxic dinoflagellates]	*Gonyaulax catenella*	Pacific coast of North America	Onset within 10 min to 4 hr, with thirst, weakness, & numbness around mouth, lips, & tongue, & in some cases at fingertips; followed by muscular incoordination, progressive paralysis, & respiratory failure. Death (reported in ∿5% of cases) within 24 hr.	Tetrahydropurine derivative ($C_{10}H_{17}N_7O_4$·2HCl & mol wt of 372) contains 2 basic groups, with pK_a's of 8.3 & 11.5. Produces progressive muscle weakness & paralysis by blocking action potentials in motor axons through ability to prevent increase in early transient ionic permeability associated with inward movement of Na^+ during excitation; also direct effect on excitable muscle membrane, heart, & circulating blood volume. Human LD ∿3 mg; mouse LD_{50} = 3.4 μg/kg.	15, 21, 22, 30, 35, 44, 45, 54-57, 59, 60, 71, 75, 87-89, 94
171		*G. polyramma*	Japan; South Africa			
172		*G. tamarensis*	Atlantic coast of North America; North Sea			
173		*Gymnodinium* sp.	South Africa			
174		*G. brevis*	Florida; Gulf of Mexico			
175		*Pyrodinium phoneum*	North Sea			

[6] Also known as Gonyaulax or mussel poisoning; saxitoxin or mytilotoxin.

Contributors: Russell, Findlay E., and Carlson, Richard W.

References

[1] Alender, C. B., and F. E. Russell. 1966. In R. A. Boolootian, ed. Physiology of Echinodermata. Interscience, New York. pp. 529-543.

[2] Alender, C. B., et al. 1965. Toxicon 3:9.

[3] Austin, L., et al. 1965. Aust. J. Exp. Biol. Med. Sci. 43:79.

[4] Banner, A. H., et al. 1963. Proc. Gulf Carib. Fish. Inst. 16th Annu. Sess., p. 84.

[5] Baxter, E. H., et al. 1968. Toxicon 6:45.

[6] Bernstein, M. E. 1969. Ibid. 7:287.

[7] Boylan, D. B., and P. J. Scheuer. 1967. Science 155(3758):52.

continued

Part III. Marine Organisms

[8] Brock, V. E. 1956. Copeia (3):195.

[9] Cameron, A. M., and R. Endean. 1966. Toxicon 4: 111.

[10] Cantacuzene, J., and A. Damboviceanu. 1934. C. R. Soc. Biol. 117:136.

[11] Carlisle, D. B. 1962. J. Mar. Biol. Ass. U.K. 42:155.

[12] Carlson, R. W. Unpublished. Univ. Southern California, Los Angeles, 1971.

[13] Carlson, R. W., et al. 1971. Toxicon 9:379.

[14] Castex, M. N. 1963. La Raya Fluvial. Castellvi, Santa Fe, Argentina.

[15] Cheymol, J., and F. Bourillet. 1966. Actual. Pharmacol. 19:1.

[16] Cheymol, J., et al. 1965. Arch. Fac. Med. Madrid 8:151.

[17] Chu, G. W. T. C., and C. E. Cutress. 1955. Hawaii Med. J. 14:403.

[18] Cleland, J. B., and R. V. Southcott. 1965. Aust. Nat. Health Med. Res. Counc. (Canberra) Spec. Rep. Ser. 12.

[19] De Laubenfels, M. W. 1932. Proc. U.S. Nat. Mus. 81:85.

[20] Endean, R., and C. Rudkin. 1965. Toxicon 2:225.

[21] Evans, M. H. 1968. Ibid. 4:289.

[22] Evans, M. H. 1970. Mar. Pollut. Bull. 1(12):186.

[23] Feigen, G. A., et al. 1966. Toxicon 4:15.

[24] Feigen, G. A., et al. 1968. Ibid. 6:17.

[25] Freeman, S. E., and R. J. Turner. 1971. Brit. J. Pharmacol. 41:154.

[26] Friess, S. L., et al. 1959. J. Pharmacol. Exp. Ther. 126:323.

[27] Friess, S. L., et al. 1960. Ann. N.Y. Acad. Sci. 90: 893.

[28] Friess, S. L., et al. 1967. Biochem. Pharmacol. 16: 1617.

[29] Ghiretti, F. 1960. Ann. N.Y. Acad. Sci. 90:726.

[30] Gibbard, J., et al. 1939. Can. J. Pub. Health 30(4): 193.

[31] Gillet, K., and F. McNeill. 1962. The Great Barrier Reef and Adjacent Isles. Coral Press, Sydney, Australia.

[32] Haavaldsen, R., and F. Fonmun. 1963. Nature (London) 199:286.

[33] Habermehl, G., and G. Volkwein. 1970. Justus Liebigs Ann. Chem. 731:53.

[34] Halstead, B. W. 1959. Dangerous Marine Animals. Cornell Maritime Press, Cambridge, Md.

[35] Halstead, B. W. 1965. Poisonous and Venomous Marine Animals of the World. U.S. Government Printing Office, Washington, D. C. v. 1.

[36] Halstead, B. W. 1967. Ibid. v. 2.

[37] Halstead, B. W. 1970. Ibid. v. 3.

[38] Halstead, B. W., and F. R. Modglin. 1958. Z. Tropenmed. Parasitol. 9:129.

[39] Halstead, B. W., et al. 1955. J. Morphol. 97:1.

[40] Halstead, B. W., et al. 1955. Trans. Amer. Microsc. Soc. 75:381.

[41] Hashimoto, Y., and T. Okaichi. 1960. Ann. N.Y. Acad. Sci. 90:667.

[42] Hessell, D. W., et al. 1960. Ibid. 90:788.

[43] Hopkins, D. G. 1964. Med. J. Aust. 1:81.

[44] Kao, C. Y. 1966. Pharmacol. Rev. 18:997.

[45] Kao, C. Y., and F. A. Fuhrman. 1967. Toxicon 5: 25.

[46] Keen, T. E. B., and H. D. Crone. 1969. Ibid. 7:55.

[47] Kohn, A. J. 1963. Venomous Poisonous Anim. Noxious Plants Pac. Reg., p. 83.

[48] Kohn, A. J., et al. 1960. Ann. N.Y. Acad. Sci. 90(3): 706.

[49] Lane, C. E. Unpublished. Univ. Miami, Coral Gables, 1971.

[50] Lane, C. E., and E. Dodge. 1958. Biol. Bull. 115: 219.

[51] Larsen, J. B., and C. E. Lane. 1966. Toxicon 4: 199.

[52] Lee, R. K. C., and H. Q. Pang. 1945. Amer. J. Trop. Med. 25:281.

[53] Maretski, A., and J. Del Casillo. 1967. Toxicon 4: 245.

[54] McFarren, E. F. 1967. Ibid. 4:294.

[55] McFarren, E. F., et al. 1956. Proc. Nat. Shellfish Ass. 47:114.

[56] McFarren, E. F., et al. 1965. Toxicon 3:111.

[57] Medcof, J. C., et al. 1947. Bull. Fish. Res. Bd. Can. 75:1.

[58] Mendes, E. G., et al. 1963. Science 139:408.

[59] Meyer, K. F. 1953. N. Engl. J. Med. 249:765, 804, 843.

[60] Meyer, K. F., et al. 1928. J. Prev. Med. 2:365.

[61] Mosher, H. W., et al. 1964. Science 144:1100.

[62] Nigrelli, R. F., and P. Zahl. 1952. Proc. Soc. Exp. Biol. Med. 81:397.

[63] Nitta, S. 1934. Yakugaku Zasshi 54:648.

[64] Pawlowsky, E. N., and A. K. Stein. 1929. Arch. Dermatol. Syph. 157:647.

[65] Pearson, R. B. Unpublished. Univ. Queensland, Brisbane, Australia, 1971.

[66] Phillips, C., and W. H. Brady. 1953. Sea Pests. Univ. Miami Press, Coral Gables.

[67] Picken, L. E. R., and R. J. Skaer. 1966. In W. J. Rees, ed. The Cnidaria and Their Evolution. Academic Press, London. pp. 19-50.

[68] Puffer, H. W. 1970. Proc. West. Pharmacol. Soc. 13: 120.

[69] Ruggieri, G. D., et al. 1960. Zoologica (New York) 45:1.

[70] Russell, F. E. 1966. Toxicon 4:65.

[71] Russell, F. E. 1965. Advan. Mar. Biol. 3:256.

continued

Part III. Marine Organisms

[72] Russell, F. E. 1969. In H. D. Graham, ed. The Safety of Foods. Avi, Westport, Conn.

[73] Russell, F. E. 1969. In W. S. Hoar and D. J. Randall, ed. Fish Physiology. Academic Press, New York. v. 3.

[74] Russell, F. E. 1970. N. Engl. J. Med. 282:753.

[75] Russell, F. E. 1971. Int. Encycl. Pharmacol. Ther., Sect. 71, 2:3.

[76] Russell, F. E., and J. A. Emery. 1960. Ann. N.Y. Acad. Sci. 90:805.

[77] Russell, F. E., and R. D. Lewis. 1956. In E. E. Buckley and N. Porges, ed. Venoms. American Association for the Advancement of Science, Washington, D.C. pp. 43-53.

[78] Russell, F. E., and P. R. Saunders, ed. 1967. Animal Toxins. Pergamon Press, Oxford.

[79] Russell, F. E., and A. van Harreveld. 1954. Arch. Int. Physiol. 62:322.

[80] Russell, F. E., et al. 1958. Amer. J. Med. Sci. 235:566.

[81] Russell, F. E., et al. 1958. Med. Arts Sci. 12:78.

[82] Saunders, P. R. 1959. Science 129(3344):272.

[83] Saunders, P. R. 1960. Ann. N.Y. Acad. Sci. 90:798.

[84] Saunders, P. R., and L. Tokes. 1961. Biochim. Biophys. Acta 52:526.

[85] Saunders, P. R., et al. 1962. Amer. J. Physiol. 203:429.

[86] Schaeffer, R. C., et al. 1971. Toxicon 9:69.

[87] Schantz, E. J. 1960. Ann. N.Y. Acad. Sci. 90:843.

[88] Schantz, E. J., et al. 1957. J. Amer. Chem. Soc. 79:5230.

[89] Schantz, E. J., et al. 1958. J. Ass. Offic. Agr. Chem. 41:160.

[90] Scheuer, P. J., et al. 1967. Science 155:1267.

[91] Simon, S. E., et al. 1964. Arch. Int. Pharmacodyn. 149:318.

[92] Skeie, E. 1962. Acta Pathol. Microbiol. Scand. 55:166.

[93] Skeie, E. 1965. Munksgaard, Kobenhaven 1.

[94] Sommer, H., et al. 1937. Arch. Pathol. 24:532.

[95] Sullivan, T. D., and R. F. Nigrelli. 1956. Proc. Amer. Ass. Cancer Res. 2:151.

[96] Sutherland, S. K., et al. 1970. Toxicon 8:249.

[97] Thompson, D. A. 1964. Science 146(3641):244.

[98] Thron, C. D., et al. 1964. Toxicol. Appl. Pharmacol. 6:182.

[99] Trethewie, E. R. 1965. Toxicon 3:55.

[100] Tsuda, K., et al. 1964. Chem. Pharm. Bull. 12:1357.

[101] Whitley, G. P. 1957. Aust. Mus. Mag. 12(5):139.

[102] Whysner, J. A., and P. R. Saunders. 1963. Toxicon 1:113.

[103] Whysner, J. A., and P. R. Saunders. 1966. Ibid. 4:177.

[104] Whyte, J. M., and R. Endean. 1962. Ibid. 1:25.

[105] Woodward, R. B. 1964. Pure Appl. Chem. 9:49.

[106] Yaffee, H. S. 1970. N. Engl. J. Med. 282:51.

82. HIGHER-PLANT TOXINS

Toxic Effects: CNS = central nervous system. Data in brackets refer to the column heading in brackets.

	Species [Distribution]	Plant Part	Toxin [Synonym]	Toxic Effects	Remarks
1	*Abrus praecatorius* [Southern Florida, India, tropics]	Seed; root less toxic	L-Abrine [N-methyltryptophan], abric acid [54]	Onset may be delayed several hours to 2 da; vomiting, diarrhea, acute gastroenteritis, chills, sometimes convulsions. [41] Rectal bleeding [26,40]; in severe cases, death from heart failure [54].	One seed chewed may be fatal to child; 60-g seed fatal to horse. [54] 0.5 ml of 1% saline extract of seed injected subcutaneously lethal to rabbit in 24 hr [38,59]. Injected L-abrine found in liver of rat [37].
2	*Aconitum napellus* [Northeastern United States, Canada, Europe, Asia [27]]	Root, leaf, flower, seed	Mainly aconine, aconitine, napelline, picraconitine; 8 other alkaloids reported. [43,56]	Tingling, burning, numbness in lips, tongue, throat; later, possibly in fingers. [43] Great restlessness, dyspnea, slow pulse, muscular weakness, incoordination, cold & livid skin, pupillary constriction or dilation [43,51]; vomiting, diarrhea, convulsions, possibly death in 0.5-8 hr from respiratory or cardiac paralysis [18,43].	Lethal dose, 1-6.5 mg; young leaves mistakenly eaten for parsley, and roots for horseradish. [43] Considered most dangerous of British plants [18].

continued

	Species [Distribution]	Plant Part	Toxin [Synonym]	Toxic Effects	Remarks
3	*Agrostemma githago* [United States, Canada, Europe [41]]	Seed	Glycosides, especially gypsogenin [githagenin]; saponin. [24]	Man: vertigo, depressed breathing, gastroenteritis, vomiting, diarrhea, headache, sharp pains in spine, difficult locomotion, sometimes coma & death [10,41]. Cattle, horse: colic, diarrhea, muscular tremors, rigidity, coma, death [41,54]. Poultry: diarrhea, weakness.	Milled seeds sometimes in wheat flour [18]; frequent ingestion of small amounts causes chronic githagism [41]; 0.1-0.45 kg/45 kg live wt fatal to stock [41].
4	*Antiaris toxicaria* [Southern Asia, East Indies [45]]	Milky sap	Complex of cardenolides, mainly α- & β-antiarin [7,9]	Skin irritation, blistering, swelling; vomiting, convulsions, death from heart failure if toxin injected into bloodstream. [9]	One of the principal arrow poisons [9,45]. Cloth made from bark causes severe itching if sap not completely removed [9].
	Astragalus spp. [Northern hemisphere [41]]				Hazard to livestock in United States; toxicity varies with species & locality; some species harmless. Horses may become addicted to grazing on locoweeds and return to them after recovery from poisoning. [3]
5	Locoweed	Fresh plant	Locoine [3, 52]	Cattle, goat, horse, sheep: depression, edema of eyelids, distorted vision causing erratic behavior [3]; weakness, loss of muscular control, inappetence, emaciation, abortion, sometimes death.	
6	Poison vetch	Fresh plant	Selenium [3, 52]	Cattle, goat, horse, sheep: constipation, emaciation, declining pulse & blood pressure, paralysis, death.	
7	Other species	Fresh plant	Unidentified toxin [3, 52]	Cattle, goat, horse, sheep: temporary collapse, incoordination of hind legs, gastroenteritis, weight loss, dyspnea, paralysis, asphyxia [52].	
8	*Atropa belladonna* [Eastern United States, Europe, Asia [41]]	Entire plant, especially root, leaf, seed	Mainly hyoscyamine & its isomer, and atropine (the racemic mixture) [12,43]. Scopolamine [hyoscine]; 10 other alkaloids reported. [55,56]	Man, acute: dry skin, mouth, throat [18]; flushing of face, cyanosis, mydraisis, nausea, vomiting, constipation, slurred speech, giddiness, stupor, coma, rapid & weak pulse, fever, death from asphyxia & heart failure [27]. Chronic: erythema, urticaria, vesicular eruptions, slurred speech, mydriasis, glaucoma, muscular tremors or twitchings; sudden withdrawal causes nausea, salivation, perspiration. [8] Cattle: mydriasis, constipation, rapid pulse, labored breathing, frenzy, paralysis [27].	Leaves are source of the drug belladonna [43]. Children & grazing animals often poisoned by eating fruit. Flesh of rabbits that have eaten plant is toxic to humans. [18]
9	*Blighia sapida* [Southern Florida, Jamaica, western tropical Africa; rare in other tropical regions]	Arils of unripe fruit, ripe & unripe seeds	Hypoglycine A [α-amino-β-(2-methylenecyclopropyl) propionic acid]2/ in unripe arils & seeds; its γ-glutamyl derivative, hypoglycine B [γ-L-glutamyl-α-amino-β-(2-methylenecyclopropyl) propionic acid] in ripe & unripe seeds [44]	Abdominal pain, recurrent vomiting, sleepiness, convulsions, coma, death in severe cases [4]. Rabbit: general hyperemia of of organs with some cellular necrosis, especially in liver & kidney. Lung congestion causing respiratory failure [16] & shock [34].	Ripe arils commonly eaten raw in Nigeria [44], and cooked in Jamaica where seasonal (Dec-Mar) outbreaks of "vomiting sickness" [4] and ∿5000 deaths between 1886-1950 [44] have resulted. Red membrane (placenta) is nontoxic [34].

1/ More powerful than digitalin [9]. 2/ Twice as toxic as hypoglycine B.

continued

	Species [Distribution]	Plant Part	Toxin [Synonym]	Toxic Effects	Remarks
10	*Cannabis sativa* [United States, Mexico, tropical America, Europe, temperate Asia [41]]	Upper leaf, flower bract, resin	Chief hallucinogenic principle is Δ^1-3,4-*trans*-tetrahydrocannabinol [12]	Man: Smoking causes exaltation, inebriety, confusion, followed by CNS depression [54]; in some subjects, large quantity causes strong psychoactive effects, which may merge into depression [28,50] and cause death from heart failure [2]. Man, dog, monkey: hypotension & hypothermia [25]. Plant has insect-repellant, piscicidal, & skin-irritating properties [14].	Various narcotic preparations ("hashish," "ganja") [14]. Seeds also yield hempseed oil for paints; plant stems yield hemp fiber for cordage & carpets. Seeds are sterilized for use in "birdseed"
11	*Cicuta* spp. [Northern temperate regions [18, 41]]	Primarily root; stem, leaf less toxic	Cicutoxin [40]	Man, other animals: abdominal pain, nausea, vomiting, diarrhea, mydriasis, labored breathing, foaming at mouth, weak & rapid pulse, epileptoid convulsions, death from respiratory failure [18,41] within 15 min or 2-3 hr [39].	Genus includes the most poisonous plants in the United States. Fatalities have resulted from mistaking roots for parsnips. [18, 41]
12	*Colchicum autumnale* [United States, Europe, North Africa [18]]	Entire plant, mainly mature corm, seed [41]	Colchicine [acetyltrimethylcolchicinic acid] [12, 43]	Man: burning in throat; 6-8 hr later a feeling of suffocation, oppression in chest, difficulty in swallowing, vomiting, diarrhea, colic, tenesmus, giddiness, weakness in legs, arthralgia, cyanosis, labored breathing, convulsions; death from respiratory exhaustion after 7-36 hr of consciousness. [53] Other animals: nausea, vomiting, colic, diarrhea, hematuria, depression, unconsciousness, paralysis, mydriasis, profuse perspiration, death in 1-3 da [53].	Widely grown in flower gardens. Arrests cellular mitosis; employed in treatment of gout. [43] Doubles chromosomes, and is therefore used in plant breeding [12].
13	*Conium maculatum* [North America, temperate South America, Europe, northern Africa, Asia [18,41]]	Fruit, especially the unripe; root, stem, leaf	Coniine, *N*-methylconiine, γ-coniceine, conhydrine, pseudoconhydrine [12]	Man: mydriasis [18], often blindness; burning & dryness in throat, muscular weakness, paralysis of extremities, death from respiratory paralysis. [27,41] Man, horse: nausea, unsuccessful efforts to vomit [41]. Cattle, horse: mydriasis [18]; grinding of teeth, inappetence, salivation, bloating, muscular weakness especially in hind legs, rapid & feeble respiration, paralysis, death from respiratory failure [27,41].	Leaves most toxic when plant is flowering; root less toxic in spring. [41] Resemblance of fruit to anise, leaves to parsley, and root to parsnips responsible for many human fatalities [18]; plant commonly fatal to livestock [10].
14	*Croton tiglium* [Africa, southern Asia, East Indies, Pacific islands [45]]	Root, leaf, bark, seed	Crotin, croton resin [43]; ricinine [56]; seed oil contains TPA [12-*O*-tetradecanoylphorbol-13-acetate] & PDD [phorbol-12,13-didecanoate] [32].	Pain in back of throat & in anus [43]; vomiting, drastic purging, possibly collapse & death [43]. Croton oil is a skin irritant, causing reddening, swelling, pustules [9]. TPA & PDD are most active tumor promoters (co-carcinogens) known [32].	Croton oil formerly a human & veterinary purgative; now used to lubricate machinery. Crushed fruits & leaves are fish poisons. Smoke from burning wood inflames eyes. [9]
15	*Datura metel; D. stramonium;* other spp. [Temperate, subtropical, tropical regions]	Entire plant, especially seeds	Scopolamine [hyoscine]; hyoscyamine & its isomer; atropine (the racemic mixture). [12]	Man: mydriasis, headache, nausea, vomiting (rarely), vertigo, thirst, dry & burning sensation in skin, loss of muscular control. Acute poisoning causes mania, convulsions, death [41]; nonfatal poisoning usually causes memory loss, mental confusion [45,53]. Cattle: mydriasis, suspension of secretions or diarrhea, rapid heart action, paralysis, death from asphyxia [27]. Swine: convulsive twitching [27].	*Datura* spp. widely grown for ornament. Children poisoned by eating seeds, eating or sucking flower; adults poisoned by ingesting seeds for intoxicating effect, taking infusion of leaves, accidentally ingesting seeds in grain, or gathering young plants with other greens. Important ritual narcotics of North & South

continued

	Species [Distribution]	Plant Part	Toxin [Synonym]	Toxic Effects	Remarks
					American Indians. Alkaloids used medicinally as sedatives, anticholinergics. [12]
16	*Delphinium* spp. [North temperate regions, especially western United States]	Leaves, seeds; to a lesser degree, roots	Delphinine [2]; delphinoidine, delphisine, staphisagroine [41]; 30 other alkaloids reported [55, 56]. *D. ajacis* contains ajacine & ajaconine [2].	Man: burning & inflammation of mouth & pharynx, lowered blood pressure, nausea, abdominal pain, labored respiration, itching, cyanosis. Cattle: uneasiness, stiffness, staggering, constipation, frothing at mouth, nausea, bloating [41]; spasms, respiratory failure [54]. Horse: less susceptible [49]. Sheep: usually graze larkspur safely; poisoned by *D. ajacis*. [2]	Second to *Astragalus* in causing livestock fatalities in United States [41]. Leaves & seeds may cause dermatitis; cultivars common in flower gardens; seeds long used in insecticide. [54]
17	*Digitalis lanata; D. purpurea* [Western United States, western Europe, other areas [27]]	Entire plant, especially leaves, seeds	*D. purpurea:* aglycones of digitoxin, gitoxin, & gitalin, mainly digitoxigenin, gitoxigenin, & gitaloxigenin [43]. *D. lanata:* Same as above, plus digoxigenin (aglycone of digoxin) & diginatigenin [12].	Anorexia, nausea, vomiting, slow & pronounced pulse in early stages [41]; cardiac arrhythmias, diarrhea, abdominal pain, headache, fatigue, malaise, drowsiness, convulsions, rapid & irregular pulse, death in severe cases [51].	Common in flower gardens [41]. Dry plants in hay have poisoned horses & cattle; fresh leaves fatal to turkeys. [18] Digoxin or digitoxin, alone or with other *Digitalis* glycosides, used in cardiovascular therapy [12].
18	*Dioscorea hispida* [Southern Asia, East Indies, Pacific islands [9]]	Entire plant, tuber	Dioscorine [45]	Discomfort, then burning, in throat; giddiness, vomiting of blood, suffocation, drowsiness, exhaustion. [9] Paralysis of CNS [45].	Often causes fatalities in Philippines [45]. Edible after slicing or grating, salting, sun-drying, repeated washing & soaking, boiling [9].
19	*Erythroxylon coca* [Northern South America, East Indies [45]]	Leaf	Cocaine; 13 other alkaloids reported. [55, 56]	Leaves: mild stimulation, suppression of hunger & thirst, chronic intoxication [22]. Cocaine: general CNS stimulation followed by depression, numbness of tongue, vertigo, nausea, delirium, convulsions, rapid then weak pulse [43]; paralysis of respiratory centers, cyanosis, often sudden death from asphyxia [51].	Leaves chewed mainly by Indians of Peru & Bolivia; also steeped for medicinal "tea." [22] Cocaine has been widely used as a local anesthetic, but is being replaced by safer drugs; abused by addicts who suffer hallucinations and often an illusion of insects crawling on their skin. [43]
20	*Euphorbia* spp. [Worldwide [41]]	Milky sap	Euphorbiosteroid [18]; euphorbon, euphorboresene [9,41]; also cyanogenetic glycosides & co-carcinogenic esters of diterpenes in some species	Man: dermatitis, eye irritation, temporary blindness [13]; swelling around mouth & eyes; burning in mouth & throat, sneezing, vomiting, diarrhea, hemorrhagic gastroenteritis [54]; fainting, death [9,41]. Other animals: blistering of skin, loss of hair, weakness, collapse, death [41].	Various species grown for ornament. Euphorbium derived from *E. resinifera* formerly used medicinally, now in paint as protectant. [9] *Euphorbia* sap mixed with arrow poison as cohesive irritant [54].

continued

	Species [Distribution]	Plant Part	Toxin [Synonym]	Toxic Effects	Remarks
21	*Helleborus niger* [United States, Europe [41]]	Rootstock, leaf	Helleborin[3/], helleborein[4/] [18, 41]; toxic alkaloids	Man: severe dermatitis in some individuals [41]; violent inflammation of mucous membranes in stomach & intestines, vomiting, dizziness, convulsions, sometimes death [43]. Cattle, horse: purging (bloody), salivation, nausea, excessive urination, intermittent pulse, death [18].	Cultivated in flower gardens; 250 g fresh root or 70 g dried root toxic to horse [11]. Use in medicine and as veterinary parasiticide abandoned as too dangerous [18].
22	*Hippomane mancinella* [Central America, West Indies, northern South America; now rare in Florida [40]]	Milky sap, leaf, fruit	Physostigmine or similar alkaloid, plus a sapogenin [35]	Man: Sap causes severe burning of skin, swelling, & blistering; may cause hemorrhage in eyes & temporary blindness. [1, 35] Eating fruit or drinking water containing leaves causes oral inflammation & severe gastroenteritis; ulceration of intestinal tract proceeds slowly. [35,41]	Sap used by Caribs as arrow poison; employed in folk medicine; smoke from burning wood highly irritating. [1] Toxicity greatly exaggerated [40].
23	*Hyoscyamus niger* [Northwestern United States, Canada [41]; Mediterranean region, Asia, Oceania]	Entire plant	Mainly hyoscyamine & scopolamine [hyoscine] [43]	See *Atropa belladonna,* entry 8	Usually avoided by animals because of unpleasant taste; children poisoned by eating seeds & pods. [41] Roots mistaken for parsnips [18]. Extract long employed as sedative; dropped from United States Pharmacopeia.
24	*Jatropha curcas* [Southern Florida, tropics [40]]	Yellow sap, seed	Curcin (a toxalbumin), purgative oil [54]	Man: burning in throat, bloating, dizziness, vomiting, diarrhea, drowsiness, possibly dysuria & mydriasis [53]; severe leg cramps, deafness [29,30]; violent purgative action often fatal to children [1]. Other animals: hemorrhagic enteritis, staggering, dull vision, mydriasis, bloating, paralysis, somnolence, convulsions, fever, shivering, coma, death in 1-3 da in acute cases [53].	Bark & sap used as fish poison [9]. Seeds yield "hell oil," formerly given as purgative, now used in soap-making & as lubricant; seeds of *J. multifida,* a tropical ornamental, often cause poisoning in children. [40]
25	*Kalmia angustifolia; K. latifolia* [Northeastern United States, Pacific Coast, Canada [10]]	All parts except wood	Acetylandromedol (a resinoid) or arbutin (a glycoside), resembling curare in action on muscles & depression of CNS [24]	Man: symptoms similar to those produced in other animals, plus pain in head, sweating, tingling of skin [10]. Other animals (usually sheep): watering of eyes & nose, salivation, frothing at mouth, vomiting, impaired vision or blindness, dizziness, convulsions [41]; muscular paralysis, respiratory failure [24].	Children poisoned by eating leaves; frequent cause of fatal poisoning of livestock, but grazed by deer; fruits toxic to birds. [21] Nectar yields bitter, toxic honey [24]. The toxin gives a blue spot at about R_f 0-50 in thin layer chromatography [57].
26	*Lathyrus sativus;* other spp. [Southern United States, southern Europe, northern Africa, Asia [41,53]]	Mature plant, seed	N-β-D-Glycopyranosyl-N-α-L-arabinosyl-α,β-diaminopropionitrile [48]	Man: pain in back [54]. Sudden weakness in legs; further ingestion may cause leg paralysis. Cattle: constipation, weak pulse, skin numbness [53]; lameness, general weakening of tendons & ligaments, tissue fragility, connective tissue malformations such as exostoses, hernias, & aneurysms [53]. Horse: dyspnea, paralysis of hindlimbs & larynx, asphyxia [18].	*L. cicera, L. clymenum,* & *L. sativus* used as food & fodder [54], but cause many cases of neurolathyrism in man & livestock [53]. *L. odoratus,* the sweet pea of flower gardens, has fatally poisoned children [33].

[3/] Neurotoxin similar to aconitine. [4/] Cardiac glycoside similar to digitalis.

continued

	Species [Distribution]	Plant Part	Toxin [Synonym]	Toxic Effects	Remarks
27	*Leucaena leucocephala* [Southern United States, tropics of both hemispheres]	Leaf, especially the immature; root, bark, seed	Mimosine [β-{N-(3-hydroxy-4-pyridone)}-α-aminopropionic acid] [23]	Cattle (when plant is major part of, or sole, diet): lack of coordination, temporary blindness; enlarged thyroid in offspring. [23] Donkey, horse, mule: alopecia of mane & tail, possible deformation or loss of hoofs; in severe cases, lameness, debility, death from hunger & thirst. Sheep: overindulgence causes shedding of fleece [23]. Swine: total alopecia, impaired vision, emaciation, various degrees of paralysis, respiratory failure [27].	High in protein; widely grown as fodder for cattle, goats, sheep. [27] Ripe seeds used as coffee substitute; unripe seeds & young leaves cooked as vegetables occasionally cause hair loss in man. In cell culture, mimosine inhibits DNA synthesis [61].
28	*Lupinus* spp. [Temperate regions [53]]	Seed most toxic, pod less, leaf least	*d*-Lupanine, sparteine; 30 other alkaloids reported. [55, 56]	Lupinosis (nonalkaloidal): liver damage, anemia, cachexia, fever, general jaundice, coma, paralysis, constipation, then hemorrhagic diarrhea; swelling of ears, eyelids, lips, nose; usually chronic, rarely acute. [18,53] Lupine poisoning (alkaloidal): frothing at mouth, dyspnea, frenzied actions, nausea, bloating, convulsions, coma, possibly death [18,53].	Deaths of many livestock, especially sheep, from ingestion of seeds in quantity [19,27]. Toxicity varies with species, season, & location [19]. Debittered & "sweet" cultivars of *L. luteus, L. albus,* etc., with methionine, valued as feedstuff [58].
29	*Manihot esculenta* [Tropics [9]]	Root, especially skin & juice; stem, mature leaf, fruit	Cyanogenetic glycosides [47]	Man, livestock: nausea; rapid, labored breathing; rapid, irregular, weak pulse. Twitching, staggering, spasms of neck & legs, convulsions, mydriasis, coma, death from respiratory paralysis. [53]	Bitter cultivars high in CN⁻, yield starch for tapioca; sweet cultivars with little of no CN⁻, widely cooked as starchy vegetables. Protein-rich leaves edible when cooked. [47]
30	*Melia azedarach* [Southern United States, tropics of both hemispheres [53]]	Leaf, bark, flower, fruit pulp	Paraisine in leaf [56]; azaridine [54] & tazettine [2] in fruit & bark; alkaloids affecting CNS [54].	Man, leaf poisoning: stomatitis, decreased urine formation, violent & bloody vomiting [53]. Fruit poisoning: nausea, vomiting, labored breathing, palpitation, paralysis [41]. Other animals (especially swine): vomiting, colic, diarrhea, labored breathing, cyanosis, convulsions or paralysis, death by asphyxia [54].	Common shade tree. Roots, leaves, bark, flowers, fruit used to stupefy fish [1]; various parts of tree used in folk medicine [54]. Lethal dose for swine: 3 g fruit/0.45 kg body wt [52].
31	*Metopium toxiferum* [Southern Florida, Bahama Islands, West Indies]	Entire plant, especially sap	Probably similar to poison ivy, but is known to affect individuals immune to poison ivy [40]	Dermatitis similar to that caused by poison ivy; blistering may continue for weeks, spreading from one area to another; may be accompanied by intense itching & burning. Severe cases require hospitalization. [40]	One of commonest native trees of southern Florida. Smoke from burning wood highly irritating. Clear, sticky sap turns black on exposure to air. [40]
32	*Nerium oleander* [Southern United States, subtropics, tropics [45]]	Root, leaf, bark, flower	Glycosides, mainly nerin & oleandrin [folinerin], resembling digitalis in action; rosagenin in bark similar to strychnine. [54]	Man: mydriasis, abdominal cramps, nausea, vomiting, slow & irregular pulse, bloody diarrhea, death from cardiac or respiratory paralysis [53]; dermatitis in sensitive individuals [1]. Other animals: symptoms similar to those produced in man, plus sweating, gnashing of teeth, groaning, sometimes polyuria [53].	Popular ornamental shrub. Often poisons grazing animals; 15-20 g fresh plant fatal to horse. Smoke from burning plant toxic; meat roasted on skewers of oleander wood, or food stirred with oleander sticks, fatally toxic. [1,27, 53] Children have died from eating flowers [29,30].

continued

	Species [Distribution]	Plant Part	Toxin [Synonym]	Toxic Effects	Remarks
33	*Papaver somnif-erum* [Southern Europe, Asia, sub-tropics, tropics]	Milky exu-date from incised unripe seed pod, dried as "opi-um" [41]	Mainly mor-phine; also codeine, the-baine, papa-verine, nar-cotine. [53] 20 other iso-quinoline al-kaloids [55, 56].	Acute: temporary euphoria followed by CNS depression, symmetrical pinpoint pu-pils, depressed respiration, cyanosis, coma, death from depression of respiration & cir-culation. Chronic (from drug abuse): varies with individual case [42]; strong psychoactive depression & psychological dependence [50].	Cultivated in flower gardens as well as for production of impor-tant narcotics, analgesics, & re-laxants [12]. Seeds commonly used in bakery products; yield edible oil. Animals rarely poi-soned by eating seed pods.[41]
34	*Phytolacca americana* [Eastern & southern United States, Europe, South Africa [54]]	Root, mature (red) stem, seed	Saponin (an emetic), res-in affecting CNS [39]	Burning & bitterness in mouth; after 2 hr or more, vomiting, purging, spasms, ab-dominal cramps, lowered pulse, impaired vision, sometimes convulsions, death from respiratory paralysis. Causes abortion in cattle. [24,27,39]	Young shoots edible if well cooked [41]. Fruit juice used as food coloring; root gathered ac-cidentally with shoots, or mis-taken for parsnips or horseradish. [41,54] Has often poisoned swine; extract formerly used me-dicinally has caused poisoning from overdoses. [54]
35	*Prunus* spp. [5] [Northern hemisphere, Orient [41]]	Leaf, espe-cially wilted; bark, seed	Hydrocyanic acid, formed by enzyme action on amygdalin & on mandelonitrile glucoside, *d* form [prunasin] [6] & *dl* form [prulaurasin] [7] [11,27]	Animals: uneasiness, staggering, rolling eyes, protruding tongue, falling, convul-sions, labored breathing, bloating. Death may occur in <1 hr. [24,41]	*P. serotina* & *P. virginiana* fre-quently fatal to livestock in United States. Toxicity of leaves varies with age, weather, & de-gree of wilting. [49]
36	*Ricinus com-munis* [Southern United States, subtropics, tropics [41]]	Seed	Ricin (a tox-albumin), ri-cinine [8], castor bean allergen [46]; purga-tive oil	Man, by ingestion: burning in mouth, throat, & stomach; vomiting, diarrhea, thirst, rapid then faint pulse, cramps of abdomen & legs, convulsions, shallow res-piration. Contact & inhalant: conjunctivi-tis, dermatitis, respiratory allergy. [46] Other animals, by ingestion: hemorrhagic enteritis, staggering, dulled vision, heart weakness, bloating, paralysis, convulsions, fever, shivering, coma, death in 1-3 da [53].	Some varieties cultivated as orna-mentals; 2-3 seeds may be fatal to a child, and 6 seeds to a horse. [54] Necklaces of castor beans have caused allergic reactions [40]; pollen has caused respira-tory allergy [33]. Seeds yield castor oil; efforts being made to detoxify protein-rich press cake for stock feed. [44]
37	*Senecio* spp. [Worldwide]	Entire plant espe-cially seed	Numerous pyrrolizidine alkaloids; most com-mon are se-neciphylline & senecio-nine. Also senecifoli-dine, isatidine, pterophine, retrorsine, platyphylline [9], scleratine. [20,54-56]	Man: abdominal pain, vomiting, ascites, en-larged liver, emaciation, bloody diarrhea; usually fatal if not treated early. [53] Grazing animals: yawning, inappetence, emaciation, staggering, colic [41]; uncon-sciousness, death from liver cirrhosis [54]. *Senecio* alkaloids have shown both tumor-genic & tumor-inhibiting activity [20].	Seeds of various species in har-vested grain have caused chronic & acute poisoning in man [54]; *Senecio* poisoning common in livestock [41].

[5] Wild cherry, peach, and apricot. [6] In *P. padus.* [7] In *P. laurocerasus.* [8] An alkaloid, less toxic than ricin. [9] Similar to atropine in action.

continued

	Species [Distribution]	Plant Part	Toxin [Synonym]	Toxic Effects	Remarks
38	*Strophanthus* spp. [Tropical America, tropical Africa [45]]	Root, bark, seed	64 cardiac glycosides & aglycones reported [54], mainly strophanthin [K-strophanthoside] [43]; also trigonelline (an alkaloid) [56].	Vomiting, slow & irregular pulse, blurred vision, delirium, death from circulatory failure of cardiac origin [51].	Arrow poisons from several species [45]. *S. sarmentosus* yields sarmentogenin, a precursor of cortisone [54]; strophanthin from *S. kombe* formerly used as cardiotonic [43].
39	*Strychnos castelnaei; S. toxifera;* other spp. [Central America, northern South America [1]]	Root, bark	>60 alkaloids, mainly toxiferines I-IV, VI-X, XII; caracurines I, III-IX. [55,56]	Haziness of vision, relaxation of facial muscles, inability to raise head, loss of muscular contraction in arms & legs [31]; muscle nerve end-plate paralysis, depressant effects on muscles of respiration, death from asphyxia [17].	Bark of *S. castelnaei* usually used as basis of curare, famed as Indian blowgun poison [36]. Tubocurarine chloride used as skeletal muscle relaxant in surgery & as anticonvulsant [12].
40	*S. nuxvomica* [Southeast Asia, East Indies [27, 45]]	Leaf, wood, bark, flower, seed	Strychnine, brucine; 6 other alkaloids reported. [55, 56] Loganin (a glycoside) [43].	Action on spinal cord causes excessive reflex irritability, followed by rapid tonic convulsions with intermissions of exhaustion & sweating; extreme muscular rigidity, asphyxia, death. Mind not affected. [29,30]	Strychnine used as rodenticide; 0.5-4 mg formerly prescribed as stimulant; 60-90 mg may be fatal. [12,43] Fruit pulp eaten by birds [29,30].
41	*Thevetia peruviana* [Southern Florida, tropics [9, 45]]	Root, milky sap, leaf, bark, seed	Thevetin [10/] [11], thevetoxin, neriifolin [54] (cardiac glycosides); also peruvoside [11/] [5].	Numbness & burning in mouth, dryness of throat, dilated pupils [11]; vomiting, diarrhea, high blood pressure, erratic heartbeat, death from asphyxia & sudden heart paralysis [45]. Sap may cause dermatitis [1].	Seeds used for suicide, homicide, & abortion, and as fish poison; 15-20 g fatal to horse. [9] In India, peruvoside is considered a promising new drug for congestive heart failure [5].
42	*Toxicodendron radicans; T. vernix* [North America [41]; introduced into South Africa [54]]	Entire plant	Urushiol [12/] [12,15]	Skin irritation, swelling, blisters, extreme discomfort, itching; sometimes fatal to children. [41] Poisoning results from contact with plant or contaminated clothing, objects, or animals [10].	Smoke from burning plant toxic; ~1,000,000 Americans suffer ivy poisoning each year. [15]
43	*Veratrum viride* [United States, Canada [10]]	Entire plant	Protoveratrine A & B, germidine, germitrine; 15 other alkaloids reported. [55, 56]	Burning in throat & esophagus [10]; vomiting, abdominal pain, diarrhea, muscular weakness, spasms, possibly convulsions, rapid pulse, shallow breathing, semiconsciousness, death from asphyxia [41]. *V. californicum* produces cyclopian malformation in offspring of sheep which ingest plant [6].	This & related species yield veratrum, an antihypertensive drug, declining in use because of undesirable side effects [43]. Veratrum alkaloids block membrane action potential [60].
44	*Zigadenus* spp. [Western hemisphere, especially United States & Brazil [41]]	Bulb, stem, leaf, flower	Zygadenine, zygacine, germidine (glycoalkaloids); 8 other glycoalkaloids reported. [55,56]	Animals: salivation, nausea, vomiting, lowered temperature, staggering, rapid breathing, convulsions, paralysis, possibly coma & death [24,49].	Frequent cause of livestock poisoning in western United States; 20-900 g fatal to 45-kg sheep. Children occasionally poisoned by eating bulb. [41]

10/ One-eighth as potent as ouabain.   11/ As potent as ouabain.   12/ A complex of 3-pentadecylcatechol and other toxic principles.

continued

Contributors: Larson, Edward, and Morton, Julia F.

References

[1] Allen, P. H. 1943. Amer. J. Trop. Med. 23(1):S1.

[2] Al-Rawi, A. 1966. Iraq Min. Agr. (Baghdad) Tech. Bull. 145.

[3] Anonymous. 1966. Pasture and Range Plants. Phillips Petroleum, Bartlesville, Okla.

[4] Arnold, L. E. 1947. Jamaica Med. Rev. 1:26.

[5] Arora, R. B., et al. 1967. Indian J. Exp. Biol. 5:31.

[6] Binns, W., et al. 1963. Amer. J. Vet. Res. 24:1164.

[7] Bisset, N. G. 1966. Lloydia 29(1):1.

[8] Brookes, V. J. 1958. Poisons: Their Properties, Chemical Identification, Symptoms, and Emergency Treatments. Ed. 2. Van Nostrand, New York.

[9] Burkill, I. H. 1966. A Dictionary of the Economic Products of the Malay Peninsula. Ministry of Agriculture and Cooperatives, Kuala Lumpur. v. 1, p. 2.

[10] Chestnut, V. K. 1898. USDA Div. Bot. Bull. 20.

[11] Chopra, R. N., et al. 1965. Poisonous Plants of India. Ed. 2. Indian Council of Agricultural Research, Delhi. v. 1.

[12] Claus, E. P., et al. 1970. Pharmacognosy. Ed. 6. Lea and Febiger, Philadelphia.

[13] Crowder, J. I., and R. R. Sexton. 1964. Arch. Ophthalmol. 72:476.

[14] Dastur, J. F. 1951. Useful Plants of India and Pakistan. Ed. 2. D. B. Taraporevala, Bombay.

[15] Dawson, C. R. 1956. Trans. N.Y. Acad. Sci., II, 18(5):427.

[16] Doughty, D. D., and E. Larson. 1960. Trop. Geogr. Med. 12:243.

[17] Fanshawe, D. B. 1950. Brit. Guiana Forest Dep. Bull. 2.

[18] Forsyth, A. A. 1954. Min. Agr. Fish. Food (London) Bull. 161.

[19] Gardner, M. R. 1964. J. Agr. West. Aust. 5(11):890.

[20] Gharbo, S. A., and A.-A. M. Habib. 1969. Lloydia 32(4):503.

[21] Graham, E. H. 1935. Carnegie Inst. Carnegie Mus. (Pittsburgh) Bot. Pam. 1.

[22] Granier-Doyeux, M. 1962. Bull. Narcot. 14(4):1.

[23] Hamilton, R. I., et al. 1968. Aust. Vet. J. 44:484.

[24] Hardin, J. W. 1966. N.C. Agr. Exp. Sta. Bull. 414.

[25] Hardman, H. F., et al. 1970. Univ. Mich. Med. Cent. J. 36:238.

[26] Hart, R. 1963. Univ. Rochester Bull. Suppl. Mater. 7(4):14.

[27] Hurst, E. 1942. Poison Plants of New South Wales. Snelling, Sydney, Australia.

[28] Kiplinger, G. F., et al. 1970. Univ. Mich. Med. Cent. J. 36:240.

[29] Kirtikar, K. R. 1893. Poisonous Plants of Bombay. Natural History Society, Bombay. v. 1.

[30] Kirtikar, K. R. 1903. Ibid. v. 2.

[31] Krantz, J. C., and C. J. Carr. 1961. The Pharmacologic Principles of Medical Practice. Ed. 5. Williams and Wilkins, Baltimore.

[32] Kreibech, G., and E. Hecker. 1970. Z. Krebsforsch. 74:448.

[33] Lampe, K. F., and R. Fagerstrom. 1968. Plant Toxicity and Dermatitis. Williams and Wilkins, Baltimore.

[34] Larson, E., et al. 1953. Quart. J. Fla. Acad. Sci. 16(3):151.

[35] Lauter, W. M., and P. A. Foote. 1955. J. Amer. Pharm. Ass. Sci. Ed. 44:361.

[36] LeCointe, P. 1947. Amazonia Brasileria III. Companhia Editoria Nacional, Sao Paulo, Brazil.

[37] Lin, J., et al. 1970. Toxicon 8:197.

[38] Misra, D. S., et al. 1966. Indian J. Exp. Biol. 4:161.

[39] Montgomery, F. H., and C. K. H. Roe. 1965. Ont. Dep. Agr. Publ. 508.

[40] Morton, J. F. 1971. Plants Poisonous to People. Hurricane House, Miami.

[41] Muenscher, W. C. 1951. Poisonous Plants of the United States. Rev. ed. Macmillan, New York.

[42] Oettingen, W. F. von. 1952. Poisoning. P. B. Hoeber, New York.

[43] Osol, A., et al. 1967. The U.S. Dispensatory and Physician's Pharmacology. Ed. 26. J. B. Lippincott, Philadelphia.

[44] Plimmer, J. R., and C. E. Seaforth. 1963. Trop. Sci. 5(3):137.

[45] Quisumbing, E. 1951. Philipp. Dep. Agr. Natur. Resour. Tech. Bull. 16.

[46] Raymond, W. D. 1961. Trop. Sci. 3(1):19.

[47] Rogers, D. J. 1963. Bull. Torrey Bot. Club 90:43.

[48] Rukmini, C. 1969. Indian J. Chem. 7(10):1062.

[49] Schmutz, E. M., et al. 1968. Livestock-poisoning Plants of Arizona. Univ. Arizona Press, Tucson.

[50] Seevers, M. H. 1970. Pharmacologist 12(2):172.

[51] Sollmann, T. H. 1957. A Manual of Pharmacology. Ed. 8. W. B. Saunders, Philadelphia.

[52] Sperry, O. E., et al. 1964. Tex. Agr. Exp. Sta. Bull. 1017.

[53] Steyn, D. G. 1934. Toxicology of Plants in South Africa. Central News Agency, Johannesburg.

[54] Watt, J. M., and M. G. Breyer-Brandwijk. 1962. Medicinal and Poisonous Plants of Southern and Eastern Africa. Ed. 2. E. and S. Livingstone, London.

[55] Willaman, J. J., and H.-L. Li. 1970. Lloydia 33(3A).

[56] Willaman, J. J., and B. C. Schubert. 1961. U.S. Dep. Agr. Tech. Bull. 1234.

[57] Clarke, E., et al. 1972. Toxicon 10:526.

[58] Gladstone, J. 1970. J. Agr. West. Aust. 11(2):26.

[59] Misra, D. S., et al. 1968. Indian J. Exp. Biol. 6:108.

[60] Ohta, M., et al. 1972. Toxicon 10:531.

[61] Tsai, X., and X. Ling. 1971. Ibid. 9:241.

Toxicology: LD_{50} = lethal dose for 50% of test subjects; i.p. = intraperitoneal; ppm = parts per million

	Species	Mycotoxin (Synonym)	Toxicology	Remarks
1	*Amanita muscaria* [1] [7]	Two isoxazole derivatives, ibotenic acid & its decarboxylation product, muscimol; sometimes also muscazone; very little muscarine	Muscimol 5 times as toxic as ibotenic acid [7]; within a few min or 1-2 hr: preliminary excitement, muscular twitching, dyspnea, diarrhea, vomiting [56]; drowsiness, followed by hallucinations lasting 3-18 hr [21]; depression, unconsciousness; rarely fatal [7]	Toxicity highly variable; fungus employed as an intoxicant in Siberia [7,55]
2	*A. phalloides* [2] [7, 56]	Chiefly aminitine; also the cyclopeptides phalloine & phalloidine	After 6-24 hr: abdominal pain [12]; vomiting, diarrhea, intense thirst, recurrent drowsiness [19]; respiratory & circulatory depression, delirium, & sometimes convulsions [38]; jaundice, hepatitis, coma, death from renal failure [45,55]	One of the most deadly fungi, mortality, ∿50%; cause of the majority of "mushroom deaths" in the United States [12,38,43]; genus contains other equally poisonous & some edible species
3	*Aspergillus flavus* [5]	Aspertoxin; *O*-methylsterigmatocystin	Toxic to chick embryos, but no evidence on toxicity in mammals [39]	Structurally related to aflatoxins (*see* entry 5); identified in culture extracts of fungi isolated from grain implicated in animal toxicity; no information on presence in foods
4	*A. nidulans; A. versicolor; Bipolaris* sp. [18,39]	Sterigmatocystin	Oral LD_{50} in rats = 120-160 mg/kg body wt; i.p. LD_{50} = 60-65 mg/kg body wt; prominent pathological finding: necrosis of the liver & renal tubules [18,36]; sarcomas induced following subcutaneous injection in rats [9]; hepatocellular carcinoma after oral dosing of 1.5-2.25 mg·rat^{-1}·da^{-1} for 1 yr [35,37]	
5	*A. flavus; A. parasiticus* [15]	Aflatoxins B_1, B_2, B_{2a}, G_1, G_2, G_{2a}, M_1, M_2	Acute poisoning: single-dose LD_{50} in susceptible species = 0.5-15 mg/kg body wt [58]; aflatoxins B_1 & G_1 lethal to all domestic & laboratory animal species (mice & sheep relatively resistant); necrosis of liver & kidney primary signs [15] Subacute poisoning: field outbreaks of toxicity observed in cattle, sheep, swine, & poultry at levels of 0.1-2.5 ppm aflatoxins in feed [15] Chronic exposure: tumors of the liver & kidneys in rats, ducks, & rainbow trout induced by aflatoxins B_1 & G_1 at levels of 0.01-1.0 ppm in feed [3,15,60]	Originally identified in peanut meal [2,40], but also found occasionally in moldy oilseeds (cottonseeds, soybeans), grains (barley, corn, millet, rice, sorghum, wheat), pulses (beans, cowpeas, peas), cassava, & yams [1, 22,59]; aflatoxins M_1 & M_2 found in milk of cattle fed aflatoxins B_1 & B_2 [15]
6	*A. ochraceous* [57]	Ochratoxins A, B, & C	LD_{50} of ochratoxin A in rats = 22 mg/kg body wt; ochratoxins B & C inactive; renal tubular necrosis primary lesion in acute poisoning of rats; no evidence of carcinogenic activity in chronic experiments [33,34]	Toxin-producing fungus isolated from wheat [57]; little information on occurrence in feed ingredients or food sources [42]; no established field cases of poisoning by toxins
7	*Claviceps purpurea* [3] [7,46]	Derivatives of *d*-lysergic acid: ergocornine, ergocristine, ergocryptine (ergokryptine), ergonovine (ergometrine), ergotamine, ergotoxine	Acute poisoning: in man & other animals, vomiting & diarrhea, respiratory, visual, & motor difficulties, convulsions, lowered blood pressure, shallow respiration, unconsciousness; during pregnancy, possibly uterine hemorrhage, abortion, peripheral gangrene Chronic, convulsive poisoning: vomiting, itching, paresthesia, analgesia of the extremities, anorexia or uncontrollable hunger, diarrhea, muscle contracture, delirium, sometimes a tabes-like complex [43,44] Gangrenous type: acute pain in the extremities; dry gangrene including the tips of the ears & tail in cattle, comb & wattles in poultry [12]	Occurs on barley, oats, rye, wheat, & other grasses; cause of many cases of poisoning (ergotism) in livestock; epidemics of "St. Anthony's Fire" caused by ergot-contaminated bread as recently as 1951 in France & India; ergot preparations valued medicinally, chiefly for their effect on the muscles of the uterus [12]

[1] Not found in the tropics; widely distributed in northern regions [56]. [2] Found in North America, Europe, East Indies, and Australia. [3] Found in North America, Europe, and Australia; mycotoxin in the sclerotium of the fungus.

continued

	Species	Mycotoxin (Synonym)	Toxicology	Remarks
8	*Cladosporium* sp. [20]	Cladosporic acids	Ingestion of toxic grain followed by alimentary toxic aleukia (ATA) in man [25,26]; after 1-3 da, clinical syndrome begins with gastroenteritis, nausea, vomiting, & diarrhea lasting up to 9 da; asymptomatic period (2 wk-2 mo) accompanied by bone marrow destruction with leukopenia & agranulocytosis; in 3rd stage (5-20 da), sepsis, agranulocytosis, fever, petechial hemorrhage, necrotic lesions of the mouth, pharynx, & esophagus; mortality, 0-80%, depending on quantity ingested; toxicosis reproduced in cats & guinea pigs, but other animal species not susceptible	Maximal toxin production by fungi at −5 to 0°C, particularly when freezing & thawing occur alternately [20]
9	*Fusarium poae; F. sporotrichioides*	Fusariogenin		
10	*F. nivale*	Butenolide [62]	Clinical signs of toxicosis (fescue toxicity) in cattle: lameness, gangrene & sloughing of hooves, & elevation of body temp [63]; toxin isolated from fungus implicated in etiology, possibly acting synergistically with T-2 toxin (*see* entry 12) [64], but etiologic role not established	Fungal involvement suggested by sporadic & seasonal occurrence of toxicity of tall fescue pastures [63]
11	*F. nivale*	Fusarenon-X [50]; nivalenol [48]	Structurally related toxins all containing 12,13-epoxytrichothec-9-ene nucleus [16] with different substituents; i.p. or oral LD_{50} for mice = 3-8 mg/kg body wt [8,23,47,50]; potent necrotizing action on skin; massive hemorrhage & ulceration of gastrointestinal tract caused by oral administration	Toxin-producing fungi isolated from batches of corn & other cereal grains implicated in animal poisoning [14]; toxin production supported by many cereal grains [51]; no information on human exposure
12	*F. tricinctum*	Diacetoxyscirpenol [8]; T-2 toxin [4]		
13	*Gibberella zeae* [54]	Zearalenone	Potent estrogenic activity; mammary hypertrophy, uterine eversion, abortion, & testicular atrophy involved in field cases [6]; in rats, uterine weight & growth rate increased by doses of 20-40 µg/animal [28]	Toxicosis involves mainly swine fed corn on which the producing fungus had grown during storage [6]; general distribution of toxin in other grains unknown
14	*Penicillium islandicum* [24,41]	Islanditoxin	Cyclic peptide, with oral LD_{50} for mice = 6.5 mg/kg body wt [29]; hemorrhagic necrosis of the liver induced in acute poisoning, cirrhosis in chronic dosing	Hepatomas caused when rats fed fungal cultures containing these 2 compounds (plus others) over long periods [53]	Originally identified in cultures of fungus isolated from toxic ("yellowed") rice; no information on occurrence of toxins in human foods, or on the relationship of food contamination to human disease [52]
15		Luteoskyrin	Lipid-soluble pigment, with oral LD_{50} = 211 mg/kg body wt [29]; necrosis of the liver induced by acute dosing, fibrosis & nodular hyperplasia by chronic dosing; other liver toxicity accompanied by mitochondrial damage		
16	*P. rubrum* [31,32]	Rubratoxin B	In rats & mice, i.p. LD_{50} = 0.2-0.4 mg/kg body wt, and in dogs & chickens, 4.5 mg/kg body wt [61]; oral LD_{50} = 400 mg/kg body wt; hemorrhagic necrosis of liver & spleen involved in acute poisoning; no evidence of carcinogenic activity after chronic dosing in rats	Although the rubratoxin-producing organism was originally isolated from toxic corn, production of toxin demonstrated only in laboratory experiments
17	*Pithomyces* [13,17]	Sporidesmin	"Facial eczema" in ruminants, reflecting photosensitization secondary to liver damage caused by the toxic agent [11,27]; in sheep, liver damage (cholangitis, biliary obstruction, fibrosis) & photosensitization caused by a single dose of 0.5 mg/kg body wt [30]; LD_{50} in sheep = 1.0-2.0 mg/kg body wt [30]; rabbits & guinea pigs sensitive, but other species resistant to the toxin [10]	Toxicosis confined to ruminants grazing on ryegrass pastures; localization & seasonal character related to growth characteristics & distribution of the fungus [49]

Contributors: Wogan, G. N.; Larson, Edward, and Morton, Julia F.

continued

References

[1] Allcroft, R., and R. B. A. Carnaghan. 1963. Chem. Ind. (London), p. 50.

[2] Allcroft, R., et al. 1961. Vet. Rec. 73:428.

[3] Ayres, J. L., et al. 1971. J. Nat. Cancer Inst. 46(3): 561.

[4] Bamburg, J. R., et al. 1968. Tetrahedron 24:3329.

[5] Burkhardt, H. J., and J. Forgacs. 1968. Ibid. 24: 717.

[6] Christensen, C. M., et al. 1965. Appl. Microbiol. 13: 653.

[7] Claus, E. P., et al. 1970. Pharmacognosy. Ed. 6. Lea and Febiger, Philadelphia.

[8] Dawkins, A. W. 1966. J. Chem. Soc. 1:116.

[9] Dickens, F., et al. 1966. Brit. J. Cancer 20:134.

[10] Dodd, D. C. 1960. N.Z. J. Agr. Res. 3:491.

[11] Done, J., et al. 1960. Res. Vet. Sci. 1:76.

[12] Forsyth, A. A. 1954. Min. Agr. Fish. Food (London) Bull. 161.

[13] Fridichsons, J., and A. Mathieson. 1962. Tetrahedron Lett. (26):1265.

[14] Gilgan, M. W., et al. 1966. Arch. Biochem. Biophys. 114:1.

[15] Goldblatt, L. A., ed. 1969. Aflatoxin: Scientific Background, Control, and Implications. Academic Press, New York.

[16] Grove, J. F. 1969. J. Chem. Soc. 21:1266.

[17] Hodges, R., et al. 1963. Chem. Ind. (London), p. 42.

[18] Holzapfel, C. W., et al. 1966. S. Afr. Med. J. 40: 1100.

[19] Hurst, E. 1942. Poison Plants of New South Wales. Snelling, Sydney, Australia.

[20] Joffe, A. Z. 1965. Mycotoxins Foodst. Proc. Symp. 1964, p. 77.

[21] Lampe, K. F., and R. Fagerstrom. 1968. Plant Toxicity and Dermatitis. Williams and Wilkins, Baltimore.

[22] Loosmore, R. M., et al. 1964. Vet. Rec. 76:64.

[23] Marasas, W. F. O., et al. 1969. Toxicol. Appl. Pharmacol. 15:471.

[24] Marumo, S. 1959. Bull. Agr. Chem. Soc. Jap. 23: 428.

[25] Mayer, C. F. 1953. Mil. Surg. 113:173.

[26] Mayer, C. F. 1953. Ibid. 113:295.

[27] McFarlane, D., et al. 1959. N.Z. J. Agr. Res. 2:194.

[28] Mirocha, C. J., et al. 1967. Appl. Microbiol. 15:497.

[29] Miyake, M., and M. Saito. 1965. Mycotoxins Foodst. Proc. Symp. 1964, p. 133.

[30] Mortimer, P. H., and A. Taylor. 1962. Res. Vet. Sci. 3:147.

[31] Moss, M. O., et al. 1968. Nature (London) 220:767.

[32] Moss, M. O., et al. 1968. Tetrahedron Lett. (5): 367.

[33] Purchase, I. F. H., and W. Nel. 1967. In R. I. Mateles and G. N. Wogan, ed. Biochemistry of Some Foodborne Microbial Toxins. Massachusetts Institute of Technology Press, Cambridge. p. 153.

[34] Purchase, I. F. H., and J. J. Theron. 1968. Food Cosmet. Toxicol. 6:479.

[35] Purchase, I. F. H., and J. J. van der Watt. 1968. Ibid. 6:555.

[36] Purchase, I. F. H., and J. J. van der Watt. 1969. Ibid. 7:135.

[37] Purchase, I. F. H., and J. J. van der Watt. 1970. Ibid. 8:289.

[38] Quisumbing, E. 1951. Philipp. Dep. Agr. Natur. Resour. Tech. Bull. 16.

[39] Rodricks, J. V. 1969. J. Agr. Food Chem. 17:457.

[40] Sargeant, K., et al. 1961. Vet. Rec. 73:1219.

[41] Shibata, S., et al. 1968. Tetrahedron Lett. (27): 3179.

[42] Shotwell, O. L., et al. 1969. Appl. Microbiol. 17: 765.

[43] Sollmann, T. H. 1957. A Manual of Pharmacology. Ed. 8. W. B. Saunders, Philadelphia.

[44] Steyn, D. G. 1934. Toxicology of Plants in South Africa. Central News Agency, South Africa.

[45] Steyn, D. G. 1966. S. Afr. Med. J. 40(14):405.

[46] Svoboda, G. H., et al. 1954. J. Amer. Pharm. Assoc. Sci. Ed. 43(5):257.

[47] Tatsuno, T., et al. 1968. Chem. Pharm. Bull. 16: 2519.

[48] Tatsuno, T., et al. 1969. Tetrahedron Lett. (33): 2823.

[49] Thornton, R. H., and J. C. Percival. 1959. Nature (London) 183:63.

[50] Ueno, Y., et al. 1969. Experientia 25:1062.

[51] Ueno, Y., et al. 1970. Chem. Pharm. Bull. 18:304.

[52] Uraguchi, K., et al. 1961. Jap. J. Exp. Med. 31:19.

[53] Uraguchi, K., et al. 1961. Ibid. 31:435.

[54] Urry, W. H., et al. 1966. Tetrahedron Lett. (45): 3109.

[55] Van der Merwe, K. J., et al. 1965. Nature (London) 205:1112.

[56] Watt, J. M., and M. G. Breyer-Brandwijk. 1962. Medicinal and Poisonous Plants of Southern and Eastern Africa. Ed. 2. E. and S. Livingstone, London.

[57] Wieland, G. 1968. Science 159:946.

[58] Wogan, G. N. 1965. Mycotoxins Foodst. Proc. Symp. 1964, p. 163.

[59] Wogan, G. N. 1968. Fed. Proc. Fed. Amer. Soc. Exp. Biol. 27:932.

[60] Wogan, G. N., and P. M. Newberne. 1967. Cancer Res. 27(1):2370.

[61] Wogan, G. N., et al. 1971. Toxicol. Appl. Pharmacol. 19:712.

[62] Yates, S. G., et al. 1967. Tetrahedron Lett. (7): 621.

[63] Yates, S. G., et al. 1969. J. Agr. Food Chem. 17: 437.

[64] Yates, S. G., et al. 1970. Appl. Microbiol. 19:103.

Data in light brackets refer to the column heading in brackets. Figures in heavy brackets are reference numbers.

	Substance (Synonym)	Industrial or Medicinal Use [Common Source of Poisoning]	Clinical Syndrome	Pathological Changes	Other Toxic Effects [Metabolic Effects]	Treatment
1	Acetylsalicylic acid (Aspirin)	Analgesic, antipyretic, anti-inflammatory, antirheumatic, & uricosuric drug [21]	Respiratory stimulation [22]	....	Gastric irritation [21] [Uncoupling of oxidative phosphorylation, interference with thyroid function [6]]	....
2	Acrylamide	Manufacture of polymers; grouting agent, flocculator	Peripheral neuropathy—sensory & motor [19]; cerebellar ataxia after acute exposure	Dying-back axonal degeneration [18], no CNS abnormality demonstrated [32]	Hyperhydrosis	....
3	Alcohols	CNS depressants, hypnotics [20]	Excitation, coma [20]	....	Alcoholism [41] [Hypoglycemia [16], hyperlipemia [35]]	Disulfiram [8,20]
4	Amphetamine	CNS stimulant, treatment of obesity [20]	Euphoria, alertness followed by reduced mental acuity [20]	....	Habituation, psychosis [21]	....
5	Arsenic	Insecticide [Accidental ingestion]	Peripheral neuropathy—sensory & motor (onset 1-3 wk after single acute exposure; insidious in chronic exposure) [25]	Axonal degeneration [25]	Acute gastroenteritis, acute dermatitis, chronic pigmentation, white line on nails	Chelating agent (probably ineffective following acute poisoning)
6	Barbiturates	Hypnotics, general anesthetics [21]	Respiratory depression, hypotension [21]	....	....	Artificial respiration [42]
7	Belladonna alkaloids, atropine	Cholinergic blocking agents [21]	Excitement, mania, hot & dry skin, dilated pupils, tachycardia [21]	....	....	Barbiturates, physostigmine, artificial respiration [20]
8	Bromides	Sedatives, antiepileptics [20]	Mental symptoms [20]	....	Skin lesions, gastrointestinal disturbances [20]	Sodium chloride, ammonium chloride [20]
9	α-Bungarotoxin	Experimental tool [43] [Snake bite—active ingredient of venom [38]]	Muscle relaxation, respiratory failure [12]	....	....	....
10	Caffeine	CNS stimulant [21]	Wakefulness, restlessness, alertness, respiratory stimulation, convulsions [21]	....	....	....
11	Carbon disulfide	Manufacture of viscose rayon	Peripheral neuropathy—sensory & motor [48], Parkinsonism, psychiatric disturbances	Axonal degeneration in dogs [1], basal ganglia degeneration in monkeys [45], vascular lesions important? [49]	[Inhibits formation of norepinephrine from dopamine]	....

continued

	Substance (Synonym)	Industrial or Medicinal Use [Common Source of Poisoning]	Clinical Syndrome	Pathological Changes	Other Toxic Effects [Metabolic Effects]	Treatment
12	Chlorinated naphthalenes—Aldrin; Dieldrin; Endrin	Insecticides	Epilepsy [26], no long-term effects	None	....	Anticonvulsants
13	Chloroform	General anesthetic [21]	Hypotension, arrhythmias, respiratory depression [21]	....	Liver damage [21]	....
14	Cocaine	Local anesthetic [21]	Tremor, convulsion, coma [21]	....	Euphoric excitement, addiction [20]	Barbiturates [21]
15	Curare	Muscle relaxant [21]	Muscle relaxation, diplopia, respiratory paralysis [21]	....	....	Neostigmine [5], edrophonium [21]
16	Cyclopropane	General anesthetic [21]	Cardiac depression & arrhythmias, respiratory depression, nausea, vomiting [21]	....	....	....
17	DDT	Insecticide—not toxic in normal use [23]	Tremors, hyperirritability, convulsions [23]	....	....	....
18	Diacetylmorphine (Heroin)	Analgesic drug [21]	Euphoria [21]	....	Addiction [21]	....
19	Diethyl ether (Ethyl ether)	General anesthetic [20]	Respiratory depression, muscle relaxation, nausea, vomiting [20]	....	Irritation of respiratory mucosa [20]	....
20	Diphenylhydantoin	Antiepileptic, antiarrhythmic drug [20]	Ataxia, tremors, nausea [21], peripheral neuropathy [37]	Cerebellar degeneration	Megaloblastic anemia, gum hyperplasia	....
21	Diphtheria exotoxin	[Infection with Corynebacterium diphtheriae]	Polyneuropathy—motor more than sensory, initially affecting cranial nerves	Segmental demyelination [30, 39]	Fever, nasopharyngeal or laryngeal lesions, myocarditis	Antitoxin
22	Disulfiram	Treatment of alcoholism	Peripheral neuropathy—sensory & motor [24]	....	[Toxic effect due to metabolism to carbon disulfide ? [9]]	....
23	Ephedrine	Sympathomimetic drug, bronchodilator, vasopressor [21]	Tachycardia, hypertension, CNS stimulation [21]	....	....	....
24	Ergot alkaloids; ergotamine	Treatment of migraine [21]	Nausea, vomiting, paresthesia [21]	Gangrene [21]	....	....
25	Gold	Treatment of rheumatoid arthritis	Polyneuropathy—pain initially, then motor more than sensory [14]	....	Fever, dermatitis [Allergic mechanism ?]	Chelating agent
26	Haloperidol	Tranquilizer	Parkinsonism	....	....	....
27	Halothane	General anesthetic [20]	Cardiac depression, muscle relaxation, respiratory depression [20]	....	....	....

continued

	Substance (Synonym)	Industrial or Medicinal Use [Common Source of Poisoning]	Clinical Syndrome	Pathological Changes	Other Toxic Effects [Metabolic Effects]	Treatment
28	Hexamethonium	Ganglionic blocking agent [20]	Hypotension, constipation, dilation of pupils, dry mouth [20]	....	Tolerance [21]	....
29	Isoniazid	Antibiotic for tuberculosis	Peripheral neuropathy—mainly sensory, later mildly motor [2]	Axonal degeneration [11]	[Combines with & inactivates pyridoxine [3]; liability to neuropathy inherited, depending on amount of detoxicating enzyme [15]]	Prevented by pyridoxine
30	Lead	[Any industrial exposure, particularly manufacture of accumulators; lead glazed pottery, old lead paint]	Encephalopathy with fits, particularly in acute poisoning in children; motor peripheral neuropathy, particularly in chronic poisoning in adults	Cerebral edema, segmental demyelination [17]	Anemia, colic, blue line on gums	Chelating agent
31	Lindane (BHC)	Insecticide—not toxic in normal use; symptoms follow heavy exposure only	Epilepsy [46]	None	....	....
32	Lysergide (LSD; lysergic acid diethylamide)	Serotonin antagonist [21]	Hallucinations [21]	Chromosomal damage [21]	Psychoses, tolerance [21]	Phenothiazines, barbiturates
33	Manganese	[Mining]	Parkinsonism	....	....	....
34	Mercury, inorganic Metal (vapor)	[Mining; any industrial exposure]	Tremor, psychosis	None	None	None
35	Salts	[Suicide]	None	....	Renal failure	Dimercaprol
36	Mercury, organic Alkyl	[Accidental ingestion of contaminated fish [33], seedgrain]	Blindness, ataxia, sensory disturbance [28]	Degeneration of cerebellum, occipital & parietal cortex [27]	Renal damage	None
37	Aryl	Fungicide, seed dressing	None	....	Renal ? & skin damage	....
38	Methadone	Analgesic [21], treatment of morphine addiction [13]	Respiratory depression [21]	....	Tolerance, addiction [21]	....
39	Methyl bromide	Fumigant	Psychosis, epilepsy, peripheral neuropathy	....	Lung damage	Anticonvulsants
40	Methyl chloride	[Refrigerators]	Drowsiness & confusion, epilepsy; symptoms may occur several hr to 3 da after exposure [40]	....	Vomiting	Anticonvulsants
41	Morphine	Analgesic [21]	Drowsiness, euphoria, respiratory depression, emesis, constipation, constriction of pupils [21]	....	Tolerance, addiction [21]	Nalorphine, levallorphan [21]

continued

	Substance (Synonym)	Industrial or Medicinal Use [Common Source of Poisoning]	Clinical Syndrome	Pathological Changes	Other Toxic Effects [Metabolic Effects]	Treatment
42	Nicotine	Ganglionic stimulant & depressant [20]	Hypotension, dilation of pupils, dry mouth, nausea, vomiting, respiratory paralysis [20]	....	Thromboangiitis, tobacco amblyopia [20]	Artificial respiration [20]
43	Nitrofurantoin	Antibiotic, mainly for urinary infections	Peripheral neuropathy—sensory & motor, particularly when renal failure present [36]	Axonal degeneration [34]	....	....
44	Organophosphates	Insecticides	Anticholinesterase effects, headache, abdominal pain & vomiting, sweating, miosis, muscular twitching	....	[Depress cholinesterase levels]	Atropine, pralidoxime
45	Neurotoxins, delayed action—e.g., triorthocresyl phosphate (TOCP)	TOCP used as plasticide [Accidental contamination of cooking oils]	Early—anticholinesterase effects; delayed (up to a week after exposure)—peripheral sensory & motor neuropathy; spastic weakness	Dying-back axonal degeneration, initially affecting primary spindle afferents; pyramidal tract degeneration [10]	....	....
46	Phenothiazines— chlorpromazine; prochlorperazine; trifluoperazine	Tranquilizers [21]	Dyskinesia [29], Parkinsonism [21]	....	Weight gain, dry mouth, constipation, hypotension, drug hypersensitivity [21]	....
47	Phentolamine	Adrenergic blocking agent; diagnosis of pheochromocytoma [21]	Tachycardia, gastric stimulation [21]	....	....	....
48	Physostigmine	Anticholinesterase, ophthalmological use [21]	Sweating, salivation, diarrhea, constriction of pupils, hypotension, bronchoconstriction, muscle fasciculation [21]	....	....	Atropine [21]
49	Procaine; lidocaine; tetracaine; dibucaine	Local anesthetic [20]	CNS stimulation & depression, peripheral cardiovascular depression, salivation, tremor, convulsion, coma [20]	....	....	Barbiturates [20]
50	Propranolol	β-Adrenergic blocking agent, antiarrhythmic drug [20]	Congestive failure, hypotension, dizziness, nausea, diarrhea, dry mouth [21]	....	[Hypoglycemia [21]]	....
51	Reserpine	Hypotensive agent	Parkinsonism	....	Depression [Depletes brain of norepinephrine, dopamine, & serotonin]	....

continued

	Substance (Synonym)	Industrial or Medicinal Use [Common Source of Poisoning]	Clinical Syndrome	Pathological Changes	Other Toxic Effects [Metabolic Effects]	Treatment
52	Streptomycin	Antibiotic	Deafness	Degeneration of cranial nerve VIII	....	....
53	Strychnine	CNS stimulant, rat poison [20]	Convulsions, asphyxia [20]	....	....	Barbiturates, mephenesin [20]
54	Succinylcholine	Muscle relaxant [21]	Muscle paralysis, apnea [21]	....	....	Curare [21]
55	Tetraethyllead	Gasoline additive	Confusion & psychosis	....	....	....
56	Tetrodotoxin	Experimental tool, experimental muscle relaxant [47] [Puffer fish—active ingredient of poison [47]]	Respiratory failure, hypotension [47]	None [47]	[None [47]]	Artificial respiration [47]
57	Thallium	Fungicide [Accidental ingestion]	Peripheral neuropathy—sensory & motor; retrobulbar neuritis, chorea, confusion	Axonal bead-like swellings, followed by degeneration [44]	Gastrointestinal in acute; alopecia	Chelating agent (probably ineffective following acute poisoning)
58	Trichlorethylene	Industrial solvent	Narcosis; cranial nerve palsies (particularly V)	Brain-stem degeneration, particularly nerve tracts & nuclei of V [7]	[Toxic agent is probably a breakdown product of trichlorethylene]	....
59	Veratrum alkaloids; protoveratrines A & B	Hypotensive drug [31]	Hypotension, bradycardia, temporary apnea, nausea, vomiting [31]	....	....	....
60	Vinca alkaloids—vincristine; vinblastine	Chemotherapeutic agent, particularly reticuloses	Peripheral neuropathy—initially sensory, severe motor later	Axonal degeneration [4]	Paralytic ileus, alopecia [Destruction of neurotubules, proliferation of neurofilaments [50]]	....

Contributors: Le Quesne, Pamela M.; Narahashi, Toshio

References

[1] Alpers, B. J., and F. H. Lewey. 1940. Arch. Neurol. Psychiat. 44:725.

[2] Biehl, J. P., and J. H. Skavlen. 1953. Amer. Rev. Tuberc. 68:296.

[3] Biehl, J. P., and R. W. Vilter. 1954. Proc. Soc. Exp. Biol. Med. 85:389.

[4] Bradley, W. G., et al. 1970. J. Neurol. Sci. 10:107.

[5] Bridenbaugh, P. O., and H. C. Churchill-Davidson. 1968. J. Amer. Med. Ass. 203:541.

[6] Brody, T. M. 1956. J. Pharmacol. Exp. Ther. 117:39.

[7] Buxton, P. H., and M. Hayward. 1967. J. Neurol. Neurosurg. Psychiat. 30:511.

[8] Cahn, S. 1970. The Treatment of Alcoholics: An Evaluative Study. Oxford Univ. Press, New York.

[9] Casier, H., and E. Merlevede. 1962. Arch. Int. Pharmacodyn. Ther. 139:165.

[10] Cavanagh, J. B. 1964. J. Pathol. Bacteriol. 87:365.

[11] Cavanagh, J. B. 1967. J. Neurol. Neurosurg. Psychiat. 30:26.

[12] Chang, C. C., and C. Y. Lee. 1963. Arch. Int. Pharmacodyn. Ther. 144:241.

[13] Dole, V. P., and M. Nyswander. 1965. J. Amer. Med. Ass. 193:646.

[14] Endtz, L. J. 1958. Rev. Neurol. 99:395.

[15] Evans, D. A. P., et al. 1960. Brit. Med. J. 2:485.

[16] Field, J. B., et al. 1963. J. Clin. Invest. 42:497.

[17] Fullerton, P. M. 1966. J. Neuropathol. Exp. Neurol. 25:214.

continued

[18] Fullerton, P. M., and J. M. Barnes. 1966. Brit. J. Ind. Med. 23:210.

[19] Garland, T. O., and M. W. H. Patterson. 1967. Brit. Med. J. 4:134.

[20] Goodman, L. S., and A. Gilman, ed. 1970. The Pharmacological Basis of Therapeutics. Ed. 4. Macmillan, New York.

[21] Goth, A. 1970. Medical Pharmacology. Ed. 5. C. V. Mosby, St. Louis.

[22] Graham, J. D. P., and W. A. Parker. 1948. Quart. J. Med. 17:153.

[23] Hayes, W. J. 1959. DDT: Human and Veterinary Medicine. Birkhäuser, Basel.

[24] Hayman, M., and P. A. Wilkins. 1956. Quart. J. Stud. Alc. 17:601.

[25] Heyman, A., et al. 1956. N. Engl. J. Med. 254:401.

[26] Hoogendam, I., et al. 1965. Arch. Environ. Health 10:441.

[27] Hunter, D., and D. S. Russell. 1954. J. Neurol. Neurosurg. Psychiat. 17:235.

[28] Hunter, D., et al. 1940. Quart. J. Med. 9:193.

[29] Hunter, R., et al. 1964. J. Neurol. Neurosurg. Psychiat. 27:219.

[30] Kaeser, H. E., and E. H. Lambert. 1962. Electroencephalogr. Clin. Neurophysiol., Suppl. 22:29.

[31] Krayer, O., and G. H. Acheson. 1946. Physiol. Rev. 26:383.

[32] Kuperman, A. S. 1958. J. Pharmacol. Exp. Ther. 123:180.

[33] Kurland, L. T., et al. 1960. World Neurol. 1:370.

[34] Lhermitte, F., et al. 1963. Presse Med. 71:767.

[35] Losowsky, M. S., et al. 1963. Amer. J. Med. 35:794.

[36] Loughridge, L. W. 1962. Lancet 2:1133.

[37] Lovelace, R. E., and S. J. Horowitz. 1968. Arch. Neurol. (Chicago) 18:69.

[38] Mebs, D., et al. 1972. Hoppe Seylers Z. Physiol. Chem. 353:243.

[39] Morgan-Hughes, J. A. 1968. J. Neurol. Sci. 7:157.

[40] Morgan Jones, A. 1942. Quart. J. Med. 11:29.

[41] National Institute on Alcohol Abuse and Alcoholism. 1972. Alcohol and Health. U.S. Government Printing Office, Washington, D.C.

[42] Nilsson, E. 1951. Acta Med. Scand., Suppl. 253.

[43] O'Brien, R. D., et al. 1972. Annu. Rev. Pharmacol. 12:19.

[44] Prick, J. J. G., et al. 1955. Thallium Poisoning. Elsevier, Amsterdam.

[45] Richter, R. 1945. J. Neuropathol. Exp. Neurol. 4:324.

[46] Schmiedeberg, J., and H. J. Wasserburger. 1953. Anz. Schaedlingsk. 26:129.

[47] Simpson, L. L., ed. 1971. Neuropoisons, Their Pathophysiological Actions. Plenum Press, New York. v. 1.

[48] Vigliani, E. C. 1954. Brit. J. Ind. Med. 11:235.

[49] Vigliani, E. C., and C. L. Cuzzullo. 1950. Med. Lav. 41:49.

[50] Wisniewski, H., et al. 1968. J. Cell Biol. 38:224.

85. BIOCIDES

Part I. Antibiotics

Mode of Action: RNA = ribonucleic acid; tRNA = transfer ribonucleic acid; mRNA = messenger ribonucleic acid; DNA = deoxyribonucleic acid; S = Svedberg unit (1 S = 0.1 ps). **Acute Toxicity: Route**—i.p. = intraperitoneal, i.v. = intravenous, s.c. = subcutaneous; LD_{50} = lethal dose for 50% of test subjects.

	Antibiotic [Synonym] (Empirical Formula)	Mode of Action	Organisms Affected	Acute Toxicity			Reference
				Animal	Route	LD_{50}, mg/kg[1]	
1	Amphotericin B	Increases cell permeability	Strains of *Leishmania brasi-*	Dog	Oral	>500	4,17,19,
2	$(C_{46}H_{73}NO_{20})$	by binding to sterols in	*liensis & L. donovani;* many	Mouse	i.p.	1640	25-27,
3		membranes of sensitive	yeasts & fungi, including		i.v.	4.5	29
4		cells, thus permitting	those responsible for many		Oral	>8000	
5		K^+ & other essential	systemic fungal infections	Rat	Oral	>1000	
		metabolites to leak	of man				
		out, causing cellular					
		death. Fungistatic or					
		fungicidal, depending on concentration & time of contact.					

[1] Unless otherwise indicated.

continued

Part I. Antibiotics

Antibiotic [Synonym] (Empirical Formula)	Mode of Action	Organisms Affected	Acute Toxicity			Reference
			Animal	Route	LD_{50}, mg/kg [1]	
6 7 Ampicillin [α-Aminobenzyl-penicillin] ($C_{16}H_{19}N_3O_4S$)	Inhibits bacterial cell-wall synthesis	*Aerobacter, Diplococcus, Escherichia, Haemophilus, Neisseria, Proteus, Salmonella, Shigella, Staphylococcus, Streptococcus*	Mouse Mouse, rat	i.v. Oral & s.c.	Nontoxic[2] Nontoxic[3]	13,22, 31
8 Bacitracin A ($C_{66}H_{103}N_{17}O_{16}S$)	Disrupts structure and/or function of bacterial cell walls	*Actinomyces, Clostridium, Corynebacterium, Diplococcus, Haemophilus, Neisseria, Staphylococcus, Streptococcus, Treponema*	....	.....	....	13,38
9 10 11 12 Benzylpenicillinic acid [Benzylpenicillin; penicillin G] ($C_{16}H_{18}N_2O_4S$)	Inhibits cell-wall formation, probably by inhibiting peptidoglycan peptidyltransferase, which is responsible for cross-linking reaction in murein synthesis. Bactericidal.	Gram-positive bacteria; *Haemophilus* & *Neisseria* spp.	Mouse	i.p. i.v. Oral s.c.	3880 1800 12,750 6000	2,13,26, 28,29, 31,34, 37-39
13 14 15 16 17 18 19 Cephalothin ($C_{16}H_{17}N_2O_6S_2$)	Inhibits cell-wall formation, probably by inhibiting peptidoglycan peptidyltransferase, which is responsible for cross-linking reaction in murein synthesis. Bactericidal.	Penicillinase-producing staphylococci, β-hemolytic streptococci, pneumococci, & sensitive strains of certain gram-negative bacteria (e.g., *Escherichia coli, Haemophilus influenzae, Klebsiella* sp., *Paracolobactrum* sp., *Proteus mirabilis*)	Mouse Rat	i.p. i.v. Oral i.p. i.v. Oral s.c.	>7000 5000 >10,000 6300 >4000 >10,000 7500	1,10,13, 35
20 21 22 23 24 25 Chloramphenicol ($C_{11}H_{12}N_5O_2Cl_2$)	Inhibits protein synthesis; binds to 50S ribosome subunit, thus inhibiting reactions after formation of aminoacyl tRNA but before formation of complete polypeptide chain on ribosome. Bacteriostatic.	Large viruses, rickettsiae; many species of gram-positive & gram-negative bacteria; blue-green algae	Mouse Rabbit Rat	i.p. i.v. Oral s.c. i.v. i.v.	1320 100-200 2640 2300-2585 117 175-278	5,12,13, 26,29, 31,34
26 27 28 29 Chlortetracycline ($C_{22}H_{23}N_2O_8Cl$)	Tetracyclines inhibit protein synthesis at a stage preceding peptide-bond formation on ribosome	*Amoeba* spp.; lymphogranuloma-psittacosis group, *Rickettsia; Aerobacter, Bacillus, Borrelia, Brucella, Clostridium, Corynebacterium, Diplococcus, Escherichia, Haemophilus, Leptospira, Mycobacterium, Mycoplasma, Neisseria, Pasteurella, Proteus, Pseudomonas, Salmonella, Shigella, Staphylococcus, Streptococcus, Treponema*	Mouse Rat	i.v. Oral i.v. Oral	134 1500 118 3000	12,13, 31,34
30 31 32 33 Colistin ($C_{45}H_{85}N_{13}O_{10}$)	Disorganizes cell membrane, causing leakage of intracellular materials & subsequent death. Bacteriostatic or bactericidal, depending on concentration & time of contact.	A wide spectrum of gram-negative bacteria; less active against gram-positive bacteria & fungi. Colistin & polymyxin are the most active inhibitors of *Pseudomonas*.	Mouse	i.p. i.v. Oral s.c.	126-220 220-400 767-2000 138-220	26,31, 41

[1] Unless otherwise indicated. [2] No lethal effect from 2 g/kg. [3] No observable toxic effects from single doses up to 5 g/kg.

continued

Part I. Antibiotics

Antibiotic [Synonym] (Empirical Formula)	Mode of Action	Organisms Affected	Acute Toxicity				Reference
			Animal	Route	LD$_{50}$, mg/kg[1]		
34 Dactinomycin [Actinomycin D; Actinomycin C$_1$] (C$_{60}$H$_{76}$N$_{12}$O$_{15}$)	Selectively inhibits RNA synthesis; binds to guanine residues in DNA, but not in RNA. Formation of stable dactinomycin-DNA complex inhibits DNA-dependent RNA synthesis by RNA nucleotidyltransferase (RNA polymerase).	Gram-positive microorganisms (e.g., *Lactobacillus*, *Streptococcus*). Highly toxic to both normal & neoplastic mammalian cells.	Mouse	i.v.	0.67-0.74		11-13, 24
35 Erythromycin 36 [Erythromycin A] 37 (C$_{37}$H$_{67}$NO$_{13}$) 38 39 40 41	Interferes with transfer of amino acids from aminoacyl tRNA to proteins on ribosomes. Bacteriostatic or bactericidal, depending on nature of organism & drug concentration.	Most gram-positive bacteria. Certain large viruses, Rickettsiales, *Haemophilus*, & *Neisseria* sensitive; also some strains of *Bordetella*, *Brucella*, *Listeria*, *Mycoplasma*, *Pasteurella*, & *Treponema*.	Mouse / Mouse / Mouse / Mouse / Rat / Rat / Rat	i.p. / i.v. / Oral / s.c. / i.v. / Oral / s.c.	490 / 426 / 2927 / 1849 / 209 / >2000 / 1442		12,13, 31,34, 36,40
42 Gentamicin[4] 43 44 45	Blocks protein synthesis by action on 30S ribosome subunit	Many species of gram-positive & gram-negative bacteria (broad spectrum)	Mouse / Mouse / Mouse / Mouse	i.p. / i.v. / Oral / s.c.	430 / 75 / >9505 / 485		9,15,30, 42,43
46 Griseofulvin 47 (C$_{17}$H$_{17}$O$_6$Cl) 48 49 50 51 52	Causes morphological distortions, and therefore possible inhibition of cell-wall synthesis. Tests on several fungi show nucleus as possible target site. Fungistatic.	Some species of *Epidermophyton*, *Microsporum*, & *Trichophyton*	Mouse / Mouse / Mouse / Mouse / Rat / Rat / Rat	i.p. / i.v. / Oral / s.c. / i.p. / i.v. / Oral	1550 / 280 / >5000 / >12,000 / 5000 / Nontoxic[5] / 10,000		3,13,14, 17,26, 31,32, 36
53 Lincomycin 54 (C$_{18}$H$_{34}$N$_2$O$_6$S) 55 56	Inhibits protein synthesis; binds to 50S ribosome	Most gram-positive bacteria; *Leptospira pomona*	Mouse / Mouse / Mouse / Rat	i.p. / Oral / s.c. / Oral	1000 / >2000 / >800 / >4000		7,16,21, 23
57 Novobiocin 58 (C$_{31}$H$_{36}$N$_2$O$_{11}$)	Directly inhibits DNA synthesis; indirectly interferes with protein synthesis & cell-wall permeability	*Bacillus*, *Corynebacterium*, *Diplococcus*, *Haemophilus*, *Listeria*, *Neisseria*, *Nocardia*, *Staphylococcus*, *Streptococcus*	Mouse / Mouse	i.v. / Oral	407 / 962 to >1000		12,31, 34,36
59 Nystatin 60 (C$_{47}$H$_{75}$NO$_{18}$) 61 62 63	Binds to sterols in cell membrane, causing leakage of K$^+$ & other cell constituents. Fungistatic or fungicidal, depending on concentration & time of contact.	Many yeasts & fungi, including those responsible for many systemic infections in man	Mouse / Mouse / Mouse / Rat / Rat	i.p. / Oral / s.c. / i.p. / Oral	30,000-50,000[6] / >12,500,000[6] / 120 / 85,000-93,000[6] / >8,340,000[6]		8,13,17, 19,26, 27,29, 31,34
64 Polymyxin B (C$_{56}$H$_{104}$N$_{16}$O$_{14}$Cl$_5$)	Affects organization & function of bacterial cell membranes	*Aerobacter*, *Brucella*, *Escherichia*, *Haemophilus*, *Pasteurella*, *Pseudomonas*, *Salmonella*, *Shigella*, *Vibrio*	Mouse	i.v.	6-9		13,31, 34

[1] Unless otherwise indicated. [4] 3 components: gentamicin C$_1$ (C$_{21}$H$_{43}$N$_5$O$_7$), gentamicin C$_{1a}$ (C$_{19}$H$_{39}$N$_5$O$_7$), and gentamicin C$_2$ (C$_{20}$H$_{41}$N$_5$O$_7$). [5] Rats have an intravenous dose of 200 mg/kg. [6] Units/kg.

continued

Part I. Antibiotics

Antibiotic [Synonym] (Empirical Formula)	Mode of Action	Organisms Affected	Acute Toxicity			Reference
			Animal	Route	LD$_{50}$, mg/kg	
65 Streptomycin 66 (C$_{21}$H$_{39}$N$_7$O$_{12}$) 67 68	Inhibits protein synthesis by binding to ribosome. Bacteriostatic or bactericidal, depending on concentration.	Some larger viruses & rickettsiae; many species of gram-positive & gram-negative bacteria, including spirochetes	Mouse	i.p. i.v. Oral s.c.	610 85 15,550 500	2,6,13, 26,29
69 Tetracycline 70 (C$_{22}$H$_{24}$N$_2$O$_8$) 71	Inhibits binding of aminoacyl tRNA to amino acid site of 70S or 30S ribosomes, but no effect on binding of tRNA to peptidyl site	Some larger viruses & rickettsiae; many species of gram-positive & gram-negative bacteria, including spirochetes (see also Chlortetracycline, entries 26-29)	Mouse	i.p. i.v. Oral	20-300[7/] 150-170[7/] >3000[7/]	18,20, 26,31, 33

[7/] Tetracycline hydrochloride.

Contributors: Meyers, Edward; Gale, George O., and Jukes, Thomas H.

References

[1] American Medical Association. 1967. New Drugs Evaluated by A. M. A. Council on Drugs. Chicago. p. 53.

[2] Bacharach, A. L., et al. 1959. J. Pharm. Pharmacol. 11:737.

[3] Bent, K. J., and R. H. Moore. 1966. Symp. Soc. Gen. Microbiol. 16:82.

[4] Borowski, E., et al. 1970. Tetrahedron Lett. (45): 3903.

[5] Brock, T. D. 1961. Bacteriol. Rev. 25:32.

[6] Brock, T. D. 1966. Symp. Soc. Gen. Microbiol. 16: 131.

[7] Chang, F. N., et al. 1966. Proc. Nat. Acad. Sci. U.S. 55:431.

[8] Chong, C. N., and R. W. Rickards. 1970. Tetrahedron Lett. (59):5145.

[9] Cooper, D. J., et al. 1969. J. Infec. Dis. 119:342.

[10] Flynn, E. H. 1966. Antimicrob. Ag. Chemother., p. 715.

[11] Foley, G. E. 1955-56. Antibiot. Annu., p. 432.

[12] Goldberg, I. H. 1965. Amer. J. Med. 39:722.

[13] Goodman, L. S., and A. Gilman, ed. 1970. The Pharmacological Basis of Therapeutics. Ed. 4. Macmillan, New York.

[14] Grove, J. F. 1963. Quart. Rev. Chem. Soc. 17:1.

[15] Hahn, F. E., and S. G. Sarre. 1969. J. Infec. Dis. 119:364.

[16] Herr, R. R., and M. E. Bergy. 1962. Antimicrob. Ag. Chemother., p. 560.

[17] Hildick-Smith, G., et al. 1964. Fungus Diseases and Their Treatment. Little, Brown; Boston.

[18] Kaul, P. N. 1961. Hindustan Antibiot. Bull. 4:59.

[19] Lampen, J. O. 1966. Symp. Soc. Gen. Microbiol. 16:111.

[20] Laskin, A. I. 1967. In D. Gottlieb and P. D. Shaw, ed. Antibiotics. J. Springer, New York. v. 1, p. 331.

[21] Lewis, C., et al. 1962. Antimicrob. Ag. Chemother., p. 570.

[22] Lynn, B. 1965. Antibiot. Chemother. (Basel) 13: 125.

[23] Magerlein, B. J., et al. 1966. Antimicrob. Ag. Chemother., p. 727.

[24] Manaker, R. A., et al. 1954-55. Antibiot. Annu., p. 853.

[25] Mechlinski, W., and C. P. Schaffner. 1970. Tetrahedron Lett. (44):3873.

[26] Newton, B. A. 1965. Annu. Rev. Microbiol. 19: 209.

[27] Oroshnik, W., and A. D. Mebane. 1963. Fortschr. Chem. Org. Naturst. 21:17.

[28] Park, J. T. 1966. Symp. Soc. Gen. Microbiol. 16:70.

[29] Porter, J. N. 1964. In P. L. Altman and D. S. Dittmer, ed. Biology Data Book. Federation of American Societies for Experimental Biology, Washington, D.C. pp. 312-325.

[30] Rinehart, K. L., Jr. 1969. J. Infec. Dis. 119:345.

[31] Schnitzer, R. J., and F. Hawking, ed. 1964. Exp. Chemother. 3.

[32] Sharpe, H. M., and E. G. Tomick. 1960. Toxicol. Appl. Pharmacol. 2:44.

[33] Siddigui, M. A. Q., and K. Hosokawa. 1969. Biochem. Biophys. Res. Commun. 36:711.

[34] Spector, W. S., ed. 1957. Handbook of Toxicology. W. B. Saunders, Philadelphia. v. 2.

[35] Spencer, J. L., et al. 1966. Antimicrob. Ag. Chemother., p. 573.

[36] Stecher, P. G., et al., ed. 1960. The Merck Index. Ed. 7. Merck, Rahway, N. J.

continued

85. BIOCIDES

Part I. Antibiotics

[37] Stewart, G. T. 1965. The Penicillin Group of Drugs. Elsevier, New York.

[38] Strominger, J. L., and D. J. Tipper. 1965. Amer. J. Med. 39:708.

[39] Strominger, J. L., et al. 1967. Fed. Proc. Fed. Amer. Soc. Exp. Biol. 26:9.

[40] Taubman, S. B., et al. 1963. Antimicrob. Ag. Chemother., p. 395.

[41] Vogler, K., and R. O. Studer. 1966. Experientia 22:345.

[42] Waitz, J. A., and M. J. Weinstein. 1969. J. Infec. Dis. 119:355.

[43] Weinstein, M. J., et al. 1963. Antimicrob. Ag. Chemother., p. 1.

Part II. Insecticides

Route: IAM = insecticide added to medium; i.v. = intravenous; i.p. = intraperitoneal; p.c. = percutaneous. Data in brackets refer to the column heading in brackets. Values in parentheses are ranges, estimate "c" (*see* Introduction).

	Insecticide [Synonym]	Mode of Action	Organisms Affected	Acute Toxicity			Reference
				Animal	Route	LD$_{50}$[1] mg/kg [ppm]	
1	Aldicarb [2-Methyl-2-(methylthio)-	Anticholinesterase	Most animals	Rat	Oral	1	10,13
2	propionaldehyde *O*-(methylcarbamoyl)oxime]			Housefly[2], ♀	Topical	5.5	10,13
3				Mosquito[3], larva	IAM	[0.16]	10,13
4	Aldrin [1,2,3,4,10,10-Hexachloro-	Contact & stomach toxicant	Most insects	Dog	Oral	(65-95)	1,2,16
5	1,4,4a,5,8,8a-hexahydro-*endo,*			Guinea pig	Oral	33	1,2,16
6	*exo*-1,4:5,8-dimethanonaphthalene]			Mouse	Oral	44	1,2,16
7				Rabbit	Oral	(50-80)	1,2,16,20
8				Rat, ♂	Oral	(38-54)	1,2,5,12,16,20
9				♀	i.v.	18	2,16
10					Oral	(46-67)	1,2,5,12,16,20
11	Allethrin [Allylrethronyl *d,l-cis,*	Neurotoxin	Selected insects	Rat, ♀	Oral	680	8,14,15
12	*trans*-chrysanthemate]			Housefly[2], ♀	Topical	8.0	13-15
13				Mosquito[3], larva	IAM	[0.14]	13-15
14	Azinphos-methyl [Guthion; *O,O*-	Anticholinesterase	Most animals	Rat, ♀	Oral	11	8,14,15
15	dimethyl *S*-(4-oxo-1,2,3-benzotriazin-3[4*H*]-ylmethyl) phosphorodithioate]			Housefly[2], ♀	Topical	2.7	13-15
16				Mosquito[3], larva	IAM	[0.025]	13-15
17	Carbaryl [Arylam; 1-naphthyl *N*-	Anticholinesterase	Insects & other invertebrates	Rat, ♂	Oral	850	8,14,15
18	methylcarbamate]			Housefly[2], ♀	Topical	900	13-15
19				Mosquito[3], larva	IAM	[1.0]	13-15
20	Carbofuran [2,2-Dimethyl-2,3-dihydrobenzofuranyl-7-*N*-methylcarbamate]	Anticholinesterase	Most animals	Rat	Oral	5	10,13
21				Housefly[2], ♀	Topical	4.6	10,13
22				Mosquito[3], larva	IAM	[0.052]	10,13
23	Chlordane [Octachlorodihydrodicyclopentadiene; Octa-Klor;	Contact & stomach toxicant; marked residual toxicity	Insects, especially locusts, grasshoppers, crickets, soil insects, cockroaches, flies, ants, & insects predatory on cotton	Goat	Oral	180	16,26
24				Mouse	Oral	430	16,23
25	Toxichlor; 1,2,4,5,6,7,8,8-octachloro-2,3,3a,4,7,7a-hexahydro-			Rat	Oral	(335-430)	5,16
26	4,7-methanoindene]			Sheep	Oral	(500-1000)	16,26
27				Chick	Oral	(220-230)	16,22

[1] Unless otherwise indicated. [2] *Musca domestica.* [3] *Culex pipiens quinquefasciatus (C. fatigans).*

continued

Part II. Insecticides

	Insecticide [Synonym]	Mode of Action	Organisms Affected	Acute Toxicity				Reference
				Animal	Route	LD_{50}[1] mg/kg [ppm]		
28	Chlorobenzilate [Ethyl 4,4'-di-	Direct con-	Mites	Mouse	Oral	4850		6,7,16
29	chlorobenzilate]	tact action				729[4]		9,16
30		for some ascaridi-		Rat	Oral	3100		6,7,16
31		ans, residual action				702[4]		9,16
32		for others; low insecticidal powers				735(682-792)[5]		9,16
33	DDT [Chlorophenothane; dico-	Peripheral	Wide range	Cat	Oral	(400-500)[6]		14-16
34	phane; Neocid; zeidane; 1,1,1-	neuro-	of insects	Dog	i.v.	∿50		14-16
35	trichloro-2,2-bis(p-chlorophe-	toxin	& other	Guinea pig	Oral	400		14-16
36	nyl)ethane]		arthropods	Horse	Oral	300		14-16
37				Monkey	Oral	200		14-16
38				Mouse	Oral	(150-400)[6]		14-16
39				Rabbit	Oral	(250-400)[6]		14-16
40				Rat	i.v.	(40-60)		14-16
41					Oral	(150-420)[6]		14-16
42				♀	Oral	118		8,14,15
43				Housefly[2], ♀	Topical	2.0		13-15
44				Mosquito[3], larva	IAM	[0.007]		13-15
45	DDVP [Dichlorvos; dimethyl 2,2-	Anticholin-	Insects &	Rat	Oral	(56-80)		8,13
46	dichlorovinyl phosphate]	esterase	other in-	Housefly[2], ♀	Topical	1.8		8,13
47				Mosquito[3], larva	IAM	[0.075]		8,13
48	Demeton[7]	Anticholin- esterase	Most animals	Rat	Oral	(2.5-6.2)		8
49	Diazinon [O,O-Diethyl O-(2-iso-	Anticholin-	Most animals	Rat, ♀	Oral	76		8,14,15
50	propyl-6-methyl-4-pyrimidyl)	esterase		Housefly[2], ♀	Topical	2.85		13-15
51	phosphorothioate]			Mosquito[3], larva	IAM	[0.086]		13-15
52	Dicapthon [O,O-Dimethyl O-	Anticholin-	Selected in-	Rat	Oral	(330-400)		8,13
53	(2-chloro-4-nitrophenyl) phos-	esterase	sects	Housefly[2], ♀	Topical	1.5		8,13
54	phorothioate]			Mosquito[3], larva	IAM	[0.027]		8,13
55	Dicofol [Kelthane; 1,1-bis(p-	....	Mites	Rat, ♂	Oral	809		18
56	chlorophenyl)-2,2,2-trichloro- ethanol]			♀	Oral	684		18
57	Dieldrin [1,2,3,4,10,10-Hexa-	Central	Household	Dog	Oral	(56-80)		1,2,14,15
58	chloro-6,7-epoxy-1,4,4a,5,6,7,8,	neuro-	& agricul-	Guinea pig	Oral	49		1,2,14,15
59	8a-octahydro-endo-exo-1,4:5,8-	toxin	tural in-	Mouse	Oral	38		1,2,14,15
60	dimethanonaphthalene]		sects, grass-	Rabbit	Oral	(45-50)		1,2,14,15
61			hoppers,	Rat	Oral	(38-87)		1,2,14,15
62			locusts,		p.c.	(60-90)		2,14,15
63			crickets,	♀	Oral	46		8,14,15
64			cotton in-	Sheep	Oral	(50-75)		1,2,14,15
65			sects, soil-	Housefly[2], ♀	Topical	0.95		13-15
66			inhibiting insects	Mosquito[3], larva	IAM	[0.0078]		13-15
67	Dilan[8]	May be	Insects, es-	Mouse	i.p.	600		16
68		similar to	pecially		Oral	∿1100		16
69		DDT &	potato	Rat	Oral	300[9]; 4000[10]		16
		methoxy- chlor	leafhopper, Mexican bean beetle, & other bean insects					

[1] Unless otherwise indicated. [2] *Musca domestica.* [3] *Culex pipiens quinquefasciatus (C. fatigans).* [4] Technical product used. [5] 25% xylene emulsion used. [6] LD_{50} dose varies with sex and type of vehicle used; young animals are particularly susceptible. [7] A mixture of Demeton-O [O,O-diethyl O-(2-ethylthioethyl) phosphorothioate] and De-meton-S [O,O-diethyl S-(2-ethylthioethyl) phosphorothioate]. [8] A mixture of the two nitroalkyl DDT analogues: Bulan [2-nitro-1,1-bis(p-chlorophenyl)butane] and Prolan [2-nitro-1,1-bis(p-chlorophenyl)propane]. [9] Bulan. [10] Pro-lan.

Part II. Insecticides

	Insecticide [Synonym]	Mode of Action	Organisms Affected	Acute Toxicity			Reference
				Animal	Route	LD_{50}[1] mg/kg [ppm]	
70	Endrin [1,2,3,4,10,10-Hexachloro-6,7-epoxy-1,4,4a,5,6,7,8,8a-octa-hydro-endo,endo-1,4:5,8-di-methanonaphthalene]	Contact & stomach toxicant; may have delayed neuro-toxic ac-tion	Insects, es-pecially Orthoptera, cotton plant in-sects, flies, & other household insects	Guinea pig	Oral	(∼16-36)	16,21
71				Monkey	Oral	∼3	16,21
72				Rabbit, ♀	Oral	(7-10)	16,21
73				Rat, ♂, young	Oral	28.8	16,21
74				adult	Oral	(40-43.4)	16,19,21
75				♀, young	Oral	16.8	16,21
76				adult	Oral	7.3	16,21
77	Fenitrothion [O,O-Dimethyl O-(3-methyl-4-nitrophenyl) phos-phorothioate]	Anticholin-esterase	Selected in-sects	Rat, ♀	Oral	500	8,14,15
78				Housefly[2], ♀	Topical	2.2	13-15
79				Mosquito[3], larva	IAM	[0.0056]	13-15
80	Fenthion [O,O-Dimethyl O-4-(methylthio-3-methylphenyl) phosphorothioate]	Anticholin-esterase	Birds; most insects	Rat	Oral	(215-245)	8,13
81				Housefly[2], ♀	Topical	2.2	8,13
82				Mosquito[3], larva	IAM	[0.0042]	8,13
83	Heptachlor [1,4,5,6,7,8,8-Hepta-chloro-3a,4,7,7a-tetrahydro-4,7-methanoindene]	Central neuro-toxin	Cotton in-sects, grass-hoppers, soil insects, onion thrips, & alfalfa weevils	Guinea pig	Oral	116	14-16
84				Mouse	Oral	68	14-16
85				Rat	p.c.	(195-250)	5,14,15,25
86				♂	Oral	(60-169)	14-16,25
87				♀	Oral	162	8,14,15
88				Chick	Oral	63	14,15,17
89				Housefly[2], ♀	Topical	2.25	13-15
90				Mosquito[3], larva	IAM	[0.0056]	13-15
91	Isodrin [1,2,3,4,10,10-Hexachloro-1,4,4a,5,8,8a-hexahydro-endo,endo-1,4:5,8-dimethanonaph-thalene]	Neuro-toxic, with de-layed ac-tion	Many in-sects, es-pecially Orthoptera and cotton, household, & soil insects	Rat, ♂, weanling	Oral	27.8(23.3-33.3)	16
92				6 mo old	Oral	42.1(35.9-49.4)	16
93				♀, weanling	Oral	16.4(12.6-21.5)	16
94				6 mo old	Oral	11.7(10.2-13.5)	16
95	Lindane [Gamma-BHC; gamma-HCH; Gammexane; γ-1,2,3,4,5,6-hexachlorocyclohexane]	Central neuro-toxin	Insects	Dog	i.v.	7.5[11]	14-16
96					Oral	(40-200)[11]	14-16
97				Guinea pig	Oral	(100-127)	14-16
98				Mouse	Oral	86	14-16
99				Rabbit	Oral	(60-200)	11,12,14-16
100				Rat	i.p.	(35-85)	14-16
101					Oral	(125-230)	3,11,12,14-16,24
102				♀	Oral	91	8,14,15
103				Housefly[2], ♀	Topical	0.9	13-15
104				Mosquito[3], larva	IAM	[0.025]	13-15
105	Malathion [O,O-Dimethyl S-(1,2-dicarboxyethyl) phosphorodi-thioate]	Anticholin-esterase	Selected in-sects	Rat, ♀	Oral	1000	8,14,15
106				Housefly[2], ♀	Topical	27.5	13-15
107				Mosquito[3], larva	IAM	[0.081]	13-15
108	Methoxychlor [2,2-Bis(p-methoxy-phenyl)-1,1,1-trichloroethane]	Peripheral neuro-toxin	Most insects	Rat, ♀	Oral	6000	8,14,15
109				Housefly[2], ♀	Topical	9.0	13-15
110				Mosquito[3], larva	IAM	[0.067]	13-15

[1] Unless otherwise indicated. [2] *Musca domestica.* [3] *Culex pipiens quinquefasciatus (C. fatigans).* [11] Lethal dose.

continued

Part II. Insecticides

Insecticide [Synonym]	Mode of Action	Organisms Affected	Acute Toxicity			Reference
			Animal	Route	LD$_{50}$ mg/kg [ppm]	
111 Mevinphos [Dimethyl 2-carbomethoxy-1-methylvinyl phosphate]	Anticholin-esterase	Most animals	Rat	Oral	(3.7-6.1)	8,13
112 Parathion [*O,O*-Diethyl *O*-(*p*-nitrophenyl) phosphorothioate]	Anticholin-esterase	Most animals	Rat, ♀	Oral	3.6	8,14,15
113			Housefly [2], ♀	Topical	0.85	13-15
114			Mosquito [3], larva	IAM	[0.0032]	13-15
115 Perthane [1,1-Dichloro-2,2-bis(*p*-ethylphenyl)ethane]	....	Many insects	Rat	Oral	9.34	4
116 Phorate [*O,O*-Diethyl *S*-(/ethyl-thio/methyl) phosphorodithioate]	Anticholin-esterase	Most animals	Rat	Oral	(1.1-2.3)	8
117 Phthalthrin [2,3,5,6-Tetrahydro-phthalimidomethyl *d,l-cis,trans-*chrysanthemate]	Neurotoxin	Selected insects	Rat, ♀	Oral	>20,000	8,14,15
118			Housefly [2], ♀	Topical	1.95	13-15
119			Mosquito [3], larva	IAM	[0.062]	13-15
120 Propoxur [2-Isopropoxyphenyl-*N*-methylcarbamate]	Anticholin-esterase	Insects & other invertebrates	Rat, ♀	Oral	104	8,14,15
121			Housefly [2], ♀	Topical	21.5	13-15
122			Mosquito [3], larva	IAM	[0.33]	13-15
123 Ronnel [*O,O*-Dimethyl *O*-(2,4,5-trichlorophenyl) phosphoro-thioate]	Anticholin-esterase	Insects & other invertebrates	Rat	Oral	(1250-2630)	8,13
124			Housefly [2], ♀	Topical	2.6	8,13
125			Mosquito [3], larva	IAM	[0.03]	8,13
126 Toxaphene [12]	Contact & stomach toxicant; neurotox-ic symptoms after direct application manifested following a latent period	Agricultur-al insects, soil insects, orthopteran pests, household insects, insects on livestock, certain phytophagous acarians	Guinea pig	Oral	69	16
127			Mouse	Oral	112	16
128			Rat	Oral	∿69	16
129 Zectran [4-Dimethylamino-3,5-xylyl *N*-methylcarbamate]	Anticholin-esterase	Selected insects; snails	Rat	Oral	(15-63)	10,13
130			Housefly [2], ♀	Topical	65	10,13
131			Mosquito [3], larva	IAM	[0.50]	10,13

[2] *Musca domestica*. [3] *Culex pipiens quinquefasciatus (C. fatigans)*. [12] Chlorinated camphene containing 67-69% chlorine.

Contributors: Metcalf, Robert L.; Fitzhugh, O. G.

References

[1] Borgmann, A. R., et al. Unpublished. Alcon Labs, Inc., Ft. Worth, Texas, 1972.

[2] Council of Europe, Working Party on Poisonous Substances in Agriculture. 1962. Agricultural Pesticides. Strasbourg, France.

[3] Dallemagne, M. J., and E. Philippot. 1948. Arch. Int. Pharmacodyn. Ther. 76:274.

[4] Finnegan, J. K., et al. 1955. Ibid. 103:404.

[5] Gaines, T. B. 1960. Toxicol. Appl. Pharmacol. 2:88.

[6] Gasser, R. 1952. Congr. Int. Phytopharm., C. R., 3e, Paris, p. 357.

[7] Gasser, R. 1952. Experientia 8:65.

[8] Hayes, W. H., Jr. 1963. Clinical Handbook on Economic Poisons. U.S. Public Health Service, Washington, D.C.

[9] Horn, J. J., et al. 1955. J. Agr. Food. Chem. 3:752.

[10] Kenaga, E. E., and W. E. Allison. 1969. Bull. Entomol. Soc. Amer. 15:85.

[11] Lehman, A. J. 1948. Ass. Food Drug Offic. U.S. Quart. Bull. 12:82.

[12] Lehman, A. J. 1951. Ibid. 15:122.

[13] Metcalf, R. L. 1968. World Health Organ. Insecticide Mimeogr. Doc. VBC/68.66.

[14] Metcalf, R. L. 1966. Kirk-Othmer Encycl. Chem. Technol., 2nd Ed., 2:677.

[15] Metcalf, R. L., et al. 1962. Destructive and Useful Insects. Ed. 4. McGraw-Hill, New York.

[16] Negherbon, W. O. 1959. Handbook of Toxicology. W. B. Saunders, Philadelphia. v. 3.

[17] Sherman, M., and E. Ross. 1961. Toxicol. Appl. Pharmacol. 3:521.

[18] Smith, R. B., Jr., et al. 1959. Ibid. 1:119.

continued

Part II. Insecticides

[19] Street, J. C., et al. 1957. Proc. Annu. Meet. West. Sect. Amer. Soc. Anim. Prod. 8(46):1.

[20] Treon, J. F., and F. P. Cleveland. 1955. J. Agr. Food Chem. 3:402.

[21] Treon, J. F., et al. 1955. Ibid. 3:842.

[22] Turner, H. F., and W. G. Eden. 1952. J. Econ. Entomol. 45:130.

[23] U.S. Food and Drug Administration. 1947. FDA Quart. Rep. 3.

[24] U.S. Public Health Service, National Communicable Disease Center. 1956. Clinical Memoranda on Economic Poisons. Atlanta, Ga.

[25] Velsicol Corporation. 1965. In Food Agr. Organ. U.N. Meet. Rep. PL-10.

[26] Welch, H. 1948. J. Econ. Entomol. 41:36.

Part III. Nematocides

Route: Resp = respiratory (inhalation of vapor). Data in brackets refer to the column heading in brackets.

| | Nematocide [Synonym] | Mode of Action | Organisms Affected | Acute Toxicity | | | Reference |
				Animal	Route	LD$_{50}$, mg/kg[1] [ppm]	
1	Carbofuran [Furadan; NIA 10242; 2,3-dihydro-2,2-dimethyl-7-benzofuranyl methylcarbamate]	Contact; systemic in plants	Nematodes; soil, sucking, rasping, & chewing insects	Guinea pig	Resp	53[2]	3,9
2				Rabbit	Dermal	10,200	
3					Oral	8-14	
4	Chloropicrin [Chlor-O-Pic; Picfume; trichloronitromethane]	Fumigant	Nematodes; soil insects; fungi; weed seeds	Rat, white	Resp	[20]	2,11
5	Dasanit [B-25141; Bay 25141; O,O-diethyl O-(p/methylsulfinyl/phenyl) phosphorothioate]	Contact	Nematodes, especially Belonolaimus spp.	Rat, white	Dermal	3-30	1,5,9, 12
6					Oral	2-11	
7	DBCP [Fumazone; Nemagon; Oxy BBC; 1,2-dibromo-3-chloropropane]	Fumigant	Nematodes	Rabbit	Dermal	1420	1,2,11, 12,14
8				Rat, white	Oral	173	
9					Resp	[100]	
10	DD mixture[3] [D-D; Vidden D]	Fumigant	Nematodes, but Trichodorus christiei reestablishes rapidly	Rat, white	Dermal	2100	3,9,11, 15
11					Oral	140	
12					Resp	[500]	
13	Dimpylate [Diazinon; Sarolex; O,O-diethyl O-(2-isopropyl-6-methyl-4-pyrimidinyl) phosphorothioate]	Contact	Nematodes, especially Belonolaimus spp.; soil insects	Rat, white	Dermal	379-1107	2,13
14					Oral	66-600	
15	Ethoprop [Prophos; Mocap; V-C 9-104; O-ethyl S,S-dipropyl phosphorodithioate]	Contact	Nematodes, especially endoparasites; soil insects	Rabbit	Dermal	26	4,5,9, 10
16				Rat, white	Oral	62	
17	Ethylene dibromide [Dowfume W-85; EDB; Soilbrom-85; 1,2-dibromoethane]	Fumigant	Nematodes, but Trichodorus christiei reestablishes rapidly	Rat, white	Oral	108-170	2,7,8
18					Resp	[200]	
19	Metham sodium [SMDC; Vapam; sodium methyldithiocarbamate]	Fumigant	Nematodes; soil insects; fungi; weed seeds	Rabbit	Dermal	800	2,11
20				Rat, white	Oral	820	
21	Methyl bromide [Brom-O-Gas; Dowfume MC-2; bromomethane]	Fumigant	Nematodes; soil insects; fungi; weed seeds	Rat, white	Resp	[200]	2,6,11
22	Telone[4]	Fumigant	Nematodes	Rat, white	Oral	250-500	2,11
23					Resp	[500]	
24	Vorlex[5]	Fumigant	Nematodes; soil insects; fungi; weed seeds	Rat, white	Oral	100	2,11

[1] Unless otherwise indicated. [2] mg/m³ of air of 75 WP (75% active ingredient as a wettable powder formulation) aerosolized. [3] Mixture of 1,3-dichloropropene, 1,2-dichloropropane, and related chlorinated hydrocarbons. [4] A mixture of 1,3-dichloropropene and related chlorinated hydrocarbons. [5] A mixture of methyl isothiocyanate and chlorinated hydrocarbons.

continued

Part III. Nematocides

Contributors: Smart, Grover C., Jr., and Dickson, Donald W.

References

[1] Brodie, B. B., and G. W. Burton. 1967. Plant Dis. Rep. 51:562.

[2] Kenaga, E. E. 1966. Bull. Entomol. Soc. Amer. 12:161.

[3] Maggenti, A. R., et al. 1970. Ibid. 54:1012.

[4] Munson, J. D., and R. E. Hill. 1970. J. Econ. Entomol. 63:1614.

[5] O'Bannon, J. H., and A. L. Taylor. 1967. Plant Dis. Rep. 51:995.

[6] Peterson, G. W. 1970. Ibid. 54:572.

[7] Radewald, J. D., et al. 1969. Ibid. 53:385.

[8] Rhoades, H. L. 1968. Ibid. 52:573.

[9] Rhoades, H. L. 1969. Ibid. 53:728.

[10] Rhoades, H. L. 1970. Ibid. 54:411.

[11] Rhoades, H. L., et al. 1966. Fla. Agr. Exp. Sta. Bull. 707.

[12] Smart, G. C., Jr. 1969. Proc. Symp. Trop. Nematol. Univ. Puerto Rico Agr. Exp. Sta., p. 68.

[13] Streu, H. T., and L. M. Vasvary. 1966. Bull. N. J. Acad. Sci. 11:17.

[14] Toung, M.-C. 1969. Plant Dis. Rep. 53:300.

[15] Youngson, C. R., and C. A. I. Goring. 1970. Ibid. 54:196.

Part IV. Fungicides

Mode of Action: DNA = deoxyribonucleic acid; tRNA = transfer ribonucleic acid; NADH = nicotinamide adenine dinucleotide (reduced form); CoA = coenzyme A. LD_{50} = lethal dose for 50% of test subjects. Data in brackets refer to the column heading in brackets.

	Fungicide [Synonym]	Mode of Action	Organisms Affected	Toxicity			Reference
				Animal	Route	LD_{50}[1] [No-Effect Level]	
1	Benomyl [Methyl 1-(butyl-carbamyl)-2-benzimidazole carbamate]	Little known other than that compound is preventive, curative, systemic fungicide & mite ovacide	Fungi & mite ova	Rat, ♂♀, fasted	Oral, acute	>10,000 mg/kg	4
2 3	Botran [2,6-Dichloro-4-nitroaniline]	Inhibits protein synthesis	Used primarily for control of *Botrytis* spp., *Rhizopus* spp., & *Sclerotinia* spp. for postharvest rots of fruits	Rat	Oral Acute In diet for 2 yr	>10,000 mg/kg [100 ppm]	28, 29
4 5 6 7	Captan [*N*-(Trichloromethyl-thio)-4-cyclo-hexene-1,2-dicarboximide]	Reacts mainly with thiols; affects many metabolic processes involving sulfhydryl-containing enzymes & cofactors. Possibly fungistatic with lower dosages; fungicidal with higher dosages or prolonged exposure.	Toxic to most fungi	Dog Rat	Oral Chronic feeding Oral Chronic feeding	Nontoxic[2] [4000 ppm] 9000 mg/kg [1000 ppm]	12-14, 18
8 9	Carboxin [5,6-Dihydro-2-methyl-1,4-oxathiin-3-carboxanilide]	Affects several aspects of metabolism including inhibition of oxidation of pyruvate & acetate and interference with nucleic acid synthesis. Systemic action in plants against fungi.	Especially Basidiomycetes	Rabbit Rat	Dermal, acute Oral, acute	>8000 mg/kg 3200 mg/kg	16, 27
10 11	Chloroneb [1,4-Dichloro-2,5-dimethoxy-benzene]	Toxicity to *Rhizoctonia solani* apparently involved with direct or indirect inhibition of DNA synthesis at the nucleotide polymerization stage	Soil-borne fungi such as *Rhizoctonia solani*, *Sclerotium rolfsii*, & *Pythium* spp.	Rat, ♂♀	Dermal, acute Oral, acute	>5000 mg/kg[3] >11,000 mg/kg	5, 22, 23

[1] Unless otherwise indicated. [2] No evidence of chronic toxicity in subjects fed 300 mg technical-grade captan·kg^{-1}·da^{-1} for 66 wk; skin may become irritated. [3] Lethal dose.

continued

Part IV. Fungicides

	Fungicide [Synonym]	Mode of Action	Organisms Affected	Toxicity			Reference
				Animal	Route	LD$_{50}$[1] [No-Effect Level]	
12	Chlorothalonil [Tetrachloro-isophthaloni-trile]	Little known other than that compound is fungicide for control of leaf spots	Turf & orna-mental fungi	Rabbit	Dermal, acute	>10,000 mg/kg	3
13				Rat	Oral, acute	>10,000 mg/kg	
14	Cycloheximide[4] [3-[2-(3,5-Di-methyl-2-oxo-cyclohexyl)-2-hydroxyethyl] glutarimide]	Inhibits incorporation of leu-cine into ribosomal protein in a cell-free system; no inter-ference with activation of amino acids or their transfer to tRNA. Resistance deter-mined by ribosomes, not by enzymes in supernatant.	Certain proto-zoa & yeasts; *Ceratocystis fagacearum; Higginsia hiemalis* (cherry leaf spot); *Ustilago tritici;* some turf diseases	Guinea pig	Oral	65 mg/kg	7,8, 20, 21
15				Monkey	Oral	60 mg/kg	
16				Mouse	Oral	133 mg/kg	
17				Rat	Oral	2.5 mg/kg	
18	Dexon [p-(Di-methylamino)-benzenediazo sodium sul-fonate]	Inhibits NADH by respiratory chain involved—basis for specificity apparently acces-sability to intromitochon-drial site involved	*Phytophthora* spp. & *Pyth-ium* spp.	Guinea pig, ♂	Oral	150 mg/kg	2, 26
19				Rat ♂♀, wean-ling	In diet for 16 wk	[80 ppm]	
20				♀	Oral	60 mg/kg	
21	Dichlone [2,3-Dichloro-1,4-naphthoqui-none]	Inactivates enzymes with functional amino or sulfhy-dryl groups. Toxicity results from concomitant inhibition of phosphorylation, certain dehydrogenases, carboxylases, & CoA.	Soil fungi; *Venturia inaequalis* (apple scab)	Dog	Chronic feeding	500 ppm[5]	12, 19
22				Rat	Oral	1.3 mg/kg	
23					Chronic feeding	1580 ppm[5]	
24	Dinocap [Kara-thane; 2-(1-methylheptyl)-4,6-dinitro-phenyl cro-tonate]	Unknown; possibly attri-butable to phenolic portion of molecule	Acarians; fungi causing pow-dery mildew In vitro: many fungi affected by 0.01% concentration	Dog	In diet for 1 yr	[50 ppm[6]]	10, 11, 15
25				Rat, ♂	Oral	980 mg/kg	
26				♀	Oral	1190 mg/kg	
27				Duck, Pekin	....	25 ppm[7]	
28	Dodine [n-Dodecylguani-dine acetate]	Mechanism of action unknown, but surfactant properties & highly basic nature of guani-dine moiety implicated. In solution at pH 5.6 dodine acetate (positively charged dodine cation) reacts rapid-ly with surface carboxyl & phosphate groups of spores. Enzyme inhibition at the cytoplasmic membrane or of intracellular reaction (rather than physical disorganization of the membrane) may explain toxicity.	In vivo: *Fusi-cladium cera-si* (peach scab); *F. ef-fusum* (pecan scale); *Higgin-sia hiemalis* (cherry leaf spot); *Venturia inaequalis* (ap-ple scab) In vitro: many other fungi	Dog	Chronic feeding	[>50 to <200 ppm]	12, 24, 25
29				Rabbit	....	Fn[8]	
30				Rat, ♂	Oral	750 mg/kg	
31					Chronic feeding	[200 ppm]	
32	Dyrene [2,4-Di-chloro-6-o-chloroanilino-s-triazine]	Mechanism of action not clar-ified	*Alternaria solani & Phytophthora infestans* (potato pathogens); *Fu-sarium* spp.; *Helminthosporium* spp. (cause diseases of turf)	Dog	Oral	7.1 g/kg	1, 12
33					Chronic feeding	[5000 to >10,000 ppm]	
34				Rat	Oral	2.7 g/kg	
35					Chronic feeding	[5000 ppm]	

1/ Unless otherwise indicated. 4/ Highly specific fungicide; inhibits some organisms with 0.1-1.0 ppm, but does not in-hibit others with 100 ppm or more. 5/ Minimal-effect level.

6/ No weight loss observed. 7/ Dose producing cataracts. 8/ Severe skin irritation.

continued

Part IV. Fungicides

Fungicide [Synonym]	Mode of Action	Organisms Affected	Toxicity		LD$_{50}$[1] [No-Effect Level]	Reference
			Animal	Route		
36 37 Ferbam[9] [Ferric dimethyldithiocarbamate]	Inhibits urease, pancreatic & malt amylases, polyphenol oxidases. A dithiocarbamate-metal-protein complex possibly formed through either sulfhydryl or carboxyl (or both) groups of the protein.	Many fungi	Rat	Oral Chronic feeding	>17 g/kg [250 ppm]	12, 17, 19
38 39 40 41 Glyodin [2-Heptadecylglyoxalidine acetate; 2-heptadecyl-2-imidazoline acetate]	Unknown	In vivo: *Venturia inaequalis* (apple scab) & other fruit-infecting pathogens In vitro: many fungi	Dog Guinea pig Rat	Chronic feeding Chronic feeding Oral Chronic feeding	[210 ppm[10]] [450 ppm[10]] 1300 mg/kg[11]; 3800 mg/kg[12] [270 ppm[10]]	12
42 43 44 Maneb [Manganous ethylenebis(dithiocarbamate)]	Mechanism of action only partially known. Inhibits metabolism, but not synthesis, of citrate, apparently through inhibition of aconitate hydrase; no detectable effect on other enzymes of the tricarboxylic acid cycle at minimal lethal doses.	In vivo: fungal pathogens of foliage, e.g., *Phytophthora infestans* (potato late blight), *Peronospora cubensis* (downy mildew of muskmelon), *Alternaria* spp. (cause leaf spot on many plants)	Dog Rat	Chronic feeding Oral Chronic feeding	[80 ppm] 6750 mg/kg [25 ppm]	12, 17, 19
45 46 Pentachloronitrobenzene [PCNB]	Unknown	*Rhizoctonia solani*	Rat	Oral In diet for 2 yr	>12 g/kg [2500 ppm]	15
47 48 Phenylmercuric acetate [PMA; Scutl]	Low concentrations of phenylmercuric salts react with both amino- & sulfhydryl-dependent enzymes; higher concentrations react with other kinds of proteins. Reactions involving organometallic compounds usually reversible in the presence of another thiol such as cysteine or glutathione.	In vitro: bacteria; fungi; algae	Rat	Oral In diet	37 mg/kg As low as 0.5 ppm[13]	6,9, 17
49 50 51 Sodium *o*-phenylphenate	Unknown	In vitro: fungi causing postharvest diseases of citrus fruits	Dog Rat	Chronic feeding Oral Chronic feeding	[20,000 ppm] 2700 mg/kg [2000 ppm]	12

[1] Unless otherwise indicated. [9] Insolubility limits use for in vitro studies. [10] 2-Heptadecylimidazoline. [11] Heptadecylimidazoline component. [12] 2-Heptadecyl-1-(hydroxy-ethylimidazoline) component. [13] Levels resulting in renal lesions [6].

Contributor: Mitchell, J. E.

References

[1] Burchfield, H. P. 1957. Contrib. Boyce Thompson Inst. 19:169.

[2] Chemagro Corporation. 1972. Dexon Technical Sheet. Kansas City, Mo.

[3] Diamond Alkali Co. 1972. Chlorothalonil Technical Research Bulletin. Cleveland, O.

[4] Du Pont de Nemours, E. I., & Co. 1972. Benomyl Technical Data Sheet. Biochemicals Dep., Wilmington, Del.

[5] Du Pont de Nemours, E. I., & Co. 1972. Chloroneb Technical Data Sheet. Biochemicals Dep., Wilmington, Del.

continued

Part IV. Fungicides

[6] Fitzhugh, O. G., et al. 1950. Arch. Ind. Hyg. Occup. Med. 2:433.

[7] Ford, J. H., et al. 1958. Plant Dis. Rep. 42:680.

[8] Greig, M. E., et al. 1959. Toxicol. Appl. Pharmacol. 1:599.

[9] Hunter, D., et al. 1940. Quart. J. Med. 9:193.

[10] Kirby, A. H. M., et al. 1958. Nature (London) 182: 1445.

[11] Larson, F. S., et al. 1959. Arch. Int. Pharmacodyn. Ther. 119:31.

[12] Lehman, A. J. 1965. Summaries of Pesticide Toxicity. Association of Food and Drug Officials, Topeka, Kan. pp. 93-111.

[13] Leukens, R. J. 1959. Phytopathology 49:339.

[14] Leukens, R. J., and H. D. Sisler. 1958. Ibid. 48: 235.

[15] Martin, H. 1961. Can. Dep. Agr. Publ. 1093.

[16] Mathrie, D. E. 1970. Phytopathology 60:671.

[17] Owens, R. G. 1963. Annu. Rev. Phytopathol. 1:77.

[18] Owens, R. G., and G. Blaac. 1960. Contrib. Boyce Thompson Inst. 20:475.

[19] Owens, R. G., and H. M. Novotny. 1958. Ibid. 19: 463.

[20] Siegel, M. R., and H. D. Sisler. 1965. Biochim. Biophys. Acta 103:558.

[21] Siegel, M. R., et al. 1966. Biochem. Pharmacol. 15: 1213.

[22] Sisler, H. D. 1969. J. Agr. Food Chem. 17:123.

[23] Sisler, H. D. 1969. Phytopathology 59:627.

[24] Somers, E., and D. J. Fisher. 1967. J. Gen. Microbiol. 48:147.

[25] Somers, E., and R. J. Pring. 1966. Ann. Appl. Biol. 58:457.

[26] Tolmsoff, W. J. 1965. Ph.D. Thesis. Univ. California, Davis (Univ. Microfilm 66-2401, Ann Arbor, Mich.).

[27] Uni-Royal Chemical Co. 1972. Carboxin Technical Data Sheet. Naugatuck, Conn.

[28] Upjohn International, Inc. 1972. Botran Technical Data Sheet. T.U.C.O. Products Division, Kalamazoo, Mich.

[29] Weber, P. J., and J. M. Ogawa. 1965. Phytopathology 55:159.

Part V. Herbicides

The metabolic pathways described refer principally to higher plants, although limited consideration is given to certain microorganisms. Data, for the most part, include work published since 1966, and should consequently be considered as supplemental to data in reference 110. Metabolic changes and other anomalous plant responses described do not necessarily all have physiological significance as a site of primary or even secondary herbicidal action. Some responses are mediated mainly by physiological concentrations, and others by herbicidal concentrations. Variations in absorption and translocation which could thus influence metabolism at the site of action have recently been discussed [17, 73, 111, 157, 165, 185] and are not considered herein. For reviews with comprehensive descriptions of the metabolism of numerous herbicides, consult references 4, 44, 48, 65, 87, 162, 185, 254, 271, and 272. **Herbicide:** Names are those assigned by the Weed Science Society of America, and chemical nomenclature is essentially according to the American Chemical Society. **Rat Oral LD$_{50}$** = acute lethal oral dose for 50% of rats tested [21]. **Mode of Action:** DNA = deoxyribonucleic acid; RNA = ribonucleic acid; mRNA = messenger ribonucleic acid; rRNA = ribosomal ribonucleic acid; tRNA = transfer ribonucleic acid; S = Svedberg unit (1 S = 0.1 ps); ADP = adenosine 5'-diphosphate; ATP = adenosine 5'-triphosphate; NADH = nicotinamide adenine dinucleotide (reduced form); NADP$^+$ and NADPH = nicotinamide adenine dinucleotide phosphate (oxidized and reduced forms, respectively). Figures in brackets are reference numbers.

	Herbicide [Synonym]	Rat Oral LD$_{50}$ mg/kg	Mode of Action	Organism Affected (Synonym)
1	Amitrole [3-	24,600	Multiple & specific sites involved	Various plants & microorganisms [101]
2	Amino-s-		Interferes with histidine metabolism in bacteria [103],	Chlorella vulgaris [224]
3	triazole]		algae [45], fungi [101], yeast [130], & higher plants	Cirsium arvense [101]
4			[101]. Imidazoleglycerophosphate, a histidine pre-	Prototheca zopfii [45]
5			cursor, accumulates in these cells. Competitively in-	Rosa hybrid [101]
6			hibits activity of imidazoleglycerolphosphate dehy-	Saccharomyces cerevisiae [130,131]
7			dratase, an enzyme of histidine biosynthesis [106, 131].	Salmonella typhimurium [103,106]
8			Interferes with purine & one-carbon metabolism, as	Cirsium arvense [101]
9			evidenced (i) by inhibited formation of imidazole	Salmonella typhimurium [38,101]
10			ring [101] in purine-requiring mutants, by reversal of	Scenedesmus [46]
11			amitrole toxicity with adenine [46,101], methionine [38], or serine, and (ii) by reduced incorporation of formate & glycine into RNA [25]	Triticum aestivum [25]

continued

	Herbicide [Synonym]	Rat Oral LD$_{50}$ mg/kg	Mode of Action	Organism Affected (Synonym)
12			Accumulation by cells inhibited by purines	*Salmonella typhimurium* [102]
13			Inhibits activity of catalase, fatty acid peroxidase, &	*Brassica hirta (Sinapis alba)* [149]
14			other enzymes [149,269] while suppressing synthe-	*Chlorella pyrenoidosa* [78]
15			sis of one of the phosphorylase isozymes [78]. In-	*Papaver somniferum* [269]
16			creases activities of phosphatase & amylase [109].	Many other species [109]
17			Transformed into many metabolites, such as β-(3-	*Avena sativa* [145]
18			amino-1,2,4-triazolyl-1)-α-alanine [146,160,	*Chlorella pyrenoidosa* [79]
19			230], or a highly phytotoxic product [145].	*Phaseolus vulgaris* [160,230]
20			Conjugates with sugars to form glycosides, as	*Pinus ponderosa* [146]
			well as with amino acids [79].	
21			Level of soluble N compounds decreased 1 da after	*Cirsium arvense* [43]
22			treatment, but increased above level in controls	*Malus pumila (Pyrus malus)* [231]
23			after 5 da. Causes decrease in protein content of	*Poteriochromonas stipitata* [210]
			light-grown leaf tissue [231].	
24			Inhibits protein synthesis in plants grown in light [25],	*Glycine max* [169]
25			but not in etiolated plants or plant parts [40,169]	*Phaseolus vulgaris* [40]
26				*Triticum aestivum* [25]
27			Reduces carbohydrate content of foliage, but not of	*Cynodon dactylon* [172]
			stolon	
28			Inhibits chloroplast development in leaves [24,87] &	*Linum usitatissimum* [12]
29			roots [12]. Chloroplasts lack normal grana, lamel-	*Lycopersicon esculentum* [197]
30			lar membranes, 70S ribosomes, DNA, & fraction I	*Ocimum basilicum* [67]
31			[22,27], but may contain super grana [67]. Auto-	*Triticum aestivum* [22,24,27]
32			radiography indicates that amitrole does not accu-	*Zea mays* [87]
			mulate in chloroplasts [197].	
33			Inhibits carotenogenesis, but not chlorophyll syn-	*Poteriochromonas* [62]
34			thesis, in etiolated wheat seedlings. Carotenoids do	*Triticum aestivum* [41]
			not provide protection for chloroplast components	
			at high light intensities. Coproporphyrin accumu-	
			lates in treated algal cells [62].	
35			No effect on phototropism, geotropism [202], &	*Ipomoea nil (Pharbitis nil)* [117]
36			photoperiodism [117]	*Triticum aestivum* [202]
37			Induces marked increases in 4 anthocyanidins	*Triticum aestivum* [26]
38	Atrazine [2-Chloro-4-(ethylamino)-6-(iso-propylamino)-s-triazine]	3080	Primary sensitive site located within chloroplasts for resistant (e.g., maize) & susceptible species. Inhibits Hill reaction & its associated noncyclic photophosphorylation; ineffective against cyclic photophosphorylation.	Numerous plants [28,220]
39			Indirectly affects respiration by inhibiting photosyn-	*Agrostemma githago* [181]
40			thesis, causing a deficit of assimilation products	*Brassica kaber* var. *pinnatifida (Sinapis arvensis)* [181]
41				*Glycine max* [219]
42				*Phaseolus vulgaris* [181]
43				*Zea mays* [219]
44			Metabolism & detoxication occur by Dechlorination—conversion of s-chlorotriazines to 2-hydroxy derivatives, e.g., hydroxyatrazine &	*Triticum aestivum* [220]
45			hydroxysimazine	*Zea mays* [220]
46			N-Dealkylation—formation of 2-chloro-4-amino-6-isopropylamino-s-triazine & 2-chloro-4-amino-6-	*Pisum sativum* [166]
47			ethylamino-s-triazine	*Sorghum bicolor (S. vulgare)* [166]

continued

Part V. Herbicides

	Herbicide [Synonym]	Rat Oral LD$_{50}$ mg/kg	Mode of Action	Organism Affected (Synonym)
48 49			Glutathione conjugation—glutathione-atrazine formation [137]	*Sorghum bicolor (S. vulgare)* [137] *Zea mays* [221]
50			At subherbicidal concentrations, *s*-triazine increases protein content & total soluble amino acids	Numerous plants [201]
51			Subherbicidal treatments stimulated activities of nitrate reductase, alanine aminotransferase (glutamic-pyruvic transaminase), α-amylase, starch pyrophosphorylase, & ATPase. Increased nitrate reductase & aminotransferase activity may enhance amino acid formation necessary for protein synthesis.	*Phaseolus vulgaris* [227]
52			Alters auxin & cytokinin metabolism of callus tissue, which may affect protein synthesis	*Glycine max* [64]
53			Causes chromosomal aberrations (aneuploidy & polyploidy) in pollen mother cells	*Sorghum bicolor (S. vulgare)* [140]
54 55			Induces swelling of fret system & disruption of granal discs; finally, rupture of grana membranes & chloroplast envelopes	*Echinochloa crusgalli* [100] *Phaseolus vulgaris* [14]
56			Increases RNA synthesis in isolated chromatin from treated plants	*Glycine max* [192]
57 58	Barban [4-Chloro-2-butynyl *m*-chloro-carbanilate]	1350	Probably acts as gene antiderepressor substance, inhibiting mRNA synthesis. Blocks gibberellin-induced amylase synthesis in aleurone layer and inhibits amino acid incorporation into protein.	*Hordeum vulgare* [268] Numerous other higher plants [153]
59 60			Resembles phytokinins by preserving chlorophyll levels in excised tissue and maintaining enzyme activity	*Avena sativa* [52] *Hordeum vulgare* [154,156]
61			Inhibits meristematic activity in roots, but increases RNA, protein, & nucleotide content of cells	*Avena fatua* [135]
62 63			Rapidly inhibits cytokinesis & cell elongation, blocking cell division at metaphase	*Nicotiana* sp. [267] *Triticum aestivum* [42]
64	Bromacil [5-Bromo-3-*sec*-butyl-6-methyl-uracil]	5200	Strongly inhibits Hill reaction in isolated chloroplasts	*Brassica rapa* [105]
65 66 67			Inhibits nitrate reductase synthesis in cauliflower & radish; activity of proteolytic enzyme appreciably reduced in squash cotyledons	*Brassica oleracea* var. *botrytis* [99] *Cucurbita maxima* [15] *Raphanus sativus* [99]
68			Inhibits cell wall formation in root cells, resulting in cells with multiple nuclei. Causes progressive swelling of grana & fret membranes in leaf tissue.	*Avena sativa* [16]
69			Interferes with cell division by preventing septation	*Escherichia coli* K-12 [116]
70 71 72			Depletes carbohydrates & alanine, but increases aspartic, glutamic, & malic acids	*Agropyron repens* [240] *Glycine max* [57] *Zea mays* [57]
73			Retards senescence, due to cytokinin-like activity	*Zea mays* [71]
74 75	Bromoxynil [3,5-Dibromo-4-hydroxybenzonitrile]	250	Stimulates ADP-limited O$_2$ utilization, but inhibits non-ADP-limited O$_2$ uptake. Relieves oligomycin-inhibited O$_2$ uptake in mitochondria. Inhibits photoreduction & noncyclic phosphorylation, but not cyclic phosphorylation, in isolated chloroplasts.	*Solanum tuberosum* [167] *Spinacia* sp. [167]
76			Inhibits proteolytic enzyme development in cotyledons of intact embryos, but not in excised cotyledons, suggesting interference with hormonal control as well as with proteinase synthesis	*Cucurbita maxima* [191]
77			Interferes with amylase development in distal halves of intact seeds during germination	*Hordeum vulgare* [189]

continued

Part V. Herbicides

	Herbicide [Synonym]	Rat Oral LD$_{50}$ mg/kg	Mode of Action	Organism Affected (Synonym)
78	Cacodylic acid [Hydroxydimethylarsine oxide]	830	Hydrolyzed by plant tissues to form Na$_3$AsO$_4$, a more phytotoxic substance (cacodylic acid itself possibly nonherbicidal)	Numerous plants [8]
79	CDAA [N,N-Diallyl-2-chloroacetamide]	750	May prevent activation of amino acids & aminoacyl tRNA formation, or interfere with the transfer of aminoacyl tRNA to the polypeptide during protein synthesis. Inhibits gibberellic acid-induced amylase production.	Numerous plants [115]
80			Blocks incorporation of malonic acid into lipids by excised hypocotyls	Sesbania exaltata [151]
81 82	Chloramben [3-Amino-2,5-dichlorobenzoic acid]	3500	Converted to N-glucosylarylamine, N-glucosylchloramben, or N-(3-carboxy-2,5-dichlorophenyl)glucosylamine metabolites by a soluble enzyme system	Abutilon theophrasti [237] Numerous other plants [76,238]
83			Interferes with amylase synthesis in distal halves of intact seeds	Hordeum vulgare [189]
84 85 86	Chlorpropham [CIPC; isopropyl m-chlorocarbanilate]	5000-7500	Stops gibberellin-induced amylase synthesis in barley aleurone layer. Prevents loss of enzyme activity. Retards senescence by blocking synthesis & transcription of a mRNA. Inhibits protein synthesis.	Glycine max [169] Hordeum vulgare [154] Zea mays [169]
87			Severely inhibits kinetin-induced cell proliferation, but not indoleacetic acid-induced cell enlargement	Nicotiana tabacum [122]
88			Inhibits mitosis, blocking cell division at metaphase and arresting anaphase; cytostatic effect	Numerous plants [97,152]
89 90			Converted to water-soluble glycoside metabolite in susceptible species	Amaranthus retroflexus [114] Polygonum lapathifolium [114]
91			Reduces epicuticular lipids by 50% in leaves following treatment of roots	Pisum sativum [235]
92			Exerts uncoupling effect on oxidative phosphorylation linked to NO$_3^-$ oxidation	Nitrobacter agilis [258]
93			Causes nuclear migration when injected into living cell of stem	Vicia faba [260]
94			Inhibits phytochrome-regulated seed germination	Numerous plants [155]
95			Inhibits both photochemical & oxidative production of ATP	Chlorella [207]
96	2,4-D [(2,4-Dichlorophenoxy)-acetic acid]	300-1000	Includes decarboxylation of acetic acid side chain, ring hydroxylation, protein or peptide conjugation, & complex formation with sugars	Numerous plants [143,270]
97			Principally metabolized to (2,4-dichlorophenoxy)propionic acid	Numerous plants [93]
98			Converted by soluble enzyme preparations to succinic acid & glyoxylate	Arthrobacter spp. [245,246]
99			Converted to inactive homologs with longer side chains; inactivated by binding to protein	Medicago sativa [55,142,198]
100			Inhibits DNA synthesis transcribed by Escherichia coli DNA nucleotidyltransferase (DNA polymerase) from pea embryo DNA & chromatin	Pisum sp. [215]
101			Subherbicidal concentrations favor production of mRNA. Inhibitory levels induce synthesis of rRNA.	Glycine max [70]
102			Increases RNA level, but not DNA level, in single-cell algae	Poteriochromonas stipitata [211]

continued

Part V. Herbicides

	Herbicide [Synonym]	Rat Oral LD$_{50}$ mg/kg	Mode of Action	Organism Affected (Synonym)
103			Activity of chromatin RNA nucleotidyltransferase (RNA polymerase) from treated plants higher than that of control plants	*Glycine max* [180]
104			Herbicidal effect may partially result from aberrant growth induced by abnormal nucleic acid metabolism	Numerous plants [95]
105			Affects activity of many plant enzymes, such as carbonate dehydratase (carbonic anhydrase) & catalase [233], which are decreased, while others, such as β-fructofuranosidase (invertase) [206], glucose-6-phosphate dehydrogenase [29], glutathione reductase [29], hemicellulase [242], & hydrolase are increased	Numerous plants [29,206,233,242]
106			Bacterial fumarate hydratase (fumarase) is particularly sensitive	*Pseudomonas fluorescens* [136]
107			Some effects due to increased ethylene production in plant tissue	Several plants [1,108]
108			Reduces indoleacetic acid content & inhibits cell elongation [250]	....
109			Prevents formation of certain amino acids from glucose & blocks incorporation of other amino acids into protein in root tips	*Pisum sativum* [126]
110			Inhibits cocarboxylase or, more likely, interferes with thioctic acid (α-lipoic acid) metabolism; affects coenzyme A activity	*Pisum sativum* [127]
111			Inhibits respiration stimulated by 2,4-dinitrophenol, indicating it affects the coupling of phosphorylation with the electron transport chain	*Brassica oleracea* [144]
112			Increases respiration as evidenced by increased cytochrome oxidase & peroxidase activities; inhibits catalase activity	*Zea mays* [158]
113			Inhibits oxidative phosphorylation	*Pisum* sp. [54]
114			Mitochondria isolated from treated plants are larger and incorporate more amino acids into protein, and more phosphate into lipids & RNA than do controls	*Glycine max* [33]
115			Inhibits stomatal opening in both light & dark	Several plants [195]
116			First causes rupture & disintegration of tonoplast, followed by pulling away of chloroplast from cell wall; enlargement & swelling of grana compartments of chloroplasts	*Nicotiana tabacum* [94,255]
117			Enhances growth of cultured, excised tissue	*Glycine max* [259]
118				*Nicotiana tabacum* [259]
119			Depresses light-induced anthocyanin synthesis in excised dark-grown plants	*Sorghum* sp. [252]
120			Interferes with photosynthesis, Hill reaction, cyclic & noncyclic phosphorylation, & stability of the chlorophyll-protein-lipid complex	*Pisum* sp. [53]
121				*Zea mays* [158]
122			Affects exudation from excised roots by inhibiting metabolic processes of root cells	Numerous plants [171]
123			Causes changes in biopotentials, affecting permeability of membrane surface	*Nitella* [88]
124	Dalapon [2,2-Dichloropropionic acid]	♂, 9330; ♀, 7570	Resistant to degradation in plants & bacteria. Blocks synthesis of pantothenic acid [39] by competitively inhibiting enzymes which form pantothenic acid from β-alanine & pantoic acid; interferes with enzyme activity by binding with protein [123].	*Agarbacterium* sp. [39,123]

continued

Part V. Herbicides

	Herbicide [Synonym]	Rat Oral LD$_{50}$ mg/kg	Mode of Action	Organism Affected (Synonym)
125			Reduces proteolytic activity of cotyledons	*Cucurbita maxima* [15]
126			Causes accumulation of flavonoid glycosides	*Lupinus* sp. [253]
127			Increases fumarate & one unidentified organic acid in germinating seedlings, and decreases aconitate, citrate, malate, & succinate	*Triticum aestivum* [184]
128	DCPA [Dimethyl tetrachloroterephthalate]	3000	Arrests cell division in root tissue while cell enlargement continues. Dinucleate cells occur in radicle tissue of corn roots, and mitosis is stopped at metaphase.	*Cynodon dactylon* [35]
129				*Zea mays* [35]
130			Causes meristematic activity in phloem & xylem parenchyma cells, resulting in a mass of undifferentiated tissue within the stele	*Lycopersicon esculentum* [9]
131	Diallate [S-(2,3-Dichloroallyl) diisopropylthiocarbamate]	395	Reduces wax formation on leaf surfaces by 50%	*Pisum sativum* [235]
132			Vapor causes, first, shoot inhibition, then, reduced mitotic activity	*Avena fatua* [20]
133				*Triticum aestivum* [20]
134	Dicamba [3,6-Dichloro-o-anisic acid]	2900 ± 800	Inhibits dipeptidase activity in cotyledons during germination	*Cucurbita maxima* [13]
135			Causes reduced levels of anthocyanins and swollen stems in treated plants	*Phaseolus vulgaris*[1] [204]
136			Alters both pattern of O$_2$ consumption & cell permeability	*Cyperus rotundus* [148]
137			Expected metabolites, 5-hydroxydicamba & 3,6-dichlorogentisic acid, were not found in roots	*Cyperus rotundus* [199]
138			Inhibits amylase production but not protein & RNA synthesis	*Zea mays* [169]
139			Severely affects root growth & germination; inhibits mitosis in seedlings	*Hordeum vulgare* [263]
140	Dichlobenil [2,6-Dichlorobenzonitrile]	3160	Uncouples oxidative phosphorylation	*Cucumis sativus* [72]
141			Does not inhibit RNA & protein synthesis in dark-grown mesocotyl sections	*Zea mays* [169]
142			Depresses CO$_2$ fixation in seedlings	*Prunus persica* [2]
143			Translocated to leaves where ∿90% evaporates from the leaf and the other 10% is converted to soluble & insoluble conjugates	*Oryza sativa* [251]
144				*Triticum aestivum* [251]
145			Inhibits dipeptidase activity of cotyledons; partially inhibits development of phytase activity	*Cucurbita maxima* [13,190]
146	Dichlormate [3,4-Dichlorobenzyl methylcarbamate]	♂, 1870; ♀, 2140	Interferes with normal carotenoid synthesis [41] resulting in photodestruction of chloroplast membranes, pigments, & 70S ribosomes [23,98]	*Triticum aestivum* [23,41,98]
147	Dinoseb [2-sec-Butyl-4,6-dinitrophenol]	5-60	Inhibits RNA nucleotidyltransferase (RNA polymerase) activity ∿20%	*Zea mays* [169]
148	Diphenamid [N,N-Dimethyl-2,2-diphenylacetamide]	970 ± 140	Degraded to N-methyl-2,2-diphenylacetamide, 2,2-diphenylacetamide, & 2,2-diphenylacetic acid in whole plants & excised shoots	*Lycopersicon esculentum* [212]
149				*Triticum aestivum* [212]
150			Metabolites given in entires 148-149 more toxic than diphenamid	*Echinochloa crusgalli* [124]
151				*Lycopersicon esculentum* [124]

[1] 3-wk-old plants, greenhouse-grown.

continued

Part V. Herbicides

	Herbicide [Synonym]	Rat Oral LD$_{50}$ mg/kg	Mode of Action	Organism Affected (Synonym)
152			Interferes with dipeptidase activity in cotyledons during germination	*Cucurbita maxima* [13]
153	Diquat [6,7-Dihydrodipyrido/1,2-a:2′,1′-c/-pyrazine-diium ion]	400-440	Phytotoxicity due to photochemically produced, stable, free radicals	Several plants [19]
154			Inhibits NADPH & cyclic ATP formation in illuminated, isolated chloroplasts	*Beta vulgaris* [262,273]
155			Interferes with O$_2$ evolution in photosynthesis by acting as a Hill-reaction oxidant	*Scenedesmus* [200]
156			Retards synthesis of bacterial chlorophyll & some proteins more than that of other cellular constituents	*Rhodospirillum rubrum* [118]
157			Causes bleaching of chlorophyll & damage to plastid membranes	*Chlorella vulgaris* [236]
158	Diuron [3-(3,4-Di-chloro-phenyl)-1,1-di-methyl-urea]	3400	Strongly inhibits the Hill reaction, stopping photosynthetic O$_2$ evolution & phosphorylation	Numerous plants [58,207]
159			Inhibits electron transport in chromatophores at a site between NADH & cytochrome *b*	*Rhodospirillum rubrum* [264]
160			Degrades by demethylation to a monomethyl derivative, and finally to an aniline derivative	*Glycine max* [241]
161				*Gossypium hirsutum* [241]
162			Converted to 3,4-dichloronitrobenzene, 3-(3,4-di-chlorophenyl)-1-methylurea, & 3,4-dichloroaniline metabolites	*Zea mays* [183]
163			Inhibits electron transport pathway of mitochondria between cytochromes *b* & *c*; stops O$_2$ uptake	*Saccharomyces cerevisiae* [91,112]
164			High application rates inhibit nitrification by blocking oxidation of NO$_2^-$ to NO$_3^-$	*Nitrobacter* sp. [56]
165			Causes reduction in no. of chloroplasts/cell & gross alterations in structure of remaining chloroplasts in light-grown *Euglena*	*Euglena gracilis* [223]
166			Causes bleaching of chlorophyll in isolated chloroplasts exposed to light. This action blocked by ascorbic acid & 2,6-dichloroindophenol (2,6-dichlorophenol-indophenol).	*Spinacia* sp. [239]
167			Increases activities of catalase, *o*-diphenol oxidase (polyphenol oxidase), & peroxidase in sensitive species	*Amaranthus retroflexus* [113]
168	DSMA [Di-sodium methane-arsonate]	1800	Extracts from treated plants indicate that the carbon-arsenic bond remains intact and is not degraded	*Cynodon dactylon* [63]
169	Endothal [7-Oxabi-cyclo/2.2.1/hep-tane-2,3-dicar-boxylic acid]	206	Interferes with RNA metabolism, resembling inhibition caused by dactinomycin (actinomycin D); inhibits development of proteolytic & dipeptidase activities in cotyledons of intact embryos	*Cucurbita maxima* [13,191]
170	EPTC [S-Ethyldi-propyl-thiocar-bamate]	1652	Mitotic poison when applied to soil, mainly inhibiting cell division in root meristem	Numerous plants [104]
171			Inhibits surface wax formation on developing leaves of cabbage & pea [81], but not stem [151] & petiole tissue [256]	*Brassica oleracea* [81]
172				*Cassia obtusifolia* [256]
173				*Pisum sativum* [81]
174				*Sesbania exaltata* [151]

continued

Part V. Herbicides

	Herbicide [Synonym]	Rat Oral LD$_{50}$ mg/kg	Mode of Action	Organism Affected (Synonym)
175	Fluometuron [1,1-Di-	♂, 8900; ♀, 7900	Pronounced cytokinin-like activity, e.g., retardation of senescence	*Zea mays* [71]
176 177 178	methyl-3-(α,α,α-tri-fluoro-*m*-tolyl)urea; 1,1-dimeth-		Degraded by demethylation, forming first demethyl-fluometuron & 3-(*m*-trifluoromethylphenyl)urea (TFMPU), then several unidentified metabolites; TFMPU & demethylfluometuron half as toxic as fluometuron	*Gossypium hirsutum* [203] *Triticum aestivum* [203] *Zea mays* [177]
179	yl-3-(*m*-tri-fluoro-methyl-phenyl)-		Does not influence respiration or light-independent reactions, but interferes with light-dependent chloro-phyll synthesis & photosynthetic O$_2$ evolution	*Chlorella pyrenoidosa* [226]
180	phenyl)-urea]		Causes increase in second-state 4-oxidation of both malate & succinate in isolated mitochondria; over-comes inhibition of oxidative phosphorylation caused by oligomycin, in-dicating an uncoupling response similar to that produced by dinitrophenol	*Glycine max* [163]
181	Ioxynil [4-Hydroxy-3,5-diiodo-benzoni-trile]	110	With succinate as substrate for mitochondrial oxida-tion, ioxynil stimulates both ADP-limited & ADP$^+$-phosphate-deficient O$_2$ utilization, circumvents oligo-mycin-inhibited respiration, and slightly stimulates mitochondrial ATPase activity	*Solanum tuberosum* [68]
182			Inhibits photosynthetic electron transport—inhibits pigment system II at low concentrations, and blocks pigment system I at higher concentrations	Numerous plants [86]
183 184 185			Causes increase in soluble amino acids and decrease in proteins [188]; inhibits protein & RNA synthesis [169]	*Glycine max* [169] *Triticum aestivum* [188] *Zea mays* [169]
186			Causes chlorophyll loss following infiltration into leaf discs	*Hordeum vulgare* [59]
187 188	Linuron [3-(3,4-Di-chloro-phenyl)-1-methoxy-1-methylurea]	1500	Degraded to 3-(3,4-dichlorophenyl)-1-methoxyurea & 3,4-dichloroaniline in light-grown plants; initial step is demethylation. Some of the herbicide is bound and not extractable.	*Glycine max* [175] *Zea mays* [175]
189			Strongly inhibits the Hill reaction	Numerous plants [58]
190	MCPA [*l*(4-Chloro-*o*-tolyl)oxy*l*-acetic acid]	700	High concentration (0.01 *M*) causes fragmentation of mitochondria into spherical granules, indicating that mitochondria attempt to improve respiration by in-creasing their surface area	*Vicia faba* [129]
191			Causes marked uncoupling of oxidative phosphoryl-ation	*Zea mays* [60]
192			Inhibits growth, respiration, & phosphate uptake	Algae [128]
193	MH [Maleic hydrazide; 1,2-dihy-dro-3,6-pyridazine-dione]	6950	Becomes fixed within the plant and is not significant-ly metabolized, being bound to the 80%-ethanol-insoluble fraction of the root tissue and possibly fixed to the cell wall	*Zea mays* [179]
194			Mutagenic to blue-green algae at pH 5, producing mutations to streptomycin resistance; not mutagenic at pH 8.0	*Anacystis* sp. [90]
195			Affects cell division but not elongation in root cells; inhibition not alleviated by cysteine, kinetin, nucleo-sides, purines, pyridoxal, pyruvate, or pyrimidines	*Zea mays* [178]

continued

Part V.   Herbicides

	Herbicide [Synonym]	Rat Oral LD$_{50}$ mg/kg	Mode of Action	Organism Affected (Synonym)
196 197 198			Depresses activities of glucose-1-phosphatase, glucose-6-phosphatase, & ATPase, but significantly stimulates activities of enzymes of sucrose biosynthesis [261]; stimulates acid phosphatase activity [11], but decreases catalase activity	*Beta vulgaris* [261] *Hordeum vulgare* [11] *Zea mays* [80]
199 200 201 202 203	Monuron [3-(*p*-Chloro-phenyl)-1,1-di-methyl-urea]	3600	*N*-Demethylated & subsequently hydrolyzed to anilines by a microsomal oxidase system from leaves	*Gossypium hirsutum* [48,75,241]
			Inhibits electron transport pathway of mitochondria between cytochromes *b* & *c*; inhibits O$_2$ uptake	*Saccharomyces cerevisiae* [91]
			Strongly inhibits the Hill reaction, inhibiting O$_2$ evolution more than ferricyanide reduction	*Pisum sativum* [120]
			Causes partial closure of stomata	*Rumex conglomeratus* [5]
			Inhibits proteolytic activity of plant homogenates—possibly forming a stable enzyme-inhibitor complex	Numerous plants [214]
204	Nitralin [4-(Methyl-sulfonyl)-2,6-dinitro-*N,N*-dipro-pylaniline]	<2000	Inhibits root growth & cell division as evidenced by marked reduction of mitotic figures, inhibition of cell-wall formation, & multinucleated cells	*Zea mays* [47,82]
205 206 207	Nitrofen [2,4-Di-chloro-phenyl *p*-nitro-phenyl ether]	1470 ± 365	Interferes with oxidative & photosynthetic phosphorylation, acting primarily as inhibitor of chloroplast noncyclic electron transport & coupled photophosphorylation; in mitochondria, acting primarily as electron transport inhibitor with malate, NADH, & succinate as substrates	*Solanum tuberosum* [170] *Spinacia oleracea* [170]
			Requires light for activation, the mechanism of action being a photobiochemical process involving chloroplast xanthophylls	*Oryza sativa* [161]
208			Acts on cell & organelle membrane permeability, the effects on photosynthesis & respiration being dependent upon permeability changes in the organelles	Several plants [194]
209 210 211 212	Paraquat [1,1'-Di-methyl-4,4'-bi-pyridin-ium ion]	150	Phytotoxicity apparently results from stable free radicals produced by photochemical reactions, making paraquat an effective contact herbicide which kills green tissue	Numerous plants [19]
			Interferes with NADP$^+$ reduction & cyclic ATP formation in illuminated chloroplasts	*Beta vulgaris* [273]
			Not appreciably metabolized by leaves, but may be degraded photolytically	Numerous plants [229]
			Causes disintegration of the plasmalemma followed by disruption of chloroplast membranes and loss of turgidity in mesophyll cells	*Prosopis juliflora* [31]
213	PCP [Penta-chloro-phenol]	27-80	Causes decrease in mitotic index as compared to controls; anomalies include disturbed metaphase, lagging chromosomes, & cytomixis	*Vicia faba* [7]

continued

Part V. Herbicides

	Herbicide [Synonym]	Rat Oral LD$_{50}$ mg/kg	Mode of Action	Organism Affected (Synonym)
214	Phenmedi-pham	8000	Degraded to N-(3-hydroxyphenyl)-methylcarbamate & m-methylaniline	Several plants [119]
215 216	[Methyl m-hydroxy-carbanilate		Causes rapid decrease of CO_2 assimilation in both resistant (sugar beet) & susceptible (mustard) plants; however, CO_2 fixation returns to normal in sugar beet	*Beta vulgaris* [133] *Brassica* sp. [133]
217	m-methyl-carbanilate]		Interferes with Hill reaction in both resistant & susceptible species; no effect on other isolated enzymes such as catalase, peroxidase, & aminotransferases (transaminases)	*Beta vulgaris* [132,257]
218	Picloram [4-Amino-3,5,6-tri-chloro-picolinic acid]	8200	Possesses properties of growth regulator & auxin, substituting as auxin source in growth of callus tissue & coleoptile sections	Several plants & plant parts [85, 121,209]
219 220			Increases nucleic acids in sensitive plants as a result of lowering activities of deoxyribonuclease & ribonucleases	*Cucumis sativus* [150] *Glycine max* [150]
221 222			Stimulates RNA & protein synthesis in soybean (sensitive species) & barley	*Glycine max* [218] *Hordeum vulgare* [218]
223 224			Causes loss of leaf movement & epinastic curvature in mesquite leaves & stems by increase in ethylene production—same symptoms as ethylene fumigation	*Acacia farnesiana* [30] *Prosopis juliflora* [30]
225			Stimulates growth in coleoptile sections at 10^{-6} to $10^{-4}M$, but inhibits at higher concentrations; extensively depresses geotropic curvature at $10^{-5}M$	*Avena sativa* [209]
226			Causes swelling of mitochondria previously contracted by ATP; does not affect succinate-induced contraction; causes no additional swelling of mitochondria in 0.3 M KCl medium	*Hordeum vulgare* [50]
227			Markedly increases water uptake by seeds	Numerous plants [37]
228			Interferes with stomatal behavior; disturbs water metabolism	Numerous plants [51]
229 230			Significantly reduces photosynthetic CO_2 fixation by excised leaves [217], and translocation of photosynthetic assimilates [139]	*Glycine max* [217] *Vitis vinifera* [139]
231 232			Causes coagulation of cytoplasm of staminal hair cells & stipular cells; stops chloroplast division	*Tradescantia* sp. [208] *Vicia faba* [208]
233			Causes cessation of apical meristematic activity; alters internal anatomy of leaves & stem	*Phaseolus vulgaris* [69]
234			May chelate free and/or bound metallic ions, leading to inhibition of activities of metal-requiring enzymes	Several plants [49]
235 236			Reduces soluble protein concentration in monocots, but no effect on dry wt; protein of treated plants is quantitatively & qualitatively changed as compared to controls (shown by electrophoretic separation of soluble proteins)	*Glycine max* [49] Monocots [32]
237 238	Prometryne [2,4-Bis-(isopropyl-amino)-6-(methyl-thio)-s-triazine]	3750	Significantly inhibits photosynthetic CO_2 fixation in excised leaves of cotton (moderately tolerant) & soybean (sensitive); no effect on dark CO_2 fixation; inhibition of the Hill reaction	*Glycine max* [225] *Gossypium hirsutum* [225]
239			Decreases second-state 3-oxidation of both malate & succinate in isolated mitochondria, which is overcome by 2,4-dinitrophenol, indicating a response similar to that of oligomycin	*Glycine max* [163,244]

continued

Part V. Herbicides

Herbicide [Synonym]	Rat Oral LD$_{50}$ mg/kg	Mode of Action	Organism Affected (Synonym)
240		Replaces either thymine or uracil in nucleic acid molecules	*Escherichia coli* [243]
241 Propachlor [2-Chloro-*N*-isopropylacet-anilide]	1200	May prevent transfer of aminoacyl tRNA to growing polypeptide chains, or inhibit activation of amino acids to form aminoacyl tRNA	*Cucurbita maxima* [61]
242		Converted to glutathione conjugate & α-glutamyl-cysteine conjugate	*Zea mays* [138]
243 Propanil [3',4'-Di-chloropro-pionanilide]	2270	Influences synthesis of both DNA & RNA and their breakdown in excised roots	*Zea mays* [18]
244 245		Hydrolyzed to 3',4'-dichloroaniline by aryl acylamidase (aryl-acylamine amidohydrolase) in rice [3,77]; aniline moiety recovered in several compounds including sugars & lignin conjugates [266]; barnyard grass unable to detoxify propanil and it accumulates to lethal proportions [265]	*Echinochloa crusgalli* [265,266] *Oryza sativa* [3,77]
246		In soils, converted to tetrachloroazobenzene [141]	....
247		Depresses nuclear division and causes some chromosomal aberrations in root cells	*Allium cepa* [196]
248		Inhibits Hill reaction (O$_2$-liberating pathway) in photosynthesis	*Spinacia* sp. [173]
249 250		Severely inhibits water uptake & transpiration of rice & barnyard grass immediately after treatment, but rice recovers after 1 hr, whereas barnyard grass does not	*Echinochloa crusgalli* [174] *Oryza sativa* [174]
251 252		Inhibits ATP incorporation into RNA, leucine incorporation into protein, and gibberellin-controlled induction of α-amylase	*Glycine max* [169] *Zea mays* [169]
253 Propham [Isopropyl carbanilate]	5000	Affects alignment & orientation of spindle apparatus in endosperm cells; does not modify morphology of microtubules & other cellular organelles	*Haemanthus katharinae* [96]
254		Produces cytostatic effect in dividing plant cells; acts as a mitotic poison	Numerous plants [152]
255		Inhibits kinetin-stimulated proliferation in pith culture, but not indoleacetic acid-induced cell enlargement	*Nicotiana tabacum* [122]
256		Inhibits Hill reaction in isolated chloroplasts	Several plants [168]
257		Blocks gibberellin-induced synthesis of amylase in aleurone	*Hordeum vulgare* [154]
258 Pyrazon [5-Amino-4-chloro-2-phenyl-3(2*H*)-pyridazi-none]	3000	Inhibits O$_2$ evolution (photosynthesis) from illuminated leaf disks	Numerous plants[2/] [66,182]
259 260 261		Blocks Hill reaction is isolated chloroplasts; does not affect respiration rates of roots & leaf disks	*Beta vulgaris* [74] *Chenopodium album* [74] *Hordeum vulgare* [107]
262		Causes chloroplasts to become spherical & swollen; inhibits grana formation; causes thylakoids to become swollen & perforated	*Phaseolus vulgaris* [10]
263		Converted in shoots of sugar beet (resistant species) to *N*-glucose pyrazon, or *N*-[2-chloro-4-phenyl-3(2*H*)-pyridazinone] glucosamine & 5-amino-4-chloro-3(2*H*)-pyridazinone (ACP)	*Beta vulgaris* [234]
264		Increases DNA & RNA in treated leaves	*Brassica oleracea* [205]

2/ 4- to 5-wk-old plants.

continued

Part V. Herbicides

	Herbicide [Synonym]	Rat Oral LD$_{50}$ mg/kg	Mode of Action	Organism Affected (Synonym)
265	Pyrichlor [2,3,5-Tri-chloro-4-pyridinol]	1200	Inhibits O$_2$ evolution & noncyclic photophosphoryla-tion in isolated chloroplasts, suggesting that pigment system II is affected	*Spinacia oleracea* [164]
266			Inhibits both coupled & uncoupled reactions, suggest-ing action on a nonphosphorylating intermediate close to the electron carrier chain	*Glycine max* [125]
267			Leads to progressive destruction of chloroplast ultra-structure, swelling of the fret system, and loss of starch; does not affect shape & size of mitochondria	*Nicotiana tabacum* [83]
268			Affects carotenoid synthesis in etiolated seedlings	*Triticum aestivum* [41]
269	Siduron [1-(2-Methyl-cyclo-hexyl)-3-phenyl-urea]	>5000	Disrupts normal nucleic acid metabolism necessary for protein & cell-wall synthesis in roots	*Hordeum vulgare* [232]
270	Simazine[3/] [2-Chloro-4,6-bis-(ethyl-amino)-s-triazine]	5000	Plants treated with subherbicidal concentrations and grown under suboptimal temperatures and low NO$_3^-$ levels show increases in total N, dry weight, & nitrate reductase activity	*Zea mays* [6,248]
271			Inhibits photosynthesis, interfering with the Hill re-action & photophosphorylation	Numerous plants [36,104]
272			Apparently affects respiration less than photosynthesis	*Oryza sativa* [247]
273			Reduces pigment content—mainly xanthophyll (lu-tein), carotene, & chlorophyll *a*—of algal suspensions	*Chlorella vulgaris* [186]
274			Disturbs normal catabolism; sugars decrease but as-partic & citric acids increase	Numerous plants [84]
275			Reduces chloroplast protein & chlorophyll content in treated plants	*Avena* sp. [228]
276	Solan [3'-Chloro-2-methyl-*p*-valeroto-luidide]	10,000	Strongly inhibits Hill reaction and cyclic & noncyclic photophosphorylation in isolated chloroplasts	*Spinacia oleracea* [222]
277	2,4,5-T[4/] [(2,4,5-Tri-chloro-phenoxy)-acetic acid]	300	Lengthens duration of mitotic cycle, primarily be-cause it extends duration of DNA synthesis	*Vicia faba* [147]
278	2,3,6-TBA [2,3,6-Tri-chloroben-zoic acid]	750-1000	Inhibits geotropic & phototropic responses in shoots. A strong auxin which resists biological breakdown in plants.	Monocots [249]
279 280	TCA [Tri-chloroace-tic acid]	5000	Reduces indoleacetic acid content of sensitive plants but not of resistant varieties; increases tryptophan content of plants, especially of sensitive varieties	*Lupinus luteus* [187] *Pisum sativum* [187]
281			Inhibits hydrolytic activity of proteolytic enzymes; may inhibit enzymes which counteract toxic effects of free NH$_3$ by converting it to amides	*Lupinus luteus* [159]

[3/] For a comprehensive review of metabolic pathways in plants, animals, and soils, consult reference 89. [4/] Addi-tional metabolic pathways for 2,4,5-T are considered in many of the references for 2,4-D (*see* entries 95-122).

continued

Part V. Herbicides

Herbicide [Synonym]	Rat Oral LD$_{50}$ mg/kg	Mode of Action	Organism Affected (Synonym)
282 Trifluralin [α,α,α-Tri-fluoro-2,6-dinitro-N,N-dipropyl-p-tolu-idine]	10,000	Suppresses synthesis of DNA, RNA, & protein in root tips, but shoots appear unaffected	*Zea mays* [213]
283 284		Concentration of $10^{-4} M$ inhibits mitochondrial respiration & oxidative phosphorylation ∿25% in sorghum (susceptible) & maize (less susceptible)	*Sorghum bicolor (S. vulgare)* [176] *Zea mays* [176]
285		May affect sulfhydryl groups, thus inhibiting cell division	*Zea mays* [216]
286		Induces changes in fatty acid components of seeds	*Glycine max* [193]
287 288		Inhibits lateral root formation; causes radial enlargement of cortical, pericycle, & endodermal root cells	*Gossypium hirsutum* [92] *Zea mays* [213]
289		Disrupts mitosis (acts as a mitotic poison); multinucleate cells may occur in meristematic regions of shoots & roots	Several plants [34,213]
290		Causes pronounced change in no. & organization of palisade cells in leaves; xylem elements appear less organized and pericycle walls extremely thick	*Glycine max* [5/] [134]

5/ 7-wk-old plants.

Contributors: Bartels, Paul G., and Hull, Herbert M.

References

[1] Abeles, F. B. 1968. Weed Sci. 16:498.

[2] Abrens, J. F., and O. A. Leonard. 1970. Abstr. Weed Soc. Amer., p. 51.

[3] Akatsuka, T., et al. 1969. Ibaraki Daigaku Nogakubu Gakujutsu Hokoku 17:45.

[4] Akhavein, A. A., and D. L. Linscott. 1968. Residue Rev. 23:97.

[5] Allaway, W. G., and T. A. Mansfield. 1967. New Phytol. 66:57.

[6] Allinson, D. W., and R. A. Peters. 1970. Agron. J. 62:246.

[7] Amer, S. M., and E. M. Ali. 1970. Cytologia 34:533.

[8] Anastasia, F. B., et al. 1970. Proc. Northeast. Weed Contr. Conf., p. 102.

[9] Anderson, J. L. 1971. Abstr. Weed Soc. Amer., p. 60.

[10] Anderson, J. L., and J. P. Schaelling. 1970. Weed Sci. 18:455.

[11] Antonielli, M. 1967. G. Bot. Ital. 73:169.

[12] Arntzen, C. J., et al. 1970. Bot. Gaz. (Chicago) 131:14.

[13] Ashton, F. M., and R. C. Tsay. 1970. Abstr. Weed Soc. Amer., p. 34.

[14] Ashton, F. M., et al. 1963. Bot. Gaz. (Chicago) 124:336.

[15] Ashton, F. M., et al. 1968. Weed Sci. 16:169.

[16] Ashton, F. M., et al. 1969. Weed Res. 9:198.

[17] Audus, L. J. 1967. Agrochimica 11:309.

[18] Baker, J. B., and T. D. Pizzalato. 1968. Plant Physiol. 43:S-3.

[19] Baldwin, B. C. 1968. Proc. Brit. Weed Contr. Conf., 9th, p. 102.

[20] Banting, J. D. 1970. Weed Sci. 18:80.

[21] Barrier, G. E., et al. 1970. Herbicide Handbook of the Weed Society of America. W. F. Humphrey, Geneva, N. Y.

[22] Bartels, P. G., and A. Hyde. 1970. Plant Physiol. 46:825.

[23] Bartels, P. G., and E. J. Pegelow, Jr. 1968. J. Cell Biol. 37:1.

[24] Bartels, P. G., and T. E. Weier. 1969. Amer. J. Bot. 56:1.

[25] Bartels, P. G., and F. T. Wolf. 1965. Physiol. Plant. 18:805.

[26] Bartels, P. G., and F. T. Wolf. 1967. Biochim. Biophys. Acta 136:166.

[27] Bartels, P. G., et al. 1967. Plant Physiol. 42:736.

[28] Barth, A., and H. J. Michel. 1969. Pharmazie 24:11.

[29] Basler, E., and G. D. Wills. 1968. Proc. Okla. Acad. Sci. 47:1.

[30] Baur, J. R., and P. W. Morgan. 1969. Plant Physiol. 44:831.

[31] Baur, J. R., et al. 1969. Weed Res. 9:81.

[32] Baur, J. R., et al. 1970. Agron. J. 62:627.

[33] Baxter, R., and J. B. Hanson. 1968. Planta 82:246.

[34] Bayer, D. E., et al. 1967. Amer. J. Bot. 54:945.

[35] Bingham, S. W. 1968. Weed Sci. 16:449.

[36] Bishop, N. I. 1962. Biochim. Biophys. Acta 57:186.

[37] Biswas, P. K., and R. L. Haynes. 1970. Physiol. Plant. 23:583.

[38] Boguslawski, J., et al. 1967. Acta Biochim. Pol. 14:133.

[39] Bounds, H. C., et al. 1969. Can. J. Microbiol. 15:1121.

[40] Brown, J. C., and M. C. Carter. 1968. Weed Sci. 16:222.

[41] Burns, E. R., et al. 1971. Plant Physiol. 47:144.

[42] Burstrom, H. G. 1968. Physiol. Plant. 21:1137.

[43] Burt, G. W., and T. J. Muzik. 1970. Ibid. 23:498.

[44] Casida, J. E., and L. Lykken. 1969. Annu. Rev. Plant Physiol. 20:607.

[45] Casselton, P. J. 1966. Physiol. Plant. 19:411.

continued

85. BIOCIDES

Part V. Herbicides

[46] Castelfranco, P., and T. Bisalputra. 1965. Amer. J. Bot. 52(3):222.

[47] Cathie, L. S. 1969. Meded. Rijksfac. Landbouwwetensch. Gent. 34:1045.

[48] Chakraborty, T. K., and V. S. Mani. 1968. Pest Articles News Sum., C, 14(14):364.

[49] Chang, I. K., and C. L. Foy. 1970. Abstr. Weed Soc. Amer., pp. 14, 35.

[50] Chang, I. K., and C. L. Foy. 1971. Weed Sci. 19:54.

[51] Chang, I. K., and C. L. Foy. 1971. Ibid. 19:58.

[52] Chesalin, G. A., and A. A. Timofeeva. 1965. Dokl. Vses. Akad. Sel'skokhoz. Nauk (6):10.

[53] Chkanikov, D. I., and V. I. Kostina. 1968. Khim. Sel. Khoz. 6:536.

[54] Chkanikov, D. I., and O. D. Mikityuk. 1970. Fiziol. Rast. 17:757.

[55] Chkanikov, D. I., et al. 1968. Khim. Sel. Khoz. 6:41.

[56] Corke, C. T., and F. R. Thompson. 1970. Can. J. Microbiol. 16:567.

[57] Couch, R. W., and D. E. Davis. 1966. Weeds 14:251.

[58] Crafts, A. S. 1961. The Chemistry and Mode of Action of Herbicides. Interscience, New York.

[59] Davies, P. J., et al. 1968. Weed Res. 8:241.

[60] Deeva, V. P. 1967. Dokl. Akad. Nauk Beloruss. SSR 11:74.

[61] Dhillon, N. S., et al. 1970. Abstr. Weed Soc. Amer., p. 62.

[62] Doerfling, P., et al. 1970. Experientia 26:728.

[63] Duble, R. L., et al. 1969. J. Agr. Food Chem. 17:1247.

[64] Ebert, E., and C. J. van Assche. 1969. Experientia 25:758.

[65] Ebert, E., and P. W. Muller. 1968. Ibid. 24:1.

[66] Eshel, Y. 1969. Weed Res. 9:167.

[67] Favali, M. A., and G. Conti. 1970. Protoplasma 70:153.

[68] Ferrari, T. E., and D. E. Moreland. 1969. Plant Physiol. 44:429.

[69] Fisher, D. A., et al. 1968. Bot. Gaz. 129:67.

[70] Fites, R. C., et al. 1969. Ibid. 130:118.

[71] Foy, C. L., and H. Hiranpradit. 1970. Abstr. Weed Soc. Amer., p. 38.

[72] Foy, C. L., and D. Penner. 1965. Weeds 13:226.

[73] Foy, C. L., et al. 1971. In P. L. Altman and D. S. Dittmer, ed. Respiration and Circulation. Federation of American Societies for Experimental Biology, Bethesda, Md. pp. 742-791.

[74] Frank, R., and C. M. Switzer. 1969. Weed Sci. 17:344.

[75] Frear, D. S. 1968. Science 162:674.

[76] Frear, D. S. 1968. Phytochemistry 7:381.

[77] Frear, D. S., and G. G. Still. 1968. Ibid. 7:913.

[78] Fredrick, J. F. 1969. Ibid. 8:725.

[79] Fredrick, J. F., and A. C. Gentile. 1967. Ann. N.Y. Acad. Sci. 144:362.

[80] Gaspar, T., and O. Cledjo. 1969. Weed Res. 9:348.

[81] Gentner, W. A. 1966. Weeds 14:27.

[82] Gentner, W. A., and L. G. Burk. 1968. Weed Sci. 16:259.

[83] Geronimo, J., and J. W. Herr. 1970. Ibid. 18:48.

[84] Gnanarethinam, J. L., and J. Carles. 1966. C. R. Soc. Biol. 160:1096.

[85] Goodin, J. R., and F. L. Beecher. 1967. Plant Physiol. 42:S-23.

[86] Gromet-Elhanan, Z. 1968. Progr. Photosyn. Res. 3:1197.

[87] Guillot-Salomon, T., et al. 1967. Bull. Soc. Fr. Physiol. Veg. 13:63.

[88] Gunar, I. I., et al. 1968. Izv. Timiryazev. Sel'skokhoz. Akad. (1):3.

[89] Gunther, F. A., and J. D. Gunther. 1970. Residue Rev. 32:413.

[90] Gupta, R. S., and H. D. Kumar. 1970. Arch. Mikrobiol. 70:330.

[91] Hachimori, A., et al. 1968. J. Biochem. (Tokyo) 64:119.

[92] Hacskaylo, J., and V. A. Amato. 1968. Weed Sci. 16:513.

[93] Hagin, R. D., et al. 1970. J. Agr. Food Chem. 18:848.

[94] Hallam, N. D. 1970. J. Exp. Bot. 21:1031.

[95] Hanson, J. B., and F. W. Slife. 1969. Residue Rev. 25:59.

[96] Helper, P. K., and W. T. Jackson. 1969. J. Cell Sci. 5:727.

[97] Herichova, A. 1968. Biologia (Bratislava) 23:536.

[98] Herrett, R. A., and R. V. Berthold. 1965. Science 149:191.

[99] Hewitt, E. J., and B. A. Notton. 1966. Biochem. J. 101:39.

[100] Hill, E. R., et al. 1968. Weed Sci. 16:377.

[101] Hilton, J. L. 1969. J. Agr. Food Chem. 17:182.

[102] Hilton, J. L., and D. D. Kaufman. 1967. Weeds 15:255.

[103] Hilton, J. L., and D. D. Kaufman. 1968. Weed Sci. 16:152.

[104] Hilton, J. L., et al. 1963. Annu. Rev. Plant Physiol. 14:353.

[105] Hilton, J. L., et al. 1964. Weeds 12:129.

[106] Hilton, J. L., et al. 1965. Arch. Biochem. Biophys. 112:544.

[107] Hilton, J. L., et al. 1969. Weed Sci. 17:541.

[108] Holm, R. E., and F. B. Abeles. 1968. Planta 78:293.

[109] Homma, S. 1969. Nippon Sanshigaku Zasshi 38:75.

[110] Hull, H. M. 1968. In P. L. Altman and D. S. Dittmer, ed. Metabolism. Federation of American Societies for Experimental Biology, Bethesda, Md. pp. 477-484.

[111] Hull, H. M. 1970. Residue Rev. 31:1.

[112] Inoue, Y., et al. 1967. Agr. Biol. Chem. 31:422.

[113] Islamov, I. 1969. Uzb. Biol. Zh. 13:22.

[114] James, C. S., and G. N. Prendeville. 1969. J. Agr. Food Chem. 17:1257.

continued

Part V. Herbicides

[115] Jaworski, E. G. 1969. Ibid. 17:165.

[116] Jones, R. J., and R. R. Hewitt. 1971. Can. J. Microbiol. 17:87.

[117] Kadman-Zahavi. A. 1970. Isr. J. Bot. 19:558.

[118] Kaneshiro, T., and G. Zweig. 1965. Appl. Microbiol. 13:939.

[119] Kassenbeer, H. 1969. Abstr. Schering Akt. Ges., p. 5.

[120] Katyurin, V. M., et al. 1969. Sov. Plant Physiol. 16:149.

[121] Kefford, N. P., and O. H. Caso. 1966. Bot. Gaz. 127:159.

[122] Keitt, G. W. 1967. Physiol. Plant. 20:1076.

[123] Kemp, T. R., et al. 1969. Weed Sci. 17:444.

[124] Kesner, C. D., and S. K. Ries. 1967. Science 155:210.

[125] Killion, D. D., and R. E. Frans. 1969. Weed Sci. 17:468.

[126] Kim, W. K., and R. G. S. Bidwell. 1967. Can. J. Bot. 45:1751.

[127] Kim, W. K., and R. G. S. Bidwell. 1967. Ibid. 45:1789.

[128] Kirkwood, R. C., and W. W. Fletcher. 1970. Weed Res. 10:3.

[129] Klasova, A. 1968. Biologia (Bratislava) 23:61.

[130] Klopotowski, T., and G. Bagdasarian. 1966. Acta Biochim. Pol. 13:153.

[131] Klopotowski, T., and A. Wiater. 1965. Arch. Biochem. Biophys. 112:562.

[132] Kötter, C. 1969. Abstr. Schering Akt. Ges., p. 15.

[133] Kötter, C., and F. Arndt. 1968. J. Int. Inst. Sugar Beet Res. 3:126.

[134] Kust, C. A., and B. E. Struckmeyer. 1971. Weed Sci. 19:147.

[135] Ladonin, V. F., and K. M. Svittser. 1967. Sov. Plant Physiol. 14:853.

[136] Lamartiniere, C. A., et al. 1969. Bull. Environ. Contam. Toxicol. 4:113.

[137] Lamoureux, G. L., et al. 1970. J. Agr. Food Chem. 18:81.

[138] Lamoureux, G. L., et al. 1971. Ibid. 19:346.

[139] Leonard, O. A., et al. 1967. Weed Res. 7:208.

[140] Liang, G. H. L., et al. 1969. Weed Sci. 17:8.

[141] Lieb, H. B., and C. C. Still. 1969. Plant Physiol. 44:1672.

[142] Linscott, D. L., et al. 1968. J. Agr. Food Chem. 16:844.

[143] Loos, M. A. 1969. Degradation of Herbicides. M. Dekker, New York.

[144] Lotlikar, P. D., et al. 1968. Weed Sci. 16:161.

[145] Lund-Höie, K. 1970. Weed Res. 10:367.

[146] Lund-Höie, K., and D. E. Bayer. 1968. Physiol. Plant. 21:196.

[147] Macleod, R. D. 1969. Chromosoma 27:327.

[148] Magalhaes, A. C., and F. M. Ashton. 1969. Weed Res. 9:48.

[149] Makovcova, O. 1967. Meded. Rijksfac. Landbouwwetensch. Gent. 32:1036.

[150] Malhotra, S. S., and J. B. Hanson. 1970. Weed Sci. 18:1.

[151] Mann, J. D., and M. Pu. 1968. Ibid. 16:197.

[152] Mann, J. D., and W. B. Starey. 1966. Cytologia 31:203.

[153] Mann, J. D., et al. 1965. Weeds 13:63.

[154] Mann, J. D., et al. 1967. Biochim. Biophys. Acta 138:133.

[155] Mann, J. D., et al. 1967. Nature (London) 213:420.

[156] Mann, J. D., et al. 1967. Plant Cell Physiol. 8:613.

[157] Martin, J. T., and B. E. Juniper. 1970. Cuticles of Plants. St. Martin, New York.

[158] Mashtakov, S. M. 1968. Sel'skokhoz. Biol. 3:64.

[159] Mashtakov, S. M., and P. A. Moshchuk. 1967. Agrokhimiya (9):80.

[160] Massini, P. 1963. Acta Bot. Neer. 12:64.

[161] Matsunaka, S. 1969. J. Agr. Food Chem. 17:171.

[162] Matsunaka, S. 1969. Residue Rev. 25:45.

[163] McDaniel, J. L., and R. E. Frans. 1969. Weed Sci. 17:192.

[164] Meikle, R. W. 1970. Ibid. 18:475.

[165] Mitchell, J. W., and P. J. Linder. 1963. Residue Rev. 2:51.

[166] Montgomery, M. L., et al. 1969. J. Agr. Food Chem. 17:1241.

[167] Moreland, D. E., and W. J. Blackmon. 1970. Weed Sci. 18:419.

[168] Moreland, D. E., and K. L. Hill. 1959. J. Agr. Food Chem. 7:832.

[169] Moreland, D. E., et al. 1969. Weed Sci. 17:556.

[170] Moreland, D. E., et al. 1970. Ibid. 18:636.

[171] Mozhaeva, L. V., and N. V. Pil'shchikova. 1968. Izv. Timiryazev. Sel'skokhoz. Akad. (6):3.

[172] Nagarajan, M., and J. S. Rao. 1970. Madras Agr. J. 57:67.

[173] Nakamura, H., and S. Matsunaka. 1969. Weed Res. (Jap.) 8:33.

[174] Nakamura, H., et al. 1968. Ibid. 7:100.

[175] Nashed, R. B., and R. D. Ilnicki. 1970. Weed Sci. 18:25.

[176] Negi, N. S., et al. 1968. Ibid. 16:83.

[177] Neptune, M. D., and H. H. Funderburk, Jr. 1969. Proc. Annu. Meet. S. Weed Sci. Soc. 22:369.

[178] Nooden, L. D. 1969. Physiol. Plant. 22:260.

[179] Nooden, L. D. 1970. Plant Physiol. 45:46.

[180] O'Brien, T. J., et al. 1968. Biochim. Biophys. Acta 169:35.

[181] Olech, K. 1966. Ann. Univ. Mariae Curie-Sklodowska E21:289.

[182] Olech, K. 1967. Ibid. E22:185.

[183] Onley, J. H., et al. 1968. J. Agr. Food Chem. 16:426.

[184] Oyolu, C., and R. C. Huffaker. 1964. Crop Sci. 4:95.

[185] Palm, C. E., et al. 1968. Weed Control. National Academy of Sciences, Washington, D.C. v. 2.

[186] Paromenskaya, I. N., and G. N. Lyalin. 1968. Sov. Plant Physiol. 15:842.

[187] Parshakova, Z. P., and S. M. Mashtakov. 1967. Dokl. Akad. Nauk Beloruss. SSR 11:271.

[188] Paxton, D., and S. Zalik. 1968. Can. J. Bot. 46:89.

continued

Part V. Herbicides

[189] Penner, D. 1968. Weed Sci. 16:519.

[190] Penner, D. 1970. Ibid. 18:360.

[191] Penner, D., and F. M. Ashton. 1968. Ibid. 16:323.

[192] Penner, D., and R. Early. 1971. Abstr. Weed Soc. Amer., p. 108.

[193] Penner, D., and W. Meggitt. 1970. Crop Sci. 10: 553.

[194] Pereira, J. F., et al. 1971. Abstr. Weed Soc. Amer., p. 61.

[195] Poskuta, J., and L. Indeka. 1968. Bull. Acad. Pol. Sci., Ser. Sci. Biol. 16:779.

[196] Prasad, I., and D. Pramer. 1969. Cytologia 34:351.

[197] Putala, E. C. 1967. Ph.D. Thesis. Univ. California, Berkeley.

[198] Rakitin, Y. V., et al. 1966. Sov. Plant Physiol. 13: 30.

[199] Ray, B., and M. Wilcox. 1969. Physiol. Plant. 22: 503.

[200] Rensen, J. J. S. van. 1969. Meded. Landbouw-hogesch. Wageningen 69:11.

[201] Ries, S. K., et al. 1967. Proc. Nat. Acad. Sci. U.S. 58:526.

[202] Roberts, J. 1968. Weed Res. 8:151.

[203] Rogers, R. L., and H. H. Funderburk. 1968. J. Agr. Food Chem. 16:434.

[204] Rogerson, A. B., and C. L. Foy. 1968. Proc. S. Weed Conf. 21:347.

[205] Romanovskaya, O. I., and G. R. Ozolinya. 1967. Agrokhimiya (8):112.

[206] Rutherford, P. P., et al. 1969. Phytochemistry 8: 1859.

[207] St. John, J. B. 1971. Abstr. Weed Soc. Amer., p. 77.

[208] Sawamura, S., and W. T. Jackson. 1968. Cytologia 33:545.

[209] Schrank, A. R. 1968. Physiol. Plant. 21:314.

[210] Schröder, I. 1970. Z. Allg. Mikrobiol. 10:295.

[211] Schröder, I., et al. 1967. Acta Biol. Med. Ger. 18: 125.

[212] Schultz, D. P., and B. G. Tweedy. 1970. Abstr. Weed Soc. Amer., p. 16.

[213] Schultz, D. P., et al. 1968. Plant Physiol. 43:265.

[214] Schwenk, P., and A. Barth. 1967. Flora (Jena) 158:285.

[215] Schwimmer, S. 1968. Plant Physiol. 43:1008.

[216] Shahied, S. I., and J. Giddens. 1970. Agron. J. 62: 306.

[217] Sharma, M. P., and W. H. Vanden Born. 1971. Can. J. Bot. 49:69.

[218] Sharma, M. P., and W. H. Vanden Born. 1971. Abstr. Weed Soc. Amer., p. 41.

[219] Shimabukuro, R. H., and V. P. Masteller. 1971. Ibid., p. 43.

[220] Shimabukuro, R. H., and H. R. Swanson. 1969. J. Agr. Food Chem. 17:199.

[221] Shimabukuro, R. H., et al. 1970. Plant Physiol. 46:103.

[222] Shirakawa, N. 1969. Zasso Kenkyu 9:11.

[223] Shneyour, A., et al. 1969. Exp. Cell Res. 58:1.

[224] Siegel, J. N., and A. C. Gentile. 1966. Plant Physiol. 41:670.

[225] Sikka, H. C., and D. E. Davis. 1969. Weed Sci. 17: 122.

[226] Sikka, H. C., and D. Pramer. 1968. Ibid. 16:296.

[227] Singh, B., and D. K. Salunkhe. 1970. Can. J. Bot. 48:2213.

[228] Singh, R. P., and S. H. West. 1967. Weeds 15:31.

[229] Slade, P. 1966. Weed Res. 6:158.

[230] Smith, L. W., et al. 1968. Weed Sci. 16:523.

[231] Solecka, M., et al. 1969. J. Amer. Soc. Hort. Sci. 94:55.

[232] Splittstoesser, W. E. 1968. Weed Sci. 16:344.

[233] Sruoginite, A. V., and A. P. Shpokene. 1968. Tr. Akad. Nauk Litov. SSR, B, 3:161.

[234] Stephenson, G. R., and S. K. Ries. 1969. Weed Sci. 17:327.

[235] Still, G. G., et al. 1970. Plant Physiol. 46:307.

[236] Stokes, D. M., et al. 1970. Aust. J. Biol. Sci. 23: 265.

[237] Stoller, E. W. 1969. Plant Physiol. 44:854.

[238] Stoller, E. W., and L. M. Wax. 1968. Weed Sci. 16: 283.

[239] Stranger, C. E., and A. P. Appleby. 1971. Abstr. Weed Soc. Amer., p. 77.

[240] Swann, C. W., and K. P. Buckholtz. 1966. Weeds 14:103.

[241] Swanson, C. R., and H. R. Swanson. 1968. Weed Sci. 16:137.

[242] Tanimoto, E., and Y. Masuda. 1968. Physiol. Plant. 21:820.

[243] Temperli, A., et al. 1966. Z. Naturforsch. 21b:903.

[244] Thompson, O. C., et al. 1969. J. Agr. Food Chem. 17:997.

[245] Tiedje, J. M., and M. Alexander. 1969. Ibid. 17: 1080.

[246] Tiedje, J. M., et al. 1969. Ibid. 17:1021.

[247] Tieszen, L. L. 1970. Plant Physiol. 46:442.

[248] Tweedy, J. A., and S. K. Ries. 1967. Ibid. 42: 280.

[249] Vanderbeek, L. C. 1967. Ann. N. Y. Acad. Sci. 144:374.

[250] Venchikova, T. A. 1969. Izv. Akad. Nauk Turkm. SSR, Ser. Biol. Nauk 2:82.

[251] Verloop, A., and W. B. Nimmo. 1970. Weed Res. 20:59.

[252] Vince, D. 1968. Planta 82:261.

[253] Volynets, A. P., et al. 1970. Dokl. Akad. Nauk SSSR 194:458.

[254] Wain, R. L. 1965. Sci. Progr. (London) 53:221.

continued

[255] White, J. A., and D. D. Hemphill. 1971. Abstr. Weed Soc. Amer., p. 13.

[256] Wilkinson, R. E., and W. S. Hardcastle. 1969. Weed Sci. 17:335.

[257] Willenbrink, J. 1969. Abstr. Schering Akt. Ges., p. 7.

[258] Winely, C. L., and C. L. San Clemente. 1971. Can. J. Microbiol. 17:47.

[259] Witham, F. H. 1968. Plant Physiol. 43:1455.

[260] Worley, J. F. 1967. Biochem. Physiol. Plant Growth Subst., 6th Int. Conf. Plant Growth Subst. Proc., p. 1635.

[261] Wort, D. J., and B. Singh. 1970. Agron. J. 62:57.

[262] Würzer, B. 1969. Naturwissenschaften 56:452.

[263] Wuu, K. D., and W. F. Grant. 1966. Can. J. Genet. Cytol. 8:481.

[264] Yamashita, J., and M. D. Kamen. 1968. Biochim. Biophys. Acta 153:848.

[265] Yih, R. Y., et al. 1968. Plant Physiol. 43:1291.

[266] Yih, R. Y., et al. 1968. Science 161:376.

[267] Yung, K. H. 1970. Plant Cell Physiol. 11:677.

[268] Yung, K. H., and J. D. Mann. 1967. Plant Physiol. 42:195.

[269] Zemanek, J., and J. Ambrozova. 1967. Biol. Plant. 9:270.

[270] Zemskaya, V. A., et al. 1969. Agrokhimiya (6): 116.

[271] Zweep, W. van der, and J. L. P. van Oorschot. 1970. Z. Pflanzenkr. (Pflanzenpathol.) Pflanzenschutz, Suppl. 5:61.

[272] Zweig, G. 1969. Residue Rev. 25:69.

[273] Zweig, G., et al. 1969. J. Agr. Food Chem. 17:176.

VII. ENVIRONMENT AND SURVIVAL

86. TEMPERATURE TOLERANCE: MAN

Part I. Heat and Cold Tolerance with and without Protective Clothing

Graph [1] shows experimentally determined ranges of human thermal tolerance for men at rest without protection, and also protected with the best clothing for each zone of heat or cold. The environments used in establishing the curves were all characterized by low air movement (less than 100 ft/min), and approximately equal air and wall temperatures. Tolerance limits for nude subjects: surface pain, above 121°C (250°F); body heat storage, from 60° to 121°C (140° to 250°F); fatigue and dehydration, from 38° to 60°C (100° to 140°F); fatigue, shivering and general body cooling, with increasing discomfort, from 18° to −18°C (65° to 0°F); and cold pain and increasing danger of frostbite, below −18°C (0°F). The improvement in thermal tolerance times caused by properly chosen clothing is depicted by the two solid curves, one in the hot zone and one in the cold. For information on zone boundaries for work in heat, and for more detailed analyses of thermal stress resistance, consult reference 2.

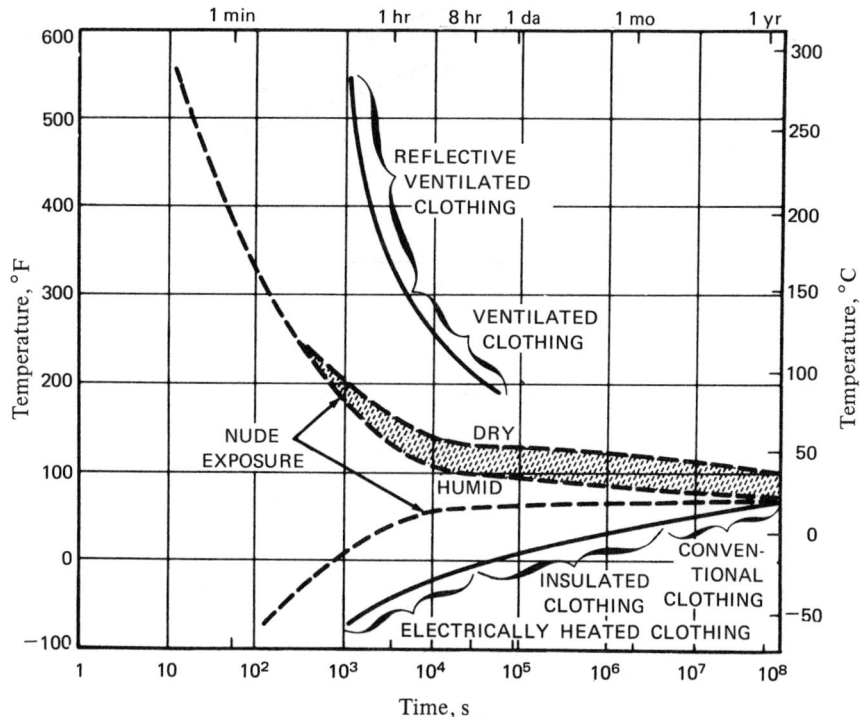

Contributor: Blockley, W. Vincent

References: [1] Blockley, W. V. 1964. NASA SP-3006:104. [2] Blockley, W. V. 1964. Ibid. 3006:108,109,111,116.

continued

Part II. Cold Tolerance with Adequate Body Insulation

Graph [1] shows cold tolerance of subjects at rest wearing adequate body insulation for increasingly cold environments. The prime limiting factor in voluntary tolerance of cold distress is the development of painfully cold feet. (The hands are more easily protected inside the clothing.) Even when the total body insulation is an impractical 5.9 clo[1]/ (close to wearing a sleeping bag), ordinary footgear limits

tolerance time at $-18°C$ ($0°F$)—shown by the vertical bar—to 77-104 minutes, which are average times for five men. The chart shows that the improvement so far achieved with insulated boots is not impressive, particularly if the socks become wet. There is a distinct risk of tissue damage when any part of the skin reaches $4°C$ ($39°F$). Most subjects refuse to continue before this point is reached.

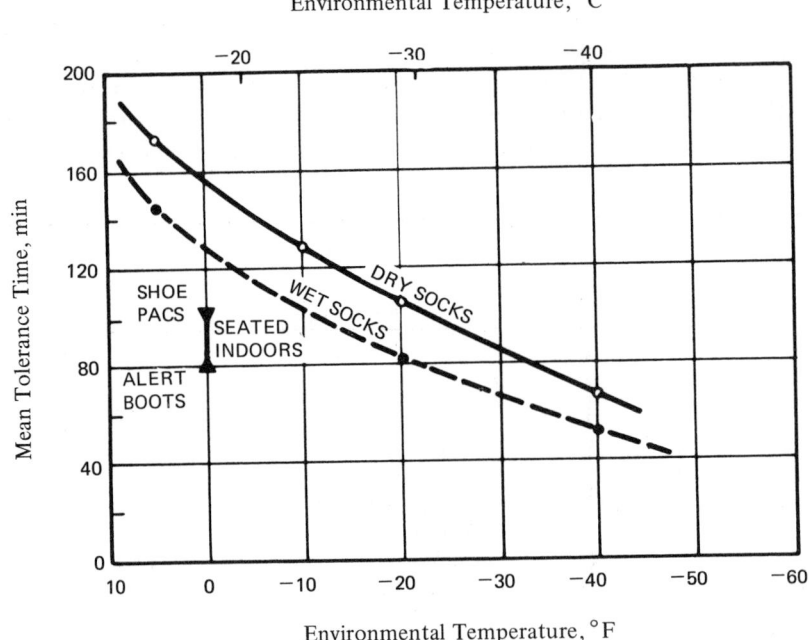

1/ 1 clo = insulation maintaining difference of $0.18°C$ for flow of 1 $kcal·m^{-2}·hr^{-1}$.

Contributor: Blockley, W. Vincent

Specific Reference
[1] Blockley, W. V. 1964. NASA SP-3006:124.

General References
[2] Carlson, L. D. 1954. Man in Cold Environment: A Study in Physiology. Univ. Washington, School Medicine, Seattle.

[3] Skrettingland, K. R., et al. 1961. Arctic Aeromed. Lab. TN-61-7.

Part III. Cold Tolerance in Water

The "voluntary tolerance, flight clothing" zone in the graph [1] shows the average results from numerous experimental

studies, including a recent one using a diver's "wet suit" in conjunction with a flight suit and long underwear. Such

continued

Part III. Cold Tolerance in Water

experiments are typically terminated when the subject declines to accept the discomfort any longer, or reaches a skin temperature below 10°C (50°F). The second limit shown, pertaining to men protected by potentially waterproof garments, reflects the fact that hands and feet cannot be adequately insulated and remain functional. Nude men in 24°C (75°F) water reach within 12 hours one of the tolerance limits: rectal temperature below 35°C (95°F), blood sugar below 60 mg/100 ml, or muscle cramps. The extent to

which real survival time would exceed these limits is difficult to predict, since injury, the equipment available, and psychological factors, such as belief in the possibility of rescue, influence tolerance. An analysis of over 25,000 personnel on ships lost at sea during 1940-44 showed that of those who reached life rafts, half died by the sixth day if the air temperature was below 5°C (41°F); survival time increased with increasing air temperature.

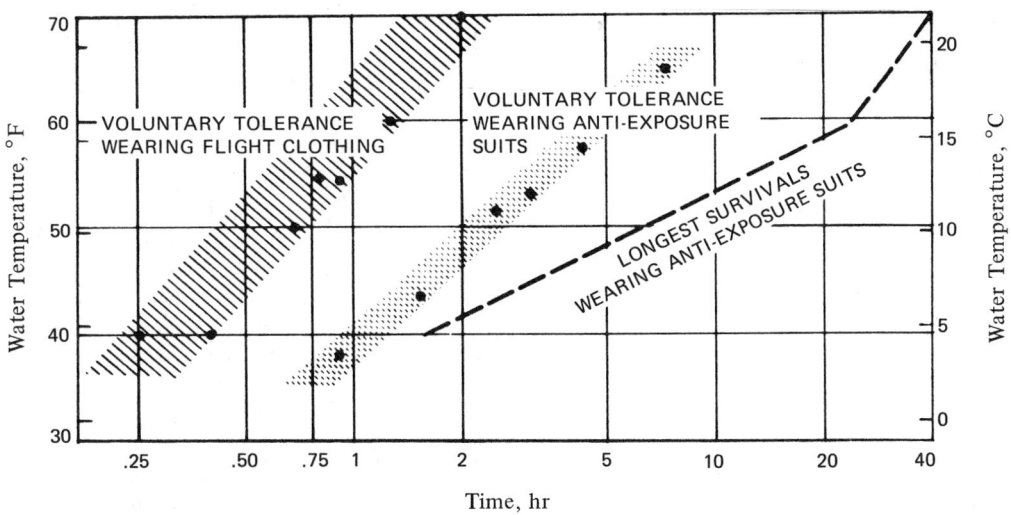

Contributor: Blockley, W. Vincent

Specific Reference

[1] Blockley, W. V. 1964. NASA SP-3006:128.

General References

[2] Beckman, E. L., and E. Reeves. 1966. Aerosp. Med. 37:1136.
[3] Hall, J. F., et al. 1953. WADC Tech. Rep. 53-323.
[4] McCance, R. A., et al. 1956. Med. Res. Counc. (Gt. Brit.) SR-291.
[5] Molnar, G. W. 1946. J. Amer. Med. Ass. 131:1046.
[6] Santa Maria, L. J., et al. 1964. Aerosp. Med. 35: 144.
[7] U.S. Navy. 1958. Med. News Lett. 31(x):12.

87. TOLERANCE TO TEMPERATURE EXTREMES: ANIMALS

Part I. Mammals

Race or breed, sex, nutritive state, and oxygen pressure also affect tolerance. Data on immersion in water are for

complete immersion of the body, except the head and neck. **Temp Measured:** B = body; Br = breast (essentially

continued

Part I. Mammals

the same as rectal and vaginal temperatures [20]); Cp = cheek pouch; I = intestinal; i.p. = intraperitoneal. **Survival:** n/n' = number survived/number tested. Data in brackets refer to the column heading in brackets. Values in parentheses are ranges, estimate "c" (*see* Introduction).

	Species (Synonym) [Specification]	Temp Measured	Ex-treme	Tolerance Limit Temp, °C [Relative Humidity, %]	Duration	Survival n/n' [%]	Remarks	Refer-ence
				Ambient Temperature				
1	*Homo sapiens*[1]	Water	Lower	15	50-70 min	6/6	Subjects nude, resting	54
2				(6-10)	45-60 min	8/8	Subjects nude, resting	8,54
3	[♂]	Air	Lower	−28.9	....	.....	Subjects walking 3.5 mi/hr	9
4		Water	Lower	4.6	53-98 min	0/7	Dachau experiments	17
5				(4-12)	30-100 min	23/35	Dachau prisoners clothed in aviator's garb including head-gear & life jacket	40
6				−1	30 min	1/10	Subjects from torpedoed vessel swimming in field ice	40
7	[Adult[2]]	Air	Upper	50 [15]	6 hr	.....	Subjects working (189 cal·m^{-2}·hr^{-1}), dressed in army jungle uniforms; wind, 2 mi/hr	48
8				50 [22]	6 hr	.....	Subjects working (188 cal·m^{-2}·hr^{-1}), dressed in shorts; wind, 2 mi/hr	48
9				50 [35]	6 hr	.....	Subjects resting, dressed in shorts; wind, 2 mi/hr	48
10	*Alopex lagopus* [3.8-5.5 kg]	Air	Lower	−30	2 hr or more	4/4	Cold chamber; metabolic rate not greatly affected	51
11				−80	1 hr	1/1	Cold chamber; subject shivering, no hypothermia	51
12	*Bos indicus* [Young ♀; 193-300 kg]	Air	Upper	40.6 [51]	∼24 hr	3/3	Controlled indoor thermal con-ditions; wind, 0.5 mi/hr; respi-ration markedly elevated	41
13	[Adult ♀]	Air	Lower	−12.8	2 wk	2/2	Controlled indoor thermal con-ditions; wind, 0.5 mi/hr	28
14	*B. taurus* Brown Swiss [Young ♀; 161-281 kg]	Air	Upper	40.6 [51]	24 hr	3/3	Controlled indoor thermal con-ditions; wind, 0.5 mi/hr; res-piration markedly elevated	29
15	Holstein & Jersey [Adult ♀]	Air	Lower	−12.8 [66]	....	.....	Controlled indoor thermal con-ditions; wind, 0.5 mi/hr	28
16	*Canis familiaris* [Adult]	Air	Upper	58 [23]	2.5 hr	1/1	Hot, dry air; nearly fatal	1
17	[27.6 kg]	Water	Lower	20	5 hr	4/4	Subjects conscious at end of test; not markedly affected	54
18				10	2-5 hr	4/4	Serious impairment approach-ing unconsciousness in 3 sub-jects, 4th not markedly affected	54

[1] In experiments on man (except for those at Dachau), tolerance limit indicates either that subjects refused to continue due to discomfort of the experiment, or that the investigator believed continuation would endanger health or life of subjects. [2] Subjects acclimated to work in heat; maintained thermal balance 2-6 hr after exposure.

continued

Part I. Mammals

Species (Synonym) [Specification]	Temp Measured	Tolerance Limit			Remarks	Reference-
		Ex-treme	Temp, °C [Relative Humidity, %]	Duration	Survival n/n' [%]	
19			0	1-5 hr	4/4	In 1 hr, serious impairment approaching unconsciousness in 3 subjects; 4th not markedly affected	54
20 Eskimo [Young; 9-15 kg]	Air	Lower	(−20 to −30)	2 hr or more	2/2	Cold chamber; metabolic rate not greatly affected	51
21 *Cavia porcellus* [Adult]	Air	Upper	44 [23]	7 hr	27/38	....	1
22 *Didelphis marsupialis virginiana* [Adult]	Air	Lower	0	Several da	.....	Almost no reduction in body temp	10
23 *Erinaceus europaeus;*	Air	Lower	(14.5-17.0)	....	.....	Body temp, 15-30°C; subjects half-wakeful	20
24 *E. europaeus roumanicus* [Adult]			4.2	231 da	6/9	Hibernating animals	32
25			(5.5-14.5)	....	.....	Body temp ∿1°C >ambient temp; subjects essentially poikilothermic	20
26			<5.5	....	.....	Body temp maintained at ∿6°C, or temp rose and animals awoke	20
27			(4.5-10)	216-396 hr	5/5	Hibernating animals	30
28 *Felis catus* [Adult]	Air	Upper	40.6 [65]	7 hr	.....	Controlled-temp room	46
29			43.3 [35]	7 hr	.....		
30			58 [23]	1.3-2.0 hr	2/2	Hot, dry air	1
31 *Lepus alleni* [Adult, 2-3 kg]	Air	Upper	51	4 hr	3/6	Body temp of survivors at end of test, 41.2-42.3°C; lethal body temp for this species, 44.9-45.9°C in 8 subjects	49
32 *Macaca mulatta (M. rhesus)* [Adult]	Air	Lower	−20	2 hr	11/11	Subjects immobilized in wooden boxes in cold chamber; mean fall in rectal temp, 3°C	11
33 *Mesocricetus auratus* [Adult]	Air	Lower	5	∿6 mo	12/12	Hibernation with periodic awakening during last 3 mo	37
34			(4-10)	60 da	10/17	Hibernation, with hypothermic periods averaging 4.4 da at 4°C and 2.4 da at 10°C	31
35			(4-6)	6 da	22/45	....	5
36	Water	Lower	3.0	1 hr	.....	Body temp 3.4°C or less at end of 1 hr	5
Mus musculus							
37 [3.5-11 mo old]	Air	Lower	(1.5-6.5)	24 hr	8/8	Cold room	55
38 [Adult]	Air	Upper	37 [20]	3 hr	16/19	....	1
39		Lower	5.8	2 mo	8/8	Cold room, with mean daily temp range of 7°C	55
40			−35	0.4 hr	0/10	Controlled-temp room; wind, 2 mi/hr	22
41 [20-27 g[3/]]	Air	Lower	−9	200 min	[50]	Acclimated to 30°C	15
42			−17	200 min	[50]	Acclimated to 20°C	15
43			−20	200 min	[50]	Acclimated to 10°C	15
44			−22	200 min	[50]	Acclimated to 0°C	15

[3/] Food and water supplied in all tests.

continued

Part I.   Mammals

	Species (Synonym) [Specification]	Temp Measured	Extreme	Temp, °C [Relative Humidity, %]	Duration	Survival n/n' [%]	Remarks	Reference
45	*Mustela rixosa* [38-70 g]	Air	Lower	−20	2 hr or more	3/3	Cold chamber; metabolic rate ∿3 times that under temperate thermal conditions	51
46	*Myotis lucifugus* [Juvenile; 4.6 g]	Air	Upper	51	>15 min	1/1	Body temp, 48°C; no ill effects	19
47	[Adult]	Air	Upper	54	Indefinite	[100]	Collected from summer roost in attic	19
48			Lower	−5	3-4 hr	12/12	Controlled-temp room; subjects restrained; unrestrained subjects survived with spontaneous rewarming after overnight exposure	44
49				−9	0-6.5 hr	31/31	Methacrylic acid (Lucite) metabolic chamber; no spontaneous arousal in 6 subjects when rewarmed	24
50	[5.2 g]	Air	Lower	0.5	>2 hr	5/5	Metabolic chamber; metabolic rate 4 times that at 2°C	21
51	[6.4 g]	Air	Upper	44	>30 min	0/5	Metabolic chamber; metabolic rate less than that at 41.5°C	21
52	*Neotoma fuscipes* [186.7 ± 33.4 g]	Air	Upper	(38-40)	7 hr	2/6	Controlled-temp chamber	34
53	*N. lepida* [110.3 ± 16.4 g]	Air	Upper	(38-40)	7 hr	3/7	Controlled-temp chamber; subjects from coastal population	34
54	[138.5 ± 14.7 g]	Air	Upper	(38-40)	7 hr	3/12	Controlled-temp chamber; subjects from desert population	34
55	*Oryctolagus cuniculus* [Adult]	Air	Upper	41 [23]	4 hr	26/31	Hot, dry atmosphere	1
56	[1767 g]	Air	Lower	−35	3.5-6.5 hr	0/9	Controlled-temp room; wind, 2 mi/hr	22
57	[3950 g]	Water	Lower	20	5 hr	4/4	Subjects conscious at end of test	54
58				10	30 min	4/4	Serious impairment, approaching unconsciousness	54
59				0	20 min	4/4		
60	*Ovis aries*, Merino [♂, castrate]	Air	Upper	43.3 [65]	7 hr	.....	Subjects survived exposure in controlled-temp room	35
61	*Peromyscus leucopus*	Air	Lower	−11	200 min	[50]	Acclimated to 30°C	15
62	[23-27 g[3/]]			−20	200 min	[50]	Acclimated to 20°C	15
63				−28	200 min	[50]	Acclimated to 10°C	15
64	*P. maniculatus* [Adult[4/]]	Air	Upper	40	6.1 hr	0/4	Winter subjects	52
65					13.3 hr	0/4	Summer subjects	52
66					13.5 hr	0/3	Acclimated to 29.5-32.5°C	52
67					13.5 hr	0/3	Acclimated to 6.0-10.5°C	52
68			Lower	−20	1.4 hr	0/3	Acclimated to 29.5-32.5°C	52
69					3.1 hr	0/4	Summer subjects	52
70					5.2 hr	0/3	Acclimated to 6.0-10.5°C	52
71					10.2 hr	0/4	Winter subjects	52

[3/] Food and water supplied in all tests.   [4/] Food and water not provided during tests.

continued

Part I. Mammals

	Species (Synonym) [Specification]	Temp Measured	Extreme	Temp, °C [Relative Humidity, %]	Duration	Survival n/n' [%]	Remarks	Reference
				Tolerance Limit				
72	[22-25 g $\underline{3/}$]	Air	Lower	−8	200 min	[50]	Acclimated to 30°C	15
73				−21	200 min	[50]	Acclimated to 20°C	15
74				−28	200 min	[50]	Acclimated to 10°C	15
75				−35	200 min	[50]	Acclimated to 0°C	15
76				−35	200 min	[50]	Acclimated to −10°C	15
77	*Procyon cancrivorus* [1.16 kg]	Air	Lower	(0-10)	2 hr or more	1/1	Cold chamber; metabolic rate ∿3 times that under temperate thermal conditions	51
	Rattus norvegicus [Adult $\underline{3/}$]							
78		Air	Upper	40 [15]	213 ± 36 min	0/16	Heated room	18
79				(42-43) [15]	144 ± 12 min	0/36		
80				45	79 ± 7 min	0/16		
81				(49-51)	31 ± 1 min	0/36		
82	[60-da-old ♂]	Air	Lower	2	8 da	[50]	....	59
83	[60-da-old ♀]	Air	Lower	2	90 da	[50]		
84	[120-da-old ♂]	Air	Lower	2	25 da	[50]		
85	[120-da-old ♀]	Air	Lower	2	60 da	[50]		
86	[227 g]	Water	Lower	20	2 hr	3/3	Subjects conscious at end of test	54
87				10	20 min	3/3	Serious impairment, approaching unconsciousness	54
88				0	10 min	3/3	Serious impairment, approaching unconsciousness	54
89	[235 g]	Air	Lower	−35	0.75-2 hr	0/12	Cold chamber; wind, 2 mi/hr	22
90	[250 g]	Air	Lower	−40	20 min	.....	Summer; estimated from O_2 consumption	16
91	[280 g]	Air	Lower	−60	20 min	.....	Winter; estimated from O_2 consumption	16
92	[280-350 g]	Air	Lower	−18	200 min	[50]	Acclimated to 30°C	15
93				−29	200 min	[50]	Acclimated to 20°C	15
94				−35	200 min	[50]	Acclimated to 10°C	15
95	Sprague-Dawley or Wistar [250-350 g]	Air	Upper	44 [50]	192 ± 26.7 min	13/13	Duration value is exposure time for body temp to reach 42°C	14
96	*Setonix brachyurus* [Adult; 2.5-4.0 kg]	Air	Upper	44 [52]	2 hr	5/5	Temp cabinet. Animals coated feet & part of tail with saliva by licking them.	6
97	*Spermophilus leucurus (Citellus leucurus)* [Adult]	Air	Upper	42.6	2 hr	5/5	Controlled-temp cabinet	23
98	*S. tridecemlineatus (C. tridecemlineatus)* [Adult]	Air	Lower	5	62-120 da	2/2	Hibernating animals	37
99				(−1.7 to +10)	....	.....	Hibernating animals	25
100	*Sus scrofa*, Berkshire [59 kg]	Air	Upper	37.8 [65]	7 hr	.....	Controlled-temp room. Subjects unable to withstand 40.6 °C at any humidity.	45

$\underline{3/}$ Food and water supplied in all tests.

continued

Part I. Mammals

Species (Synonym) [Specification]	Temp Measured	Tolerance Limit				Remarks	Reference
		Ex-treme	Temp, °C	Duration	Survival n/n' [%]		
			Body Temperature				
101 *Homo sapiens* [1] [Young ♂]	Rectal	Lower	(33-35)	>10-60 min	3/3	Subjects nude, immersed in water at 10-15°C	54
102 [♀]	Rectal	Lower	(18-20)	>2 hr	1/1	Subject inebriated. One of lowest rectal temp recorded in a human survivor.	33
103 [Adult]	Rectal	Upper	(42.2-42.8)	Short	.....	Heat stroke; death likely if exposure maintained more than a short time	50
104		Lower	27	65 min	0/7	Unanesthetized; immersed in water at 4-6°C	17
105 *Antrozous pallidus* [Adult; 19-25 g]	i.p.	Upper	43.5	1 hr	[100]	Respirometer. Survival limited to 15-20 min if body temp reached 44°C.	36
106 *Bos indicus* [Young ♀; 193-300 kg]	I	Upper	41.0	∿24 hr	3/3	Controlled indoor temp conditions; wind, 0.5 mi/hr. Respiration markedly elevated.	29
107 *B. taurus* Brown Swiss [Young ♀; 161-281 kg]	I	Upper	41.5	∿24 hr	3/3		
108 Hereford [Adult]	Rectal	Upper	(38.2-41.3)	11.5 hr	3/3	Exposure to air temp of 40°C	56
109 *Canis familiaris* [Adult]	Heart	Lower	(17.2-19.6)	....	0/50	Immersed in cold water while anesthetized	17
110	I	Upper	42.8	....	.....	Controlled-temp room. Subject developed "staggers."	47
111		Lower	20	....	7/7	Fasted, depilated, & anesthetized subjects cooled by immersion in ice water at 2-4°C. Experiment repeated once.	42
112	Liver	Lower	5	>30 min	6/8	Cooled by perfusion	27
113 *Cavia porcellus* [Adult]	I	Upper	42.7	7 hr	[50]	Ambient temp, 41°C	1
114		Lower	19.4 ± 1.2	....	14/28	Immersion in ice water	13
115 *Dasypus novemcinctus* [Adult]	I	Lower	(29-32.5)	3-6 hr	6/6	Cold room, 0-21°C	57
116 *Didelphis marsupialis virginiana* [Adult]	I	Lower	32	....	.....	Air temp, 14°C	57
117 *Erinaceus europaeus; E. europaeus roumanicus* [Adult]	Br	Lower	6.0	....	.....	Approx. lowest body temp for 20 hibernating subjects	20
118 *Felis catus* [2-8 da old]	I	Lower	(7-8)	....	9/21	Immersion in cold water	4
119 [Adult]	I	Upper	43.4	Few min	[50]	Hot, dry air	1
120 *Heterohyrax brucei* [Adult ♂; 1120-1513 g]	I	Upper	(40-41)	....	3/3	Subjects exposed to equatorial sun (ambient temp, 22.5-26.5°C); muscle tone reduced	7

[1] In experiments on man (except for those at Dachau), tolerance limit indicates either that subjects refused to continue due to discomfort of the experiment, or that the investigator believed continuation would endanger health or life of subjects.

continued

Part I. Mammals

Species (Synonym) [Specification]	Temp Measured	Tolerance Limit				Remarks	Reference
		Ex-treme	Temp, °C	Duration	Survival n/n' [%]		
121 *Lepus alleni; L. califor-*	Rectal	Upper	(41.2-43.5)	4 hr	3/6	Exposure to air temp of 51°C	49
122 *nicus* [Adult]			(43.0-44.1)	∿1-1.5 hr	9/9	Exposure to sun on hot day while restrained in wire cage	49
123			45.4	<15 min	.....		
124 *Macaca mulatta (M.*	I	Upper	44.0	....	0/1	Fully etherized subject heated in warm-air chamber (40-50°C) to 44.0°C; died shortly afterward	53
rhesus) [Adult]							
125		Lower	(12.5-14.0)	....	2/3	Fully etherized subjects cooled 3-5 hr in cold-air chamber (5-13°C); body temp of survivors remained at 14-16°C for >45 min	53
126 *Marmosa mexicana isthmica* [Adult]	I	Lower	31.5	....	.....	Air at 21.5°C	57
127 *Marmota monax* [Adult]	I	Lower	(6-14)	2-3 mo	.....	Outdoors during hibernation	43
128 *Mesocricetus auratus* [2-11 da old]	I	Lower	<1	0.5-1 hr	7/11	29 subjects immersed in cold water; majority survived	4
129 *Mus musculus* [45-109 da old]	I	Lower	(8.5-14.5)	25-65 min	20/20	Cold air, ∿−10°C	39
130 *Myotis daubentonii* [Adult]	I	Lower	(−1.6 to −0.9)	17-90 min[5/]	1/5	Chamber immersed in cold bath	26
131 *M. lucifugus* [Adult]	Rectal	Upper	42	Lengthy	[100]	Exposure to air temp of 45°C	19
132		Lower	(−4 to +10)	40-50 min	[100]	Exposure to air temp of −5 to +10°C	44
133 *M. sodalis*	I	Lower	0.55 (−0.4 to +3.0)	Several hr	6/6	Respirometer chamber	19
134 *M. yumanensis* [5-7 g]	i.p.	Upper	43.5	1 hr	[100]	Respirometer. Few subjects survived body temp of 44°C for >15-20 min.	36
135 *Ornithorhynchus ana-tinus* [693 g]	I	Upper	35.3	>17 min	1/1	Metabolic chamber in water bath at 35°C. Animal unconscious at end of test.	38
136		Lower	31.8	70 min	1/1	Metabolic chamber in water bath at 5°C. Animal very active at end of test.	38
137 *Oryctolagus cuniculus* [4-8 da old]	I	Lower	(6.2-8.1)	....	5/7	Immersed in water. Cause of 2 deaths uncertain.	4
138 [Adult]	I	Upper	43.4	Few min	[50]	Hot, dry air	1
139 *Ovis aries* [Adult]	I	Upper	41.7	....	.....	Controlled-temp room	35
140 *Phascolarctos cinereus* [Adult]	I	Lower	35	....	.....	Exposed to air temp of 7.7°C	57
141 *Rattus norvegicus* [0-17 da old]	i.p.	Lower	(3.5-9)	82 min or less	.....	41 subjects cooled in metabolic chamber immersed in cold water. Hearts stopped beating between 3 & 9°C; recovery on rewarming.	12

[5/] Duration of body temp below 0°C.

continued

Part I. Mammals

	Species (Synonym) [Specification]	Temp Measured	Tolerance Limit				Remarks	Refer-ence
			Ex-treme	Temp, °C	Duration	Survival n/n' [%]		
142	[1-9 da old]	I	Lower	(2-10)	....	.....	Subjects immersed in water. Body temp of 10°C endured for >2 hr, of 2°C for 0.5 hr.	2
143	[Adult]	I	Upper	42.5	Few min	[50]	Mean lethal temp for subjects exposed to hot, dry air	1
144	[200 g]	I	Lower	15.1	2 hr	[50]	Mean hypothermic level for subjects immersed in water at various temp	3
145	Spermophilus lateralis (Citellus lateralis) [Adult; 164 g]	I	Upper	(41.5-43.5)	4 hr	8/11	3 subjects that died shortly after 4-hr exposure to ambient temp of 37°C had rectal temp of ∿43.5°C; highest body temp in survivors was 42.5°C	58
146	S. tridecmlineatus (C. tridecemlineatus [Adult]	Cp	Upper	(41.6-42.3)	....	4/4	Heated environment	25
147			Lower	(1.3-13)	....	.....	∿50% of 49 hibernating subjects observed were in this temp range	25
148	Sus scrofa, Berkshire [59 kg]	I	Upper	41.7	....	.....	Controlled-temp room. Near limit of survival.	45
149	Tachyglossus aculeatus [2363 g]	I	Lower	(25.5-29.1)	77-102 min	3/3	Metabolic chamber in water bath at 4-8°C	38
150			Upper	(34.8-40.0)	60-71 min	2/3	Metabolic chamber in water bath at 35-37°C	38
151	Tadarida brasiliensis [Adult; 10-12 g]	i.p.	Upper	43.5	1 hr	[100]	Respirometer. Few subjects survived body temp of 44°C for >15-20 min.	36

Contributors: Sealander, John A., Jr.; Dawson, William R., and Bennett, Albert F.

References

[1] Adolph, E. F. 1947. Amer. J. Physiol. 151:564.

[2] Adolph, E. F. 1948. Ibid. 155:366.

[3] Adolph, E. F. 1948. Ibid. 155:378.

[4] Adolph, E. F. 1951. Ibid. 166:75.

[5] Adolph, E. F., and J. W. Lawrow. 1951. Ibid. 166: 62.

[6] Bartholomew, G. A. 1956. Physiol. Zool. 29:26.

[7] Bartholomew, G. A., and M. Rainy. 1971. J. Mammal. 52:81.

[8] Behnke, A. R., and C. P. Yaglou. 1951. J. Appl. Physiol. 3:591.

[9] Belding, H. S. 1949. In L. H. Newburgh, ed. Physiology of Heat Regulation and the Science of Clothing. W. B. Saunders, Philadelphia. p. 352.

[10] Britton, S. W., and W. E. Atkinson. 1938. J. Mammal. 19:94.

[11] Dugal, L.-P., and G. Fortier. 1952. J. Appl. Physiol. 5:143.

[12] Fairfield, J. 1948. Amer. J. Physiol. 155:355.

[13] Gosselin, R. E. 1949. Ibid. 157:103.

[14] Hainsworth, F. R., and E. M. Stricker. 1969. Ibid. 217:494.

[15] Hart, J. S. 1953. Can. J. Zool. 31:80.

[16] Hart, J. S., and O. Heroux. 1963. Ibid. 41:712.

[17] Hegnauer, A. H. 1959. Ann. N. Y. Acad. Sci. 80: 315.

[18] Heilbrunn, L. V., et al. 1946. Physiol. Zool. 19: 404.

[19] Henshaw, R. E., and G. E. Folk, Jr. 1966. Ibid. 39: 223.

[20] Herter, K. 1934. Z. Vergl. Physiol. 20:511.

[21] Hock, R. J. 1951. Biol. Bull. 101:289.

[22] Horvath, S. M., et al. 1948. Science 107:171.

[23] Hudson, J. W. 1962. Univ. Calif. Berkeley Publ. Zool. 64:1.

[24] Hurst, R. N., and J. E. Wiebers. 1968. J. Mammal. 49:791.

[25] Johnson, G. E. 1928. J. Exp. Zool. 50:15.

continued

Part I. Mammals

[26] Kalabuchow, N. I. 1935. Zool. Jahrb. Abt. Allg. Zool. Physiol. Tiere 55:47.

[27] Kenyon, J. R. 1961. Brit. Med. Bull. 17:43.

[28] Kibler, H. H., and S. Brody. 1950. Mo. Agr. Exp. Sta. Res. Bull. 464.

[29] Kibler, H. H., and S. Brody. 1951. Ibid. 473.

[30] Kristofferson, R., and A. Soivio. 1964. Ann. Zool. Fenn. 1:370.

[31] Kristofferson, R., and A. Soivio. 1966. Ibid. 3:66.

[32] Kristofferson, R., and A. Soivio. 1967. Ann. Acad. Sci. Fenn., A4, 122:1.

[33] Laufman, H. 1951. J. Amer. Med. Ass. 147:1201.

[34] Lee, A. K. 1963. Univ. Calif. Berkeley Publ. Zool. 64:57.

[35] Lee, D. H. K., and K. Robinson. 1941. Proc. Roy. Soc. Queensland 53:189.

[36] Licht, P., and P. Leitner. 1967. Comp. Biochem. Physiol. 22:371.

[37] Lyman, C. P. 1954. J. Mammal. 35:545.

[38] Martin, C. J. 1903. Phil. Trans. Roy. Soc. London B195:1.

[39] Meader, R. G., and C. Marshall. 1938. Yale J. Biol. Med. 10:365.

[40] Molnar, G. W. 1946. J. Amer. Med. Ass. 131:1046.

[41] Murigen, I. I. 1937. Bull. Exp. Biol. Med. (USSR) 4:100.

[42] Penrod, K. E. 1949. Amer. J. Physiol. 157:436.

[43] Rasmussen, A. T. 1915. Ibid. 39:20.

[44] Reite, O. B., and W. H. Davis. 1961. Proc. Soc. Exp. Biol. Med. 121:1212.

[45] Robinson, K., and D. H. K. Lee. 1941. Proc. Roy. Soc. Queensland 53:145.

[46] Robinson, K., and D. H. K. Lee. 1941. Ibid. 53:159.

[47] Robinson, K., and D. H. K. Lee. 1941. Ibid. 53:171.

[48] Robinson, S., et al. 1945. Amer. J. Physiol. 143:21.

[49] Schmidt-Nielsen, K., et al. 1965. Hvalradets Skr. 48:125.

[50] Schmitt, M. G. 1944. In O. Glasser, ed. Medical Physics. Year Book, Chicago. v. 1, p. 436.

[51] Scholander, P. F., et al. 1950. Biol. Bull. 99:237.

[52] Sealander, J. A. 1951. Amer. Midl. Natur. 46:257.

[53] Simpson, S. 1902. J. Physiol. (London) 28:xxxvii.

[54] Spealman, C. R. 1946. Amer. J. Physiol. 146:262.

[55] Sumner, F. B. 1913. J. Exp. Zool. 15:315.

[56] Taylor, C. R., and C. P. Lyman. 1967. Physiol. Zool. 40:280.

[57] Wislocki, G. B. 1933. Quart. Rev. Biol. 8:385.

[58] Yousef, M. K., and W. G. Bradley. 1971. Comp. Biochem. Physiol. 39A:671.

[59] Zarrow, M. X., and M. E. Denison. 1956. Amer. J. Physiol. 186:216.

Part II. Birds

Race or breed, sex, nutritive state, oxygen pressure, and acclimation also affect tolerance. **Temp** refers to air temperature unless otherwise indicated. **Survival:** n/n' = number survived/number tested; percentages indicate percent survived. Data in brackets refer to the column heading in brackets.

	Subject		Tolerance Limit				Remarks	Reference
	Species	Specification	Extreme	Temp, °C [Relative Humidity]	Duration	Survival n/n'		
1	*Anas platyrhynchos*	1023 g	Lower	−18[1]	225 hr	0/8	....	13,18
2	*A. platyrhynchos domesticus*	2016 g	Lower	−40	384 hr	1/1	....	10
3	*Bonasa umbellus*	615 g	Lower	−18[1]	185 hr	0/2	....	13,18
4	*Colinus virginianus*	164-167 g	Upper	44.4, 45.1 [10%][2]	100 min	2/2	Mean body temp for 2 subjects: 43.1, 43.4°C	20
5		167 g	Lower	−18[1]	60 hr	0/10		13,18
6	*Columba livia*	....	Upper	54-55[3]	1-3 hr	3/3	Birds selected for test on basis of having "sat especially quietly" in experiments at 50.9°C	6
7				50.9[3]	2-6 hr	100%	Body temp stabilized at 43.1 ± 0.7°C	

[1] Maintained in cold chamber. [2] Confined in small wind tunnel; air speed, 2.5 mi/hr. [3] Tested in temperature-controlled room.

continued

87. TOLERANCE TO TEMPERATURE EXTREMES: ANIMALS

Part II. Birds

	Subject		Tolerance Limit				Remarks	Reference
	Species	Specification	Extreme	Temp, °C [Relative Humidity]	Duration	Survival n/n'		
8				45 [<50%][4]	2-6 hr	100%	Body temp stabilized at 42.6 ± 0.1°C	
9		300-400 g	Lower	−40	96-120 hr	11/25	....	24
10					144 hr	4/25	....	
11				−40[5]	24 hr	4/6	....	
12	Summer[6]	343 g	Lower	−57	500 min	50%	....	14
13	Winter[6]	380 g	Lower	−57	6500 min	50%	....	14
14	*Corvus cryptoleucus*	583 g	Upper	45.8 [10%][2]	4 hr	1/1	Body temp maintained between 42.5 & 43.2°C	20
15	*Eurostopodus guttatus*	89 g	Upper	52.8 [18%][4]	2 hr	1/1	....	8
16	*Gallus gallus*[7]	....	Upper	43	2 hr	50%	....	26
17		1539-1642 g	Lower	−35[8]	3.3-29.5 hr	0/11	Cloacal temp, 21-41°C	15
18		....	Lower	20-27.8[9]	23-27 hr	5/5	Cloacal temp, 27.8-31.1°C	25
19	Embryo	0-21 da	Upper	28-43	504 hr	50%	....	9,11
20		0 da	Lower	−5	48 hr	50%	....	22
21		0-5 da	Lower	21	120 hr	50%	....	7
22		5 da	Lower	0	38 hr	49%	....	10
23		17 da	Lower	10	144 hr	60%	....	10
	Hesperiphona vespertina							
24	Summer; molting[6]	56.4 g	Lower	−48[1]	250 min	50%	....	14
25	Summer; post-molt[6]	56.0 g	Lower	−48[1]	1250 min	50%	....	14
26	Winter[6]	62.5 g	Lower	−48[1]	8300 min	50%	....	14
27	*Junco hyemalis*	21.8 g	Lower	−14[1]	37 hr	0/12	....	18
28	*Larus canus*, nestling	3-8 da	Upper	43	0.5-1.0 hr	0/3	....	2
29	*Meleagris gallopavo*	4869 g	Lower	−18[1]	324 hr	0/2	....	13,18
30	*Passer domesticus*	24.9-25.8 g	Upper	48.6 [20%][10]	2 hr	3/3	Subjects obtained at Houston, Texas	16
31				48.6 [20%][10]	2 hr	0/3	Subjects obtained at Boulder, Colorado	
32				47.6 [20%][10]	2 hr	1/3	Subjects obtained at Ann Arbor, Michigan	
33		22-25 g	Upper	44.1-45.2 [12%][2]	1 hr	4/4	Mean body temp of 4 subjects, 44.0-44.7°C	20
34		approx. 27 g	Upper	39.1	5.0-27.2 hr	0/12	....	17
35			Lower	−15.1 to −12.6[1]	6.5-16.5 hr	0/13	....	17
36	*Phalacrocorax auritus*	7.5-9 wk	Upper	45 [18%]	60 min	1/1	Temp cabinet. Body temp held below 42°C.	19
37	*Phalaenoptilus nuttallii*	....	Upper	>48	....	100%	....	4
38	*Phasianus colchicus*	1167 g	Lower	approx. −1	340 hr	0/24	Artificially propagated stock maintained in outdoor pens. 100% survival occurred in control group maintained under same temp conditions with food available.	13,18
39		1210 g	Lower	−17.2 to +4.4	35 da	13/25	Wild stock maintained in outdoor pens without shelter	13

[1] Maintained in cold chamber. [2] Confined in small wind tunnel; air speed, 2.5 mi/hr. [4] Tested in metabolism chamber. [5] Wind speed, 5 mi/hr. [6] Food and water supplied in all tests. [7] Synonym: *Gallus domesticus.* [8] Tested in cold room. [9] Water temperature. [10] Tested in walk-in incubator.

continued

Part II. Birds

Subject			Tolerance Limit				Remarks	Refer-ence
Species	Specifi-cation	Extreme	Temp, °C [Relative Humidity]	Duration	Survival n/n'		Remarks	Refer-ence
40	877 g	Lower	−24	12 da	0/13		....	12
41 *Scardafella inca*	42 g	Upper	44 [<53%][4]	110 min	10/11		Body temp of bird that died reached 47.6°C after activity	21
42 *Serinus canarius*	20 g	Lower	−35[8]	0.6 hr	0/2		....	15
43 *Struthio camelus*	63-104 kg	Upper	50	8 hr	......		Prolonged panting did not produce alkalosis	23
44 *Sturnus vulgaris*	69.8 g	Upper	24-30	1.25 da	0/4		Deprived of food & water	5
45	77.0 g	Lower	2-4	1.0 da	0/4		Deprived of food & water	5
46 Summer[6]	72.3 g	Lower	−48[1]	140 min	50%		....	14
47 Winter[6]	82.9 g	Lower	−48[1]	1100 min	50%		....	14
48 *Sula dactylatra,* hatchlings	<100 g	Upper	44-53[11]	<20 min	......		Substratum temp was 43.5-49.0°C. Air temp was 32.4-36.3°C.	3
49 *Troglodytes aedon*	....	Upper	approx. 37.9	1 hr	2/2		Deep throat temp, 45.3°C	1
50		Lower	−13.9	4 hr	1/1		....	17
51 Embryo	1-13 da	Lower	15.6-21.1 [80-90%]	20-48 hr	2/16		....	1
52 Fledgling	1-16 da	Upper	45.2	<1.5 hr	1/11		Deep throat temp, 46.6°C	1
53	1-9 da	Lower	6.4-10.1	<1 hr	5/9		Deep throat temp, 6.8-11.3°C	1
54 *Zenaidura macroura*	91-117 g	Upper	45.4 [12%][2]	2 hr	3/3		Mean body temp of 3 subjects, 43.3-43.6°C	20
55 *Zonotrichia albicollis*	26.6 g	Lower	−17[1]	16 hr	0/3		....	18

[1] Maintained in cold chamber. [2] Confined in small wind tunnel; air speed, 2.5 mi/hr. [4] Tested in metabolism chamber. [6] Food and water supplied in all tests. [8] Tested in cold room. [11] Black-bulb temperature.

Contributors: Sealander, John A., Jr.; Dawson, William R. and Bennett, Albert F.

References

[1] Baldwin, S. P., and S. C. Kendeigh. 1932. Sci. Publ. Cleveland Mus. Natur. Hist. 3.
[2] Barth, E. K. 1951. Nytt Mag. Naturvidensk. 88:213.
[3] Bartholomew, G. A. 1966. Condor 68:523.
[4] Bartholomew, G. A., et al. 1962. Ibid. 64:117.
[5] Brenner, F. J. 1965. Wilson Bull. 77:388.
[6] Calder, W. A., and K. Schmidt-Nielsen. 1967. Amer. J. Physiol. 213:883.
[7] Dareste, C. 1891. Recherches sur la production artificielle des monstruosités. Ed. 2. C. Reinwald, Paris.
[8] Dawson, W. R., and C. D. Fisher. 1969. Condor 71: 49.
[9] De la Roche, F. 1810. J. Phys. Chim. Hist. Natur. Arts 71:289.
[10] Dougherty, J. E. 1926. Amer. J. Physiol. 79:39.
[11] Edwards, C. L. 1902. Ibid. 6:351.
[12] Errington, P. E. 1939. Wilson Bull. 51:22.
[13] Gerstell, R. 1942. Pa. Game Comm. Res. Bull. 3.

[14] Hart, J. S. 1962. Physiol. Zool. 35:224.
[15] Horvath, S. M., et al. 1948. Science 107:171.
[16] Hudson, J. W., and S. L. Kimzey. 1966. Comp. Biochem. Physiol. 17:203.
[17] Kendeigh, S. C. 1934. Ecol. Monogr. 4:301.
[18] Kendeigh, S. C. 1945. J. Wildl. Manage. 9:217.
[19] Lasiewski, R. C., and G. K. Snyder. 1969. Auk 86: 529.
[20] Lasiewski, R. C., et al. 1966. Comp. Biochem. Physiol. 19:445.
[21] MacMillen, R. E., and C. H. Trost. 1967. Ibid. 20: 263.
[22] Moran, T. 1925. Proc. Roy. Soc. B98:436.
[23] Schmidt-Nielsen, K., et al. 1969. Condor 71:341.
[24] Streicher, E., et al. 1950. Amer. J. Physiol. 161:300.
[25] Sturkie, P. D. 1946. Ibid. 147:531.
[26] Yeates, N. T. M., et al. 1941. Proc. Roy. Soc. Queensland 53:105.

continued

87. TOLERANCE TO TEMPERATURE EXTREMES: ANIMALS

Part III. Reptiles and Amphibians

Sex, nutritive state, oxygen pressure, and state of acclimation also affect tolerance. **Rapid:** Quick heating or cooling (generally about 1°C/min) used to bring animals to tolerance limit. **Survival:** n/n′ = number survived/number tested; percentages indicate percent survived. Data in brackets refer to the column heading in brackets.

Order[1] & Species (Synonym)	Acclimation Temp °C[1]	Temp Measured	Extreme	Temp, °C [Rapid]	Duration hr[1]	Survival n/n′	Remarks	Reference
Reptilia								
Crocodylia								
1 *Alligator mississipiensis*	....	Cloaca	Upper	39	....	≮50%	....	16,65
2				38-39	<0.5	4/4	In water; emerged	16
3				38	....	0/2	In sun	16
4			Lower	4	....	≮50%	....	16,65
5		Body	Lower	1-3	2-4	4/7	....	65
Serpentes[2]								
6 *Coluber constrictor*	....	Cloaca	Upper	[42.4]	....	....	....	8
7 *Crotalus cerastes*	....	Cloaca	Upper	45-47	0.2	0/4	Air temp, 55°C	55
8				[41.6]	....	0/1	....	19
9		Body	Upper	[45]	....	0/4	....	55
10 *C. viridis*	....	Cloaca	Upper	[39-40]	....	....	....	8
11 *Lampropeltis getulus*	....	Cloaca	Upper	[42]	....	....	....	8
12			Lower	[−2]	....	....	....	8
13 *Masticophis flagellum flagellum (Coluber flagellum)*	....	Cloaca	Upper	42.4-44	....	0/3	....	19
14 *Natrix sipedon*	....	Air	Upper	43	24	3/3	....	48
15				39-43	24	0/4	....	48
16		Air	Lower	1.5	<24	30/30	....	48
17				−2	12	0/8	....	48
18				−5 to −2	12-24	0/10	....	48
19 *Pituophis catenifer*	....	Cloaca	Upper	[40.5]	....	....	....	8
20			Lower	[−3.0]	....	....	....	8
21 *Thamnophis radix*	....	Air	Upper	42	12-24	0/8	....	48
22				37-40	>48	15/15	....	48
23			Lower	1.5	>24	5/5	....	48
24				−3.8 to −1.5	....	8/16	Hibernating 12-18 inches deep	1
25 *T. sirtalis*	Room temp	Cloaca	Upper	38.9	4-5	10/10	Acclimated for 5 da	62
26			Lower	4.2	2-3	10/10	Acclimated for 5 da	62
Sauria[2]								
27 *Amphibolurus barbatus*	....	Body	Upper	46	0.17[3]	0/7	....	46
28 *A. ornatus*	40	Body	Upper	46	1.06[3]	0/6	....	5
29 *Anguis fragilis*	....	Air	Upper	37	1.0	5/12	High humidity	37
30 *Anniella pulchra*	....	Cloaca	Upper	[34.0]	....	....	....	8
31 *Anolis carolinensis*	....	Cloaca	Upper	[41.0-42.8]	....	0/5	....	8
32	32 (day), 20 (night)	Cloaca	Upper	42.5	0.95[3]	0/8	14 hr/da photoperiod	45

[1] Unless otherwise indicated. [2] Suborder. [3] Mean survival time.

continued

Part III. Reptiles and Amphibians

	Order & Species (Synonym)	Accli-mation Temp °C	Temp Mea-sured	Tolerance Limit				Remarks	Ref-er-ence
				Ex-treme	Temp, °C [Rapid]	Duration hr	Survival n/n'		
33		25	Body	Upper	[42.9]	....	....	16 hr/da photoperiod	39
34				Lower	[15.6]	....	....		39
35		20	Cloaca	Upper	[40.8]	....	6/6	....	18
36	A. homolechis	....	Air	Upper	29.0-38.8	....	100%	....	59
37	Cnemidophorus tigris (C. tessellatus)	....	Cloaca	Upper	[46.0]	....	0/1	....	17
38	Crotaphytus collaris	....	Cloaca	Upper	[45.0-47.5]	....	0/3	....	17
39			Body	Upper	>45	....	<50%	....	24
40	Dipsosaurus dorsalis	....	Cloaca	Upper	47-47.5	0.5	0/3	....	19
41			Body	Upper	>47	....	....		23
42	Egernia stokesii	....	Cloaca	Upper	43.5	0.27[3/]	0/5	....	46
	Eumeces obsoletus								
43	Embryo	....	Soil	Upper	42-43	0.5	2/10	....	27
44				Lower	−3	0.5	7/14	....	27
45	Hatchling	....	Air	Upper	42.7	0.5	2/2	....	27
46	Adult	....	Body	Upper	42	4-5	....	....	21
47			Air	Upper	43	0.5	2/2	....	27
48				Lower	−5 to −2.4	0.5	5/10	....	27
49	Gehyra variegata	....	Cloaca	Upper	43.5	1.21[3/]	0/7	....	46
50	Gerrhonotus multi-carinatus (Elgaria multicarinata)	....	Cloaca	Upper	[40.0-40.5]	....	....	....	8
51					[39]	....	....		44
52				Lower	[−4 to −2]	....	....		8
53	Heloderma suspectum	....	Cloaca	Upper	[44.2]	....	0/1		4
54	Lacerta agilis	....	Body	Upper	41	....	....		43
55	Phrynosoma cornu-tum	....	Body	Upper	[47.9]	....	....		57
56			....	Lower	[9.5]	....	....		57
57		27	Cloaca	Upper	[45.1]	....	....		3
58		25	Body	Upper	[47.8]	....	....	16 hr/da photoperiod	39
59				Lower	[7.7]	....	....	16 hr/da photoperiod	39
60	Pygopus baileyi	....	Cloaca	Upper	43.5	0.83[3/]	0/1	....	46
61	Sceloporus graciosus	....	Cloaca	Upper	44.8-45.5	....	0/5	....	8
62					[43.6]	....	0/39		17
63	S. occidentalis	....	Cloaca	Upper	[45.9-46.1]	....	21/21	Summer-acclimated field animals, males & females	41
64					45.5	....	7/7	In direct sunlight	41
65					45.2	....	24/24	Spring-acclimated animals	41
66		9-33.5	Body	Upper	42.8	1.5	34/34	Acclimated for 10-25 da to 16 hr photoperiod (LD16:8) and daily temp program	42
67					41.9	1.5	27/27	Acclimated for 10-25 da to 8 hr photoperiod (LD8:16) and daily temp program	42
68		25	Cloaca	Upper	44.7	....	6/6	Acclimated for 30 da, fasting	41
69			Body	Upper	[44.6]	....	....	16 hr/da photoperiod	39
70				Lower	[14.0]	....	....	16 hr/da photoperiod	39
71		7	Cloaca	Upper	45.4	....	4/4	Acclimated for 34 da	41

[3/] Mean survival time.

continued

Part III. Reptiles and Amphibians

	Order & Species (Synonym)	Accli- mation Temp °C	Temp Mea- sured	Tolerance Limit				Remarks	Ref- er- ence
				Ex- treme	Temp, °C [Rapid]	Duration hr[1]	Survival n/n'		
72	*S. occidentalis biseri- atus*	....	Air	Upper	>40	....	≮50%	....	22
73	*Uma notata*	....	Cloaca	Upper	45-51	0.2	0/5	Air temp, 55°C	55
74					45.6	....	....	Maximum observed field temp	50
75	*Uta stansburiana*	....	Cloaca	Upper	[48.3]	....	0/52	....	17
76	*U. stansburiana hes- peris*	....	Air	Upper	>41.5	....	≮50%	....	22
77	*Xantusia vigilis*	....	Cloaca	Upper	40	1.25	3/6	....	19
78			Air	Upper	38	96	≮50%	....	19
79		25	Body	Upper	[40.7]	....	....	16 hr/da photoperiod	39
80				Lower	[4.3]	....	....	16 hr/da photoperiod	48
	Chelonia								
81	*Chelydra serpentina*	....	Cloaca	Upper	39-41	<0.5	0%	Body wt, 500 g	2
82				Lower	<4	....	≮50%	Body wt, 500 g	2
83		20	Body	Upper	[39.5]	....	8/8	....	33
84	*Chrysemys picta belli*	....	Cloaca	Upper	[42.3]	....	0/3	....	8
85	*(C. belli)*	20	Body	Upper	[41.7]	....	10/10	16 hr/da photoperiod	32
86	*(C. marginata belli)*	....	Cloaca	Upper	40	0.5	≮50%	Air temp, 42.5°C; body wt, 520 g	2
87		20	Water	Upper	40.9-42.2	0.4	112/112	Acclimated for 2 wk, 16 hr/da photoperiod	33
88	*C. scripta (Pseudemys*	....	Cloaca	Upper	[41.2]	....	....	....	8
89	*scripta)*	20	Body	Upper	[41.7]	....	11/11	....	33
90	*C. scripta elegans (P.*	....	Cloaca	Upper	46	....	≮50%	....	58
91	*scripta elegans)*			Lower	1	....	≮50%	....	58
92	*Gopherus agassizii*	....	Cloaca	Upper	[39.5-43]	....	....	....	8
93					36-38	....	100%	....	25
94		20	Body	Upper	[39.0]	....	0/1	....	33
95	*Sternotherus odoratus*	20	Body	Upper	[41.1]	....	19/19	....	33
96	*Terrapene carolina*	20	Body	Upper	[42.6]	....	37/37	....	33
97	*Testudo graeca*	20	Body	Upper	[43.2]	....	1/2	....	33
98	*Trionyx spinifer*	20	Body	Upper	[41.1]	....	10/10	....	33
	Amphibia								
	Salientia								
99	*Acris crepitans*	35	Water	Upper	43.2	0.25	11/11	Acclimated for 3-7 da	26
100		4 & 14	Water	Upper	40.5	0.2-1.2	27/27	Acclimated for 3-7 da	26
	Bufo americanus								
101	Embryo	....	Water	Upper	>30	....	≮50%	....	52
102				Lower	<10	....	≮50%	....	52
103	Larva	30	Water	Upper	36.3	24	≮50%	....	29
104		25	Water	Upper	43.5	Few min	≮50%	....	20
105		15	Water	Upper	[40.5]	Few min	5/10	....	20
106		10	Water	Upper	34	24	≮50%	....	20
107	*B. bufo*	....	Water	Upper	33.5	24	5/21	....	37
108				Lower	−1	2	≮50%	....	36

[1] Unless otherwise indicated.

continued

Part III. Reptiles and Amphibians

	Order & Species (Synonym)	Accli-mation Temp °C	Temp Mea-sured	Tolerance Limit				Remarks	Ref-er-ence
				Ex-treme	Temp, °C [Rapid]	Duration hr	Survival n/n'		
109			Body	Lower	−0.5 to −0.15	0.3-2.4	⋖50%	....	35
110					−1	0.3-2.4	5/5		36
111	B. marinus	....	Water	Upper	41-42	0.4	0/24	Body wt, 30-240 g	64
112					41	0.2	7/24		63
113			Air	Lower	<10	....	⋖50%	Body wt, 30-240 g	64
114			Axilla	Lower	7	....	0/6		63
115		27	Water	Upper	41.7	96	3/3	....	11
116			Cloaca	Upper	40	0.6	20/20	Acclimated for 11 da, unfed	40
117		26	Cloaca	Upper	[39.3]	....	9/9	....	9
118				Lower	[11.0]	....		....	9
119	Crinia signifera	25	Cloaca	Upper	[33.5-35.7]	....	75/75	....	10
120	Dendrobates auratus	26	Cloaca	Upper	[34.8]	....	....	....	9
121				Lower	[9.5]	....		....	9
122	Hyla arborea	....	Cloaca	Lower	−0.5	....	⋖50%	....	38
123	H. caerulea	30	Cloaca	Upper	[39.2]	....	18/18	....	10
124	H. crucifer	....	Cloaca	Upper	[36.2]	....	6/6	Summer-acclimated field animals	7
125	Embryo	....	Water	Upper	28	....	⋖50%	....	52
126				Lower	<6	....	⋖50%	....	52
	H. regilla								
127	Larva	27	Water	Upper	36	>2	....	....	13
128	Adult	23	Cloaca	Lower	[−1.0]	....		....	9
129		5	Cloaca	Upper	[31.8]	....		....	9
130	Leptodactylus penta-	27	Body	Upper	[35]	....		....	6
131	dactylus			Lower	[4-8]	....		....	6
132	Limnodynastes tas-maniensis	25	Cloaca	Upper	[33.4]	....	12/12	....	10
133	Pseudacris triseriata	35	Water	Upper	40.4	0.2	8/8	Acclimated for 3-7 da	26
134		25	Water	Upper	39.6	0.5	11/11	Acclimated for 3-7 da	26
135		4 & 14	Water	Upper	38.5	0.2-0.8	36/36	Acclimated for 3-7 da	26
	Rana catesbeiana								
136	Embryo	....	Water	Upper	32[4]	....	⋖50%	....	53
137				Lower	15[4]	....	⋖50%	....	53
138	Larva	....	Water	Upper	36.3	....	⋖50%	Brief exposure	12
139	Adult	....	Air	Upper	>22	>24	⋖50%	High humidity	15
140		23	Body	Upper	[33.7]	....	3/3	....	11
141			Water	Upper	33.5	....	3/3	Acclimated to 23°C for approx. 3 da	11
142	R. pipiens	29	Water	Upper	35.2	....	3/3	Acclimated for approx. 2 da	11
143		15 & 25	Cloaca	Upper	38-41	0.25-0.5	690/690	Acclimated to different photoperiods	49
144		12	Water	Upper	33.7	....	3/3	Acclimated for approx. 2 da	11
145		5	Water	Upper	31.2	....	3/3	....	11
146	Embryo	....	Water	Upper	34[4]	....	⋖50%	Florida animals	54
147					28[4]	....	⋖50%	Vermont animals	54
148				Lower	11[4]	....	⋖50%	Florida animals	54
149					6[4]	....	⋖50%	Vermont animals	54

[4] Incubation temp.

continued

Part III. Reptiles and Amphibians

	Order & Species (Synonym)	Acclimation Temp °C	Temp Measured	Extreme	Temp,°C [Rapid]	Duration hr	Survival n/n'	Remarks	Reference
150	First cleavage to	....	Water	Upper	>26	21	≮50%	....	47
151	yolk plug			Lower	<4	480	≮50%	....	47
152	Adult	....	Body	Upper	[38.5]	....	1/3	Spring-acclimated animals	15
153				Lower	−1.5 to −1.8	2	0/2	....	14
154			Cloaca	Upper	[35.7]	....	10/10	Costa Rican animals	9
155				Lower	[−1.6]	....	....	Costa Rican animals	9
156		30	Cloaca	Upper	[36.5]	....	23/23	....	61
157		25	Cloaca	Upper	[38.7]	....	....	Vermont animals	31
158		23	Body	Upper	[34.5]	....	3/3	....	11
159		15	Body	Upper	[38.0]	....	....	12 hr/da photoperiod	49
160	*R. temporaria (R.*	....	Water	Upper	32.5-33.0	1	2/10	....	37
161	*fusca)*		Body	Lower	−1.4	0.3	1/1	Body wt, 20 g	65
162	Embryo	....	Water	Upper	24	....	≮50%	....	52
163				Lower	1	....	≮50%	....	52
	Scaphiopus hammondi								
164	Larva	26	Water	Upper	39	>2	....		13
165	Adult	30	Cloaca	Upper	[40.3]	....	....		9
166		23	Cloaca	Lower	[0.0]	....	....		9
167	*Xenopus laevis (X.*	....	Body	Upper	33	....	0/3	....	64
168	*dactylethra)*		Cloaca	Lower	0	....	≮50%	....	64
	Caudata								
169	*Ambystoma macula-*	10	Cloaca	Upper	34.5	0.5	12/12	Acclimated for 1 wk, unfed	28
170	*tum*	5	Water	Upper	37.6	0.5	5/5	Tested 2 da after removal from stream water at 5°C	60
171	Embryo	....	Water	Lower	>3.5	>23	≮50%	....	30
172	Juvenile	....	Water	Upper	[38.6]	....	....	Morning	56
173					[39.7]	....	....	Afternoon	56
174	Adult	....	Cloaca	Upper	[34.9]	....	....	Spring-acclimated	28
175		20	Water	Upper	[37.2]	....	4/4	Acclimated for 1 mo	30
176		15	Water	Upper	[38.4]	0.5	9/9	Acclimated for 2 wk	60
177	*A. tigrinum*	....	Cloaca	Upper	[36.7]	....	6/6		7
178	*Amphiuma means*	20	Water	Upper	[37.1]	....	3/3	Acclimated for 1 mo	30
179	*Desmognathus fuscus*	17-26	Water	Upper	[33.5]	....	....	....	66
180		20	Water	Upper	[36.3]	....	7/7	....	30
181	*Notophthalmus viri-*	17-26	Water	Upper	[37.3]	....	....	....	66
182	*descens (Diemicty-*	20	Water	Upper	[38.3]	....	10/10	....	30
	lus viridescens)								
183	*Plethodon cinereus*	20	Water	Upper	[34.7]	....	5/5	Acclimated for 1 mo	30
184	*P. glutinosus*	20	Water	Upper	[35.1]	....	7/7	....	30
185	*Salamandra salaman-*	....	Cloaca	Lower	−3 to −2	24	0/3	....	34
186	*dra (S. maculosa)*		Water	Lower	−3.5	1.1	≮50%	....	34
187	*Taricha torosa*	22	Water	Upper	36.0	0.3	0/4	....	51

Contributors: Sealander, John A., Jr.; Adolph, Edward F.; Bennett, Albert F., and Dawson, William R.

References

[1] Bailey, R. M. 1949. Ecology 30:238.
[2] Baldwin, F. M. 1925. Biol. Bull. 48:432.
[3] Ballinger, R. E., and J. Schrank. 1970. Physiol. Zool. 43:19.
[4] Bogert, C. M., and R. Martin del Campo. 1956. Bull. Amer. Mus. Natur. Hist. 109:1.
[5] Bradshaw, S. D., and A. R. Main. 1968. J. Zool. 154:193.

continued

Part III. Reptiles and Amphibians

[6] Brattstrom, B. H. 1960. Year B. Amer. Phil. Soc., p. 284.

[7] Brattstrom, B. H. 1963. Ecology 44:238.

[8] Brattstrom, B. H. 1965. Amer. Midl. Natur. 73:376.

[9] Brattstrom, B. H. 1968. Comp. Biochem. Physiol. 24:93.

[10] Brattstrom, B. H. 1970. Ibid. 35:69.

[11] Brattstrom, B. H., and P. Lawrence. 1962. Physiol. Zool. 35:148.

[12] Brett, J. R. 1944. Publ. Ont. Fish. Res. Lab. 63.

[13] Brown, H. A. 1969. Copeia, p. 138.

[14] Cameron, A. T., and T. I. Brownlee. 1913. Quart. J. Exp. Physiol. 7:115.

[15] Cameron, A. T., and T. I. Brownlee. 1915. Ibid. 9: 247.

[16] Colbert, E. H., et al. 1946. Bull. Amer. Mus. Natur. Hist. 86:327.

[17] Cole, L. C. 1943. Ecology 24:94.

[18] Corn, M. J. 1971. J. Herpetol. 5:17.

[19] Cowles, R. B., and C. M. Bogert. 1944. Bull. Amer. Mus. Natur. Hist. 83:261.

[20] Davenport, C. B., and W. E. Castle. 1895. Arch. Entwicklungsmech. Organ. 2:227.

[21] Dawson, W. R. 1960. Physiol. Zool. 33:87.

[22] Dawson, W. R., and G. A. Bartholomew. 1956. Ibid. 29:40.

[23] Dawson, W. R., and G. A. Bartholomew. 1958. Ibid. 31:100.

[24] Dawson, W. R., and J. R. Templeton. 1963. Ibid. 36:219.

[25] Dill, D. B. 1938. Life, Heat, and Altitude. Harvard Univ. Press, Cambridge.

[26] Dunlap, D. G. 1968. Physiol. Zool. 41:432.

[27] Fitch, A. V. 1964. Herpetologica 20:184.

[28] Gatz, A. J., Jr. 1971. Ibid. 27:157.

[29] Hathaway, E. S. 1927. U.S. Bur. Fish. Bull. 43:169.

[30] Hutchison, V. H. 1961. Physiol. Zool. 34:92.

[31] Hutchison, V. H., and M. R. Ferrance. 1970. Herpetologica 26:1.

[32] Hutchison, V. H., and R. J. Kosh. 1964. Ibid. 20: 233.

[33] Hutchison, V. H., et al. 1966. Ibid. 22:32.

[34] Jecklin, L. 1935. Rev. Suisse Zool. 42:731.

[35] Kalabuchov, N. 1934. Dokl. Akad. Nauk SSSR 1: 424.

[36] Kalabuchow, N. I. 1935. Zool. Jahrb. Abt. Allg. Zool. Physiol. Tiere 55:47.

[37] Kirk, R. L., and L. Hogben. 1946. J. Exp. Biol. 22: 213.

[38] Knauthe, K. 1891. Zool. Anz. 14:104.

[39] Kour, E. L., and V. H. Hutchison. 1970. Copeia, p. 219.

[40] Krakauer, T. 1970. Comp. Biochem. Physiol. 33:15.

[41] Larson, M. W. 1961. Herpetologica 17:113.

[42] Lashbrook, M. K., and R. L. Livezy. 1970. Physiol. Zool. 43:38.

[43] Liberman, S. S., and N. V. Pokrowskaia. 1943. Zool. Zh. 22:247.

[44] Licht, P. 1964. Comp. Biochem. Physiol. 13:27.

[45] Licht, P. 1968. Amer. Midl. Natur. 79:149.

[46] Licht, P., et al. 1966. Copeia, p. 162.

[47] Lillie, F. R., and F. P. Knowlton. 1897. Zool. Bull. 1:179.

[48] Lueth, F. X. 1941. Copeia, p. 125.

[49] Mahoney, J. J., and V. H. Hutchison. 1969. Oecologia (Berlin) 2:143.

[50] Mayhew, W. W. 1964. Herpetologica 20:170.

[51] McFarland, W. N. 1955. Copeia, p. 191.

[52] Moore, J. A. 1940. Amer. Natur. 74:188.

[53] Moore, J. A. 1942. Biol. Bull. 83:375.

[54] Moore, J. A. 1942. Biol. Symp. 6:189.

[55] Mosauer, W. 1936. Ecology 17:56.

[56] Pough, F. H., and R. E. Wilson. 1970. Physiol. Zool. 43:194.

[57] Prieto, A. A., and W. Whitford. 1971. Copeia, p. 498.

[58] Rodbard, S., and D. Feldman. 1946. Proc. Soc. Exp. Biol. Med. 6:43.

[59] Ruibal, R. 1961. Evolution 15:98.

[60] Sealander, J. A., and B. W. West. 1969. Herpetologica 25:122.

[61] Seibel, R. A. 1970. Ibid. 26:208.

[62] Stewart, G. R. 1965. Ibid. 21:81.

[63] Stuart, L. 1951. Copeia, p. 220.

[64] Taylor, N. B. 1931. J. Physiol. (London) 71:156.

[65] Weigmann, R. 1929. Z. Wiss. Zool. 134:641.

[66] Zweifel, R. G. 1957. Ecology 38:64.

Part IV. Fishes

Tolerance Limit: The water temperature extreme permitting 50% survival, unless otherwise indicated; geographical distribution, nutritive state, type of water, and water pressure also affected tolerance.

	Species (Synonym)	Acclimation Temp, °C	Tolerance Limit			Reference
			Extreme	Temp, °C	Duration hr	
	Osteichthyes					
1	*Carassius auratus*	37	Upper	42	14	11
2			Lower	15	14	

continued

Part IV.   Fishes

	Species (Synonym)	Acclimation Temp, °C	Tolerance Limit			Reference
			Extreme	Temp, °C	Duration hr	
3		30	Upper	38	24	5
4			Lower	9	24	
5		20	Upper	35	24	5
6			Lower	2	24	
7		10	Upper	31	24	5
8			Lower	<0	24	
9		2	Upper	28	14	11
10	*Clupea harengus* Larva	15.5[1]	Upper	23.0	24	3
11			Lower	−0.5	24	
12		10.5[1]	Upper	22.0	24	3
13			Lower	−1.0	24	
14		15[2]	Upper	23.7	24	3
15			Lower	−0.5	24	
16		8[2]	Upper	22.5	24	3
17			Lower	−1.8	24	
18	Fry	15	Upper	23.4	24	3
19			Lower	−1.0	24	
20		7.5	Upper	22.2	24	3
21			Lower	−1.8	24	
22	Yearling	15	Upper	22.9	24	3
23			Lower	−0.75	24	
24		11	Upper	22.0	24	3
25			Lower	−1.1	24	
26	*Coregonus clupeaformis*, eggs hatching	......	Upper	10	.....	22
27			Lower	0.5	.....	
28	*Cyprinus carpio*	26	Upper	35.7	24	2
29		20	Upper	31-34	24	2
30		......	Lower	−0.7[3]	24	23
31	*Esox lucius*, juvenile	30	Upper	35.5	1.1	24
32				33.5	6.7	
33		25	Upper	34.5	1.0	24
34				32.5	6.1	
35	*Fundulus heteroclitus*	......	Upper	40	2	4,18
36			Lower	1[3]	.....	
37		28	Upper	37	.....	9
38		20	Upper	34	.....	9
39			Lower	2	48	
40		14	Upper	32	.....	9
41			Lower	1	48	
42	*Gadus morhua (G. callarias)*	......	Upper	19.8-24.4[4]	.....	17
43			Lower	−2	13	8
44	Embryo	......	Upper	10	.....	20
45			Lower	−1		
46	*Gasterosteus aculeatus (G. bispinosus)*	......	Upper	31.7-33[4]	.....	17,23,26
47			Lower	−0.7[3]		
48	Embryo	......	Upper	23	.....	20
49			Lower	8		

[1] Autumn-spawned.   [2] Spring-spawned.   [3] Body temperature.   [4] Rise in temperature of 1°C/5 min until death.

continued

Part IV. Fishes

	Species (Synonym)	Acclimation Temp, °C	Tolerance Limit			Reference
			Extreme	Temp, °C	Duration hr	
50	Ictalurus punctatus (I. lacustris)	25	Upper	34	.....	14
51			Lower	6	24	
52		20	Upper	33	.....	14
53			Lower	3	24	
54		15	Upper	30	.....	14
55			Lower	0	24	
56	Lepomis macrochirus purpurescens	30	Upper	34	.....	14
57			Lower	11	24	
58		20	Upper	32	.....	14
59			Lower	5	24	
60		15	Upper	31	.....	14
61			Lower	3	24	
62	Leptocottus armatus	......	Upper	29.5	.....	21
63			Lower	0.3	0.3	10
64	Melanogrammus aeglefinus	......	Upper	18.5-22.9[4]	.....	17
65	Micropterus salmoides	30	Upper	36	.....	14
66			Lower	12	24	
67		20	Upper	33	.....	14
68			Lower	6	24	
69		10	Upper	28	24	15
70	Oncorhynchus keta	2.5	Lower	−0.1[5]	66.7	7
71	Fry	5.0	Lower	−0.5	17.5	7
72				−1.0	5.6	
73				−1.5	0.7	
74		2.5	Lower	−0.5	22.0	7
75				−1.0	7.2	
76				−1.5	0.5	
77	Juvenile	20	Upper	23.7	168	6
78			Lower	6.5	92	
79		10	Upper	22.6	168	6
80			Lower	0.5	92	
81		5	Upper	21.8	168	6
82	Yearling	5.0	Lower	−0.5	46.7	7
83				−1.0	0.5	
84				−1.5	0.2	
85	O. tshawytscha	20	Upper	25.1	168	6
86			Lower	4.5	92	
87		10	Upper	24.3	168	6
88			Lower	0.8	92	
89		5	Upper	21.5	168	6
90	Osmerus mordax	......	Upper	21.5-28.5[4]	.....	17
91	Perca flavescens	25[6]	Upper	32	.....	14
92			Lower	9	24	
93		25[7]	Upper	30	.....	14
94			Lower	4	24	
95		20	Upper	29	24	14
96			Lower	3	24	
97		10	Upper	25	24	14
98		5	Upper	21	24	14

[4] Rise in temperature of 1°C/5 min until death. [5] Ultimate incipient lethal. [6] Summer. [7] Winter.

continued

Part IV. Fishes

	Species (Synonym)	Acclimation Temp, °C	Tolerance Limit			Reference
			Extreme	Temp, °C	Duration hr	
99	*Petromyzon marinus*, prolarva & postlarva	20	Upper	34.0	1.5	19
100				31.5	8.3	
101				30.5	41.7	
102				30.0	166.7	
103		15	Upper	34.0	0.8	19
104				31.5	7.5	
105				30.0	66.7	
106				29.0	166.7	
107	*Pseudopleuronectes americanus*	......	Upper	27.9-30.6[4]	......	17
108			Lower	−2	1	
109	*Salmo clarki clarki*, yearling	20	Upper	29.9[8]	......	16
110		15	Upper	29.1[8]	......	16
111		10	Upper	27.6[8]	......	16
112	*S. gairdneri*, embryo	......	Upper	13	......	20
113			Lower	3		
	S. salar					
114	Prolarva & postlarva	......	Upper	28	1	1
115				24	65	
116				22	188	
117	Alevin	20	Upper	26	10	1
118				24	48	
119		10	Upper	25-26	2	1
120				24	12	
121		5	Upper	25	1	1
122				22.5	6	
123	Parr	......	Upper	29.8	24	25
124	Yearling	......	Upper	28.5	24	25
	S. trutta trutta					
125	Prolarva & postlarva	......	Upper	28	1	1
126				25	12	
127				23	144	
128				22	288	
129	Alevin	20	Upper	26	7	1
130				24	60	
131		10	Upper	26	1	1
132				24	48	
133	Parr	......	Upper	29.1	24	25
134	Yearling	......	Upper	26.4	24	25
135	*Salvelinus fontinalis*	23	Lower	1	48	12
136		20; 25	Upper	25	......	12
137		10	Upper	24	24	13
138		3	Upper	23	......	12
139	*Scomber scombrus*, embryo	......	Upper	21	......	20
140			Lower	11		
	Chondrichthyes					
141	*Raja ocellata (R. diaphanes)*	......	Upper	28	24	8
142				26.5-26.9[4]	......	17
143			Lower	−2	14	8
144	*Squalus acanthias*	......	Upper	28.5-29.1[4]	......	17

[4] Rise in temperature of 1°C/5 min until death.   [8] Heating rate for critical thermal maximum was 0.4°C/min.

continued

Part IV. Fishes

Contributors: Bardach, John E., Hart, J. Sanford; Brett, J. R.

References

[1] Bishai, H. M. 1960. J. Cons. Cons. Perm. Int. Explor. Mer 25(2):129.

[2] Black, E. C. 1953. J. Fish. Res. Bd. Can. 10(4):196.

[3] Blaxter, J. H. S. 1960. J. Mar. Biol. Ass. U.K. 39(3): 605.

[4] Borodin, N. A. 1934. Zool. Jahrb. Abt. Allg. Zool. Physiol. Tiere 53:313.

[5] Brett, J. R. 1944. Publ. Ont. Fish. Res. Lab. 63.

[6] Brett, J. R. 1952. J. Fish. Res. Bd. Can. 9(6):265.

[7] Brett, J. R., and D. F. Alderdice. 1958. Ibid. 15(5): 805.

[8] Britton, S. W. 1924. Amer. J. Physiol. 67:411.

[9] Doudoroff, P. 1945. Biol. Bull. 88:194.

[10] Fries, E. F. B. 1952. Copeia, p. 147.

[11] Fry, F. E. J., et al. 1942. Rev. Can. Biol. 1:50.

[12] Fry, F. E. J., et al. 1946. Publ. Ont. Fish. Res. Lab. 66.

[13] Hart, J. S. 1947. Trans. Roy. Soc. Can., V, 41:57.

[14] Hart, J. S. 1952. Publ. Ont. Fish. Res. Lab. 72.

[15] Hathaway, E. S. 1928. U.S. Bur. Fish. Bull. 43:169.

[16] Heath, W. G. 1963. Science 142:486.

[17] Huntsman, A. G., and M. I. Sparks. 1924. Contrib. Can Biol. Fish 2(6):97.

[18] Loeb, J., and H. Westeneys. 1912. J. Exp. Zool. 12:543.

[19] McCauley, R. W. 1963. J. Fish. Res. Bd. Can. 20(2): 483.

[20] Moore, J. A. 1940. Amer. Natur. 74:188.

[21] Morris, R. W. 1960. Limnol. Oceanogr. 5:175.

[22] Price, J. W. 1940. J. Gen. Physiol. 23:449.

[23] Schmidt, P. J., et al. 1936. Dokl. Akad. Nauk SSSR 3:305.

[24] Scott, D. P. 1964. J. Fish. Res. Bd. Can. 21(5): 1043.

[25] Spass, J. T. 1960. Hydrobiologia 15(1-2):78.

[26] Weigmann, R. 1936. Biol. Zentralbl. 56:301.

Part V. Aquatic Invertebrates

Data pertain to adult stages unless otherwise specified, and are meaningful only under the conditions of the experiments performed. Limits of tolerance vary with temperature and duration of acclimation; rate of temperature changes and period of time spent at a critical temperature; taxonomic strain, color, sex, size, stage of life history, age, and degree of hunger of specimens; geographic source; season collected; concentration of specimens; and salinity, gas content and pH of water. **Survival:** Percentages indicate present survived.

Class[1] & Species (Synonym)	Acclimation Temp °C	Extreme	Temp °C	Duration min[2]	Survival	Remarks	Reference
			Tolerance Limit				
Chordata							
1 Cephalochordata[3] *Branchiostoma caribaeum (B. lanceolatus)*	....	Upper	40.6	....	.....	Lethal temp	33
2 Thaliacea *Salpa maxima (S. africana)*	....	Upper	37.7	....	.....	Lethal temp	33
Echinodermata							
3 Ophiuroidea *Ophioderma brevispinum*	22.3-25.3	Upper	37.7	....	.....	Optimum temp, 30°C ±	19
4		Lower	−0.6	....	.....		
5 Asteroidea *Asterias forbesi*	....	Upper	42	9	.....	....	22
6			36	9	.....		
7			32	43	.....		
8 Echinoidea *Arbacia punctulata*	....	Upper	42	9	.....	....	40

[1] Unless otherwise indicated. [2] Unless otherwise specified. [3] Subphylum.

continued

Part V. Aquatic Invertebrates

	Class[1] & Species (Synonym)	Accli- mation Temp °C	Tolerance Limit				Remarks	Ref- er- ence
			Ex- treme	Temp °C	Duration min[2]	Sur- vival		
9				38	45			
10				37	89			
11	*Psammechinus microtuberculatus* (*Echinus microtuberculatus*)	....	Upper	39.1	....	.....	Lethal temp	33
12	*Strongylocentrotus purpuratus*	....	Upper	23.5	....	.....	From Alaska to Mexico	8
	Chaetognatha							
	Chaetognatha[4]							
13	*Sagitta elegans*	20	Upper	25.5-27.5	....	.....	Temp increased 1°C/5 min	14
	Arthropoda							
	Arachnida							
14	*Hydrachna cruenta*	....	Upper	46.2	....	.....	Lethal temp	4
	Merostomata							
15	*Limulus polyphemus*[5]	....	Lower	−1 to 0	....	.....	Heart beat stopped	1
16		30	Upper	46.2	....	.....	Optimum temp, 41°C	19
17		20	Upper	41	....	.....	Optimum temp, 38.1°C	19
18		16	Upper	41	....	.....	....	19
	Crustacea—Malacostraca[6,7]							
	Homarus americanus							
19	3rd stage zoeae	15	Upper	32.5	....	.....	....	13
20	4th stage zoeae	25	Upper	34.2	....	.....	....	13
21		15	Upper	34.0	....	.....	....	13
22	5th stage zoeae	20	Upper	34.9	....	.....	....	13
23		15	Upper	33.4	....	.....	....	13
24	Adults	25	Upper	29.5	....	.....	Salinity, 25‰; oxygen, 4.3 mg/ liter	21
25		15	Upper	28.2	....	.....		
26		5	Upper	22.1	....	.....		
27	*Orconectes rusticus*	30	Upper	36.6	24 hr	50%	Acclimated for 1 wk	25
28		22-26	Upper	36.4	12 hr	50%	Acclimated for 1-5 wk	25
29				35.6	24 hr	50%		
30		4	Upper	35.3	12 hr	50%	Acclimated for 4 da	25
31				34.8	12 hr	50%	Acclimated for 8 da	
32				34.1	12 hr	50%	Acclimated for 12 da	
33				33.5	12 hr	50%	Acclimated for 16 da	
	Uca pugnax							
34	1st stage zoeae	25	Upper	38	31 hr	50%	Salinity 35‰	27
35					1 hr	50%	Salinity 10‰	
36	Adults	23-24	Upper	44	9.5	50%	Temperate zone animals	28
37				42	38	50%		
38			Lower	3.5	7 hr	50%	....	
39		15	Upper	44	7	50%	Temperate zone animals	28
40				42	18	50%		
	U. rapax[8]							
41	1st stage zoeae	25	Upper	38	61 hr	50%	Salinity 35‰	27
42					1 hr	50%	Salinity 10‰	
43	Adults	23-24	Upper	44	11	50%	Tropical zone animals	28
44				42	40	50%		

[1] Unless otherwise indicated. [2] Unless otherwise specified. [4] Phylum. [5] For additional tables of data on *Limulus polyphemus*, consult reference 15. [6] Subclass. [7] For data on 35 oceanic species, consult reference 30. [8] For data on other *Uca* species, consult reference 29.

continued

Part V. Aquatic Invertebrates

	Class [1] & Species (Synonym)	Acclimation Temp °C	Tolerance Limit				Remarks	Reference
			Extreme	Temp °C	Duration min[2]	Survival		
45			Lower	7	35	50%	....	
46		15	Upper	42	18	50%	Tropical zone animals	28
47			Lower	7	55	50%	....	
	Crustacea—Cirripedia[6]							
48	*Lepas fascicularis*	29	Upper	42.3	....	.....	Optimum temp, 30.3-32.1°C	19
49			Lower	−2.3 to +1.4	....	.....		
	Crustacea—Copepoda[6]							
50	*Cyclops serulatus, C. vernalis, & C. viridis (Calamus serulatus, C. vernalis, C. viridis)*	....	Upper	30.0-38.5	....	.....	Critical temp for dormancy	3
	Crustacea—Branchiopoda[6]							
	Artemia salina							
51	Eggs	25	Upper	103.5	75	.....	....	12
52	Adults	....	Upper	39	....	.....	Optimum temp, 25°C	18
53	*Daphnia pulex*	....	Upper	30	....	.....	Optimum temp, 14-18°C	23
54			Lower	0	....	.....		
	Insecta							
	Aedes aegypti							
55	Eggs	....	Upper	46.7	....	50%	....	7
56	1st instar larvae	....	Upper	45.7	....	50%	....	7
57	2nd instar larvae	....	Upper	46.3	....	50%	....	7
58	3rd instar larvae	....	Upper	45.6	....	50%	....	7
59	4th instar larvae	....	Upper	45.1	....	50%	....	7
60	Pupae	....	Upper	46.7	....	50%	....	7
	Culex pipiens							
61	Eggs	....	Upper	42.5	....	50%	....	7
62	Larvae	....	Upper	40	....	.....	Lethal temp	4
63			Lower	−4	1 hr	0%	Lethal temp	17
64	1st instar	....	Upper	41.7	....	50%	....	7
65	2nd instar	....	Upper	41.8	....	50%	....	7
66	3rd instar	....	Upper	41.8	....	50%	....	7
67	4th instar	....	Upper	41.3	....	50%	....	7
68	Pupae	....	Upper	42.0	....	50%	....	7
	Annelida							
	Polychaeta							
69	*Eunice schemacephala (E. fucata)*, 4-5 da old	29	Upper	42.7	....	.....	Optimum temp 35.5°C ±	19
70			Lower	−2.3	....	.....		
	Mollusca							
	Cephalopoda							
71	*Octopus vulgaris*	....	Upper	36	....	.....	....	33
	Bivalvia							
72	*Crassostrea virginica*	24	Upper	47.5	....	50%	Rapid temp increase	9
73				41.0	....	50%	Slow temp increase	
74	*Mercenaria mercenaria (Venus mercenaria)*	15	Upper	45.2	....	.....	Temp increased 1°C/5 min	11

[1] Unless otherwise indicated. [2] Unless otherwise specified. [6] Subclass.

continued

Part V. Aquatic Invertebrates

	Class & Species (Synonym)	Accli-mation Temp °C	Tolerance Limit				Remarks	Ref-er-ence
			Ex-treme	Temp °C	Duration min[2]	Sur-vival		
75	*Mya arenaria*	15	Upper	40.6	....	.....	Temp increased 1°C/5 min	11
76	*Mytilus edulis*	15	Upper	40.8	....	.....	Temp increased 1°C/5 min	11
77	*Placopecten magellanicus (Pec-ten grandis)*	....	Upper	23.5	....	.....	Acclimated to natural water; le-thal temp rose 1°C for each 5°C increase in acclimation temp	5
78	*Spisula solidissima (Mactra sol-idissima)*	15	Upper	37.0	....	.....	Temp increased 1°C/5 min	11
	Gastropoda							
79	*Fimbria fimbria (Tethys leporina)*	....	Upper	40.5	....	.....	Lethal temp	34
80	*Ilyanassa obsoleta (Nassarius ob-soleta)*	25	Upper	39	3 hr	20%	Parasitized	32
81				39	3 hr	50%	Nonparasitized	
82				37	6 hr	55%	....	
83	*Littorina littorea*	....	Upper	46.0	....	.....	Temp increased 1°C/5 min	6
84				43.7	....	.....	Lethal temp, high tide	10
85				43.5	....	.....	Lethal temp, mid-tide	10
86				41.0	....	.....	Lethal temp, low tide	10
87	*Lymnaea stagnalis*	....	Lower	12	....	.....	Limit for growth and develop-ment	1

Platyhelminthes

	Trematoda							
88	*Himasthla quissetensis*[9]	25	Upper	41	6 hr	100%	Cercariae survived for at least 6 hr	31
89	*Zoogonus rubellus*[10]	25	Upper	39	30	.....	Cercariae died within 30 min	31
	Turbellaria							
90	*Dugesia gonocephala*	....	Upper	27.5	....	.....	Summer acclimated	2
91				25.0	....	.....	Winter acclimated	

Ctenophora

	Nuda							
92	*Beroe cucumis*	14	Upper	29.7-30.0	....	.....	Optimum temp, 18-28°C	19

Coelenterata (Cnideria)

	Anthozoa							
93	*Actinia equina*	....	Upper	43.5	....	.....	Mean lethal temp	33
94	*Porites clavaria*	....	Upper	36.4	....	.....	Mean lethal temp	20
	Scyphozoa							
95	*Aurelia aurita*	29	Upper	38.5	....	.....	Optimum temp, 29°C ±	19
96			Lower	7.7-11.8	....	.....		
97		14	Upper	30	....	.....	Optimum temp, 18-23°C	19
98			Lower	−1.4	....	.....		
99	*Cassiopea frondosa*	29	Upper	40	....	.....	Optimum temp, 32.5°C ±	19
100			Lower	8.3-9.7	....	.....		
101	*Cyanea capillata (C. arctica)*	14	Upper	26.8-28.0	....	.....	Optimum temp, 19°C ±	19
	Hydrozoa							
102	*Pennaria tiarella*	29	Lower	−2.3	....	.....	Movements ceased	19
103		22-26	Upper	34.7	....	.....	Movements ceased	19
104			Lower	−0.6	....	.....		

[2] Unless otherwise specified. [9] Parasite of bird. [10] Parasite of fish.

continued

Part V. Aquatic Invertebrates

Class [1]/ & Species (Synonym)	Accli- mation Temp °C	Tolerance Limit				Remarks	Ref- er- ence
		Ex- treme	Temp °C	Duration min[2]/	Sur- vival		
Protozoa							
Ciliatea							
105 Blepharisma lateritia	20	Upper	40.0	....	46%	....	15
106 Paramecium aurelia	....	Upper	31.5	....	.....	....	2
107	20	Upper	40.0	....	58.8%	....	15
108 Stentor coeruleus	....	Lower	2	....	.....	Ciliary movement ceased	1
109 Tetrahymena pyriformis, strain E	....	Upper	39.0	9 hr	.....	....	24
Rhizopodea							
110 Amoeba proteus	....	Upper	35.5-38.3	60	.....	....	26
Mastigophora [11]/							
111 Euglena gracilis	....	Upper	37.5-44.0	....	.....	Lethal temp	16

[1]/ Unless otherwise indicated. [2]/ Unless otherwise specified. [11]/ Superclass.

Contributors: Vernberg, F. J. and Dexter, Ralph W.

References

[1] Belehrádek, J. 1935. Protoplasma Monogr. (Berlin) 8:1.

[2] Bläsing, I. 1953. Zool. Jahrb. Abt. Allg. Zool. Physiol. Tiere 64(2):112.

[3] Coker, R. E. 1934. J. Elisha Mitchell Sci. Soc. 50: 143.

[4] Davenport, C. B., and W. E. Castle. 1895. Arch. Entwicklungsmech. Organismen 2:227.

[5] Dickie, L. M. 1958. J. Fish. Res. Bd. Can. 15:1189.

[6] Evans, R. G. 1948. J. Anim. Ecol. 17:165.

[7] Farid, M. A. 1949. Amer. J. Hyg. 49:83.

[8] Farmanfarmaian, A., and A. Giese. 1963. Physiol. Zool. 36:237.

[9] Fingerman, M., and L. D. Fairbanks. 1956. Anat. Rec. 125:636.

[10] Gowanlock, J. N., and F. R. Hayes. 1926. Contrib. Can. Biol. 3:133.

[11] Henderson, J. T. 1929. Ibid. 4:397.

[12] Hinton, H. E. 1954. Ann. Mag. Natur. Hist., Ser. 12, 7:158.

[13] Huntsman, A. G. 1924. Contrib. Can. Biol. 2:91.

[14] Huntsman, A. G., and M. I. Sparks. 1924. Ibid. 2: 95.

[15] Hutchison, R. H. 1913. J. Exp. Zool. 15:131.

[16] Jahn, T. L. 1933. Arch. Protistenk. 79:249.

[17] Luyet, B. J., and P. M. Gehenio. 1940. Biodynamica 3:33.

[18] Mathias, P. 1937. Actual. Sci. Ind. 447:1.

[19] Mayer, A. G. 1914. Carnegie Inst. Wash. Publ. 183:1.

[20] Mayer, A. G. 1918. Ibid. 252:175.

[21] McLeese, D. W. 1956. J. Fish. Res. Bd. Can. 13: 247.

[22] Orr, P. R. 1955. Physiol. Zool. 28:290.

[23] Pagliani, G. 1935. Atti Soc. Ital. Sci. Natur. Mus. Civ. Stor. Natur. Milano 74:295.

[24] Slater, J. V. 1954. Amer. Natur. 88:168.

[25] Spoor, W. A. 1955. Biol. Bull. 108:77.

[26] Thornton, F. E. 1932. Physiol. Zool. 5:246.

[27] Vernberg, F. J. 1969. Amer. Zool. 9:333.

[28] Vernberg, F. J., and R. E. Tashian. 1959. Ecology 40:589.

[29] Vernberg, F. J., and W. B. Vernberg. 1966. Oikos 18:118.

[30] Vernberg, F. J., and W. B. Vernberg. 1970. Mar. Biol. 6:26.

[31] Vernberg, W. B. 1961. Exp. Parasitol. 11:270.

[32] Vernberg, W. B., and F. J. Vernberg. 1963. Ibid. 14:330.

[33] Vernon, H. M. 1899. J. Physiol. (London) 25:131.

[34] Wright, W. R. 1927. Bull. Entomol. Res. 18:91.

88. TOLERANCE TO TEMPERATURE EXTREMES: PLANTS

Part I. Minimum and Maximum: Freshwater Algae

Almost all algae tested so far will survive water freezing to carbon dioxide freezing temperatures [1].

	Location	Month [Water Temp, °C]	Species[1]	Temp, °C Minimum	Temp, °C Maximum	Reference
1	Alaska, Barrow, USA	Aug [6]	*Spirogyra* sp.	−7[2]	40[3]	4
2			*Ulothrix zonata*	−15[4]	38[3]	4
3	Texas, USA	....	*Chlorella sorokiniana* 7-11-05[5]	15[6]; 20[7]	...	12
4				....	42	10,11
5	Yellowstone National Park, USA	....	*Anacystis thermalis*	....	42	5
6			Bacillariophyceae[8]	−11	51	5
7			*Chroococcus minutis fuscus*	....	46	5
8			*C. yellowstonensis*	....	41	5
9			Chrysophyceae[8]	....	40	5
10			*Nostoc sphaericum*	....	30	5
11			*Oscillatoria filiformis*	59	83	5
12			*O. geminata*	....	45	5
13			*Protococcus botryoides*	....	80	5
14			*Synechococcus eximius*	70	84	5
15			*S. vulcanus*	46	85	5
16			*S. vulcanus bacillarioides*	57	70	5
17			Xanthophyceae[8]	....	33	5
18	Iceland, Reykir; warm rivulet	[32]	*Cladophora fracta*	−2.5[9]	35[4]	3
19		[25]	*Oedogonium capillare*	0[4]	36[4]	3
20	Austria, Laxenburg; pond	Nov [2]	*Cladophora fracta*	−4[4]	36[4]	9
21	Austria, Ybbs River	Nov [4]	*Spirogyra* sp.	0[4]	28[4]	9
22	Japan	....	*Anabaena* sp.	....	40	8
23			*Chroococcus* sp.	....	57	8
24			*Oscillatoria amphibia*	....	50	8
25			*O. formosa*	....	50	8
26			*O. okeni*	....	44	8
27			*O. proboscidea*	....	47	8
28			*O. tenuis tergestina*	....	44	8
29	Worldwide, hot acid springs	....	*Cyanidium caldarium*	....	60	2
30	Unspecified	....	*Chlamydomonas nivalis*	−36	4	7
31			*Chlorella pyrenoidosa* Emerson	7[6]; 8[7]	...	12
32				....	29	11
33			Cryptophyceae[8]	....	40	5
34		Jan-Feb	*Nostoc kihlmani*	<−17	...	6

[1] Unless otherwise indicated. [2] Exposed 12 hr, 10% survival. [3] Exposed 30 min. [4] Exposed 12 hr, 100% survival. [5] Synonym: *C. pyrenoidosa* 7-11-05. [6] At 4700 lumens/m². [7] At 17,000 lumens/m². [8] Class. [9] Exposed 3 da.

Contributors: Biebl, Richard; Sorokin, Constantine; Allen, Mary Belle

References

[1] Allen, M. B. Unpublished. Univ. of Alaska, 1971.
[2] Ascione, R., et al. 1966. Science 153:752.
[3] Biebl, R. 1967. Botaniste 50:38.
[4] Biebl, R. 1969. Mikroskopie 25:3.
[5] Copeland, J. J. 1936. Ann. N. Y. Acad. Sci. 36:1.
[6] Höfler, K. 1951. Verh. Zool. Bot. Ges. Wien 92:234.
[7] Huber-Pestalozzi, G. 1926. In C. J. Schröter, ed. Das Pflanzenleben der Alpen. A. Raustein, Zurich. p. 942.
[8] Molisch, H. 1926. Pflanzenbiologie in Japan. G. Fischer, Jena.
[9] Schölm, H. 1968. Protoplasma 65:97.
[10] Shihira, I., and R. Krauss. 1965. *Chlorella.* Port City Press, Baltimore.
[11] Sorokin, C. 1959. Nature (London) 184:613.
[12] Sorokin, C. 1960. Biochim. Biophys. Acta 38:197.

continued

Part II. Minimum and Maximum: Marine Algae

It is important to note that from the cold waters of northern latitudes to the warm waters of southern latitudes there is an increasing heat and a decreasing cold resistance; the reverse is true from south to north. Almost all algae tested will survive water-freezing to carbon dioxide freezing temperatures. Diatoms in Arctic Ocean ice actively grow at a temperature of $-2°C$ [1]. **Location:** Latitudes are approximate. **Exposure Range:** Plants were exposed for 12 hours at various temperatures and observed 24 hours later. **Mean Water Temp:** Values are for July. Data in brackets refer to the column heading in brackets.

	Location	Exposure Range, °C	Mean Water Temp, °C	Species [1]	Temp, °C		Reference
					Minimum	Maximum	
				Intertidal [2]			
1	Alaska, Douglas Island & Juneau, USA; 58° N latitude	−20 to +36	10.2	*Chaetomorpha cannabina*	−8	30	7
2				*Enteromorpha torta*	−8	34	
3				*Pylaiella littoralis*	−8	30	
4				*Ulva lactuca*	−8	30	
5	Alaska, Izembek Lagoon, USA; 55° N latitude	−10 to +34	11.2	*Ectocarpus confervoides*	−10	30	7
6	California, Pacific Grove, USA; 36° N latitude	−10 to +34	14.3	*Cladophora trichotoma*	2	32	7
7				*Porphyra perforata*	−10	32	
8	Washington, Friday Harbor, USA; 48.5° N latitude	−20 to +30	11.6	*Cladophora trichotoma*	−8	32	7
9				*Enteromorpha linza*	−8	30	
10				*Porphyra fucicola*	−8	30	
11				*Ulva lactuca*	−8	30	
12	Puerto Rico; 16° N latitude	−10 to +45	29	*Catenella opuntia*	1	40	5
13				*Enteromorpha flexuosa*	−2	35	
14				*Polysiphonia ferulacea*	3	35	
15				*Ulva lactuca*	−2	35	
16	Greenland, Godhavn; 69° N latitude	−10 to +30	6	*Chaetomorpha linum*	−10	28	6
17	France, Roscoff, Bretagne; 48° N latitude	−8 to +35	17	*Catenella repens*	−8	30	4
18				*Cladophora rupestris*	−8	30	
19				*Enteromorpha compressa*	−8	35	
20				*Polysiphonia ferulacea*	3	35	
21				*Porphyra umbilicalis*	−8	30	
22				*Ulva lactuca*	−8	30	
23	Italy, Naples; 41° N latitude	−7 to +42	24	*Bangia fuscopurpurea*	−7	35	3
24				*Cladophora hamosa*	−7	35	
25				*C. spinulosa*	−2	35	
26				*Polysiphonia pulvinata*	−7	35	
27				*Porphyra leucosticta*	−7	30	
				Sublittoral [2]			
28	Alaska, Cape Glazenap, USA; 55° N latitude	−10 to +34	...	*Ceramium pacificum*	<2	24	7
29				*Ptilota asplenoides*	2	24	
30	Alaska, Izembek Lagoon, USA; 55° N latitude	−10 to +34	11.2	*Alaria* sp.	<2	22	7
31				*Dictyosiphon foeniculaceus*	<2	24	
32				*Rhodymenia pertusa*	<2	26	
33	California, La Jolla, USA; 33° N latitude	−7 to +32	20	*Dictyota flabellata*	4	26	7
34				*Laurencia subopposita*	4	26	
35	California, Pacific Grove, USA; 36° N latitude	−10 to +34	14.3	*Cryptopleura violacea*	1	22	7
36				*Rhodymenia pacifica*	−1.5	30	
37	Washington, Friday Harbor, USA; 48.5° N latitude	−20 to +30	11.6	*Callophyllis heanophylla*	−1	22	7
38				*Monostroma zostericola*	−1	22	
39				*Rhodymenia pertusa*	−1	22	

[1] Unless otherwise indicated. [2] Determinations by single method.

continued

Part II. Minimum and Maximum: Marine Algae

Location	Exposure Range, °C	Mean Water Temp, °C	Species [1]	Temp, °C		Reference
				Minimum	Maximum	
40 Puerto Rico; 16° N latitude	−10 to +45	29	*Anadyomena stellata*	8	32	5
41			*Ceramium nitens*	14	35	
42			*Dictyota dichotoma*	5	32	
43			*D. divaricata*	8	35	
44			*Florideae* [3,4]	5-14	32-35	
45			*Laurencia obtusa*	14	35	
46 Greenland, Godhavn; 69° N latitude	−10 to +30	6	*Laminaria saccharina*	−4	24	6
47			*Turnerella pennyi*	−4	22	
48 France, Roscoff, Bretagne; 48° N	−8 to +35	17	*Arthrocladia villosa*	−2	27	4
49 latitude			*Dictyota dichotoma*	3	27	
50			*Florideae* [3,5]	−2 to +3	27-30	
51			*Ulva olivacea*	−2	30	
52 Italy, Naples; 41° N latitude	−7 to +42	24	*Ceramium strictum*	−2	30	2
53			*Cladophora prolifera*	−2	30	
54			*Taonia atomaria*	−2	30	
Intertidal, Littoral, Sublittoral, & Unspecified [6]						
55 Connecticut, USA	....	...	*Fucus vesiculosus*	−60 to −20 [7]	....	9
56 Russia, White Sea	....	...	*Ascophyllum nodosum*	−28 [8]	39.3-41.5	8
57			*Fucus vesiculosus* [9]	−28 [10]	41.6-42.5	
58			*F. vesiculosus* [11]	−20	41.9	
59 Japan	....	...	*Bangia fuscopurpurea* [12]	−55	....	10
60			*Enteromorpha linza* [12]	−20	....	
61			*Porphyra onoi* [12]	−10	....	
62			*P. pseudolinearis* [12,13]	−70	....	
63			*P. pseudolinearis* [12,14]	−55	....	
64			*Ulothrix flacca* [12]	−25	....	

[1] Unless otherwise indicated. [3] Class. [4] Data for 9 species. [5] Data for 23 species. [6] Determinations by various methods. [7] Plants slowly cooled, then immediately removed and gradually warmed. [8] Survival 5 hr. [9] Littoral zone. [10] Survival 4-5 hr. [11] Sublittoral zone. [12] Intertidal zone. Exposed 24 hr, 50% survival. [13] Male. [14] Female.

Contributor: Biebl, Richard

References

[1] Allen, M. B. In press, 1971.
[2] Baas-Becking, L. G. M. 1930. Contributions to Marine Biology. Stanford Univ. Press, Palo Alto.
[3] Biebl, R. 1939. Jahrb. Wiss. Bot. 88:389.
[4] Biebl, R. 1958. Protoplasma 50:217.
[5] Biebl, R. 1962. Bot. Mar. 4:241.
[6] Biebl, R. 1968. Flora (Jena) B157:327.
[7] Biebl, R. 1970. Protoplasma 69:61.
[8] Feldmann, N. L., and M. I. Lutova. 1963. Cah. Biol. Mar. 4:435.
[9] Parker, J. 1960. Biol. Bull. 119:474.
[10] Terumoto, I. 1964. Teion Kagaku Seibutsu Hen 19:28.

Part III. Minimum and Maximum: Lichens

Data are for relatively sensitive and highly resistant species selected from more than 50 lichens. Temperatures without brackets are for air-dried thalli; temperatures in brackets are for fully hydrated thalli. After treatment of air-dry tissue for 30 minutes, a 50% decrease in respiration was observed.

continued

Part III. Minimum and Maximum: Lichens

	Species	Temp, °C		Refer-ence
		Minimum	Maximum	
1	*Alectoria implexa*	....	72-73	3
2	*A. ochroleuca*	....	72	3
3	*A. sarmentosa*	....	70-74	3
4	*Anaptychia leucomelaena*	[<−78]	....	1
5	*Cladonia foliacea*	....	92-96	3
6	*C. pocillum & C. pyxidata*	....	101	3
7	*C. rangiferina*	[−196[1/]]	....	2
8	*C. rangiformis*	....	99 [46.5]	3
9	*Cora pavonia*	[<−78]	....	1
10	*Lecanora melanophthalma*[2/]	[−196[3/]]	....	2
11	*Lobaria pulmonaria*	[−196[1/]]	73 [36.5]	2
12	*Parmelia cirrhata*	[<−78]	....	1
13	*Umbilicaria cylindrica*	....	95	3
14	*U. hirsuta*	....	100	3
15	*U. pustulata*	....	98 [45.5]	3
16	*U. vellea*	....	98-100 [44]	3
17	*Usnea dasypoga*	....	71-74 [<35]	3
18	*U. florida*	....	70	3
19	*Xanthoria elegans*[2/]	[−196[3/]]	....	2
20	3 other antarctic spp.	−196[1/]	....	2
21	16 Central European, Mediterranean, & tropical spp.[4/]	[<−78]	....	1

[1/] Sustains temperature of −196°C for 6 hr only when gradually frozen. [2/] Antarctic. [3/] Sustains temperature of −196°C for 6 hr when gradually decreased as well as when directly exposed to −196°C; no damage was observed in subsequent growth of test algae (isolated from frozen thalli) when compared with growth prior to treatment and with growth of untreated control group thalli. [4/] 23 species tested, not all of which are mentioned in the reference.

Contributor: Pisek, A.

References

[1] Kappen, L., and O. L. Lange. 1970. Lichenologist 4:289.

[2] Kappen, L., and O. L. Lange. 1972. Flora (Jena) 61:1.

[3] Lange, O. L. 1953. Ibid. 40:39.

Part IV. Minimum and Maximum: Mosses

	Type of Vege-tation	Location	Species (Synonym)	Temp, °C		Remarks	Ref-er-ence
				Mini-mum	Maxi-mum		
1	Mossy forest (hanging moss)	El Yunque, Puerto Rico	*Bazzania stolonifera*	−7	35	Treated in water. Exposure: min temp, 24 hr; max temp, 12 hr.	1
2			*Herberta juniperoidea (H. juniperina)*	−16	35		
3			*Metzgeria hamata*	1	32		
4			*Omphalanthus filiformis*	1	35		
5			*Plagiochila* spp.[1/]	−7	35		
6				1	32		
7			*Rhizogonium spiniforme*	−16	40		

[1/] Values are for 3 species.

continued

Part IV. Minimum and Maximum: Mosses

	Type of Vegetation	Location	Species (Synonym)	Temp, °C Minimum	Temp, °C Maximum	Remarks	Reference
8			*Thuidium urceolatum*	1	35		
9			*Trichocolea elliottii*	1	32		
10	Rain forest, wet soil	El Yunque, Puerto Rico	*Bazzania schwaneckiana*	−7	35	Treated in water. Exposure: min temp, 24 hr; max temp, 12 hr.	1
11			*Neesioscyphus* sp.	1	35		
12			*Pilotrichidium callicostatum*	−7	35		
13			*Plagiochila* spp.[1]	−7	32		
14			*Symphyogyna trivittata*	1	32		
15	Forest	Ratnapura, Ceylon	*Bryum* sp.	−7	42	Treated in water. Exposure: min temp, 24 hr; max temp, 30 min.	2
16			*Leucoloma amoenevirens*	−14	50		
17			*Schistochila commutata*	−11	42		
18	Unspecified	El Yunque, Puerto Rico	*Cyclolejeuna convexistipa*	1	32	Epiphyllic; treated in water. Exposure: min temp, 24 hr; max temp, 12 hr.	1
19		....	*Barbula fallax (B. gracilis)*	......	110-115	Least sensitive of 50 species tested; injury after 30-min heating in dry state over phosphorus pentoxide	4
20			*Ceratodon purpureus*	......	100-105		
21			*Grimmia trichophylla*	......	105-110		
22			*Pleurochaete squarrosa*	......	100-105		
23				......	105-110		
24			*Frullania dilatata*	......	70-75	Most sensitive of 50 species tested; injury after 30-min heating in dry state over phosphorus pentoxide	4
25			*Gymnomitrium obtusum*	......	65-70		
26			*Plagiothecium curvifolium*	......	65-70		
27			*P. denticulatum*	......	70-75		
28			30 species	−20 to −10	....	Min temp for majority of species tested. Exposure: 18 hr in turgescent state during winter.	3
29			30 species[2]	−30 to −20	....	Min temp for least sensitive. Exposure: 18 hr in turgescent state during winter.	3

[1] Values are for 3 species. [2] Species include *Ceratodon purpureus, Dicranum scoparium, Grimmia pulvinata, Plagiothecium denticulatum, P. undulatum.*

Contributors: Biebl, Richard; Pisek, A.

References

[1] Biebl, R. 1964. Protoplasma 59:133.
[2] Biebl, R. 1967. Flora (Jena) B157:25.
[3] Irmscher, E. 1912. Jahrb. Wiss. Bot. 50:387.
[4] Lange, O. L. 1955. Flora (Jena) 142:381.

Part V. Minimum and Maximum: Vascular Plants Other Than Fruit and Vegetable Crops

Temperature: Where two temperatures are given in a single column, the first is for summer and the second for winter, unless otherwise specified.

continued

Part V. Minimum and Maximum: Vascular Plants Other Than Fruit and Vegetable Crops

	Type of Vegetation	Location	Species (Synonym)	Plant Parts	Temp, °C		Remarks	Reference
					Minimum	Maximum		
			Pteridophyta [1] (Filices)					
1	Virgin forest	Puerto Rico	Dicranopteris pectinata (Gleichenia pectinata)	Detached fronds	2	48	Exposure: min temp, 24 hr; max temp, 30 min in nylon bags placed in water bath, then kept in damp room 5 da	1
2			Nephrolepis biserrata	Detached fronds	2	48		
3			N. rivularis	Detached fronds	−1.2	50		
4			Oleandra articulata	Detached fronds	2	48		
5			Thelypteris deltoidea (Dryopteris deltoidea)	Detached fronds	2	42		
6	Summer green fern; beech-mixed forest	Göttingen, Germany	Athyrium filix-femina	Rhizomes	−3.5; −8.5	......	Exposure: 4 hr. Injury: 10%.	3
7			Dryopteris spinulosa	Rhizomes	−3.5; −8.5	......		
8			Pteridium aquilinum	Rhizomes	−2.5 [2]	......		
9	Winter green fern	Göttingen, Germany	Dryopteris spinulosa	Fronds	−15.5; −7	48.0	Exposure: min temp, 2 hr; 1st value for Dec-Feb, 2nd value for April. Max temp, 30 min; value for Dec-Feb; only slightly different in summer. Injury: 10%.	3
10			Phyllitis scolopendrium	Fronds	−14.8; −13	47.5		
11			Polypodium vulgare	Fronds	−18.1; −13	48.5		
12			Polystichum lobatum	Fronds	−13.6; −11	48.5		
13	Mediterranean fern	Majorca	Adiantum capillus-veneris	Fronds	−1.5	......	Exposure: 2 hr; value for April. Injury: 10%.	3
14			Asplenium glandulosum	Fronds	−12	......		
15			Polypodium australe (P. serratum)	Fronds	−7.0	......		
			Pinophyta (Gymnospermae)					
16	Evergreen	New Haven, Connecticut	Abies guatemalensis [3]	Leaves or needles	−6	......	Exposure: slowly cooled, then immediately removed and gradually warmed; value for Jan	12
17			Cupressus lusitanica	Leaves or needles	−10	......		
18			Juniperus virginiana	Leaves or needles	−52	......		
19			Picea abies (P. excelsa)	Leaves or needles	−58	......		
20			Pinus palustris	Leaves or needles	−25	......		
21			P. strobus	Leaves or needles	>−189	......		
22			P. sylvestris	Leaves or needles	−62	......		
23			Tsuga canadensis	Leaves or needles	−45	......		
24		Moscow Mt., Idaho	Abies grandis	....	−15; −55 to −45	......	Exposure: slow freezing; thawing at intervals of 2-3 hr. Injury: 15-20% to mature leaves, none at 1-2°C higher temp.	11
25			Pinus ponderosa	....	−15; −60 to −50	......		
26		Innsbruck, Austria	Abies alba	Branches with mature needles	−4	46	Exposure: cooled 5°C/hr for 6-8 hr, slowly thawed; values for summer	13, 14, 16, 26

[1] On the whole, resistance to temperature, especially frost, increases with water loss. Leaves of Polypodium vulgare are totally resistant to dehydration and in dehydrated state, totally resistant to frost (−196°C) [4]. [2] Value for summer and winter. [3] Age, 1-6 yr.

continued

Part V. Minimum and Maximum: Vascular Plants Other Than Fruit and Vegetable Crops

	Type of Vegetation	Location	Species (Synonym)	Plant Parts	Temp, °C Minimum	Temp, °C Maximum	Remarks	Reference
27			*Picea abies*	Branches with mature needles	−3.5 to −8; −38	40.5; 42.5	Exposure: slow freezing; thawing at intervals of 2-3 hr. Injury: 15-20% to mature needles, none at 1-2°C higher temp.	13, 14, 16, 26
28			*Pinus cembra*	Branches with mature needles	−10; −40	......		
29			*P. mugo*	Branches with mature needles	−6; −35	......		
30			*Taxus baccata*	Branches with mature needles	−4.0; −40.0	47.5; 50.5		
31		Lago di Garda, Riva, Italy	*Cedrus atlantica & C. deodara*	Branches with mature needles	−8 to −6; −17 to −15	......	Exposure: slow freezing; thawing at intervals of 2-3 hr. Injury: 15-20% to mature leaves, none at 1-2°C higher temp.	8, 9
32			*Cupressus sempervirens*	Branches with mature needles	−9 to −7; −18 to −14	......		
33			*Pinus pinea*	....	−7 to −5; −14 to −11	......		
34		Mediterranean area	*Pinus halepensis*	Branches with mature needles	−11 to −15	......	Exposure: cooled 3-5°C/hr for 3 hr; slowly thawed; value for winter	9
35		Japan	*Picea glehnii*	Twigs or leaves	−12	......	Exposure: 24 hr	19
36			*P. mariana (Abies mariana)*	Twigs or leaves	−28	......		
37			*Pinus parviflora (P. pentaphylla)*	Twigs or leaves	−28	......		
38			*Taxus cuspidata*	Twigs or leaves	−30	......		

Magnoliophyta (Angiospermae)

	Type of Vegetation	Location	Species (Synonym)	Plant Parts	Minimum	Maximum	Remarks	Reference
39	Evergreen	New Haven, Connecticut	*Ilex opaca*	Leaves or needles	−35	......	Exposure: slowly cooled, then immediately removed and gradually warmed; value for Jan	12
40		Innsbruck, Austria	*Erica carnea*	....	−4.5; −18.5	......	Exposure: slow freezing; thawing at intervals of 2-3 hr. Injury: 15-20% to mature leaves, none at 1-2°C higher temp.	14, 16, 26
41			*Rhododendron ferrugineum*	....	−5[4]; −15[4] to −25[5]	......		
42			*R. hirsutum*	....	−5.5; −28.5	......		
43		Lago di Garda, Riva, Italy	*Arbutus unedo*	....	−8 to −5; −10	......	Exposure: slow freezing; thawing at intervals of 2-3 hr. Injury: 15-20% to mature leaves, none at 1-2°C higher temp.	8
44			*Laurus nobilis*	....	−6; −10.5 to −10.0	......	Exposure: slow freezing; thawing at intervals of 2-3 hr. Injury: 15-20% to mature leaves, none at 1-2°C higher temp.	8
45				Mature leaves	−5.0; −9.2	49.5; 50.5	....	15, 16

[4] In protected position. [5] In exposed position.

continued

Part V. Minimum and Maximum: Vascular Plants Other Than Fruit and Vegetable Crops

	Type of Vege-tation	Loca-tion	Species (Synonym)	Plant Parts	Temp, °C		Remarks	Ref-er-ence
					Mini-mum	Maxi-mum		
46			Nerium oleander	....	−6 to −4; −9	......	Exposure: slow freezing; thawing at intervals of 2-3 hr. Injury: 15-20% to mature leaves, none at 1-2°C higher temp.	8
47			Olea europaea	....	−6.5; −12	......		
48			Poncirus trifoliata	....	−16 to −12; −22 to −18	......		
49			Prunus laurocerasus (Lauro-cerasus officinalis)	Mature leaves or branches	−5.2	46.5	Min temp: value for summer; max temp: value for summer & winter	15, 16
50			Quercus ilex	....	−6; −13.5 to −13.0	......	Exposure: slow freezing; thawing at intervals of 2-3 hr. Injury: 15-20% to mature leaves, none at 1-2°C higher temp.	8
51			Trachycarpus fortunei	....	−13 to −10; −15.0 to −12.5	......		
52			Viburnum tinus	....	−7 to −4; −12 to −11	......		
53		Medi-terra-nean area	Acacia dealbata[6]	Branches with mature leaves	−6.0	......	Exposure: cooled 3-5°C/hr for 3 hr; slowly thawed; value for winter	9
54			Arbutus unedo[7]	Branches with mature leaves	−9.5	......		
55			Chamaerops humilis[7]	Branches with mature leaves	−9.0	......		
56			Eucalyptus globulus[6]	Branches with mature leaves	−3.0	......		
57			Laurus nobilis[7]	Branches with mature leaves	−9.0	......		
58			Magnolia grandiflora[6]	Branches with mature leaves	−12	......		
59			Nerium oleander[7]	Branches with mature leaves	−7.0	......		
60			Olea europaea sativa[7]	Branches with mature leaves	−10	......		
61			Quercus ilex[7]	Branches with mature leaves	−12	......		
62			Q. suber[7]	Branches with mature leaves	−7.0	......		
63			Trachycarpus fortunei[6]	Branches with mature leaves	−11	......		
64			Viburnum tinus[7]	Branches with mature leaves	−10	......		
65		Japan	Laurus nobilis	Twigs or leaves	−15	......	Exposure: 24 hr	19
66			Rhododendron brachycar-pum (R. fauriai)	Twigs or leaves	−28	......		
67	Decidu-ous	....	Robinia pseudoacacia	Bark sections with adjacent wood	−5; <−60	......	Exposure: placed in jars and cooled 1°C/hr; removed for thawing. Injury: 50% survival.	22
68		Japan	Betula tauschii	Twigs or leaves	−30	......	Exposure: 24 hr	19
69			Platanus orientalis	Twigs or leaves	−25	......		
70			Populus nigra	Twigs or leaves	−30	......		

[6] Species introduced into area. [7] Native species.

continued

Part V. Minimum and Maximum: Vascular Plants Other Than Fruit and Vegetable Crops

	Type of Vegetation	Location	Species (Synonym)	Plant Parts	Temp, °C Minimum	Temp, °C Maximum	Remarks	Reference
71			*Rosa pendulina*	Twigs or leaves	−28	......		
72			*Salix koriyanagi*	Twigs or leaves	−30	......		
73	Deciduous; beech forest	Göttingen, Germany	*Acer campestre*	Buds[8]; leaves[9]	−23; −4	......	Exposure: 2 hr; gradual temp change; 1st value for Jan, 2nd value for May. Injury: 15-20% to buds & leaves.	25
74			*Alnus glutinosa*	Buds[8]; leaves[9]	−28; −3	......		
75			*Corylus avellana*	Buds[8]; leaves[9]	−25; −3.5	......		
76			*Fagus sylvatica*	Buds[8]; leaves[9]	−22; −2.5	......		
77			*Fraxinus excelsior*	Buds[8]; leaves[9]	−27; −2	......		
78	Herbaceous	....	*Begonia decandra*	....	0.5-1	46	Exposure: min temp, 24 hr; max temp, 30 min	1
79			*Digitalis purpurea*[10]	Young leaves	−4.0	......	Exposure: 2 hr; gradual temp changes; value for May. Injury: 15-20% to unfolding leaves.	25
80			*Erica tetralix*	....	....	50.5; 47.5; 45	Exposure: 30 min; 1st value for Dec, 2nd for Mar, 3rd for May	7
81			*E. tetralix*[10]	....	−3.5	......	Exposure: 2 hr; gradual temp changes, value for May. Injury: 15-20% to unfolding leaves.	25
82			*E. tetralix*[11]	....	−20.0	......	Exposure: 2 hr; gradual temp changes; value for Dec-Feb. Injury: 15-20% to leaves or shoots.	25
83			*Impatiens parviflora*	Entire plant	....	41.5	Exposure: 30 min in saturated humidity. Injury: <50% of plants. Max temp tolerated in dry atmosphere, 2-5°C higher.	20
84			*Oxalis acetosella*	Entire plant	....	40.5		
85			*O. acetosella*[10]	....	−3.5	......	Exposure: 2 hr; gradual temp changes; value for May. Injury: 15-20% to unfolding leaves.	25
86			*O. acetosella*[12]	....	−11.5	......	Exposure: 2 hr; gradual temp changes; value for Dec-Feb. Injury: 15-20% to leaves or shoots.	25
87			*Stellaria media*	....	−2.5; −9.7	......	Exposure: slow freezing; thawing at intervals of 2-3 hr. Injury: 15-20% to mature leaves, none at 1-2°C higher temp.	26
88			*Veronica tournefortii*	....	−4.3; −10.8	......		
89		Costa Brava, Spain	*Asparagus acutifolius*[13]	Detached branches	....	53	Exposure: 30 min at temp increments of 1°C, then room temp & diffuse light	6
90	Herbaceous; xerophytic	....	*Datura stramonium*	Entire plant	....	47	Exposure: 30 min in saturated humidity. Injury: <50% of plants. Max temp tolerated in dry atmosphere, 2-5°C higher.	20
91			*Dianthus carthusianorum*	Entire plant	....	48		
92			*Festuca glauca*	Entire plant	....	50.5		
93			*Iris chamaeiris*	Entire plant	....	49.5		

[8] In January. [9] In May. [10] One of most sensitive of 28 herbaceous species tested. [11] One of least sensitive of 15 species tested. [12] One of most sensitive of 15 species tested. [13] One of least sensitive of 39 species tested.

continued

Part V. Minimum and Maximum: Vascular Plants Other Than Fruit and Vegetable Crops

Type of Vegetation	Location	Species (Synonym)	Plant Parts	Temp, °C Minimum	Temp, °C Maximum	Remarks	Reference
94		*Sedum acre*	Entire plant	....	48.5-49.5		
95		*S. reflexum*	Entire plant	....	49.5-50.0		
96		*Opuntia* sp.	....	....	63	....	10
97	Tropical	....	*Kalanchoe blossfeldiana*	Entire plant	....	44.5	Plants not flowering; leaves barely mature. Exposure: 30 min.	7
98				Entire plant	....	44.5	Plants not flowering; leaves succulent. Exposure: 30 min.	7
99				Entire plant	....	46.5	Plants flowering; leaves barely succulent. Exposure: 30 min.	7
100				Entire plant	....	49	Plants flowering; leaves succulent. Exposure: 30 min. Injury: 50% survival.	7
101	Tropical; virgin forest	Puerto Rico	*Euterpe globosa*	Branches with leaves	−1.5	50	Exposure: min temp, 24 hr; max temp, 30 min	1
102			*Guarea trichilioides*	Branches with leaves	21	48		
103	Tropical; rain forest	Mauritania & Ivory Coast, Africa	*Piptadeniastrum africanum (Piptadenia africana)*[14]	Entire plant	....	44	Tested in natural habitat. Exposure: gradual warming, 30 min at each temp increment, then returned to normal temp. Injuries studied after 3 da.	5
104	Desert	Mauritania & Ivory Coast, Africa	*Citrullus colocynthis*[14]	Entire plant	....	44	Tested in natural habitat. Exposure: gradual warming, 30 min at each temp increment, then returned to normal temp. Injuries studied after 3 da.	5
105			*Cucumis prophetarum*[14]	Entire plant	....	44-46		
106			*Phoenix dactylifera*[15]	Entire plant	....	57		
107	Hydrophytic	....	*Elodea callitrichoides*	Entire plant	....	38.5	Exposure: 30 min in saturated humidity. Injury: <50% of plants. Max temp tolerated in dry atmosphere, 2-5°C higher.	20
108			*E. canadensis*	Entire plant	....	39.0-39.5		
109			*Vallisneria spiralis*	Entire plant	....	41.5		
110	Hydrophytic	Puerto Rico	*Avicennia marina*	....	21	48	Exposure: min temp, 24 hr; max temp, 30 min	1
111			*Conocarpus erecta*	....	<−3	<54		
112	Hothouse	....	*Coleus* sp.	....	>1	......	Injury: after 28 hr at 0.3-1.0°C	24
113			*Gloxinia grandiflora*	....	1-5	......	Exposure: 24 hr in saturated humidity. Injury: none; at >24 hr, injury to leaves and later to shoot apex.	21
114			*Impatiens sultani*	....	>1	......	Injury: after 28 hr at 0.3-1.0°C	24
115	Cereals & crop plants	....	*Avena sativa*	....	−12 to −9	......	Conditioned to low temp	17, 18
116			*Brassica napus*	....	−9	......	Conditioned to low temp	17
117			*Camellia sinensis (Thea sinensis)*	....	>−10	......	Diploid & triploid varieties. Exposure: 30 min; value for Dec.	23

[14] One of most sensitive of 44 species tested. [15] One of least sensitive of 44 species tested.

continued

Part V. Minimum and Maximum: Vascular Plants Other Than Fruit and Vegetable Crops

	Type of Vegetation	Location	Species (Synonym)	Plant Parts	Temp, °C		Remarks	Reference
					Minimum	Maximum		
118				....	−15	......	Diploid & triploid varieties. Exposure: 60 min; value for Jan.	23
119			Hordeum vulgare	....	−15 to −10	......	Conditioned to low temp	17, 18
120			Linum spp.	....	−12	......	Exposure: open air; value for Jan. Injury: 50% survival.	2
121			Secale cereale	....	−25 to −15	......	Conditioned to low temp	17, 18
122			Triticum aestivum	....	−22 to −10	......		
123		Costa Brava,	Brassica fruticulosa[16]	Detached branches	....	44	Exposure: 30 min at temp increments of 1°C, then room	6
124		Spain	Ruscus aculeatus[13]	Detached branches	....	55	temp & diffuse light	

[13] One of least sensitive of 39 species tested. [16] One of most sensitive of 39 species tested.

Contributor: Pisek, A.

References

[1] Biebl, R. 1964. Protoplasma 59:133.
[2] Dillman, A. C. 1941. J. Amer. Soc. Agron. 33(9): 787.
[3] Kappen, L. 1964. Flora (Jena) 155:123.
[4] Kappen, L. 1966. Ibid. B156:427.
[5] Lange, O. L. 1959. Ibid. 147:595.
[6] Lange, O. L., and R. Lange. 1963. Ibid. 153:387.
[7] Lange, O. L., and B. Schwemmle. 1960. Planta 55: 208.
[8] Larcher, W. 1954. Ibid. 44:607.
[9] Larcher, W. 1963. Veroeff. Mus. Ferdinand. Innsbruck 43:153.
[10] McDougal, D. I., and E. B. Working. 1921. Carnegie Inst. Wash. Yearb. 20:46.
[11] Parker, J. 1955. Ecology 36:377.
[12] Parker, J. 1960. Nature (London) 187:1133.
[13] Pisek, A., and R. Kemnitzer. 1968. Flora (Jena) B157:314.
[14] Pisek, A., and R. Schiessl. 1947. Ber. Naturwiss. Med. Ver. Innsbruck 47:33.
[15] Pisek, A., et al. 1967. Flora (Jena) B157:239.
[16] Pisek, A., et al. 1968. Ibid. B158:110.
[17] Roemer, T., and W. Rudorf. 1941. Handbuch der Pflanzenzuechtung. P. Parey, Berlin.
[18] Roemer, T., and F. Scheffer. 1944. Ackerbaulehre. Ed. 2. P. Parey, Berlin.
[19] Sakai, A. 1962. Teion Kagaku Seibutsu Hen 11:1.
[20] Sapper, I. 1935. Planta 23:518.
[21] Seible, D. 1939. Beitr. Biol. Pflanz. 26:289.
[22] Siminovitch, D., and D. R. Briggs. 1953. Plant Physiol. 28:15.
[23] Simura, T. 1957. Proc. Int. Genet. Symp., 1956, p. 321.
[24] Spranger, E. 1941. Gartenbauwissenschaft 16:90.
[25] Till, O. 1956. Flora (Jena) 143:499.
[26] Ulmer, W. 1937. Jahrb. Wiss. Bot. 84:553.

Part VI. Minimum: Fruit and Vegetable Crops

	Species (Synonym)	Specifications	Temp, °C	Reference
		Fruits		
1	Ananas comosus (A. sativus)	Immature	−1.6	20
2		Ripe	−1.2	20

continued

Part VI. Minimum: Fruit and Vegetable Crops

	Species (Synonym)	Specifications	Temp, °C	Reference
3	*Carya illinoensis* 'Schley'	....	−6.9	20
4	*Castanea sativa*	Italy	−4.5	20
5	*Citrullus vulgaris*	Rind	1.8	20
6		Flesh	1.5	20
7	*Citrus limon*	....	−5.0[1]	11
8		Open flowers	0[2]	5
9		Fruitlets	−1[2]	5
10		Mature leaves & branches; Riva, Italy	−3.2[3]; −7.9[4]	16
11		Rind & flesh; California	−2.1	20
12	*C. sinensis*	Leaves	−6.7	7
13		Rind	−2.5	20
14		Flesh	−2.2	20
15	*Cocos nucifera*	Flesh	−3.6	20
16		Milk	−0.9	20
17	*Corylus avellana*	Catkins	−26 to −18	14
18		Open flowers	−36 to −20	14
19		Leaf buds	−36 to −23	14
20		Bark & wood	−36 to −31	14
21	*Cucumis melo*	Rind	2.0; 1.8[5]	20
22		Flesh	1.7; 1.7[5]	20
23	*Diospyros virginiana* 'Tanenashi'	....	−2.0	20
24	*Ficus carica* 'Mission'	Fresh; California	−2.7	20
25	*Fragaria* spp.	....	−1.2	20
26		Buds showing petals, open flowers, & fruitlets	−2[2]	5
27	*Juglans regia*	....	−6.7	20
28		Buds showing petals, open flowers, & fruitlets	−1[2]	22
29	*Malus* sp.[6]	....	−2.7	20
30	*M. pumila*	Bursting	−10.0 to −5.5	21
31		Mouse-ears	−8.3 to −3.9	21
32		Green clusters	−7.0 to −2.2	6,21
33		Pink buds	−6.0 to −1.9	6,15,21
34		Buds showing petals	−4 to −3[2]	5,22
35		Open flowers	−5.0 to −1.3	5,6,10,15,21,22
36		Petal fall	−4.5 to −2.5	6
37		Setting	−3.2 to −1.1	6,15,21
38		Fruitlets	−2.6 to −1.0	5,6,22
39		Stems, winter	−31.0 to −23.0	8
40		Roots[7]	−12 to −11	4
41		Bark, winter	−40.0 to −25.0	10
42		Bark, during flowering	−8.0 to −6.0	10
43		Trunk cortex, Jan	−29	17
44		Trunk cortex, during flowering	−8	17
45		Fall & winter varieties	−1.9	20
46		Summer varieties	−2.0	20
47	*Mangifera indica* 'Faizanson'	....	−1.2	20
48	*Musa sapientum*	Peel, immature	−1.2	20
49		Peel, mature	−1.4	20
50		Pulp, immature	−1.0	20
51		Pulp, mature	−3.3	20
52	*Olea europaea*	Fresh, green	−1.9	20

[1] Tested in winter; cooled 3-5°C/hr, exposed 3 hr to low temperature, then slowly thawed. [2] Temperature endured for 30 min. [3] Value for summer. [4] Value for winter. [5] Value for honeydew. [6] Crab apple. [7] American first-year stock and French second-year seedlings.

continued

Part VI. Minimum: Fruit and Vegetable Crops

	Species (Synonym)	Specifications	Temp, °C	Reference
53	O. europaea sativa	Nov-Feb	−12 to −10[8]	12
54		May-Aug	−7[8]	12
55	Persea americana (P. gratissima) 'Collinson'	....	−2.6	20
56	Phoenix canariensis (P. jubae)	....	−5.0[1]	11
57	Prunus spp.[9]	Buds showing petals	−4 to −1	5,15,22
58		Open flowers	−2.0 to −0.5	5,15,22
59		Setting	−2.0	15
60		Fruitlets	−1.0 to −0.5[2]	5,22
61	Prunus spp.[10]	Buds showing petals	−5 to −1	5,21,22
62		Open flowers	−3.0 to −0.5	5,21,22
63		Setting	−1.1	21
64		Fruitlets	−1.0 to −0.5[2]	5,22
65	Prunus sp.[11]	Trunk cortex, Jan	−23	17
66		Trunk cortex, during flowering	−8	17
67	P. amygdalus	Buds showing petals	−3.5 to −2.0	5,15,22
68		Open flowers	−3 to −1	5,15,22
69		Setting	−1.0	15
70		Fruitlets	−1.0 to +0.5[2]	5,22
71	P. armeniaca	Buds showing petals	−4 to −1	1,5,15,21,22
72		Burst calyx	−1.7 to −1.1	1
73		Open flowers	−3.3 to −0.5	1,5,15,21,22
74		Petal fall	−2.8 to −2.2	1
75		Withered stamens	−2.8 to −1.7	1
76		Setting	−1.1 to −0.5	15,21
77		Fruitlets	−0.5 to 0[2]	5,22
78		Trunk cortex, Jan	−21	17
79		Trunk cortex, during flowering	−7 to −6	17
80	P. avium	Buds showing petals	−4.5 to −2	5,15,22
81		Open flowers	−2	5,15,22
82		Setting	−1.0	15
83		Fruitlets	−1[2]	5,22
84		Sweet immature; California	−3.5	20
85		Sweet mature; California	−4.3	20
86		Sweet mature; eastern grown	−4.0	20
87		Roots	−10	4
88	P. persica	Before showing pink	−12.0 to −9.5	18
89		Buds showing petals	−4.0 to −3.9	5,15,18,21,22
90		Open flowers	−3.0 to −2.0	5,15,18,21,22
91		Setting	−1.1 to −1.0	15,21
92		Fruitlets	−1[2]	5,22
93		Hard ripe fruit	−1.4	20
94		Seedling roots	−11 to −10	4
95	Pyrus communis	Buds showing petals	−4 to −2	5,15,22
96		Open flowers	−2.0 to −1.5	5,15,22
97		Setting	−1.0	15
98		Fruitlets	−1[2]	5,22
99		Hard & soft ripe fruit	−2.4	20
100		Stem builders	−31.0 to −20.0	9
101		Roots[12]	−10	4

[1] Tested in winter; cooled 3-5°C/hr, exposed 3 hr to low temperature, then slowly thawed. [2] Temperature endured for 30 min. [8] Exposed 4-6 hr. [9] Plum. [10] Prune. [11] Cherry. [12] French stock.

continued

Part VI. Minimum: Fruit and Vegetable Crops

	Species (Synonym)	Specifications	Temp, °C	Reference
102	*Ribes nigrum*	Grape stage	−3.0 to −2.5 [13]	13
103		Open flowers	−4.0 to −3.0 [13]	13
104		Setting	−5.0 to −4.0 [13]	13
105	*Rubus* spp.	Buds showing petals, open flowers, & fruitlets	−2 [2]	5
106		Black varieties	−1.7	20
107		White variety	−2.0	20
108	*R. idaeus*	....	−1.2	20
109		Buds developing	−7.0	3
110		Buds showing petals, open flowers, & fruitlets	−2 [2]	5
111		Cane hardened	−18.0	3
112		Roots	−16 [14]	2
113	*Vaccinium* sp. 'Rubel'	....	−2.5	20
114	*Vitis* spp.	Buds showing petals	−1 [2]	22
115		Open flowers & fruitlets	−0.5 [2]	22
116		Roots	−15.5 to −11.0	4
117		American	−2.5	20
118	*V. vinifera*	European	−3.9	20
		Vegetables		
119	*Allium cepa*	Mild	1.1	20
120		Strong	1.0	20
121	*A. cepa* 'Yellow Strassburg'	Sets	1.4	20
122	*Apium graveolens*	....	1.3	20
123	*Asparagus officinalis*	....	1.2	20
124	*Beta vulgaris*	....	1.9-2.8	20
125			−7 to −5 [15]	19
126	*Brassica oleracea* 'Jersey Wakefield'	Early	0.4	20
127	*B. oleracea* var. *botrytis*	....	1.0	20
128	*B. oleracea* var. *capitata*	....	−11 to −8 [15]	19
129	*Capsicum* spp.	Green	1.0	20
130	*Cichorium endivia*	Broad-leaved	0.8	20
131		Curled	0.6	20
132	*C. endivia* 'Belgian'	....	0.7	20
133	*Cucumis sativus*	....	0.8	20
134	*Cucurbita maxima* 'Hubbard'	Winter	1.5	20
135	*C. pepo*	Summer, cymling	1.5	20
136	*C. pepo* 'Cocozelle'	Italian	0.6	20
137	*C. pepo* 'Connecticut Pie'	....	1.0	20
138	*Daucus carota*	....	1.3	20
139	*Ipomoea batatas*	....	1.9	20
140	*Lactuca* spp.	....	0.4	20
141	*Lactuca* sp. 'Romaine'	....	0.8	20
142	*Lycopersicon esculentum*	Mature green & ripe	0.9	20
143	*Phaseolus* spp.	....	1.3	20
144		Pods	1.0	20
145	*P. lunatus*	....	1.1	20
146		Pods	0.7	20
147	*Pisum sativum*	....	1.1	20
148		Pods	1.1	20
149	*Raphanus sativus* 'French Breakfast'	....	2.6	20
150	*Rheum rhaponticum*	....	2.0	20

[2] Temperature endured for 30 min. [13] Exposed 30 min, in cold chamber. [14] Soil temperature in open air. [15] Plants conditioned to low temperature.

continued

Part VI. Minimum: Fruit and Vegetable Crops

	Species (Synonym)	Specifications	Temp, °C	Reference
151	*Solanum melongena*	....	0.9	20
152	*S. tuberosum*	....	1.7	20
153		....	−3 to −2[15/]	19
154	*Spinacia oleracea*	....	0.9	20
155	*Taraxacum officinale*	....	1.2	20
156	*Zea mays*	Sweet; milk stage	1.7	20

[15/] Plants conditioned to low temperature.

Contributors: Modlibowska, Irena; Pisek, A.

References

[1] Atkinson, J. D. 1952. N. Z. J. Sci. Technol. A34(3): 277.

[2] Brierley, W. G., and R. H. Landon. 1946. Proc. Amer. Soc. Hort. Sci. 47:215.

[3] Brierley, W. G., and R. H. Landon. 1954. Ibid. 63: 173.

[4] Carrick, D. B. 1920. Cornell Univ. Agr. Exp. Sta. Mem. 36.

[5] Davison, J. R. 1947. Agr. Gaz. N. S. W. 58:254.

[6] Durand, R. 1962. Int. Hort. Congr., 16th, Brussels 5:190.

[7] Hendershott, C. H. 1962. Proc. Amer. Soc. Hort. Sci. 80:247.

[8] Karnatz, H. 1958. Mitt. Obstbauvers. Alten Landes 13:54.

[9] Karnatz, H. 1958. Ibid. 13:225.

[10] Kohn, H. 1959. Gartenbauwissenschaft 24:314.

[11] Larcher, W. 1963. Veroeff. Mus. Ferdinand. Innsbruck 43:153.

[12] Larcher, W. 1963. Protoplasma 57:569.

[13] Modlibowska, I. Unpublished. East Malling Research Station, Kent, England, 1964.

[14] Olden, E. J. 1953. Foren. Vaxtforadl. Frukttrad Balsgard (28-31).

[15] Perraudin, G. Unpublished. Stations Fédérales d'Essais Agricoles de Lausanne, Sous-Station en Valais, Switzerland, 1962.

[16] Pisek, A., et al. 1967. Flora (Jena) B157:239.

[17] Pisek, A. Unpublished. Univ. Innsbruck, Austria, 1971.

[18] Proebsting, E. L., and H. H. Mills. 1961. Proc. Amer. Soc. Hort. Sci. 78:104.

[19] Roemer, T., and W. Rudorf. 1941. Handbuch der Pflanzenzuechtung. P. Parey, Berlin.

[20] Wright, R. C. 1937. U.S. Dep. Agr. Circ. 447.

[21] Yakima County Extension Service. 1957. Goodfruit Grower 7(6):1.

[22] Young, F. D. 1947. U.S. Dep. Agr. Farmers Bull. 15:88.

89. LETHAL TEMPERATURES: INVERTEBRATES

Part I. Insects

Biological Zero: Temperature at which all vital processes are arrested by cold. For additional information on lethal cold, consult references 3, 57, 67, and 72; for additional information on lethal heat, consult references 28, 43, and 46. Data in brackets refer to the column heading in brackets.

	Species (Synonym)	Stage [Limits for Development °C]	Activity Range [Acclimation Temp] °C	Lethal Cold		Lethal Heat		Reference
				Temp [Biological Zero] °C	Exposure Time	Temp, °C [Relative Humidity, %]	Exposure Time	
1	*Acanthoscelides ob-*	Egg [15-35]	....	−6	7 da	....	....	34
2	*tectus*			−4	15 da	....	....	34
3				8	40 da	....	....	34
4				11	51 da	....	....	34

continued

89. LETHAL TEMPERATURES: INVERTEBRATES

Part I. Insects

Species (Synonym)	Stage [Limits for Development, °C]	Activity Range [Acclimation Temp] °C	Lethal Cold		Lethal Heat		Reference
			Temp [Biological Zero] °C	Exposure Time	Temp, °C [Relative Humidity, %]	Exposure Time	
	All stages [15-33]	....	−7	9 da	55	30 min	76
			−4	29 da	60	12 min	79
			0	34 da	....	....	34
Aedes aegypti	Larva	....	0.5	17 hr	44	1 hr	56
	4th instar	....	5	65 hr	....	....	54
Agria housei (Pseudosarcophaga affinis)	Larva	[23]	....	....	45	130 min	41
		[29]	....	....	45	200 min	41
Anagasta kuehniella (Ephestia kuehniella)	Egg	....	−24.5	....	45 [8]	6 hr	48,71
	Larva	....	−10[1]	8 da			75
			−5.8 to +8.1	....	45 [8]	30-120 min	48,71
	[8-31]	....	−19	1 da	....	....	50
			−10	6 da	....	....	50
			−7	17 da	....	....	50
	Pupa	....	−15	4 hr	....	....	4
	[11-31]	....	−19	1 da	....	....	50
			−10	5 da	....	....	50
	Adult	....	−15	50-80 min	45 [8]	5-21 min	4,48
Anthonomus grandis	Adult	13.3-35.0	−13.8 [−4 to +3]	....	50-60	....	42
Apis mellifera	Adult	....	−6	30 min	....	....	35,36
			0	100 hr	42	20 hr	35,36
			5	100 hr	48	1 hr	37
Blattella germanica	Adult	....	....	....	43-45 [0-90]	1 hr	39
		[15]	7	167 hr	....	....	24
		[25]	7	132 hr	....	....	24
		[35]	7	48 hr	....	....	24
Calliphora erythrocephala	Larva	[12]	....	....	41	1 hr	33
		[30]	....	....	41	75 min	33
	Adult	....	....	....	41 [10]	1 hr	6
			....	....	41.2	35-120 min[2]	26
			....	....	45 [100]	5 min	6
			....	....	47 [70]	5 min	6
Cimex lectularius	Adult	12-22[3]	....	....	43.5 [0-90]	1 hr	15,27
		[14-17]	4.5[4]	....	....	....	52
		[30]	7[4]	....	....	....	52
Cryptolestes ferrugineus	Adult [20-39]	....	−5.5	20 da	....	....	50
			−5	30 da	....	....	50
			0	60 da	....	....	50
Culex pipiens	Egg	15-25	5	....	....	....	11
	0-da old	....	2	18 hr	....	....	31
	1-da old	....	2	5 da	....	....	31
	Adult	....	....	....	40	45 min	45
		21.1-26.5	−17	....	44.5-45.6	....	49,70
Dermestes maculatus	Egg	....	....	....	40	6 da	63
	[17-39]	....	....	....	55	15 min	63
	Larva [17-39]	....	....	....	50	15 min	63

[1] 0% relative humidity. [2] Value is lethal dose for 50% of test subjects. [3] Temperature preferred by organism. [4] Chill-coma temperature, the low temperature below which locomotory responses to stimuli fail to occur.

continued

Part I. Insects

	Species (Synonym)	Stage [Limits for Development, °C]	Activity Range [Acclimation Temp] °C	Lethal Cold		Lethal Heat		Reference
				Temp [Biological Zero] °C	Exposure Time	Temp, °C [Relative Humidity, %]	Exposure Time	
50		Larva & adult	....	....	....	52	2.5 min	38
51		Pupa [17-39]	....	....	....	57.5	15 min	63
52		Adult	....	....	....	47	14 min	38
53			....	....	....	52.5	15 min	63
54	*Drosophila melano-gaster*	Adult	....	−8	45 min	....	....	18
55	*Dytiscus marginalis*	Egg	....	[0]	....	31	....	8
56		Larva	....	[6]	....	35	....	9
57	*Epilachna varivestis*	Pupa	....	−11	8 hr	....	....	5
58				−5.5	7 da	....	....	5
59				0.2	12 da	....	....	5
60		Adult	....	−20	1.5 hr	....	....	5
61				−6	7 hr	....	....	5
62				−3	28 da	....	....	5
63	*Galleria mellonella*	Larva [10-?]	....	−18 to −15	2 hr	46	80 min	17
64				....	....	49	40 min	17
65		Half-grown	....	....	....	45	24 hr	60
66		Full-grown	....	−10	3 hr	....	....	60
67		Adult	....	−10	1 hr	46	17 hr	60
68				8.8	30 min	....	....	58
69	*Glossina morsitans*	Puparia	....	....	....	40	10 hr	66
70				....	....	50	30 min	66
71		[8-32]	....	−11	4 hr	44	6 hr	64
72				−10	6 hr	....	....	64
73	*Hylotrupes bajulus*	Larva	....	....	....	55 [30]	75 min	7
74	*Lasioderma serri-corne*	Egg [19-38]	....	....	....	50	20 min	19
75		Larva, 3rd instar	....	10[5]	8 wk	....	....	20
76				10[6]	12 wk	....	....	20
77		[19-38]	....	7.2[7]	3 wk	....	....	20
78		Larva, 4th instar	....	44[7]	3 wk	50	25 min	21
79				10[5]	20 wk	....	....	21
80				10[6]	24 wk	....	....	21
81				7.2	5 wk	....	....	21
82		Pupa	....	5	8 wk	50	20 min	21
83		Adult	....	....	....	50	20 min	21
84		All stages	....	−15	61 hr	50	4 hr	65
85	*Melanoplus san-guinipes (M. mex-icanus)*	Egg	[22]	−30	16 hr	50	2 hr	61
86		Nymph	....	−7	48 hr	....	....	61
87		Adult	22-37	−10	24 hr	54	10 min	61
88			....	−15 to −10	1 hr	....	....	44
89	*Musca domestica*	Egg	....	....	....	43	....	25
90		Larva	....	2.5	6 da	....	....	74
91				5	8 da	....	....	74
92				[5]	....	49	....	10,32
93		Pupa	....	11	32 da	....	....	74,75
94				2.5	12 da	....	....	74,75
95			>7.8	0-1 [10]	7 da	....	....	13

5/ 55% relative humidity.   6/ 65% relative humidity.   7/ 75% relative humidity.

continued

Part I. Insects

	Species (Synonym)	Stage [Limits for Development, °C]	Activity Range [Acclimation Temp] °C	Lethal Cold		Lethal Heat		Reference
				Temp [Biological Zero] °C	Exposure Time	Temp, °C [Relative Humidity, %]	Exposure Time	
96		Adult	....	6.1[4]	30 min	....	....	58
97			>15.6[3]	0	3 da	44.6	....	29,59
98	*Myzus persicae*	Adult		−2	50 da	....	....	1
99	*Oncopeltus fasci-*	Egg	15-32	3	30 min	....	....	47,68
100	*atus*	Nymph	18-32	2.5	30 min	....	....	47,68
101		Adult	18-32	2.5	30 min	....	....	47,68
102	*Oryzaephilus suri-*	Adult [18-38]	....	−5.5	15 da	....	....	50
103	*namensis*			−5	20 da	....	....	50
104				−2	25 da	....	....	50
105	*Ostrinia nubilalis*	Larva	....	−31.7	10 min	58	11 min	81
106	*(Pyrausta nubila-*			−20	1 da	....	....	40
107	*lis)*			−15	1 da	....	....	40
108		Larva[8]	....	−20	73 da	....	....	40
109				−15	90 da	....	....	40
110	*Pediculus humanus*	Egg	....	....	....	50	30 min	16
111				....	....	52	10 min	16
112				....	....	53.5	5 min	16
113				−20	4 hr	....	....	14
114				−17.5	5 hr	....	....	14
115				−15	10 hr	....	....	14
116				−10	24 hr	....	....	14
117				−5	71 hr	....	....	14
118		Adult	....	....	....	33-38 [0-90]	24 hr	51
119				....	....	46	45 min	16
120				....	....	49.5	10 min	16
121				....	....	51.5	5 min	16
122				−20	30 min	....	....	14
123				−17.5	1 hr	....	....	14
124				−15	2 hr	....	....	14
125				−10	7.5 hr	....	....	14
126				−5	20 hr	....	....	14
127			20-39	....	....	46.5 [30-90]	1 hr	51
128	*Periplaneta ameri-*	Nymph & adult	....	−15 to −10	1 hr	....	....	44
129	*cana*	Adult	....	....	....	38 [10]	24 hr	23
130				....	....	42-45 [0-90]	1 hr	39
131			[10-15]	6.5[4]	30 min	....	....	58
132			[30]	8.7[4]	30 min	....	....	58
133	*Phormia regina*	Adult	....	−13.5	12 hr	44.5 [41]	1 hr	62
134				4.5	10 da	38	15 da	62
135	*Plodia interpunc-*	Egg	....	2.4	8 da	....	....	22
136	*tella*	Egg [15-36]	....	−16.5	4 hr	....	....	2
137				4	5 hr	....	....	2
138	*Ptinus ocellus (P. tectus)*	Larva [7-28]	....	−2	90 da	....	....	50
139	*Sitophilus grana-*	Egg	....	−5.5	10 da	....	....	50
140	*rius*			5	59 da	....	....	50
141		Developmental	....	−5.5	15 da	....	....	50
142				−2.5	30 da	....	....	50
143				0	50 da	....	....	50

[3] Temperature preferred by organism. [4] Chill-coma temperature, the low temperature below which locomotory responses to stimuli fail to occur. [8] Cold-hardy.

continued

Part I. Insects

	Species (Synonym)	Stage [Limits for Development, °C]	Activity Range [Acclimation Temp] °C	Lethal Cold		Lethal Heat		Ref-er-ence
				Temp [Biological Zero] °C	Exposure Time	Temp, °C [Relative Humidity, %]	Exposure Time	
144		Adult	....	−5	45 da	....	....	50
145				−2.5	80 da	....	....	50
146				−0.5	32-33 da	....	....	50,77
147				4.5	43 da	....	....	77
148		[12-34]	....	−12 to −10	10 da	....	....	69
149				−5 to −3	45 da	....	....	69
150	Tenebrio molitor	Larva	[18]	....	....	42	1 hr	55
151			[35]	....	....	44	1 hr	55
152		Pupa	[23]	....	....	44	1 hr	53
153		Young ♂ [10-32]	....	....	....	42.5	190 min	12
154		Older ♂ [10-32]	....	....	....	42.5	2 hr	12
155		Adult	[15]	7.4[4/]	30 min	....	....	58
156			[35]	10.8[4/]	30 min	....	....	58
157	Tribolium casta-neum	Egg	....	....	....	50	15 min	30
158				....	....	51	10 min	30
159		Larva	....	....	....	51	15 min	30
160		Pupa	....	....	....	51	10 min	30
161		Adult	....	....	....	51	15 min	30
162		All stages [21-40]	....	3	28 da	....	....	78
163	T. confusum	Egg	....	−16.5	5 hr	....	....	2
164				4	6 hr	....	....	2
165		[18-36]	....	....	....	50	15 min	30
166		Younger egg	....	10	10 da	....	....	80
167		Older egg	....	10	15 da	....	....	80
168		Larva [18-36]	....	....	....	51	10 min	30
169		Pupa [18-36]	....	....	....	50	15 min	30
170				....	....	51	10 min	30
171		Adult	....	....	....	51	10 min	30
172				−15 to −10	1 hr	50	1 hr	45
173		All stages	....	3	28 da	....	....	78
174	Trogoderma grana-rium	Egg [21-41]	....	−1	9 da	60	30 min	73
175		Larva	....	−19	15 da	....	....	50
176		Larva, 1st instar [21-41]	....	....	....	60	30 min	73
177		Larva, 5th instar [21-41]	....	....	....	60	25 min	73
178		Adult	....	....	....	60	35 min	73

[4/] Chill-coma temperature, the low temperature below which locomotory responses to stimuli fail to occur.

Contributors: Howe, R. W.; Mutchmor, John A.

References

[1] Adams, J. B. 1962. Can. J. Zool. 40:951.

[2] Adler, V. E. 1960. J. Econ. Entomol. 53:973.

[3] Asahina, E. 1969. Advan. Insect Physiol. 6:1.

[4] Atwal, A. S. 1960. Can. J. Zool. 38:131.

[5] Auclair, J. L. 1960. Ann. Entomol. Soc. Quebec 5:18.

[6] Beattie, M. V. F. 1928. Bull. Entomol. Res. 18:397.

[7] Becker, G., and I. Loebe. 1961. Anz. Schaedlingsk. 34:145.

[8] Blunck, H. 1914. Z. Wiss. Zool. 111:76.

[9] Blunck, H. 1924. Ibid. 121:171.

continued

89. LETHAL TEMPERATURES: INVERTEBRATES

Part I. Insects

[10] Bodenheimer, F. S. 1931. Z. Angew. Entomol. 18: 492.

[11] Boissezon, P. de. 1930. Bull. Soc. Zool. France 55: 255.

[12] Bowler, K. 1967. Entomol. Exp. Appl. 10:16.

[13] Bucher, G. E. 1947. Doctoral Diss. (14):42.

[14] Busvine, J. R. 1944. Bull. Entomol. Res. 35:115.

[15] Buxton, P. A. 1933. Trans. Roy. Soc. Trop. Med. Hyg. 26:325.

[16] Buxton, P. A. 1940. Brit. Med. J. (4130):341.

[17] Cantwell, G. E. 1971. Agr. Res. 19:14.

[18] Chiang, H. C., et al. 1962. Can. Entomol. 94:722.

[19] Childs, D. P. 1965. Tob. Sci. 9:55.

[20] Childs, D. P., et al. 1968. J. Econ. Entomol. 61: 992.

[21] Childs, D. P., et al. 1970. Ibid. 63:1860.

[22] Cline, L. D. 1970. Ibid. 63:1081.

[23] Cloudsley-Thompson, J. L. 1962. Entomol. Exp. Appl. 5:270.

[24] Colhoun, E. H. 1960. Ibid. 3:27.

[25] Davidson, J. 1944. J. Anim. Ecol. 13:26.

[26] Davison, F. 1971. J. Insect Physiol. 17:575.

[27] Deal, J. 1941. J. Anim. Ecol. 10:323.

[28] Dill, D. B., et al., ed. 1964. Adapt. Environ., Handb. Physiol., Sect. 4.

[29] Dove, W. E. 1916. J. Econ. Entomol. 9:528.

[30] Faria Estacio, F. L. de. 1956. Lisboa Min. Ultramar, Junta Invest. Ultramar Estud. Ensaios Doc. 18.

[31] Farid, M. A. 1949. Amer. J. Hyg. 49:83.

[32] Fay, R. W. 1939. J. Econ. Entomol. 32:851.

[33] Fraenkel, G., and H. S. Hopf. 1940. Biochem. J. 34:1085.

[34] Franssen, C. J. H. 1962. Versl. Landbouwk. Onderzoek. 67:13.

[35] Free, J. B., and Y. Spencer-Booth. 1959. Bee World 40:173.

[36] Free, J. B., and Y. Spencer-Booth. 1960. Entomol. Exp. Appl. 3:222.

[37] Free, J. B., and Y. Spencer-Booth. 1962. Ibid. 5: 249.

[38] Galichet, P. F. 1960. C. R. Acad. Agr. Fr. 46:404.

[39] Gunn, D. L., and F. B. Notley. 1936. J. Exp. Biol. 13:28.

[40] Hanec, W., and S. D. Beck. 1960. J. Insect Physiol. 5:169.

[41] House, H. L., et al. 1958. Can. J. Zool. 36:629.

[42] Hunter, W. D., and W. D. Pierce. 1912. U.S. Dep. Agr. Bur. Entomol. Bull. 114.

[43] Kiyoku, M. 1962. Okayama Daigaku Nogakubu Gakujutsu Hokoku 19:17.

[44] Knipling, E. B., and W. N. Sullivan. 1957. J. Econ. Entomol. 50:368.

[45] Knipling, E. B., and W. N. Sullivan. 1958. Ibid. 51: 344.

[46] Levins, R. 1969. Amer. Natur. 103:483.

[47] Lin, S., et al. 1954. Physiol. Zool. 27:287.

[48] Mansbridge, G. 1936. Ann. Appl. Biol. 23:803.

[49] Maslow, A. W. 1930. Arch. Schiffs. Tropenhyg. 34: 170.

[50] Mathlein, R. 1961. Medd. Statens Vaextskyddsanst. 12:83.

[51] Mellanby, K. 1932. J. Exp. Biol. 9:222.

[52] Mellanby, K. 1939. Proc. Roy. Soc. B127:473.

[53] Mellanby, K. 1954. Proc. Roy. Entomol. Soc. London C19:10.

[54] Mellanby, K. 1954. Nature (London) 173:582.

[55] Mellanby, K. 1958. Ibid. 181:1403.

[56] Mellanby, K. 1960. Bull. Entomol. Res. 50:821.

[57] Meryman, H. T., ed. 1966. Cryobiology. Academic Press, New York.

[58] Mutchmor, J. A., and A. G. Richards. 1961. J. Insect Physiol. 7:141.

[59] Nieschulz, O. 1935. Zool. Anz. 110:225.

[60] Oertel, E. 1962. Proc. Int. Congr. Entomol., 11th, Vienna, 1960, 2:532.

[61] Parker, J. R. 1930. Bull. Univ. Mont. Agr. Exp. Sta. 223.

[62] Parrish, W. D., and W. E. Bickley. 1966. J. Econ. Entomol. 59:804.

[63] Paul, C. F., et al. 1963. Indian J. Entomol. 24:167.

[64] Phelps, R. J., and P. M. Burrows. 1969. Entomol. Exp. Appl. 12:23.

[65] Popov, V. I. 1960. Nauch. Tr. Vissh Selskostop. Inst. Sofia Agron. Fak. 8:327.

[66] Potts, W. H. 1933. Bull. Entomol. Res. 24:293.

[67] Prosser, C. L., et al., ed. 1967. The Cell and Environmental Temperature. Pergamon Press, New York.

[68] Richards, A. G. 1964. Physiol. Zool. 37:199.

[69] Rodionova, L. Z. 1960. Zool. Zh. 39:1624.

[70] Rudolfs, W. 1925. J. N. Y. Entomol. Soc. 33:163.

[71] Salt, R. W. 1936. Minn. Univ. Agr. Exp. Sta. Tech. Bull. 116.

[72] Salt, R. W. 1961. Annu. Rev. Entomol. 6:55.

[73] Soliman, A. A., and A. H. Kashef. 1962. Bull. Soc. Entomol. Egypte 46:513.

[74] Sømme, L. 1961. Nor. Entomol. Tidsskr. 11:191.

[75] Sømme, L. 1968. Entomol. Exp. Appl. 11:143.

[76] Stanev, M. Ts. 1958. Nauch. Tr. Nauchnoizsled. Inst. Zasht. Rast. 1:189.

[77] Stojanovic, T. 1965. Leptop. Nauc. Rad. Poljoprivr. Fak. Novom Sadu (Jugoslav.) 9.

[78] Takashima, F., et al. 1969. Res. Bull. Plant Protect. Serv. Jap. 7:17.

[79] Vulchev, S. 1967. Rast. Zasht. (Sofia) 15(7):8.

[80] Watters, F. L. 1966. J. Stored Prod. Res. 2:81.

[81] Worthley, L. H. L., and D. J. Caffrey. 1927. U.S. Dep. Agr. Tech. Bull. 53.

continued

Part II. Parasitic Helminths and Protozoa

	Species	Stage	Environment or Host	Lethal Extreme	Lethal Exposure		Reference
					Temp, °C	Time	
			Acanthocephala				
1	*Macracanthorhynchus hirudinaceus*	Egg	Water	Heat	60	10 min	40
2	*Moniliformis dubius*	Egg	Cockroach	Heat	38	4 wk	72
			Nematoda				
3	*Ancylostoma caninum*	Egg	Culture	Cold	−10	6 hr	7
4			Fecal culture	Cold	7-9	1 wk	85
5			Water	Heat	42	....	47
6		Larva	Culture	Cold	−20	1 da	7
7	*A. duodenale*	Egg	Fecal culture	Cold	7-9	1 wk	85
8	*Ascaridia galli*	Egg	....	Heat	54	5 min	33
9			Water	Heat	43	12 hr	1
10				Cold	−12 to −8	22 hr	1
11	*Ascaris lumbricoides*	Egg	....	Heat	68	0.44 s	87
12					65	1.9 min[1]	70
13			Feces	Heat	50	Few hr	70
14			Water	Heat	60	5 min	56
15					54	5 min	54
16	*Bunostomum trigonocephalum*	Egg	Culture	Heat	40	12 hr	10
17				Cold	−12	2 hr	10
18			0.78% saline solution	Heat	50	20 min	65
19				Cold	−10	7 hr	65
20		Larva, 1st instar	Culture	Cold	−12	1 hr	10
21			Water	Heat	50	20 min	53
22		Larva, infective	100% relative humidity	Heat	45	6 hr	10
23				Cold	−12	1.5 hr	10
24			Water	Heat	50	2 hr	53
25	*Enterobius vermicularis*	Egg	....	Cold	−20	6 hr	9
26			Dry heat	Heat	56	1 min	37
27			22% relative humidity	Heat	44-48	3 hr	37
28			Moist state	Cold	−10 to −8	6 da	37
29	*Haemonchus contortus*	Egg	....	Cold	−12	2 da	76
30			Fecal culture	Heat	60	2.5 hr	76
31			Water	Heat	37.8	5 da	13
32	*Metastrongylus apri (M. elongatus)*	Egg	Water	Heat	50	3 min	92
33	*Necator americanus*	....	Fecal culture	Cold	7-9	1 wk	85
34	*Oesophagostomum columbianum*	Egg	Fecal pellet	Heat	58	15 min	20
35			Moist feces	Cold	−18	10 da	20
36		Larva, infective	100% relative humidity	Heat	50	30 min	75
37				Cold	−18	24 hr	75
38			Water	Heat	50	15 min	64
39					50	2 hr	2

[1] Estimated 100% kill.

continued

Part II. Parasitic Helminths and Protozoa

	Species	Stage	Environment or Host	Lethal Extreme	Lethal Exposure Temp, °C	Time	Reference
40	*Ostertagia circum-*	Egg	Water	Heat	45	1-2 da	25
41	*cincta*	Larva	....	Heat	45	30 min	50
42			Water	Cold	−6	2 wk	25
43	*Strongyloides pa-*	Egg	Culture	Heat	50	1 hr	63
44	*pillosus*	Larva, 1st & 2nd	Fecal culture or water	Cold	−10 to −7	12 min	62
45		instars	Water	Heat	51	1.5 min	62
46		Larva, 3rd instar	Water	Heat	56	1 min	62
47				Cold	−70	0.5 min	62
48					−21	8-10 min	62
49			Wet charcoal	Cold	−5	24 hr	91
50	*S. stercoralis*	Filaria	Charcoal culture	Heat	50	90 min	26
51	*Strongylus equi-*	Egg	....	Cold	−10 to −6	47-56 da	43
52	*nus*		Feces	Heat	45	96 da	27
53	*Toxascaris leonina*	Egg	Agar medium	Heat	40	1-3 da	57
54	*Toxocara canis*	Egg	Agar medium	Heat	40	1-3 da	57
55				Cold	−15	5 da	57
56	*Trichinella spiralis*	Larva	....	Cold	−37	2 min	29
57			Ringer's solution	Heat	60	10 min	59
58			Tissue	Cold	−34 to −18	24 hr	68
59					−33.7	310 min	6
60			Water	Heat	55	....	67
61	*Trichostrongylus colubriformis*	Egg	*Escherichia coli* culture or feces	Heat	35	12 hr	98
62		Egg, unembryonated	Fecal pellet	Heat	50	12 hr	5
63				Cold	−95	1 da	5
64		Egg, embryonated	Fecal pellet	Heat	50	12 hr	5
65				Cold	−95	2 da	5
66		Larva, 1st instar	Fecal pellet	Heat	50	12 hr	5
67				Cold	−95	4 da	5
68		Larva, 2nd instar	Fecal pellet	Heat	50	12 hr	5
69				Cold	−95	32 da	5
70		Larva, 3rd instar	Desiccated	Heat	50	16-24 da	4
71			Dry	Cold	−28	64-128 da	4
72			Fecal pellet	Cold	−95	64 da	5
73			Dry	Heat	50	64-128 da	4
74				Cold	−95	<128 da	4
75			Moist	Heat	50	4 da	4
76				Cold	−95	>128 da	4
77			Water	Heat	50	8 hr	4
78					50	12 hr	5
79				Cold	−28	64 da	4
80	*Trichuris trichiura*	Egg	Water	Heat	60	5 min	44
81					52	3 min	54
82	*Wuchereria ban-crofti*	Microfilaria	Mosquito	Cold	3.8	37 hr	58
	Platyhelminthes—Cestoda[2]						
83	*Diphyllobothrium*	Egg	Water	Heat	35	Immediate death	30
84	*latum*			Cold	−10	3 hr	30

[2] Class.

continued

Part II. Parasitic Helminths and Protozoa

	Species	Stage	Environment or Host	Lethal Extreme	Lethal Exposure Temp, °C	Time	Reference
85		Plerocercoid	Fish	Heat	50	10 min	8
86				Cold	−18	2-4 da	88
87					−10	24 hr	8
88			Hank's solution	Cold	−9	....	99
89	*Echinococcus*	Egg	0% relative humidity	Cold	4	4 hr	42
90	*granulosus*		Water	Heat	60	10 min	48,55
91	*Hymenolepis dim-*	Egg	....	Heat	50	Few min	8
92	*inuta*	Cysticercoid	Beetle	Heat	40	5 da	95
93	*H. nana*	....	Water	Heat	....	10 hr	80
94		Egg	Dry	Heat	41-42	4 hr	80
95	*Taenia pisiformis*	Egg	....	Heat	59	10 min	78
96			0% relative humidity	Cold	4	4 hr	42
97	*T. saginata*	Egg	....	Heat	59	10 min	78
98			Water	Cold	−30	16-19 da	84
99		Cysticercus	Tissue	Heat	71	5 min	9
100				Cold	−10 to −8	65 hr	73,94
101		Encapsulated	Tissue	Heat	56	14-32 min	3
102		Decapsulated	Ringer's solution	Heat	56	5 min	3
103	*T. solium*	Cysticercus	Pork	Cold	−10	4 da	94
104			Tissue	Heat	48	15 min	14
105				Cold	−35	1 hr	14
	Platyhelminthes—Trematoda[2]						
106	*Clonorchis sinensis*	Metacercaria	Fish	Heat	60	15 s	32
107	*Fasciola hepatica*	Egg	....	Heat	33-35	24 hr	93
108			Water	Cold	0	24 hr	97
109		Metacercaria	....	Heat	35	14 da	16
110				Cold	−20	12 hr	16
111					−5 to −1	>24 hr	93
112	*Paragonimus*	Metacercaria	Crabs	Heat	55	20 min	52
113	*westermanii*			Cold	−40	3 min	90
114	*Schistosoma haematobium*	Cercaria	Water	Heat	50	Immediate death	61
115				Cold	2	7 hr	61
116	*S. japonicum*	Egg	0.2% NaCl solution	Heat	55	3 min	34
117			Water	Cold	−18	18 hr	34
118	*S. mansoni*	Egg	Feces or tissue	Heat	25	1 hr	11
119		Miracidium	Water	Heat	45	30 min	12
120				Cold	−5	1 hr	12
121		Cercaria	Water	Heat	45	30 min	41,66
	Protozoa—Ciliophora[3]						
122	*Balantidium coli*	Trophozoite	Culture (PR strain)	Heat	47	15 min	86
	Protozoa—Sporozoa[3]						
123	*Babesia bigemina*	....	Blood	Heat	48	10 min	69
124	*Eimeria tenella*	Oocyst	Buffered solutions	Heat	52	2.4 min[4]	23
125			2.5% $K_2Cr_2O_7$	Heat	53	3 hr	22
126				Cold	−12	7 da	22

[2] Class. [3] Subphylum. [4] Estimated 50% kill.

continued

Part II. Parasitic Helminths and Protozoa

	Species	Stage	Environment or Host	Lethal Extreme	Lethal Exposure		Reference
					Temp, °C	Time	
127			Water	Heat	53	10 min	24
128					50	16 min[5]	18
129		Sporozoite	Cell culture	Heat	44	72 hr	83
130	Isospora rivolta	Oocyst	....	Heat	65	1 hr	81
131			2.5% $K_2Cr_2O_7$	Heat	50	4 hr	45
132	Plasmodium cathemerium	Asexual forms	Blood	Heat	50	10 min	28
133	P. vivax	Gametocyte	Mosquito	Heat	37.5	2-3 hr	82
134	Toxoplasma gondii	Proliferative	Mouse liver	Cold	−14	3 hr	51
135			Peritoneal exudate	Heat	56	5 min	71
136		Cyst	Mouse brain	Cold	−15	24 hr	35
137					−14	3 hr	51
138			Mouse brain suspension	Heat	56	10 min	35,51

Protozoa—Sarcomastigophora[3]

	Species	Stage	Environment or Host	Lethal Extreme	Lethal Exposure		Reference
139	Entamoeba coli	Trophozoite	Culture	Heat	45	3 hr	17
140				Cold	3	8-10 hr	17
141		Cyst	Water	Heat	76	5 min	15
142	E. histolytica	Trophozoite	Culture	Heat	45	1-3 hr	17
143					41-43	72 hr	77
144				Cold	−70	2 hr	49
145					3	1-4 da	17
146			Feces	Heat	37	2-5 hr	89
147		Cyst	Feces	Cold	0-6	1-4 da	79
148			Feces & physiological saline	Heat	50	5 min	79
149			Water	Heat	68	5 min	15
150					51	5 min	38
151					50	2 min	19
152					50	5 min	100
153			Ringer's solution	Cold	−15	24 hr	31
154	Giardia lamblia; Iodamoeba buetschlii	Cyst	Water	Heat	64	5 min	15
155	Leishmania brasiliensis; L. caninum; L. donovani; L. tropica	Leptomonad (culture)	Saline solution	Heat	40	30 min	74
156	Trichomonas vaginalis	Trophozoite	Culture	Heat	50	4 min	36
157			Water	Heat	50	5 min	46
158	Tritrichomonas foetus	Trophozoite	5% glycerol & semen	Cold	−79	∿20 hr after overnight pretreatment at 5°C	39
159	Trypanosoma brucei	Crithidia (culture)	Alsever's solution	Cold	−20	1 da	60
160	T. cruzi	Crithidia (culture)	Culture	Heat	47	15 min	9
161					46	15 min	21
162			Locke's solution	Heat	44.3	1 hr	96

[3] Subphylum. [5] 50% kill.

Contributor: Daly, James J.

continued

89. LETHAL TEMPERATURES: INVERTEBRATES

Part II. Parasitic Helminths and Protozoa

References

[1] Ackert, J. E. 1931. Parasitology 23:360.

[2] Agrawal, V. 1966. Trans. Amer. Microsc. Soc. 85: 99.

[3] Allen, R. W. 1947. J. Parasitol. 33:331.

[4] Anderson, F. L., and N. D. Levine. 1968. Ibid. 54: 117.

[5] Anderson, F. L., et al. 1966. Ibid. 52:713.

[6] Augustine, D. L. 1933. Amer. J. Hyg. 17:697.

[7] Balasingam, E. 1964. Can. J. Zool. 42:907.

[8] Belding, D. L. 1952. Textbook of Clinical Parasitology. Ed. 2. Appleton-Century-Crofts, New York.

[9] Belding, D. L. 1965. Ibid. Ed. 3.

[10] Belle, E. A. 1959. Can. J. Zool. 37:289.

[11] Benex, J. 1960. Bull. Soc. Pathol. Exot. 53:526.

[12] Benex, J., and R. Deschiens. 1964. Ibid. 56:987.

[13] Berberian, J. F., and J. D. Mizelle. 1957. Amer. Midl. Natur. 57:421.

[14] Biagi, F., et al. 1963. Prensa Med. Mex. 28:253.

[15] Boeck, W. C. 1921. Amer. J. Hyg. 1:365.

[16] Boray, J. C., and K. Enigk. 1964. Z. Tropenmed. Parasitol. 15:324.

[17] Cabrera, H. A., and R. J. Porter. 1958. Exp. Parasitol. 7:285.

[18] Chang, K. 1937. Amer. J. Hyg. 26:337.

[19] Chang, S. L. 1950. Ibid. 56:82.

[20] Chhabra, R. C., and K. S. Singh. 1965. Indian J. Helminthol. 17:125.

[21] Dalma, J. 1961. Rev. Fac. Med. Tucuman (Argent.) 3:291.

[22] Edgar, S. A. 1954. Trans. Amer. Microsc. Soc. 73: 237.

[23] Farmer, J. N., and E. R. Becker. 1958. Proc. Iowa Acad. Sci. 65:535.

[24] Fish, F. 1931. Science 73:292.

[25] Furman, D. P. 1944. Amer. J. Vet. Res. 5:79.

[26] Gailliard, H. 1951. Ann. Parasitol. Hum. Comp. 26: 204.

[27] Galofre, E. J., and W. A. Rosa. 1942. Publ. Univ. Buenos Aires Fac. Agron. Vet. Inst. Parasitol. Enferm. Parasitar. 2:1.

[28] Gingrich, W. D. 1941. J. Infect. Dis. 68:46.

[29] Gould, S. E., and L. J. Kaasa. 1949. Amer. J. Hyg. 49:17.

[30] Guttowa, A. 1961. Acta Parasitol. Pol. 9:371.

[31] Halpern, B., and R. E. Dolkart. 1954. Amer. J. Trop. Med. Hyg. 3:276.

[32] Hsu, H. F., and L. S. Wang. 1938. Chin. Med. J., Suppl. 2:385.

[33] Itagaki, S. 1927. Proc. World's Poult. Congr. (Ottawa), p. 339.

[34] Ito, J. 1955. Jap. J. Med. Sci. Biol. 8:175.

[35] Jacobs, L., et al. 1960. J. Parasitol. 46:11.

[36] Johnson, G., and M. H. Trussel. 1944. Proc. Soc. Exp. Biol. Med. 57:252.

[37] Jones, M. F., and L. Jacobs. 1941. Amer. J. Hyg. D33:88.

[38] Jones, M. F., and W. L. Newton. 1950. Amer. J. Trop. Med. 30:53.

[39] Joyner, L. P., and G. H. Bennett. 1956. J. Hyg. 54: 335.

[40] Kates, K. C. 1942. J. Agr. Res. 64:93.

[41] Krakower, C. A. 1940. Puerto Rican J. Pub. Health Trop. Med. 16:26.

[42] Laws, G. F. 1968. Exp. Parasitol. 22:227.

[43] Lucker, J. T. 1941. J. Agr. Res. 63:193.

[44] Magrova, E. 1967. Biologia (Bratislava) 22:364.

[45] Mahrt, J. L. 1968. J. Protozool. 15:308.

[46] Malysko, E., and J. Dziegielewska. 1966. Wiad. Parazytol. 12:491.

[47] McCoy, O. R. 1930. Amer. J. Hyg. 11:413.

[48] Meymarian, E., and C. W. Schwabe. 1962. Amer. J. Trop. Med. Hyg. 11:360.

[49] Molinari, V. 1956. Bull. Soc. Pathol. Exot. 49:254.

[50] Morgan, D. O. 1928. J. Helminthol. 6:183.

[51] Motomura, I. 1967. Trop. Med. (Nagasaki) 9:201.

[52] Nakagawa, K. 1916. J. Infect. Dis. 18:131.

[53] Narain, B. 1965. Parasitology 55:551.

[54] Nolf, L. O. 1932. Amer. J. Hyg. 16:288.

[55] Nosik, A. F. 1952. Sb. Tr. Khar'kov. Vet. Inst. 21: 304.

[56] Ogata, S. 1925. Ann. Trop. Med. Parasit. 19:301.

[57] Okoshi, S., and M. Usui. 1968. Jap. J. Vet. Sci. 30: 29.

[58] Omori, N. 1962. Endem. Dis. Bull. Nagasaki Univ. 4:1.

[59] Otto, G. F., and E. Abrams. 1939. Amer. J. Hyg. 29:115.

[60] Polge, C., and M. A. Soltys. 1957. Trans. Roy. Soc. Trop. Med. Hyg. 51:519.

[61] Porter, A. 1938. Publ. S. Afr. Inst. Med. Res. 8:1.

[62] Prasad, D. 1963. J. Zool. Soc. India 15:122.

[63] Premvati. 1963. Parasitology 53:483.

[64] Premvati, and S. S. Lal. 1961. J. Parasitol. 47:943.

[65] Premvati, G., and B. Narain. 1969. Parasitology 59: 185.

[66] Purnell, R. 1966. Ann. Trop. Med. Parasitol. 60: 182.

[67] Ranson, B. H. 1916. J. Agr. Res. 5:819.

[68] Ranson, B. H., and B. Schwartz. 1919. Ibid. 17:201.

[69] Rees, C. W. 1937. J. Parasitol. 23:175.

[70] Reyes, W. L., et al. 1963. Amer. J. Trop. Med. Hyg. 12:46.

[71] Rifaat, M. A., and T. A. Morsy. 1963. J. Trop. Med. Hyg. 66:178.

[72] Robinson, E. S., and A. W. Jones. 1971. Exp. Parasitol. 29:292.

[73] Schmey, M., and G. Bugge. 1931. Berlin. Tieraerztl. Wochenschr. 47:193.

continued

Part II. Parasitic Helminths and Protozoa

[74] Senekji, H. A., and N. Zebouni. 1941. Amer. J. Hyg. 34(c):67.

[75] Shanker, R. 1970. Indian Vet. J. 47:120.

[76] Shorb, D. A. 1944. J. Agr. Res. 69:279.

[77] Siddiqui, W. A. 1963. J. Protozool. 10:480.

[78] Silverman, P. H. 1956. Trans. Roy. Soc. Trop. Med. Hyg. 50:7.

[79] Simitch, T., et al. 1954. Arch. Inst. Pasteur Alger. 32:223.

[80] Simitch, T., et al. 1955. Ibid. 33:30.

[81] Stojanovic, B. 1949. Arh. Biol. Nauka 1:188.

[82] Stratmann-Thomas, W. K. 1940. Amer. J. Trop. Med. 20:703.

[83] Strout, R. G., and C. A. Ouellette. 1969. Exp. Parasitol. 25:324.

[84] Suvorov, V. Y. 1965. Med. Parazitol. Parazit. Bolez. 34:98.

[85] Svensson, R. 1925. Chin. Med. J. 39:667.

[86] Svensson, R. 1955. Exp. Parasitol. 4:502.

[87] Swales, W. E., and D. K. Froman. 1939. Can. J. Res. D17:169.

[88] Titova, S. D. 1955. Med. Parazitol. Parazit. Bolez. 24:255.

[89] Tsuchiya, H. 1945. Amer. J. Trop. Med. 25:277.

[90] Tsuda, M. 1959. Kiseichugaku Zasshi 8:812.

[91] Turner, J. H., and G. I. Wilson. 1961. J. Parasitol. 47:30.

[92] Ustinov, I. D. 1963. Tr. Vses. Inst. Gel'mintol. 10:56.

[93] Vasil'eva, I. 1960. Zool. Zh. 39:1478.

[94] Viljoen, N. R. 1937. Onderstepoort J. Vet. Sci. Anim. Ind. 9:337.

[95] Voge, M., and J. A. Turner. 1956. Exp. Parasitol. 5:580.

[96] Von Brand, T., et al. 1946. J. Gen. Physiol. 30:163.

[97] Wade, L. 1952. Amer. J. Vet. Res. 13:345.

[98] Wang, G.-T. 1967. Ibid. 28:1085.

[99] Wikgren, B., and P. Nikander. 1963-64. Mem. Soc. Fauna Flora Fenn. 40:189.

[100] Yorke, W., and A. R. D. Adams. 1926. Ann. Trop. Med. Parasitol. 20:317.

90. THERMAL DEATH: RICKETTSIA AND BACTERIA

Bacteria which do not form endospores are generally killed within 20 minutes when directly exposed in fluid to temperatures of 70°C or over (moist heat), but thermophiles are somewhat more resistant. Bacterial endospores may resist moist heat at 100°C for two minutes to many hours. However, no known living organism can survive compressed steam at 121°C (routine autoclaving) for 20 minutes. Inconsistencies in the values may be attributed to different experimental methods and to the fact that all of the variables, especially pH (a crucial factor), were not always reported. **Heating Menstruum:** PS = phosphate solution; PW = peptone water.

	Species (Synonym)	Heating Menstruum	Temp °C	Time min	Reference
	Rickettsiales				
1	Coxiella burneti	Egg-yolk sac	63-65	30	34
2		Milk	62-66	30	8,20
3			72.22	0.25	8,20
4		Milk[1]	62.7	30-40	18
5		Milk[2]	67.7-68.7	0.25	23
6		Milk[3]	71.6	0.25	24
7	Miyagawanella lymphogranulomatosis	....	56	10	35,43
8			60	10	20
9	Rickettsia prowazeki	....	50	30	19
10			56	30	35
11		Milk	50	15	43

	Species (Synonym)	Heating Menstruum	Temp °C	Time min	Reference
12	R. quintana	Louse excreta	60	20	44
13			100[4]	20	44
14	R. rickettsi	....	50	Few	35
	Schizomycetes				
15	Actinobacillus lignieresii	....	62	10	39,44
16	A. mallei	....	55	10	5,44
17			60	120	39
18			75	60	39
19	Actinomyces bovis	....	60	15	8,43
20			60	60	19
21			62-64	3-10	42
22	A. israelii	....	60	60	39

[1] Sealed tubes in water. [2] Naturally infected. [3] Artificially infected. [4] Dry heat.

continued

	Species (Synonym)	Heating Menstruum	Temp °C	Time min	Reference		Species (Synonym)	Heating Menstruum	Temp °C	Time min	Reference
23	Aerobacter aerogenes	....	55	30	44	64		Milk	62.2	7	30
24	A. aerogenes[5]	....	60	30	19	65		0.85% NaCl	62.5	10	1
	Bacillus anthracis					66	Clostridium botulinum, spores	....	100	300	33,44
25	Spores	....	100	5-10	8,33	67			105	100	44
26			140[4]	180	6,21	68			120	5	44
27			150[4]	60	33,44	69			180[4]	5-15	33,44
28		Boiling water	100	10	39	70		pH 7.0	100	480	8
29		Live steam	100	5-10	39	71			115	10-40	8
30		Saline suspension[6]	90	15-45	44	72		2% aminoids	100	65	10
31			95	10-25	44	73		PS, pH 7	100	330	21
32			100	5-10	44	74			105	110	39
33		Physiological NaCl solution	100	5-10	26	75			110	33	39
34		Horsehair or bristles[7]	121	15	44	76			115	11	39
35	Vegetative forms	....	54	30	39	77			120	4	21
36			55	60	44	78		Canned corn	100	105	15
37			60	30	8	79			115	15	15
38	B. subtilis, spores	....	100	15-20	14	80		Canned pears	100	30	15
39		5% Phenol	63	55-65	37	81			115	5	15
40			70	30	37	82	C. novyi	....	105	6	44
41			80	5	37	83	Spores	....	80	60	32
42		PS, pH 4.4	100	2	14	84			100	10	33
43		pH 5.6	100	7	14		C. perfringens				
44		pH 7.6	100	11	14	85	Spores	....	80[11]	10	7
45		pH 8.4	100	9-11	14	86		Suspension[6]	90	30	44
46		1% PW[8]	100	11	14	87			100	5	44
47		1% PW[9]	100	18	14	88	Spores[12]	0.85% NaCl	90	29	17
48	Bordetella pertussis	....	55	30	6,21	89	Spores[13]	PS, pH 7.0-7.12	100	10	10
49	Borrelia recurrentis	....	50	30	12	90			105	27	10
50	Brucella abortus	Milk	51.7	30	11	91	Vegetative cells	....	80	10	7
51			60	10	41	92	C. septicum, spores	....	170[4]	7	28,33
52			61.1-62.7	30	6	93			110	5	33
53			70.1	0.25	41	94	C. tetani	....	105	3-25	19
54		0.85% NaCl	62.5	10	1	95		Steam	100	40-60	5
55	B. melitensis	....	60	10	8,44	96	Spores	....	80[4]	60	39
56		Aqueous emulsion	57.5	10	16	97			100[11]	10	8
57		Milk	60	20	41	98			105	3-25	44
58			61.1-62.7	30	5	99		pH 4.1 or 10.2	100	11	8
59		0.85% NaCl	62.5	10	1	100		pH 7.2	100	29	8
60	B. suis	....	60	10	44	101		Live steam	100	5	39
61			61.6-62.7	3	5	102		0.85% NaCl	100	10-25	27
62		Milk[10]	60	20	30	103		pH 1.2	100	5	27
63			61.1	15	30	104		pH 4	100	10	27
						105		pH 7	100	30	27
						106	Corynebacterium diphtheriae	....	58	10	39,43
						107			60	10	8
						108			100	1	39
						109		Water	58	10	39
						110			100	1	39
						111		Suspension or broth culture	58	10	5,44

[4] Dry heat. [5] Strains not resistant to heat. [6] Containing 1 million spores/ml. [7] In bundles not more than 2.5 inches thick. [8] Incubated at 21-23°C. [9] Incubated at 41°C. [10] 5 × 10⁸ cells/ml. [11] Most spores destroyed at this temperature. [12] 5 strains. [13] 1 strain.

continued

	Species (Synonym)	Heating Menstruum	Temp °C	Time min	Reference
112		Milk	55-60	2	36
113		Ice cream	65.6	0.5	29
114	C. xerosis	Broth, pH 7.6	56	4	29
115	Diplococcus	....	52	10	39
116	pneumoniae		55	20	3,19
117		Broth 14/	60	30	16
118		Blood broth	56	5-7	25
119		Melted dextrose agar	60	15	16
120	Erysipelothrix rhusiopathiae (E. insidiosa)	....	55	15	39,44
121	Escherichia coli 15/	....	60	15-20	39
122	E. coli	....	55	60	44
123			60	15	44
124			60	30	6,19
125			62.5	30	41
126		Broth	61	10	14
127		Milk	57.2	>60	13
128		Whole milk	69	10	14
129		Skim milk	65	10	14
130		Whey	63	10	14
131		Cream	73	10	14
132	Fusobacterium fusiforme (Fusiformis fusiformis)	....	55	15	44
133			60	2	19
134	Haemophilus ducreyi	....	55	60	39,44
135	H. influenzae	....	50-55	30	39,44
136			55	30	19
137		Bouillon	58	10	16
138			62	2	16
139	Klebsiella pneumoniae	....	55	30	9,44
140	Lactobacillus thermophilus	....	71	30	2
141			82	2.5	2
142	Leptospira icterohemorrhagiae	....	50	10	8,44
143			50-55	30	39
144			60	0.1	8,44
145	Listeria monocytogenes	....	55	60	44
146		Milk	65	0.5-0.6	44
147			75	0.1	44
148	Mycobacterium bovis	Milk	61.1	10	16
149			62.8	6	16
150	M. marinum (M. balnei)	....	56	10	44
151	M. tuberculosis	....	58	30	6
152	sis		59	20	6
153			60	15	44

	Species (Synonym)	Heating Menstruum	Temp °C	Time min	Reference
154			60	15-20	8,19
155			65	2	6
156			71	0.33	13
157			82	<0.33	13
158			100	<0.25	13
159		Milk	55	60	13
160			60	10	13
161			62	30	39
162			65	2	13
163		Milk 16/	60	20	44
164		Ice cream	62.6	6	29
165	Neisseria gonorrhoeae	....	42	300-900	44
166			50	2-3	14
167			55	<5	39,44
168			55	5	8,9
169		Infected pus 14/	50	5	16
170	N. meningiti-	....	50	Fast	39
171	dis		55	<5	44
172			55	5	8
173		Bouillon 14/	60	1	16
174	N. meningiti- dis 17/	....	Room	180	44
175	Nocardia asteroides	....	60	60	39
176			70	5	44
177	N. madurae	....	60	5	44
178	Pasteurella multocida	....	<45	......	2,19
179			50	15	9
180		Broth	55	15	44
181	P. multocida (P. septica)	....	60	Few	44
182	Proteus vul- garis	....	55	60	3,44
183		Broth	55	60	43
184	Pseudomonas aeruginosa	Water bath	55	60	44
185	Salmonella spp. 18/	....	55	60	44
186			60	15-20	44
187	S. enteritidis	Broth, pH 7.05	60	23.8	41
188			65	13.8	41
189	S. gallinarum	Liquid whole egg	60.5-61.6	6	38
190	S. paratyphi	....	55	30	44
191			55	60	43
192			60	15-20	43
193		Broth, pH 7.05	60	8.9	41
194			65	4.3	41
195		Milk	60-62	10	41
196	S. typhi (S. typhosa)	....	56	......	8
197			60	4.3	14
198		Suspension 19/	47	120	44
199			49	48	44

14/ Sealed tubes.   15/ Most strains.   16/ Closed vessel.   17/ Dried.   18/ Most species.   19/ Containing 100,000 organisms/ml.

continued

	Species (Synonym)	Heating Menstruum	Temp °C	Time min	Reference
200			51	15	44
201			53	7	44
202			55	2.5	44
203			59	0.35	44
204		Broth	55	23.5	41
205			56	4-9	25
206			60	4.3	14
207			60	8.9	41
208		Milk	55	30	36
209			60	20	16
210			61.1-62.7	4-8	41
211		Milk[14]	60	8	16
212			63	4	16
213		Ice cream	57.2	10	29
214			62.8	3	29
215	S. typhimurium	Broth, pH 7.05	60	13.8	41
216		Liquid whole egg	58.8	3.7	45
217			60	2.6	45
218	Shigella dysenteriae	....	55	60	8,44
219		Water	60	1	41
220		Milk	60	6	41
221			60	16	31
222		Milk[14]	60	10	31
223	Staphylococcus aureus	....	60	10	19
224			60[20]	30	8
225			60	30-60	9
226			60[21]	60	21,43
227			80[22]	60	5
228			190.6-218.3[4]	30	14
229		Broth	58	10	16
230			65.6	2	40
231		Broth[14]	65	18.8	41

	Species (Synonym)	Heating Menstruum	Temp °C	Time min	Reference
		Skim milk			
232		9% serum solids	60	30	22
233		20% serum solids	60	35	22
234		14% sugar	60	30	22
235		Custard filling	88	10	14
236	Streptobacillus moniliformis (Actinobacillus muris)	Serum broth	55	30	44
237	Streptococcus faecalis	....	60	30	44
238			60	60	21
239	S. pyogenes[15]	....	60	30-60	39
240	S. pyogenes	....	54	30	8
241			55[22]	10	39
242			55	30	43,44
243			60	30	19
244		Milk	60	15	4
245			62	30	21
246		Cream	57.2	5	29
247			61.1	1	29
248		Melted dextrose agar	60	15	16
249		Dilute salt gelatin	60	10	16
250	Treponema pallidum	....	43	10	8
251			50-55	30	19
252		Suspension[23]	41	120	5
253		Infected rabbit testicular tissue	39	300	39
254			40	180	39
255			41	120	39
256			41.5	60	5,39
257	Vibrio comma	....	55	10	5,39
258			55	15	44
259			56	30	8
260	V. fetus	....	56	5	44

[4] Dry heat. [14] Sealed tubes. [15] Most strains. [20] Some strains withstand 70°C for a short time. [21] Generally killed at this temperature and time. [22] Required for some strains. [23] Infected rabbit testicular tissue.

Contributors: Dupré, Margaret V.; Frobisher, Martin

References

[1] Boak, R., and C. M. Carpenter. 1928. J. Infec. Dis. 43:327.

[2] Breed, R. S., et al., ed. 1957. Bergey's Manual of Determinative Bacteriology. Ed. 7. Williams and Wilkins, Baltimore.

[3] Bryan, A. H., et al. 1962. Bacteriology: Principles and Practice. Ed. 6. Barnes and Noble, New York.

[4] Buchanan, R. E., and E. D. Buchanan. 1951. Bacteriology. Ed. 5. Macmillan, New York.

[5] Burrows, W. 1963. Textbook of Microbiology. Ed. 18. W. B. Saunders, Philadelphia.

[6] Burrows, W., et al. 1968. Ibid. Ed. 19.

[7] Canada, J. C., et al. 1964. Appl. Microbiol. 12:273.

[8] Cruickshank, R., ed. 1960. Mackie and McCartney's Handbook of Bacteriology. Ed. 10. E. and S. Livingstone, Edinburgh.

[9] Dubos, R. J., ed. 1958. Bacterial and Mycotic Infections of Man. Ed. 3. J. B. Lippincott, Philadelphia.

[10] Esty, J. R., and K. F. Meyer. 1922. J. Infec. Dis. 31:650.

[11] Evans, A. C. 1917. J. Bacteriol. 2:185.

[12] Felsenfeld, O. 1965. Bacteriol. Rev. 29:46.

continued

[13] Foster, E. M., et al. 1957. Dairy Microbiology. Prentice-Hall, Englewood Cliffs, N. J.

[14] Frazier, W. C. 1958. Food Microbiology. McGraw-Hill, New York.

[15] Halversen, W. V., and G. L. Hays. 1936. J. Bacteriol. 32:466.

[16] Hampil, B. 1932. Quart. Rev. Biol. 7:171.

[17] Headlee, M. R. 1931. J. Infec. Dis. 48:468.

[18] Huebner, R. J., and J. A. Bell. 1951. J. Amer. Med. Ass. 145:301.

[19] Jacobs, M. B., and M. J. Gerstein. 1960. Handbook of Microbiology. Van Nostrand, Princeton, N. J.

[20] Jawetz, E., et al. 1960. Review of Medical Microbiology. Ed. 4. Lange Medical Publications, Los Altos, Calif.

[21] Joklik, W. K., and D. T. Smith, ed. 1972. Zinsser's Microbiology. Ed. 15. Appleton-Century-Crofts, New York.

[22] Kadan, R. S., et al. 1963. Appl. Microbiol. 11:45.

[23] Lennette, E. H., et al. 1952. Amer. J. Hyg. 55:246.

[24] Marmion, B. P., et al. 1951. Min. Health (Gt. Brit.) Mon. Bull. 10:119.

[25] Morton, H. E. Unpublished. Univ. Pennsylvania School of Medicine, Philadelphia, 1953.

[26] Murray, T. J. 1931. J. Infec. Dis. 48:457.

[27] Murray, T. J., and M. R. Headlee. 1931. Ibid. 48:436.

[28] Oag, R. K. 1940. J. Pathol. Bacteriol. 51:137.

[29] Oldenbusch, C., et al. 1930. Amer. J. Pub. Health 20:615.

[30] Park, S. E., et al. 1932. J. Bacteriol. 24:461.

[31] Park, W. H. 1927. Amer. J. Pub. Health 17:36.

[32] Park, W. H., and A. W. Williams. 1939. Pathogenic Microorganisms. Ed. 11. Lea and Febiger, Philadelphia.

[33] Pelczar, M. J., Jr., and R. D. Reid. 1972. Microbiology. McGraw-Hill, New York.

[34] Ransom, S. E., and R. J. Heubner. 1951. Amer. J. Hyg. 53(1):110.

[35] Rivers, T. M., and F. L. Horsfall, Jr., ed. 1959. Viral and Rickettsial Infections of Man. Ed. 3. J. B. Lippincott, Philadelphia.

[36] Rosenau, E. C. 1908. U.S. Pub. Health Serv. Hyg. Lab. Bull. 42.

[37] Russell, A. D., and M. Loosemore. 1964. Appl. Microbiol. 12:403.

[38] Schneider, M. D. 1951. Food Technol. 5:349.

[39] Smith, D. T., et al., ed. 1964. Zinsser's Microbiology. Ed. 13. Appleton-Century-Crofts, New York.

[40] Stritar, J. E. 1941. Amer. Meat Inst. 36th Annu. Conv., p. 15.

[41] Tanner, F. W. 1944. The Microbiology of Foods. Ed. 2. Garrard Press, Champaign, Ill.

[42] Waksman, S. A., and H. A. Lechevalier. 1959-62. The Actinomycetes. Williams and Wilkins, Baltimore. v. 1-3.

[43] Wilson, G. S., and A. A. Miles. 1955. Topley and Wilson's Principles of Bacteriology and Immunity. Ed. 4. Williams and Wilkins, Baltimore. v. 1 & 2.

[44] Wilson, G. S., and A. A. Miles. 1964. Ibid. Ed. 5.

[45] Winter, A. R., et al. 1946. Amer. J. Pub. Health 36:451.

91. THERMAL DEATH: NONPATHOGENIC BACTERIA

The time and temperature necessary for destruction of bacteria by heat depend on strain differences, nutritional conditions for growth (including sporulation), composition and pH of the heating menstruum, concentration and age of the cells, and the type of medium and conditions used for recovery of survivors.

Part I. D Values

D Value: Number of minutes necessary at a given temperature for 90% destruction of a population.

Species	Heating Menstruum [Sporulation Conditions]	Population no./ml or g	Temp °C	D Value	Reference
Bacillus cereus, T[1]	0.5 M NaH$_2$PO$_4$-KH$_2$PO$_4$ buffer, pH 7.0				
1	[15°C]	1 X 10^6	90	4.8	13
2	[20°C]	1 X 10^6	90	21.7	13
3	[30°C]	1 X 10^6	90	36.1	13
4	[37°C]	1 X 10^6	90	33.2	13
5	[41°C]	1 X 10^6	90	16.9	13
6	[25 µg cycloserine/ml]	1 X 10^6	90	1.4	13

[1] Vegetative cells grown in aerated medium at 30°C until committed to spore formation, then kept under specified sporulation conditions until sporulation was completed.

continued

Part I.   D Values

	Species	Heating Menstruum [Sporulation Conditions]	Population no./ml or g	Temp °C	D Value	Reference
7		[100 μg cycloserine/ml]	1×10^6	90	1.0	13
8		[200 μg cycloserine/ml]	1×10^6	90	2.7	13
9		[18 μg methicillin sodium[2]/ml]	1×10^6	90	23.2	13
10		[36 μg methicillin sodium[2]/ml]	1×10^6	90	13.0	13
11		[54 μg methicillin sodium[2]/ ml]	1×10^6	90	11.1	13
12		[108 μg methicillin sodium[2]/ml]	1×10^6	90	5.8	13
13	*B. coagulans*[3]	Citric acid-NaH$_2$PO$_4$ buffer, pH 4	3×10^5	98.9	9.5	10
14		Plus 20 ppm oil of garlic	3×10^5	98.9	7.5	10
15		Citric acid-Na$_2$HPO$_4$ buffer, pH 4	3×10^5	93.3	22.0	10
16				87.8	59.0	10
17		Plus 20 ppm allyl isothiocyanate	3×10^5	93.3	16.0	10
18				87.8	42.4	10
19		Plus 20 ppm or 50 ppm allyl isothiocyanate	3×10^5	98.9	6.0	10
20		Plus 20 ppm oil of garlic	3×10^5	93.3	21.4	10
21				87.8	53.0	10
22		Plus 50 ppm oil of garlic	3×10^5	93.3	21.0	10
23		Plus 20 ppm oil of onion	3×10^5	98.9	8.6	10
24				87.8	56.0	10
25		Tomato juice, pH 4.2	5.7×10^5	106	0.51	9
26			1×10^7	89	10.41	7
27		Plus 100 ppm CuSO$_4$	1×10^7	89	5.65	7
28		Plus 100 ppm captan[4]	1×10^7	89	7.71	7
29		Plus 100 ppm dichlone[5]	1×10^7	89	8.24	7
30		Plus 100 ppm maneb[6]	1×10^7	89	7.89	7
31		Plus 100 ppm zineb[7]	1×10^7	89	9.07	7
32		Tomato juice, pH 4.3	3×10^5	98.9	8.5	10
33				93.3	23.0	10
34				87.8	72.5	10
35		Plus 20 ppm allyl isothiocyanate	3×10^5	98.9	7.2	10
36				93.3	20.8	10
37				87.8	65.0	10
38		Plus 50 ppm allyl isothiocyanate	3×10^5	93.3	17.2	10
39				87.8	55.0	10
	B. coagulans ATCC 8038[8]	M/40 phosphate buffer, pH 7				
40		[30°C]	1×10^9	89	11.57	6
41				86	71.14	6
42		[45°C]	1×10^9	89	32.05	6
43		[Agar containing 1 ppm MnSO$_4$]	1×10^6	96.4	13.51	1
44		[Agar containing 1 ppm MnSO$_4$ + 0.2% KH$_2$PO$_4$]	1×10^6	96.4	26.80	1
45		[Agar containing no KH$_2$PO$_4$]	1×10^6	96.4	28.05	1
46		[Agar containing 0.5% KH$_2$PO$_4$]	1×10^6	96.4	8.67	1
47		[Medium containing 1 ppm MnSO$_4$ + 0.05% KH$_2$PO$_4$]	1×10^6	96.4	41.32	1
48		[Corn concoction agar, 45°C]	1×10^9	93	15.34	5
49		[Thermoacidurans agar, 45°C]	1×10^9	93	20.29	5
50		[Tomato juice agar, 30°C]	1×10^7	93	4.4	15
51			1×10^9	93	4.50-6.06	5

[2] Synonym: Celbenin®. [3] Synonym: *B. coagulans* var. *thermoacidurans*. [4] Synonym: N-trichloromethylthio-3a,4,7,7a-tetrahydrophthalimide. [5] Synonym: 2,3-dichloro-1,4-naphthoquinone. [6] Synonym: manganous ethylenebis(dithiocarbamate). [7] Synonym: zinc ethylenebis(dithiocarbamate). [8] Spores.

continued

Part I. D Values

	Species	Heating Menstruum [Sporulation Conditions]	Population no./ml or g	Temp °C	D Value	Reference
52		[Tomato juice agar, 37°C]	1×10^7	93	11.0	15
53			1×10^9	93	11.01	6
54		[Tomato juice agar, 45°C]	1×10^7	93	20.0	15
55			1×10^9	96	8.31	6
56				93	18.80	5
57		[Tomato juice agar, 45°C, pH 5]	1×10^9	93	6.11	5
58		[Tomato juice agar, 45°C, pH 6]	1×10^9	93	5.57	5
59		[Tomato juice agar, 45°C, pH 7]	1×10^9	93	5.69	5
60	*B. stearothermophilus* NCIB 8919[8/]	Water	9.7×10^5	115	18.3	2
61		2% (wt/vol) NaCl solution	9×10^5	115	14.0	2
62		4% (wt/vol) NaCl solution	9.4×10^5	115	11.3	2
63		8% (wt/vol) NaCl solution	8.8×10^5	115	10.3	2
64	*Clostridium* sp., PA 3679[8/]	Distilled water	2.5×10^2	132	0.0174	8
65				121	0.170	8
66				110	3.03	8
67			1×10^4	143	0.0054	22
68				132	0.062	22
69				121.1	0.67	22
70			2.6×10^4	132.2	0.0496	23
71				121.1	0.80	23
72				115.6	3.04	23
73				104.4	36.9	23
74			2.5×10^5	132	0.0280	8
75				121	0.369	8
76				110	8.09	8
77		Distilled water, pH 7.1	1×10^4	127	0.204	22
78		Phosphate buffer, pH 6.95	1×10^4	121.1	0.81	4
79				115.6	2.29	4
80				107.3	13.3	4
81		Phosphate buffer, pH 7.0, M/1.875	1×10^4	127	0.564	22
82		Phosphate buffer, pH 7.0, M/3.75	1×10^4	127	0.513	22
83		Phosphate buffer, pH 7.0, M/7.5	1×10^4	127	0.454	22
84		Phosphate buffer, pH 7.0, M/15	1×10^4	127	0.382	22
85				121.1	1.04	3
86				115.6	2.80	3
87				112.8	5.24	3
88			1.2×10^4	137.8	0.0288	18
89			1×10^6	137.8	0.0252	18
90				121.1	1.06	17
91			1×10^8	137.8	0.0253	18
92		Asparagus—chopped	5.6×10^2	121.1	0.49	20
93		Beans, green—strained	1×10^4	121.1	0.288	21
94		Beans, lima—chopped	6.7×10^2	121.1	1.62	20
95		Beans, snap—chopped	6.7×10^2	121.1	0.26	20
96		Beets—chopped	7.7×10^2	121.1	0.51	20
97		Carrots—chopped	6.7×10^2	121.1	0.36	20
98		Chicken soup—strained	1×10^4	121.1	0.274	21
99		Corn—chopped	6.7×10^2	121.1	2.13	20
100		Custard pudding—strained	1×10^4	121.1	0.328	21
101		Mushrooms—chopped	6.7×10^2	121.1	1.22	20

[8/] Spores.

continued

Part I. D Values

	Species	Heating Menstruum [Sporulation Conditions]	Population no./ml or g	Temp °C	D Value	Reference
102		Okra—chopped	6.7×10^2	121.1	0.34	20
103		Pea puree	2.3×10^4	121.1	1.45	14
104		Pea puree—canned	2.2×10^4	132.2	0.0945	11
105				127	0.421	11
106				121.1	1.67	11
107				115.6	5.50	11
108		Plus 3.5 ppm subtilin	2.2×10^4	132.2	0.0654	11
109				126.7	0.26	11
110				121.1	1.04	11
111				115.6	3.60	11
112		Plus 7 ppm subtilin	2.2×10^4	132.2	0.0569	11
113				126.7	0.23	11
114				121.1	0.872	11
115				115.6	2.89	11
116		Plus 14 ppm subtilin	2.2×10^4	132.2	0.0506	11
117				126.7	0.193	11
118				121.1	0.704	11
119				115.6	2.46	11
120		Peas—chopped	6.7×10^2	121.1	1.16	20
121		Peas—strained	1×10^4	143	0.0084	22
122				132	0.081	22
123				121.1	1.087	22
124				121.1	0.430	21
125		Pork—strained	1×10^4	121.1	0.508	21
126		Pumpkin—chopped	6.7×10^2	121.1	0.40	20
127		Spinach—chopped	6.7×10^2	121.1	1.05	20
128		Squash, summer—chopped	6.7×10^2	121.1	0.63	20
129		Sweet potatoes—chopped	6.7×10^2	121.1	0.85	20
130		Vegetables & bacon—strained	1×10^4	121.1	0.365	21
131		Vegetables & beef—strained	1×10^4	121.1	0.37	21
132	*Escherichia coli* ATCC 9637	Raw milk	1×10^6	80	0.00016	19
133				75.6	0.00195	19
134				57.2	1.3	19
135				51.7	28.2	19
136		Chocolate milk	1×10^6	80	0.00028	19
137				75.6	0.00265	19
138				57.2	2.6	19
139				51.7	32.2	19
140		Raw cream, 40% fat	1×10^6	80	0.00036	19
141				75.6	0.00093	19
142				57.2	3.5	19
143				51.7	34.4	19
144		Ice cream mix	1×10^6	81.1	0.00053	19
145				77.8	0.00120	19
146				57.2	5.1	19
147				51.7	39.3	19
148	*Pseudomonas fragi*	Grown in milk, 10% fat for 20 hr at 25°C, heated in milk	6.9×10^5	52	2.1	12
149		Grown and heated in cream, 20% fat	6.9×10^5	52	1.9	12
150		Grown in cream, 20% fat, heated in skim milk	6.9×10^5	52	2.0	12
151		Grown and heated in skim milk	6.9×10^5	52	3.1	12

continued

Part I. D Values

	Species	Heating Menstruum [Sporulation Conditions]	Population no./ml or g	Temp °C	D Value	Reference
152		Grown in skim milk, heated in cream, 20% fat	6.9×10^5	52	3.0	12
153	*Streptococcus faecalis* ATCC 7080	Beef pie	2×10^7	74	0.11	16
154				71	0.33	16
155				65.5	2.00	16
156				60.0	12.20	16
157		Chicken a la king	2×10^7	74	0.07	16
158				71	0.33	16
159				65.5	1.90	16
160				60	13.50	16
161		Fish cakes	2×10^7	74	0.07	16
162				71	0.33	16
163				65.5	1.90	16
164				60	11.25	16
165		Fish sticks	2×10^7	74	0.13	16
166				71	0.35	16
167				65.5	2.30	16
168				60	15.70	16
169		Lobster pie	2×10^7	74	0.0094	16
170				71	0.24	16
171				65.5	1.60	16
172				60	10.50	16
173		Tuna pie	2×10^7	74	0.07	16
174				71	0.28	16
175				65.5	1.90	16
176				60	11.25	16

Contributor: Walker, Homer W.

References

[1] Amaha, M., and Z. J. Ordal. 1957. J. Bacteriol. 74: 596.

[2] Cook, A. M., and R. J. Gilbert. 1969. J. Appl. Bacteriol. 32:96.

[3] Desrosier, N. W., and W. B. Esselen. 1951. J. Bacteriol. 61:541.

[4] Desrosier, N. W., et al. 1954. J. Milk Food Technol. 17:207.

[5] El-Bisi, H. M., and Z. J. Ordal. 1956. J. Bacteriol. 71:1.

[6] El-Bisi, H. M., and Z. J. Ordal. 1956. Ibid. 71:10.

[7] El-Bisi, H. M., et al. 1955. Food Res. 20:554.

[8] Frank, H. A., and L. L. Campbell. 1957. Appl. Microbiol. 5:243.

[9] Knock, G. G., et al. 1959. J. Sci. Food Agr. 10:337.

[10] Kosker, O., et al. 1951. Food Res. 16:510.

[11] Le Blanc, F. R., et al. 1953. Food Technol. 7:181.

[12] Luedecke, L. O., and L. G. Harmon. 1966. Appl. Microbiol. 14:716.

[13] Murrell, W. G., and A. D. Warth. 1965. Spores 3:1.

[14] O'Brien, R. T., et al. 1956. Food Technol. 10: 352.

[15] Ordal, Z. J., and R. V. Lechowich. 1958. Proc. Res. Conf. Res. Counc. Amer. Meat Inst. Found. Chicago Univ., 10th, p. 91.

[16] Ott, T. M., et al. 1961. J. Food Sci. 26:1.

[17] Pflug, I. J., and W. B. Esselen. 1953. Food Technol. 7:237.

[18] Pflug, I. J., and W. B. Esselen. 1954. Food Res. 19: 92.

[19] Read, R. B., et al. 1961. Appl. Microbiol. 9:415.

[20] Reynolds, H., et al. 1952. Food Res. 17:153.

[21] Secrist, J. L., and C. R. Stumbo. 1956. Food Technol. 10:543.

[22] Secrist, J. L., and C. R. Stumbo. 1958. Food Res. 23:51.

[23] Stumbo, C. R., et al. 1950. Food Technol. 4:321.

continued

Part II. Minutes to Kill

Time: The average number of minutes necessary to kill the bacteria at a specified temperature. Values are for spores, unless otherwise indicated.

	Species	Heating Menstruum [Sporulation Conditions]	Population no./ml or g	Temp °C	Time min	Refer-ence
1	*Bacillus cereus*	Distilled water, pH 6.9	1×10^6[1]	105	4	5
2			1×10^6[2]	105	5	5
3			1×10^6[3]	105	7	5
4			1×10^6[4]	105	8	5
5		$M/50$ phosphate buffer, pH 7.0	2×10^7	99.5	6	14
6			2×10^7[5]	53	4	14
7	*B. coagulans*[6]	Tomato juice, pH 4.4-4.5	1×10^5	121.1	0.25	3
8		Plus 1% NaCl, pH 4.3-4.4	1×10^5	121.1	0.23	3
9		Plus 2% NaCl, pH 4.2-4.3	1×10^5	121.1	0.185	3
10		Plus 4% NaCl, pH 4.2	1×10^5	121.1	0.17	3
11		Plus 8% NaCl, pH 4.0-4.1	1×10^5	121.1	0.155	3
12		Plus 0.5% acetic acid, pH 4.0-4.1	1×10^5	121.1	0.15	3
13		Plus 1% acetic acid, pH 3.9-4.0	1×10^5	121.1	0.14	3
14		Plus 2% acetic acid, pH 3.7	1×10^5	121.1	0.125	3
15		Plus 200 mg ascorbic acid/pint, pH 4.4	1×10^5	121.1	0.24	3
16		Plus 0.5% citric acid, pH 3.8-3.9	1×10^5	121.1	0.35	3
17		Plus 1% citric acid, pH 3.5-3.6	1×10^5	121.1	0.32	3
18		Plus 0.5% lactic acid, pH 3.8-3.9	1×10^5	121.1	0.25	3
19		Plus 1% lactic acid, pH 3.5-3.6	1×10^5	121.1	0.19	3
20		Plus 2% lactic acid, pH 3.1-3.2	1×10^5	121.1	0.29	3
21		Plus 2% lactic acid, pH 3.2-3.3	1×10^5	121.1	0.14	3
22		Plus 10% glucose [7], pH 4.3	1×10^5	121.1	0.27	3
23		Plus 25% glucose [7], pH 4.3	1×10^5	121.1	0.32	3
24		Plus 40% glucose [7], pH 4.3	1×10^5	121.1	0.43	3
25		Plus 50% glucose [7], pH 4.3	1×10^5	121.1	0.52	3
26		Plus 0.1% sodium benzoate, pH 4.4-4.5	1×10^5	121.1	0.22	3
27		Plus 10% sucrose, pH 4.4	1×10^5	121.1	0.20	3
28		Plus 25% sucrose, pH 4.4	1×10^5	121.1	0.22	3
29		Plus 40% sucrose, pH 4.4	1×10^5	121.1	0.26	3
30		Plus 50% sucrose, pH 4.4	1×10^5	121.1	0.28	3
31		Plus 0.1% black pepper, pH 4.4-4.5	1×10^5	121.1	0.23	3
32		Plus 1% black pepper, pH 4.5	1×10^5	121.1	0.20	3
33		Plus 0.01% oil of cloves, pH 4.4-4.5	1×10^5	121.1	0.20	3
34		Plus 0.1% oil of cloves, pH 4.4-4.5	1×10^5	121.1	0.19	3
	B. subtilis	$M/15$ phosphate buffer, pH 7				
35		[Medium containing 1% $(NH_4)_2SO_4$, at 37°C]	5×10^7	95	36	12
36		[Medium containing $CaCl_2$ or $FeCl_3$]	5×10^7	95	41	12
37		[Medium containing $MgCl_2$]	5×10^7	95	46	12
38		[Medium containing $MgSO_4$]	5×10^7[8]	95	49	12
39		[Medium containing NaCl]	5×10^7	95	38	12
40		[Medium containing Na_2HPO_4]	5×10^7[8]	95	49	12
41		[Medium containing Na_2SO_4]	5×10^7	95	38	12
42		[Medium containing 1% KCl, at 37°C]	5×10^7	95	36	12
43		[Medium containing K_2HPO_4]	5×10^7	95	46	12

[1] Spores stored in water 4 mo at 5°C. [2] Spores stored in water 4 mo at 46°C. [3] Spores stored in water 4 mo at 37°C. [4] Spores stored in water 4 mo at 20 or 30°C. [5] Cells 6-8 hr old. [6] Synonym: *B. coagulans thermoacidurans.* [7] Synonym: dextrose. [8] Spores 5 da old.

continued

Part II. Minutes to Kill

	Species	Heating Menstruum [Sporulation Conditions]	Population no./ml or g	Temp °C	Time min	Refer-ence
44		[Medium containing 0.5 mole/kg K_2SO_4]	5×10^7	95	37	12
45		[Medium containing 0.1% ammonium lactate or ammonium oxalate]	5×10^7	95	35	12
46		[Medium containing 1% amygdalin, at 37°C]	5×10^7	95	36	12
47		[Medium containing 0.1% potassium tartrate or sodium acetate]	5×10^7	95	45	12
48		[Medium containing 0.1% sodium citrate or sodium formate]	5×10^7	95	37	12
49		[0.1-1.0% glucose medium]	5×10^7	95	50	12
50				100	24	12
51		[1% glucose + 1% peptone, at 21-23°C]	5×10^7	100	14	12
52		[1% lactose medium]	5×10^7	95	51	12
53		[1% starch medium]	5×10^7	95	53	12
54		[Medium containing 0.1% asparagine]	5×10^7	95	47	12
55		[Medium containing 0.1% tyrosine]	5×10^7	95	45	12
56		[1% gelatin medium]	5×10^7	100	16	12
57		[1% isoelectric gelatin medium or 1% glucose gelatin + 1% peptone medium, at 37°C]	5×10^7	100	22	12
58		[Blood digest medium]	5×10^7	100	12	12
59				95	33	12
60		[Casein digest medium, at 41°C]	5×10^7	100	18	12
61		[1% casein digest medium]	5×10^7	95	42	12
62		[Peptic digest medium]	5×10^7	100	10	12
63				95	30	12
64		[Peptic digest medium, at 28°C]	5×10^7	95	31	12
65		[1% peptone medium, at 41°C]	5×10^7	100	18	12
66		[1% Berna peptone medium]	5×10^7	95	42	12
67		[1% Difco peptone medium]	5×10^7	95	37	12
68		[1% Parke-Davis peptone medium, at 37°C]	5×10^7	95	36	12
69		[1% Witte peptone medium]	5×10^7	95	39	12
70		[1% peptone water, at 21-23°C]	5×10^7	100	11	12
71				95	23	12
72		[1% peptone water, at 28°C]	5×10^7	95	31	12
73		[1% peptone water, at 37°C]	5×10^7	95	36	12
74		[1% Difco proteose medium]	5×10^7	95	50	12
75		[Trypic digest medium, at 28°C]	5×10^7	95	31	12
76		[Asparagus infusion medium]	5×10^7	95	47	12
77		[Bean juice infusion medium]	$5 \times 10^{7[8/]}$	95	49	12
78		[Corn juice medium or hay infusion medium]	5×10^7	95	43	12
79		[Pea juice medium]	5×10^7	95	47	12
80		[Potato infusion medium]	5×10^7	95	55	12
81		[Spinach juice infusion medium]	$5 \times 10^{7[8/]}$	95	49	12
82	*Clostridium* sp., PA 3679	Distilled water	1×10^4	121.1	4.80	7
83		Phosphate buffer, pH 7.0	1×10^2	115	19	1
84				110	45	2
85			1×10^3	105	105	11
86			4×10^3	110	39.4	11
87			1×10^4	121.1	5.40	6
88				115	28	2
89				110	65	1

[8/] Spores 5 da old.

continued

Part II. Minutes to Kill

	Species	Heating Menstruum [Sporulation Conditions]	Population no./ml or g	Temp °C	Time min	Refer- ence
90			2.5×10^4	115	16[9]	13
91				115	22[10]	13
92			6×10^5	120	4.8	11
93				115	14.4	11
94				110	38	11
95			1×10^6	110	90	1
96		Plus 18 amino acids	1×10^4	115	25	2
97		Plus 1% ovalbumin	1×10^4	115	33	2
98		Plus 2.5% ovalbumin	1×10^4	115	40	2
99		Plus 2.5% serum albumin	1×10^4	115	45	2
100		Plus 2.5% heat-denatured serum albumin, 1% serum albu- min, & 2.5-5.0% polypeptone	1×10^4	115	38	2
101		Plus 2.5% heat-denatured serum albumin, 0.25% serum albu- min, 1.0-2.5% yeast nucleic acid, 4×10^9 living cells of *Staphylococcus aureus* or *Sarcina lutea*	1×10^4	115	35	2
102		Plus 5×10^7 killed cells of *Torula utilis*/ml	1×10^4	115	40	2
103		M/15 phosphate buffer, pH 7	1×10^1	110	35	1
104			1×10^2	120	10	1
105				105	110	1
106			1×10^4	121.1	4	7
107				120	13	1
108				105	160	1
109			1×10^5	115	33	1
110				110	75	1
111			1×10^6	120	17	1
112				105	240	1
113		Asparagus brine, pH 5.03	6×10^5	110	27.4	11
114		pH 5.10	6×10^5	120	2.8	11
115				115	9.2	11
116		pH 5.43	4×10^3	110	14.1	11
117				105	55	11
118		Asparagus puree, pH 5.1	1×10^4	121.1	0.16	10
119		pH 8.0	1×10^4	121.1	4.30	10
120		canned, pH 5.65	1×10^4	121.1	2.05	7
121		frozen, pH 5.5	1×10^4	121.1	1.10	10
122		pH 6.15	1×10^4	121.1	2.10	7
123		pH 6.7	1×10^4	121.1	3.60	10
124		Green bean puree, pH 5.1	1×10^4	121.1	0.50	10
125		pH 5.9	1×10^4	121.1	2.80	10
126		pH 7.5	1×10^4	121.1	4.60	10
127		fresh, pH 5.85	1×10^4	121.1	3.20	7
128		pH 6.0	1×10^4	121.1	6.0	7
129		canned, pH 5.5	1×10^4	121.1	3.30	10
130		pH 5.7	1×10^4	121.1	4.30	7
131		Beet puree, pH 5.9	1×10^4	121.1	3.0	10
132		pH 7.1	1×10^4	121.1	7.30	10
133		Carrot puree, canned, pH 5.48	1×10^4	121.1	4.70	7
134		fresh, pH 6.18	1×10^4	121.1	5.70	7
135		Corn puree, canned, pH 6.48	1×10^4	121.1	2.30	7
136		frozen, pH 6.64	1×10^4	121.1	6.50	7
137		pH 7.20	1×10^4	121.1	2.60	7

[9] Recovery medium incubated at 37°C. [10] Recovery medium incubated at 27°C.

continued

Part II. Minutes to Kill

	Species	Heating Menstruum [Sporulation Conditions]	Population no./ml or g	Temp °C	Time min	Refer-ence
138		White corn puree, pH 4.7	1×10^4	121.1	0.08	10
139		pH 5.2	1×10^4	121.1	1.80	10
140		pH 6.1	1×10^4	121.1	5.60	10
141		pH 6.6	1×10^4	121.1	4.40	10
142		pH 6.9	1×10^4	121.1	4.70	10
143		Yellow corn puree, pH 5.0	1×10^4	121.1	0.45	10
144		pH 6.6	1×10^4	121.1	9.0	10
145		canned, pH 5.5	1×10^4	121.1	3.30	10
146		Hominy, pH 7.8	1×10^2	121.1	2.60	6
147			1×10^4	121.1	7.20	6
148		Pea puree, pH 4.7	1×10^4	121.1	0.14	10
149		pH 5.6	1×10^4	121.1	5.60	10
150		pH 7.2	1×10^4	121.1	6.80	10
151		canned, pH 3.15	1×10^4	121.1	3.15	7
152		frozen, pH 5.2	1×10^4	121.1	1.10	10
153		pH 6.75	1×10^4	121.1	4.50	7
154		pH 6.95	1×10^2	121.1	2.40	7
155			1×10^4	121.1	4.20	7
156			1×10^4	121.1	7.10	7
157		Pumpkin, pH 5.1	1×10^4	121.1	0.15	10
158		pH 7.6	1×10^4	121.1	1.70	10
159		Spinach, pH 5.39	4×10^3	105	80	11
160		puree, canned, pH 5.55	1×10^4	121.1	5.55	7
161		frozen, pH 5.60	1×10^4	121.1	5.60	7
162		pH 6.69	1×10^4	121.1	7.00	7
163		Hubbard squash puree, pH 5.73	1×10^4	121.1	1.40	7
164		Sweet potato puree, pH 5.0	1×10^4	121.1	0.33	10
165		pH 6.1	1×10^4	121.1	6.50	10
166		pH 6.9	1×10^4	121.1	4.70	10
167		frozen, pH 5.5	1×10^4	121.1	3.60	10
168	*Escherichia coli*	Raw milk, 4% fat	1×10^6[11]	72.2	0.34	9
169				70.8	0.92	9
170				68.3	2.6	9
171				65.8	6.9	9
172				63.3	16	9
173				61.4	24	9
174	*Lactobacillus* spp.	Orange concentrate, pH 3.45	3.4×10^5	65.6	1.20	8
175		Orange juice, single strength, pH 3.7	3.4×10^5	65.6	0.28	8
176	*Streptococcus fae-calis*, A6-1	Ground ham	2.5×10^8[12]	65.6	10	4
177	*S. faecalis*, ARA-10	Ground ham	1.2×10^8[12]	65.6	15	4
178	*S. faecalis*, HD-4B	Ground ham	3.3×10^7[12]	65.6	25	4
179	*S. faecalis*, HS-5	Ground ham	8.3×10^8[12]	65.6	55	4
180	*S. faecalis*, HS-6	Ground ham	8.4×10^8[12]	65.6	20	4

[11] Cells 24 hr old. [12] Cells 48 hr old.

Contributor: Walker, Homer W.

References
[1] Amaha, M. 1953. Food Res. 18:411.
[2] Amaha, M., and K. Sakaguchi. 1954. J. Bacteriol. 68:338.
[3] Anderson, E. E., et al. 1949. Food Res. 14:499.
[4] Brown, W. L., et al. 1960. Ibid. 25:345.
[5] Curran, H. R. 1935. J. Infec. Dis. 56:196.

continued

Part II. Minutes to Kill

[6] Desrosier, N. W., and W. B. Esselen. 1950. Mass. Agr. Exp. Sta. Bull. 456.

[7] Esselen, W. B., and I. J. Pflug. 1956. Food Technol. 10:557.

[8] Murdock, D. I., et al. 1953. Food Res. 18:85.

[9] Prucha, M. J., and W. J. Corbett. 1940. J. Milk Technol. 3:269.

[10] Sognefest, P., et al. 1948. Food Res. 13:400.

[11] Townsend, C. T., et al. 1938. Ibid. 3:323.

[12] Williams, O. B. 1929. J. Infec. Dis. 44:421.

[13] Williams, O. B., and J. M. Reed. 1942. Ibid. 71:225.

[14] Williams, O. B., and C. H. Zimmerman. 1951. J. Bacteriol. 61:63.

Part III. Percent Destruction of Population

Population Reduction: Percent reduction of the population at a specified temperature. Values are for cells 24 hours old, unless otherwise indicated.

	Species & Heating Menstruum	Population no./ml or g	Temp °C	Time min [1]	Population Reduction %	Reference
	Clostridium sp., PA 3679					
1	0.066 *M*	1×10^5 [2]	121.1	7.5	99.99	3
2	phosphate		115	15.3		
3	buffer, pH 7		111	37.2		
4			107	76.3		
	Escherichia coli, K1					
5	Ringer's solu-	1×10^7	55	7.39	99.9	2
6	tion	1×10^7 [3]	55	4.63		
7		1×10^7 [4]	55	1.42		
8		1×10^7 [5]	55	2.37		
9		1×10^7 [6]	55	3.09		
10		1×10^7 [7]	55	3.68		
	Micrococcus freudenreichii					
11	Milk, 4.6% fat	1×10^6	73.9	0.067	99.99	4
12			71.1	0.25		
13			68.3	1.00		
14			65.6	5.00		
15			62.8	17.50		
16			60	55.0		

	Species & Heating Menstruum	Population no./ml or g	Temp °C	Time min [1]	Population Reduction %	Reference
17	Ice cream	1×10^6	76.7	0.033	99.99	4
18	mix, pH 6.7		73.9	0.17		
19			71.1	0.75		
20			68.3	2.50		
21			65.6	15.0		
	Pseudomonas mephitica, 29					
22	Sterile skim	5.6×10^8	49.4	12.0	99.99999	1
23	milk	8.1×10^8	45.7	97.80		
24		8.3×10^8	48.0	22.58		
	P. viscosa, 3					
25	Sterile skim	5.1×10^8	50.2	11.55	99.99999	1
26	milk		48.0	17.25		
27			45.7	37.17		
	Streptococcus faecalis, ATCC 7080					
28		5×10^4 or	76.7	1.20	99.99999	5
29		1×10^5	71.1	2.68		
30			65.6	13.50		
31			60	67.20		

[1] Average. [2] Spores. [3] Cells 30 min old. [4] Cells 3 hr old. [5] Cells 6 hr old. [6] Cells 8 hr old. [7] Cells 12 hr old.

Contributor: Walker, Homer W.

References

[1] Kaufman, O. W., and R. H. Andrews. 1954. J. Dairy Sci. 37:317.

[2] Lemcke, R. M., and H. R. White. 1959. J. Appl. Bacteriol. 22:193.

[3] Reynolds, H., et al. 1952. Food Res. 17:153.

[4] Speck, M. L. 1947. J. Dairy Sci. 30:975.

[5] Webster, R. C., and W. B. Esselen. 1956. J. Milk Food Technol. 19:209.

Part I. Animal Pathogens

#	Species (Synonym) [Structure]	Thermal Death Temp °C	Time min	Ref-er-ence	#	Species (Synonym) [Structure]	Thermal Death Temp °C	Time min	Ref-er-ence
1	*Allescheria boydii*	56	60	17	22	*M. canis*	70	10	1
2	*Arthroderma benhamiae (Tricho-phyton mentagrophytes)*	75	10	1	23	*M. ferrugineum (Trichophyton fer-rugineum)*	60	10	10
3	[Spores in water]	55	5	7	24	*M. racemosum* [Spores in water]	55	5	7
4	*A. ciferri (T. georgiae)* [Spores in water]	55	5	7	25	*Nannizzia cajetani (Microsporum cookei)* [Spores in water]	55	5	7
5	*A. gloriae (T. gloriae)* [Spores in water]	55	5	7	26	*N. grubyia (M. vanbreuseghemii)* [Spores in water]	55	5	7
6	*A. uncinatum (T. ajelloi)* [Spores in water]	55	5	7	27	*N. gypsea (M. gypseum)*	60	60	17
7	*Aspergillus fumigatus* [Spores in dry air]	121.1	30	15,18	28	[Spores in water]	55	5	7
8	*A. niger* [Spores]	55	30	18	29	*N. obtusa* [Spores in water]	55	5	7
9	*Basidiobolus* spp.	50	30	3	30	*N. persicolor* [Spores in water]	55	5	7
10	*Blastomyces dermatitidis* [1/]	60	30	5,17	31	*Paracoccidioides brasiliensis* [3/]	60	60	17
11	*Candida albicans*	60	10	11	32	*Penicillium* spp.	50-60	30	18
12	*C. tropicalis*	60	10	11	33	*Phialophora verrucosa*	100	15	2
13	*Cephalosporium granulomatis*	53	5	20	34	*Pityrosporum ovale*	60	30	14
14	*Coccidioides immitis* [Arthrospores]	60	4	16	35	*Rhizopus equinus*	100	20	6
15	*Cryptococcus neoformans*	50	10	13	36	*Sporothrix schenckii (Sporotrichum schenckii)*	59	5	9
16	*Epidermophyton floccosum*	50	10	1	37	*Trichophyton megninii*	55	10	1
17	*Fonsecaea compactum*	100	15	2	38	*T. phaseoliforme* [Spores in water]	55	5	7
18	*F. pedrosoi*	100	15	2	39	*T. rubrum*	60	60	17
19	*Geotrichum candidum*	56	60	17	40	*T. schoenleinii*	60	10	19
20	*Histoplasma capsulatum* [2/]	55	15	4,8	41	*T. tonsurans*	49	10	19
21	*Microsporum audouinii*	60	60	17	42	*T. violaceum*	60	60	17

[1/] Thermal death for yeast phase, 56°C for 60 min. [2/] Thermal death for yeast phase, 60°C for 5 min [12]. [3/] Same values for yeast phase.

Contributors: Gordon, Morris A.; Beneke, E. S.; Eichbaum, Francisco W.

References

[1] Bonar, L., and A. D. Dreyer. 1932. Amer. J. Pub. Health 22:909.

[2] Carrión, A. L. 1950. Ann. N. Y. Acad. Sci. 50:1255.

[3] Coremans-Pelsencer, J. 1966. Bull. Cl. Sci. Acad. Roy. Belg., Ser. 5, 52:775.

[4] DeMonbreun, W. A. 1934. Amer. J. Trop. Med. 14:93.

[5] Denton, J. F., and A. F. Di Salvo. 1963. J. Bacteriol. 85:717.

[6] Dodge, C. W. 1935. Medical Mycology. C. V. Mosby, St. Louis.

[7] Engelhardt-Zasada, Ch., and H. Prochacki. 1971. Int. Soc. Hum. Anim. Mycol. Proc., 5th Congr., Paris. Inst. Pasteur, p. 35.

[8] Goos, R. D. 1964. Mycologia 56:662.

[9] Hektoen, L., and C. F. Perkins. 1900. J. Exp. Med. 5:77.

[10] Kadisch, E. 1930. Dermatol. Z. 57:412.

[11] Kadisch, E. 1930. Ibid. 60:48.

[12] Kao, C. J., and J. Schwarz. 1956. J. Infec. Dis. 99:219.

[13] Kligman, A. M., and F. D. Weidman. 1949. Arch. Dermatol. Syphilol. 60:726.

[14] Martin-Scott, I. 1952. Brit. J. Dermatol. 64:257.

[15] Paulo de Almeida, F. 1939. Mycologia Medica. Melhoramentos, Sao Paulo.

[16] Roessler, W. G., et al. 1946. J. Infec. Dis. 79:12.

[17] Smith, D. T., et al., ed. 1964. Zinsser's Microbiology. Ed. 13. Appleton-Century-Crofts, New York.

[18] Togashi, K. 1949. Biological Characters of Plant Pathogens: Temperature Relations. Meibundo, Tokyo.

[19] Verujsky, D. 1887. Ann. Inst. Pasteur 1:369.

[20] Weidman, F. D., and A. M. Kligman. 1945. J. Bacteriol. 50:491.

continued

Part II. Plant Pathogens

Thermal Death: Time—Death usually occurred after 10-minute exposure to the specified temperature, unless otherwise indicated. Data in brackets refer to the column heading in brackets.

	Species	Structure	Condition	Host or Medium	Thermal Death Temp, °C [Time]	Reference
1	*Alternaria brassicae*	....	.....	....	55	18
2	*A. tenuis*	Mycelia	.....	Pear fruit	47 [7 min]	4
3	*Armillaria mellea* [1]	Mycelia	.....	....	65	18
4	*Aspergillus chevalieri*	Mycelia	.....	Prune fruit	>80	11
5	*A. fischeri* var. *glaber*	Mycelia	.....	Strawberry fruit	>80	10
6	*A. mangini*	Mycelia	.....	Prune fruit	>80	11
7	*A. niger* [1]	Spores	Dry	....	69	18
8			Moist	....	62	18
9	*Botryodiplodia theobromae*	Mycelia	.....	Guava fruit	>40	17
10	*Botrytis cinerea* [2]	....	.....	....	55	18
11			.....	....	35 [few da]	19
12	*Cephalosporium sacchari*	....	Wet	Sugarcane seed	46	16
13	*Ceratocystis fimbriata*	Mycelia	.....	Sweet potato	43 [3 da]	7
14	*Ceratostomella ips*	....	.....	....	55	18
15	*C. ulmi*	Spores	.....	....	51-57	18
16	*Cercospora beticola*	....	Dry	....	95	18
17			Moist	....	46	18
18	*Cladosporium fulvum* [1]	....	Dry	....	70	18
19	*C. herbarum*	Mycelia, spores	.....	Soil	50-60	5
20	*Colletotrichum gloeosporioides*	Mycelia	.....	Mango fruit	56 [5 min]	15
21			.....	Papaya fruit	46 [20 min]	9
22	*Corticium vagum* [2]	Sclerotia	.....	....	60	18
23	*Fomes annosus*	Basidiospores	Dry	....	45 [30 min]	13
24			Wet	....	45 [1 hr]	13
25		Conidia	Dry	....	45 [2-3 hr]	13
26			Wet	....	45 [40-60 min]	13
27	*F. applanatus*	Mycelia	.....	....	65	18
28	*Fusarium oxysporum* [2]	....	.....	....	57	18
29			.....	Soil	50-60	5
30	*Gibberella zeae* [2]	Ascospores	.....	....	>65	18
31	*Glomerella gossypii*	Conidia	Dry	....	>95	18
32			Moist	....	51	18
33	*Guignardia bidwellii*	Spores	.....	....	>60	18
34	*Lentinus lepideus*	Mycelia	Dry	....	>105	18
35			Moist	....	>60	18
36	*Macrophomina phaseoli* [1]	....	.....	....	55	18
37	*Merulius lacrymans* [1]	Mycelia	.....	....	50-55	18
38	*Monilinia fructicola*	Spores	.....	....	52 [2-3 min]	14
39	*M. fructigena* [3]	....	.....	....	52	18
40	*Mycogone perniciosa*	....	.....	....	>42	18
41	*Nigrospora oryzae*	Spores	Dry	....	>67	18
42			Moist	....	56	18
43	*Ophiobolus sativus* [1]	....	.....	....	54	18
44	*Penicillium expansum*	Mycelia	.....	Pear fruit	47 [7 min]	3
45		Spores	.....	....	54 [0.5-1 hr]	1
46	*Phymatotrichum omnivorum*	....	.....	....	46-51	18

[1] Variability among different strains or in different hosts. [2] Extreme variability among different strains or in different hosts. [3] Synonym: *Sclerotinia fructigena*.

continued

Part II. Plant Pathogens

	Species	Structure	Condition	Host or Medium	Thermal Death Temp, °C [Time]	Reference
47	*Physoderma zeae-maydis*	....	.....	....	>80	18
48	*Phytophthora citrophthora*[4/]	Spores	.....	....	46	18
49	*P. infestans*[2/]	Mycelia	.....	....	45	18
50		Spores	.....	....	25	18
51	*Piricularia oryzae*	Conidia	.....	....	51	18
52		Mycelia	.....	....	52-55	18
53	*Polystictus versicolor*[1/]	Mycelia	.....	....	>70	18
54	*Pyrenophora graminea*	Conidia	.....	....	52	18
55		Mycelia	.....	....	55	18
56	*P. teres*	Conidia	.....	....	45	18
57		Mycelia	.....	....	55	18
58	*Rhizopus stolonifer*	Spores	.....	....	55	18
59		Spores—germinating	.....	....	52	14
60	*Sclerospora graminicola*	Conidia	.....	....	40	18
61		Oospores	Dry	....	118	18
62			Moist	....	53	18
63	*Sclerotinia libertiana*	Sclerotia	.....	....	60	18
64	*S. sclerotiorum*[1/]	....	.....	....	35 [few da]	19
65		Sclerotia	Moist	....	50	18
66	*Septoria lycopersici*	Spores	.....	....	53	18
67	*Stemphylium botryosum*	Mycelia, spores	.....	Soil	60	5
68	*Streptomyces fradiae*	Mycelia	.....	....	55	6
69		Spores	.....	....	>65 [20 min]	6
70	*Synchytrium endobioticum*	Sporangia	Moist	....	85	18
71	*Taphrina deformans*	Mycelia	.....	....	46	18
72	*Trametes pini*	Mycelia	.....	....	65-70	18
73	*Trichoderma viride*[5/]	Mycelia, spores	.....	Soil	55	5
74	*Ustilago avenae*	....	.....	....	45-53	18
75	*U. hordei*[1/]	....	.....	....	43-48	18
76	*U. maydis*[1,6/]	Spores	Dry	....	106	18
77			Moist	....	52	18
78			Wet	Water	100 [24 hr]	12
79	*U. nuda*[1/]	Spores	.....	....	>42	18
80	*U. scitaminea*	....	Wet	Sugarcane seed	50 [2 hr]	16
81	*U. tritici*[1/]	....	.....	....	45-48	18
82	*Verticillium* spp.	Mycelia, spores	.....	Soil	50-60	5
83	*V. albo-atrum*	Conidia	Wet	....	50	2
84	*V. malthousei*	Conidia	Wet	....	36	8
85	*Xeromyces bisporus*	....	Moist	Prunes	>80	11

[1/] Variability among different strains or in different hosts.
[2/] Extreme variability among different strains or in different hosts. [4/] Synonym: *Pythiacystis citrophthora*. [5/] Synonym: *T. lignorum*. [6/] Synonym: *U. zeae*.

Contributors: Whitehead, Marvin D.; Beneke, E. S.

References
[1] Baldy, R. W., et al. 1970. J. Bacteriol. 102:514.
[2] Bell, A. A., and J. T. Presley. 1969. Phytopathology 59:1147.
[3] Ben-Arie, R., and R. Barkai-Golan. 1969. Int. J. Appl. Radiat. Isotop. 20:687.
[4] Ben-Arie, R., and S. Guelfat-Reich. 1969. Plant Dis. Rep. 53:363.
[5] Bollen, G. J. 1969. Neth. J. Plant Pathol. 75:157.
[6] Eber, J. B., and J. I. Frea. 1970. Microbios 2:43.
[7] Daines, R. H., et al. 1962. Phytopathology 52:1138.

continued

Part II. Plant Pathogens

[8] Forer, L. B., and P. J. Wuest. 1971. Ibid. 61:128.

[9] Hunter, J. E., et al. 1969. Plant Dis. Rep. 53:279.

[10] McEvoy, I. J., and M. R. Stuart. 1970. Isr. J. Agr. Res. 9:59.

[11] Pitt, J. I., and J. H. B. Christian. 1970. Appl. Microbiol. 20:682.

[12] Ponte, J. J. da. 1967. Sydowia 21:159.

[13] Ross, E. W. 1969. Phytopathology 59:1789.

[14] Smith, W. L., Jr., et al. 1971. Ibid. 58:887.

[15] Smoot, J. J., and R. H. Segall. 1963. Plant Dis. Rep. 47:739.

[16] Srinivasan, K. V. 1971. Sugarcane Pathol. News Lett. (VI):46.

[17] Srivastava, M. P., and R. N. Tandon. 1969. Indian Phytopathol. 22:268.

[18] Togashi, K. 1949. Biological Characters of Plant Pathogens: Temperature Relations. Meibundo, Tokyo.

[19] Van Den Berg, L., and C. P. Lentz. 1968. Can. J. Bot. 46:1477.

Part III. Saprophytes

Most fungi do not grow at temperatures of 50-60°C. Those that do are known as thermophilic fungi [5], and a number of these have been reported killed, under specific conditions, by short exposure to temperatures of 50-60°C or higher. Some of the factors affecting the thermal death point are moisture, acidity, organic substances, and fungal structures.

| | Species (Synonym) | Structure | Heating Menstruum | Thermal Death | | Reference |
				Temp °C	Time min[1]	
1	*Achlya flagellata*	Zoospores[2]	Water	50	30	27
2	*Allomyces* sp.	Mitosporangia & sporophytic mycelia	....	30	<24 hr	19
3	*Alternaria* sp.	Spores	....	65-70[3]	....	16
4	*A. brassicicola*	Mycelia & spores	Culture broth	60	90	20
5	*A. solani*	Spores	Culture	50	25	34
6	*Aspergillus* sp.	Spores	Saline solution	60	90	27
7				70	60	27
8	*A. candidus*	Conidia	Milk	62.8	30	30
9	*A. flavus*	Conidia	Milk	60	30	30
10	*A. fumigatus*[4]	Conidia	Milk	68.3	30	30
11	*A. niger*	Conidia	Culture	60	10	34
12			Milk	60	30	34
13	*A. oryzae*	Conidia	Milk	54.5	30	30
14	*Byssochlamys fulva*	Ascospores	Canned blueberries	96	30	13
15			Syrup	86-88	30	24
16	*Candida* sp.	Budding cells	Soil	35-43[5]	....	18
17	*C. krusei (Torula monosa)*[4]	Budding cells	Milk	98	10	6
18	*C. lusitaniae*	Budding cells	....	45	....	32
19	*C. nivalis*	Budding cells	....	45	120	23
20	*C. norvegensis*	Budding cells	....	43	....	32
21	*C. obtusa*	Budding cells	....	41	....	32
22	*C. thomasii*	Budding cells	Water	58	10	6
23	*C. zeylandoides*	Budding cells	....	32	....	32
24	*Chaetomium* sp.	Ascospores	....	<65.5[6]	4 hr	17
25	*C. olivaceum*	Ascospores	Water	50	30	2
26	*C. thermophile* var. *coprophile*	Perithecia	Compost	67	10	4

[1] Unless otherwise specified. [2] Exclusive of mycelia. [3] Dry heat. [4] Sometimes pathogenic. [5] 93% killed. [6] Moist heat.

continued

Part III. Saprophytes

	Species (Synonym)	Structure	Heating Menstruum	Thermal Death		Refer-ence
				Temp °C	Time min [1]	
27	Circinella sp.	Spores	Milk	57.5	30	30
28	Cladosporium sp.	Spores	....	65-70[3]	....	16
29	Coprinus sp.	Mycelia	Stable manure	57+	....	26
30	Diehliomyces microsporus	Ascospores	....	<65.5[6]	4 hr	17
31	Emericella nidulans (Aspergillus nidulans)	Conidia	Milk	60	30	30
32	Entomophthora obscura	Mycelia	Culture	27	....	12
33	E. virulenta	Mycelia	Culture	36	....	12
34	Eurotium repens (Aspergillus repens)	Conidia	Cream	62.8	30	30
35	Fusarium sp.	Conidia	Milk	62.8	30	30
36	F. lini	Mycelia	....	55	10	31
37	Geotrichum sp.	Arthrospores	....	60[6]	4 hr	17
38	G. candidum (G. lactis)[4]	Arthrospores	Milk	54.5	30	30
39			Saline solution	60	<1 hr	25
40	Hansenula anomala (Saccharomyces anomalous)	Budding cells	Water	58	10	6
41	Humicola grisea var. thermoidea	Spores	Compost	72	10	10
42	H. lanuginosa	Hyphae	Yeast starch agar	60+	....	5
43	Monilia sp.	Conidia	Bread	60	20	22
44	Mucor sp.	Spores	Milk	57.5	30	30
45	M. ambiguus	Spores	Malt agar	60	....	14
46	M. geophilus	Spores	Malt agar	60	....	14
47	M. heterosporus	Spores	Malt agar	60	....	14
48	M. hiemalis	Spores	Malt agar	60	....	14
49	M. piriformis	Spores	Malt agar	60	....	14
50	M. pusillus (M. mirus)	Spores	Compost	72	10	10
51			Culture	50-57+	25	5
52	M. sphaerosporus	Spores	Malt agar	60	....	14
53	Mycoderma sp.	Budding cells	Dates with 22% water	60	....	8
54	M. vini	Budding cells	Water	58	10	6
55		Spores	Cystine agar	32	7 da	21
56	Myriococcum albomyces	Hyphae	Yeast starch agar	60+	....	5
57	Papulaspora byssina	Spores	....	65.5[6]	4 hr	17
58	P. thermophila	Bubils	Compost	61	10	9
59	Penicillium sp.	Conidia	....	65-70[3]	....	16
60		Sclerotia	Canned blueberries	90.5	16.6 hr	37
61	P. camembertii; P. citrinum; P. expansum; P. notatum; P. variabile	Conidia	Milk	54.5	30	30
62	P. chrysogenum	Conidia	Milk	57.5	30	30
63	P. digitatum	Conidia	Culture	58.5	....	1
64	P. purpurogenum; P. roqueforti	Conidia	Milk	57.2	30	30
65	Pichia membranaefaciens (Willia belgica)	Budding cells	Water	58	10	6
66	Radiomyces embreei	Hyphae	Casein-glucose	50	....	7
67	R. spectabilis	Hyphae	Casein-glucose	45	....	7
68	Rhizopus chinensis	Spores	Potato agar	52	24 hr	35
69	R. oryzae[4]	Spores	Potato agar	45.5	24 hr	35
70	R. stolonifer	Spores	....	55	30	1,28,34
71			Milk	51.7	30	30
72			Potato agar	35	24 hr	35
73	Rhodotorula mucilaginosa (Cryptococcus ludwigi)	Budding cells	Water	58	10	6

[1] Unless otherwise specified. [3] Dry heat. [4] Sometimes pathogenic. [6] Moist heat.

continued

Part III. Saprophytes

	Species (Synonym)	Structure	Heating Menstruum	Thermal Death Temp °C	Thermal Death Time min[1]	Reference
74	*Saccharomyces cerevisiae*	Budding cells	Bread	68	30	36
75			Oil	57.5	20 s	3
76			Water	58	10	6
77		Spores	Radiofrequency	76	1.5 s	14
78	*S. ellipsoideus*	Budding cells	Dates with 22% water	60	....	8
79			Oil	54.6	20 s	3
80	*Saprolegnia parasitica*	Zoospores[2]	Water	50	30	27
81	*Sartorya fumigata (Aspergillus malignus)*	Conidia	Water	100	60	15
82		Mycelia	Canned strawberries	100	12	15
83	*S. fumigata (A. fischeri)*	Conidia & mycelia	Water	80	30	15
84	*Schizophyllum commune*	Mycelia	....	48	....	11
85	*Sepedonium* sp.	Spores	Water	60	30	2
86	*Spicaria* sp.	Spores	Water	60	30	2
87	*Sporotrichum thermophile*	Conidia	Compost	65	....	10
88	*Stilbella thermophila*	Spores	Compost	62	....	10
89	*Syncephalastrum* sp.	Spores	Milk	62.8	30	30
90	*Talaromyces emersonii*	Ascospores & conidia	Ground chips	98	30	29
91	*T. thermophilus*	Conidia	Compost	68	....	10
92	*Thermomyces* sp.	Chlamydospores	Culture	65	....	33
93	*Thielaviopsis paradoxa*	Mycelia	Culture	52.5-53.5	....	1
94	*Torula* sp.	Budding cells	Dates with 22% water	60	....	8
95	*Torulopsis* sp.	Budding cells	Soil	35-43	....	18
96	*Trichoderma koningii*	Spores	....	65.5[6]	4 hr	17
97	*Trichothecium* sp.	Spores	Culture	55	10	34
98	*T. roseum (Cephalothecium roseum)*	Spores	Culture	47-48	....	1

[1] Unless otherwise specified. [2] Exclusive of mycelia. [6] Moist heat.

Contributors: Beneke, E. S.; Schmitt, John A.

References

[1] Ames, A. 1915. Phytopathology 5:11.

[2] Anderson, F. A. 1956. M. S. Thesis. Pennsylvania State Univ., University Park.

[3] Bruner, P.-O. 1953. Sv. Bryggeritidskr. 68:257, 300.

[4] Celerin, E. M., and C. L. Fergus. 1971. Mycologia 63:1030.

[5] Cooney, D. G., and R. Emerson. 1964. Thermophilic Fungi. W. H. Freeman, San Francisco.

[6] Dougherty, R. 1944. In F. W. Tanner, ed. Microbiology of Foods. Ed. 2. Garrard Press, Champaign, Ill. p. 136.

[7] Embree, R. W. 1964. Bull. Torrey Bot. Club 91(4): 263.

[8] Esau, P., and W. V. Cruess. 1933. Fruit Prod. J. Amer. Food Mfr. 12(5):144.

[9] Fergus, C. L. 1971. Mycologia 63:426.

[10] Fergus, C. L., and R. M. Amelung. 1971. Ibid. 63: 675.

[11] Gäumann, E. 1939. Angew. Bot. 21:59.

[12] Hall, I. M., and J. V. Bell. 1960. J. Insect Pathol. 2: 247.

[13] Hull, R. 1939. Ann. Appl. Biol. 26:800.

[14] Jacob, F. C., et al. 1964. Amer. J. Enol. Viticult. 15(2):69.

[15] Kavanagh, J., et al. 1963. Nature (London) 198: 1322.

[16] Kopeikovskii, V. M., and V. K. Kostenko. 1963. Izv. Vyssh. Ucheb. Zaved. Pishch. Tekhnol. 33(2):26.

[17] Lambert, E. B., and T. T. Ayer. 1957. Plant Dis. Rep. 41(4):348.

[18] Lund, A. 1956. Wallerstein Lab. Commun. 19(66): 221.

[19] Machlis, L., and J. M. Crasemann. 1956. Amer. J. Bot. 43:601.

[20] Manning, D. L. Unpublished. Ohio State Univ., Columbus, 1970.

[21] McVeigh, I., and K. Morton. 1964. Mycologia 56: 672.

[22] Morison, C. B. 1933. Cereal Chem. 10:462.

continued

Part III. Saprophytes

[23] Nash, C. H., and N. A. Sinclair. 1968. Can. J. Microbiol. 14(6):691.

[24] Olliver, M., and T. Rendle. 1934. Chem. Ind. (London) 53:166T.

[25] Parsons, J. E. 1963. M. S. Thesis. Ohio State Univ., Columbus.

[26] Perrier, A. 1929. C. R. Acad. Sci. 188:1426.

[27] Schmitt, J. A. Unpublished. Ohio State Univ., Columbus, 1971.

[28] Smith, W. C., and R. D. Bassett. 1963. Phytopathology 53:747.

[29] Tansey, M. R. 1971. Mycologia 63:537.

[30] Thom, C., and S. H. Ayers. 1916. J. Agr. Res. 6:153.

[31] Tochinai, Y. 1926. J. Coll. Agr. Hokkaido Imp. Univ. 14:171.

[32] Uden, N. van, and L. do Carmo-Dousa. 1959. Port. Acta Biol. 6(3):239.

[33] Waksman, S. A., et al. 1939. Soil Sci. 47:37.

[34] Wallace, G. I., and F. W. Tanner. 1931. Proc. Soc. Exp. Biol. Med. 28:970.

[35] Weimer, J. L., and L. L. Harter. 1923. J. Agr. Res. 24:1.

[36] Wells, E. P. 1917. Vt. Agr. Exp. Sta. Bull. 203.

[37] Williams, C. C., et al. 1951. Food Res. 6:69.

93. THERMAL INACTIVATION: ANIMAL VIRUSES

Group Classification based on Melnick [41] and Wilner [66]. **Inactivation: Unit of Measurement**—Bovine ID_{50} = bovine infectious dose (median); CIU = cell infectious unit; EI = egg infectivity; EID_{50} = egg infectious dose (median); ELD_{50} = egg lethal dose (median); If_{50} = interferon evoking dose (median); Mouse ID_{50} = mouse infectious dose (median); Mouse LD_{50} = mean lethal dose for 50% of test sub- jects (mouse); PFU = plaque-forming unit; Pock-FU = pock forming unit; TCID = tissue culture infectious dose; $TCID_{50}$ = tissue-culture infectious dose (median). For additional information on Oncornaviruses, consult reference 60. For additional information on Newcastle disease virus, consult reference 48.

	Group Classification	Virus	Temp °C	Time	Inactivation Amount $-\log_{10}$	Unit of Measurement, $\log^{-1}$	Remarks	Reference
1	Adeno-	Adeno-	4	180 da	1.5-2.0	$TCID_{50}$	....	12
2	virus	virus	4	180 da	0	$TCID_{50}$	....	51
3			20	15 da	0.5-1.0	$TCID_{50}$	....	12
4			37	10 da	1.4-2.3	$TCID_{50}$	pH 7.5; maximum stability at pH 6	51
5			50	21 min	>6.0	PFU	....	64
6			55	3 min	7.5	$TCID_{50}$	....	12
7			56	4 min	1.8-2.8	$TCID_{50}$	....	51
8	Arena-	Lym-	4	4 da	1.0-2.0	Mouse LD_{50}	Partially purified viruses; diminishing rate	6
9	virus	pho-	37	1.5 hr	2.5	Mouse LD_{50}	Partially purified virus; albumin protects	47
10		cytic chorio-meningitis	37	2 hr	2.3 to <4.7	Mouse ID_{50}	Many substances were protective	34
11	Corona-	Avian	4	140 da	4.8	ELD_{50}	1st-order kinetics for 99.99% inactivation	46
12	virus	infec-	25	11 da	4.2	ELD_{50}	1st-order kinetics for 99.99% inactivation	46
13		tious	37	28 hr	4.4	ELD_{50}	1st-order kinetics for 99.99% inactivation	46
14		bron-	50	15 min	7.0	ELD_{50}	High egg-passaged Beaudette strain	28
15		chitis	50	20 min	3.6-4.4	PFU	Slower rate after 20 min	8
16	Herpes-	Cyto-	3	61 hr	1.2	CIU	1st-order kinetics	32
17	virus	mega-	4	1 da	1.6	PFU	....	50
18		lovirus	15	22 hr	1.2	CIU	1st-order kinetics	32
19			22	2 da	1.2	PFU	No inactivation in 6 hr	50
20			23	39 hr	1.2	CIU	1st-order kinetics	32
21			36	5 hr	1.8	PFU	No inactivation in 1 hr	50

continued

	Group Classification	Virus	Temp °C	Time	Inactivation		Remarks	Reference
					Amount $-\log_{10}$	Unit of Measurement, $\log^{-1}$		
22			37	5.4 hr	1.8	CIU	1st-order kinetics	32
23			50	40 min	2.7	PFU	....	50
24		Herpes	4	8 da	3.6	PFU	Immature virus; no inactivation in 4 da	50
25		sim-	4	8 da	0.5	PFU	Mature extracellular virus	50
26		plex	22	4 da	0.8	PFU	....	50
27			36	20 hr	1.3	PFU	Serum stabilizes	50
28			50	15 min	3.5	PFU	....	50,65
29			60	15 s	<5.0	PFU	In dairy products	57
30		Pseudo-rabies	50	1 hr	0-4.5	$TCID_{50}$	Vaccine strain more stable	3
31	Meta-	Respira-	4	5 hr	3.0	PFU	Partially purified virus	52
32	myxo-	tory	4	2 da	1.0	$TCID_{50}$	Diminishing rate; maximum stability at pH 7.5	24
33	virus	syncy-	25	5 hr	1.0	PFU	Partially purified virus in distilled water	52
34		tial	25	4 da	2.8	$TCID_{50}$	No further reduction on 5th & 6th days	24
35			37	5 hr	2.7	PFU	Partially purified virus in distilled water	52
36			37	2 da	2.0	$TCID_{50}$	In cell culture fluid	24
37			50	15 min	2.7	PFU	Partially purified virus in distilled water	52
38	Oncor-	Rous	37	20 hr	1.5	Pock-FU	!st-order kinetics	13
39	navirus	sarco-	50	1 hr	1.2	Pock-FU	1st-order kinetics	13
40		ma	60	5 min	3.0	Pock-FU	1st-order kinetics	13
41	Ortho-	Influen-	45	1 hr	2.0	EID_{50}	....	33
42	myxo-	za	50	15 min	4.5	EID_{50}	....	65
43	virus		50	10 min	3.0	EID_{50}	....	33
44			55	10 min	<6.5	EID_{50}	....	45
45			56	3 min	2.0	EID_{50}	....	33
46	Papova-	Polyoma	4	56 da	0	$TCID_{50}$	....	4
47	virus		37	56 da	2.5	$TCID_{50}$	....	4
48			65	30 min	3.0	Mouse ID_{50}	....	4
49		Vacuo-	50	1 hr	0	PFU	....	62
50		lating virus (SV_{40})	56	1 hr	1.5	PFU	....	59
51	Para-	Canine	4	2 wk	0.3	PFU	In cell culture medium	5
52	myxo-	distem-	5	6 da	0.5	EID_{50}	In buffer	35
53	virus	per	25	1 da	0.3-1.4	EID_{50}	Inversely related to initial titer	35
54			37	6 hr	1.8	EID_{50}	....	35
55			37	6 hr	1.8	PFU	....	5
56			56	12 min	1.5	PFU	....	5
57		New-	38	3 hr	1.0	ELD_{50}	....	9
58		castle	43	20 min	1.0	ELD_{50}	Diminishing rate	9
59		disease	55	20 min	4.8	ELD_{50}	Diminishing rate	9
60			56	30 min	6.0-7.0	EI	Diminishing rate beyond 30 min	25
61			60	3 min	5.0	ELD_{50}	Diminishing rate	9
62		Rinder-	4	28 da	1.0-1.8	$TCID_{50}$	....	49
63		pest	36.5	1 da	1.2-2.3	$TCID_{50}$	....	49
64			56	30 min	3.0-4.4	$TCID_{50}$	....	49
65	Picorna-	Foot-	4	6 mo	1.2	Mouse LD_{50}	In cell culture fluid	15
66	virus	and-	4	6 mo	4.5	Mouse LD_{50}	In tissue suspension	15
67		mouth	4	18 wk	1.0	$TCID_{50}$	In buffer; most stable at pH 7	2
68		disease	37	2 da	3.8-6.4	PFU	In cell culture fluid	16
69			37	4 da	3.7	Mouse LD_{50}	In tissue suspension	16

continued

	Group Classification	Virus	Temp °C	Time	Inactivation		Remarks	Reference
					Amount $-\log_{10}$	Unit of Measurement, $\log^{-1}$		
70			37	2 da	2.3	$TCID_{50}$	In cell culture fluid	2
71			55	1 min	3.0	$TCID_{50}$	In buffer; slower rate after 1 min	2
72			56	5 min	1.0-4.0	PFU	In milk at pH 7.6 & 6.7	54
73			63	15 min	⋨5.0	PFU	In milk at pH 7.6 & 6.7	54
74			72	15 s	1.7-4.7	PFU	In milk at pH 7.6 & 6.7	54
75			85	4 hr	9.9	Bovine ID_{50}	Large-volume inoculum	11
76		Polio-virus	4	60 da	0.7	PFU	....	61
77			10	2 wk	⋩1.0	$TCID_{50}$	In foods	37
78			22	1 wk	0.3 to >5.6	PFU	Neutral to highly acid foods	7,27
79			37	1 da	1.0	$TCID_{50}$	Purified virus in buffer	36
80			37	5 da	1.0-1.5	PFU	Unpurified virus in buffer	36,38
81			45	10 min	2.0	PFU	Purified virus in water	1,18
82			45	1 da	3.0	$TCID_{50}$	Slower rate after 1 da	10
83			50	15 min	>5.0	PFU	1 M $MgCl_2$ stabilizes	65
84			50	15 min	0.2-7.0	PFU	pH 4; attenuated viruses more stable	63
85			60	3.5 min	>6.0	TCID	In whole egg	56
86			60	37 s	2.0	TCID	In egg white	56
87			65	1 hr	7.0-7.3	$TCID_{50}$	....	40
88			75	1 hr	7.3	$TCID_{50}$	No infectivity for types 1 & 3	40
89		Rhino-virus	20	15 da	2.0	PFU	....	10
90			50	10 min	2.2	PFU	Slower rate after 10 min	10
91			55	2 min	2.0	PFU	Slower rate after 2 min	10
92	Poxvirus	Vaccinia	6	40 da	0.4	PFU	....	55
93			26	20 da	1.5	PFU	....	55
94			37	20 da	3.1	PFU	....	55
95			50	45 min	2.7	Pock-FU	Slower rate after 45 min	30
96			50	1 hr	5.2	Pock-FU	1st-order kinetics with fresh virus stock	67
97			50	2 hr	2.8-4.2	PFU	....	43,64
98			55	15 min	5.8	Pock-FU	Slower rate after 15 min	30
99			55	30 min	5.0	Pock-FU	1st-order kinetics with fresh virus stock	67
100		Variola	50	2 hr	1.5-3.5	Pock-FU	Rate dependent on suspending fluid	23
101			56	1 hr	5.8	Pock-FU	Slower rate after 1 hr	23
102	Reovirus	Blue-tongue	4	15 da	0.2	PFU	....	29
103			25	15 da	3.0	PFU	Many substances were protective	29
104			37	1 da	5.5	PFU	Partially purified virus	29
105			37	3 da	1.3	ELD_{50}	Tissue suspension	58
106			46	4 hr	0.4	ELD_{50}	Tissue suspension	58
107			56	1 min	2.0	ELD_{50}	Tissue suspension	58
108		Reovirus	4	2 wk	1.0	PFU	....	53
109			24	2 wk	2.2	PFU	....	53
110			37	2 wk	5.6	PFU	....	53
111			50	40 min	4.2	PFU	Slower rate after 40 min	14
112			60	2 min	4.9	PFU	No inactivation beyond 2 min	14
113			60	10 s	2.0-5.0	PFU	In dairy products; diminishing rate	57
114	Rhabdo-virus	Rabies-virus	4	10 da	0.6	Mouse LD_{50}	In tissue culture medium	31
115			33	1 da	0.6	Mouse LD_{50}	In tissue culture medium	31
116		Vesicular sto-matitis	50	15 min	3.0	PFU	....	65

continued

Group Classification	Virus	Temp °C	Time	Inactivation Amount $-\log_{10}$	Unit of Measurement, $\log^{-1}$	Remarks	Reference
117 Toga-virus	Eastern equine encephalitis	50	2 hr	4.1	PFU	Partially purified virus in buffer; diminishing rate beyond 2 hr	42
118	Semliki	22	3 hr	0 to >2.0	PFU	Rate increased with dilution of medium	17
119	Forest	35	1 hr	1.1	PFU	Diluted in phosphate buffer	17
120		50	15 min	0.7-1.8	PFU	Tissue culture medium plus 10% serum	17
121	Tick-	4	2 wk	4.0	Mouse LD_{50}	Physiological saline	22
122	borne	4	2 wk	0	Mouse LD_{50}	In milk	22
123	en-	20	2 hr	4.0	Mouse LD_{50}	Physiological saline	22
124	ceph-	50	2 min	2.0 to >5.7	If_{50}	pH 7; varies with buffer	39
125	alitis	62	30 min	<2.2	Mouse LD_{50}	In milk	21
126		72	10 s	<6.7	Mouse LD_{50}	In milk	20
127	Venezuelan equine encephalitis	50	4 hr	3.2	PFU	Partially purified virus in buffer	42
128 None	Infectious hepatitis	56	30 min	....	....	3 out of 4 humans infected; serum ingested	26
129	Serum	21	3 mo	....	....	3 out of 5 humans infected; pooled plasma injected	44
130	hepa-titis	21	6 mo	....	....	1 out of 20 humans infected; 1-2 ml pooled plasma injected	44
131		60	4 hr	....	....	5 out of 10 humans infected; pooled plasma injected	44
132		60	10 hr	....	....	No humans infected	19,44

Contributors: Cliver, Dean O., and Salo, Richard J.

References

[1] Ackermann, W. W., et al. 1970. Arch. Environ. Health 21:377.

[2] Bachrach, H. L., et al. 1957. Proc. Soc. Exp. Biol. Med. 95:147.

[3] Bartha, A., et al. 1969. Acta Vet. (Budapest) 19:97.

[4] Brodsky, I., et al. 1959. J. Exp. Med. 109:439.

[5] Bussell, R. H., and D. T. Karzon. 1962. Virology 18:589.

[6] Camyre, K. P., and C. J. Pfau. 1968. J. Virol. 2:161.

[7] Cliver, D. O., et al. 1970. J. Milk Food Technol. 33:484.

[8] Coria, M. F. 1972. Appl. Microbiol. 23:281.

[9] DiGioia, G. A., et al. 1970. Ibid. 19:451.

[10] Dimmock, N. J. 1967. Virology 31:338.

[11] Dimopoullos, G. T., et al. 1959. Amer. J. Vet. Res. 20:510.

[12] Dossena, G. 1961. Ann. Sclavo 3:178.

[13] Dougherty, R. M. 1961. Virology 14:371.

[14] Engler, R. 1968. Z. Naturforsch. 23:884.

[15] Fellowes, O. N. 1962. Amer. J. Vet. Res. 23:1035.

[16] Fellowes, O. N. 1966. Appl. Microbiol. 14:206.

[17] Fleming, P. 1971. J. Gen. Virol. 13:385.

[18] Fujioka, R. S., et al. 1969. Proc. Soc. Exp. Biol. Med. 132:825.

[19] Gellis, S. S., et al. 1948. J. Clin. Invest. 27:239.

[20] Gresikova, M., et al. 1961. Acta Virol. (Prague) 5:31.

[21] Gresikova-Kohutova, M. 1959. Ibid. 3:215.

[22] Gresikova-Kohutova, M. 1959. Cesk. Epidemiol. Mikrobiol. Immunol. 8:26.

[23] Hahon, N., and E. Kozikowski. 1961. J. Bacteriol. 81:609.

[24] Hambling, M. 1964. Brit. J. Exp. Pathol. 45:647.

[25] Hanson, R. P., et al. 1949. Proc. Soc. Exp. Biol. Med. 70:283.

[26] Havens, W. P., Jr. 1945. Ibid. 58:203.

[27] Heidelbaugh, N. D., and D. J. Giron. 1969. J. Food Sci. 34:239.

[28] Hopkins, S. R. 1967. Avian Dis. 11:261.

[29] Howell, P. G., et al. 1967. Onderstepoort J. Vet. Res. 34:317.

[30] Kaplan, C. 1958. J. Gen. Microbiol. 18:58.

[31] Kissling, R. E., and D. R. Reese. 1963. J. Immunol. 91:362.

continued

[32] Krugman, R. D., and C. R. Goodheart. 1964. Virology 23:290.

[33] Lauffer, M. A., et al. 1948. Arch. Biochem. 16: 321.

[34] Lehmann-Grube, F. 1968. Arch. Gesamte Virusforsch. 23:202.

[35] Lo, J. P., et al. 1965. Acta Vet. Scand. 6:329.

[36] Lund, E. 1963. Arch. Gesamte Virusforsch. 13: 395.

[37] Lynt, R. K., Jr. 1966. Appl. Microbiol. 14:218.

[38] Majer, M., and R. Thomssen. 1965. Arch. Gesamte Virusforsch. 17:585.

[39] Mayer, V., and I. Slavik. 1967. Acta Virol. (Prague) 11:26.

[40] Medearis, D. N., Jr., et al. 1960. Proc. Soc. Exp. Biol. Med. 104:419.

[41] Melnick, J. L. 1971. Progr. Med. Virol. 13:462.

[42] Mika, L. A., et al. 1963. J. Infec. Dis. 113:195.

[43] Munyon, W., et al. 1970. J. Virol. 5:32.

[44] Murray, R., et al. 1955. N. Y. State J. Med. 55: 1145.

[45] Otte, P. 1971. Zentralbl. Bakteriol. Parasitenk. Infektionskr. Hyg. Abt. I, Orig. A217:441.

[46] Page, C. A., and C. H. Cunningham. 1962. Amer. J. Vet. Res. 23:1065.

[47] Pfau, C. J., and K. P. Camyre. 1967. Arch. Gesamte Virusforsch. 20:430.

[48] Picken, J. C., Jr. 1964. In R. P. Hanson, ed. Newcastle Disease Virus: An Evolving Pathogen. Univ. Wisconsin Press, Madison. pp. 167-188.

[49] Plowright, W., and R. D. Ferris. 1961. Arch. Gesamte Virusforsch. 11:516.

[50] Plummer, G., and B. Lewis. 1965. J. Bacteriol. 89: 671.

[51] Rafajko, R. R., and J. C. Young. 1964. Proc. Soc. Exp. Biol. Med. 116:683.

[52] Rechsteiner, J. 1969. J. Gen. Virol. 5:397.

[53] Rhim, J. S., et al. 1961. Virology 15:428.

[54] Sellers, R. F. 1969. Brit. Vet. J. 125:163.

[55] Sharp, D. G., et al. 1964. Proc. Soc. Exp. Biol. Med. 115:811.

[56] Strock, N. R., and N. N. Potter. 1972. J. Milk Food Technol. 35:247.

[57] Sullivan, R., et al. 1971. Appl. Microbiol. 22:315.

[58] Svehag, S.-E. 1963. Arch. Gesamte Virusforsch. 13: 499.

[59] Sweet, B. H., and M. R. Hilleman. 1960. Proc. Soc. Exp. Biol. Med. 105:420.

[60] Vogt, P. 1965. Advan. Virus Res. 11:293.

[61] Wallis, C., and J. L. Melnick. 1961. Tex. Rep. Biol. Med. 19:683.

[62] Wallis, C., and J. L. Melnick. 1961. Ibid. 19:701.

[63] Wallis, C., and J. L. Melnick. 1962. Proc. Soc. Exp. Biol. Med. 111:305.

[64] Wallis, C., et al. 1962. J. Immunol. 89:41.

[65] Wallis, C., et al. 1965. Virology 26:694.

[66] Wilner, B. I. 1969. A Classification of Major Groups of Human and Other Animal Viruses. Ed. 4. Burgess, Minneapolis.

[67] Woodroofe, G. M. 1960. Virology 10:379.

94. SURVIVAL TIME AND THERMAL INACTIVATION: PLANT VIRUSES

Test Species: Infecting medium was plant juice, unless otherwise indicated. **Inactivation Temp:** Temperature at which infectivity is lost in 10 minutes. For information on half-time of virus inactivation, consult reference 77.

	Virus	Plant Source of Virus	Test Species	Survival		Inactivation Temp, °C	Reference
				Temp °C	Time		
1	Alfalfa mosaic	*Pisum sativum; Vicia faba*	*Phaseolus vulgaris* 'Stringless Green Refugee'	20[1]	<5 da	70	78
2	Dolichos lablab strain	*Vicia faba minor*	*V. faba minor*	25	>24 to <48 hr	65-71	51
3	La Salle strain	*Nicotiana tabacum* 'Samsun'	*Phaseolus vulgaris; Vigna sinensis*	Room[1]	<7 da	65	15
4	Pierce strain	*Nicotiana glutinosa*	*Phaseolus vulgaris*	Room	9 da	62-65	53
5	Vein necrosis strain	*Glycine max*[2]	*Phaseolus vulgaris* 'Pinto U.I. 111'	18	>30 to <32 hr	62-64	80
6	Arabis mosaic, Forsythia strain	*Chenopodium quinoa*	*C. quinoa*	±20	>36 da	64	60

[1] Kept in dark. [2] Synonyms: *G. maxima, G. soja.*

continued

	Virus	Plant Source of Virus	Test Species	Survival		Inactivation Temp, °C	Ref-er-ence
				Temp °C	Time		
	Barley yellow dwarf						
7	M A V strain	*Avena byzantina*	*A. byzantina*[3/]	....	....	>65 to <70	31
8	R P V strain	*Avena byzantina* 'Coast Black'	*A. byzantina* 'Coast Black'[4/]	....	....	>65 to <70	31
9	Bean common mosaic	*Phaseolus vulgaris*	*P. vulgaris* 'Stringless Green Refugee'	18	>28 to <32 hr	56-58	53
10	Bean-pod mottle	*Phaseolus vulgaris*	*P. vulgaris* 'Pinto U.I. 111'	18	>62 to <93 da	70-75	81
11	Bean southern mosaic	*Phaseolus vulgaris*	*P. vulgaris* 'Pinto U.I. 111'	18	32 wk	90-95	79
12	Bean yellow mosaic	*Phaseolus vulgaris*	*P. vulgaris* 'Stringless Green Refugee'	18	>24 to <32 da	58-60	83
13	Bean yellow stipple	*Phaseolus vulgaris*	*P. vulgaris* 'Pinto U.I. 111'	18	5 da	72-75	82
14	Black locust mosaic[5/]	*Chenopodium quinoa*	*C. quinoa*	20-28	>5 to <10 da	72	62
15	Chrysanthemum stunt	....	....	25	7-8 wk	96-98	11
16	Cucumber mosaic	*Cucumis sativus*	*C. sativus* 'Beit Alpha'; *Chenopodium amaranticolor*	Room	<22 da	60-70	17
17		*Nicotiana glutinosa*	*Chenopodium amaranticolor; N. tabacum*[6/]	26-32	>3 da	>60 to <65	36
18		*Nicotiana tabacum*	*N. tabacum*[6/]	26-32	>1 to <2 da	>65 to <70	36
19	Dahlia mosaic	*Verbesina encelioides*	*V. encelioides*	18	>28 to <35 da	90	12
20	Grape infectious degeneration	*Chenopodium quinoa*	*C. quinoa*	4	>34 to <64 da	>55 to <57	75
21	Malva mosaic	*Malva parviflora*	*M. parviflora; M. alcea* var. *fastigiata*	Room	<4 da	65	44
22	Malva mosaic[7/]	*Malva neglecta*	*M. neglecta*[8/]	19-22	>72 to <120 hr	65	32
23	Melon mosaic	*Cucumis sativus* 'Beit Alpha'; *Cucurbita pepo* 'Sihi Lavan'	*Cucumis sativus* 'Beit Alpha'; *Chenopodium amaranticolor*	Room	<7 da	65	17
	Nasturtium mosaic						
24	Brazilian strain	*Tropaeolum majus*	*T. majus*	25	<24 hr	58	65
25	California strain	*Tropaeolum majus*	*T. majus*	Room	>4 da	55	34
26	Ring mosaic strain	*Chenopodium quinoa*	*C. quinoa*	Room	>11 to <18 da	>66 to <68	59
27		*Nicotiana alata*	*N. alata*	Room	5 da	>65 to <68	63
28	Ring-spot mosaic strain	*Nicotiana tabacum*	*Chenopodium quinoa*	Room	>24 to <30 da	62-65	64
29	Oat mosaic	*Avena sativa*	*A. sativa*[9/]	20	>24 to <48 hr	>44 to <46	73
30	Pea-enation mosaic	*Pisum sativum*	*P. sativum* 'Prince of Wales'	18	>72 to <96 hr	56-58	71
31	Pea mosaic	*Pisum sativum*	*P. sativum* 'Prince of Wales'	18	>48 to <72 hr	60-64	22
32	Pea mosaic-2[10/]	*Pisum sativum*	*Phaseolus vulgaris* 'Stringless Green Refugee'	18	>24 to <28 hr	58-60	83

[3/] Plant juice fed to *Macrosiphum avenae.* [4/] Plant juice fed to *Rhopalosiphum padi.* [5/] Synonym: *Robinia* mosaic virus. [6/] Infecting medium was plant juice kept in darkened drawer. [7/] Synonym: *Malva* virus 1 Ryshkow. [8/] Infecting medium was sap from frozen leaves diluted 1:1 with tap water. [9/] Infecting medium was plant juice diluted 1:1 with distilled water. [10/] Synonym: Pea mottle virus.

continued

	Virus	Plant Source of Virus	Test Species	Survival		Inactivation Temp, °C	Ref-er-ence
				Temp °C	Time		
33	Pea streak	*Pisum sativum* 'Perfected Prince of Wales'	*P. sativum* 'Perfected Prince of Wales'	18	>16 to <32 da	58-60	26
34	Pea stunt	*Pisum sativum*	*P. sativum* 'Perfected Prince of Wales'	18	>48 to <72 hr	58-60	27
35	Pelargonium, strain R5	*Nicotiana tabacum; Gomphrena globosa*	*N. tabacum; G. globosa*	20 ± 1	264 hr	75	57
36	Pelargonium leaf curl	*Nicotiana clevelandii*	*N. clevelandii*	18	>21 da	90	33
37	Populus mosaic	*Nicotiana megalo-siphon*	*N. megalosiphon* [9]	~18	<2 da	76	61
38	Potato A	*Nicandra physaloides*	*Solanum tuberosum* [11]	18	12-18 hr	44-52	43
39		*Nicotiana tabacum*	*N. tabacum*	Room	Few hr	~50	28
40	Veinbanding strain	*Nicotiana tabacum*	*N. tabacum*	Room	4 da	>58 to <60	39
41	Potato M	*Solanum tuberosum* 'Seedling E K'	*Datura metel*	20	>2 to <4 da	65-70	2
42	Potato S	*Solanum tuberosum* 'Seedling 41956'	*Nicotiana debneyi*	20	>4 to <8 da	>55 to <60	2
43	Potato X	*Nicotiana rustica*	*N. rustica*	16-20	>360 da	72	41
44		*Nicotiana tabacum* 'White Burley'	*N. tabacum* 'White Burley'	14-17	>234 da	>70 to <75	74
45	Mottle strain	*Nicotiana tabacum* 'Connecticut Havana 38'	*N. tabacum* 'Connecticut Havana 38'	Room	>28 da	>68 to <70	38
46	Ring-spot strain	*Nicotiana tabacum* 'Connecticut Havana 38'	*N. tabacum* 'Connecticut Havana 38'	Room	>28 da	>65 to <68	38
47		*Datura metel*	*D. metel*	Room	>7 da	65-70	10
48	Potato Y	*Datura metel*	*D. metel*	Room	2-3 da	55-60	10
49		*Nicotiana tabacum* [12]	*N. tabacum* [13]	20-22	>6 to <18 da	56-62	20
50		*Nicotiana tabacum*	*N. tabacum*	15	>3 to <4 da	57	23
51		*Nicotiana tabacum; Solanum tuberosum*	*N. tabacum* 'Connecticut Havana 38'	Room	<6 da	60	39
52	Necrotic strain	*Nicotiana tabacum* 'Samsun'	*N. tabacum* 'Samsun'	21-23	>50 da	62	37
53		*Nicotiana tabacum* 'White Burley'	*N. tabacum* 'White Burley'	20	70 da	61	1
54	Pepper veinbanding mosaic strain	*Nicotiana tabacum* 'Turkish'	*N. tabacum* 'Turkish' [14]	23	>10 to <15 da	61-65 [15]	67
55	Standard strain	*Nicotiana tabacum* 'White Burley'	*N. tabacum* 'White Burley'	20	12 da	61	1
56	Vein-necrosis strain	*Nicotiana tabacum* 'White Burley'	*N. tabacum* 'White Burley'	Room	>7 to <8 da	58	48
57	Potato aucuba mosaic	*Nicotiana tabacum* 'Samsun'	*Capsicum frutescens* 'Early California Wonder'	18-20 [1]	30-95 da [16]	>65 to <70 [17]	40
58		*Nicotiana tabacum*	*N. sylvestris*	15	>4 da	68	24
59		*Solanum tuberosum* 'President'	*S. nodiflorum*	15 [1]	>3 to <4 da	>63 to <65	16

[1] Kept in dark. [9] Infecting medium was plant juice diluted 1:1 with distilled water. [11] Synonym: *S. demissum*. [12] Recently infected. [13] Test plant also infected with potato X virus. [14] Infecting medium was undiluted plant juice. [15] Heated for 11 min. [16] In sap diluted 1:9 with distilled water. [17] In sap diluted 1:4 with distilled water.

continued

	Virus	Plant Source of Virus	Test Species	Survival		Inactivation Temp, °C	Reference
				Temp °C	Time		
60	Potato mop-top	*Nicotiana tabacum* 'Xanthi'	*Chenopodium amaranti-color*; *N. tabacum* 'Xanthi'	20	14 wk	80	30
61	Potato yellow dwarf	*Nicotiana rustica*	*N. rustica*	23-27	>12 hr	52	9
	Prunus						
62	Strain A	*Cucumis sativus* 'Chicago Pickling'	*C. sativus* 'Chicago Pickling' [18]	24	>6 to <9 hr	>52 to <54	25
63	Strain B	*Cucumis sativus* 'Chicago Pickling'	*C. sativus* 'Chicago Pickling'	24	>15 to <18 hr	44	25
64	Squash mosaic	*Cucumis sativus* 'Beit Alpha'	*C. sativus* 'Beit Alpha'; *Cucurbita pepo*	Room	<24 da	65	17
65	Sugar beet curly top	*Beta vulgaris*	*B. vulgaris* [19]	Room	>7 to <14 da	80	5
66	Sugar beet yellows	*Beta vulgaris*	*Chenopodium capitatum*	20-24	>12 to <24 hr	>50 to <55	6
67	Tobacco etch	*Nicotiana tabacum*	*N. tabacum* [20]	Room	12 da	58	4
68	Tobacco mosaic [21]	*Nicotiana tabacum* 'Turkish'	*N. glutinosa*; *Phaseolus vulgaris* 'Early Golden Cluster'	68	>70 da	93	54
69	Tobacco necrosis	*Nicotiana tabacum*	*Vigna sinensis* 'Black' [14]	....	....	92	55
70		*Nicotiana tabacum*	*N. tabacum* [14]; *Vigna sinensis* [14]	21	20	72	70
	Tobacco rattle						
71	Brazilian strain	*Lycopersicon esculentum*	*Chenopodium amaranti-color*; *Petunia hybrida*	25	>13 to <22 da	>78 to <80	40,66
72	Type M	*Nicotiana tabacum* 'White Burley'	*Phaseolus vulgaris* 'Prince'	20	>6 wk	85	13
73	Tobacco ring spot	*Nicotiana rustica*	*N. tabacum* [22]	26-32	>6 da	>65 to <70	36
74		*Nicotiana sylvestris*	*N. tabacum* [22]	26-32	>1 to <2 da	>60 to <65	36
75		*Nicotiana tabacum*	*N. tabacum* [22]	26-32	6 da	>65 to <70	36
76		*Phaseolus vulgaris*	*P. vulgaris* 'Henderson Bush Lima'; *P. vulgaris* 'Stringless Green Refugee'	Room	<7 da	66	53
	Tobacco streak						
77	Brazilian strain	*Nicotiana tabacum*	*N. tabacum*	Room	>12 to <24 hr	50-55	19
78		*Nicotiana tabacum* [23]	*N. tabacum* 'Turkish' [24]	Room (?)	>9 to <27 hr	>60 to <65	18
79	Canadian strain	*Nicotiana tabacum* (?)	*N. tabacum* (?)	Room	<24 hr	54	7
80	Pea strain	*Phaseolus vulgaris*	*P. vulgaris* 'Pinto U.I. 111'	22	>26 to <27 hr	64	52
81	Standard strain	*Nicotiana tabacum* 'Havana'	*N. tabacum* 'Havana'	22	>24 to <36 hr	53	35
82		*Phaseolus vulgaris* 'Pinto U.I. 78'	*P. vulgaris* 'Pinto U.I. 78'	18	<24 hr	54-56	72
83	Tomato aspermy	*Nicotiana tabacum*	*N. tabacum*	25	3-6 da	65-70	11
84	Chrysanthemum strain	*Nicotiana tabacum*	....	23-25	>12 hr	70	49
85	Tomato bunchy top	*Lycopersicon esculentum*	*Nicotiana glutinosa*	Room	>24 hr	70	45
86	Tomato bushy stunt	*Lycopersicon esculentum* (?)	*Vigna sinensis*; *Nicotiana glutinosa*	Room	>22 da	80	68

[14] Infecting medium was undiluted plant juice. [18] Infecting medium was plant juice diluted with 25 parts 0.03 *M* phosphate buffer. [19] Infecting medium was plant juice or phloem exudate. [20] Starch-iodine method. [21] For additional information, consult reference 54. [22] Infecting medium was plant juice kept in dark. [23] Sap from recovered tissue. [24] Infecting medium was plant juice in 0.02 *M* phosphate buffer plus sodium sulfite.

continued

	Virus	Plant Source of Virus	Test Species	Survival		Inactivation Temp, °C	Reference
				Temp °C	Time		
87	Tomato mosaic 25/	*Lycopersicon esculentum* 'Bonny Best'	*L. esculentum* 'Bonny Best'	Room	>138 da	85-95	76
88	Tomato ring spot	*Lycopersicon esculentum*	*L. esculentum*	3	>21 to <45 hr	>56 to <58	58
89	Beet ring-spot strain	*Nicotiana tabacum* 'White Burley'	*Phaseolus vulgaris* (?)	20	>2 to <3 wk	>63 to <66	29
90	Pelargonium strain kl	*Nicotiana tabacum; Gomphrena globosa*	*N. tabacum* 26/; *G. globosa* 26/	20 ± 1	48 hr	62	57
91	Tomato black-ring strain	*Nicotiana tabacum* (?)	*N. tabacum* (?)	Room	>7 da	>58 to <62	69
92	Tomato ring-spot strain; peach yellow-bud strain	*Petunia hybrida*	*P. hybrida* 27/	−10	>5 da	>58 to <60	14
93	Tomato spotted wilt	*Lycopersicon esculentum* 28/	*L. esculentum* 29/	21	>2 hr	>44 to <46	50
94		*Lycopersicon esculentum* 'Dwarf Champion' 28/	*L. esculentum* 'Dwarf Champion' 30/	17.2	>4.5 to <6 hr	>42 to <45	3
95		*Lycopersicon esculentum* 'Dwarf Champion' 12/	*Nicotiana tabacum* 'Blue Prior' 31/	....	....	45.5	8
96	Corcova strain	*Lycopersicon esculentum*	*L. esculentum; Nicotiana glutinosa*	21	<2.5 hr	>46 to <48	21
97	Tomato tip-blight strain	*Lycopersicon esculentum*	*L. esculentum* 14/	18	<1 hr	41.5	46
98	Watermelon mosaic Strain 1	*Cucurbita pepo* 'Caserta Bush'	*C. pepo* 'Caserta Bush' 32/	18	>16 to <32 da	60-65	56
99	Strain 2	*Cucumis sativus; Cucurbita maxima; C. pepo*	*Chenopodium amaranticolor*	20	>12 da	66	47
100	Wheat striate mosaic	*Triticum durum* 'Ramsay'	*T. durum* 'Ramsay' 33/	4	>3 to <4 da	>45 to <55	42

12/ Recently infected. 14/ Infecting medium was undiluted plant juice. 25/ Strain of tobacco mosaic virus. 26/ Infecting medium was plant juice obtained by homogenizing dehydrated tissue in 0.03 M phosphate buffer. 27/ Infecting medium was plant juice diluted 1:5 with water. 28/ Young plant recently infected. 29/ Infecting medium was untreated plant juice. 30/ Infecting medium was plant juice diluted with water. 31/ Infecting medium was sap diluted with 7 parts buffer containing sodium sulfite. 32/ Method Bos. 33/ Infecting medium was juice of leaf-hoppers, *Endria inimica.*

Contributors: Silberschmidt, Karl M.; Zaumeyer, William J.; Webb, Raymon E.

References
[1] Aubert, O. 1960. Mem. Soc. Vaudoise Sci. Natur. 12(77):153.
[2] Bagnall, R. H., et al. 1956. Wis. Agr. Exp. Sta. Res. Bull. 198.
[3] Bald, J. G., and G. Samuel. 1931. Aust. Counc. Sci. Ind. Res. Bull. 54.
[4] Bawden, F. C., and B. Kassanis. 1941. Ann. Appl. Biol. 28:107.
[5] Bennett, C. W. 1935. J. Agr. Res. 50:211.
[6] Bennett, C. W. 1960. U.S. Dep. Agr. Tech. Bull. 1218.
[7] Berkeley, G. H., and J. H. H. Phillips. 1943. Can. J. Res. C21:181.
[8] Best, R. J. 1946. Aust. J. Exp. Biol. Med. Sci. 24: 21.
[9] Black, L. M. 1938. Phytopathology 28:863.
[10] Borges, M. V. 1958. Agron. Lusitana 20(4):287.
[11] Brierley, P. 1963. Plant Dis. Rep. 47(6):445.

continued

[12] Brierley, P., and F. F. Smith. 1950. Ibid. 34(12): 363.

[13] Cadman, C. H., and B. D. Harrison. 1959. Ann. Appl. Biol. 47:542.

[14] Cadman, C. H., and R. M. Lister. 1961. Phytopathology 51:29.

[15] Cervantes, J., and R. H. Larson. 1961. Wis. Agr. Exp. Sta. Res. Bull. 229.

[16] Clinch, P. E. M., et al. 1936. Sci. Proc. Roy. Dublin Soc., N. S. 21:431.

[17] Cohen, S., and F. E. Nitzany. 1963. Phytopathology 53(2):193.

[18] Costa, A. S., and A. M. B. Carvalho. 1961. Phytopathol. Z. 42:113.

[19] Costa, A. S., et al. 1940. J. Agron. Sao Paulo 3:1.

[20] Darby, J. F., et al. 1951. Wis. Agr. Exp. Sta. Res. Bull. 177.

[21] Delle Coste, A. C., and S. Zabala. 1946. Publ. Inst. Sanid. Veg. (Buenos Aires) A17.

[22] Doolittle, S. P., and F. R. Jones. 1925. Phytopathology 15:763.

[23] Dykstra, T. P. 1939. Ibid. 29:40.

[24] Dykstra, T. P. 1939. Ibid. 29:917.

[25] Fulton, R. W. 1957. Ibid. 47:683.

[26] Hagedorn, D. J., and J. C. Walker. 1949. Ibid. 39:837.

[27] Hagedorn, D. J., and J. C. Walker. 1949. J. Agr. Res. 78:617.

[28] Hansen, H. P. 1937. Tidsskr. Planteavl 42:631.

[29] Harrison, B. D. 1957. Ann. Appl. Biol. 45:462.

[30] Harrison, B. D., and R. A. C. Jones. 1970. Ibid. 65(3):393.

[31] Heagy, J., and W. F. Rochow. 1965. Phytopathology 55:809.

[32] Hein, A. 1956. Phytopathol. Z. 28(2):205.

[33] Hollings, M. 1962. Ann. Appl. Biol. 50(2):189.

[34] Jensen, D. D. 1950. Phytopathology 40:967.

[35] Johnson, J. 1936. Ibid. 26:285.

[36] Johnson, J., and T. J. Grant. 1932. Ibid. 22:741.

[37] Klinkowski, M., and K. Schmelzer. 1957. Phytopathol. Z. 28:285.

[38] Koch, K. L. 1933. Phytopathology 23:319.

[39] Koch, K. L., and J. Johnson. 1935. Ann. Appl. Biol. 22:37.

[40] Kollmer, G. F., and R. H. Larson. 1960. Wis. Agr. Exp. Sta. Res. Bull. 223.

[41] Ladenburg, R. C., et al. 1950. Ibid. 165.

[42] Lee, P. E., and W. Bell. 1963. Can. J. Bot. 41:767.

[43] MacLachlan, D. S., et al. 1953. Wis. Agr. Exp. Sta. Res. Bull. 180.

[44] Majorana, G., and K. Silberschmidt. 1967. Ann. Fac. Agr. Univ. Bari 21:5.

[45] McClean, A. P. D. 1931. Repub. S. Afr. Dep. Agr. Tech. Serv. Sci. Bull. 100.

[46] Milbrath, J. A. 1939. Phytopathology 29:156.

[47] Molnar, A., and K. Schmelzer. 1964. Phytopathol. Z. 51(4):361.

[48] Nobrega, N. R., and K. Silberschmidt. 1944. Arq. Inst. Biol. (Sao Paulo) 15:307.

[49] Noordam, D. 1952. Tijdschr. Plantenziekten 58:121.

[50] Norris, D. O. 1946. Aust. Counc. Sci. Ind. Res. Bull. 202.

[51] Nour, M. A., and J. J. Nour. 1962. Phytopathology 52:427.

[52] Patino, G., and W. J. Zaumeyer. 1959. Ibid. 49:43.

[53] Pierce, W. H. 1934. Ibid. 24:87.

[54] Price, W. C. 1933. Ibid. 23:749.

[55] Price, W. C. 1938. Amer. J. Bot. 25:603.

[56] Regenmortel, M. H. van, et al. 1962. Phytopathol. Z. 43(3):205.

[57] Reinert, R. A., et al. 1963. Phytopathology 53:1292.

[58] Samson, R. W., and E. P. Imle. 1942. Ibid. 32:1037.

[59] Schmelzer, K. 1960. Z. Pflanzenkr. Pflanzenschutz 67:193.

[60] Schmelzer, K. 1962. Phytopathol. Z. 46(2):105.

[61] Schmelzer, K. 1966. Ibid. 55(4):317.

[62] Schmelzer, K. 1967. Ibid. 58:59.

[63] Schumann, K. 1963. Ibid. 48:135.

[64] Schwarz, R. 1958. Ibid. 33:375.

[65] Silberschmidt, K. 1953. Phytopathology 43:304.

[66] Silberschmidt, K., et al. 1967. Phytopathol. Z. 60:278.

[67] Simon, J. N. 1956. Phytopathology 46:53.

[68] Smith, K. M. 1935. Ann. Appl. Biol. 22:731.

[69] Smith, K. M. 1946. Parasitology 37:126.

[70] Smith, K. M., and J. G. Bald. 1935. Ibid. 27:231.

[71] Stubbs, M. W. 1937. Phytopathology 27:242.

[72] Thomas, R. R., and W. J. Zaumeyer. 1950. Ibid. 40:832.

[73] Toler, R. W., and T. T. Hebert. 1964. Ibid. 54(4):428.

[74] Vasudeva, R. S., and T. B. Lal. 1945. Indian J. Agr. Sci. 14:288.

[75] Vuittenez, A., and J. Kuszala. 1963. C. R. Acad. Agr. Fr. 49(10):795.

[76] Walker, M. N. 1926. Phytopathology 16:431.

[77] Yarwood, C. E., and E. S. Sylvester. 1959. Plant Dis. Rep. 43(2):125.

[78] Zaumeyer, W. J. 1938. J. Agr. Res. 56:747.

[79] Zaumeyer, W. J., and L. L. Harter. 1943. Ibid. 67:297.

[80] Zaumeyer, W. J., and G. Patino. 1960. Phytopathology 50:226.

[81] Zaumeyer, W. J., and H. R. Thomas. 1948. J. Agr. Res. 77:81.

[82] Zaumeyer, W. J., and H. R. Thomas. 1950. Phytopathology 40:847.

[83] Zaumeyer, W. J., and B. L. Wade. 1935. J. Agr. Res. 51:715.

Rectal Temperature: **Min** = minimum; **Max** = maximum. **Critical Air Temperature:** Air temperature at which the normal animal first begins to show a change in deep body temperature. **Temperature Regulating Mechanism:** + = present; − = absent. **Thermoneutrality Zone:** The air temperature at which the normal animal has the lowest metabolic rate.

	Animal	Rectal Temperature, °C			Critical Air Temperature, °C		Temperature Regulating Mechanism			Thermo-neutrality Zone, °C	Reference
		Normal	Min	Max	Low	High	Sweating	Shivering	Panting		
1	Man	37	21	44	1	32	+	+	−	24-31	4,6,16,22,23,71,76,77
2	Camel	37-38	34	41	....	......	+	+	−	....	69
3	Cat	39	19	44	....	36	−	+	+	24-28	3,25,63,64,70
4	Cattle, Brahman	38-39	..	......	1	32	−	...	+	10-27	52-54
5	Cattle, dairy	38-39	..	43	....	40	+	...	+	5-16	24,51,52,55
6	Dog	38-39	24	42	−80	42-58	−	+	+	18-25	1,26,29,47,60,72
7	Donkey	36-38	..	......	....	......	+	+	...	....	69
8	Elephant	36-37	..	......	....	......	...	...	...	....	17
9	Goat	38-39	..	......	....	40	+	...	...	20-30	7,13,14,48
10	Guinea pig	39	17	43	−15	32	−	+	+	30-31	1,3,15,27,38,70
11	Horse	38	..	......	....	......	+	...	...	....	14,49
12	Monkey[1]	37-38	19	43	....	38	+	+	−	27-30	9,18,28,33,41,42,62,70
13	Mouse, white	37	10	......	10	37	−	...	−	30-33	1,30,38
14	Rabbit	39	20	42	−29	32	−	+	+	20-30	1,8,11,13,20,36,50,57,58
15	Rat, white	37-38	15	44	−10	32	−	+	−	28-30	2,3,5,21,32,35,37,38,56,70
16	Seal, fur[2]	38	..	43-44	....	......	+	+	+	....	10
17	Seal, harbor[3]	37	..	......	−30	......	...	...	...	−10 to +30	31,46
18	Sheep	39-40	..	42	....	32	+	+	+	10	12,13,34,40,61,75
19	Swine	37-38	..	42-45	....	30	−	+	+	0-20	15,44,45,59,66,68
20	Chicken	41-42	15	47	−35	32	−	+	+	19-29	43,65,73,74
21	Ostrich	38	..	......	....	......	...	...	+	....	19
22	Pigeon	43	..	47	−85	42	−	+	+	20-30	27,39,65,67

[1] *Macaca* sp. [2] *Callorhinus* sp. [3] *Phoca* sp.

Contributor: Ederstrom, H. E.

References

[1] Adolph, E. F. 1947. Amer. J. Physiol. 151:564.

[2] Adolph, E. F. 1948. Ibid. 155:378.

[3] Adolph, E. F. 1951. Ibid. 166:75.

[4] Adolph, E. F., and G. W. Malnar. 1946. Ibid. 146: 507.

[5] Allen, S. C., and V. E. Hall. 1952. Fed. Proc. Fed. Amer. Soc. Exp. Biol. 11:4.

[6] Andersen, K. L., and B. Hellstrom. 1960. Acta Physiol. Scand. 50:88.

[7] Andersson, B. 1957. Ibid. 41:90.

[8] Ariel, I., et al. 1943. Cancer Res. 3:448.

[9] Aron, H. 1911. Philipp. J. Sci. B6:101.

[10] Bartholomew, G. A., and F. Wilke. 1956. J. Mammal. 37:327.

[11] Blair, J. R., et al. 1951. Fed. Proc. Fed. Amer. Soc. Exp. Biol. 10:15.

[12] Bligh, J. 1959. J. Physiol. (London) 146:142.

[13] Bligh, J., et al. 1971. Ibid. 212:377.

[14] Brody, S. 1945. Bioenergetics and Growth. Reinhold, New York.

[15] Brück, K., et al. 1969. Fed. Proc. Fed. Amer. Soc. Exp. Biol. 28:1035.

[16] Burton, A. C., and O. G. Edholm. 1955. Man in a Cold Environment. E. Arnold, London.

[17] Buss, I. W., and A. Wallner. 1965. J. Mammal. 46: 104.

[18] Chaffee, R. R. J., et al. 1969. Fed. Proc. Fed. Amer. Soc. Exp. Biol. 28:1029.

[19] Crawford, E. C., Jr., and K. Schmidt-Nielsen. 1967. Amer. J. Physiol. 212:347.

[20] Degerman, G., and J. H. Kihlstrom. 1964. Acta Physiol. Scand. 62:46.

[21] Depocas, F., et al. 1957. J. Appl. Physiol. 10:393.

[22] Ehrmantraut, W. R., et al. 1957. Arch. Intern. Med. 99:57.

[23] Ferris, E. B., Jr., et al. 1938. J. Clin. Invest. 17:249.

[24] Findley, J. D., and D. Robertshaw. 1965. J. Physiol. (London) 179:285.

[25] Forster, R. E., and T. B. Ferguson. 1952. Amer. J. Physiol. 169:255.

continued

[26] Galvao, P. E. 1947. Ibid. 148:478.

[27] Giaja, J., and S. Gelineo. 1933. Arch. Int. Physiol. 37:20.

[28] Guerra, F., and J. R. Brobeck. 1944. J. Pharmacol. Exp. Ther. 80:209.

[29] Hammel, H. T., et al. 1958. Amer. J. Physiol. 194:99.

[30] Hart, J. S. 1951. Can. J. Zool. 29:225.

[31] Hart, J. S., and L. Irving. 1959. Ibid. 37:447.

[32] Hart, J. S., et al. 1956. J. Appl. Physiol. 9:404.

[33] Hayward, J. N., and M. A. Baker. 1968. Amer. J. Physiol. 215:389.

[34] Hemingway, A., et al. 1966. J. Appl. Physiol. 21:1223.

[35] Heroux, O. 1962. Can. J. Biochem. 40:537.

[36] Heroux, O. 1967. Can. J. Physiol. Pharmacol. 45:451.

[37] Heroux, O., and F. Depocas. 1959. Can. J. Biochem. 37:473.

[38] Herrington, L. P. 1940. Amer. J. Physiol. 129:123.

[39] Hoffman, R. A. 1958. Ibid. 195:751.

[40] Hofmyer, H. S., et al. 1969. J. Appl. Physiol. 26:517.

[41] Hongo, T. T., and C. P. Luck. 1953. J. Physiol. (London) 122:570.

[42] Honjo, S., et al. 1963. Aerosp. Med. 35:A-123.

[43] Horvath, S. M., et al. 1948. Science 107:171.

[44] Ingram, D. L. 1965. Nature (London) 207:415.

[45] Irving, L. 1956. J. Appl. Physiol. 9:421.

[46] Irving, L., and J. S. Hart. 1957. Can. J. Zool. 35:497.

[47] Irving, L., and J. Krog. 1954. J. Appl. Physiol. 6:667.

[48] Jenkinson, D. M., and D. Robertshaw. 1971. J. Physiol. (London) 212:455.

[49] Jirka, M., and J. Kotas. 1959. Ibid. 147:74.

[50] Keeton, R. W. 1924. Amer. J. Physiol. 69:307.

[51] Kibler, H. H., and S. Brody. 1949. Mo. Agr. Exp. Sta. Res. Bull. 450.

[52] Kibler, H. H., and S. Brody. 1950. Ibid. 464.

[53] Kibler, H. H., and S. Brody. 1951. Ibid. 473.

[54] Kibler, H. H., and S. Brody. 1954. Ibid. 552.

[55] Kibler, H. H., et al. 1949. Ibid. 435.

[56] Krog, H., et al. 1955. J. Appl. Physiol. 7:349.

[57] Lee, D. H. K., et al. 1941. Proc. Roy. Soc. Queensland 53:129.

[58] Lee, R. C. 1939. J. Nutr. 18:473.

[59] Lee, R. C., et al. 1941. Ibid. 21:321.

[60] Lozinsky, E. 1924. Amer. J. Physiol. 67:388.

[61] McFarlane, W. V., et al. 1963. Nature (London) 197:270.

[62] McMurrey, J. D., et al. 1956. Surg. Gynecol. Obstet. 102:75.

[63] Perkins, J. F. 1945. Amer. J. Physiol. 145:264.

[64] Prouty, L. R. 1949. Fed. Proc. Fed. Amer. Soc. Exp. Biol. 8:128.

[65] Randall, W. C. 1943. Amer. J. Physiol. 39:56.

[66] Richards, S. A. 1970. Biol. Rev. Cambridge Phil. Soc. 45:223.

[67] Riddle, O., et al. 1934. Amer. J. Physiol. 107:333.

[68] Roos, A., et al. 1947. J. Clin. Invest. 26:505.

[69] Schmidt-Nielsen, K., et al. 1957. Amer. J. Physiol. 188:103.

[70] Scholander, P. F., et al. 1950. Biol. Bull. 99:237.

[71] Scholander, P. F., et al. 1958. J. Appl. Physiol. 12:1.

[72] Spurr, G. B., et al. 1954. Amer. J. Physiol. 178:275.

[73] Sturkie, P. D. 1946. Ibid. 147:531.

[74] Sturkie, P. D. 1954. Avian Physiology. Comstock, Ithaca.

[75] Webster, A. J. F., et al. 1969. Can. J. Physiol. Pharmacol. 47:553.

[76] Windham, C. H., et al. 1954. J. Appl. Physiol. 6:681.

[77] Winslow, C. E. A., et al. 1937. Amer. J. Physiol. 120:288.

96. RESPONSES TO CHANGES IN AMBIENT TEMPERATURE: HOMOIOTHERMIC ANIMALS

Data are for animals in resting state.

Animal	Ambient Temp °C	Variable	Response		Reference
			Single Exposure	Repeated or Continued Exposure	
		Increased Ambient Temperature			
1 Man	32	Rectal temp	Increased	Returned towards normal	6,8,16
2	34	Heat production	Increased	Returned towards normal	12,14,16
3	29	Tissue conductance	Increased sharply	Increased	5,14,16
4	28	Evaporative cooling, cutaneous	Increased sharply	Increased in relation to body temp	4,16

continued

	Animal	Ambient Temp °C	Variable	Response Single Exposure	Response Repeated or Continued Exposure	Reference
5	Cat	32	Rectal temp	Increased	....	1,16
6		38	Heat production	Increased	....	1,16
7		>32	Tissue conductance	Decreased slightly	....	1,16
8		35	Evaporative cooling, respiratory	Increased	....	1,16
9	Ox	24	Rectal temp	Increased	Returned towards normal	16
10		40	Heat production	No increase unless body temp increased	Diminished below normal	3,10,16
11		....	Tissue conductance	Increased	Remained high	16
12		25	Evaporative cooling, cutaneous	Increased	....	16
13		20	Evaporative cooling, respiratory	Increased	Increased further	2,16
14	Rat	30	Rectal temp	Increased	Decreased	16
15		31	Heat production	Increased	Decreased	16
16		....	Tissue conductance	Increased	Decreased	16
17		33	Evaporative cooling, saliva spreading	Increased	Decreased cutaneous loss	16
18	Sea	25	Rectal temp	Increased	....	18
19	lion	25	Heat production	Increased	....	7,17
20	(seal)	>21	Tissue conductance	Increased	....	7,18
21		30	Evaporative cooling	Increased	....	13
22	Chick-	30	Rectal temp	Increased	Decreased	15
23	en	35	Heat production	Increased	Decreased	15
24		....	Tissue conductance	Increased	Decreased	15
25		27	Evaporative cooling, respiratory	Increased	Decreased	15
			Decreased Ambient Temperature			
26	Man	24	Rectal temp	Increased	Returned towards normal	6,16
27		23	Heat production	Increased	Returned towards normal or remained elevated	9,11,12,16
28		<29	Tissue conductance	Decreased slowly	Decreased further	9,11,12,16
29	Cat	10	Rectal temp	Decreased	Increased	1,16
30		10	Heat production	Increased	....	1,16
31		0	Heat production	Increased	Increased	1,16
32		<32	Tissue conductance	Decreased	Decreased[1]	1,16
33	Ox	24	Rectal temp	No change	No consistent change	16
34		<18	Heat production	Increased	No change	3,10,16
35		<15	Tissue conductance	Decreased	....	16
36	Rat	<20	Rectal temp	Decreased	Increased	16
37		<25	Heat production	Increased	Increased	16
38		....	Tissue conductance	Decreased	Increased	16
39	Sea	2 (air), 20 (water)	Heat production	Increased	....	7,17
40	lion (seal)	13 (water, in winter)	Heat production	....	Increased	7,17
41		18	Tissue conductance	Decreased	Decreased further in winter	7,18
42	Chick-	−1	Rectal temp	....	Increased	15
43	en	25	Heat production	Increased	Increased	15
44		....	Tissue conductance	Decreased	Increased	15

[1] Decreased further at 10°C.

continued

Contributor: Whittow, G. Causey

References

[1] Adams, T., et al. 1970. J. Appl. Physiol. 29:852.
[2] Bianca, W. 1968. In E. S. E. Hafez, ed. Adaptation of Domestic Animals. Lea and Febiger, Philadelphia. pp. 97-118.
[3] Blaxter, K. L., and F. W. Wainman. 1960. J. Agr. Sci. 56:81.
[4] Fox, R. H., et al. 1963. J. Physiol. (London) 166: 530.
[5] Fox, R. H., et al. 1963. Ibid. 166:548.
[6] Glaser, E. M. 1949. Ibid. 110:330.
[7] Hart, J. S., and L. Irving. 1959. Can. J. Zool. 37: 447.
[8] Hellon, R. F., et al. 1956. J. Physiol. (London) 123: 559.
[9] Hong, S. K., et al. 1969. Fed. Proc. Fed. Amer. Soc. Exp. Biol. 28:1143.

[10] Johnson, H. D. 1967. Publ. Amer. Ass. Advan. Sci. 86:189.
[11] Kang, B. S., et al. 1963. J. Appl. Physiol. 18:483.
[12] Leblanc, J. 1956. Ibid. 9:395.
[13] Matsuura, D. T., and G. C. Whittow. 1972. Fed. Proc. Fed. Amer. Soc. Exp. Biol. 31:826.
[14] Piwonka, R. W., et al. 1965. J. Appl. Physiol. 20:379.
[15] Whittow, G. C. 1965. In P. D. Sturkie, ed. Avian Physiology. Ed. 2. Comstock, Ithaca. pp. 186-238, 239-271.
[16] Whittow, G. C., ed. 1971. Comparative Physiology of Thermoregulation. Academic Press, New York. v. 2.
[17] Whittow, G. C., and D. T. Matsuura. 1972. Physiologist 15:303.
[18] Whittow, G. C., et al. 1972. Physiol. Zool. 45:68.

97. TEMPERATURE REGULATION AND WATER LOSS: MAN

Occasionally average values only were available for body weight loss, sweat loss, and for changes in rectal and oral temperatures. **Subject: Attire**—Clothed subjects were usually considered to be separated from the environment by 1 clo unit (reciprocal of thermal conductance). **Body Weight Loss:** The sum of respiratory and evaporative water loss, plus in some instances water loss due to dripping sweat (not necessarily synonymous with water loss due to effective evaporative heat loss). **Activity or Condition:** met. = metabolic; $\dot{V}_{O_2}$ = basal oxygen consumption. **Site** (of temperature measurement): R = rectal; O = oral; S = skin; B = body (mean value); U = unspecified. Data in brackets refer to the column heading in brackets. Values in parentheses are ranges, estimate "c" (see Introduction).

	Subjects		Activity or Condition	Experiments, no. [Duration]	Ambient Temp, °C Dry Bulb [Wet Bulb]	Water Intake	Body Weight Loss [Sweat Loss]	Temperature		Reference
	Attire	No. [Age]						Site	Change, °C	
1	Nude	6	Climbed 2.5% grade at 2.17 km/hr, with 10 min rest each hr	11 [4 hr]	37.8 [24.8]	0	1st hr: 0.88 kg; 4th hr: 0.76 kg	R	+0.29/hr	12
2				4 [4 hr]	37.8 [24.8]	?, plus 25 g glucose	1st hr: 0.82 kg; 4th hr: 0.74 kg	R	+0.19/hr	
3				1 [4 hr]	37.8 [24.8]	?, plus 100 g glucose	1st hr: 0.77 kg; 4th hr: 0.74 kg	R	+0.32/hr	
4			Rested	6 [2.25-5.5 hr]	43.3 [32.2]	0	....	R	+0.28(+0.15 to +0.69)/hr	10
5								O	+0.29(+0.13 to +0.59)/hr	
6		5	Climbed 2.5% grade at 2.17 km/hr, with 10 min rest each hr	6 [2 hr]	35.0 [32.4]	0	1st hr: 1.17 kg; 2nd hr: 0.98 kg	R	+0.56/hr	12

continued

	Subjects		Activity or Condition	Experiments, no. [Duration]	Ambient Temp, °C Dry Bulb [Wet Bulb]	Water Intake	Body Weight Loss [Sweat Loss]	Temperature		Reference
	Attire	No. [Age]						Site	Change, °C	
7			Rested	7 [4.25-5.5 hr]	36.0 [33.3]	0	....	R	+0.18(+0.17 to +0.18)/hr	10
8								O	+0.15/hr	
9		3 [21-39 yr]	Walked 4.8 km/hr	3 [2 hr]	(21-22)	0	0.31(0.12-0.50) kg/hr	R	+0.28(0 to +0.65)/hr	2
10								S	−0.37(−0.65 to −0.20)/hr	
11					30	0	0.94(0.66-1.12) kg/hr	R	+0.49(+0.25 to +0.65)/hr	
12								S	+0.03(−0.05 to +1.00)/hr	
13			Rested	3 [2 hr]	(21-22)	0	0.10(0.06-0.16) kg/hr	R	−0.27(−0.35 to −0.15)/hr	2
14								S	−0.17(−0.45 to −0.05)/hr	
15					30	0	0.26(0.14-0.35) kg/hr	R	+0.03(−0.15 to +0.15)/hr	
16								S	−0.05(−0.035 to +0.300)/hr	
17		2	Rested for 3 hr at 20°C, then stepped at a rate of 2000 ft·lb/min in heat; wind velocity in heat 0.25-0.4 m/s. Subjects underwent exposures at each of indicated water intakes	4 [4 hr]	33.9 [32.2]	"Ad lib"	2.2 kg[1,2]; 2.9 kg[3]	R	+1[4]	17
18						Wt loss replaced	2.21 kg[1,5]; 3.4 kg[3]	R	+1[4]	
19						3-5% deficit maintained during exposure	2.45 kg[1,5]; 2.5 kg[3,5]	R	+1.3[4]	
20						5.8% deficit maintained during exposure	2.2 kg[1,5]; 1.9 kg[3,5]	R	+1.5[4]	
21			Rested; air movement <6 m/min	1 [4 hr]	(43-44) [(28-29)]	2% of body wt at start; thereafter, intake equal to total wt loss	6(3-8) g/min[1]; 5.5(3.8-6) g/min[3]	R	+1[1,6]; +1.3[3,6]	15
22	Shorts	14	Rested	16 [8.0-12.5 hr]	(43.0-43.5) [(28.0-33.0)]	0	0.48(0.40-0.59)%/hr	O	+0.11(+0.03 to +0.16)/hr	8
23								S	+0.085(+0.03 to +0.28)/hr	
24		6 [20-24 yr]	Rested	11 [12 hr]	43 [28]	0	0.48(0.43-0.54)%/hr	O	+0.10(+0.04 to +0.24)/hr	14
25		3	Stepped off 30.5-cm stool 12 times/min for 30 min at beginning of test	12 [1.83-2.67 hr]	37.8	0	1.33(1.0-1.6) kg/hr	R	+1.25(+1.10 to +1.40)/hr	9

1/ Subject 1. 2/ Water deficits averaging 2.2% body weight resulted when water was taken "ad lib." 3/ Subject 2. 4/ Means for 2 subjects; no ranges available. 5/ Replaced during exposure. 6/ Rectal temperature rose throughout heat exposure.

continued

	Subjects		Activity or Condition	Experiments, no. [Duration]	Ambient Temp, °C Dry Bulb [Wet Bulb]	Water Intake	Body Weight Loss [Sweat Loss]	Temperature		Reference
	Attire	No. [Age]						Site	Change, °C	
26		2	Stepped off 30.5-cm stool 12 times/min for 30 min at beginning of test	[2.38 hr]	33.9 [32.8]	0	1.18(0.74-1.62) kg/hr	R	+1.14(+1.05 to +1.23)/hr	9
27 28 29	Shorts, socks, & shoes	3	Worked continuously; met. heat production, 342 kcal/m²/hr; mean work rate, 65% max $\dot{V}_{O_2}$	3 [1 hr]	22 [13.8]	Equal to wt loss	98 g/hr [368 g/m²/hr]	R S B	+1.42 −1.36 0.85	5
30 31 32			Worked continuously; met. heat production, 315 kcal/m²/hr; mean work rate, 61% max $\dot{V}_{O_2}$	3 [1 hr]	22 [13.8]	0	733 g/hr [324 g/m²/hr]	R S B	+1.5 +0.57 +1.32	5
33 34 35			Worked intermittently—30 s work, 30 s rest; met. heat production, 334 kcal/m²/hr; mean work rate, 62% max $\dot{V}_{O_2}$	3 [1 hr]	22 [13.8]	0	702 g/hr [292 g/m²/hr]	R S B	+1.91 −1.56 +1.21	5
36 37 38			Worked intermittently—30 s work, 30 s rest; met. heat production, 333 kcal/m²/hr; mean work rate, 61% max $\dot{V}_{O_2}$	3 [1 hr]	22 [13.8]	Equal to wt loss	28 g/hr [321 g/m²/hr]	R S B	+1.64 −1.47 1.02	5
39	Shorts, socks, & gym shoes	2	Rested 1 hr in gradient calorimeter, worked 1 hr at 3-7.5 kcal/min, then rested 1 hr in gradient calorimeter; wind velocity, 1.25 m/s in calorimeter, & 8 cm/s in exercise room except when air temp = 35°C, velocity = 83 cm/s	[3 hr]	Calorimeter: 30, 35, or 40; during work: 20, 25, or 35 ["Low"]	0	(0.1-1.4) kg	R	+0.5/kg water loss[1]; +0.33/kg water loss[3]	16
40	Shorts, socks, & boots	15	Walked 2.17 km/hr	15 [1.5 hr]	48.9 [2.67]	0 before work; 1.2 liters during work	668 g/m²/hr	R	+1.76[7]	11
41						2 liters before work; 1.2 liters during work	685 g/m²/hr	R	+1.65[7]	
42	Regulation army dress, plus helmet, rifle, & 23.7 kg pack	22[8]	Walked 29 km over undulating gravel road	1 [6.5 hr]	(21.7-27) [(16-17.5)]; (27-31) [(19-19.2)]; (24.3-30.6) [(17.8-18)]	2.709 liters	1.958 kg[9] [4.508 liters/man]	R	+1	18

[1] Subject 1. [3] Subject 2. [7] Total temperature change in 1.5 hr. For rectal temperature and sweat rates, experimental vs. control, probability <0.005. [8] Experiments done on 5 different days with 6 men from each group [13]. Data are only for those completing the march. No information available as to which subjects marched on which days. [9] 2.9% body weight.

continued

Subjects		Activity or Condition	Experiments, no. [Duration]	Ambient Temp, °C Dry Bulb [Wet Bulb]	Water Intake	Body Weight Loss [Sweat Loss]	Temperature		Reference
Attire	No. [Age]						Site	Change, °C	
43	Regulation army dress, plus helmet, rifle, & 24.1 kg pack	18[8/]	Walked 29 km over undulating gravel road	1 [6.5 hr]	(22.5-31.1) [(17-19.2)]; (26.5-32) [(19-19.2)]	1.175 liters	3.398 kg[10/] [4.525 liters/man]	R	+1.3	18
44	Clothed	2	Climbed 4% grade at 2.17 km/hr, with 10 min rest each hr	6 [3 hr]	32.2 [29.2]	0	1st hr: 1.48 kg; 3rd hr: 1.28 kg	R	+0.39/hr	12
45		1	Walked 6.8 km/hr	[2.55 hr]	33 [15]	0.8 kg	0.58 kg/hr	U	+0.31/hr	4
46		Unspecified	Expended 190 kcal/m²/hr	14 [6 hr]	34.5(31.9-38.0) [30.9(28.9-32.9)]	Unlimited	Initial rate: +1.52(+1.12 to +1.85) kg/hr End rate: +0.78(+0.37 to +1.06) kg/hr	...	....	7
47				12 [6 hr]	44.3(40.5-50.1) [26.3(24.1-27.4)]	Unlimited	Initial rate: +1.36(+1.06 to +1.71) kg/hr End rate: +1.04(+0.76 to +1.24) kg/hr	...	....	
48	Clothed (?)	1 [20 yr]	Rested in bed	[48 hr]	21.1 [(10.1-13.1)]	700 ml/da	0.039 kg/hr	R	+0.0058/hr	3
49				[26 hr]	35 [19]	1st da: 700 ml; 2nd da: 215 ml	0.105 kg/hr	R	+0.017/hr	
50		Unspecified	Walked in rough desert terrain at ∿1.9-2.5 km/hr	29[11/] [2.5-4.7 hr]	(32.2-37.8)	0	(0.5-6.8)%	R	+0.31/1% body wt loss	1
51			Worked in hot room	27 [12 hr]	48.9	0	(0.5-9.9)%	R	+0.2/1% body wt	1
52	Unspecified	12	Expended 1.65 kcal/m²/hr for 10 min, with 20 min rest each half hr	[4/da on 6 consecutive da]	37.8 [34.4]	0.55 ml/hr	12.25 g/kg	R	38.17 in 4th hr[12/]	6
53						0.70 ml/hr	12.90 g/kg	R	38.00 in 4th hr[12/]	
54						0.85 ml/hr	13.80 g/kg	R	37.90 in 4th hr[12/]	

[8/] Experiments done on 5 different days with 6 men from each group [13]. Data are only for those completing the march. No information available as to which subjects marched on which days. [10/] 4.8% body weight. [11/] Observations. [12/] Actual rectal temperature.

Contributor: Senay, L. C., Jr.

References

[1] Adolph, E. F. 1947. Physiology of Man in the Desert. Interscience, New York.

[2] Craig, F. N. 1952. J. Appl. Physiol. 4:826.

[3] Di Giovanni, C., Jr., and N. C. Birkhead. 1964. Aerosp. Med. 35(3):225.

continued

[4] Dill, D. B. 1938. Life, Heat and Altitude. Harvard Univ. Press, Cambridge.

[5] Ekblom, B., et al. 1970. Acta Physiol. Scand. 79:475.

[6] Ellis, F. P., et al. 1954. J. Physiol. (London) 125: 61P.

[7] Gerking, S. D., and S. Robinson. 1946. Amer. J. Physiol. 147:370.

[8] Hertzman, A. B., and I. D. Ferguson. 1959. WADC Tech. Rep. 59:398.

[9] Ladell, W. S. S. 1955. J. Physiol. (London) 127:11.

[10] Lee, D. H. K., and A. G. Mulder. 1935. Ibid. 84:279.

[11] Moroff, S. V., and D. E. Bass. 1965. J. Appl. Physiol. 20(2):267.

[12] Pitts, G. C., et al. 1944. Amer. J. Physiol. 142:253.

[13] Senay, L. C., Jr. Unpublished. St. Louis Univ. School of Medicine, St. Louis, 1971.

[14] Senay, L. C., Jr., and M. L. Christensen. 1965. J. Appl. Physiol. 20(2):278.

[15] Senay, L. C., Jr., and W. van Beaumont. 1969. Pfluegers Arch. Gesamte Physiol. Menschen Tiere 312:82.

[16] Snellen, J. W. 1966. Acta Physiol. Pharmacol. Neer. 14:99.

[17] Strydom, N. B., and L. D. Holdsworth. 1968. Int. Z. Angew. Physiol. Einschl. Arbeitsphysiol. 26:95.

[18] Strydom, N. B., et al. 1966. S. Afr. Med. J. 40:539.

98. METABOLISM AND TEMPERATURE: HIBERNATING MAMMALS AND BIRDS

Hibernation in homoiotherms embraces a number of conditions, ranging from deep torpor to lethargy—from greatly reduced to slightly lowered body temperature, heart rate, and metabolic rate. Data in brackets refer to the column heading in brackets.

	Species (Synonym)	Distribution	Temperature, °C		Heart Rate beats/min [Respiratory Rate, breaths/min]	O_2 Consumption [CO_2 Production] ml/g/hr	Reference
			Ambient	Rectal[1]			
			Mammalia				
1	Cercaertus nanus	Australia	3.2-4.6	....	28	....	5
2			9.0	10.8	....	0.05	5
3			18.0	20.1	....	0.24	5
4	Cricetus cricetus	Europe	6.5	....	....	0.032	26
5	Erinaceus europaeus	Europe	....	5.5	2-6 [1-3]	....	2
6			−20 to −6	3.7	6 [3]	....	2
7			2-3	6.2-7.7	18-24	0.014-0.033	40,41
8			4	8.0-32.0	....	....	2
9			6	....	....	0.028	26
10	Glis glis	Europe	....	....	....	0.029	28
11			6	....	....	0.017	24
12			8.7	....	....	0.021	24
13			11.3	....	....	0.024	24
14	Marmosa microtarsus	South America	11	17	....	0.20	43
15	Marmota marmota	Europe	....	....	....	0.02	27
16			10	10.5	[0.35]	0.018 [0.012][2]	24
17	M. monax	Eastern North America	....	4-7	4-5	0.008-0.034	9
18			....	36.5-37.2	80-95	0.32-0.975	33,34
19	Mesocricetus auratus	Middle East	5	5-6	14-15	0.183 [0.132[3]][2]	32
20			5.5	5.5[4]	....	0.032	32
21	Microdipodops pallidus	California; Nevada	7-9	....	16.1-18.2	....	6
22	Muscardinus avellanarius	Europe	6	....	[9-10]	....	40
23			10.0	....	....	0.040	42
24			11.6	....	10-12	....	42

[1] Unless otherwise indicated. [2] Respiratory quotient may be derived from the formula CO_2/O_2. [3] Calculated. [4] Oral temperature.

continued

	Species (Synonym)	Distribution	Temperature, °C		Heart Rate beats/min [Respiratory Rate, breaths/min]	O$_2$ Consumption [CO$_2$ Production] ml/g/hr	Reference
			Ambient	Rectal[1/]			
25	*Myotis lucifugus*	North America	0.5	....	....	0.113	19
26			1.3	1.35	....	....	19
27			2.0	....	....	0.022-0.034	19
28			10	....	....	0.071	19
29			23	23.2[5/]	....	0.45	19
30	*M. myotis*	Europe	1.7	....	....	0.020 [0.009][2/]	16
31			2.5	....	....	0.051 [0.033][2/]	16
32			4-6	<10	<100	<5	35
33	*Nyctalus noctula*	Europe	4.3	....	....	0.030	25
34			20	....	....	0.403[3/] [0.314][2/]	23
35			30	....	....	0.682[3/] [0.484][2/]	23
36	*Perognathus hispidus*	Western United States	5	22.7	....	1.50	36
37			11.0-16.8	12.0-20.3	25-60 [12]	0.15	45
38	*P. longimembris*	Western North America	3	5	....	....	4
39	*Pipistrellus pipistrellus*	Europe	5	....	....	0.053 [0.038][2/]	25
40	*Plecotus auritus*	Europe	5	6.5	....	0.069 [0.049[3/]][2/]	23
41			10	10.7	....	0.094 [0.079[3/]][2/]	23
42			19.7	....	....	0.255	23
43	*Setifer setosus*	Madagascar	15	16.5	....	0.072	18
44	*Sicista betulina*	Northern Europe	4.0	7.2	....	....	43
45			16.3	18.3	....	....	43
46	*Spermophilus citellus* (*Citellus citellus*)	Northern Europe	7	7.2	....	0.015	23
47	*S. franklini (C. franklini)*	Central United States	....	3.8-4.0	2-4	....	12
48	*S. lateralis (C. lateralis)*	Western United States	−2 to +5	0.5-1.0	....	....	44
49			5.4	5.5-15.0[6/]	....	1.44	15
50			6	6.5-8.0	....	....	11
51			19-23	20-26	....	0.96-1.09[7/]	10
52	*S. mohavensis (C. mohavensis)*	Mohave Desert, California	21	21.2	....	0.10	3
53	*S. tereticaudus (C. tereticaudus)*	Western United States	....	34	....	0.70	22
54			23	34.8	....	1.20	22
55			24	25	....	....	22
56	*S. tridecemlineatus (C. tridecemlineatus)*	Central United States	....	3-10[4/]	5-20 [14-15]	0.081-0.191	1
57			5.0	5.5	3-15 [1-3]	....	30
58			6	6.5-8.0	....	....	11
59	*S. undulatus (C. undulatus)*	Alaska; Siberia	....	0.5-9.0	2.7	....	12
60			....	38.0	180	....	38
61			0	2.0	4	....	20
62			2	4.5	68 [10]	....	20
63			5.9	5.9	[6]	0.063	20
64	*Tadarida brasiliensis*	Western North America	10	....	....	0.061	17
65			20	....	....	0.175	17
66	*Tamias striatus*	Eastern United States; southeastern Canada	−7 to +32	36.0-40.3	165	1.03	46
67			11.2-23.5	20.3	23	0.06-0.74	46
68	*Tenrec ecaudatus (Centetes ecaudatus)*	Madagascar	15	15.5	....	0.043	18
69	*Thalarctos maritimus*	Arctic regions	−50 to −20	....	27	....	14

[1/] Unless otherwise indicated. [2/] Respiratory quotient may be derived from the formula CO_2/O_2. [3/] Calculated. [4/] Oral temperature. [5/] Subcutaneous temperature. [6/] Hypothalamic temperature. [7/] Respiratory quotient.

continued

Species (Synonym)	Distribution	Temperature, °C		Heart Rate beats/min [Respiratory Rate, breaths/min]	O$_2$ Consumption [CO$_2$ Production] ml/g/hr	Reference
		Ambient	Rectal[1]			
70 *Ursus americanus*[8]	North America	....	31.2	....	0.05	21
71		....	....	8-12	....	13
72		−16.5	33.0	....	....	20
73 *Zapus hudsonius*	Northern North America	....	5.0	....	0.040	37
Aves						
74 *Aeronautes saxatilis*	Western North America	5	18	....	....	8
75 *Apus apus*	Europe; Asia; Africa	19	23[9]	[8-10]	0.7 [0.31][2]	29
76 *Calypte anna*	North America	8.2	8.8	[50]	....	8
77		24	....	....	0.84	39
78 *Chordeiles minor*	Western North America	16	18	45	0.200	31
79		19.5	19.5	56	....	31
80 *Colius striatus*	Southern Africa	20	22.8	....	0.2	7
81 *Phalaenoptilus nuttallii*	Western United States	4.8	4.8	18	0.06	8
82		17	18	....	....	8
83 *Selasphorus sasin*	California	22	....	....	1.24	39

[1] Unless otherwise indicated. [2] Respiratory quotient may be derived from the formula CO$_2$/O$_2$. [8] Not a true hibernator, as indicated by the difference between ambient and rectal temperatures. [9] Proventricular temperature, taken orally.

Contributor: Simmonds, Richard C.

References

[1] Baldwin, F. M., and K. L. Johnson. 1941. J. Mammal. 22:180.

[2] Bartels, H., et al. 1969. Resp. Physiol. 7:278.

[3] Bartholomew, G. A. 1960. Bull. Harvard Mus. Comp. Zool. 124:193.

[4] Bartholomew, G. A., and T. Cade. 1957. J. Mammal. 38:60.

[5] Bartholomew, G. A., and J. W. Hudson. 1962. Physiol. Zool. 35:94.

[6] Bartholomew, G. A., and R. E. MacMillen. 1961. Ibid. 34:177.

[7] Bartholomew, G. A., and C. H. Trost. 1970. Condor 72:141.

[8] Bartholomew, G. A., et al. 1957. Ibid. 59:145.

[9] Benedict, F. G., and R. C. Lee. 1938. Carnegie Inst. Wash. Publ. 457.

[10] Bintz, G. L., et al. 1971. Comp. Biochem. Physiol. 38A:121.

[11] Burlington, R. F., and B. K. Whitten. 1971. Ibid. 38A:469.

[12] Dawe, A. R., and P. R. Morrison. 1955. Amer. Heart J. 49:367.

[13] Folk, G. E., Jr., et al. 1966. Amer. Zool. 6:583.

[14] Folk, G. E., Jr., et al. 1970. Arctic 23:130.

[15] Hammel, H. T., et al. 1968. Physiol. Zool. 41:341.

[16] Hari, P. 1909. Pfluegers Arch. Gesamte Physiol. Menschen Tiere 130:112.

[17] Herreid, C. F. 1963. J. Cell. Comp. Physiol. 61:201.

[18] Hildwein, G. 1964. C. R. Acad. Sci. 259:2009.

[19] Hock, R. J. 1951. Biol. Bull. 101:289.

[20] Hock, R. J. 1958. Cold Inj. Trans. Conf., 5th, p. 61.

[21] Hock, R. J. 1960. Bull. Harvard Mus. Comp. Zool. 124:153.

[22] Hudson, J. W. 1964. Ann. Acad. Sci. Fenn., A4, 71:219.

[23] Kayser, C. 1939. Ann. Physiol. Physicochim. Biol. 15:1087.

[24] Kayser, C. 1940. Ibid. 16:127.

[25] Kayser, C. 1950. Arch. Sci. Physiol. 4:361.

[26] Kayser, C. 1959. C. R. Soc. Biol. 153:167.

[27] Kayser, C., et al. 1967. Ibid. 161:918.

[28] Kayser, C., et al. 1969. Ibid. 163:212.

[29] Koskimies, J. 1950. Ann. Acad. Sci. Fenn., A4, 15:1.

[30] Landau, B. R., and A. R. Dawe. 1958. Amer. J. Physiol. 194:75.

[31] Lasiewski, R. C., and W. R. Dawson. 1964. Condor 66:477.

[32] Lyman, C. P. 1948. J. Exp. Zool. 109:55.

[33] Lyman, C. P. 1951. Amer. J. Physiol. 167:638.

[34] Lyman, C. P. 1958. Ibid. 194:83.

[35] Mejsnar, J., and L. Jansky. 1970. Can. J. Physiol. Pharmacol. 48:102.

[36] Morrison, P. R., and F. A. Ryser. 1962. J. Mammal. 43:529.

[37] Morrison, P. R., and F. A. Ryser. 1962. J. Cell. Comp. Physiol. 60:169.

[38] Nardone, R. M. 1955. Amer. J. Physiol. 182:364.

[39] Pearson, O. P. 1950. Condor 52:145.

[40] Saissy, J. A. 1811. Mem. Accad. Sci. Turin (2):1.

[41] Sarajas, H. S. S. 1954. Acta Physiol. Scand. 32:28.

[42] Schenk, P. 1922. Pfluegers Arch. Gesamte Physiol. Menschen Tiere 197:66.

continued

[43] Suomalainen, P. 1947. Arch. Soc. Zool. Bot. Fenn. Vanamo 2:33.

[44] Twente, J. W., and J. A. Twente. 1970. Novosibirsk (USSR), p. 181.

[45] Wang, L. C.-H., and J. W. Hudson. 1970. Comp. Biochem. Physiol. 32A:275.

[46] Wang, L. C.-H., and J. W. Hudson. 1971. Ibid. 38A: 59.

99. LIFE SPAN AND TEMPERATURE: INSECTS

Life Span: Values are means in terms of days, unless otherwise indicated. Data in brackets refer to the column heading in brackets.

	Species (Synonym) [Strain]	Sex	Temp °C	Life Span da	Reference		Species (Synonym) [Strain]	Sex	Temp °C	Life Span da	Reference
1	Aphelinus semiflavus	♂	32.2	4.9	4	19			34	187.8[3/]	
2		♀	15.6	70.4		20			36	36.7[3/]	
3			21.1	46.0		21	Musca domestica [NAIDM]	♂	27	16.9	7
4			26.7	19.9		22		♀	27	28.7	
5			29.4	16.0		23	Nasonia vitripennis (Mor-	♀	18	45.8	1
6	Bracon brevicornis (Habro-	♀	22.0	61.0[1/]	2	24	moniella vitripennis)		25	39.8	
7	bracon serinopae) [Wild]		30.0	36.0[1/]		25	[Wild]		30	35.4	
8	B. juglandis (H. juglandis)	♀	22	45[1/]	2	26	Praon palitans [Wild]	♂	26.7	9.3	4
9	[Wild]		30	28[1/]		27		♀	18.3	20.4	
10	Calliphora erythrocephala	♂	24	49.9-52.8[2/]	3	28			21.1	18.2	
11	[Wild]	♀	24	46.8-55.3[2/]		29			23.9	10.6	
12	Drosophila melanogaster	♂♀	10	120.5	6	30	Tribolium castaneum	♂	30	36.3	8
13			15	92.4		31	[Wild]	♀	30	26.4	
14			20	40.2		32	Trioxys utilis [Wild]	♂	32.2	2.6	4
15			25	28.5		33		♀	15.6	14.1	
16			30	13.6		34			21.1	14.3	
17	D. subobscura [K/B]	♂	18.5	98.6	5	35			26.7	7.9	
18			24.8	47.6		36			29.4	6.5	

[1/] Median, not mean. [2/] Mean for two different wild strains. [3/] Minutes, not days.

Contributors: Samis, H. V., and Baird, M. B.; Lints, F. A.

References

[1] Clark, A. M., and R. N. Kidwell. 1967. Exp. Gerontol. 2:79.

[2] Clark, A. M., and R. F. Smith. 1967. Ibid. 2:217.

[3] Ducoff, H. S., et al. 1970. Ibid. 5:233.

[4] Force, D. C., and P. S. Messenger. 1964. Ann. Entomol. Soc. Amer. 57:406.

[5] Hollingsworth, M. J. 1969. Exp. Gerontol. 4:49.

[6] Loeb, J., and J. H. Northrop. 1917. J. Biol. Chem. 32:103.

[7] Rockstein, M., and H. M. Lieberman. 1959. Gerontologia 3:23.

[8] Tribe, M. A. 1967. Exp. Gerontol. 2:183.

100. DIAPAUSE: MITES AND INSECTS

Diapause in mites and insects is a state of dormancy which in many species is induced by exposure to a suitable photoperiodic regimen. The diapause response can be modified by other environmental factors, such as temperature, relative humidity, and diet. **Dormant Stage:** I = instar; FI = final instar; FD = fully developed; S = small; HG = half-grown.

	Species	Dormant Stage	Diapause Duration	Reference		Species	Dormant Stage	Diapause Duration	Reference
	Arachnida—Acarina					Insecta			
1	Panonychus ulmi	Egg	....	6,25	3	Aedes hexodontus	Embryo, FD	Throughout winter	5
2	Tetranychus urticae	Adult	....	25					

continued

	Species	Dormant Stage	Diapause Duration	Reference		Species	Dormant Stage	Diapause Duration	Reference
4	*A. triseriatus*	Embryo, FD	Up to 6 mo	3	22	*Lestes sponsa*	Egg	15 wk	14
5	*Anax imperator*	Larva, FI	2-3 mo	13	23	*Locusta migratoria gallica*	Egg	14-156 da	24
6	*Antheraea pernyi*	Pupa	3-8 mo	38	24	*Malacosoma disstria*	Embryo, FD	6-9 mo	21
7	*Anthonomus grandis*	Adult	3-5 mo	18	25	*Melanoplus bivittatus*	Embryo, FD	Few weeks to many months	35
8	*Anthrenus verbasci*	Larva, S-FI	Many months	7					
9	*Apanteles glomeratus*	Prepupa	....	20	26	*M. differentialis*	Egg	Few weeks to many months	8,9
10	*Bombyx mori*	Embryo, S or HG	5-9 mo	23					
11	*Cephus cinctus*	Prepupa	6-8 mo	34	27	*Melittobia chalybii*	Larva, FI	Several months	37
12	*Ceratophyllus fasciatus*	Larva, FI	2-12 mo	2	28	*Nasonia vitripennis* [2]	Prepupa	Winter	36
					29	*Ostrinia nubilalis*	Larva, FI	6-9 mo	1
13	*Drosophila deflexa*	Larva, FI	Several months	4	30	*Pectinophora gossypiella*	Larva	Winter	28
14	*Dytiscus marginalis*	Adult [1]	....	22					
15	*Ephestia elutella*	Larva	8-9 mo	39	31	*Phaenicia sericata*	Larva, FI	Several weeks to many months	15,31
16	*Epilachna varivestis*	Adult	Throughout winter	17					
17	*Eurydema ornatum*	Adult	....	10	32	*Pieris rapae*	Pupa	6-52 wk	30
18	*Eurygaster integriceps*	Adult	....	19	33	*Popillia japonica*	Larva, 3rd I	50 da	26,27
					34	*Reduvius personatus*	Larva	Few weeks to several months	33
19	*Heliothis virescens*	Pupa	Winter, 3-8 mo	32					
20	*H. zea*	Pupa	Winter, 3-8 mo	32	35	*Samia cynthia*	Pupa	....	16
21	*Leptinotarsa decemlineata*	Adult [1]	Many months	11,12	36	*Trichogramma cacoeciae*	Larva, FI	6-8 mo	29

[1] Reproductive diapause involving corpus allatum. [2] Synonym: *Mormoniella vitripennis*.

Contributor: Hayes, D. K.

References

[1] Arbuthnott, K. D. 1949. U.S. Dep. Agr. Tech. Bull. 869.

[2] Bacot, A. 1914. J. Hyg., Plague Suppl. 3:447.

[3] Baker, F. C. 1935. Can. Entomol. 67:149.

[4] Basden, E. B. 1954. Proc. Roy. Entomol. Soc. London A29:114.

[5] Beckel, W. E. 1958. Can. J. Zool. 36:541.

[6] Blair, C. A., and J. R. Groves. 1952. J. Hort. Sci. 27:14.

[7] Blake, G. M. 1958. Bull. Entomol. Res. 49:751.

[8] Bodine, J. H. 1929. Physiol. Zool. 2:459.

[9] Bodine, J. H., et al. 1939. J. Cell. Comp. Physiol. 14:173.

[10] Bonnemaison, L. 1948. C. R. Acad. Sci. 227:985.

[11] Breitenbrecher, J. K. 1918. Carnegie Inst. Wash. Publ. 263:341.

[12] Busnel, R. G., and A. Drilhon. 1937. C. R. Soc. Biol. 124:916.

[13] Corbet, P. S. 1954. Ph.D. Thesis. Cambridge Univ., England.

[14] Corbet, P. S. 1956. Proc. Roy. Entomol. Soc. London A31:45.

[15] Cragg, J. B., and P. Cole. 1952. J. Exp. Biol. 29:600.

[16] Danilevskii, A. S. 1939. Zool. Zh. 18:1926.

[17] Douglass, J. R. 1928. J. Econ. Entomol. 21:203.

[18] Earle, N. W., et al. 1959. Ibid. 52:710.

[19] Fedetov, D. M. 1944. C. R. Acad. Sci. URSS 42:408.

[20] Geispits, K. F., and I. I. Kyao. 1953. Entomol. Obozr. 33:32.

[21] Hodson, A. C., and C. J. Weinman. 1945. Minn. Agr. Exp. Sta. Tech. Bull. 170.

[22] Joly, P. 1945. Arch. Zool. Exp. Gen. 84:49.

[23] Kogure, M. 1933. J. Dep. Agr. Kyushu Imp. Univ. Fukuoka Jap. 4:1.

[24] LeBerre, J. R. 1951. Rev. Zool. Agr. Appl. 10-12:1.

[25] Lees, A. D. 1953. Ann. Appl. Biol. 40:449.

[26] Ludwig, D. 1932. Physiol. Zool. 5:431.

[27] Ludwig, D. 1953. J. Gen. Physiol. 36:751.

[28] Lukefahr, M. J. 1964. U.S. Dep. Agr. Tech. Bull. 1304.

[29] Marchal, P. 1936. Ann. Epiphyt. Phytogenet. 2:447.

[30] Mazaki, S. 1955. Jap. J. Appl. Zool. 20:98.

[31] Melanby, K. 1938. Parasitology 30:392.

[32] Phillips, J. R., and L. D. Newsom. 1966. Ann. Entomol. Soc. Amer. 59:154.

[33] Readio, P. A. 1931. Ibid. 24:19.

continued

[34] Salt, R. W. 1947. Can. J. Res. D25:66.
[35] Salt, R. W. 1949. Ibid. D27:236.
[36] Saunders, D. S. 1967. Science 156:1126.
[37] Schmieder, R. G. 1933. Biol. Bull. 65:338.

[38] Tanaka, Y. 1950. Nippon Sanshigaku Zasshi 19: 358, 429.
[39] Waloff, N. 1949. Trans. Roy. Entomol. Soc. London 100:147.

101. DISPERSION OF SMALL ORGANISMS

Numerical Unit: Values are means.

Part I. Invertebrates

Dispersion by flight may possibly be aided by air movements.

	Species[1] (Synonym) [Means of Dispersion]	Specification	Numerical Unit					Reference
		Horizontal Dispersion						
1	*Aedes* spp. [Flight]	Miles from original case	17.5	32.5	47.5	62.5	92.5	94
2		Yellow fever cases	14.6	4.6	1.7	0.8	0.6	
3	*A. aegypti* [Flight]	Meters from release point	250	750	1250	1750	2250	101
4		Eggs in traps	1100	46	40	5	2	
5	*A. albopictus* [Flight]	Yards from release point	25	125	225	325	475	5
6		Mosquitoes recovered	92	43	2	1	2	
7	*A. communis* [Flight]	Feet from release point (subarctic)	75	300	1000	2000	5000	45
8		Mosquitoes recovered	80	24	6	1	1	
9	*A. leucocelaenus* & *Haemagogus spegazzinii* [Flight]	Kilometers from release point	1.0	2.0	2.4	2.7	4.0	12
10		Mosquitoes recovered/100 man-hr	7	1	0.5	0.2	0.1	
11	*A. polynesiensis* [Flight]	Yards from release point	0	50	100	150	.....	43
12		Mosquitoes recovered	37	8	3	0	.....	
13	*Agriotes obscurus* [Walking, running]	Feet from release point	3	9	18	36	.....	82
14		Beetles/trap	12	3	0.6	0.1	.....	
15	*Anastatus bifasciatus* [Flight]	Yards from release point	100	900	1500	3900	5100	16
16		Eggs parasitized, % (north)	23	12	9	4	3	
17		Eggs parasitized, % (south)	32	20	18	13	12	
18	*Anopheles funestus* [Flight]	Yards from riverbank	100	600	1200	2400	.....	20
19		Mosquitoes recovered	179	14	5	4	.....	
20	*A. gambiae* [Flight]	Miles from release point	0.75	1.25	1.75	2.25	.....	20
21		Mosquitoes recovered	3.0	2.2	1.6	1.2	.....	
22	*A. pseudopunctipennis* [Flight]	Miles from release point	0-1	1-2	2-3	3-4	.....	80
23		Mosquitoes recovered	16	10	8	6	.....	
24	*A. quadrimaculatus* [Flight]	Feet from source	500	1500	2500	3500	6500	84
25		Mosquitoes recovered	71	43	30	22	6	
26		Feet from mosquito source	500	1500	3500	6000	.....	84
27		Malarial infections, %	36	22	11	4	.....	
28		Miles from breeding-site reservoir	0.5	1.5	2.5	3.5	4.5	27
29		Females recovered/wk	542	79	5	3	1	
30	*Anthonomus grandis* [Flight]	Yards from overwinter area	220	440	880	1320	1760	33
31		Weevils trapped	59	43	26	17	9	
32	*A. grandis* [Crawling]	Feet from release point	25	75	125	175	225	78
33		Females recovered	17	8	3	0	0	
34		Males recovered	28	6	2	1	0	

[1] Unless otherwise indicated.

continued

Part I. Invertebrates

Species (Synonym) [Means of Dispersion]	Specification	Numerical Unit					Reference
35 Apis mellifera [Flight]	Yards from apiary	5.5	11	16.5	22	33	26
36	Honeybees found	6.1	3.9	2.6	1.6	0.3	
37	Yards from apiary	200	500	833	1126	1426	58
38	Pollination/2 yd²	8	7	7	6	6	
39	Feet from apiary	130	630	1130	1730	2330	6
40	Honeybees/2 yd²	21	13	9	7	5	
41	Yield of red clover seed, lb/2 yd²	65	34	22	14	8	
42	Colonies/acre	0	1	3	5	.....	67
43	Bees/yd²	0	0.24	0.30	0.90	.....	
44	Freshly tripped flowers	57	74	81	116	.....	
45	Average no. seeds/acre	35	45	72	117	.....	
46	Pounds honey/colony	....	103	92	73	.....	
47 Bruchus pisorum [Flight]	Feet from field margin	300	600	1400	.....	.....	53
48	Weevils found	51	50	48	.....		
49	Miles from overwinter area	1	2	3	4	5	93
50	Weevils found	13	8	5	3	2	
51 Calendra maidis [Crawling]	Rows from field margin	3	9	33	50	.....	11
52	Beetles found	19	11	3	1	.....	
53 Camnula pellucida & Melanoplus	Yards from release point	10	50	90	160	.....	81
54 sanguinipes (M. mexicanus) [Crawling2/]	Grasshoppers recovered	179	85	51	18	.....	
55 Carpocapsa pomonella [Flight]	Feet from release point	75	189	264	332	.....	90
56	Moths recovered, %	57	25	13	5	.....	
57 Catocala spp. [Flight]	Feet from release point	13	88	163	238	363	7
58	Moths recovered	8.1	3.7	2.4	1.5	0.6	
59 Chalcodermus aeneus [Unknown]	Feet from field margin	1.0	11.5	37.5	105.0	138.5	4
60	Curculios found	1.49	0.71	0.40	0.14	0.07	
61 Chionaspis furfura [Crawling]	Inches from dispersal point	1.5	7.5	13.5	19.5	22.5	40
62	Mature scales recovered	44	7	3	0.7	0.1	
63 Circulifer tenellus [Flight]	Miles from breeding area	15	35	52	105	215	31
64	Leafhoppers recovered	499	151	76	15	4	
65 Cochliomyia hominivorax [Crawling]	Inches from carcass	3	9	15	21	.....	89
66	Larvae/ft²	320	47	4	2	.....	
67 C. macellaria [Flight]	Miles from release point (rural)	0.5	2.5	4.5	6.5	9.0	73
68	Flies recovered	2.2	0.2	0.1	0.1	0.1	
69 C. macellaria3/ [Flight]	Miles from release point	0.25	1	2	3	.....	72
70	Traps containing flies, %	70	56	42	34	.....	
71 Conotrachelus nenuphar [Flight]	Feet from release point	50	136	235	335	478	86
72	Curculios recovered	48	13	5	2	1	
73 Cryptolestes pusillus (Laemophloeus	Feet from probable source	50	100	200	400	.....	24
74 minitum) [Flight]	Beetles recovered	11	8	6	3	.....	
75 Culex pipiens quinquefasciatus (C.	Miles from release point	0.2	0.5	0.75	1.0	2.5	77
76 quinquefasciatus) [Flight]	Mosquitoes recovered	47	3	2	5	1	
77 C. tarsalis [Flight]	Miles from release point	0.25	0.75	1.25	1.75	2.5	77
78	Mosquitoes recovered	13	12	2	0	1	
79 Culicoides grahamii [Flight]	Yards from edge of breeding site	5	25	50	100	150	65
80	Punkies caught biting	20	12	9	6	4	
81 C. impunctatus [Flight]	Yards from breeding center	0	28	57	80	89	49,50
82	Males recovered	1163	311	130	43	16	
83	Females recovered	832	258	135	77	58	

2/ Some flight by adults. 3/ Yellow-eyed mutant strain.

continued

Part I. Invertebrates

	Species (Synonym) [Means of Dispersion]	Specification	Numerical Unit					Reference
84	*Cylas formicarius elegantulus*	Yards from source	440	880	940	1320	1760	14
85	[Flight]	Sweet potato plants infested, %	42	32	31	26	21	
86	*Daphnia pulex* [Crawling]	Inches from source	0.5	2.5	4.5	6.5	8.5	8
87		Fleas found	21	9	5	2	0	
88	*Dendroctonus ponderosae (D. monti-*	Yards from road	2	5	10	15	.....	3
89	*colae)* [Flight]	Trees killed, %	63	26	10	1	.....	
90	*D. valens* [Flight]	Feet from source logs	50	200	400	.....	.....	37
91		Infestation	63	27	9	.....	.....	
92	*Diabrotica duodecimpunctata*	Feet from field margin	11.5	37.5	105.0	138.0	273.0	4
93	[Flight]	Beetles found	0.32	0.27	0.22	0.20	0.17	
94	*D. vittata* [Flight]	Miles from release point	0.25	0.75	1.25	1.75	.....	25
95		Beetles recovered	31	11	4	1	.....	
96	*Dissosteira longipennis & Melanoplus*	Miles from release point	25	75	125	175	225	96
97	*sanguinipes (M. mexicanus)* [Flight]	Grasshoppers recovered	8	6	6	5	5	
98	*Drosophila funebris* [Flight]	Meters from release point	5	25	45	65	75	88
99		Flies recovered	17.1	1.7	0.5	0	0.1	
100	*D. melanogaster* [Flight]	Meters from release point	5	15	25	35	45	88
101		Flies recovered	12.7	5.7	2.1	0.7	0.3	
102	*D. repleta* [Flight]	Feet from release point	25	150	250	350	450	70
103		Flies recovered	30	6	2	1	0	
104	*Empoasca fabae* [Flight]	Miles from nearest land	3	6	9	10	.....	85
105		Leafhoppers recovered	38	27	21	20	.....	
106	*Epitrix cucumeris* [Flight]	Feet from field margin	100	200	300	400	500	97
107		Injuries/tuber	9	8	7	7	6	
108	*Glossina* sp. [Flight]	Miles from release point	0	0.67	1.33	2.67	3.33	44
109		Males recovered	593	208	136	51	3	
110		Yards from thicket to host animals	25	50	100	125	.....	61
111		Flies recovered during dry season	52	13	7	2	.....	
112		Yards from thicket to host animals	80	180	230	280	.....	61
113		Flies recovered during wet season	35	14	8	3	.....	
114		Miles from fly belt	0.13	0.63	1.3	3.0	4.0	62
115		Compounds ruined, %	83	40	29	17	13	
116	*G. morsitans* [Flight]	Yards (following man)	50	1000	2000	4000	6000	87
117		Flies found	30	13	9	5	3	
118	*Grapholitha molesta* [Flight]	Feet from orchard	100	1320	2640	.....	.....	30
119		Moths recovered	194	14	8	.....	.....	
120	*Haematobia irritans* [Flight]	Yards from release point	0	100	200	300	400	41
121		Flies recovered	2087	497	290	142	64	
122	*Harmolita grandis* [Crawling]	Feet from wheat stubble	10	20	30	40	60	52
123		Infestation, %	4.2	3.0	2.1	1.4	0.5	
124	*H. grandis* [Flight]	Feet from wheat stubble	1	50	100	150	.....	52
125		Infestation, %	52	19	13	10	.....	
126	*H. tritici* [Flight]	Yards from wheat stubble	58	174	290	450	.....	13
127		Adults recovered	18	9	4	0	.....	
128	*Heliothis zea (H. armigera)* [Flight]	Rows from light traps (convergence)	1	3	5	10	.....	10
129		Plants infested, %	32.4	31.9	31.6	31.3	.....	
130	*Hippodamia convergens* [Flight]	Miles from release point	0.5	1.5	2.5	3.5	5.5	18
131		Beetles recovered	1.1	0.6	0.4	0.3	0.1	
132	*Hylurgopinus rufipes* [Flight]	Feet from source	60	120	320	579	816	99
133		Beetles/ft^2	14	9	5	2	1	

continued

Part I. Invertebrates

	Species (Synonym) [Means of Dispersion]	Specification	Numerical Unit					Reference
134	*Liriomyza pusilla* [Flight]	Feet from field margins	20	100	140	180	260	98
135		Mines/leaf	76	64	50	31	17	
136		Feet from field margins	9	100	300	400	600	98
137		Potato yield, bushels/acre	171	188	202	206	211	
138	*Littorina* sp. [Crawling]	Inches from release point	0.5	1.5	3.5	4.5	5.5	8
139		Periwinkles found	20	9.5	3.0	2.0	1.0	
140	*Lydella thompsoni (L. stabulans gri-sescens)* [Flight]	Miles from release point	0.5	1.5	2.5	4.0	6.0	2
141		European corn borers parasitized, %	27	18	13	9	5	
142	*L. thompsoni (L. stabulans grises-cens)* [Flight⁴/]	Miles from colonization point	0.5	2	4	6	.....	57
143		European corn borer larvae parasitized, %	25	16	12	9	.....	
144	*Macrosteles divisus* [Flight, crawling]	Feet from release point	50	100	150	200	.....	56
145		Leafhoppers recovered	9	3	1	1	.....	
146		Feet from release point	30	225	450	.....	.....	56
147		Days to first leafhopper recovery	4	13	16	.....	.....	
148	*M. divisus* [Flight]	Miles from nearest land	3	6	9	10	.....	85
149		Leafhoppers recovered	516	145	37	18	.....	
150	*Mayetiola destructor (Phytophaga destructor)* [Flight]	Feet from wheat field	100	400	600	.....	.....	60
151		Flies recovered	95	16	12	.....	.....	
152		Feet from hibernation	7.5	27.5	47.5	97.5	.....	48
153		Plants infested, %	44	16	9	3	.....	
154	*Melanoplus* spp.⁵/ [Crawling]	Feet from release point	100	200	300	400	.....	63
155		Grasshoppers on bare ground	17	10	6	3	.....	
156	*Merodon equestris* [Flight]	Feet from old planting	7	85	150	200	300	23
157		Plants infested, %	42	21	16	13	10	
158	*Musca domestica* [Flight]	Yards from release point	35	110	220	330	.....	100
159		Flies recovered	572	189	49	28	.....	
160		Miles from release point	0.43	1.5	2.5	3.5	4.5	74
161		Tagged flies/trap	106	12	3	2	0.3	
162	*Myzus persicae* [Flight]	Rows from field margin	10	20	40	80	160	51
163		Aphids/100 late potato leaves	3076	1700	1009	661	484	
164	*Ostrinia nubilalis (Pyrausta nubilalis)* [Flight]	Rows from light trap (convergence)	1	2	3	4	5	39
165		Plants infested, %	85	62	51	41	36	
166	*O. nubilalis (P. nubilalis)* [Crawling]	Inches from source	40	56	80	89	103	64
167		Larvae found	56	34	13	7	0	
168	*Pectinophora gossypiella* [Flight]	Feet from moth source	1000	2750	3500	6250	.....	66
169		Worms/boll	1.33	1.22	1.20	1.16	.....	
170	*Phaenicia* sp. [Flight]	Miles from release point	0.25	0.5	0.75	1	1.25	83
171		Flies recovered	31	11	5	2	0.5	
172	*P. sericata* [Flight]	Miles from release point	0.5	1	2	4	6	55
173		Flies recovered, %	1.8	1.3	0.3	0.1	0	
174	*Phormia regina* [Flight]	Miles from release point	0.8	1.4	2.4	3.4	4.4	83
175		Flies recovered	1.8	1.3	0.8	0.5	0.3	
176		Miles from release point	0.25	0.5	0.75	1.13	.....	83
177		Flies recovered	0.12	0.02	0.03	0	.....	
178		Miles from release point	0.5	2	4	7	12	55
179		Flies recovered, %	9.1	0.8	0.1	0	0	
180	*Popillia japonica* [Crawling]	Inches from release point	22	42	62	84	.....	38
181		Larvae recovered	79	44	23	7	.....	

⁴/ Except for passive transportation by other insects. ⁵/ *See also* lines 53, 54, 96, and 97.

continued

Part I. Invertebrates

	Species (Synonym) [Means of Dispersion]	Specification	Numerical Unit					Reference
182	*P. japonica* [Flight]	Feet from field margin	0	38.5	73.5	118.5	.....	102
183		Damage to corn, %	3.4	0.8	0.4	0	.....	
184	*Porthetria dispar* [Flight]	Feet from source	330	660	1320	2640	.....	15
185		Males recovered	67	22	5	3	.....	
186	*P. dispar* [Carried by wind]	Feet from source	50	150	250	350	600	9
187		Larvae found	16	4.4	1.7	1.3	0.5	
188	*Psila rosae* [Flight]	Yards inside field	1	5	15	30	50	103
189		Larval mines/100 carrots	78	48	28	15	6	
190	*P. rosae* [Burrowing in soil]	Yards from plants	15	25	35	45	.....	69
191		Pupae found	94	42	15	0	.....	
192	*Psorophora* sp. [Flight]	Miles from release point	0.5	2.5	4.5	6.5	.....	75
193		Mosquitoes recovered	8	0.6	0.2	0.1	.....	
194	*P. confinnis* [Flight]	Miles from release point	0.5	2.5	4.5	8.5	11.5	42
195		Mosquitoes/trap	3.7	1.8	1.1	0.4	0	
196	*Rhabdoscelus obscurus (Rhabdocnemis obscura)* [Flight, crawling]	Feet from release point	265	405	600	740	825	91
197		Weevils recovered	26	16	8	4	1	
198		Days to recovery	52	60	70	76	78	
199	*Rhagoletis pomonella* [Flight]	Feet from release point	0	25	50	75	100	59
200		Flies recovered	21.4	9.1	18.5	7.1	6.5	
201		Feet from release point	125	150	175	200	225	59
202		Flies recovered	7.5	15.3	7.3	5.7	7.3	
203		Feet from release point	250	....	.....	.....	.....	59
204		Flies recovered	6.4	....	.....	.....	.....	
205	*Saissetia oleae* [Air currents]	Feet from source	13	35	100	250	450	76
206		Scales recovered	564	433	293	172	92	
207	*Scolytus multistriatus* [Flight]	Feet from source	50	100	200	400	1000	92
208		Twig crotches wounded, %	39	23	14	8	3	
209	*Simulium* sp. [Flight]	Miles from release point	0.5	2.5	4.5	9.5	10.5	17
210		Flies recovered/visit	2.3	0.4	0.1	0.1	0	
211		Days between release & recovery	5	25	45	65	.....	17
212		Flies recovered	30	0	1	2	.....	
213	*Sitophilus oryza* [Probably crawling]	Feet from source	50	100	200	.....	.....	24
214		Weevils found	3	2	1	.....	.....	
215	*Tiphia vernalis* [Flight]	Feet from feeding area	7	21	49	77	90	35
216		Eggs/larva	3.7	3.2	2.7	2.5	2.4	
217	*Trichogramma minutum* [Flight, crawling]	Meters for female to move	1	2	3	4	5	32
218		Days to move	2.0	2.2	2.3	3.7	3.5	
219		Meters for female to move	6	7	8	9	.....	32
220		Days to move	4.0	2.0	4.0	2.0	.....	
221	*Trioza tripunctata* [Crawling]	Rows from field margin	1	3	5	10	20	68
222		Psyllids/10 bushes	10	5	4	3	1	
223	*Tyloderma fragariae* [Crawling]	Yards from release point	25	50	68	125	300	79
224		Borers recovered	6	6	5	4	1	

Vertical Dispersion

	Species (Synonym) [Means of Dispersion]	Specification	Numerical Unit					Reference
225	*Anopheles* spp. [Flight]	Altitude, meters (in forest)	0	5	10	15	.....	19
226	*A. darlingi*	Distribution, %	51	18	19	12	.....	
227	*A. mediopunctatus*	Distribution, %	3	3	39	55	.....	
228	*A. shannoni*	Distribution, %	4	7.3	24.6	64.1	.....	

continued

Part I. Invertebrates

	Species[1] (Synonym) [Means of Dispersion]	Specification	Numerical Unit					Reference
229	*A. quadrimaculatus* [Flight]	Altitude, feet	1.5	3.0	6.0	7.5	.....	28
230		Mosquitoes recovered	14	11	5	3	.....	
231	*Circulifer tenellus* [Flight]	Altitude, feet	8	16	25	.....	.....	21
232		Leafhoppers recovered	276	211	169	.....	.....	
233		Altitude, feet	2.5	15	32	.....	.....	54
234		Parasitism, %	3.5	1.7	0.9	.....	.....	
235	*Drosophila* sp. [Flight]	Altitude, miles	0.54	2.04	3.54	5.04	.....	71
236		Flies recovered	413	237	164	117	.....	
237	*Epitrix hirtipennis* [Flight]	Altitude, feet	12	19	23	.....	.....	1
238		Beetles recovered	96	41	18	.....	.....	
239		Altitude, feet	1.5	5	14	19	23	22
240		Beetles recovered/da	56	32	12	6	2	
241	*Heterodera rostochiensis* [Wind currents]	Height of trap, inches	22	38	55	.....	.....	95
242		Viable cysts	2077	435	103	.....	.....	
243	*Ostrinia nubilalis (Pyrausta nubilalis)* [Flight]	Altitude, feet	5	10	15	.....	.....	34
244		Moths recovered	914	547	332	.....	.....	
245	*Psallus seriatus* [Flight, air currents]	Altitude, feet	5.5	11.5	17.5	23.5	.....	22
246		Fleahoppers recovered	61	97	156	250	.....	
247	Aphididae [6] [Flight]	Altitude, feet	50	250	500	1000	1500	46
248		Aphids recovered	270	228	75	21	6	
249	Aphididae [6,7] [Flight]	Altitude, feet	10	150	500	1000	.....	47
250		Aphids recovered/hr	18	3	2	1	.....	
251	Aphididae [6] [Air currents, flight]	Altitude, miles	0.17	1.15	2.17	2.68	.....	71
252		Aphids recovered	2	9	16	19	.....	
253	Coccinellidae [5] [Flight]	Altitude, miles	0.17	1.15	2.17	2.68	.....	71
254		Beetles recovered	40	13	4	1	.....	
255	Insects, miscellaneous [Flight, air currents]	Altitude, feet	10	177	277	.....	.....	29
256		Insects recovered/ft^3 air	239	51	21	.....	.....	
257		Altitude, feet	20	1000	4000	7000	.....	36
258		Insects recovered/10 min flight	26	6	3	1	.....	

[1] Unless otherwise indicated. [5] *See also* lines 53, 54, 96, and 97. [6] Family. [7] 7-hr collection.

Contributor: Wolfenbarger, D. O.

References

[1] Annand, P. N. 1932. U.S. Dep. Agr. Circ. 244.
[2] Baker, W. A., et al. 1949. U.S. Dep. Agr. Tech. Bull. 983.
[3] Bedard, W. D. Unpublished.
[4] Bissell, T. L. 1939. J. Econ. Entomol. 32:546.
[5] Bonnet, D. D., and D. J. Worcester. 1946. Amer. J. Trop. Med. 26:465.
[6] Braun, E., et al. 1953. Sci. Agr. 33:437.
[7] Brower, A. E. 1930. Entomol. News 41:10, 44.
[8] Brownlee, J. 1911. Proc. Roy. Soc. Edinburgh. B31:262.
[9] Burgess, A. F. 1913. U.S. Dep. Agr. Bur. Entomol. Bull. 119.
[10] Carruth, L. A., and T. W. Kerr. 1937. J. Econ. Entomol. 30:297.
[11] Cartwright, O. L. 1929. S. C. Agr. Exp. Sta. Bull. 257.
[12] Causey, O. R., and H. W. Kumm. 1948. Amer. J. Trop. Med. 28:469.
[13] Chamberlain, T. R. 1941. U.S. Dep. Agr. Tech. Bull. 784.
[14] Cockerham, K. L., et al. 1954. La. Agr. Exp. Sta. Tech. Bull. 483.
[15] Collins, C. W., and S. F. Potts. 1932. U.S. Dep. Agr. Tech. Bull. 336.
[16] Crossman, S. S. 1917. J. Econ. Entomol. 10:177.
[17] Dalmat, H. T., and C. L. Gibson. 1952. Ann. Entomol. Soc. Amer. 45:605.
[18] Davidson, W. M. 1925. Trans. Amer. Entomol. Soc. 50:163.

continued

101. DISPERSION OF SMALL ORGANISMS

Part I. Invertebrates

[19] Deane, L. M., et al. 1953. Folia Clin. Biol. 20:101.
[20] De Meillon, B. 1934. Publ. S. Afr. Inst. Med. Res. 6:195.
[21] Dobzhansky, T., and S. Wright. 1943. Genetics 28:304.
[22] Dominick, C. B. 1943. Va. Agr. Exp. Sta. Bull. 355.
[23] Doucette, C. F. 1942. U.S. Dep. Agr. Tech. Bull. 809.
[24] Douglas, W. A. 1941. U.S. Dep. Agr. Circ. 602.
[25] Dudley, J. E., and E. M. Searls. 1923. J. Econ. Entomol. 16:363.
[26] Echert, J. E. 1933. J. Agr. Res. 47:257.
[27] Eyles, D. E., et al. 1945. Pub. Health Rep. 60:1265.
[28] Ficht, G. A., and T. E. Heinton. 1941. J. Econ. Entomol. 34:599.
[29] Freeman, J. A. 1938. Nature (London) 142:153.
[30] Frost, S. W. 1933. Pa. Agr. Exp. Sta. Bull. 301.
[31] Fulton, R. A., and V. E. Romney. 1940. J. Agr. Res. 61:737.
[32] Fye, R. E., and J. J. Jensen. 1969. J. Econ. Entomol. 62:1291.
[33] Gaines, J. C. 1932. Ibid. 25:1181.
[34] Gaines, J. C., and K. P. Ewing. 1938. Ibid. 31:674.
[35] Gardner, T. R. 1938. Ibid. 31:204.
[36] Glick, P. A. 1939. U.S. Dep. Agr. Tech. Bull. 673.
[37] Graham, S. A. 1922. Rep. Minn. State Entomol. 19:15.
[38] Hawley, I. M. 1934. J. Econ. Entomol. 27:503.
[39] Hervey, G. E. R., and C. E. Palm. 1935. Ibid. 28:670.
[40] Hill, C. B. 1952. Va. Agr. Exp. Sta. Tech. Bull. 119.
[41] Hoelscher, C. E., et al. 1968. J. Econ. Entomol. 61:370.
[42] Horsfall, W. R. 1942. Arkansas Agr. Exp. Sta. Bull. 427.
[43] Jachowski, L. A., Jr. 1954. Amer. J. Hyg. 60:186.
[44] Jackson, C. H. N. 1940. Ann. Eugen. 10:332.
[45] Jenkins, D. W., and C. C. Hassett. 1951. Can. J. Zool. 29:178.
[46] Johnson, C. G. 1951. Sci. Progr. (London) 39(153):41.
[47] Johnson, C. G., and H. L. Penman. 1951. Nature (London) 168:337.
[48] Jones, E. T. Unpublished.
[49] Kettle, D. S. 1951. Bull. Entomol. Res. 42:239.
[50] Kettle, D. S. 1951. Proc. Roy. Entomol. Soc. London A26:59.
[51] Klostermeyer, C. E. 1953. Wash. Agr. Exp. Sta. Tech. Bull. 9.
[52] Larrimer, W. H., and A. L. Ford. 1919. J. Econ. Entomol. 12:417.
[53] Larson, A. O., et al. 1933. Ibid. 26:1063.
[54] Lawson, F. R., et al. 1951. U.S. Dep. Agr. Tech. Bull. 1030.
[55] Lindquist, A. W., et al. 1951. J. Econ. Entomol. 44:397.
[56] Linn, M. B. 1940. N.Y. Agr. Exp. Sta. Ithaca Bull. 742.
[57] MacCreary, D., and P. L. Rice. 1949. Ann. Entomol. Soc. Amer. 42:141.
[58] MacVicar, R. M., et al. 1952. Sci. Agr. 32:67.
[59] Maxwell, C. W. 1968. J. Econ. Entomol. 61:103.
[60] McCullough, J. W. 1917. Ibid. 10:162.
[61] Moggridge, J. Y. 1949. Bull. Entomol. Res. 40:43.
[62] Morris, K. R. S. 1952. Ibid. 43:375.
[63] Munro, J. A., and H. S. Telford. 1942. N. Dak. Agr. Exp. Sta. Bull. 309.
[64] Neiswander, C. R., and J. R. Savage. 1931. J. Econ. Entomol. 23:389.
[65] Nicholas, W. L. 1953. Ann. Trop. Med. Parasitol. 47:309.
[66] Ohlendorf, W. 1926. U.S. Dep. Agr. Bull. 1374.
[67] Pankiw, P., et al. 1956. Can. J. Agr. Sci. 36:114.
[68] Peterson, A. 1923. N. J. Agr. Exp. Sta. Bull. 378.
[69] Petherbridge, F. R., and D. W. Wright. 1943. Ann. Appl. Biol. 30:348.
[70] Pimental, D., and R. W. Fay. 1955. J. Econ. Entomol. 48:19.
[71] Profft, J. 1939. Arb. Physiol. Angew. Entomol. Berlin-Dahlem 6:119.
[72] Quarterman, K. D., et al. 1949. Amer. J. Med. 29:973.
[73] Quarterman, K. D., et al. 1954. J. Econ. Entomol. 47:405.
[74] Quarterman, K. D., et al. 1954. Ibid. 47:413.
[75] Quarterman, K. D., et al. 1955. Ibid. 48:30.
[76] Quayle, H. J. 1911. Ibid. 4:301.
[77] Reeves, W. C., et al. 1948. Mosquito News 8:61.
[78] Reinhard, H. J., and F. L. Thomas. 1933. Tex. Agr. Exp. Sta. Bull. 475.
[79] Richter, P. O. 1939. Ky. Agr. Exp. Sta. Bull. 389.
[80] Rickard, E. R. 1928. Bol. Inst. Clin. Quir. (Buenos Aires) 4:133.
[81] Riegert, P. W., et al. 1954. Can. Entomol. 86:223.
[82] Roebuck, A., et al. 1947. Ann. Appl. Biol. 34:186.
[83] Schoof, H. F., and R. R. Siverly. 1954. J. Econ. Entomol. 47:830
[84] Smith, G. E., et al. 1941. Amer. J. Hyg. 34(C):102.
[85] Stearns, L. A., and D. MacCreary. 1938. J. Econ. Entomol. 31:226.
[86] Steiner, H. M., and H. N. Worthley. 1941. Ibid. 34:249.
[87] Swynnerton, C. F. M. 1936. Trans. Roy. Entomol. Soc. London 84:1.
[88] Timofeeff-Ressovsky, N. V., and E. A. Timofeeff-Ressovsky. 1940. Z. Indukt. Abstamm. Vererbungsl. 79:44.
[89] Travis, B. V., et al. 1940. J. Econ. Entomol. 33:847.
[90] Van Leeuwen, E. R. 1940. Ibid. 33:162.
[91] Van Zwaluwenburg, R. H., and J. S. Rosa. 1940. Hawaii. Plant. Rec. 44:3.
[92] Wadley, F. M., and D. O. Wolfenbarger. 1944. J. Agr. Res. 69:299.
[93] Wakeland, C. R. 1934. J. Econ. Entomol. 28:981.
[94] Walcott, A. M., et al. 1937. Amer. J. Trop. Med. 17:677.

continued

Part I. Invertebrates

[95] White, J. H. 1953. Nature (London) 172:686.

[96] Willis, H. R. 1939. J. Econ. Entomol. 32:401.

[97] Wolfenbarger, D. O. 1940. Ann. Entomol. Soc. Amer. 33:391.

[98] Wolfenbarger, D. O. 1948. Fla. Entomol. 31:15.

[99] Wolfenbarger, D. O. Unpublished. Univ. Florida, Agricultural Research and Education Center, Homestead, 1971.

[100] Wolfinsohn, M. 1953. Bull. Res. Counc. Isr. B3: 263.

[101] Wolfinsohn, M., and R. Galun. 1953. Ibid. B2: 433.

[102] Woodside, A. M. 1954. J. Econ. Entomol. 47:349.

[103] Wright, D. W., and D. G. Ashby. 1946. Ann. Appl. Biol. 33:69.

Part II. Viruses, Bacteria, and Fungi

	Disease (Organism) [Means of Dispersion]	Specification	Numerical Unit					Reference
			Horizontal Dispersion					
	Viruses							
1	Beet mosaic [Black bean aphid]	Yards from steckling bed	220	1320	1560	....	.....	30
2		Plants infected, %	95	57	12	....	.....	
3	Cabbage mosaic [Cabbage aphid]	Miles from fields	0.06	0.6	7	22	.....	29
4		Plants infected, %	72.6	46.9	1.0	0.3	.....	
5	Celery mosaic [Insects]	Feet from harborer plants	3	15	28	75	120	42
6		Diseased plants, %	100	89	52	28	16	
7	Cucumber cucurbit mosaic [Insects]	Yards from harborer plant	1	140	225	350	500	7
8		Days to first symptoms	17	42	45	47	49	
9	Eastern X-disease of peach [Certain	Feet from chokecherry plants	50	125	187.5	....	.....	39
10	leafhoppers]	Plants infected, %	89.3	8.1	0.6	....	.....	
11	Mild streak of black raspberry [Pre-	Feet from wild brambles	25	75	125	....	.....	15
12	sumably insects]	Infections, %	66	29	12	....	.....	
13	Potato calico [Insects]	Rows from source	1	2	3	4	5	28
14		Diseased plants	26	27	22	17	13	
15	Potato leaf roll [Insects]	Inches from inoculum source	18	36	54	72	90	11
16		Plants infected	41	25	17	12	8	
17	Potato leaf roll [Aphids]	Rows from infected plants	1	2	3	4	6	21
18		Diseased plants, %	21	12	7	5	1	
19	Potato mosaic [Insects]	Rows from diseased plants	1	2	3	4	6	20
20		Diseased plants, %	36	24	18	13	6	
21	Potato yellow dwarf [Six-spotted	Feet from old meadow	1	30	60	90	135	9
22	leafhopper]	Diseased plants, %	80	23	14	9	4	
23	Rugose mosaic of potato [Insects]	Inches from inoculum source	18	36	54	72	90	11
24		Infected plants	37.7	11.7	5.3	3.4	2.8	
25	Severe streak of raspberry [Insects]	Rows from wild brambles	3	8	13	18	23	5
26		Diseased plants	165	86	47	21	1	
27	Sudden death of clove [Probably	Tree intervals	1	2	3	4	5	23
28	insects]	Life expectancy, months	30.2	38.1	42.7	46.0	48.5	
29	Sugar beet curly top [Beet leafhop-	Miles from breeding ground	57	265	315	385	430	32
30	per]	Diseased plants, %	100	15	10	4	1	
31	Tristeza disease of citrus [Aphids]	Feet from inoculum source	5	25	45	65	73	1
32		Diseased trees, %	51	29	20	15	13	
33	Wheat streak mosaic [Eriophyd	Yards from source	0	8	18	50	90	33
34	mite]	Plants infected, %	64	26	19	10	5	

continued

Part II. Viruses, Bacteria, and Fungi

Disease (Organism) [Means of Dispersion]	Specification	Numerical Unit					Reference
35 (Bacteria colonies on sea-water me-	Miles from land (over water)	5	80	275	....	.....	45
36 dium) [Air currents]	Bacterial colonies	41	58	65	....	.....	
37	Miles from sea (over land)	0	0.06	0.25	0.50	1.0	45
38	Bacterial colonies	548	292	215	177	138	
Fungi							
39 (Air-borne spores) [Wind]	Degrees north of equator	57°30′	64°20′	68°55′	71°5′	.....	26
40	Fungus colonies on plate	3.61	0.49	0.48	0.72	.....	
41 Beet downy mildew (Peronospora	Meters from seed plants	10	150	1000	....	.....	13
42 sp.) [Wind]	Plants injured, %	28	8	1	....	.....	
43 Blossom infection (Monilinia laxa[1])	Feet from center of nearest source row	22	44	66	88	.....	44
[Air currents]							
44	Blossom infection, %	55.7	39.1	29.3	22.4	.....	
45 (Bovista plumbea) [Air currents]	Meters from release point	5	10	15	20	.....	38
46	Spores recovered	912	323	165	102	.....	
47 Cedar and apple rust (Gymnospor-	Yards from infected trees	0	55	110	220	440	17
48 angium sp.) [Air currents]	Leaf infections	64	40	33	26	19	
49 Chestnut blight (Endothia parasit-	Feet from spore source	27	85	180	266	.....	12
50 ica) [Air currents]	Ascospores found	23	11	8	8	.....	
51 Crown rust of oats (Puccinia coro-	Feet from inoculum source	3	5	7.7	10.3	13	43
52 nata) [Wind]	Infections, %	92.9	53.4	35	19.5	0.7	
53 Damping-off disease [Mycelial	Centimeters from inoculum source	1	3	5	7	9	2
54 growth]	Plants damped-off, %	100	8	7	16	0	
55 Downy mildew (Pseudoperonospora	Feet from spore source	10	50	100	200	400	19
56 humilis) [Air currents]	Leaves infected, %	26	16	12	7	3	
57 Dutch elm disease (Ceratostomella	Feet from inoculum source	200	600	1000	1800	2300	18
58 ulmi) [Elm bark beetles]	Diseased trees/acre	2.52	0.38	0.47	0	0.09	
59	Feet from inoculum source	12.5	62.5	300	575	.....	4
60	Diseased trees	27	19	11	8	.....	
61 Eyespot disease of wheat (Helmin-	Meters from spore source	5	20	50	70	.....	24
62 thosporium sacchari) [Water]	Culms with eye spots, %	26.7	12.3	7.7	6.4	.....	
63 Hollyhock rust [Gravity & ejection]	Spore dispersal, millimeters	0.15	0.35	0.55	0.75	.....	3
64	Spores/0.1 mm²	14.6	7.6	2.2	0.1	.....	
65 Leaf spots on tulips [Raindrop	Centimeters from conidia source	15.2	34.6	58.0	79.8	102.0	41
66 splash & wind]	Lesions/plant	31.6	20.1	12.9	8.5	5.1	
67 Loose smut of wheat (Ustilago tri-	Meters from spore source	2	4	24	80	.....	25
68 tici) [Air currents]	Smutted heads	241	234	114	0	.....	
69 Maize rust (Puccinia sorghi) [Wind]	Kilometers from spore source	0.5	2.5	4.5	6.5	.....	46
70	Plants attacked, %	100	3	0.3	0	.....	
71 Onion mildew (Peronospora de-	Feet from onion sets	120	780	1750	2000	.....	22
72 structor) [Air currents]	Lesions/100 ft row	1138	98	1	0	.....	
73 Potato late blight (Phytophthora in-	Centimeters from edge of infective group	30	90	150	210	270	10
74 festans) [Wind]	Plants infected, %	89	63	43	22	5	
75 Powdery mildew on barley (Ery-	Meters from source	1.5	3.5	5.5	7.5	8.5	27
76 siphe graminis) [Wind]	Plants affected, %	99	84	76	70	68	
77 Stem rust (Puccinia graminis)	Feet from barberry hedge	15	125	225	325	425	16
78 [Wind]	Grass infected, %	100	41	5	1	0.5	
79 Stem rust on rye (P. graminis se-	Meters from source plant	50	300	1000	3000	.....	8
80 calis) [Wind]	Yield (grams)/100 ears	47.6	92.3	122.3	149.7	.....	

[1] Synonym: Sclerotinia laxa.

continued

Part II. Viruses, Bacteria, and Fungi

Disease (Organism) [Means of Dispersion]	Specification	Numerical Unit					Reference
81 (*Tilletia tritici*) [Air currents]	Meters from release point	5	10	15	20	.....	38
82	Spores recovered	800	168	49	30	.....	
83 Tobacco blue mold (*Peronospora*	Yards from source	0	4	8	12	.....	40
84 *tabacina*) [Wind]	Plant lesions/1000 in.^{2} of field	140	8	1	0.5	.....	
85 Wheat stem rust (*Puccinia graminis*)	Miles from known source	200	360	580	740	940	36
86 [Air currents]	Spores collected	13,092	10,768	8883	7920	6975	
87 White pine blister rust (*Cronartium*	Feet from gooseberry bush	50	150	350	450	650	35
88 *ribicola*) [Air currents]	Diseased trees, %	75	55	40	36	29	
Vertical Dispersion							
89 (Bacteria, miscellaneous) [Air cur-	Altitude, feet	1500	6000	12,000	15,000	.....	31
90 rents]	Bacteria	113	48	15	5	.....	
Fungi							
91 Azalea flower spot (*Ovulinia azal-*	Inches above ground	4	10	18	48	.....	34
92 *eae*) [Air currents, water drip]	Infections	42	28	17	0	.....	
93 Onion mildew (*Peronospora de-*	Altitude, feet	100	200	700	1200	.....	22
94 *structor*) [Air currents]	Spores/ft^3 air	32	102	451	801	.....	
95 Wheat stem rust (*Puccinia graminis*)	Feet above barberry bushes	1000	2000	7000	12,000	.....	37
96 [Air currents]	Aeciospores recovered	19	14	5	1	.....	
97	Altitude, feet	1000	5000	10,000	14,000	.....	6
98	Uredospores	48,200	7730	144	40	.....	
99	Elevation, meters	30	400	600	800	.....	14
100	Spores/cm^2/min	1458	490	339	231	.....	

Contributor: Wolfenbarger, D. O.

References

[1] Bitancourt, A. A., and A. J. Rodriguez. 1948. Arq. Inst. Biol. (Sao Paulo) 18:313.

[2] Blair, I. D. 1943. Ann. Appl. Biol. 30:118.

[3] Buller, A. H. 1924. Researches on Fungi. Longmans, Green; London. v. 3, p. 533.

[4] Collins, D. L., et al. 1940. N.Y. Agr. Exp. Sta. Ithaca Bull. 740.

[5] Cooley, L. M. 1936. N.Y. State Agr. Exp. Sta. (Geneva) Bull. 665.

[6] Craigie, J. H. 1945. Sci. Agr. 25:285.

[7] Doolittle, S. P. 1925. J. Agr. Res. 31:1.

[8] Fisher, H. 1950. Z. Pflanzenkr. (Pflanzenpathol.) Pflanzenschutz 57:1.

[9] Frampton, V. L. 1942. Phytopathology 32:799.

[10] Gregory, P. H. 1945. Trans. Brit. Mycol. Soc. 28:26.

[11] Gregory, P. H., and D. R. Read. 1949. Ann. Appl. Biol. 36:475.

[12] Heald, F. D., et al. 1915. J. Agr. Res. 3:493.

[13] Höchapfel, H. 1950. Nachrichtenbl. Deut. Pflanzenschutzdienstes (Braunschweig) 2:124.

[14] Hubert, K. 1932. Fortschr. Landwirt. 7:195.

[15] Jeffers, W. F., and W. W. Woods. 1948. Phytopathology 38:222.

[16] Johnson, A. G., and J. G. Dickson. 1919. Wis. Agr. Exp. Sta. Bull. 304.

[17] Jones, L. R., and E. T. Bartholomew. 1915. Ibid. 257.

[18] Liming, O. N., et al. 1951. Phytopathology 41: 146.

[19] Magie, R. O. 1942. N.Y. Agr. Exp. Sta. Geneva Tech. Bull. 267.

[20] Murphy, P. A. 1921. Can. Exp. Farms Bull. 44.

[21] Murphy, P. A., and E. J. Worthley. 1920. Phytopathology 10:407.

[22] Newhall, A. G. 1938. Ibid. 28:257.

[23] Nutman, F. J., and F. M. L. Sheffield. 1949. Ann. Appl. Biol. 36:419.

[24] Oort, A. J. P. 1936. Tijdschr. Plantenziekten 42: 179.

[25] Oort, A. J. P. 1940. Ibid. 46:1.

[26] Pady, S. M., et al. 1950. Phytopathology 40:632.

[27] Pape, H., and B. Rademacher. 1934. Angew. Bot. 16:115.

[28] Porter, D. R. 1935. Hilgardia 9:383.

[29] Pound, G. S. 1946. Phytopathology 36:1035.

[30] Pound, G. S. 1947. J. Agr. Res. 75:31.

continued

Part II. Viruses, Bacteria, and Fungi

[31] Proctor, B. E. 1934. Proc. Amer. Acad. Arts Sci. 69:315.

[32] Romney, V. E. 1939. U.S. Dep. Agr. Circ. 518.

[33] Slykhuis, J. T. 1955. Phytopathology 45:116.

[34] Smith, F. F., and F. Weiss. 1942. U.S. Dep. Agr. Tech. Bull. 798.

[35] Snell, W. H. 1941. J. Forest. 39:537.

[36] Stakman, E. C., and L. M. Hamilton. 1939. Plant Dis. Rep., Suppl. 117:69.

[37] Stakman, E. C., et al. 1923. J. Agr. Res. 24:599.

[38] Stepanov, K. M. 1935. Tr. Zashch. Rast., II, 8:1.

[39] Stoddard, E. M. 1947. Conn. Agr. Exp. Sta. New Haven Bull. 506.

[40] Waggoner, P. E., and G. S. Taylor. 1955. Plant Dis. Rep. 39:79.

[41] Wallace, E. R. 1934. Holland County (Engl.) Counc. Bulb Res. Subcomm. Rep. 1933, p. 37.

[42] Wellman, F. L. 1935. Phytopathology 25:289.

[43] Wilson, E. E., and G. A. Baker. 1946. Ibid. 36: 418.

[44] Wilson, E. E., and G. A. Baker. 1946. J. Agr. Res. 72:301.

[45] Zobell, C. E. 1942. Publ. Amer. Ass. Advan. Sci. 17:55.

[46] Zogg, H. 1949. Phytopathol. Z. 15:143.

Part III. Pollen and Seeds

	Species [Means of Dispersion]	Specification	Numerical Unit					Reference
		Horizontal Dispersion						
1	*Abies alba* [Air currents]	Yards from seed trees	55	165	275	....	....	19
2		Seedlings/acre	22	9	3	....	....	
3	*Agropyron cristatum* [Wind]	Rods from field	5	15	25	....	....	23
4		Pollen grains	72	29	10	....	....	
5	*A. intermedium* [Wind]	Rods from field	5	12	25	....	....	23
6		Pollen grains	44	17	4	....	....	
7	*Beta* sp. [Wind]	Meters from seed fields	0	300	500	800	....	21
8		Pollen grains/cm²	11,613	1941	1075	278	....	
9		Feet from contaminant	2.3	20.7	43.2	73.2	....	4
10		Hybrids, %	5.6	0.3	0.2	0	....	
11	*Brassica rapa* [Insects]	Feet from contaminating plants	1	4.5	15	30	42.5	5
12		Proportion of hybrid seed, %	42	14	2.5	0.2	0.5	
13	*Bromus* sp. [Wind]	Rods from field	5	15	25	40	60	23
14		Pollen grains	146	41	21	10	4	
15	*Cedrus atlantica* [Wind]	Feet from source tree	40	120	240	325	700	38
16		Pollen grains	189	116	71	51	0.1	
17	*C. libani* [Wind]	Feet from source tree	15	75	135	195	....	38
18		Pollen grains	127	62	37	22	....	
19	*Citrullus vulgaris* [Honeybees]	Feet from field margin	50	250	450	650	....	28
20		Melons/acre	734	653	623	605	....	
21	*Dactylis* sp. [Wind]	Meters from field	0	200	400	600	800	21
22		Pollen grains/cm²	3096	447	172	120	86	
23	*Daucus carota* [Honeybees]	Rows from pollen source	1	2	3	4	5	12
24		Seed yield, lb/2 yr	31.0	27.3	27.9	24.6	22.7	
25		Rows from pollen source	6	7	8	9	10	12
26		Seed yield, lb/2 yr	24.6	22.8	21.1	20.6	19.7	
27	*Fraxinus* sp. [Wind]	Feet from source tree	25	50	150	400	....	38
28		Pollen grains	2545	1008	141	29	....	

continued

Part III. Pollen and Seeds

	Species [Means of Dispersion]	Specification	Numerical Unit					Reference
29	*Gilia* sp. [Insects]	Feet from white flowers	25	250	2640	10,560	....	11
30		Hybridization of blue-flowered plants, frequency	100	74	44	26	....	
31	*Gossypium* sp. [Wind]	Feet from marker plants	7.5	20.0	32.5	45.0	70.0	1
32		Natural crossing, %	29.7	8.5	9.1	5.1	0.8	
33		Feet from red cotton	16	35	51	99	189	15
34		Hybrids, %	6.9	3.0	2.0	0.9	0.3	
35	*G. arboreum* [Bumblebees]	Feet from flowers dusted with methylene blue	1.5	4.5	7.5	10.5	13.5	33
36		Flowers with dye particles, %	60	49	44	40	38	
37	*G. hirsutum* [Bumblebees]	Feet from flowers dusted with methylene blue	1.5	4.5	7.5	10.5	13.5	33
38		Flowers with dye particles, %	94	85	81	78	76	
39	*G. hirsutum* [Wind or insects]	Feet from contaminant	1	10	100	700	1800	29
40		Cross-pollination, %	18	12	7	3	1	
41	*Helianthus* sp. [Honeybees]	Feet from apiary	0	200	600	1000	....	13
42		Seed yield, lb/acre	1285	981	918	889	....	
43	*Juglans regia* [Air currents]	Feet from pollen source	60	150	500	1000	1600	8
44		Pollen grains/mm²/da	4	2.8	1.4	0.6	0	
45	*Juniperus scopularum* [Air currents]	Yards from seed source	22	44	66	88	....	19
46		Seedlings/acre	5588	259	192	0	....	
47	*Leontodon* sp. [Wind]	Feet from source	2	12	20	28	32	6
48		Seeds	184	21	9	3.5	1	
49	*Lolium* sp. [Wind]	Meters from ryegrass field	0	200	500	700	900	21
50		Pollen grains/cm²	4045	1053	535	345	204	
51	*L. perenne* [Wind]	Centimeters from rough clone contaminant	40	120	200	280	....	35
52		Rough plants, %	40.1	13.8	7.2	3.8	....	
53	*Lycopersicon esculentum* [Air currents; insects (?)]	Feet from contaminant	6	18	30	42	54	10
54		Cross-pollination, %	1.1	0.6	0.4	0.2	0.1	
55	*Malus pumila* [Wind]	Feet from source tree	0	165	330	....	....	38
56		Pollen grains	13	2	0.9	....	....	
57		Feet from pollen source	8	19	42	....	....	30
58		Fruit set/100 blossom spurs	52	34	18	....	....	
59	*M. pumila* [Honeybees]	Yards from bee colonies	25	50	75	100	150	20
60		Fruit set, %	7	7	6	6	4	
61	*Oryza sativa* [Dehiscence & wind]	Centimeters from pollen source	25	50	100	150	200	31
62		Pollen grains	22	9	3	1	0.4	
63	*Panicum virgatum* [Wind]	Rods from field	5	15	25	40	60	23
64		Switchgrass pollen	27	7	4	2	0.5	
65	*Parthenium argentatum* [Wind]	Yards from guayule plants	100	400	850	1200	....	14
66		Pollen grains/in.²	89	49	27	17	....	
67	*Paspalum notatum* [Wind]	Rods from albino seedling isolation blocks	0	5	10	15	....	18
68		Albinos, %	14.0	19.3	21.7	23.0	....	
69	*Pennisetum glaucum* [Wind]	Yards from release point	4	50	200	400	....	18
70		Pollen, %	100.0	8.9	0.8	0.4	....	
71	*Persea* sp. [Honeybees]	Feet from apiary	125	562.5	1062.5	....	....	37
72		Fruit yield (bushels)/tree	2.38	1.26	0.94	....	....	

continued

Part III. Pollen and Seeds

	Species [Means of Dispersion]	Specification	Numerical Unit					Ref- er- ence
73	*P. americana,* 'Taylor' [Insects]	Rows from nearest reciprocating variety	1	2	3	4	5	36
74		Fruit set/tree	61	54	50	47	44	
75	*Phaseolus limensis* [Insects]	Feet from pollen parent	2.5	5.0	7.5	12.5	32.5	2
76		Natural hybrids, %	7.3	2.7	1.4	0.6	0.5	
77	*P. lunatus* [Air currents; insects (?)]	Yards from kidney beans	1	2	3	5	9	3
78		Cross-pollination, %	5	4	3	2	1	
79	*P. vulgaris* [Air currents; insects (?)]	Yards from sieva beans	1	2	3	5	9	3
80		Cross-pollination, %	9	7	6	4	3	
81	*Phleum pratense* [Wind]	Meters from timothy field	0	100	200	300	500	21
82		Pollen grains/cm²	2613	781	505	343	140	
83	*Picea* sp. [Wind]	Feet from source tree	0	165	330	....	....	38
84		Pollen grains	9.7	0.1	0.7	....	....	
85	*P. mariana* [Air currents]	Feet from seed trees	10	80	160	240	....	34
86		Seedlings/acre	71,260	47,180	18,540	0	....	
87	*Pinus* spp. [Air currents]	Yards from seed tree stand	3344	6248	8426	....	....	25
88		Seedlings/acre	2002	991	507	....	....	
89	*P. cembroides* [Wind]	Feet from source tree	10	75	150	225	300	38
90		Pollen grains	8479	462	86	38	52	
91	*P. echinata*	Miles from forest margin	0	0.1	0.2	0.25	....	7
92		Pollen, %	100	17	15	10	....	
93	*P. monticola* [Air currents]	Yards from seed source	22	44	66	88	....	19
94		Seedlings/acre	616	177	57	9	....	
95	*Populus* sp. [Wind]	Feet from source tree	50	500	1400	3200	4200	38
96		Pollen grains	107	86	76	69	66	
97	*P. deltoides* [Wind]	Feet from source tree	25	250	500	1550	3550	38
98		Pollen grains	115	62	46	20	0.3	
99	*Pseudotsuga taxifolia* [Air currents]	Feet from seed trees	2	4	6	8	....	19
100		Seedlings/acre	304	170	91	35	....	
101	*Raphanus* sp. [Insects]	Feet from contaminant	1	3	4.2	5.0	6.4	5
102		Contamination, %	11.9	5.3	1.7	1.6	0.7	
103	*R. sativus* [Wind or insects]	Feet from contaminant	1	95	191	335	420	9
104		Cross-pollination, %	75	18	10	3	0	
105	*Secale cereale* [Wind]	Rods from rye field	5	15	25	40	60	23
106		Pollen grains	453	232	124	52	11	
107		Meters from rye field	100	300	500	700	....	21
108		Pollen grains/cm²	4181	2579	1834	1343	....	
109	*S. cereale* [Air currents]	Feet from pollen source	0.3	2.6	5.2	7.9	10.5	32
110		Cross-pollination, %	24	18	13	9	7	
111	*Trifolium hybridum* [Honeybee]	Yards from bee colonies	5.5	440	880	....	....	26
112		Seeds/head	38	33	32	....	....	
113		Miles from bee yards	0.125	0.625	1.5	2.5	....	16
114		Seed yield, lb/acre	128	95	77	67	....	
115	*T. pratense* [Honeybees]	Yards from apiary	200	500	833	1126	1426	24
116		Seed yield, lb/acre	166	149	139	133	128	
117	*T. repens* [Honeybees]	Miles from bee yards	0.125	0.625	1.5	2.5	....	16
118		Seed yield, lb/acre	206	102	46	13	....	
119	*Tsuga heterophylla* [Air currents]	Yards from seed trees	22	44	66	88	....	19
120		Seedlings/acre	1434	169	101	0	....	
121	*Ulmus* sp. [Wind]	Feet from source tree	500	1100	2700	5500	....	38
122		Pollen grains	115	152	12	8	....	

continued

Part III. Pollen and Seeds

Species [Means of Dispersion]	Specification	Numerical Unit					Reference
123 *Zea mays* [Wind]	Rods from field	5	15	25	40	60	23
124	Pollen grains	18	6	3	2	0.8	
125	Feet from pollen source	10	30	50	70	....	4
126	Pollen grains	7330	341	121	30	....	
127	Feet from pollen source	4	16	28	40	44	17
128	Seed set	256	197	122	75	31	
129	Feet from contaminating plants	13	29	45	61	77	4
130	Hybridization, %	7	6	3	1.3	0.3	
131	Rods north of contaminating field	5	25	60	100	....	22
132	Outcrossed seeds, %	16.5	0.8	0.2	0.2	....	
Vertical Dispersion							
133 *Beta vulgaris* [Air currents]	Altitude, feet	1000	2000	3000	4000	....	27
134	Pollen grains	56	26	14	9	....	

Contributor: Wolfenbarger, D. O.

References

[1] Afzal, M., and A. H. Khan. 1950. Agron. J. 42:89.

[2] Allard, R. W. 1954. Proc. Amer. Soc. Hort. Sci. 64:410.

[3] Barrons, K. 1938. Ibid. 36:637.

[4] Bateman, A. J. 1947. Heredity 1:235.

[5] Bateman, A. J. 1947. J. Genet. 48:257.

[6] Brownlee, J. 1911. Proc. Roy. Soc. Edinburgh B31:262.

[7] Buell, M. F. 1947. J. Elisha Mitchell Sci. Soc. 63:163.

[8] Crane, H. L., et al. 1937. Yearb. Agr. (U.S. Dep. Agr.), p. 827.

[9] Crane, M. B., and K. Mather. 1943. Ann. Appl. Biol. 30:301.

[10] Currence, T. M., and J. M. Jenkins, Jr. 1942. Proc. Amer. Soc. Hort. Sci. 41:273.

[11] Epling, C., and T. Dobzhansky. 1942. Genetics 27:317.

[12] Franklin, D. F. 1971. Arkansas Agr. Ext. Serv. Misc. Publ. 127:112.

[13] Furgala, B. 1954. Glean. Bee Cult. 82:532.

[14] Gardner, E. J. 1946. J. Amer. Soc. Agron. 38:264.

[15] Green, J. M., and M. D. Jones. 1953. Agron. J. 45:366.

[16] Harrison, C. M., et al. 1945. Mich. Agr. Exp. Sta. Quart. Bull. 28:85.

[17] Haskell, G., and P. Dow. 1951. Emp. J. Exp. Agr. 19:45.

[18] Hodgson, H. J. 1949. Agron. J. 41:337.

[19] Hoffmann, J. V. 1911. J. Agr. Res. 11:1.

[20] Hutson, R. 1926. N.J. Agr. Exp. Sta. Bull. 434.

[21] Jensen, I., and H. Bøgh. 1941. Tidsskr. Planteavl 46:238.

[22] Jones, M. D., and J. S. Brooks. 1950. Okla. Agr. Exp. Sta. Tech. Bull. 38.

[23] Jones, M. D., and L. C. Newell. 1946. Nebr. Agr. Exp. Sta. Res. Bull. 148.

[24] MacVicar, R. M., et al. 1952. Sci. Agr. 32:67.

[25] McQuilken, W. E. 1940. Ecology 21:135.

[26] Megee, C. R., and R. H. Kelty. 1932. Mich. Agr. Exp. Sta. Quart. Bull. 14:271.

[27] Meier, F. C., and F. Artschwager. 1938. Science 88:507.

[28] Parris, G. K., and J. D. Haynie. 1950. Fla. Dep. Agr. Tallahassee Bull. 135:45.

[29] Pope, O. A., et al. 1944. J. Agr. Res. 68:347.

[30] Roberts, R. H. 1945. Proc. Amer. Soc. Hort. Sci. 46:87.

[31] Rodrigo, P. A. 1925. Philipp. Agr. 14:155.

[32] Roemer, T. 1931. Z. Zuecht. A17:14.

[33] Stephens, S. G., and M. D. Finkner. 1953. Econ. Bot. 7:257.

[34] U.S. Department of Agriculture Forest Service. 1939. Lake States Forest Exp. Sta. Tech. Note 147.

[35] Wit, F. 1952. Euphytica 1:95.

[36] Wolfe, H. S., et al. 1934. Fla. Agr. Exp. Sta. Bull. 272.

[37] Wolfenbarger, D. O. 1954. Fla. Agr. Exp. Sta. Annu. Rep. 1953, p. 290.

[38] Wright, J. W. 1952. U.S. Dep. Agr. Forest Serv. Northeast. Forest Exp. Sta. Pap. 46.

With good management, and if other factors are favorable, many of the plants listed will grow and develop satisfactorily outside the pH range specified. Field crops and veg- etables generally are not as sensitive to soil pH as are flowers and shrubs.

	Species (Synonym)	pH	Reference
	Pinophyta[1]		
1	*Abies* spp.	4.5-6.5	6
2	*Chamaecyparis thyoides*	4.5-6.0	6
3	*Ginkgo biloba*	5.5-7.0	6
4	*Juniperus* spp.	5.5-7.5	6
5	*J. virginiana*	5.0-8.0	6
6	*Larix* spp.	4.5-7.5	6
7	*Picea* spp.	4.5-6.5	6
8	*Pinus* spp.	4.5-6.5	6
9	*P. palustris*	4.5-6.0	6
10	*Pseudotsuga taxifolia*	5.0-6.5	6
11	*Taxodium distichum*	6.0-7.5	6
12	*Taxus* spp.	5.0-7.5	6
13	*Thuja occidentalis*	6.0-7.5	4
14	*Tsuga canadensis*	4.5-6.0	6
	Magnoliophyta[2]—Magnoliopsida[3]		
15	*Acacia* spp.	6.5-8.0	6
16	*Acer* spp.	5.5-7.5	6
17	*Aesculus hippocastanum*	5.5-7.0	6
18	*Alnus* spp.	6.0-7.5	6
19	*Althaea* spp.	6.0-8.0	5
20	*Antirrhinum majus*	6.0-7.5	4
21	*Apium graveolens dulce*	6.0-7.5	6
22	*Arachis hypogaea*	5.0-6.5	4
23	*Beta vulgaris*	≮6.2	1
24	*Betula lenta*	4.5-6.0	6
25	*Brassica napus* var. *napobrassica (B. napobrassica)*	≮5.5	1
26	*B. nigra*	6.0-7.5	4
27	*B. oleracea* var. *botrytis*	5.5-7.5	5
28	*B. oleracea* var. *capitata*	6.0-7.5	4
29	*B. oleracea* var. *gemmifera*	6.0-7.5	5
30	*B. oleracea* var. *italica*	6.0-7.0	4
31	*B. rapa*	5.5-7.0	5
32	*Camellia japonica*	4.5-6.0	6
33	*Cannabis sativa*	6.0-7.5	6
34	*Capsicum frutescens*	5.5-7.0	5
35	*Carya ovata*	6.0-6.5	6
36	*Castanea dentata*	4.5-6.5	6
37	*Catalpa* spp.	6.0-7.5	6
38	*Celtis* spp.	6.0-7.5	6
39	*Centaurea nigra*	5.0-7.5	2
40	*Chrysanthemum morifolium*	6.0-7.5	4
41	*Citrullus vulgaris*	5.0-6.5	6
42	*Citrus limon*	6.0-7.6	4

	Species (Synonym)	pH	Reference
43	*C. sinensis*	6.0-7.5	4
44	*Coleus blumei*	6.0-7.5	6
45	*Cornus florida*	5.0-6.5	6
46	*Cucumis melo*	6.0-8.0	5
47	*C. sativus*	5.5-7.0	4
48	*Cucurbita pepo*	5.5-7.0	5
49	*Dahlia* spp.	6.0-8.0	5
50	*Datura stramonium*	6.0-7.5	4
51	*Daucus carota*	5.5-7.0	4
52	*Delphinium* spp.	6.0-8.0	5
53	*Dianthus caryophyllus*	6.0-7.5	4
54	*Eucalyptus* spp.	6.5-8.0	6
55	*Fagopyrum sagittatum (F. esculentum)*	5.5-7.0	3,4
56	*Fagus grandifolia*	5.0-6.5	4
57	*F. sylvatica*	6.0-7.5	6
58	*Fragaria* spp.	5.0-6.5	4,5
59	*Gardenia jasminoides*	5.0-7.0	5
60	*Gleditsia triacanthos*	6.0-7.5	6
61	*Glycine max (G. soja)*	6.0-7.5	6
62	*Gossypium hirsutum*	5.0-6.5	6
63	*Helianthus annuus*	6.0-7.5	3,4
64	*Hibiscus esculentus*	6.0-7.5	5
65	*Ilex aquifolium*	5.0-6.5	6
66	*I. opaca*	4.5-6.0	6
67	*Impatiens balsamina*	6.0-7.5	4
68	*Ipomoea batatas*	5.0-6.5	6
69	*Juglans* spp.	6.0-7.5	6
70	*Lactuca sativa*	6.0-7.5	6
71	*Lathyrus montanus*	5.0-7.0	2
72	*Lespedeza* spp.	5.0-6.5	5
73	*Ligustrum* spp.	6.0-7.5	5
74	*Linum usitatissimum*	5.0-7.0	3,4
75	*Liquidambar styraciflua*	5.0-6.5	6
76	*Liriodendron tulipifera*	5.5-7.5	6
77	*Lotus corniculatus*	5.0-7.5	2
78	*Lycopersicon esculentum*	5.5-7.5	4
79	*Magnolia grandiflora*	5.0-7.0	5
80	*Malus pumila*	5.0-6.5	4
81	*Matthiola incana*	6.0-7.5	4
82	*Medicago sativa*	≮6.6	1
83	*Melilotus alba*	6.0-7.5	6
84	*Nicotiana tabacum*	5.5-7.5	3,4
85	*Nyssa sylvatica*	4.5-6.0	6
86	*Oenothera biennis*	6.0-8.0	4
87	*Ostrya virginiana*	6.0-7.0	6
88	*Petroselinum crispum (P. hortense)*	5.0-7.0	4
89	*Petunia* spp.	6.0-8.0	5

[1] Synonym: Gymnospermae. [2] Synonym: Angiospermae. [3] Synonym: Dicotyledoneae.

continued

	Species (Synonym)	pH	Reference		Species (Synonym)	pH	Reference
90	*Phaseolus vulgaris*	6.0-7.5	3,4	120	*Viola* spp.	6.0-7.5	5
91	*Pisum sativum*	6.0-8.0	5	121	*Vitis* spp.	6.0-8.0	5
92	*Platanus* spp.	5.5-7.5	6	122	*Zinnia* spp.	6.0-8.0	5
93	*Populus* spp.	5.5-7.5	6		Magnoliophyta[2]—Liliopsida[6]		
94	*P. tremuloides*	4.5-5.5	6				
95	*Prunus persica (Amygdalus persica)*	6.0-7.5	4	123	*Agrostis alba*	5.0-6.5	6
96	*Pyrus communis*	6.0-7.5	4	124	*A. tenuis*	4.0-6.0	2
97	*Quercus alba*	6.0-8.0	5	125	*Allium cepa*	6.0-7.5	6
98	*Q. coccinea*	4.5-6.5	6	126	*Ananas comosus*	4.5-6.0	5
99	*Q. falcata*	4.5-5.0	6	127	*Asparagus officinalis*	6.0-8.0	4
100	*Q. robur*	6.0-7.5	6	128	*Avena sativa*	5.0-6.8	1
101	*Q. velutina*	4.5-6.5	6	129	*Cynodon dactylon*	5.5-7.5	5
102	*Raphanus sativus*	5.5-7.0	6	130	*Dactylis glomerata*	4.5-8.0	2
103	*Rhododendron obtusum* var. *amoenum*	4.5-6.0	4	131	*Festuca rubra*	4.0-8.0	2
104	*Ricinus communis*	6.0-7.5	4	132	*Gladiolus* spp.	6.0-8.0	5
105	*Robinia* spp.	5.5-7.5	6	133	*Hordeum vulgare*	<5.7	1
106	*Rosa* sp.	5.5-7.0	4	134	*Iris* spp.	6.0-8.0	5
107	*Rubus* spp.[4]	6.0-8.0	5	135	*Lilium longiflorum*	6.0-7.0	4
108	*Rumex acetosa*	4.0-7.5	2	136	*Oryza sativa*	5.0-6.5	3,4
109	*Salix* spp.	5.5-7.5	6	137	*Paspalum dilatatum*	6.0-7.0	5
110	*Solanum tuberosum*	5.4-6.7	1	138	*Phleum pratense*	6.0-8.0	4
111	*Spinacia oleracea*	6.0-7.5	4	139	*Poa pratensis*	5.5-7.5	3,4
112	*Tilia* spp.	6.0-7.5	6	140	*Saccharum officinarum*	6.0-8.0	3,4
113	*Trifolium pratense*	6.0-7.5	3,4	141	*Secale cereale*	5.0-7.0	3,4
114	*T. repens*	6.0-6.8	1	142	*Sorghum bicolor (S. vulgare)*	5.5-7.5	3,4
115	*Tropaeolum majus*	5.5-7.5	4	143	*Tradescantia virginiana*	5.0-7.5	4
116	*Ulmus americana*	6.0-8.0	5	144	*Triticum aestivum*	5.5-7.5	3
117	*Vaccinium* spp.	4.5-6.0	6	145	*Tulipa gesneriana*	6.0-7.5	6
118	*Vicia* spp.	5.5-7.5	5	146	*Zea mays*	5.5-7.5	3,4
119	*Vigna* spp.[5]	5.5-7.5	5				

[2] Synonym: Angiospermae. [4] Most species. [5] Many species. [6] Synonym: Monocotyledoneae.

Contributors: Welch, C. D.; Larsen, Sigurd

References
[1] Blood, J. W. 1957. J. Sci. Food Agr. 8:645.

[2] Grime, J. P., and J. G. Hodgson. 1969. In I. H. Rorison, ed. Ecological Aspects of the Mineral Nutrition of Plants. B. H. Blackwell, Oxford. pp. 67-99.

[3] Ignatieff, V. 1949. FAO Agr. Stud. 9:108.

[4] Spurway, C. H. 1941. Mich. Agr. Exp. Sta. Spec. Bull. 306.

[5] Welch, C. D. 1964. Tex. Agr. Exp. Sta. Leafl. L-164.

[6] Wherry, E. T. Unpublished. Univ. Pennsylvania, Philadelphia, 1965.

103. EFFECT OF LIGHT ON DEVELOPMENT: MAGNOLIOPHYTES

Part I. Wavelengths for Maximum Response

Response & Species	Wavelength nm	Energy Level erg/cm^2	Duration of Exposure	Reference		Response & Species	Wavelength nm	Energy Level erg/cm^2	Duration of Exposure	Reference
Promotion of germination					3	*Lactuca sativa* 'Grand Rapids'	660	24,000	4 min	3
1 *Arabidopsis thaliana*	650	35,000	....	21						
2 *Billbergia elegans*	660	6000	4 min	6	4	*Lepidium densiflorum*	660	6000	8 min	24

continued

103. EFFECT OF LIGHT ON DEVELOPMENT: MAGNOLIOPHYTES

Part I. Wavelengths for Maximum Response

	Response & Species	Wavelength nm	Energy Level erg/cm²	Duration of Exposure	Reference
5	*L. virginicum*	650	1.4×10^6	8 min	24
6	*Pinus virginiana*	660	6000	1 hr	25
7	*Wittrockia superba*	660	6000	4 min	6
	Inhibition of germination				
8	*Lactuca sativa* 'Grand Rapids'	735	70,000	4 min	3
9	*L. sativa* 'Great Lakes'	735	2400	12 hr	10
10	*Lamium amplexicaule*	730	2	Continuous	10
11	*Lepidium densiflorum*	735	1.8×10^5	8 min	24
12	*L. virginicum*	735	30,000	8 min	24
13	*Lycopersicon esculentum* 'Ace'	740	1500	4 min, cyclic	13
	Promotion of stem elongation				
14	*Phaseolus vulgaris* 'Pinto'	735	7000	5 min	8
	Inhibition of stem elongation				
15	*Hordeum vulgare* 'Colsess I'	660	1.3×10^5	1.5 min	2
16	*Lactuca sativa* 'Grand	430	500	15 hr	15
17	Rapids'	730	1100	8 hr	9
18	*Phaseolus vulgaris*	660	1.5×10^6	2 min	5
	Promotion of leaf expansion				
19	*Phaseolus vulgaris*	650	1.5×10^6	2 min	5
20	*Pisum sativum* 'Little Marvel'	650	160	4 min	18

	Response & Species	Wavelength nm	Energy Level erg/cm²	Duration of Exposure	Reference
	Promotion of flowering				
21	*Hordeum vulgare* 'Wintex'	650	60,000	2 min	1
22	*Hyoscyamus niger*	650	1500	6 min	19
23	*Ipomoea nil*[1]	650	3.2×10^5	2 min	16
24	*Xanthium pensylvanicum*	735	3.0×10^5	4 min	4
	Inhibition of flowering				
25	*Chenopodium rubrum*	630	2000	4 min	11
26	*Glycine max* 'Biloxi'	630	50,000	30 s	17
27	*Ipomoea nil*[1]	740	1.2×10^6	2 min	16
28	*Xanthium pensylvan-*	630	40,000	5 min	17
29	*icum*	750	7500	90 min	14
	Twining				
30	*Cuscuta indecora*	483	1300	5 hr	12
31		730	1700	5 hr	12
	Pigment synthesis				
32	*Lycopersicon esculentum*[2]	640	6000	20 s	20
	Anthocyanin synthesis				
33	*Brassica oleracea* var. *capitata*[3]	690	5×10^7	4 hr	22
34	*B. rapa*[3]	725	1.0×10^8	8 hr	22
35	*Malus* sp. 'Arkansas'[4]	650	2.6×10^8	16 hr	23
36	*Sorghum bicolor*[5] 'Wheatland Milo'[3]	470	3×10^7	4 hr	7

[1] Synonym: *Pharbitis nil.* [2] Fruit. [3] Seedling. [4] Fruit skin. [5] Synonym: *S. vulgare.*

Contributor: Downs, R. J.

References

[1] Borthwick, H. A., et al. 1949. Bot. Gaz. 110:103.

[2] Borthwick, H. A., et al. 1951. Ibid. 113:95.

[3] Borthwick, H. A., et al. 1952. Proc. Nat. Acad. Sci. U.S. 38:662.

[4] Borthwick, H. A., et al. 1952. Ibid. 38:929.

[5] Downs, R. J. 1955. Plant Physiol. 30:468.

[6] Downs, R. J. 1964. Phyton (Buenos Aires) 21:1.

[7] Downs, R. J., and H. W. Siegelman. 1963. Plant Physiol. 38:25.

[8] Downs, R. J., et al. 1957. Bot. Gaz. 118:199.

[9] Evans, L. T., et al. 1965. Planta 64:201.

[10] Hendricks, S. B., et al. 1959. Bot. Gaz. 121:1.

[11] Kasperbauer, M. J., et al. 1963. Ibid. 124:444.

[12] Lane, H. C., and M. J. Kasperbauer. 1965. Plant Physiol. 40:109.

[13] Mancinelli, A. L. 1966. Bot. Gaz. 127:1.

[14] Mancinelli, A. L., and R. J. Downs. 1967. Plant Physiol. 42:95.

[15] Mohr, H., and M. Wehrung. 1960. Planta 55:438.

[16] Nakayama, S., et al. 1960. Bot. Gaz. 121:237.

[17] Parker, M. W., et al. 1946. Ibid. 108:1.

[18] Parker, M. W., et al. 1949. Amer. J. Bot. 36:194.

[19] Parker, M. W., et al. 1950. Bot. Gaz. 111:242.

[20] Piringer, A. A., and P. H. Heinze. 1954. Plant Physiol. 29:467.

[21] Shropshire, W., et al. 1961. Plant Cell Physiol. 2:63.

[22] Siegelman, H. W., and S. Hendricks. 1957. Plant Physiol. 32:393.

[23] Siegelman, H. W., and S. Hendricks. 1958. Ibid. 33:185.

[24] Toole, E. H., et al. 1955. Ibid. 30:15.

[25] Toole, V. K., et al. 1961. Ibid. 36:285.

continued

Part II. Light Exposure for Development

Development: Flowering—DF = days to flowering, PMF = percent of plants with male flowers, NFP = number of flowers per plant, DM = days to maturity.

	Species	Light Exposure hr	Development Flowering	Height cm	Reference		Species	Light Exposure hr	Development Flowering	Height cm	Reference
1	Acer rubrum	8	....	4	3	45	Hydrangea macrophyl-	8	....	18	12
2		12	....	10		46	la 'Engels White'	12	....	33	
3		14	....	42		47		14	....	45	
4		16	....	45		48		16	....	53	
5	Bougainvillea glabra	10	45 DF	95	1	49		24	....	80	
6		12	57 DF	100		50	Malus sylvestris[1]	8	1 PMF	180	11
7		13	151 DF	150		51	'Jonathan'	12	26 PMF	230	
8		14	153 DF	125		52		14	10 PMF	220	
9	Cannabis sativa	8	36 PMF	....	2	53		16	31 PMF	272	
10		11	44 PMF	....		54	Petunia hybrida 'Bal-	8	>65 DF	....	10
11		14	0 PMF	....		55	lerina'	9	>65 DF	....	
12	Chrysanthemum mori-	10	61 DF	55	1	56		10	>65 DF	....	
13	folium 'Lucifer'	12	69 DF	55		57		12	56 DF	....	
14		13	82 DF	62		58		14	47 DF	....	
15		14	173 DF	120		59		16	41 DF	....	
16	Coffea arabica	8	106 DF	....	9	60	Pinus virginiana	8	....	7	7
17		12	120 DF	....		61		12	....	12	
18		14	>315 DF	....		62		14	....	23	
19		18	>315 DF	....		63		16	....	49	
20		24	>315 DF	....		64	Rudbeckia newmanii	10	>102 DF	....	1
21	Cosmos sulphureus	10	27 DF	70	1	65		12	>102 DF	....	
22	'Klondyke'	12	33 DF	80		66		13	89 DF	48	
23		13	35 DF	85		67		14	82 DF	62	
24		14	135 DF	200		68	Silphium trifoliatum	10	107 DF	95	1
25	C. sulphureus 'Orange	10	34 DF	75	1	69		12	110 DF	140	
26	Flare'	12	34 DF	75		70		13	70 DF	162	
27		13	40 DF	78		71		14	89 DF	190	
28		14	42 DF	88		72	Solidago altissima	10	81 DF	58	1
29	Fragaria ananassa 'Cli-	11	5.0 NFP	....	5	73		12	92 DF	55	
30	max'	13	3.0 NFP	....		74		13	102 DF	80	
31		15	0 NFP	....		75		14	141 DF	140	
32		17	0 NFP	....		76	Triodia flava	10	38 DF	65	1
33	F. ananassa 'Mastodon'	11	0.8 NFP	....	5	77		12	44 DF	62	
34		13	0.3 NFP	....		78		13	47 DF	75	
35		15	11.0 NFP	....		79		14	56 DF	62	
36		17	20.0 NFP	....		80	Viburnum burkwoodii	8	....	49	6
37	F. ananassa 'Red Rich'	11	4.6 NFP	....	5	81		12	....	60	
38		13	5.0 NFP	....		82		14	....	81	
39		15	14.0 NFP	....		83		16	....	113	
40		17	23.0 NFP	....		84	Weigela florida var.	8	....	21	4
41	Glycine max 'Lee'	13	63 DM	....	8	85	varigata	12	....	71	
42		13.5	90 DM	....		86		14	....	254	
43		14	109 DM	....		87		16	....	478	
44		14.5	123 DM	....							

[1] Synonym: Pyrus malus.

continued

103. EFFECT OF LIGHT ON DEVELOPMENT: MAGNOLIOPHYTES

Part II. Light Exposure for Development

Contributor: Downs, R. J.

References
[1] Allard, H. A., and W. W. Garner. 1940. U.S. Dep. Agr. Tech. Bull. 727.
[2] Borthwick, H. A., and N. H. Scully. 1954. Bot. Gaz. 116:14.
[3] Downs, R. J., and H. A. Borthwick. 1956. Ibid. 117: 310.
[4] Downs, R. J., and H. A. Borthwick. 1956. Proc. Amer. Soc. Hort. Sci. 68:518.
[5] Downs, R. J., and A. A. Piringer. 1955. Ibid. 66:234.
[6] Downs, R. J., and A. A. Piringer. 1958. Ibid. 72:511.
[7] Downs, R. J., and A. A. Piringer. 1958. Forest Sci. 4:185.
[8] Johnson, H. W., et al. 1960. Bot. Gaz. 122:77.
[9] Piringer, A. A., and H. A. Borthwick. 1955. Turrialba 5:72.
[10] Piringer, A. A., and H. M. Cathey. 1960. Proc. Amer. Soc. Hort. Sci. 76:649.
[11] Piringer, A. A., and R. J. Downs. 1959. Ibid. 73:9.
[12] Piringer, A. A., and N. W. Stuart. 1955. Ibid. 65: 446.

104. PHOTOPERIOD, WITH TEMPERATURE INTERACTIONS, FOR FLOWERING: MAGNOLIOPHYTES

Photoperiodic Classification: Varietal differences account for multiple classifications, with the more common classification being given priority in listing. **Light Period:** ">" should be interpreted as the specified number of hours or more; "<" as "up to" the specified number of hours. **Temperature Interactions:** Th = photoperiodic response occurring at relatively high temperatures (plants may also flower at other day lengths at lower temperatures), or reproductive development promoted by high temperatures during photoperiodic induction; Tl = photoperiodic response occurring at relatively low temperatures (plants may also flower at other day lengths at higher temperatures), or reproductive development promoted by low temperatures during photoperiodic induction; Tp = thermoperiodic (i.e., development affected by alternation of temperature between day and night periods); Tq = quantitative effect of temperature on critical day length (i.e., an increase in temperature lowers the minimum limits for long-day plants and raises the maximum limits for short-day plants), or on the degree of photoperiodic response; Va = vernalization not essential but promotes reproductive development; Ve = vernalization essential for complete reproductive development—or other low-temperature preconditioning of embryo plants, seedlings, buds, or plants—prior to photoperiodic induction; Vo = vernalization not effective. For additional information, consult references 6, 28, 39, 55, 67, 75, 97, 98, and 114. Data in brackets refer to the column heading in brackets.

	Species (Synonym)	Photoperiodic Classification [Light Period, hr]	Temperature Interactions	Reference
		Magnoliopsida (Dicotyledoneae)		
1	*Anagallis arvensis*	Long day required [>12]	Va (vernalization allows GO strain to flower on 8-hr photoperiods)	21,22,25
2	*Antirrhinum majus*	Long day favorable	Th; Tl (day neutral)	66,86
3	*Arabidopsis thaliana*	Long day required [>5]	Va ; Ve (depends on variety)	62-64,76,77
4	*Beta saccharifera*	Long day required	Tl (7-9°C); Ve	40
5	*B. vulgaris*	Long day favorable	Th; Tl (long day required)	91
6	*Brassica campestris*	Long day favorable	Th	41
7	*B. hirta (Sinapis alba)*	Long day favorable [>9.5]	Va	12,13,24,102
8	*Bryophyllum crenatum*	Long days then short days required [>12.5, long day; <11.5, short day]	Tl (short days)	33,89
9	*B. daigremontianum*	Long days then short days required [>12, long day; >12, short day]	Tl (10°C will replace short days required)	116
10	*Cannabis sativa*	Short day favorable to required in some races[1] [<20]	....	18,78

[1] Quantitative for primordium development, qualitative for opening of flower.

continued

	Species (Synonym)	Photoperiodic Classification [Light Period, hr]	Temperature Interactions	Reference
11	*Capsicum frutescens*	Day neutral; short day favorable	Tp	26,32,35
12	*Cestrum nocturnum*	Long days then short days required [>12, long days; <12.5, short days]	Tl (short days); Th (long days)	47,94,95
13	*Chenopodium rubrum*	Short day required [<10 to 12]	....	29
14	*Chrysanthemum maximum*	Long day required	....	91
15	*C. morifolium*, cultivars	Short day required to day neutral	No vernalization to absolute required	23,79,85,99
16	*Cucumis sativus*	Day neutral	....	30,108
17	*Daucus carota*	Day neutral	Ve (4-10°C)	96
18	*Digitalis purpurea*	Long day favorable	Ve	8
19	*Fagopyrum sagittatum (F. esculentum)*	Day neutral	....	7,44
20	*Fragaria chiloensis*	Short day required [<10][2/]	Tq	31,53
21	*F. chiloensis* 'Everbearing'	Long day favorable; day neutral	....	31
22	*Glycine max (G. soja)*	Short day required to favorable	Th; Tq	17,43,81
23	*G. max* 'Biloxi'	Short day required [<14]	....	14-16
24	*Gossypium hirsutum*	Day neutral; short day favorable	Tq	11,61
25	*Helianthus annuus*	Short day favorable; day neutral	....	34
26	*Hyoscyamus niger*, annual	Long day required [>11]	Vo	56
27	biennial	Long day required [>11]	Ve	56
28	*Ilex aquifolium*	Day neutral	....	90
29	*Ipomoea batatas*	Short day required; short day favorable	....	71
30	*I. nil (Pharbitis nil)*	Short day favorable [<9, mature plant]	Tl[3/]; Va	57,60,106
31	*Kalanchoe blossfeldiana*	Short day required [<12]	Tl	51,52,93,100
32	*Lactuca sativa*	Long day favorable	Th; Tl (day neutral)	9,20,107
33	*Lycopersicon esculentum*	Day neutral; long day favorable; short day favorable	Tp	1,113
34	*Medicago sativa*	Long day favorable	Th; Tl (day neutral)[4/]	92
35	*Nicotiana tabacum*	Day neutral[2/]	....	43
36	*N. tabacum* 'Havana'	Long day favorable	....	91
37	*N. tabacum* 'Maryland mammoth'	Short day required [<14]	Th; <13°C (day neutral)	43,92
38	*Oenothera biennis*	Long day favorable	Tl	92
	Perilla frutescens var. *crispa*			
39	Red-leaf variety	Short day required [<14]	Th	4,42,74,80,111
40	Green-leaf variety	Short day required [<8]	Th	4,42,74,80,111
41	*Phaseolus vulgaris*	Day neutral; short day required[5/]	....	5,43,71
42	*Phlox paniculata*	Long day required	Th	91
43	*Pisum sativum*	Day neutral; long day favorable	....	10,92
44	*Raphanus sativus*	Long day required	....	43,84
45	*Rhododendron* sp.	Day neutral	....	90
46	*Silene armeria* "N" strain	Short day required [<12.5]	Th; Tq; Va (5°C)	65,68,69,112
47	*Solanum tuberosum*	Long day favorable; short day favorable; day neutral	....	2,44,58
	Trifolium pratense			
48	'American Medium'	Long day favorable [>9]	....	109
49	'English Montgomery'	Long day required [>12]	....	109
50	*Vicia faba*	Day neutral	Va	37
51	*Xanthium strumarium*	Short day required [<15.5]	....	46,48-50

[2/] Data applicable to most varieties. [3/] At 15°C will flower under continuous illumination. [4/] Vegetative on warm nights. [5/] Photoperiod influences fruit development, but floral initiation is not affected.

continued

Species (Synonym)	Photoperiodic Classification [Light Period, hr]	Temperature Interactions	Reference
Liliopsida (Monocotyledoneae)			
52 *Allium cepa*	Long day favorable; short day favorable; day neutral	Tl	44,70,101
53 *Avena sativa*	Long day required [>9]	Tl; Va (winter varieties); Vo (spring varieties)	44
54 *Hordeum vulgare*, spring	Long day favorable	Vo	88,91
55 winter	Long day required [>12]	Va (7-9°C)	19,44
56 *Lemna perpusilla* strain 6746	Short day favorable [<14]	....	54,59,87,110
Lolium temulentum			
57 Strain Ba 6137-21	Long day required [>14]	Vo	27,36,38,83
58 Strain *Ceres*	Long day required [>9]	Vo	27,36,38,83
59 *Oryza sativa*, summer	Day neutral	Th	103,104
60 winter	Short day required [<12]	Th	103,104
61 *Phleum pratense*	Long day required [>12]	....	3,109
62 *Poa pratensis*	Long day favorable	Th; Tl (day neutral or short day favorable); Ve	3,82,92,105
63 *Triticum aestivum*, spring	Long day favorable	Vo	44,72,73,115
64 winter	Long day favorable [>12]2/	Va	44,72,73,115
65 *Zea mays*	Day neutral; short day required	....	45,71

2/ Data applicable to most varieties.

Contributor: Papenfuss, Herbert D.

References

[1] Adams, J. 1924. Amer. J. Bot. 11:229.
[2] Allard, H. A. 1938. J. Agr. Res. 57:775.
[3] Allard, H. A., and M. W. Evans. 1941. Ibid. 62:193.
[4] Allard, H. A., and W. W. Garver. 1940. U.S. Dep. Agr. Tech. Bull. 727.
[5] Allard, H. A., and W. J. Zaumeyer. 1944. Ibid. 867.
[6] Altman, P. L., and D. S. Dittmer, ed. 1962. Growth, including Reproduction and Morphological Development. Federation of American Societies for Experimental Biology. Washington, D.C. pp. 501-506.
[7] Arthur, J. M., and J. D. Guthrie. 1927. Mem. Hort. Soc. N.Y. 3:73.
[8] Arthur, J. M., and E. K. Harvill. 1941. Contrib. Boyce Thompson Inst. 12:111.
[9] Arthur, J. M., et al. 1930. Amer. J. Bot. 16:338.
[10] Aso, K., and U. Muari. 1924. Nogaku Kai Ho 254: 31.
[11] Berkeley, E. E. 1931. Ann. Mo. Bot. Gard. 18:573.
[12] Bernier, G. 1963. Naturwissenschaften 50:101.
[13] Bernier, G. 1964. Acad. Roy. Belg. Cl. Sci. Mem. Collect. 16:1.
[14] Blaney, L. T., and K. C. Hamner. 1957. Bot. Gaz. 119:10.
[15] Borthwick, H. A., and M. W. Parker. 1938. Ibid. 100:245.
[16] Borthwick, H. A., and M. W. Parker. 1939. Ibid. 100:374.
[17] Borthwick, H. A., and M. W. Parker. 1939. Ibid. 101:341.
[18] Borthwick, H. A., and N. J. Scully. 1954. Ibid. 116: 14.
[19] Borthwick, H. A., et al. 1941. Ibid. 103:326.
[20] Bremer, A. H. 1931. Gartenbauwissenschaft 4:469.
[21] Brulfert, J. 1965. Rev. Gen. Bot. 72:641.
[22] Brulfert, J., and P. Chouard. 1961. C. R. Acad. Sci. 253:179.
[23] Cathey, H. M. 1957. Proc. Amer. Soc. Hort. Sci. 69:485.
[24] Chailahjan, M. Ch. 1946. C. R. (Dokl.) Acad. Sci. URSS 54:735.
[25] Chouard, P. 1947. Bull. Soc. Bot. Fr. 93:399.
[26] Cochran, H. L. 1936. Cornell Univ. Agr. Exp. Sta. Mem. 190.
[27] Cooper, J. P. 1960. Ann. Bot. (London), N.S. 24: 232.
[28] Cooper, J. P. 1960. Herb. Abstr. 30:71.
[29] Cumming, B. G. 1959. Nature (London) 184:1044.
[30] Danielson, L. L. 1944. Plant Physiol. 19:638.
[31] Darrow, G. M., and G. F. Waldo. 1934. U.S. Dep. Agr. Tech. Bull. 453.
[32] Dorland, R. E., and F. W. Went. 1947. Amer. J. Bot. 34:393.
[33] Dostal, R. 1966. Advan. Front. Plant Sci. 14:11.
[34] Dyer, H. J., et al. 1959. Bot. Gaz. 121:50.
[35] Eguchi, T. 1937. Proc. Imp. Acad. (Tokyo) 13:332.
[36] Evans, L. T. 1958. Nature (London) 182:197.
[37] Evans, L. T. 1959. Ann. Bot. (London), N.S. 23: 521.

continued

[38] Evans, L. T. 1960. J. Agr. Sci. 54:410.

[39] Evans, L. T., ed. 1969. The Induction of Flowering. Cornell Univ. Press, Ithaca.

[40] Fife, J. M., and C. Price. 1953. Plant Physiol. 28:475.

[41] Friend, D. J. C., and V. A. Helson. 1966. Science 153:1115.

[42] Funke, G. L. 1943. Rec. Trav. Bot. Neer. 40:392.

[43] Garner, W. W., and H. A. Allard. 1920. J. Agr. Res. 18:553.

[44] Garner, W. W., and H. A. Allard. 1923. Ibid. 23:871.

[45] Gerhard, E. 1940. J. Landwirt. 87:161.

[46] Gilbert, B. E. 1926. Bot. Gaz. 81:1.

[47] Griesel, W. O. 1966. Plant Physiol. 41(Suppl.):xxvii.

[48] Hamner, K. C. 1938. Bot. Gaz. 99:615.

[49] Hamner, K. C. 1960. Cold Spring Harbor Symp. Quant. Biol. 25:269.

[50] Hamner, K. C., and J. Bonner. 1938. Bot. Gaz. 100:388.

[51] Harder, R., and G. Gümmer. 1947. Planta 35:88.

[52] Harder, R., and H. von Witsch. 1940. Gartenbau-wissenschaft 15:226.

[53] Hartman, H. T. 1947. Plant Physiol. 22:407.

[54] Hillman, W. S. 1958. Nature (London) 181:1275.

[55] Hillman, W. S. 1962. The Physiology of Flowering. Rinehart and Winston, New York.

[56] Hsu, J. C. S., and K. C. Hamner. 1967. Plant Physiol. 42:725.

[57] Imamura, S., et al. 1966. Shokubutsugaku Zasshi 79:714.

[58] Jones, H. A., and H. A. Borthwick. 1938. Amer. Potato J. 15:331.

[59] Kandeler, R. 1955. Z. Bot. 43:61.

[60] Kimura, K., and A. Takimoto. 1963. Shokubutsu-gaku Zasshi 76:67.

[61] Konstantinov, P. N. 1938. U.S. Dep. Agr. Off. Exp. Exp. Sta. Rec. 78:170.

[62] Laibach, F. 1943. Naturwissenschaften 31:246.

[63] Laibach, F. 1949. Ber. Deut. Bot. Ges. 62:27.

[64] Laibach, F. 1951. Beitr. Biol. Pflanz. 28:173.

[65] Lang, A. 1965. Handb. Pflanzenphysiol. 15(1):1380.

[66] Laurie, A. 1930. Proc. Amer. Soc. Hort. Sci. 27:319.

[67] Leopold, A. C. 1951. Quart. Rev. Biol. 26:247.

[68] Liverman, J. L. 1952. Ph.D. Thesis. California Institute of Technology, Pasadena.

[69] Liverman, J. L., and A. Lang. 1951. Abstr. W. Soc. Natur. Annu. Meet.

[70] Magruder, R., and H. A. Allard. 1937. J. Agr. Res. 54:719.

[71] McClelland, T. B. 1928. Ibid. 37:603.

[72] McKinney, H. H., and W. J. Sandow. 1933. J. Hered. 24:169.

[73] McKinney, H. H., and W. J. Sandow. 1935. J. Agr. Res. 51:621.

[74] Moshkov, B. S. 1939. C. R. (Dokl.) Acad. Sci. URSS 22:456.

[75] Murneek, A. E., and R. O. Whyte, ed. 1948. Vernalization and Photoperiodism. Chronica Botanica, Waltham, Mass.

[76] Napp-Zinn, K. 1960. Planta 54:409.

[77] Napp-Zinn, K. 1960. Ibid. 54:445.

[78] Nelson, C. H. 1944. Plant Physiol. 19:294.

[79] Okada, M. 1957. Engei Gakkai Zasshi 26:59.

[80] Pagorleckii, B. K. 1953. Agrobiologiya 3:153.

[81] Parker, M. W., and H. A. Borthwick. 1943. Bot. Gaz. 104:612.

[82] Peterson, M. L., and W. E. Loomis. 1949. Plant Physiol. 24:31.

[83] Peterson, M. L., et al. 1961. Crop Sci. 1:17.

[84] Plitt, T. M. 1932. Plant Physiol. 7:337.

[85] Post, K. 1942. Cornell Univ. Agr. Exp. Sta. Bull. 787.

[86] Post, K., and C. L. Weddle. 1940. Proc. Amer. Soc. Hort. Sci. 37:1037.

[87] Purves, W. K. 1961. Planta 56:684.

[88] Purvis, O. N. 1934. Ann. Bot. (London) 48:919.

[89] Resende, F. 1952. Port. Acta Biol. A3:318.

[90] Roberts, R. H. Unpublished.

[91] Roberts, R. H., and B. E. Struckmeyer. 1938. J. Agr. Res. 56:633.

[92] Roberts, R. H., and B. E. Struckmeyer. 1939. Ibid. 59:699.

[93] Rünger, W. 1959. Planta 53:602.

[94] Sachs, R. M. 1956. Plant Physiol. 31:185.

[95] Sachs, R. M. 1956. Ibid. 31:430.

[96] Sakr, E. S., and H. C. Thompson. 1942. Proc. Amer. Soc. Hort. Sci. 41:343.

[97] Salisbury, F. B. 1963. The Flowering Process. Macmillan, New York.

[98] Samygin, G. A. 1946. Tr. Inst. Fiziol. Rast. Akad. Nauk SSSR 3:129.

[99] Schwabe, W. W. 1950. J. Exp. Bot. 1:329.

[100] Schwabe, W. W. 1956. Ann. Bot. (London), N.S. 20:1.

[101] Scully, N. J., et al. 1945. Bot. Gaz. 107:52.

[102] Sen, B., and S. C. Chakravarti. 1942. Nature (London) 149:139.

[103] Sircar, S. M. 1946. Proc. Nat. Inst. Sci. India 12:191.

[104] Sircar, S. M., and B. Pariji. 1945. Nature (London) 155:395.

[105] Sprague, V. G. 1948. J. Amer. Soc. Agron. 40:144.

[106] Takimoto, A. 1966. Shokubutsugaku Zasshi 79:474.

[107] Thompson, H. C., and J. E. Knott. 1933. Proc. Amer. Soc. Hort. Sci. 30:507.

[108] Tiedjens, V. A. 1928. J. Agr. Res. 36:721.

[109] Tincker, M. A. H. 1925. Ann. Bot. (London) 39:721.

[110] Umemura, K., and Y. Oata. 1965. Plant Cell Physiol. 6:793.

[111] Wellensiek, S. J. 1959. Proc. Kon. Ned. Akad. Wetensch. C62:195.

[112] Wellensiek, S. J. 1966. Z. Pflanzenphysiol. 54:377.

[113] Went, F. W. 1945. Amer. J. Bot. 32:469.

[114] Withrow, R. B., ed. 1959. Publ. Amer. Ass. Advan. Sci. 55.

[115] Wort, D. J. 1941. Bot. Gaz. 102:725.

[116] Zeewart, J. A. D., and A. Lang. 1962. Planta 58:531.

105. VERNALIZATION REQUIREMENT FOR FLOWERING: MAGNOLIOPHYTES

Since only a few plants are capable of initiating flowers at the near-freezing temperatures optimal for vernalization, the effects of cold treatment do not become measurable until after the treated plant material has been removed from the low vernalization temperature to conditions which usually include higher temperature and a long photoperiod. Determination of the effectiveness of vernalization must, therefore, be indirect. The effects increase with increasing exposure to cold, until a maximum duration (saturation) is reached, beyond which there is little or no further observable promotion of flowering. Temperatures from −8 to +10°C are effective. Conditions before or after cold treatment may influence the nature and length of the vernalization requirement. **Vernalization Requirement:** Abs = absolute—plants are unable to flower without cold treatment; Quant = quantitative—flowering is accelerated by vernalization, but will eventually occur without it. **Vernalization Duration:** Number of days required for "saturation," unless otherwise indicated. Data in brackets refer to the column heading in brackets.

| | Species (Synonym) | Stage Treated | Vernalization | | | Requirement After Vernalization | Remarks | Reference |
			Require- ment	Duration da	Temp, °C [Optimum]			
1	*Agropyron cristatum*	Plants	Abs (?)	....	....	Long day	Medium vernalization requirement	28,74
2	*Agrostis alba*	Plants	Quant	30-90 [1]	0-2	Long day	....	27
3	*Althaea rosea*, biennial	Plants	Abs	169 [2]	Outside	Long day	....	33
4	*Anthriscus cerefolium*	Seeds, plants	Quant	2-4	5-15 [9]	Long day	....	62
5	*Apium graveolens*	Plants [3]	Abs	15-30	4.5-10	Long day, 15.6-21.0°C	....	77,78
6	*Arabidopsis thaliana*	Seeds	Quant	56	−3.5 to +5 [2]	Day length >4 hr, 20-25°C	Duration depends on day length after completion of cold treatment	10,58
7	*Avena sativa*, winter	Seeds	Quant	30-45 [1]	0-2	Long day	Summer varieties do not respond to vernalization	31,75
8	*Beta vulgaris*	Seeds	Abs	>80	0-5	≮15-hr day, 12-16°C	....	31,81
9		Plants	Abs	35-50 [1]	0-2	≮15-hr day, 12-16°C	Plants 4 mo old	15,21
10		Beets	Abs	20-60 [1]	5-6	≮15-hr day, 12-16°C	....	21,22,31
11	*Brassica campestris*	Plants [3]	Abs	....	....	....	Biennial strains only	14,45
12	*B. hirta (Sinapis alba)*	Seeds	Quant	10-30	0-3	Short (8-9 hr) or long day	Very small effect, except when grown in short day after vernalization	24,34, 51,72
13	*B. juncea*	Seeds	Quant	42	2-7	....	....	73
14	*B. napus* var. *napobrassica* (*B. napobrassica*)	Plants, seeds	Abs	56-71 [1]	3	Long day or continuous light, ∼20°C	....	29,81
15	*B. nigra (Sinapis nigra)*	Seeds	Quant	45 [4]	2	Long day	Small effect	72
16	*B. oleracea* var. *botrytis*	Plants	Abs	14 [4]	5	Long day	....	69
17	*B. oleracea* var. *capitata*	Seeds	Abs	28 [4]	0	Long day	Most varieties cannot be seed-vernalized	25
18		Plants	Abs	60	4.4	Long day	Some varieties do not require vernalization	78
19	*B. oleracea gongylodes*	Plants	Abs	60-90	5	Long day	....	29,31
20	*B. pekinensis*	Plants	Quant	16	4.4	≮16-hr photoperiod	Will flower in short day (8 hr), but much later than in long day	52
21	*B. rapa (B. rapa esculenta)*	Seeds	Abs	10-50 [1]	0-5	Long day	....	29,31

[1] Requirement depends on strain or variety. [2] November 1 - April 18. [3] Seeds cannot be vernalized. [4] Not specified whether indicated exposure time results in "saturation".

continued

	Species (Synonym)	Stage Treated	Vernalization			Requirement After Vernalization	Remarks	Reference
			Require-ment	Duration da	Temp, °C [Optimum]			
22	*Bromus inermis*	Plants	Abs; quant[1]	30-90[1]	0-2	Long day (>14-15 hr)	....	32,60,74
23	*Campanula longestyla*	Plants[3]	....	....	Winter, outside	>14-hr day, 18-22°C	No vernalization re-quired for photo-period <16 hr	53
24	*C. medium*	Plants[3], 21 wk old	Abs	>70	5	Photoperiod <16 hr	....	85
25		Plants[3], 29 wk old	Abs	49	5	Photoperiod <16 hr		
26	*Centaurea cyanus*	Seeds	Quant	21[4]	0-4	....	Small effect in short day (8 hr); no vernal-ization required when grown in continuous light	51
27		Plants	Quant	10[4]	0-4	....		
28	*Chondrilla juncea*	Plants	Quant	56	10.5	Slight pref-erence for long day	....	8
	Chrysanthemum morifolium							
29	'Magnet'	Plants	Quant	28	4.4	Long day	Small effect	79
30	'Shuokan'	Plants	Abs[5]	28	3	Short or long day	Low temperature was applied during the night only	33
31	'Sunbeam'	Plants	Quant	20	5-7	Short or long day	Large effect	70
32	*Cicer arietinum*	Seeds	Quant	30-40	2	....	Small effect	72,86
33	*Cichorium endivia*	Seeds	Quant	28	1-2.5	16-hr day	....	29,35,36
34	*C. intybus*	Seeds, plants	Abs	42	5	Long day	....	37
35	*Coreopsis grandiflora*	Plants	Abs	42	3	Long day	....	44
36	*Dactylis glomerata*	Plants[3]	Abs; quant[1]	30-90	0-2	Long day	Low-altitude mediter-ranean strains require little or no cold	20,28,32
37	*Daucus carota*	Plants	Abs	40-65[1]	4	Long day	....	12,29,78
38		Roots	Abs	60	4	Long day	....	14,78
39	*Delphinium ajacis*	Seeds	Quant	30	3	Short day	Small effect when plants grown in 14- to 18-hr photoperiod	40
40	*D. consolida*	Seeds	Quant	20[4]	0-4	8-hr day or continuous light, 15°C	Effect reliable but small	51
41		Young plants	Quant	10[4]	0-4	8-hr day or continuous light, 15°C		
42	*Dianthus barbatus*	Plants[3]	Abs	42-63	5	Short or long day	....	82
43	*D. caryophyllus*	Plants	Quant	21	2-5	Short or long day, <10°C	Plants with 9-11 pairs of leaves at beginning of treatment	5
44	*D. purpurea*	Plants[3]	Abs	100	4-10	Long day or continuous light	....	3,12,29

[1] Requirement depends on strain or variety. [3] Seeds cannot be vernalized. [4] Not specified whether indicated exposure time results in "saturation". [5] More or less.

continued

Species (Synonym)	Stage Treated	Vernalization			Requirement After Vernalization	Remarks	Reference
		Require-ment	Duration da	Temp, °C [Optimum]			
45 Euphorbia lathyris	Plants	Abs	....	....	Short or long day	....	14
46 Festuca elatior	Plants	Quant	30-90 [1]	0-2	16-hr day	Some varieties have a more or less absolute cold requirement, others do not require cold	6,27,32
47 F. elatior arundinaceae	Plants	Quant	>14	5	Long day	....	76
48 F. rubra	....	Abs	....	....	18-hr day	Long vernalization requirement	28,74
49 Geum urbanum	Plants[3]	Abs	42-70	0-4	Short or long day	42 wk required for induction of apical meristem	14,83
50 Helianthus annuus	Seeds	Quant	35	2	14- to 16-hr day	No effect with 17- to 18-hr day	72
Hordeum vulgare							
51 Winter	Seeds	Quant	20-40 [1]	[0-3]	Long day	....	29,31,
52 Spring	Seeds	Quant	0-15 [1]	[6-8]	Long day	Effect very slight, if any	67,86
53 Hyoscyamus niger, biennial	Plants[1]	Abs	42-105	3-10	Long day	Effect of vernalization increases up to 105 days	48
54 Lactuca sativa	Seeds	Quant	10-20 [1]	2-5	Long day	....	31,41,86
55 Lathyrus odoratus	Seeds	Quant	40	2	Long day	Small effect	72
56 Lens culinaris (L. esculenta)	Seeds	Quant	10-12	6-10	....	Small effect	31,86
57 Linum austriacum	Plants	Abs	60	3-4	Long day	....	23
58 Lolium multiflorum	Seeds	Quant	30-40	0-5	Long day, continuous light	Some varieties show little or no response	18,47
59 Biennial	Seeds	Abs[5]	35-42	0-5	Long day, continuous light	....	18
60 L. perenne	Seeds	Abs; quant[1]	0-90 [1]	0-5	Long day, continuous light	....	16,17
61 Lunaria annua (L. biennis)	Plants	Abs	77-84	5	Short (8 hr) or long (16 hr) day	Plants 11 wk old at beginning of treatment. Younger plants less sensitive. Plants >6 wk cannot be vernalized.	64,84
62 Lupinus angustifolius	Seeds	Quant	14-21	2-5	Long day	....	31,56, 68,72
63 Matthiola incana	Plants	Quant	10-20	10	Long day	Plants with 10 leaves at beginning of cold treatment	46,65
64 Medicago tribuloides	Seeds	Quant	7-35	3.3	14.5-hr day	....	2
65 Melilotus alba	Plants[3]	Abs	30	3-7	13- to 16-hr day	Plants with 3 leaves at beginning of treatment. Will flower without vernalization in 17- to 24-hr photoperiod.	39,42

[1] Requirement depends on strain or variety. [3] Seeds cannot be vernalized. [5] More or less.

continued

Species (Synonym)	Stage Treated	Vernalization			Requirement After Vernalization	Remarks	Reference
		Requirement	Duration da	Temp, °C [Optimum]			
66 *M. officinalis*	Plants	Abs (?)	28	4.4-10	13- to 16-hr day	Will flower without vernalization in 17- to 24-hr photoperiod	42,47
67 *Myosotis alpestris*	....	Abs	....	....	16-hr day	....	57
68 *Oenothera biennis*	Plants[3/]	Abs	30[4/]	4	Long day or continuous light	....	12,29
69 *O. parviflora*	Plants[3/]	Abs	30-60	3-4	Long day	....	13,14
70 *Oryza sativa*	Seeds	Quant	14-21	3-12	....	Small effect	9,86
71 *Papaver somniferum*	Seeds	Quant	35	2-3	Long day	Small effect	49
72 *Phalaris canariensis*	Seeds	Quant	42	0-5	Long day or continuous light, 10-21°C	Small effect	18
73 *P. tuberosa*	Plants	Abs; quant[1/]	0-100[1/]	2.5-4	Long day	Seedlings with 3 leaves at beginning of cold treatment. Seeds may be vernalized but are less responsive.	43,56
74 *Phleum pratense*	Seeds	Quant	0-35	0-5	Long day	Small effect	18,19
75 *Pisum sativum*	Seeds	Quant	20-30	2-7	Short (8 hr) or long (16 hr) day	Small but significant effect; not all varieties respond to vernalization	4,14,38, 72
76 *P. sativum arvense*	Seeds	Quant	27-40	0-2	Long day	....	54,72
77 *Poa pratensis*	Plants	Abs	≮62	Winter, outside	Long day (16 hr)	....	28,50
78 *Raphanus sativus*	Seeds	Quant	10-46[1/]	0-5	Long day	Small effect. Seedlings respond in same manner.	72,87
79 *Ricinus communis*	Seeds	Quant	28-40	2	Long day	Very slight effect	71
80 *Salvia horminum*	Seeds	Quant	39[4/]	2	Long day	....	72
Secale cereale							
81 Winter	Seeds	Quant	30-50[1/]	[0-5]	Long day	....	31
82 Petkus winter	Seeds	Quant	49	[1]	Long day	....	29,66
83 Spring	Seeds	Quant	0-14	[0-5]	Long day	Usually small effect	31
84 *Sorghum vulgare sudanense*	Seeds	Quant	....	−2 to +5	....	Small effect	63
85 *Spinacia oleracea*	Seeds	Quant	10-15[1/]	2-8	Long day	Cold treatment lowers critical day length	29,41,61
86 'Nobel'	Seeds	Quant	56	[2-5]	Photoperiod 10-12 hr or longer	No vernalization required for photoperiod ≮18 hr	80
87 *Tagetes patula*	Seeds	Quant	36[4/]	2	Long day	Very small effect	72
88 *Trifolium incarnatum*	Seeds	Quant	40[4/]	0	Long day	....	54
89 *T. pratense*	Seeds	Quant	10-40[1/]	3-8	Long day	....	9,55
90 *T. repens*	Seeds	Abs; quant[1/]	15-30	3	Long day	....	7
91 *T. subterraneum*	Plants	Quant	14-35[1/]	7	Long day or continuous light	....	26
92 'Dwalganup', early	Seeds	Quant	14-21	3.3	Short or long day	....	1
93 'Tallarook', late	Seeds	Abs	42-56	3.3	≮14.5-hr day, 20-24°C	....	1

[1/] Requirement depends on strain or variety. [3/] Seeds cannot be vernalized. [4/] Not specified whether indicated exposure time results in "saturation".

continued

	Species (Synonym)	Stage Treated	Vernalization			Requirement After Vernalization	Remarks	Reference
			Require-ment	Duration da	Temp, °C [Optimum]			
	Triticum aestivum							
94	Winter	Seeds	Quant	40-70[1]	0-3	Long day	....	29-31,55
95	Spring	Seeds	Quant	0-14[1]	0-8	Long day	Small effect	
96	*Vicia faba*	Seeds	Quant	35-40	2	....	Relatively small effect	72
97	*Viola hirta*	....	Abs	....	....	Short day	Production of chasmo-gamic flowers	11
98	*V. odorata*	....	Quant	....	....	Short day	Production of chasmo-gamic flowers	11
99	*Zea mays*	Seeds	Quant	34	3	....	Small effect	59

[1] Requirement depends on strain or variety.

Contributor: Ketellapper, H. J.

References

[1] Aitken, Y. 1955. Aust. J. Agr. Res. 6:212.
[2] Aitken, Y. 1955. Ibid. 6:258.
[3] Arthur, J. M., and E. K. Harvill. 1941. Contrib. Boyce Thompson Inst. 12:111.
[4] Barber, H. N. 1959. Heredity 13:33.
[5] Blake, J. 1956. Bull. Soc. Fr. Physiol. Veg. 2:169.
[6] Bommer, D. 1959. Z. Acker - Pflanzenbau 109:95.
[7] Cairns, D. 1941. N. Z. J. Sci. Technol. A22:279.
[8] Caso, O. H., and N. P. Kefford. 1968. Aust. J. Biol. Sci. 21:883.
[9] Chadwick, D., et al., ed. 1935. Imp. Bur. Plant Genet. (Cambridge) Bull. 17.
[10] Chintraruck, Bh., and H. J. Ketellapper. 1969. Plant Cell Physiol. 10:271.
[11] Chouard, P. 1948. C. R. Acad. Sci. 266:1831.
[12] Chouard, P. 1951. Bull. Soc. Bot. Fr. 98:11.
[13] Chouard, P. 1956. Bull. Soc. Fr. Physiol. Veg. 2:125.
[14] Chouard, P. 1960. Ann. Rev. Plant Physiol. 11:191.
[15] Chroboczek, E. 1934. Cornell Univ. Agr. Exp. Sta. Mem. 154.
[16] Cooper, J. P. 1954. Congr. Int. Bot. Rapp. Commun. 8(11/12):356.
[17] Cooper, J. P. 1956. J. Agr. Sci. 47:262.
[18] Cooper, J. P. 1957. Ibid. 49:361.
[19] Cooper, J. P. 1958. J. Brit. Grassl. Soc. 13:81.
[20] Cooper, J. P. 1963. In L. T. Evans, ed. Environmental Control of Plant Growth. Academic Press, New York. p. 381.
[21] Curth, P. 1955. Zuechter 25:176.
[22] Curth, P. 1962. Z. Pflanzenzuecht. 47:254.
[23] Dang, K. D., and F. Chodat. 1958. Experientia 14:68.
[24] Denffer, D. von 1939. Jahrb. Wiss. Bot. 88:759.
[25] Dikshit, N. N., and U. P. Singh. 1952. Current Sci. (India) 21:249.
[26] Evans, L. T. 1959. Aust. J. Agr. Res. 10:1.
[27] Fedorov, A. K. 1960. Izv. Akad. Nauk SSSR, Ser. Biol. (5):775.
[28] Gardner, F. P., and W. E. Loomis. 1953. Plant Physiol. 28:201.
[29] Gassner, G. 1918. Z. Bot. 10:417.
[30] Gassner, G. 1953. Zuechter 23:193.
[31] Hänsel, H. 1953. Z. Pflanzenzuecht. 32:233.
[32] Hanson, A. A., and V. Sprague. 1953. Agron. J. 45:248.
[33] Harada, H. 1962. Rev. Gen. Bot. 69:201.
[34] Harder, R., and I. Störmer. 1936. Landwirt. Jahrb. 83:401.
[35] Harrington, J. F., et al. 1956. Science 125:601.
[36] Harrington, J. F., et al. 1959. Neth. J. Agr. Sci. 7:68.
[37] Hartmann, T. A. 1956. Proc. Kon. Ned. Akad. Wetensch. C59:677.
[38] Highkin, H. R. 1956. Plant Physiol. 31:399.
[39] Johnson, I. J. 1933. Sci. Agr. 113:746.
[40] Junges, W. 1957. Planta 49:11.
[41] Junges, W. 1959. Z. Pflanzenzuecht. 41:103.
[42] Kasperbauer, M. J., et al. 1962. Plant Physiol. 37:165.
[43] Ketellapper, H. J. 1960. Ecology 41:298.
[44] Ketellapper, H. J., and A. Barbaro. 1966. Phyton 25:33.
[45] Kloen, D. 1954. Congr. Int. Bot. Rapp. Commun. 8(11/12):291.
[46] Kohl, H. C. 1958. Proc. Amer. Soc. Hort. Sci. 72:481.
[47] Koreisa, J. V. 1935. Herb. Rev. 3:94.
[48] Lang, A. 1951. Zuechter 21:241.
[49] Lecat, P. 1955. C. R. Acad. Sci. 241:1984.
[50] Lindsey, K. E., and M. L. Peterson. 1962. Crop. Sci. 2:71.
[51] Listowski, A., and A. Jasmanowicz. 1961. Rocz. Nauk Roln. A83:695.
[52] Lorentz, O. A. 1946. Proc. Amer. Soc. Hort. Sci. 47:309.
[53] Mathon, C. C. 1960. Bull. Soc. Bot. Fr. 107:92.
[54] McKee, R. 1935. U.S. Dep. Agr. Circ. 377.
[55] McKinney, H. H. 1940. Bot. Rev. 6:25.
[56] McWilliam, J. R. 1968. Aust. J. Biol. Sci. 21:395.
[57] Michniewicz, M., and A. Lang. 1962. Planta 58:549.
[58] Napp-Zinn, K. 1957. Ibid. 50:177.

continued

[59] Napp-Zinn, K. 1961. Encycl. Plant Physiol. (Berlin) 16:24.

[60] Newell, L. C. 1951. Agron. J. 43:417.

[61] Parlevliet, J. E. 1967. Meded. Landbouwhogesch. Wageningen 67(2):1.

[62] Parlevliet, J. E. 1967. Z. Pflanzenphysiol. 58:76.

[63] Peregudov, W. I. 1960. Tr. Stavropol. Sel'skokhoz. Inst. 9:3.

[64] Pierik, R. L. M. 1967. Meded. Landbouwhogesch. Wageningen 67(6):1.

[65] Post, K. 1936. Proc. Amer. Soc. Hort. Sci. 33:649.

[66] Purvis, O. N., and F. G. Gregory. 1937. Ann. Bot. (London), N.S. 1:569.

[67] Razumov, V. I., and M. I. Smirnova. 1934. Tr. Prikl. Bot. Genet. Selek., Ser. 15, 4:47.

[68] Rudorf, W. 1958. In H. Kappert and W. Rudorf, ed. Grundlagen der Pflanzenzuechtung. P. Parey, Berlin. v. 1., p. 225.

[69] Sadik, S., and J. L. Ozbun. 1968. Plant Physiol. 43: 1696.

[70] Schwabe, W. W. 1950. J. Exp. Bot. 1:329.

[71] Séchet, J. 1953. Bull. Soc. Bot. Fr. 100:44.

[72] Séchet, J. 1953. Botaniste 37:1.

[73] Sen, B., and J. C. Chakravarti. 1942. Indian J. Agr. Sci. 8:245.

[74] Stepanov, V. N. 1958. Izv. Timiryazev. Sel'skokhoz. Akad. (2):7.

[75] Taylor, J. W., and F. A. Coffman. 1938. J. Amer. Soc. Agron. 30:1010.

[76] Templeton, W. C. 1960. Diss. Abstr. 21:20.

[77] Thompson, H. C. 1944. Proc. Amer. Soc. Hort. Sci. 45:425.

[78] Thompson, H. C. 1953. In W. E. Loomis, ed. Growth and Differentiation in Plants. Iowa State College Press, Ames. p. 179.

[79] Vince, D. 1955. J. Hort. Sci. 30:34.

[80] Vlitos, A. J., and W. Meudt. 1955. Contrib. Boyce Thompson Inst. 18:159.

[81] Voss, J. 1940. Zuechter 12:34, 73.

[82] Waterschoot, H. F. 1960. Proc. Kon. Ned. Akad. Wetensch. C60:318.

[83] Weber, M.-R. 1956. Bull. Soc. Fr. Physiol. Veg. 2: 169.

[84] Wellensiek, S. J. 1958. Proc. Kon. Ned. Akad. Wetensch. C61:561.

[85] Wellensiek, S. J. 1960. Meded. Landbouwhogesch. Wageningen 60(7):1.

[86] Whyte, R. O. 1946. Crop Production and Environment. Faber and Faber, London.

[87] Yamamoto, K. 1933. Sapporo Norin Gakkaiho 25: 260.

106. FACTORS AFFECTING TRANSPIRATION RATES: MAGNOLIOPHYTES

Part I. Various Conditions

For information on plant canopy loss per land area, or consumptive use of water by crops, and on stomatal resistance to water loss, consult reference 7. Water loss was calculated from data obtained in experiments using intact plants growing in specified substrates or leaves. **Specifications:** AR = air rate; AT = air temperature; cal = calorie; EA = evaporation from atmometer; LI = light intensity; LT = leaf temperature; REF = radiation energy flux; RH = relative humidity; SD = saturation deficit of air; SM = soil moisture; ST = soil temperature; VPD = vapor pressure deficit; WHC = soil moisture in percent of water-holding capacity on dryweight basis. **Water Loss:** dm = decimeter. Data in brackets refer to the column heading in brackets.

	Species	Plant Part [Substrate]	Specifications		Water Loss mg/dm^2 leaf surface/hr	Reference
			Constants	Variables		
1	*Ananas comosus* 'Smooth Cayenne'	Intact plant [Nutrient solution]	4 plants per treatment, 2 mo old; LI = 35,000 lux	AT (day & night) = 25°C EA (1430 hr) = 3.1 ml/hr	271[1]/	13
2				(0030 hr) = 1.4 ml/hr	50[2]/	
3				AT (day) = 25°C EA (1530 hr) = 3.0 ml/hr	140[1]/	
4				At (night) = 15°C EA (0030 hr) = 0.9 ml/hr	75[2]/	
5	*Avena sativa* 'Wintok'	Intact plant [Sand]	12 plants, 6-wk-old seedlings; high RH	SM = 5%	83	11
6				10%	1208	
7				15%	1542	
8	*Citrus aurantium*	Intact plant [Soil]	3 plants, 5-6 mo old; AT = 25°C	SM = 7%	474	1
9				9%	636	
10				11%	654	
11				17%	780	

[1]/ Minimum. [2]/ Maximum.

continued

Part I. Various Conditions

	Species	Plant Part [Substrate]	Constants	Variables	Water Loss mg/dm² leaf surface/hr	Reference
12	*C. jambhiri*	Intact plant	1 plant; AT = 30.0; RH = 35%	LI = 100,000 lux; WHC = 100	236	5
13		[Nutrient	1 plant; AT = 30.0; RH = 35%	LI = 0 lux	18	4
14		solution]		26,100 lux	155	
15	*C. limet*	Intact plant	3 plants, 5-6 mo old; AT =	SM = 7%	294	1
16	*tioides*	[Soil]	25°C	9%	630	
17				11%	900	
18				17%	1338	
19	*Gossypium*	Intact plant	2 plants, 8-12 wk old; LI =	SD = 4.4 mbars, in light	400	6
20	*barbadense*	[Nutrient	24,000 lux; REF = 0.2 cal/	in dark	30	
21		solution]	cm²/min	SD = 13.7 mbars, in light	380	
22				in dark	50	
23				SD = 27.9 mbars, in light	570	
24				in dark	60	
25			1 plant; AT = 30.0; RH = 35%	LI = 2800 lux	67	4
26				61,200 lux	290	
27	*Helianthus*	Intact plant	1 plant; AT = 30.0; RH = 35%	LI = 7000 lux	270	4
28	*annuus*	[Nutrient solution]		65,600 lux	576	
29	*Hordeum*	Intact plant	12 plants, 6-wk-old seedlings;	SM = 5%	125	11
30	*vulgare*	[Sand]	high RH	10%	792	
31	'Chase'			15%	958	
32	'Delta'	Detached	6-8 leaves, 13-da-old plants;	1000 hr	2200	9
33		leaves	AT = 30°C[2]/ day, 15°C[1]/	1200 hr	2600	
34			night; RH = 40-60%; LI =	1400 hr	2600	
35			10,800-27,000 lux	1800 hr	900	
36	*Musa acuminata*	Detached leaves	40-45 of youngest leaves, 1-m-tall plants; AT = 25°C; RH = 32%; LI = 35,000 lux	....	1100	2
37	*Phaseolus*	Intact plant	1 plant; AT = 30.0; RH = 35%	LI = 55,000 lux; WHC = 100	450	5
38	*vulgaris*	[Nutrient	1 plant; AT = 30.0; RH = 35%	LI = 2200 lux	90	4
39		solution]		53,300 lux	321	
40	'Black Valentine'	Intact leaves	12-48 leaves, 20-da-old plants; AT = 23°C; LI =	Shaded	76	8
41			13,200 lux; AR = 160 cm³/min	Unshaded	109	
42			12-48 leaves, 20-da-old	AR = 160 cm³/min	107	8
43			plants; AT = 23°C; LI = 13,200 lux; in light only	AR = 260 cm³/min	150	
44	*Prosopis*	Intact plant	2 plants, 9-wk-old seedlings	ST = 13°C	104	12
45	*glandulosa*	[Soil]		21°C	204	
46	var. *glan*			29°C	242	
47	*dulosa*			38°C	225	
48				AT = 18.0°C; RH = 77%; VPD = 3.5 mm Hg	96	
49				AT = 22.0°C; RH = 63%; VPD = 7.3 mm Hg	208	

[1]/ Minimum. [2]/ Maximum.

continued

Part I. Various Conditions

	Species	Plant Part [Substrate]	Specifications — Constants	Specifications — Variables	Water Loss mg/dm² leaf surface/hr	Reference
50				AT = 29.5°C; RH = 50%; VPD = 14.5 mm Hg	217	
51				AT = 32.0°C; RH = 41%; VPD = 20.9 mm Hg	250	
52	*Sorghum vul-*	Intact	Several plants with 6-8 leaves;	LT = 15°C	18	3
53	*gare su-*	leaves	REF = 0.06 cal/cm²/min	20°C	24	
54	*danense*			25°C	30	
55	'Greenleaf'			30°C	42	
56				35°C	48	
57			REF = 0.46 cal/cm²/min	LT = 25°C	84	3
58				30°C	102	
59				35°C	132	
60	*Triticum*	Intact	Several plants with 6-8 leaves;	LT = 15°C	60	3
61	*aestivum*	leaves	REF = 0.06 cal/cm²/min	20°C	72	
62	'Gabo'			25°C	78	
63				30°C	84	
64				35°C	96	
65			REF = 0.46 cal/cm²/min	LT = 20°C	72	3
66				25°C	90	
67				30°C	102	
68				35°C	132	
69	'Red Chief'	Intact plant [Sand]	12 plants, 6-wk-old seedlings; high RH	SM = 5%	125	11
70				10%	833	
71				15%	1250	
72	*Vitis vinifera* 'Sultana'	Detached leaves	6 leaves from fruiting plants; AT = 20°C	....	473	10

Contributors: Krauss, Beatrice; Ehrler, W. L.

References
[1] Bielorai, H., and K. Mendel. 1969. J. Amer. Soc. Hort. Sci. 94:203.
[2] Brun, W. A. 1965. Plant Physiol. 40:800.
[3] Downes, R. W. 1970. Aust. J. Biol. Sci. 23:777.
[4] Ehrler, W. L. 1968. Plant Physiol. 43:208.
[5] Ehrler, W. L. Unpublished. U.S. Dep. of Agriculture, Soil and Water Conservation Research Division, Phoenix, Ariz., 1971.
[6] Ehrler, W. L., et al. 1966. Plant Physiol. 41:72.
[7] Ekern, P. C. 1971. Eos, Trans. Amer. Geophys. Union 52:286.
[8] Koontz, H. V., and R. E. Foote. 1966. Physiol. Plant. 19:316.
[9] Livnè, A., and Y. Vaadia. 1965. Ibid. 18:660.
[10] Prossingham, J. V., et al. 1967. Aust. J. Biol. Sci. 20:1151.
[11] Salim, M. H., and G. W. Todd. 1965. Agron. J. 57: 595.
[12] Wendt, C. W., et al. 1968. Ibid. 60:383.
[13] Yader, R. C. 1969. M.S. Thesis. Univ. Hawaii, Honolulu. pp. 51, 53.

Part II. Variation in Soil Conditions for Corn

	Soil Condition	Leaf Area per Plant cm²	Dry Matter per Plant, g — Grain	Dry Matter per Plant, g — Total	Water Loss per Plant kg	Water Loss per Gram Grain, g[1]	Water Loss per Gram Dry Matter, g[1]
				8-Year Average			
1	No manure added Infertile	2948	17	72	38.3	8263	531

[1] Average of ratios.

continued

Part II. Variation in Soil Conditions for Corn

	Soil Condition	Leaf Area per Plant cm^2	Dry Matter per Plant, g		Water Loss		
			Grain	Total	per Plant kg	per Gram Grain, g[1]	per Gram Dry Matter, g[1]
2	Intermediate	3573	26	108	51.3	2485	489
3	Fertile	4612	74	213	77.4	1100	368
	2.4 lb manure added/plant						
4	Infertile	4244	52	180	68.9	1518	396
5	Intermediate	4638	67	207	75.6	1173	369
6	Fertile	5121	108	289	94.2	879	327
	3-Year Average						
7	Too dry (50% saturation)	6818	165	379	105.5	695	289
8	Favorable (70% saturation)	8056	230	522	162.2	716	316
9	Too wet (95% saturation)	7031	168	422	142.1	854	341

[1] Average of ratios.

Contributor: Kiesselbach, T. A.

Reference: Kiesselbach, T. A. 1929. Proc. Int. Congr. Plant Sci., 1926, 1:87.

Part III. Diurnal Variation for Corn

	Time	Temp °C	Mean Relative Humidity %	Light Intensity Energy cal/cm^2/min	Soil Matric Potential[1] coulomb		Transpiration Rate[2]	
					Wet Treatment	Dry Treatment	Wet Treatment	Dry Treatment
1	0800 hr	22.0	53	.....	..	..	.....	.....
2	0900 hr	22.0	53	0.41	5	..	10.9	9.4
3	1000 hr	24.5	57	0.71	8	6	22.0	14.5
4	1100 hr	24.5	57	0.93	8	8	29.4	22.5
5	1200 hr	28.0	55	1.05	12	12	33.2	24.1
6	1300 hr	28.0	55	1.11	20	17	34.6	25.8
7	1400 hr	28.7	54	1.10	2	25	36.5	27.0
8	1500 hr	28.7	54	0.88	1	36	30.0	25.3
9	1600 hr	29.3	44	0.62	4	45	26.4	22.4

[1] Measured by tensiometer. [2] Values are means for three plants.

Contributor: Ehrler, W. L.

Reference: Ehrler, W. L. Unpublished. U.S. Dep. of Agriculture, Soil and Water Conservation Research Division, Phoenix, Ariz., 1971.

continued

Part IV. Annual Variation

For information on transpiration under greenhouse conditions, consult reference 2. **Average:** Values are comparable only when derived from data for identical years. Figures in brackets are reference numbers.

	Species	Seasonal Water Requirement, $g^{1/}$						
		1912	1913	1914	1915	1916	1917	Average
	South Dakota [1]							
1	*Medicago sativa*	735	735	1038	696	673	866	790
2	*Setaria italica*	239	293	311	171	233	278	254
3	*Sorghum vulgare sudanense*	....	....	......	272	314	344	310
4	*Triticum aestivum*	463	436	528	333	352	487	433
	Colorado [3]							
5	*Amaranthus retroflexus*	....	320	306	229	340	307	300
6	*Avena sativa*	449	617	615	445	809	636	595
7	*Bouteloua gracilis*	....	....	389	312	336	290	332
8	*Gossypium hirsutum*	488	657	574	443	612	522	549
9	*Hordeum vulgare*	443	....	501	404	664	522	507
10	*Medicago sativa*	657	834	890	695	1047	822	824
11	*Secale cereale*	....	....	622	469	800	625	629
12	*Setaria italica*	187	286	295	202	367	284	273
13	*Sorghum vulgare sudanense*	....	....	394	260	426	378	365
14	*Triticum aestivum*	394	496	518	405	636	471	487
15	*Vigna sinensis*	.....	571	659	413	767	481	578
16	*Zea mays*	280	399	368	253	495	346	357

$1/$ Ratio of weight of water transpired to weight of dry matter produced.

Contributor: Bailey, Lowell F.

References
[1] Dillman, A. C. 1931. J. Agr. Res. 42:187.
[2] Ehrler, W. L. 1969. In C. C. Hoff and M. L. Riedesel, ed. Physiological Systems in Semi-Arid Environments. Univ. New Mexico Press, Albuquerque. pp. 239-247.
[3] Shantz, H. L., and L. N. Piemeisel. 1927. J. Agr. Res. 34:1093.

Part V. Solutes

Data are for plants grown in a nonsaline nutrient solution and then placed in a medium containing a soluble salt or other osmotically active solute. Decreased transpiration normally follows, caused by (i) decreased leaf area resulting from the effect of salinity on plant growth, and (ii) decreased transpiration per unit leaf area possibly resulting from stomatal closure. Stomatal resistance has often been assumed to depend on pressure (turgor) potential of leaves; therefore, values are given for both control and experimental plants. Measurements of pressure potential and transpiration rate were made at the end of the period of exposure to the solute. **Transpiration Rate** is expressed on a unit leaf area basis, unless otherwise indicated. Data in brackets refer to the column heading in brackets.

continued

106. FACTORS AFFECTING TRANSPIRATION RATES: MAGNOLIOPHYTES

Part V. Solutes

	Species (Synonym) [Stage]	Solute	Duration of Exposure	Temp, °C [Relative Humidity]	Osmotic Potential of Root Medium bar	Leaf Pressure Potential, bar Control Plants	Experimental Plants	Transpiration Rate % of Control	Reference
1	*Allium cepa* [4-leaf stage]	NaCl & CaCl$_2$	7 da	16.5 [80%]	−5.4	8.8	2.2	65 ± 12	3
2				33 [35%]	−5.4	4.8	−3.1	24 ± 5	
3	*Capsicum frutescens* 'California Wonder' [8- to 10-leaf stage]	Polyethylene glycol 400[1]	1 da	25 [70%]	−5.4	...	.....	64	7
4			7 da	25 [70%]	−5.4	5.1	7.3	67	
5		600[1]	1 da	25 [70%]	−5.4	...	.....	64	
6			7 da	25 [70%]	−5.4	5.1	4.9	81	
7		1000[1]	1 da	25 [70%]	−5.4	...	.....	63	
8			7 da	25 [70%]	−5.4	5.1	4.1	77	
9		1540[1]	1 da	25 [70%]	−5.0	...	.....	67	
10			7 da	25 [70%]	−5.0	5.1	4.3	73	
11		4000[2]	1 da	25 [70%]	−5.1	...	.....	44	
12			7 da	25 [70%]	−5.1	5.1	5.7	71	
13		Polyethylene glycol 400 or NaCl[3]	3 da	25 [70%]	−5.0	...	.....	68	6
14	*Gossypium hirsutum* [6- to 10-leaf stage]	NaCl	3 wk	32 ± 1[4], 26 ± 1[5]	−3.5	2.5	3.5	120[6]	1
15					−6.5	2.5	5.0	120[6]	
16			4 wk	32 ± 1[4], 26 ± 1[5]	−8.5	2.5	6.0	130[6]	
17		NaCl & CaCl$_2$	5 da	16.5 [80%]	−8.5	9.7	10.8	64 ± 8	3
18				33 [35%]	−5.4	8.0	9.0	72 ± 6	
19					−8.5	6.0	8.0	66 ± 6	
20	[8.5-wk-old stage]	NaCl	4 wk	25 [75%]	−6	5.7	6.0	93[7]	4
21					−12	5.7	3.3	85[7]	
22	[10-wk-old stage]	NaCl	Gradually salinized 15 da, steady 7 da	26[8], 38[9] [25%]	−15	...	.....	78[10]	5
23					−15	9.6	8.7	130[10,11]	
24				26[8], 38[9] [90%]	−15	...	.....	79[10]	
25	[18-wk-old stage]	NaCl	11 wk	26[8], 38[9] [90%]	−15	7.9	6.2	100[10,11]	5
26	*Hordeum vulgare* [6-wk-old stage]	NaCl	Gradually salinized 20 da	25 [75%]	0[12], −15[12]	8	9	89[13]	8
27					−15	8	13	98[13]	
28	*Lycopersicon esculentum* 'Grosse Lisse' [5-leaf stage]	Mannitol	6 hr	25 ± 2 [68-80%]	−5	8	6	24	13
29					−10	8	0	16	
30			21 hr	25 ± 2 [68-80%]	−5	...	.....	67	
31					−10	...	.....	18	
32			28 hr	25 ± 2 [68-80%]	−5	8	7	63	
33					−10	8	0	13	

[1] Untreated commercial product. [2] Dissolved in water and passed through deionizing resin to remove toxic material. [3] No signfiicant difference in response of plants to the two solutes was recorded. [4] Day. [5] Night. [6] Unusually high transpiration rates may have resulted from lower than normal pressure potentials for the control leaves. [7] Light intensity = 5mW·cm^{-2}. [8] Minimum. [9] Maximum.

[10] Growing in sunlit chambers. [11] Leaf area of control plants was nearly 10 times that of salinized plants. Shading undoubtedly lowered transpiration rate per unit leaf area for control treatment. [12] Roots split equally between solutions of two osmotic potentials. [13] Transpiration rate expressed as water loss per unit top weight; light intensity = 32,000 lux.

continued

106. FACTORS AFFECTING TRANSPIRATION RATES: MAGNOLIOPHYTES

Part V. Solutes

Species (Synonym) [Stage]	Solute	Duration of Exposure	Temp, °C [Relative Humidity]	Osmotic Potential of Root Medium bar	Leaf Pressure Potential, bar Control Plants	Experimental Plants	Transpiration Rate % of Control	Reference
34	NaCl	6 hr	25 ± 2 [68-80%]	−5	8	6	28	13
35				−10	8	3	19	
36		21 hr	25 ± 2 [68-80%]	−5	...	.....	92	
37				−10	...	.....	65	
38		28 hr	25 ± 2 [68-80%]	−5	8	12	90	
39				−10	8	12	67	
40 [30- to 50-cm stage]	Concentrated Hoagland solution	40 hr	27	−2.5	...	.....	55[14]	12
41				−5	...	.....	36[14]	
42 *Nicotiana tabacum* [60-da-old stage]	None (ABA)[15]	2 da	....	0	...	.....	61[16]	11
43	None (kinetin)[15]	2 da	....	0	...	.....	132[16]	11
44	Mannitol	2 da	....	−3.6	...	.....	75[16]	11
45	NaCl	2 da	....	−3.6	...	.....	69[16]	11
46	NaCl (ABA)[15]	2 da	....	−3.6	...	.....	48[16]	11
47	NaCl (kinetin)[15]	2 da	....	−3.6	...	.....	100[16]	11
Phaseolus vulgaris								
48 [2-leaf stage]	NaCl & CaCl$_2$	5 da	16.5 [80%]	−3.4	5.7	6.7	70 ± 9	3
49			33 [35%]	−3.4	...	.....	56 ± 8	
50 [4 da after primary leaves unfold]	NaCl	1 da	20 [65%]	−1.2	4.2	2.8	96[17]	2
51				−2.5	4.2	2.6	92[17]	
52				−3.8	4.2	1.1	59[17]	
53				−5.0	4.2	1.0	34[17]	
54 [8-leaf stage[18]]	NaCl & CaCl$_2$	3 wk	25[4], 15[5] [40%]	−2.1	...	.....	87	9
55				−3.7	...	.....	74	
56				−5.2	...	.....	47	
57			25[4], 15[5] [96%]	−2.1	...	.....	94	
58				−3.7	...	.....	96	
59				−5.2	...	.....	89	
60 [2-wk-old stage]	NaCl	Gradually salinized 3 da, steady 4 da	25 [65%]	−1.4	...	.....	89[19]	10
61				−2.1	...	.....	83[19]	
62				−2.8	...	.....	72[19]	
63				−3.5	...	.....	61[19]	
64 [3-wk-old stage]	NaCl	2 wk	25 [75%]	−2	6.9	5.9	77[7]	4
65				−4	6.9	5.7	49[7]	
66 [5-wk-old stage]	NaCl	16 da	25 [70-85%]	0[12], −4[12]	2.3	5.3	99[19]	8
67				0[12], −4[12]	2.3	2.2	60[19]	

[4] Day. [5] Night. [7] Light intensity = 5mW·cm^{-2}. [12] Roots split equally between solutions of two osmotic potentials. [14] Light intensity = 30,000 lux. [15] Leaves sprayed three successive times with $4 \times 10^{-5} M$ dl-abscisic acid (ABA), or $5 \times 10^{-5} M$ kinetin, prior to transpiration measurements. [16] Transpiration measurements made on detached leaves with petioles in water, at 2300 lux light intensity. [17] Light intensity = 4×10^4 erg·cm^{-2}·s^{-1}. [18] Data are averages for treatment at three root temperatures—13, 21, and 30°C. [19] Light intensity = 13,000 lux.

Contributors: Rawlins, Stephen L.; Janes, Byron E.

References
[1] Boyer, J. S. 1965. Plant Physiol. 40:229.
[2] Brouwer, R. 1963. Acta Bot. Neer. 12:248.
[3] Gale, J., et al. 1967. Physiol. Plant. 20:408.
[4] Hoffman, G. J., and C. J. Phene. 1971. Trans. Amer. Soc. Agr. Eng. 14:1103.
[5] Hoffman, G. J., et al. 1971. Agron. J. 63:822.

continued

[6] Janes, B. E. 1968. Physiol. Plant. 21:334.

[7] Janes, B. E. Unpublished. Univ. Connecticut, Dep. Plant Science, Storrs, 1971.

[8] Kirkham, M. B., et al. 1969. Plant Physiol. 44:1378.

[9] Lunt, O. R., et al. 1960. Trans. Int. Congr. Soil Sci., 7th, 1:560.

[10] Meiri, A., and A. Poljakoff-Mayber. 1970. Soil Sci. 109:26.

[11] Mizraki, Y., et al. 1970. Plant Physiol. 46:169.

[12] Oertli, J. J., and W. F. Richardson. 1968. Soil Sci. 105:177.

[13] Slatyer, R. O. 1961. Aust. J. Biol. Sci. 14:519.

107.   FACTORS AFFECTING OSMOTIC POTENTIAL: PLANTS

Part I.   Root Solutions

Osmotic potential (OP) is a major component of total water potential which governs water movement into and through plants. Osmotic potential (negative values, since solutes decrease the potential of water at a given temperature and pressure) has an absolute value equal to the positive pressure which would have to be applied to the solution to prevent net entry of water into the solution when the solution is separated from pure water by a semipermeable membrane.

Values are for osmotic potential at 25°C. **Growth Conditions:** min = minimum. **Method:** C = cryoscopy; V = vapor pressure osmometry; T = thermocouple psychrometry. **Osmotic Adjustment Ratio** $= \dfrac{\Delta OP_{plant\ sap}}{\Delta OP_{root\ medium}}$, where ΔOP refers to the change in osmotic potential of plant sap and root medium due to added osmotic agents. Data in brackets refer to the column heading in brackets.

	Species [Growth Conditions]	Age or Stage [Time at Minimum OP]	ΔOP of Medium	Method	Plant Part	OP of Base Nutrient Medium [OP of Control Plant Sap], bars	Osmotic Agent & Level, bars [Time After Addition]	Osmotic Adjustment Ratio	Reference
	Long-Term Effects or Steady-State Responses								
1	*Beta vulgaris* [Outdoor sand culture]	189 da [170 da]	−0.24 to −1.17 bars every 2 da	C	Mature leaf blade	−0.8 [−15.0]	Mixed Cl⁻ salts		3
2							−2.0	1.52	
							−5.9	1.40	
3							Mixed SO_4^{2-} salts		3
4							−1.2	1.99	
5							−3.1	1.21	
							−4.8	0.75	
6	*Capsicum frutescens* [Greenhouse; water culture]	Seedling stage + 53-70 da [42-53 da]	−1 bar every 2 da	C	Young leaf	−0.4 [−9.8]	NaCl[1]		1
7							−1.1	0.39	
8							−2.2	1.05	
9							−3.3	0.80	
							−4.4	0.85	
10					Mature leaf	−0.4 [−10.9]	NaCl[1]		1
11							−1.1	0.10	
12							−2.2	1.15	
13							−3.3	0.93	
							−4.4	1.15	
14					Root[2]	−0.4 [−5.3]	NaCl[1]		1
15							−1.1	1.28	
16							−2.2	1.04	
17							−3.3	0.78	
							−4.4	0.70	

[1] Added to base nutrient solution.   [2] OP values are for rinsed roots, corrected for dilution by water in free space.

continued

Part I.   Root Solutions

	Species [Growth Conditions]	Age or Stage [Time at Minimum OP]	ΔOP of Medium	Meth-od	Plant Part	OP of Base Nutrient Medium [OP of Control Plant Sap], bars	Osmotic Agent & Level, bars [Time After Addition]	Osmotic Adjust-ment Ratio	Ref-er-ence
	Gossypium hirsutum 'Acala'								
18	[Outdoor sand	203 da [184	−0.24 to −1.17	C	Mature	−0.8 [−15.4]	Mixed Cl⁻ salts		3
	culture]	da]	bars every 2		leaves		−2.0	0.20	
19			da		on main		−5.9	0.34	
					stalk		Mixed SO_4^{2-} salts		3
20							−1.2	−0.12	
21							−3.1	0.00	
22							−4.8	0.19	
	[Greenhouse;	6- to 10-	−1.67 bars	V	Leaf	−0.4 [−11.1[3]]	NaCl + CaCl₂[4]		4
23	water culture]	leaf stage	every 2 or				−5	1.08	
		+ 11 da	3 da			−0.4 [−12.2[5]]	NaCl + CaCl₂[4]		4
24		[5 da]					−8.1	1.00	
						−0.4 [−13.1[3]]	NaCl + CaCl₂[4]		4
25							−8.1	1.04	
	'Acala 4-42' [Wa-	38 da [9-16	−1.5 to −3.0	C	Leaf	−0.4 [−11.3]	NaCl[1]		1
26	ter culture,	da]	bars every 2-				−3.3	0.76	
27	25°C; 12,912		4 da				−6.6	1.28	
28	lux; 40% rel-						−13.3	0.92	
	ative humidity]				Stem	−0.4 [−11.4]	NaCl[1]		1
29							−3.3	0.40	
30							−6.6	0.60	
31							−13.3	1.01	
					Root[6]	−0.4 [−5.4]	NaCl[1]		1
32							−3.3	1.30	
33							−6.6	1.05	
34							−13.3	0.98	
	Hedysarum carno-	2-leaf stage	−0.77 bars/da	?	Top	−0.6 [−9.3]	NaCl		5
35	*sum* [Green-	[3 da]					−13.8	2.8[7]	
	house; water	2-leaf stage	−0.77 bars/da	?	Top	−0.6 [−9.3]	NaCl		5
36	culture 30-	[11 da]					−4.6	2.3	
37	34°C max,						−9.3	4.8[7]	
	18-20°C min]		−2.32 bars/da	?	Top	−0.6 [−9.3]	NaCl		5
38							−4.6	1.4	
39							−9.3	1.4	
40							−13.8	1.4	
41							−18.4	2.1[7]	
			−4.64 bars/da	?	Top	−0.6 [−9.3]	NaCl		5
42							−4.6	1.0	
43							−9.3	1.0	
44							−13.8	1.4	
45							−18.4	1.4	
	Hordeum vulgare	79 da [60	−0.24 to −1.17	C	Entire	−0.8 [−11.1]	Mixed Cl⁻ salts		3
46	[Outdoor sand	da]	bars every 2		plant		−2.0	2.72	
47	culture]		da				−5.9	2.02	

[1] Added to base nutrient solution. [3] After plants were exposed to 30°C for 24 hr. [4] Isosmotic mixture. [5] After plants were exposed to 16.5°C for 24 hr. [6] OP values are for unrinsed roots, uncorrected for dilution of cell sap by free-space solution. [7] Not at steady state, but still increasing steadily.

continued

Part I. Root Solutions

	Species [Growth Conditions]	Age or Stage [Time at Minimum OP]	ΔOP of Medium	Method	Plant Part	OP of Base Nutrient Medium [OP of Control Plant Sap], bars	Osmotic Agent & Level, bars [Time After Addition]	Osmotic Adjustment Ratio	Reference
48							Mixed SO$_4^{2-}$ salts −1.2	7.41	3
49							−3.1	2.42	
50							−4.8	2.04	
51	*Lycopersicon esculentum* [Outdoor sand culture]	145 da [126 da]	−0.24 to −1.17 bars every 2 da	C	Mature leaflet	−0.8 [−10.5]	Mixed Cl$^-$ salts −2.0	0.74	3
52							−5.9	0.88	
53							Mixed SO$_4^{2-}$ salts −1.2	1.11	3
54							−3.1	0.95	
55							−4.8	0.97	
56	*Medicago sativa* [Outdoor sand culture]	83-176 da [64-157 da][8/]	−0.24 to −1.17 bars every 2 da	C	All stems & leaves	−0.8 [−15.2]	Mixed Cl$^-$ salts −2.0	0.80	3
57							−5.9	0.52	
58							Mixed SO$_4^{2-}$ salts −1.2	0.11	3
59							−3.1	0.48	
60							−4.8	0.72	
61	*Phaseolus vulgaris* 'Pinto' [Greenhouse; water culture]	Cordate leaf stage + 11 da [5 da]	−1.67 bars every 2 or 3 da	V	Leaf	−0.4 [−8.2[5/]]	NaCl + CaCl$_2$[4/] −3	1.03	4
62						−0.4 [−10.0[3/]]	NaCl + CaCl$_2$[4/] −3	0.93	4
63	*Solanum tuberosum* 'Russet Burbank' [Greenhouse; water culture]	15-cm plants + 41-81 da [41 da]	−0.25 bar/da for 4-40 da	V	2.5-cm tip of 3rd youngest leaf	−0.3 [−7.3]	Polyethylene glycol (Carbowax 1540) −8.4 to 0	0.79[9/]	8
64	*Sorghum bicolor*[10/] 'Dwarf Milo' [Outdoor sand culture]	90 da [71 da]	−0.24 to −1.17 bars every 2 da	C	6th leaf from base	−0.8 [−12.2]	Mixed Cl$^-$ salts −2.0	1.26	3
65							−5.9	1.29	
66							Mixed SO$_4^{2-}$ salts −1.2	1.32	3
67							−3.1	0.78	
68							−4.8	0.64	

Short-Term Effects & Time Course Studies

69	*Agropyron desertorum* [Greenhouse; water culture]	48 da [1 da]	−0.25 to −2.0 bars/da	V	3-4 tillers, basal portion	∼−0.5 [−13.1]	Polyethylene glycol (Carbowax 1540) −10 to −1.25	0.81[11/]	8
70			−2.5 bars/da	V	3-4 tillers, basal portion	∼−0.5 [−13.1]	CaCl$_2$ + NaCl[12/] −10.7	0.81	8
71	*Allium cepa* 'White Globe' [Greenhouse; water culture]	4-leaf stage + 6-14 da [1-8 da]	−1.67 bars every 2 or 3 da	V	Leaf	−0.4 [−7.7 to −6.4]	CaCl$_2$ + NaCl[4/] −5 [1 da]	0.34	4
72							[2 da]	0.26	
73							[4 da]	0.38	
74							[8 da]	0.48	

[3/] After plants were exposed to 30°C for 24 hr. [4/] Isosmotic mixture. [5/] After plants were exposed to 16.5°C for 24 hr. [8/] Average for 3 harvests at 83, 126, and 176 da after 64, 107, and 157 da at minimum OP. [9/] From regression coefficient over a range of OP values from −8.7 to −0.3 bars. [10/] Synonym: *S. vulgare.* [11/] Average value. [12/] Equimolar mixture.

continued

Part I. Root Solutions

	Species [Growth Conditions]	Age or Stage [Time at Minimum OP]	ΔOP of Medium	Method	Plant Part	OP of Base Nutrient Medium [OP of Control Plant Sap], bars	Osmotic Agent & Level, bars [Time After Addition]	Osmotic Adjustment Ratio	Reference
	Capsicum frutescens [Greenhouse; water culture]	50-56 da [1 & 2 da]	−1.5 bars every 2 da	C	Leaf Young	−0.4 [−8.8]	NaCl		2
75							−1.7, −3.3, & −5.0 [1 da]	0.35 [11]	
76							[2 da]	0.75 [11]	
					Mature	−0.4 [−9.1]	NaCl		2
77							−1.7, −3.3, & −5.0 [1 da]	0.35 [11]	
78							[2 da]	0.51 [11]	
					Stem & petiole	−0.4 [−9.6]	NaCl		2
79							−1.7, −3.3, & −5.0 [1 da]	0.57 [11]	
80							[2 da]	0.87 [11]	
					Root[2]	−0.4 [−7.3]	NaCl		2
81							−1.7, −3.3, & −5.0 [1 da]	0.39 [11]	
82							[2 da]	0.59 [11]	
	'California Wonder' [26.5°C; 21,520-26,900 lux; 70% relative humidity]	∿29-34 da [1-5 da]	−5 bars abruptly	C	Leaf (?)	−0.5 [−12.4 to −10.2][13]	Polyethylene glycol 400[13]		6
83							−5 [1 da]	0.22	
84							[2 da]	0.49	
85							[3 da]	0.47	
86							[4 da]	0.59	
87							[5 da]	0.64	
		∿30-34 da [1 da]	−1 bar every 8 or 24 hr	C	Leaf & petiole	−0.5 [−10.0][13]	Polyethylene glycol 400[13]		6
88							−2.0	0.70	
89							−4.7	0.51	
90							−6.3	0.49	
91							−6.8	0.53	
92							−9.8	0.42	
							NaCl[13]		6
93							−2.3	0.61	
94							−4.2	0.31	
95							−6.0	0.35	
96							−6.9	0.46	
97							−10.1	0.53	
					Stem	−0.5 [−9.1][13]	Polyethylene glycol 400[13]		6
98							−2.0	0.70	
99							−4.7	0.38	
100							−6.3	0.43	
101							−6.8	0.49	
102							−9.8	0.61	

[2] OP values are for rinsed roots, corrected for dilution by water in free space. [11] Average value. [13] No indication whether OP values reported were for 0° or 25°C; values for solutions and plant saps given as reported.

continued

Part I. Root Solutions

	Species [Growth Conditions]	Age or Stage [Time at Minimum OP]	ΔOP of Medium	Method	Plant Part	OP of Base Nutrient Medium [OP of Control Plant Sap], bars	Osmotic Agent & Level, bars [Time After Addition]	Osmotic Adjustment Ratio	Reference
103							NaCl[13]		6
							−2.3	0.87	
104							−4.2	0.52	
105							−6.0	0.55	
106							−6.9	0.75	
107							−10.1	0.82	
					Root	−0.5 [−5.7][13]	Polyethylene glycol 400 [13]		6
108							−2.0	1.05	
109							−4.7	0.19	
110							−6.3	0.38	
111							−6.8	0.43	
112							−9.8	0.40	
							NaCl[13]		6
113							−2.3	1.09	
114							−4.2	0.45	
115							−6.0	0.55	
116							−6.9	0.55	
117							−10.1	0.52	
	Hordeum vulgare	25-45 da	−1 bar, then	T	Tip of 2nd	−0.4 [−26.0 to	NaCl[14]		7
118	'Liberty' [25°C;	[1 da]	−3 bars		leaf from	−17.1]	−1.0	0.00	
119	32,280 lux; 70-		every 3 or		top		−3.0	2.13	
120	85% relative hu-		4 da				−6.0	2.72	
121	midity]						−9.0	2.66	
122							−12.0	2.50	
123							−15.0	1.76	
							NaCl[15]		7
124							−1.0	1.10	
125							−3.0	0.13	
126							−6.0	0.95	
127							−9.0	0.83	
128							−12.0	0.32	
129							−15.0	0.63	
130	*Lycopersicon es-*	5-leaf stage	−5 bars im-	C	Leaf	−0.8 [−10.9][13]	KNO₃, −5[13]	0.75	9
131	*culentum* 'Grosse	[1 da]	posed abrupt-				NaCl, −5[13]	1.11	
132	Lisse' [Green-		ly				Mannitol, −5[13]	0.29	
133	house; water cul-						Sucrose, −5[13]	0.82	
134	ture]		−10 bars im-	C	Leaf	−0.8 [−10.9][13]	KNO₃, −10[13]	0.93	9
135			posed abrupt-				NaCl, −10[13]	0.98	
136			ly				Mannitol, −10[13]	0.93	
137							Sucrose, −10[13]	0.86	
	Phaseolus vulgaris	17-37 da	Two 2-bar in-	T	Tip of	−0.4 [−11.5 to	NaCl[14]		7
138	'Blue Lake' [25°C;	[1-16 da]	crements, 2		leaflet,	−9.3]	−2 [1 da]	0.95	
139	32,280 lux; 70-		da apart		youngest		−4 [1 da]	0.52	
140	85% relative hu-				expanded		[4 da]	1.27	
141	midity]				trifoliate		[6 da]	1.20	

[13] No indication whether OP values reported were for 0° or 25°C; values for solutions and plant saps given as reported. [14] All roots were at indicated OP. [15] Half of roots were at indicated OP; rest of roots were at −0.4 bars.

continued

Part I. Root Solutions

	Species [Growth Conditions]	Age or Stage [Time at Minimum OP]	ΔOP of Medium	Method	Plant Part	OP of Base Nutrient Medium [OP of Control Plant Sap], bars	Osmotic Agent & Level, bars [Time After Addition]	Osmotic Adjustment Ratio	Reference
142							[8 da]	1.47	
143							[10 da]	1.27	
144							[14 da]	1.15	
145							[16 da]	1.37	
							NaCl[15]		7
146							−2 [1 da]	0.70	
147							−4 [1 da]	0.22	
148							[4 da]	0.50	
149							[6 da]	0.20	
150							[8 da]	0.22	
151							[10 da]	0.27	
152							[14 da]	0.40	
153							[16 da]	0.92	
154	'Dwarf Red' [Greenhouse; water culture]	15-21 da [1 da]	−1 bar every 2 da	C	Cordate leaf	−0.4 [−8.7]	NaCl −1.1, −2.2, & −3.3 [1 da]	0.56[11]	2
155					1st trifoliate	−0.4 [−8.1]	NaCl −1.1, −2.2, & −3.3 [1 da]	0.97[11]	2
156					Stem & petiole	−0.4 [−8.4]	NaCl −1.1, −2.2, & −3.3 [1 da]	0.67[11]	2
157					Root[2]	−0.4 [−6.4]	NaCl −1.1, −2.2, & −3.3 [1 da]	0.83[11]	2
158		Cordate leaf stage [0.5-1 da]	−1.1 bars at 1800 hr	C	Leaf	−0.4 [−8.2 to −7.8]	CaCl$_2$ −1.1 [12 hr]	0.3	2
159							[18 hr]	1.4	
160							[1 da]	0.5	
161							NaCl −1.1 [12 hr]	0.3	2
162							[18 hr]	1.1	
163							[1 da]	0.9	
164							Na$_2$SO$_4$ −1.1 [12 hr]	0.5	2
165							[18 hr]	1.3	
166							[1 da]	0.8	
167					Stem	−0.4 [−7.8 to −7.3]	CaCl$_2$ −1.1 [12 hr]	0.3	2
168							[18 hr]	0.7	
169							[1 da]	0.3	
170							NaCl −1.1 [12 hr]	0.5	2
171							[18 hr]	0.7	
172							[1 da]	1.1	
173							Na$_2$SO$_4$ −1.1 [12 hr]	0.1	2
174							[18 hr]	0.5	
175							[1 da]	0.8	

[2] OP values are for rinsed roots, corrected for dilution by water in free space. [11] Average value. [15] Half of roots were at indicated OP; rest of roots were at −0.4 bars.

continued

Part I. Root Solutions

	Species [Growth Conditions]	Age or Stage [Time at Minimum OP]	ΔOP of Medium	Method	Plant Part	OP of Base Nutrient Medium [OP of Control Plant Sap], bars	Osmotic Agent & Level, bars [Time After Addition]	Osmotic Adjustment Ratio	Reference
176					Root[2]	−0.4 [−7.0 to −6.5]	CaCl$_2$ −1.1 [12 hr]	0.5	2
177							[18 hr]	0.7	
178							[1 da]	0.8	
179							NaCl −1.1 [12 hr]	1.0	2
180							[18 hr]	0.6	
181							[1 da]	1.0	
182							Na$_2$SO$_4$ −1.1 [12 hr]	1.0	2
183							[18 hr]	0.6	
184							[1 da]	0.9	
	'Wade' [26.5°C; 21,520-26,900 lux; 70% relative humidity]	∿30-34 da [1 da]	−1 bar every 8 or 24 hr	C	Leaf & petiole	−0.5 [−8.2][13]	Polyethylene glycol 400 [13]		6
185							−2.8	0.57	
186							−3.3	1.00	
187							−4.9	0.14	
188							NaCl[13] −2.4	0.33	6
189							−3.5	1.50	
190							−4.6	0.33	
					Stem	−0.5 [−9.5][13]	Polyethylene glycol 400 [13]		6
191							−2.8	0.76	
192							−3.3	1.43	
193							−4.9	0.30	
194							NaCl[13] −2.4	0.83	6
195							−3.5	1.09	
196							−4.6	0.63	
					Root	−0.5 [−5.7][13]	Polyethylene glycol 400 [13]		6
197							−2.8	0.67	
198							−3.3	0.67	
199							−4.9	0.25	
200							NaCl[13] −2.4	0.72	6
201							−3.5	0.87	
202							−4.6	0.26	
203	*Solanum tuberosum* 'Russet Burbank' [Greenhouse; water culture]	15 cm tall + 6 da [1 da]	−0.25 to −2.0 bars/da	V	2.5-cm tip of 3rd youngest leaf	∿−0.5 [−6.3]	Polyethylene glycol (Carbowax 1540) −10 to −1.25	0.93[11]	8
204			−2.5 bars/da	V	2.5-cm tip of 3rd youngest leaf	∿−0.5 [−6.3]	CaCl$_2$ + NaCl[12] −12	1.03	8

[2] OP values are for rinsed roots, corrected for dilution by water in free space. [11] Average value. [12] Equimolar mixture. [13] No indication whether OP values reported were for 0° or 25°C; values for solutions and plant saps given as reported.

continued

107. FACTORS AFFECTING OSMOTIC POTENTIAL: PLANTS

Part I. Root Solutions

Contributor: Bernstein, Leon

References
[1] Bernstein, L. 1961. Amer. J. Bot. 48:909.
[2] Bernstein, L. 1963. Ibid. 50:360.
[3] Eaton, F. M. 1942. J. Agr. Res. 64:357.
[4] Gale, J., et al. 1967. Physiol. Plant. 20:408.
[5] Hamza, M. 1969. C. R. Acad. Sci. 268:1925.

[6] Janes, B. E. 1966. Soil Sci. 101:180.
[7] Kirkham, M. B., et al. 1969. Plant Physiol. 44:1378.
[8] Ruf, R. H., et al. 1963. Soil Sci. 96:326.
[9] Slatyer, R. O. 1961. Aust. J. Biol. Sci. 14:519.

Part II. Other Factors

Osmotic potential is an important parameter in the water relationship of plant cells and whole plants. A complete knowledge of plant-water relationships also requires information on other relevant parameters, hydrostatic pressure differences and matric potential. For a detailed discussion of plant-water relationships, consult references 10 and 29.

Most available experimental techniques for the measurement of osmotic potential have inherent errors [28]; it is therefore advisable to consult the literature source accompanying each value. **Experimental Condition:** TPW = time from perianth wilt.

	Factor	Species (Synonym)	Plant Part	Experimental Condition	Osmotic Potential atm	Reference
	Species variation					
1	Vascular	*Atriplex* sp.	Leaf	....	−30.8	27
2		*Beta vulgaris*	Leaf	....	−11.3	27
3		*Capsicum frutescens*	Leaf	....	−12.8	27
4		*Ceratonia siliqua*	Leaf	....	−17.8	27
5		*Citrus sinensis*	Fruit	....	−14.8 to −9.9	16
6		*Helianthus annuus*	Leaf	....	−5.3	27
7		*Hordeum vulgare*	Leaf	....	−7.2	27
8		*Medicago sativa*	Leaf	....	−12.6	27
9		*Nicotiana tabacum*	Leaf	....	−7.6	27
10		*Phaseolus vulgaris*	Leaf	....	−6.5	27
11		*Pisum sativum*	Leaf	....	−6.1	27
12		*Solanum tuberosum*	Leaf	....	−9.3	27
13		*Triticum aestivum (T. vulgare)*	Leaf	....	−12.0	27
14		*Zea mays*	Leaf	....	−13.6	27
15	Halophytic	*Avicennia marina (A. nitida)*	Leaf	....	−46.5	26
16		*Batis maritima*	Leaf	....	−60.5	26
17		*Glaux maritima*	....	....	−14.6	3
18		*Laguncularia racemosa*	Leaf	....	−50.0	26
19		*Salicornia europaea*	....	....	−39.7	3
20		*S. virginica (S. pacifica)*	Leaf	....	−53.5	26
21		*Spartina patens*	....	....	−20.9	3
22		*Triglochin maritima*	....	....	−24.6	3
23	Nonvascular	*Ascophyllum nodosum*	....	....	−23.9	2
24		*Cladophora graminea*	....	....	−40.0	7
25		*Fucus vesiculosus*	....	....	−26.5	11
26		*Laminaria cloustonii*	....	....	−32.2 to −27.9	18
27		*Neurospora crassa*	....	....	−18.0	23
28		*Nitella clavata*	....	....	−6.0 to −5.2	4
29		*Spyridia filamentosa*	....	....	−29.0	21

continued

Part II. Other Factors

	Factor	Species (Synonym)	Plant Part	Experimental Condition	Osmotic Potential atm	Reference
30	Cell variation	*Allium cepa*	Guard cells	Cells open	−17.8	22
31				Cells closed	−15.1	22
32			Epidermal cells[1]	Cells open	−5.4	22
33				Cells closed	−5.4	22
34		*Chrysanthemum morifolium*	Guard cells	Stomatal aperture: 0	−28.1 ± 4.1	5
35			Epidermis	Stomatal aperture: 0	−20.4 ± 2.8	5
36			Guard cells	Stomatal aperture: 2 μ	−20.1 ± 0.8	5
37			Epidermis	Stomatal aperture: 2 μ	−14.1 ± 2.4	5
38			Guard cells	Stomatal aperture: 4 μ	−29.9 ± 3.5	5
39			Epidermis	Stomatal aperture: 4 μ	−13.2 ± 1.0	5
40			Guard cells	Stomatal aperture: 6 μ	−38.0 ± 4.0	5
41			Epidermis	Stomatal aperture: 6 μ	−14.3 ± 2.2	5
42		*Clivia* sp.	Endosperm	TPW, 63 da	−6.7	24
43				74 da	−6.2	24
44				105 da	−6.7	24
45			Ovule	TPW, 17 da	−9.1	24
46			Central vacuole	TPW, 44 da	−10.3	24
47		*Commelina communis*	Guard cells	0.1-0.6 M mannitol	−5.2	36
48				0.3 M KCl + 0.01 M CaCl₂	−18.8	36
49		*Cyclamen* sp.	Guard cells	Cells open	−21.4	35
50				Cells closed	−12.9	35
51			Epidermal cells[1]	Cells open	−11.1	35
52				Cells closed	−11.1	35
53		*Haemanthus katharinae*	Endosperm	TPW, 36 da	−4.3	24
54				49 da	−3.6	24
55				88 da	−7.2	24
56			Ovule	TPW, 1 da	−3.6	24
57			Central vacuole	TPW, 15 da	−8.6	24
58				36 da	−5.5	24
59		*Iresine* sp.	Guard cells	Cells open	−39.5	35
60				Cells closed	−21.7	35
61			Epidermal cells[1]	Cells open	−11.4	35
62				Cells closed	−11.4	35
63		*Pelargonium hortorum*	Guard cells	Stomatal aperture: 0	−12.5 ± 1.7	5
64			Epidermis	Stomatal aperture: 0	−11.8 ± 1.6	5
65			Guard cells	Stomatal aperture: 2 μ	−22.7 ± 4.2	5
66			Epidermis	Stomatal aperture: 2 μ	−13.8 ± 2.3	5
67			Guard cells	Stomatal aperture: 4 μ	−29.1 ± 2.9	5
68			Epidermis	Stomatal aperture: 4 μ	−13.3 ± 1.2	5
69			Guard cells	Stomatal aperture: 6 μ	−29.6 ± 5.7	5
70			Epidermis	Stomatal aperture: 6 μ	−9.5 ± 1.5	5
71			Leaf sap	....	−6.5 ± 0.3	5
72			Epidermal sap	....	−13.1 ± 0.9[2]; −15.0 ± 0.03[3]	5
73		*Rumex patientia*	Guard cells	Cells open	−18.8	25
74				Cells closed	−12.8	25
75			Epidermal cells[1]	Cells open	−13.8	25
76				Cells closed	−12.8	25

[1] Adjacent to guard cells. [2] Measured by melting point method. [3] Measured by plasmolysis method.

continued

Part II. Other Factors

	Factor	Species (Synonym)	Plant Part	Experimental Condition	Osmotic Potential atm	Reference
77		*Veronica beccabunga*	Guard cells	Cells open	−24.7	22
78				Cells closed	−19.7	22
79			Epidermal cells[1]	Cells open	−11.8	22
80				Cells closed	−11.8	22
81		*Vicia faba*	Guard cells	Cells open	−10.2	30
82				Cells closed	−8.6	30
83			Epidermal cells[1]	Cells open	−5.7	30
84				Cells closed	−4.1	30
85	Exudate variation	*Brassica hirta (Sinapis alba)*	Root	Xylem exudate; medium contained 1.0 mM KCl + 0.1 mM CaCl$_2$	−0.45	9
86		*Cucurbita pepo*	Stem	Phloem exudate	−10.5 to −8.2	15
87		*Salix viminalis*	Stem	Phloem exudate	−11.0 to −3.5	34
88		*Yucca flaccida*	....	Phloem exudate	−14.8 to −12.7	32
89		*Zea mays*	Root	Xylem exudate; medium contained 50 mM KCl + 0.1 mM CaCl$_2$	−2.6	8
90				Xylem exudate; medium contained 0.1 mM KCl + 0.1 mM CaCl$_2$	−1.1	8
91				Xylem exudate; medium contained 25 mM K$_2$SO$_4$ + 0.1 mM CaSO$_4$	−1.5	1
92				Xylem exudate; medium contained 0.1 mM K$_2$SO$_4$ + 0.1 mM CaSO$_4$	−0.8	1
93	Temperature	*Potamogeton crispus*	Leaf	0°C	−14.3	12
94				13°C	−13.5	12
95				23°C	−12.7	12
96				30°C	−11.1	12
97	Height on tree	*Juglans nigra*	Leaf	2.4 cm	−16.8	13
98				6.3 cm	−17.8	13
99				9.7 cm	−18.2	13
100				11.6 cm	−17.2	13
101				13.4 cm	−18.3	13
102				15.9 cm	−18.3	13
103	Diurnal variation	*Ambrosia trifida*	Top leaf	0600 hr	−12.5	14
104				1000 hr	−15.3	14
105				1400 hr	−17.4	14
106				1700 hr	−16.5	14
107				2000 hr	−16.3	14
108		*Andropogon scoparius*	Shoot	1200 hr	−23.0	31
109				1400 hr	−25.0	31
110				1600 hr	−27.0	31
111				1800 hr	−26.0	31
112				2000 hr	−21.5	31
113				2200 hr	−20.5	31
114				0400 hr	−21.0	31
115				0800 hr	−25.0	31

[1] Adjacent to guard cells.

continued

Part II. Other Factors

	Factor	Species (Synonym)	Plant Part	Experimental Condition	Osmotic Potential atm	Reference
116	Seasonal varia-	*Linnaea borealis*	....	October	−19.6	20
117	tion			December	−25.0	20
118				March	−25.6	20
119				May	−14.3	20
120		*Pinus cembra,* young	....	Summer	−18 to −16	33
121				Early winter, some snow cover	−24.5	33
122				Winter, some snow cover	−30.0	33
123				Winter, complete snow cover	−22.5	33
124				Winter, no snow cover	−42.0	33
125		mature	....	Winter	−28.0	33
126		*Populus tremuloides*	....	October	−15.0	20
127				December	−16.8	20
128				January	−16.2	20
129				February	−13.7	20
130				March	−17.0	20
131				May	−10.6	20
132	Osmotic poten-	*Gossypium hirsutum*	Leaf	Growth medium, 0 atm	−9.8	6
133	tial of growth			−3 atm	−9.1	6
134	medium 4/			−6 atm	−11.5	6
135				−12 atm	−8.9	6
136			Stem	Growth medium, 0 atm	−9.9	6
137				−3 atm	−8.1	6
138				−6 atm	−7.5	6
139				−12 atm	−10.0	6
140			Root	Growth medium, 0 atm	−4.5	6
141				−3 atm	−5.4	6
142				−6 atm	−4.8	6
143				−12 atm	−4.3	6
144		*Lolium perenne*	Leaf	Growth medium, −0.4 atm	−11.5	19
145				−1.1 atm	−11.8	19
146				−3.3 atm	−10.5	19
147				−5.3 atm	−9.0	19
148				−7.9 atm	−6.9	19
149		*Phaseolus vulgaris* 'Dwarf Red'	Leaf	Growth medium, −1.3 atm	−10.4; −9.1	17
150				−2.3 atm	−11.6; −9.3	17
151				−3.3 atm	−12.6; −9.3	17
152		*Zea mays*	Leaf	Growth medium, −0.4 atm	−7.1	19
153				−1.1 atm	−6.7	19
154				−3.3 atm	−5.1	19
155				−5.3 atm	−3.6	19
156				−7.9 atm	−4.5	19

4/ **Osmotic Potential** value is difference in osmotic potential between growth medium and plant part.

Contributor: Collins, J. C.

References

[1] Anderson, W. P., and J. C. Collins. 1969. J. Exp. Bot. 20:72.

[2] Atkins, W. R. G. 1916. Some Recent Researches in Plant Physiology. Whittaker, London. p. 179.

[3] Baron, W. M. H. 1968. Physiological Aspects of Water and Plant Life. W. Heinemann, London.

[4] Barr, C. E., and T. C. Broyer. 1964. Plant Physiol. 39:48.

continued

[5] Bearce, B. C., and H. C. Kohl. 1970. Ibid. 46: 5151.

[6] Bernstein, L. 1961. Amer. J. Bot. 48:909.

[7] Blinks, L. R. 1951. In G. M. Smith, ed. Manual of Phycology. Chronica Botanica, Waltham, Mass. pp. 263-291.

[8] Collins, J. C. Unpublished. Univ. East Anglia, Norwich, England, 1969.

[9] Collins, J. C., and A. P. Kerrigan. Unpublished. Univ. East Anglia, Norwich, England, 1972.

[10] Dainty, J. 1969. In M. B. Wilkins, ed. Physiology of Plant Growth and Development. McGraw-Hill, New York. pp. 419-452.

[11] Dognon. A. 1921. C. R. Soc. Biol. 85:112.

[12] Gamma, H. 1932. Protoplasma 16:489.

[13] Harris, J. A., et al. 1917. Bull. Torrey Bot. Club 44:267.

[14] Herrick, E. M. 1933. Amer. J. Bot. 20:18.

[15] Jarvis, P. Unpublished. Grassland Research Institute, Berkshire, England, 1969.

[16] Kaufmann, M. B. 1970. Plant Physiol. 46:146.

[17] Lagerwerff, J. V., and H. E. Eagle. 1961. Ibid. 36: 472.

[18] Lapicque, L. 1921. C. R. Soc. Biol. 85:206.

[19] Lawlor, D. W. 1969. J. Exp. Bot. 20:895.

[20] Lewis, F. J., and G. M. Tuttle. 1920. Ann. Bot. (London) 34:405.

[21] Mosebach, G. 1936. Beitr. Biol. Pflanz. 24:113.

[22] Mouravieff, I. 1958. Bull. Soc. Bot. Fr. 105:467.

[23] Robertson, N. F., and S. R. H. Rizvi. 1968. Ann. Bot. (London), N. S. 32:279.

[24] Ryczkowski, M. 1961. Z. Pflanzenphysiol. 61:422.

[25] Sayre, J. D. 1926. Ohio J. Sci. 26:233.

[26] Scholander, P. F., et al. 1964. Proc. Nat. Acad. Sci. U.S. 52:119.

[27] Shimsi, D. S., and A. L. Livne. 1967. Ann. Bot. (London), N. S. 31:505.

[28] Slatyer, R. O. 1966. Protoplasma 62:34.

[29] Slatyer, R. O. 1967. Plant Water Relationships. Academic Press, New York.

[30] Stalfelt, M. G. 1967. Physiol. Plant. 20:634.

[31] Stoddart, L. A. 1935. Plant Physiol. 10:661.

[32] Tammes, T. M. L., and J. van Die. 1964. Acta Bot. Neer. 7:76.

[33] Tranquillini, W. 1963. In A. J. Rutter and F. H. Whitehead, ed. Water Relations of Plants. Blackwell Scientific Publications, London.

[34] Weatherley, P. E., et al. 1959. J. Exp. Bot. 10:1.

[35] Wiggans, R. G. 1921. Amer. J. Bot. 8:30.

[36] Willmer, C. M., and T. A. Mansfield. 1970. New Phytol. 69:639.

108. MAXIMUM PERMISSIBLE EXPOSURE TO IONIZING RADIATION: MAN

The unit of dose equivalent is the rem. Number of rems = no. of rad $\times$ QF $\times$ n. (Rad = unit of absorbed dose and corresponds to 100 ergs/g of medium. QF = quality factor, which is the ratio of the absorbed dose, in rads, from reference X rays to the absorbed dose, in rads, from the given radiation field required to produce the same effect as the reference X rays. Reference X rays in most cases have been those from 200-250 kilovolt X-ray sources, or γ radiation from cobalt-60. n = relative damage factor.) *Abbreviations:* ICRP = International Commission on Radiological Protection; NCRP = National Council on Radiation Protection; MPC = maximum permissible concentration of a radionuclide.

Part I. Quality Factors

The values for heavy charged particles are given in terms of linear energy transfer in water. **Energy:** MeV = million electron volts, one electron volt being the energy given to an electron when it is accelerated through a potential difference of one volt (1.602×10^{-12} ergs). For additional information, consult reference 1.

Energy, MeV[1]	Quality Factor	Reference	Energy, MeV[1]		Quality Factor	Reference	Energy, MeV[1]		Quality Factor	Reference
Neutrons			6	2.5	8		Heavy Charged Particles			
			7	5.0	7					
1 Thermal	3	4	8	10	6.5		13	>3.5[3]	1	3
2 10^{-4}	2		9	50	5	2	14	3.5-7.0[3]	1-2	
3 0.02	5		10	400	3.5		15	7.0-23[3]	2-5	
4 0.1	8		11	1000	∿2.5-∿10[2]	5	16	23-53[3]	5-10	
5 0.5	10		12	2000	∿2.5-∿10[2]		17	53-175[3]	10-20	

[1] Unless otherwise indicated. [2] Approximately 10 at surface; approximately 2.5 at depth of 20-30 cm. [3] KeV/μ, or thousand electron volts per micron; 1 KeV/μ = 10 MeV/cm.

continued

108. MAXIMUM PERMISSIBLE EXPOSURE TO IONIZING RADIATION: MAN

Part I. Quality Factors

Contributor: Morgan, Karl Z.

References

[1] Morgan, K. Z., and J. E. Turner. 1972. Amer. Inst. Phys. Handb., ch. 8:291.
[2] Neufeld, J., et al. 1966. Health Phys. 12:227.
[3] RBE (Relative Biological Effectiveness) Committee. 1963. Ibid. 9:357.
[4] U.S. National Bureau of Standards. 1957. Nat. Bur. Stand. (U.S.) Handb. 63.
[5] Wright, H. A., et al. 1969. Health Phys. 16:13.

Part II. Dose Equivalent to Body Organs

Values are the recommended maximum permissible doses of ionizing radiation to the various organs of the body, and are in addition to doses from medical and background exposure. The values apply to both internal and external exposure. **Integrated:** The limit in rems for a person occupationally exposed is the age in years minus 18, then multiplied by 5. For detailed information regarding dose commitment, planned special exposures, simultaneous exposure to several organs, exposures for pregnant women, emergency exposures, etc., consult the appended references.

| Body Organ | Limits for Persons Occupationally Exposed | | | | Limits for Persons Not Occupationally Exposed | |
| | ICRP Recommendations | | NCRP Recommendations | | ICRP Recommended | NCRP Recommended |
	Annual[1] rem/yr	Integrated rem	Annual[2] rem/yr	Integrated rem	Annual, rem/yr	Annual, rem/yr
1 Whole body	5	5(age −18)	5[3]	5(age −18)	0.5	0.5
2 Gonads	5	5(age −18)	5[3]	5(age −18)	0.5	0.5
3 Red bone marrow	5	5(age −18)	5[3]	5(age −18)	0.5	0.5
4 Lenses of eyes	15	....	5[3]	5(age −18)	1.5	0.5
5 Fetus	..	....	...	....	...	0.5
6 Skin	30	....	15	....	3	...
7 Bone	30	....	15	....	3	...
8 Thyroid	30	....	15	....	3	...
9 Hands	75	....	75	....	7.5	...
10 Forearms	75	....	30	....	7.5	...
11 Hands, feet, forearms, & ankles	75	....	...	....	7.5	...
12 All other body organs	15	....	15	....	1.5	...

[1] For exceptions to these limits, *see* Part III. [2] The NCRP specifies that in exceptional cases one-third of these values may be received in a 13-wk period provided the limit for integrated dose equivalent is not exceeded. [3] In exceptional cases, well-distributed in time, the NCRP permits these dose limits to be increased to 15 rem/yr.

Contributor: Morgan, Karl Z.

General References

[1] International Commission on Radiological Protection. 1965. Int. Comm. Radiol. Prot. Publ. 9.
[2] Morgan, K. Z., and J. E. Turner. 1967. Principles of Radiation Protection. J. Wiley, New York.
[3] National Council on Radiation Protection. 1971. Nat. Counc. Radiat. Prot. Rep. 39.

continued

Part III. Occupational Exposure Levels

Values are those recommended by the International Commission on Radiological Protection, and represent the maximum permissible exposures when the dose is distributed as indicated. Exposure levels are given in terms of the permissible annual dose equivalent, which is the dose rate reached after 50 years of exposure at the maximum permissible concentration of a radionuclide. **Equivalence of Rapid & Slow Intake of Radionuclides: ELT** = equivalent length of time. **Symbol:** R = permissible annual dose equivalent.

Exposure Category	Limits for Exposure [1] to Radionuclides with Short Effective Half-Lives		Resulting Dose Integrated over 50-Yr Period [2]	Equivalence of Rapid & Slow Intake of Radionuclides	
	rem/13 wk	rem/yr	rem/50 yr	13 Wk Dose	ELT at MPC yr
1 Routine exposures	R/4	R	R/4	1 × MPC	1/4
2 Single exposures or exposures on a 13-wk basis [3]	R/2	R	R/2	2 × MPC	1/2
3 Planned special exposures [4]	2 R	2 R	2 R	8 × MPC	2
4 Planned special once-in-a-lifetime exposures [4]	5 R	5 R	5 R	20 × MPC	5

[1] External and/or internal exposure. [2] Actually, for external and/or internal exposures to radionuclides of any effective half-life, the period would be $50 + \tau/2$ yr, where τ is the period in years of intake of the radionuclide. [3] The ICRP states it would be undesirable to repeat this quarterly dose equivalent of R/2 at close intervals, but in order to provide flexibility it is permitted on infrequent occasions to receive 2 R in one year to the total body, gonads, or red bone marrow, provided the accumulated dose equivalent limit of 5(age-18) rem is not exceeded. Ordinarily the annual dose equivalent R should not be exceeded for any of the body organs. [4] Planned special exposures are not permitted if a single exposure in excess of R/2 has been received in the previous 12 mo, or if at any time the worker has received an abnormal exposure in excess of 5 R. Planned special exposures are not permitted to women of reproductive capacity. They are not permitted to total body, gonads, or red bone marrow if the accumulated dose equivalent limit of 5(age-18) will be exceeded as a consequence.

Contributor: Morgan, Karl Z.

Reference: Morgan, K. Z., and J. E. Turner. 1967. Principles of Radiation Protection. J. Wiley, New York.

Part IV. Concentrations of Radionuclides in Air, Water, and Bodies of Radiation Workers

Values are for radionuclides ingested (in water) or inhaled (in air). In applying the values of maximum permissible concentration (MPC), care must be exercised not to exceed the maximum permissible annual dose commitment. This corresponds to the body intake at the MPC for one year, and for a long-lived radionuclide corresponds to the build-up of a very small fraction of the maximum permissible body burden in one year. **Radionuclide: Z**—atomic number; **Type of Decay**—α = alpha particle, β^- = negatron, β^+ = positron, γ = gamma ray, e^- = internal conversion electron, ϵ = orbital electron capture, SF = spontaneous fission; **s or i**—s = soluble compounds, i = insoluble compounds. **q:** The maximum permissible burden in the total body resulting from exposure of the reference man (a typical adult worker in the United States) at the MPC for 50 years. When these burdens are reached in the total body, the dose rates to the critical body organs correspond to the dose limits given in Part I for the person occupationally exposed. **Maximum Permissible Concentrations of Radionuclides: Critical Organ**—that organ receiving the radiation dose that results in the greatest damage to the body; GI = gastrointestinal tract, (S) = stomach, (SI) = small intestine, (ULI) = upper large intestine, (LLI) = lower large intestine. *Abbreviation:* μCi = microcurie, one millionth of a curie, or 3.7×10^4 disintegrations per second.

Radionuclide					Maximum Permissible Concentrations of Radionuclides					
Z	Symbol & Mass No.	Type of Decay	s or i	q [1] μCi	In Water			In Air		
					Critical Organ	40-hr wk μCi/ml	168-hr wk μCi/ml	Critical Organ	40-hr wk μCi/ml	168-hr wk μCi/ml
1 1	^{3}H (HTO or ^{3}H$_2$O)	β^-	s	1000	Body tissue	0.1	0.03	Body tissue	5×10^{-6}	2×10^{-6}

[1] When other footnote numbers appear in this column, q pertains only to the critical organ specified in the footnote.

continued

Part IV. Concentrations of Radionuclides in Air, Water, and Bodies of Radiation Workers

| | | Radionuclide | | | | Maximum Permissible Concentrations of Radionuclides | | | | | |
| | | | | | | In Water | | | In Air | | |
	Z	Symbol & Mass No.	Type of Decay	s or i	q[1] μCi	Critical Organ	40-hr wk μCi/ml	168-hr wk μCi/ml	Critical Organ	40-hr wk μCi/ml	168-hr wk μCi/ml
2	4	^{7}Be	ϵ, γ	s	600	GI (LLI)	0.05	0.02	Total body	6×10^{-6}	2×10^{-6}
3				i	....	GI (LLI)	0.05	0.02	Lung	10^{-6}	4×10^{-7}
4	6	^{14}C (CO$_2$)	β^-	s	300	Fat	0.02	8×10^{-3}	Fat	4×10^{-6}	10^{-6}
5	9	^{18}F	β^+	s	20	GI (SI)	0.02	8×10^{-3}	GI (SI)	5×10^{-6}	2×10^{-6}
6				i	....	GI (ULI)	0.01	5×10^{-3}	GI (ULI)	3×10^{-6}	9×10^{-7}
7	11	^{22}Na	β^+, γ	s	10	Total body	10^{-3}	4×10^{-4}	Total body	2×10^{-7}	6×10^{-8}
8				i	....	GI (LLI)	9×10^{-4}	3×10^{-4}	Lung	9×10^{-9}	3×10^{-9}
9	11	^{24}Na	β^-, γ	s	7	GI (SI)	6×10^{-3}	2×10^{-3}	GI (SI)	10^{-6}	4×10^{-7}
10				i	....	GI (LLI)	8×10^{-4}	3×10^{-4}	GI (LLI)	10^{-7}	5×10^{-8}
11	14	^{31}Si	β^-, γ	s	$10^{[2]}$	GI (S)	0.03	9×10^{-3}	GI (S)	6×10^{-6}	2×10^{-6}
12				i	....	GI (ULI)	6×10^{-3}	2×10^{-3}	GI (ULI)	10^{-6}	3×10^{-7}
13	15	^{32}P	β^-	s	6	Bone	5×10^{-4}	2×10^{-4}	Bone	7×10^{-8}	2×10^{-8}
14				i	....	GI (LLI)	7×10^{-4}	2×10^{-4}	Lung	8×10^{-8}	3×10^{-8}
15	16	^{35}S	β^-	s	90	Testis	2×10^{-3}	6×10^{-4}	Testis	3×10^{-7}	9×10^{-8}
16				i	....	GI (LLI)	8×10^{-3}	3×10^{-3}	Lung	3×10^{-7}	9×10^{-8}
17	17	^{36}Cl	β^-	s	80	Total body	2×10^{-3}	8×10^{-4}	Total body	4×10^{-7}	10^{-7}
18				i	....	GI (LLI)	2×10^{-3}	6×10^{-4}	Lung	2×10^{-8}	8×10^{-9}
19	17	^{38}Cl	β^-, γ	s	9	GI (S)	0.01	4×10^{-3}	GI (S)	3×10^{-6}	9×10^{-7}
20				i	....	GI (S)	0.01	4×10^{-3}	GI (S)	2×10^{-6}	7×10^{-7}
21	19	^{42}K	β^-, γ	s	10	GI (S)	9×10^{-3}	3×10^{-3}	GI (S)	2×10^{-6}	7×10^{-7}
22				i	....	GI (LLI)	6×10^{-4}	2×10^{-4}	GI (LLI)	10^{-7}	4×10^{-8}
23	20	^{45}Ca	β^-	s	30	Bone	3×10^{-4}	9×10^{-5}	Bone	3×10^{-8}	10^{-8}
24				i	....	GI (LLI)	5×10^{-3}	2×10^{-3}	Lung	10^{-7}	4×10^{-8}
25	20	^{47}Ca	β^-, γ	s	5	Bone	10^{-3}	5×10^{-4}	Bone	2×10^{-7}	6×10^{-8}
26				i	....	GI (LLI)	10^{-3}	3×10^{-4}	GI (LLI)	2×10^{-7} [3]	6×10^{-8} [3]
27	21	^{46}Sc	β^-, γ	s	$10^{[4]}$	GI (LLI)	10^{-3}	4×10^{-4}	Liver	2×10^{-7} [5]	8×10^{-8} [5]
28				i	....	GI (LLI)	10^{-3}	4×10^{-4}	Lung	2×10^{-8}	8×10^{-9}
29	21	^{47}Sc	β^-, γ	s	$50^{[4]}$	GI (LLI)	3×10^{-3}	9×10^{-4}	GI (LLI)	6×10^{-7}	2×10^{-7}
30				i	....	GI (LLI)	3×10^{-3}	9×10^{-4}	GI (LLI)	5×10^{-7}	2×10^{-7}
31	21	^{48}Sc	β^-, γ	s	$9^{[4]}$	GI (LLI)	8×10^{-4}	3×10^{-4}	GI (LLI)	2×10^{-7}	6×10^{-8}
32				i	....	GI (LLI)	8×10^{-4}	3×10^{-4}	GI (LLI)	10^{-7}	5×10^{-8}
33	23	^{48}V	$\beta^+, \epsilon, \gamma$	s	$8^{[6]}$	GI (LLI)	9×10^{-4}	3×10^{-4}	GI (LLI)	2×10^{-7}	6×10^{-8}
34				i	....	GI (LLI)	8×10^{-4}	3×10^{-4}	Lung	6×10^{-8}	2×10^{-8}
35	24	^{51}Cr	ϵ, γ	s	800	GI (LLI)	0.05	0.02	Total body	10^{-5} [5]	4×10^{-6} [5]
36				i	....	GI (LLI)	0.05	0.02	Lung	2×10^{-6}	8×10^{-7}
37	25	^{52}Mn	$\beta^+, \epsilon, \gamma$	s	$5^{[7]}$	GI (LLI)	10^{-3}	3×10^{-4}	GI (LLI)	2×10^{-7}	7×10^{-8}
38				i	....	GI (LLI)	9×10^{-4}	3×10^{-4}	Lung	10^{-7}	5×10^{-8} [5]
39	25	^{54}Mn	ϵ, γ	s	$20^{[4]}$	GI (LLI)	4×10^{-3}	10^{-3}	Liver	4×10^{-7}	10^{-7}
40				i	....	GI (LLI)	3×10^{-3}	10^{-3}	Lung	4×10^{-8}	10^{-8}
41	25	^{56}Mn	β^-, γ	s	$2^{[7]}$	GI (LLI)	4×10^{-3}	10^{-3}	GI (LLI)	8×10^{-7}	3×10^{-7}
42				i	....	GI (LLI)	3×10^{-3}	10^{-3}	GI (LLI)	5×10^{-7}	2×10^{-7}
43	26	^{55}Fe	ϵ	s	1000	Spleen	0.02	8×10^{-3}	Spleen	9×10^{-7}	3×10^{-7}
44				i	....	GI (LLI)	0.07	0.02	Lung	10^{-6}	3×10^{-7}

[1] When other footnote numbers appear in this column, q pertains only to the critical organ specified in the footnote. [2] For lung. [3] Also for lung. [4] For liver. [5] Also for lower large intestine. [6] For kidney. [7] For pancreas.

continued

Part IV. Concentrations of Radionuclides in Air, Water, and Bodies of Radiation Workers

	Z	Symbol & Mass No.	Type of Decay	s or i	q[1] µCi	In Water Critical Organ	In Water 40-hr wk µCi/ml	In Water 168-hr wk µCi/ml	In Air Critical Organ	In Air 40-hr wk µCi/ml	In Air 168-hr wk µCi/ml
45	26	^{59}Fe	β^-,γ	s	20[8]	GI (LLI)	2×10^{-3}	6×10^{-4}	Spleen	10^{-7}	5×10^{-8}
46				i	....	GI (LLI)	2×10^{-3}	5×10^{-4}	Lung	5×10^{-8}	2×10^{-8}
47	27	^{57}Co	ϵ,γ,e^-	s	200	GI (LLI)	0.02	5×10^{-3}	GI (LLI)	3×10^{-6}	10^{-6}
48				i	....	GI (LLI)	0.01	4×10^{-3}	Lung	2×10^{-7}	6×10^{-8}
49	27	^{58m}Co	β^+,ϵ,γ	s	200	GI (LLI)	0.08	0.03	GI (LLI)	2×10^{-5}	6×10^{-6}
50				i	....	GI (LLI)	0.06	0.02	Lung	9×10^{-6}	3×10^{-6}
51	27	^{58}Co	β^+,ϵ	s	30	GI (LLI)	4×10^{-3}	10^{-3}	GI (LLI)	8×10^{-7}	3×10^{-7} [9]
52				i	....	GI (LLI)	3×10^{-3}	9×10^{-4}	Lung	5×10^{-8}	2×10^{-8}
53	27	^{60}Co	β^-,γ	s	10	GI (LLI)	10^{-3}	5×10^{-4}	GI (LLI)	3×10^{-7}	10^{-7} [9]
54				i	....	GI (LLI)	10^{-3}	3×10^{-4}	Lung	9×10^{-9}	3×10^{-9}
55	28	^{59}Ni	ϵ	s	1000	Bone	6×10^{-3}	2×10^{-3}	Bone	5×10^{-7}	2×10^{-7}
56				i	....	GI (LLI)	0.06	0.02	Lung	8×10^{-7}	3×10^{-7}
57	28	^{63}Ni	β^-	s	200	Bone	8×10^{-4}	3×10^{-4}	Bone	6×10^{-8}	2×10^{-8}
58				i	....	GI (LLI)	0.02	7×10^{-3}	Lung	3×10^{-7}	10^{-7}
59	28	^{65}Ni	β^-,γ	s	4[10]	GI (ULI)	4×10^{-3}	10^{-3}	GI (ULI)	9×10^{-7}	3×10^{-7}
60				i	....	GI (ULI)	3×10^{-3}	10^{-3}	GI (ULI)	5×10^{-7}	2×10^{-7}
61	29	^{64}Cu	β^-,β^+,ϵ	s	10[8]	GI (LLI)	0.01	3×10^{-3}	GI (LLI)	2×10^{-6}	7×10^{-7}
62				i	....	GI (LLI)	6×10^{-3}	2×10^{-3}	GI (LLI)	10^{-6}	4×10^{-7}
63	30	^{65}Zn	β^+,ϵ,γ	s	60	Total body	3×10^{-3}	10^{-3} [11,12]	Total body	10^{-7} [11,12]	4×10^{-8} [11]
64				i	....	GI (LLI)	5×10^{-3}	2×10^{-3}	Lung	6×10^{-8}	2×10^{-8}
65	30	^{69m}Zn	γ,e^-,β^-	s	0.7[13]	GI (LLI)	2×10^{-3}	7×10^{-4}	Prostate	4×10^{-7} [5]	10^{-7}
66				i	....	GI (LLI)	2×10^{-3}	6×10^{-4}	GI (LLI)	3×10^{-7}	10^{-7}
67	30	^{69}Zn	β^-	s	0.8[13]	GI (S)	0.05	0.02	Prostate	7×10^{-6}	2×10^{-6}
68				i	....	GI (S)	0.05	0.02	GI (S)	9×10^{-6}	3×10^{-6}
69	31	^{72}Ga	β^-,γ	s	5[6]	GI (LLI)	10^{-3}	4×10^{-4}	GI (LLI)	2×10^{-7}	8×10^{-8}
70				i	....	GI (LLI)	10^{-3}	4×10^{-4}	GI (LLI)	2×10^{-7}	6×10^{-8}
71	32	^{71}Ge	ϵ	s	100[6]	GI (LLI)	0.05	0.02	GI (LLI)	10^{-5}	4×10^{-6}
72				i	....	GI (LLI)	0.05	0.02	Lung	6×10^{-6}	2×10^{-6}
73	33	^{73}As	ϵ,γ	s	300	GI (LLI)	0.01	5×10^{-3}	Total body	2×10^{-6}	7×10^{-7}
74				i	....	GI (LLI)	0.01	5×10^{-3}	Lung	4×10^{-7}	10^{-7}
75	33	^{74}As	$\beta^-,\beta^+,\epsilon,\gamma$	s	40	GI (LLI)	2×10^{-3}	5×10^{-4}	GI (LLI)	3×10^{-7}	10^{-7}
76				i	....	GI (LLI)	2×10^{-3}	5×10^{-4}	Lung	10^{-7}	4×10^{-8}
77	33	^{76}As	β^-,γ	s	20	GI (LLI)	6×10^{-4}	2×10^{-4}	GI (LLI)	10^{-7}	4×10^{-8}
78				i	....	GI (LLI)	6×10^{-4}	2×10^{-4}	GI (LLI)	10^{-7}	3×10^{-8}
79	33	^{77}As	β^-,γ	s	80	GI (LLI)	2×10^{-3}	8×10^{-4}	GI (LLI)	5×10^{-7}	2×10^{-7}
80				i	....	GI (LLI)	2×10^{-3}	8×10^{-4}	GI (LLI)	4×10^{-7}	10^{-7}
81	34	^{75}Se	ϵ,γ	s	90	Kidney	9×10^{-3}	3×10^{-3} [9]	Kidney	10^{-6} [9]	4×10^{-7}
82				i	....	GI (LLI)	8×10^{-3}	3×10^{-3}	Lung	10^{-7}	4×10^{-8}
83	35	^{82}Br	β^-,γ	s	10	Total body	8×10^{-3} [14]	3×10^{-3} [14]	Total body	10^{-6}	4×10^{-7}
84				i	....	GI (LLI)	10^{-3}	4×10^{-4}	GI (LLI)	2×10^{-7}	6×10^{-8}
85	37	^{86}Rb	β^-,γ	s	30	Pancreas	2×10^{-3} [9]	7×10^{-4} [9]	Pancreas	3×10^{-7} [9]	10^{-7} [9,12]
86				i	....	GI (LLI)	7×10^{-4}	2×10^{-4}	Lung	7×10^{-8}	2×10^{-8}
87	37	^{87}Rb	β^-	s	200[4,9]	Pancreas	3×10^{-3}	10^{-3}	Pancreas	5×10^{-7}	2×10^{-7} [9,12]
88				i	....	GI (LLI)	5×10^{-3}	2×10^{-3}	Lung	7×10^{-8}	2×10^{-8}

[1] When other footnote numbers appear in this column, q pertains only to the critical organ specified in the footnote. [4] For liver. [5] Also for lower large intestine. [6] For kidney. [8] For spleen. [9] Also for total body. [10] For bone. [11] Also for prostate. [12] Also for liver. [13] For prostate. [14] Also for small intestine.

continued

Part IV. Concentrations of Radionuclides in Air, Water, and Bodies of Radiation Workers

	Z	Symbol & Mass No.	Type of Decay	s or i	$q^{1/}$ μCi	In Water Critical Organ	In Water 40-hr wk μCi/ml	In Water 168-hr wk μCi/ml	In Air Critical Organ	In Air 40-hr wk μCi/ml	In Air 168-hr wk μCi/ml
89	38	^{85m}Sr	ϵ, γ	s	50	GI (SI)	0.2	0.07	GI (SI)	4×10^{-5}	10^{-5}
90				i	....	GI (SI)	0.2	0.07	GI (SI)	3×10^{-5}	10^{-5}
91	38	^{85}Sr	ϵ, γ	s	60	Total body	3×10^{-3}	10^{-3}	Total body	2×10^{-7}	8×10^{-8}
92				i	....	GI (LLI)	5×10^{-3}	2×10^{-3}	Lung	10^{-7}	4×10^{-8}
93	38	^{89}Sr	β^-	s	4	Bone	3×10^{-4}	10^{-4}	Bone	3×10^{-8}	10^{-8}
94				i	....	GI (LLI)	8×10^{-4}	3×10^{-4}	Lung	4×10^{-8}	10^{-8}
95	38	^{90}Sr	β^-	s	2	Bone	10^{-5}	4×10^{-6}	Bone	10^{-9}	4×10^{-10}
96				i	....	GI (LLI)	10^{-3}	4×10^{-4}	Lung	5×10^{-9}	2×10^{-9}
97	38	^{91}Sr	β^-, γ	s	$3^{10/}$	GI (LLI)	2×10^{-3}	7×10^{-4}	GI (LLI)	4×10^{-7}	2×10^{-7}
98				i	....	GI (LLI)	10^{-3}	5×10^{-4}	GI (LLI)	3×10^{-7}	9×10^{-8}
99	38	^{92}Sr	β^-, γ	s	$2^{10/}$	GI (ULI)	2×10^{-3}	7×10^{-4}	GI (ULI)	4×10^{-7}	2×10^{-7}
100				i	....	GI (ULI)	2×10^{-3}	6×10^{-4}	GI (ULI)	3×10^{-7}	10^{-7}
101	39	^{90}Y	β^-	s	$3^{10/}$	GI (LLI)	6×10^{-4}	2×10^{-4}	GI (LLI)	10^{-7}	4×10^{-8}
102				i	....	GI (LLI)	6×10^{-4}	2×10^{-4}	GI (LLI)	10^{-7}	3×10^{-8}
103	39	^{91m}Y	β^-, γ	s	$5^{10/}$	GI (SI)	0.01	0.03	GI (SI)	2×10^{-5}	8×10^{-6}
104				i	....	GI (SI)	0.01	0.03	GI (SI)	2×10^{-5}	6×10^{-6}
105	39	^{91}Y	β^-, γ	s	$5^{10/}$	GI (LLI)	8×10^{-4}	3×10^{-4}	Bone	4×10^{-8}	10^{-8}
106				i	....	GI (LLI)	8×10^{-4}	3×10^{-4}	Lung	3×10^{-8}	10^{-8}
107	39	^{92}Y	β^-, γ	s	$2^{10/}$	GI (ULI)	2×10^{-3}	6×10^{-4}	GI (ULI)	4×10^{-7}	10^{-7}
108				i	....	GI (ULI)	2×10^{-3}	6×10^{-4}	GI (ULI)	3×10^{-7}	10^{-7}
109	39	^{93}Y	β^-, γ, e^-	s	$2^{10/}$	GI (LLI)	8×10^{-4}	3×10^{-4}	GI (LLI)	2×10^{-7}	6×10^{-8}
110				i	....	GI (LLI)	8×10^{-4}	3×10^{-4}	GI (LLI)	10^{-7}	5×10^{-8}
111	40	^{93}Zr	β^-, γ, e^-	s	$100^{10/}$	GI (LLI)	0.02	8×10^{-3}	Bone	10^{-7}	4×10^{-8}
112				i	....	GI (LLI)	0.02	8×10^{-3}	Lung	3×10^{-7}	10^{-7}
113	40	^{95}Zr	β^-, γ, e^-	s	20	GI (LLI)	2×10^{-3}	6×10^{-4}	Total body	10^{-7}	4×10^{-8}
114				i	....	GI (LLI)	2×10^{-3}	6×10^{-4}	Lung	3×10^{-8}	10^{-8}
115	40	^{97}Zr	β^-, γ	s	$5^{10/}$	GI (LLI)	5×10^{-4}	2×10^{-4}	GI (LLI)	10^{-7}	4×10^{-8}
116				i	....	GI (LLI)	5×10^{-4}	2×10^{-4}	GI (LLI)	9×10^{-8}	3×10^{-8}
117	41	^{93m}Nb	γ, e^-	s	$200^{10/}$	GI (LLI)	0.01	4×10^{-3}	Bone	10^{-7}	4×10^{-8}
118				i	....	GI (LLI)	0.01	4×10^{-3}	Lung	2×10^{-7}	5×10^{-8}
119	41	^{95}Nb	β^-, γ	s	40	GI (LLI)	3×10^{-3}	10^{-3}	Total body	5×10^{-7}	$2 \times 10^{-7}\ ^{5/}$
120				i	....	GI (LLI)	3×10^{-3}	10^{-3}	Lung	10^{-7}	3×10^{-8}
121	41	^{97}Nb	β^-, γ	s	$10^{10/}$	GI (ULI)	0.03	9×10^{-3}	GI (ULI)	6×10^{-6}	2×10^{-6}
122				i	....	GI (ULI)	0.03	9×10^{-3}	GI (ULI)	5×10^{-6}	2×10^{-6}
123	42	^{99}Mo	β^-, γ	s	8	Kidney	5×10^{-3}	$2 \times 10^{-3}\ ^{5/}$	Kidney	7×10^{-7}	3×10^{-7}
124				i	....	GI (LLI)	10^{-3}	4×10^{-4}	GI (LLI)	2×10^{-7}	7×10^{-8}
125	43	^{96m}Tc	ϵ, γ, e^-	s	$60^{6/}$	GI (LLI)	0.4	0.1	GI (LLI)	8×10^{-5}	3×10^{-5}
126				i	....	GI (LLI)	0.3	0.1	Lung	3×10^{-5}	10^{-5}
127	43	^{96}Tc	ϵ, γ	s	$10^{6/}$	GI (LLI)	3×10^{-3}	10^{-3}	GI (LLI)	6×10^{-7}	2×10^{-7}
128				i	....	GI (LLI)	10^{-3}	5×10^{-4}	GI (LLI)	2×10^{-7}	8×10^{-8}
129	43	^{97m}Tc	ϵ, γ, e^-	s	$20^{6/}$	GI (LLI)	0.01	4×10^{-3}	GI (LLI)	2×10^{-6}	8×10^{-7}
130				i	....	GI (LLI)	5×10^{-3}	2×10^{-3}	Lung	2×10^{-7}	5×10^{-8}
131	43	^{97}Tc	ϵ	s	$60^{6/}$	GI (LLI)	0.05	0.02	Kidney	$10^{-5}\ ^{5/}$	$4 \times 10^{-6}\ ^{5/}$
132				i	....	GI (LLI)	0.02	8×10^{-3}	Lung	3×10^{-7}	10^{-7}
133	43	^{99m}Tc	β^-, γ	s	200	GI (ULI)	0.2	0.06	GI (ULI)	4×10^{-5}	10^{-5}
134				i	....	GI (ULI)	0.08	0.03	GI (ULI)	10^{-5}	5×10^{-6}

$^{1/}$ When other footnote numbers appear in this column, q pertains only to the critical organ specified in the footnote. $^{5/}$ Also for lower large intestine. $^{6/}$ For kidney. $^{10/}$ For bone.

continued

Part IV.   Concentrations of Radionuclides in Air, Water, and Bodies of Radiation Workers

	Z	Symbol & Mass No.	Type of Decay	s or i	q[1] μCi	In Water Critical Organ	In Water 40-hr wk μCi/ml	In Water 168-hr wk μCi/ml	In Air Critical Organ	In Air 40-hr wk μCi/ml	In Air 168-hr wk μCi/ml
135	43	^{99}Tc	β^-	s	10[6]	GI (LLI)	0.01	3×10^{-3}	GI (LLI)	2×10^{-6}	7×10^{-7}
136				i	....	GI (LLI)	5×10^{-3}	2×10^{-3}	Lung	6×10^{-8}	2×10^{-8}
137	44	^{97}Ru	ϵ, γ, e^-	s	30[6]	GI (LLI)	0.01	4×10^{-3}	GI (LLI)	2×10^{-6}	8×10^{-7}
138				i	....	GI (LLI)	0.01	3×10^{-3}	GI (LLI)	2×10^{-6} [3]	6×10^{-7}
139	44	^{103}Ru	β^-, γ, e^-	s	20[6]	GI (LLI)	2×10^{-3}	8×10^{-4}	GI (LLI)	5×10^{-7}	2×10^{-7}
140				i	....	....	....	....	Lung	8×10^{-8}	3×10^{-8}
141	44	^{105}Ru	β^-, γ, e^-	s	2[6]	GI (ULI)	3×10^{-3}	10^{-3}	GI (ULI)	7×10^{-7}	2×10^{-7}
142				i	....	GI (ULI)	3×10^{-3}	10^{-3}	GI (ULI)	5×10^{-7}	2×10^{-7}
143	44	^{106}Ru	β^-, γ	s	3[6]	GI (LLI)	4×10^{-4}	10^{-4}	GI (LLI)	8×10^{-8}	3×10^{-8}
144				i	....	GI (LLI)	3×10^{-4}	10^{-4}	Lung	6×10^{-9}	2×10^{-9}
145	45	^{103m}Rh	γ, e^-	s	200[6,8]	GI (S)	0.4	0.1	GI (S)	8×10^{-5}	3×10^{-5}
146				i	....	GI (S)	0.3	0.1	GI (S)	6×10^{-5}	2×10^{-5}
147	45	^{105}Rh	β^-, γ	s	40[6]	GI (LLI)	4×10^{-3}	10^{-3}	GI (LLI)	8×10^{-7}	3×10^{-7}
148				i	....	GI (LLI)	3×10^{-3}	10^{-3}	GI (LLI)	5×10^{-7}	2×10^{-7}
149	46	^{103}Pd	ϵ, γ, e^-	s	20[6]	GI (LLI)	0.01	3×10^{-3}	Kidney	10^{-6}	5×10^{-7}
150				i	....	GI (LLI)	8×10^{-3}	3×10^{-3}	Lung	7×10^{-7}	3×10^{-7}
151	46	^{109}Pd	β^-, γ, e^-	s	7[6]	GI (LLI)	3×10^{-3}	9×10^{-4}	GI (LLI)	6×10^{-7}	2×10^{-7}
152				i	....	GI (LLI)	2×10^{-3}	7×10^{-4}	GI (LLI)	4×10^{-7}	10^{-7}
153	47	^{105}Ag	ϵ, γ	s	30[6]	GI (LLI)	3×10^{-3}	10^{-3}	GI (LLI)	6×10^{-7}	2×10^{-7}
154				i	....	GI (LLI)	3×10^{-3}	10^{-3}	Lung	8×10^{-8}	3×10^{-8}
155	47	^{110m}Ag	β^-, γ	s	10[6]	GI (LLI)	9×10^{-4}	3×10^{-4}	GI (LLI)	2×10^{-7}	7×10^{-8}
156				i	....	GI (LLI)	9×10^{-4}	3×10^{-4}	Lung	10^{-8}	3×10^{-9}
157	47	^{111}Ag	β^-, γ	s	20[6]	GI (LLI)	10^{-3}	4×10^{-4}	GI (LLI)	3×10^{-7}	10^{-7}
158				i	....	GI (LLI)	10^{-3}	4×10^{-4}	GI (LLI)	2×10^{-7}	8×10^{-8}
159	48	^{109}Cd	ϵ, γ, e^-	s	20[4,6]	GI (LLI)	5×10^{-3}	2×10^{-3}	Liver	5×10^{-8}	2×10^{-8} [15]
160				i	....	GI (LLI)	5×10^{-3}	2×10^{-3}	Lung	7×10^{-8}	3×10^{-8}
161	48	^{115m}Cd	β^-, γ, e^-	s	3[4]	GI (LLI)	7×10^{-4}	3×10^{-4}	Liver	4×10^{-8} [15]	10^{-8}
162				i	....	GI (LLI)	7×10^{-4}	3×10^{-4}	Lung	4×10^{-8}	10^{-8}
163	48	^{115}Cd	β^-, γ, e^-	s	3[4]	GI (LLI)	10^{-3}	3×10^{-4}	GI (LLI)	2×10^{-7}	8×10^{-8}
164				i	....	GI (LLI)	10^{-3}	4×10^{-4}	GI (LLI)	2×10^{-7}	6×10^{-8}
165	49	^{113m}In	γ, e^-	s	30[6]	GI (ULI)	0.04	0.01	GI (ULI)	8×10^{-6}	3×10^{-6}
166				i	....	GI (ULI)	0.04	0.01	GI (ULI)	7×10^{-6}	2×10^{-6}
167	49	^{114m}In	$\beta^-, \epsilon, \gamma, e^-$	s	6[6]	GI (LLI)	5×10^{-4}	2×10^{-4}	Kidney	10^{-7} [5,16]	4×10^{-8} [5,16]
168				i	....	GI (LLI)	5×10^{-4}	2×10^{-4}	Lung	2×10^{-8}	7×10^{-9}
169	49	^{115m}In	β^-, γ, e^-	s	30[6,8]	GI (ULI)	0.01	4×10^{-3}	GI (ULI)	2×10^{-6}	8×10^{-7}
170				i	....	GI (ULI)	0.01	4×10^{-3}	GI (ULI)	2×10^{-6}	6×10^{-7}
171	49	^{115}In	β^-	s	30[6]	GI (LLI)	3×10^{-3}	9×10^{-4}	Kidney	2×10^{-7}	9×10^{-8}
172				i	....	GI (LLI)	3×10^{-3}	9×10^{-4}	Lung	3×10^{-8}	10^{-8}
173	50	^{113}Sn	ϵ, γ, e^-	s	30[10]	GI (LLI)	2×10^{-3}	9×10^{-4}	Bone	4×10^{-7}	10^{-7}
174				i	....	GI (LLI)	2×10^{-3}	8×10^{-4}	Lung	5×10^{-8}	2×10^{-8}
175	50	^{125}Sn	β^-, γ, e^-	s	7[10]	GI (LLI)	5×10^{-4}	2×10^{-4}	GI (LLI)	10^{-7}	4×10^{-8}
176				i	....	GI (LLI)	5×10^{-4}	2×10^{-4}	Lung	8×10^{-8}	3×10^{-8}
177	51	^{122}Sb	β^-, γ	s	20	GI (LLI)	8×10^{-4}	3×10^{-4}	GI (LLI)	2×10^{-7}	6×10^{-8}
178				i	....	GI (LLI)	8×10^{-4}	3×10^{-4}	GI (LLI)	10^{-7}	5×10^{-8}
179	51	^{124}Sb	β^-, γ	s	10	GI (LLI)	7×10^{-4}	2×10^{-4}	GI (LLI)	2×10^{-7} [9]	5×10^{-8}
180				i	....	GI (LLI)	7×10^{-4}	2×10^{-4}	Lung	2×10^{-8}	7×10^{-9}
181	51	^{125}Sb	β^-, γ, e^-	s	40[2]	GI (LLI)	3×10^{-3}	10^{-3}	Lung	5×10^{-7}	2×10^{-7} [17]
182				i	....	GI (LLI)	3×10^{-3}	10^{-3}	Lung	3×10^{-8}	9×10^{-9}

[1] When other footnote numbers appear in this column, q pertains only to the critical organ specified in the footnote. [2] For lung. [3] Also for lung. [4] For liver. [5] Also for lower large intestine. [6] For kidney. [8] For spleen. [9] Also for total body. [10] For bone. [15] Also for kidney. [16] Also for spleen. [17] Also for bone.

continued

Part IV. Concentrations of Radionuclides in Air, Water, and Bodies of Radiation Workers

		Radionuclide				Maximum Permissible Concentrations of Radionuclides					
						In Water			In Air		
	Z	Symbol & Mass No.	Type of Decay	s or i	q [1] μCi	Critical Organ	40-hr wk μCi/ml	168-hr wk μCi/ml	Critical Organ	40-hr wk μCi/ml	168-hr wk μCi/ml
183	52	^{125m}Te	γ, e$^-$	s	20 [18]	Kidney	5×10^{-3} [5]	2×10^{-3} [5,19]	Kidney	4×10^{-7}	10^{-7}
184				i	....	GI (LLI)	3×10^{-3}	10^{-3}	Lung	10^{-7}	4×10^{-8}
185	52	^{127m}Te	β^-, γ, e$^-$	s	7 [18]	Kidney	2×10^{-3} [5,19]	6×10^{-4}	Kidney	10^{-7} [19]	5×10^{-8} [19]
186				i	....	GI (LLI)	2×10^{-3}	5×10^{-4}	Lung	4×10^{-8}	10^{-8}
187	52	^{127}Te	β^-	s	20 [6]	GI (LLI)	8×10^{-3}	3×10^{-3}	GI (LLI)	2×10^{-6}	6×10^{-7}
188				i	....	GI (LLI)	5×10^{-3}	2×10^{-3}	GI (LLI)	9×10^{-7}	3×10^{-7}
189	52	^{129m}Te	β^-, γ, e$^-$	s	3 [6,18]	GI (LLI)	10^{-3} [15,19]	3×10^{-4}	Kidney	8×10^{-8}	3×10^{-8} [19]
190				i	....	GI (LLI)	6×10^{-4}	2×10^{-4}	Lung	3×10^{-8}	10^{-8}
191	52	^{129}Te	β^-, γ, e$^-$	s	5 [6]	GI (S)	0.02	8×10^{-3}	GI (S)	5×10^{-6}	2×10^{-6}
192				i	....	GI (ULI)	0.02	8×10^{-3}	GI (ULI)	4×10^{-6}	10^{-6}
193	52	^{131m}Te	β^-, γ, e$^-$	s	4 [6]	GI (LLI)	2×10^{-3}	6×10^{-4}	GI (LLI)	4×10^{-7}	10^{-7}
194				i	....	GI (LLI)	10^{-3}	4×10^{-4}	GI (LLI)	2×10^{-7}	6×10^{-8}
195	52	^{132}Te	β^-, γ, e$^-$	s	3 [6]	GI (LLI)	9×10^{-4}	3×10^{-4}	GI (LLI)	2×10^{-7}	7×10^{-8}
196				i	....	GI (LLI)	6×10^{-4}	2×10^{-4}	GI (LLI)	10^{-7}	4×10^{-8}
197	53	^{126}I	β^-, ϵ, γ	s	1	Thyroid	5×10^{-5}	2×10^{-5}	Thyroid	8×10^{-9}	3×10^{-9}
198				i	....	GI (LLI)	3×10^{-3}	9×10^{-4}	Lung	3×10^{-7}	10^{-7}
199	53	^{129}I	β^-, γ, e$^-$	s	3	Thyroid	10^{-5}	4×10^{-6}	Thyroid	2×10^{-9}	6×10^{-10}
200				i	....	GI (LLI)	6×10^{-3}	2×10^{-3}	Lung	7×10^{-8}	2×10^{-8}
201	53	^{131}I	β^-, γ, e$^-$	s	0.7	Thyroid	6×10^{-5}	2×10^{-5}	Thyroid	9×10^{-9}	3×10^{-9}
202				i	....	GI (LLI)	2×10^{-3}	6×10^{-4}	GI (LLI)	3×10^{-7} [3]	10^{-7} [3]
203	53	^{132}I	β^-, γ, e$^-$	s	0.3	Thyroid	2×10^{-3}	6×10^{-4}	Thyroid	2×10^{-7}	8×10^{-8}
204				i	....	GI (ULI)	5×10^{-3}	2×10^{-3}	GI (ULI)	9×10^{-7}	3×10^{-7}
205	53	^{133}I	β^-, γ, e$^-$	s	0.3	Thyroid	2×10^{-4}	7×10^{-5}	Thyroid	3×10^{-8}	10^{-8}
206				i	....	GI (LLI)	10^{-3}	4×10^{-4}	GI (LLI)	2×10^{-7}	7×10^{-8}
207	53	^{134}I	β^-, γ	s	0.2	Thyroid	4×10^{-3}	10^{-3}	Thyroid	5×10^{-7}	2×10^{-7}
208				i	....	GI (S)	0.02	6×10^{-3}	GI (S)	3×10^{-6}	10^{-6}
209	53	^{135}I	β^-, γ, e$^-$	s	0.3	Thyroid	7×10^{-4}	2×10^{-4}	Thyroid	10^{-7}	4×10^{-8}
210				i	....	GI (LLI)	2×10^{-3}	7×10^{-4}	GI (LLI)	4×10^{-7}	10^{-7}
211	55	^{131}Cs	ϵ	s	700	Total body	0.07	0.02	Total body	10^{-5} [12]	4×10^{-6} [12]
212				i	....	GI (LLI)	0.03	9×10^{-3}	Lung	3×10^{-6}	10^{-6}
213	55	^{134m}Cs	β^-, γ, e$^-$	s	100 [6,9]	GI (S)	0.02	0.06	GI (S)	4×10^{-5}	10^{-5}
214				i	....	GI (ULI)	0.03	0.01	GI (ULI)	6×10^{-6}	2×10^{-6}
215	55	^{134}Cs	β^-, γ	s	20	Total body	3×10^{-4}	9×10^{-5}	Total body	4×10^{-8}	10^{-8}
216				i	....	GI (LLI)	10^{-3}	4×10^{-4}	Lung	10^{-8}	4×10^{-9}
217	55	^{135}Cs	β^-	s	200	Liver	3×10^{-3}	10^{-3} [9,16]	Liver	5×10^{-7} [16]	2×10^{-7} [9,16]
218				i	....	GI (LLI)	7×10^{-3}	2×10^{-3}	Lung	9×10^{-8}	3×10^{-8}
219	55	^{136}Cs	β^-, γ	s	30	Total body	2×10^{-3}	9×10^{-4}	Total body	4×10^{-7}	10^{-7}
220				i	....	GI (LLI)	2×10^{-3}	6×10^{-4}	Lung	2×10^{-7}	6×10^{-8}
221	55	^{137}Cs	β^-, γ, e$^-$	s	30	Total body	4×10^{-4}	2×10^{-4} [20]	Total body	6×10^{-8}	2×10^{-8}
222				i	....	GI (LLI)	10^{-3}	4×10^{-4}	Lung	10^{-8}	5×10^{-9}
223	56	^{131}Ba	ϵ, γ	s	50	GI (LLI)	5×10^{-3}	2×10^{-3}	GI (LLI)	10^{-6}	4×10^{-7}
224				i	....	GI (LLI)	5×10^{-3}	2×10^{-3}	Lung	4×10^{-7}	10^{-7}
225	56	^{140}Ba	β^-, γ	s	4 [10]	GI (LLI)	8×10^{-4}	3×10^{-4}	Bone	10^{-7}	4×10^{-8}
226				i	....	GI (LLI)	7×10^{-4}	2×10^{-4}	Lung	4×10^{-8}	10^{-8}

[1] When other footnote numbers appear in this column, q pertains only to the critical organ specified in the footnote. [3] Also for lung. [5] Also for lower large intestine. [6] For kidney. [9] Also for total body. [10] For bone. [12] Also for liver. [15] Also for kidney. [16] Also for spleen. [18] For testis. [19] Also for testis. [20] Also for liver, spleen, and muscle.

continued

Part IV. Concentrations of Radionuclides in Air, Water, and Bodies of Radiation Workers

	Z	Symbol & Mass No.	Type of Decay	s or i	q[1] μCi	In Water Critical Organ	In Water 40-hr wk μCi/ml	In Water 168-hr wk μCi/ml	In Air Critical Organ	In Air 40-hr wk μCi/ml	In Air 168-hr wk μCi/ml
227	57	^{140}La	β^-, γ	s	9[4]	GI (LLI)	7×10^{-4}	2×10^{-4}	GI (LLI)	2×10^{-7}	5×10^{-8}
228				i	....	GI (LLI)	7×10^{-4}	2×10^{-4}	GI (LLI)	10^{-7}	4×10^{-8}
229	58	^{141}Ce	β^-, γ	s	30[4]	GI (LLI)	3×10^{-3}	9×10^{-4}	Liver	4×10^{-7}	2×10^{-7} [5,17]
230				i	....	GI (LLI)	3×10^{-3}	9×10^{-4}	Lung	2×10^{-7}	5×10^{-8}
231	58	^{143}Ce	β^-, γ	s	7[4]	GI (LLI)	10^{-3}	4×10^{-4}	GI (LLI)	3×10^{-7}	9×10^{-8}
232				i	....	GI (LLI)	10^{-3}	4×10^{-4}	GI (LLI)	2×10^{-7}	7×10^{-8}
233	58	^{144}Ce	α, β^-, γ	s	5[10]	GI (LLI)	3×10^{-4}	10^{-4}	Bone	10^{-8} [12]	3×10^{-9}
234				i	....	GI (LLI)	3×10^{-4}	10^{-4}	Lung	6×10^{-9}	2×10^{-9}
235	59	^{142}Pr	β^-, γ	s	7[10]	GI (LLI)	9×10^{-4}	3×10^{-4}	GI (LLI)	2×10^{-7}	7×10^{-8}
236				i	....	GI (LLI)	9×10^{-4}	3×10^{-4}	GI (LLI)	2×10^{-7}	5×10^{-8}
237	59	^{143}Pr	β^-	s	20[4,10]	GI (LLI)	10^{-3}	5×10^{-4}	GI (LLI)	3×10^{-7}	10^{-7}
238				i	....	GI (LLI)	10^{-3}	5×10^{-4}	Lung	2×10^{-7}	6×10^{-8}
239	60	^{144}Nd	α	s	0.1	Bone	2×10^{-3} [5]	7×10^{-4}	Bone	8×10^{-11}	3×10^{-11}
240				i	....	GI (LLI)	2×10^{-3}	8×10^{-4}	Lung	3×10^{-10}	10^{-10}
241	60	^{147}Nd	α, β^-, γ	s	10[4]	GI (LLI)	2×10^{-3}	6×10^{-4}	Liver	4×10^{-7} [5]	10^{-7} [5]
242				i	....	GI (LLI)	2×10^{-3}	6×10^{-4}	Lung	2×10^{-7}	8×10^{-8}
243	60	^{149}Nd	β^-, γ	s	3[4]	GI (LLI)	8×10^{-3}	3×10^{-3}	GI (LLI)	2×10^{-6}	6×10^{-7}
244				i	....	GI (ULI)	8×10^{-3}	3×10^{-3}	GI (ULI)	10^{-6}	5×10^{-7}
245	61	^{147}Pm	α, β^-	s	60[10]	GI (LLI)	6×10^{-3}	2×10^{-3}	Bone	6×10^{-8}	2×10^{-8}
246				i	....	GI (LLI)	6×10^{-3}	2×10^{-3}	Lung	10^{-7}	3×10^{-8}
247	61	^{149}Pm	β^-, γ	s	20[10]	GI (LLI)	10^{-3}	4×10^{-4}	GI (LLI)	3×10^{-7}	10^{-7}
248				i	....	GI (LLI)	10^{-3}	4×10^{-4}	GI (LLI)	2×10^{-7}	8×10^{-8}
249	62	^{147}Sm	α	s	0.1	Bone	2×10^{-3} [5]	6×10^{-4}	Bone	7×10^{-11}	2×10^{-11}
250				i	....	GI (LLI)	2×10^{-3}	7×10^{-4}	Lung	3×10^{-10}	9×10^{-11}
251	62	^{151}Sm	β^-, γ	s	100[10]	GI (LLI)	0.01	4×10^{-3}	Bone	6×10^{-8}	2×10^{-8}
252				i	....	GI (LLI)	0.01	4×10^{-3}	Lung	10^{-7}	5×10^{-8}
253	62	^{153}Sm	β^-, γ	s	20[4]	GI (LLI)	2×10^{-3}	8×10^{-4}	GI (LLI)	5×10^{-7}	2×10^{-7}
254				i	....	GI (LLI)	2×10^{-3}	8×10^{-4}	GI (LLI)	4×10^{-7}	10^{-7}
255	63	^{152}Eu (9.2 hr)	$\beta^-, \epsilon, \gamma$	s	8[4]	GI (LLI)	2×10^{-3}	6×10^{-4}	GI (LLI)	4×10^{-7}	10^{-7}
256				i	....	GI (LLI)	2×10^{-3}	6×10^{-4}	GI (LLI)	3×10^{-7}	10^{-7}
257	63	^{152}Eu (13 yr)	$\beta^-, \epsilon, \gamma$	s	20[6]	GI (LLI)	2×10^{-3}	8×10^{-4}	Kidney	10^{-8}	4×10^{-9}
258				i	....	GI (LLI)	2×10^{-3}	8×10^{-4}	Lung	2×10^{-8}	6×10^{-9}
259	63	^{154}Eu	$\beta^-, \epsilon, \gamma$	s	5[6,10]	GI (LLI)	6×10^{-4}	2×10^{-4}	Kidney	4×10^{-9} [17]	10^{-9} [17]
260				i	....	GI (LLI)	6×10^{-4}	2×10^{-4}	Lung	7×10^{-9}	2×10^{-9}
261	63	^{155}Eu	β^-, γ	s	70[6]	GI (LLI)	6×10^{-3}	2×10^{-3}	Kidney	9×10^{-8}	3×10^{-8} [17]
262				i	....	GI (LLI)	6×10^{-3}	2×10^{-3}	Lung	7×10^{-8}	3×10^{-8}
263	64	^{153}Gd	ϵ, γ, e^-	s	90[10]	GI (LLI)	6×10^{-3}	2×10^{-3}	Bone	2×10^{-7}	8×10^{-8}
264				i	....	GI (LLI)	6×10^{-3}	2×10^{-3}	Lung	9×10^{-8}	3×10^{-8}
265	64	^{159}Gd	β^-, γ	s	20[10]	GI (LLI)	2×10^{-3}	8×10^{-4}	GI (LLI)	5×10^{-7}	2×10^{-7}
266				i	....	GI (LLI)	2×10^{-3}	8×10^{-4}	GI (LLI)	4×10^{-7}	10^{-7}
267	65	^{160}Tb	β^-, γ	s	20[6,9,10]	GI (LLI)	10^{-3}	4×10^{-4}	Bone	10^{-7} [9,15]	3×10^{-8}
268				i	....	GI (LLI)	10^{-3}	4×10^{-4}	Lung	3×10^{-8}	10^{-8}
269	66	^{165}Dy	β^-, γ	s	10[10]	GI (ULI)	0.01	4×10^{-3}	GI (ULI)	3×10^{-6}	9×10^{-7}
270				i	....	GI (ULI)	0.01	4×10^{-3}	GI (ULI)	2×10^{-6}	7×10^{-7}
271	66	^{166}Dy	β^-, γ, e^-	s	5[10]	GI (LLI)	10^{-3}	4×10^{-4}	GI (LLI)	2×10^{-7}	8×10^{-8}
272				i	....	GI (LLI)	10^{-3}	4×10^{-4}	GI (LLI)	2×10^{-7}	7×10^{-8}
273	67	^{166}Ho	β^-, γ, e^-	s	5[10]	GI (LLI)	9×10^{-4}	3×10^{-4}	GI (LLI)	2×10^{-7}	7×10^{-8}
274				i	....	GI (LLI)	9×10^{-4}	3×10^{-4}	GI (LLI)	2×10^{-7}	6×10^{-8}

[1] When other footnote numbers appear in this column, q pertains only to the critical organ specified in the footnote. [4] For liver. [5] Also for lower large intestine. [6] For kidney. [9] Also for total body. [10] For bone. [12] Also for liver. [15] Also for kidney. [17] Also for bone.

continued

Part IV. Concentrations of Radionuclides in Air, Water, and Bodies of Radiation Workers

		Radionuclide				Maximum Permissible Concentrations of Radionuclides					
				s	q [1]	In Water			In Air		
	Z	Symbol & Mass No.	Type of Decay	or i	μCi	Critical Organ	40-hr wk μCi/ml	168-hr wk μCi/ml	Critical Organ	40-hr wk μCi/ml	168-hr wk μCi/ml
275	68	^{169}Er	β^-, γ	s	30 [10]	GI (LLI)	3×10^{-3}	9×10^{-4}	GI (LLI)	6×10^{-7}	2×10^{-7}
276				i	....	GI (LLI)	3×10^{-3}	9×10^{-4}	Lung	4×10^{-7}	10^{-7}
277	68	^{171}Er	β^-, γ, e^-	s	9 [10]	GI (ULI)	3×10^{-3}	10^{-3}	GI (ULI)	7×10^{-7}	2×10^{-7}
278				i	....	GI (ULI)	3×10^{-3}	10^{-3}	GI (ULI)	6×10^{-7}	2×10^{-7}
279	69	^{170}Tm	$\beta^-, \epsilon, \gamma, e^-$	s	9 [10]	GI (LLI)	10^{-3}	5×10^{-4}	Bone	4×10^{-8}	10^{-8}
280				i	....	GI (LLI)	10^{-3}	5×10^{-4}	Lung	3×10^{-8}	10^{-8}
281	69	^{171}Tm	β^-	s	90 [10]	GI (LLI)	0.01	5×10^{-3}	Bone	10^{-7}	4×10^{-8}
282				i	....	GI (LLI)	0.01	5×10^{-3}	Lung	2×10^{-7}	8×10^{-8}
283	70	^{175}Yb	β^-, γ	s	30 [6,10]	GI (LLI)	3×10^{-3}	10^{-3}	GI (LLI)	7×10^{-7}	2×10^{-7}
284				i	....	GI (LLI)	3×10^{-3}	10^{-3}	GI (LLI)	6×10^{-7}	2×10^{-7}
285	71	^{177}Lu	β^-, γ	s	20 [10]	GI (LLI)	3×10^{-3}	10^{-3}	GI (LLI)	6×10^{-7}	2×10^{-7}
286				i	....	GI (LLI)	3×10^{-3}	10^{-3}	GI (LLI)	5×10^{-7}	2×10^{-7}
287	72	^{181}Hf	β^-, γ	s	4 [8]	GI (LLI)	2×10^{-3}	7×10^{-4}	Spleen	4×10^{-8}	10^{-8}
288				i	....	GI (LLI)	2×10^{-3}	7×10^{-4}	Lung	7×10^{-8}	3×10^{-8}
289	73	^{182}Ta	β^-, γ	s	7 [4]	GI (LLI)	10^{-3}	4×10^{-4}	Liver	4×10^{-8}	10^{-8}
290				i	....	GI (LLI)	10^{-3}	4×10^{-4}	Lung	2×10^{-8}	7×10^{-9}
291	74	^{181}W	ϵ, γ	s	70 [4]	GI (LLI)	0.01	4×10^{-3}	GI (LLI)	2×10^{-6}	8×10^{-7}
292				i	....	GI (LLI)	0.01	3×10^{-3}	Lung	10^{-7}	4×10^{-8}
293	74	^{185}W	β^-	s	30 [10]	GI (LLI)	4×10^{-3}	10^{-3}	GI (LLI)	8×10^{-7}	3×10^{-7}
294				i	....	GI (LLI)	3×10^{-3}	10^{-3}	Lung	10^{-7}	4×10^{-8}
295	74	^{187}W	β^-, γ	s	30 [4,9]	GI (LLI)	2×10^{-3}	7×10^{-4}	GI (LLI)	4×10^{-7}	2×10^{-7}
296				i	....	GI (LLI)	2×10^{-3}	6×10^{-4}	GI (LLI)	3×10^{-7}	10^{-7}
297	75	^{183}Re	ϵ, γ	s	80	GI (LLI)	0.02 [9]	6×10^{-3}	Total body	3×10^{-6}	9×10^{-7}
298				i	....	GI (LLI)	8×10^{-3}	3×10^{-3}	Lung	2×10^{-7}	5×10^{-8}
299	75	^{186}Re	β^-, γ	s	20 [21]	GI (LLI)	3×10^{-3}	9×10^{-4}	GI (LLI)	6×10^{-7}	2×10^{-7}
300				i	....	GI (LLI)	10^{-3}	5×10^{-4}	GI (LLI)	2×10^{-7}	8×10^{-8}
301	75	^{187}Re	β^-	s	300 [22]	GI (LLI)	0.07	0.03 [23]	Skin	9×10^{-6}	3×10^{-6}
302				i	....	GI (LLI)	0.04	0.02	Lung	5×10^{-7}	2×10^{-7}
303	75	^{188}Re	β^-, γ	s	7 [21]	GI (LLI)	2×10^{-3}	6×10^{-4}	GI (LLI)	4×10^{-7}	10^{-7}
304				i	....	GI (LLI)	9×10^{-4}	3×10^{-4}	GI (LLI)	2×10^{-7}	6×10^{-8}
305	76	^{185}Os	ϵ, γ, e^-	s	8 [6]	GI (LLI)	2×10^{-3}	7×10^{-4}	GI (LLI)	5×10^{-7}	2×10^{-7}
306				i	....	GI (LLI)	2×10^{-3}	7×10^{-4}	Lung	5×10^{-8}	2×10^{-8}
307	76	^{191m}Os	β^-, γ, e^-	s	100 [6]	GI (LLI)	0.07	0.03	GI (LLI)	2×10^{-5}	6×10^{-6}
308				i	....	GI (LLI)	0.07	0.02	Lung	9×10^{-6}	3×10^{-6}
309	76	^{191}Os	β^-, γ, e^-	s	20 [6]	GI (LLI)	5×10^{-3}	2×10^{-3}	GI (LLI)	10^{-6}	4×10^{-7}
310				i	....	GI (LLI)	5×10^{-3}	2×10^{-3}	Lung	4×10^{-7}	10^{-7}
311	76	^{193}Os	β^-	s	10 [6]	GI (LLI)	2×10^{-3}	6×10^{-4}	GI (LLI)	4×10^{-7}	10^{-7}
312				i	....	GI (LLI)	2×10^{-3}	5×10^{-4}	GI (LLI)	3×10^{-7}	9×10^{-8}
313	77	^{190}Ir	ϵ, γ	s	40 [4,6,8]	GI (LLI)	6×10^{-3}	2×10^{-3}	GI (LLI)	10^{-6}	4×10^{-7}
314				i	....	GI (LLI)	5×10^{-3}	2×10^{-3}	Lung	4×10^{-7}	10^{-7}
315	77	^{192}Ir	β^-, γ	s	6 [6]	GI (LLI)	10^{-3}	4×10^{-4}	Kidney	10^{-7} [16]	4×10^{-8}
316				i	....	GI (LLI)	10^{-3}	4×10^{-4}	Lung	3×10^{-8}	9×10^{-9}
317	77	^{194}Ir	β^-	s	7 [6]	GI (LLI)	10^{-3}	3×10^{-4}	GI (LLI)	2×10^{-7}	8×10^{-8}
318				i	....	GI (LLI)	9×10^{-4}	3×10^{-4}	GI (LLI)	2×10^{-7}	5×10^{-8}
319	78	^{191}Pt	ϵ, γ	s	10 [6]	GI (LLI)	4×10^{-3}	10^{-3}	GI (LLI)	8×10^{-7}	3×10^{-7}
320				i	....	GI (LLI)	3×10^{-3}	10^{-3}	GI (LLI)	6×10^{-7}	2×10^{-7}

[1] When other footnote numbers appear in this column, q pertains only to the critical organ specified in the footnote. [4] For liver. [6] For kidney. [8] For spleen. [9] Also for total body. [10] For bone. [16] Also for spleen. [21] For thyroid. [22] For skin. [23] Also for skin.

continued

Part IV. Concentrations of Radionuclides in Air, Water, and Bodies of Radiation Workers

	Z	Symbol & Mass No.	Type of Decay	s or i	q[1]/ μCi	In Water Critical Organ	In Water 40-hr wk μCi/ml	In Water 168-hr wk μCi/ml	In Air Critical Organ	In Air 40-hr wk μCi/ml	In Air 168-hr wk μCi/ml
321	78	^{193m}Pt	ϵ, γ	s	100[6]/	GI (LLI)	0.03	0.01	GI (LLI)	7×10^{-6}	2×10^{-6}
322				i	....	GI (LLI)	0.03	0.01	GI (LLI)	5×10^{-6}	2×10^{-6}
323	78	^{193}Pt	ϵ	s	70	Kidney	0.03	9×10^{-3}	Kidney	10^{-6}	4×10^{-7}
324				i	....	GI (LLI)	0.05	0.02	Lung	3×10^{-7}	10^{-7}
325	78	^{197m}Pt	β^-, γ, e^-	s	5[6]/	GI (ULI)	0.03	0.01	GI (ULI)	6×10^{-6}	2×10^{-6}
326				i	....	GI (ULI)	0.03	9×10^{-3}	GI (ULI)	5×10^{-6}	2×10^{-6}
327	78	^{197}Pt	β^-, γ	s	10[6]/	GI (LLI)	4×10^{-3}	10^{-3}	GI (LLI)	8×10^{-7}	3×10^{-7}
328				i	....	GI (LLI)	3×10^{-3}	10^{-3}	GI (LLI)	6×10^{-7}	2×10^{-7}
329	79	^{196}Au	β^-, γ, e^-	s	40	GI (LLI)	5×10^{-3}	2×10^{-3}	GI (LLI)	10^{-6}	4×10^{-7}
330				i	....	GI (LLI)	4×10^{-3}	10^{-3}	Lung	6×10^{-7}	2×10^{-7}
331	79	^{198}Au	β^-, γ	s	20[6]/	GI (LLI)	2×10^{-3}	5×10^{-4}	GI (LLI)	3×10^{-7}	10^{-7}
332				i	....	GI (LLI)	10^{-3}	5×10^{-4}	GI (LLI)	2×10^{-7}	8×10^{-8}
333	79	^{199}Au	β^-, γ	s	70[6]/	GI (LLI)	5×10^{-3}	2×10^{-3}	GI (LLI)	10^{-6}	4×10^{-7}
334				i	....	GI (LLI)	4×10^{-3}	2×10^{-3}	GI (LLI)	8×10^{-7}	3×10^{-7}
335	80	^{197m}Hg	ϵ, γ, e^-	s	4	Kidney	6×10^{-3}	2×10^{-3}	Kidney	7×10^{-7}	3×10^{-7}
336				i	....	GI (LLI)	5×10^{-3}	2×10^{-3}	GI (LLI)	8×10^{-7}	3×10^{-7}
337	80	^{197}Hg	ϵ, γ, e^-	s	20	Kidney	9×10^{-3}	3×10^{-3}	Kidney	10^{-6}	4×10^{-7}
338				i	....	GI (LLI)	0.01	5×10^{-3}	GI (LLI)	3×10^{-6}	9×10^{-7}
339	80	^{203}Hg	β^-, γ, e^-	s	4	Kidney	5×10^{-4}	2×10^{-4}	Kidney	7×10^{-8}	2×10^{-8}
340				i	....	GI (LLI)	3×10^{-3}	10^{-3}	Lung	10^{-7}	4×10^{-8}
341	81	^{200}Tl	ϵ, γ	s	40[6]/	GI (LLI)	0.01	4×10^{-3}	GI (LLI)	3×10^{-6}	9×10^{-7}
342				i	....	GI (LLI)	7×10^{-3}	2×10^{-3}	GI (LLI)	10^{-6}	4×10^{-7}
343	81	^{201}Tl	ϵ, γ, e^-	s	40[6]/	GI (LLI)	9×10^{-3}	3×10^{-3}	GI (LLI)	2×10^{-6}	7×10^{-7}
344				i	....	GI (LLI)	5×10^{-3}	2×10^{-3}	GI (LLI)	9×10^{-7}	3×10^{-7}
345	81	^{202}Tl	ϵ, γ, e^-	s	20[6]/	GI (LLI)	4×10^{-3}	10^{-3}	GI (LLI)	8×10^{-7}	3×10^{-7}
346				i	....	GI (LLI)	2×10^{-3}	7×10^{-4}	Lung	2×10^{-7}	8×10^{-8}
347	81	^{204}Tl	β^-	s	10[6]/	GI (LLI)	3×10^{-3}	10^{-3}	Kidney	6×10^{-7}	2×10^{-7} [5]/
348				i	....	GI (LLI)	2×10^{-3}	6×10^{-4}	Lung	3×10^{-8}	9×10^{-9}
349	82	^{203}Pb	ϵ, γ	s	30[6]/	GI (LLI)	0.01	4×10^{-3}	GI (LLI)	3×10^{-6}	9×10^{-7}
350				i	....	GI (LLI)	0.01	4×10^{-3}	GI (LLI)	2×10^{-6}	6×10^{-7}
351	82	^{210}Pb	α, β^-, γ	s	0.4	Kidney	4×10^{-6} [9]/	10^{-6} [9]/	Kidney	10^{-10}	4×10^{-11}
352				i	....	GI (LLI)	5×10^{-3}	2×10^{-3}	Lung	2×10^{-10}	8×10^{-11}
353	82	^{212}Pb	$\alpha, \beta^-, \gamma, e^-$	s	0.02	Kidney	6×10^{-4} [5]/	2×10^{-4} [5]/	Kidney	2×10^{-8}	6×10^{-9}
354				i	....	GI (LLI)	5×10^{-4}	2×10^{-4}	Lung	2×10^{-8}	7×10^{-9}
355	83	^{206}Bi	ϵ, γ	s	1[6]/	GI (LLI)	10^{-3}	4×10^{-4}	Kidney	2×10^{-7} [5]/	6×10^{-8}
356				i	....	GI (LLI)	10^{-3}	4×10^{-4}	Lung	10^{-7}	5×10^{-8}
357	83	^{207}Bi	ϵ, γ	s	2[6]/	GI (LLI)	2×10^{-3}	6×10^{-4}	Kidney	2×10^{-7}	6×10^{-8}
358				i	....	GI (LLI)	2×10^{-3}	6×10^{-4}	Lung	10^{-8}	5×10^{-9}
359	83	^{210}Bi	α, β^-	s	0.04[6]/	GI (LLI)	10^{-3}	4×10^{-4}	Kidney	6×10^{-9}	2×10^{-9}
360				i	....	GI (LLI)	10^{-3}	4×10^{-4}	Lung	6×10^{-9}	2×10^{-9}
361	83	^{212}Bi	α, β^-, γ	s	0.01[6]/	GI (S)	0.01	4×10^{-3}	Kidney	10^{-7}	3×10^{-8}
362				i	....	GI (S)	0.01	4×10^{-3}	Lung	2×10^{-7}	7×10^{-8}
363	84	^{210}Po	α	s	0.03	Spleen	2×10^{-5} [15]/	7×10^{-6}	Spleen	5×10^{-10} [15]/	2×10^{-10} [15]/
364				i	....	GI (LLI)	8×10^{-4}	3×10^{-4}	Lung	2×10^{-10}	7×10^{-11}
365	85	^{211}At	α, ϵ, γ	s	0.02[24]/	Thyroid	5×10^{-5} [25]/	2×10^{-5} [25]/	Thyroid	7×10^{-9} [25]/	2×10^{-9}
366				i	....	GI (ULI)	2×10^{-3}	7×10^{-4}	Lung	3×10^{-8}	10^{-8}

[1]/ When other footnote numbers appear in this column, q pertains only to the critical organ specified in the footnote. [5]/ Also for lower large intestine. [6]/ For kidney. [9]/ Also for total body. [15]/ Also for kidney. [24]/ For ovary. [25]/ Also for ovary.

continued

Part IV. Concentrations of Radionuclides in Air, Water, and Bodies of Radiation Workers

	Z	Symbol & Mass No.	Type of Decay	s or i	q [1] μCi	In Water — Critical Organ	In Water — 40-hr wk μCi/ml	In Water — 168-hr wk μCi/ml	In Air — Critical Organ	In Air — 40-hr wk μCi/ml	In Air — 168-hr wk μCi/ml
367	86	^{220}Rn [26]	α, β⁻, γ, e⁻	..	....	....	....	....	Lung	3×10^{-7}	10^{-7}
368	86	^{222}Rn [26]	α, β⁻, γ	..	....	....	....	....	Lung	3×10^{-8}	10^{-8}
369	88	^{223}Ra	α, β⁻, γ	s	0.05	Bone	2×10^{-5}	7×10^{-6}	Bone	2×10^{-9}	6×10^{-10}
370				i	....	GI (LLI)	10^{-4}	4×10^{-5}	Lung	2×10^{-10}	8×10^{-11}
371	88	^{224}Ra	α, β⁻, γ, e⁻	s	0.06	Bone	7×10^{-5}	2×10^{-5}	Bone	5×10^{-9}	2×10^{-9}
372				i	....	GI (LLI)	2×10^{-4}	5×10^{-5}	Lung	7×10^{-10}	2×10^{-10}
373	88	^{226}Ra	α, β⁻, γ	s	0.1	Bone	4×10^{-7}	10^{-7}	Bone	3×10^{-11}	10^{-11}
374				i	....	GI (LLI)	9×10^{-4}	3×10^{-4}	Lung	5×10^{-11}	2×10^{-11}
375	88	^{228}Ra	α, β⁻, γ, e⁻	s	0.06	Bone	8×10^{-7}	3×10^{-7}	Bone	7×10^{-11}	2×10^{-11}
376				i	....	GI (LLI)	7×10^{-4}	3×10^{-4}	Lung	4×10^{-11}	10^{-11}
377	89	^{227}Ac	α, β⁻, γ	s	0.03	Bone	6×10^{-5}	2×10^{-5}	Bone	2×10^{-12}	8×10^{-13}
378				i	....	GI (LLI)	9×10^{-3}	3×10^{-3}	Lung	3×10^{-11}	9×10^{-12}
379	89	^{228}Ac	α, β⁻, γ, e⁻	s	0.04 [10]	GI (ULI)	3×10^{-3}	9×10^{-4}	Liver	8×10^{-8}	3×10^{-8} [17]
380				i	....	GI (ULI)	3×10^{-3}	9×10^{-4}	Lung	2×10^{-8}	6×10^{-9}
381	90	^{227}Th	α, β⁻, γ	s	0.02 [10]	GI (LLI)	5×10^{-4}	2×10^{-4}	Bone	3×10^{-10}	10^{-10}
382				i	....	GI (LLI)	5×10^{-4}	2×10^{-4}	Lung	2×10^{-10}	6×10^{-11}
383	90	^{228}Th	α, β⁻, γ, e⁻	s	0.02	Bone	2×10^{-4}	7×10^{-5}	Bone	9×10^{-12}	3×10^{-12}
384				i	....	GI (LLI)	4×10^{-4}	10^{-4}	Lung	6×10^{-12}	2×10^{-12}
385	90	^{230}Th	α, γ	s	0.05	Bone	5×10^{-5}	2×10^{-5}	Bone	2×10^{-12}	8×10^{-13}
386				i	....	GI (LLI)	9×10^{-4}	3×10^{-4}	GI (LLI)	10^{-11}	3×10^{-12}
387	90	^{231}Th	α, β⁻, γ	s	30 [10]	GI (LLI)	7×10^{-3}	2×10^{-3}	GI (LLI)	10^{-6}	5×10^{-7}
388				i	....	GI (LLI)	7×10^{-3}	2×10^{-3}	GI (LLI)	10^{-6}	4×10^{-7}
389	90	^{232}Th	α, β⁻, γ, e⁻	s	0.04	Bone	5×10^{-5}	2×10^{-5}	Bone	2×10^{-12} [27]	7×10^{-13} [27]
390				i	....	GI (LLI)	10^{-3}	4×10^{-4}	Lung	10^{-11}	4×10^{-12}
391	90	^{234}Th	β⁻, γ	s	4 [10]	GI (LLI)	5×10^{-4}	2×10^{-4}	Bone	6×10^{-8}	2×10^{-8}
392				i	....	GI (LLI)	5×10^{-4}	2×10^{-4}	Lung	3×10^{-8}	10^{-8}
393	90	Th-nat [28]	α, β⁻, γ, e⁻	s	0.01	Bone	3×10^{-5}	10^{-5}	Bone	2×10^{-12} [27]	6×10^{-13} [27]
394				i	....	GI (LLI)	3×10^{-4}	10^{-4}	Lung	4×10^{-12}	10^{-12}
395	91	^{230}Pa	α, β⁻, ε, γ	s	0.07 [10]	GI (LLI)	7×10^{-3}	2×10^{-3}	Bone	2×10^{-9}	6×10^{-10}
396				i	....	GI (LLI)	7×10^{-3}	2×10^{-3}	Lung	8×10^{-10}	3×10^{-10}
397	91	^{231}Pa	α, β⁻, γ	s	0.02	Bone	3×10^{-5}	9×10^{-6}	Bone	10^{-12}	4×10^{-13}
398				i	....	GI (LLI)	8×10^{-4}	3×10^{-4}	Lung	10^{-10}	4×10^{-11}
399	91	^{233}Pa	β⁻, γ	s	40 [6]	GI (LLI)	4×10^{-3}	10^{-3}	Kidney	6×10^{-7}	2×10^{-7}
400				i	....	GI (LLI)	3×10^{-3}	10^{-3}	Lung	2×10^{-7}	6×10^{-8}
401	92	^{230}U	α, β⁻, γ	s	0.01	Kidney	7×10^{-5}	2×10^{-5}	Kidney	3×10^{-10}	10^{-10}
402				i	....	GI (LLI)	10^{-4}	5×10^{-5}	Lung	10^{-10}	4×10^{-11}
403	92	^{232}U	α, β⁻, γ, e⁻	s	0.01	Bone	2×10^{-5}	8×10^{-6}	Bone	10^{-10}	3×10^{-11}
404				i	....	GI (LLI)	8×10^{-4}	3×10^{-4}	Lung	3×10^{-11}	9×10^{-12}
405	92	^{233}U	α, γ	s	0.5	Bone	10^{-4}	4×10^{-5}	Bone	5×10^{-10}	2×10^{-10}
406				i	....	GI (LLI)	9×10^{-4}	3×10^{-4}	Lung	10^{-10}	4×10^{-11}
407	92	^{234}U	α, γ	s	0.05	Bone	10^{-4}	4×10^{-5}	Bone	6×10^{-10}	2×10^{-10}
408				i	....	GI (LLI)	9×10^{-4}	3×10^{-4}	Lung	10^{-10}	4×10^{-11}
409	92	^{235}U	α, β⁻, γ	s	0.03	Kidney	10^{-4} [17]	4×10^{-5}	Kidney	5×10^{-10}	2×10^{-10} [17]
410				i	....	GI (LLI)	8×10^{-4}	3×10^{-4}	Lung	10^{-10}	4×10^{-11}

[1] When other footnote numbers appear in this column, q pertains only to the critical organ specified in the footnote. [6] For kidney. [10] For bone. [17] Also for bone. [26] The daughter elements of ^{220}Rn and ^{222}Rn are assumed present to the extent that they occur in unfiltered air; for all other isotopes the daughter elements are not considered as part of the intake, and if present they must be considered on the basis of rules for mixtures. [27] Provisional values. [28] In the case of natural thorium the double curie is used, i.e., 3.7×10^{10} disintegrations/s of ^{232}Th plus 3.7×10^{10} disintegrations/s of ^{228}Th.

continued

Part IV. Concentrations of Radionuclides in Air, Water, and Bodies of Radiation Workers

	Z	Symbol & Mass No.	Type of Decay	s or i	q[1] μCi	In Water Critical Organ	In Water 40-hr wk μCi/ml	In Water 168-hr wk μCi/ml	In Air Critical Organ	In Air 40-hr wk μCi/ml	In Air 168-hr wk μCi/ml
411	92	^{236}U	α, γ	s	0.06	Bone	10^{-4}	5×10^{-5}	Bone	6×10^{-10}	2×10^{-10}
412				i	....	GI (LLI)	10^{-3}	3×10^{-4}	Lung	10^{-10}	4×10^{-11}
413	92	^{238}U	α, γ, e^-	s	0.005	Kidney	2×10^{-5}	6×10^{-6}	Kidney	7×10^{-11}	3×10^{-11}
414				i	....	GI (LLI)	10^{-3}	4×10^{-4}	Lung	10^{-10}	5×10^{-11}
415	92	U-nat [29]	$\alpha, \beta^-, \gamma, e^-$	s	0.005	Kidney	2×10^{-5}	6×10^{-6}	Kidney	7×10^{-11}	3×10^{-11}
416				i	....	GI (LLI)	5×10^{-4}	2×10^{-4}	Lung	6×10^{-11}	2×10^{-11}
417	92	^{240}U + ^{240}Np	$\alpha, \beta^-, \gamma, e^-$	s	4 [6]	GI (LLI)	10^{-3}	3×10^{-4}	GI (LLI)	2×10^{-7}	8×10^{-8}
418				i	....	GI (LLI)	10^{-3}	3×10^{-4}	GI (LLI)	2×10^{-7}	6×10^{-8}
419	93	^{237}Np	α, β^-, γ	s	0.06	Bone	9×10^{-5}	3×10^{-5}	Bone	4×10^{-12}	10^{-12}
420				i	....	GI (LLI)	9×10^{-4}	3×10^{-4}	Lung	10^{-10}	4×10^{-11}
421	93	^{239}Np	α, β^-, γ	s	30 [10]	GI (LLI)	4×10^{-3}	10^{-3}	GI (LLI)	8×10^{-7}	3×10^{-7}
422				i	....	GI (LLI)	4×10^{-3}	10^{-3}	GI (LLI)	7×10^{-7}	2×10^{-7}
423	94	^{238}Pu	α, γ	s	0.04	Bone	10^{-4}	5×10^{-5}	Bone	2×10^{-12}	7×10^{-13}
424				i	....	GI (LLI)	8×10^{-4}	3×10^{-4}	Lung	3×10^{-11}	10^{-11}
425	94	^{239}Pu	α, γ	s	0.04	Bone	10^{-4}	5×10^{-5}	Bone	2×10^{-12}	6×10^{-13}
426				i	....	GI (LLI)	8×10^{-4}	3×10^{-4}	Lung	4×10^{-11}	10^{-11}
427	94	^{240}Pu	α, γ	s	0.04	Bone	10^{-4}	5×10^{-5}	Bone	2×10^{-12}	6×10^{-13}
428				i	....	GI (LLI)	8×10^{-4}	3×10^{-4}	Lung	4×10^{-11}	10^{-11}
429	94	^{241}Pu	α, β^-, γ	s	0.9	Bone	7×10^{-3}	2×10^{-3}	Bone	9×10^{-11}	3×10^{-11}
430				i	....	GI (LLI)	0.04	0.01	Lung	4×10^{-8}	10^{-8}
431	94	^{242}Pu	α	s	0.05	Bone	10^{-4}	5×10^{-5}	Bone	2×10^{-12}	6×10^{-13}
432				i	....	GI (LLI)	9×10^{-4}	3×10^{-4}	Lung	4×10^{-11}	10^{-11}
433	94	^{243}Pu	$\alpha, \beta^-, \gamma, e^-$	s	7 [10]	GI (ULI)	0.01	3×10^{-3}	GI (ULI)	2×10^{-6}	6×10^{-7}
434				i	....	GI (ULI)	0.01	3×10^{-3}	GI (ULI)	2×10^{-6}	8×10^{-7}
435	94	^{244}Pu	$\alpha, \beta^-, \gamma, e^-$ (99.7%), SF (0.3%)	s	0.04	Bone	10^{-4}	4×10^{-5}	Bone	2×10^{-12}	6×10^{-13}
436				i	....	GI (LLI)	3×10^{-4}	10^{-4}	Lung	3×10^{-11}	10^{-11}
437	95	^{241}Am	α, γ	s	0.1	Kidney	10^{-4} [17]	4×10^{-5}	Kidney	6×10^{-12} [17]	2×10^{-12} [17]
438				i	....	GI (LLI)	8×10^{-4}	3×10^{-4}	Lung	10^{-10}	4×10^{-11}
439	95	^{242m}Am	$\alpha, \beta^-, \gamma, \epsilon, e^-$	s	0.07	Bone	10^{-4} [15]	4×10^{-5}	Bone	6×10^{-12} [15]	2×10^{-12} [15]
440				i	....	GI (LLI)	3×10^{-3}	9×10^{-4}	Lung	3×10^{-10}	9×10^{-11}
441	95	^{242}Am	$\alpha, \beta^-, \gamma, \epsilon, e^-$	s	0.06 [4]	GI (LLI)	4×10^{-3}	10^{-3}	Liver	4×10^{-8}	10^{-8}
442				i	....	GI (LLI)	4×10^{-3}	10^{-3}	Lung	5×10^{-8}	2×10^{-8}
443	95	^{243}Am	α, β^-, γ	s	0.05	Bone	10^{-4} [15]	4×10^{-5}	Bone	6×10^{-12} [15]	2×10^{-12} [15]
444				i	....	GI (LLI)	8×10^{-4}	3×10^{-4}	Lung	10^{-10}	4×10^{-11}
445	95	^{244}Am	$\alpha, \beta^-, \gamma, e^-$	s	0.2 [6,10]	GI (SI)	0.1	0.05	Bone	4×10^{-6} [15]	10^{-6} [15]
446				i	....	GI (SI)	0.1	0.05	GI (SI)	2×10^{-5} [3]	8×10^{-6} [3]
447	96	^{242}Cm	α, γ	s	0.05 [4]	GI (LLI)	7×10^{-4}	2×10^{-4}	Liver	10^{-10}	4×10^{-11}
448				i	....	GI (LLI)	7×10^{-4}	2×10^{-4}	Lung	2×10^{-10}	6×10^{-11}
449	96	^{243}Cm	α, γ	s	0.09	Bone	10^{-4}	5×10^{-5}	Bone	6×10^{-12}	2×10^{-12}
450				i	....	GI (LLI)	7×10^{-4}	2×10^{-4}	Lung	10^{-10}	3×10^{-11}
451	96	^{244}Cm	α, γ	s	0.1	Bone	2×10^{-4}	7×10^{-5}	Bone	9×10^{-12}	3×10^{-12}
452				i	....	GI (LLI)	8×10^{-4}	3×10^{-4}	Lung	10^{-10}	3×10^{-11}
453	96	^{245}Cm	α, β^-, γ	s	0.04	Bone	10^{-4}	4×10^{-5}	Bone	5×10^{-12}	2×10^{-12}
454				i	....	GI (LLI)	8×10^{-4}	3×10^{-4}	Lung	10^{-10}	4×10^{-11}
455	96	^{246}Cm	α	s	0.05	Bone	10^{-4}	4×10^{-5}	Bone	5×10^{-12}	2×10^{-12}
456				i	....	GI (LLI)	8×10^{-4}	3×10^{-4}	Lung	10^{-10}	4×10^{-11}

[1] When other footnote numbers appear in this column, **q** pertains only to the critical organ specified in the footnote. [3] Also for lung. [4] For liver. [6] For kidney. [10] For bone. [15] Also for kidney. [17] Also for bone. [29] In the case of natural uranium the double curie used corresponds to 3.7×10^{10} disintegrations/s of ^{238}U plus 3.7×10^{10} disintegrations/s of ^{235}U plus 1.7×10^9 disintegrations/s of ^{235}U.

continued

Part IV. Concentrations of Radionuclides in Air, Water, and Bodies of Radiation Workers

	Z	Symbol & Mass No.	Type of Decay	s or i	$q\underline{1/}$ μCi	In Water Critical Organ	40-hr wk μCi/ml	168-hr wk μCi/ml	In Air Critical Organ	40-hr wk μCi/ml	168-hr wk μCi/ml
457	96	^{247}Cm	$\alpha, \beta^-, \gamma, e^-$	s	0.04	Bone	10^{-4}	4×10^{-5}	Bone	5×10^{-12}	2×10^{-12}
458				i	....	GI (LLI)	6×10^{-4}	2×10^{-4}	Lung	10^{-10}	4×10^{-11}
459	96	^{248}Cm	α (89%),	s	0.005	Bone	10^{-5}	4×10^{-6}	Bone	6×10^{-13}	2×10^{-13}
460			SF (11%)	i	....	GI (LLI)	4×10^{-5}	10^{-5}	Lung	10^{-11}	4×10^{-12}
461	96	^{249}Cm	$\alpha, \beta^-, \gamma, e^-$	s	$1\underline{10/}$	GI (S)	0.06	0.02	Bone	$10^{-5}\underline{30/}$	4×10^{-6}
462				i	....	GI (S)	0.06	0.02	GI (S)	10^{-5}	4×10^{-6}
463	97	^{249}Bk	α, β^-, γ	s	$0.7\underline{10/}$	GI (LLI)	0.02	6×10^{-3}	Bone	9×10^{-10}	3×10^{-10}
464				i	....	GI (LLI)	0.02	6×10^{-3}	Lung	10^{-7}	4×10^{-8}
465	97	^{250}Bk	$\alpha, \beta^-, \gamma, e^-$	s	$0.05\underline{10/}$	GI (ULI)	6×10^{-3}	2×10^{-3}	Bone	10^{-7}	5×10^{-8}
466				i	....	GI (ULI)	6×10^{-3}	2×10^{-3}	GI (ULI)	10^{-6}	4×10^{-7}
467	98	^{249}Cf	α, γ	s	0.04	Bone	10^{-4}	4×10^{-5}	Bone	2×10^{-12}	5×10^{-13}
468				i	....	GI (LLI)	7×10^{-4}	2×10^{-4}	Lung	10^{-10}	3×10^{-11}
469	98	^{250}Cf	α	s	0.04	Bone	4×10^{-4}	10^{-4}	Bone	5×10^{-12}	2×10^{-12}
470				i	....	GI (LLI)	7×10^{-4}	3×10^{-4}	Lung	10^{-10}	3×10^{-11}
471	98	^{251}Cf	α, γ	s	0.04	Bone	10^{-4}	4×10^{-5}	Bone	2×10^{-12}	6×10^{-13}
472				i	....	GI (LLI)	8×10^{-4}	3×10^{-4}	Lung	10^{-10}	3×10^{-11}
473	98	^{252}Cf	α, γ, SF	s	$0.01\underline{10/}$	GI (LLI)	2×10^{-4}	7×10^{-5}	Bone	6×10^{-12}	2×10^{-12}
474				i	....	GI (LLI)	2×10^{-4}	7×10^{-5}	Lung	3×10^{-11}	10^{-11}
475	98	^{253}Cf	$\alpha, \beta^-, \gamma, e^-$	s	$0.04\underline{10/}$	GI (LLI)	4×10^{-3}	10^{-3}	Bone	8×10^{-10}	3×10^{-10}
476				i	....	GI (LLI)	4×10^{-3}	10^{-3}	Lung	8×10^{-10}	3×10^{-10}
477	98	^{254}Cf	SF	s	$0.0007\underline{10/}$	GI (LLI)	4×10^{-6}	10^{-6}	Bone	5×10^{-12}	2×10^{-12}
478				i	....	GI (LLI)	4×10^{-6}	10^{-6}	Lung	5×10^{-12}	2×10^{-12}
479	99	^{253}Es	$\alpha, \beta^-, \gamma, e^-$	s	$0.04\underline{10/}$	GI (LLI)	7×10^{-4}	2×10^{-4}	Bone	8×10^{-10}	3×10^{-10}
480				i	....	GI (LLI)	7×10^{-4}	2×10^{-4}	Lung	6×10^{-10}	2×10^{-10}
481	99	^{254m}Es	$\alpha, \beta^-, \gamma, e^-$	s	$0.02\underline{10/}$	GI (LLI)	5×10^{-4}	2×10^{-4}	Bone	5×10^{-9}	2×10^{-9}
482				i	....	GI (LLI)	5×10^{-4}	2×10^{-4}	Lung	6×10^{-9}	2×10^{-9}
483	99	^{254}Es	$\alpha, \beta^-, \gamma, e^-$	s	$0.02\underline{10/}$	GI (LLI)	4×10^{-4}	10^{-4}	Bone	2×10^{-11}	6×10^{-12}
484				i	....	GI (LLI)	4×10^{-4}	10^{-4}	Lung	10^{-10}	4×10^{-11}
485	99	^{255}Es	α, β^-, γ	s	$0.04\underline{10/}$	GI (LLI)	8×10^{-4}	3×10^{-4}	Bone	5×10^{-10}	2×10^{-10}
486				i	....	GI (LLI)	8×10^{-4}	3×10^{-4}	Lung	4×10^{-10}	10^{-10}
487	100	^{254}Fm	α, γ, e^-	s	$0.02\underline{10/}$	GI (ULI)	4×10^{-3}	10^{-3}	Bone	6×10^{-8}	2×10^{-8}
488			(99.9448%),	i	....	GI (ULI)	4×10^{-3}	10^{-3}	Lung	7×10^{-8}	2×10^{-8}
			SF (5.52 × 10^{-2}%)								
489	100	^{255}Fm	α, γ	s	$0.04\underline{10/}$	GI (LLI)	10^{-3}	3×10^{-4}	Bone	2×10^{-8}	6×10^{-9}
490				i	....	GI (LLI)	10^{-3}	3×10^{-4}	Lung	10^{-8}	4×10^{-9}
491	100	^{256}Fm	SF	s	$0.0008\underline{10/}$	GI (ULI)	3×10^{-5}	9×10^{-6}	Bone	3×10^{-9}	10^{-9}
492				i	....	GI (ULI)	3×10^{-5}	9×10^{-6}	Lung	2×10^{-9}	6×10^{-10}

$\underline{1/}$ When other footnote numbers appear in this column, q pertains only to the critical organ specified in the footnote.
$\underline{10/}$ For bone. $\underline{30/}$ Also for stomach.

Contributor: Morgan, Karl Z.

General References
[1] International Commission on Radiological Protection. 1959. Int. Comm. Radiol. Prot. Publ. 2.
[2] Ibid. 4, 1964.
[3] National Council on Radiation Protection. 1963. Nat. Counc. Radiat. Prot. Rep. 22:80.

Radiation: **Type**—MeV = million electron volts, one electron volt being the energy given to an electron when it is accelerated through a potential difference of one volt and corresponding to 1.602×10^{-12} ergs (1 MeV $= 1.6 \times 10^{-6}$ ergs); kV = kilovolt, a unit of electrical potential equal to 1000 volts; **Exposure**—μCi = microcurie, one millionth of a curie or 3.7×10^{4} disintegrations per second; mCi = millicurie, one thousandth of a curie or 3.7×10^{7} disintegrations per second; rem = roentgen equivalent man, the quantity of radiation absorbed in tissue which gives the same observable effect as one rep of X or gamma rays (rep = roentgen equivalent physical, the equivalent of 93 ergs per gram energy absorption). **Late Effects: Latent Period**—For acute exposures equal to the elapsed time between exposure and appearance of the effect; for chronic exposures equal to the elapsed time between the beginning of the exposure and

appearance of the effect; **Incidence**—n/n' = number of individuals affected/number exposed; percentages gives percent of individuals affected. **Remarks**: RBE = relative biological effectiveness, i.e., the ratio of absorbed dose, in rads, of a test radiation required to produce a given biological effect to the amount of a standard radiation, such as 250 kV X rays or gamma rays from Cobalt-60, required to produce the same effect. *Abbreviations & Symbols:* rad = unit of absorbed dose, 1 rad corresponding to 100 ergs per gram of absorber; R = roentgen, the quantity of X- or γ-radiation such that the associated corpuscular emission per 0.001293 gram of air produces, in air, ions carrying one electrostatic unit of electrical charge of either sign. For further information on the quantitative estimation of late effects of all kinds of external radiation in a variety of mammalian systems, consult reference 67.

	Radiation		Late Effects			Remarks	Reference
	Type	Exposure	Latent Period	Effects	Incidence n/n' [1]	Remarks	
				Man			
1	A-bomb	In utero	Through 16 yr	No change in risk of leukemia or other cancers	....	....	25
2			Through 5 yr	Microcephaly	33/169	Most sensitive at 7 to 15 wk of gestation	42
3				Mental retardation	15/169		
4		Epilation dose	2 yr	Radiation-induced cataracts	10 cases	These cases are ones which affected vision; many more detectable only with slit lamp	12,13,27
5		>50 rads	Through 21 yr	Increased incidence of cancers of lung, thyroid, & mammary gland	....	....	69,70,74
6		20-300+ rads	Through 21 yr	Increased incidence of leukemia; dose-dependent increase.	Approx. 2-3 cases per 10^6 man rads per year	Dose-response curve shows no threshold and does not differ significantly from linearity. Risk declining from peak incidence 6-7 yr after the bomb. RBE for neutron component appears to be 5.	24
	Neutrons, from cyclotron						
7	0-20 MeV	10-135 neutrons	2-10 yr	Cataracts, minimal	3/10	....	1
8				slight to moderate	4/10		
9				severe	3/10		
10		Chronic exposure	....	Mild epilation	2 cases	....	1
11				No blood changes	....		
12	16 MeV	Epilation dose	2 yr	Radiation-induced cataracts	2 cases	....	12,13
13		400-500 neutrons	2 mo-5 yr	Severe epidermolytic reaction, skin atrophy & fibrosis, persistent ulcerations, and diminished repair by normal tissues	13/16	....	59
14				Radiation osteitis; severe gastrointestinal reactions	....		

[1] Unless otherwise specified.

continued

	Radiation		Late Effects			Remarks	Reference
	Type	Exposure	Latent Period	Effects	Incidence n/n' [1]		
15	Radium, residual	Ingested 0.02-180 µg	1-48 yr	Radiation osteitis; osteomyelitis; loss of teeth; susceptibility to fractures; deafness; arthritis. Dose-dependent increase.	....	Shape of dose-response curve is unknown.	3,40,71, 72
16		>(10-20) µg	1-48 yr	Marrow hyperplasia & jaw necrosis	13 cases	....	3,40,71, 72
17		1000-10,000 rads locally over a variable time	>3 yr	Osteosarcomas & carcinomas proportional to dose	....	Shape of dose-response curve is non-linear	21
18		10,000-23,200 rads locally over a variable time	>3 yr	Reduced incidence of osteosarcomas & carcinomas	....	Shape of dose-response curve is non-linear	21
19		Treatment 1-10 µg	20-30 yr	Changes in bone density	24/24	....	38
20				Distortion of Haversian canals & normal bone configuration	7/24		
21				Edentia & mandibular lesions	3/19		
22				Aseptic necrosis of bone	7/24		
23				Fibrosarcoma & honeycombed teeth	....		
24		4-10 µg	20-30 yr	Dental changes	100%	....	38
25	Uranium ores	Variable doses	>10 yr	Four-fold increase in risk of respiratory neoplasms; 35-fold difference between lowest & highest cumulative exposure categories	....	....	68
26	X rays	Variable doses	Months to years	Skin atrophy, telangiectasis, sclerosis, pigmentation, alopecia, altered vasomotion, diminished sweat & sebaceous function, loss of cutaneous ridges & fingerprints, ulcers & keratoses, malignancies, hyperkeratotic warty growths, deformed & brittle nails, loss of nails, & subungual hyperkeratoses	....	....	4,23,43, 71,72
27		In utero, diagnostic	Up to 10 yr after birth	Cancer	572 deaths per 10^6 man rads in first 10 yr of life	Incidence a linear function of dose	58
28		To the head, therapeutic 200-6900 rads	Variable	Radiation-induced cataracts	....	Single exposure threshold = 200 rads; multiple exposure threshold (given at intervals of 3 wk or more) = 400 rads	41
29		1200-6900 rads	Variable	Radiation-induced cataracts	100%	....	41
30		500-3000 rads to spine for treatment of ankylosing spondylitis	Variable	Leukemia increased by a factor of 9.5; cancer of heavily irradiated sites increased by factor of 1.6	....	....	15

[1] Unless otherwise specified.

continued

	Radiation		Late Effects			Remarks	Refer-ence
	Type	Exposure	Latent Period	Effect	Incidence n/n' [1]		
31		500-1050 R in 1 to 3 fractions to induce artificial menopause	3-24 yr	Risk of leukemia increased by a factor of 4.6; no demonstrable increase in risk of other malignancies	....	....	20
32		Radiologists exposed occupationally; dose unknown	Lifetime	Life expectancy: No effect to 4.8 yr reduction, depending upon baseline population used for comparison	....	Present radiological practices have effectively corrected the risks where demonstrated	14,54, 55,73
				Dog			
33	Neutrons	0.012-0.11 neutrons/da	1 yr	Reduction of lymphocytes only observed effect	....	....	6
34		1.7 neutrons/da	1 yr	Mucoid conjunctivitis, keratoconjunctivitis, & corneal opacities; reduced size of spleen & testes, increased incidence of infection; hypoplasia of bone marrow & regional lymph nodes; hemorrhage of lymph nodes, heart, stomach, & small bowel; reduction of all formed elements of blood	....	....	6
35	Fast	150 neutrons	2 yr	Destruction and chronic inflammation of cornea; no cataracts	....	....	45
36		900 neutrons	2 yr	Radiation-induced cataracts	60-75%	....	45
37	X rays	0.06-0.12 R/da	1+ yr	Spermatogenesis: no effect	....	....	11
38		0.6 R/da	4 mo to years	Sperm count 0-6% of normal	....	....	11
39		3 R/da to total of 375 R	1 yr	Permanent sterility	....	More effective than 375 R given as single exposure	11
40		100-2000 R to the testes	10-25 wk	Aspermia or oligospermia, followed by recovery	....	....	11
41		100-300 R	Lifetime	Decrease in life expectancy; 9%/100 R	....	....	2
42		100 R, in fractions	Lifetime	Decrease in life expectancy	....	No effect of fractionation	2
43		300 R, in fractions	Lifetime	Life expectancy decreased as time between fractions decreased; no serious impairment of reproductive ability	....	....	2
44		300-375 R	10-25 wk	Aspermia or oligospermia followed by recovery	....	....	11
45		750 R, as weekly 50 R fractions	Lifetime	Complete sterilization of female pups	....	....	2
				Guinea Pig			
	Gamma rays From cobalt-60	Repeated daily					
46		15 R/da	106 da	Death	50%	....	61
47		30 R/da	63 da	Death	50%	....	61
48		60 R/da	41 da	Death	50%	....	61
49		90 R/da	20 da	Death	50%	....	61
50		120 R/da	18 da	Death	50%	....	61

[1] Unless otherwise specified.

continued

	Radiation		Late Effects			Remarks	Refer-ence
	Type	Exposure	Latent Period	Effects	Incidence n/n' [1]		
51	From radium	Repeated daily 0.11 R/da	38 mo	Death	75%	Survival not different from controls	26,39,60
52		1.1 R/da to total of 1050 R	32 mo	Death	50%	....	26,39,60
53		2.2 R/da to total of 2100 R	32 mo	Death	50%	....	26,39,60
54		4.4 R/da to total of 2400 R	18 mo	Death	50%	....	26,39,60
55			....	Significant reduction of formed elements of blood	....		
56		8.8 R/da to total of 2300 R	5 mo	Death	50%	....	26,39,60
57			....	Significant reduction of formed elements of blood	....		
				Mouse			
58	Carbon-14, in sodium formate	100 μCi, injected intraperitoneally during gestation	Lifetime	Normal offspring: no effect on weight, longevity, or tumor incidence; no effect on fertility	....	....	57
59	Gamma rays From cobalt-60	5-200,000 R/da in daily 8 hr periods	Lifetime	Life expectancy decreased	....	Effects not a monotonic function of dose rate; multiple physiological systems involved	53
60	From A-bomb	Single	Lifetime	Life expectancy decreased	....	Effect is a curvilinear function of radiation dose	63
61	Neutrons, fast	Single	....	Radiation-induced cataracts	....	RBE = 2	18
62			....	Radiation-induced cataracts	....	RBE = 4	16,51,52
63		3 neutrons	....	Radiation-induced cataracts	....	RBE = 6	16,51,52
64		Repeated daily	Lifetime	Life expectancy decreased	....	RBE = 10	46
65		3-50 rems	Lifetime	Life expectancy decreased	....	RBE may vary with dose rate	44
66		0.004-11.4 neutrons/da	....	Life expectancy decreased; carcinogenesis	....	RBE is dose-rate dependent, 10 at lowest and 3 at highest rate	65,66
67	Plutonium-239	Intravenous injections 3.1-15.6 μCi/kg	190-250 da	Osteogenic sarcoma	....	....	10,71,72
68		>6.1 μCi/kg	Lifetime	Marked decrease in life expectancy	....	....	10,71,72
69	Protons, 60 MeV	Single	<2 mo	Radiation-induced cataracts	....	RBE = 1	19
70	Radium-226	Intravenous injections 12+ μCi/kg	250-300 da	Osteogenic sarcoma	....	....	10,71,72
71		50+ μCi/kg	....	Marked decrease in life expectancy; debilitation and increased incidence of infection	....	....	10,71,72

[1] Unless otherwise specified.

continued

	Radiation		Late Effects			Remarks	Reference
	Type	Exposure	Latent Period	Effects	Incidence n/n' [1]		
72	Strontium-89	Single injections 2.5 μCi/g	200 da	Bone tumors	....	....	9
73		5.0 μCi/g	150 da	Bone tumors	....	....	9
74		Monthly injections 0.05 μCi/g	500 da	Bone tumors	....	....	9
75		0.10 μCi/g	425 da	Bone tumors	....	....	9
76		0.20 μCi/g	350 da	Bone tumors	....	....	9
77		0.50 μCi/g	220 da	Bone tumors	....	....	9
78		1.00 μCi/g	160 da	Bone tumors	....	....	9
79	Strontium-90	Single injections	200+ da	Carcinomas of mucous membranes of head	....	Incidence increased in proportion to dose	47
80		0.2 μCi/g	225-250 da	Leukemia induction	....	Incidence highest at this dose; at higher doses other problems remove animals before leukemia can develop	49
81			200+ da	Osteosarcoma induction	1/17	....	48
82		0.4 μCi/g	200+ da	Osteosarcoma induction	11/27	....	48
83		0.8 μCi/g	200+ da	Osteosarcoma induction	41/46	....	48
84		1.6 μCi/g	200+ da	Osteosarcoma induction	42/50	....	48
85	X rays	Single	Lifetime	Reduction of life expectancy	....	Dose-response curve is linear; slope = 5.44% loss/100 rads. Sensitivity to radiation-induced life shortening decreases with advancing age, but fetus is less sensitive than adult.	29,36,64
86				Carcinogenesis	....	Fetus less sensitive to radiation carcinogenesis than adult	64
87		Fractionated	Lifetime	Reduction of life expectancy	....	Daily exposures are only half as deleterious as acute exposures in reducing life expectancy. Shape of dose-response curve varies with mouse strain studied.	75
88	X rays or gamma rays	External	Lifetime	Increase in occurrence of all spontaneously occurring cancers	....	....	62
				Rat			
89	Cerium-144	1-3 mCi/kg	200 da	Osteogenic sarcoma; liver atrophy with ascites and jaundice	....	....	37
90	Gamma rays, from cobalt-60	20 R/da	332 da	Death	50%	....	61
91		60 R/da	236 da	Death	50%	....	61
92		70 R/da	72 da	Death	50%	....	61
93		80 R/da	53 da	Death	50%	....	61
94		90 R/da	48 da	Death	50%	....	61
95		120 R/da	38 da	Death	50%	....	61
96	Neutrons	0.012-0.06 neutrons/da	1 yr	No effect	....	....	6
97		0.11 neutrons/da	1 yr	Neoplasms, including leukemias, increased by a factor of 3	....	....	6
98		1.7 neutrons/da	1 yr	Increased infection and neoplasms; bilateral cataracts; hypoplasia of spleen; atrophy of testes & ovarian follicles; reduction of formed elements of blood	....	....	6

[1] Unless otherwise specified.

continued

	Radiation		Late Effects			Remarks	Reference
	Type	Exposure	Latent Period	Effects	Incidence n/n' [1]	Remarks	Reference
99	Fast	220 rads, in single exposure	Lifetime	Reduction of life expectancy	....	Effect decreases as age at exposure increases	28
100	Phosphorus-32	External	4-5 mo	Tumors; lens opacities; skin injuries	....	....	76
101	Plutonium-239	3.1-6.2 mCi/g	300-400 da	Osteogenic sarcoma	....	....	7
102		4.6-7.8 mCi/g	3-7 mo	Non-neoplastic bone injury, with some evidence of recovery; destruction of germ cells in both sexes	....	....	7
103	Praseodymium-144	1-3 mCi/kg	200 da	Osteogenic sarcoma; liver atrophy with ascites and jaundice	....	....	37
104	Radium	0.125 μCi/g	5 mo	Damage to epiphyseal cartilage, overgrowth with atypical bone, & loss of normal bone cells; destruction of ovary	....	....	7,71,72
105		0.5 μCi/g	3-6 mo	Blood vessel injury; ovarian injury	....	....	7,71,72
106		0.6 μCi/g	5 mo	Damage to epiphyseal cartilage & bone; atypical bone formed	....	....	7,71,72
107	Strontium-89	Single or monthly injections	Variable	Osteogenic carcinomas	....	Appearance earlier with higher doses; single injections less effective	50
	Strontium-90	In drinking water for 10-30 da					
108		33 μCi, total dose	425 da	1 μCi in skeleton	....	....	22
109			380 da	No increase in neoplasms	....		
110		650 μCi, total dose	346 da	2 μCi in skeleton	....	....	22
111			372 da	Leukemias	7×10^{-4}/rat/wk		
112				Osteogenic sarcomas	0		
113		790 μCi, total dose	117 da	11 μCi in skeleton	....	....	22
114			254 da	Leukemias	1.7×10^{-3}/rat/wk		
115				Osteogenic sarcomas	7.6×10^{-3}/rat/wk		
116		464 μCi, total dose	40 da	33 μCi in skeleton	....	....	22
117			106 da	Leukemias	0		
118				Osteogenic sarcomas	0.023/rat/wk		
119	Thorium oxide	0.3 ml	14 mo	Fibroblastic tumors	14/60	....	23
120		2.5 ml	10-17 mo	Sarcomas	33/50	....	23
121		5.0 ml	10-17 mo	Sarcomas	50/50	....	23
122	X rays, 250 kV	25-400 R to whole body	10.5-11 mo	Mammary gland neoplasia	....	Dose-incidence curve linear; above 400R incidence erratic	8
123		1-6 exposures 0 R	Lifetime	Normal life expectancy = 28.6 mo	....	....	35
124		120 R	Lifetime	Life expectancy reduced to 24.9 mo	....	....	35

[1] Unless otherwise specified.

continued

Radiation		Late Effects			Remarks	Refer-ence
Type	Exposure	Latent Period	Effects	Incidence n/n' 1/		
125		240 R	Lifetime	Life expectancy reduced to 23.1 mo	....	....	35
126		480 R	Lifetime	Life expectancy reduced to 20.2 mo	....	....	35
127		400 R to in-tact or cas-trated ♂	16 mo	Mammary cancers, mostly fibrosarcomas	50%	....	56
128		400 R to ♀	10 mo	Mammary cancers, one or more per rat	79%	....	17
129		1000 R to whole body	Through 500 da	Decrease in life expectancy; nephrosclerosis; generalized arteriosclerosis, hypertension, thrombocy-topenia, anemia; earlier onset of neoplasms	....	Animals breathing 5% oxy-gen	5,30-34

1/ Unless otherwise specified.

Contributor: Yuhas, John M.

References

[1] Abelson, P. H., and P. G. Kruger. 1949. Science 110:655.
[2] Andersen, A., and L. Rosenblatt. 1968. U.S. At. Energy Comm. Rep. TID-4500:11.1.
[3] Aub, J. C., et al. 1952. Medicine (Baltimore) 31:221.
[4] Behrens, C. F., et al. 1959. Atomic Medicine. Ed. 3. Williams and Wilkins, Baltimore.
[5] Bennett, L. R., et al. 1953. Radiology 61:411.
[6] Blair, H. A., ed. 1954. Nat. Nucl. Energy Ser. Div. VI-2.
[7] Bloom, W., ed. 1948. Ibid. IV-22I.
[8] Bond, V. P., et al. 1960. Radiat. Res. 12:276.
[9] Brues, A. M. 1949. J. Clin. Invest. 28:1286.
[10] Brues, A. M. 1953. Nucl. Sci. Abstr. 7:10.
[11] Casarett, G. W., and H. A. Eddy. 1968. Proc. Symp. Dose Rate Mammal. Radiat. Biol., p. 14.1.
[12] Cogan, D. C., et al. 1949. Science 110:654.
[13] Cogan, D. C., et al. 1952. Arch. Ophthalmol. 47:55.
[14] Court Brown, W., and R. Doll. 1958. Brit. Med. J. 2:181.
[15] Court Brown, W., and R. Doll. 1965. Ibid. 2:1327.
[16] Cronkite, E. P., and V. P. Bond. 1956. Annu. Rev. Physiol. 18:483.
[17] Cronkite, E. P., et al. 1960. Radiat. Res. 12:81.
[18] Darden, E. B., Jr., et al. 1967. Int. J. Radiat. Biol. 12:435.
[19] Darden, E. B., Jr., et al. 1970. Radiat. Res. 43:598.
[20] Doll, R., and P. Smith. 1968. Brit. J. Radiol. 42:362.
[21] Evans, R. D., et al. 1969. In C. W. Mays, et al., ed. Delayed Effects of Bone Seeking Radionuclides. Univ. Utah Press, Salt Lake City. pp. 157-194.
[22] Harris, R. J. C., ed. 1963. Cellular Basis and Aetiology of Late Somatic Effects of Ionizing Radiation. Academic Press, New York.
[23] Hueper, W. C. 1942. Occupational Tumors and Allied Diseases. C. C. Thomas, Springfield, Ill.

[24] Ishimaru, T., et al. 1971. Radiat. Res. 45:216.
[25] Jablon, S., and H. Kato. 1970. Lancet 1:1000.
[26] Jacobson, L. O., and E. H. Marks. 1947. Radiology 49:286.
[27] Jammet, H. 1963. Ann. Ocul. 196:329.
[28] Jones, D., and D. Kimeldorf. 1964. Radiat. Res. 22:112.
[29] Kohn, H. I., and P. Guttman. 1963. Ibid. 18:348.
[30] Lamson, B. G., et al. 1957. Arch. Pathol. 64:505.
[31] Lamson, B. G., et al. 1958. Ibid. 66:311.
[32] Lamson, B. G., et al. 1958. Ibid. 66:322.
[33] Lamson, B. G., et al. 1959. J. Nat. Cancer Inst. 22:1059.
[34] Lamson, B. G., et al. 1962. Radiat. Res. 16:54.
[35] Lamson, B. G., et al. 1963. Ibid. 18:255.
[36] Lindop, R., and J. Rotblat. 1961. Proc. Roy. Soc. B154:332.
[37] Lisco, H., et al. 1947. Radiology 49:361.
[38] Looney, W. B. 1951. Nucl. Sci. Abstr. 5:6046.
[39] Lorenz, E., et al. 1947. Radiology 49:274.
[40] Martland, H. S. 1931. Amer. J. Cancer 15:2435.
[41] Merriam, G., and E. Focht. 1957. Amer. J. Roentgenol. Radium Ther. 77:759.
[42] Miller, R. W. 1956. Pediatrics 18:1.
[43] Mohs, F. E. 1952. J. Amer. Dent. Ass. 45:160.
[44] Mole, R. 1961. Int. J. Radiat. Biol. 3:493.
[45] Moses, C., et al. 1953. Arch. Ophthalmol. 50:609.
[46] Neary, G. J., et al. 1952. Int. J. Radiat. Biol. 4:239.
[47] Nilsson, A. 1968. Acta Radiol. Ther. Phys. Biol. 7:27.
[48] Nilsson, A. 1970. Ibid. 9:155.
[49] Nilsson, A. 1971. Ibid. 10:115.
[50] Prosser, C. L., et al. 1947. Radiology 49:299.
[51] Riley, E. F., et al. 1954. Radiat. Res. 1:556.
[52] Riley, E. F., et al. 1955. Ibid. 3:342.
[53] Sacher, G., and D. Grahn. 1964. J. Nat. Cancer Inst. 32:277.

continued

[54] Seltzer, W., and P. Sartwell. 1958. J. Amer. Med. Ass. 166:585.

[55] Seltzer, W., and P. Sartwell. 1965. Amer. J. Epidemiol. 81:2.

[56] Shellabarger, C., et al. 1960. Radiat. Res. 12:94.

[57] Simpson, L., and L. L. Bennett, Jr. 1962. Ibid. 17:145.

[58] Stewart, A., and G. Kneale. 1970. Lancet 1:1185.

[59] Stone, R. S. 1948. Amer. J. Roentgenol. Radium Ther. 59:771.

[60] Stone, R. S., ed. 1951. Nat. Nucl. Energy Ser. Div. IV-20.

[61] Thomson, J. F., et al. 1953. Amer. J. Roentgenol. Radium Ther. 69:830.

[62] Upton, A. C. 1967. Cancer Res. 24:1861.

[63] Upton, A. C., et al. 1960. Ibid. 20:1.

[64] Upton, A. C., et al. 1964. Biological Effects of Neutron and Proton Irradiations. International Atomic Energy Agency, Vienna. v. 2, pp. 337-344.

[65] Upton, A. C., et al. 1967. Radiat. Res. 32:493.

[66] Upton, A. C., et al. 1970. Ibid. 41:467.

[67] Van Cleave, C. D. 1968. U.S. At. Energy Comm. TID Rep. 24310.

[68] Wagoner, J., et al. 1965. N. Engl. J. Med. 273:181.

[69] Wanebo, C. K., et al. 1968. Amer. Rev. Resp. Dis. 98:778.

[70] Wanebo, C. K., et al. 1968. N. Engl. J. Med. 279:667.

[71] Warren, S. 1942. Arch. Pathol. 34:443, 562, 749, 917, 1070.

[72] Warren, S. 1943. Ibid. 35:121.

[73] Warren, S., and O. Lombard. 1966. Arch. Environ. Health 13:415.

[74] Wood, J. W., et al. 1969. Amer. J. Epidemiol. 89:4.

[75] Yuhas, J. M. 1969. J. Gerontol. 24:451.

[76] Zirkle, R., ed. 1951. Nat. Nucl. Energy Ser. Div. IV-22E.

110. CHARACTERISTICS AND COMPOSITION OF THE ATMOSPHERE

For additional information, consult General References appended to Part I.

Part I. Regional Temperatures

Thermal Region	Altitude	Temperature	Remarks
1 Tropo-sphere	Earth's surface to 11 km (average). The top (tropopause) is lowest (8-10 km) in Arctic in winter, highest (16-18 km) in tropics.	From value at earth's surface, temperature decreases with increasing altitudes ∿6.5°C/km (average rate) to a minimum temperature of ∿−57°C at tropopause (minimum may vary from ∿−80°C in Tropics to ∿−47°C in Arctic in summer)	Region of greatest water vapor, turbulence, vertical air currents, & weather. Altitude & temperature vary with latitude, season, & weather regime. Typical or average mid-latitude values are given with major variations.
2 Strato-sphere	∿ 11-20 km	Region is almost isothermal, ∿−57°C	Turbulence generally suppressed in stratosphere. Temperature rise due to absorption of ultraviolet & X radiation and ozone formation in the high stratosphere. Altitude & temperature undergo latitudinal, seasonal, & short term variations.
3	∿20-50 km	Temperature increases slowly with altitude to a maximum of ∿0°C at top of stratosphere (stratopause)	
4 Meso-sphere	∿50-80 km	From ∿0°C at stratopause, temperature decreases with increasing altitude to a minimum of ∿−90°C at top of mesosphere (mesopause)	Region of decreasing temperature. Altitude & temperature undergo latitudinal, seasonal, & short-term variations.
5 Thermo-sphere	∿80 km to upper limit of atmosphere	From ∿−90°C at mesopause, temperature increases with altitude up to ∿200 km. Above 200 km, varies widely from ∿600°C during quiet sun to perhaps 2000°C during maximum sunspot activity.	Region of large temperature variations depending upon solar activity

continued

Part I.   Regional Temperatures

	Thermal Region	Altitude	Temperature	Remarks
6	Exo-sphere	∿550 km	Temperature cannot be defined in usual way because of large mean free path of particles	Region of outer fringe of atmosphere where particles escape from atmosphere to space
Ref-er-ence	[2]	[1-3,5]	[1,3-5]	[1,3-5]

Contributors: Caskey, James E., Jr.; Hitchcock, Fred A.

Specific References

[1] Campen, C. S., Jr., et al., ed. 1960. U.S. Air Force Handb. Geophys., ch. 1,2.

[2] Sawyer, J. S. 1963. Quart. J. Roy. Meteorol. Soc. 89:156.

[3] U.S. Committee on Extension to the Standard Atmosphere. 1962. U.S. Standard Atmosphere, 1962. National Aeronautics and Space Administration, U.S. Air Force, and U.S. Weather Bureau, Washington, D.C.

[4] U.S. Committee on Extension to the Standard Atmosphere. 1966. U.S. Standard Atmosphere Supplements, 1966. Environmental Science Services Administration, National Aeronautics and Space Administration, and U.S. Air Force, Washington, D.C.

[5] Valley, S. L., ed. 1965. U.S. Air Force Handb. Geophys. Space Environ., ch. 2.

General References

[6] Huschke, R. E., ed. 1959. Glossary of Meteorology. American Meteorological Society, Boston.

[7] Johnson, J. C. 1954. Physical Meteorology. J. Wiley, New York.

[8] Kuiper, G. P. 1951. The Atmosphere of the Earth and Planets. Univ. Chicago Press, Chicago.

[9] Letestu, S., ed. 1966. International Meteorological Tables. World Meteorological Organization, Geneva, Switzerland.

[10] List, R. J., ed. 1951. Smithsonian Meteorological Tables. 6th Rev. ed. Smithsonian Institution, Washington, D.C.

[11] Malone, T. F., ed. 1951. Compendium of Meteorology. American Meteorological Society, Boston.

Part II.   Ionospheric Characteristics

The ionosphere, embedded in the high stratosphere and mesosphere, is that portion of the atmosphere sufficiently ionized to affect radio-wave propagation. Ionization is produced by absorption of solar ultraviolet, corpuscular, and X-radiation by atmospheric gases. The temperatures of the ionospheric layers are equivalent to the temperatures of the corresponding thermal regions (high stratosphere and mesosphere) in which they are embedded (*see* Part I).

	Region	Altitude of Maximum Ionization	Characteristics		Region	Altitude of Maximum Ionization	Characteristics
1	D layer	∿90 km	Region of ionosphere with lowest electron density. Altitude of maximum ionization may decrease considerably during solar flares.	3	F$_1$ layer	∿200 km[1/]	Region absent at night
2	E layer	∿110 km	Electron density in daytime is approx. 10 times that of D region, and is much higher than nighttime value	4	F$_2$ layer	∿300 km	Region of ionosphere with highest but erratic electron density; electron density varies with sunspot activity; daytime value is approx. 10 times the nighttime value
				Ref-er-ence	[2]	[1,2]	[1,2]

[1/] In daytime.

Contributors: Caskey, James E., Jr.; Hitchcock, Fred A.

continued

Part II. Ionospheric Characteristics

References

[1] Campen, C. S., Jr., et al., ed. 1960. U.S. Air Force Handb. Geophys., ch. 15.

[2] Valley, S. L., ed. 1965. U.S. Air Force Handb. Geophys. Space Environ., ch. 12.

Part III. Chemical Composition of Dry Air

Data are given only for the troposphere. In the high stratosphere, the mesosphere, and the thermosphere, interacting physical and chemical processes (e.g., diffusive separation of gases, radiative detachment, radiative recombination, ionization, dissociative recombination, and chemical reactions) produce highly unstable or variable concentrations.

	Classification	Component	Concentration ppm[1]		Classification	Component	Concentration ppm[1]
1	Substances found	Nitrogen	$78.084 \pm 0.004\%$[2]	10	Substances found	Carbon dioxide	330 ± 10
2	in greater non-	Oxygen	$20.946 \pm 0.002\%$[2]	11	in variable con-	Methane	2.0
3	variable con-	Argon	$0.934 \pm 0.001\%$[2]	12	centrations;	Sulfur dioxide	0-1
	centrations			13	may depart sig-	Ozone[3]	0-0.07 (summer)
4	Substances found	Neon	18.18 ± 0.04	14	nificantly from		0-0.02 (winter)
5	in lesser non-	Helium	5.24 ± 0.004	15	normal from	Nitrogen dioxide	0-0.02
6	variable con-	Krypton	1.14 ± 0.01	16	time to time	Iodine	0-0.01
7	centrations	Nitrous oxide	0.5 ± 0.1	17	and place to	Ammonia	Trace
8		Hydrogen	0.5	18	place	Carbon monoxide	0 to trace
9		Xenon	0.087 ± 0.001				

[1] Unless otherwise specified. [2] Percentages are percent by volume. [3] Formed in stratosphere, but small amounts are transported downward, and some reach the earth's surface.

Contributors: Caskey, James E., Jr.; Hitchcock, Fred A.

References

[1] Glueckauf, E. 1951. Compendium of Meteorology. American Meteorological Society, Boston. p. 6.

[2] U.S. Committee on Extension of the Standard Atmosphere. 1962. U.S. Standard Atmosphere, 1962. National Aeronautics and Space Administration, U.S. Air Force, and U.S. Weather Bureau, Washington, D.C.

[3] Valley, S. L., ed. 1965. U.S. Air Force Handb. Geophys. Space Environ., ch. 6.

111. CHEMISTRY OF AIR POLLUTANTS

Part I. Classification

	Major Classes	Subclasses	Typical Members
1	Inorganic gases	Oxides of nitrogen	Nitrogen dioxide, nitric oxide
2		Oxides of sulfur	Sulfur dioxide, sulfuric acid
3		Other inorganics	Ammonia, carbon monoxide, chlorine, hydrogen fluoride, hydrogen sulfide, ozone
4	Organic gases	Hydrocarbons	Benzene, butadiene, butene, ethylene, isooctane, methane
5		Aldehydes, ketones	Acetone, formaldehyde
6		Other organics	Acids, alcohols, chlorinated hydrocarbons, peroxyacyl nitrates, polynuclear aromatics
7	Aerosols	Solid particulate matter	Dusts, smoke
8		Liquid particulates	Fumes, oil mists, polymeric reaction-products

continued

Part I. Classification

Contributor: Haagen-Smit, A. J.

Reference: Weisburd, M. I., and S. Smith Griswold. 1962. Pub. Health Serv. Publ. 937.

Part II. Products

	Reactant	Reaction	Product
	General Reactions		
1	Sulfur dioxide, oxygen (+ catalysts)	$SO_2 + O \rightarrow SO_3 \rightarrow H_2SO_4$	Sulfuric acid, sulfates, aerosols
2	Olefins, sulfur dioxide, oxides of nitrogen, oxygen and sunlight	$SO_2 + ROO\cdot \rightarrow RO\cdot + SO_3$	Sulfuric acid, aerosols
3	Styrene, halogens, sunlight	$C_6H_5CH{=}CH_2 + Cl_2$	Eye irritant
4	Nitric oxide, oxygen	$2NO + O_2 \rightarrow 2NO_2$	Nitrogen dioxide (slow reaction)
	Photolysis [1]		
5	Nitrogen dioxide	$NO_2 \rightarrow NO + O$	Nitric oxide, atomic oxygen (main primary reaction)
6	Aldehydes	$RCHO \rightarrow R\cdot + H\dot{C}O$	Alkyl, formyl
7	Ketones	$R_1R_2CO \rightarrow R\cdot + R\dot{C}O$	Alkyl, acyl
8	Alkyl nitrites	$RONO \rightarrow RO\cdot + NO \rightarrow R\cdot + NO_2$	Alkyl, alkoxyl, nitric oxide, & nitrogen dioxide
9	Nitrous acid	$HNO_2 \rightarrow HO\cdot + NO \rightarrow H + NO_2$	Hydroxyl radical, atomic hydrogen, nitric oxide, & nitrogen dioxide
	Thermal Reactions [1]		
10	Ozone, olefins	$O_3 + R_2C{=}CR_2 \rightarrow R\cdot, RO\cdot, ROO\cdot$	Alkyl, alkoxyl, peroxyalkyl
11	Atomic oxygen, hydrocarbons	$O + RH \rightarrow R\cdot + HO\cdot$	Alkyl, hydroxyl
12	Atomic oxygen, aldehydes	$O + RCHO \rightarrow R\dot{C}O + HO\cdot$	Acyl, hydroxyl
	Organic Chain Reactions		
13	Alkyl, oxygen	$R\cdot + O_2 \rightarrow ROO\cdot$	Peroxyalkyl
14	Peroxyalkyl, oxygen	$ROO\cdot + O_2 \rightarrow RO\cdot + O_3$	Alkoxyl, ozone
15	Alkoxyl, hydrocarbons	$RO\cdot + RH \rightarrow ROH + R\cdot$	Alkyl, alcohol
16	Peroxyalkyl, hydrocarbons	$ROO\cdot + RH \rightarrow ROOH + R\cdot$	Alkyl, hydroperoxide
17	Hydroxyl, hydrocarbons	$HO\cdot + RH \rightarrow R\cdot + H_2O$	Alkyl, water
	Consumption of Free Radicals & Reactive Intermediates		
18	Peroxyalkyl, nitric oxide	$RO\dot{O} + NO \rightarrow ROONO \rightarrow R\dot{O} + NO_2$	Alkoxyl, nitrogen dioxide
19	Peroxyalkyl, olefins	$RO\dot{O} + {:}C{=}C{:} \rightarrow ROO\text{-}C\text{-}C\cdot$	Polymers
20	Peroxyalkyl, nitrogen dioxide	$RO\dot{O} + NO_2 \rightarrow ROONO_2$	Peroxyalkyl nitrate
21	Peroxyalkyl, sulfur dioxide	$RO\dot{O} + SO_2 \rightarrow SO_3 + RO\cdot$	Sulfur trioxide, alkoxyl
22	Peroxyacyl, nitrogen dioxide	$R(CO)O\dot{O} + NO_2 \rightarrow R(CO)OONO_2$	Peroxyacyl nitrate
23	Alkoxyl	$2RCH_2\dot{O} \rightarrow RCH_2OH + RCHO$	Alcohol, aldehyde
24	Alkoxyl, nitric oxide	$R\dot{O} + NO \rightarrow RONO$	Alkyl nitrite
25	Alkyl, hydroxyl	$R\cdot + H\dot{O} \rightarrow ROH$	Alcohol
26	Atomic oxygen, oxygen	$O + O_2 \rightarrow O_3$	Ozone
27	Atomic oxygen, sulfur dioxide	$O + SO_2 \rightarrow SO_3$	Sulfur trioxide
28	Ozone, olefins	$O_3 + {:}C{=}C{:} \rightarrow RCHO$	Aldehydes, ketones, ozonides
29	Ozone, nitric oxide	$O_3 + NO \rightarrow NO_2 + O_2$	Nitrogen dioxide, oxygen
30	Ozone, nitrogen dioxide	$O_3 + NO_2 \rightarrow N_2O_5 \rightarrow HNO_3$	Nitric acid

[1] Generation of free radicals and reactive intermediates.

continued

111. CHEMISTRY OF AIR POLLUTANTS

Part II. Products

Contributor: Haagen-Smit, A. J.

General References

[1] Haagen-Smit, A. J. 1962. In A. Stern, ed. Air Pollution. Academic Press, New York. v. 1, p. 41.

[2] Johnstone, H. F., and D. R. Coughanowr. 1958. Ind. Eng. Chem. 50:1169.

[3] Katz, M. 1961. World Health Organ. Monogr. Ser. 46:97.

[4] Leighton, P. A. 1961. Photochemistry of Air Pollution. Academic Press, New York.

[5] Renzetti, N. A., and G. J. Doyle. 1959. J. Air Pollut. Contr. Ass. 8:293.

[6] Renzetti, N. A., and G. J. Doyle. 1960. Int. J. Air Pollut. 2:327.

[7] Saltzman, B. E. 1958. Ind. Eng. Chem. 50:4, 677.

[8] Shuck, E. A., et al. 1960. Air Pollut. Found. Rep. 31.

[9] Wayne, L. G. 1962. Los Angeles County Air Pollut. Contr. Dist. Tech. Progr. Rep. 3.

112. EMISSION OF AIR POLLUTANTS

Abbreviation: gal = gallon.

Part I. Motor Vehicles and Gasoline Evaporation

Values must be used with great care, as emission quantities vary appreciably with engine type, driving conditions, speed, temperature, and other factors. **Pollutant:** (HC) = hydrocarbons.

	Emission Source	Pollutant	Quantity Emitted [1]	Reference
	Motor vehicles			
1	Gasoline engines (blow-by emissions included, but not evaporation losses)	Ammonia	2 lb/1000 gal	6
2		Carbon monoxide	80 g/mile	6
3		Nitrogen oxides	4.6 g/mile	6
4		Aldehydes	0.36 g/mile	5
5		Benzo[*a*]pyrene	0.3 g/1000 gal	4
6		Hydrocarbons	12.0 g/mile	6
7		Organic acids	0.13 g/mile	5
8		Particulates	0.6 g/mile	6
9	Diesel engines	Carbon monoxide	225 lb/1000 gal	6
10		Nitrogen oxides	370 lb/1000 gal	6
11		Sulfur dioxide	27 lb/1000 gal	6
12		Aldehydes	3 lb/1000 gal	6
13		Hydrocarbons	37 lb/1000 gal	6
14		Organic acids	3 lb/1000 gal	6
15		Particulates	13 lb/1000 gal	6
	Gasoline evaporation loss [2]			
16	Storage tanks (refinery & bulk terminal) [3]	Gasoline (HC)	49.6 lb/1000 gal; 0.82% of vol	6
17	Filling tank vehicles [4]	Gasoline (HC)	8.25 lb/1000 gal; 0.14% of vol	1

[1] All factors representative of age and population mix of autos and trucks in 1970 using 1975 federal testing procedure. [2] Volume loss calculated assuming an average gasoline specific gravity of 0.73. [3] Loss calculated assuming 75% floating roof tanks and 25% cone roof tanks. [4] Loss calculated assuming 50% splash fill and 50% submerged fill.

continued

Part I. Motor Vehicles and Gasoline Evaporation

	Emission Source	Pollutant	Quantity Emitted[1]	Reference
18	Filling station tanks[4]	Gasoline (HC)	9.4 lb/1000 gal; 0.154% of vol	2
19	Splash fill	Gasoline (HC)	11.5 lb/1000 gal	2
20	Submerged fill	Gasoline (HC)	7.3 lb/1000 gal	2
21	Filling automobile tanks	Gasoline (HC)	11.6 lb/1000 gal; 0.19% of vol	3
22	Automobile gas tanks & carburetors	Gasoline (HC)	3.6 g/mile	6

[1] All factors representative of age and population mix of autos and trucks in 1970 using 1975 federal testing procedure. [4] Loss calculated assuming 50% splash fill and 50% submerged fill.

Contributor: Southerland, James H.

References

[1] American Petroleum Institute. 1959. Amer. Petrol. Inst. Statist. Bull. 2514.

[2] Chass, R. L., et al. 1963. J. Air Pollut. Contr. Ass. 13(11):524.

[3] MacKnight, R. A., et al. 1959. Emissions from Underground Gasoline Storage Tanks. Los Angeles County Air Pollution Control District, Los Angeles.

[4] Magill, P. L., and R. W. Benoliel. 1952. Ind. Eng. Chem. 44:1347.

[5] U.S. Environmental Protection Agency. 1972. Compilation of Air Pollutant Emission Factors. Nat. Air Pollut. Contr. Admin. Publ. AP-42(Feb.).

[6] U.S. Environmental Protection Agency. 1973. Ibid. AP-42(Apr.).

Part II. Open Burning, Incineration, Combustion, and Selected Industrial Processes

Emission of pollutants from open burning, incineration, combustion, and selected industrial processes are difficult to quantify because of the wide range of material burned, burning conditions, process variations, etc. **Emission Source:** Btu = British thermal unit. **Emission Factors by**

Pollutant: Sulfur Oxides—S indicates weight percent sulfur in fuel; **Particulates**—A indicates weight percent ash in coal. *Abbreviations:* Neg. indicates negligible amount produced; hp = horse power. Values in parentheses are ranges, estimate "c" (*see* Introduction).

	Emission Source (uncontrolled basis)	Emission Factors by Pollutant									Reference
		Ammonia lb/ton	Carbon Monoxide lb/ton	Nitrogen Oxides lb/ton	Sulfur Oxides lb/ton	Aldehydes lb/ton	Benzo[a]-pyrene mg/ton	Hydrocarbons lb/ton	Organic Acids lb/ton	Particulates lb/ton	
	Refuse, open burning										2,5
1	Automobile components	....	125	4	Neg.	.....	....	30	.....	100	
2	Municipal refuse	....	85	6	1	.....	80	30	.....	16	
	Horticultural refuse										
3	Agricultural field burning	....	100	2	Neg.	.....	270	20	.....	17	
4	Landscape refuse & pruning	....	60	2	Neg.	.....	....	20	.....	17	
5	Wood	....	50	2	Neg.	.....	....	4	.....	17	
	Refuse, incinerator burning Municipal										3,5
6	Multiple chamber, uncontrolled	0.3	35	3	2.5	.....	....	1.5	0.6	30(8-70)	
7	With settling chamber & water spray system	....	35	3	2.5	.....	....	1.5	.....	14(3-35)	

continued

Part II. Open Burning, Incineration, Combustion, and Selected Industrial Processes

	Emission Source (uncontrolled basis)	Ammonia lb/ton	Carbon Monoxide lb/ton	Nitrogen Oxides lb/ton	Sulfur Oxides lb/ton	Aldehydes lb/ton	Benzo[a]pyrene mg/ton	Hydrocarbons lb/ton	Organic Acids lb/ton	Particulates lb/ton	Reference
	Industrial & commercial										2,5
8	Single chamber	....	20(4-200)	2	2.5	.....	400	15(0.5-50)	.....	15(4-31)	
9	Multiple chamber	....	10(1-25)	3	2.5	0.3	1200	3(0.3-20)	.....	7(4-8)	
10	Controlled air	....	Neg.	10	1.5	.....	....	....	.....	1.4(0.7-2)	
	Flue-fed										3,5
11	Original	0.4	20	3	0.5	.....	....	15(2-40)	22	30(7-70)	
12	Modified	....	10	10	0.5	.....	....	3(0.3-20)	.....	6(1-10)	
	Domestic, single chamber										5
13	Without primary burner	....	300	1	0.5	.....	....	100	.....	35	
14	With primary burner	....	Neg.	2	0.5	.....	....	2	.....	7	
	Natural gas combustion[1]										1,4,5
15	Power plant	....	0.67	23.5	0.023	0.116	....	0.04	0.155	0.58	
16	Industrial process boilers	0.0013	0.015	4.65-8.9[2]	0.023	0.116	0.8	1.55	0.27	0.697	
17	Domestic & commercial	....	0.77	3.1-4.7[3]	0.023	0.39	....	0.3	0.039	0.736	
	Coal combustion										
	Utility & large industrial (>100 × 10^6 Btu/hr input)										4,5
	Pulverized										
18	General	....	1	18	38 S	0.005	....	0.3	.....	16 A	
19	Wet bottom	....	1	30	38 S	0.005	....	0.3	.....	13 A	
20	Dry bottom	....	1	18	38 S	0.005	....	0.3	.....	17 A	
21	Cyclone	....	1	55	38 S	0.005	....	0.3	.....	2 A	
	General industrial & large commercial (10-100 × 10^6 Btu/hr input)										4,5
22	Stoker	....	2	15	38 S	0.005	....	1	.....	13 A	
	Commercial & domestic (<10 × 10^6 Btu/hr input)										4,5
23	Hand-fired units	....	90	3	38 S	0.005	....	20	.....	20	
24	Stoker	....	10	6	38 S	0.005	....	3	.....	2 A	
	Fuel oil combustion										4,5
25	Power plant	....	0.76	26.7	40.3 S	0.25	....	0.51	.....	2	
	Industrial & commercial										
26	Distillate oil[4]	....	1.1	11.3-22.7[5]	40.8 S	0.57	....	0.85	.....	4.25	
27	Residual oil[6]	....	1.0	10.2-20.3[5]	40.8 S	0.25	....	0.76	.....	5.8	
28	Domestic	....	1.42	3.4	40.8 S	0.57	....	0.85	.....	2.84	
29	Asphaltic concrete	....	....	.....	......	.....	....	....	.....	15.0	5
	Copper smelters										5
30	Roasting	....	....	....	60[7]	.....	....	....	.....	45[7]	
31	Smelting[8]	....	....	....	320[7]	.....	....	....	.....	20[7]	
32	Converting	....	....	....	870[7]	.....	....	....	.....	60[7]	
33	Refining	....	....	....	......	.....	....	....	.....	10[7]	

[1] Natural gas density, 0.05165 lb/ft³. [2] Use 4.65 for boilers of <500 hp, and 8.9 for boilers of >7500 hp. [3] Use 3.1 for domestic units, and 4.7 for commercial units. [4] Distillate oil density, 7.05 lb/gal. [5] Use lower number for tangentially fired units, and higher number for horizontally fired units. [6] Residual oil density, 7.88 lb/gal. [7] Values are pounds of pollutant per ton of ore. [8] Reverberatory furnace.

continued

Part II.   Open Burning, Incineration, Combustion, and Selected Industrial Processes

	Emission Source (uncontrolled basis)	Ammo-nia lb/ton	Carbon Mon-oxide lb/ton	Nitro-gen Oxides lb/ton	Sulfur Oxides lb/ton	Alde-hydes lb/ton	Benzo[a]-pyrene mg/ton	Hydro-carbons lb/ton	Or-ganic Acids lb/ton	Partic-ulates lb/ton	Refer-ence
	Iron & steel										5
34	Blast furnace	....	1750	....	......	.....	....	....	.....	150	
35	Sintering	....	44	....	......	.....	....	....	.....	42	
36	Open hearth furnace	....	....	....	......	.....	....	....	.....	13	
37	Electric furnace	....	18	....	......	.....	....	....	.....	10	
38	Basic oxygen furnace	....	139	....	......	.....	....	....	.....	51	
39	Metallurgical coke	....	1.3 [9]	....	4.0 [9]	.....	....	4.2 [9]	.....	3.5 [9]	5
40	Nitric acid	....	....	52.5	......	.....	....	....	.....	....	5
	Portland cement										5
41	Kilns	....	....	2.6	10.2	.....	....	....	.....	237	
42	Dryers, grinders, etc.	....	....	....	......	.....	....	....	.....	64	
43	Sulfuric acid	....	....	....	40	.....	....	....	.....	....	5
	Wood pulping										5
44	Recovery boilers & direct contact evaporators	....	60 [10]	....	5.0 [10]	.....	....	....	.....	151 [10]	
45	Smelt dissolving tank	....	....	....	......	.....	....	....	.....	2.0 [10]	
46	Lime kiln	....	10 [10]	....	......	.....	....	....	.....	45 [10]	
47	Fluidized-bed calciner	....	....	....	......	.....	....	....	.....	72 [10]	

[9] Values are pounds of pollutant per ton of coal.   [10] Values are pounds of pollutant per ton of pulp.

Contributor: Southerland, James H.

References

[1] American Gas Association Laboratories. Unpublished. Cleveland, O., 1970.

[2] Hangebrauck, R. P., et al. 1964. J. Air Pollut. Contr. Ass. 14:267.

[3] U.S. Environmental Protection Agency. 1968. Nat. Air Pollut. Contr. Admin. Publ. 999-AP-42.

[4] U.S. Environmental Protection Agency. 1972. Ibid. AP-42(Feb.).

[5] U.S. Environmental Protection Agency. 1973. Ibid. AP-42(Apr.).

113.   EFFECTS OF AIR POLLUTANT INHALATION ON MAMMALS

The following tables are based, in part, on particulate and gaseous pollutants currently monitored at both urban and nonurban sites under the National Air Surveillance Networks, U.S. Environmental Protection Agency ([34,64,70] in Part I). **Exposure:** i.t. = intratracheal. **Effects:** ↑ = increase, ↓ = decrease; conc = concentration; LC_{50} = lethal concentration for 50% of inoculated group, EC_{50} = effective concentration for 50% of inoculated dose; AP = alkaline phosphatase; BANA = benzoylarginine-β-naphthyl-amide; BCG = bacille Calmette-Guérin; Bp = benzo[a]-pyrene; BUN = blood urea nitrogen; LDH = lactic dehydrogenase; PBI = protein-bound iodine; SGOT = serum glutamic oxaloacetic transaminase; SGPT = serum glutamic pyruvic transaminase; SPF = specific pathogen free; TCID = tissue culture infective dose. Data in brackets refer to the column heading in brackets. Values in parentheses are ranges, estimate "c" (*see* Introduction).

continued

Part I. Particulate Air Pollutants

	Substance	Animal	Dose	Exposure [Duration]	Particle Size μ	Effects	Reference
	Antimony compounds						
1	Antimony tri-hydride	Mice	100 mg/m³	Single	....	Death in 105 min	63
2	Antimony tri-oxide	24 guinea pigs	45.4 mg/m³	2-3 hr/da [33-609 da]	....	16% mortality; extensive pneumonitis, fatty degeneration of liver, hypertrophy of lymphoid follicles of spleen	18
3		44 guinea pigs, 190-660 g	2.0-30.0 mg/m³	5-10 min	∿1.46	85.9% total retention, 69.8% retention in upper respiratory tract, 67% alveolar retention, & 15% alveolar deposition; upper respiratory tract clearance essentially complete in 2-6 hr	34
4		20 rabbits	89 mg/m³	100 hr/mo [10 mo]	0.6	Endogenous lipid pneumonia	24
5		50 rats	100-125 mg/m³	100 hr/mo [14.5 mo]	0.6	Endogenous lipid pneumonia	24
6		55 rats	1320 mg/m³	2 or 12 hr	<1	Small dust aggregates on alveoli & alveolar sac surfaces at 2-12 hr; atelectasis & ↑ in cellularity of alveolar walls from leukocytic exudate at 12 hr; ↑ in extracellular dust conc in proximal racemi, particularly in evaginating alveoli at 3 da	15
7	Antimony tri-sulfide	4 dogs	5.4(5.32-5.55) mg/m³	7 hr/da, 5 da/wk [7-10 wk]	⋙2	No deaths; ECG at 10 wk suggested some myocardial injury; slight swelling of myocardial fibers in all dogs several wk after exposure	10
8		6 rabbits	5.6 mg/m³	7 hr/da, 5 da/wk [6 wk]	⋙2	No deaths; ECG indicated slight to moderate myocardial damage, T waves especially affected; marked cardiac dilatation, myocardium flabby, fibers swollen	10
9		5 rabbits	27.8 mg/m³	7 hr/da, 5 da/wk [6 wk]	⋙2	No deaths; ECG indicated myocardial changes or coronary inadequacy; slight to moderate parenchymatous changes in myocardium; parenchymatous degeneration in liver & tubular epithelium of kidneys	10
10		10 rats, Wistar	3.07 mg/m³	7 hr/da, 5 da/wk [6 wk]	⋙2	No deaths; elevation of RS-T segments in all ECG leads just before end of experiment; mild congestion & focal areas of hemorrhage in lungs; slight to moderate cardiac hyperemia, parenchymatous degeneration of myocardium	10
	Arsenic compounds						
11	Arsenic trioxide	Rats, albino	1.3 μg/m³	3 mo	....	No toxic effects	48
12			4.9 μg/m³	3 mo	....	↓ in blood-SH groups, ↓ in chronaxial ratio of antagonistic muscles; significant ↑ in As content of liver & lung; pericellular edema in brain, inflammatory infiltrate in bronchi & fatty degeneration of liver	48

continued

	Substance	Animal	Dose	Exposure [Duration]	Particle Size μ	Effects	Reference
13			60.7 μg/m^3	3 mo	....	Effects same as 4.9 μg/m^3. Also, eosinophilia, ↓ in blood cholinesterase, ↑ in blood pyruvic acid; ↑ in latent period of conditioned reflexes; significant ↑ in As content of brain; neuron plasmolysis & karyolysis of pyramidal tract	48
14	& fly ash (1% As)	85 mice, hairless, 20 g	179.4 ± 35.6 μg As/m^3 & 16.2 ± 3.1 mg ash/m^3	6 hr/da, 5 da/wk [1-6 wk]	<10	Arsenic content, μg/100 g, at 2 wk: liver, 63-85; kidney, 173-208; marked ↓ in As in both organs after 4-6 wk; As content of skin 89-172 μg/100 g from 1-6 wk	7
	Asbestos compounds						
15	Amosite	60 hamsters, Syrian Golden, 2 mo	2.5 mg in saline	Intratracheal injection 1/wk [6 wk]	Length 17.6	39/60 deaths in 11 mo; mild interstitial fibrosis & adenomatoid foci in lungs; occasional calcified pleural plaques; no respiratory tract tumors	61
16		20-28 rats, Osborne-Mendel, 12 wk	10 mg in 1:1 beeswax:tricaprylin	Single intrapulmonary injection	....	No pulmonary neoplasms or keratinizing metaplasia at 24 mo	62
17		Rats, Charles River	∿50 mg/m^3	16 hr/wk [11 mo]	Diameter 0.2-0.5; length 3-5	↓ in proline hydroxylase & advanced fibrosis in lung	16
18	African, hammer-milled	24 rats, Wistar, 168-308 g	41.4 mg/m^3	7 hr/da, 5 da/wk [6.5 wk]	Length 80% 1-6	Dust retention at 56 da in lungs: SPF rats, 2.78 mg; standard rats, 2.08 mg; free & phagocytized dust in respiratory bronchioles of both groups with less free dust in standard rats with bronchitis; tissue reaction progressive & more severe than that observed with chrysotile—see entries 20 & 22	15
19	Chrysotile	Rats, Charles River	∿50 mg/m^3	16 hr/wk [15 mo]	Diameter 0.2; length 6-15	↓ in proline hydroxylase & advanced fibrosis in lungs	16
20	African, hammer-milled	24 rats, Wistar, 168-308 g	37.2 mg/m^3	7 hr/da, 5 da/wk [6.5 wk]	Length 79% 1-6	Dust retention at 56 da in lungs: SPF rats, 0.40 mg; standard rats, 0.54 mg; less free dust in standard rats with bronchitis; tissue reaction nonprogressive but more severe than that observed with Canadian chrysotile—see entry 22	15
21	Canadian	20-28 rats, Osborne-Mendel, 12 wk	10 mg in 1:1 beeswax:tricaprylin	Single intrapulmonary injection	....	No pulmonary neoplasms or keratinizing metaplasia at 24 mo	62
22		24 rats, Wistar, 168-308 g	71.0 mg/m^3	7 hr/da, 5 da/wk [6.5 wk]	Length 90% 1-6	Dust retention at 56 da in lungs: SPF rats, 0.34 mg; standard rats, 0.66 mg; little free dust in standard rats with bronchitis; tissue reaction nonprogressive	15

continued

Part I. Particulate Air Pollutants

	Substance	Animal	Dose	Exposure [Duration]	Particle Size μ	Effects	Reference
23	Hammer-milled (contains 245.5 μg Ni, 21.6 μg Co, 153.3 μg Cr, & 37.9 μg Pb/mg)	64 rats, albino	3.5 mg in H_2O	1-6 i.t. injections	....	30% survivors at 16 mo; 16% survivors after <16 mo showed primary pulmonary adenocarcinomas with no evidence of metastases; tumors not related to degree of asbestosis (fibrosis) which was multifocal	26
24		131 rats, albino (70/131 received single i.t. injection of 0.05 ml 5% NaOH to impair dust clearance)	86(42-146) mg/m³	6 hr/da, 5 da/wk [62 wk]	....	Survivors at 16 mo: 44% with, 67% without, impairment. Survivors after <16 mo: 48% with, & 24% without, impairment showed primary malignant tumors with no evidence of metastases; tumors primarily adenocarcinomas; fibrosarcoma, squamous cell carcinoma & mesothelioma also observed; tumors not related to degree of asbestosis (fibrosis).	26
25	Harsh	9 hamsters, Syrian Golden, 2 mo	1.25 mg in human plasma	Intratracheal injection 1/wk [24 wk]	Length 36	5/9 deaths in 173-404 da; mild interstitial fibrosis & adenomatoid foci in lungs; occasional calcified pleural plaques; no respiratory tract tumors	61
26	Natural	20 rats, Sprague-Dawley, 200-225 g (injected with 1 μCi ³H-thymidine/g body wt)	2.6 mg in saline	Single i.t. injection	....	Bronchiolitis obliterans with proliferation of connective tissue & fibrosis of respiratory tissue; no pleural lesions; significant, reversible ↑ in ³H-labelled mesothelial cells at 5-7 da—*see* entry 30	11
27	Ball- & hammer-milled	55 rats	24.5 mg in H_2O	Intratracheal injection	Diameter 0.05-1	Proliferative polypoid fibroblastic inflammation in smaller bronchi & bronchioles, respiratory bronchioles & alveolar ducts at 4 da with subsequent collagenization; 90 & 100% mortality at 18 & 24 mo, respectively—*see* entry 29	27
28	Soft	60 hamsters, Syrian Golden, 2 mo	0.5-1.25 mg in saline	Intratracheal injection 1/wk [10 wk]	Length 30	35/60 deaths in 6 mo; mild interstitial fibrosis & adenomatoid foci in lungs; occasional calcified pleural plaques; no respiratory tract tumors	61
29	Synthetic	55 rats	59 mg in H_2O	Intratracheal injection	Diameter 0.02-0.04; length 0.08 to <1.0	Proliferative polypoid fibroblastic inflammation in respiratory bronchioles & alveolar ducts at 4 da ascribed to high local conc of magnesium silicate; no subsequent collagenization; recovery in 6 mo; 22 & 53% mortality at 12 & 24 mo, respectively	27
30		16 rats, Sprague-Dawley, 200-225 g (injected with 1 μCi ³H-thymidine/g body wt)	2.6 mg in saline	Single i.t. injection	....	No pulmonary or pleural lesions; marked ↓ in ³H-labelled mesothelial cells at 14 da	11

continued

Part I. Particulate Air Pollutants

	Substance	Animal	Dose	Exposure [Duration]	Particle Size μ	Effects	Reference
31	Crocidolite	Rats, Charles River, $\sim$250 g	4 mg in saline	Single i.t. injection	Diameter 0.4-0.5; length 3-6	Significant ↑ in lung proline hydroxylase at 1-3 wk	16
32		Rats, Charles River	$\sim$50 mg/m^3	16 hr/wk [14 mo]	Diameter 0.4-0.5; length 3-6	↓ in proline hydroxylase & advanced fibrosis in lung	16
33		20-28 rats, Osborne-Mendel, 12 wk	10 mg in 1:1 beeswax:tricaprylin	Single intrapulmonary injection	....	No pulmonary neoplasms or keratinizing metaplasia at 24 mo	62
34		25-30 rats, Osborne-Mendel, 12 wk	40 mg in gelatin on fibrous glass pledgets	Implanted over left lung & pericardium	Diameter <1; length 10-100	80% deaths; 74% mesothelial sarcomas at 23 mo with no lymphatic or vascular metastases; 50% sarcomas derived from pericardial mesothelium; 100% extensive fibrosis	62
35	African, hammer-milled	24 rats, Wistar, 130-167 g	49.5 ± 12.1 mg/m^3	7.5 hr/da, 5 da/wk [6 wk]	....	2.53 mg dust retention in lungs at 58 da; most dust concentrated in alveolar phagocytes in respiratory bronchioles at 1-58 da	69
36	Rhodesian	20-28 rats, Osborne-Mendel, 12 wk	10 mg in 1:1 beeswax:tricaprylin	Single intrapulmonary injection	....	No pulmonary neoplasms or keratinizing metaplasia at 24 mo	62
37	Benzo[a]pyrene	30 hamsters, Syrian Golden	100 μg in gelatin	Intratracheal injection 1/wk [60 wk]	....	No metaplasia or neoplasia in respiratory tract	51
38	& channel black, calcined	68 rats	0.1 mg Bp & 10 mg channel black in casein	Intratracheal injection 1/10 da [60 da]	....	Preneoplastic lesions including diffuse hyperplasia & proliferation of bronchial epithelium, focal planocellular metaplasia of bronchiolar epithelium & adenomas after 1 mo; moderately pronounced fibrosis, consolidation & atelectasis after 3-4.5 mo; death from inflammatory changes	42
39		Rats, 100 g	5 mg Bp & 5 mg channel black in casein	Single i.t. injection	Channel black 128 Å	Significant ↓ in Bp elimination rate in 5th wk; complete elimination in $\sim$50 da	41
40	& thermal decomposition black	Rats, 100 g	5 mg Bp, 1.25 mg channel black, & 3.75 mg thermal decomposition black to casein	Single i.t. injection	Channel black 128 Å; thermal decomposition black 3047 Å	↑ in Bp elimination rate; complete elimination in 35 da	41
41	& thermal decomposition black	Rats, 100 g	5 mg Bp & 5 mg thermal decomposition black in casein	Single i.t. injection	Thermal decomposition black 3047 Å	Complete Bp elimination in 23 da	41
42	& hematite	Hamsters, Syrian Golden	3.0-3.4 mg Bp & hematite (48.3% Bp) in saline	Single or repeated i.t. injections 1/wk [10 wk]	....	Bp retention rates similar for both single- & repeated-dose regimens	51

continued

Part I. Particulate Air Pollutants

	Substance	Animal	Dose	Exposure [Duration]	Particle Size μ	Effects	Reference
43		Hamsters, Syrian Golden, young adult	2 mg Bp & 2-10 mg hematite in saline	Single i.t. injection	Hematite ⩾8.2- ⩾17.5	Constant Bp retention rate at 1-21 da with increasing hematite doses; ↓ in % Bp retention with decreasing particle size with 2 mg each Bp & hematite	50,53
44		60 hamsters, Syrian Golden, 11 wk	3 mg Bp & 3 mg hematite in saline	Intratracheal injection 1/wk [15 wk]	Hematite 90% <1.0; 48% <0.25	Survivors at 16 wk: 37% ♂, 70% ♀; 100% nonmetastatic bronchogenic carcinoma, including squamous cell carcinoma, anaplastic carcinoma & adenocarcinoma, in proximal bronchi of both ♂ & ♀ surviving >16 wk	54
45		99 hamsters, Syrian Golden, 90-95 g (46/99 received 5000 IU vitamin A palmitate orally twice/wk for lifespan)	3 mg Bp & 3 mg hematite in saline	Intratracheal injection 1/wk [10 wk]	Hematite 94% <1.0	32 & 11% respiratory tract tumors & 21 & 2% squamous tumors (primarily tracheal & bronchial carcinomas) with Bp, & Bp plus supplemental vitamin A regimens, respectively; ↑ in forestomach papillomas with Bp; marked ↓ in forestomach papillomas with no malignancies in Bp plus vitamin A group	52
46		4-8 hamsters/group, Syrian Golden, 100-120 g	5 mg Bp & 5 mg hematite in saline	1, 4, or 10 i.t. injections	Hematite 93% <5; 68% <1 (wt)	Focal squamous metaplasia of tracheobronchial epithelium by light microscopy 48 hr after multiple (4 or 10) doses only; adjacent areas of epithelial hyperplasia after 10 doses revealed widened intercellular spaces & abnormal desmosomes; many lysosomes, few mitochondria, & numerous tonofilaments in cytoplasm; polylobulated nuclei; enlarged, pleomorphic nucleoli with electron microscopy	29
47	& sulfur dioxide	Hamsters; 21 rats	10 mg Bp/m³ & 3.5 ppm SO₂	1 hr/da, 5 da/wk [494 exposure days]	....	Mortality in 98 wk: rats, 12.2%; hamsters, 100%; 2/21 pulmonary squamous cell carcinomas with renal metastases in rats at 655-698 da; 1/21 advanced pulmonary squamous metaplasia at 623 da; no significant pathology in hamsters	32
48		Hamsters; 21 rats	10 ppm SO₂ & 10 mg Bp/m³ & 3.5 ppm SO₂	1 hr/da, 5 da/wk [534 exposure days] & 1 hr/da, 5 da/wk [494 exposure days]	....	Mortality in 98 wk: rats, 38.1%; hamsters, 100%; 5/21 pulmonary squamous cell carcinomas with renal metastases in rats at 547-794 da; 2/21 advanced pulmonary squamous metaplasia at 485-715 da; no significant pathology in hamsters	32

continued

Part I. Particulate Air Pollutants

	Substance	Animal	Dose	Exposure [Duration]	Particle Size μ	Effects	Reference
	Beryllium compounds						
49	Beryllium fluoride	5 cats, young adult	0.97 mg/m³ in H_2O	6 hr/da [207 da]	0.61(0.33-0.94)	No deaths; lung damage	66
50		6 cats, young adult	10 mg/m³ in H_2O	6 hr/da [3 wk]	0.63(0.52-0.74)	No deaths	66
51		14 dogs, young adult	0.97 mg/m³ in H_2O	6 hr/da [207 da]	0.61(0.33-0.94)	3 deaths; suspected macrocytic anemia	Consolidation, emphysema, & slight edema in lungs; Be tended to accumulate in lungs, pulmonary lymph nodes, liver, skeleton, & bone marrow	66
52		6 dogs, young adult	10 mg/m³ in H_2O	6 hr/da [3 wk]	0.63(0.52-0.74)	1 deaths; 3 sacrificed moribund	
53		6 dogs, young adult; 3 rabbits	2.2(2.0-2.4) mg/m³ in H_2O	6 hr/da [23 wk]	....	↓ in RBC count & Hb levels; ↑ in mean corpuscular volume consistent with macrocytic anemia	64
54		20 guinea pigs, young adult	10 mg/m³ in H_2O	6 hr/da [3 wk]	0.63(0.52-0.74)	7 deaths	66
55		20 mice, young adult	10 mg/m³ in H_2O	6 hr/da [3 wk]	0.63(0.52-0.74)	6 deaths	66
56		4 monkeys, rhesus	27 µg (5.2 µg Be)/ft³ in H_2O	6 hr/da [7-16 da]	....	2/4 deaths after 13-16 exposures from pneumonitis; pulmonary emphysema, edema, granulomas (2/4), & fibrosis; marked alveolar hyperplasia (4/4) & slight to moderate metaplasia (4/4) of alveoli, & bronchial & bronchiolar epithelium; marked lymph node hyperplasia (4/4); multiple extrapulmonary lesions	56
57		10 rabbits, young adult	0.97 mg/m³ in H_2O	6 hr/da [207 da]	0.61(0.33-0.94)	No deaths; suspected macrocytic anemia; lung damage	66
58			10 mg/m³ in H_2O	6 hr/da [3 wk]	0.63(0.52-0.74)	1 death; suspected macrocytic anemia; lung damage	66
59		120 rats, young adult	0.97 mg/m³ in H_2O	6 hr/da [207 da]	0.61(0.33-0.94)	73 deaths; minimal lung lesions	66
60		40 rats, young & old adult	10 mg/m³ in H_2O	6 hr/da [3 wk]	0.63(0.52-0.74)	7 deaths; minimal lung lesions	66
61	Berryllium oxide	6 dogs, Beagle, 7.3-10.8 kg	120(40-300) mg/m³	20 min	....	4/6 Be-containing granulomas in lungs at 30 mo with no excess collagen formation	40
62		65 rats	39.57 µg/liter	1-5 hr/da [1-35 hr]	0.285(0.11-1.25)	Large amounts of dust (>24 mg Be/100 g) in lungs at >1 yr; little tendency for Be to be redistributed from lungs to other tissues; fibrous tissue proliferation from 35 da to >1 yr but no granulomatous inflammation in lungs	19

continued

Part I. Particulate Air Pollutants

	Substance	Animal	Dose	Exposure [Duration]	Particle Size μ	Effects	Reference
63		2 cats; 10 dogs; 20 guinea pigs, mixed English; 2 monkeys, rhesus; 9 rabbits, New Zealand; 90 rats, Wistar; (all young adults)	10 & 82 mg/m³ in H₂O (special grade of BeO)	6 hr/da, 5 da/wk [15-40 da]	0.47-0.59	68% mortality in rats exposed to 82 mg/m³ for 15 da; all other treated animals survived	28
64			83 mg/m³ in H₂O (refractory grade GC of BeO)	6 hr/da, 5 da/wk [60 da]	1.13	All animals survived	
65			84-86 mg/m³ in H₂O (fluorescent grade of BeO)	6 hr/da, 5 da/wk [10-17.5 da]	<1.0	5% mortality in rats exposed to 87 mg/m³ for 10 da; all other treated animals survived	
66			88 mg/m³ in H₂O (refractory grade SP of BeO)	6 hr/da, 5 da/wk [10 da]	0.71	All animals survived	
67	Calcined	6 dogs, Beagle; 5 monkeys, cynamolgus (all adults)	3.3-4.4 mg Be/m³	3 × 30 min/mo [2 yr]	....	Significant Be levels in lungs with higher conc present in monkeys; no histological or ultrastructural pulmonary changes; no changes in air-blood barrier thickness or capillary-alveolar surface area ratio	13
68		30 guinea pigs, 360-400 g	2 mg in saline	Single i.t. injection	1-5	Pulmonary edema in all treated animals at 15 da; peribronchial lymphoid hyperplasia at 15-60 da in animals receiving BeO calcined at 500 or 1100°C only; no specific pulmonary reaction at 30-60 da with BeO calcined at 1600°C	12
69	Beryllium phosphate	4 monkeys, rhesus	66 µg [5.6 µg Be]/ft³	6 hr/da [30 da]	....	1/4 deaths at 75 da from pneumonitis; pulmonary emphysema & fibrosis; minimal extrapulmonary lesions	56
70	Beryllium sulfate	4 cats, young adult	0.95 mg [0.04 mg Be]/m³ in H₂O	6 hr/da [100 da]	0.25	No deaths; 20% body wt loss. µg Be/g fresh tissue from 4 sacrificed animals: lung, 0.08; liver, 0.02; kidney, 0.01; spleen, 0.01.	65
71		5 cats, young adult	10 mg [0.43 mg Be]/m³ in H₂O	6 hr/da [95 da]	1.5	1 death; no change in body wt	65
72			47 mg [2 mg Be]/m³ in H₂O	6 hr/da [51 da]	0.96	4 deaths; 43% body wt loss	65

The effects for items 63-66 also note: Damage in lungs only; dust particles in peribronchial & perivascular tissues, as well as in alveoli & phagocytes; inflammation, edema, & thickening of alveolar walls; bronchial epithelial desquamation & hyperplasia

continued

Part I. Particulate Air Pollutants

	Substance	Animal	Dose	Exposure [Duration]	Particle Size μ	Effects	Reference
73		12 dogs	3.6-4.0 mg/m³ in H₂O	6 hr/da [2 mo]	....	↓ in RBC count & Hb levels; ↑ in mean corpuscular volume consistent with macrocytic anemia; spontaneous recovery from anemia after 3.5-4 mo	64
74		5 dogs, young adult	0.95 mg [0.04 mg Be]/m³ in H₂O	6 hr/da [100 da]	0.25	No deaths; 10% body wt loss. μg Be/g fresh tissue from 5 sacrificed animals: lung, 0.6; pulmonary lymph nodes, 0.7; liver, 0.01; kidney, 0.003; spleen, 0.01.	65
75			10 mg [0.43 mg Be]/m³ in H₂O	6 hr/da [95 da]	1.5	No deaths; 11% body wt loss; leukocytosis. μg Be/g fresh tissue from 4 sacrificed animals: lung, 4; pulmonary lymph nodes, 2; liver, 1.8; kidney, 0.8; spleen, 0.004; femur, 0.8	
76			47 mg [2 mg Be]/m³ in H₂O	6 hr/da [51 da]	0.96	4 deaths; 4% body wt loss; leukocytosis	
77		20 guinea pigs, 400-600 g	0.95 mg [0.04 mg Be]/m³ in H₂O	6 hr/da [100 da]	0.25	No deaths; 18% body wt gain	65
78		34 guinea pigs, 400-600 g	10 mg [0.43 mg Be]/m³ in H₂O	6 hr/da [95 da]	1.5	2 deaths; 100% body wt gain	65
79		12 guinea pigs, 400-600 g	47 mg [2 mg Be]/m³ in H₂O	6 hr/da [51 da]	0.96	7 deaths; 37% body wt gain	65
80		10 guinea pigs, 400-600 g	100 mg [4.3 mg Be]/m³ in H₂O	6 hr/da [14 da]	1.1	3 deaths; 2% body wt loss	65
81		83 hamsters	0.95 mg [0.04 mg Be]/m³ in H₂O	6 hr/da [100 da]	0.25	No deaths; no change in body wt	65
82		10 hamsters	47 mg [2 mg Be]/m³ in H₂O	6 hr/da [51 da]	0.96	5 deaths; 18% body wt loss	65

The bracketed note spanning rows 74–76 Effects column: "Reversible macrocytic anemia after 3-8 wk; significant changes in phospholipid & free cholesterol of whole RBC; tendency to hypoalbuminemia & hyperglobulinemia; acute inflammatory response in lung, with erosion & proliferation of bronchial epithelium"

continued

Part I. Particulate Air Pollutants

	Substance	Animal	Dose	Exposure [Duration]	Particle Size μ	Effects	Reference
83			100 mg [4.3 mg Be]/m³ in H₂O	6 hr/da [14 da]	1.1	2 deaths; 8% body wt loss	65
84		38 mice	47 mg [2 mg Be]/m³ in H₂O	6 hr/da [51 da]	0.96	4 deaths; 6% body wt loss	65
85			100 mg [4.3 mg Be]/m³ in H₂O	6 hr/da [14 da]	1.1	No deaths; 13% body wt loss	65
86		2 monkeys	0.95 mg [0.04 mg Be]/m³ in H₂O	6 hr/da [100 da]	0.25	No deaths; 10% body wt gain. μg Be/g fresh tissue from 2 sacrificed animals: lung, 1.2; pulmonary lymph nodes, 1.3; liver, 0.5; kidney, 0.01; spleen, 0.1.	65
87		5 monkeys	10 mg [0.43 mg Be]/m³ in H₂O	6 hr/da [95 da]	1.5	No deaths; 31% body wt loss	65
88		1 monkey	47 mg [2 mg Be]/m³ in H₂O	6 hr/da [51 da]	0.96	1 death; 25% body wt loss	65
89		4 monkeys, rhesus	66 μg [5.6 μg Be]/ft³ in H₂O	6 hr/da [7 da]	....	1/4 deaths at 52 da from pneumonitis; pulmonary emphysema, granulomas (1/4 at 6 mo), fibrosis; desquamation of bronchial & bronchiolar epithelium; marked lymph node hyperplasia (2/4); minimal extrapulmonary lesions	56
90		23 rabbits, 2.6-4.0 kg	0.95 mg [0.04 mg Be]/m³ in H₂O	6 hr/da [100 da]	0.25	No deaths; 15% body wt gain. μg Be/g fresh tissue from 5 sacrificed animals: lung, 1.6; pulmonary lymph nodes, 0; liver, 0.004; kidney, 0.003; spleen, 0.01.	65
91		24 rabbits, 2.6-4.0 kg	10 mg [0.43 mg Be]/m³ in H₂O	6 hr/da [95 da]	1.5	2 deaths; no change in body wt; leukocytosis	65
92		10 rabbits, 2.6-4.0 kg	47 mg [2 mg Be]/m³ in H₂O	6 hr/da [51 da]	0.96	1 death; 7% body wt gain; leukocytosis	65
93		3 rabbits, 2.6-4.0 kg	100 mg [4.3 mg Be]/m³ in H₂O	6 hr/da [14 da]	1.1	No deaths; no change in body wt; leukocytosis	65
94		20 rats, 250-280 g	0.95 mg [0.04 mg Be]/m³ in H₂O	6 hr/da [100 da]	0.25	No deaths; 20% body wt gain	65
95		40 rats	4 mg/m³ in H₂O	6 hr/da [23 wk]	....	↓ in RBC count; ↑ in mean corpuscular volume consistent with macrocytic anemia	64
96		47 rats, 250-280 g	10 mg [0.43 mg Be]/m³ in H₂O	6 hr/da [95 da]	1.5	23 deaths; 28% body wt gain; leukocytosis; inhalation of HF vapor (8 mg/m³) doubles toxicity of BeSO₄ poisoning	65

continued

Part I. Particulate Air Pollutants

	Substance	Animal	Dose	Exposure [Duration]	Particle Size μ	Effects	Ref-er-ence
97		15 rats, 250-280 g	47 mg [2 mg Be]/m^3 in H$_2$O	6 hr/da [51 da]	0.96	13 deaths; no change in body wt; leukocytosis	65
98		10 rats, 250-280 g	100 mg [4.3 mg Be]/m^3 in H$_2$O	6 hr/da [14 da]	1.1	10 deaths; 2% body wt loss; leukocytosis	65
99		150 rats, Sprague-Dawley, 6 wk	34.2 μg Be/m^3	7 hr/da, 5 da/wk [72 wk]	0.12	↑ in mortality in ♀ only; 100% alveolar adenocarcinomas at 13 mo; ↑ in Be content in lungs with conc plateau at 36 wk; significant ↓ in Be content of excised tumors compared to nonmalignant tissue; maximum Be levels in tracheobronchial lymph nodes at 36-52 wk with greater Be deposition in ♂	43,44
100		136 rats, Wistar & Sherman, 140-210 g	12 μg [1 μg Be]/ft^3 in H$_2$O	8 hr/da, 5.5 da/wk [6 mo]	....	46 deaths. Apparent effect on lung tissue: stimulation of epithelial cell proliferation without connective tissue reaction; foam-cell clustering; focal mural infiltration; lobular septal cell proliferation; peribronchial alveolar wall epithelization; granulomatosis & neoplasia.	57
101	Beryllium chloride (10%), be-ryllium fluoride (40%), & beryllium oxide (50%) in rocket exhaust	2 dogs, Beagle, 8.1-10.8 kg	115 mg Be/m^3	20 min	<1 to >5	3.9-5.5 μg Be/g wet lung at 3 yr; Be (<0.05-1 μ) deposited in histiocytic lysosomes in septal interstitium in association with collagen bundles & ↑ in numbers of septal capillaries	46
	Cadmium compounds						
102	Cadmium chloride	93 dogs	0.32 mg/liter	30 min	....	Dose = LC$_{90}$. Immediately after exposure: occasional vomiting, bradycardia, rapid asthmatic-respiration; ∿50% mortality within 24 hr, with rapidly developing hemoconcentration & anoxemia due to pulmonary edema (treatment with 2,3-dimercapto-1-propanol reduced mortality ∿50%). Pulmonary lesions at 1-3 mo consisted of occasional emphysema & scarring only. No significant changes in organs other than lungs after exposure, although large amount of CdCl$_2$ is distributed throughout body, particularly kidneys; fraction remaining in lungs is fixed & persists at least 15 wk.	30
103		∿73 mice	0.1 (0.085-0.17) mg/liter	21-36 min	<2	Overall retention, 10.5-23%; 12 hr after exposure, Cd content ↓ in lungs, ↑ in liver & kidneys, then stays constant; contents of gut & gut wash pass through maximum at 2 & 6 hr, respectively	38
104		200 rats, Sprague-Dawley	125 (78-164) mg-min/m^3	1/2 wk [24 wk]	....	Well tolerated with single or repeated exposure	35
105		100 rats, Sprague-Dawley	∿250 mg-min/m^3	15 min	....	1 death; acute pulmonary edema within 24 hr, peak within 3 da; proliferative interstitial pneumonitis from 3-10 da; permanent lung damage, with perivascular & peribronchial fibrosis	35

continued

Part I. Particulate Air Pollutants

	Substance	Animal	Dose	Exposure [Duration]	Particle Size μ	Effects	Reference
106	Cadmium oxide fumes	10 dogs	4(3-7) mg/m^3	6 hr/da, 5 da/wk [1102 hr]	98%<3	Cd dust deposition in mg/100 g tissue: lungs, 2.6; liver, 2.6; kidneys, 5.7; lesser amounts in bones & teeth (no color change). No demonstrable gross or microscopic changes in lungs, liver, or kidneys. Blood levels, 0.07 mg/100 g; urine levels, 0.13 mg/100 g; no change in Hb, Hct, RBC, WBC, BUN, alkaline phosphatase, or sulfobromophthalein retention.	39
107		14 dogs	3100-10,000 mg-min/m^3	10-20 min	0.3-0.5	LC$_{50}$ = ~4000 mg-min/m^3 — Resting animals: approx 11% CdO fumes retained in lungs of guinea pigs, mice, monkeys, rabbits, & rats; conc in dried tissues proportional to dosage; LC$_{50}$ = 1 mg/100 g for rats, <10 mg/100 g for monkeys	5
108		100 guinea pigs, 300 g	640-6450 mg-min/m^3	13-30 min	0.3-0.5	LC$_{50}$ = ~3500 mg-min/m^3	
109		30 mice, 15 g	660-1130 mg-min/m^3	15 min	0.3-0.5	LC$_{50}$ = <700 mg-min/m^3	
110		34 monkeys	4500-28,200 mg-min/m^3	10-30 min	0.3-0.5	LC$_{50}$ = ~15,000 mg-min/m^3	
111		53 rabbits	640-3690 mg-min/m^3	13-30 min	0.3-0.5	LC$_{50}$ = ~2500 mg-min/m^3	
112		160 rats, Sprague-Dawley, 250-300 g	150-1300 mg-min/m^3	10-15 min	0.3-0.5	LC$_{50}$ = 500 mg-min/m^3	
113		136 rats, Sprague-Dawley	~500 mg-min/m^3	Single	>0.5	.... Acute pulmonary edema within 24 hr, peak within 3 da; proliferative interstitial pneumonitis in 3-10 da; permanent lung damage, with perivascular & peribronchial fibrosis	35
114		244 rats, Sprague-Dawley	800-1000 mg-min/m^3	Single	>0.5	Dose = LC$_{50}$	
115	Cadmium sulfide	10 dogs	4.0-(3-7) mg/m^3	6 hr/da, 5 da/wk [895 hr]	98%<3	Cd dust deposition in mg/100 g tissue: lungs, 3.6; kidney, 1.1; liver, 0.33; urine, 0.05; blood, 0.03. No change in Hb, Hct, RBC, WBC, BUN, alkaline phosphatase, or sulfobromophthalein retention; no dose-related pathology.	39
116	Carbon black	Guinea pigs (exposed to aerosol of killed, radioactive ^{35}S, & viable Escherichia coli for 10 min)	15 mg/m^3	6 hr/da, 5 da/wk [4 wk]	0.5-2.8	Significant ↓ in clearance rate of viable bacteria; no significant change in clearance rate of radioactive bacteria 3 hr after exposure	49
117		6 mice/group BALB/c, 3 mo (exposed to airborne, killed E. coli for 60 min)	558 ± 154 µg/m^3	100 hr/wk [102-192 da]	1.8-2.2	↓ in antibody-forming cells in serum & spleen, & ↑ in antibody-forming cells in mediastinal lymph nodes at 102-135 da; effects on immunosuppression in serum & spleen at 192 da more pronounced than those observed with SO$_2$—see entry 224 in Part II	70

continued

Part I. Particulate Air Pollutants

	Substance	Animal	Dose	Exposure [Duration]	Particle Size μ	Effects	Reference
118		10 rabbits	1.6-2.3 mg/m³	45 min	<1.0	No effects on tracheal ciliary velocity	14
119		150 rats, 150-200 g	6.1 Coh units[1]	2 wk	....	C particles alone & with excess positive & negative ions caused improved learning rates	8
120		18 rats, Long Evans Hooded, 328 g	4.0 mg/m³	16 da	2.2	Significant ↑ in phospholipid & phosphatidyl choline content in dry lung; no significant change in total lipid content, or surface-tension area & pressure-volume measurements	45
121	& SO₂	10 rabbits	2.5-2.9 mg/m³	45 min	....	Significant ↓ in tracheal ciliary velocity; effects similar to SO₂ alone	14
122	& uranium dioxide	20 rats	1.5 mg carbon & 5 µg UO₂ in H₂O	Single	6.0	Simultaneous administration of inert dust accelerated expulsion of biologically active particles from lungs when particles were present in very small amounts; greater lung burden increased phagocytic response	31
123	Chromium dust, mixed	Mice, A, Swiss, & C57BL, 8-10 wk	0.01-4.6 mg/m³	4 hr/da, 5 da/wk [52 wk]	0.8(0.5-5.0)	No bronchiogenic carcinoma; no ↑ in incidence of benign lung tumors, although spontaneous tumors appeared earlier in dust-exposed mice	4
124		185 rats, Wistar & McCollum, 2-4 mo	0.01-4.6 mg/m³	4 hr/da, 5 da/wk [101 wk]	0.8(0.5-5.0)		
125	Cobalt metal & tungsten carbide (1:3)	20 guinea pigs, ∿600 g	8800-10,600 parts/cm³ & 2800 parts/cm³	8 hr/da [20 da] rested 5 da, then 8 hr/da [15 da]	0.5-2.0	15 deaths; acute pulmonary consolidation; slight diffuse dust pigmentation only in survivors at 6-19 mo	17
	Lead compounds						
126	Lead chloride	Rats, Wistar-derived, 300 g	100 µg/m³	12 hr/da, 6 da/wk [2 wk]	0.17	No significant change in number of alveolar macrophages recovered after respiratory tract lavage at 2 wk; no histopathologic inflammatory response	9
127	Lead sesqui-oxide	Rats, Wistar-derived, 300 g	150 µg/m³	24 hr/da [<2 wk]	0.15	Significant ↓ in number of alveolar macrophages recovered after respiratory tract lavage at <2 wk; no histopathologic inflammatory response	9
	Molybdenum compounds						
128	Calcium molybdate	24 guinea pigs, young adult	4.5 mg Mo/ft³	1 hr/da [26 da]	....	20.8% deaths; no clinical signs of toxicity	20
129	Molybdenite	25 guinea pigs, young adult	3.1 mg Mo/ft³	1 hr /da [24 da]	....	4.2% deaths; ↑ in respiratory rate; 3.9 mg Mo/10 g lung	20
130	Molybdenum trioxide dust	51 guinea pigs, young adult	5.8 mg Mo/ft³	1 hr/da [24 da]	....	51% deaths; eye & nose irritation, loss of appetite & body wt, diarrhea, some muscular incoordination, alopecia; greatest conc of Mo in kidneys & bones; spleen & lung content high	20

[1] Coh unit = $\dfrac{\text{optical density} \times 100}{\text{linear feet of air sampled through filter}}$.

continued

Part I. Particulate Air Pollutants

	Substance	Animal	Dose	Exposure [Duration]	Particle Size μ	Effects	Reference
131	Molybdenum trioxide fumes	Guinea pigs, young adult	1.5 mg Mo/ft^3	1 hr/da [25 da]	....	No deaths	20
132			5.4 mg Mo/ft^3	1 hr/da [25 da]	....	8.3% deaths	20
	Nickel compounds						
133	Nickel chloride	Rats, Wistar-derived, 300 g	109 μg/m^3	12 hr/da, 6 da/wk [2 wk]	0.32	No significant change in number of alveolar macrophages recovered after respiratory tract lavage at 2 wk; hyperplastic bronchial epithelium & peribronchial lymphocytic infiltration; less abundant alveolar macrophages than with NiO—see entry 134	9
134	Nickel monoxide	Rats, Wistar-derived, 300 g	120 μg/m^3	12 hr/da, 6 da/wk [2-6 wk]	0.25	Significant ↑ in number of alveolar macrophages recovered after respiratory tract lavage at 2-6 wk; significant macrophage accumulation in alveolar walls at <2 wk, & hypersecretion in bronchial epithelium	9
135	Silica	∿60 rats, Wistar, 150 g[2]	438 ± 115 μg lung deposition	6 hr	0.72	Significant ↑ in SiO$_2$ retention in lungs after 25 da post-exposure cooling, or 2-30 da pre-exposure & 25 da post-exposure cooling; significant ↓ in suprarenal ascorbic acid content after 2-30 da pre-exposure & 25 da post-exposure cooling only	15
136		24 rats, Wistar, 168-308 g	35.7 mg/m^3	7 hr/da, 5 da/wk [6.5 wk]	90% <2.8; 52% 1-1.4	Dust retention at 56 da in lungs: SPF rats, 3.5 mg; standard rats, 3.2 mg; granulomatous, mainly peribronchiolar, silica reaction in lungs at 56 da; no exacerbation of bronchitis in standard rats	15
137	Amorphous[3]	10-20 rats/group, ∿200 g	40 mg in saline	Single i.t. injection	0.10	Nonprogressive histopathologic lung lesions greater than those observed with quartz to 1 mo; ↑ in lung wt at 1 mo & lymph node wt to 8 mo; SiO$_2$ retention <20% in lungs & ∿5% in lymph nodes at 8 mo	15
138	81% pure	Rats, Sprague-Dawley, ∿200 g[2]	20 mg in saline	Single i.t. injection	<5	Significant ↑ in pulmonary SiO$_2$ content in BCG-treated rats at 60 da; less SiO$_2$ retention in lungs than with calcined amorphous silica—see entry 140	21
139	85% pure[4]	10-20 rats/group, ∿200 g	40 mg in saline	Single i.t. injection	0.05	Nonprogressive histopathologic lung lesions similar to those observed with quartz at 1 mo; ↑ in wt & collagen content of lungs at 1 mo, & lymph nodes to 8 mo; SiO$_2$ retention ∿30% in lungs & ∿ 5% in lymph nodes at 8 mo	15

[2] Injected intravenously with 5 mg BCG (Gothenburg substrain) 2 mo prior to treatment. [3] Produced by combustion of silicon-halogen in hydrogen gas. [4] Obtained from smoke of ferrosilicon smelting furnace.

continued

Part I. Particulate Air Pollutants

	Substance	Animal	Dose	Exposure [Duration]	Particle Size μ	Effects	Reference
140	86% pure, calcined	Rats, Sprague-Dawley, ~200 g[2/]	20 mg in saline	Single i.t. injection	....	Significant ↑ in SiO_2 content of lungs & significant ↓ in SiO_2 content of hilar lymph nodes in BCG-treated rats at 60 da; less SiO_2 retention in lungs than with quartz	21
141	97.6% pure	50 guinea pigs, 260-730 g	1.5 mg/ft³	8 hr/da [1-24 mo]	~0.02	No deaths; reversible periductal & peribronchiolar intra-alveolar giant cell accumulations; residual sequelae: emphysema, mural fibrosis, bronchiolar & ductal stenosis	55
142		10 rabbits, New Zealand, 1.1-1.6 kg	1.5 mg/ft³	8 hr/da [1-2 mo]	~0.02	Progressive functional incapacitation & Hct elevation, both possibly due to combined effect of pulmonary vascular obstruction & emphysema; right & left ventricular pressure elevation	59
143		65 rats, Sprague-Dawley, 200-240 g	1.5 mg/ft³	8 hr/da [6-12 mo]	~0.02	Most deaths from pulmonary vascular obstruction, with pulmonary insufficiency due to emphysema; most rats removed from dust after 6-mo exposure rapidly recovered	58
	Silica compounds Kieselguhr						
144	81% silica	10-20 rats/group, ~200 g	40 mg in saline	Single i.t. injection	<5	Nonprogressive histopathologic lung lesions similar to those observed with quartz at 1-2 mo; SiO_2 retention ~20% in lungs & ~2% in lymph nodes at 8 mo	15
145	86% silica, calcined	10-20 rats/group, ~200 g	40 mg in saline	Single i.t. injection	....	Nonprogressive histopathologic lung lesions similar to those observed with quartz at 1-2 mo; progressive ↑ in lymph node wt from 1-4 mo & collagen content to 8 mo; SiO_2 retention ~25% in lungs & <5% in lymph nodes at 8 mo	15
146	Quartz	20 rats, Wistar	30 mg in H_2O	Single i.t. injection	0.47	Leukocytic infiltration of alveoli, particularly evaginating alveoli of respiratory bronchioles & alveolar ducts with some proliferation of alveolar cells at 24 hr; obliteration of evaginating alveoli by avascular fibroblastic tissue at 4 da; obliteration of some respiratory bronchioles & alveolar ducts at 7 da	15
147	84 ± 6% silica	20 guinea pigs, 20 hamsters, 20 rats[5/]	280 mg/m³	6 hr/da, 5 da/wk [115 da]	2.36	Few silicotic nodules in lungs of all species with & without impairment at ~11 mo; atelectasis, desquamative pneumonitis, & alveolar proteinosis localized in nonemphysematous regions; abundant quartz dust in macrophages; dust, lipid material, & cholesterol needles in proteinaceous material; severity of lesions in guinea pigs >hamsters >rats	22,23

[2/] Injected intravenously with 5 mg BCG (Gothenburg substrain) 2 mo prior to treatment. [5/] 50% of each species with experimental emphysema from i.t. injection of 1-6 mg papain 4 mo prior to exposure.

continued

Part I. Particulate Air Pollutants

	Substance	Animal	Dose	Exposure [Duration]	Particle Size μ	Effects	Reference
148	98.3% silica	10-20 rats/ group, ∿200 g	40 mg in saline	Single i.t. injection	1.2	Progressive histopathologic lung lesions to 8 mo; progressive ↑ in lung & lymph node wt & collagen content to 8 mo; SiO_2 retention >50% in lungs & >10% in lymph nodes at 8 mo	15
149	100% silica	Rats, Sprague-Dawley, ∿200 g[6]	20 mg in saline	Single i.t. injection	1.2	Significant ↑ in SiO_2 content of lungs & significant ↓ in SiO_2 content of hilar lymph nodes in BCG-treated rats at 60 da when compared to animals receiving quartz alone	21
150	Dörentrup, 98.9% silica	94 rats, 160-180 g[7]	20 mg/m³	5 hr/da, 5 da/wk [30-120 da]	....	SiO_2 retention at 120 da in mediastinal lymph nodes of rats: 572 µg; with P204 treatment, 24 µg	15
151		90 rats, 160-180 g[8]	30 mg/m³	5 hr/da [3 da]	....	SiO_2 clearance at 3 mo from lungs of rats: 43.7%; with P204 treatment, 62.3%. SiO_2 retention at 3 mo in mediastinal lymph nodes, 16 µg; with P204 treatment, 1 µg.	15
152	& titanium dioxide	30 rats, 160-180 g	20 mg quartz/ m³ & 80 mg TiO_2/m³	5 hr/da [9 da]	TiO_2 0.05-0.2	TiO_2 & SiO_2 dust clearance at 4 mo from lungs: 19.2%; with TiO_2 alone, 66.4%. TiO_2 & SiO_2 retention at 4 mo in mediastinal lymph nodes, 196 µg; with TiO_2 alone, 11 µg.	15
153	Quartz-glass, 85% silica	10-20 rats/ group, ∿200 g	40 mg in saline	Single i.t. injection	0.30	Progressive histopathologic lung lesions similar to those observed with quartz to 4 mo; ↑ in lung wt at 1-2 mo & lung collagen content; ↑ in lymph node wt from 1-4 mo & collagen content; SiO_2 retention ∿20% in lungs & >5% in lymph nodes at 8 mo	15
154	Silicate, aluminium	15 hamsters	3.5 mg in H_2O	Single i.t. injection	Diameter 2.0; length <15	Alveolar infiltration by macrophages & giant cells at 2.5 mo; "asbestos" bodies demonstrated	25
	Sulfur compounds						
155	Ammonium sulfate	6 guinea pigs, 200-300 g	1.0 mg/m³ in H_2O	1 hr	0.29 (wt)	Significant ↑ (∿29%) in pulmonary flow resistance	1
156	Ferric sulfate	15 guinea pigs	1 mg/m³ in H_2O	1 hr	....	Significant ↑ (77%) in pulmonary flow resistance	2
157	Zinc sulfate	7 guinea pigs, 200-300 g	0.91 mg/m³ in H_2O	1 hr	0.29 (wt)	Significant ↑ (∿40%) in pulmonary flow resistance	1
158	Zinc ammonium sulfate	4 cats (paralyzed with gallamine triethiodide, vagotomized, & artificially ventilated)	40-50 mg SO_4/ m³ in H_2O	3 min	∿85% <0.25	6% ↑ in pulmonary flow resistance & 25% ↓ in pulmonary compliance similar, but less severe, than that observed with histamine phosphate aerosol; effects prevented by prior i.v. infusion of isoproterenol (32 µg·kg⁻¹·min⁻¹)	15

[6] Injected i.v. with 5 mg BCG (Gothenburg substrain) 2 mo prior to treatment. [7] 48/94 received s.c. injections of 0.01 mg polyvinylpyridin-N-oxide (P204) every 5 da.

[8] 40/90 received 10 s.c. injections of 0.01 mg P204 within 56 da after exposure.

continued

Part I. Particulate Air Pollutants

	Substance	Animal	Dose	Exposure [Duration]	Particle Size μ	Effects	Reference
159		42 guinea pigs, 200-300 g	0.25, 0.5, 1.1, & 1.8 mg/m³ in H_2O	1 hr	0.29 (wt)	Significant ↑ (∿22, 40, 82, & 130%, respectively) in pulmonary flow resistance	1
160		21 guinea pigs, 200-300 g	1.5 & 2.43 mg/m³ in H_2O	1 hr	0.51 (wt)	Significant ↑ (∿43 & 68%, respectively) in pulmonary flow resistance	1
161		10 guinea pigs, 200-300 g	1.4 mg/m³ in H_2O	1 hr	0.74 (wt)	Significant ↑ (∿21%) in pulmonary flow resistance	1
162		11 guinea pigs 200-300 g	1.1 & 3.6 mg/m³ in H_2O	1 hr	1.4 (wt)	Significant ↑ (∿32%) in pulmonary flow resistance at 3.6 mg/m³ only	1
163	& sulfur dioxide	10 guinea pigs, 200-300 g	0.25 mg/m³ zinc ammonium sulfate in H_2O & 2.5 ppm SO_2	1 hr	0.29 (wt)	Significant ↑ (∿60%) in pulmonary flow resistance; ∿20% ↑ in resistance with 0.25 mg/m³ zinc ammonium sulfate (0.29 μ) or 2.6 ppm SO_2 alone	1
164	Sulfuric acid aerosol	8 dogs, Beagle, 7-17 kg (4/8 with pulmonary impairment from prior NO_2 exposure of 26 ppm × 191 da)	889 ± 296 μg/m³	21 hr/da [620 da]	90% <0.5	Significant ↓ in mean single-breath CO diffusion capacity, residual volume, & net lung volume (inflated & deflated) in impaired & unimpaired dogs; significant ↑ in total expiratory resistance in both groups; high N_2 washout values (impaired); significant ↓ in lung wt (both groups) & heart wt (unimpaired); nonspecific histopathologic changes	33
165		∿68 guinea pigs, 1-2 mo	8-16 mg/m³ in H_2O	72 hr	∿1	No deaths	3
166		64 guinea pigs, 1-2 mo	18 mg/m³ in H_2O	8 hr	∿1	Dose = LC_{50}; pulmonary hemorrhage & edema; focal congestion of suprarenals & spleen; survivors showed hilar consolidation, pneumonic changes, & fibrosis in lungs	3
167		38 guinea pigs, 12-18 mo	50 mg/m³ in H_2O	8 hr	∿1		
168		Guinea pigs, 200-250 g	∿28.8 mg/m³ in H_2O, 20°C	8 hr	2.7	Dose = LC_{50}; at higher conc, onset of death delayed; more extensive pulmonary damage, including bronchial desquamation, hemorrhagic consolidation, edema, emphysema	36
169			∿49.0 mg/m³ in H_2O, 0°C	8 hr	0.8		
170			∿60.9 mg/m³ in H_2O, 20°C	8 hr	0.8	Dose = LC_{50}; rapid death with bronchospasm & resultant emphysema	36
171		32 guinea pigs, 200-250 g	12.4(4-22.4) ppm	8 hr/da [8 da]	....	No significant difference in mortality rates with or without prior exposure; no tolerance indicated	37
172		152 guinea pigs, old & young	2.2(1.15-3.0) mg/m³ in H_2O	24 hr/da [18-140 da]	0.4	Young animals: slight edema in trachea & larynx; ↓ in mucus in major bronchus; ↓ in lymphocytes in pulmonary lymphatic channels. Older animals also showed ↑ in mucus at tracheal bifurcation.	67

continued

Part I. Particulate Air Pollutants

	Substance	Animal	Dose	Exposure [Duration]	Particle Size μ	Effects	Reference
173			2.97(2.1-4.14) mg/m³ in H₂O	24 hr/da [18-140 da]	0.9	Most active particle size physiologically; slight edema; ↑ in desquamated epithelium in minor bronchi; some reduction in lymphocytes in pulmonary lymphatics; ↓ in bronchial mucus	67
174			10.79(1.46-26.5) mg/m³ in H₂O	24 hr/da [18-140 da]	0.6	↓ in lymphatic-histiocytic foci in lungs; ↓ in mucus of bronchi & larynx; less spasm of minor bronchi	67
175		35 guinea pigs	0.47(0.087-1.61) mg/liter in H₂O	7 hr	93-99% <2	3/3 died within 2.75 hr at 0.087 mg/liter	68
176		5 mice	0.38 mg/liter in H₂O	7 hr/da [5 da]	93-99% <2	4 died	
177		70 mice	0.47(0.087-1.61) mg/liter in H₂O	7 hr	93-99% <2	2/5 died after 3.5 hr at 0.55 mg/liter	
178		2 rabbits	0.38 mg/liter in H₂O	7 hr/da [5 da]	93-99% <2	All died	
179		2 rabbits	1.47 mg/liter in H₂O	3.5 hr	93-99% <2	1 died	
180		2 rats	0.38 mg/liter in H₂O	7 hr/da [2 da]	93-99% <2	All died	
181		2 rats	0.7 mg/liter in H₂O	7 hr	93-99% <2	All died	

Effects (bracketed for rows 175-181): Nasal irritation & respiratory distress; degenerative changes of respiratory tract epithelium, pulmonary hyperemia & edema, & in some cases focal pulmonary hemorrhages; areas of atelectasis & emphysema. No deaths at lower dosages for same exposure period.

	Substance	Animal	Dose	Exposure [Duration]	Particle Size μ	Effects	Reference
	Vanadium compounds						
182	Vanadium trioxide	5 dogs; 10 rabbits; rats	40-70 mg/m³	2 hr/da [9-12 mo]	....	Moderate hypochromic anemia, 33% ↓ in WBC; ↓ in blood albumin:γ-globulin ratio, 50% ↓ in blood ascorbic acid (vitamin C); ↑ in blood chloride; ∿30% ↓ in serum SH groups; ↑ in serum amino & nucleic acid content; marked inhibition of respiration in liver & brain; suppurative bronchitis, septic bronchopneumonia, emphysema, moderate interstitial pulmonary sclerosis; greatest deposition in liver, kidneys, & lungs	47
183	Vanadium pentoxide	5 dogs; 10 rabbits; rats	8-18 mg/m³	2 hr/da [9-12 mo]	....	Biochemical & morphological changes similar to those observed with V₂O₃. Also, bronchospasm & 30% ↓ in urinary 5-hydroxyindoleacetic acid.	47
184		12 rabbits, 1.7-2.5 kg	0.02-0.04 mg/liter	1 hr/da [5-8 mo]	78% ⋗5	Chronic inflammatory changes in tracheal & nasal mucosa; emphysema & bronchopneumonic foci with some particles of V dust; probable reduction in elasticity of lungs; small round-cell infiltrates in liver (probably infectious); pyelonephritic changes in kidneys; varying increased V content in examined organs, except intestines	60

continued

Part I. Particulate Air Pollutants

	Substance	Animal	Dose	Exposure [Duration]	Particle Size μ	Effects	Reference
185		13 rabbits, 2-2.5 kg	0.6 mg/liter (3.9-45.0 ct[9]/ total dose)	40-60 min/ da [1-3 da]	78% ⊁ 5	Dyspnea, copious nasal discharge, conjunctival irritation. Severe cases: apathy, anorexia, & emaciation; death in 1-2 da preceded by rales & accelerated breathing. Marked acute laryngotracheitis, acute bronchopneumonic changes with accumulation of V dust; fatty degeneration of liver; increased, varying content of V in most examined organs; occasional enteritis.	60
186		6 rabbits, 2-2.5 kg	10.9 ct[9]/ total dose	2 da	78% ⊁ 5	Slight chronic tracheitis; atelectatic, apparently postpneumonic, lung areas, & emphysema at 26 da. No dust in some lungs although high V content in examined organs, including lungs.	60
187	Zinc oxide	132 rats, Wistar, 250 g	0.4-0.6 mg/m³	10-120 min	0.7-1.6	16 total deaths; 15 deaths with UV-irradiated (50 s) ZnO. Marked hypothermia; ↑ in hypothermia with irradiated ZnO; 1.1-5.5 mg Zn/g lung in sacrificed animals.	6

[9]/ Ct = concentration × time.

Contributors: Palm, Paul E., Nick, M. Susan, Arnold, Elsie P., and Platz, Barbara B.

References

[1] Amdur, M. O., and M. Corn. 1963. Amer. Ind. Hyg. Ass. J. 24(4):326.

[2] Amdur, M. O., and D. Underhill. 1968. Arch. Environ. Health 16:460.

[3] Amdur, M. O., et al. 1952. Arch. Ind. Hyg. Occup. Med. 5:318.

[4] Baetjer, A. M., et al. 1959. Arch. Ind. Health 20:124.

[5] Barrett, H. M., et al. 1947. J. Ind. Hyg. Toxicol. 29:279.

[6] Beeckmans, J. M., and J. R. Brown. 1963. Arch. Environ. Health 7:346.

[7] Bencko, V., and K. Symon. 1970. Atmos. Environ. 4(2):157.

[8] Bevilacqua, D. M., and C. W. LaBelle. 1963. Amer. Ind. Hyg. Ass. J. 24:448.

[9] Bingham, E., et al. 1972. Arch. Environ. Health 25(6):406.

[10] Brieger, H., et al. 1954. Ind. Med. Surg. 23:521.

[11] Bryks, S., and F. D. Bertalanffy. 1971. Arch. Environ. Health 23(6):469.

[12] Chiappino, G., et al. 1969. Arch. Pathol. 87(2):131.

[13] Conradi, C., et al. 1971. Arch. Environ. Health 23(5):348.

[14] Dalhamn, T., and L. Strandberg. 1963. Int. J. Air Water Pollut. 7:517.

[15] Davies, C. N., ed. 1967. Inhaled Particles and Vapours. Pergamon Press, London. v. 2.

[16] Davis, H. V., and A. L. Reeves. 1971. Amer. Ind. Hyg. Ass. J. 32(9):599.

[17] Delahant, A. B. 1955. Arch. Ind. Health 12:116.

[18] Dernehl, C. U., et al. 1945. J. Ind. Hyg. Toxicol. 27:256.

[19] Dutra, F. R., et al. 1951. Arch. Ind. Hyg. Occup. Med. 4:65.

[20] Fairhall, L. T., et al. 1945. Pub. Health Bull. 293.

[21] Göthe, C.-J., and A. Swensson. 1970. Arch. Environ. Health 20(5):579.

[22] Gross, P., and R. T. P. deTreville. 1968. Ibid. 17(5):720.

[23] Gross, P., and R. T. P. deTreville. 1969. Ibid. 18(3):340.

[24] Gross, P., et al. 1955. Arch. Ind. Health 11:479.

[25] Gross, P., et al. 1967. Amer. Ind. Hyg. Ass. J. 28(6):541.

[26] Gross, P., et al. 1967. Arch. Environ. Health 15(3):343.

[27] Gross, P., et al. 1970. Amer. Ind. Hyg. Ass. J. 31(2):125.

[28] Hall, R. H., et al. 1950. Arch. Ind. Hyg. Occup. Med. 2:25.

[29] Harris, C. C., et al. 1971. Cancer Res. 31:1977.

[30] Harrison, H. E., et al. 1947. J. Ind. Hyg. Toxicol. 29:302.

[31] LaBelle, C. W., and H. Brieger. 1959. Arch. Ind. Health 20:100.

[32] Laskin, S., et al. 1970. At. Energy Comm. Symp. Ser. 18:321.

[33] Lewis, T. R., et al. 1973. Arch. Environ. Health 26:16.

continued

Part I. Particulate Air Pollutants

[34] Palm, P. E., et al. 1956. Arch. Ind. Health 13:355.

[35] Paterson, J. C. 1947. J. Ind. Hyg. Toxicol. 29:294.

[36] Pattle, R. E., and H. Cullumbine. 1956. Brit. Med. J. 2:913.

[37] Pattle, R. E., et al. 1956. J. Pathol. Bacteriol. 72:219.

[38] Potts, E. M., et al. 1950. Arch. Ind. Hyg. Occup. Med. 2:175.

[39] Princi, F., and I. F. Greever. 1950. Ibid. 1:651.

[40] Prine, J. R., et al. 1966. Amer. J. Clin. Pathol. 45:448.

[41] Pylev, L. N. 1967. Gig. Sanit. 32(5):174.

[42] Pylev, L. N. 1969. Ibid. 34(2):283.

[43] Reeves, A. L., and A. J. Vorwald. 1967. Cancer Res. 27(1):446.

[44] Reeves, A. L., et al. 1967. Ibid. 27(1):439.

[45] Rhoades, R. A. 1972. Life Sci. 11(1):33.

[46] Robinson, F. R., et al. 1968. Arch. Environ. Health 17(2):193.

[47] Roshchin, I. V. 1967. Gig. Sanit. 32(6):345.

[48] Rozenshtein, I. S. 1970. Ibid. 35(1-3):16.

[49] Rylander, R. 1969. Arch. Environ. Health 18:551.

[50] Saffiotti, U. 1970. At. Energy Comm. Symp. Ser. 18:27.

[51] Saffiotti, U., et al. 1965. J. Air Pollut. Contr. Ass. 15(1):23.

[52] Saffiotti, U., et al. 1967. Cancer 20(5):857.

[53] Saffiotti, U., et al. 1967. Proc. Amer. Ass. Cancer Res. 8(Abstr.):57.

[54] Saffiotti, U., et al. 1968. Cancer Res. 28:104.

[55] Schepers, G. W. H. 1957. Arch. Ind. Health 16:203.

[56] Schepers, G. W. H. 1964. Ind. Med. Surg. 33:1.

[57] Schepers, G. W. H., et al. 1957. Arch. Ind. Health 15:32.

[58] Schepers, G. W. H., et al. 1957. Ibid. 16:125.

[59] Schepers, G. W. H., et al. 1957. Ibid. 16:280.

[60] Sjöberg, S.-G. 1950. Acta Med. Scand., Suppl. 238.

[61] Smith, W. E., et al. 1965. Ann. N.Y. Acad. Sci. 132(1):456.

[62] Stanton, M. F., et al. 1969. Amer. Ind. Hyg. Ass. J. 30(3):236.

[63] Stock, A., and O. Guttman. 1904. Ber. Deut. Chem. Ges. 37:885.

[64] Stokinger, H. E., and C. A. Straud. 1951. J. Lab. Clin. Med. 38:173.

[65] Stokinger, H. E., et al. 1950. Arch. Ind. Hyg. Occup. Med. 1:379.

[66] Stokinger, H. E., et al. 1953. Ibid. 8:493.

[67] Thomas, M. D., et al. 1958. Arch. Ind. Health 17:70.

[68] Treon, J. F., et al. 1950. Arch. Ind. Hyg. Occup. Med. 2:716.

[69] Wagner, J. C., and J. W. Skidmore. 1965. Ann. N.Y. Acad. Sci. 132(1):77.

[70] Zarkower, A. 1972. Arch. Environ. Health 25(1):45.

Part II. Gas and Vapor Air Pollutants

	Substance	Animal	Dose	Exposure [Duration]	Effects	Reference
1	Acrolein vapor	4 dogs, Beagle, 9.9 kg; 30 guinea pigs, Princeton or Hartley-derived, 392 g; 17 monkeys, squirrel, 751 g; 30 rats, Sprague-Dawley-derived, 254 g	0.22 ppm in C_2H_5OH	24 hr/da [90 da]	No hematologic changes; nonspecific inflammatory changes in lungs, heart, liver, & kidneys in dogs, guinea pigs, & monkeys; emphysema, acute congestion, vacuolization of bronchiolar epithelium, & bronchiolar constriction in 2/4 dogs; focal subcapsular hemorrhage of spleen in 2/4 dogs	69
2		2 dogs, 11 kg; 15 guinea pigs, 436 g; 7-9 monkeys, 581 g; 15 rats, 326 g (strains of all species same as in entry 1)	0.7 ppm in C_2H_5OH	8 hr/da, 5 da/wk [6 wk]	No hematologic or biochemical changes; chronic inflammatory changes & occasional emphysema in lungs of all animals, particularly dogs & monkeys	69
3			1.8 ppm in C_2H_5OH	24 hr/da [90 da]	Severe eye & nasal irritation in dogs & monkeys; no hematologic changes; nonspecific inflammatory changes in brain, lungs, heart, liver, & kidneys in all animals; tracheal squamous metaplasia (9/9) & basal cell hyperplasia (6/9) in monkeys; bronchopneumonia in dogs	69

continued

Part II. Gas and Vapor Air Pollutants

	Substance	Animal	Dose	Exposure [Duration]	Effects	Reference
4			3.7 ppm in C_2H_5OH	8 hr/da, 5 da/wk [6 wk]	Severe eye & nasal irritation in dogs & monkeys; no hematologic or biochemical changes; nonspecific inflammatory changes in lungs, liver, & kidneys of all animals; squamous metaplasia & basal cell hyperplasia of trachea in dogs & monkeys; necrotizing bronchitis & bronchiolitis (bronchiolitis obliterans) with squamous metaplasia of lungs in 7/9 monkeys; bronchopneumonia in dogs	69
5		4 guinea pigs, 228 g	10.5 ppm	6 hr	2 deaths; acute emphysema, focal edema & congestion; inflammatory cell infiltration; acute desquamating tracheitis in 1/4	91
6		4-15 mice	0.4 ppm	6 hr	Voluntary running activity depressed $\sim$50%	78
7		20 mice, $\sim$20 g	10.5 ppm	6 hr	9 deaths delayed >24 hr after exposure; consolidation, congestion, & inflammatory cell reaction in lungs	91
8		4 rabbits, 3 mo	0.6 ppm	4 hr/da [30 da]	No ophthalmologic or biochemical effects; no effect on 6-phosphogluconate dehydrogenase, malate dehydrogenase, glucose-6-phosphate dehydrogenase, or LDH	59,74
9			2 ppm	4 hr	No ophthalmologic or biochemical effects	
10		2 rabbits, 2.6-2.8 kg	10.5 ppm	6 hr	1 death; generalized emphysema, focal congestion & consolidation in lungs; focal desquamation of bronchial epithelium	91
11		20 rats, albino	0.03 mg/m^3 (0.013 ppm[1]/)	Long-term	No effects in normal or silicotic rats	103
12		10 rats, albino, 90-130 g	0.15 mg/m^3 (0.065 ppm[1]/)	$\sim$60 da	No effects except for significant ↑ in fluorescent leukocyte counts at 24 hr	54
13			0.51 mg/m^3 (0.22 ppm[1]/)	$\sim$60 da	Significant, reversible ↑ in fluorescent leukocyte counts at 1 wk; significant, reversible ↓ in whole blood cholinesterase at 5 wk & urinary coproporphyrin at 7 wk; loss of conditioned reflexes to sound & light; proliferation of bronchial epithelium & marked eosinophilic infiltration of bronchial wall	
14		20 rats, albino (10/20 with chronic pulmonary insufficiency from i.t. injection of 50 mg SiO$_2$ 2.5 mo prior to exposure)	0.74 mg/m^3 (0.32 ppm[1]/)	Long-term	Significant change in blood cholinesterase in silicotic rats, & after starvation in normal rats; ↑ in urinary 17-ketosteroids in both groups; significant ↓ in adrenal ascorbic acid (vitamin C) content in silicotic rats; change in chronaxy of antagonistic muscles in both groups	103
15		22 rats, Holtzman, 200-300 g	8 ppm	4 hr	Marked ↑ in liver AP & tyrosine-α-ketoglutarate transaminase 2 hr after exposure; little effect on hepatic enzymes with adrenalectomy; moderate pulmonary edema & inflammation	79
16		6 rats, Sprague-Dawley-derived, 200-300 g[2]/	3.9 ppm	4 hr/da [9 da]	Significant ↓ in liver:body wt ratio; no significant change in liver AP	81

[1]/ Calculated by contributors. [2]/ Fasted during 24-hr interval between beginning of exposure and sacrifice.

continued

Part II. Gas and Vapor Air Pollutants

	Substance	Animal	Dose	Exposure [Duration]	Effects	Reference
17		15 rats, Sprague-Dawley-derived, 200-300 g[2/]	4.4 ppm	2, 4, & 8 hr	Significant ↓ (35%) in liver AP after 2-hr exposure; significant ↑ (222%) in liver AP after 8-hr exposure concomitant with dyspnea & nasal irritation; no significant change in lung & kidney AP	81
18		20 rats, Sprague-Dawley-derived, 200-300 g[2/]	6.4 ppm	4 hr	Significant ↑ in liver:body wt ratio, liver AP, & adrenal wt; no significant change in lung: body wt ratio, or lung & serum AP	81
19		20 rats, Sprague-Dawley-derived, 200-300 g[2/]	12 ppm	4 hr	36 & 72% ↓ in serum & lung AP, respectively	81
20	Acrolein aerosol	20 guinea pigs	1.1-1.3 mg-min/m³ × 10⁵	≫10 hr	7 deaths; initial hyperactivity & eye irritation followed by slow & deep respiration; edematous & hemorrhagic lungs; enlarged liver & fluid in peritoneal cavity (particle size, 0.7 μ)	99
21		12 guinea pigs, 250 g (intact & tracheotomized)	17 ppm	1 hr	Significant ↑ in pulmonary flow resistance & tidal volume, significant ↓ in respiration rate & minute volume, & no change in pulmonary compliance in intact animals only (particle size, 0.55-2.2 μ)	25
22		50 mice	0.6-0.7 mg-min/m³ × 10⁵	≫10 hr	Pulmonary edema & hemorrhage, pleural effusion; dose near fatal (particle size, 0.7 μ)	99
23		5 rabbits	1.2-1.4 mg-min/m³ × 10⁵	≫10 hr	3 deaths; initial hyperactivity & eye irritation followed by slow & deep respiration; convulsions prior to death (particle size, 0.7 μ)	99
24	Ammonia	10 rats	102.3 ppm	5 hr/da, 5 da/wk [60 da]	No significant effect on rate of tracheal ciliary activity; 60% moderate to severe damage of tracheal mucosa	22
25		49 rats, Sprague-Dawley & Long-Evans-derived	262 mg/m³ (377 ppm[1/])	24 hr/da [90 da]	No deaths; 25% mild nasal irritation; 8% slight leukocytosis, nonspecific circulatory & degenerative changes in lungs & kidneys	20
26	& carbon, activated	10 rats	118.7 ppm NH₃ & 3.4 mg carbon/m³	5 hr/da, 5 da/wk [60 da]	Significant ↓ in rate of tracheal ciliary activity; 80% severe damage of tracheal mucosa (carbon particle size: 95%, <3 μ; 65%, <1 μ)	22
27	Carbon monoxide	6 dogs	100 ppm	5.75 hr/da [11 wk]	Gait, postural & position reflex disturbances; ECG characteristic of anoxia & necrosis of single heart muscle fibers in 4/6; "anoxic necrosis" of cerebral cortex (white matter), globus pallidus, & brain stem 3 mo after final dose	65
28		11 dogs	1800-2200 ppm	9 hr	Death within 7-8 hr by slow asphyxia; collapse, convulsions, apparent unconsciousness in ∿2 hr	128

[1/] Calculated by contributors. [2/] Fasted during 24-hr interval between beginning of exposure and sacrifice.

continued

Part II. Gas and Vapor Air Pollutants

	Substance	Animal	Dose	Exposure [Duration]	Effects	Reference
29		17 dogs	1800-2200 ppm followed by 1300-1800 ppm	2 hr 15 hr	Unconsciousness & slow, shallow respiration after 13-19 hr; death by slow asphyxia within 12-16 hr; severe edema in brain after 16 hr; focal myelin degeneration in survivors after 16-165 da	128
30		3-4 dogs	6000 ppm	20-30 min	All deaths by rapid asphyxia within 20-30 min; hyperglycemia, hyperuricemia, ↓ in plasma CO_2 & plasma CO_2 capacity, ↑ in H^+ conc; severe perivascular & perineuronal edema, particularly in corpus striatum, cortex, & dorsal motor nucleus of vagus nerve; neurons extensively damaged; some petechial hemorrhages	128
31		5 guinea pigs	650 ppm	4 hr	↓ in respiratory rate after 4 hr; slight ↓ in tidal volume; both reversible within 45 min	82
32		4-15 mice	250 ppm	6 hr	Voluntary running activity depressed ∿50%	78
33		Mice	4000 ppm	45-50 min	Lethal concentration	31
34		18 monkeys, cynamolgus, 1.6-3.0 kg	19.9 or 65.5 ppm	22-23 hr/da, 7 da/wk [104 wk]	No hematologic or biochemical changes, no change in blood P_{O_2}, P_{CO_2} or pH, & no histopathologic lesions at either level; slight to moderate ↑ in blood carboxyhemoglobin at 4-12 wk & significant ↑ in ♀ heart:body wt ratios at both levels	34
35		Rats, adult	5000 ppm	30 min	Lethal conc; unconsciousness, spasmodic contraction of muscles at death	77
36		10 rats, albino	1 mg/m³ (0.87 ppm [1])	72 da	Estrous cycle changes in exposed ♀ & progeny of first generation of exposed ♀; significant changes in % fertilization of exposed ♀, duration of pregnancy, & wt gain of progeny; increased litter size & pituitary function	71
37		Rats, albino, 110-160 g (2/3 groups received rations supplemented with 25 or 50 μg pyridoxine)	22 mg/m³ (19.2 ppm [1])	5-6 hr/da, 6 da/wk [105-114 da]	↑ in blood carboxyhemoglobin after 6-hr exposure with or without supplemental vitamin B_6; no significant effect on urinary 4-pyridoxic or xanthurenic acid in either group; ↓ in vitamin-B_6 content in tissues of rats without supplement	85
38		5-9 rats, albino, 150-300 g	100 ppm	96 hr	No effect on sleep or waking patterns	19
39		9 rats, albino, 150-300 g	250 ppm	96 hr	↓ in proportion of REM (rapid eye movement) phases in total sleep time; frequent daytime wakefulness & relaxation phases; recovery within 24 hr	19
40		35 rats, Holtzman, 160-180 g	50 ppm	5 hr/da, 5 da/wk [12 wk]	Significant ↑ in Zn & Mg in nuclear fraction of homogenized liver at 6-12 wk, & 3 & 12 wk, respectively. Significant ↓ in Cu in nuclear & mitochondrial fractions at 12 wk; in Mg in mitochondrial & supernatant fractions at 3 & 6 wk; & in Co & Fe in all three fractions at 6 & 9 wk, & 3 & 6 wk, respectively.	73

[1] Calculated by contributors.

continued

Part II. Gas and Vapor Air Pollutants

	Substance	Animal	Dose	Exposure [Duration]	Effects	Reference
41	& nitric oxide, nitrogen dioxide, or ozone	160 mice, Swiss, 10 wk; 96 rats, Sprague-Dawley, 10 wk	120 ppm CO & 3 ppm NO, 3 ppm NO_2, or 0.75 ppm O_3	7 hr	No significant enhancement of blood carboxyhemoglobin with addition of NO, NO_2, or O_3	68
42	& ammonia & nitrogen dioxide	50 rats, albino, 75-80 g (2/3 groups received rations supplemented with 25 or 50 μg pyridoxine)	21 mg CO/m^3 (18.3 ppm[1]), 16 mg NH_3/m^3 (23 ppm[1]), & 2.1 mg NO_2/m^3 (1.1 ppm[1])	6 hr/da, 1 da/wk [139-149 da]	61% body wt loss, 54% dermal ulceration, & significant ↓ in vitamin-B_6 content of tissues in rats without supplement; significant ↓ in SGPT & SGOT, & urinary vitamin B_6 & 4-pyridoxic acid in nonsupplemented rats & rats treated with 25 μg pyridoxine only	86
43	Formaldehyde vapor	2 dogs, Beagle; 15 guinea pigs, Princeton-derived; 3 monkeys, squirrel; 3 rabbits, New Zealand; 15 rats, Sprague-Dawley & Long-Evans-derived	4.6 ± 0.4 mg/m^3 (3.75 ± 0.33 ppm[1])	24 hr/da [90 da]	No hematologic changes; interstitial inflammation in lungs of all species; focal chronic inflammation in heart & kidneys of guinea pigs & rats only	20
44		5 guinea pigs[3]	1.0-1.5 mg/liter (815-1222 ppm[1])	5 hr/da [10 da]	Significant ↑ in ^{35}S-methionine uptake by blood proteins; no significant change in ^{35}S-methionine uptake by renal & hepatic proteins	93
45		60 mice, C3H	0.05 mg/liter (40.8 ppm[1])	1 hr/da, 3 da/wk [35 wk]	Basal cell hyperplasia & stratification of epithelium of trachea & major bronchi; no lung tumors	60
46			0.10 mg/liter (81.5 ppm[1])	1 hr/da, 3 da/wk [35 wk]	Basal cell hyperplasia, stratification of epithelium, & squamous cell metaplasia of trachea & major bronchi; no lung tumors	60
47			0.20 mg/liter (163 ppm[1])	1 hr/da, 3 da/wk [11 da]	15 deaths	60
48		72 rats	0.6-1.3 mg/liter (489-1060 ppm[1])	30 min	LC_{50} = 1 mg/liter (815 ppm[1]). Deaths from 6 hr-15 da; marked eye & nasal irritation; respiratory distress (up to 2 wk); lung hemorrhages, intraalveolar & perivascular edema.	104
49		Rats, albino	0.012 mg/m^3 (0.01 ppm[1]) & 1.0 mg/m^3 (0.82 ppm[1])	Gestation period	Significant ↓ in ascorbic acid content of maternal liver & whole embryo, & in DNA content of maternal & fetal livers at both conc	47
50		11 rats, albino, 230-250 g[3]	3.5 mg/m^3 (2.85 ppm[1]), 1.05 mg/m^3 (0.86 ppm[1]), & 0.6-0.7 mg/m^3 (0.49-0.57 ppm[1])	10 da each [30 da]	Significant ↓ in ^{35}S-methionine uptake by hepatic proteins after 20 & 30 da; no significant change in ^{35}S-methionine uptake in blood, kidneys, or testes	93
51	Formaldehyde aerosol	20 guinea pigs	~20 mg/m^3	≥10 hr	7 deaths after exposure (particle size, 0.7 μ)	99

[1] Calculated by contributors. [3] Injected with 1 ml ^{35}S-methionine solution (2 × 10^5 cpm/ml) after exposure period.

continued

Part II. Gas and Vapor Air Pollutants

	Substance	Animal	Dose	Exposure [Duration]	Effects	Reference
52		Guinea pigs, 250 g (intact & tracheotomized)	50 ppm	1 hr	Qualitatively similar results to those described for acrolein aerosol—*see* entry 21 (particle size, 0.55-2.2 μ)	25
53		100 mice	∿20 mg/m^3	≫10 hr	4 deaths during exposure, 13 deaths after exposure; eye irritation, followed by slow, deep respiration; convulsions prior to death (particle size, 0.7 μ)	99
54		5 rabbits	∿20 mg/m^3	≫10 hr	3 deaths after exposure; edema & hemorrhage in lungs of all animals (particle size, 0.7 μ)	99
55	Hydrogen fluoride	Guinea pigs, Hartley, 340-360 g	4300 ppm	15 min	Dose = LC$_{50}$; eye & nasal irritation; respiratory distress, body wt loss, general weakness; nasal mucosa necrosis with acute inflammation; selective necrosis of renal tubules; hepatocellular intracytoplasmic globules; dermal collagen changes with acute inflammation; possible myeloid hyperplasia of bone marrow	96
56		Rats, Wistar, 100-120 g	1310 ppm	60 min		
57			2040 ppm	30 min		
58			2690 ppm	15 min		
59			4970 ppm	5 min		
60	Nickel carbonyl	5 dogs	0.2-1.0 mg/liter (28.6-143 ppm[1/])	30 min	∿99% Ni excreted in urine within 6 da	118
61		32 rats	0.6 mg/liter (80 ppm) in C$_2$H$_5$OH:(C$_2$H$_5$)$_2$O	30 min	↑ in Ni content of NaCl-precipitable fraction of lung & liver RNA; ↓ in Ni content in NaCl-soluble fraction of lung RNA; no change in NaCl-soluble liver RNA	112
62		27 rats, Sprague-Dawley, 140-170 g[4/]	0.2 mg Ni/liter (80 ppm Ni(CO)$_4$[1/])	15 min	Significant ↓ (84 & 42%) in benzpyrene hydroxylase in lung & liver, respectively, at 52 hr in induced rats	113
63		75 rats, Wistar	0.17-0.5 mg/liter (24.3-71.5 ppm[1/]) in C$_2$H$_5$OH	30 min	LC$_{50}$ = 0.24 mg/liter (34.3 ppm[1/]) ⎱ Severe pulmonary congestion & edema in immediate deaths & extensive pneumonitis in survivors ≮1 da; severe necrosis of pericentral hepatocytes; focal necrosis, reticular degeneration, accumulated megakaryocytes in spleen; tubular epithelial degeneration & less glomerular degeneration in kidneys	62
64		6 rats, Wistar	0.38 (0.083-0.54) mg/liter [54.3(11.9-77.2) ppm[1/]] in C$_2$H$_5$OH	30 min/da [10 exposures, 48 da]	1/6 deaths; ∿50% increase in Hb ⎰	
65		137 rats, Wistar, 145-273 g	0.03-0.06 mg/liter (4.3-8.6 ppm[1/]) in C$_2$H$_5$OH:(C$_2$H$_5$)$_2$O	30 min/da, 3 da/wk [12 mo]	↑ in mortality; significant ↑ in heart, lung, & adrenal wt; extensive pulmonary lesions; squamous metaplasia of bronchial epithelium	111
66		>64 rats, Wistar, 200-250 g	0.03 mg/liter (4 ppm) in C$_2$H$_5$OH:(C$_2$H$_5$)$_2$O	30 min/da, 3 da/wk [12 or 26 mo]	25 & 88% mortality at 1 & 2 yr, respectively; pulmonary squamous cell carcinoma or adenocarcinoma with metastases in 3 rats after 24-26 mo	109,110

[1/] Calculated by contributors. [4/] 19/27 received 15 mg recrystallized phenothiazine by gastric intubation to induce benzpyrene hydroxylase 4 hr after exposure.

continued

Part II. Gas and Vapor Air Pollutants

	Substance	Animal	Dose	Exposure [Duration]	Effects	Reference
67				30 min/da, 3 da/wk [12 mo]	Significant ↑ in Ni (63 µg/g) content of total lung parenchyma with ↑ in Ni localization in nuclear, mitochondrial, microsomal, & supernatant fractions; significant ↑ in Ni:N$_2$ & Ni:RNA ratios in nuclear & mitochondrial lung fractions; significant ↑ in Ni in microsomal & supernatant liver fractions	114
68		345 rats, Wistar, 200-250 g[5/]	0.6 mg/liter (80 ppm) in C$_2$H$_5$OH: (C$_2$H$_5$)$_2$O	30 min	Mortality in 3 wk: 72%; with sodium diethyldithiocarbamate (Dithiocarb), 0. 1/60 survivors with sodium diethyldithiocarbamate showed pulmonary anaplastic carcinoma with metastases in liver & spleen at 26 mo; 1/71 survivors without sodium diethyldithiocarbamate showed pulmonary adenocarcinoma with metastases in liver & spleen at 24 mo.	109,110
69			0.6 mg/liter (80 ppm) in C$_2$H$_5$OH: (C$_2$H$_5$)$_2$O	30 min	Significant ↑ in Ni (91 µg/g) content of total lung parenchyma with ↑ in Ni localization in microsomal & supernatant fractions; significant ↑ in Ni:N$_2$ & Ni:RNA ratios in microsomal & supernatant lung fractions; significant ↑ in Ni in microsomal & supernatant liver fractions	114
	Nitrogen oxide compounds					
70	Nitric oxide	Guinea pigs	16 or 50 ppm	⪢4 hr	No significant change in respiratory rate or tidal volume	82
71		4 guinea pigs, New England strain A	15 ppm	24 hr/da [12-13 wk]	Significant ↑ in ratio of LDH-positive alveolar cells to alveoli with minimal lung lesions (slight chronic inflammation & congestion)	101
72		Mice, albino	⪢310 ppm	8 hr	No deaths	92
73			350 ppm	8 hr	100% mortality during exposure	92
74	Nitrogen dioxide	11 cats	220-270 ppm	∼90 min	No deaths	127
75		3 cats	330-370 ppm	60-90 min	2 deaths; ↑ in blood methemoglobin	127
76		2 dogs/group	39 or 53 ppm	60 min	Respiratory distress & eye irritation at 50% rat LC$_{50}$; no histopathologic lesions	16
77			52 or 85 ppm	15 min		
78			125 or 164 ppm	5 min		
79		9 guinea pigs	5 ppm	10-20 min	Pulmonary flow resistance nearly doubled after 10-min exposure	117
80		20 guinea pigs	5.2 & 6.5 ppm	4 hr	Reversible ↑ in respiratory rate after 180 & 135 min for respective doses; ↓ in tidal volume	82
81		14 guinea pigs	9 & 13 ppm	2 hr	Reversible ↑ in respiratory rate after 90 & 60 min for respective doses; ↓ in tidal volume	82
82		12 guinea pigs	10 ppm	24 hr/da [1-6 wk]	Significant ↑ in ratio of type 2 pneumonocytes to total alveolar wall cells; type 2 pneumonocytes showed fine, closely-packed lamellae in cytosomes & intracellular lipid bodies with electron microscopy; lipid bodies also present within macrophages & alveolar lumina; ↑ in histiocytic desquamation; interstitial & intra-alveolar aggregates of neutrophils; interstitial plasma cell infiltrates	129

[5/] 60/345 received single s.c. injection of 50 mg/kg sodium diethyldithiocarbamate trihydrate (Dithiocarb) 15 min after exposure.

continued

Part II. Gas and Vapor Air Pollutants

	Substance	Animal	Dose	Exposure [Duration]	Effects	Reference
83		7 guinea pigs, 270-370 g	80 ppm	3 hr	Significant ↑ in lung water content; no significant change in total lung histamine content, or lung histamine content per g parenchyma or per 100-g body wt	115
84		24 guinea pigs, albino	2 ppm	24 hr/da [1-3 wk]	Significant ↑ in ratio of LDH-positive alveolar wall cells to alveoli, primarily in right lower lobe; 21% focal bronchopneumonia	102
85		50 guinea pigs, Hartley, 250 g	10 ppm	24 hr/da [12-45 da]	Significant ↑ in O₂ consumption, LDH & aldolase in lung at 26-45 da, 12-45 da, & 12 & 32 da, respectively; significant ↓ in lung aldolase at 26 da; significant ↑ in O₂ consumption in kidney, LDH in kidney & liver, & aldolase in liver; significant ↓ in aldolase in kidney	12
86		18 guinea pigs, Hartley, 250 g	15 ppm	22-23 hr/da [26-40 da]	Significant ↑ in anaerobic LDH isoenzyme in lung at 26-40 da; significant ↓ in aerobic LDH isoenzyme in lung at 26-40 da; no significant change in anaerobic or aerobic LDH isoenzymes in liver & kidney	11
87		48 hamsters, Syrian, 80-100 g	45-55 ppm	21-23 hr/da [10 wk]	15/48 deaths; hyperplasia & hypertrophy of epithelium of terminal & respiratory bronchioles & proximal alveolar ducts with extensive inflammatory infiltrate, pulmonary edema, but no emphysema at 10 wk; minimal epithelial hypertrophy only at 14 wk; significant ↑ in ratios of lung:body wt & right ventricle:body wt at 10 wk only	63
88		9 mice	3.7 ppm	6 hr	No depression of voluntary activity during exposure	82
89		42 mice	16(7.7-20.9) ppm	6 hr	Voluntary running activity depressed ~50%	78,82
90		Mice, Swiss, 21 g (30-110/group challenged with airborne, viable *Klebsiella pneumoniae*, type A, strain A-D)	0.5 ppm	6, 18, or 24 hr/da, 7 da/wk [1-12 mo]	Significant ↑ in mortality in all groups (6, 18, & 24 hr/da) challenged with *K. pneumoniae* after 6 mo & in 24 hr/da group only after 24 mo; significant ↑ in aerobic LDH isoenzymes in serum in 24 hr/da group only after 12 mo; significant ↓ in clearance rate of *K. pneumoniae* in 6 & 18 hr/da groups after 12 mo & in 24 hr/da group after 6-12 mo	36
91		Mice, Swiss, 21 g (30-110/group challenged with airborne, viable *K. pneumoniae*, type A, strain A-D)	0.5 ppm	6, 18, or 24 hr/da, 7 da/wk [1-12 mo]	Alveolar expansion & septal breakage after 3 mo, & alveolar expansion with bronchiolar obstruction, moderate to severe pneumonitis & severe bronchitis after 6 mo in all 3 groups (6, 18, & 24 hr/da) without *K. pneumoniae* challenge; severe pneumonitis & alveolar expansion with proliferative bronchiolitis & airway blockage in exposed groups with challenge	8

continued

Part II.   Gas and Vapor Air Pollutants

	Substance	Animal	Dose	Exposure [Duration]	Effects	Reference
92		6 monkeys, squirrel, 0.5-0.8 kg (challenged with 10^4-10^5 airborne, viable *K. pneumoniae* within 1 hr after exposure)	10 or 15 ppm	2 hr	1/2 ↓ in tidal volume at 11-17 da after 15 ppm NO_2 & *K. pneumoniae;* ↑ in anaerobic LDH isoenzyme in serum at 2 hr-2 da after 10 ppm NO_2 with or without challenge; alveolar expansion with septal breaks after 10 & 15 ppm NO_2 with or without challenge; interstitial lymphocytic infiltrate after 15 ppm NO_2 only; proliferative bronchiolar epithelium after 10 & 15 ppm NO_2 with challenge	57
93		10 monkeys, squirrel, 0.6-0.8 kg[6]	5 ppm	24 hr/da [2 mo]	28% mortality 3-10 da after bacterial challenge; 33% mortality 5 da after viral challenge; ↓ in minute volume after bacterial challenge, & marked ↓ in tidal volume after viral challenge; 5/7 reduced lung clearance after bacterial challenge	58
94		4 rabbits, 3-4 mo	18-20 ppm	4 hr/da [34 da]	No ophthalmologic or biochemical effects; no effect on 6-phosphogluconate dehydrogenase, malate dehydrogenase, glucose-6-phosphate dehydrogenase, or LDH	59,74
95		21 rabbits, albino, 0.5-1 yr	8-12 ppm	24 hr/da [3-4 mo]	Hyperplasia & hypertrophy of bronchial epithelium, particularly in terminal bronchioles, with associated infiltration of macrophages; ↑ in size & elastic content of alveolar ducts & peripheral alveoli; apparent alveolar disruption; changes in bronchiolar epithelium only reversible within 1 wk-1 mo	56
96		5 rabbits, 2.2-2.7 kg	315 ppm	15 min	Dose = LC_{50}; severe respiratory distress, eye irritation; focal accumulations of intra-alveolar macrophages, some proliferation of alveolar epithelial lining with inflammatory infiltrate in survivors at 7-21 da	16
97		6 rabbits, New Zealand, 1-3 kg	50 ppm	3 hr	No significant change in benzpyrene hydroxylase in tracheobronchial mucosa at 30 min-7 da after exposure	87
98		Rabbits, New Zealand, 1.8-2.2 kg[7]	15 ppm	3 hr	Alveolar macrophages showed ↓ in resistance to rabbitpox virus challenge in vitro at 48 hr, and marked ↓ in phagocytic capacity for killed BCG vaccine	1
99		Rabbits, New Zealand, 1.8-2.2 kg	25 ppm	3 hr	Alveolar macrophages inoculated in vitro with 0.1 TCID parainfluenza 3 virus per cell showed significant ↓ in interferon production at 48 hr	122
100		10 rats/group, 100-120 g	115 ppm	60 min	Dose = LC_{50}	16
101			162 ppm	30 min		
102			201 ppm	15 min		
103			416 ppm	5 min	Dose = LC_{50}; death within 30 min-3 da; severe respiratory distress & eye irritation; 10-15% body wt loss; darkened areas & occasional purulent nodules in lungs	16

6/ Challenged with 4-9 × 10^5 airborne, viable *K. pneumoniae,* type A, strain A-D, within 1 hr after exposure, or with single i.t. injection of 0.5 ml 10 × $10^{6.4}$ mouse LD_{50} units type A influenza virus, strain PR-8, 24 hr prior to exposure.
7/ Received i.t. injection of 5 × 10^5 TCID parainfluenza 3 virus prior to exposure.

continued

Part II. Gas and Vapor Air Pollutants

	Substance	Animal	Dose	Exposure [Duration]	Effects	Reference
104		10 rats/group 200-300 g	88 ppm	240 min	Dose = LC_{50}; death from pulmonary edema	52
105			163 ppm	60 min		
106			174 ppm	30 min		
107			420 ppm	15 min		
108		32 rats, Long Evans Hooded, 230-330 g	6 ppm	∿23 hr/da, 7 da/wk [6 wk]	Significant, reversible ↓ in mucociliary transport as measured by first edge time & 20% transport time; slight ↑ in interstitial edema & vascular congestion	46
109		9 rats, Sprague-Dawley	0.8 ppm	24 hr/da [∿2.7 yr]	Sustained ↑ (∿20%) in respiratory rate; slight widening & ↑ in elastic content of alveolar ducts	43
110		18 rats, Sprague-Dawley, ∿100 g	0.8 ppm & 2 ± 1 ppm	24 hr/da [69 da] or 24 hr/da [∿2 yr]	Sustained tachypnea; no changes in pulmonary flow resistance or dynamic compliance; significant ↑ (∿20%) in absolute lung wt; hypertrophy, ↓ in number of cilia & cytoplasmic extensions, & cytoplasmic crystalloid inclusion bodies in epithelium of terminal bronchioles with light & electron microscopy	44
111		6 rats, Sprague-Dawley, 175-200 g	0.9-1.1 ppm	4 hr	Peroxidative changes in lung lipids indicative of diene conjugation with maximum effect at 24-48 hr	120
112		Rats, Sprague-Dawley, 175-200 g (4/group received 10 mg/da dl-α-tocopherol orally for ⩤3 da)	0.94-1.25 ppm	4 hr/da [6 da]	↑ in peroxidative changes in lung lipids indicative of diene conjugation in rats without vitamin E supplement; vitamin E supplement only partially effective in preventing changes	120
113		4 rats, Wistar, weanling	0.15-0.5 ppm	23 hr/da [2-6 wk]	No changes in lung, liver, or body wt; ↑ in aspartic acid excretion in urine; no ↑ in glutamic acid excretion	94
114		Rats, Wistar, 100 g	2 ppm	24 hr/da [1 da-12 mo]	Hypertrophy of ciliated epithelium in terminal bronchioles with cloudy swelling & loss of cilia at 72 hr; focal epithelial hyperplasia at 7 da with recovery apparent from 14-21 da; focal cytoplasmic swelling of type 1 cells in peripheral alveoli at 21 da with electron microscopy	105
115		Rats, Wistar, 100 g	17 ppm	24 hr/da [<1-43 da]	Loss of cilia in epithelium of terminal bronchioles at 4-8 hr to 3 wk; hypertrophy of ciliated & nonciliated epithelium & multilayered focal hyperplasia with prevalent mitotic figures at 48 hr; cytoplasmic disruption of type 1 cells in proximal alveoli with fibrin-covered basement membrane at 4 hr; interstitial fibrin & edema with macrophage infiltration at 8-24 hr; increased mitotic activity with ↑ in cells with altered structure at 24-48 hr; abundant subendothelial swelling at 7 da with light & electron microscopy	105

continued

Part II. Gas and Vapor Air Pollutants

	Substance	Animal	Dose	Exposure [Duration]	Effects	Reference
116		36 rats, Wistar, 100 g (received i.p. injection of 100-300 μCi ^{3}H-thymidine 1 hr prior to sacrifice)	2 ppm	24 hr/da [$\geqslant$360 da]	Marked $\uparrow$ in labeling index from $\uparrow$ in ^{3}H-labeled type 2 alveolar cells in peripheral alveoli from 24 hr to 7 da with no $\uparrow$ in tissue cellularity; no significant change in labeling indexes of cells lining terminal bronchioles or alveoli at distal ends of terminal bronchioles	39
117		36 rats, Wistar, 100 g (received i.p. injection of 100-300 μCi ^{3}H-thymidine 1 hr prior to sacrifice)	17 ppm	24 hr/da [$\geqslant$7 da]	Marked $\uparrow$ in labeling indexes of terminal bronchioles with concomitant $\uparrow$ in mitotic figures at 24-168 hr; of alveoli at distal ends of terminal bronchioles with $\uparrow$ in cellularity from type 2 alveolar cells at 48 hr-7 da; & of type 2 alveolar cells in peripheral alveoli with no $\uparrow$ in cellularity at 24 hr-7 da	39
118	& ferric oxide	22 dogs, Beagle, 7.0-13.3 kg	26 ppm NO_2 & 922 ± 535 μg Fe_2O_3/m^3	21 hr/da [193 da]	Significant $\uparrow$ in total pulmonary resistance	66
119	Nitrogen oxides	11 dogs, Beagle	0.24 mg NO/m^3 (0.20 ppm[1/]) & 0.94-1.88 mg NO_2/m^3 (0.41-0.82 ppm[1/])	16 hr/da [5 yr]	9% significant $\uparrow$ in mean & diastolic systemic arterial pressure at 5 yr; 9% bradycardia by static ECG & 9% premature ventricular contractions by exercise ECG at 5 yr; 18% mild to moderate left ventricular hypertrophy, & 9% possible mild right ventricular hypertrophy by VCG at 5 yr	9
120		11 dogs, Beagle	1.84-2.45 mg NO/m^3 (1.50-2.0 ppm[1/]) & 0.38 mg NO_2/m^3 (0.17 ppm[1/])	16 hr/da [5 yr]	9% significant $\uparrow$ in systolic pulmonary arterial pressure at 5 yr; 27% probable right ventricular hypertrophy, old posterior myocardial infarction or left axis deviation by VCG at 5 yr	9
121	Nitrogen pentoxide	Rats, albino, 110-160 g (2/3 groups received rations supplemented with 25 or 50 μg pyridoxine)	2.6 mg/m^3 (0.59 ppm[1/])	5-6 hr/da, 6 da/wk [104-113 da]	$\uparrow$ in blood methemoglobin at 6 hr not affected by supplemental vitamin B_6; significant $\downarrow$ in urinary 4-pyridoxic acid & tissue vitamin-B_6 content, & significant $\uparrow$ in urinary xanthurenic acid in rats without supplement only	85
122	Nitro-olefin vapors 2-Nitro-2-butene	30 guinea pigs, English, 300-400 g	0.37-1.14 ppm	2 hr	Significant $\uparrow$ in pulmonary flow resistance with $\downarrow$ in C-chain lengths; significant $\uparrow$ in tidal volume; $\downarrow$ in respiratory rate at >0.37 ppm; all effects reversible	80
123		4 guinea pigs; 10 mice, Swiss; 2 rabbits, albino; 10 rats, CFN	10-20 ppm, 50% & 90% RH	6 hr/da, 5 da/wk [3 mo]	Nose & eye irritation, dyspnea, cyanosis, peripheral vasodilation, incoordination, loss of equilibrium, anorexia, $\downarrow$ in growth; congestion, hyperemia, edema, occasional pseudomembrane formation in nasal passages, larynx, trachea, & bronchi; congestion, hepatization, hemorrhage, consolidation, emphysema, atelectasis, abscess formation, fibrosis in lungs; clinical signs & gross pathologic changes more severe with 50% RH than with 90% RH; death in >50% after <24 exposures at 20 ppm; 2-nitro-2-butene & 4-nitro-4-nonene less toxic to rats than 3-nitro-3-hexene at 10 ppm	28,70

[1/] Calculated by contributors.

continued

Part II. Gas and Vapor Air Pollutants

	Substance	Animal	Dose	Exposure [Duration]	Effects	Reference
124		34 mice, Rolfsmeyer, 25-35 g	0.23-1.10 ppm	6 hr	Significant, reversible depression in voluntary activity during exposure; no ↓ in effect with increasing C-chain length	80
125	2-Nitro-2-heptene	1 rabbit; 6 rats, Osborne-Mendel	26-308 ppm, 47% & 92% RH	5 hr	No mortality in rabbits at ≯72, & in rats at ≯135 ppm; hydrothorax at 92% RH	26
126	2-Nitro-2-hexene	1 rabbit; 6 rats, Osborne-Mendel	152-515 ppm, 47% & 92% RH	5 hr	No mortality at 152 ppm; hydrothorax at 92% RH	26
127	3-Nitro-3-hexene	4 dogs, Beagle	0.2 ppm	6 hr/da, 5 da/wk [18 mo]	Eye & respiratory tract irritation; no significant hematologic, biochemical (PBI, cholesterol, BUN), or pathologic changes	28
128		4 dogs, Beagle; 200 rats, CFN	1.0 & 2.0 ppm, 60% RH	36-42 mo	Eye & respiratory tract irritation in dogs; no significant hematologic or biochemical (PBI, cholesterol, BUN) changes in dogs; in rats, primary pulmonary adenocarcinoma in 6/100 at 1.0 ppm, and 11/100 at 2.0 ppm; interstitial pneumonitis with atelectasis, abscess formation & emphysema in rats at both conc	28
129		5 guinea pigs, English, 300-400 g	0.74-1.56 ppm	2 hr	Significant ↑ in pulmonary flow resistance & tidal volume; ↓ in respiratory rate; all effects reversible	80
130		4 guinea pigs; 10 mice, Swiss; 2 rabbits, albino; 10 rats, CFN	10-20 ppm, 50% & 90% RH	6 hr/da, 5 da/wk [3 mo]	Similar effects to those observed with 2-nitro-2-butene—see entry 123	28,70
131		14 mice, Rolfsmeyer, 25-35 g	0.32-0.77 ppm	6 hr	Significant, reversible depression of voluntary activity during exposure; no ↓ in effect with increasing C-chain length	80
132		40 mice, Swiss Webster, weanling	0.2 ± 0.07 ppm, 49 & 90% RH	6 hr/da, 5 da/wk [15 mo]	No eye or respiratory tract irritation; 5/40 (4/20 at 49% RH) primary pulmonary adenocarcinoma; bacterial infection considered possible contributing factor	27
133	2-Nitro-2-nonene	4 guinea pigs; 10 mice, Swiss; 2 rabbits, albino; 10 rats, CFN	20 ppm, 50% & 90% RH	6 hr/da, 5 da/wk [3 mo]	No deaths; minimal clinical signs; gross pathological changes similar to those observed with 2-nitro-2-butene—see entry 123; less toxic than 2-nitro-2-butene & 4-nitro-4-nonene at 20 ppm	28
134		1 rabbit; 6 rats, Osborne-Mendel	43-64 ppm, 47% & 92% RH	5 hr	No mortality; hydrothorax at 92% RH	26
135	3-Nitro-3-nonene	1 rabbit; 6 rats, Osborne-Mendel	10-89 ppm, 47% & 92% RH	5 hr	No mortality in rabbits at 89 ppm & in rats at 10 ppm; hydrothorax at 92% RH	26
136	4-Nitro-4-nonene	5 guinea pigs, English, 300-400 g	1.02-4.31 ppm	2 hr	No significant ↑ in pulmonary flow resistance at <1.82 ppm; significant ↑ in tidal volume; ↓ in respiratory rate; all effects reversible	80
137		4 guinea pigs; 10 mice, Swiss; 2 rabbits, albino; 10 rats, CFN	10-20 ppm, 50% & 90% RH	6 hr/da, 5 da/wk [3 mo]	Similar effects to those observed with 2-nitro-2-butene—see entry 123	28

continued

Part II. Gas and Vapor Air Pollutants

	Substance	Animal	Dose	Exposure [Duration]	Effects	Reference
138		19 mice, Rolfs-meyer, 25-35 g	0.41-2.82 ppm	6 hr	Significant, reversible depression in voluntary activity during exposure; no ↓ in effect with increasing C-chain length	80
139	2-Nitro-2-octene	1 rabbit; 6 rats, Osborne-Mendel	44-141 ppm, 47% & 92% RH	5 hr	No mortality at 44 ppm, hydrothorax at 92% RH	26
140	3-Nitro-2-octene	1 rabbit; 6 rats, Osborne-Mendel	19-54 ppm, 47% & 92% RH	5 hr	No mortality at ≯47 ppm; hydrothorax at 92% RH	26
141	3-Nitro-3-octene	1 rabbit; 6 rats, Osborne-Mendel	72-268 ppm, 47% & 92% RH	5 hr	No mortality at 72 ppm; hydrothorax at 92% RH	26
142	2-Nitro-2-pentene	1 rabbit; 6 rats, Osborne-Mendel	55-344 ppm, 47% & 92% RH	5 hr	No mortality at 55 ppm; lung hemorrhages & hydrothorax; more severe hydrothorax at 92% RH	26
143	3-Nitro-3-pentene	1 rabbit; 6 rats, Osborne-Mendel	268-468 ppm, 47% & 92% RH	5 hr	100% mortality; lung hemorrhages & hydrothorax; more severe hydrothorax at 92% RH	26
144	Ozone	14 cats, anesthetized	34.5 ppm (wt)	3 hr	Dose = LC_{50}	75
145		<44 dogs	1 ppm	8, 16, or 24 hr/da [18 mo]	Significant ↑ in monoamine oxidase in cerebral cortex for 8 & 16 hr/da groups only; significant ↓ in monoamine oxidase in cerebral cortex, & in catechol methyltransferase in parietal & occipital cortex at 24 hr/da; no significant change in catecholamines in cerebral cortex, cholinesterase in parietal & occipital cortex, or 5'-nucleotidase & adenosine triphosphatases in parietal cortex at 24 hr/da	121
146		12 guinea pigs, 270-370 g	1 ppm	3 hr	Significant ↑ in lung water content; significant ↓ in total lung histamine content	115
147		15 guinea pigs, random bred, 300-400 g	0.34-0.68 ppm	2 hr	Significant ↑ in respiratory rate & ↓ in tidal volume at all conc; all effects reversible	82
148		14 guinea pigs, random bred, 300-400 g	1.08 or 1.35 ppm	2 hr	Reversible ↑ in pulmonary flow resistance, ↓ in tidal volume; preexposure increased tolerance to edema formation & lethal effects of O_3, but did not affect changes in pulmonary function	82
149		14 guinea pigs, 300 g; hamsters, 57 g	1.06(0.75-1.24) ppm	6 hr/da [268 da]	Chronic bronchiolitis, bronchiolar wall fibrosis, emphysema	108
150		63 guinea pigs, U. of Chicago, 2-3 mo	51.7 ppm (wt)	3 hr	Dose = LC_{50}; lung hemorrhages at ≮6 ppm for 18 hr	75
151		Hamsters	1 ppm	6 hr	Concurrent exposure to O_3 & exercise fatal to rats & mice, but not to hamsters at otherwise lethal levels	107
152		Hamsters	10.5 ppm	4 hr	Dose = LC_{50}	106,116
153		Mice	1 ppm	4 hr	Mild pulmonary edema with leukocytic infiltrate at 20 hr after exposure	100
154			1.56(0.53-1.86) ppm	∼3 wk	Significant ↓ in spontaneous activity	10

continued

Part II. Gas and Vapor Air Pollutants

	Substance	Animal	Dose	Exposure [Duration]	Effects	Reference
155			3.2 ppm	4 hr	Acute pulmonary edema at 4 hr with recovery apparent at $>$20 hr after exposure; leukocytic infiltration of alveoli with maximum effect at $\sim$20 hr; desquamation of bronchiolar epithelium corresponding to degree of edema	100
156		10 mice, 22 g	1.06(0.75-1.24) ppm	6 hr/da [268 da]	Chronic bronchitis, bronchiolar wall fibrosis, mild emphysema	108
157		18 mice, young, 19 g	4 ppm	4 hr	61% mortality	106
158		18 mice, old, 35 g	4 ppm	4 hr	5.6% mortality	106
159		39 mice, 25-35 g	0.2-0.5 ppm	6 hr	$\downarrow$ in voluntary activity during exposure	82
160		20 mice, 30-35 g	5 ppm	4 hr	20% $\uparrow$ in lung wet wt 2-4 hr after exposure; no change in lung dry wt; 22% $\uparrow$ in wet wt & 25% $\uparrow$ in dry wt in spleen; no wt change in liver	100
161		Mice, albino, 22 g	3.6-4.1 ppm	4 hr	LC_{50} = 3.8 ppm	106
162		22 mice, albino, 22-26 g	8 ppm	4 hr	Significant $\uparrow$ in RBC acetylcholinesterase; no significant change in RBC glutathione	50
163		Mice, Hamilton, 17-32 g	0.1-4.4 ppm followed by 8.0-9.2 ppm challenge dose after 4 da	4 hr	Preexposure to $\not<$0.3 ppm protects against mortality & edema formation; degree of protection dose-related; marked protection from 4-102 da after $\sim$2 ppm	72
164		32 mice, IVAN-NMRI, 4 da-8 wk	0.65-1.3 ppm	6-7 hr/da [1-2 da]	Partial disruption of endothelial lining cells in small alveolar capillaries with lamellar inclusions in alveolar corner cells with electron microscopy after 7-hr exposure; progressive endothelial disruption extending into interstitium with associated edema-like foci, & focal fragmentation of basement membrane & epithelium; degeneration of mitochondria, endoplasmic reticulum & inclusions in alveolar corner cells after 13-hr exposure; severity of lesions greater in 4-da- to 1-mo-old mice	7
165		50 mice, Swiss	4.6 ppm	4 hr	Dose = LC_{50}	116
166		62 mice, Swiss Webster, 18-25 g	3-10 ppm	4 hr	EC_{50} for production of pulmonary edema = 3.66 ppm	119
167		$\sim$40 mice, Swiss Webster, 18-25 g[8]	1.5 ppm	4 hr/da, 5 da/wk [2 mo]	No histopathologic upper respiratory tract or lung lesions with O_3 alone; no significant histopathologic changes in mice with induced tuberculosis	119
168		32 mice, Swiss Webster, 18-20 mo (received i.p. injection of 100-300 μCi ^{3}H-thymidine 1 hr prior to sacrifice)	0.5 & 1.2 ppm	6 hr	9.4% mortality; significant, reversible $\downarrow$ in number of ^{3}H-labeled peripheral alveolar cells at 1-72 hr; no histopathologic pulmonary changes	38
169		32 mice, Swiss Webster, 18-20 mo (received i.p. injection of 100-300 μCi ^{3}H-thymidine 1 hr prior to sacrifice)	2.5 & 3.5 ppm	6 hr	28% mortality; significant, reversible $\downarrow$ in number of ^{3}H-labeled peripheral alveolar cells at 1-72 hr; pulmonary edema at $\not>$48 hr & alveolar cell proliferation at 48 & 72 hr	38

[8] $\sim$20/40 received i.v. injection of 0.20 mg H, 37 Rv *Mycobacterium tuberculosis*.

continued

Part II. Gas and Vapor Air Pollutants

	Substance	Animal	Dose	Exposure [Duration]	Effects	Reference
170		220 mice, Swiss Webster, 25 g (100/220 received daily s.c. injections of 0.2 mg desmosterol)	4.5 ppm	2 hr/da, every 3 da [≥75 da]	Survivors: pulmonary adenomas, 6.6%; with demosterol, 8.0%; epithelial alterations, including hyperplasia, squamous metaplasia, & adenoma, 23%; with demosterol, 38%; mild to severe inflammatory changes, including acute bronchitis, bronchiolitis, & bronchopneumonia, 91%; with demosterol, 82%	127
171		Mice; rats, 150-400 g	1 ppm	6 hr	Concurrent exposure to O_3 & exercise (15 min/hr) fatal at otherwise non-lethal levels; challenge dose demonstrated tolerance lasting 4-6 wk, depending on duration of prior exposure	107
172		8 rabbits	10 ppm	1 hr/wk [6 wk]	↑ in precipitin titers after 3-4 wk; maximum effect 10 da after final exposure. O_3 reacts with tissue protein to form an antigenic structure which gives rise to antibodies.	100
173		4 rabbits, 3 mo	2 ppm	4 hr/da [25 da]	No ophthalmologic or biochemical effects; no effect on 6-phosphogluconate dehydrogenase, malate dehydrogenase, glucose-6-phosphate dehydrogenase, & LDH	59,74
174		5 rabbits, 6 mo-2 yr	10 ppm	1 hr	Marked, reversible ↑ in Mg-activated serum AP at ≥24 hr after exposure & at 3-6 da	100
175		8 rabbits, 1.2-1.7 kg	8-45 ppm increased with time	1 hr/wk [49 wk]	Shallow respiration with marked, reversible ↓ in O_2 consumption & tidal volume; return to preexposure levels at ≤2 da after first exposure, & at 5 hr after tenth exposure; no edema tolerance demonstrated; ↓ in functional compensation at ≤4 mo corresponding to appearance of bronchiolar fibrosis	100
176		96 rabbits, albino, 1-2 kg	1-7 ppm	3 hr	Significant ↑ in % heterophilic leukocytes in total cells recovered following pulmonary lavage, with maximum effect at 3 ppm & persistent effect (to 24 hr) at 5 ppm; ↓ in % alveolar macrophages	18
177		6 rabbits/group, albino, 1-2 kg[9]	0.67-9.5 ppm	3 hr	Significant ↓ in phagocytic activity of alveolar macrophages recovered by pulmonary lavage with maximum effect at 4 ppm	18
178		20 rabbits, Flemish giant, 5.5-6.4 kg (unilateral pulmonary exposure by means of bronchial catheterization)	12 ± 1 ppm	3.5-4.5 hr	Significant ↑ in percent wet wt of exposed lung	4
179		7 rabbits, Flemish giant, 5.5-6.4 kg (unilateral pulmonary exposure by means of bronchial catheterization)	12 ± 1 ppm	6 ± 1 hr	100% mortality; bilateral ↑ in respiratory rate after 0.5-3.5 exposure hr; ↑ in pulmonary flow resistance after 1 hr; bilateral ↓ in dynamic compliance; bilateral ↓ in tidal volume & ↑ in minute volume to 3.5 hr followed by ↓ in both volumes; ↓ in inspiratory & expiratory flow rates after 3.5 hr; all effects unilateral unless otherwise noted	4

[9] Received i.t. injection of 1 ml concentrated culture of *Streptococcus* (Lancefield C) 30 min prior to sacrifice.

continued

Part II. Gas and Vapor Air Pollutants

	Substance	Animal	Dose	Exposure [Duration]	Effects	Reference
180		6 rabbits, New Zealand, ∿1 kg	0.4 ppm	6 hr/da, 5 da/wk [6 mo]	Significant, reversible ↑ in serum trypsin inhibitor capacity after first 6-hr exposure	88
181		6 rabbits, New Zealand, ∿1 kg	0.4 ppm	6 hr/da, 5 da/wk [10 mo]	Significant ↑ in total serum trypsin-protein esterase at 165 da; chronic pulmonary inflammation, emphysema with thickened intra-alveolar septa, & thickening of walls of small pulmonary arteries	89,90
182		6 rabbits, New Zealand, 2.5-3.5 lb	0.5 or 2.0 ppm	8 hr/da [7 da]	↑ in % polymorphonuclear leukocytes & ↓ in absolute number of alveolar macrophages in total cells recovered following pulmonary lavage; significant ↑ in osmotic fragility of macrophages at 2 ppm only; no detectable lipid peroxidation in macrophages	33
183		8 rabbits, New Zealand, 1-3 kg	0.75 ppm	3 hr	Significant, reversible ↓ (27%) in benzpyrene hydroxylase in tracheobronchial mucosa at 0.5-24 hr	87
184		38 rabbits, New Zealand, 3 mo	36 ppm (wt)	3 hr	Dose = LC_{50}	75
185		75 rats	2 ppm	3 hr	Marked, reversible ↓ in O_2 consumption & minute volume at 8-12 hr after exposure, & in tidal volume at 8 to >12 hr corresponding to time of maximum edema; moderate ↓ in O_2 consumption & ↑ in tidal & minute volumes at <50 hr	100
186		45 rats, 144-380 g	1 ppm	6 hr/da [200 da]	Mortality: no exercise, 0; with exercise (15 min/hr), 55%; death from pulmonary edema & hemorrhage; three factors tend to augment injurious response: youth, drinking water with 10% C_2H_5OH, & respiratory infection	106
187		10 rats, 150 g	9.2 ppm	45 min	Marked, reversible ↑ in lung RNA content at 1-2 hr & 6-24 hr after exposure; reversible ↑ in lung DNA at 4-6 hr; no change in lung RNA:DNA ratio; pulmonary edema & acute inflammatory reaction at <2 hr. Slight, reversible ↑ in liver RNA at 1 hr; reversible ↓ in liver RNA & DNA at 6 hr & 2-4 hr, respectively; marked, reversible ↑ in liver RNA:DNA ratio at 1-4 hr; no pathological liver changes.	100
188		8 rats, 200-400 g	6 ppm	4 hr	Mortality in rats: no preexposure, 75%; with preexposure (1.7 ppm × 4 hr), 25%. ↓ in lung AP & 5'-nucleotidase in both groups with greater AP ↓ in rats without preexposure.	100
189		30 rats, albino, 80-100 g	0.021 mg/m³ (0.011 ppm[1/])	24 hr/da [93 da]	No toxic effects	35
190			0.11 mg/m³ (0.056 ppm[1/])	24 hr/da [93 da]	Significant ↓ in body wt at <3 wk & O_2 consumption at <6 wk; reversible ↓ in blood cholinesterase at 2.5-3.5 mo; significant ↑ in urinary 17-ketosteroids at <6 wk; significant, reversible ↓ in adrenal ascorbic acid content at 3-3.5 mo	35
191		Rats, albino, 250 g	3.6-6.4 ppm	4 hr	LC_{50} = 4.8 ppm	106

[1/] Calculated by contributors.

continued

Part II.   Gas and Vapor Air Pollutants

	Substance	Animal	Dose	Exposure [Duration]	Effects	Reference
192		16 rats, CFN, 200-395 g	33 ppm	1 hr	No significant change in ratio of reduced to oxidized NADP (nadide phosphate) in tracheal epithelium; no histopathologic tracheal lesions	83,84
193		30 rats, Charles River-derived, 145 g[10/]	15 ppm	8 hr	Significant ↑ in survival time of PABA-treated rats	51
194		24 rats, Sprague-Dawley, weanling[11/]	0.5 ppm	24 hr/da [2-6 wk]	No respiratory distress or body wt loss; 4% mortality, pulmonary edema & ↑ in % wet lung wt in vitamin E-depleted rats at 6 wk; significant ↑ in arachidonic & palmitic acids & significant ↓ in oleic & linoleic acids in lung tissue lipids in vitamin E-depleted rats at 6 wk; little change in lung lavage lipids	95
195		Rats, Sprague-Dawley, 2 mo[12/]	0.7 ± 0.15 ppm	24 hr/da [5 da]	Significant ↑ in cathepsins A, B, & D, & in β-N-acetylglucosaminidase in lung fractions; no significant change in cathepsin C; no significant difference in specific enzyme activities in vitamin E-deficient & supplemented rats	30
196		Rats, Sprague-Dawley, 2 mo[13/]	0.79 ± 0.14 ppm	24 hr/da [7 da]	Significant ↑ in acid phosphatase, β-N-acetylglucosaminidase, cathepsins C & D, & BANA-amidohydrolase in lung fractions; ↑ in specific enzyme activities greater than that observed at 0.7 ± 0.15 ppm— see entry 195; no significant difference between vitamin-E-deficient & supplemented rats	30
197		15 rats, Sprague-Dawley, 10 wk	0.8-1.5 ppm	6 hr/da, 4 da/wk [130 da]	Significant ↑ in urine pH & significant ↓ in titratable acidity at ≮91-98 da; no significant changes in creatinine, uric acid, or amino acid excretion	55
198		24 rats, Sprague-Dawley, weanling[11/]	1.0 ppm	24 hr/da [8.2-18.5 da]	LC$_{50}$ = 1.0 ppm × 8.2 da for vitamin E-depleted rats & 1.0 ppm × 18.5 da for supplemented rats; severe respiratory distress & significant ↑ in arachidonate at 9 da in lung tissue lipids in both groups	95
199		6-10 rats, Wistar, 2-3 mo	2.4 ppm	4 hr/da, 5 da/wk [32, 64, & 160 hr]	No pulmonary lesions at 32 hr; 10 & 20% slight pulmonary hemorrhage & edema at 64 hr; 10% slight pulmonary edema at 160 hr	76
200				6 hr/da, 4 da/wk [32 hr]	100% slight pulmonary edema; 10% slight or moderate, & 80% severe pulmonary hemorrhage	76
201				6 hr/da, 4 da/wk [64 hr]	100% slight pulmonary edema; 40% slight & 30% moderate pulmonary hemorrhage	76
202				6 hr/da, 4 da/wk [176 hr]	100% slight pulmonary edema; 67% slight pulmonary hemorrhage	76

[10/] 15/30 received intraperitoneal injection of 2 ml 20 mM p-aminobenzoic acid for 3 da prior to exposure. [11/] 12/24 fed vitamin E-free diet, or diet containing 100 mg dl-α-tocopherol acetate/kg diet for 4 wk prior to exposure. [12/] 1/5 groups received vitamin E-deficient diet; 4/5 groups received diet containing 10.5, 45, 150 or 1500 mg dl-α-tocopherol acetate/kg diet for 4 wk prior to exposure. [13/] 1/3 groups received vitamin E-deficient diet; 2/3 groups received diet containing 10.5 or 45 mg dl-α-tocopherol acetate/kg diet for 4 wk prior to exposure.

continued

Part II. Gas and Vapor Air Pollutants

	Substance	Animal	Dose	Exposure [Duration]	Effects	Reference
203		Rats, Wistar, adult	1 ppm	6 hr	Marked tolerance to subsequent exposure to O_3; tolerance lasted >4.5 wk	116
204		55 rats, Wistar, adult	2.5-12.6 ppm	4 hr	LC_{50} = 5.5 ppm; pulmonary edema & hemorrhage, congested liver, dark adrenals	116
205	& nitrogen oxide measured as NO_2	126 mice	<0.5, 1, 2.5, 4, 6, & 8 ppm O_3 & <0.2, 1-2, 4-5, 10, 17, 35, & 500 ppm NO_x	4 hr	No ↑ in degree of mortality in mice exposed to increasing conc of both gases, compared to O_3 alone	106
206		36 rats, 200-250 g	6(1.2-10.7) ppm O_3 & 8(1.8-27.0) ppm NO_2	4 hr	Mild respiratory distress at lower conc. Acute gasping dyspnea; pulmonary edema, emphysema, & some congestion of alveolar capillaries but no hemorrhage at higher conc.	29
207	Peroxyacetyl nitrate	54 mice, C57BL, 75-149 g	2.8-8.6 ppm	6 hr	6- & 24-hr activity$_{50}$ (conc required to produce 50% ↓ in voluntary running activity) = 4.5 & 4.0 ppm, respectively; effects reversible in 2-4 da	15
208		177 mice, A-strain, 98-114 da	97-145 ppm	2 hr	LC_{50} (4 wk) = 106 ppm; delayed mortality at lower doses; ↑ in susceptibility with age & higher exposure temp	14
209	Sulfur dioxide	20 cats, adult, 2-5 kg (tracheotomized & artificially ventilated; SO_2 administered via tracheal cannula)	19 ± 5.9 ppm	30 min	10% significant ↑ in pulmonary flow resistance; 30% significant ↓ in lung compliance; all effects reversible; no morphological changes in airway caliber	21
210		8 dogs, 17-30 kg (tracheotomized & artificially ventilated; SO_2 administered via tracheal cannula)	0.01-1.0% = 100-10,000 ppm	0.1-6 min	Bronchoconstriction limited to exposed lobe; bronchoconstriction associated with bronchial arterial vasodilatation; response unaffected by acute or chronic denervation	17
211		Dogs, anesthetized, 9-25 kg	200-850 ppm	1-4 min/da [<2 da]	Pulmonary vasoconstriction; bronchoconstriction, preceded & followed by bronchodilatation; ↑ in pulmonary arterial blood pressure; ↓ in myocardial force of contraction accompanied by bradycardia; systemic shock; 86.02% SO_2 absorbed	98
212		8 dogs, Beagle, 7-17 kg (4/8 with pulmonary impairment from prior NO_2 exposure of 26 ppm × 191 da)	5.1 ± 0.4 ppm	21 hr/da [620 da]	Significant ↑ in single-breath N_2 washout in unimpaired dogs; no body wt or hematologic changes; nonspecific histopathologic pulmonary changes	67
213		Guinea pigs	1-100 ppm	10 min	↑ (15-80%) in pulmonary flow resistance	41
214			34 ppm	1 hr	Reversible ↑ in pulmonary flow resistance, with maximum effect at 20 min; reversible ↓ in dynamic lung compliance; effects not potentiated by propanolol but reduced by atropine & tracheostomy	32
215		Guinea pigs (exposed to aerosol of killed, radioactive ^{35}S, & viable *Escherichia coli* for 10 min	10.4 ppm	6 hr/da, 5 da/wk [4 wk]	No significant change in clearance rates of viable & radioactive bacteria at 3 hr after exposure	97

continued

984

Part II. Gas and Vapor Air Pollutants

	Substance	Animal	Dose	Exposure [Duration]	Effects	Reference
216		10 guinea pigs	5 ppm	20 min	Reversible ↑ in pulmonary flow resistance; more rapid ↓ in inspiratory than expiratory resistance during recovery	117
217		8 guinea pigs	112 ppm	113 hr	50% mortality at 54-113 hr	124,125
218		Guinea pigs, 250 g (intact & tracheotomized)	5 ppm	1 hr	Qualitatively similar results to those described for acrolein aerosol—see entry 21	25
219		12 guinea pigs, 0.55-1.2 kg	6.8-17.3 ppm	1 hr	Reversible ↓ (<5-38%) in tidal volume in 10/12; dose-related ↑ in Hb; no effect on blood inorganic sulfur conc	64
220		6 guinea pigs, 270-370 g	50 ppm	3 hr	No significant effect on lung water or histamine content	115
221		120 guinea pigs, Hartley, 180 ± 50 g	5.72 ppm	22 hr/da, 7 da/wk [52 wk]	No effects on survival rate or pulmonary function; no hematologic or biochemical changes; ↓ in spontaneous pulmonary & tracheal lesions; increased cytoplasmic vacuolization in liver	2
222		36 hamsters, Syrian Golden (18/36 with experimental emphysema from aerosol of 3% papain for 2-4 hr, 17-20 da prior to exposure)	650 ppm	4 hr/da, 5 da/wk [19-74 da]	No significant changes in lung compliance, recoil (at 20-60% volumes) or resistance in impaired or unimpaired animals; bronchial dilatation, dysplasia of bronchial epithelium, & loss of cilia & hypersecretion in trachea in both groups; dilatation of alveolar ducts & focal emphysema in unimpaired animals	49
223		12 mice	109 ppm	238 hr	Lethargy, rhinitis, conjunctivitis, dyspnea	124,125
224		6 mice/group, BALB/c, 3 mo (exposed to airborne, killed *E. coli* for 60 min)	2.01 ± 0.21 ppm	∼100 hr/wk [102-192 da]	Reversible ↑ in number of antibody-forming cells in mediastinal lymph nodes at 102-192 da; hyperplasia of mediastinal nodes at 102 da; significant, reversible ↑ in number of antibody-forming cells in spleen & in agglutinin titers in serum at 135-192 da	130
225		Mice, Ha/ICR, 40-44 da (1/2 groups with experimental, mild upper respiratory tract, bacterial infection)	9.6-11.2 ppm	24 hr/da [≯3 da]	Moderate rhinitis with focal necrosis & desquamation of respiratory epithelium in nasomaxillary turbinates in mice without infection at 48-72 hr; severe necrotizing rhinitis with loss of cilia & ulceration of respiratory epithelium in mice with infection at 24-72 hr; minimal tracheal & pulmonary lesions in both groups	45
226		30 mice, Swiss Webster, 27-33 g[14/]	19.3-27 ppm	∼24 hr/da [≯7 da]	Significant ↑ (∼12%) in amounts of pneumonia at ≮19.3 ppm; mild congestion & thickening of alveolar walls after 7 da at 27 ppm	40
227		9 monkeys, cynamolgus, 1.6-3.0 kg	1.28 ± 0.42 ppm	24 hr/da [78 wk]	No deaths; no significant changes in body wt, pulmonary function, arterial O_2 & CO_2 tension, or arterial pH; no hematologic & biochemical effects, or pulmonary lesions; minimal submucosal lymphoid hyperplasia in trachea in 2/9	3

[14/] Experimentally infected with aerosol of 100 MID_{50} A2/Japan 305/57 influenza virus prior to exposure.

continued

Part II. Gas and Vapor Air Pollutants

	Substance	Animal	Dose	Exposure [Duration]	Effects	Reference
228			10.0 ppm	22 hr/da, 7 da/wk [104 wk]	No effects on body wt, or on dynamic lung compliance, N_2 washout or CO diffusion capacity; no hematologic or biochemical changes; no histopathologic lesions in lungs or trachea	13
229		20 rabbits	74-239 ppm	45 min	Significant ↓ in tracheal ciliary velocity	24
230		Rabbits, 3-4 mo	6 ppm	4 hr/da [32 da]	No ophthalmologic or biochemical effects; no effect on 6-phosphogluconate dehydrogenase, malate dehydrogenase, glucose-6-phosphate dehydrogenase, or LDH	59,74
231			10 ppm	4 hr		
232		4 rabbits, New Zealand, 3-4 kg	23.5 ± 5 ppm	14 or 62 hr	Marked ↑ in exogenous cyanolytic sulfite in plasma after 14 exposure hr; ↑ greater after 62 exposure hr; decay to one-half immediate post-exposure values after 5-30 hr; cyanolytic sulfite conc less in corresponding serum sample; no change in plasma free sulfite	53
233		15 rats	0.1 mg/m^3 (0.04 ppm[1])	96 da	Significant, reversible ↓ in latent period of unconditioned defensive reflex at 3 mo; reversible EEG changes in response to rhythmic illumination at 3-5 mo; splenic congestion at 3-6.5 mo	37
234		15 rats	0.5 mg/m^3 (0.19 ppm[1])	96 da	Significant, reversible ↓ in RBC & blood nucleic acids, & urinary coproporphyrin & vitamin B_1 & B_2 after >1.5 mo; significant, reversible ↑ in WBC with altered fluorescence after 0.5 mo, in urinary ascorbic acid (vitamin C) after >1.5 mo, & in latent period of unconditioned defensive reflex after 2 mo; EEG changes in response to rhythmic illumination at 3 mo with partial recovery at 5 mo; inflammatory changes in trachea, bronchi, & lungs; slight dystrophic changes in liver, heart, & kidneys; splenic congestion; reversible changes in large neurons of pons, medulla oblongata, & cerebellum	37
235		9 rats	10 ppm	6 hr/da [10 wk]	Hypertrophy & hypersecretion of tracheal mucosa; marked ↓ in mucus flow rate; no change in ciliary frequency; ↑ in ciliary density in tracheal mucosa, & severe edema, vascularization & collagen degeneration of lamina propria with light & electron microscopy at 10 & 14 wk	23
236		10 rats, albino	0.15 mg/m^3 (0.06 ppm[1])	72 da	↑ in litter size of exposed ♀	71
237		10 rats, albino	4 mg/m^3 (1.53 ppm[1])	72 da	Estrous cycle changes in exposed ♀ & progeny of first generation of exposed ♀; significant changes in % fertilization of exposed ♀, duration of pregnancy, & wt gain of progeny; ↑ in litter size	71
238	& carbon, amorphous	24 guinea pigs (exposed to aerosol of killed, radioactive ^{35}S, & viable *Escherichia coli* for 10 min)	10.4 ppm SO$_2$ & 15 mg carbon/m^3	6 hr/da, 5 da/wk [4 wk]	12.5% mortality after two exposure wk; localized pulmonary infections in survivors; significant ↓ in clearance rates of viable & radioactive bacteria 3 hr after exposure—*see* entries 116 in Part I & 215 (carbon particle size, 0.5-2.8 μ)	97

[1] Calculated by contributors.

continued

Part II. Gas and Vapor Air Pollutants

	Substance	Animal	Dose	Exposure [Duration]	Effects	Reference
239		6 mice/group, BALB/c, 3 mo (exposed to airborne, killed *E. coli* for 60 min)	2.02 ± 0.18 ppm SO_2 & 554 ± 142 μg carbon/m^3	$\sim$100 hr/wk [102-192 da]	Reversible ↑ in number of antibody-forming cells in mediastinal lymph nodes at 102-192 da; reversible hyperplasia of mediastinal nodes at 102-192 da; ↓ in number of antibody-forming cells in serum & spleen at ⋖102 da; effects on immunosuppression in serum & spleen at 192 da more pronounced than with SO_2 alone—*see* entries 224, & in Part I entry 117 (carbon particle size, 1.8-2.2 μ)	130
240		10 rabbits	116-121 ppm SO_2 & 2.5-2.9 mg carbon/m^3	45 min	Significant ↓ in tracheal ciliary velocity; effect similar to that observed with SO_2 alone	24
241	& carbon monoxide	10 rats, albino	4 mg SO_2/m^3 (1.53 ppm[1]/) & 2 mg CO/m^3 (1.75 ppm[1]/)	72 da	More severe estrous cycle changes in exposed ♀ & progeny of first generation of exposed ♀ than with SO_2 or CO alone; significant changes in % fertilization of exposed ♀, duration of pregnancy, & wt gain of progeny; ↑ in number of stillbirths, litter size, & pituitary function; inflammatory changes & ↓ in number of primordial follicles in ovaries—*see* entries 36 & 237	71
242	& fly ash	100 guinea pigs, Hartley; 9 monkeys, cynamolgus	5.0 ppm SO_2 & 0.5 mg fly ash/m^3	22 hr/da, 7 da/wk [52-78 wk]	No effects on survival, body wt, or pulmonary function; no hematologic or biochemical changes; no histopathologic changes attributable to exposure	61
243	& graphite, synthetic	Rats, Sprague-Dawley	1.0 ppm SO_2 & 1.0 mg graphite/m^3	12 hr/da, 7 da/wk [119 da]	No change in ciliary frequency of tracheal mucosa at ⋗119 da (graphite particle size, 1.5 μ)	42
244			3.0 ppm SO_2 & 1.0 mg graphite/m^3	12 hr/da, 7 da/wk [109 da]	No significant change in ratio of dust-laden cells to total number of alveolar cells at 56 & 109 da with SO_2 & graphite compared to graphite alone (graphite particle size, 1.5 μ)	42
	& inorganic salts, insoluble					
245	& calcium sulfate	9 monkeys, cynamolgus, 1.6-3.0 kg	10.0 ppm SO_2 & 10.0 mg CaSO$_4$/m^3	22 hr/da, 7 da/wk [104 wk]	No effects on body wt, or dynamic lung compliance, N$_2$ washout or CO diffusion capacity; no hematologic or biochemical changes; no histopathologic lesions in lungs or trachea (CaSO$_4$ particle size, 1.28 μ)	13
246	& ferric oxide fumes	11 guinea pigs	$\sim$2.5 ppm SO_2 & $\sim$8-10 mg Fe$_2$O$_3$/m^3	1 hr	↑ ($\sim$10%) in pulmonary flow resistance; no potentiation of response to SO_2 alone (Fe$_2$O$_3$ particle size, 0.07 μ)	5
247	& manganese dioxide	10 guinea pigs	$\sim$2.5 ppm SO_2 & $\sim$8-10 mg MnO$_2$/m^3	1 hr	↑ ($\sim$10%) in pulmonary flow resistance; no potentiation of response to SO_2 alone (MnO$_2$ particle size, ⋗5 μ)	5
	& inorganic salts, soluble					
248	& ammonium thiocyanate	15 guinea pigs	1.9 ppm SO_2 & 10 mg NH$_4$SCN/m^3 in H$_2$O	1 hr	↑ ($\sim$105%) in pulmonary flow resistance; potentiation ratio (response to SO_2 & aerosol/SO_2 alone) at 1 hr = 5.0 (NH$_4$SCN particle size, ⋗0.1 μ)	5

[1]/ Calculated by contributors.

continued

Part II. Gas and Vapor Air Pollutants

	Substance	Animal	Dose	Exposure [Duration]	Effects	Reference
249	& ferrous sulfate	20 guinea pigs	$\sim$1.5 ppm SO_2 & 1 ± 0.12 mg Fe/m^3 in H_2O	1 hr	↑ ($\sim$55%) in pulmonary flow resistance; potentiation evident after 10 min (Fe particle size, 0.07 μ)	5
250	& manganese chloride	18 guinea pigs	1 to $\sim$4.5 ppm SO_2 & 1 ± 0.15 mg Mn/m^3 in H_2O	1 hr	↑ ($\sim$50%) in pulmonary flow resistance; potentiation evident after 10 min (Mn particle size, 0.09 μ)	5
251	& potassium chloride	16 guinea pigs	2.3 ppm SO_2 & 10 mg KCl/m^3 in H_2O	1 hr	↑ ($\sim$70%) in pulmonary flow resistance; potentiation ratio at 1 hr = 3.3 (KCl particle size, ≯0.1 μ)	5
252	& sodium chloride	20 cats, adult, 2-5 kg (tracheotomized & artificially ventilated; SO_2 administered via tracheal cannula)	17.9 ± 8.3 ppm SO_2 & 10 ± 0.2 mg $NaCl/m^3$	0.5 hr	No significant changes in pulmonary flow resistance; 15% significant, reversible ↓ in lung compliance; no morphological changes in airway caliber—see entry 209 (NaCl particle size, 0.25 μ)	21
253		10 guinea pigs	2 ppm SO_2 & 10 mg $NaCl/m^3$ in H_2O	1 hr	↑ ($\sim$50%) in pulmonary flow resistance; potentiation ratio at 1 hr = 2.4; no potentiation after 10 min; resistance remained elevated 5 hr after exposure with 10 mg $NaCl/m^3$ × 0.5 hr (NaCl particle size, ≯0.1 μ)	5
254		59 hamsters, Syrian Golden, 25-49 da (39/59 received single i.t. injection of influenza D virus, strain PR8, prior to exposure or after 76 exposure da)	650 ppm SO_2	3 hr/da, 5 da/wk [≯10 mo]	Dilatation of bronchi & alveolar ducts, & slight focal emphysema after 7 exposures; slight dysplasia of bronchial epithelium, congestion & atelectasis after 29-32 exposures; desquamation of tracheal epithelium, possible hyperplasia of bronchial epithelium & focal alveolar bronchiolization after 42-63 exposures; no additive histopathology with SO_2 & virus (NaCl particle size, 0.0004 μ)	48
255	& sodium ortho-vanadate	23 guinea pigs	$\sim$2 ppm SO_2 & 0.7 ± 0.08 mg V/m^3 in H_2O	1 hr	↑ ($\sim$60%) in pulmonary flow resistance; potentiation evident after 10 min (V particle size, 0.05 μ)	5
256	& sulfuric acid aerosol	11 dogs, Beagle	0.14 mg SO_2/m^3 (0.05 ppm [L/]) & 0.10 mg H_2SO_4/m^3	16 hr/da [5 yr]	9% significant ↑ in systemic arterial pressure at 5 yr; 9% left axis deviation & mid-precordial elevated P-waves at 5 yr, & persistent ST-segment elevations at 4.5-5 yr by static ECG; 18% premature ventricular contractions at 5 yr by exercise ECG; 9% possible mild left ventricular hypertrophy at 4.5-5 yr by VCG	9
257		12 dogs, Beagle	0.5 ppm SO_2 & 100 μg H_2SO_4/m^3	16 hr/da, 7 da/wk [18 mo]	No significant effects on total pulmonary flow resistance, dynamic lung compliance, or single-breath CO diffusion capacity at 18 mo	123
258		8 dogs, Beagle, 7-17 kg (4/8 with pulmonary impairment from prior NO_2 exposure of 26 ppm × 191 da)	5.1 ± 0.4 ppm SO_2 & 904 ± 288 μg H_2SO_4/m^3	21 hr/da [620 da]	High N_2 washout values in unimpaired dogs; otherwise, results similar to those observed with H_2SO_4 alone—see entry 164 in Part I (H_2SO_4 particle size, 90% <0.5 μ)	67

[L/] Calculated by contributors.

continued

Part II. Gas and Vapor Air Pollutants

	Substance	Animal	Dose	Exposure [Duration]	Effects	Reference
259	^{35}S-sulfur dioxide	Dogs, tracheotomized	1-150 ppm	20-40 min	↑ (50-125%) in pulmonary flow resistance after 10 s of exposure; no change in lung compliance	6

Contributors: Palm, Paul E., Nick, M. Susan, Arnold, Elsie P., and Platz, Barbara B.

References

[1] Acton, J. D., and Q. N. Myrvik. 1972. Arch. Environ. Health 24(1):48.

[2] Alarie, Y., et al. 1970. Ibid. 21(6):769.

[3] Alarie, Y., et al. 1972. Ibid. 24(2):115.

[4] Alpert, S. M., and T. R. Lewis. 1971. Ibid. 23(6):451.

[5] Amdur, M. O., and D. Underhill. 1968. Ibid. 16:460.

[6] Balchum, O. J., et al. 1959. Fed. Proc. Fed. Amer. Soc. Exp. Biol. 18:20.

[7] Bils, R. F. 1970. Arch. Environ. Health 20(4):468.

[8] Blair, W. H., et al. 1969. Ibid. 18(2):186.

[9] Bloch, W. N., Jr., et al. 1972. Ibid. 24(5):342.

[10] Boche, R. D., and J. J. Quilligan, Jr. 1960. Science 131:1733.

[11] Buckley, R. D., and O. J. Balchum. 1967. Arch. Environ. Health 14(3):424.

[12] Buckley, R. D., and O. J. Balchum. 1967. Ibid. 14(5):687.

[13] Busey, W. M., et al. 1971. Hazleton Lab. (Vienna, Va.) Rep. 71-29.

[14] Campbell, K. I., et al. 1967. Arch. Environ. Health 15:739.

[15] Campbell, K. I., et al. 1970. Ibid. 20(1):22.

[16] Carson, T. R., et al. 1962. Amer. Ind. Hyg. Ass. J. 23:457.

[17] Cho, Y. W., et al. 1968. Arch. Environ. Health 16(5):651.

[18] Coffin, D. L., et al. 1968. Ibid. 16(5):633.

[19] Colmant, H. J. 1972. Staub 32(4):32.

[20] Coon, R. A., et al. 1970. Toxicol. Appl. Pharmacol. 16:646.

[21] Corn, M., et al. 1972. Arch. Environ. Health 24(4):248.

[22] Dalhamn, T., and L. Reid. 1967. In C. N. Davies, ed. Inhaled Particles and Vapours. Pergamon Press, London. v. 2.

[23] Dalhamn, T., and J. Rhodin. 1956. Brit. J. Ind. Med. 13:110.

[24] Dalhamn, T., and L. Strandberg. 1963. Int. J. Air Water Pollut. 7:517.

[25] Davis, T. R. A., et al. 1967. Arch. Environ. Health 15(4):412.

[26] Deichmann, W. B., et al. 1958. Arch. Ind. Health 18:312.

[27] Deichmann, W. B., et al. 1963. Toxicol. Appl. Pharmacol. 5:445.

[28] Deichmann, W. B., et al. 1965. Ind. Med. Surg. 34(10):800.

[29] Diggle, W. M., and J. C. Gage. 1955. Brit. J. Ind. Med. 12:60.

[30] Dillard, C. J., et al. 1972. Arch. Environ. Health 25(6):426.

[31] Douglas, C. G., and J. S. Haldane. 1912. J. Physiol. (London) 44:275.

[32] Douglas, J. S., et al. 1969. Arch. Environ. Health 18(4):627.

[33] Dowell, A. R., et al. 1970. Ibid. 21(2):121.

[34] Eckardt, R. E., et al. 1972. Ibid. 25(6):381.

[35] Eglite, M. E. 1968. Gig. Sanit. 31(1-3):18.

[36] Ehrlich, R., and M. C. Henry. 1968. Arch. Environ. Health 17(6):860.

[37] Elfimova, E. V., and M. I. Gusev. 1969. Gig. Sanit. 34(1-3):161.

[38] Evans, M. J., et al. 1971. Arch. Environ. Health 22(4):450.

[39] Evans, M. J., et al. 1972. Ibid. 24(3):180.

[40] Fairchild, G. A., et al. 1972. Ibid. 25(3):174.

[41] Frank, N. R., et al. 1962. J. Appl. Physiol. 17:252.

[42] Fraser, D. A., et al. 1968. J. Air Pollut. Contr. Ass. 18(12):821.

[43] Freeman, G., et al. 1966. Arch. Environ. Health 13(4):454.

[44] Freeman, G., et al. 1968. Ibid. 17(2):181.

[45] Giddens, W. E., Jr., and G. A. Fairchild. 1972. Ibid. 25(3):166.

[46] Giordano, A. M., and P. E. Morrow. 1972. Ibid. 25(6):443.

[47] Gofmekler, V. A., et al. 1968. Gig. Sanit. 33(7-9):112.

[48] Goldring, I. P., et al. 1967. Arch. Environ. Health 15(2):167.

[49] Goldring, I. P., et al. 1970. Ibid. 21(1):32.

[50] Goldstein, B. D., et al. 1968. Ibid. 16:648.

[51] Goldstein, B. D., et al. 1972. Ibid. 24(4):243.

[52] Gray, E. L., et al. 1954. Arch. Ind. Hyg. Occup. Med. 10:418.

[53] Gunnison, A. F., and A. W. Benton. 1971. Arch. Environ. Health 22(3):381.

[54] Gusev, M. I., et al. 1966. Gig. Sanit. 31(1):8.

[55] Hathaway, J. A., and R. E. Terrill. 1962. Amer. Ind. Hyg. Ass. J. 23:392.

[56] Haydon, G. B., et al. 1967. Amer. Rev. Resp. Dis. 95:797.

[57] Henry, M. C., et al. 1969. Arch. Environ. Health 18(4):580.

[58] Henry, M. C., et al. 1970. Ibid. 20(5):566.

[59] Hine, C. H., et al. 1960. J. Air Pollut. Contr. Ass. 10:17.

[60] Horton, A. W., et al. 1963. J. Nat. Cancer Inst. 30:31.

continued

[61] Kantz, R. J., II, et al. 1972. Toxicol. Appl. Pharmacol. 22(2):320.

[62] Kincaid, J. F., et al. 1953. Arch. Ind. Hyg. Occup. Med. 8:48.

[63] Kleinerman, J., and C. R. Cowdrey. 1968. Yale J. Biol. Med. 40:579.

[64] Lee, S. D., and R. M. Danner. 1966. Arch. Environ. Health 12(5):583.

[65] Lewey, F. H., and D. L. Drabkin. 1944. Amer. J. Med. Sci. 208:502.

[66] Lewis, T. R., et al. 1969. Arch. Environ. Health 18(4):596.

[67] Lewis, T. R., et al. 1973. Ibid. 26:16.

[68] Lutmer, R. F., et al. 1967. Atmos. Environ. 1(1):45.

[69] Lyon, J. P., et al. 1970. Toxicol. Appl. Pharmacol. 17:726.

[70] MacDonald, W. E., et al. 1963. Amer. Ind. Hyg. Ass. J. 24:539.

[71] Mamatsashvili, M. I. 1970. Gig. Sanit. 35(5):277.

[72] Matzen, R. N. 1957. Amer. J. Physiol. 190:84.

[73] Mazaleski, S. C., et al. 1970. Amer. Ind. Hyg. Ass. J. 31(2):183.

[74] Mettier, J. S. R., et al. 1960. Arch. Ind. Health 21:13.

[75] Mittler, S., et al. 1956. Ind. Med. Surg. 25:301.

[76] Mittler, S., et al. 1957. Arch. Ind. Health 15:191.

[77] Moss, R. H., et al. 1951. Arch. Ind. Hyg. Occup. Med. 4:53.

[78] Murphy, S. D. 1964. J. Air Pollut. Contr. Ass. 14:303.

[79] Murphy, S. D. 1965. Toxicol. Appl. Pharmacol. 7:833.

[80] Murphy, S. D., et al. 1963. Ibid. 5:319.

[81] Murphy, S. D., et al. 1964. Ibid. 6:520.

[82] Murphy, S. D., et al. 1964. Amer. Ind. Hyg. Ass. J. 25:246.

[83] Nasr, A. N. M., et al. 1971. Arch. Environ. Health 22(5):538.

[84] Nasr, A. N. M., et al. 1971. Ibid. 22(5):545.

[85] Nizhegorodov, V. M., and Ya. L. Markhotskii. 1969. Gig. Sanit. 34(4-6):272.

[86] Nizhegorodov, V. M., and Ya. L. Markhotskii. 1971. Ibid. 36(1-3):137.

[87] Palmer, M. S., et al. 1972. Arch. Environ. Health 25(6):439.

[88] P'an, A. Y. S., and Z. Jegier. 1971. Ibid. 23(3):215.

[89] P'an, A. Y. S., and Z. Jegier. 1972. Ibid. 24(4):233.

[90] P'an, A. Y. S., et al. 1972. Ibid. 24(4):229.

[91] Pattle, R. E., et al. 1957. Brit. J. Ind. Med. 14:47.

[92] Pflesser, G. 1936. Naunyn Schmiedebergs Arch. Exp. Pathol. Pharmakol. 181:145.

[93] Rapoport, K. A., et al. 1968. Gig. Sanit. 33(10-12):64.

[94] Ripperton, L. A., and D. R. Johnston. 1959. Amer. Ind. Hyg. Ass. J. 20:324.

[95] Roehm, J. N., et al. 1972. Arch. Environ. Health 24(4):237.

[96] Rosenholtz, M. J., et al. 1963. Amer. Ind. Hyg. Ass. J. 24:253.

[97] Rylander, R. 1969. Arch. Environ. Health 18:551.

[98] Salem, H., and D. M. Aviado. 1961. Ibid. 2:56.

[99] Salem, H., and H. Cullumbine. 1960. Toxicol. Appl. Pharmacol. 2:183.

[100] Scheel, L. D., et al. 1959. J. Appl. Physiol. 14:67.

[101] Sherwin, R. P., et al. 1967. Amer. Rev. Resp. Dis. 96:319.

[102] Sherwin, R. P., et al. 1972. Arch. Environ. Health 24(1):43.

[103] Sinkuvene, D. S. 1970. Gig. Sanit. 35(3):325.

[104] Skog, E. 1950. Acta Pharmacol. Toxicol. 6:299.

[105] Stephens, R. J., et al. 1972. Arch. Environ. Health 24(3):160.

[106] Stokinger, H. E. 1957. Arch. Ind. Health 15:181.

[107] Stokinger, H. E., et al. 1956. Ibid. 14:158.

[108] Stokinger, H. E., et al. 1957. Ibid. 16:514.

[109] Sunderman, F. W. 1966. Lung Tumours Anim. Proc. 3rd Quadren. Int. Conf. Cancer, Perugia, 1965, p. 551.

[110] Sunderman, F. W., and A. J. Donnelly. 1965. Amer. J. Pathol. 46:1027.

[111] Sunderman, F. W., et al. 1957. Arch. Ind. Health 16:480.

[112] Sunderman, F. W., Jr. 1963. Amer. J. Clin. Pathol. 39(6):549.

[113] Sunderman, F. W., Jr. 1967. Cancer Res. 27(1):950.

[114] Sunderman, F. W., Jr., and F. W. Sunderman. 1963. Amer. J. Clin. Pathol. 40(6):563.

[115] Suzuki, T. 1969. Bull. Tokyo Med. Dent. Univ. 16(2):99.

[116] Svirbely, J. L., and B. E. Saltzman. 1957. Arch. Ind. Health 15:111.

[117] Swann, H. E., et al. 1965. Arch. Environ. Health 10:24.

[118] Tedeschi, R. E., and F. W. Sunderman. 1957. Arch. Ind. Health 16:486.

[119] Thienes, C. H., et al. 1965. Amer. Ind. Hyg. Ass. J. 26(3):255.

[120] Thomas, H. V., et al. 1968. Science 159:532.

[121] Trams, E. G., et al. 1972. Arch. Environ. Health 24(3):153.

[122] Valand, S. B., et al. 1970. Ibid. 20(3):303.

[123] Vaughan, T. R., Jr., et al. 1969. Ibid. 19(1):45.

[124] Weedon, F. R., et al. 1939. Contrib. Boyce Thompson Inst. 10:281.

[125] Weedon, F. R., et al. 1940. Ibid. 11:365.

[126] Werthamer, S., et al. 1970. Arch. Environ. Health 20(1):16.

[127] Wirth, W. 1930. Naunyn Schmiedebergs Arch. Exp. Pathol. Pharmkol. 157:264.

[128] Yant, W. P., et al. 1934. Pub. Health Bull. 211.

[129] Yuen, T. G. H., and R. P. Sherwin. 1971. Ibid. 22(1):178.

[130] Zarkower, A. 1972. Ibid. 25(1):45.

continued

Part III. Complex Air Pollutants

	Substance	Animal	Dose	Exposure [Duration]	Effects	Reference
	Diesel engine exhaust fumes, running conditions					
1	Light load, 1600 rpm	10 guinea pigs; 40 mice; 4 rabbits	$<10\ \mu g\ V/m^3$, 0.056% CO, 23 ppm NO_2, 46 ppm NO_x, 16 ppm CHO as HCHO, 74 mg particulates/m^3, 6 ft visibility, 7.5 s lacrimation time	5 hr	No mortality; fumes more irritating, but less toxic than with increased load; moderate to severe capillary & venous congestion in lungs; desquamation of tracheal epithelium	16
2		10 guinea pigs; 50 mice; 2 rabbits	$<10\ \mu g\ V/m^3$, 0.056% CO, 23 ppm NO_2, 46 ppm NO_x, 16 ppm CHO as HCHO, 74 mg particulates/m^3, 6 ft visibility, 7.5 s lacrimation time	7 hr	Mortality at 7 hr: guinea pigs, 100%; rabbits, 50%; mice, 20%. At 14 hr, rabbits, 100%; mice, 90%. Pulmonary congestion, consolidation, edema & emphysema; moderate to severe tracheitis with focal desquamation of epithelial lining.	16
3		50 mice; 2 rabbits	$<10\ \mu g\ V/m^3$, 0.056% CO, 23 ppm NO_2, 46 ppm NO_x, 16 ppm CHO as HCHO, 74 mg particulates/m^3, 6 ft visibility, 7.5 s lacrimation time	14 hr		
4	Increased load, 1600 rpm	10 guinea pigs; 40 mice; 4 rabbits	$<10\ \mu g\ V/m^3$, 0.041% CO, 51 ppm NO_2, 209 ppm NO_x, 6.0 ppm, CHO as HCHO, 122 mg particulates/m^3, 1-6 ft visibility, 10 s lacrimation time	5 hr	Mortality: guinea pigs, 90%; rabbits, 48%; mice, 0. Fumes irritating; pulmonary congestion, edema & emphysema; minimal tracheal & bronchial changes except for bronchopneumonia in 3 animals.	16
5	Increased load with worn fuel injector, 1600 rpm	10 guinea pigs; 40 mice; 4 rabbits	$<10\ \mu g\ V/m^3$, 0.038% CO, 43 ppm NO_2, 174 ppm NO_x, 6.4 ppm CHO as HCHO, 53 mg particulates/m^3, 6 ft visibility, 20 s lacrimation time	5 hr	Mortality: guinea pigs, 60%; mice, 2%; rabbits, 0. Fumes practically nonirritating; pulmonary congestion, edema & emphysema; minimal tracheal & bronchial changes.	16
6	Increased fuel-to-air ratio; light load, 1600 rpm	10 guinea pigs; 40 mice; 4 rabbits	$<10\ \mu g\ V/m^3$, 0.17% CO, 12 ppm NO_2, 44 ppm NO_x, 155 ppm CHO as HCHO, 1070 mg particulates/m^3, 0.5 ft visibility, 5.5 s lacrimation time	5 hr	100% mortality in all species; fumes violently irritating particularly to eyes; severe pulmonary hemorrhage & congestion; marked tracheal & bronchial lesions, with loss of epithelial lining	16
	Gasoline engine exhaust fumes					
7	Auto exhaust, irradiated	12 dogs, Beagle	100 ppm CO, 24-30 ppm HC, 0.1 ppm NO, 0.5-1.0 ppm NO_2, 0.2-0.4 ppm O_x	16 hr/da, 7 da/wk [18 mo]	At 18 mo, no significant effects on total expiratory flow resistance, dynamic lung compliance, or single-breath CO diffusion capacity	17

continued

Part III. Complex Air Pollutants

	Substance	Animal	Dose	Exposure [Duration]	Effects	Reference
8		11 dogs, Beagle	114.5 mg CO/m^3 (100 ppm[1/]), 15.7-19.6 mg HC/m^3, 0.12 mg NO/m^3 (0.1 ppm[1/]), 0.94-1.88 mg NO_2/m^3 (0.4-0.8 ppm[1/]), 0.39-0.78 mg O_x/m^3	16 hr/da [5 yr]	9% significant ↑ in mean systemic arterial pressure at 5 yr; 9% bradycardia at 4.5-5 yr by static ECG; 27% possible mild right ventricular hypertrophy at 5 yr by VCG	2
9		5-10 guinea pigs, English	Air:exhaust ratio, 1140:1	4-6 hr	↑ (26%) in pulmonary flow resistance & (6%) in tidal volume; ↓ (17%) in respiratory rate	14
10			Air:exhaust ratio, 360:1	4-6 hr	↑ (29%) in pulmonary flow resistance & (12%) in tidal volume; ↓ (20%) in respiratory rate	14
11			Air:exhaust ratio, 150:1	4-6 hr	↑ (113%) in pulmonary flow resistance & (25%) in tidal volume; ↓ (33%) in respiratory rate	14
12		Guinea pigs, Hartley, NIH; hamsters, Syrian Golden; mice, Swiss, A/J, C57BL, LAF_1, 101; rats, Sprague-Dawley; 5-8 wk	≯20-100 ppm CO, ≯6-36 ppm HC, ≯0.4-2.0 ppm NO, ≯0.3-1.9 ppm NO_2, ≯0.2-1.0 ppm O_3	7 da/wk [23 mo]	No effect on mortality or body wt; initial ↓ followed by ↑ in voluntary running activity at >23 da in mice; significant ↑ in WBC at >14 mo in mice; no pulmonary function changes in mice or guinea pigs; significant ↓ in total lipids/g dry lung wt at higher exhaust conc in rats; significant ↓ in fertility & litter survival rate in mice; no significant difference in incidence of pulmonary inflammatory changes, including vasculitis, bronchitis, & pneumonia, or neoplasia in mice exposed to raw or irradiated exhaust—*see* entry 26	9
13		Mice, A/J, C57BL, LAF_1, 4 mo	0.1-0.47 μg organic Pb/m^3 & 3.2-15.6 μg inorganic Pb/m^3	24 hr/da [15 mo]	Significant ↑ in bone lead content at >9.6 μg Pb/m^3—*see* entry 28	13
14		600 mice, LAF_1, 12-13 wk	≯100 ppm CO, ≯45 ppm HC, ≯3.5 ppm NO, ≯1.5 ppm NO_2, ≯1.5 ppm O_x	16 hr/da [46 da premating & 7-10 da postnatal]	Significant ↑ in nonpregnancy average in ♀ with preexposure of ♂; significant ↓ in implantation scars & number of pups/litter with preexposure of ♂ or ♀; significant ↑ in 1-8 da mortality of exposed nursling mice; significant ↑ in 8-21 da mortality of exposed or unexposed nursling mice sired by preexposed ♂	12
15		Rats, Sprague-Dawley	Air:exhaust ratio, 150:1[2/]	6 hr	No change in lung & serum AP, SGOT, & cholinesterase; blood carboxyhemoglobin levels greater than those observed with raw exhaust	15
16		180 mice, Swiss, 5 wk[3/]	12-25 ppm CO & 0.08-0.14 ppm O_x	4 hr	No significant ↑ in mortality from streptococcal pneumonia	9
17			≮25 ppm CO & 0.15-0.41 ppm O_x	4 hr	Significant ↑ in mortality from streptococcal pneumonia	9
18		5-10 mice, Taconic	Air:exhaust ratio, 1140:1	4-6 hr	Voluntary running activity decreased 16%	14
19			Air:exhaust ratio, 360:1	4-6 hr	Voluntary running activity decreased 43%	14

[1/] Calculated by contributors. [2/] 290 ppm CO, 0.78 ppm total O_x, 5.5 ppm NO_2, 1.0 ppm NO, 1.93 ppm HCHO, 0.17 ppm acrolein, 8.9 μg olefin/liter, $HCl:NO_x$ = 3:4.

[3/] Received aerosol of *Streptococcus* (Lancefield C) for 0.5 hr after exhaust exposure.

continued

Part III. Complex Air Pollutants

	Substance	Animal	Dose	Exposure [Duration]	Effects	Reference
20			Air:exhaust ratio, 150:1	4-6 hr	Voluntary running activity decreased 77%	14
21		4-15 mice, Taconic	2700 ppm total gas	6 hr	Voluntary running activity decreased ∿50%	14
22	& sulfur oxides	12 dogs, Beagle	CO, HC, NO, NO₂, & O_x [4/]; 0.5 ppm SO₂; 100 μg H₂SO₄ aerosol/ m³	16 hr/da, 7 da/wk [18 mo]	At 18 mo, no significant effects on total expiratory flow resistance, dynamic lung compliance or single-breath CO diffusion capacity	17
23		11 dogs, Beagle	CO, HC, NO, NO₂, & O_x [5/]; 0.14 mg SO₂/m³ (0.05 ppm [1/]); 0.10 mg H₂SO₄ aerosol/m³	16 hr/da [5 yr]	9% significant ↑ in mean & systolic pulmonary arterial pressure at 5 yr; 9% possible intraventricular conduction defect at 5 yr by VCG; 9% myocardial infarction at 5 yr	2
24	Auto exhaust, raw	12 dogs, Beagle	100 ppm CO, 24-30 ppm HC, 1.5-2.0 ppm NO, 0.1 ppm NO₂	16 hr/da, 7 da/wk [18 mo]	At 18 mo, no significant effects on total expiratory flow resistance, dynamic lung compliance, or single-breath CO diffusion capacity	17
25		11 dogs, Beagle	114.5 mg CO/m³ (100 ppm [1/]), 15.7-19.6 mg HC/m³, 1.84-2.45 mg NO/m³ (1.5-2.0 ppm [1/]), 0.19 mg NO₂/m³ (0.1 ppm [1/])	16 hr/da [5 yr]	9% significant ↑ in mean pulmonary arterial pressure at 5 yr; no changes in static or exercise ECG, or VCG at 4.5-5 yr	2
26		Guinea pigs, Hartley, NIH; hamsters, Syrian Golden; mice, Swiss, A/J, C57BL, LAF₁, 101; rats, Sprague-Dawley; 5-8 wk	⪢20-100 ppm CO & ⪢0.4-2.0 ppm NO	7 da/wk [23 mo]	No effect on mortality or body wt; initial ↓ followed by ↑ in voluntary running activity at >10 da in mice; no hematologic or pulmonary function changes in mice &/or guinea pigs; significant ↓ in total lipids/g dry lung wt at higher exhaust conc in rats; no significant difference in incidence of pulmonary inflammatory changes, including vasculitis, bronchitis, & pneumonia, or neoplasia in mice exposed to raw or irradiated exhaust—*see entry 12*	9
27		4-15 mice	7100 ppm total gas	6 hr	Voluntary running activity decreased ∿50%	14
28		Mice, A/J, C57BL, LAF₁, 4 mo	0.08-0.47 μg organic Pb/m³ & 2.6-15.6 μg inorganic Pb/m³	24 hr/da [15 mo]	Significant ↑ in bone lead content at >9.6 μg Pb/m³; bone lead content greater than that observed with irradiated exhaust—*see entry 13*	13
29	& sulfur oxides	5 dogs, tracheotomized	CO, HC, NO, & NO₂ [6/]; 0.5 ppm SO₂; 100 μg H₂SO₄ aerosol/m³	10 min-2 hr	100% removal of SO₂, 90% of NO₂, & 73% of NO in upper airway; no removal of CO & hydrocarbons	17
30		12 dogs, Beagle	CO, HC, NO, & NO₂ [6/]; 0.5 ppm SO₂; 100 μg H₂SO₄ aerosol/m³	16 hr/da, 7 da/wk [18 mo]	At 18 mo, no significant effects on total expiratory flow resistance, dynamic lung compliance or single-breath CO diffusion capacity	17

[1/] Calculated by contributors. [4/] Concentrations same as those listed in entry 7. [5/] Concentrations same as those listed in entry 8. [6/] Concentrations same as those listed in entry 24.

continued

Part III.   Complex Air Pollutants

	Substance	Animal	Dose	Exposure [Duration]	Effects	Reference
31		11 dogs, Beagle	CO, HC, NO, & NO$_2$[7]; 0.14 mg SO$_2$/m^3 (0.05 ppm[1]); 0.10 mg H$_2$SO$_4$ aerosol/ m^3	16 hr/da [5 yr]	9% significant ↑ in mean pulmonary arterial & wedge pressures at 5 yr; 9% bradycardia at 5 yr by static ECG; 9% mild to moderate left ventricular hypertrophy, 9% possible mild right ventricular hypertrophy, & 9% possible myocardial infarction at 5 yr by VCG	2
	Smog Natural photochemical					
32	Greater Los Angeles, Calif.	Guinea pigs, Hartley; hamsters, Syrian Golden; rabbits; rats, Charles River	9.1-13.5 ppm CO, 1.9-14.4 pphm NO, 4.4-7.7 pphm NO$_2$, 3.0-5.0 ppm O$_x$[8]	24 hr/da [∿2-3 yr]	No effects on survival rate of rats; significant, reversible ↑ in total expiratory flow resistance in guinea pigs, particularly in older animals, with oxidant levels >0.5 ppm; significant ↓ in SGOT in rabbits after 2-3 yr; significant, reversible ↑ in urinary 17-ketogenic steroids in guinea pigs subjected to 8-wk stress period; no ↑ in incidence of histopathologic lesions except for cardiac & pulmonary calcinosis in guinea pigs	18
33		Mice, 5-21 mo	>0.4 ppm O$_x$	2-3 hr	Epithelial & endothelial swelling, disruption of epithelial cytoplasm with lamellar inclusions & few normal mitochondria present in lungs of 9- & 21-mo-old mice; effects reversible after 14 hr in 9-mo-old mice only; no ultrastructural lung lesions in 5-mo-old mice	1
34		Mice, A, A/J, C57BL, 6 wk	9.1-13.5 ppm CO, 1.9-14.4 pphm NO, 4.4-7.7 pphm NO$_2$, 3.0-5.0 ppm O$_x$[8]	24 hr/da [∿2 yr]	Significant ↑ in single & multiple pulmonary adenomas in A/J mice which died; significant ↓ in pulmonary adenomas in killed, 6-19 mo old A/J mice; no significant ↑ in pulmonary adenomas in killed A-strain mice (2 colonies) or C57 mice which died (1 colony). Significant ↑ in interstitial pneumonitis in A/J mice which died or were killed, & in severe interstitial pneumonitis & acute bronchopneumonia in C57 mice >8 mo old which died; severe interstitial pneumonitis associated occasionally with ↑ in squamous metaplasia & bronchiolar-alveolar cell hyperplasia in C57 mice.	7,8, 18
35	Riverside, Calif.	Guinea pigs, Hartley; mice, A, A/J, C57BL; rabbits, New Zealand; rats, Charles River	1.7 ppm CO, 2.4 ppm HC, 1.5 pphm NO, 1.9 pphm NO$_2$, 5.7 pphm O$_x$[9]	24 hr/da [∿2.5 yr]	Significant ↓ in survival rate of ♂ C57 mice only; no significant effect on total expiratory flow resistance in guinea pigs; significant ↓ in SGOT in rabbits & in lung AP in rats, particularly in older rats; slight ↑ in pulmonary adenomas in A-strain mice <23 mo old; no significant ↑ in pulmonary adenomas in A/J or C57 mice; pulmonary calcinosis in guinea pigs	5

[1] Calculated by contributors. [7] Concentrations same as those listed in entry 25. [8] Range of average concentrations for overall measuring period, 1962-67, for 4 stations located in central Los Angeles, Azusa, Burbank, and near Hollywood Freeway. [9] Average annual concentrations for overall measuring period, 1965-67.

continued

Part III. Complex Air Pollutants

	Substance	Animal	Dose	Exposure [Duration]	Effects	Reference
36		24 mice, C57BL, 113 da	2.7-4.4 ppm HC, 3.7 pphm NO$_2$, 6.2-23.9 pphm O$_x$	24 hr/da [13 mo]	Difference in monthly average voluntary running activity highly correlated to average & peak O$_x$ conc, & to temperature	4
37	Synthetic Gasoline vapor, ozonized	14 mice, C57BL/6, young adult	0.51 ppm O$_3$ (1.71 ppm O$_x$)	∿3 wk	Very slight but significant ↓ in spontaneous activity	3
38		30 mice, C57BL, 12 wk	1.25 ppm O$_x$	24 hr/da [19 wk]	Significant ↓ in fertility & litter rates, & in litter size of exposed ♀; significant ↑ in mortality of exposed offspring at <21 da	10
39		405 mice, C57BL	....	92 wk	Significant ↑ in pulmonary tumors, which frequently displayed characteristics of neoplastic growth, at 71-85 wk; marked hyperplasia & metaplasia of bronchial & bronchiolar epithelium at ≪19 wk; significant ↓ in spontaneous extrapulmonary tumors	11
40		Rabbits	1-3 ppm	1 hr/da [24 hr]	Complete cessation of ciliary activity after 24 exposure hr; return to 35% normal activity in 48 hr	6
41	Photochemical (irradiated mixture of 8 ppm propylene, 2.8 ppm NO, 25 ppm CO, & water vapor)	Mice, 8-20 mo	>0.4 ppm O$_x$	3 hr	Slight epithelial & endothelial swelling, & vacuolization of epithelial cytoplasm with lamellar inclusions & few normal mitochondria present in lungs of 8-mo-old mice; similar but more severe & irreversible lung lesions in 15- & 20-mo-old mice	1

Contributors: Palm, Paul E., Nick, M. Susan, Arnold, Elsie P., and Platz, Barbara B.

References

[1] Bils, R. F. 1968. J. Air Pollut. Contr. Ass. 18(5): 313.

[2] Bloch, W. N., Jr., et al. 1972. Arch. Environ. Health 24(5):342.

[3] Boche, R. D., and J. J. Quilligan, Jr. 1960. Science 131:1733.

[4] Emik, L. O., and R. L. Plata. 1969. Arch. Environ. Health 18(4):574.

[5] Emik, L. O., et al. 1971. Ibid. 23(5):335.

[6] Falk, H. L., and P. Kotin. 1957. J. Air Pollut. Contr. Ass. 7:12.

[7] Gardner, M. B. 1966. Arch. Environ. Health 12(3): 305.

[8] Gardner, M. B., et al. 1970. Ibid. 20:310.

[9] Hueter, F. G., et al. 1966. Ibid. 12(5):553.

[10] Kotin, P., and M. Thomas. 1957. Arch. Ind. Health 16:411.

[11] Kotin, P., et al. 1958. Cancer 11:473.

[12] Lewis, T. R., et al. 1967. Arch. Environ. Health 15(1):26.

[13] Lutmer, R. F., et al. 1967. Atmos. Environ. 1(5): 585.

[14] Murphy, S. D. 1964. J. Air Pollut. Contr. Ass. 14: 303.

[15] Murphy, S. D., et al. 1963. Arch. Environ. Health 7:66.

[16] Pattle, R. E., et al. 1957. Brit. J. Ind. Med. 14:47.

[17] Vaughan, T. R., Jr., et al. 1969. Arch. Environ. Health 19(1):45.

[18] Wayne, L. G., and L. A. Chambers. 1968. Ibid. 16: 871.

Location: G = greenhouse, F = field, Fo = forest. **Concentration:** Many concentrations were given in mg/m³, but were changed to ppm at 25°C and 760 mm Hg, by a conversion factor of 1 mg/m³ = 0.382 ppm SO₂ (by volume). **Duration:** Gr. sn. = growing season. **Fumigation:** A = artificial, N = natural. **Effects:** ↑ = increase, ↓ = decrease. Data in brackets refer to the column heading in brackets. Values in parentheses are ranges, estimate "c" (*see* Introduction).

Part I. Herbaceous Plants

Species (Synonym) [Developmental Stage]	Location [G or F]	Sulfur Dioxide Exposure Concentration ppm Mean [Max]	Duration [Fumigation]	Environmental Conditions	Effects	Reference
		Cereals, Grains, and Grasses				
1 *Agrostis palustris*	New Jersey, USA	0.75	6 hr [A]	27-30°C, 50% RH	Slight leaf discoloration	3
2 'Penncross'	[G]	0.85 or 1.80	6 hr [A]	27-30°C, 50% RH	Severe leaf discoloration	
3 *Avena* sp.	Garson, Ont. [F]	0.17 [0.63 for 1 hr]	8 hr [N]	....	Leaf injury	5
4 *A. sativa*	New York, USA	0.66	5 hr [A]	....	Slight leaf discoloration	31
5	[G]	1.35	1 hr [A]	29-30°C, 53% RH	16.7% leaf injury	33
6		1.35	1 hr [A]	29-30°C, 76% RH	5.7% leaf injury	33
7		1.35	2 hr [A]	29-30°C, 51% RH	9.4% leaf injury	33
8		1.35	2 hr [A]	29-30°C, 73% RH	13.3% leaf injury	33
9 [8 cm]	Hatersheim, W. Ger. [F]	2.0	5 × 30 min/da for 7 da [A]	....	Slight leaf injury	30
10			150 min/da for 7 da [A]	....	Severe leaf injury on 3rd da	
A. sativa [1]						11
11 [2-leaf or shooting]	W. Ger. [F (pots)]	0.69	20 hr [A]	....	No ↓ in yield wt	
12 [Shooting or grain formation]	W. Ger. [F (pots)]	0.96	38 hr [A]	....	No ↓ in yield wt	
A. sativa [2]						11
13 [Cotyledon]	W. Ger. [F (pots)]	1.77	8 hr [A]	....	14% ↓ in yield wt	
14 [Cotyledon or small rosette]	W. Ger. [F (pots)]	0.88	16 hr [A]	....	No ↓ in yield wt	
15 [Large rosette]	W. Ger. [F (pots)]	0.88	16 hr [A]	....	10% ↓ in yield wt	
16 [Grain formation]	W. Ger. [F (pots)]	0.88	16 hr [A]	....	26% ↓ in yield wt	
A. sativa						
17 'Banner'	Stroh, Wash. [F]	0.315 [0.93 for 0.3 hr]	17.3 hr [N]	....	No leaf injury	20
18 'Clintland 64'	USA [G]	≮0.5	4 hr [A]	29 ± 3°C, ∿60% RH; ∿11,000 lux	Slight leaf necrosis	29
19 [21 da old]	USA [G]	≮0.05	4 hr [A]	29 ± 3°C, ∿60% RH; ∿11,000 lux; ≮0.05 ppm NO₂	Slight to moderate leaf necrosis, leaf chlorosis	29
20 'Flaming Gold'	Biersdorf, W. Ger. [F (pots)]	0.01	Gr. sn. [N]	HF & HCl in air	Slight leaf necrosis; no ↓ in straw yield wt, yield quality, & growth ht	12, 13
21		0.05	Gr. sn. [N]	HF & HCl in air	Severe leaf necrosis; ↓ in grain & straw yield wt, & in yield quality	
22 *Cynodon dactylon* 'Kansas P-16'	New Jersey, USA [G]	0.75, 0.85, or 1.80	6 hr [A]	27-30°C, 50% RH	No leaf injury	3

[1] Mixed culture with common vetch. [2] Mixed culture with charlock.

continued

Part I. Herbaceous Plants

	Species (Synonym) [Developmental Stage]	Location [G or F]	Sulfur Dioxide Exposure		Environmental Conditions	Effects	Reference
			Concentration ppm Mean [Max]	Duration [Fumigation]			
23	*Festuca rubra* 'Highlight' & 'Pennlawn'	New Jersey, USA [G]	0.75	6 hr [A]	27-30°C, 50% RH	Slight to moderate leaf discoloration	3
24			0.85 or 1.80	6 hr [A]	27-30°C, 50% RH	Severe leaf discoloration	
25	*Hordeum* sp. [15-20 cm]	Czechoslovakia [G]	0.369	2.3 hr [A]	Noon	No leaf injury on 1st da, slight leaf necrosis on 2nd da	7
26			0.761	3.75 hr [A]	Noon	No leaf injury on 1st da, leaf necrosis moderate on 2nd da, severe on 3rd	
27			0.90	2 hr 25 min [A]	Evening	No leaf necrosis on 1st-4th da	
28			1.00	2.5 hr [A]	Morning	No leaf injury on 1st da, moderate leaf necrosis on 4th da	
29			1.34	2.3 hr [A]	Noon	No leaf injury on 2nd da, moderate leaf necrosis on 4th da	
30	*Hordeum* sp.[1] [15-20 cm]	Czechoslovakia [G]	0.138	4 hr [A]	Noon	No leaf injury on 1st da, slight leaf necrosis on 4th da	7
31			0.715	3 hr [A]	Noon	Leaf necrosis slight on 1st da, moderate on 2nd da	
32	'Hannchen'	Stroh, Wash. [F]	0.315 [0.93 for 0.3 hr]	17.3 hr [N]	....	Slight leaf injury	20
33	*Hordeum* sp. 'Hannchen' [19 da after seeding]	Summerland, B.C. [F]	2.38	3 hr [A]	41-50% RH, sunshine	5.7% SO_2 absorption, 20% leaf injury	20
34	*H. vulgare*	USA [G]	0.66	5 hr [A]	....	Slight leaf injury	31
35		Hatersheim, W. Ger. [F]	0.3-0.6 (pretreatment concn for prime exposure; (*see* entry 40)	18-333 hr [A]	....	No leaf injury	30
36			1.2-2.0	2-4 hr [A]	....	Slight to moderate leaf injury	
37			1.2-2.0 (prime exposure; no pretreatment)	2-4 hr [A]	....	Moderate to severe leaf injury	
38		New York, USA [G]	1.35	1 hr [A]	29-30°C, 53% RH	16.7% leaf injury	33
39					29-30°C, 76% RH	18.2% leaf injury	
40				2 hr [A]	29-30°C, 51% RH	47.4% leaf injury	
41					29-30°C, 73% RH	69.4% leaf injury	
42	[3-leaf]	UK [G]	0.004-0.06	[A]	10-20% RH (stomata closed)	0.0005-0.006 cm/s absorption of $^{35}SO_2$ [3]	23
43					30-40% RH (stomata closed)	0.0016-0.0023 cm/s absorption of $^{35}SO_2$ [3]	
44					50-60% RH (stomata closed)	0.007-0.013 cm/s absorption of $^{35}SO_2$ [3]	

[1] Mixed culture with common vetch. [3] Velocity of deposition = $\dfrac{SO_2 \text{ absorbed/cm}^2 \text{ leaf surface}}{\text{dosage } SO_2 \text{ in air } (\mu g \cdot s \cdot cm^{-3})}$.

continued

Part I. Herbaceous Plants

#	Species (Synonym) [Developmental Stage]	Location [G or F]	Sulfur Dioxide Exposure Concentration ppm Mean [Max]	Duration [Fumigation]	Environmental Conditions	Effects	Reference
45					70-75% RH (stomata closed)	0.005-0.021 cm/s absorption of $^{35}SO_2$ [3]	
46					75-80% RH (stomata closed)	0.007-0.026 cm/s absorption of $^{35}SO_2$ [3]	
47					80-85% RH (stomata closed)	0.007-0.014 cm/s absorption of $^{35}SO_2$ [3]	
48					80-85% RH (stomata open)	0.037-0.089 cm/s absorption of $^{35}SO_2$ [3]	
49					85-90% RH (stomata open)	0.045-0.13 cm/s absorption of $^{35}SO_2$ [3]	
50					90-95% RH (stomata open)	0.11-0.25 cm/s absorption of $^{35}SO_2$ [3]	
51	*Lolium multiflorum* [4] [1st or 2nd crop]	W. Ger. [F]	0.76	9 or 25 hr [A]	....	No ↓ in yield wt	11
52	[2nd crop]	W. Ger. [F]	0.38	25 hr [A]	....	No ↓ in yield wt	
53			1.92	10 hr [A]	....	34% ↓ in yield wt	
54	*L. multiflorum* [5]	Biersdorf, W. Ger. [F (pots)]	0.46 or 0.92	8 hr/da for 6 da [A]	....	No ↓ in yield wt	11
55	*L. perenne*	Biersdorf, W. Ger. [F (pots)]	0.96	12 hr [A]	....	26% ↓ in yield wt	11
56	*L. perenne* [4]	Biersdorf, W. Ger. [F (pots)]	0.96	12 hr [A]	....	12% ↓ in yield wt	11
57	*L. perenne* 'Aberystwyth S23' [Seed to 3-leaf]	Manchester, UK [G]	0.01-0.06 [0.2 for 24 hr]	46-81 da [N]	Smoke; high & low fertility soil	No leaf injury, ↓ in yield wt	2
58	'Lamora'	New Jersey, USA [G]	0.85	6 hr [A]	27-30°C, 50% RH	Slight leaf discoloration	3
59	'Manhattan'	New Jersey, USA [G]	0.85	6 hr [A]	27-30°C, 50% RH	No leaf injury	3
60	'Lamora' & 'Manhattan'	New Jersey, USA [G]	0.75 or 1.80	6 hr [A]	27-30°C, 50% RH	Slight leaf discoloration	3
61	*Phleum* sp.	Rayside, Ont. [F]	0.21 [0.66 for 1 hr]	8 hr [N]	....	Leaf injury	5
62	*Poa annua*	New Jersey, USA [G]	0.75 or 0.85	6 hr [A]	27-30°C, 50% RH	Slight leaf discoloration	3
63			1.80	6 hr [A]	27-30°C, 50% RH	Moderate leaf discoloration	
64	*P. nipponica*	Japan [G]	1.0	2 hr/da for 40 da [A]	9-19°C, 59-92% RH	No leaf injury	18
65	*P. pratensis* 'Delta' & 'Merion'	New Jersey, USA [G]	0.75	6 hr [A]	27-30°C, 50% RH	No leaf injury	3
66			0.85	6 hr [A]	27-30°C, 50% RH	Slight to moderate leaf discoloration	
67			1.80	6 hr [A]	27-30°C, 50% RH	Moderate leaf discoloration	

[3] Velocity of deposition = $\dfrac{SO_2 \text{ absorbed/cm}^2 \text{ leaf surface}}{\text{dosage } SO_2 \text{ in air } (\mu g \cdot s \cdot cm^{-3})}$. [4] Mixed culture with red clover. [5] Mixed culture with crimson clover and hairy vetch.

continued

Part I. Herbaceous Plants

Species (Synonym) [Developmental Stage]	Location [G or F]	Sulfur Dioxide Exposure		Environmental Conditions	Effects	Reference
		Concentration ppm Mean [Max]	Duration [Fumigation]			
68 *Secale cereale*	USA [G]	0.66	5 hr [A]	....	Slight leaf discoloration	31
69	New York, USA [G]	1.35	1 hr [A]	29-30°C, 53% RH	5.9% leaf injury	33
70				29-30°C, 76% RH	No leaf injury	
71			2 hr [A]	29-30°C, 51% RH	11.9% leaf injury	
72				29-30°C, 73% RH	3.6% leaf injury	
73	W. Ger. [F]	0.77	8 hr/da for 10 da [A]	....	No ↓ in yield wt	11
74 *S. cereale* [1,6]	W. Ger. [F]	0.58	8 hr/da for 10 da [A]	....	No ↓ in yield wt	11
75 'Lochows Petkuser' [6]	Biersdorf, W. Ger. [F (pots)]	0.05	Gr. sn. [N]	HF & HCl in air	Leaf necrosis, ↓ in grain yield wt & yield quality	12, 13
76		0.06-0.104	Gr. sn. [N]	HF & HCl in air	4-6% severe leaf necrosis, ↓ in straw yield wt & yield quality	
77 'Prolific'	Stroh, Wash. [F]	0.315 [0.93 for 0.3 hr]	17.3 hr [N]	....	Severe leaf injury	20
78 *Sorghum* sp.	New York, USA [F or G]	0.5	4-8 hr [A]	....	No leaf injury	34
79 *Triticum* sp.	Hatersheim, W. Ger. [F]	2.0	3 hr [A]	....	Severe leaf injury	30
80 'Little Club' [Adult]	USA [G]	0.19-0.46	3-24 hr/da [A]	20-23°C; light 16 hr/da	No leaf injury, no fruiting injury, no change in yield wt	24
81 'Marquis'	Stroh, Wash. [F]	0.315 [0.93 for 0.3 hr]	17.3 hr [N]	....	Moderate leaf injury	20
82 *T. aestivum (T. sativum)*	New York, USA [G]	1.35	1 hr [A]	29-30°C, 53% RH	4.2% leaf injury	33
83				29-30°C, 76% RH	No leaf injury	
84			2 hr [A]	29-30°C, 51% RH	No leaf injury	
85				29-30°C, 73% RH	3.1% leaf injury	
86 'Heines VII' [6]	Biersdorf, W. Ger. [F (pots)]	0.015	Gr. sn. [N]	....	15% ↓ in grain yield wt	13
87		0.024	Gr. sn. [N]	HF & HCl in air	No change in grain yield wt	12, 13
88		0.050	Gr. sn. [N]	....	No change in grain yield wt	13
89		0.051	Gr. sn. [N]	HF & HCl in air	↓ in grain yield wt	12, 13
90 'Heines Koga II' [7]	Biersdorf, W. Ger. [F (pots)]	0.050	Gr. sn. [N]	....	↓ in grain & straw yield wt	13
91 *T. durum* 'Pellissier'	Stroh, Wash. [F]	0.315 [0.93 for 0.3 hr]	17.3 hr [N]	....	Moderate to severe leaf necrosis	20
92 *Zea* sp. [8]	W. Ger. [F]	0.77	24 hr [A]	....	No ↓ in yield	11
93 *Z. mays* [9]	New York, USA [F or G]	0.5	4-8 hr [A]	....	No leaf injury	34
94 'Badische Landmais' [15-20 cm]	Giessen, W. Ger. [G]	0.08, 0.19, or 0.38	12 hr/da for 13 da [A]	22.4°C, 58% RH; soil S deficient	No leaf injury, ↑ in yield wt & growth ht	6
95		0.56	12 hr/da for 13 da [A]	22.4°C, 58% RH; soil S deficient	Leaf necrosis, ↑ in yield wt & growth ht	

[1] Mixed culture with common vetch. [6] Winter variety. [7] Summer variety. [8] Mixed culture with sunflower, field peas, and common vetch. [9] Field and sweet.

continued

Part I. Herbaceous Plants

	Species (Synonym) [Developmental Stage]	Location [G or F]	Sulfur Dioxide Exposure		Environmental Conditions	Effects	Reference
			Concentration ppm Mean [Max]	Duration [Fumigation]			
96	*Zoysia japonica* 'Common' & 'Meyer'	New Jersey, USA [G]	0.75, 0.85, or 1.80	6 hr [A]	27-30°C, 50% RH	No leaf injury	3
	Rush						15
97	[Early growth]	Japan [G]	0.3	2 da [A]	....	Slight leaf discoloration	
98	[Middle growth]	Japan [G]	0.3	2 da [A]	....	Moderate leaf discoloration	
99	[Late growth]	Japan [G]	0.3	2 da [A]	....	Severe leaf discoloration	
Vegetables and Dicotyledonous Crops							
100	*Allium cepa*	[G]	1.0	2 hr/da for 40 da [A]	9-19°C, 59-92% RH	Moderate leaf injury	18
101	*Apium graveolens*	Rayside, Ont. [F]	0.29 [0.87 for 1 hr]	8 hr [N]	....	Leaf injury	5
102		[F or G]	0.5	4-8 hr [A]	....	Severe leaf injury	34
103	*Arachis hypogaea* 'Starr' [28 da old]	[G]	0.02-0.03	4-5 hr [A]	21 ± 1°C, 54 ± 3% RH; 21,520 lux for 14 hr; 0.008-0.01 ppm O$_3$; surplus moisture	Slight leaf chlorosis	1
104			0.05-0.12	4-14 hr [A]	21 ± 1°C, 54 ± 3% RH; 21,520 lux for 14 hr; inadequate to surplus moisture	Slight leaf discoloration	
105	*Beta vulgaris*	Garson, Ont. [F]	0.33 [1.31 for 1 hr]	8 hr [N]	....	Leaf injury	5
106		Japan [G]	1.0	2 hr/da for 40 da [A]	9-19°C, 59-92% RH	No leaf injury	18
107	'Criewen Yellow'	Biersdorf, W. Ger. [F (pots)]	0.05	Gr. sn. [N]	HF & HCl in air	Leaf necrosis & chlorosis, ↓ in root yield wt [10], no ↓ in leaf yield quality	12, 13
108			0.124-0.159	Gr. sn. [N]	HF & HCl in air	Leaf necrosis & chlorosis, ↓ in root yield wt & leaf yield quality [10]	
109	*B. vulgaris* var. *cicla*	Rayside, Ont. [F]	0.27 [0.88 for 1 hr]	8 hr [N]	....	Leaf injury	5
110	*Brassica campestris*	Japan [G]	1.0	2 hr/da for 40 da [A]	9-19°C, 59-92% RH	No leaf injury	18
	B. kaber [11]						11
111	[Cotyledon]	W. Ger. [F (pots)]	0.86	16 hr [A]	....	13% ↓ in yield wt	
112			1.73	8 hr [A]	....	69% ↓ in yield wt	
113	[Small rosette]	W. Ger. [F (pots)]	0.86	16 hr [A]	....	11% ↓ in yield wt	
114	[Large rosette]	W. Ger. [F (pots)]	0.86	16 hr [A]	....	No ↓ in yield wt	
115	[Grain formation]	W. Ger. [F (pots)]	0.86	16 hr [A]	....	6% ↓ in yield wt	
116	*B. napus* 'Liho' [7]	Biersdorf, W. Ger. [F (pots)]	0.124	Gr. sn. [N]	....	↓ in seed yield wt, no ↓ in yield quality	13
117	*B. oleracea*	Garson, Ont. [F]	0.45 [0.94 for 1 hr]	8 hr [N]	....	Leaf injury	5
118		Japan [G]	1.0	2 hr/da for 40 da [A]	9-19°C, 59-92% RH	No leaf injury	18

[7] Summer variety. [10] Second annual crop only. [11] Mixed culture with oats.

continued

Part I. Herbaceous Plants

Species (Synonym) [Developmental Stage]	Location [G or F]	Sulfur Dioxide Exposure		Environmental Conditions	Effects	Reference
		Concentration ppm Mean [Max]	Duration [Fumigation]			
119 *B. rapa*	Garson, Ont. [F]	0.23 [1.31 for 1 hr]	8 hr [N]	....	Leaf injury	5
120	New York, USA [G]	1.0	1-2 hr [A]	....	Leaf discoloration & necrosis	31
121 *Cichorium* sp.	New York, USA [F or G]	0.1-1.0	2-8 hr [A]	....	Moderate to severe leaf injury	34
122 *C. endivia*	New York, USA [G]	1.0	1-2 hr [A]	....	Leaf discoloration & necrosis	31
123 *Cucumis sativus*	Garson, Ont. [F]	0.25 [1.08 for 1 hr]	8 hr [N]	....	Leaf injury	5
124	[F or G]	0.5	4-8 hr [A]	....	Severe leaf injury	34
125 *Cucurbita* sp.	Rayside, Ont. [F]	0.38 [0.64 for 1 hr]	8 hr [N]	....	Leaf injury	5
126 *C. pepo*	[F or G]	0.5	4-8 hr [A]	....	Severe leaf injury	34
127 *Daucus carota*	Garson, Ont. [F]	0.25 [1.08 for 1 hr]	8 hr [N]	....	Leaf injury	5
128 'Nantaise'	Biersdorf, W. Ger. [F (pots)]	0.104	Gr. sn. [N]	HF & HCl in air	No ↑ or ↓ in carotene, ↓ in yield wt	12, 13
129 *Fagopyrum* sp.	Morgan, Ont. [F]	0.15 [0.56 for 1 hr]	8 hr [N]	Hot, humid (rain), cloudy	Leaf injury	5
130 *F. sagittatum (F. esculentum)*	Tokyo, Japan [G]	0.065	13 da [A]	0-120 ppm Na_2SO_4 added	↓ in yield wt, no ↓ in growth ht	9
131			21 da [A]	....	1.77% S content on 21st da, leaf injury on 16th	
132		0.13 or 0.26	13 da [A]	0-120 ppm Na_2SO_4 added	↓ in yield wt, ↓ in growth ht	
133		0.13	21 da [A]	....	2.74% S content on 21st da, leaf injury on 3rd	
134		0.26	12 da [A]	....	2.38% S content on 12th da, leaf injury on 1st	
135	Japan [G]	0.065	30 or 50 da [A]	....	1.2% S content, leaf injury	8
136		0.13	30 or 50 da [A]	....	0.75% S content, leaf injury	
137		0.26	30 or 50 da [A]	....	0.6% S content, leaf injury	
138	New York, USA [G]	0.4	7 hr [A]	....	Leaf discoloration	31, 32
139		0.98	6 hr [A]	....	100% leaf injury (discoloration, necrosis, withering) [12/]; slight leaf injury (discoloration, necrosis, withering) [13/]	33
140		1.44	5 hr [A]	....	No leaf injury [13/]	33
141	New York, USA [F or G]	0.5	4-8 hr [A]	....	Severe leaf injury	34
142 *Fragaria* sp. (*F. grandiflora*)	Japan [G]	1.0	2 hr/da for 40 da [A]	9-19°C, 59-92% RH	No leaf injury	18

[12/] After 4-6 hours in turgid plants. [13/] After 5-6 hours in wilted plants.

continued

114. EFFECTS OF LOW CONCENTRATIONS OF SULFUR DIOXIDE ON VASCULAR PLANTS

Part I. Herbaceous Plants

	Species (Synonym) [Developmental Stage]	Location [G or F]	Sulfur Dioxide Exposure		Environmental Conditions	Effects	Reference
			Concentration ppm Mean [Max]	Duration [Fumigation]			
143	*Glycine max* 'Hark' [21 da old]	USA [G]	0.05	4 hr [A]	$29 \pm 3°C$, ∿60% RH; ∿11,000 lux; 0.05 ppm NO_2	Slight leaf discoloration	29
144	'Kino'	Hereford, Ariz. [F]	>0.5-6.0	30 min [A]	....	Leaf necrosis 0.66% ↓ in yield quality/1% leaf necrosis	4
145	*Gossypium* sp.	Bayonne, N.J. & Staten Island, N.Y. [G]	0.28-0.32	10 wk [N]	0.09-0.11 ppm O_3	Leaf injury & chlorosis, ↓ in growth ht	14
146	*G. hirsutum*	[F or G]	0.5	4-8 hr [A]	....	Slight leaf injury	34
147	*Humulus lupulus* [Leaves, 5-10 sets]	Czechoslovakia [G]	0.14	4 hr [A]	Noon	No leaf injury on 1st da, moderate to severe leaf necrosis on 4th da	7
148			0.88	2 hr 25 min [A]	Evening	No leaf necrosis on 4th da	
149			1.20	3 hr [A]	Noon	Moderate leaf necrosis on 4th da	
150	*Ipomoea batatas*	[F or G]	0.5	4-8 hr [A]	....	Severe leaf injury	34
151	*Lactuca sativa*	Rayside, Ont. [F]	0.38 [0.64 for 1 hr]	8 hr [N]	....	Leaf injury	5
152		Japan [G]	1.0	2 hr/da for 40 da [A]	9-19°C, 59-92% RH	No leaf injury	18
153	var. *longifolia*	New York, USA [G]	1.0	1-2 hr [A]	....	Leaf discoloration & necrosis	31
154	*Lycopersicon esculentum*	Rayside, Ont. [F]	0.38 [0.64 for 1 hr]	8 hr [N]	....	Leaf injury	5
155		[F or G]	0.5	4-8 hr [A]	....	Severe leaf injury	34
156		New York, USA [G]	0.66	5 hr [A]	....	Leaf discoloration & necrosis	31
157			1.13	6 hr [A]	....	Moderate leaf injury (discoloration, necrosis) 12/; no leaf injury 13/	33
158	'Bonnie Best' [2nd internode]	....	0.5	4 hr/da, 2 da/wk [A]	....	Chloroplast disintegration	22
159	'Rhineland's Glory' [F (pots)]	Biersdorf, W. Ger.	0.061	Gr. sn. [N]	....	Severe leaf injury, no fruit injury, no ↓ in plant top yield wt	13
160			0.104	Gr. sn. [N]	HF & HCl in air	↓ in fruit maturation time & plant top yield wt, severe leaf injury, no fruit injury	12, 13
161	'Roma VF' [4-6 leaves]	USA [G]	0.5	4 hr [A]	$29 \pm 3°C$, ∿60% RH; ∿11,000 lux	No leaf injury	29
162	*Medicago sativa*	Logan, Utah [F]	0.141	1078 hr [A]	....	1% ↓ in photosynthesis, ↑ in respiration, slight leaf chlorosis	26

12/ After 4-6 hours in turgid plants.   13/ After 5-6 hours in wilted plants.

continued

Part I. Herbaceous Plants

Species (Synonym) [Developmental Stage]	Location [G or F]	Sulfur Dioxide Exposure Concentration ppm Mean [Max]	Duration [Fumigation]	Environmental Conditions	Effects	Reference
163		0.188	628 hr [A]	....	10[14]-27%[15] ↓ in pho-	
164		0.291	336 hr [A]	....	tosynthesis, ↑ in S content, 1.2% leaf necrosis[15], 5.5% leaf chlorosis[15]	
165	Stroh, Wash. [F]	0.220 [0.58 for 0.3 hr]	∿6.6 hr [N]	....	Slight leaf injury	20
166	New York, USA [F or G]	0.5	4-8 hr [A]	....	Severe leaf injury	34
167 [1 yr old]	Logan, Utah [F]	0.236	528 hr [A]	....	11-16% ↓ in photosynthesis, ↓ in chlorophyll, 1.7% leaf necrosis on 18th da, 7% leaf chlorosis & leaf abscission on 8th	26
168 [3rd crop]	Salt Lake City, Utah [F]	1.0	78 min [A]	....	No ↑ or ↓ in photosynthesis, 30% SO$_2$ absorption, no leaf necrosis	26
169		1.90	70 min [A]	....	88% ↓ in photosynthesis[16], 47% SO$_2$ absorption, 14% leaf necrosis	
170 'Eifel'	Biersdorf, W. Ger. [F (pots)]	0.009	Gr. sn. [N]	HF & HCl in air	Slight leaf discoloration & necrosis, no ↓ in yield wt or quality	12, 13
171		0.051	Gr. sn. [N]	HF & HCl in air	Severe leaf necrosis, 10% mortality, ↓ in yield wt[17] & quality	
172		0.104	Gr. sn. [N]	HF & HCl in air	Severe leaf necrosis, 20% mortality, ↓ in yield wt[18] & quality	
173 'Grimm'	New York, USA [G]	0.154	402 hr [A]	21(16-31)°C, 62 (30-77)% RH; 0-0.79 g S/100 ml nutrient solution added to soil	↑ in yield wt	21
174		0.2	25 da [A]	....	No leaf injury	31, 32
175		0.4	7 hr [A]	....	Leaf discoloration	
176	Summerland, B.C. [F]	0.10	504 hr [A]	46-100% RH	No leaf injury	20
177		0.14	525 hr [A]	30-95% RH	3% leaf injury	
178		0.20	87.5 hr [A]	58(32-87)% RH	No leaf injury	
179		0.30	66.5 hr [A]	54(34-80)% RH	Slight leaf injury	
180 [Cotyledon]	Summerland, B.C. [F]	0.52	67.5 hr [A]	30-75% RH	No leaf injury	20
181 'Province 34886' [3rd crop]	Logan, Utah [F]	0.430	240 min/da for 3 da [A]	....	↓ in photosynthesis[16], 0.2% leaf necrosis, 2.3% leaf chlorosis on 13th da	26
182		0.573	253 min/da for 3 da [A]	....	↓ in photosynthesis[16], slight leaf necrosis, slight leaf chlorosis on 13th da	
183 'Turkestan' [Transplanted seedling]	Utah [F]	0.1	6-7 hr/da, 6 da/wk Gr. sn. [A]	pH 5.5, low nutrient, no S added	↓ in transpiration, ↑ in CO$_2$ assimilation & yield wt, no ↑ or ↓ in chlorophyll	27

[14] First exposure period. [15] Second exposure period. [16] During exposure period only; recovery post treatment. [17] First of two crops only. [18] Two of two crops.

continued

Part I. Herbaceous Plants

	Species (Synonym) [Developmental Stage]	Location [G or F]	Sulfur Dioxide Exposure		Environmental Conditions	Effects	Reference
			Concentration ppm Mean [Max]	Duration [Fumigation]			
184					pH 5.5 or 7.0, normal nutrient, no S added	↓ in transpiration, ↑ in chlorophyll & yield wt	
185					pH 5.5, high nutrient, no S added	↓ in transpiration; ↑ in CO_2 assimilation, chlorophyll, yield wt	
186					pH 5.5, low nutrient, 0.8 ppm S added	↓ in transpiration, no ↑ or ↓ in chlorophyll, slight ↑ in yield wt	
187					pH 5.5, high nutrient, 0.8 ppm S added	↓ in transpiration, no ↑ or ↓ in chlorophyll & yield wt	
188					pH 5.5, low or high nutrient, 1.5 ppm S added	No ↑ or ↓ in transpiration, chlorophyll, yield wt	
189					pH 5.5, normal nutrient, 5 ppm S added	No ↑ or ↓ in transpiration & yield wt, slight ↓ in chlorophyll	
190					pH 7.0, normal nutrient, 5 ppm S added	No ↑ or ↓ in transpiration, chlorophyll, yield wt	
191					pH 5.5, low or high nutrient, 10 ppm S added	No ↑ or ↓ in transpiration, CO_2 assimilation, chlorophyll; slight ↓ in yield wt	
192					pH 5.5 or 7.0, normal nutrient, 90 ppm S added	No ↑ or ↓ in transpiration & yield wt, slight ↓ in chlorophyll	
193	'Turkestan 19301' [1 yr old]	Logan, Utah [F]	0.656	159 min [A]	....	↓ in photosynthesis [16], 78.5% SO_2 absorption, 0.1% leaf necrosis	26
194			0.69	170 min [A]	....	18% ↓ in photosynthesis [16], 80.7% SO_2 absorption, 0.1% leaf necrosis	
195			0.811	64 min [A]	....	44% ↓ in photosynthesis [16], 82.7% SO_2 absorption, slight leaf necrosis, 2.0% leaf chlorosis	
196			0.840	60 min [A]	....	25% ↓ in photosynthesis [16], 77.2% SO_2 absorption, slight leaf necrosis	
197			0.900	18 min [A]	....	13% ↓ in photosynthesis [16], 81% SO_2 absorption, 0.5% leaf necrosis, 5.0% leaf chlorosis	
198			0.978	253 min [A]	....	38% ↓ in photosynthesis [16], ↓ in respiration [16], 15.5% SO_2 absorption, 3-12% leaf necrosis & 11-14% leaf chlorosis on 36th da	

[16] During exposure period only; recovery post treatment.

continued

Part I. Herbaceous Plants

	Species (Synonym) [Developmental Stage]	Location [G or F]	Sulfur Dioxide Exposure		Environmental Conditions	Effects	Reference
			Concentration ppm Mean [Max]	Duration [Fumigation]			
199			1.26	14 min [A]	....	13% ↓ in photosynthesis [16], ∿140 mg SO_2 absorption	
200	'Turkestan 19303' [3rd crop]	Logan, Utah [F]	0.348	4.25 hr/da for 41 da [A]	Partial drought	4-6% ↓ in photosynthesis [16], slight leaf chlorosis	26
201	'Turkestan 19315'	Logan, Utah [F]	0.417	251 min/da for 5 da [A]	....	9% ↓ in photosynthesis [16], slight leaf necrosis, slight leaf chlorosis on 15th da	26
202	Melilotus sp.	[F or G]	0.5	4-8 hr [A]	....	Severe leaf injury	34
203	M. alba	New York, USA [G]	0.66	5 hr [A]	....	Leaf discoloration & necrosis	31
204			0.82	6 hr [A]	....	Moderate leaf injury (discoloration, necrosis) [12]; no leaf injury [13]	33
205			1.44	6 hr [A]	....	No leaf injury	33
206	Nicotiana sp.	Japan [G]	1.0	2 hr/da for 7 da [A]	25°C, 76-90% RH; 20-80 kilolux	↓ in photosynthesis, leaf injury	25
207	'Ksandy' [15-20 cm]	Giessen, W. Ger. [G]	0.38	12 hr/da for 9 da [A]	24°C, 58% RH; soil S deficient	No leaf injury, ↑ in yield wt & growth ht	6
208			0.56	12 hr/da for 9 da [A]	24°C, 58% RH; soil S deficient	No leaf injury, ↑ in yield wt & growth ht	
209	'W3'	Bayonne, N.J. & Staten Island, N.Y. [G]	0.28-0.32	10 wk [N]	0.09-0.11 ppm O_3	Leaf injury & chlorosis, ↓ in growth ht	14
210	N. glutinosa [10-12 wk old]	[G]	1.25	4 hr [A]	27°C, 70-80% RH; 10,760 lux	18% leaf necrosis	10
211	N. rustica var. brasilia [10-12 wk old]	[G]	1.25	4 hr [A]	27°C, 70-80% RH; 10,760 lux	76% leaf necrosis	10
212	N. tabacum 'Basma Serez'	Ottawa, Ont. [G]	1.0	⊁12 hr [A]	20 ± 1°C, >70% RH; 21,520 lux	1-25 % leaf injury in 10-12 hr	16
213	'Bel W3' [4-6 leaves]	USA [G]	0.5	4 hr [A]	29°C, ∿60% RH; ∿11,000 lux	No leaf injury	29
214	'Burley 21,37,49' & 'Ky 9'	Beltsville, Md. [G]	0.75	4 hr [A]	....	No leaf injury	19
215			1.25	4 hr [A]	....	2-10% leaf injury	
216	'Samsun' & 'Samsun (NN)' [10-12 wk old]	[G]	1.25	4 hr [A]	27°C, 70-80% RH; 10,760 lux	20-28% leaf necrosis	10
217	'Turkish'	[F or G]	0.5	4-8 hr [A]	....	Severe leaf injury	34
218	'White Gold'	Ottawa, Ont. [G]	0.20	⊁12 hr [A]	20 ± 1°C, >70% RH; 21,520 lux	No leaf injury	16
219			1.0	⊁10 hr [A]	20 ± 1°C, >70% RH; 21,520 lux	∿1% leaf injury in 10 hr	
220				⊁13 hr [A]	20 ± 1°C, >70% RH; 21,520 lux	2-30% leaf injury in 11-13 hr	

[12] After 4-6 hours in turgid plants. [13] After 5-6 hours in wilted plants. [16] During exposure period only; recovery post treatment.

continued

Part I. Herbaceous Plants

	Species (Synonym) [Developmental Stage]	Location [G or F]	Sulfur Dioxide Exposure Concentration ppm Mean [Max]	Duration [Fumigation]	Environmental Conditions	Effects	Reference
221	'Xanthi' [10-12 wk old]	[G]	1.25	4 hr [A]	27°C, 70-80% RH; 10,760 lux	17% leaf necrosis	10
222	*Phaseolus vulgaris*	Bayonne, N.J. & Staten Island, N.Y. [G]	0.28-0.32	10 wk [N]	0.09-0.11 ppm O_3	Leaf injury & chlorosis, ↓ in growth ht	14
223	[1-2 pair true leaves]	Czechoslovakia [G]	0.14	4 hr [A]	Noon	Leaf necrosis slight on 1st da, moderate on 4th da	7
224			0.70	3 hr [A]	Noon	Severe leaf necrosis on 1st-3rd da	
225	[21 da old]	USA [G]	0.05	4 hr [A]	29 ± 3°C, ∿60% RH; ∿11,000 lux; 0.05 ppm NO_2	2% leaf discoloration	29
226			0.5	4 hr [A]	29 ± 3°C, ∿60% RH; ∿11,000 lux	No leaf injury	
227	[1st trifoliate leaves]	[G]	0.5	4 hr/da, 2 da/wk [A]	....	Disintegration of chloroplasts, leaf injury (discoloration, necrosis, withering)	22
228	*Piper nigrum*	New York, USA [G]	1.0	5 hr [A]	....	Leaf discoloration & necrosis	31
229	*Pisum* sp.	Garson, Ont. [F]	0.17 [0.63 for 1 hr]	8 hr [N]	....	Leaf injury	5
230	*Raphanus* sp.	Rayside, Ont. [F]	0.38 [0.64 for 1 hr]	8 hr [N]	....	Leaf injury	5
231	'Siletta' [15-20 cm]	Giessen, W. Ger. [G]	0.08, 0.19, or 0.38	12 hr/da for 20 da [A]	13°C, 77% RH; soil S deficient	No leaf injury, ↑ in yield wt, slight ↓ in growth ht	6
232			0.56	12 hr/da for 20 da [A]	13°C, 77% RH; soil S deficient	Leaf necrosis, ↑ in yield wt, slight ↓ in growth ht	
	Raphanus sativus 'Cherry Belle'						
233	[21 da old]	USA [G]	0.05	4 hr [A]	29 ± 3°C, ∿60% RH; ∿11,000 lux; 0.05 ppm NO_2	Slight leaf necrosis and/or chlorosis	29
234			0.5	4 hr [A]	29 ± 3°C, ∿60% RH; ∿11,000 lux	No leaf injury	
235	[3-4 da after seeding]	USA [G]	0.05	8 hr/da, 5 da/wk for 5 wk [A]	20°C, 80% RH; 10,760 lux for 12 hr/da	No leaf injury, 17-30% ↓ in root yield wt, no ↓ in leaf yield wt	28
236	*Rheum* sp.	Garson, Ont. [F]	0.17 [0.63 for 1 hr]	8 hr [N]	....	Leaf injury	5
237	*Ricinus communis*	New York, USA [G]	1.0	5 hr [A]	....	Leaf discoloration & necrosis	31
238	*Solanum melongena*	[F or G]	0.5	4-8 hr [A]	....	Severe leaf injury	34
239		USA [G]	1.0	1-2 hr [A]	....	Leaf discoloration & necrosis	31
240	*S. tuberosum* 'Lori'	Biersdorf, W. Ger. [F (pots)]	0.015	Gr. sn. [N]	....	No ↑ or ↓ in vitamin C, ↓ in tuber yield wt	13

continued

Part I. Herbaceous Plants

Species (Synonym) [Developmental Stage]	Location [G or F]	Sulfur Dioxide Exposure		Environmental Conditions	Effects	Reference
		Concentration ppm Mean [Max]	Duration [Fumigation]			
241		0.024	Gr. sn. [N]	HF & HCl in air	No ↑ or ↓ in vitamin C, no ↓ in tuber yield wt	12, 13
242		0.05	Gr. sn. [N]	HF & HCl in air	No ↑ or ↓ in vitamin C, ↓ in tuber yield wt	
243	*Spinacia oleracea*	Garson, Ont. [F]	0.34 [1.34 for 1 hr]	8 hr [N]		Leaf injury	5
244		Japan [G]	1.0	2 hr/da for 40 da [A]	9-19°C, 59-92% RH	Moderate leaf injury	18
245	'Matador'	Biersdorf, W. Ger. [F (pots)]	0.009	Gr. sn. [N]	HF & HCl in air	Slight leaf necrosis, no ↓ in leaf yield wt [19]	12, 13
246			0.104	Gr. sn. [N]	HF & HCl in air	Leaf necrosis, ↓ in leaf yield wt [19]	
247	*Trifolium incarnatum* [20]	W. Ger. [F (pots)]	0.45	8 hr/da for 6 da [A]		96% ↓ in yield wt	11
248			0.90	4-8 hr/da for 3-6 da [A]		98.8-99.9% ↓ in yield wt	
249	*T. pratense*	Garson, Ont. [F]	0.17 [0.60 for 1 hr]	8 hr [N]	Warm, sunny	Leaf injury	5
250		[F (pots)]	0.94	12 hr [A]		54% ↓ in yield wt	11
251	*T. pratense* [21] [1st crop]	W. Ger. [F]	0.38	9 hr [A]		No ↓ in yield wt	11
252			0.75	9 hr [A]		14% ↓ in yield wt	
253			1.88	9 hr [A]		64% ↓ in yield wt	
254	[2nd crop]	W. Ger. [F]	0.38	25 hr [A]		33% ↓ in yield wt	11
255			0.75	25 hr [A]		25% ↓ in yield wt	
256			1.88	10 hr [A]		97% ↓ in yield wt	
257	'Eifel'	Biersdorf, W. Ger. [F (pots)]	0.009	Gr. sn. [N]	HF & HCl in air	Slight leaf necrosis, no ↓ in yield quality	12, 13
258			0.051	Gr. sn. [N]	HF & HCl in air	25% leaf necrosis, ↓ in yield wt [22] & quality	
259	*Vicia* sp.	New York, USA [F or G]	0.5	4-8 hr [A]		Moderate leaf injury	34
260	*V. faba* [5 wk old]	UK [G]	0.25	24 hr/da for 3 da [A]	20°C, 50-55% RH; 10,000 lux for 12 hr	↑ in stomata opening	17
261	*V. faba* [23]	W. Ger. [F]	0.38	48 hr [A]		36% ↓ in yield wt	11
262			0.75	24 hr [A]		63% ↓ in yield wt	

[19] Three of three crops. [20] Mixed culture with Italian ryegrass and hairy vetch. [21] Mixed culture with Italian ryegrass. [22] Second of two crops only. [23] Mixed culture with lupine, field peas, and common vetch.

Contributors: Palm, Paul E., Nick, M. Susan, Arnold, Elsie P., and Platz, Barbara B.

References

[1] Applegate, H. G., and L. C. Durrant. 1969. Environ. Sci. Technol. 3(8):759.

[2] Bleasdale, J. K. A. 1952. Nature (London) 169:376.

[3] Brennan, E., and P. M. Halisky. 1970. Phytopathology 60(11):1544.

[4] Davis, C. R. 1972. J. Air Pollut. Contr. Ass. 22(12):964.

[5] Dreisinger, B. R. 1965. Proc. Air Pollut. Contr. Ass. Annu. Meet., 58th, Toronto.

[6] Faller, N., et al. 1970. Plant Soil 33:177.

[7] Fiala, V., and P. Hautke. 1962. Rostlinna Vyroba (Prague) 8:1043.

[8] Fujiwara, T. 1968. Nippon Shokubutsu Byori Gakkaiho 34(5):336.

continued

Part I. Herbaceous Plants

[9] Fujiwara, T. 1970. Ibid. 36:127.

[10] Grosso, J. J., et al. 1971. Phytopathology 61:945.

[11] Guderian, R. 1967. Schriftenr. Landesanst. Immiss. Bodennutzungschutz Landes Nordrhein-Westfalen 4: 80.

[12] Guderian, R., and H. Stratmann. 1962. Forschungsber. Landes Nordrhein-Westfalen 1118:5.

[13] Guderian, R., and H. Stratmann. 1968. Ibid. 1920:3.

[14] Hindawi, I. J. 1968. J. Air Pollut. Contr. Ass. 18(5): 307.

[15] Kamada, S., et al. 1969. Taiki Osen Kenkyu 4(1): 131.

[16] Macdowall, F. D. H., and A. F. W. Cole. 1971. Atmos. Environ. 5:553.

[17] Majernik, O., and T. A. Mansfield. 1970. Nature (London) 227(5256):377.

[18] Matsushima, J., and M. Harada. 1964. Mie Daigaku Nogakubu Gakujutsu Hokoku 30:11.

[19] Menser, H. A., and G. H. Hodges. 1970. Agron. J. 62(2):265.

[20] National Research Council of Canada. 1939. Nat. Res. Counc. Can. Publ. 815.

[21] Setterstrom, C., et al. 1938. Contrib. Boyce Thompson Inst. 9:179.

[22] Solberg, R. A., and D. F. Adams. 1956. Amer. J. Bot. 43:755.

[23] Spedding, D. J. 1969. Nature (London) 224:1229.

[24] Swain, R. E., and A. B. Johnson. 1936. Ind. Eng. Chem. Ind. Ed. 28(1):42.

[25] Taniyama, T., and H. Arikado. 1968. Nippon Sakumotsu Gakkai Kiji 37:366.

[26] Thomas, M. D., and G. R. Hill. 1937. Plant Physiol. 12:309.

[27] Thomas, M. D., et al. 1943. Ibid. 18:345.

[28] Tingey, D. T., et al. 1971. J. Amer. Soc. Hort. Sci. 96(3):369.

[29] Tingey, D. T., et al. 1971. Phytopathology 61: 1506.

[30] Zahn, R. 1970. Staub 30(4):162.

[31] Zimmerman, P. W. 1952. Proc. Gov. Conf. Exhib. Atmos. Pollut., Trenton, p. 23.

[32] Zimmerman, P. W. 1952. Proc. U.S. Tech. Conf. Air Pollut., p. 127.

[33] Zimmerman, P. W., and W. Crocker. 1934. Contrib. Boyce Thompson Inst. 6:455.

[34] Zimmerman, P. W., and A. E. Hitchcock. 1956. Ibid. 18(6):263.

Part II. Ligneous Plants

Species (Synonym) [Developmental Stage]	Location [G, F, or Fo]	Sulfur Dioxide Exposure		Environmental Conditions	Effects	Reference
		Concentration ppm Mean [Max]	Duration [Fumigation]			
Deciduous Trees						
1 *Acer* sp.	Skead, Ont. [F]	0.045 [3.64 for 0.5 hr]	Gr. sn. [N]	....	No leaf injury	4
2 *A. platanoides* [3-yr old, new leaves]	New York, USA [G]	0.50	24 hr/da for 30 da [A]	16-27°C, 70% RH; ≯26,900 lux for 15 hr	Leaf chlorosis on 12th da	26
3		0.75	8 hr/da for 30 da [A]	16-27°C, 70% RH; ≯26,900 lux for 15 hr	Leaf chlorosis on 21st da	
4		2.0	4-8 hr [A]	27°C, 70% RH; 26,900 lux	0 to <10% leaf necrosis	
5 *A. pseudoplatanus* [Shrub]	UK [F]	0.83	3 hr [A]	17-23°C, 55-80% RH	20% leaf necrosis	24, 25
6		1.40	3 hr [A]	17-23°C, 55-80% RH	80% leaf necrosis	
7 *Alnus* sp.	Rayside, Ont. [F]	0.21 [0.46 for 1 hr]	8 hr [N]	....	Leaf injury	4
8 *A. incana*	Stangenberg, E. Ger. [Fo]	0.08 [>0.46]	Gr. sn. [N]	8°C; As, HCl, & smoke in air	↓ in chlorophyll, ht, & circumference; severe leaf injury (discoloration, necrosis)	19

continued

Part II. Ligneous Plants

Species (Synonym) [Developmental Stage]	Location [G, F, or Fo]	Sulfur Dioxide Exposure		Environmental Conditions	Effects	Reference
		Concentration ppm Mean [Max]	Duration [Fumigation]			
9 Betula alba	Callum, Ont. & Kukagami, Ont. [F]	0.19-0.21 [0.46-0.48 for 1 hr]	8 hr [N]	....	Leaf injury	4
10 B. papyrifera	Marble, Wash. [F]	0.153 [0.41 for 0.3 hr]	46 hr [N]	Humid	No leaf injury	17
11	Stroh, Wash. [F]	0.254 [0.67 for 0.3 hr]	47.6 hr [N]	Humid	Severe leaf injury	
12 Cephalanthus sp.	[F or G]	0.50	4-8 hr [A]	....	Severe leaf injury	27
13 Citrus aurantium	[G]	1.0	2 hr/da for 21 da [A]	....	No ↑ or ↓ in S, Ca, K, Mg	14
14 [2 yr old]	[G]	1.0	2 hr/da for 34 da [A]	....	↑ in S, N, Ca; ↓ in K & total linear growth; 0.7% leaf abscission	14, 16
15 [2-3 yr old]	[G]	1.0	2 hr/da for 40 da [A]	9-19°C, 59-92% RH	No leaf injury, ↓ in total linear growth	15
16 C. hassaku	[G]	1.0	2 hr/da for 13 da [A]	....	↑ in S, no leaf abscission	14
17 [2 yr old]	[G]	1.0	2 hr/da for 24-34 da [A]	....	↑ in S, slight ↑ in N & Ca, 0-2.3% leaf abscission, ↓ in total linear growth	14, 16
18 [2-3 yr old]	[G]	1.0	2 hr/da for 40 da [A]	9-19°C, 59-92% RH	No leaf injury, ↓ in total linear growth	15
19 C. nobilis var. unshiu 19 [2 yr old]	[G]	1.0	2 hr/da for 13 da [A]	....	↑ in S, no ↑ or ↓ in N, slight ↑ in Ca, 0.9% leaf abscission	14, 16
20			2 hr/da for 24 da [A]	....	↑ in S, no ↑ or ↓ in N, slight ↓ in Ca, 2.2% leaf abscission	16
21			2 hr/da for 34 da [A]	....	↑ in S, N, & total linear growth; no ↑ or ↓ in Ca; ↓ in K; 15.8% leaf abscission	14, 16
22 [2-3 yr old]	[G]	1.0	2 hr/da for 40 da [A]	9-19°C, 59-92% RH	No leaf injury, ↑ in total linear growth	15
23 Corylus sp.	Burwash, Ont. [Fo]	0.23 [1.14 for 1 hr]	8 hr [N]	....	Leaf injury	4
24 C. cornuta 'California' [Small transplant]	Wenatchee, Wash. [F]	0.53-0.60	7 hr [A]	....	No leaf injury	22
25		1.00-1.08	4-6 hr [A]	....	10-38% leaf necrosis	
26 Crataegus sp.	New Jersey, USA [F]	0.1	3 da [N]	Fog	Leaf necrosis	1
27 Fagus silvatica	UK [F]	0.77-1.01	3 hr [A]	17-23°C, 55-80% RH	10-20% leaf necrosis	24
28 [3 yr old]	Biersdorf, W. Ger. [F (pots)]	0.015-0.024	Gr. sn. [N]	HF & HCl in air	↓ in shoot & cross-sectional area growth	5,6
29 Ginkgo biloba [3 yr old, new leaves]	New York, USA [G]	0.25-0.50	24 hr/da for 30 da [A]	16-27°C, 70% RH; ≥26,900 lux for 15 hr	Leaf chlorosis on 30th da (at 0.25 ppm) & on 14th da (at 0.50 ppm)	26

continued

Part II. Ligneous Plants

Species (Synonym) [Developmental Stage]	Location [G, F, or Fo]	Sulfur Dioxide Exposure Concentration ppm Mean [Max]	Duration [Fumigation]	Environmental Conditions	Effects	Reference
30		0.75	8 hr/da for 30 da [A]	16-27°C, 70% RH; ⊅26,900 lux for 15 hr	Leaf chlorosis on 30th da	
31		2.0	8 hr [A]	27°C, 70% RH; 26,900 lux	No leaf necrosis	
32 *Malus* sp.	Marble, Wash. [F]	0.280 [0.93 for 0.3 hr]	21 hr [N]	....	No leaf injury	17
33	Stroh, Wash. [F]	0.437 [2.11 for 0.3 hr]	24 hr [N]	....	Severe leaf injury	17
34	[F or G]	0.50	4-8 hr [A]	....	Slight leaf injury	27
35 'Manks Codlin'	UK [F]	0.48	6 hr [A]	21-28°C, 45-66% RH	60% leaf necrosis	24
36 *M. sylvestris (M.*	Biersdorf, W. Ger.	0.015-0.024	Gr. sn. [N]	HF & HCl in air	↓ in yield quality	5,6
37 *communis)* 'Elli-sons Orange' [1] [2 yr old]	[F (pots)]	0.05	Gr. sn. [N]	HF & HCl in air	↓ in yield wt & in thickness growth	
38 *Platanus acerifolia, P. occidentalis, P. orientalis, P. racemosa, P. wrightii* [Seedlings]	Beltsville, Md. [G]	2.0	3 hr [A]		Leaf discoloration & abscission	21
39 *Populus grandidentata*	Skead, Ont. [F]	0.20 [0.66 for 1 hr]	8 hr [N]		Leaf injury	4
40 *P. tremuloides*	St. Charles, Ont. [F]	0.13 [0.42 for 1 hr]	8 hr [N]	Humid (drizzle)	Leaf injury	4
41 *Prunus* sp.	[F or G]	0.5	4-8 hr [A]	....	Severe leaf injury	27
42 'Burbank'	[F or G]	0.1-1.0	2-8 hr [A]	....	Moderate leaf injury	
43 'Moorpark'	[F or G]	0.50	4-8 hr [A]	....	Slight leaf injury	
44 *P. avium* 'Primavera' [2]	Biersdorf, W. Ger. [F (pots)]	0.061	Gr. sn. [N]	....	↓ in shoot growth	6
45		0.104	Gr. sn. [N]	HF & HCl in air	75% leaf necrosis, ↓ in shoot growth	5,6
46		0.05	Gr. sn. [N]	HF & HCl in air	↓ in thickness growth	5,6
P. cerasus 'Grosse Lange Loth' [3]						5,6
47 [1 yr old]	Biersdorf, W. Ger. [F (pots)]	0.050	Gr. sn. [N]	HF & HCl in air	↓ in shoot growth	
48 [2 yr old]	Biersdorf, W. Ger. [F (pots)]	0.024	Gr. sn. [N]	HF & HCl in air	↓ in yield wt	
49 *P. domestica* [4]	Biersdorf, W. Ger. [F (pots)]	0.061-0.104	Gr. sn. [N]	HF & HCl in air	↓ in fruit yield wt, quality, & in thickness growth	5,6
50 [2 yr old]	Biersdorf, W. Ger. [F (pots)]	0.05	Gr. sn. [N]	HF & HCl in air	↓ in thickness	
51 *P. padus* [Shrub]	UK	0.64	3 hr [A]	....	No leaf injury	24,
52		1.43	3 hr [A]	17-23°C, 55-80% RH	30% leaf necrosis	25
53 *Pyrus* sp. [5]	New Jersey, USA [F]	0.10	3 da [N]	Fog	Leaf necrosis	1
54 *Pyrus* sp. 'Zwijndrechtse wijnpeern' [6]	UK [F]	0.50	6 hr [A]	20-25°C, 58-80% RH	40% leaf necrosis	24

[1] Grafted on EM IX. [2] Grafted on F12/1. [3] Grafted on *P. mahaleb*. [4] Grafted on St. Julien. [5] Crab apple. [6] Pear.

continued

Part II. Ligneous Plants

	Species (Synonym) [Developmental Stage]	Location [G, F, or Fo]	Sulfur Dioxide Exposure		Environmental Conditions	Effects	Reference
			Concentration ppm Mean [Max]	Duration [Fumigation]			
55	'Chojura' [Blossom before, during, or after pollination]	Chiba, Japan [G]	0.2	8 hr [A]	....	No ↓ in fruit yield no.	13
56	[Blossom before pollination]	Chiba, Japan [G]	0.5	8 hr [A]	....	Slight ↓ in fruit yield no.	13
57	[Blossom during pollination]	Chiba, Japan [G]	0.5	8 hr [A]	....	Severe ↓ in fruit yield no.	13
58	[Blossom after pollination]	Chiba, Japan [G]	0.5	8 hr [A]	....	Moderate ↓ in fruit yield no.	13
59	'Conference'	UK [F]	0.51	6 hr [A]	18-24°C, 63-77% RH	50% leaf necrosis	24
60	'Legipont'	UK [F]	0.48	6 hr [A]	15-22°C, 55-96% RH	40% leaf necrosis [z/]	24
61	Quercus palustris [3 yr old, new leaves]	New York, USA [G]	0.25-0.50	24 hr/da for 30 da [A]	16-27°C, 70% RH; ⋗26,900 lux for 15 hr	No leaf injury or leaf chlorosis on 30th da	26
62			0.75	8 hr/da for 30 da [A]	16-27°C, 70% RH; ⋗26,900 lux for 15 hr	No leaf injury	
63	Q. robur (Q. pedunculata) [3 yr old]	Biersdorf, W. Ger. [F (pots)]	0.024	Gr. sn. [N]	HF & HCl in air	↓ in thickness & cross-sectional area growth	5,6
64			0.051	Gr. sn. [N]	HF & HCl in air	↓ in shoot growth	
65	Salix sp.	Kukagami, Ont. [Fo]	0.30 [0.41 for 1 hr]	8 hr [N]	....	Leaf injury	4
66	Solanum pseudocapsicum	New York, USA [F or G]	0.5	4-8 hr [A]	....	No leaf injury	27
67	Sorbus aucuparia [Shrub]	UK [F]	0.54-1.43	3 hr [A]	17-23°C, 55-80% RH	30-100% leaf necrosis	24, 25
68	Ulmus americana [Seedling]	Beltsville, Md. [G]	2.0	3 hr [A]	24°C, 85% RH	Leaf injury (discoloration, necrosis, withering, abscission)	21
69	U. parvifolia [3 yr old, new leaves]	New York, USA [G]	0.25	24 hr/da for 30 da [A]	16-27°C, 70% RH; ⋗26,900 lux for 15 hr	Leaf chlorosis on 11th da, leaf necrosis on 30th da	26
70			0.75	8 hr/da for 30 da [A]	16-27°C, 70% RH; ⋗26,900 lux for 15 hr	Leaf chlorosis on 5th da	
71			2.0	2-8 hr [A]	27°C, 70% RH; 26,900 lux	10-90% leaf necrosis	
			Conifers				
72	Larix sp.	Kukagami, Ont. [Fo]	0.26 [0.41 for 1 hr]	8 hr [N]	....	Leaf injury	4
73	L. dahurica [Adult]	Stangenberg, E. Ger. [Fo]	0.08 [>0.45]	Gr. sn. [N]	7.7°C; As, HCl, & smoke in air	↓ in chlorophyll, ht, & circumference; moderate leaf discoloration; leaf necrosis	19

z/ Leaves wet from rain for 3-5 hr.

continued

Part II. Ligneous Plants

| Species (Synonym) [Developmental Stage] | Location [G, F, or Fo] | Sulfur Dioxide Exposure | | Environmental Conditions | Effects | Reference |
		Concentration ppm Mean [Max]	Duration [Fumigation]			
74 *L. decidua* [Adult]	Stangenberg, E. Ger. [Fo]	0.08 [>0.45]	Gr. sn. [N]	7.7°C; As, HCl, & smoke in air	↓ in chlorophyll, ht, & circumference; severe leaf discoloration; leaf necrosis; 100% mortality	19
75 *(L. europaea)* [3 yr old]	Biersdorf, W. Ger. [F (pots)]	0.015-0.024	Gr. sn. [N]	HF & HCl in air	↓ in thickness & cross-sectional area growth	5,6
76		0.05	Gr. sn. [N]	HF & HCl in air	↓ in shoot growth	
77 *L. leptolepis*[8] [2 yr old]	Tharandt, E. Ger. [G]	0.34	20-50 hr [A]	12-21°C, 56-94% RH	Slight to moderate leaf discoloration	23
78		0.73	10 hr [A]	12-21°C, 56-94% RH	Severe leaf discoloration	
79 *L. occidentalis*	Marble, Wash. [Fo]	0.153 [0.41]	46 hr [N]	....	No leaf injury	17
80	Stroh, Wash. [Fo]	0.254 [0.67]	47.6 hr [N]	....	Leaf discoloration & necrosis	17
81 [Small transplant]	Wenatchee, Wash.	0.53-0.60	7 hr [A]	....	35% leaf necrosis	22
82	[F]	1.00-1.08	5 hr [A]	....	86% leaf necrosis	
83 [4- to 6-yr-old transplant]	Summerland, B.C.	0.30	8 hr [A]	28°C, 67% RH	Slight leaf injury	17
84	[F]	0.41	83 hr [A]	15-35°C, 40-95% RH	Severe leaf injury	
85 *Picea* sp.	Vykmanov, Krusne Hory, Czechoslovakia [Fo]	0.031 [>0.188 (1.2%)[9]]	8 mo avg [N]	....	↑ in S, leaf injury	12
86	Skead, Ont. [Fo]	0.045 [3.64 for 0.5 hr]	10 yr Gr. sn. [N]	....	No leaf injury	4
87 [Normal grafting]	Munich, W. Ger. [G]	0.75	15 da [A]	....	Severe leaf injury (discoloration, withering, abscission)	20
88 [2-3 yr old, resistant scion]	Munich, W. Ger.	0.75	15 da [A]	....	Leaf discoloration	20
89	[G]	1.5	6 da [A]	....		
90 [Adult]	Celna, Krusne Hory, Czechoslovakia [Fo]	0.021 [>0.188 (0.1%)[9]]	8 mo avg [N]	....	Moderate leaf injury	11, 12
91	Behanky, Krusne Hory, Czechoslovakia [Fo]	0.052 [0.376 (0.43%)[9]]	12 mo avg [N]	....	Severe leaf injury	12
92	Mnisek, Krusne Hory, Czechoslovakia [Fo]	0.055 [0.376 (0.53%)[9]]	3 mo avg [N]	....	Severe leaf injury, mortality	11, 12
93 [80 yr old]	Moscow, USSR [Fo]	0.11-0.41	[N]	0.27-1.87 mg/m³ NO_x	↑ in S, ↓ in N, mortality	18
94 *P. abies* [Adult]	Stangenberg, E. Ger. [Fo]	0.08 [>0.45]	Gr. sn. [N]	7.7°C; As, HCl, & smoke in air	↓ in chlorophyll, ht, & circumference; severe leaf discoloration; leaf necrosis; 57-90% mortality	19
95 *P. abies (P. excelsa)* [4 yr old]	Biersdorf, W. Ger. [F (pots)]	0.024	Gr. sn. [N]	HF & HCl in air	↓ in thickness growth	5,6

[8] Hybrid of *L. leptolepis* '219' and '101' with *L. decidua*. [9] Percent of total measurement time.

continued

Part II. Ligneous Plants

	Species (Synonym) [Developmental Stage]	Location [G, F, or Fo]	Sulfur Dioxide Exposure		Environmental Conditions	Effects	Reference
			Concentration ppm Mean [Max]	Duration [Fumigation]			
96	*Pinus* sp.	Skead, Ont. [Fo]	0.33 [0.66 for 1 hr]	8 hr [N]	....	Leaf injury	4
97		St. Charles, Ont. [Fo]	0.20 [0.52 for 1 hr]	8 hr [N]	Warm, fog, cloudy	Leaf injury	
98	*P. banksiana* [Adult]	Stangenberg, E. Ger. [Fo]	0.08 [>0.45]	Gr. sn. [N]	7.7°C; As, HCl, & smoke in air	↓ in chlorophyll, ht, & circumference; severe leaf discoloration; leaf necrosis; 57-90% mortality	19
99	*P. contorta* [Small transplant]	Wenatchee, Wash. [F]	0.53-0.60	7 hr [A]	....	No leaf injury	22
100			1.00-1.08	4-7 hr [A]	....	No leaf injury in 5 hr, 33% leaf necrosis in 4 hr & 50% in 7 hr	22
101	[6-yr-old transplant]	Summerland, B.C. [F]	0.46	736 hr [A]	−11 to +30°C, 40-100% RH	↑ in S, trace leaf injury	17
102	*P. monticola* [Small transplant]	Wenatchee, Wash. [F]	1.00-1.08	7 hr [A]	....	33% leaf necrosis	22
103	*P. ponderosa*	Summerland, B.C. [Fo]	0.25	450 hr [A]	0.5-36°C, 12-94% RH	↑ in S, no leaf injury	17
104			0.50	336 hr [A]	−3 to +12°C, 32-98% RH	↑ in S, trace leaf injury	
105			0.50	1008 hr [A]	2-30°C, 12-94% RH	↑ in S, no leaf injury	
106			0.75	147 hr [A]	−3 to +9°C, 32-95% RH, cloudy	↑ in S, 55% leaf injury	
107			1.50	150 hr [A]	11-35°C, 13-91% RH	10% leaf necrosis in 143 hr	
108	[1-yr-old seedling]	Summerland, B.C. [F (pots)]	0.29	44.5 hr [A]	9-36°C, 33-78% RH	Slight leaf injury	17
109	[1- to 2-yr-old seedling]	Summerland, B.C. [F (pots)]	0.50	6 hr [A]	24.5-36°C, 45-65% RH	No leaf injury	17
110	[2-yr-old seedling]	Summerland, B.C. [F (pots)]	0.29	44.5 hr [A]	9-36°C, 33-78% RH	No leaf injury	17
111	[6-yr-old seedling]	Summerland, B.C. [F]	0.33	48 hr [A]	15-34°C, 35-70% RH	No leaf injury	17
112			0.39	6.2 hr [A]	21-30°C, 65-85% RH	Slight leaf necrosis	
113			0.44	18.7 hr [A]	4.5-30°C, 55-85% RH	No leaf injury	
114	[Small transplant]	Wenatchee, Wash. [F]	1.00-1.08	5 or 7 hr [A]	....	0-20% leaf necrosis	22
115	[3- to 10-yr-old transplant]	Summerland, B.C. [F]	0.46	736 hr [A]	−10 to +30°C, 40-100% RH	↑ in S, 20% leaf injury	17
116	*P. resinosa*	Skead, Ont. [Fo]	0.30 [0.78 for 1 hr]	8 hr [N]	....	Leaf injury	4
117	*P. strobus*	Penage, Ont. [Fo]	0.21 [0.45 for 1 hr]	8 hr [N]	Humid (rain), cloudy	Leaf injury	4
118	[65-85 yr old]	Emerald Lake, Ont. [Fo]	0.008 [0.63 for 0.5 hr]	10 yr Gr. sn. [N]	....	2.5-37.5% leaf injury [10], slight ↑ in mortality	4,9, 10

[10] First- and second-year needles, respectively.

continued

Part II. Ligneous Plants

	Species (Synonym) [Developmental Stage]	Location [G, F, or Fo]	Sulfur Dioxide Exposure		Environmental Conditions	Effects	Reference
			Concentration ppm Mean [Max]	Duration [Fumigation]			
119		Lake Panache, Ont. [Fo]	0.009 [1.32 for 0.5 hr]	10 yr Gr. sn. [N]	....	6.0-43.2% leaf injury [10]	
120		Portage Bay, Ont. [Fo]	0.017 [1.24 for 0.5 hr]	10 yr Gr. sn. [N]	....	21-77% leaf injury [10], leaf abscission, slight ↑ in mortality, ↓ in yield volume	
121		West Bay, Ont. [Fo]	0.045 [3.64 for 0.5 hr]	10 yr Gr. sn. [N]	....	38-96% leaf injury [10], leaf abscission, slight ↑ in mortality, ↓ in yield volume	
122	[New needles, sensitive scion [11]]	[G]	0.05	1-3 hr [A]	26-31°C, 80% RH, sunlight	Slight leaf discoloration in 1-2 hr, slight leaf necrosis in 3 hr	2
123	[New needles, resistant scion [11]]	[G]	0.08	2 hr [A]	26-31°C, 80% RH, sunlight	No leaf injury	
124	[Sensitive scion [11]]	Pisgah, N.C. [F (pots)]	0.06-0.12	Gr. sn. [N]	0.1-0.4 pphm O_3	Leaf necrosis	3
125	[Resistant scion [11]]	Pisgah, N.C. [F (pots)]	0.06-0.12	Gr. sn. [N]	0.1-0.4 pphm O_3	No leaf injury	3
126	[10 yr old, sensitive]	Pisgah, N.C. [Fo]	0.06-0.12	Gr. sn. [N]	0.1-0.4 pphm O_3	Leaf necrosis & chlorosis, ↓ in needle growth	3
127	[10 yr old, resistant]	Pisgah, N.C. [Fo]	0.06-0.12	Gr. sn. [N]	0.1-0.4 pphm O_3	Trace leaf injury	3
128	[3 yr old, diseased]	[G]	0.05	12 hr/da for 30 da [A]	....	No leaf injury	7
129	P. sylvestris	W. Ger. [Fo]	0.03	Gr. sn. [N]	....	Slight leaf injury	8
130	[3 yr old]	Biersdorf, W. Ger. [F (pots)]	0.015-0.024	Gr. sn. [N]	HF & HCl in air	↓ in cross-sectional area growth	5,6
131			0.05	Gr. sn. [N]	HF & HCl in air	↓ in cross-sectional area growth	
132	[Adult]	W. Ger. [Fo]	0.07	Gr. sn. [N]	....	Severe leaf injury, mortality	8
133		Stangenberg, E. Ger. [Fo]	0.08 [>0.45]	Gr. sn. [N]	7.7°C; As, HCl, & smoke in air	↓ in chlorophyll, ht, & circumference; severe leaf discoloration; leaf necrosis; 45% mortality	19
134	Pseudotsuga menziesii [Small transplant]	Wenatchee, Wash. [F]	1.00-1.08	4-7 hr [A]	....	0-1% leaf injury in 4, 5, or 7 hr; 66% leaf necrosis in 6 hr	22
135	P. taxifolia	Summerland, B.C. [Fo]	0.25	450 hr [A]	−0.5 to +36°C, 14-94% RH	↑ in S, no leaf injury	17
136			0.50	52 daylight hr [A]	9-32°C, 12-92% RH, mostly cloudy	↑ in S, no leaf injury	
137			0.75	72 hr [A]	5-15°C, 38-72% RH, cloudy	↑ in S, no leaf injury	
138			1.50	150 hr [A]	11-35°C, 13-91% RH	↑ in S, slight leaf injury in 143 hr	
139	[2-yr-old seedling]	Summerland, B.C. [F (pots)]	0.29	44.5 hr [A]	10-36°C, 33-78% RH	Slight injury	

[10] First- and second-year needles, respectively. [11] Grafted on 2-year-old rootstock.

continued

114. EFFECTS OF LOW CONCENTRATIONS OF SULFUR DIOXIDE ON VASCULAR PLANTS

Part II. Ligneous Plants

Species (Synonym) [Developmental Stage]	Location [G, F, or Fo]	Sulfur Dioxide Exposure		Environmental Conditions	Effects	Reference
		Concentration ppm Mean [Max]	Duration [Fumigation]			
140 [3-yr-old transplant]	Summerland, B.C. [F]	0.22	1656 hr [A]	−10 to +24°C, 41-100% RH	↑ in S, no leaf injury	
141 *Taxus* sp.	[F or G]	0.5	4-8 hr [A]	....	Moderate leaf injury	27

Contributors: Palm, Paul E., Nick, M. Susan, Arnold, Elsie P., and Platz, Barbara B.

References

[1] Brennan, E., et al. 1967. Plant Dis. Rep. 51(10): 850.
[2] Costonis, A. C. 1970. Phytopathology 60(6):994.
[3] Costonis, A. C. 1971. Ibid. 61:717.
[4] Dreisinger, B. R. 1965. Proc. Air Pollut. Contr. Ass. Annu. Meet., 58th, Toronto.
[5] Guderian, R., and H. Stratmann. 1962. Forschungsber. Landes Nordrhein-Westfalen 1118:5.
[6] Guderian, R., and H. Stratmann. 1968. Ibid. 1920:3.
[7] Jaeger, J., and W. Banfield. 1970. Phytopathology 60:575.
[8] Knabe, W., and K. H. Günther. 1971. Allg. Forstz. 26(24):503, 513.
[9] Linzon, S. N. 1958. Rep. Ont. Dep. Lands Forest. and Ont. Dep. Mines, p. 1.
[10] Linzon, S. N. 1971. J. Air Pollut. Contr. Ass. 21(2): 81.
[11] Materna, J. 1966. Proc. Conf. Eff. Ind. Emiss. Forest. Janske Lazne, Czech., p. III-1.
[12] Materna, J., et al. 1969. Ochr. Ovzdusi 1(6):84.
[13] Matsuoka, Y., et al. 1969. Taiki Osen Kenkyu 4(1): 130.
[14] Matsushima, J. 1969. Proc. Int. Citrus Symp., 1st, 2:733.
[15] Matsushima, J., and M. Harada. 1964. Mie Daigaku Nogakubu Gakujutsu Hokoku 30:11.
[16] Matsushima, J., and M. Harada. 1966. Engei Gakkai Zasshi 35(3):241.
[17] National Research Council of Canada. 1939. Nat. Res. Counc. Can. Publ. 815.
[18] Popov, B. V. 1969. Amer. Inst. Crop Ecol. Surv. USSR Air Pollut. Lit. 2:60.
[19] Ranft, H. 1966. Proc. Conf. Eff. Ind. Emiss. Forest. Janske Lazne, Czech., p. XV-1.
[20] Rohmeder, E., et al. 1962. Forstwiss. Zentralbl. 81: 321.
[21] Santamour, F. S., Jr. 1969. Plant Dis. Rep. 53(6): 482.
[22] Scheffer, T. C., and G. G. Hedgcock. 1955. U.S. Forest Serv. Tech. Bull. 1117.
[23] Schönbach, H., et al. 1965. Zuechter 34(10):312.
[24] Spierings, F. 1967. Atmos. Environ. 1(3):205.
[25] Spierings, F. 1967. Int. Union Forest Res. Organ., 14th Congr., Sect. 24 Pap., p. 567.
[26] Temple, P. J. 1972. J. Air Pollut. Contr. Ass. 22(4): 271.
[27] Zimmerman, P. W., and A. E. Hitchcock. 1956. Contrib. Boyce Thompson Inst. 18(6):263.

115. CIRCADIAN RHYTHMS IN CONTINUOUS LIGHT AND DARKNESS: VERTEBRATES

Data are for free-running periods in constant darkness (indicated by 0 light intensity), or in constant light [32]. Values represent means, or ranges when hyphenated.

Species (Synonym)	No. of Animals	Consecutive Days Recorded	Rhythm	Light Intensity lux	Circadian Period hr	Reference
			Mammalia			
1 *Ammospermophilus leucurus*	3	Weeks	Locomotor activity	0	24.2	41
2	3	Weeks	Locomotor activity	0	24.3	42
3	3	Weeks	Locomotor activity	11	24.2	42
4	1	Weeks	Locomotor activity	11	24.2	41

continued

	Species (Synonym)	No. of Animals	Consecutive Days Recorded	Rhythm	Light Intensity lux	Circadian Period hr	Reference
5	*Apodemus flavicollis*	1	Weeks	Locomotor activity	16.5	24.3[1]	17
6		1	Weeks	Locomotor activity	16.5	25.2[2]	17
7	*Clethrionomys glareolus*	14	10 da	Total activity	0	23.8	31
8	*C. rufocanus*	22	15 da	Total activity	0	24.0	31
9	*C. rutilus*	3	Weeks	Locomotor activity	0	24.1	41,42
10		1	Weeks	Locomotor activity	17	24.7	41,42
11		1	Weeks	Locomotor activity	178	24.8	41,42
12	*Dicrostonyx groenlandicus*	1	Weeks	Locomotor activity	0	23.6	41
13		1	Weeks	Locomotor activity	0	24.7	42
14		1	Weeks	Locomotor activity	16	24.8	42
15	*Eptesicus fuscus*	1	Weeks	Total activity at 22°C	0	21.1	38
16	*Eutamias umbrinus*	1	Weeks	Locomotor activity	0	24.2	41,42
17	*Glaucomys volans*	18	Weeks	Locomotor activity	0	23.9	14
18		1	7 da	Locomotor activity	1	24.3	14
19		1	10 da	Locomotor activity	55	24.4	14
20	*Glis glis*	1	Weeks	Locomotor activity	0.6	26.5	34
21		12	Weeks	Locomotor activity	10	23.5	35
22	*Lemmus trimucronatus*	1	Weeks	Locomotor activity	0	24.8	41,42
23		1	Weeks	Locomotor activity	65	24.6	41,42
24		42	5 da	Suprarenal secretion	∼17	23.8	3
25	*Mesocricetus auratus*	1	Weeks	Locomotor activity	0	23.9	33
26		1	Weeks	Locomotor activity	0	24.3	33
27		4	Weeks	Locomotor activity	0	<24	15
28		7	Weeks	Locomotor activity	0	>24	15
29		2	6 da	Locomotor activity	0.05	24.2	6
30		2	6 da	Locomotor activity	0.5	24.4	6
31		2	6 da	Locomotor activity	22	24.5	6
32		60	5 da	Suprarenal oxygen quotient	1000	23.5	2
33		24	10 da	Suprarenal steroid secretions	∼100	24.3	1
34	*Microtus nivalis wagneri (M. wagneri)*	3	Weeks	Total activity	0	23.5	44
35	*M. oeconomus*	3	Weeks	Locomotor activity	0	24.3	41,42
36		1	Weeks	Locomotor activity	35	25.0	41,42
37		1	Weeks	Locomotor activity	220	25.3	41,42
38	*Mus musculus*	1	Weeks	Total activity	0	24.3	4
39		1	10 da	Total activity	0	23.2	6
40		1	4 da	Total activity	0	23.5	6
41		3	6 da	Total activity	0.5	23.5	5
42		5	Weeks	Total activity	11	24.1	43
43		5	6 da	Total activity	135	26.1	6
44		3	6 da	Locomotor activity	0.5	24.3	5
45		3	6 da	Locomotor activity	33	25.0	5
46		3	6 da	Locomotor activity	100	25.5	5
47		3	6 da	Locomotor activity	330	26.0	5
48		6	Weeks	Body temperature	Blinded	23.2	19
49	*Myotis lucifugus*	1	18 da	Total activity at 22°C	0	21.5	38
50		16	4 da	Body temperature	0	25[3]	27
51		7	21 da	Body temperature	0	22[4]	27
52	*Perognathus intermedius*	6	7 da	Locomotor activity	0	23.8	39
53		6	7 da	Locomotor activity	5	25.0	39
54	*P. longimembris*	4	12 da	Total activity	0	22.7-23.1	20

[1] "Night". [2] "Day". [3] Winter. [4] Summer.

continued

	Species (Synonym)	No. of Animals	Consecutive Days Recorded	Rhythm	Light Intensity lux	Circadian Period hr	Reference
55	*Peromyscus leucopus*	11	10 da	Total activity	0	24.3	40
56		7	Weeks	Locomotor activity	0	24.1	37
57		1	Weeks	Locomotor activity	0	23.2	37
58		1	15 da	Locomotor activity	0	22.8	37
59		1	10 da	Locomotor activity	0	24.7	37
60		1	10 da	Locomotor activity	15	24.5	37
61		1	Weeks	Locomotor activity	30	24.6	25
62		1	Weeks	Locomotor activity	75	24.8	25
63		1	Weeks	Locomotor activity	270	25.3	25
64	*P. maniculatus*	5	Weeks	Locomotor activity	0	24.2	41,42
65		1	Weeks	Locomotor activity	118	24.8	41,42
66		4	Weeks	Locomotor activity	925	25.5	41,42
67	*Rattus norvegicus*	4	16 da	Locomotor activity	0	24.8	18
68		1	Weeks	Locomotor activity	11	24.8	10
69		1	Weeks	Locomotor activity	11	25.3	11
70		4	14 da	Locomotor activity	56	25.5	18
71		6	Days	Hepatic tyrosine transaminase	0	>24	9
72		6	Days	Hepatic tyrosine transaminase	500-700	>24	9
73	*Rhinolophus ferrumequinum*	2	Weeks	Flight activity	0	21.8	13
74	*Spermophilus columbianus (Ci-*	4	Weeks	Locomotor activity	0	25.1	41
75	*tellus columbianus)*	2	Weeks	Locomotor activity	0	25.1	42
76		2	Weeks	Locomotor activity	∿11	24.5	41
77		1	Weeks	Locomotor activity	∿11	24.5	42
78		5	Weeks	Locomotor activity	∿17	23.8	41
79		4	Weeks	Locomotor activity	∿151-167	23.8	42
80	*S. richardsonii (C. richardsonii)*	1	Weeks	Locomotor activity	183	25.1	42
81	*S. undulatus (C. undulatus)*	4	Weeks	Locomotor activity	0	24.3	41
82		4	Weeks	Locomotor activity	0	24.3	42
83		2	Weeks	Locomotor activity	0.4-11	24.0	42
84		2	Weeks	Locomotor activity	5	24.0	41
85		5	Weeks	Locomotor activity	97-161	23.9	42
86		5	Weeks	Locomotor activity	144	23.9	41
87		1	Weeks	Locomotor activity	270	24.1	42
88		1	Weeks	Locomotor activity	280	23.9	41
89	*Tamias striatus*	3	12 da	Locomotor activity	0	25.3	38
90		3	20 da	Locomotor activity	5	23.4	38
	Aves						
91	*Carduelis chloris*	5	13-14 da	Total activity	70	23.9	8
92		5	13-24 da	Total activity	750	22.6	8
93	*Carpodacus mexicanus*	3	Weeks	Total activity at 20°C	0.02	22.2	16
94		3	Weeks	Total activity at 6°C	0.02	22.7	16
95		3	11 da	Total activity at 20°C	0.1	22.9	16
96		3	10 da	Total activity at 2°C	0.1	23.2	16
97	*Fringilla coelebs*	6	7 da	Total activity	0.2	24.2	8
98		10	9 da	Total activity	0.5	24.6	7
99		4	Weeks	Total activity	0.5	24.8	5
100		4	Weeks	Total activity	2	24.6	5
101		1	Weeks	Total activity	5-7	25.0	26
102		4	Weeks	Total activity	9	23.2	5
103		6	7 da	Total activity	10	23.2	8
104		4♂	7 da	Total activity	14	23.1	36
105		4♀	7 da	Total activity	14	26.0	36
106		4	Weeks	Total activity	135	22.0	5

continued

	Species (Synonym)	No. of Animals	Consecutive Days Recorded	Rhythm	Light Intensity lux	Circadian Period hr	Reference
107	*Passer domesticus*	53	Weeks	Total activity	0.1 (blinded)	24.6	28
108		2	Weeks	Total activity	10	24.5	30
109		2	Weeks	Total activity	11	24.2	30
110	*Pheucticus ludovicianus*	1	21 da	Total activity	10	25.0	30
111	*Pipilo erythrophthalmus*	1	Weeks	Total activity	11	24.8	30
112	*Sturnus vulgaris*	3	Weeks	Total activity	0.7	24.3	22
113		3	Weeks	Total activity	8	23.5	22
114		3	Weeks	Total activity	72	22.5	22
	Lower Vertebrates						
115	*Lacerta sicula*	3	Weeks	Total activity at 25°C	9	24.8	21
116		8	16 da	Total activity	20	∿23.5	23
117		16	16 da	Total activity	22-41	23.5	24
118		3	Weeks	Total activity	100	24.3	21
119		3	Weeks	Total activity	835	23.5	21
120	*Micropterus salmoides*	6	4 da	Swimming	0	>24	12
121	*Phoxinus phoxinus*	12	6 da	Total activity	200	27	29
122		12	6 da	Total activity	1000	29	29

Contributors: Andrews, Richard V.; Folk, G. Edgar, Jr.

References

[1] Andrews, R. V. 1968. Comp. Biochem. Physiol. 26: 179.
[2] Andrews, R. V., and G. E. Folk. 1964. Ibid. 11:393.
[3] Andrews, R. V., et al. 1968. Acta Endocrinol. 59: 36.
[4] Aschoff, J. 1952. Pfluegers Arch. Gesamte Physiol. Menschen Tiere 255:189.
[5] Aschoff, J. 1960. Cold Spring Harbor Symp. Quant. Biol. 25:11.
[6] Aschoff, J. 1962. Handb. Zool. 8(30).
[7] Aschoff, J., and R. Wever. 1966. Comp. Biochem. Physiol. 18:397.
[8] Aschoff, J., et al. 1968. Z. Vergl. Physiol. 58:307.
[9] Black, I. E., and J. Axelrod. 1968. Proc. Nat. Acad. Sci. U.S. 61:1287.
[10] Brown, F. A., Jr., and E. D. Terracini. 1959. Proc. Soc. Exp. Biol. Med. 101:457.
[11] Brown, F. A., Jr., et al. 1956. Amer. J. Physiol. 184: 491.
[12] Davis, R. E. 1964. Anim. Behav. 12:272.
[13] De Coursey, G., and P. De Coursey. 1964. Biol. Bull. 126:14.
[14] De Coursey, P. 1961. Z. Vergl. Physiol. 44:331.
[15] De Coursey, P. 1964. J. Cell. Comp. Physiol. 63: 189.
[16] Enright, J. T. 1966. Comp. Biochem. Physiol. 18: 463.
[17] Erkinaro, E. 1969. Z. Vergl. Physiol. 64:407.
[18] Folk, G. E., Jr. 1959. Proc. Iowa Acad. Sci. 66:399.
[19] Halberg, F., et al. 1959. Publ. Amer. Ass. Advan. Sci. 55:803.
[20] Hayden, P., and R. G. Lindberg. 1969. Science 164: 1289.
[21] Hoffmann, K. 1957. Naturwissenschaften 44:359.
[22] Hoffmann, K. 1960. Z. Vergl. Physiol. 43:544.
[23] Hoffmann, K. 1968. Ibid. 58:225.
[24] Hoffmann, K. 1969. Ibid. 62:93.
[25] Johnson, M. 1939. J. Exp. Zool. 82:315.
[26] Lohman, M. 1967. Comp. Biochem. Physiol. 22: 289.
[27] Menaker, M. 1961. J. Cell. Comp. Physiol. 57:81.
[28] Menaker, M. 1968. Proc. Nat. Acad. Sci. U.S. 59: 414.
[29] Mueller, K. 1968. Naturwissenschaften 55:140.
[30] Palmer, J. D. 1964. Comp. Biochem. Physiol. 12: 273.
[31] Pearson, A. M. 1962. Ann. Zool. Soc. Zool. Bot. Fenn. Vanamo 24:1.
[32] Pittendrigh, C. S. 1960. Cold Spring Harbor Symp. Quant. Biol. 25:159.
[33] Pittendrigh, C. S., and V. G. Bruce. 1957. Rhythmic and Synthetic Processes in Growth. Princeton Univ. Press, Princeton.
[34] Pohl, H. 1965. Naturwissenschaften 52:269.
[35] Pohl, H. 1968. Z. Vergl. Physiol. 58:364.
[36] Pohl, H. 1970. Ibid. 66:141.
[37] Rawson, K. S. 1959. Publ. Amer. Ass. Advan. Sci. 55:791.
[38] Rawson, K. S. 1960. Cold Spring Harbor Symp. Quant. Biol. 25:105.
[39] Stewart, M. C., and W. G. Reeder. 1968. Physiol. Zool. 41:149.
[40] Suter, R. B., and K. S. Rawson. 1968. Science 160: 1011.
[41] Swade, R. H. 1964. Diss. Abstr. 25(3):2117.
[42] Swade, R. H., and C. S. Pittendrigh. 1967. Amer. Natur. 101:431.
[43] Terracini, E. D., and F. A. Brown, Jr. 1962. Physiol. Zool. 35:27.
[44] Wolf, E. 1930. Z. Vergl. Physiol. 11:321.

Light:Dark Cycles are considered to entrain when the period equals the sum of the hours in light and in darkness. Under constant conditions of light and temperature, the period of a circadian rhythm is seldom exactly 24 hours. However, these rhythms may be "entrained" or synchronized to an oscillation in an external signal, such as light or temperature fluctuations, provided the period of the entraining cycle is not too different from the natural period of the rhythm in question. Values in parentheses are ranges, estimate "c" unless otherwise indicated (see Introduction).

| | Class & Species (Synonym) | Rhythm | Constant Conditions | | | Light:Dark Cycles | | Reference |
			Light Intensity lux [L]	Temp °C	Circadian Period hr	Which Entrain hr	Which Do Not Entrain hr	
				Arthropoda				
	Crustacea							
1	Callinectes sapidus	Color change	0	16	24	....	....	26
2	Cambarus sp.	Eye pigment migration	....	7	24	....	....	63
3				21	24			
4	C. virilis	Activity	0	13.5	(22.9-24.9)	....	....	55
5	Carcinus maenas	Activity	Light	15	12.3	15.5:8.0	....	7
6					24.6	13.3:12.5		7
7			110-750	(15-26)	24.83	....		47
8	Ligia baudiniana	Color change	0	....	24.0	10:8	....	40
9	Orchestoidea benedicti	Activity	0	20	24.3	....	....	48
10			440	20	25.1			
11	O. californiana	Activity	0	20	24.2	....	....	48
12			440	20	(24.9-25.1)			
13	O. corniculata	Activity	0	20	24.4	12:12	....	48
14			440	20	(24.4-24.7)			
15	Orconectes pellucidus	Activity & O₂ consumption	0	13	26-34	....	....	36
16	Uca maracoani	Color change	Dim red light	(28.5-29.5)	24.0	....	....	4
17	U. pugnax	Color change	Dim light	6	24	....	16:16	10
18				16	24		1:23	11
19				26	24		....	9
20			....	....	....	....	6:6	62
21							6:18	
22							18:6	
23	U. rapax	Color change	Dim red light	(28.5-29.5)	25.3	....	....	4
	Insecta							
24	Antheraea pernyi	Eclosion	0	26	22	....	....	60
25	Anthia venator	Activity	....	....	....	9:9	....	18
26	Apis mellifera	Feeding	1000	27	(21.4-26.0)	....	....	5
27	Byrsotria fumigata	Activity	0	(24-26)[b]	(23.78-24.62)[b]	....	....	54
28			Very low	(24-26)[b]	25.45			
29			250	(24-26)[b]	24.47			
30	Carausius morosus	Activity	....	....	....	8:8	....	37
31	(Dixippus morosus)					14:14		
32	Chaoborus spp.	Activity	0	....	24	....	....	41
33	Drosophila sp.	Eclosion	0	10	30	....	....	15
34				20	24			
35				30	22			
36	D. pseudoobscura	Eclosion	0	10	24.7(23.8-25.6)[b]	....	....	65
37				16	24.5			49
38				20 ± 0.2	(21.1-25.5)			33

[L] Unless otherwise specified.

continued

	Class & Species (Synonym)	Rhythm	Constant Conditions			Light:Dark Cycles		Ref-er-ence
			Light Intensity lux [1]	Temp °C	Circadian Period hr	Which Entrain hr	Which Do Not Entrain hr	
39				20	24.0(23.6-24.4)[b]			65
40				21	24.0			49
41				26	24.0			49
42				28	23.7(23.0-24.3)[b]			65
43			....	....	....	4.6:9.2	....	12
44						9.2:4.6		12
45						5.2:10.5		12
46						10.5:5.2		12
47						6.1:12.2		12
48						12.2:6.1		12
49						2:22 [2]		50
50						6:18 [2]		50
51						8:16		12
52						10:14 [2]		50
53						11:13 [2]		50
54						14:10		50
55						16:8		12
56	*Geotrupes stercorosus* (*G. silvaticus*)	Activity	0	....	(24.0-24.7)	....	....	28
57	*Gryllus campestris*	Activity	0	20	24	9:9	....	19
58	*Iridomyrmex humilis*, ♂ only	Activity	0	25	(24.5-25.0)	....	....	44
59	*Leucophaea maderae*	Activity	0	....	23.84 ± 0.24	....	....	43
60				20	25.10	....		12
61				(24-26)[b]	(23.23-24.17)[b]	....		54
62				25	24.40	....		12
63				30	24.28	....		12
64			0.5	....	24.32 ± 0.27	....		43
65			2	(24-26)[b]	24.18	11:11		12
66			7	(24-26)[b]	24.75	13:13		12
67			250	(24-26)[b]	(24.00-24.53)[b]	23:1		54
68	*Pectinophora gossypiella*	Oviposition	20	26	22.40	6:18	....	1
69						14:10		
70						18:6		
71	*Periplaneta americana*	Activity	0	(24-26)[b]	(23.75-23.83)	....	....	54
72			Dim light	18	(24-25)	....	....	16
73				(19-20)	(24.2-24.6)[b]			
74				(22-23)	(24.3-24.7)[b]			
75				(27-28)	(24.4-25.6)[b]			
76				29	(24.4-26.5)[b]			
77				31	(24-27)			
78			....	....	....	9:9	....	17
79	*Pseudosmittia arenaria*	Eclosion	....	18	....	9:9	3:3	52
80						12:12	6:6	
81						14:14	21:21	
82						18:18	24:24	
83						....	27:27	
84						....	36:36	

[1] Unless otherwise specified. [2] Also entrained by skeleton photoperiods (15 min light at beginning of each light and each dark period only).

continued

	Class & Species (Synonym)	Rhythm	Constant Conditions			Light:Dark Cycles		Ref-er-ence
			Light Intensity lux [1]	Temp °C	Circadian Period hr	Which Entrain hr	Which Do Not Entrain hr	
85	*Tenebrio molitor*	Activity	0.01	20	23.25	....	....	42
86				25	24.2(23.35-24.55)			
87				35	23.35			
88			2	25	25.5(25.0-26.4)			
89			100	25	26.1(25.05-27.2)			
90	*Velia currens*	Activity	0	....	24.2	....	....	53
91			0.5-700	....	(26-27.5)			
92		Sun orienta-tion	....	....	....	4:4	....	6
93						14:14		
94	*Veromessor andrei*	Activity	10	23 ± 0.5	23.9 ± 1.2	....	....	45

Mollusca

	Class & Species (Synonym)	Rhythm	Light Intensity lux	Temp °C	Circadian Period hr	Which Entrain hr	Which Do Not Entrain hr	Ref-er-ence
	Gastropoda							
95	*Aplysia californica*	Electrical out-put of cell no. 3 of iso-lated parieto-visceral ganglion	Low	11.2 ± 0.3	26.9	12:12	....	1
96				11.4 ± 0.4	28.6			
97			Medium	11.3	21.4			
98				15.3	25.2			
99		Impulse rate of isolated eye	0	15	27.5	....	....	34

Coelenterata [3]

	Class & Species (Synonym)	Rhythm	Light Intensity lux	Temp °C	Circadian Period hr	Which Entrain hr	Which Do Not Entrain hr	Ref-er-ence
	Anthozoa							
100	*Cavernularia obesa*	Expansion of colony	0	10	24 or 48	9:9	0.5:0.5	46
101				20	24	9:15	1.5:1.5	
102				30	24	15:9	3:3	
103						15:15	6:6	
104						....	3:21	
105						....	21:3	
106						....	18:18	
107						....	21:21	
108						....	24:24	

Protozoa

	Class & Species (Synonym)	Rhythm	Light Intensity lux	Temp °C	Circadian Period hr	Which Entrain hr	Which Do Not Entrain hr	Ref-er-ence
	Ciliata							
109	*Paramecium aurelia*, syngen 3, stock 37[P]	Mating	0	12	24.0	....	....	38
110				17	22.1 ± 1.3			
111				22	19.5			
112			300-1500	17	22.2 ± 0.5			
113	*P. bursaria*	Mating	0	17	(22-23)	....	....	24
114				25	(22-23)			
115				29.5	(22-23)			
116	*P. bursaria* T5 & 25b	Cell division	Light	18.5	24	10:10	24:24	61
117				25	24	14:14	....	
118				30	24	6:6[4]	....	
119						9:9[4]	....	
120	*P. multimicronuclea-tum*, syngen 2	Mating type reversal	0	22	<24	8:16	....	2
121			<250	22	<24			
122		Cell division	60	14	∼24	....	....	3

[1] Unless otherwise specified. [3] Synonym: Cnidaria. [4] Strain 25b only.

continued

	Class & Species (Synonym)	Rhythm	Constant Conditions			Light:Dark Cycles		
			Light Intensity lux	Temp °C	Circadian Period hr	Which Entrain hr	Which Do Not Entrain hr	Ref-er-ence
123	*Tetrahymena pyrifor-*	Cell division	15	28.5	∿21	....	3:3	64
124	*mis* W					....	6:6	
	Mastigophora							
125	*Chlamydomonas rein-*	Phototaxis	0, except test light	18	<24	....	....	13
126	*hardi*			22	∿24			
127				28	<24			
128		Cell division	4000-5000	23 ± 0.5	∿24	....	....	13
129	*Dissodinium lunula* (*Pyrocystis lunula*)	Chloroplast movement	0	22	∿24	....	....	59
130	*Euglena gracilis*	Phototaxis	0, except test light	16.7	26.2	3:3	2:10	14
131				18.5	25.5	8:8	....	14
132				23.0	23.5	2:22	....	14
133				25	23.7	....	....	25
134				26.0	23.8	4:20	....	14
135				33.0	23.2	6:18	....	14
136						8:16	....	14
137						10:14	....	14
138						12:12	....	14
139						14:10	....	14
140						16:8	....	14
141						18:6	....	14
142						20:4	....	14
143						24:24	....	14
144			....	(16-18)	....	8:8	....	51
145		Cell division	8000, random L:D[5/]	25	27.5	....	....	23
146			¼:½ da	25 ± 0.5	29.2	10:14		22
147			½:1 da	25 ± 0.5	25.6	18:10		22
148			1:2 da	25 ± 0.5	26.0	3:6:3:12		22
149			1:3 da	25 ± 0.5	24.4	4:4:4:12		22
150			2:4 da	25 ± 0.5	27.0	....		22
151			5:5 da	25 ± 0.5	32.4	....		22
152			8:8 da	25 ± 0.5	33.0	....		22
153	Autotrophic	Motility	0, except test light	(15-35)	23.5 ± 0.3	....	....	8
154				25	23.0 ± 0.3	8:8	2:2	56
155						10:10	4:4	56
156						12:12	6:6	56
157						14:14	....	56
158						18:18	Entrain	56
159						24:24	....	56
160		Cell division	800	25 ± 0.5	24.2 ± 0.45	14:10	....	20,
161						16:8		21
162	Mixotrophic	Motility	0, except test light	20	24.0	....	....	8
163				25	24.8 ± 0.9	8:8		56
164						12:12		56
165						14:14		56
166						18:18		56
167						24:24		56
168					24.8	....		8
169				30	25.5	....		8
170				35	27.5	....		8

[5/] 8 hr light in 24 hr.

continued

| | Class & Species (Synonym) | Rhythm | Constant Conditions | | | Light:Dark Cycles | | Ref-er-ence |
			Light Intensity lux	Temp °C	Circadian Period hr	Which Entrain hr	Which Do Not Entrain hr	
171	Colorless mutant	Motility	0, except test light	20	24.2 ± 0.7	....	....	39
172				27	24.6 ± 1.6			
173	Mutant PyZUL	Cell division	0	19	22.4	....	....	35
174			5000	19	23.0			
175	*Gonyaulax polyedra*	Luminescent capacity	0	21	23.0	....	....	58
176					24.4			
177			1000	(15.5-17.5)	22.8	7:7	....	31
178				(15.7-16.1)	(21.3-23.7)	6[6/]:6	6[7/]:6	30
179				(18.5-19.5)	(22.2-24.0)	8:8	24:24	32
180				21	24.4	12:12	....	58
181				(21.5-22.5)	(23.4-27.2)	16:16	....	58
182				(22.6-23.6)	(24.4-27.0)	....	....	58
183				23.7	25.7	....	....	58
184				(26.1-27.1)	26.8	....	....	58
185				(26.1-27.5)	(25.5-27.5)	....	....	58
186				(30-34)	25.5	....	....	58
187			1200	21	(23.0-26.0)	....	....	58
188			3800	21	(21.3-24.3)	....	....	58
189			6800	21	(21.2-22.8)	....	....	58
190		Luminescent glow	0	18	22.9	....	....	57
191				20	23.3			57
192				25	24.7			57
193			1200	24	24.8			29
194		Photosynthesis capacity	1100	(24.7-25.3)	26	....	....	29
195		Cell division	0	18	22.8	....	....	57
196				25	24.8	....	....	
197			1000	18.5	23.9	8:8	20:4	
198				19	22.5	18:6	....	
199				25	25.4	....	....	
200	*Gyrodinium dorsum*	Activation of phototaxis by red light	0	Room temp	∼24	....	....	27

6/ Light intensity, 8000 lux during light period.   7/ Light intensity, 2000 lux during light period.

Contributor: Sweeney, Beatrice M.

References

[1] Aschoff, J., ed. 1965. Circadian Clocks. North-Holland, Amsterdam.

[2] Barnett, A. 1966. J. Cell. Physiol. 67:239.

[3] Barnett, A. 1969. Science 164:1417.

[4] Barnwell, F. H. 1963. Biol. Bull. 125:399.

[5] Bennett, M. F., and M. Renner. 1963. Ibid. 125:416.

[6] Birukow, G. 1960. Cold Spring Harbor Symp. Quant. Biol. 25:403.

[7] Blume, J., et al. 1962. Biol. Zentralbl. 81:569.

[8] Brinkmann, K. 1966. Planta 70:344.

[9] Brown, F. A., Jr., and G. C. Stephens. 1951. Biol. Bull. 100:71.

[10] Brown, F. A., Jr., and H. M. Webb. 1948. Physiol. Zool. 21:371.

[11] Brown, F. A., Jr., and H. M. Webb. 1949. Ibid. 22:136.

[12] Bruce, V. G. 1960. Cold Spring Harbor Symp. Quant. Biol. 25:29.

[13] Bruce, V. G. 1970. J. Protozool. 17:328.

continued

[14] Bruce, V. G., and C. S. Pittendrigh. 1956. Proc. Nat. Acad. Sci. U.S. 42:676.

[15] Bünning, E. 1935. Ber. Deut. Bot. Ges. 53:594.

[16] Bünning, E. 1958. Biol. Zentralbl. 77:141.

[17] Cloudsley-Thompson, J. L. 1953. Ann. Mag. Natur. Hist., Ser. 12, 6:705.

[18] Cloudsley-Thompson, J. L. 1956. Ibid. 9:305.

[19] Cloudsley-Thompson, J. L. 1958. J. Insect Physiol. 2:275.

[20] Edmunds, L. N., Jr. 1965. J. Cell. Comp. Physiol. 66:147.

[21] Edmunds, L. N., Jr. 1966. J. Cell. Physiol. 67:35.

[22] Edmunds, L. N., Jr., and R. Funch. 1969. Planta 87:134.

[23] Edmunds, L. N., Jr., and R. Funch. 1969. Science 165:500.

[24] Ehret, C. F. 1959. Fed. Proc. Fed. Amer. Soc. Exp. Biol. 18:1232.

[25] Feldman, J. F. 1967. Proc. Nat. Acad. Sci. U.S. 57: 1080.

[26] Fingerman, M. 1955. Biol. Bull. 109:255.

[27] Forward, R. B., Jr., and D. Davenport. 1970. Planta 92:259.

[28] Geisler, M. 1961. Z. Tierpsychol. 18:389.

[29] Hastings, J. W. 1964. In A. Giese, ed. Phytophysiology. Academic Press, New York. v. 1, p. 333.

[30] Hastings, J. W., and B. M. Sweeney. 1957. Proc. Nat. Acad. Sci. U.S. 43:804.

[31] Hastings, J. W., and B. M. Sweeney. 1958. Biol. Bull. 115:440.

[32] Hastings, J. W., and B. M. Sweeney. 1959. Publ. Amer. Ass. Advan. Sci. 55:567.

[33] Honegger, H.-W. 1967. Z. Vergl. Physiol. 57:244.

[34] Jacklet, J. W. 1969. Science 164:562.

[35] Jarrett, R. M., and L. N. Edmunds, Jr. 1970. Ibid. 167:1730.

[36] Jegla, T. C., and T. L. Poulson. 1968. J. Exp. Zool. 168:273.

[37] Kalmus, H. 1938. Z. Vergl. Physiol. 25:494.

[38] Karakashian, M. W. 1968. J. Cell. Physiol. 71:197.

[39] Kirschstein, M. 1969. Planta 85:126.

[40] Kleitman, N. 1940. Biol. Bull. 78:403.

[41] La Row, E. J. 1968. Limnol. Oceanogr. 13:250.

[42] Lohmann, M. 1964. Z. Vergl. Physiol. 49:341.

[43] Lohmann, M. 1967. Experientia 23:788.

[44] McCluskey, E. S. 1963. Physiol. Zool. 36:273.

[45] McCluskey, E. S. 1967. Comp. Biochem. Physiol. 23:665.

[46] Mori, S. 1957. Publ. Seto Mar. Biol. Lab. 6:79.

[47] Naylor, E. 1960. J. Exp. Biol. 37:481.

[48] Osbeck, B. L. 1970. M.A. Thesis. Univ. California, Santa Barbara.

[49] Pittendrigh, C. S. 1954. Proc. Nat. Acad. Sci. U.S. 40:1018.

[50] Pittendrigh, C. S. 1966. Z. Pflanzenphysiol. 54:275.

[51] Pohl, R. 1948. Z. Naturforsch. 3b:367.

[52] Remmert, H. 1955. Z. Vergl. Physiol. 37:338.

[53] Rensing, L. 1961. Ibid. 44:292.

[54] Roberts, S. K. de F. 1960. Physiol. Zool. 30:70.

[55] Roberts, T. W. 1944. Ecol. Monogr. 14:359.

[56] Schnabel, G. 1968. Planta 81:49.

[57] Sweeney, B. M., and J. W. Hastings. 1958. J. Protozool. 5:217.

[58] Sweeney, B. M., and J. W. Hastings. 1960. Cold Spring Harbor Symp. Quant. Biol. 25:87.

[59] Swift, E., and W. R. Taylor. 1967. J. Phycol. 3:77.

[60] Truman, J. W. 1971. Proc. Nat. Acad. Sci. U.S. 68: 595.

[61] Volm, M. 1964. Z. Vergl. Physiol. 48:157.

[62] Webb, H. M. 1950. Physiol. Zool. 23:316.

[63] Welsch, J. H. 1941. J. Exp. Zool. 86:35.

[64] Wille, J. J., Jr., and C. F. Ehret. 1968. J. Protozool. 15:785.

[65] Zimmerman, W. F., et al. 1968. J. Insect Physiol. 14:669.

117. PERIOD OF CIRCADIAN RHYTHMS IN LIGHT AND DARKNESS: PLANTS

Abbreviations & Symbols: D = dark; L or W = white light; B = blue radiation; G = green radiation; R = red radiation; FR = far-red radiation; L_n = natural light; L_f or l_f = fluorescent light; L_i = incandescent light; $\lessdot$ = not less than; W = watt. Repeated capital letters (LL, DD, etc.) indicate continuous exposure for prolonged period; lower case letters in italics (*l, d, r,* etc.) indicate irradiation or dark period interrupting, or interposed between, sustained periods of darkness or light. Superscript numbers attached to abbreviations (L^1, L^2, etc.) identify different treatments in a sequence. **Treatment:** Experimental conditions to which rhythm is causally related. **First Min** and **First Max:** Time of first minimum, and/or first maximum, in rhythm after change or start of treatment unless specified by "ex" [e.g., "exr,D(−)" = first minimum in darkness after end of red light interruption]. **Period of Rhythm:** 0 = no evident

continued

rhythm. **No. of Observations:** Number of minima or maxima recorded. A summary of the German literature has not been included. Data in brackets refer to the column heading in brackets. For additional information on biological rhythms, consult references 4, 11, 13, and 14 in Part I.

Part I. Fungi and Algae

	Species	Rhythm [Measurement]	Pretreatment	Treatment	First Min (−) First Max (+) hr	Period of Rhythm hr [No. of Observations]	Remarks	Reference
				Prolonged Darkness				
1	*Daldinia concentrica*	Ascospore discharge [Periodicity]	2 cycles, ∿16 hr L_n:8 hr D	DD	(+) 15	22-23 [4]	....	10
2	*Euglena gracilis*	Motility & phototaxis	Daily, 12 hr L_f, 4-*W*:12 hr D	DD	(+) 17-18	∿24 [2]	No rhythm in LL. Tungsten test lamp used 30 min in every 2 hr to measure response. Daily 12 hr L:12 hr D: Max approx. midpoint of L.	2
3			DD	DD	....	0	Test lamp on only 10 min every 2 hr. Rhythm dependent on energy of photosynthesis.	2
4 5	*Gonyaulax polyedra*	Luminescent capacity [Total light emitted in 1 min when air bubbled through suspension]	LL$_f$ (cool white), 7530 lux, 25°C	DD, 25°C	exLL,DD(−)13 (+)4	23.5 [3] 22.7 [4]	Luminescence dim during light; ∿60 times brighter in darkness	15
6 7			Cycles, 7 hr L, 8610 lux:7 hr D; 21°C	7 hr L:DD; 21°C	exL,DD(−)21 (+)7	23.0 [2] 23.5 [3]	During pretreatment, rhythm was entrained to 14-hr cycles	6
8 9			Daily, 12 hr L, 8610 lux:12 hr D	12 hr L, 8610 lux:DD; 21°C	exL,DD(−)20 (+)6-7	24-25 [3-4] 24-25 [4]	....	6,7
10 11 12		Luminescent glow of undisturbed cells	Daily, 12 hr L: 12 hr D; 20°C	DD, 18°C DD, 20°C DD, 25°C	....	22.9 23.9 24.7	Period increased with increased temp. Glow at about time of max cell division.	16
13 14 15	*Neurospora crassa*	Mycelial growth & reproduction [Relative density of mycelium & zonation of reproduction]	2 mo DD, 26 ± 1°C, 45 ± 5% RH	DD, 15°C DD, 25°C DD, 35°C	....	∿24 24 24	Light for subculturing, 11 lux. Time of subculturing, not time of day, set rhythm.	1
16	Strain 21863	Zonation of mycelial growth [Relative density of mycelium]	48 hr LL after culture tubes inoculated	7 da DD, 4 different rotations with respect to the earth	....	∿24 (all 4 treatments)	Experiments at South Pole. Revolutions varied with respect to earth (therefore to sun): (i) stationary (control), once per 24 hr; (ii) counter to earth's rotation, once per 6 da (once per 28.8 hr with respect to sun); (iii) counter to earth's rotation, once per 24 hr (no rotation with respect to sun); (iv) revolution in direction of earth's rotation (once per 20 hr with respect to sun).	5

continued

Part I. Fungi and Algae

	Species	Rhythm [Measurement]	Pretreatment	Treatment	First Min (−) First Max (+) hr	Period of Rhythm hr [No. of Observations]	Remarks	Reference
					Darkness Interrupted by Light			
17	*Gony-*	Luminescent	Daily, 12 hr L,	12 hr L, 8610	ex*l*, 485 lux		2.5 hr *l*: Max effect	6
18	*aulax*	capacity	8610 lux:12 hr	lux:DD, 2.5 hr *l*	(−)11.5	26 [2]	reached at ∿8610 lux;	
	poly-	[Total light	D	of either 485,	(+)24.5	25.5 [2]	phase shift obtained	
19	*edra*	emitted in 1		1290, 2150,	ex*l*, 1290 lux		with 2.5 hr *l* was 2.5 hr	
20		min when air		4090, or 6460	(−)11.5	25.5 [3]	at 2150 lux, 5.5 hr at	
		bubbled		lux at 6th hr;	(+)24.5	24.7 [3]	3230 lux, 8.25 hr at	
21		through sus-		21°C	ex*l*, 2150 lux		4300 lux, 11 hr at 6460	
22		pension]			(−)9.5	25 [3]	lux, 11.3 hr at 7530 lux,	
					(+)23.5	23.5 [2]	11.5 hr at 8610 lux.	
23					ex*l*, 4090 lux		(*See also* lines 8 & 9:	
24					(−)8.5	21 [3]	with 12 hr L:DD, max	
					(+)16.5	21 [3]	in 6-7 hr exL)	
25					ex*l*, 6460 lux			
26					(−)6.5	25 [3]		
					(+)14.5	21 [3]		
27				12 hr L, 8610 lux:D (varied):3 hr *l*, 15,060 lux: DD	....	....	Sensitivity to L greater in 1st 12 hr of D than in 2nd 12 hr of D; continued rhythmically. 3 hr *l* pre max in luminescence: Phase delay (i.e., ex*l*,DD(+) = >24 hr). 3 hr *l* post max in luminescence: Phase advance (i.e., ex*l*,DD(+) = <24 hr).	6
28				12 hr L, 8610 lux:2.5 hr D:6 hr *l*, 15,060 lux: DD	ex*l*(−)21 (+)31	29 [3] 18 [4]	*l* before max lumines- cence caused phase de- lay	7
29								
30				12 hr L, 8610 lux:6 hr D:6 hr *l*, 15,060 lux:DD	ex*l*(−)22 (+)12	23 [3] 24 [4]	*l* at or after max lumi- nescence caused phase advance	7
31								
					Prolonged Light			
32	*Daldinia con- centri- ca*	Ascospore discharge [Periodicity]	7 cycles, 12 hr D:12 hr L, 1080 lux	LL_f, 1080 lux	....	22 [2]	LL: Rhythm faded after 2 da. 12 hr D:12 hr L: Max discharge at or near end of L.	10
33	*Gony-*	Cell division	Daily, 12 hr L,	LL, 2150 lux,	(−)42	25.4 [8]	In daily LD, cells/ml	16
34	*aulax*	[Pairs/100	9680 lux:12 hr	25°C	(+)12	25.4 [9]	showing stepwise in-	
35	*poly-*	cells (per-	D	LL,2150 lux,	(−)16	22.5 [8]	crease with time; max	
36	*edra*	cent paired		18°C	(+)43	23.9 [9]	just at end of D. In LL,	
		cells in sus- pension)]					cells/ml showed linear increase with time (periodicity lost).	
37		Luminescent	7 hr L, 8610 lux:	LL, 2475 lux,	(−)7	25.5 [3]	During pretreatment,	6
38		capacity [Total light emitted in 1 min when air bubbled through suspension]	7 hr D; 21°C	21°C	(+)17	24.5 [3]	rhythm was entrained to 14-hr cycles	

continued

Part I. Fungi and Algae

	Species	Rhythm [Measurement]	Pretreatment	Treatment	First Min (−) First Max (+) hr	Period of Rhythm hr [No. of Observations]	Remarks	Reference
39 40		Photosynthesis activity & capacity [Measured by incorporation of $^{14}CO_2$]	Daily, 12 hr L_f (cool white), 9680 lux, ~26°C:12 hr D, ~23°C	LL_f (cool white), 1180 lux, 25°C (aliquots from LL, 1180 lux, incubated with tracer for 15 min at L_f 10,330 lux)	(−)44 [1] (+)32 [1]	~24 [2] ~28 [2]	Photosynthetic capacity: Cells maintained in dim light, but tracer incorporation measured at intensity saturating for photosynthesis. Photosynthetic activity: In dim light (1180 lux), no rhythm observed after 1st small max. Cultures exLD, in LL (10,330 lux): No rhythm. In daily LD, max rate of photosynthesis occurred at 8th hr of 12 hr L. Results show that rhythm in photosynthesis [9], luminescence [6], & cell division [16] have a common controlling mechanism. (Luminescence ratio of max to min, ~60)	9
				Changes in Light Intensity				
41 42	Gonyaulax polyedra	Luminescent capacity [Total light emitted in 1 min when air bubbled through suspension]	12 hr L^1:12 hr D: 12 hr L^1_f (cool white), 8610 lux; 21°C	LL^2_f (cool white), 1290 lux, 21°C	(−)20 (+)10	24.5 [5] 24.5 [6]	exL^1,DD: Period was 24.5 hr	6
43 44				LL^3_f (cool white), 4090 lux, 21°C	(−)18 (+)7	22.0 [5] 22.8 [6]		
45 46				LL^4_f (cool white), 7320 lux, 21°C	(−)16 (+)4	20.8 [6] 22.4 [6]		
47 48			1 yr LL^1_f, 8610 lux	LL^2, 968 lux, 21°C	(−)23 (+)13	25.7 [5] 24.2 [5]	....	6
49	Neurospora crassa	Mycelial growth & reproduction [Relative density of mycelium & zonation of reproduction]	20 mo LL^1_f (cool white), 1080 lux	LL^2, 54-2152 lux, 26 ± 1°C, 45 ± 5% RH	....	~24	LL^2 of 54, 108, 160, 270, 540, 1080, & 2150 lux resulted in same no. of bands, but 538 lux or more caused less regular bands and less growth	1
				Photoperiodic Cycles Other Than 24 Hours				
50	Daldinia concentrica	Ascospore discharge [Periodicity]	2 cycles, ~16 hr L_n:8 hr D	8 cycles, 6 hr L:6 hr D	exL,D(+)5	12 [2] [8]	Period was 22-23 hr in DD	10
51 52	Gonyaulax polyedra	Luminescent capacity [Total light emitted in 1 min when air bubbled through suspension]	7 hr L, 8610 lux: 7 hr D; 21°C	7 hr L, 8610 lux: 7 hr D	L(−)7 D(+)6	[4] [4]	Rhythm entrained to 14-hr cycles. 7 hr L:DD: Period was 23.5 hr in DD	6
53			6 hr L:6 hr D	7 cycles, 6 hr L, 2150 lux:6 hr D	Each cycle, exL(+)4	12 [2]; 24	Largest max every 24 hr. Transfer to LL, 2150 lux: 1st max at 16 hr exD. Period was ~24 hr in LL or DD after cycle.	7
54				7 cycles, 6 hr L, 8610 lux:6 hr D	Each cycle, exL(+)5	12 [2]	Amplitude (max) greater at 8610 than at 2150 lux. Max every 12 hr (i.e., entrainment or repetitive resetting).	

1/ Graph started at 16 hr exD,LL [9]; ratio of max to min, ~5. 2/ Rhythm entrained by L:D cycles other than 24 hr.

continued

Part I. Fungi and Algae

	Species	Rhythm [Measurement]	Pretreatment	Treatment	First Min (−) First Max (+) hr	Period of Rhythm hr [No. of Observations]	Remarks	Reference
55				6 cycles, 6 hr L, 2150 lux:8 hr D	Each cycle, exL(+)6	16 [2/]	Transfer to LL, 2150 lux: 1st max at 13 hr exD. Period was ∿24 hr in LL or DD after cycle.	
56				3 cycles, 16 hr L, 2150 lux:16 hr D	Successive max exL: 5, 5, 15	32 [2/]	Transfer to LL, 2150 lux: 1st max at 17 hr exD. Period reverted to ∿24 hr in LL or DD after cycle.	
					Irradiation by Different Parts of Spectrum			
57	*Gony-*	Luminescent	2 cycles, 12 hr	Control: DD at	exL,DD(−)20	∿24 [3]	Phase shift in monochro-	8
58	*aulax*	capacity	L_f (cool white),	60th hr	(+)5	∿24 [4]	matic *l* dependent on	
59	*poly-*	[Total light	10,760 lux:12	3 hr monochro-	(+) measured	....	wavelength (λ). Max ef-	
	edra	emitted in 1	hr D; then 12	matic *l*, different	ex monochro-		fectiveness at 475 nm	
		min when air	hr L at 48th-	λ & intensities,	matic *l*		*(b),* & a lesser peak at	
60		bubbled	60th hr; 22°C	at 66th-69th hr;	(+) measured	∿24	650 nm *(r).* Ineffective	
		through sus-		otherwise DD	ex monochro-		λ: 350 nm, ∿550 nm,	
		pension]			matic *l* + 24		& >700 nm. Phase shift—at	
					hr D		475 nm, 7.3 ergs·mm^{-2}·s^{-1}: 6.5- hr shift; at 650 nm, 6.5 ergs·	

mm^{-2}·s^{-1}: 4-hr shift (i.e., energy of 0.95 *(b)* & 3.1 *(r)* ergs·mm^{-2}· s^{-1} were required for 2-hr shift). Effectiveness spectrum roughly corresponded with absorption spectrum of whole cells, to which chlorophyll *a* & *c,* & peridinin contributed gross features. No evidence of reversal of *r* effect by *fr* (730 nm).

	Species	Rhythm [Measurement]	Pretreatment	Treatment	First Min (−) First Max (+) hr	Period of Rhythm hr	Remarks	Reference
61	*Neuros-*	Zonation of	40 hr L_f (cool	RR, 24 ± 0.5°C	(−)....	22.0 [5]	Illumination specified	12
62	*pora*	mycelial	white), 14-*W*		(+)....	22.5 [6]	only as "red lamp"	
63	*crassa*	growth		RR, 31 ± 1°C	(−)13	22.2 [5]		
64		[Relative density of mycelium]			(+)9	25.5 [4]		
	Pilobo-	Spore ejec-	6-7 da, 12 hr L_f	RR; *l* (white),	ex 9th hr *l*		Phase reset by 1/2000-s *l*	3
65	*lus*	tion	(white):12 hr	1/2000-s flash,	(−)26	24 [3]	at 21st hr, but tran-	
66	*sphae-*		D; 25°C	at 9th or 21st	(+)6	25.7 [4]	sients occurred.	
	rospo-			hr; 25°C	ex 21st hr *l*		Rhythm faded after 4-6	
67	*rus*				(−)15	25 [4]	da RR. Plants 6-7 da	
68					(+)4	29 [5]	old at time of measure- ment.	

2/ Rhythm entrained by L:D cycles other than 24 hr.

Contributor: Cumming, Bruce G.

References

[1] Bianchi, D. E. 1964. J. Gen. Microbiol. 35:437.

[2] Bruce, V. G., and C. S. Pittendrigh. 1956. Proc. Nat. Acad. Sci. U.S. 42:676.

[3] Bruce, V. G., et al. 1960. Science 131:728.

[4] Bünning, E. 1964. The Physiological Clock. Academic Press, New York.

[5] Hamner, K. C., et al. 1962. Nature (London) 195: 476.

[6] Hastings, J. W., and B. M. Sweeney. 1958. Biol. Bull. 115:440.

[7] Hastings, J. W., and B. M. Sweeney. 1959. Publ. Amer. Ass. Advan. Sci. 55:567.

continued

Part I. Fungi and Algae

[8] Hastings, J. W., and B. M. Sweeney. 1960. J. Gen. Physiol. 43:697.

[9] Hastings, J. W., et al. 1961. Ibid. 45:69.

[10] Ingold, C. T., and V. J. Cox. 1955. Ann. Bot. (London), N.S. 19:201.

[11] Long Island Biological Association. 1960. Cold Spring Harbor Symp. Quant. Biol. 25.

[12] Pittendrigh, C. S., et al. 1959. Nature (London) 184:169.

[13] Sollberger, A. 1965. Biological Rhythm Research. Elsevier, Amsterdam.

[14] Sweeney, B. M. 1963. Annu. Rev. Plant Physiol. 14:411.

[15] Sweeney, B. M., and J. W. Hastings. 1957. J. Cell. Comp. Physiol. 49:115.

[16] Sweeney, B. M., and J. W. Hastings. 1958. J. Protozool. 5:217.

Part II. Magnoliophytes

Species (Synonym)	Rhythm [Measurement]	Pretreatment	Treatment	First Min (−) First Max (+) hr	Period of Rhythm hr [No. of Observations]	Remarks	Reference
			Prolonged Darkness				
1 2 *Avena sativa* 'Victory'	CO_2 output of coleoptiles [Increase in rate as percent of initial rate]	56 hr R[1/]	DD, 26°C	(−)35 (+)17	[1] 23 [2]	Rhythm measured at time of causative treatment	2
3 4 5	Growth rate of coleoptiles (intact seedlings) [mm/hr]	30-56 hr R[1/]	DD, 22-24.5°C	ex 30 hr R,DD (+)42 ex>30 hr R,DD (−)3-4 (+)16-18	24 [2] 22-32 [2-3] 24 [2]	1 min infrared when photographed. Rhythm measured at time of causative treatment.	1
6 7 *Cestrum nocturnum*	Opening & closing of flowers on intact plants & cut stalks	Greenhouse, 23°C (day), 17°C (night)	DD, 23°C	Start of 23°C (−)0-5 (+)7-16	24 [4] 25 [3]	Flowers opened at start of 23°C. Min: Flowers closed. Max: Flowers open. Rhythm measured at time of causative treatment.	21
8 9	Odor production of flowers on intact plants & of cut flowers	Greenhouse, 23°C (day), 17°C (night)	DD, 17°C	(−).... (+)....	26.5 [3] 27 [4]	Odor produced when flowers open, absent when closed. Rhythm measured at time of causative treatment.	21
10 11 *Kalanchoe daigremontiana*	CO_2 output of excised leaves[2/] [μg $CO_2 \cdot hr^{-1} \cdot g^{-1}$ fresh wt, initially in CO_2-free air]	Cycles, 8 hr L_i, 3000 lux:16 hr D	ex 8 hr L: DD; 26°C	(−)9 (+)16	22 [4] 24 [4]	Rhythm measured at time of causative treatment	25
12 13 *K. fedtschenkoi*	CO_2 output of excised leaves[2/] [μg $CO_2 \cdot hr^{-1} \cdot g^{-1}$ fresh wt, initially in CO_2-free air]	Cycles, 8 hr L_i, 3000 lux:16 hr D	ex 8 hr L: DD; 26°C	(−)9 ± 1 (+)16.5 ± 0.5	22 [5] 22 [5]	ex 8 hr L:DD (control): Normal max at ∼18, 41, 63, & 84 hr. Similar results obtained using 1-cm pieces of mesocotyl. Rhythm measured at time of causative treatment.	25

[1/] Pretreatment commenced at start of germination. [2/] 2-3 mo old from 2-yr-old plants.

continued

Part II. Magnoliophytes

	Species (Synonym)	Rhythm [Measurement]	Pretreatment	Treatment	First Min (−) First Max (+) hr	Period of Rhythm hr [No. of Observations]	Remarks	Reference
14				ex 16 hr L:	(−)10	21.5 [5]	Rhythm measured at time	
15				DD; 26°C	(+)18	22 [5]	of causative treatment	
16		CO_2 fixation of excised leaves [μg CO_2 fixed· $hr^{-1} \cdot g^{-1}$ fresh wt, compared with μg CO_2 output·$hr^{-1} \cdot g^{-1}$ fresh wt; initially in CO_2-free air]	Leaves excised at 1500 hr from plants in short days of 8 hr L (incandescent + mercury vapor):16 hr D	DD started with normal D (1600 hr), 26°C	CO_2 fixation (−)21 (+)9 CO_2 output (−)8 (+)17	24 [3] 25 [3] 23.5 [3] 24 [3]	Fixation measured by feeding $^{14}CO_2$ at 2-hr intervals. At end of feeding period, tubes were cleared of $^{14}CO_2$ and leaves killed. CO_2 fixation: 1st max was earlier with leaves from long day (16 hr L) than those from short day (8 hr L), i.e., transient period decreased as duration of previous light increased. Rhythm measured at time of causative treatment.	24
17								
18								
19								
20				16 hr L^2, 3000 lux (starting 1600 hr): DD; 26°C	CO_2 fixation exL^2,DD (−)12 (+)7 CO_2 output exL^2,DD	22 [4] 23 [4]		
21								
22					(−)6	24 [4]		
23					(+)18	23 [4]		
24			Leaves excised at 1500 hr from plants in long days of 16 hr L (incandescent + mercury vapor):8 hr D	DD started with normal D (1600 hr), 26°C	CO_2 fixation (−)18 (+)7 CO_2 output (−)8 (+)19	24 [2] 22 [2] 22 [2] 22 [2]	Rhythm measured at time of causative treatment	
25								
26								
27								
28				16 hr L^2, 3000 lux (starting 1600 hr): DD; 26°C	CO_2 fixation exL^2,DD (−)18 (+)7 CO_2 output exL^2,DD	24 [4] 24 [4]	Rhythm measured at time of causative treatment	
29								
30					(−)4	25 [4]		
31					(+)20	23 [4]		
32	K. pinnata	CO_2 output of excised leaves[2] [μg $CO_2 \cdot hr^{-1} \cdot g^{-1}$ fresh wt, initially in CO_2-free air]	Cycles, 8 hr L_i, 3000 lux:16 hr D	ex 8 hr L: DD; 26°C	(−)12 (+)20	24 [3] 23 [3]	Rhythm measured at time of causative treatment	25
33								
34	Phaseolus vulgaris	Leaf movement on plants with 1 (1st triplicate) leaf	L_f (cool white), 21°C, 60 ± 10% RH	DD, 21°C, 60 ± 10% RH; 5 different rotations	....	Same rhythm with all rotations (period not reported)	Experiments at South Pole. Revolutions varied with respect to earth (therefore to sun). Turntable rotated once in 1, 2, 4, 6, & 8 da. During DD, infrared radiation used for time-lapse photography. Rhythm measured at time of causative treatment.	15

2/ 2-3 mo old from 1-yr-old plants.

continued

Part II. Magnoliophytes

	Species (Synonym)	Rhythm [Measurement]	Pretreatment	Treatment	First Min (−) First Max (+) hr	Period of Rhythm hr [No. of Observations]	Remarks	Reference
35 36	*Secale cereale*	Growth rate of coleoptiles (intact seedlings) [mm/hr]	30 hr R[L]	DD, 26°C	(−)23 (+)19	36 [2] 20 [2]	2nd min & 3rd max poorly defined. Rhythm measured at time of causative treatment.	2
37 38			46 hr R[L]	DD, 26°C	(−)9 (+)3	23 [2] 16.5 [3]	2nd min poorly defined. Rhythm measured at time of causative treatment.	
39 40	*Triticum aestivum* 'Eclipse'	Growth rate of coleoptiles (intact seedlings) [mm/hr]	50 hr R	DD, 26°C	(−)4 (+)18	25 [2] 29 [2]	Plants 50 hr old at start of DD. Rhythm measured at time of causative treatment.	2
41 42			49.5 hr D: 0.5 hr L, 20 lux	DD, 26°C	(−)4 (+)21	24.5 [3] 21 [2]		
43 44			49.5 hr D: 0.5 hr L, 120 lux	DD, 26°C	(−)3 (+)22	26 [2] 18 [2]		
				Darkness Interrupted by Light				
45 46	*Avena sativa* 'Victory'	Growth rate of coleoptiles (intact seedlings) [mm/hr]	30 hr R[L]	18 hr D:5 min r:DD; 24.5 ± 1.5°C	exr,DD(−)9 (+)22	25 [2] 24 [2]	r at normal 1st max: No shift in phase	1
47 48 49 50			50 hr R[L]	17 hr D:7 hr r:DD; 21 ± 1°C	exR,D(−)5 (+)16 exr,DD(−)8 (+)18	[1] [1] [1] [1]	r at normal 1st max: Slight if any shift in phase	
51 52 53 54			50 hr R[L]	23 hr D:7 hr r:DD; 22 ± 1°C	exR,D(−)0 (+)17 exr,DD(−)7 (+)17	.... [1] [1] [1]	r between normal 1st & 2nd max: Rhythm phased by end of 7 hr r (similar results with 12 hr r for seedlings 42 hr old)	
55 56	*Chenopodium botrys*	Germination [Percent, 24 da after seeds wetted]	DD (19 da imbibition period), 25°C	0-96 hr R, 25°C	exDD,R(−)13 (+)10	Main periods: 16, 10.5, 26.5	DD: No germination. R, 0.012 g·cal⁻¹·cm⁻²·min⁻¹, from 20-W gro-lux: 1 layer red cinemoid. (Raw data showed correlation coefficient of 0.9 to harmonic curve based on period of 143, 16, 10.5, & 26.5.) Germination in white light (L_f or L_f + L_i) also indicated periods of <24 hr & frequently one of 8-12 hr. Suggestive evidence for rhythm in other species [4], but only relatively short light periods were used.	8-10
57 58	*Glycine max*	[No. of nodes flowering on 7-wk-old plants]	L_n + L_i (540 lux) to 0200 hr (20-hr da)[3]	7 cycles, 8 hr L_f, 15,800 ergs/cm²:40 hr D, 30 min l once every D (plants 3 wk old)	exL,l(−)8 (+)19	25 [2] 20 [2]	Min and max revealed by l. (1st min and 1st max refer to clock hour of l.) Effect of 3 & 30 min l essentially similar. Plants in 2nd trifoliate leaf stage when treated.	5

[L] Pretreatment commenced at start of germination. [3] Post-treatment: Same as pretreatment.

continued

Part II. Magnoliophytes

	Species (Synonym)	Rhythm [Measurement]	Pretreatment	Treatment	First Min (−) First Max (+) hr	Period of Rhythm hr [No. of Observations]	Remarks	Reference
59 60		[No. of nodes flowering]	$L_n + L_i$ (320 lux) (20-hr da)[3/]	7 cycles, 8 hr L_f (cool white), 27-30°C:64 hr D, 4 hr l once every D	exL,l(−)18 (+)28	16 [4] 14 [3]	Min & max revealed by l. (1st min & 1st max refer to clock hour of l.) Phase may be altered by l, but interactions with multiple light treatments difficult to resolve. Plants in 3rd trifoliate leaf stage when treated.	7
61 62	*Kalanchoe blossfeldiana*	[No. of plants flowering, or no. of flowers/plant]	Daily, 18 hr L, greenhouse (plants 15 wk old at end of pretreatment)	9 cycles, 11.5 hr L_n:60.5 hr D, 60 min l_f (1080 lux) once every D	exL,l(−)0-12 (+)21	24 [3] 21 [2]	Min & max revealed by l. (1st min & 1st max refer to clock hour of l.) Controls (no l) did not flower. Flowering assessed when plants were 17 wk old. Plant height was positively correlated with flowering.	6
63 64		[No. of flowers/plant]	$L_n + L_i$ (long day)	10 cycles, 12 hr L_n:60 hr D, 30 min l_i once every D	exL,l(−)8 (+)20	20 [3] 24 [3]	Periods measured 7 wk after causative treatment (1st min & 1st max refer to clock hour of l)	22
65 66	*K. fedtschenkoi*	CO_2 output of excised leaves [μg $CO_2 \cdot hr^{-1} \cdot g^{-1}$ fresh wt]	Cycles, 8 hr L:16 hr D	7 hr D:3 hr l_i, 3000 lux: DD	ex$l$(−)10 (+)18	22 [3] 20 [2]	l before normal 1st max: Phase completely reset. Rhythm measured at time of causative treatment.	26
67 68				13 hr D:5 hr l_i, 3000 lux: DD	ex$l$(−)12 (+)26	24.5 [3] 23 [3]	l at crest of 1st max: Phase not reset. 1 or 3 hr l, or 3 hr l at crest of 2nd max (35 hr D): Phase not reset. Rhythm measured at time of causative treatment.	
69 70				33 hr D:1 hr l_i, 3000 lux: DD	ex$l$(−)2 (+)10	14 [2] 20 [2]	l between normal 1st & 2nd max: Phase not completely reset. 1 hr l, 20,000 lux: Phase not reset. Rhythm measured at time of causative treatment.	
71 72				31 hr D:3 hr l_i, 3000 lux: DD	ex$l$(−)10 (+)22	22.5 [3] 18 [2]	l between normal 1st & 2nd max: Phase reset by end of 3 hr l, also with 6 hr l. Rhythm measured at time of causative treatment.	
73 74				28 hr D:6 hr l, 25 lux:DD	ex$l$(−)7 (+)18	23 [3] 23 [2]	l between normal 1st & 2nd max: Phase reset by end of 6 hr l, 8-3000 lux. l, 2 lux: Phase not completely reset; delay equals duration of l (e.g., 6 hr l delays phase by 6 hr). Rhythm measured at time of causative treatment.	

3/ Post-treatment: Same as pretreatment.

continued

Part II. Magnoliophytes

	Species (Synonym)	Rhythm [Measurement]	Pretreatment	Treatment	First Min (−) First Max (+) hr	Period of Rhythm hr [No. of Observations]	Remarks	Reference
					Prolonged Light			
75 76	Cestrum nocturnum	Opening & closing of flowers [Flowers on intact plants]	Greenhouse, 23°C (day), 17°C (night)	LL, 5380 lux, 17°C	(−).... (+)....	29 [4] 27.7 [4]	Min: Flowers closed. Max: Flowers open. Rhythm measured at time of causative treatment.	21
77 78		[Flowers on cut stalks]	Greenhouse, 23°C (day), 17°C (night)	LL, 5380 lux, 14°C	(−).... (+)....	30.3 [4] 31 [5]	Rhythm measured at time of causative treatment	
79 80		Odor production of flowers [Intact plants & cut stalks]	Greenhouse, 23°C (day), 17°C (night)	14 hr D:LL; 17°C	exD,LL (−)17-25 (+)1-6	27 [4] 28 [5]	Odor produced when flowers open, absent when closed. Rhythm persisted 7-8 da. Rhythm measured at time of causative treatment.	21
81 82	Helianthus annuus	Negative exudation (water intake by cut stump of stem) [ml/min]	L_n (greenhouse)	LL, 28 ± 1°C (plants 12 wk old)	(−) near noon (+) near midnight	∼24 ∼24	Stem intact below cut surface, with roots in soil. Max negative exudation corresponded to time of min positive exudation.	14
83 84 85	'Russian Mammoth'	Positive exudation from stump of stem [ml/hr per plant]	....	LL_i, 1080-3230 lux, 20°C, 70% RH	Hr after decapitation (−)24 (+)12	∼24 [3] ∼24 [4]	Stem intact below cut surface, with roots in Hoagland's solution. L_n (greenhouse): Min near midnight, max near noon. LL: Rhythm phased by decapitation (done 12 hr apart).	13
				LL, 108 lux, 15 & 30°C	....	∼24		
86				LL_i (ruby), low intensity, 25°C, 90% RH	....	∼24		
	Phaseolus vulgaris [4/]	Leaf movement on plants with 2 primary leaves	....	LL_f, 27 ± 0.5°C			Rhythm continued in LL for not less than 4 wk	17
87				1080 lux	....	26		
88				4840 lux	....	26		
89				7530 lux	....	26		
90				10,220 lux	....	26		
					Changes in Light Intensity			
91 92 93 94 95 96	Kalanchoe fedtschenkoi	CO_2 output of excised leaves [μg $CO_2 \cdot hr^{-1} \cdot g^{-1}$ fresh wt]	Cycles, 8 hr L:16 hr D	21 hr L^1, 3000 lux, changing to 2000 lux by 23rd hr, to 1000 lux by 25th hr, & to DD by 27th hr	From start of L^1 (−)1st, 0-24 2nd, 48 3rd, 72 (+)1st, 32 2nd, 56 3rd, 78	24 24 24 23 23 23	Gradual changes in light intensity (3000-0 lux): 80% reduction of intensity necessary to initiate rhythm, rather than reduction below a critical value	26

4/ Pinto bean.

continued

Part II. Magnoliophytes

	Species (Synonym)	Rhythm [Measurement]	Pretreatment	Treatment	First Min (−) First Max (+) hr	Period of Rhythm hr [No. of Observations]	Remarks	Reference
97 98 99 100				13 hr L¹, 3000 lux:15 hr L², 100 lux:DD	exL¹,L²(−)1 (+)4 exL²,DD(−)7 (+)2	[1] [1] 22.5 [3] 18 [2]	Sharp changes in light intensity. Effect of given intensity depended on intensity of preceding illumination. Phase reset by ending of L¹.	
				Light Interrupted by Darkness				
101 102	Chenopodium rubrum '374'	Flowering [Percent of total no. of plants flowering; assessed on ∿5th, 6th, & 7th da from end of d]	LL¹ + 36 hr L¹_f (cool white), 37,660 lux, pre d 5/	0-72 hr d in 3-hr increments, 20°C	(−)0-6 (+)15	30 [3] 30 [2]	∿16-hr interval between end of 1st min & start of 2nd min. Constant light: No flowering. Germination procedure (seeds on moist filter paper in petri dishes): LL¹_f (cool white)—12 hr (8610 lux) at 32.5°C, & 12 hr (4300 lux) at 10°C— for 2.5 da:36 hr L¹ at higher intensity, pre d. Constant temp of 20 ± 1°C from start of L¹.	11
103 104			LL¹ + 30, 36, or 42 hr L¹, 37,660 lux, pre d 6/	0-72 hr d in 3-hr increments, 20°C	(−)0-6 (+)15	28.5 [3] 27 [2]		
105 106			LL¹ + 36 hr L¹_i, 7530 lux, pre d 7/	0-72 hr d in 3-hr increments, 20°C	(−)0-6 (+)15	25.5 [3] 24 [2]	Shorter period than when L_f preceded d; maxima narrower	
107 108			LL¹ + 36 hr L¹_i, 10,760 lux, pre d 8/	0-96 hr d in 3-hr increments, 20°C	(−)0-9 (+)15	25 [4] 28 [4]	3rd & 4th max showed decreasing amplitude	
109	Ipomoea nil (Pharbitis nil) 'Violet'	Flowering [No. of flower buds initiated ∿2 wk after causative (d) treatment]	LL_f (cool white), 4300 lux, 20 ± 1°C 9/	0-72 hr d¹, 18°C 10/	....	....	LL: No flowering. d¹, 15°C: No flowering. Linear increase of flowering with increased length of d. Germination procedure: Plants in H₂SO₄ 30 min; washed in running water 16 hr; on sand 24 hr, ∿23°C; in soil 24-30 hr, 30-32°C; LL_f, 4304 lux, 20 ± 1°C.	23
110 111				8 hr d:8 hr l: 0-72 hr d¹; 18°C 10/	exl,d¹(−)54 (+)42	[1] ∿28 [2]	8 hr d:8 or 12 hr l: May influence rhythm shown in exl,d¹. Min & max not well defined in d¹.	
112 113				8 hr d:12 hr l: 0-72 hr d¹; 18°C 10/	exl,d¹(−)52 (+)42	[1] ∿26 [2]		
114 115 116 117 118 119	Kalanchoe fedtschenkoi	CO₂ output of excised leaves [µg CO₂·hr⁻¹·g⁻¹ fresh wt]	Cycles, 8 hr L:16 hr D	13 hr L¹, 3000 lux:15 hr L², 100 lux:6 hr d: L², 100 lux	exL¹,L²(−)1 (+).... exL²,d(−).... (+)0 exd,LL²(−)1 (+)3	[1] .... [1] 17 [2] 29 [2]	Short dark period inserted at crest of max (phase set by ending of L¹). Phase shifted by 6 hr d, and also by 3rd d inserted at similar time.	26

5/ Post-treatment: LL_i, 7530 lux. 6/ Post-treatment: LL²_i, 10,760 lux. 7/ Post-treatment: 36 hr L¹_i, 7530 lux. 8/ Post-treatment: 36 hr L¹_i, 10,760 lux. 9/ Post-treatment: ≮24 hr L_f (cool white), 4300 lux, 20°C; then 18 hr daily photoperiods, L_n + L_i (540 lux), 15-35°C. 10/ d¹ started 4 da after planting.

continued

Part II. Magnoliophytes

	Species (Synonym)	Rhythm [Measurement]	Pretreatment	Treatment	First Min (−) First Max (+) hr	Period of Rhythm hr [No. of Observations]	Remarks	Reference
120				13 hr L^1,	exL^1,L^2(−)0	....	Dark period at crest of max.	
121				3000 lux:15	(+)13	[1]	Phase completely reset by	
122				hr L^2, 100	exL^2,d(−)6	[1]	9 hr d.	
123				lux:9 hr d:	(+)0	....		
124				LL^2, 100 lux	ex$d$,LL^2(−)21	22 [2]		
125					(+)9	22 [3]		
126				13 hr L^1,	exL^1,L^2(−)8	[1]	Dark period in trough (min)	
127				3000 lux:25	(+)3	13 [2]	between 1st & 2nd max.	
128				hr L^2, 100	exL^2,d(−)4	[1]	Phase not reset by 6 hr d,	
129				lux:6 hr d:	(+)2	[1]	nor 3 hr d, but DD start-	
130				LL^2, 100 lux	ex$d$,LL^2(−)18	22 [2]	ing at similar time reset	
131					(+)7	23 [2]	phase.	
132 133	Musa acuminata 'Gros Michel'	Stomatal opening [Time from start of L to 1st increase in rate of photosynthesis & transpiration]	36 hr L_f (white), 20,440 lux, 30°C, 90-95% RH[3/]	4-72 hr D in 4-hr increments (plants 4-5 mo old)	Length of D (−)12 (+)28	24 ± <2 [3] 24 ± <2 [2]	Rhythm measured after dark period. 12-72 hr D: Gradual increase in stomatal opening time. 36 hr L pre D & increased intensity of 10,760-32,280 lux: Stomatal opening slightly faster after D, but difference less than that due to change of D. No rhythm if temp was 16 or 21°C during D.	3
134 135	Xanthium orientale (X. pensylvanicum)	Stomatal opening [Min to 75% opening]	Cycles, 8 hr L_n + 8 hr L (f + i) 900 lux:8 hr D[11/]	5 hr L_n:3 hr L, 15,000 lux:5-48 hr d (plants 6-7 wk old)	Length of d (−)16 (+)28	24 ± 3 [2] [1]	Small amount of opening in d, but less than in L after D. 1-23 hr d (night): Fastest opening in subsequent L was after 9-16 hr d.	18

Photoperiodic Cycles Other Than 24 Hours

	Species (Synonym)	Rhythm [Measurement]	Pretreatment	Treatment	First Min (−) First Max (+) hr	Period of Rhythm hr [No. of Observations]	Remarks	Reference
136 137	Glycine max	Flowering [No. of nodes flowering]	L_n + L_i (380 lux) to 0200 hr (20-hr da)[3/]	7 cycles, 8 hr L_f (cool white), 12,910 lux:16-64 hr D in 6-hr increments	exL,D(−)28 (+)16	24 [2] 24	Plants had 3rd trifoliate leaf at time of rhythm. Flowering measured 5 wk later.	19, 20
138 139	Hyoscyamus niger	Flowering [Days to stem elongation]	L_n (up to 80,700 lux), 13-43°C:D, 10-17°C	Cycles, 9 hr L, 8070 lux:3-63 hr D	exL,D(−)27 (+)3	[1] 30 [2]	Treatment cycles for 9-50 da. D in 6-hr increments, i.e., cycles of different lengths.	12
140 141				Cycles, 10 hr L, 9900 lux, 22 ± 1°C:8-50 hr D, 21 ± 1°C	exL,D(−)27 (+)8	[1] 28 [2]		
142 143				Cycles, 12 hr L:6-48 hr D	exL,D(−)30 (+)6	[1] 26 [2]		
144 145	Kalanchoe fedtschenkoi	CO_2 output of excised leaves [μg $CO_2 \cdot hr^{-1} \cdot g^{-1}$ fresh wt]	L_n + incandescent & mercury vapor lamps	2 cycles, 16 hr L, 3000 lux:16 hr D:16 hr L; then DD; 26°C	exL,D(+)11 exL,DD(+)17.5	∽33[12/] Initially 25	Stimulating cycles with periods of >24 hr, 26°C: CO_2 output increased in D, decreased in L_i. After DD onset, normal period at	27

[3/] Post-treatment: Same as pretreatment. [11/] Post-treatment: LL, 15,000 lux. [12/] Rhythm entrained by L:D cycles of more than 24 hr.

continued

Part II.   Magnoliophytes

	Species (Synonym)	Rhythm [Measurement]	Pretreatment	Treatment	First Min (−) First Max (+) hr	Period of Rhythm hr [No. of Observations]	Remarks	Reference
146 147			(leaves excised at 1500 hr)	3 cycles, 20 hr D:20 hr L, 25 lux L; then DD; 26°C	.... exL,DD(+)15.5	∿38-39 [12]	26°C was 22.4 ± 0.4 hr. Rhythm measured at time of causative treatment.	
148 149				3 cycles, 24 hr D:24 hr L, 25 lux; then DD; 26°C	.... exL,DD(+)15	∿46-48 [12]		
150				10 cycles, 3 hr D:3 hr L, 500 lux; then DD; 26°C	(+) 21 hr from start of 1st 3 hr D	23-24	3 hr D:3 hr L, 500 lux: No entrainment. Rhythm measured at time of causative treatment.	
151 152				12 cycles, 3 hr D:3 hr L, 1000 lux; then DD; 26°C	(+) middle of each 3 hr L exL,DD(+)16	∿6 [13] 23	Stimulating cycles with periods of <24 hr, 26°C. Rate of CO_2 output increased in L, decreased in D. Rhythm measured at time of causative treatment.	
153				5 cycles, 6 hr D:6 hr L, 100 lux; then DD; 26°C	(+) 13 hr from start of 1st 6 hr D	23	No entrainment. DD started at crest of 3rd max. Rhythm measured at time of causative treatment.	
154 155				5 cycles, 6 hr D:6 hr L, 500 lux; then DD; 26°C	.... exL,DD(+)18	12 [13] 23	Stimulating cycles with periods of <24 hr, 26°C. When DD started at different times of day, transient still ∿18 hr, i.e., basic oscillating system entrained by D:L cycles. Rhythm measured at time of causative treatment.	
156 157 158 159		CO_2 fixation of excised leaves [μg CO_2 fixed· hr^{-1}·g^{-1} fresh wt, compared with μg CO_2 output·hr^{-1}· g^{-1} fresh wt]	Short days	DD 6 hr earlier than normal D (i.e., 2 hr L pre DD)	CO_2 fixation (−)3 (+)13 CO_2 output (−)15 (+)23	22 [3] 26 [2] 19 [2] 27 [2]	Early D onset did not alter phase of rhythm for CO_2 fixation or output. 2 hr L acted as interruption to preceding D at peak of CO_2 output. Rhythm measured at time of causative treatment.	24
160 161 162 163			Long days	DD 6 hr earlier than normal D (i.e., 10 hr L pre DD)	CO_2 fixation (−)3 (+)11 CO_2 output (−)10 (+)17	21 [3] 20 [3] 22 [3] 26 [2]	Phase shift induced by 10 hr L between max of CO_2 output; phase set by end of L. Rhythm measured at time of causative treatment.	

[12] Rhythm entrained by L:D cycles of more than 24 hr.   [13] Rhythm entrained by L:D cycles of less than 24 hr.

continued

Part II. Magnoliophytes

	Species (Synonym)	Rhythm [Measurement]	Pretreatment	Treatment	First Min (−) First Max (+) hr	Period of Rhythm hr [No. of Observations]	Remarks	Reference
				Irradiation by Different Parts of Spectrum				
164	*Bauhinia*	Leaf move-	Daily, 12 hr	Control, ∿12	exL,DD(−)1	16.5 irregu-	6 hr extra radiation before	16
165	*monandra*	ment [Change	L_n until 1st	hr L:DD	(+)20	lar [5]	normal 12 hr L (control):	
166		in angle of	leaf had	from 1800		14.3 irregu-	Leaves exposed to W	
		blade to peti-	been ex-	hr		lar [4]	closed 4.5 hr earlier than	
		ole]	panded 1	D at 1800-	....	Not mea-	leaves on control; those	
			wk	2400 hr:6		sured	exposed to R, 2 hr earlier;	
				hr W, B, G, R, or FR			those exposed to B, G, &	
				at 2400-0600 hr: W			FR showed virtually no	
				at 0600-1800 hr			effect	
167	*Chenopo-*	Flowering	LL^1 + 36 hr	0-96 hr d, 5	exL^1,d(−)0-6	[1]	r, 0.85 mW/cm²: Reduced	11
168	*dium ru-*	[Percent of	L^1_i (10,760	min r at 6th	(+)....	....	1st & 2nd max, and elimi-	
169	*brum*	total no. of	lux) pre fr	hr d	exr,d(−)6-9	24 [3]	nated normal 3rd max.	
170	'374'	plants (plants	and/or d [14]		(+)15	27 [2]	exr,d, (−) or (+), refers to	
		4 da old at					length of d extending be-	
		time of caus-					yond r. 6th hr of d = skotophile phase.	
171		ative treat-		0-96 hr d, 5	exL^1,d(−)0-9	[1]	r, 0.85 mW/cm²: Normal	
172		ment; 8-14 da		min r at	(+)....	....	1st, 2nd, & 3rd max. 12th	
173		old when		12th hr d	exr,d(−)15	30 [4]	hr of d: Start of photo-	
174		flowering was			(+)3	30 [3]	phile phase.	
175		assessed)]		7 min fr pre	....	0	fr, 1.68 mW/cm² pre d:	
				0-96 hr d			Converted active form of	
							phytochrome P_{fr} to P_r, and prevented flowering.	
							Rhythmic display of flowering when fr = 0.	
176				7 min fr pre	exfr,d(−)0-6	[1]	r, 0.85 mW/cm²: Partially	
177				0-96 hr d; 5	(+)....	....	restored normal 2nd max,	
178				min r at 6th	exr,d(−)6-36	58 (mid-	but not 1st or 3rd; r in-	
				hr d		points)[2]	creased amount of P_{fr}	
179					(+)33	[1]		
180				7 min fr pre	exfr,d(−)0-12	[1]	r, 0.85 mW/cm²: Restored	
181				0-96 hr d; 5	(+)....	....	normal 2nd & 3rd max,	
182				min r at	exr,d(−)6-36	36 (mid-	but amplitude was less	
				12th hr d		points)[2]	than for control (fr = 0)	
183					(+)30	33 [2]		
184			L^1_i, 7530	3-72 hr d, 2	exL^1,r(−)3-6	27 [3]	r = 0.85 mW/cm². Width	
185			lux [3]	min r once	(+)15	33 [2]	(area) of max greater with	
				during d			L_f than with L_i (7530-	
186			L^1_f, 37,660	3-72 hr d, 2	exL^1,r(−)6	30 [3]	10,760 lux) pre d.	
187			lux [15]	min r once	(+)27	18 [2]		
				during d				
188			LL^1 + 36 hr	7 min fr pre	exL^1,r(−)9	30 [2]	r = 0.85 mW/cm²; fr = 1.68	
189			L^1_f (37,660	48 hr; 5 min	(+)24	[1]	mW/cm². r restored 1st	
			lux) pre	r once dur-			max, but amplitude &	
			fr [16]	ing d			width were less than for	
							control (fr = 0).	

[3] Post-treatment: Same as pretreatment. [14] Post-treatment: LL^1_i, 10,760 lux. [15] LL^1_i, 7530 lux. [16] Post-treatment: LL_i, 10,760 lux.

continued

Part II.   Magnoliophytes

	Species (Synonym)	Rhythm [Measurement]	Pretreatment	Treatment	First Min (−) First Max (+) hr	Period of Rhythm hr [No. of Observations]	Remarks	Reference
190				72 hr d, 10 s fr once during d	....	0	fr converted phytochrome P_{fr} to P_r. Very low per-cent flowering when fr imposed in 1st 40 hr d; gradual increase from 40th to 72nd hr d. No clear rhythm.	
191 192			30, 36, or 42 hr L^1f, 37,660 lux	72 hr d, 4 min r once during d	$exL^1,r(-)36$ (+)24	29 [3] 24 [2]	r, 0.85 mW/cm²: Phase set by transition from L^1 to d. No significant effect on phase when L^1 period varied by 12 hr, i.e., maxima at similar times in d.	
193 194			$LL^1 + 36$ hr L^1f (32,280 lux) pre d [16]	0-24 hr d	$exL^1,d(-)0-6$ (+)10-18	[2] [1]	16-hr interval between end of 1st min & start of 2nd min	
195 196				8 min r pre 0-24 hr d	$exr,d(-)0-8$ (+)12-16	[2] [1]	r, 0.23 mW/cm²: 12-hr interval between end of 1st min & start of 2nd min	
197 198				2 s fr pre 0-24 hr d	$exfr,d(-)0-6$ (+)10-18	[2] [1]	fr, 0.23 mW/cm²: 18-hr interval between end of 1st min & start of 2nd min	
199 200				50 s fr pre 0-24 hr d	$exfr,d(-)0-6$ (+)10-16	[2] [1]	fr, 0.23 mW/cm²: 12-hr interval between end of 1st min & start of 2nd min	
201 202				250 s fr pre 0-24 hr d	$exfr,d(-)0-6$ (+)10-12	[2] [1]	fr, 0.23 mW/cm²: 8-hr interval, i.e., increasing length of fr decreased length of d required for flowering	
203 204	*Glycine max*	Flowering [No. of nodes flowering on 7-wk-old plants]	$L_n + L_i$ (377 lux) to 0200 hr (12-hr da) [3] (plants 3 wk old)	7 cycles, 8 hr L:40 hr d, 30 min r once during d	$exL,r(-)8$ (+)19	24 [2] 20	r, 12,000 ergs/cm²: Effect of 3 & 30 min r essentially similar	5
205				7 cycles, 8 hr L:40 hr d, 30 min fr once during d	....	0	fr, 15,500 ergs/cm²: Effect of 3 & 30 min fr essentially similar	
206 207 208 209	*Kalanchoe fed-tschenkoi*	CO_2 output [μg $CO_2 \cdot$hr⁻¹· g⁻¹$ fresh wt]	Cycles, 8 hr L:16 hr D	28 hr D:6 hr r:DD	D(−)10 (+)20 $exr,DD(-)6$ (+)14	[1] [1] 23 [3] 24 [2]	r, 850 ergs·cm⁻²·s⁻¹ (>565 nm), at min between 1st & 2nd max: Phase completely reset (similarly with r, 3620 ergs·cm⁻²·s⁻¹)	26
210 211 212 213				28 hr D:6 hr b:DD	D(−)10 (+)21 $exb,DD(-)24$ (+)9	[1] [1] 24 [1] 24 [2]	b, 10,860 ergs·cm⁻²·s⁻¹ (>525 nm), or 1960 ergs· cm⁻²·s⁻¹: Phase not reset. 6 hr g, 6880 ergs·cm⁻²·s⁻¹ (475-575 nm): Phase delay of 4.5 hr.	
214 215 216				68 hr R:BB	R .... $exR,BB(-)4$ (+)11	24 [4] 25 [4]	R, 16,430 ergs·cm⁻²·s⁻¹: No rhythm. BB, 10,860 ergs·cm⁻²·s⁻¹: Started rhythm.	

[3] Post-treatment: Same as pretreatment.   [16] Post-treatment: LL$_i$, 10,760 lux.

continued

117. PERIOD OF CIRCADIAN RHYTHMS IN LIGHT AND DARKNESS: PLANTS

Part II. Magnoliophytes

Contributor: Cumming, Bruce G.

References

[1] Ball, N. G., and I. J. Dyke. 1954. J. Exp. Bot. 5: 421.

[2] Ball, N. G., et al. 1957. Ibid. 8:339.

[3] Brun, W. A. 1962. Physiol. Plant. 15:623.

[4] Bünning, E., et al. 1955. Ber. Deut. Bot. Ges. 68:41.

[5] Carpenter, B. H., and K. C. Hamner. 1963. Plant Physiol. 38:698.

[6] Carr, D. J. 1952. Z. Naturforsch. 76:570.

[7] Coulter, M. W., and K. C. Hamner. 1964. Plant Physiol. 49:848.

[8] Cumming, B. G. 1963. Can. J. Bot. 41:1211.

[9] Cumming, B. G. 1963. Int. Symp. Physiol. Ecol. Biochem. Germination, Greifswald, 1963, A II(1).

[10] Cumming, B. G. Unpublished. Univ. Western Ontario, Dep. Botany, London, Canada, 1966.

[11] Cumming, B. G., et al. 1965. Can. J. Bot. 48:825.

[12] Finn, J. C., and K. C. Hamner. 1960. Plant Physiol. 35:982.

[13] Grossenbacher, K. A. 1939. Amer. J. Bot. 26:107.

[14] Hagan, R. M. 1949. Plant Physiol. 24:441.

[15] Hamner, K. C., et al. 1962. Nature (London) 195: 476.

[16] Holdsworth, M. B. 1960. J. Exp. Bot. 11:40.

[17] Hoshizaki, T., and K. C. Hamner. 1964. Science 144:1240.

[18] Mansfield, T. A., and O. V. S. Heath. 1963. J. Exp. Bot. 14:334.

[19] Nanda, K. K., and K. C. Hamner. 1958. Bot. Gaz. 120:14.

[20] Nanda, K. K., and K. C. Hamner. 1959. Planta 53: 53.

[21] Overland, L. 1960. Amer. J. Bot. 47:378.

[22] Schwabe, W. W. 1955. Physiol. Plant. 8:263.

[23] Takimoto, A., and K. C. Hamner. 1964. Plant Physiol. 39:1024.

[24] Warren, D. M., and M. B. Wilkins. 1961. Nature (London) 191:686.

[25] Wilkins, M. B. 1959. J. Exp. Bot. 10:377.

[26] Wilkins, M. B. 1960. Ibid. 11:269.

[27] Wilkins, M. B. 1962. Plant Physiol. 37:735.

118. CIRCADIAN RHYTHMS OF PHYSIOLOGICAL VARIABLES AS REFLECTED IN RAT BIOASSAY

Mean Value: Occurrence is given in clock hours in local time, unless otherwise specified.

Part I. Daily Variation

Persistent daily fluctuations occur in a large number of physiological variables of the living organism. These rhythms in man or experimental animal are necessarily reflected in chemical- and bio-assay, because values of measured functions vary from hour to hour over the total time span of the rhythm. The natural ecological light:dark cycle, or the artificial one of the laboratory, has been demonstrated to be a dominant synchronizer of these rhythms, especially in the experimental animal. Once a rhythm has been demonstrated for a particular function, the times of occurrence of certain discrete phases of this same rhythm can be predicted with reasonable confidence for other animals of the same species when exposed to similar intervals of sampling and to a light:dark cycle identical to that initially used to establish the rhythm. Recent evidence indicates that, in spite of rigid control, there may be some phase shift in certain physiological variables which seem to be associated with seasonal changes. Sampling at the same time of day is often used to eliminate or minimize the variations due to rhythmicity, but this procedure can only assure that sampling was made in a trough or peak period, or somewhere between. This assurance is still dependent on a previously established rhythm determined under conditions of controlled light: dark with frequent intervals of sampling. **Physiological Variable:** mu = milliunits; meq = milliequivalents; ng = nanograms, or 10^{-9} gram; nmole = nanomoles, or 10^{-9} mole; pmole = picomoles, or 10^{-12} mole; μmole = micromoles, or 10^{-6} mole. **Method or Specification:** RD = rapid decapitation; SMA-12/60 = Technicon's Sequential Multiple Analyzer 12/60; CT = cardiac tap; JT = jugular tap. **Duration** shows beginning and end of the light phase of the light:dark cycle to which the animals were subjected, and is expressed in clock hours in local time, unless otherwise specified. Data in brackets refer to the column heading in brackets.

	Physiological Variable	Method or Specification	Duration	Sex	Sampling Interval hr	Mean Value 24-hr	Mean Value Lowest [Occurrence]	Mean Value Highest [Occurrence]	Reference
				Whole Blood					
1	Leukocytes, no./μl blood	RD	0400-1800	♂	4	5800	3000 [1900]	8000 [1100]	13
2				♀	4	4847	3500 [1900]	6000 [1100]	13
3		Tail blood	0600-1800	♂	2	14,956	9815 [1700]	20,140 [0900]	35

continued

Part I. Daily Variation

	Physiological Variable	Method or Specification	Duration	Sex	Sampling Interval hr	Mean Value 24-hr	Mean Value Lowest [Occurrence]	Mean Value Highest [Occurrence]	Reference
4	Eosinophils, no./μl blood	Tail blood	0600-1800	♂	2	294	193 [2200]	471 [0900]	35
5	Lymphocytes, no./μl blood	Tail blood	0600-1800	♂	2	12,186	8439 [1800]	15,923 [0900]	35
6	Neutrophils, no./μl blood	Tail blood	0600-1800	♂	2	2216	1371 [1900]	3488 [0900]	35
7	Reticulocytes, % RBC	RD	0600-1800	♀	4	....	∼3 [1600]	∼5.5 [0800]	11
8	Coagulation time, s	Tail blood[1]	0600-1800	♂	2	263	230 [0900]	325 [0100]	46
9	Glucose, mg/100 ml	Nonfasting; tail blood; method of Somogyi & Shaffer	0600-1800	♂	2	103	93 [0830]	115 [2330]	36
10	Urea nitrogen, mg/100 ml	Ether; JT; SMA-12/60	0600-1800	♂	2	20.8	17 [1500]	24 [0700]	49
	Plasma								
11	Adrenocorticotropic hormone, μg corticosterone produced/5 min	RD	0700-1900	♂	4	0.032	0.02 [0500]	0.07 [1700]	41
12				♀	4	0.044	0.01 [0500]	0.08 [1700]	41
13	Cholesterol, mg/100 ml	Pentobarbital Na; CT; method of Hauang, et al.	0600-1800	♂	2	69	44 [0000]	114 [1200]	49
14	Copper, μg/100 ml	Pentobarbital Na; CT; method of Stoner & Dasler	0600-1800	♂	2	103	84 [0700]	126 [0300]	57
15	Corticosteroid, μg/100 ml	RD	Natural daylight, Jan 31	♂	4	10	5 [0400]	16 [1600]	25
16	Corticosterone, μg/100 ml	Pentobarbital Na; CT	0600-1800	♂	2	24.9	13 [0530]	37 [1130]	45
17			0800-2000	♂	6	18	7 [0300]	31 [2200]	33
18				♀	6	23	13 [0400]	33 [2200]	33
19		RD	0400-1800	♂	4	10	6 [2300]	21 [1500]	13
20				♀	4	30	11 [0300]	58 [1800]	13
21			0700-1900	♂	4	7.8	1.1 [0500]	19 [1700]	30
22				♀	4	21	3.1 [1100]	35.4 [2300]	30
23	Mucoprotein, mg/100 ml	Pentobarbital Na; CT	0600-1900	♂	1	5.7	4.9 [1900]	6.9 [0500]	52
24	Phosphorus, inorganic, mg/100 ml	Pentobarbital Na; CT	0600-1900	♂	1	7.0	6.4 [0600]	7.7 [1600]	58
	Protein, g/100 ml								
25	Total	Pentobarbital Na; CT	0400-1800	♂	1	6.6	6.1 [1800]	7.1 [0100]	52
26	Albumin	Pentobarbital Na; CT	0400-1800	♂	1	3.1	2.8 [1500]	3.4 [0800]	52
27	α_1-Globulin	Pentobarbital Na; CT	0400-1800	♂	1	0.65	0.58 [1800]	0.74 [0100]	52
28	α_2-Globulin	Pentobarbital Na; CT	0400-1800	♂	1	0.71	0.62 [1800]	0.79 [0500]	52
29	β-Globulin	Pentobarbital Na; CT	0400-1800	♂	1	0.93	0.87 [2000]	1.10 [0100]	52
30	γ-Globulin	Pentobarbital Na; CT	0400-1800	♂	1	1.21	1.05 [2000]	1.34 [0200]	52
31	Sulfur, inorganic, mg/100 ml	Pentobarbital Na; CT	0600-1900	♂	1	....	1.8 [2200]	3.6 [1400]	58
32	Tyrosine, mg/100 ml	RD	0730-1630	♂	3	....	∼1.3 [1100]	2.0 [0500]	12
33	Uric acid, mg/100 ml	Pentobarbital Na; CT; method of Winzler	0600-1800	♂	1	1.5	1.0 [0300]	2.4 [1500]	49
	Serum								
34	Albumin, g/100 ml	RD; SMA-12/60	0600-1800	♂	2	2.8	2.7 [1100]	2.9 [0500]	49

[1] A fine, siliconized glass rod was drawn through blood until appearance of firm, beaded clot.

continued

Part I. Daily Variation

	Physiological Variable	Method or Specification	Duration	Sex	Sampling Interval hr	Mean Value			Reference
						24-hr	Lowest [Occurrence]	Highest [Occurrence]	
35	Alkaline phosphatase, mu/ml	RD; SMA-12/60	0600-1800	♂	2	195	133 [1700]	235 [0700]	49
36	Aspartate transaminase[2/], mu/ml	RD; SMA-12/60	0600-1800	♂	2	118	103 [1700]	132 [0700]	49
37	Calcium, mg/100 ml	RD; SMA-12/60	0600-1800	♂	2	9.3	8.9 [1100]	9.6 [0900]	49
38	Chloride, meq/liter	RD; flame photometer	0600-1800	♂	2	119	113 [1500]	140 [0700]	49
39	Copper, μg/100 ml	Pentobarbital Na; CT; neutron activation	0600-1800	♂	1	156	133 [0600]	256 [2000]	53
40	Corticosterone, μg/100 ml	RD	0600-1800	♂	4	14	4 [1200]	30 [1600]	24
41	Histamine	RD	0800-2000	♂	6	0.82	0.51 [1200]	1.07 [1800]	19
42	Interstitial cell-stimulating hormone[3/], ng/ml	RD; radioimmuno-assay	0400-1800	♂	3	0.80	0.46 [1100]	1.12 [2000]	17
43	L-Lactate dehydrogenase, mu/ml	RD; SMA-12/60	0600-1800	♂	2	201	117 [2300]	256 [0900]	49
44	Luteotropin[4/], ng/ml	RD; radioimmuno-assay	0400-1800	♂	3	12.7	8.12 [0800]	18.6 [2300]	17
45	Manganese, μg/100 ml	Neutron activation	0600-1800	♂	1	1.2	0.6 [0800]	2.4 [1600]	53
46	Phosphorus, inorganic, mg/100 ml	RD; SMA-12/60	0600-1800	♂	2	6.4	5.9 [2300]	6.9 [1700]	49
47	Potassium, meq/liter	RD; flame photometer	0600-1800	♂	2	3.7	3.5 [1300]	3.9 [2300]	49
48	Protein, total, g/100 ml	RD; SMA-12/60	0600-1800	♂	2	6.7	6.4 [1100]	6.9 [2100]	49
49	Serotonin[5/], μg/ml	RD	0400-1800	♀	2	1.46	1.19 [1100]	1.92 [2000]	54
50			0600-1800	♂	2	1.15	0.96 [1000]	1.28 [2200]	54
51	Sodium, meq/liter	RD; flame photometer	0600-1800	♂	2	154	150 [1500]	171 [0700]	49
52	Uric acid, mg/100 ml	RD; SMA-12/60	0600-1800	♂	2	0.92	0.36 [2300]	1.5 [1300]	49
	Whole Brain								
53	Acetylcholine, nmole/g brain	RD	12 hr light, 12 hr dark[6/]	♂	4	....	∿20 [after 6 hr of darkness]	∿40 [after 2 hr of darkness]	27
54	Norepinephrine, μg/g brain	RD	0600-1800	♂	1	0.25	0.22 [1230]	0.29 [1830]	50
55	Serotonin[5/], μg/g brain	RD	0600-1800	♂	1	0.68	0.62 [2130]	0.73 [1130]	50
	Caudate Nucleus								
56	Histamine, μg/g	RD	0800-2000	♂	6	1.87	1.17 [1800]	2.29 [0600]	18
57	Norepinephrine, μg/g	RD	0800-2000	♂	6	1.00	0.85 [1800]	1.25 [0600]	18
58	Serotonin[5/], μg/g	RD	0800-2000	♂	6	0.34	0.26 [0600]	0.43 [1800]	18
	Midbrain								
59	Histamine, μg/g	RD	0800-2000	♂	6	0.04	0.032 [0000]	0.063 [0600]	18
60	Serotonin[5/], μg/g	RD	0800-2000	♂	6	0.19	0.18 [1200]	0.21 [1800]	18
	Pineal Body								
61	Acetylserotonin methyl-transferase[7/] activity, pmole melatonin[8/]/mg pineal body	RD	1800-0400	♂	4	....	∿10 [1800]	∿34 [0000]	2

[2/] Synonyms: Serum glutamic oxalacetic transaminase; SGOT. [3/] Synonym: Luteinizing hormone. [4/] Synonym: Prolactin. [5/] Synonym: 5-Hydroxytryptamine. [6/] No clock hours given. [7/] Synonyms: Hydroxyindole-O-methyltransferase; HIOMT. [8/] Melatonin carrying radioactive carbon-14.

continued

Part I. Daily Variation

	Physiological Variable	Method or Specification	Duration	Sex	Sampling Interval hr	Mean Value			Reference
						24-hr	Lowest [Occurrence]	Highest [Occurrence]	
62	5-Hydroxyindoleacetic acid, ng/pineal body	RD	1800-0400	♀	4	....	∿4 [1900]	∿18 [0100]	39
63	Melatonin, ng/pineal body	RD	1800-0400	♂	4	....	∿1.3 [0430]	∿3.0 [1830]	39
64	Norepinephrine, ng/pineal body	Cervical dislocation	1900-0700	♀	6	....	3.8 [1900]	10.6 [0700]	59
65	Serotonin 5/, ng/pineal body	RD	1800-0400	♂	4	....	∿10 [2200]	∿90 [1200]	38
66	Tyrosine hydroxylase, pmole pyrocatechols·pineal body^{-1}·hr^{-1}	....	1900-0700	♂♀	4	....	∿3.0 [1500]	∿16 [0300]	34
	Heart								
67	Clostridiopeptidase A 9/ activity, μmole of product·g^{-1}·15 min^{-1}	RD	0600-1800	♂	4	1.9	1.5 [1300]	1.9 [1000]	14
	Hypophysis								
68	Adrenocorticotropic hormone, μg corticosterone/100 ml plasma	RD	0400-1800	♂	6	34	25.2 [0300]	48.7 [2300]	41
69				♀	6	33.4	26.6 [2300]	45.3 [1700]	41
70	Luteotropin 4/, IU/mg	....	0600-1800	♀	2	....	∿0.03 [2000]	∿0.11 [1600]	10
71	Thyrotropin 10/, U.S.P. mu/hypophysis	Pentobarbital Na	0800-2000	♀	3	....	∿120 [1300]	∿240 [0800]	3
	Liver								
72	ATP citrate lyase 11/, μmole·g liver^{-1}·hr^{-1}	RD	0900-2100	♂	6	....	40 [0900]	83 [2100]	37
73	Clostridiopeptidase A 9/, μmole·g^{-1}·15 min^{-1}	RD	0600-1800	♂	4	4.5	3.3 [1900]	5.9 [1000]	14
74	Glucose-6-phosphate dehydrogenase, μmole·g liver^{-1}·hr^{-1}	RD	0900-2100	♂	6	....	100 [0900]	300 [2100]	37
75	Hexokinase, μmole·g liver^{-1}·hr^{-1}	RD	0900-2100	♂	6	....	12 [0900]	33 [0300]	37
76	p-Nitroanisole O-demethylating activity, μmole p-nitrophenol·100 mg protein^{-1}·40 min^{-1}	RD	0630-2000	♂	4	....	0.55 [0200]	0.91	40
77	Ornithine-ketoacid transaminase, μmole·g liver^{-1}·hr^{-1}	RD	....	♂	6	....	220 [0300]	350 [1500]	37
78	Tyrosine transaminase μmole·g liver^{-1}·hr^{-1}	Cervical dislocation	0500-1900	♂	6	....	35 [1500]	110 [0000]	4
79		RD	0500-1700	♂	3	....	9 [1100]	40 [2100]	8
80			0900-2100	♂	6	....	60 [1500]	120 [0900]	37
81	μmole·g liver^{-1}·min^{-1}	RD	0730-1630	♂	3	....	2.9 [1200]	6.0 [0000]	21
82	mmole p-hydroxyphenylpyruvic acid·mg biuret protein^{-1}·min^{-1}	RD	0730-1830	♂	4	....	3.3 [0800]	16.7 [2200]	56

4/ Synonym: Prolactin. 5/ Synonym: 5-Hydroxytryptamine. 9/ Synonym: Collagenase. 10/ Synonyms: Thyroid-stimulating hormone; TSH. 11/ Synonym: Citrate-cleavage enzyme.

continued

Part I. Daily Variation

	Physiological Variable	Method or Specification	Duration	Sex	Sampling Interval hr	Mean Value 24-hr	Mean Value Lowest [Occurrence]	Mean Value Highest [Occurrence]	Reference
83	% of highest value	RD	0700-1900	♂	3	....	∿45 [1100]	∿95 [2200]	20
	Tryptophan oxygenase [12]								
84	μmole·g liver^{-1}·hr^{-1}	RD	0900-2100	♂	6	....	∿2.7 [0300]	∿7 [1500]	37
85	μmole·g liver^{-1}·min^{-1}	RD	0700-1900	♂	3	....	9.8 [1100]	25.6 [2000]	20
86	μmole kynurenine·g liver^{-1}·hr^{-1}	RD	0600-1800	♂	3	....	∿2 [0900]	∿7 [0000]	28
	Glycogen								
87	mg in liver/g body wt	Animals fasted 24 hr	....	♂	2	0.06	0.03 [1600]	0.14 [0200]	1
88	% of	Nonfasting	....	♂	4	3.1	1.9 [1600]	4.7 [0400]	15

Mitotic Index

	Physiological Variable	Method or Specification	Duration	Sex	Sampling Interval hr	Mean Value 24-hr	Mean Value Lowest [Occurrence]	Mean Value Highest [Occurrence]	Reference
89	Bone marrow, % cells dividing	....	0600-1800	♀	4	10	4.5 [1800]	13 [1000]	11
90	Connective tissue, skin, % cells dividing	....	0900-1800	♂	1	3.4	2.1 [1900]	5.3 [1000]	7
	Epidermis								
91	Abdominal wall, mitoses/ 1000 fields	....	....	♂	2	38	20 [2000]	73 [0900]	5
92	Pinna, mitoses/1000 cells	....	0600-1800	♂	1	8	4 [2200]	15 [0300]	44
	Epithelium								
	Cornea								
93	Mitoses/1000 cells	....	0350-2047 [13]	♂	2	4.5	0.3 [2200]	11 [0800]	32
94			0600-1800	♂	1	11	3.7 [2000]	15 [1200]	48
95					2	15	2.1 [1900]	25 [0930]	47
96						12	4.1 [1900]	22 [0700]	55
97	Mitoses/100 fields	....	Natural light of Moscow in March	♂	6	200	76 [2000]	362 [0800]	31
98	Duodenum, mitoses/ 1000 cells	....	0600-1800	♂	1	18.5	15 [0300]	23 [1200]	49
99	Esophagus, mitoses/ 1000 cells	....	....	♂	3	6	0.8 [1900]	13 [0700]	16
100	Lens [14], mitoses/100 cells	....	Natural L:D cycle; month not given	♂	6	165	108 [1800]	228 [0600]	43
	Stomach								
101	Mucous neck cells, mitoses/1000 cells	....	0600-1800	♂♀	2	3.6	1.8 [0000]	6 [0800]	9
102	Surface cells, mitoses/ 1000 cells	....	0600-1800	♂♀	2	4.9	3.4 [0000]	7 [0800]	9
103	Tongue, mitoses/1000 cells	....	0600-1800	♂	1	9.1	4 [2300]	16 [1100]	22
104	Kidney cortex, mitoses/ 1000 fields	....	....	♂	2	....	20 [2000]	60 [1300]	6
105	Liver, normal [15], mitoses/ 1000 cells	....	0600-1600	♂	2	4.9	2.2 [2000]	13 [0800]	29
106	Submandibular gland, mitoses/500 fields	....	....	♂	2	29.4	11.5 [0200]	39.7 [1000]	6

[12] Synonym: Tryptophan pyrrolase. [13] Natural light in Germany, from sunrise to sunset. [14] 3-mo-old rats. [15] 24-da-old rats.

continued

Part I. Daily Variation

	Physiological Variable	Method or Specification	Duration	Sex	Sampling Interval hr	Mean Value			Reference
						24-hr	Lowest [Occurrence]	Highest [Occurrence]	
107	Tooth enamel organ, mitoses/1000 cells	....	0600-1800	♂	1	6.5	4 [1800]	8 [0900]	23
	Suprarenal Gland								
108	Ascorbic acid, μg/100 ml	RD	Natural light in Germany in June	♂	4	357	251 [1600]	410 [0800]	42
109			0700-1900	♂	4	429	356 [2200]	482 [1700]	30
110	Corticosterone, μg/g	RD	0400-1800	♂	6	9.6	1.8 [0500]	26.3 [1700]	41
111				♀	6	14.6	4.9 [0500]	32.0 [1700]	41
112	Epinephrine, μg/g	RD	0600-1800	♂	1	590	430 [2200]	910 [1000]	51
113	Pantothenate, ng/mg fat-free wet wt	RD	0600-1800	♂	4	140	∿120[16] [1500]	∿160 [0000]	24
	Temperature								
114	Temperature, rectal, °C	Thermistor bridge circuit	0600-1800	♂	1.33	38.2	37.8 [0600]	38.7 [1900]	26

[16]/ Value estimated from graph.

Contributors: Scheving, Lawrence E., and Pauly, John E.

References

[1] Agren, G., et al. 1931. Biochem. J. 25:777.

[2] Axelrod, J., et al. 1965. J. Biol. Chem. 240:949.

[3] Bakke, J. K., and N. Lawrence. 1965. Metab. Clin. Exp. 14:841.

[4] Black, I. B., and J. Axelrod. 1968. Proc. Nat. Acad. Sci. 61:1287.

[5] Blumenfeld, C. M. 1939. Science 90:446.

[6] Blumenfeld, C. M. 1942. Arch. Pathol. 33:770.

[7] Chu, C. H. U. 1960. Anat. Rec. 138:11.

[8] Civen, M., et al. 1967. Science 157:1563.

[9] Clark, R. H., and B. L. Baker. 1962. Proc. Soc. Exp. Biol. Med. 111:311.

[10] Clark, R. H., and B. L. Baker. 1964. Science 143: 375.

[11] Clark, R. H., and D. R. Korst. 1969. Ibid. 166:236.

[12] Coburn, S., et al. 1968. Proc. Soc. Exp. Biol. Med. 129:338.

[13] Critchlow, V., et al. 1963. Amer. J. Physiol. 205: 807.

[14] Cutroneo, K. E., and G. C. Fuller. 1971. Life Sci. 10:395.

[15] Deuel, H. J., et al. 1938. J. Biol. Chem. 123:257.

[16] Dobrokhotov, V. N., and A. G. Kurdyumova. 1962. Bull. Exp. Biol. Med. (USSR) 52(8):81.

[17] Dunn, J. D., et al. 1972. Endocrinology 90:29.

[18] Friedman, A., and C. A. Walker. 1968. J. Physiol. (London) 197:77.

[19] Friedman, A., and C. A. Walker. 1969. Ibid. 202: 133.

[20] Fuller, R. W. 1970. Proc. Soc. Exp. Biol. Med. 133: 620.

[21] Fuller, R. W., and H. D. Snoddy. 1968. Science 159:738.

[22] Gasser, R. F., et al. 1972. J. Cell. Physiol. 80:437.

[23] Gasser, R. F., et al. 1972. J. Dent. Res. 51:740.

[24] Glick, D., et al. 1961. Amer. J. Physiol. 200:811.

[25] Guillemin, R., et al. 1959. Proc. Soc. Exp. Biol. Med. 101:394.

[26] Halberg, F., et al. 1954. Amer. J. Physiol. 177:361.

[27] Hanin, I., et al. 1970. Science 170:341.

[28] Hardeland, R., and L. Rensing. 1968. Nature (London) 219:619.

[29] Jackson, B. 1959. Anat. Rec. 134:365.

[30] König, A., and U. Eggers. 1970. J. Interdiscipl. Cycle Res. 1:95.

[31] Kosichenko, L. P. 1959. Bull. Exp. Biol. Med. (USSR) 49:617.

[32] Mayersbach, H. von, and L. E. Scheving. Unpublished. Medizinische Hochschule, Hannover, Germany, 1971.

[33] McCarthy, J. L., et al. 1960. Proc. Soc. Exp. Biol. Med. 104:787.

[34] McGeer, E. G., and P. L. McGeer. 1966. Science 153:73.

[35] Pauly, J. E., and L. E. Scheving. 1965. Anat. Rec. 153:1.

[36] Pauly, J. E., and L. E. Scheving. 1967. Amer. J. Anat. 120:627.

continued

Part I. Daily Variation

[37] Potter, V. R., et al. 1966. Advan. Enzyme Res. 4: 247.

[38] Quay, W. B. 1963. Gen. Comp. Endocrinol. 3:473.

[39] Quay, W. B. 1964. Proc. Soc. Exp. Biol. Med. 115: 710.

[40] Radzialowski, F. M., and W. F. Bousquet. 1967. Life Sci. 6:2545.

[41] Retiene, K., et al. 1968. Acta Endocrinol. 57:615.

[42] Rinne, U. K., and O. Kytomaki. 1961. Experientia 17:512.

[43] Sallmann, L. V., and P. Grimes. 1967. Invest. Ophthalmol. 5:560.

[44] Scheving, L. E., and J. E. Pauly. 1960. Acta Anat. 43:337.

[45] Scheving, L. E., and J. E. Pauly. 1966. Amer. J. Physiol. 210:1112.

[46] Scheving, L. E., and J. E. Pauly. 1967. Anat. Rec. 157:657.

[47] Scheving, L. E., and J. E. Pauly. 1967. Cell. Aspects Biorhythms Symp. Rhythmic Res., 1965, p. 167.

[48] Scheving, L. E., and J. E. Pauly. 1967. J. Cell Biol. 32:677.

[49] Scheving, L. E., and J. E. Pauly. Unpublished. Univ. Arkansas, School Medicine, Little Rock, 1970.

[50] Scheving, L. E., et al. 1968. Amer. J. Physiol. 214: 166.

[51] Scheving, L. E., et al. 1968. Ibid. 215:799.

[52] Scheving, L. E., et al. 1968. Ibid. 215:1096.

[53] Scheving, L. E., et al. 1968. Tex. Rep. Biol. Med. 26:341.

[54] Scheving, L. E., et al. 1972. Amer. J. Physiol. 222(2): 252.

[55] Scheving, L. E., et al. 1972. Proc. 2nd Int. Symp. Exp. Clin. Chronobiol., 1969 (in press).

[56] Shambaugh, G. E., et al. 1967. Endocrinology 81: 811.

[57] Stoner, R. E., and L. E. Scheving. Unpublished. Univ. Arkansas, School Medicine, Little Rock, 1971.

[58] Tsai, T. H., et al. 1970. Jap. J. Physiol. 20:12.

[59] Wurtman, R. J., and J. Axelrod. 1966. Life Sci. 5: 665.

Part II. Seasonal Variation

Despite the standardization of nutrition, temperature, and relative humidity, circadian rhythms vary significantly with the season of the year and the sex of the animal. The data are for nonfasting rats on a standard diet, acclimated to $23 \pm 1°C$ and 40% relative humidity. Unless otherwise indicated, the animals were subjected to a normal daylight regimen in a light:dark cycle from 1000 to 1000. Sampling interval was two hours. Data in brackets refer to the column heading in brackets.

| | Physiological Variable | Method | Time of Year | Sex | Mean Value | | | Reference |
					24-hr	Lowest [Occurrence]	Highest [Occurrence]	
1	Blood glucose, mg/ml	Fermentative determination by glucose oxidase-peroxidase (Notaidine) method	Jan	♂	0.78	0.66 [1000]	0.89 [2200]	6
2				♀	0.81	0.75 [1200]	0.90 [1400]	
3			Mar	♂	1.10	0.97 [0200]	1.25 [2200]	
4				♀	1.05	0.94 [1400]	1.12 [1000 & 1200]	
5			May	♂	0.87	0.68 [2000]	1.03 [0200]	
6				♀	0.83	0.67 [0000]	1.01 [1200]	
7			July	♂	1.08	0.99 [1000]	1.19 [0400]	
8				♀	1.04	0.94 [2200]	1.11 [1400]	
9			Sept	♂	0.88	0.62 [1800]	1.02 [1600]	
10				♀	0.88	0.71 [1200]	1.02 [0800]	
11			Nov	♂	0.95	0.86 [0800]	1.04 [2000]	
12				♀	0.89	0.78 [0800]	0.96 [1800]	
13	Brain DNA, mg/100 mg wet wt	Schmidt-Thanhauser method	Nov	♂	0.114	0.091 [0800]	0.136 [1800]	3,4
14				♀	0.125	0.087 [2200]	0.133 [0400]	
15	RNA, mg/100 mg wet wt	Schmidt-Thanhauser method	Nov	♂	0.36	0.28 [1600]	0.46 [2200]	3,4
16				♀	0.35	0.23 [0800]	0.49 [0400]	

continued

Part II.   Seasonal Variation

	Physiological Variable	Method	Time of Year	Sex	Mean Value			Ref-er-ence
					24-hr	Lowest [Occurrence]	Highest [Occurrence]	
	Liver							
17	Esterase, mean activity of	Titrimetric method of	Mar	♂	37.78	28.60 [0600]	45.10 [1000]	9
18	liver homogenate (20%)	Willstätter-Memmen; ml of		♀	27.01	23.38 [1600]	31.01 [1200]	
19		0.1 N NaOH consumed per	May	♂	34.37	26.74 [0800]	39.04 [1800]	
20		100 mg fresh liver per 15		♀	27.33	20.21 [1000]	34.37 [0200]	
21		min	July	♂	31.80	22.34 [0000]	44.15 [2000]	
22				♀	25.27	19.81 [0600]	35.07 [1800]	
23	Glycogen, mg/100 mg wet	KOH extraction (Pflüger)-	Jan	♂	2.42	0.52 [1000]	5.02 [1600]	5,7,
24	wt	Anthron modified		♀	2.65	0.90 [1200]	5.16 [1600]	8
25			Mar	♂	1.02	0.35 [2000]	1.78 [1200]	
26				♀	1.03	0.13 [1800]	2.20 [1000]	
27			May	♂	2.34	1.38 [2200]	3.07 [0800]	
28				♀	2.14	0.99 [1800]	3.42 [1000]	
29			July	♂	2.37	1.27 [2000]	3.43 [0600 & 0800]	
30				♀	2.07	0.98 [1600]	3.53 [0800]	
31			Sept	♂	3.05	0.73 [1800]	5.76 [0400]	
32				♀	2.41	0.51 [2000]	3.72 [1000]	
33			Nov	♂	2.26	0.89 [0000]	3.92 [0800]	
34				♀	2.10	0.72 [2200]	3.38 [0800]	
35	Nucleic acid, mg/100 mg	0.5 N PCA ultraviolet mea-	Jan	♂	0.94	0.83 [1000]	1.00 [1600]	1,2
36	wet wt	surements; modified methods		♂[1]	1.17	1.10 [2230]	1.34 [2030]	
37	DNA, mg/100 mg wet	Diphenylamine reaction	Jan	♂	0.175	0.117 [1000]	0.201 [2000]	1,2
38	wt			♂[1]	0.220	0.158 [1230]	0.275 [0230]	
39		Schmidt-Thanhauser method	Jan	♂	0.177	0.128 [2000]	0.237 [0000]	3,4
40				♀	0.167	0.126 [0200]	0.222 [1600]	
41				♂[2]	0.185	0.152 [1200]	0.239 [0200]	
42				♀[2]	0.178	0.127 [0200]	0.225 [1600]	
43				♂[3]	0.187	0.134 [1200]	0.211 [0000]	
44				♀[3]	0.178	0.130 [0200]	0.226 [1600]	
45			Nov	♂	0.190	0.152 [1000]	0.219 [0400]	
46				♀	0.197	0.163 [2000]	0.230 [0400]	
47	RNA, mg/100 mg wet	Orcinol reaction	Jan	♂	0.69	0.51 [0800]	0.87 [0000]	1,2
48	wt			♂[1]	1.00	0.91 [1230]	1.12 [0600]	
49		Schmidt-Thanhauser method	Jan	♂	0.51	0.32 [1200]	0.73 [0800]	3,4
50				♀	0.51	0.32 [0600]	0.67 [1200]	
51				♂[2]	0.53	0.41 [1200]	0.75 [0800]	
52				♀[2]	0.52	0.29 [0600]	0.68 [1200]	
53				♂[3]	0.53	0.40 [1200]	0.75 [0800]	
54				♀[3]	0.48	0.35 [0800]	0.73 [0400]	
55			Nov	♂	0.41	0.33 [1200]	0.64 [0600]	
56				♀	0.44	0.34 [1400]	0.65 [0400]	
	Spleen							
57	DNA, mg/100 mg wet wt	Schmidt-Thanhauser method	Jan	♂	0.153	0.110 [1200]	0.210 [1800]	3,4
58				♀	0.226	0.148 [1400]	0.293 [2000]	
59				♂[2]	0.165	0.117 [1200]	0.202 [0000]	
60				♀[2]	0.228	0.171 [0000]	0.393 [2000]	
61				♂[3]	0.175	0.128 [1200]	0.214 [2200]	
62				♀[3]	0.205	0.143 [0000]	0.258 [1200]	

[1] Animals subjected to a normal daylight regimen in a light:dark cycle from 1000 to 0800; intraperitoneal injection of 2.5 μCi H³-thymidine/g body wt, 1 hr before death. [2] 12 hr light:12 hr dark. [3] 24 hr dark.

continued

118. CIRCADIAN RHYTHMS OF PHYSIOLOGICAL VARIABLES AS REFLECTED IN RAT BIOASSAY

Part II. Seasonal Variation

	Physiological Variable	Method	Time of Year	Sex	24-hr	Mean Value Lowest [Occurrence]	Mean Value Highest [Occurrence]	Reference
63	RNA, mg/100 mg wet wt	Schmidt-Thanhauser method	Jan	♂	0.30	0.26 [0800]	0.38 [1800]	3,4
64				♀	0.32	0.23 [1800]	0.41 [0400]	
65				♂2/	0.28	0.24 [0800]	0.36 [2000]	
66				♀2/	0.32	0.27 [1800]	0.39 [0800]	
67				♂3/	0.30	0.25 [0800]	0.35 [2200]	
68				♀3/	0.31	0.24 [2000]	0.40 [0800]	

2/ 12 hr light:12 hr dark. 3/ 24 hr dark.

Contributor: von Mayersbach, H.

References

[1] Eling, W. 1967. Cell. Aspects Biorhythms Symp. Rhythmic Res., 1965, p. 105

[2] Eling, W., and H. von Mayersbach. 1966. Int. Kongr. Anat., Wiesbaden, 1965, p. 154.

[3] Horvath, G. 1964. Int. Kongr. Histo-Cytochem., 2nd, p. 137.

[4] Horvath, G., and H. von Mayersbach. Unpublished. Univ. of Nijmegen, Institute of Cytology and Histology, Netherlands, 1966.

[5] Leske, R. 1964. Int. Kongr. Histo-Cytochem., 2nd, p. 139.

[6] Leske, R. Unpublished. Univ. of Nijmegen, Institute of Cytology and Histology, Netherlands, 1966.

[7] Mayersbach, H. von, and R. Leske. 1963. Acta Morphol. Acad. Sci. Hung. 12:33.

[8] Mayersbach, H. von, and R. Leske. 1966. Acta Morphol. Neer. Scand. 6(4):343.

[9] Yap, P. H. K., and H. von Mayersbach. 1965. Histochemie 5:297.

119. PHOTOPERIODIC CONTROL MECHANISMS: HOMOIOTHERMIC ANIMALS

Data are only for selected cases of photoperiodic control mechanisms in which it appears evident that the mechanism is important in the control of the function in the natural existence of the species. Hence, most of the species listed are from middle and high latitudes.

Part I. Mammals

No known functions in mammals are obligately controlled by length of day under natural conditions; rather, photoperiodic mechanisms appear to be involved in the more precise timing of functions that would otherwise occur (in a less precisely timed manner) because of crude endogenous periodicities. **Long-Day Effects** and **Short-Day Effects** therefore refer only to natural day lengths. Several species of mammals have been transported to areas of different photoperiodic regimes, often including transequatorial movements. Many of these species have synchronized their breeding seasons to the new regime, indicative of a photoperiodic component involved with the timing of these functions. The importance of photoperiod has not been experimentally tested in most of these mammals and they are therefore not included in the table; for review and discussion consult references 4 and 38.

	Species	Long-Day Effects	Short-Day Effects	Reference
1	*Capra* sp.1/	....	Accelerates induction of estrus	10,20,54
2	*Cervus nippon*	Promotes onset of antler growth	....	23
3	*Dicrostonyx groenlandicus*	....	Induces autumn molt & development of winter pelage	30
4	*Equus* sp.2/	Accelerates onset of estrus	....	15,42
5	*Erinaceus europaeus*	Induces spermatogenesis	....	1

1/ Domestic goat. 2/ Domestic horse.

continued

Part I. Mammals

	Species	Long-Day Effects	Short-Day Effects	Reference
6	*Felis* sp.[3/]	Accelerates onset of estrus	....	17,51
7	*Glaucomys volans*	....	Promotes testicular descent	41
8	*Glis glis*	Defers hibernation	Advances autumn molt & development of winter pelage; promotes fat deposition	33
9	*Lepus americanus*	Induces estrus & testicular development; induces spring molt & summer pelage	Induces autumn molt & winter pelage	37
10	*L. brachyurus*	Inhibits autumn molt & winter pelage	Delays onset of spring molt	46
11	*L. timidus*	Induces spring molt & summer pelage	....	43
12	*Martes americana*	Decreases period of delayed implantation	....	47
13	*M. zibellina*	Decreases period of delayed implantation	....	7
14	*Mesocricetus auratus*	Maintains spermatogenesis in adults	Enhances testicular atrophy	22,29
15	*Microtus agrestis*	Accelerates ovarian & testicular development	Delays estrus & spermatogenesis	2,14,16
16	*M. arvalis*	Accelerates onset of breeding behavior & gonadal development	Delays sexual maturation & gonadal development	35,36,39
17	*M. montanus*	Enhances reproductive success; accelerates onset of sub-adult & adult molt	Delays sexual maturity	48-50
18	*Mustela* sp.[4/]	Induces estrus & testicular development	....	8,9,18,19, 26,28
19	*M. vison*	Induces estrus, decreases period of delayed implantation; stimulates spermatogenesis	Induces development of winter pelage	13,31,32,47
20	*Odocoileus virginianus*	Accelerates onset of rut; advances spring molt, accelerates growth & shedding of antlers	....	21
21	*Ovis* sp.[5/]	....	Induces estrus, increases spermatogenic activity; accelerates hair growth	24,25,27,40, 44,45,53
22	*Procyon lotor*	Accelerates estrus & seasonal reproductive activity of males	....	11
23	*Sciurus vulgaris*	Accelerates seasonal reproductive activity	....	52
24	*Sylvilagus transitionalis*	Induces female reproductive activity & spermatogenesis	....	12
25	*Vulpes fulva*	....	Induces development of female reproductive system & spermatogenesis; accelerates autumn molt & development of winter pelage	3,5,6,34

[3/] Domestic cat. [4/] Ferret. [5/] Domestic sheep.

Contributors: Farner, Donald S., Lewis, R. A., and Darden, T. R.

References

[1] Allanson, M., and R. Deanesly. 1934. Proc. Roy. Soc. B116:170.

[2] Baker, J. R., and R. M. Ranson. 1932. Ibid. B110:313.

[3] Bassett, C. F. 1946. Ann. N. Y. Acad. Sci. 48:239.

[4] Bedford, Duke of, and F. H. A. Marshall. 1942. Proc. Roy. Soc. B130:396.

[5] Belyaev, D. K. 1950. Zh. Obshch. Biol. 11:39.

[6] Belyaev, D. K., and L. G. Utkin. 1949. Karakulevod. Zverovod. 2:53.

[7] Belyaev, D. K., et al. 1951. Zh. Obshch. Biol. 12:260.

[8] Bissonnette, T. H. 1932. Proc. Roy. Soc. B110:322.

[9] Bissonnette, T. H. 1935. J. Exp. Zool. 71:341.

continued

Part I. Mammals

[10] Bissonnette, T. H. 1941. Physiol. Zool. 14:379.

[11] Bissonnette, T. H., and A. G. Csech. 1939. Ecology 20:156.

[12] Bissonnette, T. H., and A. G. Csech. 1939. Biol. Bull. 77:364.

[13] Bissonnette, T. H., and E. Wilson. 1939. Science 89:418.

[14] Breed, W. G., and J. R. Clarke. 1970. J. Reprod. Fert. 23:189.

[15] Burkhardt, J. 1947. J. Agr. Sci. 37:64.

[16] Clarke, J. R., and J. P. Kennedy. 1967. Gen. Comp. Endocrinol. 8:474.

[17] Dawson, A. B. 1941. Endocrinology 28:907.

[18] Donovan, B. T. 1967. J. Endocrinol. 39:105.

[19] Donovan, B. T., and G. W. Harris. 1956. J. Physiol. (London) 131:102.

[20] Eaton, O. N., and V. L. Simmons. 1953. U.S. Dep. Agr. Circ. 933.

[21] French, C. E., et al. 1960. J. Mammal. 41:23.

[22] Gaston, S., and M. Menaker. 1967. Science 158:925.

[23] Goss, R. J. 1969. J. Exp. Zool. 170:311.

[24] Hafez, E. S. E. 1952. J. Agr. Sci. 42:232.

[25] Hart, D. S. 1950. Ibid. 40:143.

[26] Hart, D. S. 1951. J. Exp. Biol. 28:1.

[27] Hart, D. S. 1961. J. Agr. Sci. 56:235.

[28] Harvey, N. E., and W. V. MacFarlane. 1958. Aust. J. Biol. Sci. 11:187.

[29] Hoffman, R. A., et al. 1965. Comp. Biochem. Physiol. 15:525.

[30] Jacobson, W. F. 1965. Master's Thesis. Washington State Univ., Pullman.

[31] Khronopulo, N. P. 1956. Priroda (USSR) 45:108.

[32] Khronopulo, N. P., and L. P. Drozdova. 1957. Zool. Zh. 36:938.

[33] König, C. 1960. Z. Morphol. Oekol. Tiere 48:545.

[34] Kuznyetsov, G. A. 1949. Sov. Zootekh. (1):108.

[35] Lecyk, M. 1962. Zool. Pol. 12:189.

[36] Lecyk, M. 1963. Ibid. 13:77.

[37] Lyman, C. P. 1943. Bull. Mus. Comp. Zool. Harvard Coll. 93:391.

[38] Marshall, F. H. A. 1937. Proc. Roy. Soc. B122:413.

[39] Martinet, L. 1970. Colloq. Int. Cent. Nat. Rech. Sci., Paris. 172.

[40] Means, T. M., et al. 1959. J. Anim. Sci. 18:1388.

[41] Muul, I. 1969. J. Mammal. 50:542.

[42] Nishikawa, Y. 1959. Studies on Reproduction in Horses. Japan Racing Association, Tokyo.

[43] Novikov, B. G., and G. I. Blagodatskaya. 1948. Dokl. Akad. Nauk SSSR 61:577.

[44] Ortavant, R. 1959. Ann. Zootech. 8:271.

[45] Ortavant, R., et al. 1964. Ann. N. Y. Acad. Sci. 117:157.

[46] Otsu, S. 1967. Jap. J. Appl. Entomol. Zool. 11:37.

[47] Pearson, O. P., and R. K. Enders. 1944. J. Exp. Zool. 95:21.

[48] Pinter, A. J. 1968. Amer. J. Physiol. 215:461.

[49] Pinter, A. J. 1968. Ibid. 215:828.

[50] Pinter, A. J., and N. C. Negus. 1965. Ibid. 208:633.

[51] Scott, P., and M. Lloyd-Jacob. 1959. Nature (London) 184:2022.

[52] Woitkewitsch, A. A. 1945. C. R. Acad. Sci. URSS 47:71.

[53] Yeates, N. T. M. 1949. J. Agr. Sci. 39:1.

[54] Yoshioka, Z., et al. 1951. Bull. Nat. Inst. Agr. Sci. (Tokyo) G1:101.

Part II. Birds

The known photoperiodic controls among birds involve long-day effects. The only common short-day effect is the elimination of photorefractoriness in many species. Domestic forms of *Gallus gallus,* although frequently described as photoperiodic, probably in the strict sense should not be so classified [86, 126]; however, the rate of sexual maturation of young and the rate of egg-laying can be increased somewhat with long daily photoperiods [23, 58-60, 76, 131, 132]. There is, however, circumstantial evidence that certain low-latitude (but not equatorial) natural populations of *Gallus gallus* have a photoperiodic element in the control of reproduction [4, 15, 20, 71, 128]. Gonadal Development: Rate of gonadal development as a function of day length is regular, but nonlinear in such a manner as to make the terms "minimum effective" and "optimum" day lengths meaningless; + in the ♂ column generally indicates that full testicular development and spermatogenesis can be induced by long daily photoperiods; + in the ♀ column, in most instances, refers only to partial ovarian development, for vitellogenesis and the culminative stages of ovarian development involve other mechanisms in most birds. Molt: PrN = prenuptial molt; PoN = postnuptial molt (which follows, in most cases, the period of photoperiodically induced gonadal development and therefore is regarded as an indirect long-day effect). Vernal Traits: SC = secondary sex characteristics; MB = migratory behavior; FD = fat deposition.

Species	Gonadal Development		Molt	Vernal Traits	Reference
	♂	♀			
1 *Agelaius phoeniceus*	+	+	....	SC	91
2 *A. tricolor*	+	+	....	SC	91
3 *Anas acuta*	+	0	....	....	92

continued

Part II. Birds

	Species	Gonadal Development ♂	Gonadal Development ♀	Molt	Vernal Traits	Reference
4	*A. platyrhynchos* [1]	+	+	PrN; PoN	....	7,8,65,84,114,125
5	*Bonasa umbellus*	+	+	....	....	17,18
6	*Carduelis carduelis*	+	+	....	....	118,121,122
7	*C. spinus*	+	....	....	....	106
8	*Carpodacus mexicanus*	+	....	....	....	36-38
9	*Chloris chloris*	+	....	....	....	21,24,85,106
10	*Colinus virginianus*	+	+	....	....	5,11,17,48-50
11	*Columba oenas*	+	....	....	....	70
12	*C. palumbus*	+	+	....	....	69,70
13	*Corvus brachyrhynchos*	+	+	....	....	102
14	*Coturnix coturnix* [2]	+	+	....	....	35,72,88,116
15	*Dolichonyx oryzivorus*	+	....	Molt?	MB?; FD?	25,27-29
16	*Erithacus rubecula*	+	+	PoN	MB	77,94,95,103,104,106
17	*Fringilla coelebs*	+	+	....	FD	14,24,57,106-108
18	*F. montifringilla*	+	+	....	MB; FD	24,66,67,106-108
19	*Hylocichla guttata*	+	....	....	....	1
20	*Junco hyemalis hyemalis*	+	+	PoN	MB; FD	26,30,31,42,43,99-101,127,136,138,140
21	*J. oreganus montanus*	+	+?	....	....	134
22	*J. oreganus oreganus*	+	....	....	MB; FD	134
23	*J. oreganus pinosus*	+	....	....	....	134
24	*J. oreganus shufeldti*	+	....	....	MB; FD	134
25	*J. oreganus thurberi*	+	....	....	MB; FD	134
26	*Lagopus lagopus*	+	−	PrN; PoN [3]	....	41,87
27	*Leucosticte atrata*	+	....	....	MB	46
28	*L. tephrocotis littoralis*	+	....	....	MB	46
29	*L. tephrocotis tephrocotis*	+	....	....	MB	46
30	*Meleagris gallopavo* [4]	+?	+	....	....	2,39,40,74,89,111,133
31	*Molothrus ater*	+	....	....	....	78
32	*M. ater obscurus*	+	+	....	....	90
33	*Parus major*	+	+	....	....	113
34	*Passer domesticus*	+	+	PoN	....	6,75,79,93,96-98,106,117-120
35	*P. montanus montanus*	+	+	....	....	118
36	*P. montanus saturatus*	+	+	....	....	73
37	*Passerella iliaca*	+	+	....	MB; FD	127,135
38	*Phasianus colchicus*	+	+	PrN	SC	11-13,17,18
39	*Phoenicurus phoenicurus*	+	....	PoN	....	103,105
40	*Prunella modularis*	+	+	....	....	118
41	*Pyrrhula pyrrhula*	+	+	....	....	118
42	*Serinus canarius* [5]	+	+	....	....	123,124
43	*S. canarius* [6]	+	+	PoN	SC	51-54,56,112,115
44	*Spizella arborea*	+	+?	....	MB; FD	44,101,130
45	*Streptopelia turtur*	+	+	....	....	68
46	*Sturnella magna*	+	....	....	....	22
47	*Sturnus vulgaris*	+	+	....	SC	9,10,16,64,109,110
48	*Zenaidura macroura carolinensis*	+	+	....	....	19
49	*Zonotrichia albicollis*	+	+	PrN?; PoN	MB; FD	26,31,42,62,137,139
50	*Z. atricapilla*	+	+	PrN?; PoN	FD	63,80-82,135
51	*Z. leucophrys gambelii*	+	+	PrN; PoN	MB; FD	32-34,45,47,61
52	*Z. leucophrys leucophrys*	+	+	PoN?	MB; FD	31,138

[1] Domestic and wild. [2] Japanese quail. [3] White winter plumage is apparently induced by short days. [4] Although it seems highly probable that there is an important photoperiodic component in the control of the annual gonadal cycles of this species, available data are concerned almost exclusively with the rate of sexual maturation of young birds which can be accelerated by long daily photoperiods. [5] Serin. [6] Domestic canary.

continued

Part II.   Birds

Species	Gonadal Development		Molt	Vernal Traits	Reference
	♂	♀			
53 Z. leucophrys nuttalli	+	+	....	....	63,135
54 Z. leucophrys pugetensis	+	+	PoN	MB; FD	3,63,135
55 Z. querula	+	....	....	....	129
56 Zosterops japonica japonica [7]	+	+	PoN	....	55,83

[7] Synonym: *Zonotrichia palpebrosa japonica.*

Contributors: Farner, Donald S., Lewis, R. A., and Darden, T. R.

References

[1] Annan, O. 1963. Auk 80:166.

[2] Asmundson, V. S., and B. D. Moses. 1950. Poult. Sci. 29:34.

[3] Bailey, R. E. 1950. Condor 52:247.

[4] Baker, E. C. S. 1928. The Fauna of British India: Birds. Ed. 2. Taylor and Francis, London. v. 5.

[5] Baldini, J. T., et al. 1954. Ibid. 33:1282.

[6] Bartholomew, G. A. 1949. Bull. Mus. Comp. Zool. Harvard Coll. 101:433.

[7] Benoit, J. 1936. Bull. Biol. Fr. Belg. 70:487.

[8] Benoit, J., and I. Assenmacher. 1953. Arch. Anat. Microsc. Morphol. Exp. 42:334.

[9] Bissonnette, T. H. 1931. J. Exp. Zool. 58:281.

[10] Bissonnette, T. H. 1931. Physiol. Zool. 4:542.

[11] Bissonnette, T. H., and A. G. Csech. 1936. Science 83:392.

[12] Bissonnette, T. H., and A. G. Csech. 1936. Bird-Band. 7:108.

[13] Bissonnette, T. H., and A. G. Csech. 1938. Amer. Natur. 71:525.

[14] Brisson, P., and L. Vaugien. 1957. C. R. Acad. Sci. 245:364.

[15] Bump, G., and W. H. Bohl. 1961. U.S. Fish Wildl. Serv. Spec. Sci. Rep. 62.

[16] Burger, J. W. 1949. Wilson Bull. 61:211.

[17] Clark, L. B., et al. 1936. Science 83:268.

[18] Clark, L. B., et al. 1937. Ibid. 85:339.

[19] Cole, L. J. 1933. Auk 50:284.

[20] Collias, N. E., et al. 1966. Anim. Behav. 14:550.

[21] Damsté, P. H. 1947. J. Exp. Biol. 24:20.

[22] Darst, P. H. 1967. Auk 84:265.

[23] Dobie, J. B., et al. 1946. Wash. Agr. Exp. Sta. Bull. 471.

[24] Dolnik, V. R. 1963. Dokl. Akad. Nauk SSSR 149:191.

[25] Engels, W. L. 1959. Publ. Amer. Ass. Advan. Sci. 55:759.

[26] Engels, W. L. 1961. Biol. Bull. 120:140.

[27] Engels, W. L. 1962. Ibid. 123:94.

[28] Engels, W. L. 1962. Ibid. 123:542.

[29] Engels, W. L. 1964. Auk 81:95.

[30] Engels, W. L., and C. E. Jenner. 1956. Biol. Bull. 110:129.

[31] Eyster, B. 1954. Ecol. Monogr. 24:1.

[32] Farner, D. S., and A. C. Wilson. 1957. Biol. Bull. 113:254.

[33] Farner, D. S., et al. 1953. Ibid. 105:434.

[34] Farner, D. S., et al. 1966. Ibid. 130:67.

[35] Follett, B. K., and D. S. Farner. 1966. Gen. Comp. Endocrinol. 7:111.

[36] Hamner, W. M. 1963. Science 142:1294.

[37] Hamner, W. M. 1966. Gen. Comp. Endocrinol. 7:224.

[38] Hamner, W. M. 1968. Ecology 49:211.

[39] Harper, J. A., and J. E. Parker. 1957. Poult. Sci. 36:967.

[40] Harper, J. A., and J. E. Parker. 1962. Ibid. 41:493.

[41] Höst, P. 1942. Auk 59:388.

[42] Jenner, C. E., and W. L. Engels. 1952. Biol. Bull. 103:345.

[43] Johnston, D. W. 1962. Auk 79:387.

[44] Kendeigh, S. C., et al. 1960. Anim. Behav. 8:180.

[45] King, J. R. 1961. Physiol. Zool. 34:145.

[46] King, J. R., and E. E. Wales. 1965. Ibid. 38:49.

[47] King, J. R., and D. S. Farner. 1956. Proc. Soc. Exp. Biol. Med. 93:354.

[48] Kirkpatrick, C. M. 1955. Physiol. Zool. 28:255.

[49] Kirkpatrick, C. M. 1959. Publ. Amer. Ass. Advan. Sci. 55:751.

[50] Kirkpatrick, C. M., and A. C. Leopold. 1952. Science 116:280.

[51] Kobayashi, H. 1953. Annot. Zool. Jap. 26:156.

[52] Kobayashi, H. 1954. Ibid. 27:19.

[53] Kobayashi, H. 1954. Ibid. 27:63.

[54] Kobayashi, H. 1954. Ibid. 27:128.

[55] Kobayashi, H. 1954. Endocrinol. Jap. 1:51.

[56] Kobayashi, H. 1957. Annot. Zool. Jap. 30:8.

[57] Koch, H. J., and A. F. de Bont. 1954. Ann. Soc. Roy. Zool. Belg. 82:143.

[58] Lamoreaux, W. F. 1943. J. Exp. Zool. 94:73.

[59] Larionov, W. T. 1941. C. R. Acad. Sci. URSS 30:374.

[60] Larionov, W. T. 1941. Ibid. 32:227.

[61] Laws, D. F. 1961. Z. Zellforsch. Mikrosk. Anat. 54:275.

[62] Lesher, S. W., and S. C. Kendeigh. 1941. Wilson Bull. 53:169.

continued

Part II. Birds

[63] Lewis, R. A. 1971. Ph.D. Thesis. Univ. Washington, Seattle.

[64] Lloyd, J. A. 1965. Physiol. Zool. 38:121.

[65] Lofts, B., and C. J. F. Coombs. 1965. J. Zool. 146: 44.

[66] Lofts, B., and A. J. Marshall. 1960. Ibis 102:209.

[67] Lofts, B., and A. J. Marshall. 1961. Ibid. 103:189.

[68] Lofts, B., et al. 1967. Ibid. 109:337.

[69] Lofts, B., et al. 1967. Ibid. 109:352.

[70] Lofts, B., et al. 1967. J. Zool. 151:17.

[71] Marshall, F. H. A. 1956. In A. S. Parkes, ed. Marshall's Physiology of Reproduction. Ed. 3. Longmans, Green; London. p. 1.

[72] Mather, F. B., and W. O. Wilson. 1964. Poult. Sci. 43:860.

[73] Matsui, T. 1966. Endocrinol. Jap. 13:23.

[74] McGillivray, D. B., and I. L. Kosin. 1965. Northwest Sci. 31:1.

[75] Menaker, M., and H. Keatts. 1968. Proc. Nat. Acad. Sci. U.S. 60:146.

[76] Mérat, P. 1960. Ann. Zootech. 9:241.

[77] Merkel, F. W. 1961. Vogelwarte 21:156.

[78] Middleton, A. L. A., and D. M. Scott. 1965. Auk 82:504.

[79] Middleton, J. 1965. Physiol. Zool. 38:255.

[80] Miller, A. H. 1948. J. Exp. Zool. 109:1.

[81] Miller, A. H. 1951. Auk 68:380.

[82] Miller, A. H. 1954. Condor 56:13.

[83] Miyazaki, H. 1934. Sci. Rep. Tohoku Univ., Ser. 4, 9:183.

[84] Mori, N., et al. 1953. Shiga Agr. Coll. Sci. Rep. 3:36.

[85] Murton, R. K., et al. 1970. J. Zool. 161:125.

[86] Nalbandov, A. V. 1970. Colloq. Int. Cent. Nat. Rech. Sci. 172.

[87] Novikov, B. G., and G. I. Blagodatskaya. 1948. Dokl. Akad. Nauk SSSR 61:577.

[88] Oishi, T., et al. 1966. Environ. Contr. Biol. 3:37.

[89] Olsen, M. W., and S. J. Marsden. 1952. Poult. Sci. 31:715.

[90] Payne, R. B. 1967. Condor 69:289.

[91] Payne, R. B. 1969. Univ. Calif. Berkeley Publ. Zool. 90:1.

[92] Phillips, R. E., and A. Van Tienhoven. 1960. J. Endocrinol. 21:253.

[93] Polikarpova, E. 1940. C. R. Acad. Sci. URSS 26:91.

[94] Putzig, P. 1937. Vogelzug 8:116.

[95] Putzig, P. 1938. Ibid. 9:189.

[96] Riley, G. M. 1936. Proc. Soc. Exp. Biol. Med. 34: 331.

[97] Riley, G. M., and E. Witschi. 1938. Endocrinology 23:618.

[98] Ringoen, A. R., and A. Kirschbaum. 1939. J. Exp. Zool. 80:173.

[99] Rowan, W. 1925. Nature (London) 115:494.

[100] Rowan, W. 1926. Proc. Boston Soc. Natur. Hist. 38:147.

[101] Rowan, W. 1929. Ibid. 39:151.

[102] Rowan, W. 1932. Proc. Nat. Acad. Sci. U.S. 18: 639.

[103] Schildmacher, H. 1937. Vogelzug 8:107.

[104] Schildmacher, H. 1938. Ibid. 9:146.

[105] Schildmacher, H. 1938. Biol. Zentralbl. 58:464.

[106] Schildmacher, H. 1963. Ibid. 82:31.

[107] Schildmacher, H., and L. Steubing. 1952. Ibid. 71:272.

[108] Schildmacher, H., et al. 1968. Z. Zellforsch. Mikrosk. Anat. 91:604.

[109] Schwab, R. G. 1970. Condor 72:466.

[110] Schwab, R. G., and D. F. Lott. 1969. J. Exp. Zool. 171:39.

[111] Scott, H. M., and L. F. Payne. 1937. Poult. Sci. 16:90.

[112] Steel, E., and R. A. Hinde. 1966. J. Zool. 149:1.

[113] Suomalainen, H. 1938. Ornis Fenn. 14:108.

[114] Svetozarov, E., and G. Straich. 1938. C. R. Acad. Sci. URSS 20:327.

[115] Takewaki, K., and H. Mori. 1944. J. Fac. Sci. Imp. Univ. Tokyo, IV, 6:547.

[116] Tanaka, K., et al. 1965. Poult. Sci. 44:662.

[117] Threadgold, L. T. 1960. Physiol. Zool. 33:190.

[118] Vaugien, L. 1948. Bull. Biol. Fr. Belg. 82:166.

[119] Vaugien, L. 1951. Bull. Soc. Zool. Fr. 77:395.

[120] Vaugien, L. 1955. Bull. Biol. Fr. Belg. 89:218.

[121] Vaugien, L. 1956. C. R. Acad. Sci. 242:2253.

[122] Vaugien, L. 1956. Ibid. 243:444.

[123] Vaugien, L. 1957. Ibid. 245:205.

[124] Vaugien, L. 1957. Ibid. 245:1268.

[125] Walton, A. 1937. J. Exp. Biol. 14:440.

[126] Warren, D. C., and H. M. Scott. 1936. J. Exp. Zool. 72:137.

[127] Weise, C. M. 1962. Auk 79:161.

[128] Whistler, H., and N. B. Kinnear. 1949. Popular Handbook of Indian Birds. Ed. 4. Gurney and Jackson, London.

[129] Wilson, F. E. 1968. Auk 85:415.

[130] Wilson, F. E., and G. R. Hands. 1968. Z. Zellforsch. Mikrosk. Anat. 89:303.

[131] Wilson, W. O., and H. Abplanalp. 1956. Poult. Sci. 35:532.

[132] Wilson, W. O., et al. 1964. Ibid. 43:1187.

[133] Wilson, W. O., et al. 1967. Ibid. 46:46.

[134] Wolfson, A. 1942. Condor 44:237.

[135] Wolfson, A. 1945. Ibid. 47:95.

[136] Wolfson, A. 1952. J. Exp. Zool. 121:311.

[137] Wolfson, A. 1953. Condor 55:187.

[138] Wolfson, A. 1954. J. Exp. Zool. 125:353.

[139] Wolfson, A. 1958. Ibid. 139:349.

[140] Wolfson, A. 1959. Physiol. Zool. 32:160.

VIII. PARASITISM

120. EXTERNAL PARASITES: LABORATORY ANIMALS

For information on external parasites of the cat, dog, and fowl, *see* Table 122. Data in brackets refer to the column heading in brackets.

	Species [Distribution]	Incidence in Laboratory	Host	Effect on Host	Reference
				Acarina	
1	*Cheyletiella parasitivorax* [Probably worldwide]	Common in some regions	Rabbit	Alopecia, scaling, hyperemia, pruritus, serous exudation	4,5,8,37,45
2	*Chirodiscoides caviae* [Probably worldwide]	Common	Guinea pig	Usually, no effect; sometimes pruritus, alopecia	8,23,27,28
3	*Myobia musculi* [Worldwide]	Common	Mouse	Dermatitis, alopecia, pruritus, self-inflicted trauma, secondary amyloidosis	13,15,17-19, 21,24
4	*Myocoptes musculinus* [Worldwide]	Common	Mouse	Pruritus, erythema, traumatic dermatitis, alopecia	4,17,18,20, 24,47
5	*Notoedres* sp. [North America, Europe]	Uncommon	Hamster	Scaly dermatitis, especially of ears	3,26,46
6	*N. cati* [Probably worldwide]	Uncommon	Rabbit	Scaly dermatitis of ears, face, & neck	1,6,40
7	*N. muris* [Probably worldwide]	Common in some colonies, absent in others	Rat	Wartlike, horny excrescences, especially on ears	9,10,14,28, 48
8	*Ornithonyssus bacoti* [Worldwide]	Rare in some colonies, common in others	Rat	Debility, anemia, decreased reproduction, death	4,31,39,49
9	*Psorergates simplex* [Possibly worldwide]	Common	Mouse	Caseous nodules in skin, dermal cysts; sometimes scabby dermatitis of ears	7,11,13,15, 16
10	*Psoroptes cuniculi* [Worldwide]	Common	Rabbit	Ear canker; sometimes otitis media, meningitis, death	1,4,28,34,42
11	*Radfordia affinis* [Worldwide]	Common	Mouse	Effects unknown; probably dermatitis, alopecia, pruritus	13,15,17,18, 24
12	*R. ensifera* [Worldwide]	Common	Rat	Self-inflicted dermal trauma	4,24,44
13	*Sarcoptes scabiei* [Worldwide]	Common	Rabbit	Scabby dermatitis	1,4,6,10,22
14	*Trichoecius romboutsi* [United States, Europe]	Unknown	Mouse	Effects unknown	12,13,15
				Siphonaptera	
15	*Leptopsylla segnis* [Worldwide]	Rare	Mouse	Acts as intermediate host for *Hymenolepis diminuta* & *H. nana*	25,49
16	*Nosopsyllus fasciatus* [Worldwide]	Rare	Rat	Acts as intermediate host for *Hymenolepis diminuta* & *H. nana*	25,49
17	*Spilopsyllus cuniculi* [Europe]	Rare	Rabbit	Dermal irritation, especially of ears; also vector of myxoma virus	2,30,43
18	*Xenopsylla cheopis* [Worldwide]	Uncommon	Rat	Acts as intermediate host for *Hymenolepis diminuta* & *H. nana*	25,30,49
				Mallophaga & Anoplura	
19	*Gliricola porcelli* [Worldwide]	Common in some colonies, rare or absent in others	Guinea pig	Irritation, alopecia	35,41

continued

	Species [Distribution]	Incidence in Laboratory	Host	Effect on Host	Reference
20	*Gyropus ovalis* [Worldwide]	Common in some colonies, rare or absent in others	Guinea pig	Irritation, alopecia	35,41
21	*Haemodipsus ventricosus* [Worldwide]	Rare	Rabbit	Pruritus, alopeacia, retarded growth	33
22	*Pedicinus eurygaster* [Tropical Asia]	Rare	Monkey	Effects unknown	32
23	*Polyplax serrata* [Worldwide]	Common in some colonies, rare or absent in others	Mouse	Pruritus, anemia. Vector of *Eperythrozoon coccoides.*	13,25,36,38
24	*P. spinulosa* [Worldwide]	Common in some colonies, rare or absent in others	Rat	Pruritus, anemia. Vector of *Rickettsia typhi* & *Haemobartonella muris.*	29,38

Contributor: Flynn, Robert J.

References

[1] Adams, C. E., et al. 1967. Univ. Fed. Anim. Welfare Handb. Care Manage. Lab. Anim., p. 396.

[2] Allan, R. M., and P. L. Shanks. 1955. Nature (London) 175:692.

[3] Baies, A., et al. 1968. Z. Versuchstierk. 10:251.

[4] Baker, E. W., et al. 1956. A Manual of Parasitic Mites of Medical or Economic Importance. National Pest Control Association, New York.

[5] Barr, A. R. 1955. J. Parasitol. 41:323.

[6] Camin, J. H., and W. M. Rogoff. 1952. S. Dak. Agr. Exp. Sta. Bull. 10.

[7] Cook, R. 1956. Brit. Vet. J. 112:22.

[8] Deoras, P. J., and K. K. Patel. 1960. Indian J. Entomol. 22:7.

[9] Fain, A. 1965. Acarologia 7:321.

[10] Fain, A. 1968. Acta Zool. Pathol. Antverp. 47:3.

[11] Fain, A., et al. 1966. Acarologia 8:251.

[12] Fain, A., et al. 1970. Acta Zool. Pathol. Antverp. 50:67.

[13] Flynn, R. J. 1955. Proc. Anim. Care Panel 6:75.

[14] Flynn, R. J. 1960. Ibid. 10:69.

[15] Flynn, R. J. 1963. Ibid. 13:111.

[16] Flynn, R. J., and B. Jaroslow. 1956. J. Parasitol. 42:49.

[17] Flynn, R. J., et al. 1965. Lab. Anim. Care 15: 440.

[18] Fukui, M., et al. 1961. Jikken Dobutsu 10:83.

[19] Galton, M. 1963. Amer. J. Pathol. 43:855.

[20] Gambles, M. R. 1952. Brit. Vet. J. 108:194.

[21] Haakh, U. 1958. Z. Tropenmed. Parasitol. 9:75.

[22] Hagen, K. W., Jr., and E. E. Lund. 1962. U.S. Dep. Agr. Publ. 45-53.

[23] Harrison, I. R., and M. M. Daykin. 1965. J. Inst. Anim. Tech. 16:69.

[24] Heine, W. 1962. Z. Versuchstierk. 2:1.

[25] Heston, W. E. 1941. Biology of the Laboratory Mouse. Blakiston, Philadelphia.

[26] Hindle, E. 1947. In A. N. Worden, ed. The Care and Management of Laboratory Animals. Williams and Wilkins, Baltimore. pp. 196-202.

[27] Hirst, S. 1917. Ann. Mag. Natur. Hist. Ser. 8, 20: 431.

[28] Hirst, S. 1922. Brit. Mus. (Natur. Hist.) Econ. Ser. (13).

[29] Holmes, D. T. 1959. Diss. Abstr. 19:2693.

[30] Hopkins, G. H. E., and M. Rothschild. 1953. An Illustrated Catalogue of the Rothschild Collection of Fleas (Siphonaptera) in the British Museum (Natural History). British Museum (Natural History), London. v. 1.

[31] Keefe, T. J., et al. 1964. Lab. Anim. Care 14:366.

[32] Kuhn, H.-J., and H. W. Ludwig. 1967. Z. Zool. Syst. Evolutionsforsch. 5:144.

[33] Litvishko, N. T., et al. 1965. Veterinariya 42:87.

[34] Lund, E. E. 1951. Small Stock Mag. 35:18.

[35] Meyer, K. F., and B. Eddie. 1952. Proc. Anim. Care Panel 2:23.

[36] Murray, M. D. 1961. Aust. J. Zool. 9:1.

[37] Mykytowycz, R. 1957. Commonw. Sci. Ind. Res. Organ. Wildl. Res. 2:164.

[38] Oldham, J. N. 1967. In E. Cotchin and F. J. C. Roe, ed. Pathology of Laboratory Rats and Mice. Blackwell Scientific Publications, Oxford. pp. 641-679.

[39] Olson, T. A., and R. G. Dahms. 1946. J. Parasitol. 32:56.

[40] Osborne, H. G. 1947. Can. J. Comp. Med. 11:144.

[41] Paterson, J. S. 1967. Univ. Fed. Anim. Welfare Handb. Care Manage. Lab. Anim., p. 241.

[42] Poole, C. M., et al. 1967. Annu. Rep. Argonne Nat. Lab. Biol. Med. Res. Div. ANL-7409.

[43] Rothschild, M. 1960. Entomol. Mon. Mag. 96:106.

[44] Skidmore, L. V. 1934. Can. Entomol. 66:110.

[45] Strasser, H. 1963. Kleintier Prax. 8:212.

continued

[46] Wantland, W. W. 1968. In R. A. Hoffman, et al., ed. The Golden Hamster; Its Biology and Use in Medical Research. Iowa State Univ. Press, Ames. pp. 171-183.

[47] Watson, D. P. 1961. Parasitology 51:373.

[48] Watson, D. P. 1962. Acarologia 4:64.

[49] Yunker, C. E. 1964. Lab. Anim. Care 14:455.

121. INTERNAL PARASITES: LABORATORY ANIMALS

For information on internal parasites of cat, dog, fowl, and large domestic animals, *see* Table 230. Data in brackets refer to the column heading in brackets.

Species [Distribution]	Incidence in Laboratory	Host [Localization]	Effect on Host	Reference
		Nematoda		
1 *Abbreviata caucasica* [Asia]	Probably common	Monkey [Esophagus, stomach, duodenum]	Usually inapparent; probably gastritis	41,69,83
2 *Aspiculuris tetraptera* [Worldwide]	Common	Mouse [Cecum]	Usually nonpathogenic; heavy infections sometimes cause impaction, intussusception, rectal prolapse	15,59,70,78
3 *Capillaria hepatica* [Worldwide]	Uncommon	Monkey [Liver]	Usually inapparent; sometimes cirrhosis of liver, death	20,43,69
4 *Graphidium strigosum* [North America, Europe, Australia]	Uncommon	Rabbit [Stomach]	Usually inapparent; sometimes diarrhea, anemia, emaciation, death	1,11,13,54
5 *Nochtia nochti* [Asia]	Uncommon	Monkey [Stomach]	Gastric tumors	7,18,69
6 *Oesophagostomum aculeatum* [United States, Asia]	Common	Monkey [Colon]	Nodules in wall of colon; sometimes adhesions	66,69
7 *O. apiostomum* [United States, Asia]	Common	Monkey [Colon; rarely omentum]	Nodules in wall of colon; sometimes adhesions	20,69,82
8 *O. bifurcum* [United States, Asia]	Common	Monkey [Colon]	Nodules in wall of colon; sometimes adhesions	5,18,19,66
9 *Paraspidodera uncinata* [Worldwide]	Common in some regions, uncommon in others	Guinea pig [Cecum, colon]	Usually considered nonpathogenic; possibly causes weight loss, debility, diarrhea	22,23,59,63
10 *Passalurus ambiguus* [Worldwide]	Common	Rabbit [Cecum, colon]	Apparently nonpathogenic	11,44,59,62
11 *Physaloptera tumefaciens* [Asia]	Common	Monkey [Stomach]	Gastritis (hemorrhagic, ulcerative)	41,70
12 *Streptopharagus pigmentatus* [United States, Asia]	Common	Monkey [Stomach]	Unknown	18
13 *Syphacia muris* [Worldwide]	Common	Rat [Cecum, colon]	Apparently nonpathogenic except in heavy infections	2,28,59,70
14 *S. obvelata* [Worldwide]	Common	Hamster [Cecum, colon]	Usually nonpathogenic; heavy infections sometimes cause impaction, intussusception, rectal prolapse	34,78
15		Mouse [Cecum, colon]	Usually nonpathogenic; heavy infections sometimes cause impaction, intussusception, rectal prolapse	15,28,59,70,78

continued

	Species [Distribution]	Incidence in Laboratory	Host [Localization]	Effect on Host	Reference
16	*Ternidens deminutus* [United States, Asia]	Uncommon	Monkey [Cecum, colon]	Anemia, intestinal nodules	18,20,41, 66,69,80
17	*Trichostrongylus colubriformis* [Worldwide]	Common	Monkey [Small intestine]	Unknown; probably diarrhea & eosinophilia in heavy infections	40,66,80
18	*Trichuris trichiura* [Worldwide]	Common in specimens obtained from natural habitat	Monkey [Cecum, colon]	Usually inapparent; heavy infections probably cause mild to severe colitis	18,20,66, 68,69
			Platyhelminthes—Cestoda		
19	*Bertiella studeri* [Asia]	Common in specimens obtained from natural habitat	Monkey [Small intestine]	Unknown	66,80
20	*Hymenolepis diminuta* [Worldwide]	Uncommon	Hamster [Intestine]	Usually inapparent; sometimes enteritis	21,65,78
21			Mouse [Intestine]	Usually inapparent; sometimes enteritis	70,78
22			Rat [Intestine]	Usually inapparent; sometimes enteritis	70
23	*H. nana* [Worldwide]	Common	Hamster [Intestine]	Retarded growth, weight loss, obstruction, enteritis	71,77,78
24			Mouse [Intestine]	Retarded growth, weight loss	25,26,33, 59,70, 72,78
25			Rat [Intestine]	Retarded growth, weight loss	64,70
26	*Taenia pisiformis* [Worldwide]	Common in some colonies, absent in others	Rabbit [Liver, peritoneal cavity]	Abdominal distension, lethargy, weight loss, liver damage	1,11,39, 52,53,62
27	*T. taeniaeformis* [Worldwide]	Common in some colonies, absent in others	Mouse [Liver]	Hepatic cyst formation; sometimes hepatic sarcoma	12,29,31, 56,59
28			Rat [Liver]	Hepatic cyst formation; sometimes hepatic sarcoma	31,56,70
			Platyhelminthes—Trematoda		
29	*Gastrodiscoides hominis* [Asia]	Common	Monkey [Cecum, colon]	Mucous diarrhea, mild enteritis	18,24
			Protozoa		
30	*Balantidium coli* [Worldwide]	Common	Monkey [Cecum, colon]	Usually nonpathogenic; sometimes enteritis	17,20,85
31	*Eimeria caviae* [Worldwide]	Common in some colonies, absent in others	Guinea pig [Large intestine]	Usually none; sometimes diarrhea, hemorrhagic enteritis, death	35,42,59
32	*E. falciformis* [Worldwide]	Common in some colonies, absent in others	Mouse [Intestine]	Anorexia, diarrhea, catarrhal enteritis; sometimes death	8,9,42,59
33	*E. irresidua* [Worldwide]	Common	Rabbit [Small intestine]	Hemorrhagic diarrhea, severe enteritis	27,32,57, 59
34	*E. magna* [Worldwide]	Common	Rabbit [Jejunum, ileum; sometimes cecum]	Anorexia, weight loss, mucous diarrhea, severe enteritis; sometimes death	27,32,57, 59
35	*E. media* [Worldwide]	Common	Rabbit [Small intestine]	Moderate diarrhea, enteritis	32,57,59, 60

continued

	Species [Distribution]	Incidence in Laboratory	Host [Localization]	Effect on Host	Reference
36	*E. perforans* [Worldwide]	Common	Rabbit [Small intestine]	Mild diarrhea, slight enteritis	27,32,57, 59
37	*E. stiedai* [Worldwide]	Common	Rabbit [Liver]	Anorexia, weight loss, abdominal distension, hepatomegaly; sometimes diarrhea, jaundice, death	3,57,59, 73-75
38	*Entamoeba histolytica* [Worldwide]	Common in specimens obtained from natural habitat	Monkey [Cecum, colon]	Diarrhea (sometimes hemorrhagic), colitis (sometimes ulcerative)	66,69,82, 85
39	*Giardia lamblia* [Worldwide]	Unknown; probably common in specimens obtained from natural habitat	Monkey [Anterior small intestine]	Possibly enteritis	66,69
40	*Hexamita muris* [Worldwide]	Common in some colonies, absent in others	Mouse [Small intestine, cecum]	Associated with duodenitis	25,48
41	*Klossiella cobayae* [Worldwide]	Common in some colonies, absent in others	Guinea pig [Kidneys, other organs]	Usually none; sometimes nephritis	6,10
42	*K. muris* [Worldwide]	Common in some colonies, absent in others	Mouse [Kidneys]	Usually none; sometimes necrotic foci in kidneys	58,76,84
43	*Nosema cuniculi* [Worldwide]	Common in some colonies, uncommon in others	Mouse [Brain, kidneys, other tissues]	Usually inapparent; sometimes encephalitis, nephritis, death	30,36,46
44			Rabbit [Brain, kidneys, other tissues]	Usually inapparent; sometimes encephalitis, nephritis, death	46,50
45	*Toxoplasma gondii* [Worldwide]	Common in some colonies; uncommon or absent in others	Guinea pig [Various tissues]	Inapparent to fatal disease with varied signs	45,47,49, 67,81
46			Mouse [Various tissues]	Inapparent to fatal disease with varied signs	4,16,51, 55
47			Rabbit [Various tissues]	Inapparent to fatal disease with varied signs	37,49,61, 79
48			Rat [Various tissues]	Inapparent to fatal disease with varied signs	14,38

Contributor: Flynn, Robert J.

References

[1] Adams, C. E., et al. 1967. Univ. Fed. Anim. Welfare Handb. Care Manage. Lab. Anim., p. 396.

[2] Bell, D. P. 1968. Lab. Anim. 2:1.

[3] Berson-Organisciak, J. 1966. Zwierz. Lab. 2:215.

[4] Beverley, J. K. A. 1959. Nature (London) 183: 1348.

[5] Bingham, G. A., and M. M. Rabstein. 1964. Lab. Anim. Care 14:357.

[6] Bonciu, C., et al. 1957. Arch. Roum. Pathol. Exp. Microbiol. 16:131.

[7] Bonne, C., and J. H. Sandground. 1939. Amer. J. Cancer 37:173.

[8] Breza, M., and V. Jurasek. 1959. Vet. Cas. 8:372.

[9] Cordero del Campillo, M. 1959. Rev. Iber. Parasitol. 19:351.

[10] Cossel, L. 1958. Schweiz. Z. Allg. Pathol. Bakteriol. 21:62.

[11] Cushnie, G. H. 1954. J. Anim. Tech. Ass. 5:22.

[12] Duffill, M. L., and M. F. Lyon. 1960. Ibid. 10:148.

[13] Enigk, K. 1938. Z. Parasitenk. 10:386.

[14] Eyles, D. E. 1952. J. Parasitol. 38:226.

[15] Flynn, R. J., et al. 1965. Lab. Anim. Care 15:440.

[16] Gibson, C. L., and D. E. Eyles. 1957. Amer. J. Trop. Med. Hyg. 6:990.

continued

[17] Gisler, D. B., et al. 1960. Ann. N. Y. Acad. Sci. 85:758.

[18] Graham, G. L. 1960. Ann. N. Y. Acad. Sci. 85:842.

[19] Guilloud, N. B., et al. 1965. Lab. Anim. Care 15:354.

[20] Habermann, R. T., and F. P. Williams, Jr. 1957. Amer. J. Vet. Res. 18:419.

[21] Handler, A. H. 1965. In W. E. Ribelin and J. R. McCoy, ed. The Pathology of Laboratory Animals. C. C. Thomas, Springfield, Ill. pp. 210-240.

[22] Herlich, H. 1959. J. Parasitol. 45:586.

[23] Herlich, H., and C. F. Dixon. 1965. Ibid. 51:300.

[24] Herman, L. H. 1967. Vet. Med. 62:355.

[25] Heston, W. E. 1941. In G. D. Snell, ed. Biology of the Laboratory Mouse. Blakiston, Philadelphia. pp. 349-379.

[26] Heyneman, D. 1961. Nature (London) 191:297.

[27] Horton-Smith, C. 1958. Vet. Rec. 70:256.

[28] Hussey, K. L. 1957. J. Parasitol. 43:555.

[29] Innes, J. R. M. 1967. In E. Cotchin and F. J. C. Roe, ed. Pathology of Laboratory Rats and Mice. Blackwell Scientific Publications, Oxford. pp. 229-257.

[30] Innes, J. R. M., et al. 1962. J. Neuropathol. Exp. Neurol. 21:519.

[31] Jones, T. C. 1967. In E. Cotchin and F. J. C. Roe, ed. Pathology of Laboratory Rats and Mice. Blackwell Scientific Publications, Oxford. pp. 1-23.

[32] Kheisin, E. M. 1957. Tr. Leningrad. Obshchest. Estestvoispyt. 73:150.

[33] King, V. M., and G. E. Cosgrove. 1963. Lab. Anim. Care 13:46.

[34] Kirschenblatt, I. D. 1949. Uch. Zap. Leningrad. Gos. Univ. Ser. Biol. Nauk 19:110.

[35] Kleeberg, H. H., and W. Steenken, Jr. 1963. J. S. Afr. Vet. Med. Ass. 34:49.

[36] Koller, L. D. 1969. J. Amer. Vet. Med. Ass. 155:1108.

[37] Lainson, R. 1955. Ann. Trop. Med. Parasitol. 49:384.

[38] Lainson, R. 1957. Trans. Roy. Soc. Trop. Med. Hyg. 51:111.

[39] Lapage, G. 1968. Veterinary Parasitology. Oliver and Boyd, London.

[40] Lapin, B. A., and L. A. Yakovleva. 1960. Comparative Pathology in Monkeys. C. C. Thomas, Springfield, Ill.

[41] Levine, N. D. 1968. Nematode Parasites of Domestic Animals and of Man. Burgess, Minneapolis.

[42] Levine, N. D., and V. Ivens. 1965. Ill. Biol. Monogr. 33.

[43] Lubinsky, G. 1956. Can. J. Comp. Med. 20:457.

[44] Lund, E. E. 1950. J. Parasitol. 36:13.

[45] Makstenieks, O., and J. D. Verlinde. 1957. Doc. Med. Geogr. Trop. 9:213.

[46] Malberbe, H., and V. Munday. 1958. J. S. Afr. Vet. Med. Ass. 29:241.

[47] Mariani, G. 1940. Riv. Biol. Colon. Rome 3:47.

[48] Meshorer, A. 1969. Lab. Anim. Care 19:33.

[49] Miller, L. T., and H. A. Feldman. 1953. J. Infect. Dis. 92:118.

[50] Møller, T. 1968. Z. Versuchstierk. 10:27.

[51] Mooser, H. 1950. Schweiz. Med. Wochenschr. 80:1399.

[52] Morgan, B. B., and P. A. Hawkins. 1949. Veterinary Helminthology. Burgess, Minneapolis.

[53] Napier, R. A. N. 1963. In W. Lane-Petter, ed. Animals for Research. Principles of Breeding and Management. Academic Press, New York. pp. 323-364.

[54] Neveu-Lemaire, M. 1936. Traite d'helminthologie medicale et veterinaire. Vigot, Paris.

[55] Nicolau, S., and G. Balmus. 1934. C. R. Soc. Biol. 115:959.

[56] Olivier, L. 1962. J. Parasitol. 48:373.

[57] Ostler, D. C. 1961. Vet. Rec. 73:1237.

[58] Otto, H. 1957. Z. Pathol. 68:41.

[59] Owen, D. 1968. Lab. Anim. Cent. News Lett. 35:7.

[60] Pellérdy, L., and A. Babos. 1953. Acta Vet. Hung. 3:173.

[61] Perrin, T. L. 1943. Arch. Pathol. 36:559.

[62] Poole, C. M., et al. 1967. Annu. Rep. Argonne Nat. Lab. Biol. Med. Res. Div. ANL-7409.

[63] Porter, D. A., and G. F. Otto. 1934. J. Parasitol. 20:323.

[64] Ratcliffe, H. L. 1949. In E. J. Farris and J. Q. Griffith, ed. The Rat in Laboratory Investigation. J. B. Lippincott, Philadelphia. pp. 502-514.

[65] Read, C. P. 1951. J. Parasitol. 37:324.

[66] Reardon, L. V., and B. F. Rininger. 1968. Lab. Anim. Care 18:577.

[67] Rodaniche, E. C., and T. Pinzon. 1949. J. Parasitol. 35:152.

[68] Rowland, E., and J. G. Vandenbergh. 1965. Ibid. 51:294.

[69] Ruch, T. C. 1959. Diseases of Laboratory Primates. W. B. Saunders, Philadelphia.

[70] Sasa, M., et al. 1962. In R. J. C. Harris, ed. The Problem of Laboratory Animal Disease. Academic Press, New York. pp. 195-214.

[71] Sheffield, F. W., and E. Beveridge. 1962. Nature (London) 196:294.

[72] Simmons, M. L., et al. 1964. Lab. Anim. Care 14:326.

[73] Smetana, H. 1933. Arch. Pathol. 15:175.

[74] Smetana, H. 1933. Ibid. 15:330.

[75] Smetana, H. 1933. Ibid. 15:516.

[76] Smith, T., and H. P. Johnson. 1902. J. Exp. Med. 6:303.

[77] Soave, O. A. 1963. J. Amer. Vet. Med. Ass. 142:285.

[78] Stone, W. B., and R. D. Manwell. 1966. Pub. Health Rep. 81:647.

[79] Szemeredi, G. 1968. Magy. Allatorv. Lapja 23:176.

[80] Valerio, D. A., et al. 1969. *Macaca mulatta:* Management of a Laboratory Breeding Colony. Academic Press, New York.

[81] Varela, G., et al. 1953. Rev. Inst. Salubr. Enferm. Trop. (Mex.) 13:217.

[82] Vickers, J. H. 1969. Ann. N. Y. Acad. Sci. 162:659.

[83] Yamaguti, S. 1961. In S. Yamaguti. Systema Helminthum. Interscience, New York. v. 3.

[84] Yang, Y. H., and H. C. Grice. 1964. Can. J. Comp. Med. Vet. Sci. 28:63.

[85] Young, R. J., et al. 1957. Proc. Anim. Care Panel 7:67.

For information on arthropod parasites of laboratory animals, *see* Table 120. Data in brackets refer to the column heading in brackets.

Species [Distribution]	Free Stage Location	Host [Stage & Location in Host]	Effect on Host	Reference
Pentastomida				
1 *Linguatula serrata* [Worldwide]	Eggs on vegetation	Mammals [Eggs expelled from respiratory tract; adults in nasal passages; eggs hatch in alimentary tract of herbivores, larvae & nymphs develop in mesenteries, adults in carnivores]	Severe irritation & blockage of nasal passages	13
Arachnida				
2 *Amblyomma* spp.[1] [North, Central, & South America]	Eggs in soil; unfed larvae, nymphs, & adults on grass	Cattle, dog, goat, horse, sheep; occasionally, birds [External; on host only while feeding]	Damage to hide, milk reduction. Larvae, nymphs, & adults are bloodsuckers. Vector of organisms causing Rocky Mt. spotted fever & Q fever.	1,15
3 *Argas persicus* [Warm & temperate semiarid regions of world]	All stages in crevices, cracks of housing, under bark of trees	Domestic fowl; occasionally wild birds [External; on host only while feeding]	Anemia, leg weakness, egg reduction; occasionally death. Nymphs & adults are bloodsuckers. Vector of organisms causing fowl spirochetosis & spiroplasmosis.	1,3,15
4 *Bdellonyssus sylviarum* [United States, Canada, Mexico, Europe, South Africa]	Eggs on feathers & in nests; other stages on surroundings	Poultry, wild birds [External; on body & feathers]	Skin lesions, egg reduction, retarded growth, anemia. Larvae, nymphs, & adults are bloodsuckers. Harbors neurotropic viruses.	1,3,12
5 *Boophilus annulatus*[1] [North America]	Eggs on soil; unfed larvae on grass	Principally ungulates [External]	Damage to hide, milk reduction. Larvae, nymphs, & adults are bloodsuckers. Vector of organism causing Texas cattle fever.	1,2,13, 15
6 *Demodex canis* [Worldwide]	None	Dog [Eggs, nymphs, & adults in hair follicles & sebaceous glands]	Follicle inflammation, mange, thickened skin, alopecia, emaciation; sometimes death.	1,13
7 *Dermacentor andersoni*[1] [Western North America]	Eggs on soil; unfed larvae, nymphs, & adults on vegetation	Most mammals [External; larvae & nymphs on most small animals; adults usually on larger animals during feeding period]	Paralysis, particularly in sheep. Larvae, nymphs, & adults are bloodsuckers. Vector of organisms causing Rocky Mt. spotted fever, tularemia, equine encephalomyelitis, & anaplasmosis.	12,13, 15
8 *D. variabilis*[1] [North America]	Eggs on soil; unfed larvae, nymphs, & adults on vegetation until host is found	Principally dog; other domestic & wild animals [External; on host only while feeding; larvae & nymphs mainly attack rodents & other small animals]	Skin damage. Larvae, nymphs, & adults are bloodsuckers. Vector of organisms causing bovine anaplasmosis, tularemia, & Rocky Mt. spotted fever.	1,7,12, 15

[1] Known to be a parasite of man.

continued

	Species [Distribution]	Free Stage Location	Host [Stage & Location in Host]	Effect on Host	Reference
9	Dermanyssus gallinae [Worldwide]	Eggs, nonfeeding larvae, nymphs, & adults in crevices of coops & roosts	Poultry, other birds [External]	Egg reduction, retarded growth, anemia; sometimes death. Larvae, nymphs, & adults are bloodsuckers; nocturnal feeders only. Vector of organisms causing spirochetosis & fowl cholera.	1,3
10	Eutrombicula alfreddugesi [1] [North & South America, West Indies]	Active forms in grasses, shrubs, brambles	Domestic & wild vertebrates [External; only larvae parasitic]	Irritation due to toxins; sometimes death in small poultry	1,7,13, 15
11	Ixodes ricinus scapularis [Principally Europe & Asia Minor]	Eggs on soil; larvae, nymphs, & adults on grass & shrubbery until host is found	Principally, dog; other domestic & wild animals [External; on host only while feeding; adults on head & neck of dog, on flank & leg and under tail of other animals]	Anemia. Larvae & nymphs are bloodsuckers in ear & on eyelids & head; rarely on body. Vector of louping ill virus, tick-borne fever virus of sheep, & organisms causing cattle redwater fever & bovine piroplasmosis.	1,13
12	Knemidokoptes mutans [Worldwide]	None	Chicken, turkey, other domestic birds [External; all active stages in tunnels between scales of feet, legs, neck, & comb]	Inflammation & keratinization between scales of feet & legs; also lameness	1,3,13
13	Otobius megnini [1] [United States, Mexico, South America, South Africa]	Eggs on ground and in cracks; adults & unattached larvae in outbuildings	Domestic animals [Inside ears]	Ear inflammation, anorexia, dullness; sometimes death. Larvae & nymphs are bloodsuckers.	1,13,15
14	Otodectes cyanotis [Worldwide]	None	Cat, dog, ferret [All stages in ears; sometimes external]	Inflammation, ear scabs, head-shaking, scratching, droopy ears with discharge; in severe cases, epileptiform fits	1,13
15	Psoroptes equi ovis [2] [Worldwide]	None	Sheep [External; all active stages on skin around edges of lesions]	Scabbing; wool loss (from biting & scratching); emaciation; sometimes death	1,13,15
16	Rhipicephalus sanguineus [1] [Worldwide]	Active forms near habitat of dog	Principally dog [External]	Larvae, nymphs, & adults are bloodsuckers. Vector of organisms causing canine piroplasmosis & cattle gall sickness.	1,13
17	Sarcoptes scabiei [1] [Worldwide]	None	Most mammals, sheep (on head) [External; all active stages in skin tunnels]	Scratching, papules, vesicles, keratinization, alopecia, mange, emaciation; sometimes death	1,13,15
18	Trombicula akamushi [1] [Japan, China, New Guinea]	All stages except larval on grass & other vegetation in flood plains	Mammals, birds [Only larvae parasitic]	Irritation due to toxins. Transmits scrub typhus.	7,12

[1] Known to be a parasite of man.   [2] Other varieties infest various domestic animals.

continued

Species [Distribution]	Free Stage Location	Host [Stage & Location in Host]	Effect on Host	Reference
		Insecta		
19	*Aedes aegypti*[1] [Worldwide]	Eggs near water (survive long periods of drying in soil); larvae & pupae in water in man-made containers	Warm-blooded animals [External; where hair or feathers are thinnest]	Adult ♀ are bloodsuckers. Vector of yellow fever virus, dengue virus, & other arboviruses.	7,10, 12,13
20	*Anopheles* spp.[1] [Worldwide]	Immature stages in water	Warm-blooded animals [External; where hair or feathers are thinnest]	Adult ♀ are bloodsuckers. Vector of certain arboviruses & organisms causing dog heartworm, malaria, & filariasis.	7,10, 12
21	*Bovicola bovis* [North & South America, Europe, Africa, Australia]	None	Cattle [External; eggs on hair; nymphs & adults feed on skin]	Reduced vigor, irritation, scaly skin	13,15
22	*Chrysops* spp.[1] [Worldwide]	Eggs on emergent aquatic vegetation; larvae & pupae in mud	Domestic & wild animals	Only ♀ are bloodsuckers. Vector of *Loa loa* & organisms causing tularemia, surra, & anthrax.	5,13
23	*Cimex lectularius*[1] [Worldwide]	All stages in cracks, crevices, similar hiding places	Man, domestic animals, poultry [External]	Skin irritation & welts. Nymphs & adults are bloodsuckers. Not proven to be a vector of disease.	13,15
24	*Cochliomyia hominivorax*[1,3] [Tropical & subtropical areas of western hemisphere]	Pupae in soil; adults in pastures	Warm-blooded animals, including livestock, wild mammals, cat, & dog [Eggs deposited on edges of wounds]	Infection & extension of wounds; untreated host usually dies from parasitism or secondary infection	4,9,11, 12
25	*Ctenocephalides canis* [Worldwide]; *C. felis*[1] [Worldwide]	Immature forms associated with nest or sleeping area of host; adults on ground part of time	Cat, dog, swine, other animals [External]	Coat damage (from biting & scratching). Adults are bloodsuckers. Both species vectors of dog & dwarf tapeworm, heartworm, & organism causing epidemic typhus.	6,13,19
26	*Cuclotogaster heterographus* [Worldwide]	None	Chicken, partridge [External; eggs on feathers; nymphs & adults on skin & feathers of head & neck]	Irritation	3,13
27	*Culex pipiens quinquefasciatus*[4] [Worldwide from 60°N to 50°S latitude]	Immature forms in stagnant water, ponds, ditches	Warm-blooded animals, especially birds [External; where hair or feathers are thinnest]	Adult ♀ are bloodsuckers. Vector of fowlpox virus, arboviruses, & organisms causing avian malaria & filariasis.	3,10,13
28	*Dermatobia hominis*[1] [Tropical North America; West Indies; South America]	Eggs, glued to other arthropods, hatch when suitable host is reached	Cattle, dog, horse, mule, wild animals [Larvae leave transporting arthropod on contact with host and penetrate skin]	Boil-like skin lesions, milk reduction, damage to hide, decreased growth rate; possibly death in young animals	8,11, 12,16

[1] Known to be a parasite of man. [3] Obligatory parasite. The adult stage of *C. macellaria* resembles *C. hominivorax* in appearance, but differs by being a secondary invader (facultative parasite), and by breeding in carcasses; larvae occasionally infest wool or necrotic wounds. The primary screwworm in Africa, India, and the East Indies is *Chrysomya bezziana*. [4] Including *C. quinquefasciatus*.

continued

	Species [Distribution]	Free Stage Location	Host [Stage & Location in Host]	Effect on Host	Reference
29	*Echidnophaga gallinacea* [1] [Worldwide, especially in warm climates]	Immature forms associated with nest or sleeping area of host	Domestic animals, rodents, poultry [External; skin, comb, wattles, & around eyes & ears]	Anemia; sometimes death. Adults are bloodsuckers.	3,15,18
30	*Gasterophilus intestinalis* [5] [Worldwide]	Pupae in soil; adults attack hosts only in daytime	Ass, horse, mule; rarely other animals [Eggs on foreleg fetlock hairs; larvae (maggots) in mouth, pharynx, & stomach]	Extension & infection of wounds	9,11,16
31	*Glossina* spp. [1] [Central Africa]	Larvae pass from ♀ when ready to pupate in soil; adults in vegetation	Cattle, big game animals [External]	Adults are bloodsuckers. Vector of organisms causing cattle & horse nangana, & sleeping sickness in man.	7,12, 13,19
32	*Haematobia irritans* [Worldwide]	Eggs & larvae (maggots) in fresh dung; pupae in dung or soil	Cattle, other animals [External]	Weight loss, milk reduction. Adults are bloodsuckers.	12,13, 15,19
33	*Haematopinus eurysternus* [6] [Worldwide]	None	Cattle [External; eggs on shaft or at base of hairs]	Hair damage (from rubbing); stunting; milk reduction. Nymphs & adults are bloodsuckers.	13,15
34	*H. suis* [Worldwide]	None	Swine [External; eggs on hair]	Dermatitis, skin sores, retarded growth. Nymphs & adults are bloodsuckers. Vector of swine pox virus.	15
35	*Hybomitra* spp. [7] [Worldwide]	Eggs on vegetation near water; larvae & pupae in mud	Most warm-blooded animals [External]	Only ♀ bite. Vector of organisms causing anthrax, surra, anaplasmosis, & hog cholera	7,12, 15,17
36	*Hypoderma lineatum* [8] [Americas; Europe; northern Asia; India]	Pupae in soil; adults in pastures	Cattle; rarely horse [Eggs on hair of legs & body; larger larvae form tumor under skin of back]	Skin perforation, hide & flesh damage, milk reduction	4,13,15
37	*Melophagus ovinus* [Most parts of world]	Larvae retained in ♀ until mature	Sheep; occasionally goat [External; pupae attached to wool]	Anemia; stained wool (damaged from rubbing). Adults are bloodsuckers.	13,15, 19
38	*Menacanthus stramineus* [Worldwide]	None	All domestic fowl [External; eggs attached to feathers; nymphs & adults on skin around vent]	Scabbing of skin, wasting, egg reduction	3,13
39	*Menopon gallinae* [Worldwide]	None	Chicken, guinea fowl [External; eggs, nymphs, & adults feed on scales, scabs, & feathers]	Scaling, scabbing	3,13
40	*Musca domestica* [Worldwide]	Immature forms in manure & decayed matter; adults in buildings	Any larger animal with lesions or secretions [External; adults accidentally ingested by host]	Adults cause decreased productivity of livestock. Vector of several tapeworm species; mechanical vector of many bacterial & protozoan pathogens & helminth eggs.	3,13,15

[1] Known to be a parasite of man. [5] *G. haemorrhoidalis* and *G. nasalis* are similar in many respects to *G. intestinalis*. [6] Information also applicable to *Linognathus vituli*. [7] Information also applicable to *Tabanus* spp. [8] Information also applicable to *H. bovis*.

continued

	Species [Distribution]	Free Stage Location	Host [Stage & Location in Host]	Effect on Host	Reference
41	*Oestrus ovis* [Worldwide]	Pupae on ground; adults in warm corners & crevices	Sheep; rarely goat [Larvae in nasal cavities & sinuses]	Mucosal irritation, nasal discharge, emaciation; sometimes death	9,13,15
42	*Phaenicia sericata* [Worldwide, except South America & Pacific islands]	Pupae in soil	Sheep, goat, other animals [Eggs on flesh & soiled wool; larvae on skin and in wounds]	Invades wounds and causes suppuration	8,13
43	*Phlebotomus papatasii* [1] [Mediterranean region; southern Europe; Asia]	Immature forms in dark, moist places & manure	Warm-blooded animals [External]	Swelling at site of bite. Adults are nocturnal bloodsuckers. Vector of organisms causing pappataci fever.	13,15, 18
44	*Phormia regina* [9] [Worldwide]	Pupae in soil; adults in pastures	Sheep, other mammals [Eggs in hair or wool; larvae in wounds; eggs & larvae also in carcasses]	Extension & infection of wounds	8,15
45	*Pulex irritans* [1] [Worldwide]	Eggs, larvae, & pupae in soil; adults on ground part of time	Man, dog, swine, other animals [External]	Irritation; poor condition; coat damage (from biting & scratching). Adults are bloodsuckers. Vector of organisms causing plague & murine typhus.	12,13, 19
46	*Simulium* spp. [10] [Worldwide in temperate to subarctic climates]	Immature forms on stones in running streams	All warm-blooded animals [External; on bare parts of head, body, & legs; under wings]	Red swelling, vesicles, anemia, toxemia, death. Adults are bloodsuckers. *S. occidentale* & *S. slossonae* are vectors of organism causing turkey leucocytozoan disease; some species are vectors of organism causing onchocerciasis in man & cattle.	3,13, 15,18
47	*Stomoxys calcitrans* [1] [Worldwide]	Immature forms in manure & other moist organic waste	Most mammals & birds [External]	Weight loss, milk reduction. Adults are bloodsuckers. Vector of poultry tapeworms, filariae, & spiruroids; mechanical vector of organisms causing surra & tularemia.	3,8,13
48	*Triatoma* spp. [1,11] [Worldwide]	All stages commonly found in, or close to, rodent nests or habitats	Domestic animals, poultry, wild animals with dens; usually wood rat [External]	Swelling, anemia, anaphylactic symptoms. Nymphs & adults are bloodsuckers. Vector of organism causing Chagas' disease.	7,12, 14,15, 19
49	*Trichodectes canis* [Worldwide]	None	Dog [External; eggs on hair; nymphs & adults feed on skin]	Scaly skin (from rubbing & scratching)	13

[1] Known to be a parasite of man. [9] *Chrysomya chloropyga* is similar to *P. regina* in its parasitism of sheep. [10] *S. arcticum, S. occidentale, S. ornatum,* and *S. vittalum* are the important blackfly pests of livestock. [11] 16 *Triatoma* species are found primarily in the western hemisphere.

continued

	Species [Distribution]	Free Stage Location	Host [Stage & Location in Host]	Effect on Host	Reference
50	*Wohlfahrtia vigil*[1] [North America]	Pupae on ground	Guinea pig, mink, rabbit, young of domestic & wild animals [Larvae in wounds]	Mild to extensive subcutaneous pustular lesions; occasionally death	8,13
51	*Xenopsylla cheopis*[1] [Worldwide]	Eggs & larvae in rodent nests	Man, rodents	Irritation due to bites. Vector of organisms causing plague & murine typhus; intermediate host of cestodes.	7,12

[1] Known to be a parasite of man.

Contributors: Edgar, S. Allen, and Hays, Kirby L.; Furman, Deane P.

References

[1] Baker, E. W., and G. W. Wharton. 1952. An Introduction to Acarology. Macmillan, New York.

[2] Belding, D. L. 1952. Textbook of Clinical Parasitology. Ed. 2. Appleton-Century-Crofts, New York.

[3] Biester, H. E., and L. H. Schwarte. 1965. Diseases of Poultry. Ed. 5. Iowa State Univ. Press, Ames.

[4] Bishopp, F. C., et al. 1926. U.S. Dep. Agr. Farmers Bull. 857.

[5] Dickmans, G. 1945. Amer. J. Vet. Res. 6:211.

[6] Ewing, H. E. 1929. A Manual of External Parasites. C. C. Thomas, Springfield, Ill.

[7] Gordon, R. M., and M. M. J. Lavoipierre. 1962. Entomology for Students of Medicine. Blackwell, Oxford.

[8] Hall, D. G. 1948. The Blowflies of North America. Thomas Say Foundation, Baltimore.

[9] Herms, W. B. 1961. Medical Entomology. Ed. 5. Macmillan, New York.

[10] Horsfall, W. R. 1955. Mosquitoes, Their Bionomics and Relation to Disease. Ronald Press, New York.

[11] James, M. T. 1947. U.S. Dep. Agr. Misc. Publ. 631.

[12] James, M. T., and R. F. Harwood. 1970. Herms' Medical Entomology. Ed. 6. Macmillan, New York.

[13] Lapage, G., ed. 1962. Monnig's Veterinary Helminthology and Entomology. Ed. 5. Williams and Wilkins, Baltimore.

[14] Matheson, R. 1950. Medical Entomology. Ed. 2. Comstock, Ithaca.

[15] Metcalf, C. L., and W. P. Flint. 1962. Destructive and Useful Insects: Their Habits and Control. Ed. 4. McGraw-Hill, New York.

[16] Neel, W. W. 1954. J. Econ. Entomol. 47(3):540.

[17] Oldroyd, H. 1954. The Horse-Flies of the Ethiopian Region. British Museum, London. v. 2.

[18] Patton, W. S., and K. M. Evans. 1929. Insects, Ticks, Mites and Venomous Animals. H. R. Grubb, Croydon, England. pt. 1.

[19] Smart, J. 1956. A Handbook for the Identification of Insects of Medical Importance. Ed. 3. British Museum, London.

123. ARTHROPOD PESTS: PLANTS AND PLANT PRODUCTS

Data in brackets refer to the column heading in brackets.

	Species (Synonym) [Distribution]	Destructive Stage	Host	Destructive Activity
	Arachnida			
1	*Eriophyes pyri* [All pear-growing regions]	Adult; immature	Apple, pear	Sucking causes blisters on undersides of leaves & on fruit
2	*Rhizoglyphus echinopus* [North America, Europe, Asia]	Adult; immature	Onion, ornamental bulbs	Bores into bulbs
3	*Steneotarsonemus pallidus* [North America, Europe]	Adult; immature	Greenhouse ornamentals, strawberry	Sucks plant juices; distorts buds & leaves

continued

	Species (Synonym) [Distribution]	Destructive Stage	Host	Destructive Activity
4	*Tetranychus urticae (T. telarius)* [United States, Europe, Africa, Asia, Australia]	Adult; immature	Cultivated plants	Sucks plant juices, causing loss of vigor and dropping of leaves
	Crustacea			
5	*Porcellio laevis* [Worldwide]	Adult; immature	Ornamentals, vegetables	Chews roots & growths near ground
	Insecta			
6	*Acanthoscelides obtectus* [Worldwide]	Larva	Bean, cowpea, pea	Devours inside of bean in field & in storage
7	*Agriotes* spp. [1] [Worldwide]	Larva	Cereal, forage, & truck crops	Devours or bores into seeds & roots
8	*Alabama argillacea* [North & South America]	Larva	Cotton only	Devours leaves
9	*Altica* spp. [2] [Worldwide]	Adult; larva [3]	Vegetable crops	Larva often feeds on roots; adult makes holes in leaves
10	*Amphibolips confluenta* [Worldwide]	Larva	Oak	Causes galls on oak leaves
11	*Anabrus simplex* [Western United States]	Adult; nymph	Grain, hay, many plants	Devours grain, hay, & leaves of plants
12	*Anagasta kuehniella (Ephestia kuehniella)* [Worldwide]	Larva	Mill products	Destroys grain products
13	*Anasa tristis* [North & Central America]	Adult; nymph	Squash, other Cucurbitaceae	Sap-sucking causes plants to wilt and die
14	*Anthonomus grandis* [Southern United States; Mexico]	Adult; larva	Cotton	Destroys buds; devours squares & bolls
15	*Aphis pomi* [North America]	Adult; nymph	Apple	Causes stunting [4]
16	*Aspidiotus perniciosis (Quadraspidiotus perniciosis)* [Worldwide]	Adult; nymph	Deciduous fruit trees, ornamentals	Secreted toxins cause wilting, kill infested trees, and discolor fruit
17	*Blissus leucopterus* [North America]	Adult; nymph	Corn, grains, grasses	Sap-sucking causes plants to wilt and die
18	*Cephus pygmaeus* [Northeastern United States; Europe; Near East]	Larva	Barley, rye, timothy, wheat, other grasses	Mines stems, causing breakage
19	*Ceratitis capitata* [South America; Mediterranean region; South Africa; western Australia; Hawaii]	Adult; larva	Fruits, vegetables	Larva burrows through fruit; adult makes egg punctures
20	*Choristoneura fumiferana* [Northern United States; Canada]	Larva	Fir, hemlock, larch, spruce, white pine	Causes partial to complete defoliation
21	*Chrysobothris femorata* [United States, fruit-growing areas of Canada]	Larva	Fruit trees, many shade trees	Bores into trunks of weakened trees & branches; kills tree
22	*Cladius isomerus* [Worldwide]	Larva	Rosebush	Skeletonizes plant; causes browning of leaves
23	*Coccus hesperidum* [Worldwide in greenhouses [5]]	Adult; nymph	Citrus plants, ornamentals	Sap-sucking causes plants to die back [4]
24	*Conotrachelus nenuphar* [Eastern United States; Canada]	Adult; larva	Apple, cherry, peach, plum, deciduous stone fruits	Larva feeds within and destroys fruit; adult punctures fruit
25	*Corythuca arcuata* [Worldwide]	....	Various trees & shrubs	Causes speckling of leaves & stunting
26	*Dendroctonus frontalis* [Southeastern & southern United States; Mexico; Central America]	Adult; larva	Pine; rarely spruce	Bores into bark & cambial region; may girdle & kill tree

[1] Information also valid for *Horistonotus* spp., *Limonius* spp., and *Melanotus* spp. [2] Information also valid for *Phyllotreta* spp. [3] Overwinters as adult. [4] Honeydew formed or excreted. [5] A subtropical species.

continued

	Species (Synonym) [Distribution]	Destructive Stage	Host	Destructive Activity
27	*D. ponderosae (D. monticolae)* [Western United States]	Adult; larva	Limber, lodgepole, ponderosa, sugar, western, & white bark pines	Bores into bark & cambial region; may girdle & kill tree
28	*Diabrotica undecimpunctata* [North America]	Adult; larva	Corn, Cucurbitaceae, grasses, weeds, other plants	Larva feeds on roots; adult devours foliage [6]
29	*Diprion hercyniae* [Northeastern United States; Canada; Europe]	Larva	Spruce	Devours leaves
30	*Drosophila melanogaster* [Worldwide]	Larva	Ripe or decaying fruit	Breeds in ripe fruit
31	*Empoasca fabae* [North & South America]	Adult; nymph	Alfalfa, bean, celery, potato, other plants	Leaf-sucking causes wilting & drying of leaves, & stunting [7]
32	*Epicauta vittata* [Worldwide]	Adult	Legumes, potato	Devours plants
33	*Epilachna varivestis* [United States, Mexico]	Adult; larva	Bean, cowpea, soybean, other legumes	Devours leaves, pods, & stems
34	*Epitrix hirtipennis* [Worldwide]	Adult	Tobacco	Devours leaves, especially those of young plants
35	*Eriosoma lanigerum* [North & South America, Europe, South Africa, Asia, Australia]	Adult; nymph	Apple, elm, pear	Branch- & root-sucking causes deformed twigs, knotty roots, & stunting
36	*Forficula auricularia* [Worldwide]	Adult; nymph	Growing plants, stored grain, decayed vegetation	Chewing
37	*Gryllus* spp. [North, Central, & South America]	Adult; nymph	Cotton, hay crops, flax	Devours cotton, hay, & flax plants
38	*Harmolita tritici* [Eastern & central United States]	Larva	Wheat, some grasses	Causes gall in wheat; causes breaking of stems
39	*Heliothis zea* [Worldwide]	Larva	Alfalfa, corn, cotton, tomato, other plants	Bores into and feeds on bolls, ears, & buds; stunts plants; reduces yield
40	*Hylemya antiqua* [North America, Europe]	Larva	Onion, garlic	Mines out bulbs [4]
41	*Hyphantria cunea* [United States, southern Canada]	Larva	Broad-leaved fruit, shade, & nut trees	Webs branches; devours foliage
42	*Lampetia equestris* [North America, Europe]	Larva	Narcissus, other bulbs	Bores into bulbs
43	*Lasius alienus americanus* [United States]	Adult	Corn	Symbiotic with aphids attacking corn roots
44	*Laspeyresia pomonella (Carpocapsa pomonella)* [Apple-growing regions of North & South America, Europe, South Africa, Asia, & southern Australia]	Larva	Apple, apricot, pear, quince, walnut, similar fruits	Bores into and destroys fruit or reduces its value
45	*Lepisma saccharina* [Worldwide]	Adult; nymph	Starchy substances	Devours bookbindings, fabrics, & wallpaper
46	*Leptinotarsa decemlineata* [North America, Europe]	Adult; larva	Eggplant, potato, tobacco, tomato, other Solanaceae	Devours leaves; terminates growth
47	*Liposcelis divinatorius* [Worldwide]	Adult; immature	Cereals, vegetables	Contaminates food; destroys bookbindings
48	*Lygus lineolaris* [North America]	Adult; nymph	Many plants, trees	Leaf-sucking & toxins cause bud-dropping, distortion, & stunting
49	*Magicicada septendecim* [Eastern & southern United States]	Adult	Many deciduous trees & shrubs	Oviposition punctures injure or kill twigs

[4] Honeydew formed or excreted. [6] Also vector of organisms causing bacterial wilt of Cucurbitaceae and viral yellow disease of asters. [7] Also vector of organism causing hopperburn disease.

continued

	Species (Synonym) [Distribution]	Destructive Stage	Host	Destructive Activity
50	*Malacosoma disstria* [North America]	Larva	Ash, aspen, basswood, birch, gum, oak, sugar maple, other trees	Defoliates trees in summer
51	*Manduca quinquemaculata (Protoparce quinquemaculata)* [North & South America, Europe, Hawaii]	Larva	Tobacco, tomato, other Solanaceae	Devours foliage
52	*Mayetiola destructor (Phytophaga destructor)* [North America, Europe, Asia, New Zealand]	Larva	Barley, rye, wheat	Feeds on stems, causing breaking & stunting
53	*Megachile latimanus* [Worldwide]	Adult	Various trees	Cuts off leaf fragments for nests
54	*Melanoplus femur-rubrum* [Worldwide]	Adult; nymph	Hay crops in range & pasture	Devours grasses, hay, & vegetation
55	*Microcentrum rhombifolium* [North America]	Adult; nymph	Many broad-leaved trees & shrubs	Chews leaves
56	*Murgantia histrionica* [Southern United States; Mexico; Central America]	Adult; nymph	Cabbage & related crops, other plants	Sap-sucking causes plants to wilt, brown, & die
57	*Myzus persicae* [Warm regions of the world]	Adult; nymph	Many trees & shrubs; tobacco in North America	Leaf-sucking causes curling & distortion of leaves[4]
58	*Oryzaephilus surinamensis* [Worldwide]	Adult; larva	Grain, grain products, dried fruit	Infests & devours grain, grain products, & dried fruit
59	*Oscinella frit* [North America, Europe, Asia]	Larva	Cereals, grasses	Bores into stems; eats central shoots
60	*Ostrinia nubilalis (Pyrausta nubilalis)* [Eastern United States; Europe; Asia]	Larva	Main host, corn; many vegetables, weeds, ornamentals	Bores into stalks & ears, causing breakage & reduced yield & quality
61	*Paleacrita vernata* [Eastern United States; Canada]	Larva	Fruit & shade trees	Defoliates trees in spring
62	*Pectinophora gossypiella* [Southern United States; South America; Europe; Africa; Asia; Australia]	Larva	Cotton, okra, other Malvaceae	Bores into and feeds on squares & bolls, cutting fiber and reducing yield
63	*Peridroma saucia* [Worldwide]	Larva	Many plants	Cuts down seedlings & transplants
64	*Philaenus spumarius (P. leucophthalmus)* [Eastern United States]	Nymph	Hay crops, legumes	Sap-sucking causes wilting, stunting, & reduced forage yield
65	*Phyllophaga* spp.[8] [North America]	Adult; larva	Most plants	Larva devours roots & underground parts; adult defoliates trees
66	*Phylloxera vitifoliae* [North America; Europe]	Adult; nymph	Grape vines	Root- & leaf-sucking causes galls & eventually death of vines
67	*Pieris rapae* [North America, Europe, Asia, Australia]	Larva	Cabbage, other Brassicaceae	Devours foliage
68	*Planococcus citri (Pseudococcus citri)* [Tropical & subtropical areas]	Adult; nymph	Citrus plants, ornamental plants	Sap-sucking causes plants to die back[4]
69	*Plodia interpunctella* [Worldwide]	Larva	Grain, grain products, dried fruit, nuts	Destroys & webs grain & grain products; infests fruits & nuts
70	*Plutella maculipennis* [Worldwide]	Larva	All Brassicaceae	Eats small holes in outer leaves
71	*Popillia japonica* [Eastern United States; China; Japan]	Adult; larva	Fruit trees, grasses, ornamentals, vegetables	Destroys blossoms, foliage, fruit, & turf
72	*Porthetria dispar* [Northeastern United States; Europe; Asia]	Larva	Most deciduous & evergreen trees & shrubs	Devours leaves

[4] Honeydew formed or excreted. [8] Other important June beetles belong to *Melolontha* spp. and *Polyphylla* spp.

continued

	Species (Synonym) [Distribution]	Destructive Stage	Host	Destructive Activity
73	*Pseudaletia unipuncta* [Worldwide]	Larva	Grains, grasses, some legumes	Devours foliage
74	*Psila rosae* [Northern North America; Europe]	Larva	Carrots, celery, parsley	Bores into and eats fibrous roots
75	*Psylla pyricola* [United States, Europe]	Adult; nymph	Pear	Leaf-sucking causes leaf-drop [4]
76	*Ramosia tipuliformis* [North America, Europe, Asia, Australia]	Larva	Black elder, currant, gooseberry, sumac	Burrows through canes
77	*Reticulitermes flavipes* [Eastern United States]	Adult; nymph	Wood, dead wood, cellulose products	Riddles, weakens, and destroys wood & cellulose materials
78	*Rhagoletis pomonella* [Northeastern & north-central United States]	Larva	Apple, blueberry	Bores into and destroys fruit
79	*Sanninoidea exitiosa* [All peach-growing areas]	Larva	Peach & other stone-fruit trees	Bores into trunk & roots at ground level, girdles tree trunk, and kills tree
80	*Saperda candida* [Eastern United States; Canada]	Larva	Apple, pear, & quince trees	Bores into trunk
81	*Schistocerca gregaria* [2] [North Africa, Arabia, India, Iran]	Adult; nymph	Many plants	Chews leaves
82	*Sitophilus oryza* [Worldwide]	Adult; larva	Stored grains; cereal products	Larva grows in kernels and destroys stored grains
83	*Sminthurus viridis* [Europe, Australia]	Adult; immature	Legumes	Surface feeding causes scorching of leaves
84	*Spissistilus festinus* [Worldwide]	Adult	Alfalfa	Stunts plant
85	*Tenebrio molitor* [Worldwide]	Larva	Grain products, refuse	Destroys grain & grain products
86	*Tenebroides mauritanicus* [Worldwide]	Adult; larva	Stored grain, grain products	Infests & destroys grain & grain products
87	*Thermobia domestica* [Worldwide]	Adult; nymph	Starchy substances	Devours bookbindings, fabrics, & wallpaper
88	*Thrips tabaci* [North & South America, Europe, South Africa, Asia, Australia]	Adult; larva; nymph	Bean, cabbage, cotton, onion, tomato	Sap-sucking causes leaves & buds to pucker and silver
89	*Thyridopteryx ephemeraeformis* [Eastern United States]	Larva	Cedar, other trees; ornamentals, especially arborvitae & junipers	Devours foliage
90	*Trialeurodes vaporariorum* [Worldwide]	Nymph	Most plants	Leaf-sucking causes wilting [4]
91	*Tribolium confusum* [Worldwide]	Adult; larva	Flour, grain products	Infests and contaminates flour, mixes, & bread
	Symphyla			
92	*Scutigerella immaculata* [North & South America, Europe, Africa]	Adult; immature	Vegetables & ornamentals	Chews tender plants & rootlets
	Diplopoda			
93	*Julus heserus* [Worldwide]	Adult; immature	Ornamentals, vegetables	Chews young stems & roots

[4] Honeydew formed or excreted. [2] Has a migratory phase.

Contributors: Allen, William W.; Pless, Charles D.

continued

General References

[1] Baker, W. L. 1972. U.S. Dep. Agr. Misc. Publ. 1175.

[2] Davidson, R. H., and L. M. Peairs. 1966. Insect Pests of Farm, Garden and Orchard. Ed. 6. J. Wiley, New York.

[3] Essig, E. O. 1958. Insects and Mites of Western North America. Macmillan, New York.

[4] Imms, A. D., et al. 1960. A General Textbook of Entomology. Ed. 9. E. P. Dutton, New York.

[5] Little, W. A. 1963. General and Applied Entomology. Ed. 2. Harper and Row, New York.

[6] Mallis, A. 1969. Handbook of Pest Control. Ed. 5. MacNair-Dorland, New York.

[7] Metcalf, C. L., and W. P. Flint. 1962. Destructive and Useful Insects, Their Habits and Control. Ed. 4. McGraw-Hill, New York.

[8] Pfadt, R. E. 1971. Fundamentals of Applied Entomology. Ed. 2. Macmillan, New York.

[9] U.S. Department of Agriculture. 1952. Insects. Yearbook of Agriculture. U.S. Government Printing Office, Washington, D. C.

124. NEMATODE PARASITES: PLANTS

Most of the nematode parasites of plants are found in close association with the roots or in the upper 16 inches of soil formerly occupied by the roots. In general, plant nematodes can be distinguished from saprophagous or predacious forms, also found in the soil, by the presence of a protrusile spear or stylet used to puncture and feed on plant cells. The soil is not the only habitat for nematodes; some live in freshwater rivers, lakes, and ponds, others live only in the ocean, and many are parasites of man and other animals. (For information on nematode parasites of mammals and birds, *see* Tables 239 and 230.) **Distribution:** Information on geographic distribution of plant parasitic nematodes is fragmentary and incomplete, even for the best-known species; undoubtedly, distribution is far wider than indicated. **Feeding Habits:** Feeding habits and particular tissues attacked vary with the species, host plant, and stage of development of both host and parasite. **Host:** Species of nematodes within a given genus vary in ability to attack plants; some have a rather wide host range, while others are highly host-specific, attacking only one or two crop plants. **Effect on Host:** Symptoms of nematode damage are often difficult to distinguish from those caused by other organisms or by poor growing conditions; hence, it is important in making a diagnosis to find the nematode in the diseased tissue or soil adjacent to the roots of the affected plant. Data in brackets refer to the column heading in brackets.

	Species [Distribution]	Feeding Habits	Host	Effect on Host [Control]	Reference
1	*Anguina* spp. [North America, Europe, South Africa]	Larva: ectoparasite around growing point. Adult: endoparasite of flower primordia & leaves.	Several *Agrostis* spp., other grasses & cereals	Abnormal flowers which develop into galls & leaf galls [Crop rotation; planting of gall-free seed]	24,27
2	*A. tritici* [Southern Atlantic states of United States; Europe; Egypt; southern & eastern Asia; Australia]	Larva: ectoparasite around growing point. Adult: endoparasite of flower primordia.	Emmer, rye, spelt, wheat	Stunted plants, distorted foliage, & galls instead of seed [Planting of gall-free seed (galls may be removed by salt-brine floatation or mechanical separators)]	21,38
3	*Aphelenchoides besseyi* [Southeastern United States (Maryland to Texas), Union of Soviet Socialist Republics, Africa, India, Bangladesh, Ceylon, Taiwan, Japan, Indonesia]	Vagrant ectoparasite of buds & growing point between young developing leaves	Rice, strawberry	Small, crinkled, distorted foliage on strawberry, and white tips on rice leaves [Setting of new beds with uninfested plants; hot-water treatment or methyl bromide fumigation for rice seed]	16,25
4	*A. fragariae* [Massachusetts, Connecticut, Delaware, & Maryland (United States); Europe]	Vagrant ectoparasite of buds between young developing leaves	Strawberry	Small, crinkled, distorted foliage [Setting of new beds with uninfested plants]	4,16

continued

	Species [Distribution]	Feeding Habits	Host	Effect on Host [Control]	Reference
5	*A. ritzema-bosi* [North America, Europe]	Vagrant endoparasite of buds & foliage	About 50 plants, including chrysanthemum, larkspur, phlox, strawberry, verbena, & zinnia	Crinkled, distorted leaves & leaf spots [Hot-water treatment of dormant plants; parathion sprays]	26,28, 30,54
6	*Belonolaimus longicaudatus* [Southern Atlantic states of United States]	Vagrant ectoparasite of root tips, sides of succulent roots, & other underground parts	Bean, beet, cabbage, celery, citrus, corn, cotton, cowpea, grass, millet, okra, onion, peanut, pine seedling, soybean, strawberry	Devitalized root tips & root lesions, causing many short, stubby-branched roots & severely stunted plants [Soil fumigation]	18,20, 35,43
7	*Criconemoides* spp. [Widespread]	Semi-sedentary ectoparasite of roots & other underground parts	Many plants; reported as injuring peach trees & peanut roots, pods, & pegs	Small lesions, stunting of plant [Soil fumigation]	14,41, 44
8	*Ditylenchus destructor* [Idaho & Wisconsin (United States), Prince Edward Island (Canada), Europe]	Vagrant endoparasite of tubers and to some extent of roots	Carrot, iris, potato, sweet potato, tulip	Destruction of tuber tissues, causing sunken areas, followed by rot [Crop rotation; planting of clean seed; soil fumigation]	3,61
9	*D. dipsaci* [Widespread in temperate zones]	Vagrant endoparasite of leaves, stems, bulbs, occasionally roots	Over 300 plants, including alfalfa, clover, hyacinth, iris, narcissus, oats, onion, phlox, & potato	Twisting, wrinkling, & distortion of stems & flowers, and necrosis & destruction of bulb tissues [Hot-water treatment of bulbs & corms; crop rotation; field sanitation; methyl bromide fumigation of infected onion & clover seeds; planting of resistant varieties]	23,46, 56
10	*Dolichodorus heterocephalus* [Florida, Georgia, North Carolina, & Michigan (United States)]	Vagrant ectoparasite of root tips & sides of succulent roots	Bean, celery, Chinese water chestnut, corn, tomato, many other plants growing in wet locations	Devitalized root tips, small lesions on sides of roots; sometimes extensive root destruction [Soil fumigation]	48,58
11	*Helicotylenchus* spp.[1] [Widespread in subtropical & tropical regions]	Vagrant ectoparasite; occasionally endoparasite of roots & other underground parts	Many plants, including banana, bean, cotton, cowpea, grass, pineapple, soybean, & ornamentals	Stunting of plant from retarded root growth; lesions may occur [Soil fumigation]	8,33, 45,55
12	*Heterodera glycines* [Midwestern & southern United States; China; Japan]	Sedentary parasite of roots; internal in early stages, external as adult	Adzuki bean, annual lespedeza, kidney bean, snap bean, soybean, vetch	General stunting of plant, decrease in size of root system; causes disease known as "yellow dwarf" in China & Japan [Crop rotation; soil fumigation; planting of resistant varieties]	10,36, 66
13	*H. rostochiensis* [Long Island & Steuben County, New York (United States); Newfoundland (Canada); Panama; Peru; Bolivia; Chile; Argentina; Europe; Union of Soviet Socialist Republics; Canary Islands; Algeria; India]	Sedentary parasite; internal in early stages, becoming largely external as adult; attacks roots & other underground parts	Potato, tomato, several other Solanaceae	Stunting of plant, decrease in size of root system; often causes increase in number of small branch rootlets [Crop rotation; soil fumigation; planting of resistant varieties]	15,42

[1] Information also valid for *Rotylenchus* spp. and *Scutellonema* spp.

continued

Species [Distribution]	Feeding Habits	Host	Effect on Host [Control]	Reference
14 *H. schachtii* [United States, Canada, Europe, Australia]	Sedentary parasite of roots & other underground parts; internal in early stages; external as adult	Over 100 plants, including broccoli, cabbage, cauliflower, kale, mangel-wurzel, mustard, rutabaga, sugar beet, table beet, & turnip	Stunting of plant, overall decrease in size of root system; often increase in number of small branch rootlets [Crop rotation; soil fumigation with dichloropropene-dichloropropane mixture before planting for sugar beet]	22,31, 49,62
15 *Hirschmanniella oryzae* [2] [Louisiana & Texas (United States), Venezuela, Nigeria, Japan, Indonesia, rice-growing areas of southeastern Asia]	Vagrant endoparasite of roots	Rice, various grasses, related Liliopsida	Root lesions, destruction of cortex & root hairs; in Indonesia, associated with "mentek," a rice root rot [Crop rotation, prolonged dry fallow of rice paddies]	1,59, 64,65
16 *Hoplolaimus galeatus* [North America, Europe, Philippines]	Vagrant internal, or partly external, parasite of roots	Many plants, including corn, cotton, pine tree, sugarcane, & St. Augustine &other lawn grasses	Lesions leading to complete destruction and sloughing off of cortex [Soil fumigation]	37
17 *Meloidogyne* spp. [Worldwide; most common in warm climates]	Sedentary endoparasite of roots & other underground parts	Over 2000 plants; hosts of individual species more restricted	Swellings, galls; often local necrosis of tissues; also causes increase, or reduction, of branch rootlets [Annual crops: rotation & fumigation; planting of resistant varieties; hot-water treatment of bulbs, corms, & tubers]	12,13, 32
18 *Paratylenchus* spp. [North America; British Isles; Netherlands; western Africa; Hawaii]	Vagrant ectoparasite of roots & other underground parts	Many plants, including alfalfa, cabbage, celery, cowpea, cucumber, fig tree,oats, okra, pineapple, radish, & wheat	Stunting of plants from root injury & retarded root growth [Fumigation somewhat effective]	39,40, 63
19 *Pratylenchus* spp. [Worldwide]	Vagrant endoparasite of roots & tubers [3]	Many plants, including alfalfa, corn, cotton, small grains, strawberry, tobacco, trees, & shrubs	Small brown root lesions; "brown root rot" of tobacco [Crop rotation for tobacco; row fumigation with dichloropropene-dichloropropane mixture]	53
20 *Radopholus similis* [Florida & Louisiana (United States), West Indies, Central America, Brazil, Peru, India, Taiwan (Formosa), Philippines, Indonesia, Hawaii]	Vagrant endoparasite of roots	About 50 plants, including avocado, banana, black pepper, canna, citrus, coffee, rice, sugarcane, & tea	Root lesions & disintegration [Hot-water treatment of infected citrus nursery stock; pulling of affected trees, followed by soil fumigation]	5,9,11, 57
21 *Rhadinaphelenchus cocophilus* [4] [West Indies, Honduras, Panama, Colombia, Venezuela, British Guiana]	Vagrant endoparasite of leaf petioles, trunk (near periphery), & roots	Coconut, date, & oil palms	Disintegration of trunk tissues (causing "red ring") and of root cortex [Phytosanitation]	7,29, 47

[2] Synonym: *Radopholus oryzae.* [3] All *Pratylenchus* spp. are root parasites except *P. mahogani* and *P. scribneri* observed, respectively, in diseased mahogany wood and potato tubers. [4] Synonym: *Aphelenchoides cocophilus.*

continued

Species [Distribution]	Feeding Habits	Host	Effect on Host [Control]	Reference
22	*Rotylenchulus reniformis* [Florida, Georgia, & Louisiana (United States); Puerto Rico; Colombia; Peru; Venezuela; Africa; Pakistan; Hawaii]	♀: sedentary, partly external parasite of roots	Many plants, including bean, beet, carrot, corn, cotton, cowpea, cucumber, eggplant, lettuce, okra, pea, pineapple, radish, & squash	Extensive root necrosis; collapse of cortical cells surrounding feeding sites [Soil fumigation; crop rotation]	6,51
23	*Trichodorus* spp. [Widespread; important in southern California & southeastern United States, Nicaragua, & Tunisia]	Vagrant ectoparasite of root tips	Many plants, including beet, cabbage, cauliflower, celery, chayote, corn, cotton, & fig	Devitalized root tips, causing numerous short, stubby branch rootlets [Soil fumigation]	19
24	*Tylenchorhynchus* spp. [Widespread]	Mostly external, occasionally internal, vagrant parasite of roots	Many plants, including azalea, cotton, oats, sugarcane, tobacco, & wheat	Stunting of plant [Soil fumigation]	34,50
25	*Tylenchulus semipenetrans* [Most citrus fruit-growing regions; Florida, Texas, & California, (United States); southern Europe]	♀: sedentary, partly external parasite of roots	Most *Citrus* species & closely related genera; olive	Extensive necrosis, discoloration of cortex of small roots [Planting of uninfected stock on clean land]	2,60
26	*Xiphinema* spp. [Worldwide]	Vagrant ectoparasite of root tips & sides of succulent roots	Many plants, shrubs, & trees, including clove, corn, laurel oak, oats, pecan, rose, strawberry, & some grasses	Devitalized root tips, necrosis of small roots, gall-like swellings, clusters of small, stubby branches [Soil fumigation]	17,52, 63

Contributors: Sasser, J. N.; Christie, Jesse R.

References

[1] Atkins, J. G., et al. 1955. Plant Dis. Rep. 39(3): 221.

[2] Baines, R. C., et al. 1948. Phytopathology 38(11): 912.

[3] Baker, A. D. 1946. Sci. Agr. 26(3):138.

[4] Ballard, E., and G. S. Peren. 1923. J. Pomol. Hort. Sci. 3:142.

[5] Birchfield, W. 1954. Proc. Fla. State Hort. Soc. 67: 94.

[6] Birchfield, W. 1962. Phytopathology 52:862.

[7] Blair, G. P. 1969. Commonw. Bur. Helminthol. (Gt. Brit.) Tech. Commun. 40:99.

[8] Blake, C. D. 1969. Ibid. 40:109.

[9] Bragdon, K. E., and R. W. Hanks. 1954. Proc. Fla. State Hort. Soc. 67:83.

[10] Brim, C. A., and J. P. Ross. 1966. Phytopathology 56:451.

continued

125. HELMINTH AND PROTOZOAN

For information on helminth and protozoan

Part I.

Species	Distribution	Reservoir Host of Definitive Stage	Vector, or Obligate Host Other than Man	Infective Stage
		Nematoda		
1 *Ancylostoma braziliense*	Limited distribution in warm climates	Cat, dog	None	3rd-stage larva

1/ By direct or indirect contact with body excreta containing parasite.

[11] Brooks, T. L. 1954. Proc. Fla. State Hort. Soc. 67: 81.

[12] Buhrer, E. M. 1938. Plant Dis. Rep. 22(12):216.

[13] Buhrer, E. M., et al. 1933. Ibid. 17(7):64.

[14] Chitwood, B. G. 1949. Proc. Helminthol. Soc. Wash. D.C. 16(1):6.

[15] Chitwood, B. G., and E. M. Buhrer. 1946. Phytopathology 36(3):180.

[16] Christie, J. R. 1943. U.S. Dep. Agr. Circ. 681.

[17] Christie, J. R. 1952. Proc. Soil Sci. Soc. Fla. 12:30.

[18] Christie, J. R. 1953. Down Earth 9(1):8.

[19] Christie, J. R., and V. G. Perry. 1951. Science 113: 491.

[20] Christie, J. R., et al. 1952. Phytopathology 42(4): 173.

[21] Chu, V. M. 1945. Ibid. 35(5):288.

[22] Corder, M. N., et al. 1936. Plant Dis. Rep. 20(3):38.

[23] Courtney, W. D. 1948. Proc. Bulb Grow. Short Course, p. 7.

[24] Courtney, W. D., and H. B. Howell. 1952. Plant Dis. Rep. 36(3):75.

[25] Cralley, E. M. 1952. Arkansas Farm Res. 1(1):5.

[26] Crossman, L., and J. R. Christie. 1936. Plant Dis. Rep. 20(10):155.

[27] Crossman, L., and J. R. Christie. 1937. Ibid. 21(9): 144.

[28] Dimock, A. W., and C. H. Ford. 1950. Phytopathology 40(1):7.

[29] Fenwick, D. W. 1969. Commonw. Bur. Helminthol. (Gt. Brit.) Tech. Commun. 40:89.

[30] Franklin, M. T. 1950. Ann. Appl. Biol. 37(1):1.

[31] Franklin, M. T. 1951. The Cyst-Forming Species of *Heterodera.* Commonwealth Bureau of Agricultural Parasitology, Farnham Royal, England.

[32] Garriss, H. R. 1953. N.C. Agr. Ext. Serv. Circ. 374.

[33] Golden, A. M. 1954. Phytopathology 44(7):389.

[34] Graham, T. W. 1954. Ibid. 44(6):332.

[35] Holderman, Q. L. 1955. Plant Dis. Rep. 39(1):5.

[36] Ichinohe, M. 1955. Hokkaido Nogyo Shikenjo Hokoku 48:1.

[37] Krusberg, L. R., and J. N. Sasser. 1955. Proc. Annu. Conv. Ass. S. Agr. Work., 52nd, p. 143.

[38] Leukel, R. W. 1929. U.S. Dep. Agr. Farmers Bull. 1607.

[39] Linford, M. B., et al. 1949. Pac. Sci. 3(2):111.

[40] Lownsbery, B. F., et al. 1952. Phytopathology 42(12):651.

[41] Machmer, J. H. 1953. Plant Dis. Rep. 37(3):156.

[42] Mai, W. F., and B. Lear. 1953. Cornell Univ. Agr. Ext. Bull. 870.

[43] Miller, L. I. 1952. Phytopathology 42(9):470.

[44] Minton, N. A., and D. K. Bell. 1969. J. Nematol. 1:349.

[45] Minz, G., et al. 1960. Ktavim 10(3/4):147.

[46] National Institute of Agricultural Botany, Seed Production Committee. 1951. Nat. Inst. Agr. Bot. (Gt. Brit.) Seed Notes 38.

[47] Nowell, W. 1919. West Indian Bull. 17(4):189.

[48] Perry, V. G. 1953. Proc. Helminthol. Soc. Wash. D.C. 20(1):21.

[49] Raski, D. J. 1950. Phytopathology 40(2):135.

[50] Reynolds, H. W., and M. M. Evans. 1953. Plant Dis. Rep. 37(11):540.

[51] Roman, J. 1964. J. Agr. Univ. P. R. 48:162.

[52] Schindler, A. F. 1954. Phytopathology 44(7):389.

[53] Sher, S. A., and M. W. Allen. 1953. Univ. Calif. Berkeley Publ. Zool. 57(6):441.

[54] Staniland, L. N. 1947. Agriculture (London) 54(6): 278.

[55] Steiner, G. 1938. J. Agr. Res. 56(1):1.

[56] Steiner, G., and E. M. Buhrer. 1932. Plant Dis. Rep. 16(8):76.

[57] Suit, R. F. 1954. Proc. Fla. State Hort. Soc. 67:85.

[58] Tarjan, A. C. 1952. Phytopathology 42(2):114.

[59] Taylor, A. L. 1969. Commonw. Bur. Helminthol. (Gt. Brit.) Tech. Commun. 40:264.

[60] Thomas, E. E. 1923. Calif. Agr. Exp. Sta. Tech. Pap. 2.

[61] Thorne, G. 1945. Proc. Helminthol. Soc. Wash. D.C. 12(2):27.

[62] Thorne, G. 1952. U.S. Dep. Agr. Farmers Bull. 2054.

[63] Thorne, G., and M. W. Allen. 1950. Proc. Helminthol. Soc. Wash. D.C. 17(1):27.

[64] Vecht, J. van der. 1953. Contrib. Gen. Agr. Res. Sta. (Bogor) 137.

[65] Vecht, J. van der, and B. H. H. Bergman. 1952. Ibid. 131.

[66] Winstead, N. N., et al. 1955. Plant Dis. Rep. 39(1): 9.

PARASITES: MAMMALS AND BIRDS

parasites of laboratory animals, *see* Table 121.

Man

Portal of Infection	Parasitism in Man			Identification of Parasite	
	Immature Stage	Definitive Stages			
		Primary Site	Secondary Site		
Nematoda					
Skin [1/]	Larva migrates under skin	In tunnels in skin	None	Larvae in cutaneous tunnels	1

continued

	Species	Distribution	Reservoir Host of Definitive Stage	Vector, or Obligate Host Other than Man	Infective Stage
2	A. duodenale	Tropical & subtropical United States, Europe, Africa, & Asia; western South America	None	None	3rd-stage larva
3	Ascaris lumbricoides	Worldwide; more common in warm climates	None	None	Fully embryonated egg
4	Brugia malayi	Warm climates in Asia	Cat, monkey	Armigeres; Mansonia	3rd-stage larva
5	Dracunculus medinensis	Warm climates of eastern hemisphere	Fur-bearing mammals	Cyclops	3rd-stage larva in Cyclops
6	Enterobius vermicularis	Worldwide; common in children	None	None	Fully embryonated egg
7	Loa loa	Tropical Africa	None	Chrysops	3rd-stage larva
8	Necator americanus	Warm climates	None	None	3rd-stage larva
9	Onchocerca volvulus	Mexico; Guatemala; eastern Venezuela; Surinam (Dutch Guiana) (?); tropical Africa	None	Simulium	3rd-stage larva
10	Strongyloides stercoralis	Warm, moist climates	Chimpanzee, dog	None	3rd-stage larva
11	Trichinella spiralis	Worldwide; common in United States	Bear, swine, walrus	None	Larva encysted in swine muscle
12	Trichuris trichiura	Warm, moist climates	Ape, monkey	None	Fully embryonated egg
13	Wuchereria bancrofti	Warm climates	None	Aedes, Culex, & other mosquito genera	3rd-stage larva
	Platyhelminthes—Cestoda				
14	Diphyllobothrium latum	North temperate & subarctic zones; lakes in Argentina & Chile	Bear, cat, dog	Freshwater fish; Cyclops; Diaptomus	Sparganum larva in fish flesh
15	Echinococcus granulosus	Worldwide; common in southern South America	Dog, wild Canidae	Cattle, sheep, & swine, alternating with dog	Egg in excreta of dog
16	E. multilocularis	Northern temperate zones	Wild Canidae	Small rodents	Egg in excreta of wild Canidae
17	Hymenolepis diminuta	Warm & temperate climates	Mouse, rat	Grain beetle, meal moth, rodent flea	Larva in hemocoel of insect
18	H. nana	Warm & temperate climates	Mouse, rat	None; may develop in grain beetle	Egg
19	Taenia saginata	Worldwide	None	Cattle	Cysticercus larva in beef
20	T. solium	Worldwide	None	Swine	Cysticercus larva in pork; egg

1/ By direct or indirect contact with body excreta containing parasite. 2/ From proboscis of insect vector at time of skin puncture to obtain blood or tissue juice from host. 3/ From infected food or contaminated water tak-

Man

Parasitism in Man				Identification of Parasite	
Portal of Infection	Immature Stage	Definitive Stages			
		Primary Site	Secondary Site		
Skin [1]	Larva migrates from skin via blood & lungs to epiglottis & gastrointestinal tract	Attached to small intestine	None	Egg in feces	2
Mouth [1]	Larva migrates	Lumen of small intestine	Various viscera	Egg in feces	3
Skin [2]	Larva migrates in lymphatics	Lymphatics of lower trunk	Lymphatics of upper trunk	Microfilaria (sheathed) in peripheral blood (nocturnal & subperiodic)	4
Mouth [3]	In viscera	Gravid ♀ migrates to skin	None known	Gravid ♀ in ruptured skin blister	5
Mouth [1]	In transit down small intestine	Cecum & vermiform appendix	♀ genital tract; perianal folds	Egg or adult in anal swab or anus	6
Skin [2]	Migrates in subcutaneous tissues	Migrates in subcutaneous tissues	Orbit & conjunctiva of eye	Microfilaria (sheathed) in blood (diurnal)	7
Skin [1]	Larva migrates	Attached to small intestine	None	Egg in feces	8
Skin [2]	Larva in skin (may invade eye tissues)	Adult in subcutaneous nodules; larva in skin (may invade eye tissues)	Eye tissues	Microfilaria (sheathed) in skin biopsy	9
Skin [1]	Larva migrates	Within intestinal mucosa	Lungs	Larva in feces or duodenal aspirate	10
Mouth [3]	Enters duodenal mucosa	In duodenal mucosa	Larva migrates & encysts in striated muscle	Larva in compressed or digested muscle	11
Mouth [1]	In transit down small intestine	Attached to cecum & vermiform appendix	Colon, rectum	Egg in feces	12
Skin [2]	Larva migrates in lymphatics	Lymphatics of lower trunk & legs	Lymphatics of upper trunk	Microfilaria (sheathed) in blood (usually nocturnal)	13
Platyhelminthes—Cestoda					
Mouth	Develops in small intestine	Attached to small intestine	None known	Egg in feces	14
Mouth	Develops in liver & lungs	Hydatid cysts in viscera	None	Hydatid cyst with scolex during aspiration or exploratory operation	15
Mouth	Develops unconfined in liver	Hydatid cysts in liver	None	Hydatid cyst with scolex during postmortem examination	16
Mouth	Develops in duodenum & rest of small intestine	Attached to small intestine	None	Proglottid or egg in feces	17
Mouth	Develops in duodenal villi	Attached to small intestine	None known	Egg in feces	18
Mouth	Develops in small intestine	Attached to small intestine	None known	Proglottid or egg in feces	19
Mouth	Develops in small intestine	Attached to small intestine	Cysticercus larva in various organs & tissues	Proglottid or egg in feces	20

en into mouth.

continued

	Species	Distribution	Reservoir Host of Definitive Stage	Vector, or Obligate Host Other than Man	Infective Stage
			Platyhelminthes—Trematoda		
21	*Clonorchis sinensis*	Sino-Japanese & Indo-Chinese areas	Many fish-eating mammals	Freshwater fishes; snail	Larva encysted in flesh of freshwater fish
22	*Fasciola hepatica*	Sheep-raising countries	Herbivores	Snail, moist vegetation	Larva encysted on water plants
23	*Fasciolopsis buski*	Oriental countries	Swine	Snail, water plants	Larva encysted on water plants
24	*Paragonimus kellicotti, P. westermanii*	Northern South America; southwest Pacific islands; Sino-Japanese areas	Cat, dog, swine, other animals	Crab, crayfish, snail, sputum of man	Larva encysted in soft tissues of crabs & crayfish
25	*Schistosoma haematobium*	Southern Portugal, Africa, Near East, Middle East	Baboon, gerbil	Bulinid snail	Cercaria free in freshwater
26	*S. japonicum*	China, Taiwan (Formosa), Japan, Philippines; Celebes	Many mammals	Oncomelaniid snail	Cercaria free in freshwater
27	*S. mansoni*	West Indies, Venezuela, Guianas, Brazil, Africa, Arabia	Baboon; rarely monkey	Planorbid snail	Cercaria free in freshwater
			Protozoa		
28	*Balantidium coli*	Worldwide; most common in warm climates	Monkey (?), swine	None	Mature cyst
29	*Entamoeba histolytica*	Worldwide; most common in warm climates	Dog; monkey; rat (?)	None	Four-nucleate cyst
30	*Giardia lamblia*	Worldwide; most common in warm climates	None (?)	None	Four-nucleate cyst
31	*Leishmania brasiliensis*	Western hemisphere, from southern Mexico to northern Argentina	Dog; possibly other mammals	*Phlebotomus*	Promastigote
32	*L. donovani*	South America, Mediterranean area, Africa, India, China	Dog, rodents	*Phlebotomus*	Promastigote
33	*L. tropica*	North Africa, Near East, Middle East, western India	Dog, rodents	*Phlebotomus*	Promastigote
34	*Plasmodium falciparum; P. malariae; P. vivax*	Temperate or warm climates	None	*Anopheles*	Sporozoite
35	*Toxoplasma gondii*	Worldwide	Many mammals & birds	None known	Trophozoite; oocyst
36	*Trichomonas vaginalis*	Worldwide; relatively common in ♂ & ♀	None	None	Trophozoite (only stage known)
37	*Trypanosoma cruzi*	Western hemisphere from United States to northern Argentina	Many mammals	Triatomid bug	Metacyclic trypanosome

[1] By direct or indirect contact with body excreta containing parasite. [2] From proboscis of insect vector at time of skin puncture to obtain blood or tissue juice from host. [3] From infected food or contaminated water taken into

Man

Portal of Infection	Parasitism in Man			Identification of Parasite	
	Immature Stage	Definitive Stages			
		Primary Site	Secondary Site		
Platyhelminthes—Trematoda					
Mouth[3]	In transit from duodenum to bile ducts	Distal bile ducts	Pancreatic ducts (rare)	Egg in feces	21
Mouth[3]	In transit from duodenum to bile ducts	Proximal bile ducts	Brain, lungs; abdominal wall (?)	Egg in feces	22
Mouth[3]	Develops in duodenum & jejunum	Attached to duodenum & jejunum	None	Egg in feces	23
Mouth[3]	In transit from duodenum to lungs	Lungs, near bronchioles	Abdominal viscera, brain	Egg in sputum or feces	24
Skin[4]	Migrates in blood vessels	Vesical venous plexus	CNS, lungs, pelvic organs, rectum	Egg in urine or feces	25
Skin[4]	Migrates in blood vessels	Mesenteric venules	Brain, lungs, liver	Egg in feces	26
Skin[4]	Migrates in blood vessels	Mesenteric venules	Brain, lungs, liver	Egg in feces	27
Protozoa					
Mouth[1]	None described	Large intestine	None	Trophozoite or cyst in feces	28
Mouth[1]	None described	Large intestine	Other viscera, skin	Trophozoite or cyst in feces, visceral abscesses, or skin abscesses	29
Mouth[1]	None described	Duodenal crypts	Gallbladder (?)	Trophozoite or cyst in feces	30
Skin[2]	None described	Skin	Mucous membranes	Amastigote stage in reticuloendothelial cells, skin, or viscera; promastigote stage in culture	31
Skin[2]	None described	Skin	Reticuloendothelium (fundamental)	Amastigote stage in reticuloendothelial cells, skin, or viscera; promastigote stage in culture	32
Skin[2]	None described	Skin	Mucous membranes (rare)	Amastigote stage in reticuloendothelial cells, skin, or viscera; promastigote stage in culture	33
Skin[2]	Schizont in hepatic parenchyma	Exoerythrocytic foci	Erythrocytes	Trophozoite, schizont, or gametocyte in blood	34
Unknown	None known	Reticuloendothelium; many parenchymal cells	Brain; retina	Trophozoite, pseudocyst, or cyst in focal areas of necrosis	35
Vulva[1]; urethra	None described	Vaginal fold	Bladder	Trophozoite in urine or vaginal smear	36
Skin[5]; conjunctiva	None described	Skin	In tissues & blood	Trypomastigote stage in tissues or blood; amastigote stage in macrophage; epimastigote stage in culture	37

mouth. [4] In contact with infested water. [5] From feces of insect vector while feeding on blood or tissue juice of host.

continued

	Species	Distribution	Reservoir Host of Definitive Stage	Vector, or Obligate Host Other than Man	Infective Stage
38	*T. gambiense*	Western & central Africa	Cattle (?)	*Glossina*	Metacyclic trypanosome
39	*T. rhodesiense*	Central & eastern Africa	Mammals, wild game	*Glossina*	Metacyclic trypanosome

2/ From proboscis of insect vector at time of skin puncture to obtain blood or tissue juice from host.

Contributor: Faust, Ernest Carroll

General References

[1] Faust, E. C., et al. 1968. Animal Agents and Vectors of Human Disease. Ed. 3. Lea and Febiger, Philadelphia.

[2] Faust, E. C., et al. 1970. Craig and Faust's Clinical Parasitology. Ed. 8. Lea and Febiger, Philadelphia.

Part II. Homoiothermic Animals Other Than Man

Data in brackets refer to the column heading in brackets.

	Species [Distribution]	Intermediate Host [Reservoir Host]	Definitive Host [Primary Location in Definitive Host]	Effect on Host
	Acanthocephala			
1	*Macracanthorhynchus hirudinaceus* [Worldwide]	*Cotinis; Phyllophaga*	Swine [Small intestine]	Formation of nodules
	Nematoda			
2	*Ancylostoma caninum* [Worldwide]	None	Coyote, dog, fox; also man [Small intestine]	Anemia, emaciation, skin reactions
3	*A. tubaeforme* [Worldwide]	None	Cat [Small intestine]	Anemia, emaciation, skin reactions
4	*Ascaridia galli* [Worldwide]	None	Chicken, goose, guinea fowl, turkey, wild birds [Small intestine]	Emaciation
5	*Ascaris suum* [1] [Worldwide]	None	Swine; also man [Small intestine]	Pneumonia, abdominal discomfort & obstruction, emaciation
6	*Dictyocaulus filaria* [Worldwide]	None	Goat, sheep, some wild ruminants [Bronchi, bronchioles]	Catarrhal inflammation, cough, emaciation
7	*D. viviparus* [Worldwide]	None	Cattle, deer [Bronchi, bronchioles]	Catarrhal inflammation, cough, emaciation
8	*Dirofilaria immitis* [Worldwide]	*Aedes; Anopheles; Culex*	Cat, coyote, dog, fox, wolf [Heart, pulmonary artery; microfilaria in blood]	Emaciation, cough, edema, dyspnea
9	*Haemonchus contortus* [Worldwide]	None	Goat, sheep, other ruminants [Abomasum]	Anemia, emaciation

continued

Man

| Parasitism in Man | | | | Identification of Parasite | |
| Portal of Infection | Immature Stage | Definitive Stages | | | |
		Primary Site	Secondary Site		
Skin[2/]	None described	Skin	CNS, blood, lymph nodes	Trypomastigote stage in blood, gland juice, or spinal fluid	38
Skin[2/]	None described	Skin	CNS, blood, lymph nodes	Trypomastigote stage in blood, gland juice, or spinal fluid	39

Part II. Homoiothermic Animals Other Than Man

	Species [Distribution]	Intermediate Host [Reservoir Host]	Definitive Host [Primary Location in Definitive Host]	Effect on Host
10	*H. placei* [Worldwide]	None	Cattle [Abomasum]	Anemia, emaciation
11	*Heterakis gallinarum* [Worldwide]	None	Chicken, guinea fowl, pheasant, quail, turkey, other birds [Cecum]	None. Egg carries *Histomonas*.
12	*Metastrongylus apri* [Worldwide]	*Eisenia; Lumbricus;* several other genera of earthworms	Swine [Bronchi, bronchioles]	Bronchitis, pneumonia. Vector of swine influenza virus.
13	*Oesophagostomum columbianum* [Worldwide]	None	Antelope, goat, sheep [Large intestine; larva in nodules throughout intestine]	Diarrhea, emaciation, nodules in the intestine
14	*Ostertagia circumcincta* [Worldwide]	None	Goat, sheep [Abomasum]	Anemia, emaciation
15	*O. ostertagi* [Worldwide]	None	Cattle, sheep; rarely horse [Abomasum]	Anemia, edema, emaciation
16	*Parascaris equorum* [Worldwide]	None	Horse, other Equidae [Small intestine]	Pneumonia, digestive disturbances, emaciation
17	*Strongyloides stercoralis* [Worldwide in warm climates]	None	Cat, dog, fox; also man [Mucosa of small intestine]	Diarrhea
18	*Strongylus vulgaris* [Worldwide]	None	Horse, other Equidae [Large intestine]	Anemia, edema, digestive disturbances, emaciation; larva forms aneurysms in anterior mesenteric arteries
19	*Toxascaris leonina* [Worldwide]	None	Cat, dog, fox, wild Canidae & Felidae [Small intestine]	Emaciation, poor growth
20	*Toxocara canis* [Worldwide]	None	Coyote, dog, fox [Small intestine]	Emaciation, poor growth

continued

Part II. Homoiothermic Animals Other Than Man

	Species [Distribution]	Intermediate Host [Reservoir Host]	Definitive Host [Primary Location in Definitive Host]	Effect on Host
21	*Trichinella spiralis* [Worldwide]	Same individual is both definitive & intermediate host [Swine, for infection of man]	Badger, rat, swine, many other mammals including man [Small intestine; larva in muscles]	Trichinosis, toxemia, muscle pains
22	*Trichostrongylus axei* [Worldwide]	None	Cattle, deer, goat, horse, sheep, swine; also man [Abomasum]	Emaciation
23	*T. colubriformis* [Worldwide]	None	Antelope, camel, cattle, goat, sheep [Small intestine]	Emaciation
24	*Trichuris suis* [Worldwide]	None	Swine [Cecum]	Emaciation, toxemia
25	*T. vulpis* [Worldwide]	None	Dog, fox [Cecum]	Emaciation, low-grade inflammation
	Platyhelminthes—Cestoda			
26	*Diphyllobothrium latum*[2/] [North America, Argentina, Chile, Europe, Siberia, Manchuria, Japan, Australia]	First: *Cyclops; Diaptomus.* Second: fish.	Cat, dog, fox, polar bear, other fish-eating mammals; also man [Small intestine]	Toxemia, anemia
27	*Dipylidium caninum* [Worldwide]	*Ctenocephalides; Pulex; Trichodectes*	Cat, dog, fox, wolf, other carnivores [Small intestine]	Enteritis, anal pruritus
28	*Echinococcus granulosus* [North & South America; Iceland; Europe; Africa; northern Asia; Australia]	Camel, cow, dog, goat, horse, monkey, mouse, rabbit, sheep, rodents; also man	Dog, fox, wolf, other Canidae [Small intestine]	Slight (if any) enteritis. Hydatid cysts in liver, lungs, & other organs of intermediate host cause serious damage.
29	*Hymenolepis carioca* [Worldwide]	*Anisotarsus, Aphodius,* & many other beetles; *Stomoxys*	Chicken, quail, turkey [Small intestine]	Slight damage
30	*Moniezia expansa* [Worldwide]	*Galumna; Oribatula;* other grass mites	Cattle, goat, sheep, other ruminants [Small intestine]	Emaciation
31	*Raillietina cesticillus* [Worldwide]	Several genera of ground & dung beetles	Chicken, pheasant, quail, turkey, wild Galliformes [Small intestine]	Slight damage
32	*Taenia pisiformis* [Worldwide]	Hare, rabbit, rat, squirrel	Cat, coyote, dog, fox, wolf [Small intestine]	Slight (if any) enteritis; anal pruritus
	Platyhelminthes—Trematoda			
33	*Fasciola hepatica* [Worldwide]	*Galba*[3/]; *Lymnaea; Pseudosuccinea;* other freshwater snails [Wild rabbit, for infection of ruminants]	Cat, dog, elephant, hare, horse, kangaroo, rabbit, rodents, swine; cattle, goat, sheep, & other ruminants; also man [Proximal bile duct]	Necrosis & cirrhosis of liver, calcification of bile ducts
34	*Nanophyetus salmincola* [Pacific Northwest (United States)]	First: *Goniobasis.* Second: usually *Salmo;* also, *Oncorhynchus, Salvelinus.*	Coyote, dog, fox, lynx, mink, raccoon [Small intestine]	Enteritis. Acts as vector of *Neorickettsia helminthoeca,* the cause of "salmon poisoning."

[2/] Synonym: *Dibothriocephalus latus.* [3/] Synonym: *Fossaria.*

continued

Part II. Homoiothermic Animals Other Than Man

	Species [Distribution]	Intermediate Host [Reservoir Host]	Definitive Host [Primary Location in Definitive Host]	Effect on Host
		Protozoa		
35	*Babesia bigemina* [North, Central, & South America; Europe; Africa; Asia; Australia; Pacific islands]	*Boophilus; Rhipicephalus*	Cattle [Erythrocytes]	Fever, anemia, hemoglobinuria, Texas fever
36	*Balantidium coli* [Worldwide]	None	Monkey, swine; also man [Large intestine]	Secondarily invades mucosa
37	*Eimeria ahsata* [Worldwide]	None	Goat, sheep [Cells of the small intestine]	Diarrhea, emaciation
38	*E. necatrix* [Worldwide]	None	Chicken [Cells of the small intestine]	Hemorrhagic enteritis
39	*E. tenella* [Worldwide]	None	Chicken [Cecal cells]	Hemorrhagic enteritis
40	*E. zuernii* [Worldwide]	None	Cattle [Intestinal cells]	Enteritis, hemorrhagic dysentery
41	*Histomonas meleagridis* [Worldwide]	None[4] [Chicken & wild Galliformes, for infection of turkey]	Chicken, partridge, peafowl, pheasant, quail, ruffed grouse [Cecum, liver]	Enterohepatitis, necrosis, ulceration
42	*Iodamoeba buetschlii* [Worldwide]	None	Monkey; swine; also man [Large intestine]	None
43	*Isospora bigemina* [Worldwide]	None	Cat, dog, fox, mink [Cells of the small intestine]	Diarrhea
44	*Leishmania donovani* [Central & South America, Balkan states, Russia, Mediterranean basin, Africa, Near East, India, China]	*Phlebotomus* [Dog, for infection of man]	Dog; also man [Reticulo-endothelial system]	Kala azar, reticuloendotheliosis, splenomegaly
45	*L. tropica* [Russia, Mediterranean basin, Africa, Near East, India]	*Phlebotomus* [Dog, gerbil, & *Rhombomys*, for infection of man]	Dog, gerbil; also man [Cutaneous tissues]	Oriental sore; skin ulcer
46	*Theileria annulata* [Southern Europe; Africa; Asia]	*Hyalomma*	Cattle, zebu [Lymphocytes, erythrocytes]	Fever, anemia, emaciation
47	*Toxoplasma gondii* [Worldwide]	None [Cat]	Cat, dog, rodents, other mammals including man, birds [Endothelial cells, leukocytes; in cat, intestinal cells only [6]]	Chorioretinitis, cerebral calcification, pneumonia
48	*Trichomonas gallinae* [Worldwide]	None	Chicken, dove, hawk, pigeon, turkey [Crop, esophagus]	Caseous nodules, necrosis
49	*Tritrichomonas foetus* [Worldwide]	None	Cattle [Genital system, preputial & penile membranes, uterus]	Contaminated semen, irregularity of estrus, abortion, macerated fetus
50	*Trypanosoma brucei* [Africa]	*Glossina* [Equidae, wild ruminants]	Cat, dog, Equidae, ruminants, swine [Blood]	Nagana, anemia, emaciation, edema
51	*T. cruzi* [Southwestern United States, Central & South America]	*Panstrongylus; Rhodnius; Triatoma* [Wild animals]	Armadillo, monkey, opossum, raccoon, wood rat; also man [Blood, myocardium, other tissues]	Chagas' disease; tissue destruction

[4] Transmitted in eggs of *Heterakis gallinarum*.　[5] *T. gondii* has a sexual stage in cat intestinal cells where it produces oocysts indistinguishable from those of *Isospora bigemina*.

continued

Part II. Homoiothermic Animals Other Than Man

	Species [Distribution]	Intermediate Host [Reservoir Host]	Definitive Host [Primary Location in Definitive Host]	Effect on Host
52	*T. evansi* [Central & South America; southeastern Europe; Africa; Asia]	*Stomoxys; Tabanus* [Wild animals]	Cat, dog, elephant, Equidae, ruminants, swine [Blood]	Surra, urticaria, edema, emaciation

Contributor: Levine, Norman D.

General References

[1] Chandler, A. C., and C. P. Read. 1961. Introduction to Parasitology. Ed. 10. J. Wiley, New York.

[2] Levine, N. D. 1968. Nematode Parasites of Domestic Animals and of Man. Burgess, Minneapolis.

[3] Levine, N. D. 1971. Protozoan Parasites of Domestic Animals and of Man. Ed. 2. Burgess, Minneapolis.

[4] Soulsby, E. J. L. 1968. Helminths, Arthropods and Protozoa of Domesticated Animals. Ed. 6. Williams and Wilkins, Baltimore.

[5] Wardle, R. A., and J. A. McLeod. 1952. The Zoology of Tapeworms. Univ. Minnesota Press, Minneapolis.

126. VIRAL DISEASES: ANIMALS

Data in brackets refer to the column heading in brackets.

	Virus Group [Examples]	Morphology [Nucleic Acid]	Natural Host [Natural Transmission]	Remarks
1	Poxviruses[1] [Cowpox, molluscum contagiosum, myxoma, smallpox, vaccinia]	Brick-shaped, 300 × 200 × 100 nm; core, lateral bodies, envelope [Double-stranded DNA]	Many mammals & birds [Respiratory; direct contact; mechanically, by arthropod vectors]	Cytoplasmic multiplication, often with inclusion-body formation. Common group antigen; lipoprotein hemagglutinin.
2	Herpesviruses[2] [Cytomegalovirus, herpes simplex, varicella-zoster]	Spherical, 150-200 nm; icosahedral symmetry, 162 capsomeres; envelope [Double-stranded DNA]	Nearly all vertebrates [Respiratory; salivary; congenital; autogenous]	Intranuclear multiplication with inclusion-body formation. Ether- & heat-sensitive; acid labile.
3	Adenoviruses[3] [Viruses causing acute respiratory disease & conjunctivitis]	Icosahedral, 70-80 nm; 252 capsomeres; no envelope, 12 projecting fibers [Double-stranded DNA]	Man & many other mammals; some birds [Usually respiratory in man; also in feces & urine]	Intranuclear multiplication with inclusion-body formation. Common group antigen; hemagglutinin usually present. Type 3 oncogenic in hamsters.
4	Papovaviruses [Mouse polyoma, Shope papilloma, virus causing warts]	Icosahedral, 45-55 nm; 72 capsomeres; no envelope [Supercoiled, cyclic, double-stranded DNA]	Many mammals [Direct contact; respiratory]	Initial intranuclear multiplication. Acid-, heat-, & ether-stable.
5	Diplornaviruses [African horse sickness, bluetongue, Colorado tick fever]	Icosahedral, 54-75 nm; 32, 42, or 92 capsomeres; no envelope [Double-stranded RNA]	Many mammals & birds [Respiratory; in feces; by arthropod vectors]	Cytoplasmic multiplication in close relation to mitotic apparatus. Common group antigen. Heat- & acid-resistant.
6	Leukoviruses [Bittner virus; Rous sarcoma; viruses causing leukemias of cats, mice, & birds]	Spherical, 80-120 nm; envelope with projections [Single-stranded RNA]	Cat, mouse, chicken, related birds [Some through milk, others unknown; widespread in environment]	Some Rous sarcoma strains require helper viruses for maturation.

[1] 3 major groups, many types. [2] 2 major groups, many types. [3] About 50 types.

continued

	Virus Group [Examples]	Morphology [Nucleic Acid]	Natural Host [Natural Transmission]	Remarks
7	Paramyxoviruses [Mumps, Newcastle disease, parainfluenza]	Slightly larger & more pleomorphic than myxoviruses (see entry 13) [Single-stranded RNA]	Primates, cattle, horse, fowl [Respiratory]	Serologically distinct from influenza viruses; some produce hemolysin; other properties similar
8	Pseudomyxoviruses [Canine distemper, measles, respiratory syncytial, Rinderpest]	Similar to paramyxoviruses (see entry 7) [Single-stranded RNA]	Man & other primates, cattle, dog, swine, wild ungulates & carnivores [Respiratory]	Giant cells & cytoplasmic inclusion bodies often produced. Lack neuraminidase. Hemolysin & hemagglutinins varible.
9	Rhabdoviruses [Egtved virus, rabies, vesicular stomatitis]	Bullet-shaped, 180 X 75 nm; helical symmetry; envelope [Single-stranded RNA]	Egtved virus—trout. Rabies—all mammals, some birds. Vesicular stomatitis—most mammals [Rabies—chiefly by bite; others—by contact]	Rabies—distinctive intracyloplasmic inclusion body
10	Togaviruses 4/ [Dengue, equine encephalitis, rubella; St. Louis encephalitis; yellow fever]	Spherical, 30-75 nm; icosahedral symmetry; envelope [Single-stranded RNA]	Man, many other mammals, many birds, some reptiles [Arthropod vectors, usually mosquito or tick; occasionally, respiratory or in feces]	Cytoplasmic multiplication in both vertebrates & arthropods. Hemagglutinin usually present. Most cause encephalitis in suckling mouse; rubella teratogenic in man.
11	Rhinoviruses 5/ [Viruses causing acute upper respiratory infections]	Similar to enteroviruses (see entry 12), except virion is more dense [Single-stranded RNA]	Man, cattle, horse [Usually respiratory]	Inactivated at pH 3-5; heat-resistant
12	Enteroviruses 6/ [Coxsackievirus, encephalomyocarditis, poliovirus]	Spherical, 20-30 nm; icosahedral symmetry, 32 or 42 capsomeres; no envelope [Single-stranded RNA]	Man & other primates, cattle, mouse, swine; rarely birds [Usually in feces]	Cytoplasmic multiplication. No common group antigen. Stable at pH 3.0-5.0; ether-resistant
13	Myxoviruses 7/ [Influenza]	Spherical or filamentous, 80-120 nm; helical symmetry; envelope with projections [Single-stranded RNA]	Man, horse, swine, fowl [Respiratory]	Hemagglutinin & neuraminidase in envelope. Inactivated by heat, acid, & ether.
14	Coronaviruses [Mouse hepatitis; viruses causing acute upper respiratory infection]	Pleomorphic, 80-160 nm; electron-transparent core; envelope with projections [RNA]	Man, mouse, fowl [Respiratory; in feces]	Hemagglutinins not demontrated
15	Tacaribe-LCM viruses [Lymphocytic choriomeningitis; viruses causing South American hemorrhagic fevers]	Similar to togaviruses (see entry 10), but larger—up to 110 nm [Single-stranded RNA]	Man, rodents [Arthropod vectors; in feces & urine]	Virion contains granules similar to ribosomes.
16	Human hepatitis viruses [Infectious hepatitis, serum hepatitis]	Spherical, 25 nm [(?)]	Man [Infectious hepatitis—in feces. Serum hepatitis—blood & blood products; contaminated needles]	Difficult to inactivate by disinfectants. Strongly heat-resistant; ether-resistant.
17	"Slow" viruses [Kuru, scrapie]	Extremely small particles, 16-26 nm; morphology uncertain [(?)]	Man, sheep [Ingestion of infected tissue; contact (?); oral secretions (?)]	Extremely high resistance to heat, disinfectants, & irradiation. Prolonged incubation period & clinical course.

4/ Formerly arboviruses, in part; about 20 groups, 200 types. 5/ 100 serotypes. 6/ About 7 groups, 150 serotypes. 7/ 3 major types, many serotypes.

continued

Virus Group [Examples]	Morphology [Nucleic Acid]	Natural Host [Natural Transmission]	Remarks
18 Nuclear polyhedral viruses[8/] [Silkworm jaundice]	Rod-shaped, 250-350 nm long; helical symmetry [DNA]	Larvae of Diptera, Hymenoptera, & Lepidoptera [Ingestion of virus in dead or disintegrated larvae or excreta; infected insects, scavengers, & predators]	Large polyhedral inclusions in nucleus contain protein & RNA, and surround the virion.
19 Nonoccluded insect viruses[9/] [Iridescent viruses; virus causing sacbrood of bees]	Hexagonal, 130-200 nm; icosahedral symmetry [DNA or RNA]	Larvae of Diptera, Hymenoptera, & Lepidoptera [Presumably, similar to polyhedral viruses (see entry 18)]	Virus particles form crystalloid aggregations in cytoplasm of host cells.

[8/] Many types. [9/] About 15 types.

Contributor: Minton, Sherman A., Jr.

General References

[1] Andrews, C. H., and H. G. Pereira. 1967. Viruses of Vertebrates. Williams and Wilkins, Baltimore.
[2] Fenner, F., and D. O. White. 1970. Medical Virology. Academic Press, New York.
[3] Melnick, J. L. 1970. Progr. Med. Virol. 12:337.
[4] Vago, C., and M. Bergoin. 1968. Advan. Virus Res. 13:247.
[5] Wilner, B. I. 1969. A Classification of the Major Groups of Human and Other Animal Viruses. Ed. 4. Burgess, Minneapolis.

127. VIRAL DISEASES: PLANTS

Major virus groups and members are from reference 3 and 6. Cryptograms are from references 1, 3, 5, and 6. (For complete descriptions of the cryptogram and its symbols, consult references 2 and 6.) Data on distribution, effect on host, and characteristics of some viruses are from reference 4.

CRYPTOGRAM

A cryptogram consists of four pairs of symbols as defined below. Asterisk (*) indicates property of virus is unknown; parentheses indicate enclosed information is doubtful or unconfirmed.

First pair: Type of nucleic acid/Strandedness of nucleic acid
 Symbols for type of nucleic acid
  R = ribonucleic acid (RNA)
  D = deoxyribonucleic acid (DNA)
 Symbols for strandedness
  1 = single-stranded
  2 = double-stranded
Second pair: Molecular weight of nucleic acid (in millions)/Percentage of nucleic acid in infective particle
When different pieces of the genome occur together in one type of particle, the symbol Σ indicates the total molecular weight of the pieces in the particle
When the pieces occur in different particles, the composition of each particle is listed separately (e.g., 2.3/5 + 0.9/5)
Third pair: Outline of particle/Outline of "nucleocapsid" (the nucleic acid plus the protein most closely in contact with it)
 Symbols for both properties

S = essentially spherical
E = elongated with parallel sides, ends not rounded
U = elongated with parallel sides, end(s) rounded
Fourth pair: Kinds of host infected/Kinds of vector
Symbols for kinds of host
 I = invertebrate
 S = seed plant
Symbols for kinds of vector
 Ac = mite & tick (Acarina, Arachnida)
 Ap = aphid (Aphididae, Homoptera, Insecta)
 Au = leaf-, plant-, or treehopper (Auchenorrhyncha, Homoptera, Insecta)
 Cc = mealybug (Coccidae, Homoptera, Insecta)
 Cl = beetle (Coleoptera, Insecta)
 Di = fly, mosquito (Diptera, Insecta)
 Fu = fungus (Chytridiales & Plasmodiophorales, Fungi)
 Ne = nematode (Nematoda)
 Th = thrips (Thysanoptera, Insecta)
 O = spread without a vector via contaminated environment

---

continued

Virus Group & Virus	Cryptogram	Distribution	Effect on Host	Major Characteristics of Group or Virus Particle
Tobravirus group Type member 1 Tobacco rattle virus [1]	R/1:2.3/5 + 0.9/5: E/E:S/Ne	United States, Brazil, Europe, Japan	Corky ringspot in potato tubers, stem-streaking of tobacco	Tubular particles with 2 main nucleoprotein components: (i) 180-210 nm long, 300 S, infective; (ii) shorter, 155-243 S, length characteristic of the isolate. Thermal inactivation point ∿70-80°C. Longevity in vitro = months. Nematode (*Trichodorus* spp.) vectors.
Other member 2 Pea early-browning virus	R/*:*/* + */*:E/E: S/Ne	Britain, Netherlands	Necrosis of leaflets & stipules, necrotic streaking of pea stems	
Tobamovirus group Type member 3 Tobacco mosaic virus	R/1:2/5:E/E:S/O	Worldwide	Chlorotic mottling, leaf distortion in tobacco	Tubular particles, 300 nm, 190 S. Thermal inactivation point >90°C. Longevity in vitro = years. Efficient natural vectors unknown.
Other members 4 Cucumber green mottle mosaic virus	R/1:*/5:E/E:S/O	England, India	Light & dark green mottling & distortion of leaves, stunting of cucumber plant	
5 Odontoglossum ringspot virus	R/1:*/5:E/E:S/*	United States, Japan	Ringspots partially or wholly necrotic, with pale green or yellow central portion on orchid leaves	
6 Ribgrass mosaic virus	R/1:*/5:E/E:S/*	United States, Romania	Chlorotic streaks along veins, systemic chlorotic mottling in plantain	
7 Sammon's opuntia virus	R/1:*/5:E/E:S/*	United States	Chlorotic ringspotting of cactus pads	
8 Sunn hemp mosaic virus	R/1:*/5:E/E:S/*	Africa, India	Mottling, stunting, malformation of *Crotalaria* leaves	
Potexvirus group Type member 9 Potato virus X	R/1:2.1/6:E/E: S/O(Fu)	Worldwide	Mild mosaic in potato, necrosis & mottling in tobacco	Flexuous rods, 480-580 nm long, 118 S. Thermal inactivation point = 65-75°C. Longevity in vitro = months.
Other members 10 Cactus virus X	R/*:2.1/*:E/E:S/O	United States, Europe	Symptomless in cactus	
11 Clover yellow mosaic virus	*/*:*/*:E/E:S/*	United States	Irregular chlorotic flecks, streaks, & distortion of clover leaves, plants stunted	
12 Hydrangea ringspot virus	*/*:*/*:E/E:S/*	United States	Dieback with stem lesions, leaf chlorosis in hydrangea	

[1] Very wide host range.

continued

	Virus Group & Virus	Cryptogram	Distribution	Effect on Host	Major Characteristics of Group or Virus Particle
13	White clover mosaic virus[2/]	R/1:2.4/5:E/E: S/O(Ap)	North America, Europe, New Zealand	Mild mosaic in clovers	
	Possible members				
14	Artichoke curly dwarf virus	*/*:*/*:E/E:S/*	United States	Leaf-curling & dwarfing of artichoke	
15	Cymbidium mosaic virus	R/1:*/6:E/E:S/*	Worldwide	Mosaic & necrosis in orchids	
16	Narcissus mosaic virus	R/*:*/*:E/E:S/*	Great Britain, Netherlands	Mosaic symptoms in *Narcissus*	
17	Papaya mosaic virus	R/1:2.2/7:E/E:S/*	United States, Venezuela	Leaf mosaic & stunting in papaya	
18	Potato acuba mosaic virus	*/*:*/*:E/E:S/Ap	Worldwide	Bright yellow mottle on potato leaves	
	Carlavirus group Type member				
19	Carnation latent virus	R/1:*/6:E/E:S/Ap	Worldwide	Symptomless in carnation	Flexuous rods, 620-690 nm long. Thermal inactivation point = 55-70°C. Longevity in vitro = a few days. Aphid vectors.
	Other members				
20	Chrysanthemum virus	*/*:*/*:E/E:S/Ap	Great Britain, Netherlands	Leaves with mosaic & ringspots, plants stunted	
21	Pea streak virus	*/*:*/5.4:E/E:S/Ap	United States, Europe	Streaking of leaf veins, stems, & petioles of pea	
22	Potato virus M	*/*:*/*:E/E:S/Ap	Worldwide	Interveinal mosaic & leaf-rolling in potato	
23	Potato virus S	*/*:*/*:E/E:S/Ap	Worldwide	Few or no symptoms in potato	
24	Red clover vein mosaic virus	R/1:*/6:E/E:S/Ap	North & South America, Europe	Mosaic, streaking, & stunting in various legumes	
	Possible members				
25	Freesia mosaic virus	*/*:*/*:E/E:S/Ap	Europe	Vary from none to prominent mosaic	
26	Hop latent virus	*/*:*/*:E/E:S/Ap	Europe	Vary from none to a mosaic, depending on variety	
27	Poplar mosaic virus	*/*:*/*:E/E:S/*	Europe	Yellow patches on leaves	
	Potyvirus group Type member				
28	Potato virus Y	*/*:*/*:E/E:S/Ap	Worldwide	Systemic necrotic symptoms in potato leaves	Flexuous rods, 720-800 nm long. Thermal inactivation point = 50-60°C. Longevity in vitro = a few days. Aphid vectors.
	Other members				
29	Bean common mosaic virus	*/*:*/*:E/E:S/Ap	Worldwide	Mottling, leaf distortion, stunting in bean	
30	Bean yellow mosaic virus	*/*:*/*:E/E:S/Ap	Worldwide	Systemic yellowish mosaic, leaf-curling, & stunting in bean	

2/ Infects many legumes.

continued

	Virus Group & Virus	Cryptogram	Distribution	Effect on Host	Major Characteristics of Group or Virus Particle
31	Beet mosaic virus	*/*:*/*:E/E:S/Ap	Worldwide	Vein-clearing in young leaves, mosaic in older leaves of sugar beet, plants stunted	
32	Clover yellow vein virus	*/*:*/*:E/E:S/Ap	Great Britain	Mild veinal yellowing in leaves, plants stunted	
33	Cowpea aphid-borne mosaic virus	*/*:*/*:E/E:S/Ap	United States, Europe	Mosaic, plants stunted	
34	Henbane mosaic virus	*/*:*/*:E/E:S/Ap	Great Britain	Dark green mottling in leaves, plants stunted	
35	Pea mosaic virus	*/*:*/*:E/E:S/Ap	Worldwide	Leaf-mottling & stunting in garden pea	
36	Potato virus A	*/*:*/*:E/E:S/Ap	Worldwide	Mild mosaic in potato leaves	
37	Soybean mosaic virus	*/*:*/*:E/E:S/Ap	Worldwide	Mottling & stunting of soybean	
38	Tobacco etch virus	R/1:*/5:E/E:S/Ap	North & South America	Necrosis & necrotic mottling of tobacco & pepper	
39	Watermelon mosaic virus (South Africa)	*/*:*/*:E/E:S/Ap	Africa	Severe mosaic of leaves, plants stunted	
	Possible members				
40	Celery mosaic virus	*/*:*/*:E/E:S/Ap	United States, Great Britain, Germany	Green mottling in leaves, plants stunted	
41	Cocksfoot streak virus	*/*:*/*:E/E:S/Ap	Europe	Green or yellow streaks in leaves	
42	Iris mosaic virus	*/*:*/*:E/E:S/Ap	United States, Europe	Light mosaic on leaves, plants stunted	
43	Lettuce mosaic virus	*/*:*/*:E/E:S/Ap	Worldwide	Mottling of lettuce leaves	
44	Malva vein clearing virus	*/*:*/*:E/E:S/Ap	United States, South America, Europe	Bright yellow patches on leaves, whitish vein-clearing	
45	Narcissus yellow stripe virus	*/*:*/*:E/E:S/Ap	United States, Europe	Mottling & striping of leaves, plants stunted	
46	Papaya ringspot virus	*/*:*/*:E/E:S/Ap	United States	Yellow rings in fruit, mosaic on leaves, plants stunted	
47	Parsnip mosaic virus	*/*:*/*:E/E:S/Ap	Great Britain	Mosaic on leaves	
48	Plum pox virus	*/*:*/*:E/E:S/Ap	Europe, Union of Soviet Socialist Republics	Mottled, curled leaves, pitted fruit	
49	Sugarcane mosaic virus	*/*:*/*:E/E:S/Ap	United States, Italy, Romania, Yugoslavia, Union of Soviet Socialist Republics, India	Chlorotic mottling, stunting, & blotches on maize, millet, & sorghum	

continued

	Virus Group & Virus	Cryptogram	Distribution	Effect on Host	Major Characteristics of Group or Virus Particle
50	Tulip mosaic virus [3]	*/*:*/*:E/E:S/Ap	Worldwide	Leaf-mottling & irregular stripes of color in tulip flowers	
51	Turnip mosaic virus	*/*:*/*:E/E:S/Ap	Worldwide	Mosaic & stunting in turnip, mottling & necrosis in cabbage	
52	Watermelon mosaic virus 1 (United States)	*/*:*/*:E/E:S/Ap	United States	Severe mosaic, stunting in watermelon	
53	Cucumovirus group Type member Cucumber mosaic virus [1]	R/1:1/18:S/S:S/Ap	Worldwide	Mosaic in cucumber & tobacco	Isometric particles, 30 nm diameter, 98 S. Thermal inactivation point = 60-70°C. Longevity in vitro = a few days. Aphid vectors.
54	Other member Tomato aspermy virus	*/*:*/*:S/S:S/Ap	United States, Europe	Tomato plants stunted, fruit seedless	
55	Possible member Peanut stunt virus	R/(1):1/16:S/S:S/Ap	United States	Severe dwarfing of leaves & petioles, chlorosis in peanut	
56	Tymovirus group Type member Turnip yellow mosaic virus	R/1:1.9/37:S/S:S/Cl	Europe	Mosaic in Chinese cabbage & other *Brassica* spp.	Isometric particles, 30 nm diameter, 110 S; accessory particles of 50 S are empty protein shells. Thermal inactivation point usually = 70-90°C. Longevity in vitro = a few weeks. Beetle vectors.
57	Other members Andean potato latent virus	R/1:2.0/36:S/S:S/*	South America	Symptomless in potato	
58	Belladonna mottle virus	R/1:2.0/37:S/S:S/Cl	Europe	Mild mottling, leaf deformation	
59	Cacao yellow mosaic virus	R/1:*/38:S/S:S/*	Africa	Chlorotic leaf mottling in cacao	
60	Dulcamara mottle virus	R/1:*/37:S/S:S/Cl	Great Britain	Mild mottling, leaf distortion	
61	Eggplant mosaic virus	*/*:*/35:S/S:S/Cl	South America	Dark green vein-banding, mottling	
62	Ononis yellow mosaic virus	R/1:*/34:S/S:S/*	Great Britain	Bright yellow mottling in *Ononis*	
63	Wild cucumber mosaic virus	R/1:2.4/35:S/S:S/Cl	United States	Leaf distortion, mosaic, & stunting in cucumber	
64	Comovirus group Type member Cowpea mosaic virus	R/1:1.5/24 + 2.6/33: S/S:S/Cl	United States, South America, Africa	Severe mosaic & distortion of cowpea leaves	3 components, ∿30 nm diameter; 115, 95, & 55 S; both containing RNA (115 & 95 S) required for infection. Thermal inactivation point = 60-80°C. Longevity in vitro = 1 or a few weeks. Beetle vectors.
65	Other members Bean pod mottle virus	R/1:1.5/30+2.5/37: S/S:S/Cl	United States	Local lesions in some beans, systemic mottling in others	
66	Broad bean stain virus	R/1:*/35:S/S:S/Cl	Europe, Africa	Chlorotic mottling of broad bean leaves	

[1] Very wide host range. [3] Synonym: tulip breaking virus.

continued

	Virus Group & Virus	Cryptogram	Distribution	Effect on Host	Major Characteristics of Group or Virus Particle
67	Radish mosaic virus	R/*:*/*:S/S:S/Cl	United States	Chlorotic lesions between leaf veins, yellow-green mottling	
68	Red clover mottle virus	R/1:*/35:S/S:S/*	Great Britain	Mottling, chlorotic rings & spots, leaf distortion	
69	Squash mosaic virus	R/1:2.4/35:S/S:S/Cl	North & South America	Range from symptomless to severe mosaic & leaf deformation in Cucurbitaceae	
70	Possible member True broad bean mosaic virus	R/1:2.8/35:S/S:S/*	Europe, Africa	Systemic mottling, mosaic in leaves, stem necrosis in broad bean	
71	Nepovirus group Type member Tobacco ringspot virus [1]	R/1:2.2/40:S/S:S/Ne	North America, Europe, Australia	Ringspot in tobacco, mosaic in cucumber	Isometric particles, 30 nm diameter; 3 components, 125 (infective), 95, & 50 S. Thermal inactivation point = 55-70°C. Longevity in vitro = a few days or weeks. Seedborne. Nematode (*Xiphinema* spp. & *Longidorus* spp.) vectors.
72	Other members Arabis mosaic virus	R/1:*/41:S/S:S/Ne	Europe	Dwarfing & yellowing of raspberry, mosaic in strawberry	
73	Grapevine fanleaf virus	*/*:*/*:S/S:S/Ne	Worldwide	Leaf distortion, yellow mosaic in grape	
74	Raspberry ringspot virus	R/*:*/43:S/S:S/Ne	Europe	Ringspots in raspberry & strawberry	
75	Strawberry latent ringspot virus [4]	*/*:*/*:S/S:S/Ne	Europe	Symptomless in many hosts, including strawberry & raspberry	
76	Tomato black ring virus [4]	R/*:*/38:S/S:S/Ne	Europe	Ringspot in tomato & bean	
77	Tomato ringspot virus	R/1:2.3/40:S/S:S/Ne	United States	Mosaic & ringspot in tomato & tobacco	
78	Possible member Cherry leaf roll virus	*/*:*/*:S/S:S/Ne	Europe	Upward rolling of leaves, stunting in cherry	
79	Bromovirus group Type member Brome mosaic virus [5]	R/1:1/22:S/S:S/Ne	United States, Europe	Mild mosaic	Isometric particles, 25 nm diameter, 85 S. Thermal inactivation point = 70-95°C. Longevity in vitro variable.
80	Other members Broad bean mottle virus	R/1:1.1/22:S/S:S/*	Great Britain	Mottling, necrosis, & crinkling of leaves, plants stunted	
81	Cowpea chlorotic mottle virus	R/1:1.1/24:S/S:S/Cl	United States	Yellow mosaic in cowpea	

[1] Very wide host range. [4] Wide host range. [5] Wide host range in Liliopsida.

continued

	Virus Group & Virus	Cryptogram	Distribution	Effect on Host	Major Characteristics of Group or Virus Particle
	Tombusvirus group Type member				Isometric particles, 30 nm diameter, 140 S. Thermal inactivation point = 85-90°C. Longevity in vitro = a few weeks.
82	Tomato bushy stunt virus	R/1:1.5/17:S/S:S/*	Canada, Great Britain	Yellow leaves, plants stunted	
	Other members				
83	Artichoke mottle crinkle virus	*/*:*/*:S/S:S/*	Europe	Mosaic on distorted leaves	
84	Carnation Italian ring-spot virus	*/*:*/*:S/S:S/*	United States, Europe	Chlorotic spots & rings, followed by recovery	
85	Pelargonium leaf curl virus	R/1:1.5/17:S/S:S/*	United States, Europe	Yellow spots, necrosis, leaf distortion in geranium	
86	Petunia asteroid mosaic virus	*/*:*/*:S/S:S/*	Italy	Mosaic, leaf distortion, necrosis	
	Caulimovirus group Type member				Isometric particles, 50 nm diameter, 250 S. Thermal inactivation point = 75-80°C. Longevity in vitro = a few days. Aphid vectors.
87	Cauliflower mosaic virus	D/2:4.5/16:S/S:S/Ap	Worldwide	Mosaic & mottling in Brassicaceae	
	Other members				
88	Carnation etched ring virus	*/*:*/*:S/S:S/*	United States, Europe	Whitish, necrotic flecks & rings on leaves	
89	Dahlia mosaic virus	(D)/*:*/(16):S/S:S/Ap	Worldwide	Mosaic, vein-banding, & stunting in dahlia	
	Ungrouped viruses				
90	Alfalfa mosaic virus [1]	R/1:1.3/18 + 1.1/18 + 0.9/18: U/U:S/Ap	Worldwide	Mosaic in alfalfa, calico in potato	3 bacilliform particles, 18 × 58 nm, 18 × 48 nm, & 18 × 36 nm, and 1 spheroidal particle
91	Apple chlorotic leafspot virus	R/1:*/*:E/E:S/*	Worldwide	Chlorotic spots, leaf distortion, stunting of apple	Flexuous rods, 600 × 12 nm
92	Apple stem grooving virus	*/*:*/*:E/E:S/*	Worldwide	Elongated grooves in apple wood, swelling at graft union	Flexuous rods, 600-700 × 12 nm
93	Apple tulare mosaic virus	R/*:*/12:S/S:S/*	California	Severe mosaic in apple leaves	Isometric particles, 33 nm
94	Barley stripe mosaic virus	R/1:1/4.5:E/E:S/*	Worldwide	Mottling, streaking, & stunting of barley & wheat	Tubular particles, 125 × 18 nm
95	Barley yellow dwarf virus	*/*:*/*:S/S:S/Ap	Worldwide	Stunting & chlorosis of barley & many Liliopsida	Isometric particles, 20-24 nm
96	Beet curly top virus	*/*:*/*:*/*:S/Au	North America	Leaf distortion, vein-clearing, stunted plants	....
97	Beet yellows virus	*/*:*/*:E/E:S/Ap	Worldwide	Yellow thickened leaves, necrotic spots	Flexuous rods, 1250 nm
98	Cacao swollen shoot virus	*/*:*/*:U/*:S/Cc	Africa	Chlorotic leaf symptoms, swellings on stems & tap roots	Bacilliform particles, 121-130 × 28 nm

[1] Very wide host range.

continued

	Virus Group & Virus	Cryptogram	Distribution	Effect on Host	Major Characteristics of Group or Virus Particle
99	Carnation mottle virus	R/1:*/(20):S/S/S/*	Worldwide	Mild mottling & stunting in carnation	Isometric particles, 28 nm
100	Carnation ringspot virus	R/1:1.4/20:S/S/S/Ne	Worldwide	Mosaic, ringspot, & stunting in carnation	Isometric particles, 30 nm
101	Carrot mosaic virus	*/*:*/*:E/E:S/Ap	Europe	Curling of carrot leaves, mosaic	Flexuous rods, 752 nm
102	Citrus tristeza virus	*/*:*/*:E/E:S/Ap	Worldwide	Leaf distortion, pitted stems; plants wilt & die	Flexuous rods, 2000 X 12 nm
103	Citrus variegation virus	*/*:*/*:S/S/S/Ap	Worldwide	Leaves distorted with yellowish blotches	Isometric particles, 30 nm
104	Cocksfoot mottle virus	R/1:1/25:S/S/S/Cl	Great Britain	Mottling in cocksfoot, barley, & wheat	Isometric particles, 30 nm
105	Desmodium yellow mottle virus	R/*:*/*:S/S/S/*	United States	Yellow mottling in *Desmodium,* chlorotic mottling in bean	Isometric particles, 30 nm
106	Lettuce necrotic yellows virus	R/*:*/*:U/E:S,I/Ap	Australia	Yellowing, severe stunting in lettuce	Bacilliform particles, 227 X 66 nm
107	Oat blue dwarf virus	*/*:*/*:S/S/S/Au	North America	Stunting, stiffening of oat leaves, general blue-green appearance of plant	Isometric particles, 30 nm
108	Pea enation mosaic virus	R/1:(1.3)/29:S/S: S/Ap	Worldwide	Translucent spots, mosaic in leaves of garden pea, enations on veins on underside of leaves	Isometric particles, 28 nm
109	Potato leaf roll virus	*/*:*/*:S/S/S,(I)/Ap	Worldwide	Stunting & rolling of potato leaves	Isometric particles, 24 nm
110	Potato yellow dwarf virus[4]	R/1:4.3/0.4:U/E: S,I/Au	North America	Yellowing & necrosis in potato	Bacilliform particles, 380 X 75 nm
111	Prune dwarf virus	*/*:*/*:S/S/S/*	Worldwide	Narrow, leathery leaves in prune	Isometric particles, 22 nm
112	Prunus necrotic ringspot virus	R/*:*/16:S/S/S/*	Worldwide	Necrotic ringspot in *Prunus,* mosaic in rose	Isometric particles, 23 nm
113	Rice dwarf virus	R/2:10/11:S/S:S,I/Au	Japan, Philippines	Streaking & stunting in rice, oats, & wheat	Isometric particles, 70 nm
114	Rice hoja blanca	*/*:*/*:S/S/S/Au	Central & South America	White stripes, mottling, & stunting of rice, rye, & wheat	Isometric particles, 42 nm
115	Rose mosaic virus	*/*:*/*:S/S/S/*	North America	Chlorotic areas, ring patterns on rose leaves	Isometric particles, 25 nm
116	Satellite virus[6]	R/1:0.4/20:S/S/S/Fu	North America, Europe	....	Isometric particles, 17 nm
117	Soil-borne oat mosaic virus	*/*:*/*:E/E:S/*	United States, Great Britain	Mottling, stunting of oats	Flexuous rods, 700-750 X 12 nm
118	Soil-borne wheat mosaic virus	*/*:*/*:E/E:S/(Fu)	North America, Italy, Japan	Streaking, stunting of wheat & barley	Tubular particles, 160-270 X 20 nm
119	Southern bean mosaic virus	R/1:1.4/21:S/S/S/Cl	North & South America, Europe, Africa	Severe mosaic in beans	Isometric particles, 28 nm

[4] Wide host range. [6] Multiplies only in plants infected with tobacco necrosis virus (*see* entry 121).

continued

	Virus Group & Virus	Cryptogram	Distribution	Effect on Host	Major Characteristics of Group or Virus Particle
120	Sowbane mosaic virus	R/1:1.3/17:S/S:S/Di	North America, Europe	Leaf-mottling, or latent in *Chenopodium*	Isometric particles, 26 nm
121	Tobacco necrosis virus[1/]	R/1:1.5/19:S/S:S/Fu	Worldwide	Extensive necrosis on tobacco & bean	Isometric particles, 26 nm diameter
122	Tobacco streak virus[1/]	*/*:*/*:S/S:S/*	North & South America, Europe, New Zealand	Necrosis in tobacco, followed by recovery	Isometric particles, 28 nm
123	Tomato spotted wilt virus[1/]	R/(1):*/*:S/S:S/Th	Worldwide	Chlorosis & necrosis on leaves, stunted plants	Isometric particles, 70-80 nm diameter
124	Western celery mosaic virus	*/*:*/*:E/E:S/Ap	United States	Vein-clearing, mottling, & twisting of leaflets	Flexuous rods, 784 nm
125	Wheat streak mosaic virus	R/1:2.8/*:E/E:S/Ac	Worldwide	Severe mosaic & stunting in wheat	Flexuous rods, 700 × 15 nm
126	Wheat striate mosaic virus	R/1:*/5:U/U:S,I/Au	North America	Fine chlorotic streaking, stunting of wheat, barley, & oats	Tubular particles, 270 × 65 nm
127	Wound tumor virus	R/2:Σ16/22:S/S:S, I/Au	United States	Veins enlarged, root & stem tumors in clover	Isometric particles, 70 nm

[1/] Very wide host range.

Contributor: Scott, Howard A.

References

[1] Commonwealth Mycological Institute, and the Association of Applied Biologists. 1970-72. Descriptions of Plant Viruses. Kew, Surrey, England.
[2] Gibbs, A. J. 1969. Advan. Virus Res. 14:320.
[3] Harrison, B. D., et al. 1971. Virology 45:356.
[4] Imperial Mycological Institute. 1922-70. Rev. Appl. Mycol. 1-49.
[5] Martyn, E. B., ed. 1968. Commonw. Mycol. Inst. Phytopathol. Pap. 9.
[6] Wildy, P. 1971. Monogr. Virol. 5.

128. RICKETTSIAL PARASITES: MAMMALS AND BIRDS

Host: Animals are listed in order of decreasing susceptibility.

	Species	Principal Host	Disease or Disorder	Method of Transmission
1	*Anaplasma centrale*	Cattle	Benign anaplasmosis	Tick to host
2	*A. marginale*	Cattle	Malignant anaplasmosis	Tick to host
3	*A. ovis*	Sheep, goat	Ovine anaplasmosis	Tick to host
4	*Bartonella bacilliformis*	Man	Oroya fever, verruga	Sand fly to man
5	*Chlamydia psittaci*	Mammals, birds	Psittacosis & ornithosis	Contact; contaminated swimming pools
6	*C. trachomatis*	Man, ape, monkey	Trachoma, lymphogranuloma venereum, inclusion conjunctivitis	Contact
7	*Colesiota conjunctivae*	Sheep, cattle, goat	Ophthalmia (keratoconjunctivitis)	Nonbiting flies

continued

	Species	Principal Host	Disease or Disorder	Method of Transmission
8	*Cowdria ruminantium*	Goat, sheep, cattle	Heartwater	Tick feces; tick to host
9	*Coxiella burneti*	Man, cattle, sheep, goat	Q fever	Tick feces; host to tick to host; contact with domestic animals; inhalation of infected dust; contaminated milk
10	*Ehrlichia canis*	Dog	Canine rickettsiosis	Tick to host
11	*E. phagocytophila*	Sheep, cattle	Ovine & bovine rickettsiosis	Tick to host
12	*Eperythrozoon suis*	Swine	Icteroanemia	Arthropods suspected
13	*Haemobartonella* spp.	Rodents, cat	Haemobartonellosis, anemia	Flea to host
14	*Neorickettsia helminthoeca*	Dog, fox, coyote	Salmon poisoning	Intestinal parasitic fluke to canines
15	*Rickettsia akari*	Man, mouse	Rickettsial pox	Mite to man or mouse
16	*R. australis*	Man, certain marsupials	Queensland tick typhus	Tick to man or marsupials
17	*R. conori*	Man; dog as reservoir	Boutonneuse (Marseilles, Mediterranean) fever	Tick to man
18	*R. prowazeki*	Man	Epidemic (classic Old World) typhus	Louse feces; man to louse to man
19	*R. rickettsi*	Man, rabbit, squirrel	Rocky Mountain spotted fever	Tick to man
20	*R. sibirica*	Man, rodents	Siberian tick typhus	Tick to host
21	*R. tsutsugamushi*	Man, rodents, monkey	Tsutsugamushi fever (scrub typhus)	Mite to man
22	*R. typhi*	Man, rodents	Murine or endemic typhus	Flea feces; rat to flea to man
23	*Ricolesia bovis*	Cattle; mouse (laboratory)	Infectious conjunctivitis	Contact
24	*R. conjunctivae*	Fowl	One form of ocular roup	Possibly contact
25	*Rochalimaea quintana*	Man	Trench fever	Body louse to man

Contributor: Philip, Cornelius B.

General References

[1] Breed, R. S., et al., ed. 1957. Bergey's Manual of Determinative Bacteriology. Ed. 7. Williams and Wilkins, Baltimore.

[2] Horsfall, F. L., Jr., and H. Tamm, ed. 1965. Viral and Rickettsial Infections of Man. Ed. 4. J. B. Lippincott, Philadelphia.

[3] Hull, T. G. 1963. Diseases Transmitted from Animals to Man. Ed. 5. C. C. Thomas, Springfield, Ill.

[4] Merchant, I. A., and R. A. Packer. 1967. Veterinary Bacteriology and Virology. Ed. 7. Iowa State Univ. Press, Ames.

[5] Moulton, F. R., ed. 1948. Rickettsial Dis. Man Symp. Amer. Ass. Advan. Sci., Boston, 1946.

[6] Philip, C. B. 1950. In R. L. Pullen, ed. Communicable Diseases. Lea and Febiger, Philadelphia. p. 781.

129. BACTERIAL PARASITES: MAMMALS AND BIRDS

Host: Animals are listed in order of decreasing susceptibility.

	Species (Synonym)	Host	Disease & Clinical Manifestation	Method of Transmission
1	*Actinobacillus equuli*	Horse	Joint infection in foals, purulent nephritis, abortion	Endogenous—commensal of intestinal tract; possibly in utero
2	*A. lignieresii*	Cattle, sheep	Actinobacillosis (wooden tongue)	Endogenous—commensal of alimentary tract; wound infection
3		Swine	Mastitis	Following injury by nursing piglets

continued

	Species (Synonym)	Host	Disease & Clinical Manifestation	Method of Transmission
4	*A. mallei*	Horse	Glanders (farcy)	Contact; contaminated food & water; wound infection; droplet infection
5		Man	Glanders	
6	*Actinomyces bovis*	Cattle, man	Actinomycosis, lumpy jaw	Endogenous—commensal of buccal cavity
7		Swine	Mastitis	
8		Horse	Fistulous withers, poll evil	
9	*A. israelii*	Man	Actinomycosis—cervicofacial & pulmonary abscesses	Endogenous—commensal of buccal cavity
10	*Bacillus anthracis*	Man	Anthrax—malignant pustule; woolsorter's disease (pneumonia)	Contact with contaminated soil or animal by-products; inhalation or ingestion of spores
11		Cattle, sheep, horse	Anthrax—acute septicemia	Contaminated food & water; soil-borne
12		Swine	Anthrax—acute pharyngitis	
13	*Bacteroides nodosus (Fusiformis nodosus)*	Sheep	Contagious footrot	Contact with infected animals; fomites
14	*Bordetella bronchiseptica*	Dog	Upper respiratory infections	Endogenous—commensal of upper respiratory tract
15		Swine	Bronchopneumonia, atrophic rhinitis	Contact; contaminated food & water
16		Rabbit	Snuffles	Endogenous
17		Man	Pertussis-like	Unknown—possibly animal contact
18	*B. parapertussis*	Man	Pertussis-like	Commensal of respiratory tract; droplet infection
19	*B. pertussis*	Man	Whooping cough	Commensal of respiratory tract; droplet infection
20	*Borrelia anserina*	Fowl	Spirochetosis	Arthropod vector (tick); possibly feces-borne
21	*B. recurrentis*	Man	Epidemic relapsing fever	Arthropod vector (louse, tick)
22	*B. vincentii*	Man	Associated with *Fusobacterium fusiforme* in Vincent's angina; fusospirochetal infection	Endogenous—commensal of buccal cavity
23	*Brucella abortus*	Cattle	Brucellosis, contagious abortion (undulant fever, Bang's disease)	Contaminated food & water; contact with infected animals or their secretions
24		Man	Brucellosis (undulant fever)	Animal contact; airborne; ingestion
25	*B. canis*	Dog	Brucellosis (undulant fever), abortion	Infectious secretions or sexual contact
26		Man	Brucellosis (undulant fever)	Contact with infected animals
27	*B. melitensis*	Goat, sheep, man	Brucellosis (undulant fever)	Contaminated food & water; contact with infected animals or their secretions
28	*B. suis*	Swine	Brucellosis (undulant fever), abortion	Contaminated food, water or soil
29		Man	Brucellosis (undulant fever)	Contact with infected animals or their secretions
30	*Clostridium botulinum*	Horse, man, fowl, sheep, cattle	Botulism; limberneck of fowl	Ingestion of food containing preformed toxin
31	*C. feseri (C. chauvoei)*	Cattle	Blackleg, symptomatic anthrax	Saprophyte of soil; wound infection; commensal of intestinal tract
32	*C. haemolyticum*	Cattle, sheep	Bacillary hemoglobinuria	Not known—possibly ingestion
33	*C. novyi*	Cattle	Gas gangrene	Soil-borne; wound infection
34		Sheep	Gas gangrene, infectious necrotic hepatitis, black disease	Wound infection; associated with liver fluke
35		Man	Gas gangrene	Soil-borne; wound infection
36	*C. perfringens*	Sheep	Enterotoxemia, lamb dysentery	Endogenous—commensal of intestinal tract; contaminated feed & water
37		Man	Gas gangrene, uterine infections	Soil-borne; wound infection; commensal of intestinal tract
38			Food poisoning	Ingestion of contaminated food

continued

Species (Synonym)	Host	Disease & Clinical Manifestation	Method of Transmission
39	Cattle, swine	Hemorrhagic enteritis	Endogenous—commensal of intestinal tract; contaminated food & water
40 C. septicum	Cattle, sheep, swine, horse	Malignant edema	Soil-borne; wound infection; commensal of intestinal tract
41	Man	Gas gangrene	
42 C. tetani	Horse, sheep, cattle, man	Tetanus	Wound infection; soil-borne
43 Corynebacterium diphtheriae	Man	Diphtheria	Droplet from active cases or carrier
44 C. equi	Horse	Bronchopneumonia of foals; metritis & abortion	Endogenous—commensal of upper respiratory tract; possibly soil-borne
45	Swine	Granulomatous lesions of cervical & pharyngeal lymph nodes	
46 C. pseudotuberculosis	Sheep	Caseous lymphadenitis	Wound infection; contaminated food & water
47	Horse	Ulcerative lymphangitis	Direct contact; wound infection
48	Cattle	Suppurative skin lesions	Direct contact; wound infection
49 C. pyogenes	Cattle, swine, sheep	Purulent infections, arthritis, mastitis	Endogenous—commensal of skin & mucous membranes; wound infection
50 C. renale	Cattle, swine	Pyelonephritis, cystitis	Contact with infected urine; commensal of mucous membranes of genital tract
51 Enterobacter aerogenes	Man	Urinary tract infections	Endogenous—commensal of intestinal tract
52	Cattle	Mastitis	
53 Erysipelothrix insidiosa (E. rhusiopathiae)	Swine, turkey	Erysipelas	Soil-borne; ingestion of contaminated food & water
54	Man	Erysipeloid	Wound infection; animal contact
55 Escherichia coli	Cattle	Calf scours; mastitis	Endogenous—commensal of intestinal tract of man & animals; feces-borne
56	Swine	Edema disease	
57	Chicken	Air sacculitis, coligranuloma	
58	Dog	Urogenital infections	
59	Man	Urogenital infections; peritonitis; infantile diarrhea; cholecystitis; bacteremia	
60 Francisella tularensis	Lagomorphs, rodents, man	Tularemia	Contact with infected animals; arthropod vector (deerfly); contaminated food & water; inhalation
61 Fusobacterium fusiforme	Man	Associated with Vincent's angina; fusospirochetal infection	Endogenous—commensal of buccal cavity (upper respiratory & alimentary tract)
62 Haemophilus ducreyi	Man	Chancroid (soft chancre)	Venereal infection
63 H. gallinarum	Chicken	Infectious coryza	Commensal of respiratory tract; droplet infection; contaminated food & water
64 H. influenzae	Man	Acute respiratory infections, otitis media, meningitis, bacteremia, endocarditis	Droplet infection; endogenous—commensal of respiratory tract
65 H. suis	Swine	Secondary in swine influenza; Glasser's disease	Endogenous—commensal of respiratory tract; contact with infected animals
66 Herellea vaginicola	Man	Urethritis, bronchitis, septicemia, endocarditis	Endogenous—commensal of upper digestive & respiratory tracts, skin & mucous
67	Dog	Septicemia	membranes; soil-borne
68 Klebsiella pneumoniae	Man	Urinary tract infections, pneumonia, peritonitis	Endogenous—commensal of upper respiratory tract
69	Horse	Cervicitis, metritis	Not known
70	Cattle	Mastitis	Contaminated milking equipment

continued

	Species (Synonym)	Host	Disease & Clinical Manifestation	Method of Transmission
71	*Leptospira canicola*	Dog, man	Leptospirosis	Contact with contaminated food & water; contact with infected urine or animal tissues
72	*L. icterohemorrhagiae*	Rat, man, dog	Leptospirosis	Contact with contaminated food & water; contact with infected urine or animal tissues
73	*L. pomona*	Cattle, swine, man	Leptospirosis	Contact with contaminated food & water; contact with infected urine or animal tissues
74	*Listeria monocytogenes*	Cattle, sheep	Listeriosis, meningoencephalitis, liver abscesses	Possibly ingestion of silage or vegetation; commensal of upper respiratory tract
75		Chicken	Myocardial & liver abscesses	
76		Man	Genital infections with abortion & stillbirth; meningitis; septicemia	Not known—possibly contact with infected animals
77	*Mima polymorpha*	Man	Urethritis, bronchitis, septicemia, endocarditis	Endogenous—commensal of upper digestive & respiratory tracts, skin & mucous membranes; soil-borne
78		Horse, cattle	Abortion	
79	*Moraxella bovis*	Cattle	Infectious keratitis, pinkeye	Contaminated dust; insects
80	*M. lacunata*	Man	Subacute infectious conjunctivitis	Endogenous—commensal of mucous membranes
81	*Mycobacterium avium*	Turkey, chicken	Tuberculosis of intestine, spleen, & liver	Contaminated food & water
82		Swine	Principally tuberculosis of lymph nodes	Contact with infected fowl
83	*M. bovis*	Cattle	Tuberculosis	Contact; contaminated food, water, & milk
84		Swine	Tuberculosis	Ingestion
85		Man	Tuberculosis	Contact with infected animals; contaminated milk
86	*M. leprae*	Man	Leprosy	Probably contact
87	*M. paratuberculosis*	Cattle, sheep, goat	Johne's disease	Contaminated food & water
88	*M. tuberculosis*	Man	Tuberculosis	Droplet infection; direct contact
89	*Mycoplasma gallisepticum*	Chicken	Chronic respiratory disease, air-sac infections	Commensal of upper respiratory tract; droplet infection; direct contact
90		Turkey	Sinusitis	
91	*M. hyopneumoniae*	Swine	Enzootic pneumonia	Airborne; droplet infection
92	*M. hyorhinis*	Swine	Polyserositis	Commensal of upper respiratory tract
93	*M. mycoides*	Cattle	Contagious bovine pleuropneumonia	Droplet infection; ingestion; direct contact
94	*M. pneumoniae*	Man	Primary atypical pneumonia	Droplet infection
95	*Neisseria gonorrhoeae*	Man	Gonorrhea, neonatal ophthalmia	Venereal infection; contact during parturition
96	*N. meningitidis*	Man	Epidemic cerebrospinal meningitis	Endogenous—commensal of upper respiratory tract; droplet infection
97	*Nocardia asteroides*	Dog, cat	Nocardiosis—infection of subcutaneous tissue & pleural cavity	Endogenous—commensal of skin; wound infection; inhalation
98		Man	Nocardiosis—lung abscesses, meningitis, mycetoma	
99		Cattle	Mastitis	Soil-borne; during milking
100	*Pasteurella haemolytica*	Sheep	Septicemia in lambs; mastitis	Endogenous—commensal of upper respiratory tract; droplet infection; contaminated food & water
101		Cattle	Pneumonia, hemorrhagic septicemia	

continued

	Species (Synonym)	Host	Disease & Clinical Manifestation	Method of Transmission
102	*P. multocida*	Cattle, sheep	Hemorrhagic septicemia, pneumonia	Endogenous—commensal of upper respiratory tract; contact with animals
103		Fowl	Fowl cholera	
104		Swine	Enzootic pneumonia	
105		Rabbit	Snuffles	
106		Man	Chronic pulmonary infections, cellulitis, joint infections	Contact with animals; animal bite
107	*Pseudomonas aeruginosa*	Dog	Otitis externa, dermatitis	Commensal of skin; found in water, soil, sewage, & contaminated laboratory equipment
108		Man	Otitis, wound infections (burns), urinary tract infections, sinusitis, meningitis	
109		Cattle	Mastitis	
110		Horse	Abortion	
111	*P. pseudomallei*	Sheep, goat, swine, horse	Abscesses of lungs, liver, & spleen	Soil-borne
112		Man	Melioidosis, pseudoglanders	Wound infection; possibly ingestion or inhalation; soil-borne
113	*Proteus mirabilis*	Man	Urogenital tract infections, sinusitis, otitis, wound infections	Endogenous—commensal of intestinal tract
114		Dog	Urinary tract infections, pyoderma	
115	*Salmonella choleraesuis*	Swine	Necrotic enteritis	Contaminated food & water
116		Man	Septicemia, abscesses, gastroenteritis	
117	*S. enteritidis*	Cattle	Gastroenteritis in calves	Contaminated food & water
118		Man	Gastroenteritis	
119	*S. gallinarum*	Fowl	Fowl typhoid	Contaminated food & water
120	*S. paratyphi*	Man	Paratyphoid fever	Contaminated food & water
121	*S. pullorum*	Chicken	Bacillary white diarrhea	Contaminated food & water
122	*S. typhimurium* [1]	Domestic animals, man	Gastroenteritis	Contaminated food & water
123	*S. typhi (S. typhosa)*	Man	Typhoid fever	Contaminated food & water
124	*Shigella dysenteriae*	Man	Shigellosis, bacillary dysentery	Contaminated food, water, & inanimate objects; flies
125	*S. flexneri*	Man	Shigellosis, bacillary dysentery	Contaminated food, water, & inanimate objects; flies
126	*S. sonnei*	Man	Shigellosis, bacillary dysentery	Contaminated food, water, & inanimate objects; flies
127	*Sphaerophorus necrophorus*	Man	Urinary tract & puerperal infections, appendicitis, ulcerative colitis	Endogenous—commensal of buccal & genital mucous membranes & intestinal tracts; associated with unsanitary conditions
128		Cattle	Calf diphtheria, liver abscesses, footrot	
129		Sheep, goat	Secondary in lip & leg ulceration; footrot	
130		Horse	Gangrenous dermatitis	
131	*Spirillum minus*	Man	Rat-bite fever	Rat bite
132	*Staphylococcus aureus*	Man	Boils, furuncles, osteomyelitis, pneumonia, meningitis	Endogenous—commensal of skin & mucous membranes; wound infection
133			Food poisoning	Ingestion of preformed toxin
134		Cattle	Mastitis	Endogenous—commensal of skin & mucous membranes; wound infection
135		Fowl	Purulent synovitis	
136		Dog	Pyoderma, otitis externa	
137	*Streptobacillus moniliformis*	Man	Rat-bite fever	Rat bite
138			Haverhill fever	Ingestion
139		Rat	Otitis media	Endogenous—commensal of buccal cavity

[1] A natural pathogen for all warm-blooded animals.

continued

	Species (Synonym)	Host	Disease & Clinical Manifestation	Method of Transmission
140	*Streptococcus agalactiae*	Cattle	Mastitis	Obligate parasite of mammary glands; contact with contaminated milking equipment
141	*S. dysgalactiae*	Cattle	Mastitis	Contact with contaminated milking equipment
142	*S. equi*	Horse	Strangles; genital infection in mare	Droplet infection; contaminated food & water
143	*S. faecalis*	Man	Endocarditis	Endogenous—commensal of intestinal tract
144	*S. mitis*	Man	Septicemia, endocarditis, sinusitis	Endogenous—commensal of upper respiratory tract; other viridans streptococci also involved
145	*S. pneumoniae* (*Diplococcus pneumoniae*)	Man	Lobar pneumonia, otitis, meningitis	Endogenous—commensal of upper respiratory tract; droplet infection
146	*S. pyogenes*	Man	Tonsillitis, scarlet fever, otitis media, erysipelas, puerperal infections	Contact with infected individual or carrier
147			Rheumatic fever, glomerulonephritis	Complications following *S. pyogenes* infections
148		Cattle	Mastitis	Contact with infected man
149	*S. uberis*	Cattle	Mastitis	Contact with contaminated milking equipment
150	*S. zooepidemicus*	Domestic animals	Suppurative lesions	Animal pyogenes; contact with infected animals
151	*Treponema pallidum*	Man	Syphilis	Sexual contact
152	*Vibrio cholerae*	Man	Cholera	Contaminated food & water; carriers
153	*V. fetus*	Cattle, sheep	Abortion, sterility	Venereal infection; ingestion
154		Man	Abortion; septicemia with endocarditis; thrombophlebitis	Not known, possibly contact with infected animals
155	*V. jejuni*	Cattle	Dysentery	Feces-borne; contaminated food & water
156	*Yersinia enterocolitica*	Man	Pseudotuberculosis, colitis	Feces-borne; contaminated food & water
157		Rodents	Pseudotuberculosis	
158	*Y. pestis*	Rodents	Sylvatic plague	Fleas
159		Man, rat	Bubonic or pneumonic plague	Fleas; droplet infection
160	*Y. pseudotuberculosis*	Rodents, turkey, swine, man	Pseudotuberculosis	Contaminated food & water

Contributor: Kowalski, Joseph J.

General References

[1] Bailey, W. R., and E. G. Scott. 1970. Diagnostic Microbiology. Ed. 3. C. V. Mosby, St. Louis.

[2] Bodily, H. L., et al., ed. 1970. Diagnostic Procedures for Bacterial, Mycotic and Parasitic Infections. Ed. 5. American Public Health Association, New York.

[3] Bruner, D. W., and J. H. Gillespie, ed. 1965. Hagan's Infectious Diseases of Domestic Animals. Ed. 5. Comstock, Ithaca.

[4] Carter, G. R. 1970. Outline of Veterinary Bacteriology and Mycology. Lucas Brothers, Columbia, Mo.

[5] Dubos, R. J., and J. G. Hirsch, ed. 1965. Bacterial and Mycotic Infections of Man. Ed. 4. J. B. Lippincott, Philadelphia.

[6] Merchant, I. A., and R. A. Packer. 1967. Veterinary Bacteriology and Virology. Ed. 7. Iowa State Univ. Press, Ames

[7] Wintrobe, M. M., et al., ed. 1970. Harrison's Principles of Internal Medicine. Ed. 6. McGraw-Hill, New York.

	Host Plant	Pathogen	Disease		Host Plant	Pathogen	Disease
1	*Acer* spp.	*Pseudomonas aceris*	Leaf spot	41	*Helianthus annuus*	*Pseudomonas helianthi*	Leaf spot
2		*Xanthomonas acernea*	Leaf blight				
3	*Allium cepa*	*Pseudomonas alliicola*	Scale rot	42	*Hordeum vulgare*	*Pseudomonas coronafaciens*	Halo blight
4		*P. cepacia*	Sour skin				
5		*P. cichorii, P. marginalis*	Soft rot	43		*P. striafaciens*	Stripe blight
6		*Xanthomonas striaformans*	Stripe	44		*Xanthomonas translucens*	Blight
7	*Antirrhinum majus*	*Xanthomonas antirrhini*	Leaf spot	45	*Ipomoea batatas*	*Pseudomonas viciae*	Leaf & stem spot
				46		*Streptomyces ipomoea*	Soil rot
8	*Avena sativa*	*Pseudomonas coronafaciens*	Halo blight	47	*Iris* spp.	*Pseudomonas cichorii*	Leaf blight
				48		*P. iridicola*	Leaf blight
9		*P. coronafaciens atropurpurea*	Purple spot	49		*P. marginata*	Corm scab & leaf blight
10		*P. striafaciens*	Stripe blight	50		*Xanthomonas tardicrescens*	Leaf blight
11		*Xanthomonas translucens*	Blight				
12	*Beta vulgaris*	*Corynebacterium betae*	Silvering disease	51	*Juglans* spp.	*Erwinia nigrifluens*	Bark canker
13		*Erwinia scabiegena*	Blister scab	52		*Xanthomonas juglandis*	Blight
14		*Pseudomonas aptata*	Blight	53	*Lactuca sativa*	*Pseudomonas cichorii*	Head rot
15		*P. wieringae*	Ring rot	54		*P. marginalis*	Marginal blight & rot
16		*Xanthomonas beticola*	Bacterial pocket				
17	*Capsicum frutescens*	*Erwinia carotovora*	Soft rot	55		*P. viridilivida*	Leaf spot & wilt
18		*Pseudomonas solanacearum*	Wilt	56		*Xanthomonas vitians*	Wilt & rot
19		*Xanthomonas vesicatoria*	Bacterial spot	57	*Lycopersicon esculentum*	*Corynebacterium michiganense*	Canker
20	*Chrysanthemum* spp.	*Erwinia chrysanthemi*	Bacterial blight	58		*Pseudomonas gardneri*	Fruit spot & scab
21		*Pseudomonas cichorii*	Leaf spot	59		*P. solanacearum*	Wilt
				60		*P. tomato*	Bacterial speck
22	*Citrus* spp.	*Erwinia citrimaculans*	Fruit spot	61		*Xanthomonas vesicatoria*	Bacterial spot
23		*Pseudomonas syringae*	Blast & black pit	62	*Malus pumila*	*Agrobacterium rhizogenes*	Hairy root
24		*Xanthomonas citri*	Canker	63		*A. tumefaciens*	Crown gall
25	*Cucumis sativus*	*Erwinia tracheiphila*	Wilt	64		*Erwinia amylovora*	Fire blight
26		*Pseudomonas lachrymans*	Angular leaf spot	65		*Pseudomonas melophthora*	Brown rot of fruit
27		*Xanthomonas cucurbitae*	Leaf spot				
28	*Cucurbita pepo*	*Erwinia tracheiphila*	Wilt	66		*P. papulans*	Blister spot
				67		*P. pomi*	Fruit rot
29	*Daucus carota*	*Erwinia carotovora*	Soft rot	68		*P. syringae*	Twig blight
30		*Xanthomonas carotae*	Leaf blight	69	*Medicago sativa*	*Corynebacterium insidiosum*	Wilt
31	*Fragaria virginiana*	*Corynebacterium fascians*	Cauliflower disease				
				70		*Pseudomonas medicaginis*	Stem blight
32		*Xanthomonas fragariae*	Angular leaf spot	71		*Xanthomonas alfalfae*	Leaf spot
33	*Fraxinus* spp.	*Pseudomonas savastanoi fraxini*	Canker	72	*Nicotiana* spp.	*Erwinia aroideae*	Hollow stalk & barn rot
34	*Gladiolus* spp.	*Pseudomonas gladioli*	Rot	73		*Pseudomonas angulata*	Angular leaf spot
35		*P. marginata*	Corm scab & leaf blight	74		*P. mellea*	Rust
				75		*P. polycolor*	Leaf spot & wet rot
36		*Xanthomonas gummisudans*	Blight				
				76		*P. pseudozoogloeae*	Black rust
37	*Glycine max* [1]	*Pseudomonas glycinea*	Blight	77		*P. solanacearum*	Granville disease
38		*P. tabaci*	Wildfire	78		*P. tabaci*	Wildfire
39		*Xanthomonas phaseoli sojensis*	Pustule	79		*Xanthomonas heterocea*	Rust
40	*Gossypium* spp.	*Xanthomonas malvacearum*	Angular leaf spot	80	*Oryza sativa*	*Pseudomonas oryzicola*	Leaf spot
				81		*P. setariae*	Stripe
				82		*Xanthomonas oryzae*	Blight
				83		*X. oryzicola*	Leaf streak

[1] Synonym: *G. soja*.

continued

1099

	Host Plant	Pathogen	Disease		Host Plant	Pathogen	Disease
84	*Pastinaca sativa*	*Pseudomonas pastinacae*	Brown rot	122		*Pseudomonas solanacearum*	Brown rot
85	*Phaseolus vulgaris*	*Corynebacterium flaccumfaciens*	Wilt	123		*Streptomyces scabies*	Scab
				124	*Trifolium*	*Pseudomonas cichorii*	Leaf blight
86		*Pseudomonas flectens*	Pod twist	125	spp.	*P. radiciperda*	Root rot
87		*P. phaseolicola*	Halo blight	126		*P. stizolobii*	Leaf spot
88		*P. stizolobii*	Leaf spot	127		*P. syringae*	Leaf spot & blight
89		*P. syringae*	Lilac blight				
90		*P. viridiflava*	Leaf spot & blight	128	*Triticum*	*Corynebacterium iranicum*	Spike blight
				129	*aestivum*	*C. tritici*	Spike blight
91		*Xanthomonas phaseoli*	Common blight	130		*Pseudomonas atrofaciens*	Basal glume rot
92		*X. phaseoli fuscans*	Fuscous blight	131		*Xanthomonas translucens* f. sp. *undulosa*	Black chaff
93	*Phleum pratense*	*Xanthomonas translucens* f. sp. *phlei-pratensis*	Streak	132	*Ulmus*	*Erwinia nimipressuralis*	Wetwood
94	*Pisum sativum*	*Pseudomonas pisi*	Blight	133	spp.	*Pseudomonas lignicola*	Black streak of wood
95		*Xanthomonas pisi*	Stem rot	134		*P. ulmi*	Leaf spot
96	*Populus* spp.	*Corynebacterium humiferum*	Wetwood	135	*Vicia faba*	*Pseudomonas fabae*	Blight
97		*Pseudomonas rimaefaciens*	Canker	136		*P. viciae*	Leaf & stem spot
98	*Prunus domestica*	*Pseudomonas morsprunorum*	Canker & leaf spot	137	*Vitis* spp.	*Agrobacterium tumefaciens*	Crown gall
99		*P. syringae*	Canker & gummosis	138		*Erwinia vitivora*	Blight
100		*Xanthomonas pruni*	Fruit & leaf spot	139		*Xanthomonas vitis-carnosae*	Leaf spot
101	*Prunus persica*	*Pseudomonas syringae*	Canker & gummosis	140	*Zea mays*	*Erwinia carotovora* f. sp. *zeae*	Stalk rot & leaf blight
102		*Xanthomonas pruni*	Leaf & fruit spot	141		*E. dissolvens*	Stalk rot
103	*Pyrus communis*	*Erwinia amylovora*	Fire blight	142		*Pseudomonas alboprecipitans*	Leaf blight & stalk rot
104		*Pseudomonas barkeri*	Blossom blight	143		*P. andropogonis*	Stripe
105		*P. nectarophila*	Blossom blight	144		*P. desaiana*	Stinking rot
106		*P. syringae*	Twig & blossom blight	145		*P. lapsa*	Leaf & stalk rot
				146		*Xanthomonas stewartii*	Wilt
107	*Raphanus*	*Pseudomonas maculicola*	Black spot	147	Fleshy	*Erwinia aroideae*	Soft rots
108	*sativus*	*Xanthomonas campestris*	Black rot	148	vegetable[2]	*E. atroseptica*	Soft rots
109		*X. vesicatoria raphani*	Leaf spot	149		*E. carotovora*	Soft rots
110	*Rheum rhaponticum*	*Erwinia rhapontici*	Crown rot	150		*E. chrysanthemi*	Soft rots
111		*Pseudomonas marginalis*	Soft rot	151	Numerous plants	*Agrobacterium tumefaciens*	Crown gall
112	*Ribes aureum*	*Pseudomonas ribicola*	Leaf spot	152		*Corynebacterium fascians*	Fasciation
113	*Rosa multiflora*	*Agrobacterium rhizogenes*	Hairy root	153		*Erwinia amylovora*	Fire blight of Rosaceae
114		*A. tumefaciens*	Crown gall				
115		*Pseudomonas syringae*	Leaf spot & blast	154		*E. aroideae*	Soft rot
116	*Salix* spp.	*Erwinia salicis*	Watermark	155		*E. atroseptica*	Soft rot
117		*Pseudomonas saliciperda*	Blight	156		*E. carotovora*	Soft rot
118	*Solanum tuberosum*	*Bacillus* spp.	Storage rots	157		*E. chrysanthemi*	Soft rot
119		*Corynebacterium sepedonicum*	Ring rot	158		*Pseudomonas solanacearum*	Brown rot
120		*Erwinia* spp.	Soft rot	159		*P. syringae*	Blight
121		*E. atroseptica*	Blackleg				

[2] Plants of tuberous and fleshy roots.

Contributor: Dickey, Robert S.

continued

General References

[1] Anderson, H. W. 1956. Diseases of Fruit Crops. McGraw-Hill, New York.

[2] Breed, R. S., et al., ed. 1957. Bergey's Manual of Determinative Bacteriology. Ed. 7. Williams and Wilkins, Baltimore.

[3] Chupp, C., and A. F. Sherf. 1960. Vegetable Diseases and Their Control. Ronald Press, New York.

[4] Dickson, J. G. 1956. Diseases of Field Crops. Ed. 2. McGraw-Hill, New York.

[5] Dowson, W. J. 1957. Plant Diseases Due to Bacteria. Ed. 2. Cambridge Univ. Press, London.

[6] Elliott, C. 1951. Manual of Bacterial Plant Pathogens. Ed. 2. Chronica Botanica, Waltham, Mass.

[7] U.S. Department of Agriculture. 1953. Plant Diseases. Yearbook of Agriculture. U.S. Government Printing Office, Washington, D.C.

131. FUNGAL PARASITES: PLANTS

Part I. Field, Fruit, and Vegetable Crops

This list of the important fungal pathogens represents contributions by plant pathologists from more than 20 states. Most vegetable crops have damping-off and root rot caused by *Phytophthora* spp., *Pythium* spp., or *Rhizoctonia solani*.

	Host Plant	Pathogen (Synonym)	Disease (Synonym)
1	*Agrostis pa-*	*Agaricus campestris (Psalliota campestris)*	Fairy ring
2	*lustris*	*Colletotrichum graminicola*	Anthracnose
3		*Corticium fusiforme*	Red thread
4		*Erysiphe graminis*	Powdery mildew
5		*Fusarium nivale*	Fusarium patch (pink snow mold)
6		*Gloeocercospora sorghi*	Copper spot
7		*Helminthosporium sorokinianum (H. sativum)*	Melting-out (fading-out); leaf spot
8		*H. vagans*	Leaf spot; crown rot
9		*Lepiota morgani*	Fairy ring
10		*Lycoperdon* spp.	Fairy ring
11		*Marasmius oreades*	Fairy ring
12		*Ophiobolus graminis*	Ophiobolus patch
13		*Puccinia coronata*	Crown rust
14		*P. graminis*	Stem rust
15		*Pythium aphanidermatum*	Spot blight (grease spot)
16		*Rhizoctonia solani*	Brown patch
17		*Sclerotium homoeocarpa*	Dollar spot
18		*Typhula incarnata (T. itoana)*	Typhula blight (gray snow mold)
19		*Urocystis agropyri*	Flag smut
20		*Ustilago striiformis*	Stripe smut
21	*Allium* spp.	*Alternaria porri*	Purple blotch
22		*Aspergillus niger*	Black mold
23		*Botrytis* spp.	Neck rot; blast
24		*Fusarium* spp.	Root rot; basal bulb rot
25		*Peronospora destructor*	Downy mildew
26		*Pyrenochaeta terrestris*	Pink rot
27		*Sclerotium cepivorum*	White rot
28		*Urocystis cepulae*	Smut
29	*Apium gra-*	*Cercospora apii*	Early blight
30	*veolens*	*Fusarium oxysporum*	Fusarium yellows; root rot
31		*Phoma apiicola*	Phoma rot; crown rot
32		*Sclerotinia sclerotiorum*	Pink rot
33		*Septoria apiicola; S. apii-graveolentis*	Late blight

continued

Part I. Field, Fruit, and Vegetable Crops

	Host Plant	Pathogen (Synonym)	Disease (Synonym)
34	*Arachis hy-*	*Aspergillus niger*	Aspergillus crown rot
35	*pogaea*	*Cercospora arachidicola; C. personata*	Cercospora leaf spot
36		*Pellicularia rolfsii*	Stem rot
37		*Pythium myriotylum*	Pod rot; root rot
38		*Rhizoctonia solani*	Rhizoctonia disease
39		*Rhizopus arrhizus; R. oryzae; R. stolonifer*	Seed & seedling rot
40		*Thielaviopsis basicola*	Black-hull disease
41		*Verticillium albo-atrum*	Verticillium wilt
42	*Asparagus*	*Botrytis cinerea*	Gray mold; shoot blight; spear rot
43	*officinalis*	*Fusarium oxysporum*	Fusarium wilt; root rot; transit rot
44		*Puccinia asparagi*	Rust
45	*Avena sativa*	*Colletotrichum graminicola*	Leaf spot
46		*Erysiphe graminis*	Powdery mildew
47		*Fusarium* spp.[1]	Scab; foot rot; seedling blight; head mold
48		*Helminthosporium avenae*	Leaf blotch; black stem
49		*Leptosphaeria avenaria*	Black stem
50		*Puccinia coronata*	Crown rust
51		*P. graminis*	Stem rust
52		*Ustilago avenae*	Loose smut
53		*U. kolleri*	Covered smut
54	*Beta vulgaris*	*Aphanomyces cochlioides*	Black root; tip rot
55		*Cercospora beticola*	Cercospora leaf spot
56		*Fusarium* spp.	Root rot; storage rot; fusarium yellows
57		*Peronospora schachtii*	Downy mildew
58		*Phoma betae*	Storage rot
59		*Phymatotrichum omnivorum*	Texas root rot
60		*Pleospora betae*	Leaf spot; root rot
61		*Pythium aphanidermatum*	Root rot of mature sugar beet
62		*Rhizoctonia solani*	Root rot
63		*Rhizopus stolonifer*	Storage rot
64		*Uromyces betae*	Rust
65	*Brassica*	*Alternaria brassicae*	Alternaria leaf spot
66	*oleracea*	*Fusarium oxysporum*	Cabbage yellows
67		*Mycosphaerella brassicicola*	Ring spot
68		*Peronospora parasitica*	Downy mildew
69		*Phoma lingam*	Blackleg
70		*Plasmodiophora brassicae*	Club root
71		*Rhizoctonia* spp.	Wire stem
72		*Sclerotinia sclerotiorum*	White blight
73	*Capsicum*	*Phytophthora capsici*	Root rot (chile wilt)
74	*annuum*	*Verticillium albo-atrum*	Verticillium wilt
75	*Capsicum*	*Alternaria* spp.	Black fruit rot
76	*frutescens*	*Botrytis cinerea*	Gray mold; fruit rot; stem canker
77		*Cercospora capsici*	Frogeye leaf spot
78		*Fusarium oxysporum; F. vasinfectum*	Fusarium wilt
79		*Gloeosporium piperatum*	Anthracnose
80		*Phytophthora capsici*	Phytophthora blight
81		*Sclerotium rolfsii*	Southern stem blight
82	*Carthamus*	*Alternaria* spp.	Leaf spot
83	*tinctorius*	*Fusarium oxysporum* f. sp. *carthami*	Fusarium wilt
84		*Gloeosporium carthami*	Anthracnose; blight
85		*Phytophthora drechsleri*	Root rot
86		*Puccinia carthami*	Rust
87		*Sclerotinia sclerotiorum*	Rot; wilt
88		*Septoria carthami*	Leaf spot

[1] Also the perfect states *Gibberella roseum* and *G. zeae*.

continued

Part I. Field, Fruit, and Vegetable Crops

	Host Plant	Pathogen (Synonym)	Disease (Synonym)
89	*Carya illino-*	*Cercospora fusca*	Brown leaf spot
90	*ensis*	*Cladosporium effusum*	Scab
91		*Elsinoe randii*	Nursery blight
92		*Gnomonia caryae*	Liver spot
93		*G. nerviseda*	Vein spot
94		*Microsphaera alni*	Powdery mildew
95		*Mycosphaerella caryigena*	Downy spot
96		*M. dendroides*	Leaf blotch
97		*Pellicularia koleroga*	Thread blight
98		*Phymatotrichum omnivorum*	Texas root rot
99	*Citrus* spp.	*Cercospora citri-grisea*	Greasy spot
100		*Clitocybe tabescens*	Root rot
101		*Diaporthe citri*	Melanose; stem-end rot
102		*Diplodia natalensis*	Twig & branch dieback; stem-end rot
103		*Elsinoe fawcettii*	Scab
104		*Glomerella cingulata*	Withertip; anthracnose
105		*Peronospora parasitica*	Downy mildew
106		*Phytophthora citrophthora*	Foot rot; brown rot
107	Cucurbita-	*Alternaria cucumerina*	Leaf blight
108	ceae	*Choanephora cucurbitorum*	Blossom blight; fruit rot
109		*Cladosporium cucumerinum*	Scab
110		*Colletotrichum lagenarium*[2/]	Anthracnose
111		*Erysiphe cichoracearum*	Powdery mildew
112		*Fusarium oxysporum* f. sp. *cucumerinum; F. oxysporum* f. sp. *melonis; F. oxysporum* f. sp. *niveum*	Fusarium wilt of cucumber, muskmelon, & watermelon, respectively
113		*F. solani*	Fusarium root rot
114		*Mycosphaerella citrullina*	Gummy stem blight
115		*M. melonis*	Black rot
116		*Pseudoperonospora cubensis*	Downy mildew
117		*Sclerotinia sclerotiorum*	Stem & fruit rot
118		*Sclerotium rolfsii*	Southern blight
119		*Verticillium albo-atrum*	Verticillium wilt
120	*Cynodon*	*Helminthosporium cynodontis*	Leaf blotch
121	*dactylon*	*Rhizoctonia solani*	Brown patch
122		*Sclerotinia homoeocarpa*	Dollar spot
123		*Ustilago cynodontis*	Inflorescence smut
124	*Daucus ca-*	*Alternaria dauci*	Leaf blight
125	*rota*	*A. radicina*	Black rot
126		*Botrytis cinerea*	Gray mold of stored carrot
127		*Cercospora carotae*	Blight; leaf spot
128		*Fusarium* sp.	Crown cavity; spot rots
129		*Pythium* sp.	Crown cavity; spot rots
130		*Rhizoctonia* sp.	Crown cavity; spot rots
131		*R. carotae*	Crater rot
132		*Rhizopus* spp.	Storage rot
133		*Sclerotinia sclerotiorum*	Watery soft rot
134	*Eremochloa*	*Colletotrichum graminicola*	Anthracnose
135	*ophiuroi-*	*Curvularia* sp.	Fading-out
136	*des*	*Helminthosporium* sp.	Leaf spot
137		*Rhizoctonia solani*	Brown patch
138	*Festuca* spp.	*Colletotrichum graminicola*	Anthracnose
139		*Corticium fusiforme*	Red thread
140		*Curvularia* sp.	Fading-out
141		*Fusarium roseum;* other *Fusarium* spp.	Fusarium blight; root rot
142		*Helminthosporium dictyoides; H. sorokinianum (H. sativum)*	Leaf spot; root rot (melting-out)
143		*Puccinia coronata*	Crown rust

[2/] Pathogen of plants other than squash.

continued

Part I. Field, Fruit, and Vegetable Crops

	Host Plant	Pathogen (Synonym)	Disease (Synonym)
144		*P. graminis*	Stem rust
145		*Rhizoctonia solani*	Brown patch
146		*Sclerotinia homoeocarpa*	Dollar spot
147	*Ficus carica*	*Fusarium moniliforme*	Endosepsis
148		*Phomopsis cinerescens*[3]	Canker
149	*Fragaria chi-*	*Botrytis cinerea*	Gray mold rot
150	*loensis*	*Colletotrichum fragariae*	Anthracnose
151		*Dendrophoma obscurans*	Leaf blight; stem-end rot
152		*Diplocarpon earliana*	Leaf scorch
153		*Fusarium* spp.	Rot
154		*Gnomonia fragariae*	Leaf & stem blight
155		*Mycosphaerella fragariae*	Leaf spot
156		*Phytophthora fragariae*	Red stele
157		*Pythium* spp.	Replant disease
158		*Rhizoctonia solani*	Black rot; leak
159		*Sphaerotheca humuli*	Powdery mildew
160		*Verticillium albo-atrum*	Verticillium wilt
161	*Glycine*	*Cephalosporium gregatum*	Brown stem rot
162	*max*[4]	*Cercospora kikuchii*	Purple stain
163		*C. sojina*	Frogeye leaf spot
164		*Colletotrichum truncatum*	Anthracnose
165		*Corynespora cassiicola*	Target spot
166		*Diaporthe phaseolorum sojae*	Pod & stem blight; stem canker
167		*Fusarium orthoceras*	Fusarium root rot
168		*F. oxysporum* f. *tracheiphilum*	Fusarium wilt
169		*Glomerella glycines*	Anthracnose
170		*Macrophomina phaseoli*	Charcoal rot (ashy stem blight)
171		*Pellicularia rolfsii*	Southern wilt; southern blight
172		*Peronospora manshurica*	Downy mildew
173		*Phyllosticta sojicola*	Leaf spot
174		*Phymatotrichum ominivorum*	Phymatotrichum root rot
175		*Phytophthora megasperma*	Phytophthora root rot
176		*Pythium ultimum*	Root rot
177		*Rhizoctonia solani*	Root & stem rot
178		*Sclerotinia sclerotiorum*	Stem rot
179		*Sclerotium rolfsii*	Sclerotial blight (southern blight)
180		*Septoria glycines*	Brown spot
181		*Thielaviopsis basicola*	Thielaviopsis root rot
182	*Gossypium*	*Fusarium oxysporum* f. *vasinfectum*	Fusarium wilt
183	spp.	*Glomerella gossypii*	Anthracnose; boll rot
184		*Pellicularia filamentosa*	"Sore shin" seedling stem canker
185		*Phymatotrichum omnivorum*	Root rot
186		*Phytophthora parasitica*	Rust
187		*Puccinia stakmanii*	Rust
188		*Pythium* spp.	Seedling disease
189		*Rhizoctonia solani*	Seedling disease
190		*Thielaviopsis basicola*	Damping-off; root rot
191		*Verticillium albo-atrum*	Verticillium wilt
192	*Helianthus*	*Erysiphe cichoracearum*	Powdery mildew
193	spp.	*Plasmopara halstedii*	Downy mildew
194		*Uromyces junci*	Rust
195	*Hordeum*	*Cephalosporium gramineum*	Cephalosporium stripe
196	*vulgare*	*Claviceps purpurea*	Eyespot
197		*Erysiphe graminis*	Powdery mildew
198		*Fusarium graminearum (Gibberella zeae)*	Fusarium blight (scab)
199		*F. roseum*	Foot rot

[3] Conidial stage of *Diaporthe*. [4] Synonym: *G. soja.*

continued

Part I. Field, Fruit, and Vegetable Crops

	Host Plant	Pathogen (Synonym)	Disease (Synonym)
200		*Helminthosporium gramineum*	Stripe disease
201		*H. sorokinianum (H. sativum)*	Spot blotch; root rot; foot rot; kernel blight
202		*Puccinia graminis*	Stem rust
203		*P. hordei*	Leaf rust
204		*Pyrenophora teres*	Net blotch
205		*Rhynchosporium secalis*	Scald
206		*Septoria avenae* f. sp. *triticia*	Septoria blotch
207		*Typhula incarnata (T. itoana)*	Snow mold
208		*Ustilago hordei*	Covered smut
209		*U. nigra*	Black or semiloose smut
210		*U. nuda*	Loose smut
211	*Ipomoea ba-*	*Botrytis cinerea*	Gray mold rot of sprouts; storage rot
212	*tatas*	*Diplodia tubericola*	Java black rot
213		*Endoconidiophora fimbriata*	Black rot
214		*Fusarium oxysporum* f. *batatas*	Wilt; stem rot
215		*F. solani*	Surface rot
216		*Monilochaetes infuscans*	Scurf
217		*Mucor racemosus*	Storage rot
218		*Plenodomus destruens*	Foot rot
219		*Rhizopus stolonifer*	Soft rot
220		*Sclerotium rolfsii*	Southern blight; circular spot
221		*Streptomyces ipomoea*	Pox (soil rot)
222	*Juglans re-*	*Armillaria mellea*	Root rot
223	*gia*	*Ascochyta juglandis*	Ring spot
224		*Dothiorella gregaria*	Dieback; black sap
225		*Exosporina fawcetti*	Branch wilt; canker
226		*Gnomonia leptostyla*	Leaf blotch
227		*Hendersonula toruloidea*	Branch wilt
228		*Phytophthora cactorum*	Crown rot
229	*Lactuca sa-*	*Alternaria sonchi*	Leaf spot
230	*tiva*	*Botrytis cinerea*	Gray mold
231		*Bremia lactucae*	Downy mildew
232		*Marssonina panattoniana*	Anthracnose
233		*Pellicularia filamentosa*	Bottom rot
234		*Rhizoctonia solani*	Bottom rot
235		*Sclerotinia sclerotiorum*	Damping-off; drop; watery soft rot
236	*Linum usi-*	*Fusarium oxysporum*	Fusarium wilt
237	*tatissimum*	*Melampsora lini*	Rust
238		*Mycosphaerella linorum*	Pasmo
239	*Lycopersi-*	*Alternaria solani*	Early blight
240	*con escu-*	*A. tenuis*	Black mold
241	*lentum*	*Botrytis cinerea* [5]	Gray mold
242		*Cladosporium fulvum*	Leaf mold
243		*Colletotrichum phomoides*	Anthracnose
244		*Fusarium oxysporum* f. *lycopersici*	Fusarium wilt
245		*Pellicularia filamentosa*	Soil rot
246		*Phoma destructiva*	Phoma rot
247		*Phytophthora capsici*	Root rot
248		*P. infestans*	Late blight
249		*P. parasitica*	Buckeye rot
250		*Rhizoctonia solani*	Collar rot; soil rot
251		*Rhizopus stolonifer*	Rhizopus ripe rot
252		*Sclerotinia sclerotiorum*	Stem rot
253		*Sclerotium rolfsii*	Southern stem blight
254		*Septoria lycopersici*	Leaf spot
255		*Stemphylium solani*	Gray leaf spot
256		*Verticillium albo-atrum*	Verticillium wilt

[5] Pathogens of plants grown in greenhouses.

continued

Part I. Field, Fruit, and Vegetable Crops

	Host Plant	Pathogen (Synonym)	Disease (Synonym)
257	*Malus pu-*	*Alternaria mali*	Soft rot
258	*mila*	*A. tenuis*	Core rot
259		*Armillaria mellea*	Root rot
260		*Botryosphaeria ribis*	Botryosphaeria canker; fruit rot
261		*Botrytis cinerea;* other *Botrytis* spp.	Gray mold rot; soft rot
262		*Clitocybe tabescens*	Root rot
263		*Corticium galactinum*	White root rot
264		*Cytospora* spp.	Cankers
265		*Fomes* spp.	Wood rot
266		*Gloeodes pomigena*	Sooty blotch
267		*Glomerella cingulata*	Bitter rot of fruit; stem canker
268		*Gymnosporangium clavipes*	Quince rust
269		*G. globosum*	Hawthorn rust
270		*G. juniperi*	Rust
271		*Helminthosporium papulosum*	Black pox
272		*Hydnum* spp.	Wood rot
273		*Lenzites* spp.	Wood rot
274		*Leptothyrium pomi*	Flyspeck
275		*Microthyriella rubi*	Flyspeck
276		*Monilinia fructicola*	Brown rot
277		*Nectria galligena*	European canker
278		*Neofabraea malicorticis*	Anthracnose; black spot canker
279		*Nummularia discreta*	Blister canker
280		*Penicillium expansum;* other *Penicillium* spp.	Blue mold rot
281		*Phyllosticta solitaria*	Blotch; leaf spot; canker
282		*Phytophthora cactorum*	Collar rot; fruit rot
283		*Physalospora obtusa*	Frogeye; black rot
284		*Podosphaera leucotricha*	Powdery mildew
285		*Polyporus* spp.	Wood rot
286		*Poria stereums*	Wood rot
287		*Rhizopus nigricans*	Ripe fruit rot
288		*Sclerotium rolfsii*	Southern stem blight
289		*Sphaeropsis malorum*	Black rot; canker
290		*Stereum purpureum*	Silver leaf
291		*Venturia inaequalis*	Scab
292		*Xylaria mali*	Black root rot
293	*Mangifera indica*	*Colletotrichum gloeosporioides*	Anthracnose
294	*Medicago*	*Aphanomyces euteiches*	Root rot
295	*sativa*	*Cercospora medicaginis*	Leaf spot
296		*C. zebrina*	Summer black stem; leaf spot
297		*Colletotrichum trifolii;* other *Colletotrichum* spp.	Anthracnose
298		*Fusarium* spp.	Wilt; fusarium root rot
299		*Leptosphaeria pratensis*	Leaf spot; crown & root rot
300		*Leptosphaerulina briosiana (Pseudoplea briosiana)*	Leaf spot
301		*L. trifolii (P. trifolii)*	Brown leaf spot
302		*Peronospora trifoliorum*	Downy mildew
303		*Phoma herbarum medicaginis*	Spring black stem
304		*Phymatotrichum omnivorum*	Phymatotrichum root rot
305		*Phytophthora erythroseptica*	Crown & root rot
306		*P. megasperma*	Phytophthora root rot; water mold
307		*Pseudopeziza jonesii*	Yellow leaf blotch
308		*P. medicaginis*	Common leaf spot
309		*Pythium ultimum*	Seedling & root rot
310		*Rhizoctonia solani;* other *Rhizoctonia* spp.	Seedling & root rot
311		*Sclerotinia trifoliorum*	Crown & root rot
312		*Stagonospora meliloti*	Crown & root rot

continued

Part I. Field, Fruit, and Vegetable Crops

	Host Plant	Pathogen (Synonym)	Disease (Synonym)
313		*Stemphylium botryosum*	Stemphylium leaf spot
314		*Uromyces striatus*	Rust
315		*Urophlyctis alfalfae*	Crown wart
316	*Mentha car-*	*Phoma strasseri*	Black stem rot
317	*diaca; M.*	*Puccinia menthae*	Mint rust
318	*piperita;*	*Septoria menthicola*	Leaf spot
319	*M. spicata*	*Sphaceloma menthae*	Anthracnose (leopard spot)
320		*Verticillium albo-atrum*	Verticillium wilt
321	*Nicotiana*	*Alternaria alternata; A. longipes*	Brown spot
322	*tabacum*	*Cercospora nicotianae*	Frogeye leaf spot
323		*Colletotrichum* sp.	Anthracnose
324		*Fusarium oxysporum*	Fusarium wilt
325		*Macrophomina phaseoli*	Charcoal rot
326		*Pellicularia filamentosa*	Stem canker
327		*Peronospora tabacina*	Blue mold
328		*Phytophthora parasitica*	Black shank
329		*Pythium debaryanum*	Damping-off
330		*Rhizoctonia solani*	"Sore shin"; damping-off
331		*Thielaviopsis basicola*	Black rot
332	*Olea euro-*	*Verticillium albo-atrum*	Verticillium wilt
	paea		
333	*Oryza sativa*	*Achlya klebsiana*	Seedling blight
334		*Cercospora oryzae*	Cercospora spot
335		*Cochliobolus miyabeanus*	Helminthosporium blight
336		*Helminthosporium oryzae*	Seedling blight; brown spot on leaves
337		*Leptosphaeria salvinii*	Stem rot
338		*Magnaporthe salvinii*	Stem rot
339		*Pythium* spp.	Pythium seedling blight
340		*Piricularia oryzae*	Blast
341		*Rhizoctonia oryzae*	Sheath blight
342		*Tilletia horrida*	Kernel smut
343	*Pastinaca sa-*	*Itersonilia perplexans*	Crown rot
	tiva		
344	*Persea amer-*	*Botryosphaeria ribis*	Branch canker; fruit rot
345	*icana*	*Cercospora purpurea*	Cercospora spot or blotch
346		*Diplodia theobromae*	Stem-end rot
347		*Oidium* spp.	Powdery mildew
348		*Phomopsis* spp.[3/]	Stem-end rot
349		*Phytophthora cinnamomi*	Phytophthora root rot
350		*Sphaceloma perseae*	Scab[6/]
351		*Verticillium albo-atrum*	Verticillium wilt
352	*Phaseolus*	*Botrytis cinerea*	Gray mold rot
353	*vulgaris*	*Colletotrichum lindemuthianum*	Anthracnose
354		*Erysiphe polygoni*	Powdery mildew
355		*Fusarium phaseoli; F. solani*	Root rot
356		*Macrophomina phaseoli*	Ashy stem blight; charcoal rot; leaf spot; root rot
357		*Pellicularia filamentosa*	Stem rot
358		*Rhizoctonia solani*	Root & stem rot; stem canker
359		*Sclerotinia sclerotiorum*	White mold (sclerotinia wilt)
360		*Sclerotium rolfsii*	Southern blight
361		*Uromyces phaseoli*	Rust
362	*Pisum sati-*	*Aphanomyces euteiches*	Root rot
363	*vum*	*Ascochyta pinodella*	Ascochyta foot rot
364		*A. pisi*	Ascochyta leaf & pod spot
365		*Botrytis cinerea*	Gray mold rot
366		*Colletotrichum pisi*	Anthracnose

[3/] Conidial stage of *Diaporthe.* [6/] Fruit and foliage disease.

continued

Part I. Field, Fruit, and Vegetable Crops

	Host Plant	Pathogen (Synonym)	Disease (Synonym)
367		*Erysiphe polygoni*	Powdery mildew
368		*Fusarium oxysporum* f. sp. *pisi* 1	Fusarium wilt
369		*F. oxysporum* f. sp. *pisi* 2	Near wilt
370		*F. solani*	Root rot
371		*Mycosphaerella pinodes*	Mycosphaerella blight
372		*Peronospora viciae*	Downy mildew
373		*Pythium* sp.	Root rot
374	*Poa praten-*	*Ascochyta graminicola*	Leaf spot
375	*sis*	*Colletotrichum graminicola*	Anthracnose
376		*Erysiphe graminis*	Powdery mildew
377		*Fusarium nivale*	Fusarium patch (pink snow mold)
378		*F. roseum*	Fusarium blight
379		*Helminthosporium sorokinianum (H. sativum)*	Melting-out (fading-out); leaf spot
380		*H. vagans*	Leaf & sheath rot
381		*Puccinia graminis; P. poa-sudeticae*	Stem rust
382		*P. rubigo-vera*	Orange leaf rust
383		*Rhizoctonia solani*	Brown patch
384		*Sclerotinia homoeocarpa*	Dollar spot
385		*Septoria macropoda*	Leaf blotch
386		*Typhula incarnata (T. itoana)*	Typhula blight (gray snow mold)
387		*Urocystis agropyri*	Flag smut
388		*Uromyces dactylidis*	Leaf rust
389		*Ustilago striiformis*	Stripe smut
390		*U. trebouxii*	Loose smut
391	*Prunus*	*Armillaria mellea*	Root rot
392	*amygdalus*	*Ceratocystis fimbriata*	Ceratocystis canker (mallet wound canker)
393		*Coryneum carpophilum*	Blight; shot hole
394		*Cytospora* sp.	Canker
395		*Monilinia fructicola*	American brown rot
396		*M. laxa*	European brown rot; blossom blight
397		*Phytophthora* spp.	Crown rot
398		*Rhizopus stolonifer*	Hull rot
399	*Prunus ar-*	*Eutypa armeniacae*	Limb dieback
400	*meniaca*	*Monilinia fructicola*	American brown rot
401		*M. laxa*	European brown rot; blossom blight
402	*Prunus avi-*	*Botrytis cinerea*	Blossom blight
403	*um*	*Coccomyces hiemalis*	Leaf spot
404		*Cytosphora* sp.	Cytospora canker
405		*Monilinia fructicola; M. laxa*	Brown rot
406		*Phytophthora* spp.	Crown rot
407		*Sphaerotheca pannosa*	Powdery mildew
408		*Verticillium albo-atrum*	Verticillium wilt
409	*Prunus cer-*	*Botrytis cinerea*	Blossom blight
410	*asus*	*Coccomyces hiemalis*	Leaf spot
411		*Monilinia fructicola; M. laxa*	Brown rot
412		*Podosphaera oxyacanthae*	Powdery mildew
413		*Verticillium albo-atrum*	Verticillium wilt
414	*Prunus do-*	*Armillaria mellea*	Root rot
415	*mestica*	*Cladosporium carpophilum*	Scab
416		*Coccomyces prunophorae*	Leaf spot
417		*Cytospora* sp.	Canker
418		*Dibotryon morbosum*	Black knot
419		*Monilinia fructicola*	Brown rot; blossom blight
420		*M. laxa*	European brown rot; blossom & twig blight
421		*Podosphaera oxyacanthae*	Powdery mildew
422		*Taphrina communis; T. pruni; T. rhizipes*	Plum pockets
423		*Tranzschelia pruni-spinosae*	Rust

continued

Part I. Field, Fruit, and Vegetable Crops

	Host Plant	Pathogen (Synonym)	Disease (Synonym)
424	*Prunus per-*	*Armillaria mellea*	Root rot
425	*sica*	*Botrytis cinerea*	Gray mold; fruit rot
426		*Cladosporium carpophilum*	Peach scab
427		*Clitocybe tabescens*	Root rot
428		*Coryneum carpophilum*	Coryneum blight; shot hole
429		*Cytospora cincta*	Cytospora canker
430		*C. leucostoma*	Cytospora canker
431		*Fusicoccum amygdali*	Canker
432		*Glomerella cingulata*	Ripe rot; twig blight
433		*Monilinia fructicola*	American brown rot; twig canker
434		*M. laxa*	European brown rot
435		*Penicillium expansum*	Soft rot
436		*Phytophthora cactorum*	Crown rot; sprinkler rot
437		*Rhizopus nigricans*	Fruit rot
438		*Sphaerotheca pannosa*	Powdery mildew
439		*Taphrina deformans*	Leaf curl
440		*Tranzschelia discolor*	Rust
441		*Valsa cincta; V. leucostoma*	Perennial canker; dieback
442		*Verticillium albo-atrum*	Verticillium wilt
443	*Psidium* spp.	*Glomerella cingulata*	Glomerella disease
444	*Pyrus com-*	*Botryosphaeria ribis*	Black rot
445	*munis*	*Botrytis cinerea*	Gray mold
446		*Clitocybe tabescens*	Root rot
447		*Fabraea maculata*	Leaf spot
448		*Gloeodes pomigena*	Sooty blotch
449		*Leptothyrium pomi*	Flyspeck
450		*Monilinia fructicola*	Brown rot
451		*Neofabraea malicorticis*	Black spot canker
452		*N. perennans*	Perennial canker
453		*Penicillium expansum*	Fruit rot
454		*Physalospora obtusa*	Black rot
455		*Phytophthora cactorum*	Crown rot
456		*Podosphaera leucotricha*	Powdery mildew
457		*Rhizopus nigricans*	Fruit rot
458		*Venturia pirina*	Scab
459	*Rubus*	*Botrytis cinerea*	Gray mold; fruit rot
460	spp. [7]	*Cercosporella rubi*	Double blossom; rosette
461		*Cylindrosporium rubi*	Leaf & cane spot
462		*Didymella applanata*	Spur blight; gray bark
463		*Elsinoe veneta*	Anthracnose
464		*Gymnoconia peckiana*	Orange rust
465		*Kuehneola uredinis*	Yellow rust; cane rust
466		*Kunkelia nitens*[8]	Orange rust
467		*Leptosphaeria coniothyrium*	Cane blight
468		*Mycosphaerella rubi*	Leaf & cane spot
469		*Phragmidium rubi-idaei*	Leaf rust
470		*Phyllactinia guttata*	Powdery mildew
471		*Rhizopus nigricans*	Black mold of fruit
472		*Sphaerotheca humuli*	Powdery mildew
473		*Verticillium albo-atrum*	Verticillium wilt
474	*Saccharum*	*Ceratostomella adiposum; C. paradoxa*	Pineapple disease
475	*officina-*	*Colletotrichum graminicola (C. falcatum)*	Red rot
476	*rum*	*Cytospora sacchari*	Sheath rot
477		*Melanconium sacchari*	Rind disease
478		*Phytophthora erythroseptica*	Seed piece rot
479		*Pythium* spp.	Root rot

[7] Blackberries and raspberries.   [8] Short cycle.

continued

Part I. Field, Fruit, and Vegetable Crops

	Host Plant	Pathogen (Synonym)	Disease (Synonym)
480	*Secale cere-*	*Claviceps purpurea*	Ergot
481	*ale*	*Colletotrichum graminicola*	Anthracnose
482		*Helminthosporium sorokinianum (H. sativum)*	Leaf blotch
483		*Puccinia graminis*	Leaf rust; stem rust
484		*P. rubigo-vera secalis*	Leaf rust
485		*Septoria secalis*	Septoria leaf blotch
486		*Ustilago tritici*	Loose smut
487	*Setaria itali-*	*Sclerospora graminicola*	Downy mildew
488	*ca*	*S. macrospora*	Mildew (crazy top)
489		*Sphacelotheca neglecta*	Head smut
490	*Solanum*	*Phomopsis vexans* [3]	Blight; fruit rot
491	*melongena*	*Verticillium albo-atrum*	Verticillium wilt
492	*Solanum tu-*	*Alternaria solani*	Early blight
493	*berosum*	*Botrytis cinerea*	Stem & leaf blight
494		*Colletotrichum atramentarium*	Black rot
495		*Fusarium* spp.	Wilt; tuber rot
496		*Helminthosporium solani*	Silver scurf
497		*Oospora pustulans*	Skin spot
498		*Pellicularia filamentosa*	Rhizoctonia disease
499		*Phytophthora erythroseptica*	Pink rot
500		*P. infestans*	Late blight
501		*Pythium debaryanum*	Leak (watery rot) [9]
502		*Rhizoctonia solani*	Blotch scurf; stem rot
503		*Sclerotinia sclerotiorum*	White mold
504		*Sclerotium rolfsii*	Southern blight
505		*Spondylocladium atrovirens*	Silver scurf
506		*Streptomyces scabies*	Common scab
507		*Verticillium albo-atrum*	Verticillium wilt
508	*Sorghum bi-*	*Ascochyta sorghina*	Rough leaf spot
509	*color*	*Cercospora sorghi*	Gray leaf spot
510		*Colletotrichum graminicola*	Anthracnose
511		*Gibberella fujikuroi* [10]	Stalk rot
512		*Gloeocercospora sorghi*	Zonate leaf spot
513		*Macrophomina phaseoli*	Charcoal rot
514		*Periconia circinata*	Milo disease
515		*Puccinia purpurea*	Rust
516		*Ramulispora sorghi*	Sooty stripe
517		*Sclerospora sorghi*	Downy mildew
518		*Sphacelotheca cruenta*	Loose smut
519		*S. reiliana*	Head smut
520		*S. sorghi*	Covered kernel smut
521	*Sorghum*	*Gibberella fujikuroi* [10]	Stalk rot
522	*halepense*	*Helminthosporium turcicum*	Leaf blight
523		*Macrophomina phaseoli*	Charcoal rot
524		*Sphacelotheca reiliana*	Head smut
525		*S. sorghi*	Covered kernel smut
526	*Sorghum*	*Colletotrichum* sp.	Stalk rot
527	*vulgare*	*Gibberella fujikuroi* [10]	Stalk rot
528		*Macrophomina phaseoli*	Charcoal rot
529		*Puccinia purpurea*	Leaf rust
530		*Sphacelotheca sorghi*	Covered kernel smut
531	*Spinacia ol-*	*Albugo occidentalis*	White rust
532	*eracea*	*Peronospora effusa*	Downy mildew
533	*Trifolium*	*Cercospora zebrina*	Leaf spot
534	spp.	*Colletotrichum trifolii*	Southern anthracnose
535		*Cymadothea trifolii* [11]	Sooty blotch

[3] Conidial stage of *Diaporthe*.   [9] Tuber disease.   [10] Also the imperfect stage *Fusarium moniliforme*.   [11] Pathogen of red, white, and alsike clover.

continued

Part I. Field, Fruit, and Vegetable Crops

	Host Plant	Pathogen (Synonym)	Disease (Synonym)
536		*Erysiphe polygoni* [12]	Powdery mildew
537		*Fusarium* spp.	Stolon rot
538		*Kabatiella caulivora* [13]	Northern anthracnose
539		*Leptodiscus terrestris*	Root rot
540		*Pellicularia filamentosa*	Stolon rot
541		*Phoma herbarum medicaginis*	Black stem
542		*Pseudopeziza trifolii*	Leaf spot
543		*Pseudoplea trifolii*	Leaf spot
544		*Sclerotinia trifoliorum*	Crown & root rot
545		*Stemphylium sarcinaeforme*	Target leaf spot
546		*Uromyces trifolii*	Rust
547	*Triticum*	*Cephalosporium gramineum*	Cephalosporium stripe
548	*aestivum*	*Cercosporella herpotrichoides*	Foot rot; eyespot
549		*Claviceps purpurea*	Ergot
550		*Colletotrichum graminicola*	Anthracnose
551		*Erysiphe graminis*	Powdery mildew
552		*Fusarium culmorum*	Root & culm rot
553		*F. graminearum (Gibberella zeae)*	Fusarium blight (scab)
554		*F. roseum*	Foot rot
555		*Helminthosporium sorokinianum (H. sativum)*	Crown & root rot
556		*Leptosphaeria nodorum*	Glume blotch; node canker
557		*Olpidium brassicae*	Root rot
558		*Ophiobolus graminis*	Take-all
559		*Puccinia glumarum*	Stripe rust
560		*P. graminis*	Stem rust
561		*P. rubigo-vera*	Leaf rust
562		*Pyrenophora trichostoma*	Leaf spot
563		*Pythium* spp.	Root rot
564		*Septoria avenae* f. sp. *triticia*	Septoria blotch
565		*S. nodorum*	Glume blotch
566		*S. tritici*	Leaf blotch
567		*Tilletia brevifaciens*	Dwarf bunt
568		*T. caries; T. foetida*	Bunt (stinking smut)
569		*Urocystis tritici*	Flag smut
570		*Ustilago tritici*	Loose smut
571	*Vaccinium*	*Botryosphaeria corticis*	Stem canker
572	*corymbo-*	*Botrytis cinerea*	Twig & blossom blight
573	*sum*	*Fusicoccum putrefaciens*	Fusicoccum canker
574		*Gloeosporium* sp.	Anthracnose
575		*Microsphaera penicillata vacinii*	Powdery mildew
576		*Monilinia vaccinii-corymbosi*	Mummy berry
577		*Phytophthora cinnamomi*	Phytophthora root rot
578	*Vitis* spp.	*Armillaria mellea*	Root rot
579		*Botrytis cinerea*	Fruit rot
580		*Cryptosporella viticola*	Dead-arm
581		*Elsinoe ampelina*	Anthracnose
582		*Guignardia bidwellii*	Black rot
583		*Mycosphaerella angulata*	Angular leaf spot
584		*Penicillium* spp.	Blue mold rot
585		*Phomopsis viticola* [3]	Dead-arm
586		*Plasmopara viticola*	Downy mildew
587		*Uncinula necator*	Powdery mildew

[3] Conidial stage of *Diaporthe*. [12] Pathogen of red clover. [13] Pathogen of crimson and red clover.

continued

Part I. Field, Fruit, and Vegetable Crops

	Host Plant	Pathogen (Synonym)	Disease (Synonym)
588	*Xanthosoma sagittifolium*	*Fusarium solani; F. stolonifer*	Rot on rhizomes
589	*Zea mays*	*Aspergillus* spp.	Ear & kernel mold (rot)
590		*Cochliobolus heterostrophus (Helminthosporium maydis)*	Southern corn leaf blight; seedling blight
591		*Colletotrichum graminicola*	Anthracnose
592		*Diplodia macrospora; D. zeae*	Stalk rot; dry ear rot
593		*D. maydis*	Seedling blight
594		*Fusarium graminearum (Gibberella zeae)*	Stalk rot; red ear rot; seedling blight; root rot
595		*F. moniliforme (G. fujikuroi)*	Seedling blight; stalk rot; pink ear rot
596		*F. moniliforme* var. *subglutinans*	Stalk rot
597		*Helminthosporium carbonum*	Charred ear; seedling blight
598		*H. turcicum* [14]	Northern corn leaf blight; seedling blight
599		*Kabatiella zeae*	Eyespot
600		*Macrophomina phaseoli*	Charcoal rot; stalk rot
601		*Monascus purpureus*	Silage mold
602		*Nigrospora oryzae*	Cob rot; dry ear rot
603		*Penicillium* spp.	Blue mold; kernel rot; mold
604		*Phyllosticta maydis (Mycosphaerella zea-maydis)*	Yellow leaf blight
605		*Physalospora zeae*	Gray ear rot
606		*Physoderma maydis*	Brown leaf spot
607		*Puccinia polysora*	Southern corn rust
608		*P. sorghi*	Rust
609		*Pythium arrhenomanes; P. debaryanum*	Seedling blight
610		*P. graminicola; P. ultimum*	Root & stalk rot
611		*Sclerophthora macrospora*	Mildew (crazy top)
612		*Sclerospora graminicola*	Downy mildew
613		*Sphacelotheca reiliana*	Head smut
614		*Trichoderma* spp.	Seed & kernel rot
615		*Ustilago maydis*	Smut; head smut

[14] Perfect stage.

Contributors: Andersen, Axel L., and Jones, Alan L.

General References

[1] Anderson, H. W. 1956. Diseases of Fruit Crops. McGraw-Hill, New York.

[2] Chupp, C., and A. F. Sherf. 1960. Vegetable Diseases and Their Control. Ronald Press, New York.

[3] U.S. Department of Agriculture, Agricultural Research Service, Crops Research Division. 1960. U.S. Dep. Agr. Handb. 165.

[4] Westcott, C. 1950. Plant Disease Handbook. Van Nostrand, Princeton.

continued

Part II. Forest Trees

	Host Tree	Pathogen (Synonym)	Disease
1	*Abies* spp.	*Aleurodiscus amorphus*	Aleurodiscus canker
2		*Armillaria mellea*	Shoestring root rot
3		*Coniophora puteana*	Brown butt rot
4		*Corticium galactinum*	White stringy rot
5		*Cylindrocladium scoparium*	Cylindrocladium blight
6		*Echinodontium tinctorium*	Indian paint rot
7		*Flammula alnicola*	Yellow stringy butt rot
8		*Fomes annosus*	Annosus root rot
9		*F. pini*	Red ring rot
10		*F. pinicola*	Pinicola trunk rot
11		*Herpotrichia nigra*	Brown felt snow mold
12		*Isthmiella faullii*	Needle cast
13		*Lirula abietis-concoloris; L. mirabilis; L. punctata*	Needle cast
14		*Macrophoma parca*	Needle blight
15		*Melampsora abieti-capraearum*	Needle rust
16		*Melampsorella caryophyllacearum*	Yellow witches'-broom
17		*Milesia* spp.	Needle rust
18		*Odontia bicolor*	White stringy rot
19		*Pholiota adiposa*	Pholiota rot
20		*Phytophthora cinnamomi*	Seedling blight
21		*Polyporus balsameus; P. sulphureus*	Brown butt rot
22		*P. schweinitzii*	Brown cubical butt rot
23		*P. tomentosus*	Tomentose root rot
24		*Poria subacida*	Poria root rot
25		*P. weirii*	Laminated root rot
26		*Potebniamyces balsamicola*	Needle blight
27		*Pucciniastrum epilobii; P. goeppertianum; P. vaccini*	Needle rust
28		*Rehmiellopsis balsameae*	Shoot blight
29		*Rhizina undulata*	Seedling burn blight
30		*Sarcotrochila balsameae*	Needle blight
31		*Scleroderris abieticola*	Scleroderris canker
32		*Sclerotinia kerneri*	Cone abortion
33		*Sclerotium bataticola*	Seedling blight
34		*Stereum chailletii*	Stereum butt rot
35		*S. sanguinolentum*	Red heartrot
36		*Uredinopsis pteridis;* other *Uredinopsis* spp.	Needle rust
37		*Valsa kunzei*	Stem canker
38		*Virgella robusta*	Needle cast
39	*Acer* spp.	*Armillaria mellea*	Shoestring root rot
40		*Ceratocystis coerulescens*	Sapstreak disease
41		*Ciboria acerina*	Seed abortion
42		*Cristulariella depraedens; C. pyramidalis*	Bull's-eye spot
43		*Cytospora chrysosperma*	Cytospora canker
44		*Daedalea unicolor*	Canker-rot
45		*Eutypella parasitica*	Eutypella canker
46		*Fomes applanatus*	White trunk rot
47		*F. connatus*	Moss-fungus rot
48		*F. igniarius*	Igniarius trunk rot
49		*Fusarium solani*	Stem canker
50		*Gloeosporium* spp.	Anthracnose
51		*Hydnum* spp.	Soft wet trunk rot
52		*Hypoxylon mammatum*	Stem lesion
53		*Illosporium maculicola*	Big-leaf brown spot
54		*Macrophomina phaseoli*	Seedling blight
55		*Microsphaera alni*	Powdery mildew
56		*Monochaetia monochaeta*	Large leaf spot

continued

Part II. Forest Trees

	Host Tree	Pathogen (Synonym)	Disease
57		*Nectria galligena*	Nectria canker
58		*Phyllactinia guttata*	Powdery mildew
59		*Phyllosticta minima*	Leaf spot
60		*Phymatotrichum omnivorum*	Texas root rot
61		*Phytophthora cactorum*	Bleeding canker
62		*Pleurotus ostreatus*	White trunk rot
63		*Polyporus dryadeus*	Weeping conk rot
64		*P. glomeratus*	Canker-rot
65		*P. spraguei*	Root & butt rot
66		*P. sulphureus*	Brown butt rot
67		*Rhizoctonia* spp.	Seedling root rot
68		*Rhytisma acerinum; R. punctatum*	Tar spot
69		*Schizoxylon microsporum*	Schizoxylon canker
70		*Septoria aceris*	Septoria spot
71		*Stereum murrayi*	Canker-rot
72		*Taphrina* spp.	Leaf blister
73		*Uncinula* spp.	Powdery mildew
74		*Ustulina vulgaris*	Basal rot
75		*Venturia acerina*	Leaf spot
76		*Verticillium albo-atrum; V. dahliae*	Verticillium wilt
77	*Alnus rubra*	*Cercosporella alni*	Leaf spot
78		*Didymosphaeria oregonensis*	Stem canker
79		*Hypoxylon fuscum*	Canker
80		*Melampsoridium alni*	Leaf rust
81		*Nectria galligena*	Nectria canker
82		*Taphrina japonica*	Leaf blister
83	*Betula* spp.	*Armillaria mellea*	Shoestring root rot
84		*Ciboria betulae*	Cone abortion
85		*Daedalea unicolor*	Canker-rot
86		*Diaporthe allegheniensis*	Stem canker
87		*Fomes applanatus; F. robustus*	White trunk rot
88		*F. connatus*	Moss-fungus rot
89		*F. fomentarius*	Trunk rot
90		*F. igniarius*	Igniarius trunk rot
91		*Gloeosporium betulinum*	Anthracnose
92		*Melampsoridium betulinum*	Leaf rust
93		*Melanconium betulinum*	Branch blight
94		*Microsphaera alni*	Powdery mildew
95		*Nectria galligena*	Nectria canker
96		*Phyllactinia guttata*	Powdery mildew
97		*Phytophthora cactorum*	Bleeding canker
98		*Poria laevigata*	Poria trunk rot
99		*P. obliqua*	Clinker rot
100		*Stereum murrayi*	Canker-rot
101		*Taphrina* spp.	Leaf curl; brooming
102		*Ustulina vulgaris*	White butt rot
103	*Carya* spp.	*Armillaria mellea*	Shoestring root rot
104		*Cercospora* spp.	Leaf spot
105		*Cladosporium effusum*	Pecan scab
106		*Elsinoe randii*	Nursery blight
107		*Fomes applanatus; F. connatus; F. igniarius*	Trunk rot
108		*Gnomonia* spp.	Leaf blight
109		*Hydnum erinaceus*	Soft trunk rot
110		*Microsphaera alni*	Powdery mildew
111		*Microstroma juglandis*	White mold
112		*Mycosphaerella dendroides*	Leaf blotch

continued

Part II. Forest Trees

	Host Tree	Pathogen (Synonym)	Disease
113		*Phyllactinia guttata*	Powdery mildew
114		*Pleurotus ostreatus*	Mushroom trunk rot
115		*Poria andersonii; P. spiculosa*	Canker-rot
116	*Fagus grandifolia*	*Corticium vellereum*	Heartrot
117		*Fomes applanatus*	White heartrot
118		*F. igniarius*	Igniarius trunk rot
119		*Hydnum* spp.	Soft wet rot
120		*Nectria coccinea* var. *faginata*	Beech bark disease
121		*N. galligena*	Nectria canker
122		*Phytophthora cactorum*	Bleeding canker
123		*Polyporus glomeratus*	Canker-rot
124		*Poria laevigata*	Poria trunk rot
125		*Scorias spongiosa*	Sooty mold
126		*Stereum murrayi*	Canker-rot
127	*Fraxinus* spp.	*Cytospora pruinosa (Cytophoma pruinosa)*	Branch canker
128		*Fomes fraxinophilus*	Ash heartrot
129		*F. geotropus*	White heartrot
130		*F. igniarius*	Igniarius trunk rot
131		*Gloeosporium aridum*	Anthracnose
132		*Lentinus tigrinus*	Southern mushroom heartrot
133		*Mycosphaerella effigurata; M. fraxinicola*	Common leaf spot
134	*Juglans nigra*	*Gnomonia leptostyla*	Anthracnose
135		*Microstroma juglandis*	Downy leaf spot
136		*Nectria galligena*	Nectria canker
137		*Pellicularia rolfsii*	Southern root rot
138		*Phytophthora cactorum; P. cinnamomi*	Seedling root rot
139	*Juniperus* spp.	*Cercospora sequoiae* var. *juniperi*	Needle blight
140		*Daedalea juniperina*	Daedaloid heartrot
141		*Fomes annosus*	Annosus root rot
142		*F. juniperinus*	Juniper heartrot
143		*Gymnosporangium* spp.	Leaf & twig galls
144		*Phomopsis juniperovora*	Phomopsis blight
145		*Poria cocos*	Tuckahoe
146		*Stigmina juniperina*	Needle blight
147	*Larix* spp.	*Dasyscypha willkommii*	Larch canker
148		*Fomes officinalis*	Quinine fungus
149		*F. pini*	Red heart
150		*Hypodermella laricis*	Needle cast
151		*Melampsora paradoxa*	Needle rust
152		*Meria laricis*	Needle blight
153		*Phomopsis pseudotsugae* [1]	Carrot top
154		*Rhizina undulata*	Burn-mold blight
155		*Valsa kunzei*	Stem canker
156	*Liquidambar styraciflua*	*Actinopelte dryina*	Small leaf spot
157		*Botryosphaeria ribis*	Stem canker
158		*Cercospora liquidambaris*	Leaf spot
159		*Lentinus tigrinus*	Heartrot
160		*Pleurotus ostreatus*	Heartrot
161		*Polyporus lucidus; P. ulmarius*	Heartrot
162	*Liriodendron tulipifera*	*Armillaria mellea*	Shoestring butt rot
163		*Collybia velutipes*	Heartrot
164		*Cylindrocladium scoparium*	Seedling black rot
165		*Erysiphe liriodendri*	Powdery mildew
166		*Fusarium solani*	Fusarium canker
167		*Hydnum erinaceus*	Heartrot

[1] Conidial stage of *Diaporthe*.

continued

Part II. Forest Trees

	Host Tree	Pathogen (Synonym)	Disease
168		*Nectria magnoliae*	Nectria canker
169		*Pleurotus ostreatus*	Heartrot
170	*Picea* spp.	*Armillaria mellea*	Shoestring root rot
171		*Chrysomyxa arctostaphyli*	Yellow witches'-broom
172		*Coniophora cerebella*	Butt rot
173		*Cylindrocladium scoparium*	Seedling rot
174		*Flammula alnicola*	Heartrot
175		*Fomes pini*	Red heart
176		*F. pinicola*	Heartrot
177		*Herpotrichia nigra*	Brown felt snow mold
178		*Lophodermium macrosporum*	Tar-spot needle cast
179		*Phacidium infestans*	Snow blight
180		*Phytophthora cinnamomi*	Damping-off
181		*Polyporus schweinitzii*	Brown cubical butt rot
182		*P. tomentosus*	Butt rot
183		*Pythium* spp.	Damping-off
184		*Rhizoctonia solani*	Seedling root rot
185		*Stereum sanguinolentum*	Red heart
186		*Valsa kunzei*	Stem canker
187	*Pinus monticola*	*Armillaria mellea*	Shoestring root rot
188		*Cronartium ribicola*	Blister rust
189		*Fomes pini*	Red ring rot
190		*Lecanosticta acicola*	Brown spot
191		*Lophodermella arcuata*	Needle cast
192		*Neopeckia coulteri*	Snow mold
193		*Polyporus schweinitzii; P. tomentosus*	Butt rot
194		*Poria subacida*	Stringy butt rot
195		*Rhizina undulata*	Burn mold blight
196	*Pinus ponderosa*	*Atropellis arizonica; A. piniphila*	Atropellis canker
197		*Cronartium comandrae; C. comptoniae*	Stem rust
198		*Dothistroma pini*	Dothistroma needle blight
199		*Elytroderma deformans*	Red needle blight
200		*Fomes annosus*	Annosus root rot
201		*F. pini*	Red ring rot
202		*Hymenochaete berkeleyi (Veluticeps berkeleyi)*	Brown cubical heartrot
203		*Peridermium filamentosum; P. harknessii; P. stalactiforme*	Stem rust
204		*Polyporus anceps*	Western red rot
205		*P. schweinitzii*	Brown cubical butt rot
206		*P. volvatus*	Heartrot
207		*Pythium* spp.	Damping-off
208		*Rhizoctonia solani*	Seedling root rot
209		*Sclerotium bataticola*	Seedling root rot
210		*Verticicladiella wagenerii*	Root stain disease
211	*Pinus strobus*	*Armillaria mellea*	Shoestring root rot
212		*Atropellis pinicola*	Atropellis canker
213		*Bifusella linearis*	Needle tar spot
214		*Caliciopsis pinea*	Twig canker
215		*Cronartium ribicola*	Blister rust
216		*Cylindrocladium scoparium*	Seedling blight
217		*Fomes annosus*	Annosus root rot
218		*F. pini*	Redheart
219		*Meloderma desmazierii (Hypodermella desmazierii)*	Gray needle blight
220		*Phytophthora cinnamomi*	Damping-off
221		*Polyporus schweinitzii*	Brown cubical butt rot
222		*P. tomentosus*	Root & butt rot
223		*Pythium* spp.	Damping-off

continued

Part II. Forest Trees

	Host Tree	Pathogen (Synonym)	Disease
225		*Rhizoctonia solani*	Seedling root rot
226		*Scirrhia acicola*	Brown-spot needle blight
227		*Stereum sanguinolentum*	Heartrot
228	*Pinus taeda*	*Cronartium fusiforme; C. quercuum*	Stem rust
229		*Fomes annosus*	Annosus root rot
230		*F. pini*	Red heart
231		*Fusarium* spp.	Damping-off
232		*Phytophthora cinnamomi*	Damping-off
233		*Ploioderma lethale*	Needle cast
234		*Polyporus schweinitzii*	Brown cubical butt rot
235		*Rhizoctonia solani*	Seedling root rot
236		*Scirrhia acicola*	Needle blight
237		*Sclerotium bataticola*	Damping-off
238	*Populus* spp.	*Armillaria mellea*	Shoestring root rot
239		*Cenangium singulare*	Sooty bark canker
240		*Ceratocystis tremulo-aurea*	Black canker
241		*Ciborinia bifrons*	Leaf tar spot
242		*Cryptochaete polygonia*	Brown-stain heartrot
243		*Cytospora chrysosperma*	Cytospora canker
244		*Dothichiza populea*	Dothichiza canker
245		*Fomes igniarius*	White heartrot
246		*Hypoxylon mammatum*	Hypoxylon canker
247		*Marssonina populi*	Small leaf spot
248		*Melampsora abietis-canadensis; M. medusae*	Leaf rust
249		*Septoria musiva*	Leaf blight; canker
250		*Taphrina johansonii*	Catkin blight
251		*Valsa nivea; V. sordida*	Valsa canker
252		*Venturia macularis*	Leaf blight
253	*Prunus serotina*	*Armillaria mellea*	Shoestring root rot
254		*Dibotryon morbosum*	Black knot
255		*Monilinia fructicola; M. rhododendri*	Seedling blight; fruit rot
256		*Nectria galligena*	Nectria canker
257		*Polyporus berkeleyi; P. spraguei*	Root & trunk rot
258		*Poria laevigata*	Poria trunk rot
259		*Taphrina farlowii; T. pruni-spinosae*	Leaf curl
260		*Valsa leucostoma*	Valsa canker
261	*Pseudotsuga menziesii*	*Armillaria mellea*	Shoestring root rot
262		*Echinodontium tinctorium*	Indian paint fungus
263		*Fomes officinalis*	Quinine fungus
264		*F. pini*	Red ring rot
265		*Fusarium oxysporum*	Damping-off
266		*Herpotrichia nigra*	Brown felt snow mold
267		*Phacidium abietis*	Snow mold
268		*Phaeocryptopus gaeumannii*	Needle blight
269		*Phytophthora cinnamomi*	Root rot
270		*Polyporus schweinitzii*	Brown cubical butt rot
271		*Poria weirii*	Laminated root rot
272		*Pythium* spp.	Damping-off
273		*Rhabdocline pseudotsugae*	Needle blight
274		*Rhizoctonia solani*	Seedling root rot
275		*Rosellinia herpotrichioides*	Smothering disease
276		*Sclerotium bataticola*	Seedling root rot
277		*Valsa kunzei*	Stem canker
278	*Quercus* spp.	*Actinopelte dryina*	Small leaf spot
279		*Armillaria mellea*	Shoestring root rot
280		*Ceratocystis fagacearum*	Oak wilt

continued

Part II. Forest Trees

	Host Tree	Pathogen (Synonym)	Disease
281		*Corticium galactinum*	White heart rot
282		*Cronartium fusiforme; C. quercuum*	Pine stem rust
283		*C. strobilinum*	Cone rust
284		*Elsinoe quercus-falcatae*	Red-oak spot anthracnose
285		*Endothia parasitica*	Chestnut blight
286		*Fomes everhartii*	Heartrot
287		*Fusarium solani*	Seedling blight
288		*Gnomonia veneta*	Oak anthracnose
289		*Hydnum erinaceus*	Soft heartrot
290		*Hypoxylon atropunctatum*	Southern stem blight
291		*Irpex mollis*	Canker-rot
292		*Microsphaera alni*	Powdery mildew
293		*Nectria galligena*	Nectria canker
294		*Phyllactinia guttata*	Powdery mildew
295		*Phymatotrichum omnivorum*	Texas root rot
296		*Physalospora glandicola*	Twig blight
297		*Pleurotus ostreatus*	Heartrot
298		*Polyporus berkeleyi; P. lucidus; P. sulphureus*	Butt rot
299		*P. dryadeus*	Weeping butt fungus
300		*P. dryophilus*	Heartrot
301		*P. hispidus*	Canker-rot
302		*Poria andersonii*	Heartrot
303		*P. spiculosa*	Canker-rot
304		*Sphaerotheca lanestris*	Sooty brooming disease
305		*Stereum frustulosum*	Heartrot
306		*S. gausapatum*	Sprout heartrot
307		*S. murrayi*	Canker-rot
308		*S. subpileatum*	Honeycomb heartrot
309		*Strumella coryneoidea*	Strumella canker
310		*Taphrina caerulascens*	Leaf blister
311	*Salix* spp.	*Botryosphaeria ribis*	Black stem canker
312		*Cryptodiaporthe salicina*	Stem canker
313		*Cytospora chrysosperma*	Cytospora canker
314		*Fomes applanatus; F. fraxinophilus; F. igniarius; F. robustus*	Heartrot
315		*Fusicladium saliciperdum*	Willow scab
316		*Lentinus tigrinus*	Heartrot
317		*Melampsora epitea; M. paradoxa*	Leaf rust
318		*Phyllactinia guttata*	Powdery mildew
319		*Phymatotrichum omnivorum*	Texas root rot
320		*Physalospora miyabeana*	Willow scab
321		*Polyporus glomeratus; P. hispidus; P. lucidus*	Heartrot
322		*Trametes suaveolens*	Heartrot
323		*Uncinula salicis*	Powdery mildew
324	*Sequoia sempervirens*	*Botrytis cinerea*	Gray mold blight
325		*Poria albipellucida*	White ring rot
326		*P. sequoiae*	Brown pocket rot
327		*P. versipora*	Heartrot
328		*Sclerotium bataticola*	Seedling blight
329	*Thuja plicata*	*Armillaria mellea*	Shoestring root rot
330		*Didymascella thujina*	Cedar leaf blight
331		*Fomes pini; F. pinicola; F. roseus*	Heartrot
332		*Herpotrichia nigra*	Brown felt snow mold

continued

Part II. Forest Trees

	Host Tree	Pathogen (Synonym)	Disease
333		*Poria asiatica; P. ferruginosa*	Heartrot
334		*P. weirii*	Laminated root rot
335	*Tsuga canadensis*	*Fabrella tsugae*	Needle blight
336		*Fomes annosus*	Annosus root rot
337		*Melampsora farlowii*	Twig & cone rust
338		*Phacidiopycnis pseudotsugae*	Bleeding canker
339		*Polyporus schweinitzii*	Brown cubical butt rot
340		*Poria subacida*	Stringy butt rot
341		*Pucciniastrum hydrangae; P. vaccini*	Needle rust
342		*Pythium debaryanum*	Damping-off
343		*Rhizoctonia solani*	Seedling root rot
344		*Stereum sanguinolentum*	Red heartrot
345	*Tsuga heterophylla*	*Fabrella tsugae*	Needle blight
346		*Flammula alnicola*	Butt rot
347		*Fomes annosus*	Annosus root & trunk rot
348		*F. pini*	Red ring rot
349		*F. robustus*	White heartrot
350		*Herpotrichia nigra*	Brown felt snow mold
351		*Melampsora epitea*	Leaf rust
352		*Pholiota adiposa*	Mushroom heartrot
353		*Polyporus oregonensis*	Trunk rot
354		*P. schweinitzii*	Brown cubical butt rot
355		*Poria subacida*	Stringy heartrot
356		*P. weirii*	Laminated root rot
357		*Stereum sanguinolentum*	Red heartrot
358	*Ulmus americana*	*Ceratocystis ulmi*	Dutch elm disease
359		*Chalaropsis thielavioides*	Root rot
360		*Dothiorella ulmi*	Dothiorella wilt
361		*Fomes applanatus; F. connatus; F. everhartii; F. igniarius*	Heartrot
362		*Gnomonia ulmea*	Elm leaf spot
363		*Microsphaera alni*	Powdery mildew
364		*Phyllactinia guttata*	Powdery mildew
365		*Phymatotrichum omnivorum*	Texas root rot
366		*Phytophthora cactorum*	Bleeding canker
367		*P. inflata*	Pit canker
368		*Pleurotus ostreatus; P. ulmarius*	Heartrot
369		*Polyporus dryadeus; P. lucidus; P. sulphureus*	Heartrot
370		*Pythium ultimum*	Damping-off
371		*Rhizoctonia crocorum*	Violet root rot
372		*Verticillium albo-atrum; V. dahliae*	Verticillium wilt

Contributor: Hepting, George H.

General References

[1] Baxter, D. V. 1952. Pathology in Forest Practice. Ed. 2. J. Wiley, New York.

[2] Boyce, J. S. 1961. Forest Pathology. Ed. 3. McGraw-Hill, New York.

[3] Browne, F. G. 1968. Pests and Diseases of Forest Plantation Trees. Clarendon Press, Oxford.

[4] Davidson, A. G., and R. M. Prentice, ed. 1967. Can. Dep. Forest Rural Develop. Publ. 1180.

[5] Food and Agriculture Organization of the United Nations. 1964. FAO Forest Pests 64 (Dis. Widely-Planted Forest Trees).

[6] Hepting, G. H. 1970. J. Forest. 68:78.

[7] Hepting, G. H. 1971. U.S. Dep. Agr. Agr. Handb. 386.

[8] Hepting, G. H., and G. M. Jemison. 1958. U.S. Forest Serv. Forest Resour. Rep. 14:183.

[9] Marshall, R. P., and A. M. Waterman. 1948. U.S. Dep. Agr. Farmers Bull. 1987.

[10] Pirone, P. P. 1970. Diseases and Pests of Ornamental Plants. Ed. 4. Ronald Press, New York.

[11] Westcott, C. 1960. Plant Disease Handbook. Van Nostrand, New York.

Part I. Superficial Mycoses

	Species	Disease	Natural Occurrence in Animals Other than Man	Microscopic Appearance in Man			Microscopic Appearance of Culture on Sabouraud's Agar
				Skin	Nail	Hair	
1	*Cladosporium werneckii*	Tinea nigra	None	Pigmented (light brown to dark green), branching, septate hyphae; may develop closely septate swollen cells, chlamydospores, & budding cells	None	None	Colonies at first yeast-like; pigmented hyphae produce blastospores laterally, & 1-3 septate conidia in clusters or in short chains from apiculi or short conidiophores
2	*Epidermophyton floccosum*	Tinea pedis; T. cruris; T. unguium	None	Abundant-branching, septate hyphae; may segment into chains of arthrospores	Branching, septate hyphae; may segment into chains of arthrospores	None	Macroconidia abundant, clavate, 2-6 cells, blunt-tipped, smooth thin walls; occur in clusters of 2 or 3; no microconidia; abundant chlamydospores
3	*Malassezia furfur*	Tinea versicolor	None	Clusters of spherical, thick-walled, budding cells, 3-8 μ, & short irregular hyphae	None	None	Oil-enriched medium allows growth of spherical budding cells of *Pityrosporum orbiculare*
4	*Microsporum audouinii*	Tinea capitis; T. corporis	Dog (rare); monkey (rare)	Branching, septate hyphae; may segment into chains of arthrospores	None	Ectothrix; sheath of small spores, 2-3 μ	Hyphae with chlamydospores; microconidia infrequent, clavate, 2.5-4 × 3-6 μ; macroconidia rare, rudimentary, ill-formed
5	*M. canis*	Tinea capitis; T. corporis; T. barbae; T. unguium	Cat [1]; chinchilla; dog [1]; horse; monkey	Branching, septate hyphae; may segment into chains of arthrospores	Rare	Ectothrix; sheath of small spores, 2-3 μ	Macroconidia numerous, 8-15 cells, spindle-shaped, thick rough walls, 8-15 × 40-150 μ; microconidia few, clavate, 2-4 × 3-6 μ
6	*M. distortum*	Tinea capitis (rare); T. corporis (rare)	Dog; monkey [1]	Branching, septate hyphae; may segment into chains of arthrospores	None	Ectothrix; sheath of small arthrospores, 2-3 μ	Macroconidia numerous, rough thick walls, distorted in shape, 4-14 × 3-40 μ; microconidia numerous, pyriform, 2-4 × 3-6 μ
7	*M. ferrugineum* [2]	Tinea capitis; T. corporis	None	Branching, septate hyphae; may segment into chains of arthrospores	None	Ectothrix; sheath of small arthrospores, 2-3 μ	Mycelium with occasional hyphal swellings; arthrospores, chlamydospores, rare macroconidia
8	*M. gypseum* [3]	Tinea capitis; T. corporis; T. barbae	Cat; dog [1]; horse [1]; monkey; mouse; rat	Branching, septate hyphae; may segment into chains of arthrospores	None	May be sheath of small arthrospores, 2-3 μ; more commonly large-spored ectothrix, 5-8 μ; invasion most frequently limited to hyphae inside hair	Macroconidia numerous, 4-6 cells, ellipsoid; thin rough walls, 8-12 × 30-50 μ; microconidia few, clavate
9	*M. nanum* [3]	Tinea capitis (rare); T. corporis (rare)	Swine	Branching, septate hyphae; may segment into chains of arthrospores	None	Septate & nonseptate hyphae, air bubbles inside hair	Macroconidia numerous, 2-3 cells, pyriform-to-elliptical, thin walls, finely echinulate or smooth, 5-7 × 12-18 μ; microconidia few

[1] The more common host. [2] Synonym: *Trichophyton ferrugineum.* [3] A common saprophyte in soil.

continued

Part I. Superficial Mycoses

	Species	Disease	Natural Occurrence in Animals Other than Man	Microscopic Appearance in Man			Microscopic Appearance of Culture on Sabouraud's Agar
				Skin	Nail	Hair	
10	*Piedraia hortai*	Black piedra	Primates	None	None	Nodule on hair shaft consists of brown, dichotomously branched, closely septate hyphae, 4-8 μ diameter, & asci containing 2-8 ascospores	Dark, thick-walled, closely septate hyphae; chlamydospores, asci, ascospores
11	*Trichophyton concentricum*	Tinea imbricata	None	Abundant-branching, septate hyphae; may segment into chains of arthrospores	None	None	Branching, septate, irregular hyphae, with chlamydospores, pectinate hyphae, favic chandeliers
12	*T. gallinae*	Tinea capitis (rare); T. corporis (rare)	Dog; poultry [1]; wild birds	Branching, septate hyphae (rare)	None	Ectothrix; chains of large spores, 4-8 μ	Macroconidia usually numerous, 2-10 cells, clavate, smooth & slightly thickened walls; microconidia few, small, pyriform-to-elongate
13	*Trichophyton mentagrophytes* [4]	Tinea pedis; T. cruris; T. corporis; T. capitis; T. unguium; T. barbae	Many domestic & wild animals	Branching, septate hyphae; may segment into chains of arthrospores	Branching, septate hyphae; may segment into chains of arthrospores	Ectothrix; chains of small arthrospores, 3-5 μ	Microconidia numerous, subspherical-to-pyriform, in terminal clusters or singly along hyphae; macroconidia clavate, 2-5 cells, thick walls, 4-6 × 10-50 μ
14	*T. rubrum*	Tinea pedis; T. capitis (rare); T. unguium; T. cruris; T. corporis; T. barbae	None	Branching, septate hyphae; may segment into chains of arthrospores	Branching, septate hyphae; may segment into chains of arthrospores	Ectothrix; chains of large arthrospores, approximately 5 μ	Microconidia numerous, singly along hyphae & in clusters; macroconidia infrequent, pencil-shaped thin walls, 4-6 × 10-50 μ
15	*T. schoenleinii*	Favus; Tinea capitis; T. corporis; T. unguium	Dog (rare)	Abundant hyphae; may segment into chains of arthrospores throughout cellular debris of scutulum	Branching, septate hyphae; may segment into chains of arthrospores	Hyphae, occasional arthrospore, numerous air bubbles inside hair	Irregular hyphae; chlamydospores; hyphal swellings; pectinate hyphae; favic chandeliers
16	*T. tonsurans*	Tinea capitis; T. corporis; T. unguium	None	Branching, septate hyphae; may segment into chains of arthrospores	Rare	Endothrix; large spores in chains, 4-7.5 μ	Microconidia numerous, clavate along sides of hyphae or on short conidiophores; spore-bearing hyphae stain poorly with Lacto-Phenol Cotton Blue; numerous chlamydospores; thin, smooth-walled macroconidia rare
17	*T. verrucosum*	Tinea corporis; T. capitis; T. barbae	Cattle [1]; dog (rare); donkey; goat; horse; sheep	Branching, septate hyphae; may segment into chains of arthrospores	None	Ectothrix; chains of large arthrospores, 5-10 μ	Irregular hyphae, abundant chlamydospores; best growth at 37°C
18	*T. violaceum*	Tinea capitis; T. corporis; T. barbae; T. unguium	Cattle (rare)	Branching, septate hyphae; may segment into chains of arthrospores	Branching, septate hyphae; may segment into chains of arthrospores	Endothrix; chains of large arthrospores, 4-7.5 μ	Irregular hyphae, abundant chlamydospores & hyphal swellings; microconidia rare

[1] The more common host. [4] Synonym: *Trichophyton interdigitale*.

continued

Part I. Superficial Mycoses

	Species	Disease	Natural Occurrence in Animals Other than Man	Microscopic Appearance in Man			Microscopic Appearance of Culture on Sabouraud's Agar
				Skin	Nail	Hair	
19	*Trichosporon beigelii* [5]	White piedra	Monkey	None	None	Nodule on hair shaft consists of hyphae which segment into spherical-to-rectangular cells, 2-8 μ; budding cells present	Hyphae segment into rectangular-to-spherical arthrospores; budding cells present

[5] Synonym: *T. cutaneum.*

Contributors: Halde, Carlyn; Friedman, Lorraine; Georg, Lucille K.

General References

[1] Ainsworth, G. C. 1952. Medical Mycology. I. Pitman, New York.

[2] Ainsworth, G. C., and P. K. C. Austwich. 1958. Commonw. Agr. Bur. (Gt. Brit.) Anim. Health Rev. Ser. 6.

[3] Beneke, E. S., and A. L. Rogers. 1971. Medical Mycology Manual. Ed. 3. Burgess, Minneapolis.

[4] Blair, J. E., et al. 1970. Manual of Clinical Microbiology. American Society for Microbiology, Bethesda, Md.

[5] Bodily, H. L., et al. 1970. Diagnostic Procedures for Bacterial, Mycotic, and Parasitic Infections. Ed. 5. American Public Health Association, Washington, D.C.

[6] Conant, N. F. 1971. Manual of Clinical Mycology. Ed. 3. W. B. Saunders, Philadelphia.

[7] Emmons, C. W., et al. 1970. Medical Mycology. Ed. 2. Lea and Febiger, Philadelphia.

[8] Georg, L. K. 1970. Diagnostic Procedures for the Isolation and Identification of the Etiologic Agents of Actinomycosis. Proc. Int. Symp. Mycoses, Pan-Amer. Health Org. Sci. Publ. 205.

[9] Jungerman, P. F., and R. M. Schwartzman. 1972. Veterinary Medical Mycology. Lea and Febiger, Philadelphia.

[10] Wilson, J. W., and O. A. Plunkett. 1965. The Fungous Diseases of Man. Univ. California Press, Berkeley.

Part II. Deep Mycoses

	Species	Disease Produced (Synonym)	Geographical Distribution	Animals Frequently Infected	Saprophytic Occurrence	Organs & Tissues of Man Frequently Attacked	Susceptible Laboratory Animals [L]	Microscopic Appearance		
								In Human Tissue	Of Culture at 25°C	Of Culture at 37°C
1	*Absidia corymbifera; A. ramosa; Basidiobolus haptosporus; B. meristosporus; Rhizopus arrhizus; R. oryzae*	Phycomycosis	Worldwide	Birds, cattle, dog, horse, swine	Soil	Lung, brain, eye, intestinal tract, sinus, cutaneous & subcutaneous tissues	Diabetic rabbit & rat	Nonseptate, coenocytic hyphae, 6-15 μ in width	Broad, coenocytic mycelium with sporangiophores	Similar to growth at 25°C
2	*Actinomyces israelii; A. naes-*	Actinomycosis	Worldwide	None	Man (oral cavity)	Cervicofacial region, lung, bone, cecum, appendix, liver	Hamster, mouse	Granules of filamentous, branching, gram-positive hyphae, 1 μ	Poor or no growth	Microaerophilic-to-anaerobic, filamentous, branching,

[L] Animals are often difficult or even impossible to infect because of the great variation in susceptibility.

continued

Part II. Deep Mycoses

Species	Disease Produced (Synonym)	Geographical Distribution	Animals Frequently Infected	Saprophytic Occurrence	Organs & Tissues of Man Frequently Attacked	Susceptible Laboratory Animals [1]	Microscopic Appearance		
							In Human Tissue	Of Culture at 25°C	Of Culture at 37°C
lundi; Arachnia propionica [2]							or less in width; club-shaped accretions on tips of hyphae may be present		gram-positive hyphae, 1 μ or less in width; organism grows only on enriched media [3]
3 Allescheria boydii [4]; Cephalosporium spp.; Leptosphaeria senegalensis; Madurella spp.; Phialophora jeanselmei; Pyrenochaeta romeroi	Eumycotic mycetoma (maduro-mycosis; Madura foot)	Worldwide; more frequent in tropics	Dog	Soil— A. boydii & Phialophora jeanselmei occur as saprophytes	Feet, hands, cutaneous & subcutaneous tissue, bone	Mouse— infected by A. boydii	Oval, irregular-shaped granules, 0.5-2 mm, made up of segmented, branched, hyaline or brown hyphae, 2-5 μ diameter, & chlamydospores	A. boydii: mycelium with oval-to-pyriform conidia, 5-7 X 8-10 μ, borne singly at ends of long conidiophores; dark brown, thin-walled perithecia, 50-200 μ diameter, containing evanescent asci & elliptical ascospores; coremia occasionally present	Similar to cultures at 25°C
4 Aspergillus spp.	Aspergillosis	Worldwide	Birds, cattle	Grain, soil	Ear, sinus, orbit, vagina, lung, brain	Birds, guinea pig, mouse, rabbit	Branching, septate hyphae	Conidiophore forms vesicle at tip; surface covered with sterigmata bearing long chains of conidia	Similar to growth at 25°C
5 Blastomyces dermatitidis	Blastomycosis (North American blastomycosis)	United States, Canada, Mexico, Africa	Dog, horse	Soil (probable) [5]	Lung, skin, bone	Guinea pig, mouse	Single-budding, thick-walled cells, 8-15 μ	Mycelium with oval-to-pyriform conidia, 3-5 μ, on conidiophores or attached directly to hyphae	Similar to forms observed in tissue
6 Candida albicans; other Candida spp.	Candidiasis (moniliasis; thrush; mycotic vulvovaginitis)	Worldwide	Cattle, young swine, poultry	Soil, water, man (oral cavity)	Mucous membranes, nail, skin, bronchus, lung, vagina, blood (as septicemias)	Guinea pig, mouse, rabbit, rat	Oval-to-spherical budding cells, 2-4 μ; frequently hyphae which may show clusters of blastospores attached at septations	Oval-to-spherical, single-budding cells, 2-4 μ; pseudohyphae & hyphae; clusters of budding cells often at septations; thick-walled chlamydospores, 6-9 μ, on special medium (C. albicans)	On Sabouraud similar to growth at 25°C; "germ-tubes" in serum & egg albumin (C. albicans & C. stellatoidea)
7 Coccidioides immitis	Coccidioidomycosis (coccidioidal granuloma; valley fever;	Arid, southwestern United States;	Many mammals, especially cattle, dog, horse, monkey, rodents, sheep	Soil	Lung, skin, bone, meninges	Guinea pig, hamster, mouse, other rodents	Thick-walled spherules, 20-60 μ, containing endospores, 2-5 μ	Mycelium with arthrospores, 2.5-3 X 3-4 μ, alternating with empty cells	Similar to growth at 25°C; under special conditions with special media, tissue spherules obtained in vitro

[1] Animals are often difficult or even impossible to infect because of the great variation in susceptibility. [2] Synonym: Actinomyces propionica. [3] Such as beef heart infusion agar at pH 6.8-7.5. [4] Imperfect form, Monosporium apiospermum. [5] Reported, but not confirmed.

continued

Part II. Deep Mycoses

Species	Disease Produced (Synonym)	Geographical Distribution	Animals Frequently Infected	Saprophytic Occurrence	Organs & Tissues of Man Frequently Attacked	Susceptible Laboratory Animals[1]	Microscopic Appearance		
							In Human Tissue	Of Culture at 25°C	Of Culture at 37°C
	Posada-Wernicke's disease)	Mexico; Central America; Chaco region & arid areas of northern South America							
8 *Cryptococcus neoformans*	Cryptococcosis (torulosis; European blastomycosis; Busse-Buschke disease)	Worldwide	Cat, cattle, dog, horse, monkey	Soil, bird droppings	Central nervous system, lung, skin, bone	Mouse, rat	Spherical, single-budding, thin-walled cells, 5-20 μ, usually surrounded by wide gelatinous capsule	Similar to cells seen in tissue[6]; abortive hyphae may be seen on primary isolation	Similar to cells seen in tissue; may or may not have large capsules
9 *Fonsecaea pedrosoi*[2]; *Phialophora verrucosa*; *Cladosporium carrionii*	Chromomycosis (chromoblastomycosis; verrucous dermatitis)	Worldwide; more frequent in tropics	None	Soil, wood	Usually on lower extremities, cutaneous & subcutaneous tissue, lymphatics	Mouse, rat	Single or clustered spherical, thick-walled, dark brown cells, 6-12 μ; multiply by splitting, not budding	Three types of sporulation: *Phialophora*—conidia borne within a terminal cuplike structure on flask-shaped conidiophore; *Cladosporium*—conidia in branching chains arising terminally from conidiophore; *Acrotheca*—conidia borne acropleurogenously on swollen, clublike conidiophore	Similar to cultures at 25°C
10 *Histoplasma capsulatum*	Histoplasmosis (reticuloendothelial cytomycosis; Darling's disease)	At least 30 countries	Most mammals, especially bat, cat, cattle, dog, horse, mouse, rat, skunk	Soil, especially from avian & bat habitats	Lung, liver, spleen, lymph nodes, mucous membranes, suprarenal, kidney	Guinea pig, hamster, mouse	Intracellular, oval, budding cells, 1-5 μ	Delicate mycelium with thin-walled, subspherical-to-pyriform conidia, 2-5 μ; & thick-walled, tuberculated conidia, 8-20 μ	Similar to cells observed in tissue, if grown on enriched medium
11 *Nocardia asteroides*	Nocardiosis	Worldwide	Cat, cattle, dog	Soil	Lung, brain, kidney, heart, spleen, liver	Guinea pig, mouse	Delicate, branched, gram-positive hyphae, 0.5-1 μ diameter; partially acid-fast	Branching hyphae, 0.5-1 μ diameter; break up readily into bacillary or coccoid forms	Similar to cultures at 25°C

[1] Animals are often difficult or even impossible to infect because of the great variation in susceptibility. [6] Some strains may be weakly encapsulated in vitro, giving culture a different gross appearance resembling many of the common yeasts. [2] Synonym: *Phialophora pedrosoi*.

continued

Part II. Deep Mycoses

	Species	Disease Produced (Synonym)	Geographical Distribution	Animals Frequently Infected	Saprophytic Occurrence	Organs & Tissues of Man Frequently Attacked	Susceptible Laboratory Animals[1]	Microscopic Appearance		
								In Human Tissue	Of Culture at 25°C	Of Culture at 37°C
12	*N. asteroides; N. brasiliensis; N. madurae*[8], *N. pelletieri*[8], *Streptomyces somaliensis*	Actinomycotic mycetoma	Worldwide; more frequent in tropics	None	Soil	Skin, subcutaneous tissue, bone, usually on lower extremities	Guinea pig, mouse	Granules of gram-positive, branching hyphae, 0.5-1 μ diameter	Branching hyphae, 0.5-1 μ diameter; spherical conidia, 0.5-1 μ, sparse to absent	Similar to cultures at 25°C
13	*Paracoccidioides brasiliensis*	Paracoccidioidomycosis (paracoccidioidal granuloma; Lutz-Splendore-Almeida disease; South American blastomycosis)	South & Central America, Mexico	None	Soil (probable)[5]	Mouth, lung, lymph nodes, gastrointestinal tract	Guinea pig, hamster, mouse	Multiple-budding, thick-walled cells, 10-60 μ	Mycelium with rare oval conidia, 3-5 μ	Similar to forms observed in tissue
14	*Rhinosporidium seeberi*	Rhinosporidiosis	Worldwide; most frequent in India & Ceylon	Cattle, horse, mule	None	Mucous membranes, nose, eye, vagina, penis, skin	None	Thick-walled spherule, 50-350 μ, with pore, containing up to 16,000 endospores, 7-9 μ	No culture method available	No culture method available
15	*Sporothrix schenckii*[9]	Sporotrichosis	Worldwide	Cat, cattle, dog, horse, mouse, mule, rat	Mine timbers, soil, plants	Hands, feet, cutaneous & subcutaneous tissue, lymphatics	Hamster, mouse, rat	Rarely seen without special stains; gram-positive, cigar-shaped or spherical-to-oval, usually intracellular, budding cells, 3-5 μ; asteroid forms rare	Delicate hyphae, 2 μ in width, pyriform-to-spherical conidia, 2-4 X 2-6 μ, borne in clusters on lateral branches or laterally along hyphae	Cigar-shaped, spherical or oval budding cells; must be grown on enriched medium

[1] Animals are often difficult or even impossible to infect because of the great variation in susceptibility. [5] Reported, but not confirmed. [8] Recently placed in the new genus *Actinomadura*. [9] Synonym: *Sporotrichum schenckii*.

Contributors: Halde, Carlyn; Georg, Lucille K.; Friedman, Lorraine

General References

[1] Ainsworth, G. C. 1952. Medical Mycology. I. Pitman, New York.
[2] Ainsworth, G. C., and P. K. C. Austwich. 1958. Commonw. Agr. Bur. (Gt. Brit.) Anim. Health Rev. Ser. 6.
[3] Beneke, E. S., and A. L. Rogers. 1971. Medical Mycology Manual. Ed. 3. Burgess, Minneapolis.
[4] Blair, J. E., et al. 1970. Manual of Clinical Microbiology. American Society for Microbiology, Bethesda, Md.
[5] Bodily, H. L., et al. 1970. Diagnostic Procedures for Bacterial, Mycotic, and Parasitic Infections. Ed. 5. American Public Health Association, Washington, D.C.

[6] Conant, N. F. 1971. Manual of Clinical Mycology. Ed. 3. W. B. Saunders, Philadelphia.
[7] Emmons, C. W., et al. 1970. Medical Mycology. Ed. 2. Lea and Febiger, Philadelphia.
[8] Georg, L. K. 1970. Diagnostic Procedures for the Isolation and Identification of the Etiologic Agents of Actinomycosis. Proc. Int. Symp. Mycoses, Pan-Amer. Health Org. Sci. Publ. 205.
[9] Jungerman, P. F., and R. M. Schwartzman. 1972. Veterinary Medical Mycology. Lea and Febiger, Philadelphia.
[10] Wilson, J. W., and O. A. Plunkett. 1965. The Fungous Diseases of Man. Univ. California Press, Berkeley.

IX. SENSORY AND NEURO- BIOLOGY

133. BRAIN: MAN

LATERAL SURFACE OF THE BRAIN

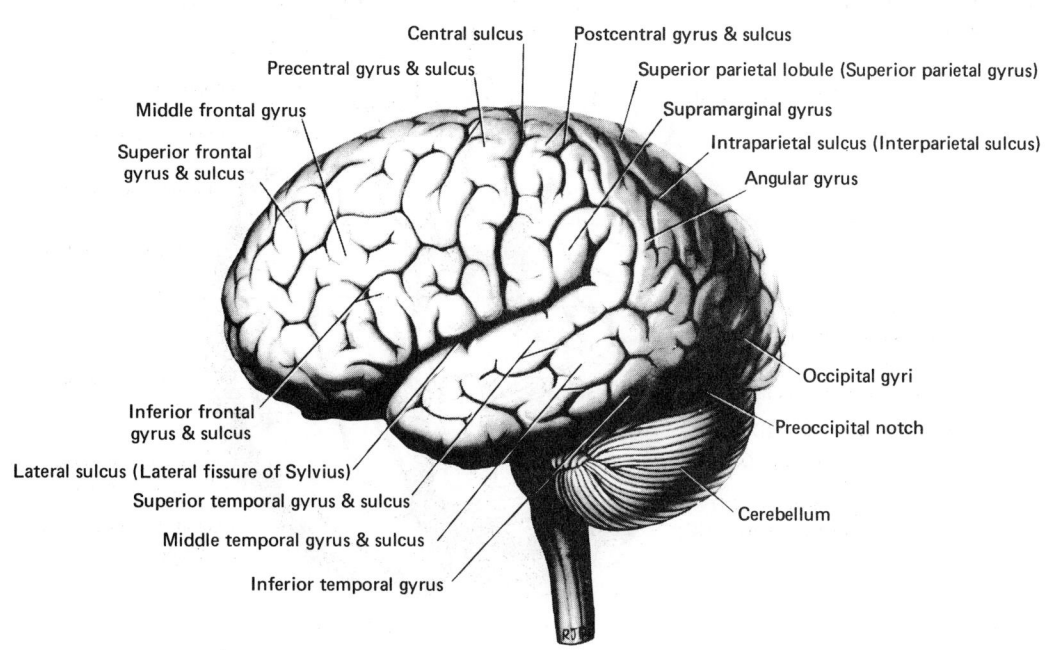

Central sulcus
Precentral gyrus & sulcus
Middle frontal gyrus
Superior frontal gyrus & sulcus
Postcentral gyrus & sulcus
Superior parietal lobule (Superior parietal gyrus)
Supramarginal gyrus
Intraparietal sulcus (Interparietal sulcus)
Angular gyrus
Occipital gyri
Preoccipital notch
Inferior frontal gyrus & sulcus
Lateral sulcus (Lateral fissure of Sylvius)
Superior temporal gyrus & sulcus
Middle temporal gyrus & sulcus
Inferior temporal gyrus
Cerebellum

MEDIAN SAGITTAL SECTION OF THE BRAIN

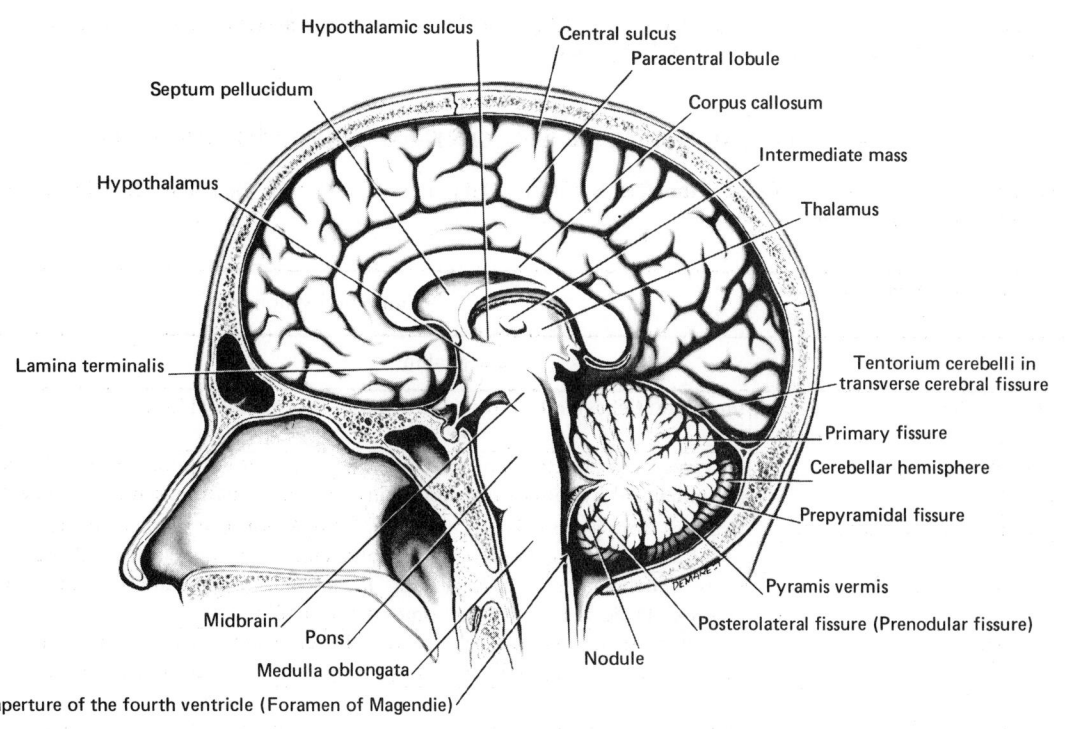

Hypothalamic sulcus
Septum pellucidum
Hypothalamus
Central sulcus
Paracentral lobule
Corpus callosum
Intermediate mass
Thalamus
Lamina terminalis
Tentorium cerebelli in transverse cerebral fissure
Primary fissure
Cerebellar hemisphere
Prepyramidal fissure
Pyramis vermis
Posterolateral fissure (Prenodular fissure)
Midbrain
Pons
Medulla oblongata
Nodule
Median aperture of the fourth ventricle (Foramen of Magendie)

continued

BASAL SURFACE OF THE BRAIN AND ROOTS OF THE CRANIAL NERVES (I-XII)

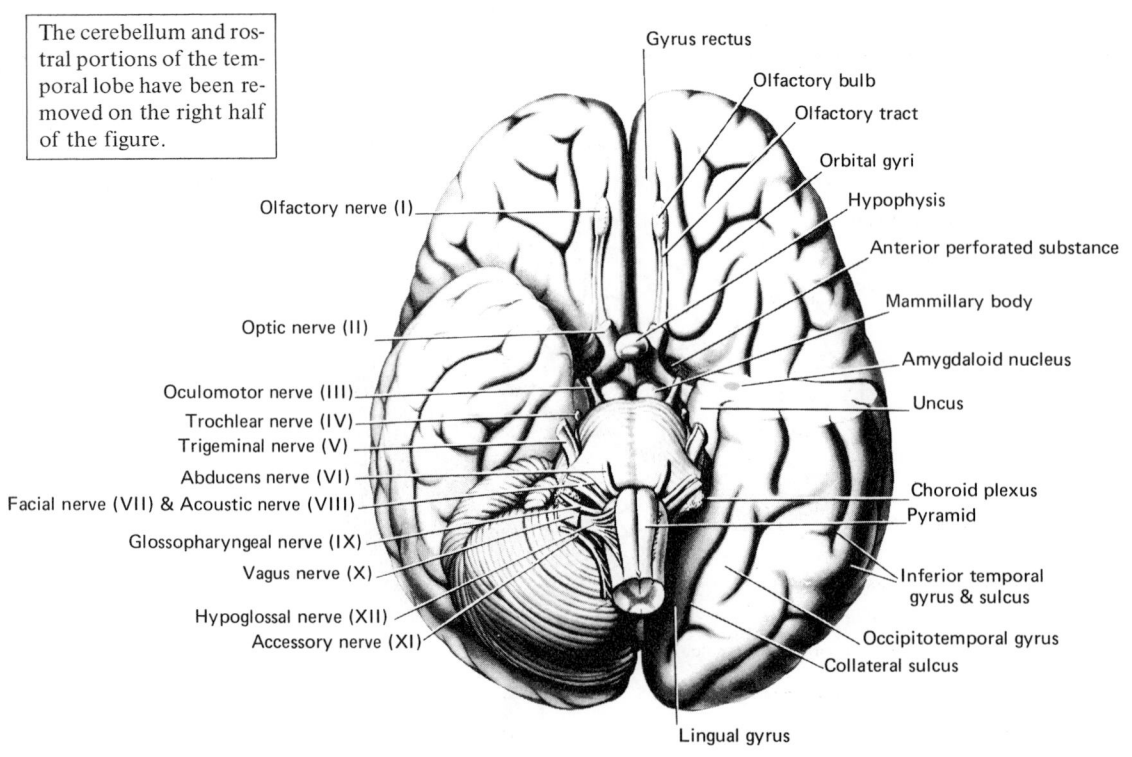

The cerebellum and ros-
tral portions of the tem-
poral lobe have been re-
moved on the right half
of the figure.

Gyrus rectus

Olfactory bulb

Olfactory tract

Orbital gyri

Hypophysis

Anterior perforated substance

Mammillary body

Amygdaloid nucleus

Uncus

Choroid plexus

Pyramid

Inferior temporal
gyrus & sulcus

Occipitotemporal gyrus

Collateral sulcus

Lingual gyrus

Olfactory nerve (I)

Optic nerve (II)

Oculomotor nerve (III)

Trochlear nerve (IV)

Trigeminal nerve (V)

Abducens nerve (VI)

Facial nerve (VII) & Acoustic nerve (VIII)

Glossopharyngeal nerve (IX)

Vagus nerve (X)

Hypoglossal nerve (XII)

Accessory nerve (XI)

Contributor: Angevine, Jay B., Jr.

Reference: Noback, C. R., and R. J. Demarest. 1967. The Human Nervous System; Basic Elements of Structure and Function. McGraw-Hill, Blakiston Division, New York.

134. PRINCIPAL NUCLEI OF THE CENTRAL NERVOUS SYSTEM: MAN

Function: C = cervical; L = lumbar; T = thoracic; S = sacral. Data in brackets refer to the column heading in brackets. Data in this table also apply to most mammals.

Structure & Nuclear Mass [Synonym]		Function [Synonym]
	Telencephalon	
1	Amygdaloid body [Amygdala; amygdaloid complex]	Important behavioral center in medial aspect of temporal lobe
2	Basolateral group	Receives from olfactory cortex & probably adjacent temporal cortex. Projects via stria terminalis to hypothalamus, midbrain reticular formation, septal nuclei, & thalamus.
3	Corticomedial gray group	Receives from hypothalamus, olfactory bulb, & nonspecific thalamic nuclei. Projects to contralateral, amygdaloid body, hypothalamus, septal nuclei, & thalamus.
4	Caudate nucleus [1]	Largest nuclear mass in basal ganglia. Receives topographical projection from cerebral cortex. Projects to outer portion of globus pallidus.
5	Claustrum	Nuclear mass internal to insular cortex. Receives input from insular cortex. Projects to caudate nucleus & putamen, and probably functions with basal ganglia.

[1] Caudate nucleus, globus pallidus, and putamen form the basal ganglia—an important part of the involuntary motor system.

continued

Structure & Nuclear Mass [Synonym]	Function [Synonym]
6 Globus pallidus [1]	Provides efferent connections for basal ganglia. Receives input from caudate nucleus & putamen—efferent fiber originates from inner portion and projects to centromedian nucleus, red nucleus, substantia nigra, subthalamus, & ventral anterior nucleus.
7 Putamen [1]	Large nuclear mass in basal ganglia. Receives topographical projection from cerebral cortex. Projects to outer portion of globus pallidus.
8 Septal nuclei [Septum pellucidum]	Important in setting behavior patterns. Receives input from amygdaloid body, hippocampus, hypothalamus, & cortex. Projects to amygdaloid body, anterior nuclei, hippocampus, hypothalamus, & trigonum habenulae [habenula].

Diencephalon

Epithalamus	
9 Trigonum habenulae [Habenula]	Receives input from amygdaloid body & septum pellucidum via stria medullaris. Projects to midbrain reticular formation (habenulopeduncular tract)
Thalamus	
10 Anterior nuclei	Interposed between cingulate cortex, hypothalamus, intralaminar nuclei, and midbrain reticular formation
11 Dorsomedial nucleus	Receives input from hypothalamus & intralaminar nuclei. Projects to frontal association cortex, areas 9, 10, 11, & 12.
12 Intralaminar nuclei—centromedian nucleus, etc.	Receives input from reticular formation & spinothalamics. Projects to probably all thalamic nuclei, but especially anterior, dorsomedial, & ventral anterior nuclei, & to basal ganglia.
13 Lateral geniculate nucleus	Thalamic site of terminations for optic nerve. Projects to visual cortex, area 17. Visual fields project topographically into nucleus.
14 Lateral posterior nucleus	Input? Projects to superior parietal lobule.
15 Medial geniculate nucleus	Tonotopically organized. Thalamic terminal for ascending auditory fibers from cochlear nuclei, inferior colliculi, & superior olive. Projects to auditory cortex, area 41.
16 Midline nucleus	Input & connections probably similar to intralaminar nuclei (entry 12)
17 Nucleus ventralis lateralis [Ventral lateral nucleus]	Somatopically organized. Receives principal input from nucleus dentatus [dentate nucleus] of cerebellum & red nucleus (dentato-rubrothalamic tract). Projects onto motor cortex, areas 4 & 6.
18 Nucleus ventralis posterolateralis [Ventral posterior lateral nucleus]	Somatopically organized. Receives cutaneous & proprioceptive information via medial lemniscus & spinothalamics. Projects extremity region, neck, & trunk onto sensory cortex, areas 1, 2, & 3.
19 Nucleus ventralis posteromedialis [Ventral posterior medial nucleus]	Somatopically organized. Receives cutaneous & proprioceptive information via gustatory & secondary trigeminal fibers. Projects to head region in sensory cortex, areas 1, 2, & 3.
20 Pulvinar	Receives connections from all ascending fiber systems (Cerebellum, medial lemniscus, reticular formation, & spinothalamics). Projects to inferior parietal lobule & posterior temporal lobe.
21 Reticular nucleus	Input & connections probably similar to intralaminar nuclei (entry 12)
22 Ventral anterior nucleus	Important link between globus pallidus & premotor cortex (area 6)
Subthalamus	
23 Subthalamic nucleus	Relay nucleus between globus pallidus, red nucleus, & ventral anterior nucleus
24 Zona incerta	Receives input from inner portion of globus pallidus. Fasciculus lenticularis runs through nucleus.
Hypothalamus	
25 Anterior nucleus	Possible role in integrating parasympathetic function
26 Dorsomedial nucleus	Projection?; function?.
27 Mammillary body	Receives large input from hippocampus via fornix. Projects to tegmental nuclei of midbrain (mammillotegmental tract) & anterior nuclei of thalamus (mammillothalamic tract); projects to & from dorsal & ventral tegmental nuclei via mammillary peduncle.
28 Periventricular area [Paraventricular nucleus]	Nucleus of neurosecretory cells which project to pars nervosa [posterior pituitary gland], and which synthesize & secrete oxytocin (hypothalamo-hypophyseal system)

[1] Caudate nucleus, globus pallidus, and putamen form the basal ganglia—an important part of the involuntary motor system.

continued

	Structure & Nuclear Mass [Synonym]	Function [Synonym]
29	Posterior nucleus, including lateral area	Possible role in integration of sympathetic functions, temperature elevating behavior (peripheral vasoconstriction, etc.), & sensation of hunger & feeding behavior. Possible efferent projection to brain stem.
30	Preoptic area	Possible function in temperature regulation. Theorized projection to tuberal [arcuate] nucleus (hypophysiotropic center); regulates changes in hypophysiotropic activity, resulting in sexual cycling.
31	Supraoptic nucleus	Nucleus of neurosecretory cells which project to pars nervosa [posterior pituitary gland], and which synthesize and secrete vasopressin (hypothalamo-hypophyseal system).
32	Tuberal nucleus [Arcuate nucleus]	Hypophysiotropic center containing presumed neurosecretory cells which project to capillaries of median eminence, and which are thought to synthesize and secrete regulating factors that excite or inhibit the various functions of anterior lobe of hypophysis [anterior pituitary]
33	Ventromedial nucleus	Presumed "satiety center" inhibiting feeding behavior

Midbrain

	Structure & Nuclear Mass [Synonym]	Function [Synonym]
34	Cranial nerves Nucleus of mesencephalic root of trigeminal nerve (V)	(See entry 55)
35	Oculomotor nerve (III) Edinger-Westphal nucleus	Preganglionic parasympathetic innervation to ganglia which control intrinsic muscles that constrict pupil and accommodate lens
36	Motor nucleus	Innervates extrinsic muscles of eye—inferior oblique, inferior, medial, & superior rectus; levator palpebrae superior
37	Trochlear nerve (IV)	Innervates superior oblique—an extrinsic eye muscle
38	Inferior colliculus	Receives primary, secondary, & tertiary auditory afferents from relay nuclei at all levels that contribute to lateral lemniscus. Projects chiefly to ipsilateral medial geniculate body via macroscopically visible inferior quadrigeminal brachium.
39	Pretectal area	Center for light reflex constriction of pupil in bright light
40	Red nucleus	Connects to spinal cord, subthalamus, & thalamus; forms the "indirect" corticospinal tract—the rubrospinal tract. Receives direct topographic projection from cerebellum & cerebral cortex.
41	Reticular formation, midbrain	Especially important region of reticular formation, owing to direct connection with diencephalon. Receives input from medulla oblongata, pons, & spinal cord. Projects to anterior, intralaminar & midline thalamic nuclei, & hypothalamus.
42	Substantia nigra	Mesencephalic nucleus of basal ganglia neurons contains dopamine; depletion of dopamine produces Parkinson's disease. Projects to globus pallidus, caudate & putamen.

Pons

	Structure & Nuclear Mass [Synonym]	Function [Synonym]
43	Cranial nerves Abducens nerve (VI)	Innervates lateral rectus muscle—an extrinsic eye muscle
44	Acoustic nerve (VIII) [Auditory nerve] Cochlear nuclei—anterior & posterior ventral, & dorsal	Receives primary auditory afferents & inhibitory afferents of olivo-cochlear bundle. Projects to contralateral nuclei of lateral lemniscus, contralateral nucleus of the trapezoid body, & superior olive. Lateral lemniscus, which consists in large part of cochlear nuclear efferents, ends primarily in inferior colliculus.
45	Nucleus of lateral lemniscus	Receives secondary & tertiary auditory afferents. Projects in ipsilateral lateral lemniscus, chiefly to inferior colliculus.
46	Nucleus of trapezoid body	Receives afferents from contralateral cochlear nucleus. Projects in lateral lemniscus chiefly to inferior colliculus.
47	Superior olive [Superior olivary complex]	Receives afferent fibers & collaterals bilaterally from cochlear nuclei. Projects in ipsilateral lateral lemniscus to inferior colliculus and directly to abducens nucleus. Probable source of olivo-cochlear bundle.
48	Facial nerve (VII) Motor nucleus	Innervates muscles for facial expression, & extrinsic & intrinsic ear muscles

continued

	Structure & Nuclear Mass [Synonym]	Function [Synonym]
49	Solitarius nucleus [Solitary nucleus]	Secondary cell bodies for gustatory sensations from taste buds in anterior two-thirds of tongue
50	Superior salivatory nucleus	Preganglionic parasympathetic to ganglia for glands in nose & palate, & lacrimal, sublingual, & submandibular [submaxillary] glands
	Trigeminal nerve (V)	
51	Center for lateral gaze [Parabducens nucleus]	In reticular formation adjacent to nucleus of abducens nerve (VI). Axons of nucleus run contralaterally in medial longitudinal fasciculus to connect abducens nucleus of one side to oculomotor nucleus of other side, providing conjugate lateral eye movement.
52	Descending nucleus	Conveys pain & temperature from face, meninges, orbit, & anterior two-thirds of tongue. Axons of nucleus join medial lemniscus and ascend to contralateral nucleus ventralis posteromedialis of thalamus.
53	Main sensory nucleus	Secondary cell bodies for tactile information from face, orbit, & anterior two-thirds of tongue. Axons of nucleus join medial lemniscus and run bilaterally to nucleus ventralis posteromedialis of thalamus.
54	Motor nucleus	Innervates anterior belly of digastric muscle, & muscles for mastication, tensor tympani, tensor veli palatini, & tympanum
55	Nucleus of mesencephalic root of trigeminal nerve (V)	Found also in midbrain. Only primary cell body within central nervous system. It conveys proprioceptive information from all head muscles into nuclei of trigeminal nerve (V).
56	Vestibular nerve (VIII)—inferior, medial, superior nuclei	(*See* entries 68, 70, & 71)
57	Pontine gray matter	Receives input from all lobes of cerebral hemisphere. Projects contralaterally to cerebellar hemisphere.
58	Reticular formation, pontine—nucleus pontis caudalis & nucleus pontis oralis	Forms reticulomesencephalic & reticulothalamic connections
	Medulla Oblongata	
	Cranial nerves	
	Glossopharyngeal nerve (IX)	
59	Ambiguus nucleus	Initiates glutination by innervation of stylopharyngeus muscles
60	Inferior salivatory nucleus	Preganglionic to otic ganglion for secretion from parotid gland
61	Solitarius nucleus	Secondary cell bodies for gustatory sensations from taste buds in posterior one-third of tongue; sensation conveyed bilaterally via medial lemniscus to nucleus ventralis posteromedialis of thalamus. Secondary cell bodies for visceral sensations from carotid body, palate, & posterior one-third of tongue.
62	Hypoglossal nerve (XII)	Innervates intrinsic musculature of tongue
	Trigeminal nerve (V)	
63	Descending nucleus	Secondary cell bodies for pain & temperature from face, meninges, orbit, teeth, & anterior two-thirds of tongue. Axons of nucleus ascend contralaterally in medial lemniscus to nucleus ventralis posteromedialis of thalamus.
	Vagus	
64	Ambiguus nucleus	Innervates muscles in larynx & pharynx responsible for glutination & phonation
65	Dorsal motor nucleus	Preganglionic parasympathetic innervation to smooth muscles in heart, blood vessels, trachea, bronchi, esophagus, stomach, intestines to lower half of colon.
66	Solitarius nucleus [Solitary nucleus]	Secondary cell bodies for gustatory sensations from taste buds in epiglottis & pharynx; sensations conveyed bilaterally via medial lemniscus to nucleus ventralis posteromedialis of thalamus. Secondary cell bodies for visceral sensations from aortic body, larynx, pharynx, thorax, & abdomen.
67	Vestibular nerve (VIII)	Balance & position of body in space
68	Inferior nucleus	Found also in pons. Secondary cell bodies receive primary axons from all canals, saccule, & utricle. Projects to cerebellum (flocculus, nucleus fastigii, nodule, & uvula), cranial nerves (abducens (VI), oculomotor (III), & trochlear (IV)), & spinal cord.

continued

	Structure & Nuclear Mass [Synonym]	Function [Synonym]
69	Lateral nucleus	Secondary cell bodies receive primary axons from all canals, saccule, & utricle; also receive input from cerebellum. Form vestibulospinal tract which projects ipsilaterally to spinal cord.
70	Medial nucleus	Found also in pons. Secondary cell bodies receive primary axons from all canals, saccule, & utricle. Project bilaterally to nuclei of abducens cranial nerve (VI) and contralaterally to oculomotor (III) & trochlear (IV) cranial nerves; project to flocculonodular lobes of cerebellum.
71	Superior nucleus	Found in pons. Secondary cell bodies receive primary axons from anterior, lateral, & posterior canals of labyrinth. Project ipsilaterally to cranial nuclei of abducens (VI), oculomotor (III), & trochlear (IV) nerves via medial longitudinal fasciculus. Correlate eye movement with position of head in space.
72	External cuneate nucleus	Secondary cell bodies for unconscious proprioceptive information from cervical regions & upper extremity to brain stem & cerebellum
73	Inferior olive [Inferior olivary complex]—dorsal & medial accessory nuclei, & principal nucleus	Receives input from reticular formation. Projects contralaterally to cerebellum. Function?
74	Nucleus cuneatus [Cuneatus]	Secondary cell bodies for conscious proprioceptive & tactile sensations from structures innervated by cervical & upper thoracic levels. Join medial lemniscus and run contralaterally to nucleus ventralis posteromedialis of thalamus.
75	Nucleus gracilis [Gracilis]	Secondary cell bodies for conscious proprioceptive & tactile sensations from structures innervated by lower thoracic, lumbar, & sacral levels. Axons of nucleus join medial lemniscus and run contralaterally to nucleus ventralis posteromedialis of thalamus.
	Reticular formation, medullary	
76	Lateral reticular nuclei	Receives afferent information from spinal cord & projects fibers to midbrain & cerebellum.
77	Reticularis gigantocellularis	Forms projections from medullary reticular formation to pontine & midbrain reticular formations; forms lateral reticulospinal tract
	Spinal Cord	
	Dorsal horn	
78	Dorsal nucleus [Nucleus of Clark]	Secondary cell bodies for unconscious proprioceptive information from thorax, abdomen, & lower extremities to cerebellum, forming posterior spinocerebellar tract; levels C_8-L_3
79	Nucleus posteromarginalis	Secondary neurons & interneurons [internuncial neurons] for somatic sensations; all levels
80	Nucleus proprius dorsalis	Secondary neurons & interneurons [internuncial neurons] for somatic sensations; all levels
81	Secondary visceral gray matter	Secondary neurons & interneurons [internuncial neurons] for visceral sensations; levels T_1-L_1
82	Substantia gelatinosa	Interneurons [internuncial neurons]; all levels
	Lateral horn	
83	Intermediolateral nucleus	Preganglionics for sympathetic nervous system; levels T_1-L_3
	Ventral horn	
84	Accessory cranial nerve (XI)	Innervates sternocleidomastoid & trapezius muscles; levels C_1-C_5
85	Dorsolateral nucleus	Innervates muscles in forearm & hand; levels C_4-C_8. Innervates muscles in leg & foot; levels L_2-S_2.
86	Phrenic nucleus	Innervates muscles in diaphragm; levels C_3-C_7
87	Ventrolateral nucleus	Innervates muscles in shoulder girdle & upper arm; levels C_4-C_8. Innervates muscles in pelvis & thigh; levels L_2-S_2.
88	Ventromedial nucleus	Innervates axial musculature; all levels

Contributors: Jacobson, Stanley, and Brawer, James R.

continued

134. PRINCIPAL NUCLEI OF THE CENTRAL NERVOUS SYSTEM: MAN

General References

[1] Brodal, A. 1969. Neurological Anatomy. Oxford Univ. Press, New York.

[2] Crosby, E. C., et al. 1962. Correlative Anatomy of the Nervous System. Macmillan, New York.

[3] Curtis, B. A., et al. 1972. An Introduction to the Neurosciences. W. B. Saunders, Philadelphia.

[4] Haymaker, W., et al. 1969. The Hypothalamus. C. C. Thomas, Springfield, Ill.

[5] Truex, R. C., and M. B. Carpenter. 1969. Human Neuroanatomy. Williams and Wilkins, Baltimore.

135. COMPARATIVE ANATOMY OF THE CENTRAL NERVOUS SYSTEM: MAJOR INVERTEBRATE PHYLA

Data are for the central nervous system of adult animals in selected classes of major invertebrate phyla. Where a central nervous system proper is difficult or impossible to demarcate naturally, the entire nervous system has been described; where a central nervous system is unambiguously recognizable, peripheral systems have generally not been considered. In certain aberrant forms, particularly those highly modified in association with parasitism, description of the basic anatomical arrangement may not remain valid due to great reduction and simplification of the adult central nervous system. When special neuropiles occur, there is typically a concomitant differentiation within the cortex of nerve cell bodies involving increased numbers and/or distinctive groupings of cell bodies. The question of homologies of particular neuropile areas among different groups [6,7] has not been discussed in the table below, as the relationship of structures given the same name in different groups, e.g., corpora pedunculata, has not always been critically established. Unless otherwise specified, data are from references 2 and 6. **Anatomical Features: Centralization**—degree of centralization, rated from 0 (lowest) to +++ (highest). **Histological Features: Predominant Neuron Type**—Bp = bipolar, Mp = multipolar, Up = unipolar. Data in brackets refer to the column heading in brackets.

Phylum & Class	Anatomical Features			Histological Features		Remarks
	Symmetry [Centralization]	Basic Arrangement	Predominant Neuron Type	Basic Organization	Special Neuropiles	
1 Porifera	Presence of true nerve cells not established; for a discussion of sponge cells exhibiting some neuronal characteristics, consult reference 8.					
Coelenterata						
2 Hydrozoa	Radial[1] [0 to +]	Dispersed subepithelial networks of neurons, often more than one net per individual; condensed nerve rings associated with velum of medusae	Bp, Mp	....	None	True synapses widespread, but some nerve nets described as syncytial; epithelial non-neuronal conduction in addition to conduction through nerve nets; ganglia, with cortex of nerve cell bodies around a fibrous core resembling simple neuropile, may occur in hydrozoan medusae (consult reference 9)
3 Scyphozoa	Radial [0 to +]	Dispersed subepithelial networks of neurons, often more than one net per individual; clusters of neurons associated with marginal sense organs; nerve rings rarely present	Bp, Mp	....	None	
4 Anthozoa	Radial, with elements of bilaterality [0]	Dispersed subepithelial networks of neurons, often more than one net per individual, with some regional differences in cell form & density	Bp, Mp	....	None	

[1] In colonial forms, may not apply to the most highly modified zooids.

continued

	Phylum & Class	Anatomical Features		Predominant Neuron Type	Histological Features		Remarks
		Symmetry [Centralization]	Basic Arrangement		Basic Organization	Special Neuropiles	
5	Platyhelminthes Turbellaria	Bilateral [+ to ++]	Extensive nerve plexuses, including a variable number of longitudinal cords & transverse commissures, with dense nerve plates in one group; anterior brain variable from well differentiated to barely distinguishable or possibly absent	Bp, Mp, Up	Brain, commonly with rind of nerve cell bodies surrounding central neuropile	No specialization of neuropile known, but distinctive cell aggregates, including globuli cells, present in some forms	Fairly distinct brain sheath in some forms
6	Trematoda	Bilateral [+]	Plexus generally dominated by 3 pairs of longitudinal cords & their commissures; brain present	Bp, Mp	No clear rind, as nerve cell bodies are scattered in the neuropile	None known	....
7	Cestoda	Bilateral, with signs of radiality [+]	Simple brain in scolex from which a variable number of longitudinal nerve cords run through strobila; transverse commissures in each proglottid	Bp, Mp	Insufficiently known	....	Reports of giant fibers
8	Nematoda	Bilateral, with signs of radiality [++]	Circumenteric nerve ring associated with several small ganglia	Up	Nerve ring fibrous & nearly devoid of nerve cell bodies; ganglia associated with ring basically comprise clusters of nerve cell bodies without neuropile	None known	Constant neuron number established for some forms; extensive neuronal anastomoses said to occur
9	Mollusca Polyplacophora	Bilateral [++]	2 pairs of longitudinal cords with numerous transverse commissures, of which the supraesophageal represents the brain	Up	Cortex of nerve cell bodies surrounding fibrous core in cords	None known	Many nerve cell bodies scattered along nerve trunks & in other tissues
10	Gastropoda	Bilateral, modified by torsion about vertical axis; bilaterality secondarily restored in varying degrees [++ to +++]	5 or more pairs of ganglia variably interconnected by longitudinal connectives & transverse commissures; cerebral, buccal, and pedal ganglia (pedal sometimes represented by longitudinal cords) symmetric & paired; remainder in loop twisted into figure-8 by torsion, often secondarily rearranged; various degrees of cephalization present; cerebral ganglia represent brain	Up	Cortex of nerve cell bodies surrounding neuropile in well demarcated ganglia; connectives & commissures largely free of nerve cell bodies	Dense neuropiles occur associated with numerous tiny nerve cell bodies, notably in brain & tentacular ganglia of certain Pulmonata	Giant nerve cell bodies common in Opisthobranchia, some reaching almost 1 mm in diameter (smaller and rarer in Pulmonata); however, axons rarely exceed 50 μ

continued

	Phylum & Class	Anatomical Features			Histological Features		Remarks
		Symmetry [Centralization]	Basic Arrangement	Predominant Neuron Type	Basic Organization	Special Neuropiles	
11	Bivalvia	Bilateral [+++]	3 pairs of ganglia interconnected by longitudinal connectives & transverse commissures; various degrees of fusion present; cerebral ganglia represent brain	Up	Cortex of nerve cell bodies surrounding neuropile; connectives & commissures generally cell-free	Series of special glomeruli known in visceroparietal ganglion of *Pecten*	....
12	Cephalopoda	Bilateral [+++]	Central ganglia concentrated into mass encircling esophagus, typically forming highly elaborate brain; highly differentiated optic lobes lateral to brain typically present	Up	Cortex of nerve cell bodies & interior neuropile common, but central islands of cells also present in both brain & optic lobes	Brain typically includes elaborate arrangement of diverse, highly differentiated neuropiles; optic lobes typically highly patterned & elaborately layered	Includes largest & most complex invertebrate brains, on a par with some vertebrate brains; *Octopus* brain plus optic lobes estimated to contain about 10^8 nerve cells; giant nerve cell bodies & axons occur; nerve cords in arms particularly massive in Octopoda; peripheral ganglia common
13	Annelida Polychaeta	Bilateral [+++]	Brain located dorsal to gut; connected by circumenteric connectives to elongate ventral chain of segmental ganglia	Up	Cortex of nerve cell bodies; central neuropile region	Several often present in brain; sizable corpora pedunculata, with large masses of small nerve cell bodies, can occur	Brain typically more highly differentiated in errant than in sedentary forms; giant fibers in some forms
14	Oligochaeta	Bilateral [+++]	Brain located dorsal to gut; connected by circumenteric connectives to elongate ventral chain of segmental ganglia	Up	Cortex of nerve cell bodies; central neuropile region	None known	Giant fibers in some forms
15	Hirudinea	Bilateral [+++]	Brain located dorsal to gut; connected by circumenteric connectives to elongate ventral chain of segmental ganglia	Up	Cortex of nerve cell bodies grouped into distinct packets; central neuropile region	None known	Giant glial cells occur (consult reference 5), as do giant nerve cell bodies
16	Arthropoda Diplopoda	Bilateral [+++]	Brain located dorsal to gut; connected circumenterically to elongate ventral chain of segmental ganglia	Up	Cortex of nerve cell bodies; central neuropile region	Several specialized neuropiles present in brain, including distinct corpora pedunculata	....

continued

	Phylum & Class	Anatomical Features		Predominant Neuron Type	Histological Features		Remarks
		Symmetry [Centralization]	Basic Arrangement		Basic Organization	Special Neuropiles	
17	Chilopoda	Bilateral [+++]	Brain located dorsal to gut; connected circumenterically to elongate ventral chain of segmental ganglia	Up	Cortex of nerve cell bodies; central neuropile region	Several specialized neuropiles present in brain, including distinct corpora pedunculata	....
18	Insecta	Bilateral [+++]	Brain, with laterally attached optic lobes, located dorsal to gut; connected circumenterically to ventral chain of segmental ganglia; ventral chain elongate to highly condensed	Up	Cortex of nerve cell bodies; central neuropile region	Several normally present in brain and optic lobes; optic neuropiles & corpora pedunculata sometimes massive & extremely highly organized	Giant fibers in some forms
19	Crustacea	Bilateral [+++]	Brain, often with attached optic lobes, located dorsal to gut; connected circumenterically to ventral chain of segmental ganglia; ventral chain elongate to highly condensed	Up	Cortex of nerve cell bodies; central neuropile region	Typically several in brain & optic lobes (where present); optic neuropiles particularly can be massive & extremely highly organized	Giant fibers in some forms; true myelin sheaths known
20	Merostomata	Bilateral [+++]	Brain located dorsal to gut; fused with highly condensed ventral chain of segmental ganglia to form mass which encircles gut	Up	Cortex of nerve cell bodies elaborately infolded in portions of brain; central neuropile region	Several present in brain, including optic neuropiles and elaborately folded corpora pedunculata	Corpora pedunculata enormous, comprising 80% of brain volume
21	Arachnida	Bilateral [+++]	Brain located dorsal to gut; connected with ventral chain of segmental ganglia; ventral chain elongate to highly condensed, frequently fused with brain to form single mass which encircles gut	Up	Cortex of nerve cell bodies; central neuropile region	Several typically present in brain, among them optic neuropiles; neuropiles can be highly organized, e.g., corpora pedunculata (consult reference 1)	Giant fibers in scorpions
22	Echinodermata	Radial, sometimes bilaterally superimposed [+]	Nerve ring(s) giving rise typically to 5 radial nerve cords; peripheral plexus(es)	Bp, Mp, Up	Layering in ring & radial cords of Asteroidea has been described; areas of true neuropile known in Asteroidea and Echinoidea (consult reference 3)	None known	Morphological differences among groups known, but details inadequately understood (e.g., elements interpreted as "ribbon axons" in the tube feet of starfish now recognized as processes of muscle cells (consult reference 4)

continued

	Phylum & Class	Symmetry [Central- ization]	Anatomical Features		Predom- inant Neuron Type	Histological Features		Remarks
			Basic Arrangement			Basic Organization	Special Neuropiles	
23	Chordata Ascidi- acea	Bilateral [+++]	Single ganglion, "cerebral" ganglion, located dorsal to pharynx between siphons		Up	Cortex of nerve cell bodies; central neuropile region	None known	Larval stage posses- ses elongated dor- sal nerve cord ho- mologous with neural tube of Vertebrata

Contributors: Weiss, Mitchell J., and Palka, John

References

[1] Babu, K. S. 1965. Zool. Jahrb. Abt. Anat. Ontog. Tiere 82:1.

[2] Bullock, T. H., and G. A. Horridge. 1965. Structure and Function in the Nervous Systems of Inverte- brates. W. H. Freeman, San Francisco. v. 1-2.

[3] Cobb, J. L. S. 1970. Z. Zellforsch. Mikrosk. Anat. 108:457.

[4] Cobb, J. L. S., and M. S. Laverack. 1967. Symp. Zool. Soc. London 20:25.

[5] Coggeshall, R. E., and D. W. Fawcett. 1964. J. Neu- rophysiol. 27:229.

[6] Hanström, B. 1928. Vergleichende Anatomie des Nervensystems der wirbellosen Tiere. J. Springer, Berlin.

[7] Holmgren, N. 1916. Kgl. Sv. Vetenskapsakad. Handl. 56:1.

[8] Lentz, T. L. 1968. Primitive Nervous Systems. Yale Univ. Press, New Haven.

[9] Mackie, G. O. 1971. Actas Simp. Int. Zoofilog. (Salamanca), p. 269.

136. SYSTEMS OF THE CEREBELLUM: MAMMALS

Data are applicable to most mammals.

Part I. Afferent

	Tract	Origin [Via Cerebellar Peduncle]	Distribution	Impulses Transmitted
1	Dorsal spinocere- bellar	Dorsal nucleus (Clarke's column) from T1 to L2 [Inferior]	Chiefly uncrossed to vermis & inter- mediate part of anterior lobe & pyramis; some fibers to tuber, uvu- la, & medial part of paramedian lobule	Proprioceptive from muscles & joints; exteroceptive from skin of trunk, hind- limb & tail
2	Ventral spinocer- ebellar	"Border cells" in ventral horn of spinal cord [Superior]	Crossed & uncrossed to vermis of an- terior lobe	Visceral afferents(?); pro- prioceptive & exterocep- tive from all parts of body
3	Cuneocerebellar	Lateral cuneate nucleus [Inferior]	Uncrossed to vermis & intermediate part of anterior lobe; some fibers to uvula, tuber & pyramis	Proprioceptive & exterocep- tive from upper limb & neck
4	Arcuatocerebellar	Arcuate nuclei [Inferior]	Bilateral; distribution unknown	Unknown
5	Olivocerebellar	All parts of inferior olive & part of accessory olive [Inferior]	Chiefly crossed to all parts of cortex & all intracerebellar nuclei; partly uncrossed to nucleus fastigii	From higher nuclei, cerebral cortex, & all levels of spi- nal cord

continued

136. SYSTEMS OF THE CEREBELLUM: MAMMALS

Part I.   Afferent

	Tract	Origin [Via Cerebellar Peduncle]	Distribution	Impulses Transmitted
6	Pontocerebellar	All parts of pontine nuclei [Middle]	Chiefly crossed to all parts of cortex except flocculonodular lobe; partly uncrossed to vermis	From spinal cord, all four lobes of cerebrum, & other centers
7	Reticulocerebellar	Lateral reticular nucleus [Inferior]	Uncrossed to entire cerebellar cortex	From all levels of spinal cord and from higher levels
8		Paramedian reticular nucleus [Inferior]	More than half uncrossed to anterior lobe; some to pyramis, uvula, & nucleus fastigii	From higher levels, including cerebral cortex
9	Vestibulocerebellar	Vestibular nuclei, chiefly medial & inferior (descending); some direct vestibular root fibers [Inferior]	Crossed & uncrossed secondary fibers to flocculonodular lobe, and some uvula & nucleus fastigii; uncrossed primary fibers to same areas	Vestibular through 8th cranial nerve
10	Perihypoglosso-cerebellar	Nucleus of Roller, nucleus prepositus, & nucleus intercalatus [Inferior]	More than half uncrossed to anterior lobe; some to pyramis, uvula, & nucleus fastigii	Unknown
11	Tectocerebellar	Superior & inferior colliculi [Superior]	Bilateral, probably to declive, folium, & tuber	Auditory & visual
12	Rubrocerebellar	Caudal two-thirds of red nucleus [Superior]	More than half crossed, chiefly to nucleus dentatus; some to nucleus fastigii	Cerebral?
13	Trigeminocerebellar	Direct sensory fibers; secondary fibers from trigeminal nucleus [Inferior?]	Bilateral to caudal anterior lobe, simplex, & dentate nucleus	Tactile & proprioceptive impulses from face & jaw
14	Lateral cervical cerebellar	Lateral cervical nucleus in C1 & 2 [Inferior]	Unknown	From all levels of spinal cord

Contributor: Snider, Ray S.

General References

[1] Dow, R. S., and G. Moruzzi. 1958. The Physiology and Pathology of the Cerebellum. Univ. Minnesota Press, Minneapolis.

[2] Eccles, J. C., et al. 1968. Exp. Brain Res. 6:195.

[3] Jansen, J., and A. Brodal. 1954. Aspects of Cerebellar Anatomy. J. G. Tanum, Oslo.

[4] Larsell, O. 1967. Comparative Anatomy and Histology of the Cerebellum. Univ. Minnesota Press, Minneapolis.

[5] Oscarsson, O. 1965. Physiol. Rev. 45:495.

[6] Precht, W., and R. Llinas. 1969. Neurobiology of Cerebellar Evolution and Development. American Medical Association, Chicago. pp. 677-702.

[7] Snider, R. S. 1967. Progr. Brain Res. 25:322.

[8] Snider, R. S., and A. Stowell. 1944. J. Neurophysiol. 7:331.

Part II.   Efferent

	Efferent Fiber	Location of Cells of Origin	Nucleus of Termination
1	Inferior cerebellar peduncle	Purkinje cells of vermis & flocculus	Uncrossed to all vestibular nuclei
2	Uncinate fasciculus—inferior cerebellar peduncle	Nucleus fastigii	Crossed to dorsolateral reticular formation, all vestibular nuclei, cervical spinal cord(?), & motor nuclei of 5th, 7th, & 12th(?) cranial nerves

continued

Part II. Efferent

	Efferent Fiber	Location of Cells of Origin	Nucleus of Termination
3	Fastigiobulbar—inferior cerebellar peduncle	Nucleus fastigii	Uncrossed to vestibular nuclei, paramedian reticular nucleus, & dorsolateral reticular formation
4	Descending superior cerebellar peduncle	Nucleus interpositus, nucleus dentatus	Crossed to tegmental nucleus of the pons, ventromedial reticular formation, olivary nuclei, & cervical spinal cord(?)
5	Accessory superior cerebellar peduncle	Nucleus fastigii	Bilateral to nucleus of 3rd cranial nerve, superior & inferior colliculus, red nucleus, periaqueductal gray, nucleus parafascicularis, & nucleus centralis medialis
6	Ascending superior cerebellar peduncle	Nucleus dentatus, nucleus interpositus, nucleus fastigii	Bilateral (mainly contralateral) to midbrain tegmentum, nucleus ventralis lateralis, nucleus ventralis posterior, red nucleus, H_1 of Forel, nucleus geniculatus medialis, zona incerta, nucleus lateralis posterior, pulvinar medialis(?), reticular nucleus, nucleus centromedianus, globus pallidus(?), nucleus centralis lateralis, & nucleus ventralis anterior

Contributor: Snider, Ray S.

General References

[1] Cohen, D., et al. 1958. J. Comp. Neurol. 109:233.
[2] Dow, R. S., and G. Moruzzi. 1958. The Physiology and Pathology of the Cerebellum. Univ. Minnesota Press, Minneapolis.
[3] Hassler, R. 1950. Deut. Z. Nervenheilk. 163:629.
[4] Jansen, J., and A. Brodal. 1954. Aspects of Cerebellar Anatomy. J. G. Tanum, Oslo.
[5] Larsell, O. 1967. Comparative Anatomy and Histology of the Cerebellum. Univ. Minnesota Press, Minneapolis.
[6] Snider, R. S. 1967. Progr. Brain Res. 25:322.
[7] Snider, R. S., et al. 1970. Int. J. Neurol. 7:141.
[8] Whiteside, J. A., and R. S. Snider. 1953. J. Neurophysiol. 16:397.

137. REFLEX ACTIONS OF SPINAL MOTOR NEURONS: MAMMALS

The diagram illustrates some of the principal reflex pathways from afferent fibers of peripheral nerves to spinal motor neurons. *Symbols:* (+) = excitatory actions; (−) = inhibitory actions.

NERVE ACTIONS

Group Ia afferents form primary endings on muscle spindles. These afferents include both those supplying the same muscle as that innervated by the motor neuron (homonymous), and those supplying synergistic (heteronymous) muscles.

Group Ib afferents provide nerve terminals of Golgi tendon organs.

Flexion reflex afferents (FRA) innervate several kinds of cutaneous, muscular, and joint receptors. These afferents include fibers of a variety of sizes in cutaneous nerves, as well as groups II and III muscle afferents and the high-threshold joint afferents.

Type A interneurons are monosynaptically excited by group Ia afferents.

Type B interneurons are monosynaptically excited by group Ib afferents.

Type C interneurons are monosynaptically excited by cutaneous afferents. (However, flexion and crossed

continued

extension reflexes would additionally involve other interneurons not belonging to Type C, and which would be polysynaptically activated by FRA.)

Renshaw cells are inhibitory interneurons activated by the recurrent collaterals of motor axons. Their stron-gest effects are on motor neurons belonging to the same motor neuron pool; therefore, the recurrent inhibitory pathway may be considered in part a negative feedback loop.

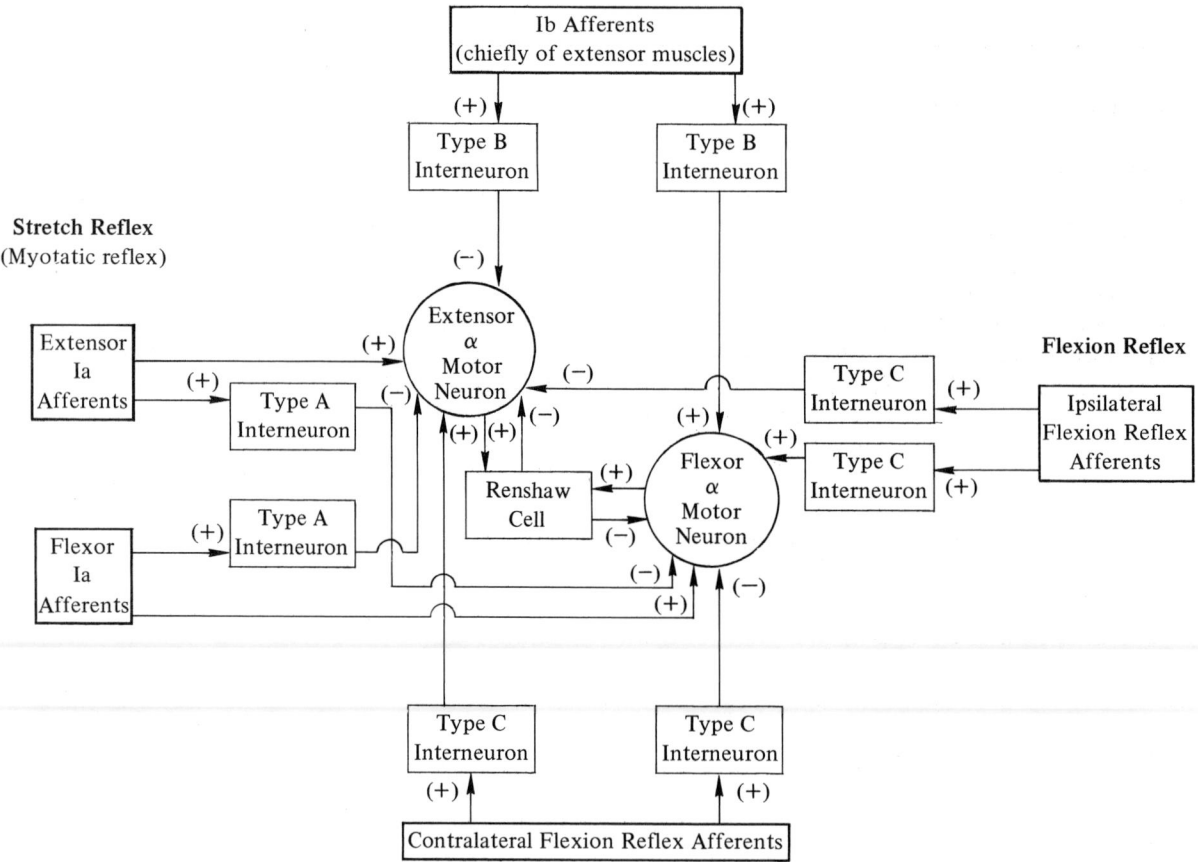

Tension Feedback
(Clasp knife reflex)

Stretch Reflex
(Myotatic reflex)

Flexion Reflex

Crossed Extension Reflex

continued

137. REFLEX ACTIONS OF SPINAL MOTOR NEURONS: MAMMALS

Contributor: Willis, W. D., Jr.

General References

[1] Araki, T., et al. 1960. J. Physiol. (London) 154: 354.

[2] Eccles, J. C. 1964. The Physiology of Synapses. Academic Press, New York.

[3] Eccles, J. C., et al. 1956. J. Neurophysiol. 19:75.

[4] Eccles, J. C., et al. 1957. J. Physiol. (London) 136: 527.

[5] Eccles, J. C., et al. 1957. Ibid. 137:22.

[6] Eccles, J. C., et al. 1957. Ibid. 138:227.

[7] Eccles, J. C., et al. 1960. Ibid. 154:89.

[8] Eccles, J. C., et al. 1961. Ibid. 159:479.

[9] Eccles, R. M., and A. Lundberg. 1958. Ibid. 144: 271.

[10] Eccles, R. M., and A. Lundberg. 1958. Acta Physiol. Scand. 43:204.

[11] Eccles, R. M., and A. Lundberg. 1959. Arch. Ital. Biol. 97:199.

[12] Granit, R. 1950. J. Neurophysiol. 13:351.

[13] Granit, R. 1952. Ibid. 15:269.

[14] Holmqvist, B. 1961. Acta Physiol. Scand., Suppl. 181.

[15] Houk, J., and E. Henneman. 1967. Brain Res. 5: 433.

[16] Hultborn, H., et al. 1971. J. Physiol. (London) 215: 613.

[17] Hunt, C. C. 1952. Ibid. 117:359.

[18] Hunt, C. C. 1954. J. Gen. Physiol. 38:117.

[19] Laporte, Y., and D. P. C. Lloyd. 1952. Amer. J. Physiol. 169:609.

[20] Lloyd, D. P. C. 1941. J. Neurophysiol. 4:184.

[21] Lloyd, D. P. C. 1943. Ibid. 6:111.

[22] Lloyd, D. P. C. 1943. Ibid. 6:317.

[23] Lloyd, D. P. C. 1946. Ibid. 9:421.

[24] Lloyd, D. P. C. 1946. Ibid. 9:439.

[25] Lloyd, D. P. C., and H. T. Chang. 1948. Ibid. 11: 199.

[26] Paintal, A. S. 1961. J. Physiol. (London) 156:498.

[27] Renshaw, B. 1940. J. Neurophysiol. 3:373.

[28] Renshaw, B. 1941. Ibid. 4:167.

[29] Renshaw, B. 1946. Ibid. 9:191.

[30] Rexed, B., and P. Therman. 1948. Ibid. 11:133.

[31] Sherrington, C. S. 1906. The Integrative Action of the Nervous System. Yale Univ. Press, New Haven.

[32] Sherrington, C. S. 1910. J. Physiol. (London) 28: 121.

[33] Wilson, V. J., et al. 1960. J. Neurophysiol. 23:144.

138. CLASSIFICATION OF AFFERENT DORSAL ROOT FIBERS: MAMMALS

Each neuron cell body of a spinal dorsal root ganglion has a dorsal root nerve fiber with a sensory process in peripheral tissue and a central process with synaptic terminals on second-order neurons in the spinal cord or medulla. (Cranial nerves are not considered here. In reference to cranial and other special nerves, it must be recognized that in spite of similarities in characteristics of sensory elements in different body regions, there are important regional differences of types and frequencies of occurrence.) These data are based on evidence documenting systematic differences in functional characteristics of sensory receptors associated with various body tissues. In a number of instances, the differences in function can be correlated with (i) the tissue in which the terminals of the sensory nerve fibers are located, (ii) the structures intimately associated with the sensory terminal (e.g., muscle spindle, Pacinian corpuscle), and (iii) the configuration and properties of the nerve terminal. In some cases, however, particularly for the thin dorsal root fibers, definitive information on structure of functionally definable elements is not available.

Part I. Primary Receptor Units with Dorsal Root Fibers

The classification of elements as various kinds of mechanoreceptors, thermoreceptors, or nociceptors is based on observations that *maximal* response from a given type is induced by one of the following: (i) mechanically induced deformation of a particular structure or tissue produced by stimuli with a given set of dynamic properties, (ii) thermally induced changes with a given set of dynamic features, or (iii) only those stimuli that directly threaten tissue integrity (i.e., are likely to or do cause damage). It should be understood that the indicated specificity of sensory receptors is

continued

in no sense absolute; many dorsal root elements are activated by both mechanical and thermal stimuli; nevertheless, even the most intense "nonspecific" stimulus ordinarily evokes only a small response relative to that produced by

a moderate level of the "specific" stimulus. Any classification of receptor function must consider the kinds of stimuli that an animal and its evolutionary forebears may encounter during natural history; this provides additional cri-

	Terminal Location, & Type	Distinguishing Characteristics	Conduction Velocity [Fiber Diameter]	Receptive Field	Most Effective Stimulus
	Cutaneous Mechanical	Displacement & velocity signaling			
1		Type II	45-65 m/s	Small area (2 mm²) of hairy skin	Skin deformation & stretch
2		Slowly adapting glabrous (probably 2 types)	55-60 m/s[1]; 40-70 m/s[2] [∿10 μ]	Small area of glabrous skin	Skin indentation
3		Type I (dome)	50-70 m/s[2] [∿10 μ]	Sensitive region in hairy skin associated with a 150- to 300 μ-diameter domelike structure	Dome indentation
		Primary velocity signaling			
4		G₂ hair	35-65 m/s	Guard hairs over area of 40-120 mm²[3]	Hair movement & skin indentation
5		Field	40-70 m/s	Area (100-350 mm²) of hairy skin[3]	Skin indentation
6		Quickly or rapidly adapting (RA) receptor of glabrous skin	46-78 m/s	Area of glabrous skin	
7		D hair	16-27 m/s	Associated with down (fine) & guard hairs over area of 20-60 mm²[3]	Hair movement & skin indentation
8		C mechanoreceptors	0.7-1.1 m/s[3]	Area (2-6 mm²) in hairy skin only	Slowly moving skin distortion
		Transient (rapid change) signaling			
9		P.C. (Pacinian)— phasic, tap	50-76 m/s	Punctate focus at threshold	Mechanical transients
10		G₁ hair receptors	66-83 m/s	Guard hairs over area of 130-350 mm²[3]	High velocity hair movement & skin indentation
11	Thermal	Cool signaling	0.7-1.2 m/s[2]; 5-15 m/s[5]	Usually 1 small area (<1 mm²)	<0.2°C change evokes detectable response

[1] For rhesus monkey. [2] For cat. [3] For cat lower leg; fields are larger on thigh, smaller on foot and toes. [4] Unmyelinated.

Units with Dorsal Root Fibers

teria for subdivision determined by dynamic features or stimulus magnitudes. **Stimulus:** Any agent or act producing a reproducible and consistent response. **Adaptation** describes decreased response to a constant stimulus of the most effective type, unless otherwise specified. Data in brackets refer to the column heading in brackets.

Terminal Morphology	Dynamic Characteristics	Adaptation	Additional Features	Reference	
Ruffini end organs	....	Slow adaptation to maintained indentation	Regular discharge; some show background activity; responsive to cooling	22-24,28	1
One type probably associated with Merkel's cells in epithelial pegs; another may be related to Ruffini end organs	....	Slow adaptation to maintained indentation	Some have irregular and others regular discharge; weakly responsive to cooling	38,43-45, 56,61	2
Pinkus' haarschiebe (dome); nerve endings associated with Merkel's cells	Prominent phasic component	Relatively slow adaptation to maintained dome deflection	Irregular discharge; some respond weakly to sudden cooling; not excited by skin stretch	22,28,40	3
Unknown—specialized termination at base of hairs (?)	Responsive to relatively slow as well as rapid movements	Rapid adaptation to maintained hair deflection	Ordinarily, not excited by cooling	17,22,51, 56	4
Unknown			Many weakly responsive to cooling	22,56	5
Encapsulated fibers of dermal papillae; possibly Meissner's corpuscles	Responsive to range of velocities from relatively slow to rapid	Variable adaptation to maintained skin deflection	....	38,43-45 61	6
Specialized terminations at hair bases (?)	Responsive to slow as well as rapid mechanical disturbance	Rapid adaptation to maintained hair displacement	Low mechanical threshold; responsive to sudden cooling; some have low-level background activity	17,22,51, 56	7
Unknown	Uniquely poor response to higher velocity stimuli	....	Responsive to sudden cooling	13,36	8
Multilamellated corpuscles	Responsive to higher derivatives of stimulus velocity (acceleration, jerk)	....	Some very responsive to vibration (150-500 Hz) & other transients	32,33,44, 48,61, 62	9
Specialized terminations around hair follicles	....	Rapid adaptation to maintained deflection	High velocity requirement may suggest high threshold	17,22,51, 56	10
Bulblike endings in basal epidermis layer	Phasic & static excitatory responses to cooling	....	Inhibited by warming; poorly responsive to mechanical stimuli; tendency for bursts of discharge	2,29,39, 42,56	11

5/ For primate.

continued

	Terminal Location, & Type	Distinguishing Characteristics	Conduction Velocity [Fiber Diameter]	Receptive Field	Most Effective Stimulus
12		Warm signaling	∿1 m/s	Small area	<0.2°C change evokes detectable response
13	Noxious	Mechanical signaling	5-35 m/s[2,5]	Several (3-20) small spots on either hairy or glabrous skin	Noxious mechanical events
14		Thermal & mechanical signaling	3-7 m/s[5]	....	Noxious heat & mechanical events
15			0.9-2.5 m/s[2]	5- to 20-mm² area	Noxious mechanical events; prolonged cold
16		Polymodal signaling	0.3-1.1 m/s[2,4,5]	Small area (1-4 mm²)	Noxious temperatures, marked mechanically produced distortion & irritant chemicals
	Subcutaneous Somatic[6]				
	Mechanical	Displacement & velocity (muscle length or tension) signaling			
17		Secondary muscle spindle (group II)	30-72 m/s [5-12 μ][2]	Muscle body	Changes in length of a muscle spindle
18		Primary muscle spindle (group Ia)	72-120 m/s [12-20 μ][2]	Muscle body	Changes in length of a muscle spindle
19		Tendon organ (group Ib)	72-120 m/s [12-20 μ][2]	Ends of extrafusal muscle bundles, muscle tendons	Changes in muscle tension at physiological muscle lengths
20		Group III (some group II)	5-40 m/s [∿2-7 μ]	Usually fascial surface of muscle	Pressure on muscle
21		Transient (rapid change) signaling	50-90 m/s	1 punctate focus in fascia or interosseus membrane	....
	Noxious	Mechanical &/or chemical signaling			
22		Group III	5-25 m/s [∿2-5 μ][2]	....	Strong pressure to body of muscle—unphysiological muscle lengths
23		Group IV	0.4-1.6 m/s [<2 μ]	Body of muscle	Strong direct pressure & temperature extremes (for muscle)

[2] For cat. [4] Unmyelinated. [5] For primate. [6] In muscles, tendons, fasciae, and interosseus membranes. [7] Tension in ten

Units with Dorsal Root Fibers

Terminal Morphology	Dynamic Characteristics	Adaptation	Additional Features	Refer-ence	
....	Phasic & static excitatory responses to heating	....	Inhibited by cooling; poorly responsive to mechanical stimuli	29,39,42	12
Unknown—"free" nerve endings (?)	Prominent phasic response	Intermediate to maintained distortion	Distinguish effectively between skin pressure produced by sharp & blunt objects	21,56	13
Unknown—"free" nerve endings (?)	Both phasic & static reponses	....	Weakly responsive to cold	41	14
Unknown—"free" nerve endings (?)	....	Intermediate to maintained mechanical distortion; slow adaptation to steady temperature	Sparse response to both mechanical & cold stimuli	11,35	15
Unknown—"free" nerve endings (?)	Both phasic & static responses	Intermediate to mechanical & heat excitation, although often develop background discharge after heat	Sensitize (lower threshold) after excitation by heat or after mechanically caused tissue damage; most common C-fiber unit in distal limb nerves	11,35,42	16
Muscle spindle secondary ending (coils & spray)	Little dynamic component	Slow adaptation to constant length	Static gamma motor (fusimotor) control	4,6,15,50	17
Muscle spindle annulospiral & associated spray endings of Ia fibers	Prominent dynamic component can dominate under fusimotor control	More rapid adaptation to position than secondary ending	Static & dynamic gamma motor (fusimotor) control	6,12,15, 49,50	18
Golgi "spray"	Increased discharge to tension$^{7/}$ changes	Slow adaptation to constant tension$^{7/}$	Responsive to muscle contraction & to marked elongation	6,31,49	19
Unknown	....	Slow adaptation to direct pressure	Some responsive to vigorous contractions & marked elongation	9,55	20
Terminal axon encased in lamella—Pacinian & (?) paciniform corpuscle	Behave like acceleration or "jerk" detectors	....	Some very responsive to vibration (150-500 Hz) & other transients	33,62	21
Unknown	....	Intermediate to slow adaptation to elongation	Responsive to injection of hypertonic solutions	9,55	22
Unknown—unencapsulated (free) endings (?)	....	Slow adaptation to maintained pressure	Some excited by contraction in presence of ischemia	8,37	23

dinous or fascial structures of muscle.

continued

	Terminal Location, & Type	Distinguishing Characteristics	Conduction Velocity [Fiber Diameter]	Receptive Field	Most Effective Stimulus
	Somatic, articular[8]				
	Mechanical	Displacement & velocity signaling			
24		Slowly adapting type 1 (tendon organ)	Large diameter—$>10\,\mu$	....	Joint movement
25		Slowly adapting type 2	Medium diameter—~ 7-$10\,\mu$	....	Joint movement
26		Velocity signaling— "phasic"	50-80 m/s	....	Joint bending & twisting
27		Transient (rapid change) signaling	50-80 m/s [~ 10-$15\,\mu$]	....	Transient components of joint movement
28	Noxious	Mechanical signaling	<30 m/s	....	Overextension of joint
	Visceral[10]				
	Mechanical	Position (static) & movement (dynamic) detectors			
		Cardiovascular			
29		Renal arterial	Not reported[11]	Renal or interlobar arterial wall	Arterial wall distortion
30		Coronary arterial	10-25 m/s[11]	Coronary arterial tree	Increased pressure
31		Suprarenal gland arterial	Unknown[12]	Small suprarenal arteries (?)	Changes in systemic arterial pressure
32		Renal venous	6-45 m/s	Probably veins of renal medulla	Renal venous pressure
		Other locations			
33		Pleural, thoracic, inlet, mediastinal	Unknown[12]	Pleura of lung base and/or interstitial tissue of mediastinum	Tracheal-pleural movement or distortion
34		Urinary bladder	Unknown[12]	Bladder wall	Tension within muscular coats
		Movement & transient detectors			
		Cardiovascular			
35		Pericardial	~ 20 m/s	Parietal pericardium	Mechanical distortion
36		Pulse-activated	24-70 m/s	Larger branches of thoracic & abdominal aorta	Pulse pressure
		Other locations			
37		Small intestine movement detectors	1.5-22 m/s [2-8 μ]	Mesentery & serosa of small intestine	Distortion of mesenteric-serosal relations

[8] In articular tissue.   [9] Directional response refers to asymmetry in activity evoked by movements from and toward an intermediate (null) position; a typically minimum discharge appears in the null position.   [10] Knowledge of the categories of afferent (receptor) neurons from these structures is less complete than for the afferent elements of cutaneous and skeletomuscular regions. The accumulated evidence, particularly from recent investigations, suggests that receptive structures are specialized in functional characteristics according to the visceral organ or deep tissue in which they are situated.   The listings presented should only be taken as indications of the classes of reasonably documented

Units with Dorsal Root Fibers

Terminal Morphology	Dynamic Characteristics	Adaptation	Additional Features	Reference	
Golgi spray	Phasic & static components to response	Slow to position	Show directional sensitivity [9]	3,14,16, 59	24
Ruffini-type ending	Prominent dynamic phase	Slow to position	Show directional sensitivity [9]	14,16,20, 59	25
Unknown	Large velocity (dynamic) & small position (static) components	Intermediate to position	....	20	26
Paciniform ending (?)	Excited by mechanical transients	....	No directional sensitivity	14,16,20	27
Unknown	....	Slow adaptation to maintained position	Afterdischarge common	20	28
Nerve terminal coils in adventitia (?)	Dynamic responsiveness not prominent	Slow adaptation to arterial pressure	....	1,53	29
Bush or tree-shaped endings in arterial adventitia	Dynamic response present	Slow to intermediate adaptation for maintained pressure	....	1,19	30
Unknown	Dynamic (phasic) & slowly adapting (static) discharge related to arterial pressure changes		....	54	31
Unknown	Responsive to level of venous pressure	Slow adaptation to steady venous pressure	Responsive to ureteral pressure above physiological range	5,7,57	32
Unknown	Movement sensitive	Slow adaptation to position	Some responsive to disturbance of trachea & mediastinum with respiration; fibers in rami of stellate ganglion	30	33
Unknown	Phasic response during isotonic stretch	Slow adaptation to isometric contraction	Discharge related to pressure	34	34
Unknown	Phasic response only to cardiac events	Rapid adaptation to pericardial distension	....	58	35
Pacinian or paciniform corpuscles (?)	Responsive to higher derivatives of mechanical displacement	Rapid adaptation to steady arterial pressure	Sensitized by epinephrine; follow sinusoidal vibration to >300 Hz	25-27,46	36
Unknown—bare "swollen" terminals (?)	Prominent phasic response	Intermediate to rapid adaptation to maintained displacement	Discharge with peristaltic activity; responsive to distension under special condition	10	37

receptor types. Cardiovascular and respiratory receptive units of the classical type, such as the baroreceptors and chemoreceptors of the carotid bifurcation and aortic arch and pulmonary stretch receptors, have afferent fibers in the vagus and glossopharyngeal nerves with supraspinal central distributions. Central interactions between such cranial nerve receptors and the spinal elements listed here are poorly understood. (Note that the innervation of much oral, facial, and visceral sensory structures stems from nerve fibers of the cranial nerves.) [11] Myelinated. [12] Probably myelinated.

continued

	Terminal Location, & Type	Distinguishing Characteristics	Conduction Velocity [Fiber Diameter]	Receptive Field	Most Effective Stimulus
38	Noxious	Cardiac	9-25 m/s	Myocardium	Myocardial ischemia
39		Intestinal, mesenteric, peritoneal	[Small] 3,10/	Intestine & coverings; parietal peritoneum	Rough handling of bowel; chemical agents
40		Bladder	Unknown	....	Bladder pressure >60 cm H_2O

3/ Unmyelinated. 10/ Myelinated.

Contributors: Perl, E. R., and P. R. Burgess

References

[1] Abraham, A. 1969. Microscopic Innervation of the Heart and Blood Vessels in Vertebrates Including Man. Pergamon Press, Oxford.

[2] Andres, K. H. 1971. Proc. Int. Union Physiol. Sci. 8:136.

[3] Andrew, B. L. 1954. J. Physiol. (London) 123:241.

[4] Appleberg, B., et al. 1966. Ibid. 185:160.

[5] Aström, A., and J. Crafoord. 1967. Acta Physiol. Scand. 70:10.

[6] Barker, D. 1962. Symp. Muscle Receptors (Hong Kong) 1961, p. 227.

[7] Beacham, W. S., and D. L. Kunze. 1969. J. Physiol. (London) 201:73.

[8] Bessou, P., and Y. Laporte. 1958. C. R. Soc. Biol. 152:1587.

[9] Bessou, P., and Y. Laporte. 1961. Arch. Ital. Biol. 99:293.

[10] Bessou, P., and E. R. Perl. 1966. J. Physiol. (London) 182:404.

[11] Bessou, P., and E. R. Perl. 1969. J. Neurophysiol. 32:1025.

[12] Bessou, P., et al. 1968. J. Physiol. (London) 196:47.

[13] Bessou, P., et al. 1971. J. Neurophysiol. 34:116.

[14] Boyd, I. A. 1954. J. Physiol. (London) 124:476.

[15] Boyd, I. A. 1962. Phil. Trans. Roy. Soc. London B245:81.

[16] Boyd, I. A., and T. D. M. Roberts. 1953. J. Physiol. (London) 122:38.

[17] Brown, A. G., and A. Iggo. 1967. Ibid. 193:707.

[18] Brown, A. M. 1967. Ibid. 190:35.

[19] Brown, A. M., and A. Malliani. 1971. Ibid. 212:685.

[20] Burgess, P. R., and F. J. Clark. 1969. Ibid. 203:317.

[21] Burgess, P. R., and E. R. Perl. 1967. Ibid. 190:541.

[22] Burgess, P. R., et al. 1968. J. Neurophysiol. 31:833.

[23] Chambers, M. R., and A. Iggo. 1967. J. Physiol. (London) 192:26.

[24] Chambers, M. R., et al. In press, 1972.

[25] Gernandt, B., and Y. Zotterman. 1946. Acta Physiol. Scand. 12:56.

[26] Gray, J. A. B., and J. L. Malcolm. 1948. J. Physiol. (London) 108:43.

[27] Gruhzit, C. C., et al. 1954. J. Pharmacol. Exp. Ther. 112:138.

[28] Harrington, T., and M. M. Merzenich. 1970. Exp. Brain Res. 10:251.

Part II. Cross-sectional Diameter, Conduction Velocity, and Correlated Functional Characteristics

Table derived from data in Part I. **Nerve Distribution:** Nerves are not distributed solely to the indicated body structures or regions. Such "pure" distributions do not exist; a fraction of afferent fibers in every seemingly definitive distribution may stem from other tissues (e.g., joint afferent units in cutaneous or muscle nerve, muscle afferent fibers in joint and some visceral nerves).

	Fiber Type (Group)	Cross-sectional Diameter	Conduction Velocity	Nerve Distribution			
				Cutaneous	Muscle	Articular	Visceral
1	Aαβ (Group I)	12-20 μ	72-120 m/s	G_1 hair; .... RA of glabrous skin	Ia (spindle); Ib (tendon organ); Pacinian-like	Ia (spindle); tendon-organ type	....
2	(Group II)	6-12 μ	30-72 m/s	Type I (dome); type II, Pacinian-like; SA of glabrous skin; G_2 hair; field	II (spindle); II pressure; articular receptors; Pacinian-like receptors; Ruffini-type	"Phasic"	Pacinian-like

continued

DORSAL ROOT FIBERS: MAMMALS

Units with Dorsal Root Fibers

Terminal Morphology	Dynamic Characteristics	Adaptation	Additional Features	Refer- ence	
Unknown	Dynamic & relatively slowly adapting phases of response to cessation of coronary circulation		Afterdischarge; fibers in rami of stellate ganglion	18,19	38
Unknown	Unknown	Unknown	Responsive to bradykinin	25,47	39
Unknown	Both dynamic & static responses	Slow adaptation to maintained pressure	Evokes spinal reflex increasing blood pressure	34,52,60	40

[29] Hensel, H., et al. 1960. J. Physiol. (London) 153: 113.
[30] Holmes, R., and R. W. Torrance. 1959. Quart. J. Exp. Physiol. 44:271.
[31] Houk, J., and E. Henneman. 1967. J. Neurophysiol. 30:466.
[32] Hunt, C. C. 1960. J. Physiol. (London) 155:175.
[33] Hunt, C. C., and A. K. McIntyre. 1960. Ibid. 153: 74.
[34] Iggo, A. 1955. Ibid. 128:593.
[35] Iggo, A. 1959. Quart. J. Exp. Physiol. 44:362.
[36] Iggo, A. 1960. J. Physiol. (London) 152:337.
[37] Iggo, A. 1961. Ibid. 155:52.
[38] Iggo, A. 1963. Acta Neuroveg. 24:225.
[39] Iggo, A. 1969. J. Physiol. (London) 200:403.
[40] Iggo, A., and A. R. Muir. 1969. Ibid. 200:763.
[41] Iggo, A., and H. Ogawa. 1971. Ibid. 216:77.
[42] Iriuchijima, J., and Y. Zotterman. 1960. Acta Physiol. Scand. 49:267.
[43] Jänig, W. 1971. Brain Res. 28:217.
[44] Jänig, W., et al. 1968. Exp. Brain Res. 6:100.
[45] Knibestöl, M., and A. B. Vallbo. 1970. Acta Physiol. Scand. 80:178.

[46] Leitner, J.-M., and E. R. Perl. 1964. J. Physiol. (London) 175:254.
[47] Lim, R. K. S. 1970. Annu. Rev. Physiol. 32:269.
[48] Lynn, B. 1969. J. Physiol. (London) 201:765.
[49] Matthews, B. H. C. 1933. Ibid. 78:1.
[50] Matthews, P. B. C. 1964. Physiol. Rev. 44:219.
[51] Merzenich, M. M., and T. Harrington. 1969. Exp. Brain Res. 9:236.
[52] Mukherjee, S. R. 1957. J. Physiol. (London) 138: 300.
[53] Niijima, A. 1971. Ibid. 219:477.
[54] Niijima, A., and D. L. Winter. 1968. Science 159: 434.
[55] Paintal, A. S. 1960. J. Physiol. (London) 152:250.
[56] Perl, E. R. 1968. Ibid. 197:593.
[57] Pines, Iu. L. 1960. Fiziol. Zh. SSSR im. I. M. Sechenova 47:1380.
[58] Ruckebusch, Y. 1961. C. R. Soc. Biol. 155:525.
[59] Skoglund, S. 1956. Acta Physiol. Scand., Suppl. 124.
[60] Talaat, M. 1937. J. Physiol. (London) 89:1.
[61] Talbot, W. H., et al. 1968. J. Neurophysiol. 31:301.
[62] Zimmermann, M. Unpublished. Univ. Heidelberg, Germany, 1972.

Part II. Cross-sectional Diameter, Conduction Velocity, and Correlated Functional Characteristics

	Fiber Type (Group)	Cross-sectional Diameter	Conduction Velocity	Nerve Distribution			
				Cutaneous	Muscle	Articular	Visceral
3	Aδ (Group III)	2-6 μ	4-30 m/s	D hair; mechanical nociceptors, cold [1]	III pressure; nociceptors	Nociceptors	Vascular pressure receptors; visceral movement or tension receptors; nociceptors
4	C (Group IV)	0.3-1.5 μ	0.4-2.0 m/s	Cold [2]; warm; C mechanoreceptors; polymodal nociceptors	Nociceptors; others (?)	(?)	Nociceptors

[1] Primate and cat glabrous. [2] Cat.

Contributors: Perl, E. R., and Burgess, P. R.

The autonomic nervous system consists of the visceral efferent nerve fibers supplying all smooth muscle, cardiac muscle, and all glandular cells except the cells of endocrine glands, of sebaceous glands, and of certain exocrine glands of the gastrointestinal tract which are primarily under hormonal control. (Despite sharing of peripheral pathways, visceral afferent fibers are, by definition, excluded from the purely efferent autonomic nervous system.) The system is divisible, for both anatomic and physiologic reasons, into a sympathetic portion (thoracolumbar outflow) and a parasympathetic portion (craniosacral outflow). In the characteristic neuronal arrangement of both parts of the system, a preganglionic fiber and a postganglionic fiber extend in series from the central nervous system to the neuroeffector organ (smooth muscle, cardiac muscle, or glandular tissue); the exception to this arrangement is in the innervation of the suprarenal medulla, where only the preganglionic neurons exist. The chemical mediator released at the neuroeffector junction of the sympathetic portion is norepinephrine, and for the parasympathetic portion it is acetylcholine. The cells of the suprarenal medulla, being homologous to postganglionic sympathetic neurons, secrete epinephrine rather than norepinephrine.

The outflow of the sympathetic portion utilizes the paired sympathetic trunks, in the ganglia of which occur many of the synapses between pre- and postganglionic fibers. Other such synapses occur in collateral ganglia, e.g., the celiac (C.G.), superior mesenteric (S.M.G.), and inferior mesenteric (I.M.G.), situated in front of the vertebral column and thus located some distance from their effector organs. The preganglionic cell bodies reside in the lateral (intermediolateral cell) column of the spinal cord from the first thoracic (T1) to the second lumbar (L2) segments, inclusively; their fibers emerge through the ventral roots. The outflow of the parasympathetic portion utilizes four cranial nerves—oculomotor (III), facial (VII), glossopharyngeal (IX), and vagus (X)—plus the second, third, and fourth sacral nerves from the corresponding sacral segments (S2, S3, S4) of the spinal cord. The synapses between pre- and postganglionic fibers occur in peripheral ganglia which are closely associated with the effector organ being supplied; therefore the postganglionic fibers are relatively short. Most, but not all, viscera receive both sympathetic and parasympathetic fibers, the two sets in a contraposed, but coordinated, functional relationship within each organ.

Various areas of the cerebral cortex send fibers to the hypothalamus (a regulating and coordinating station for visceral efferent neural activity) which in turn sends descending fibers to the preganglionic sympathetic and parasympathetic cells in the brain stem and spinal cord. Connections between the visceral efferent outflow and visceral afferent fibers at spinal-cord and brain-stem levels form reflex arcs. Visceral afferent fibers are also relayed upward to the thalamus and cortex for higher level input and control over the autonomic nervous system.

Contributor: Johnson, Robert J.

General References

[1] Appenzeller, O. 1970. The Autonomic Nervous System: An Introduction to Basic and Clinical Concepts. American Elsevier, New York.

[2] Kuntz, A. 1953. The Autonomic Nervous System. Lea and Febiger, Philadelphia.

[3] Mitchell, G. A. G. 1953. Anatomy of the Autonomic Nervous System. E. and S. Livingstone, Edinburgh.

[4] Mitchell, G. A. G. 1956. Cardiovascular Innervation. E. and S. Livingstone, Edinburgh.

[5] Pick, J. 1970. The Autonomic Nervous System: Morphological, Comparative, Clinical and Surgical Aspects. J. B. Lippincott, Philadelphia.

[6] Potts, T. K. 1925. J. Anat. 59:129.

[7] White, J. C., et al. 1952. The Autonomic Nervous System. Macmillan, New York.

continued

Sympathetic (thoracolumbar) Outflow   Parasympathetic (craniosacral) Outflow

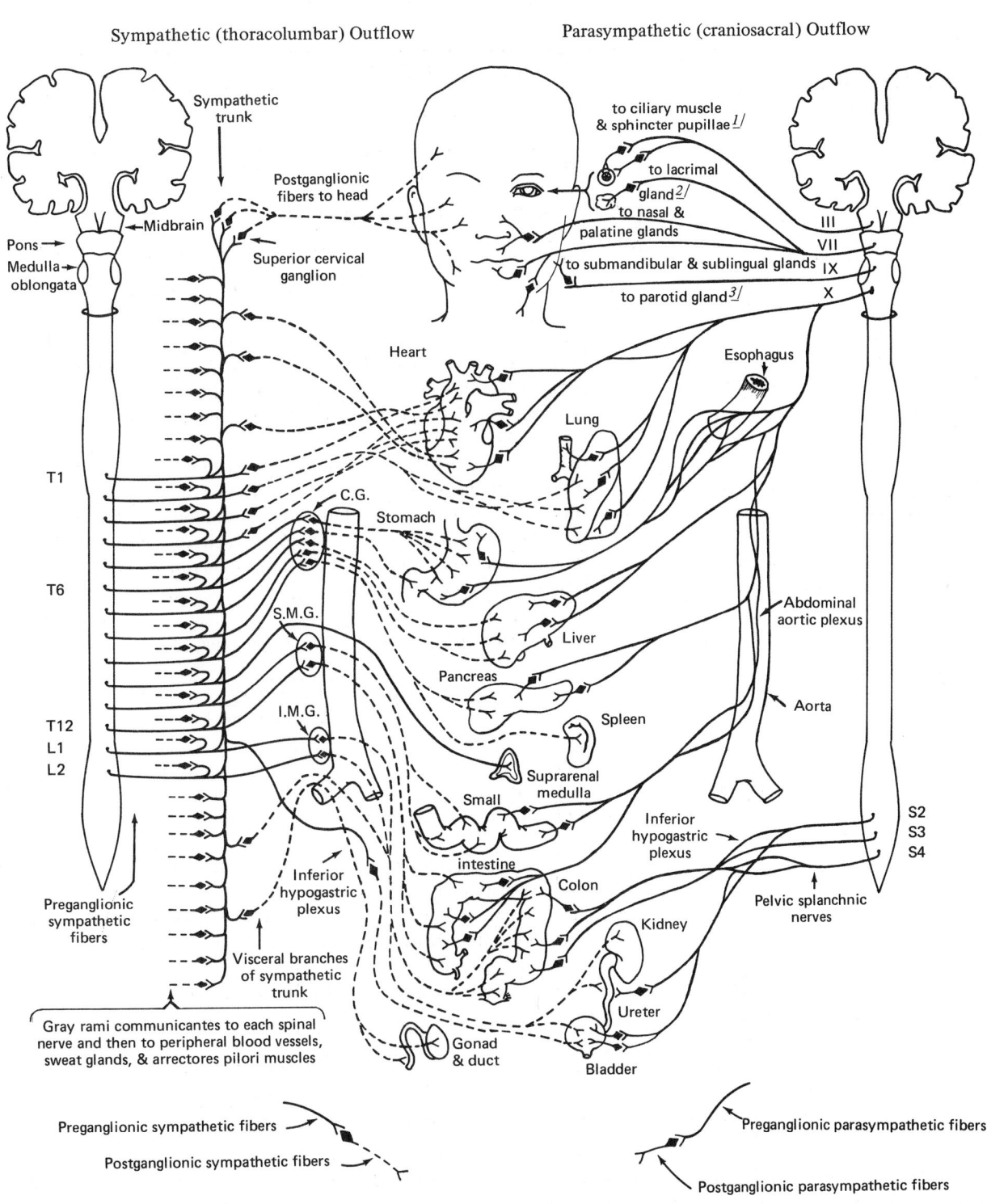

1/ Via the ciliary ganglion. 2/ Via the pterygopalatine ganglion. 3/ Via the otic ganglion.

Response: Symbols [+] to [+++] are an approximate indication of the importance of adrenergic and cholinergic nerve activity in the control of the effector organs; [±] indicates equivocal response.

Effector Organ or structure	Tissue	Adrenergic Impulses		Response to Cholinergic Impulses
		Receptor Type	Response	
1 Brain	Cerebral blood vessels	α	Slight constriction	Dilation [1]
2 Eye	Ciliary muscle	β	Relaxation for far vision (slight effect)	Contraction for near vision [+++]
3	Radial muscle of iris	α	Contraction (mydriasis) [++]	....
4	Sphincter muscle of iris	....	....	Contraction (miosis) [+++]
5 Lacrimal glands	....	....	....	Secretion [+++]
6 Salivary glands	....	α	Thick, viscous secretion [2] [+]	Profuse, watery secretion [+++]
7	Blood vessels	α	Constriction [+++]	Dilation [++]
8 Stomach	....	β	Decreased motility & tone (usually) [3] [+]	Increased motility & tone [+++]
9	....	....	Inhibited secretion?	Stimulated secretion [+++]
10	Sphincters	α	Contraction (usually) [+]	Relaxation (usually) [+]
11 Intestinal tract	....	α-β	Decreased motility & tone [3] [+]	Increased motility & tone [+++]
12	....	....	Inhibited secretion?	Stimulated secretion [++]
13	Sphincters	α	Contraction (usually) [+]	Relaxation (usually) [+]
14 Abdominal viscera	Blood vessels	α, β	Constriction [+++]; dilation [4] [+]	....
15 Liver	....	....	Glycogenolysis	....
16 Gallbladder & ducts	....	....	Relaxation [+]	Contraction [+]
17 Pancreas	Acini	....	....	Secretion [++]
18 Nasopharyngeal glands	....	....	....	Secretion [++]
19 Lung	Bronchial glands	....	Inhibition?	Stimulation [+++]
20	Bronchial muscles	β	Relaxation [+]	Contraction [++]
21	Pulmonary blood vessels	α, β	Constriction [+]; dilation [+]	Dilation [1]
22 Heart	Atria	β	Increased contractility & conduction velocity [++]	Decreased contractility; usually increased conduction velocity [++]
23	Ventricles	β	Increased contractility, conduction velocity, automaticity, & rate of idiopathic pacemakers [+++]	Some investigators claim slight decrease in contractility
24	Atrioventricular node & conduction system	....	Increased conduction velocity [++]	Decreased conduction velocity; atrioventricular block [+++]
25	Sinoatrial node	β	Increased heart rate [++]	Decreased heart rate; vagal arrest [+++]
26	Coronary blood vessels	α, β	Constriction [+]; dilation [5] [+++]	Dilation [±]

[1] Physiological significance of cholinergic vasodilation is questionable. [2] Parotid glands lack adrenergic innervation. [3] Response may result mainly from inhibition of parasympathetic ganglionic transmission by postganglionic adrenergic fibers terminating on ganglion cells of Auerbach's plexus. [4] At the usual concentrations of circulating epinephrine physiologically released, β-receptor response (vasodilation) predominates in blood vessels of skeletal muscle and liver; α-receptor response (vasoconstriction) predominates in blood vessels of other abdominal viscera. [5] Predominates in situ due to indirect effects.

continued

	Effector Organ or Structure	Tissue	Adrenergic Impulses			Response to Cholinergic Impulses
			Receptor Type	Response		
27	Spleen	Capsule	α	Contraction [+++]		....
28	Ureter	....	....	Increased motility & tone (usually)		Increased motility & tone?
29	Urinary bladder	Detrusor	β	Relaxation (usually) [+]		Contraction [+++]
30		Trigone & sphincter	α	Contraction [++]		Relaxation [++]
31	Uterus	....	α, β	Variable [6]		Variable [6]
32	Sex organs, ♂	....	....	Ejaculation [+++]		Erection [+++]
33	Skeletal muscle	Blood vessels	α, β	Constriction [++]; dilation [7] [++]		Dilation [8] [+]
34	Skin	Pilomotor muscles	α	Contraction [++]		....
35		Sweat glands	α	Slight, localized secretion [9] [+]		Generalized secretion [+++]
36	Plus mucosa	Blood vessels	α	Constriction [+++]		Dilation [1]
37	Suprarenal gland	Medulla	....	....		Secretion of epinephrine & norepinephrine

[1] Physiological significance of cholinergic vasodilation is questionable. [6] Depends on stage of menstrual cycle, amounts of circulating estrogen and progesterone, and other factors; responses of pregnant uterus differ from those of nonpregnant uterus. [7] At the usual concentrations of circulating epinephrine physiologically released, β-receptor response (vasodilation) predominates in blood vessels of skeletal muscle. [8] In skeletal muscle, vasodilation caused by the sympathetic cholinergic system is not involved in most physiological responses. [9] "Adrenergic sweating" on palms of hands and some other sites.

Contributor: Goodman, Louis S.

General Reference: Koelle, G. B. 1970. In L. S. Goodman and A. Gilman, ed. The Pharmacological Basis of Therapeutics. Ed. 4. Macmillan, New York. pp. 406-407.

141. SYNAPTIC TRANSMITTER COMPOUNDS: VERTEBRATES AND INVERTEBRATES

The principal mediators of information transfer between adjacent cells, both within the nervous system and from the nervous system to the rest of the organism, are a limited number of specialized chemical substances called transmitter compounds, or neurotransmitters. These chemical substances exert their physiological actions during the process of synaptic transmission. When an action potential arrives at the most distal processes of a neuron (the presynaptic nerve terminals), a transmitter compound is secreted. This substance diffuses across a synaptic cleft and combines with specific receptor sites in the postsynaptic membrane. Highly selective ionic permeability channels are opened or closed as a result of this combination. The ions that move, the charge they carry, and the direction of their movement determine whether the effect is excitatory or inhibitory. The direction of movement, in turn, is determined by the transmembrane concentration of the ion and the electrical potential difference across the cell membrane. (The point at which there is no net movement of ion in either direction is called the equilibrium potential, and can be defined by the Nernst equation, reference 58 in Part I.) At different synaptic junctions, the ions involved, singly or in combination, are K^+, Na^+, Ca^{2+}, and Cl^-. A specific inactivation mechanism removes the transmitter compound from its sites of action. The entire process is brief, and can be repeated up to several hundred times a second for short periods.

The following tables present biochemical, physiological, and anatomical data on the best-known or suspected neurotransmitter compounds. Data are included from both vertebrate and invertebrate nervous systems. The material included in the tables is meant to be representative, not exhaustive; the only compounds considered in any detail are acetylcholine, norepinephrine, γ-aminobutyric acid, dopamine, glutamic acid, serotonin, and glycine. Part I is a compilation of useful synaptic-transmitter-compound methodology. The other parts of the table include synaptic models (for acetylcholine, norepinephrine, and γ-aminobutyric acid) which list key references related to synaptic events; biochemical data on the enzymes and other proteins concerned with transmitter accumulation, binding, release, physiological effect, and inactivation; and lastly the physiological events recorded at synaptic junctions where the compound is suggested as a neurotransmitter. It should be noted that reference to transmitter actions as excitatory or inhibitory has

continued

been avoided—rather, ionic conductance changes are described. (Since any and all transmitters can be either "excitatory" or "inhibitory" depending on the membrane potential and ionic gradients—*see* above—the terms are not used.) Unavoidably some of the data in this table, which was completed in June 1971, will overlap information provided in other tables.

For information on prostaglandins, which are difficult to map due to present assay methods, consult references 14, 42, and 89 in Part I; for information on histamine, which may act as a neurotransmitter, consult references 14, 24, 39, 75, and 89 in Part I. In the BIOCHEMISTRY portions of the table, **Commission No.** = the number assigned an enzyme by the International Union of Biochemistry on the Nomenclature and Classification of Enzymes ([46] in Part I).

Part I. Methodology

	Determination	Material (Synonym) or Method	Reference
		Acetylcholine	
1	Ultrastructure	Clear, round, 500-Å vesicles	101
2		Freeze-etched 750-Å vesicles	82
	Affinity labels		
3	Cholinesterase	Diisopropyl fluorophosphate; pralidoxime iodide (pyridine-2-aldoxime methiodide)	96,97,100
4	Receptor	Acetylcholine analogs	55
5		α-Bungarotoxin	17,65,72
6		p-(Trimethylammonium) benzene diazonium difluoroborate	16
7		d-Tubocurarine	118
	Fractionation		
8	Synaptosomes	Central nervous system	38,120
9		Peripheral nervous system	120,122
10	Vesicles	Central nervous system	120,121
11		*Torpedo*	47,48
12	Uptake (presynaptic)	Choline	18,26
	Biochemistry		
13	Bioassay	Cat blood pressure; isolated rat duodenum; *Mya arenaria* heart; leech dorsal muscle	64
14		Guinea-pig & rat tracheae	53
15	Enzyme assay	Acetylcholinesterase	93,95
16		Choline acetyltransferase	70
17	Isolation-quantitation of acetylcholine	Dipicrylamine concentration	94
18	Histochemistry	Cholinesterase	21,56,57,62
19		Immunofluorescence	8
	Representative pharmacology		
20	Presynaptic	Botulinus toxin	4,27
21		Ca^{2+}; Mg^{2+}; hemicholinium; hexamethonium	89
22	Postsynaptic	Snake venoms	65,66,108
23	Muscarinic: agonists	Acetylcholine; acetyl-β-methylcholine; muscarine; pilocarpine	89
24	antagonists	Atropine; scopolamine (hyoscine)	89
25	Nicotinic: agonists	Acetylcholine; carbachol; nicotine	89
26	antagonists	Dihydro-β-erythroidine; hexamethonium; d-tubocurarine	89
27	Anticholinesterase	Edrophonium bromide	9,59
28		Neostigmine	32
29		Organophosphates; 284C51	30,89,100
30		Physostigmine (eserine)	89
		Catecholamines	
31	Ultrastructure	Dense-core vesicles	12,34,45,52,101
32	Induced terminal degeneration	6-Hydroxydopamine	13,110,112,113
33	Affinity labels: α-receptor	Dihydroquinoline derivatives	5-7,69

continued

1154

Part I.   Methodology

	Determination	Material (Synonym) or Method	Reference
	Fractionation		
34	Synaptosomes	Central nervous system	51,120
35	Vesicles	Central nervous system	120
36		Sympathetic nerve	10,25,92
37		Suprarenal gland	44
38	Uptake (presynaptic)	Hydroxyphenylethanolamine	49,125
39	Uptake inhibition	Cocaine	49
40		Imipramine	89
	Biochemistry		
41	Bioassay	Guinea-pig trachea	53
42		Rat blood pressure	22,116
43		Rat uterus	116
44	Enzyme assay	Dopamine-β-hydroxylase	74,119
45		Tyrosine hydroxylase	78,79,102
46	Isolation-quantitation of cate-	Isotope derivative	29,73
47	cholamines	Trihydroxyindole	2
48	Histochemistry	Ammoniacal silver, electron microscopic & light microscopic	109
49		Dichromate, electron microscopic	126-129
50		Formaldehyde-induced fluorescence	11,19,31
		Immunofluorescence & microcomplement fixation	
51		Chromogranin	3,60
52		Dopamine-β-hydroxylase	25,35,41,60,67
	Representative pharmacology		
53	Presynaptic: release	Ca^{2+}; Mg^{2+}	89
54	uptake block	Cocaine; desipramine; metaraminol	49
55	depletion	Reserpine	23,49
56		Tetrabenazine	23,91
57		Specific for norepinephrine	20
58	Postsynaptic: agonists	Dopamine; epinephrine; isoprenaline; norepinephrine	89
	antagonists		
59	α-receptor	Phenoxybenzamine; phentolamine	89
60	β-receptor	Dichlorisoproterenol; pronethalol; propranolol	89
		γ-Aminobutyric Acid	
61	Ultrastructure	Clear vesicles	84,88
62		Flat vesicles	77,111
63	Fractionation: synaptosomes	Central nervous system	51,81
64	Uptake	Schwann & endoneurial tissues	50,84
	Biochemistry		
65	Bioassay	Crustacean neuromuscular junction	106
66		Crustacean stretch receptor	28
67	Enzyme assay	γ-Aminobutyric acid; oxoglutarate aminotransferase (α-ketoglutaric acid transaminase)	40
68		Glutamate decarboxylase	40
69		Succinate semialdehyde dehydrogenase	90
70	Isolation-quantitation of GABA	Enzyme-cycling assay	87
71	Histochemistry	γ-Aminobutyric acid; oxoglutarate aminotransferase (α-ketoglutaric acid transaminase); succinate semialdehyde dehydrogenase	114,115
	Representative pharmacology		
72	Presynaptic release	Ca^{2+}	85
73	Postsynaptic antagonist	Picrotoxin	83,107

continued

Part I. Methodology

	Determination	Material (Synonym) or Method	Reference
		Glutamic Acid	
74	Ultrastructure	Clear, round vesicles	77,84,88,111
75	Uptake	Glutamate	50
	Biochemistry		
76	Bioassay	Crustacean & insect neuromuscular junctions	63
77	Enzyme assay	γ-Aminobutyric acid; oxoglutarate aminotransferase (α-ketoglutaric acid transaminase)	40
78		Glutamate decarboxylase	40
79		Succinate semialdehyde dehydrogenase	90
80	Isolation-quantitation of glutamate	Enzymatic assay	86
81	Histochemistry	γ-Aminobutyric acid; oxoglutarate aminotransferase (α-ketoglutaric acid transaminase); succinate semialdehyde dehydrogenase	114,115
82	Representative pharmacology: postsynaptic	Potentiation by aspartate	63,89
		Serotonin (5-Hydroxytryptamine)	
83	Ultrastructure	Small dense-core vesicles	45,126
84	Affinity label: receptor	Serotonin (5-Hydroxytryptamine) + NaBH$_4$	1
85	Fractionation: synaptosomes & vesicles	Central nervous system	120
86	Uptake (presynaptic)	Serotonin (5-Hydroxytryptamine)	36,37
87	Uptake inhibition	Imipramine	89
	Biochemistry		
88	Bioassay	Guinea-pig trachea	53
89	Enzyme assay	Tryptophan hydroxylase	68
90	Isolation-quantitation of serotonin	Fluorimetry	103,117,123,124
91	Histochemistry	Dichromate	126,127
92		Formaldehyde-induced fluorescence	19,31
	Representative pharmacology		
93	Presynaptic: release	Ca^{2+}; Mg^{2+}	15
94	depletion	p-Chlorophenylalanine [1]; reserpine; tetrabenazine	23,54,61,91,105
95	Postsynaptic antagonists	2-Bromolysergic acid diethylamide; lysergide (lysergic acid diethylamide); methysergide	89
		Glycine	
96	Representative pharmacology: postsynaptic antagonist	Strychnine	98
		General [2]	
97	Pathway mapping	Chromatolysis	101
98		Golgi impregnation	76
99		Nauta silver method	33,80
100		Procion yellow dye injection	104
101	Potential neurotransmitters	Rapid radiochemical screening	43
102	Visualization of synaptic boutons & axonal varicosities	Living & fixed preparations	71

[1] Selective. [2] For information on iontophoretic techniques, consult references 14, 89, and 99.

Contributors: Wallace, B. G., Sargent, P. B., Mains, R. E., and Kravitz, E. A.

continued

141. SYNAPTIC TRANSMITTER COMPOUNDS: VERTEBRATES AND INVERTEBRATES

Part I. Methodology

References

[1] Alivisatos, S. G. A., et al. 1971. Science 171:809.

[2] Anton, A. H., and D. F. Sayre. 1962. J. Pharmacol. Exp. Ther. 138:360.

[3] Banks, P., and K. B. Helle. 1971. Phil. Trans. Roy. Soc. London B261:305.

[4] Beers, W. H., and E. Reich. 1969. J. Biol. Chem. 244:4473.

[5] Belleau, B., and H. Tani. 1969. Can. J. Pharm. Sci. 4:14.

[6] Belleau, B., et al. 1968. J. Amer. Chem. Soc. 90: 823.

[7] Belleau, B., et al. 1969. Biochem. Pharmacol. 18: 1039.

[8] Benda, P., et al. 1970. Nature (London) 225:1149.

[9] Berman, J. D., and M. Young. 1971. Proc. Nat. Acad. Sci. U.S. 68:395.

[10] Bisby, M. A., and M. Fillenz. 1971. J. Physiol. (London) 215:163.

[11] Björklund, A., et al. 1968. J. Histochem. Cytochem. 16:263.

[12] Bloom, F. E. 1970. Int. Rev. Neurobiol. 13:27.

[13] Bloom, F. E., et al. 1969. Science 166:1284.

[14] Bradley, P. B. 1968. Int. Rev. Neurobiol. 11:1.

[15] Bülbring, E., and M. D. Gershon. 1967. J. Physiol. (London) 192:823.

[16] Changeux, J. P., et al. 1969. J. Gen. Physiol. 54: 225S.

[17] Changeux, J. P., et al. 1970. Proc. Nat. Acad. Sci. U.S. 67:1241.

[18] Collier, B., and F. C. MacIntosh. 1969. Can. J. Physiol. Pharmacol. 147:127.

[19] Corrodi, H., and G. Jonsson. 1967. J. Histochem. Cytochem. 15:65.

[20] Corrodi, H., et al. 1970. Eur. J. Pharmacol. 12: 145.

[21] Couteaux, R. 1955. Int. Rev. Cytol. 4:335.

[22] Crawford, T. B. B., and A. S. Outschoorn. 1951. Brit. J. Pharmacol. 6:8.

[23] Dahlström, A. 1967. Acta Physiol. Scand. 69:167.

[24] Dale, H. H. 1966. Handb. Exp. Pharmacol. 18:xxvi.

[25] De Potter, W. P. 1971. Phil. Trans. Roy. Soc. London B261:313.

[26] Diamond, I., and E. P. Kennedy. 1969. J. Biol. Chem. 244:3258.

[27] Duchen, L. W., and S. J. Strich. 1968. Quart. J. Exp. Physiol. 53:84.

[28] Edwards, C., and S. W. Kuffler. 1959. J. Neurochem. 4:19.

[29] Engleman, K., et al. 1968. Amer. J. Med. Sci. 255: 259.

[30] Eränkö, O., and N. Teräväinen. 1967. J. Histochem. Cytochem. 15:399.

[31] Falck, B., and C. Owman. 1965. Acta Univ. Lund., Sect. 2, 7:1.

[32] Fatt, P., and B. Katz. 1951. J. Physiol. (London) 115:320.

[33] Fink, R. P., and L. Heimer. 1967. Brain Res. 4:369.

[34] Geffen, L. B., and B. G. Livett. 1971. Physiol. Rev. 51:98.

[35] Geffen, L. B., et al. 1969. J. Physiol. (London) 204: 593.

[36] Gershon, M. D., and L. L. Ross. 1966. Ibid. 186: 451.

[37] Gershon, M. D., and L. L. Ross. 1966. Ibid. 186: 477.

[38] Gray, F. G., and V. P. Whittaker. 1962. J. Anat. 96:79.

[39] Green, J. P. 1964. Fed. Proc. Fed. Amer. Soc. Exp. Biol. 23:1095.

[40] Hall, Z. W., et al. 1970. J. Cell Biol. 46:290.

[41] Hartman, B. K., and S. Udenfriend. 1970. Mol. Pharmacol. 6:85.

[42] Hedqvist, P., et al. 1970. Acta Physiol. Scand. 79: 19A.

[43] Hildebrand, J. G., et al. 1971. J. Neurobiol. 2:231.

[44] Hillarp, N. A., et al. 1953. Acta Physiol. Scand. 29: 251.

[45] Hökfelt, T. 1968. Z. Zellforsch. Mikrosk. Anat. 91: 1.

[46] International Union of Biochemistry. 1965. Enzyme Nomenclature. Elsevier, New York.

[47] Israel, M., et al. 1968. C. R. Acad. Sci. 266(D):273.

[48] Israel, M., et al. 1970. J. Neurochem. 17:1441.

[49] Iversen, L. L. 1967. The Uptake and Storage of Noradrenaline in Sympathetic Nerves. Univ. Press, Cambridge, England.

[50] Iversen, L. L., and E. A. Kravitz. 1968. J. Neurochem. 15:609.

[51] Iversen, L. L., and S. H. Snyder. 1968. Nature (London) 220:796.

[52] Iwayama, T., and J. B. Furness. 1971. J. Cell Biol. 48:699.

[53] Jamieson, D. 1962. Brit. J. Pharmacol. 19:286.

[54] Jéquier, E., et al. 1967. Mol. Pharmacol. 3:274.

[55] Karlin, A. 1969. J. Gen. Physiol. 54:245s.

[56] Karnovsky, M. J. 1964. J. Cell Biol. 23:217.

[57] Kasa, P., and B. Csillik. 1966. J. Neurochem. 13: 1345.

[58] Katz, B. 1966. Nerve, Muscle and Synapse. McGraw-Hill, New York. pp. 48-57.

[59] Katz, B., and S. Thesleff. 1957. Brit. J. Pharmacol. 12:260.

[60] Kirshner, N., and A. G. Kirshner. 1971. Phil. Trans. Roy. Soc. London B261:279.

[61] Koe, B. K., and A. Weissman. 1966. J. Pharmacol. Exp. Ther. 154:499.

[62] Koelle, G. B., and C. G. Gromadzki. 1966. J. Histochem. Cytochem. 14:443.

continued

Part I. Methodology

[63] Kravitz, E. A., et al. 1970. Int. Meet. Neurobiol., 5th, Proc., p. 85.

[64] Krnjević, K., and J. F. Mitchell. 1961. J. Physiol. (London) 155:246.

[65] Lee, C. Y., et al. 1967. Nature (London) 215:1177.

[66] Lester, H. A. 1970. Ibid. 227:727.

[67] Livett, B. G., et al. 1971. Phil. Trans. Roy. Soc. London B261:359.

[68] Lovenberg, W., et al. 1967. Science 155:217.

[69] Martel, R. R., et al. 1969. Can. J. Physiol. Pharmacol. 47:909.

[70] McCaman, R. E., and J. M. Hunt. 1965. J. Neurochem. 12:253.

[71] McMahan, U. J., and S. W. Kuffler. 1971. Proc. Roy. Soc. B177:485.

[72] Miledi, R., et al. 1971. Nature (London) 229:554.

[73] Molinoff, P. B., et al. 1969. J. Pharmacol. Exp. Ther. 170:253.

[74] Molinoff, P., et al. 1970. Proc. Nat. Acad. Sci. U.S. 66:453.

[75] Monnier, M., et al. 1970. Int. Rev. Neurobiol. 12:265.

[76] Morest, D. K., and R. R. Morest. 1966. Amer. J. Anat. 118:811.

[77] Nadol, J. B., and A. J. D. de Lorenzo. 1968. J. Comp. Neurol. 132:419.

[78] Nagatsu, T., and T. Yamamoto. 1968. Experientia 24:1183.

[79] Nagatsu, T., et al. 1964. Anal. Biochem. 9:122.

[80] Nauta, W. J. H., and P. A. Gygax. 1954. Stain Technol. 29:91.

[81] Neal, M. J., and L. L. Iversen. 1969. J. Neurochem. 16:1245.

[82] Nickel, E., and L. T. Potter. 1970. Brain Res. 23:95.

[83] Obata, K., et al. 1970. Exp. Brain Res. 11:327.

[84] Orkand, P. M., and E. A. Kravitz. 1971. J. Cell Biol. 49:75.

[85] Otsuka, M., et al. 1966. Proc. Nat. Acad. Sci. U.S. 56:1110.

[86] Otsuka, M., et al. 1967. J. Neurophysiol. 30:725.

[87] Otsuka, M., et al. 1971. J. Neurochem. 18:287.

[88] Peterson, R. P., and F. A. Pepe. 1961. J. Biophys. Biochem. Cytol. 11:157.

[89] Phillis, J. W. 1970. The Pharmacology of Synapses. Pergamon Press, New York.

[90] Pitts, F. N., Jr., et al. 1965. J. Neurochem. 12:93.

[91] Pletscher, A., et al. 1968. Advan. Pharmacol. 6B:55.

[92] Potter, L. T. 1966. Pharmacol. Rev. 18:439.

[93] Potter, L. T. 1967. J. Pharmacol. Exp. Ther. 156:500.

[94] Potter, L. T. 1970. J. Physiol. (London) 206:145.

[95] Reed, D. J., et al. 1966. Anal. Biochem. 16:59.

[96] Rodgers, A. W., and E. A. Barnard. 1969. J. Cell Biol. 41:686.

[97] Rodgers, A. W., et al. 1969. Ibid. 41:665.

[98] Roper, S., et al. 1969. Nature (London) 223:1168.

[99] Salmoiraghi, G. C., and C. N. Stefanis. 1967. Int. Rev. Neurobiol. 10:1.

[100] Salpeter, M. M. 1967. J. Cell Biol. 32:379.

[101] Schmitt, F. O., ed. 1970. The Neurosciences: Second Study Program. Rockefeller Univ. Press, New York.

[102] Shiman, R., et al. 1971. J. Biol. Chem. 246:1330.

[103] Snyder, S. H., et al. 1965. Biochem. Pharmacol. 14:831.

[104] Stretton, A. O. W., and E. A. Kravitz. 1968. Science 162:132.

[105] Tagliamonte, H., et al. 1971. Nature (London) New Biol. 229:125.

[106] Takeuchi, A., and N. Takeuchi. 1967. J. Physiol. (London) 191:575.

[107] Takeuchi, A., and N. Takeuchi. 1969. Ibid. 205:377.

[108] Tamiya, N., et al. 1967. In F. E. Russell and P. R. Saunders, ed. Animal Toxins. Pergamon Press, New York. pp. 249-258.

[109] Tramezzani, J. H., et al. 1964. J. Histochem. Cytochem. 12:890.

[110] Tranzer, J. D., and H. Thoenen. 1968. Experientia 24:155.

[111] Uchizono, K. 1967. Nature (London) 214:833.

[112] Uretsky, N. J., and L. L. Iversen. 1969. Ibid. 221:557.

[113] Uretsky, N. J., and L. L. Iversen. 1970. J. Neurochem. 17:269.

[114] Van Gelder, N. M. 1965. Ibid. 12:231.

[115] Van Gelder, N. M. 1965. Ibid. 12:239.

[116] Vogt, M. 1952. Brit. J. Pharmacol. 7:325.

[117] Von Redlich, D., and D. Glick. 1969. Anal. Biochem. 29:167.

[118] Waser, P. G. 1967. Ann. N.Y. Acad. Sci. 144:737.

[119] Weinshilboum, R. M., et al. 1971. Nature (London) New Biol. 230:287.

[120] Whittaker, V. P. 1965. Progr. Biophys. Mol. Biol. 15:39.

[121] Whittaker, V. P., and M. N. Sheridan. 1965. J. Neurochem. 12:363.

[122] Whittaker, V. P., et al. 1964. Biochem. J. 90:293.

[123] Wise, C. D. 1967. Anal. Biochem. 18:94.

[124] Wise, C. D. 1967. Ibid. 20:369.

[125] Wolfe, D. E., et al. 1962. Science 138:440.

[126] Wood, J. G. 1965. Tex. Rep. Biol. Med. 23:828.

[127] Wood, J. G. 1966. Nature (London) 209:1131.

[128] Wood, J. G., and R. J. Barrnett. 1964. J. Histochem. Cytochem. 12:197.

[129] Woods, R. I. 1969. J. Physiol. (London) 203:35P.

continued

Part II.   Acetylcholine

SYNAPTIC MODEL

Abbreviations: ACh = acetylcholine; AChE = acetylcholinesterase; ChAc = choline acetyltransferase; CoA = coenzyme A.

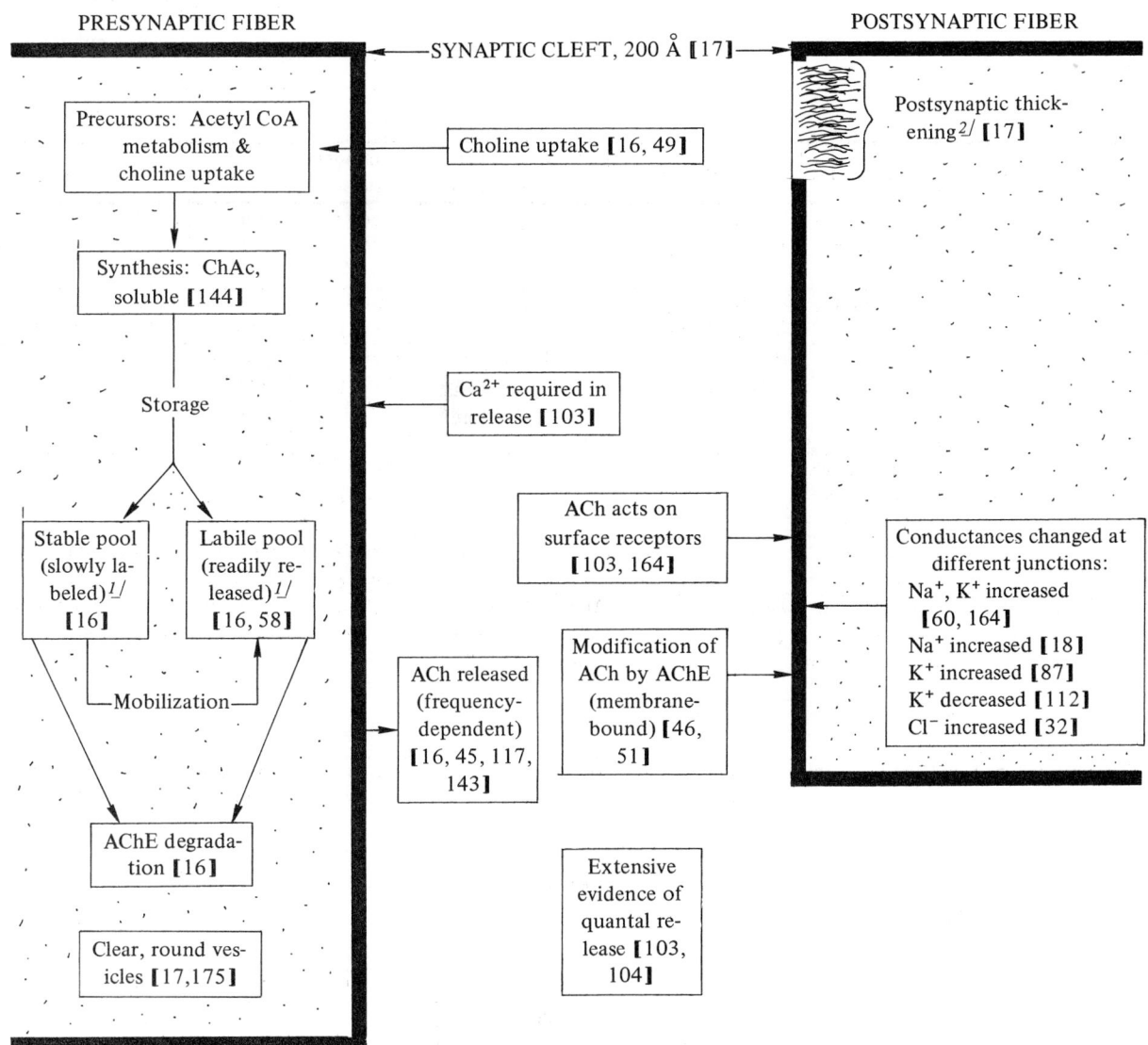

$\underline{1}$/ Control of stable and labile pools is achieved by acetylcholine inhibition [97] or mass action [144]. $\underline{2}$/ As "fuzz" in electron microscope.

continued

Abbreviations and Symbols: ACh = acetylcholine; AChE = acetylcholinesterase; CoA = coenzyme A; I_{50} = concentra- tion giving 50% inhibition; K_{eq} = equilibrium constant; K_I = dissociation constant of the enzyme-inhibitor complex;

Acetyl coenzyme A + Choline ———— Choline acetyl

Acetylcholine + Receptor ⇌ Ace

Acetylcholine + Water ———— Acetylcholinester

Enzyme [Commission No.]	Tissue Source	Molecular Weight [1]	Subcellular Localization	Cofactors + Cosubstrates	Ionic Requirements
1 Choline acetyl-transferase [2.3.1.6]	Ox brain, striate nucleus	Gel filtration: 65,000 [76]	Soluble [76]	Acetyl CoA [76]	Optimum pH: 7.5-10 [76]
2	Rabbit brain	Sedimentation: 67,000 [24]	....	Acetyl CoA [119]	Optimum pH: 7.4; salt activation: 25-300 mM [119]
3	Rat brain	Gel filtration: 50,000 [144]	Soluble, cytoplasmic; concentrated in synaptosomes [144]	Acetyl CoA [144]	Optimum pH: 7.0-7.7; salt activation: 100-300 mM [144]
4	*Torpedo marmorata* electric organ	....	....	Acetyl CoA [132]	Optimum pH: 7.4 [132]
5	Squid head ganglia	....	....	Acetyl CoA [145]	Optimum ionic strength: 0.26 M [145]
6 Acetylcholine receptor	*Electrophorus electricus* electric organ	Dialysis: >50,000 [30]	Membrane in "microsacs"; solubilized [3] by 1% deoxycholate [30, 31,56]	....	Optimum pH: 7.4; optimum ionic strength: 0.2 M [31, 56]
7	*Torpedo marmorata* electric organ	SDS-G-200 column: 88,000 & 180,000 [125]	Membrane [55,57, 125,136]; solubilized [3] by 1.5% Triton X-100 [125]	....	Optimum pH: 7.4; optimum ionic strength: 0.2 M [57]
8	*Musca domestica* brain	Failed to enter 7% polyacrylamide gel at pH 8.9 [54]	Soluble; 100,000 g supernatant after 60 min [54]	....	....

[1] Unless otherwise specified. [2] Unit = μmoles substrate reacted·min^{-1}·mg protein^{-1}. [3] Solubilized = not pelleting upon high-speed centrifugation.

tylcholine

ISTRY

K_m = Michaelis constant, measured in moles per liter; SH = sulfhydryl. Figures in heavy brackets are reference num-

bers. Data in light brackets refer to the column heading in brackets.

---

transferase [2.3.1.6] ⟶ Acetylcholine + Coenzyme A

tylcholine-Receptor

ase [3.1.1.7] ⟶ Choline + Acetic acid

Inhibitors	Apparent Michaelis Constant	Specific Activity, units [1,2]	Other Characteristics	
ACh: >50 mM; CoA: K_I = 16 μM [76]	Choline: 750 μM; acetyl CoA: 10 μM [76]	0.12 [76]	Migrated slightly toward anode in 7.5% polyacrylamide gel at pH 9.5; K_{eq} = 41 [76]	1
....	Choline: 500 μM; acetyl CoA: 10 μM [119]	....	....	2
ACh: 10-100 mM [97]; choline & acetyl CoA: >100 μM [144]; CuSO$_4$: 10^{-5} M [144]; salt: >300 mM [144]; SH reagents [144]	Choline: 16 μM; acetyl CoA: 7.7 μM [144]	0.73 [144]	Soluble in 0.15 M KCl, pH 7.0; adsorbs reversibly to particles at low ionic strength; K_{eq} = 514 ± 207 [144]	3
O$_2$; SH reagents (due to SH-catalyzed destruction of acetyl CoA) [132]	....	....	....	4
Irreversible salt inactivation: >0.3 M [145]	Choline: 1.7 mM; acetyl CoA: 0.12 mM [145]	1.8 [145]	....	5
α-Bungarotoxin [31]; gallamine triethiodide: K_I = 2.2 × 10^{-6} M [30]; hexamethonium: K_I = 3.5 × 10^{-4} M [30]; d-tubocurarine: 10^{-5} M [31], K_I = 2.6 × 10^{-5} M [30]. In vivo: SH reagents [102].	Carbachol: 1.8 × 10^{-5} M [30]; decamethonium: 8 × 10^{-7} M [30]; muscarone: 6 × 10^{-8} M [56]; phenyltrimethylammonia: 1.3 × 10^{-5} M [30]	2 × 10^{-11} mole muscarone bound/g tissue [56]; 2.6 × 10^{-8} mole α-bungarotoxin bound/g protein [31]	Phospholipoprotein is sensitive to trypsin, chymotrypsin, & phospholipase C [56]	6
α-Bungarotoxin (irreversible) [125]; decamethonium: K_m = 4.3 × 10^{-7} M [55]; nicotine: K_m = 2.5 × 10^{-6} M [55]; d-tubocurarine: K_m = 7.2 × 10^{-7} M [55]	ACh: 1.1 × 10^{-6} M [136]; ACh 1: 8 × 10^{-9} M [57]; ACh 2: 6.8 × 10^{-8} M [57]; muscarone: 7 × 10^{-6} M [136], 7 × 10^{-7} M [137]	1 × 10^{-10} mole ACh 1 bound/g tissue [57]; 8 × 10^{-10} mole ACh 2 bound/g tissue [57]; 1 × 10^{-9} mole muscarone bound/g tissue [137]	Phospholipoprotein is sensitive to trypsin, chymotrypsin, & phospholipase C [57,136]; organophosphates at 10^{-4} M block AChE without affecting receptor agonist binding [57]	7
....	Atropine: 2.1 × 10^{-6} M [55]; decamethonium: 1.9 × 10^{-7} M [55]; muscarone: 2.4 × 10^{-6} M [54]; nicotine: 3.2 × 10^{-6} M [55]	7 × 10^{-10} mole muscarone bound/g protein [54]	Protein is sensitive to trypsin & chymotrypsin; insensitive to lipase, leucine aminopeptidase, mucopeptide glucohydrolase (lysozyme), and phospholipases A, C, & D [54]	8

continued

	Enzyme [Commission No.]	Tissue Source	Molecular Weight [1]	Subcellular Localization	Cofactors + Cosubstrates	Ionic Requirements
9	Acetylcholinesterase [3.1.1.7]	Calf caudate nucleus	....	....	....	Optimum pH: 8.0; 2.5-fold activation at 0.085 M NaCl [90]
10		Rat brain	Protease solubilization, gel filtration: 100,000 [86]; Triton solubilization, gel filtration: 500,000 [43]	Membrane [86]	....	....
11		*Electrophorus electricus* electric organ	Gel filtration (polymer?): 250,000 [115]	Predominantly on innervated membrane (immunofluorescence) [14]	....	Optimum pH: 8; salt effects [160]

[1] Unless otherwise specified. [2] Unit = μmoles substrate reacted·min^{-1}·mg protein^{-1}.

Vertebrates: Acetylcholine depresses, and as often increases, firing rates in the hypothalamus, deep cerebellar nuclei, medulla, pons, and spinal cord interneurons [22, 140]. Acetylcholine and choline acetyltransferase occur in large amounts in the olfactory bulb (where acetylcholine is inhibitory to most cells) [82,142] and in the retina, but are absent in the optic nerve [82]. Acetylcholinesterase, largely membrane-bound, may be a good marker of acetylcholine synapses [3], and is also found in noncholinergic axons [59,92]; acetylcholinesterase can be liberated from intact neuromuscular junctions with collagenase [79]. Acetylcholinesterase in blood has no direct neural function [161].

Invertebrates: Tunicates—Acetylcholine, which causes muscle contractions, appears to be released following nerve

	Subjects	Site of Synaptic Junction	Presynaptic Accumulation of ACh	Neurally Evoked Release of ACh	Ca^{2+} in Release	ACh Evoked Ionic Conductance Changes[1]	
						Externally Applied	Neurally Released
				Central Nervous System			
1	Vertebrates	From reticular formation & layer 6 (polymorph cells) of cortex onto layer 5 (pyramidal cells) of cortex	Intracellular AChE in endings onto pyramidal cells [111,159]; ChAc assays of reticular formation [159]	....	....	Very long latency (5-30 s), long duration of excitation (1-3 min) [109]; decreased K$^+$ [112]	....
2		Spinal motor neuron collaterals onto Renshaw cells	Motor neurons make ACh [81]	Frequency-dependent [113]	No effect of Ca^{2+} observed [113]	Increased firing rate [52]	....
3		Mossy fibers & cerebellar Golgi cells onto cerebellar granular cells	Intracellular AChE in endings onto granule cells [111,159]	By massive electrical stimulation of cerebellum [22]	....	Increased firing rate [98,120]	....

[1] Undivided column indicates similar action of externally applied and neurally released substance.

Inhibitors	Apparent Michaelis Constant	Specific Activity, units [1,2]	Other Characteristics	
ACh: $K_I = 1 \times 10^{-2}$ M; diisopropyl-fluorophosphate: $I_{50} = 1.8 \times 10^{-6}$ M; physostigmine (eserine): 10^{-6} M [90]	ACh: 1.4×10^{-4} M [90]	18.5 [90]	Calf caudate nucleus is a particularly rich source of AChE [111]	9
Decamethonium: $K_I \sim 1.4 \times 10^{-6}$ M [86]	ACh: 1×10^{-4} M [86]	0.4 [43]	Not solubilized by 1 M NaCl [43]	10
Edrophonium bromide: $K_I = 10^{-6}$ M [15]	....	1.25×10^4 [115]; 5×10^3 [14]	Homogenous, crystallized; amino acid analysis: 3-5 amino sugars [115]	11

stimulation [67]. Atropine and *d*-tubocurarine block [67]; physostigmine (eserine) does not potentiate [66]. Muscle tissue contains acetylcholine [66], but little acetylcholinesterase [155]. *Limulus*—Acetylcholinesterase occurs in cardiac and ventral nerves [162]. Crustaceans—Acetylcholine occurs in [68,84], and can stimulate [121,177], sensory neurons. Squids—Acetylcholine, acetylcholinesterase, and choline acetyltransferase are enriched at the synaptic region of the giant synapse of the stellate ganglion [23].

Abbreviations: ACh = acetylcholine; AChE = acetylcholinesterase; ChAc = choline acetyltransferase; ? = specific site unknown.

Inactivation of ACh [1]		Morphology		Pharmacology of ACh [1]		Remarks	
Externally Applied	Neurally Released	Light Microscope	Electron Microscope	Externally Applied	Neurally Released		
Central Nervous System							
AChE [110-112]		AChE stain concentrated at synapses [111, 116]	....	Muscarinic receptor; physostigmine (eserine) potentiates [110,112]	Two-thirds of cells showed no change in conductance; possible technical problem [112]		1
AChE [44]		Golgi stains of collaterals [139]	General view of spinal cord [139]	Nicotinic receptor; dihydro-β-erythroidine blocks, physostigmine (eserine) potentiates [44]	....		2
AChE [120]		AChE in glomeruli [161]	....	Nicotinic receptor; physostigmine (eserine) potentiates [120]	....		3

continued

	Subjects	Site of Synaptic Junction	Presynaptic Accumulation of ACh	Neurally Evoked Release of ACh	Ca²⁺ in Release	ACh Evoked Ionic Conductance Changes [1]	
						Externally Applied	Neurally Released
4		Ascending, activating system of the reticular formation onto? date nucleus [2] [111], cerebellum [161], cortex [158,159], & lateral geniculate body [47]; ChAc assay [159]	Intracellular AChE paths to hippocampus [116], cau-	Arousal triples ACh output from cortex [22]; release due to stimulation of visual cortex by reticular formation [40]	....	Spontaneously active cells accelerated [22]	....
5	Invertebrates Insects	Synaptic transmission in ganglia: ? onto ?	....	....	Probable [156]	Depolarization [25,156,180]	
6	Gastropods	? onto D & H cells	....	....	....	H cells: Cl⁻ increases [32,106,152, 165]	
						D cells: Na⁺ increases [33,152]	....
7	*Aplysia*	L10 cell onto L1-L6 & R15 cells in abdominal ganglion	ChAc in L10 cell [74]	....	....	L1-L6 cells: Cl⁻ increases [18,100] R15 cell: Na⁺ increases [18,100]	
				Peripheral Nervous System			
8	Vertebrates	Spinal motor neuron onto skeletal muscle	Denervation [81]; direct measurement [143]; choline requirement [143]	Frequency-dependent [45, 108,143]	Required during presynaptic depolarization [105, 108]	K⁺ & Na⁺ increase; ΔNa conductance/ΔK conductance = 1.3 [60, 164]	
9		Autonomic preganglionic cells onto autonomic postganglionic cells	Denervation & direct analysis [81]; synthesis [84]; choline requirement [16]	Frequency-dependent [16, 82]	Required [16, 103]	Depolarization; increased conductance; Cl⁻ not involved; same reversal potential [48,53,107]	
10		Parasympathetic postganglionic cells onto gland or muscle, especially heart	Denervation [82]	Frog heart experiment [117]; frequency-dependent [82]	Required [167]	K⁺ increase; slower time course than skeletal muscle [87,167]	

[1] Undivided column indicates similar action of externally applied and neurally released substance. [2] Choline acetyltrans

Inactivation of ACh [1/]		Morphology		Pharmacology of ACh [1/]		Remarks	
Externally Applied	Neurally Released	Light Microscope	Electron Microscope	Externally Applied	Neurally Released		
AChE [116]		AChE [161]	....	Muscarinic receptor; physostigmine (eserine) potentiates [116]		....	4
....	AChE [180]	Whole ganglion [36,37]	Whole ganglion [36]	Receptor: atropine & *d*-tubocurarine block [25, 156]		ACh & AChE in nerve cord [38,123,166, 180]	5
				Receptor: methacholine (acetyl-β-methylcholine) & nicotine mimic [156]	Receptor: hexamethonium blocks [156], physostigmine (eserine) potentiates [180]		
....	....	....	....	H cells, receptor: atropine & *d*-tubocurarine block [32,165,172]; D cells, receptor: hexamethonium & *d*-tubocurarine block [72,73,172]		....	6
....	....	Whole ganglion [18,35,70,99]	Whole ganglion [35]	Receptor: *d*-tubocurarine blocks [99,100,171]		AChE in all ganglion cells [75]	7
Peripheral Nervous System							
AChE [51]		AChE concentrated at end plate [42]	AChE [46]; junctional folds, postsynaptic thickenings [17]; clear, 500-Å vesicles [17]	Nicotinic receptor; 1-25 µg curare/ml blocks [94]; Mg^{2+} competes with Ca^{2+} in release [103]; 1 µg physostigmine (eserine)/ml prolongs action [51]; hemicholinium blocks choline uptake [143]		Clear demonstration of quantal transmission [104]; botulinus toxin blocks release [63]	8
Sympathetic system: AChE [16]. Parasympathetic system: unclear; possibly AChE [48]		Comparisons of living & fixed synaptic boutons [122]		Nicotinic receptor; 1-20 µg curare/ml blocks [142]; Mg^{2+} competes with Ca^{2+} in release [16]; 1 µg physostigmine (eserine)/ml prolongs action in sympathetic ganglia [16]; ACh desensitizes [16,80]		External choline essential for maintenance of stores [39]; ACh sensitivity very discretely localized at synapses [80]	9
		AChE concentrated at synapses [42]; wiring scheme [91]	Postsynaptic thickenings; clear, 500-Å vesicles [135]				
AChE [88]		AChE concentrated at end plate [42]	Postsynaptic thickenings; clear, 500-Å vesicles [135]	Receptor: 1-10 µg atropine/ml blocks [82], 1 µg physostigmine (eserine)/ml potentiates [167]		....	10

ferase content is high in caudate nucleus [82].

continued

	Subjects	Site of Synaptic Junction	Presynaptic Accumulation of ACh	Neurally Evoked Release of ACh	Ca^{2+} in Release	ACh Evoked Ionic Conductance Changes [1]	
						Externally Applied	Neurally Released
11	Eel & marine ray	Electric nerve onto electric organ	Nerve endings rich in ACh & ChAc [89]; ACh depleted upon stimulation [29]	Yes [61,62]	....	K^+ & Na^+ increase [83,101,153, 154,176]; ΔNa permeability/ΔK permeability = 5 [102]	
12	Invertebrates Echinoderms, crustaceans, insects, & annelids	? onto heart	....	....	....	Stimulated heartbeat in crustaceans [174], insects [124,127], & annelids [149]; threshold = $10^{-10} - 10^{-9}$ g/ml. Slowed heartbeat in echinoderms; threshold = 10^{-14} g/ml [147]	....
13	Echinoderms, onychophorans, annelids, sipunculids, cephalopods, lamellibranchs, gastropods, & nematodes	Motor neuron onto muscle	....	In annelids [11]	....	Contractures in echinoderms [19], onychophorans [69], annelids [8,173], sipunculids [148], cephalopods [9], lamellibranchs [168], gastropods [50, 78,85,93], & nematodes [13, 28]	....
14	Crayfish, annelids, & mollusks	? onto ? in intestinal tract preparations	....	In annelids [5]	....	Contractures in crayfish [64,65, 96], annelids [131,178], & mollusks [77,141]; threshold = $10^{-8} - 10^{-6}$ g/ml in crayfish & mollusks, and 10^{-11} g/ml in annelids	....
15	Mollusks	Cardioregulator nerve onto heart	....	In cephalopods [71] & clams [27,146]	....	Spontaneous activity depressed; threshold = 10^{-12} g/ml [41,140, 146]	

[1] Undivided column indicates similar action of externally applied and neurally released substance.

Inactivation of ACh [1]		Morphology		Pharmacology of ACh [1]		Remarks	
Externally Applied	Neurally Released	Light Microscope	Electron Microscope	Externally Applied	Neurally Released		
AChE [61,62]		Distinct innervated & non-innervated faces of plaques [4, 133,154]	Single population of clear, 750-Å vesicles [89,134,157]	Receptor: *d*-tubocurarine blocks; physostigmine (eserine) potentiates [1,61, 62,154,176]; affinity labels [102]		Rich source of AChE [62,133]	11
AChE in echinoderms [147] & annelids [149]	……………	Complex in crustaceans [174] & insects [126, 129]	Two populations of granules in insects [95]	Receptor: in echinoderms—nicotine mimics, physostigmine (eserine) potentiates [146]; in insects—atropine, hexamethonium, & *d*-tubocurarine block [127]; in annelids—atropine blocks, nicotine mimics, & physostigmine (eserine) potentiates [149]	Receptor: in crustaceans—physostigmine (eserine) potentiates [174]	Dopamine, epinephrine, norepinephrine, & serotonin (5-hydroxytryptamine) in insects affect myocardium only [128]; details of connections unclear	12
AChE in echinoderms [10,19], onychophorans [69], annelids [11,12, 20,179], sipunculids [148], lamellibranchs [169], & gastropods [93]		AChE stain in lamellibranchs [21] & nematodes [114]	Postsynaptic thickenings; 500-Å cleft; clear, 500-Å vesicles; 600- to 1200-Å dense-core vesicles [150,151]	Nicotinic receptor; in echinoderms [10,19], onychophorans [69], annelids [12,179], & sipunculids [148], physostigmine (eserine) potentiates; in sipunculids, atropine blocks [148]; in nematodes, neostigmine & *d*-tubocurarine block [13, 28]. Receptor: in lamellibranch—atropine, benzoquinonium, hexamethonium, & methantheline block [26,168,169]. Nicotinic receptor: in echinoderms [19], & gastropods [93]—*d*-tubocurarine blocks; in lamellibranchs [26] & gastropods [93]—physostigmine (eserine) potentiates	Receptor: in lamellibranchs—methantheline blocks [170]	ACh & AChE in muscle of echinoderms [6,7, 138], annelids [6,7], lamellibranchs [169], & nematodes [114]; AChE in muscle of onychophorans [69]; γ-aminobutyric acid & epinephrine have ACh-like effect in gastropods [50]; denervation supersensitivity in cephalopods [9]	13
AChE in annelids [5,131] AChE in crayfish [64] & mollusks [141]	……………	Gut innervation in annelids [130]	…………………	Receptor: in annelids—atropine blocks [5,131, 178]; physostigmine (eserine) potentiates [5,131]. Receptor: in crayfish— atropine blocks [64,65]; in mollusks—benzoquinonium blocks [141]; in crayfish & mollusks—physostigmine (eserine) potentiates [64,141]	……………	AChE in annelids [131]	14
AChE [146]		…………………	…………………	Receptor: benzoquinonium blocks [34,118, 163]; physostigmine (eserine) potentiates [146]		Details of connections unknown; two-heart experiment [146]	15

continued

141. SYNAPTIC TRANSMITTER COMPOUNDS: VERTEBRATES AND INVERTEBRATES

Part II.   Acetylcholine

Contributors: Wallace, B. G., Sargent, P. B., Mains, R. E., and Kravitz, E. A.

References

[1] Albe-Fessard, D., and C. Chagas. 1951. C. R. Soc. Biol. 145:248.

[2] Albers, R. W., and G. J. Koval. 1961. Biochim. Biophys. Acta 52:29.

[3] Aldridge, W. N., and M. K. Johnson. 1959. Biochem. J. 73:270.

[4] Altamirano, M., et al. 1953. J. Gen. Physiol. 37:91.

[5] Ambache, N., et al. 1945. J. Exp. Biol. 21:46.

[6] Bacq, Z. M. 1935. Arch. Int. Physiol. 42:24.

[7] Bacq, Z. M. 1935. Ibid. 42:47.

[8] Bacq, Z. M. 1937. Ibid. 44:174.

[9] Bacq, Z. M. 1939. Ibid. 49:16.

[10] Bacq, Z. M. 1939. Ibid. 49:25.

[11] Bacq, Z. M., and G. Coppée. 1937. Ibid. 45:310.

[12] Bacq, Z. M., and G. Coppée. 1937. C. R. Soc. Biol. 124:1244.

[13] Baldwin, E., and V. Moyle. 1949. Brit. J. Pharmacol. 4:145.

[14] Benda, P., et al. 1970. Nature (London) 225:1149.

[15] Berman, J. D., and M. Young. 1971. Proc. Nat. Acad. Sci. U.S. 68:395.

[16] Birks, R., and F. C. MacIntosh. 1961. Can. J. Biochem. Physiol. 39:787.

[17] Birks, R., et al. 1960. J. Physiol. (London) 150:134.

[18] Blankenship, J. E., et al. 1971. J. Neurophysiol. 34:76.

[19] Boltt, R. E., and D. W. Ewer. 1963. J. Exp. Biol. 40:727.

[20] Botsford, E. F. 1941. Biol. Bull. 80:299.

[21] Bowden, J. 1958. Int. Rev. Cytol. 7:295.

[22] Bradley, P. B. 1968. Int. Rev. Neurobiol. 11:1.

[23] Bryant, S. H., and M. Brzin. 1966. J. Cell. Physiol. 68:107.

[24] Bull, G., et al. 1964. Nature (London) 201:1326.

[25] Callec, J. J., and J. Boistel. 1967. C. R. Soc. Biol. 161:442.

[26] Cambridge, G. W., et al. 1959. J. Physiol. (London) 148:451.

[27] Carroll, P. R., and L. B. Cobbin. 1968. Aust. J. Exp. Biol. Med. Sci. 46:P23.

[28] Castillo, J. del, et al. 1963. Arch. Int. Physiol. Biochim. 71:741.

[29] Chagas, C. 1952. C. R. Acad. Sci. 234:663.

[30] Changeux, J. P., et al. 1970. Ibid. 270(D):2864.

[31] Changeux, J. P., et al. 1970. Proc. Nat. Acad. Sci. U.S. 67:1241.

[32] Chiarandini, D. J., and H. M. Gerschenfeld. 1967. Science 156:1595.

[33] Chiarandini, D. J., et al. 1967. Ibid. 156:1597.

[34] Chong, G. C., and J. W. Phillis. 1965. Brit. J. Pharmacol. 25:481.

[35] Coggeshall, R. E. 1967. J. Neurophysiol. 30:1263.

[36] Cohen, M. J. 1970. In F. O. Schmitt, ed. The Neurosciences: Second Study Program. Rockefeller University Press, New York. pp. 798-812.

[37] Cohen, M. J., and J. W. Jacklet. 1967. Phil. Trans. Roy. Soc. London B252:561.

[38] Colhoun, E. H. 1958. J. Insect Physiol. 2:108.

[39] Collier, B., and F. C. MacIntosh. 1969. Can. J. Physiol. Pharmacol. 147:127.

[40] Collier, B., and J. F. Mitchell. 1966. J. Physiol. (London) 184:239.

[41] Cottrell, G. A., et al. 1968. Comp. Biochem. Physiol. 27:787.

[42] Couteaux, R. 1958. Exp. Cell Res., Suppl. 5:294.

[43] Crone, H. D. 1971. J. Neurochem. 18:489.

[44] Curtis, D. R., and R. M. Eccles. 1958. J. Physiol. (London) 141:446.

[45] Dale, H. H., et al. 1936. Ibid. 86:353.

[46] Davis, R., and G. B. Koelle. 1967. J. Cell Biol. 34:157.

[47] Defenu, G., et al. 1967. Exp. Neurol. 17:203.

[48] Dennis, M. J., et al. 1971. Proc. Roy. Soc. B177:509.

[49] Diamond, I., and E. P. Kennedy. 1969. J. Biol. Chem. 244:3258.

[50] Duncan, C. J. 1964. Z. Vergl. Physiol. 48:295.

[51] Eccles, J. C., et al. 1942. J. Neurophysiol. 5:211.

[52] Eccles, J. C., et al. 1954. J. Physiol. (London) 126:524.

[53] Eccles, R. M. 1955. Ibid. 130:572.

[54] Eldefrawi, A. T., and R. D. O'Brien. 1970. J. Neurochem. 17:1287.

[55] Eldefrawi, M. E., et al. 1971. Mol. Pharmacol. 7:104.

[56] Eldefrawi, M. E., et al. 1971. Proc. Nat. Acad. Sci. U.S. 68:1047.

[57] Eldefrawi, M. E., et al. 1971. Science 173:338.

[58] Elmqvist, D., and D. M. J. Quastel. 1965. J. Physiol. (London) 178:505.

[59] Eränkö, O. 1966. Pharmacol. Rev. 18:353.

[60] Fatt, P., and B. Katz. 1951. J. Physiol. (London) 115:320.

[61] Feldberg, W., and A. Fessard. 1942. Ibid. 101:200.

continued

Part II. Acetylcholine

[62] Feldberg, W., et al. 1940. Ibid. 97:3P.

[63] Fex, S., et al. 1966. Ibid. 184:872.

[64] Florey, E. 1954. Z. Vergl. Physiol. 36:1.

[65] Florey, E. 1956. J. Physiol. (London) 156:1.

[66] Florey, E. 1963. Comp. Biochem. Physiol. 8:327.

[67] Florey, E. 1967. Ibid. 22:617.

[68] Florey, E., and M. A. Biederman. 1959. J. Gen. Physiol. 43:509.

[69] Florey, E., and E. Florey. 1965. Comp. Biochem. Physiol. 15:125.

[70] Frazier, W. T., et al. 1967. J. Neurophysiol. 30: 1288.

[71] Fredericq, H., and Z. M. Bacq. 1940. Arch. Int. Physiol. 50:169.

[72] Gerschenfeld, H. M., and E. Stefani. 1968. Advan. Pharmacol. 6B:369.

[73] Gerschenfeld, H. M., et al. 1967. Nature (London) 213:358.

[74] Giller, E., and J. H. Schwartz. 1971. J. Neurophysiol. 34:93.

[75] Giller, E., and J. H. Schwartz. 1971. Ibid. 34:108.

[76] Glover, V. A. S., and L. T. Potter. 1971. J. Neurochem. 18:571.

[77] Greenberg, M. J. 1966. Science 154:1015.

[78] Groat, W. C. de, and R. W. Ryall. 1967. Exp. Brain Res. 3(4):299.

[79] Hall, Z. W., and R. B. Kelly. 1971. Nature (London) New Biol. 232:62.

[80] Harris, A. J., et al. 1971. Proc. Roy. Soc. B177: 541.

[81] Hebb, C. O. 1957. Physiol. Rev. 37:196.

[82] Hebb, C. O. 1963. Handb. Exp. Pharmacol. 15: 55.

[83] Higman, H. B., et al. 1964. Biochim. Biophys. Acta 79:138.

[84] Hildebrand, J. G., et al. 1971. J. Neurobiol. 2: 231.

[85] Hill, R. B. 1958. Biol. Bull. 115:471.

[86] Ho, I. K., and G. L. Ellman. 1969. J. Neurochem. 16:1505.

[87] Hutter, O. F. 1961. In E. Florey, ed. Nervous Inhibition. Pergamon Press, New York. pp. 114-123.

[88] Hutter, O. F., and W. Trautwein. 1956. J. Gen. Physiol. 39:715.

[89] Israel, M., and J. Gautron. 1969. In S. Barondes, ed. Cellular Dynamics of the Neuron. Academic Press, New York. pp. 137-152.

[90] Jackson, R. L., and M. H. Aprison. 1966. J. Neurochem. 13:1351.

[91] Jacobowitz, D. 1967. J. Pharmacol. Exp. Ther. 158: 227.

[92] Jacobowitz, D., and G. B. Koelle. 1965. Ibid. 148: 225.

[93] Jaeger, C. P. 1962. Comp. Biochem. Physiol. 7: 63.

[94] Jenkinson, D. H. 1960. J. Physiol. (London) 152: 309.

[95] Johnson, B. 1966. J. Insect Physiol. 12:645.

[96] Jones, H. C. 1962. J. Physiol. (London) 164:295.

[97] Kaita, A. A., and A. M. Goldberg. 1969. J. Neurochem. 16:1185.

[98] Kanai, T., and J. C. Szerb. 1965. Nature (London) 205:80.

[99] Kandel, E. R., et al. 1967. J. Neurophysiol. 30: 1352.

[100] Kandel, E. R., et al. 1967. Science 155:346.

[101] Karlin, A. 1967. Proc. Nat. Acad. Sci. U.S. 58: 1162.

[102] Karlin, A. 1969. J. Gen. Physiol. 54:245s.

[103] Katz, B. 1971. Science 173:123.

[104] Katz, B., and R. Miledi. 1965. J. Physiol. (London) 181:656.

[105] Katz, B., and R. Miledi. 1967. Ibid. 189:535.

[106] Kerkut, G. A., and R. C. Thomas. 1964. Comp. Biochem. Physiol. 11:199.

[107] Kobayashi, H., and B. Libet. 1970. J. Physiol. (London) 208:353.

[108] Krnjević, K., and J. F. Mitchell. 1961. Ibid. 155: 246.

[109] Krnjević, K., and J. W. Phillis. 1963. Ibid. 166: 296.

[110] Krnjević, K., and J. W. Phillis. 1963. Ibid. 166: 328.

[111] Krnjević, K., and A. Silver. 1965. J. Anat. 99: 711.

[112] Krnjević, K., et al. 1971. J. Physiol. (London) 215:247.

[113] Kuno, M., and P. Rudomin. 1966. Ibid. 187: 177.

[114] Lee, D. L. 1962. Parasitology 52:241.

[115] Leuzinger, W., and A. L. Baker. 1967. Proc. Nat. Acad. Sci. U.S. 57:446.

[116] Lewis, P. R., et al. 1967. J. Physiol. (London) 191:215.

[117] Loewi, O. 1921. Pfluegers Arch. Gesamte Physiol. Menschen Tiere 189:239.

[118] Loveland, R. E. 1963. Comp. Biochem. Physiol. 9:95.

[119] McCaman, R. E., and J. M. Hunt. 1965. J. Neurochem. 12:253.

[120] McCance, I., and J. W. Phillis. 1968. Int. J. Neuropharmacol. 7:447.

continued

Part II.   Acetylcholine

[121] McLennan, H., and D. H. York. 1966. Comp. Biochem. Physiol. 17:327.

[122] McMahan, U. J., and S. W. Kuffler. 1971. Proc. Roy. Soc. B177:485.

[123] Means, O. W. 1942. J. Cell. Comp. Physiol. 20:319.

[124] Metcalf, R. L., et al. 1964. J. Insect Physiol. 10:353.

[125] Miledi, R., et al. 1971. Nature (London) 229:554.

[126] Miller, T. 1968. J. Insect Physiol. 14:1265.

[127] Miller, T. 1968. Ibid. 14:1713.

[128] Miller, T., and R. L. Metcalf. 1968. Ibid. 14:383.

[129] Miller, T., and W. W. Thomson. 1968. Ibid. 14:1099.

[130] Millott, N. 1943. Proc. Roy. Soc. B131:271.

[131] Millott, N. 1943. Ibid. B131:362.

[132] Morris, D. 1967. J. Neurochem. 14:19.

[133] Nachmansohn, D. 1955. Harvey Lect. 49:57.

[134] Nickel, E., and L. T. Potter. 1971. Phil. Trans. Roy. Soc. London B261:383.

[135] Nilsson, E., and B. Sporrong. 1970. Z. Zellforsch. Mikrosk. Anat. 111:404.

[136] O'Brien, R. D., and L. P. Gilmore. 1969. Proc. Nat. Acad. Sci. U.S. 63:496.

[137] O'Brien, R. D., et al. 1970. Ibid. 65:438.

[138] Pentreath, V. H., and G. A. Cottrell. 1968. Comp. Biochem. Physiol. 27:775.

[139] Peters, A., et al. 1970. The Fine Structure of the Nervous System. Harper and Row, New York.

[140] Phillis, J. W. 1966. Comp. Biochem. Physiol. 17:719.

[141] Phillis, J. W. 1966. Ibid. 17:909.

[142] Phillis, J. W. 1970. The Pharamcology of Synapses. Pergamon Press, New York.

[143] Potter, L. T. 1970. J. Physiol. (London) 206:145.

[144] Potter, L. T., et al. 1968. J. Biol. Chem. 243:3864.

[145] Prince, A. K. 1967. Proc. Nat. Acad. Sci. U.S. 57:1117.

[146] Prosser, C. L. 1940. Biol. Bull. 78:92.

[147] Prosser, C. L., and C. L. Judson. 1952. Ibid. 102:249.

[148] Prosser, C. L., and C. E. Melton. 1954. J. Cell. Comp. Physiol. 44:255.

[149] Prosser, C. L., and G. L. Zimmerman. 1943. Physiol. Zool. 16:77.

[150] Reger, J. F. 1965. Z. Zellforsch. Mikrosk. Anat. 67:196.

[151] Rosenbluth, J. 1965. J. Cell Biol. 26:579.

[152] Sato, M., et al. 1968. J. Gen. Physiol. 51:321.

[153] Schoffeniels, E. 1959. Ann. N.Y. Acad. Sci. 81:285.

[154] Schoffeniels, E., and D. Nachmansohn. 1957. Biochim. Biophys. Acta 26:1.

[155] Scudder, C. L., and A. G. Karczmar. 1966. Comp. Biochem. Physiol. 17:553.

[156] Shankland, D. L., et al. 1971. J. Neurobiol. 2:247.

[157] Sheridan, M. N. 1965. J. Cell Biol. 24:129.

[158] Shute, C. C. D., and P. R. Lewis. 1967. Brain 90:497.

[159] Shute, C. C. D., and P. R. Lewis. 1965. Nature (London) 205:242.

[160] Silman, H. I., and A. Karlin. 1967. Proc. Nat. Acad. Sci. U.S. 58:1664.

[161] Silver, A. 1967. Int. Rev. Neurobiol. 10:57.

[162] Smith, C. C., and D. Glick. 1939. Biol. Bull. 77:321.

[163] S.-Rózsa, K., and I. Zs.-Nagy. 1967. Comp. Biochem. Physiol. 23:373.

[164] Takeuchi, A., and N. Takeuchi. 1960. J. Physiol. (London) 154:52.

[165] Tauc, L., and H. M. Gerschenfeld. 1962. J. Neurophysiol. 25:236.

[166] Tobias, J. M., et al. 1946. J. Cell. Comp. Physiol. 28:159.

[167] Trautwein, W., and J. Dudel. 1958. Pfluegers Arch. Gesamte Physiol. Menschen Tiere 266:324.

[168] Twarog, B. M. 1954. J. Cell. Comp. Physiol. 44:141.

[169] Twarog, B. M. 1960. J. Physiol. (London) 152:236.

[170] Twarog, B. M. 1967. Ibid. 192:857.

[171] Wachtel, H., and E. R. Kandel. 1967. Science 158:1206.

[172] Walker, R. J., and A. Hedges. 1967. Comp. Biochem. Physiol. 23:977.

[173] Walker, R. J., et al. 1968. Ibid. 24:987.

[174] Welsh, J. H. 1939. Physiol. Zool. 12:231.

[175] Whittaker, V. P. 1965. Progr. Biophys. Mol. Biol. 15:39.

[176] Whittam, R., and M. Guinnebault. 1960. Biochim. Biophys. Acta 45:336.

[177] Wiersma, C. A. G., et al. 1953. J. Exp. Biol. 30:136.

[178] Wu, K. S. 1939. Ibid. 16:184.

[179] Wu, K. S. 1939. Ibid. 16:251.

[180] Yamasaki, T., and T. Narahashi. 1960. J. Insect Physiol. 4:1.

continued

Part III.   Norepinephrine

SYNAPTIC MODEL

Abbreviation: NE = norepinephrine.

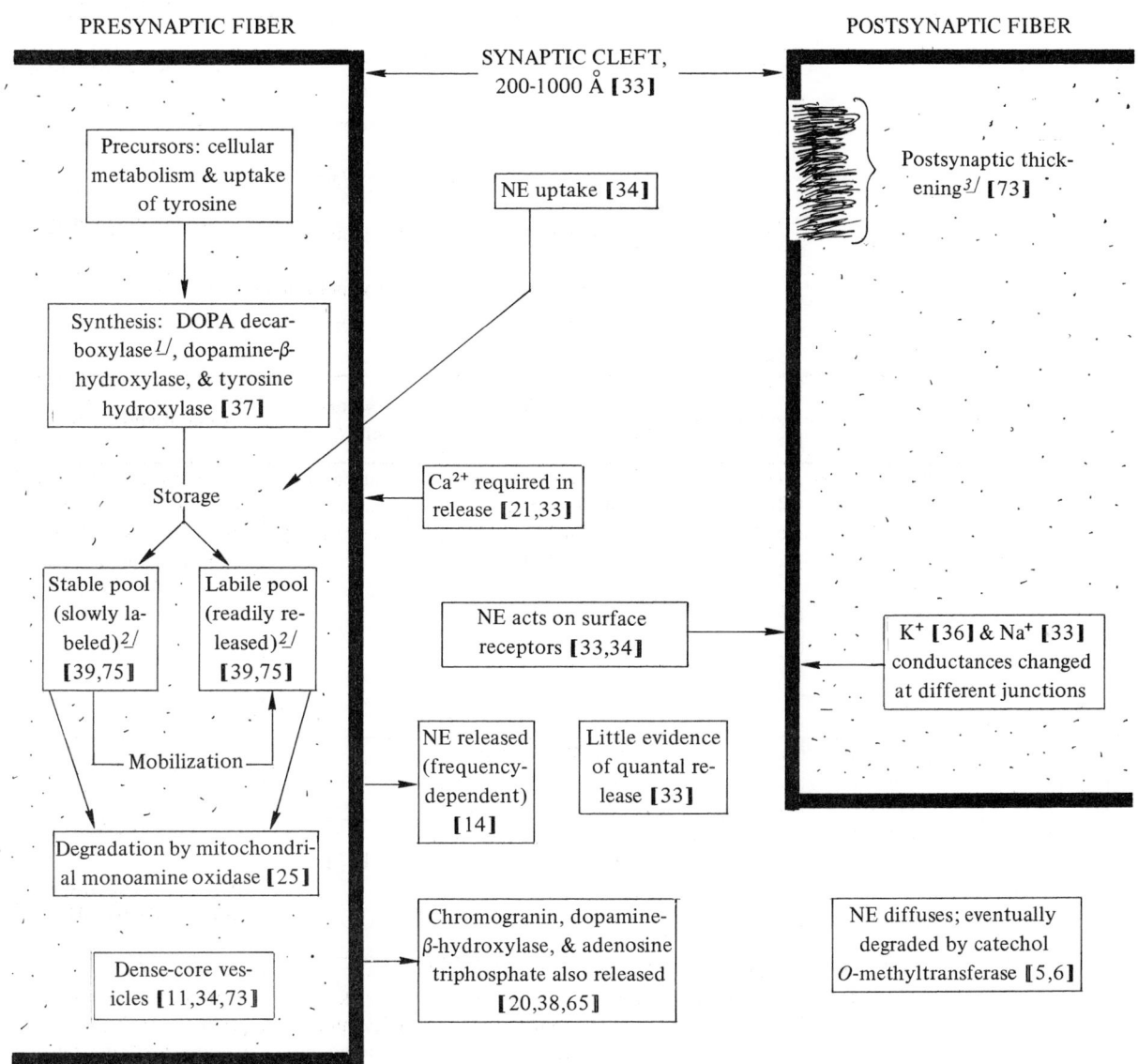

---

[1]/ Synonym: aromatic amino acid decarboxylase. [2]/ Control of stable and labile pools may be achieved by norepinephrine feedback onto tyrosine hydroxylase [68]. [3]/ As "fuzz" in electron microscope.

continued

Abbreviations and Symbols: DMPH$_4$ = 2-amino-4-hydroxy-6,7-dimethyltetrahydropteridine; FAD = flavin-adenine dinucleotide; I$_{50}$ = concentration giving 50% inhibition; K$_I$ = dissociation constant of the enzyme-inhibitor complex; K$_m$ = Michaelis constant, measured in moles per liter; NE = norepinephrine; s$_{20,w}$ = sedimentation coefficient, corrected

Tyrosine + Oxygen + Reduced pteridine	Tyrosine hydroxyl
DOPA ——————— DOPA decarboxylase [4.1.1.26] —————→ Dopamine +	
Dopamine + Oxygen + Ascorbate	Dopamine-β-hydroxyl
Norepinephrine + Oxygen + Water	Monoamine oxidase
Norepinephrine + S-Adenosylmethionine	Phenylethanolamine
Norepinephrine + S-Adenosylmethionine	Catechol O-methyl

	Enzyme (Synonym) [Commission No.]	Tissue Source	Molecular Weight	Subcellular Localization	Cofactors + Cosubstrates	Ionic Requirements
1	Tyrosine hydroxylase	Cattle suprarenal medulla	Gel filtration: 40,000 [48, 62]	Membrane of sedimented, lysed vesicles; some activity is soluble [44,48, 62]	O$_2$, reduced pteridine (tetrahydrobiopterin) [62]	Optimum pH: 6.2; sensitive to H$_2$O$_2$ oxidation; assay requires Fe^{2+} or catalase [44,62]
2		Cattle caudate nucleus	....	Activity associated with synaptic vesicle fraction [26]	....	....
3	DOPA decarboxylase (Aromatic amino acid decarboxylase) [4.1.1.26]	Guinea-pig kidney	....	Soluble, cytoplasmic [55]	Pyridoxal 5-phosphate [55]	Optimum pH: 9; optimum pH for DOPA: 6.7-7.0 (due to catechol instability at alkaline pH) [55]
4	Dopamine-β-hydroxylase [1.14.2.1]	Cattle suprarenal gland	Sedimentation: 290,000 [27]	Intravesicular; both soluble and associated with vesicle membrane [37]	Ascorbate, fumarate, & O$_2$ [40]; 2 moles enzyme-bound Cu^{2+}/mole enzyme [27]	Optimum pH: 5.5; inactivation by substrate-produced H$_2$O$_2$; assay requires Fe^{2+} or catalase [22,28]
5	Monoamine oxidase [1.4.3.4]	Cattle brain	Gel filtration: 400,000 [29]	....	Enzyme-bound FAD [29]	Optimum pH: 8.6 [29]

$^{1/}$ Unit = μmoles substrate reacted·min^{-1}·mg protein^{-1}. $^{2/}$ Solubilized = not pelleting upon high-speed centrifugation.

epinephrine

to water at 20°C; SH = sulfhydryl; Tris = tris(hydroxy-methyl)-aminomethane. Figures in heavy brackets are reference numbers. Data in light brackets refer to the column heading in brackets.

$\xrightarrow{\text{ase}}$ DOPA + Water + Oxidized pteridine

Carbon dioxide

$\xrightarrow{\text{ase [1.14.2.1]}}$ Norepinephrine + Water + Dehydroascorbate

$\xrightarrow{\text{[1.4.3.4]}}$ 3,4-Dihydroxymandelic acid + Ammonia + Hydrogen peroxide

$\xrightarrow{\text{N-methyltransferase}}$ Epinephrine + S-Adenosylhomocysteine

$\xrightarrow{\text{transferase [2.1.1.6]}}$ Normetanephrine + S-Adenosylhomocysteine

Inhibitors	Apparent Michaelis Constant	Specific Activity units[L]	Other Characteristics	
3,4-Dihydroxyphenylpropyl-acetamide (H22154): $K_I = 1 \times 10^{-5}$ M [48,72]; catechols: $K_I \sim 3 \times 10^{-3}$ M [72]; o-phenanthroline: $K_I = (1\text{-}2) \times 10^{-6}$ M [62]; tyrosine: $K_I \sim 1 \times 10^{-4}$ M [62]; 3-iodo-L-tyrosine: $K_I = 3.9 \times 10^{-7}$ M [72]; α-methyl-L-tyrosine: $K_I = 2.5 \times 10^{-5}$ M [72]; 3-iodo-α-methyl-DL-tyrosine: $K_I = 1.8 \times 10^{-7}$ M [72]; NE: $K_I \sim 1 \times 10^{-3}$ M [68]; SH reagents [48]; endogenous inhibitors [44, 62]; α,α'-dipyridyl [44]	Tyrosine: 7 μM [62]; DMPH$_4$: 5×10^{-4} M [44]	9×10^{-2} [62]	Solubilized[2] by chymotryptic digestion, tryptic digestion, & vigorous homogenization [44,48, 62]; with natural cofactor, soluble enzyme hydroxylates phenylalanine twice as fast as tyrosine [62]. Highly specific: tyramine, D-tyrosine, DL-m-tyrosine, & L-tryptophan not hydroxylated [44].	1
....	....	....	....	2
α-Methyl DOPA: $K_m = 5.4 \times 10^{-4}$ M [55]; α-hydrazino-3,4-dihydroxyphenylpropionic acid: $I_{50} = 2 \times 10^{-7}$ M [50]; SH reagents [55]	DOPA: 4×10^{-4} M; 5-hydroxytryptophan: 2×10^{-5} M [55]	0.55 [71]	Stereospecific; requires L-isomer. Nonspecific; most aromatic amino acids decarboxylated. Benzene effects. [55]	3
Benzylhydrazine: 10^{-5} – 10^{-4} M [16]; KCN: 6×10^{-4} M [40]. Cu chelators: CO (not reversed by light) [28], colchiceines [28], diethyldithiocarbamate [28], disulfiram (reduced to diethyldithiocarbamate by ascorbate) [28], SH compounds [45], & tropolones [28]	Dopamine: 5.8×10^{-3} M [71]	0.83 [71]; 3.25 [27]	Released on stimulation; $s_{20,w} = 8.9$. [37] Nonspecific; hydroxylates a variety of phenylethylamine derivatives. [16] No detectable free SH in homogenous preparation [27].	4
Kynuramine: $>2 \times 10^{-4}$ M; pargyline & pheniprazine: 10^{-6} M; SH reagents [29]	Kynuramine: $(6\text{-}7) \times 10^{-5}$ M [29]	....	Solubilized with 1% (vol/vol) non-ion NS-210; precipitates on concentration; diamines (histamine) not deaminated [29]	5

continued

Enzyme (Synonym) [Commission No.]	Tissue Source	Molecular Weight	Subcellular Localization	Cofactors + Cosubstrates	Ionic Requirements
6	Cattle kidney	Sedimentation: 290,000 [25]	....	Flavoenzyme [25]	Optimum pH: 8.5 [25]
7	Cattle liver	....	Mitochondrial membrane [77]	2 moles enzyme/ mole enzyme-bound FAD [77]	Optimum pH: 9.3 [46]
8	Rabbit brain	....	....	....	....
9	Rat brain	Gel filtration: 230,000 [79]	Mitochondrial [79]	....	Optimum $pH_{I,II,III}$: 8.3; optimum pH_{IV}: 7.4 [79]
10	Rat liver	Gel filtration: 290,000 [78]	Outer mitochondrial membrane [61,78]	....	Optimum pH: 7.2 [74]
11	Swine brain	Gel filtration: 102,000; apparent lipid complex or tetramer: 435,000 [70]	Mitochondrial [70]	Enzyme-bound FAD [70]	Optimum pH: 7.2 [70]
12 Phenylethanolamine N-methyltransferase	Mammal suprarenal medulla, rabbit heart, rabbit & rat brain	....	Soluble [3,5]	S-Adenosylmethionine [3,5]	Optimum pH_{Tris}: 8-9; optimum $pH_{phosphate}$: 7.5-8.2 [3,5]
13	Rabbit lung & *Bufo marinus* parotid gland	....	....	S-Adenosylmethionine [4,5]	Optimum pH_{Tris}: 8.5; optimum $pH_{phosphate}$: 8.0 [4,5]
14 Catechol O-methyltransferase [2.1.1.6]	Mouse brain	....	....	S-Adenosylmethionine [56]	....
15	Rat liver	24,000 [2]	Soluble [5,7]	S-Adenosylmethionine [5,7]	Absolute requirement for divalent cation; Co^{2+} & Mn^{2+} most effective [5,7]
16 Chromogranin A	Cattle suprarenal medulla	Sedimentation & osmometry: 80,000. Dimer: identical 40,000 subunits [38,66,67]	Soluble, intravesicular; accounts for 30-40% of soluble protein released by granule lysis [37,66,67]	In isolated vesicles, 10,000 nmoles epinephrine/mg chromogranin [19]	....

1/ Unit = μmoles substrate reacted·min^{-1}·mg protein^{-1}. 2/ Solubilized = not pelleting upon high-speed centrifugation.

epinephrine

Inhibitors	Apparent Michaelis Constant	Specific Activity units[1]	Other Characteristics	
Carbostyril (2-hydroxyquin-oline): $K_I = 2.5 \times 10^{-4}$ M; SH reagents [25]	....	....	Nonspecific; oxidizes most primary, secondary, & tertiary monoamines; diamines (histamine) not oxidized [25]	6
Hydralazine: $K_I = 2 \times 10^{-5}$ M; tranylcypromine: $K_I = 5.8 \times 10^{-8}$ M; SH reagents. [30] Irreversible inhibitors: nial-amide [46], pargyline, & phenylhydrazine [30]. Metal chelators: cuprizone & α,α'-dipyridyl [77].	....	....	Solubilized[2] by 1.5% Triton X-100 [77]	7
Dopamine (3-hydroxytyra-mine): $I_{50} = 14$ mM; ipro-niazid: 10^{-5} M; pargyline & pheniprazine: 10^{-6} M; serotonin (5-hydroxytryptamine): $I_{50} = 8$ mM; tranylcypromine: 10^{-7} M [41]	....	....	....	8
Harmaline: $K_I = 10^{-5} - 10^{-4}$ M; M & B 9302; $K_I = 10^{-7} - 5 \times 10^{-6}$ M; pargyline: $K_I = 5 \times 10^{-8}$ M [79]	Kynuramine: 5×10^{-5} M [79]	$0\text{-}6 \times 10^{-3}$; varies with form & substrate [79]	Solubilized[2] by 1.5% (vol/vol) Triton X-100 in presence of 10^{-3} M benzylamine; 4 characteristically different forms can be separated reproducibly by polyacrylamide gel electrophoresis [79]	9
Salts: PO_4^{2-} inhibits more than NO_3^-; NO_3^- inhibits more than Cl^- [74]	....	....	Multiple forms (?) [74]; solubilized[2] by sonication with cholic acid [78]	10
Iproniazid: 2.5×10^{-4} M; tyramine: $K_I = 1.7 \times 10^{-2}$ M; SH reagents. Metal chelators: 4,5-dihydroxy-m-benzenedisulfonic acid disodium salt & o-phenanthroline—$I_{50} = 10^{-3}$ M. [70]	Tyramine: 1.2×10^{-4} M [70]	2.8 [70]	Stable to brief exposure to pH 3 at 4°C; solubilized[2] by sonication, freezing, & thawing repeated 6 times [70]	11
SH reagents [3,5]	Normetanephrine: 5×10^{-5} M [3,5]	....	Nonspecific; will methylate phenylethanolamine & phenylisopropanolamine derivatives; absolute requirement for β-hydroxyl. Not strictly stereospecific; L-isomers preferred. Will add second N-methyl group. [3,5]	12
Chlorpromazine: $I_{50} = 2.5 \times 10^{-5}$ M; imipramine: $I_{50} = 1 \times 10^{-4}$ M [4,5]	Serotonin (5-hydroxytryptamine): 9×10^{-4} M [4,5]	....	Nonspecific; many aromatic amine derivatives methylated [4,5]	13
4-Isopropyltropolone: $I_{50} = 2.6 \times 10^{-6}$ M; methyl ester of gallic acid: $I_{50} = 1 \times 10^{-6}$ M; 4-methyl-tropolone: $I_{50} = 2.1 \times 10^{-6}$ M [56]	....	....	....	14
Catechols, pyrogallol [5]; SH reagents, tropolones [7]	Epinephrine: 1.2×10^{-4} M [7]	1.1 [2]	Nonspecific; all catechols O-methylated, mainly at $meta$ position; no stereospecificity; no monophenol activity. [5,7] Specific antibody inhibits activity [2].	15
....	Non-enzymatic; NE binding accounted for only 1-2% of vesicle catecholamine [37,66,67]		Homogenous; polar/nonpolar amino acid ratio high; asp + glu = 32% residue, lys + arg = 14% residue, pro = 9% residue; isoelectric pH = 4.5 [37,66,67]	16

continued

Enzyme (Synonym) [Commission No.]	Tissue Source	Molecular Weight	Subcellular Localization	Cofactors + Cosubstrates	Ionic Requirements
17	Sympathetic nerve	....	Vesicular; both soluble & bound [9, 19]	In isolated vesicles, 4000 nmoles NE/mg chromogranin [9,19]	....

[1]/ Unit = μmoles substrate reacted$\cdot$min$^{-1}\cdot$mg protein^{-1}.

In vertebrates, norepinephrine can be taken up avidly by regions of the brain having little or no norepinephrine [80]. In the brain stem, hypothalamus, medulla, pons, cerebral cortex, and caudate nucleus, norepinephrine has a slow, mixed action [13,49]; in the hippocampus and dorsal column of the spinal cord, norepinephrine apparently has only an inhibitory action [13]. Norepinephrine inhibits motor neurons and Renshaw cells rapidly ($\sim$1-s latency) [17,58], but is exclusively excitatory in Deiters' nucleus (see Part IV, γ-Aminobutyric Acid Physiology) [13]. Norepinephrine,

Subjects	Site of Synaptic Junction	Presynaptic Accumulation of NE	Neurally Evoked Release of NE	Ca^{2+} in Release	NE Evoked Ionic Conductance Changes [1]/	
					Externally Applied	Neurally Released
			Central Nervous System			
1	Vertebrates	Lateral olfactory tract onto mitral cell	NE fluorescence [18]	By massive electrical stimulation [8]	Sensitive [8]	Mitral cell activity depressed [59]	
2		? onto cerebellar Purkinje cell	NE fluorescence; electron microscopic autoradiography [1,12]	....	....	Spontaneous activity depressed [32, 64]	
						Conductance decrease produces hyperpolarization [64]	....
3		? onto neurosecretory cells of supraoptic nucleus	Fluorescence [15,54]	....	....	Spontaneous activity depressed [10]	....
			Peripheral Nervous System			
4	Vertebrates	Sympathetic postganglionic cells onto gland or muscle	Fluorescence [35]; chemical analysis of ganglion	Frequency-dependent [14]; stoichiometric amounts of ATP, dopamine β-hydroxylase,	Required [14]	Increased K$^+$ [33,36]	
						At other junctions, NE depolarizes; ions unknown [33]	....

[1]/ Undivided column indicates similar action of externally applied and neurally released substance. [2]/ Synonym: desmeth

epinephrine

Inhibitors	Apparent Michaelis Constant	Specific Activity units[L]	Other Characteristics	
....	....	....	Biochemically similar to suprarenal chromogranin; cross-reacts immunologically with anti-supra-renal chromogranin A sera [9,19]	17

dopamine, and serotonin have little or no effect on barbituated animals [17]. For information on norepinephrine contraction and fluorescent innervation in the hindgut of crayfish and annelids, consult references 23, 24, 42, 43, 52, 57, and 76.

Abbreviations: cAMP = cyclic adenosine 3′,5′-monophosphate; ATP = adenosine 5′-triphosphate; NE = norepinephrine; ? = specific site unknown.

Inactivation of NE[L]		Morphology		Pharmacology of NE[L]		Remarks	
Externally Applied	Neurally Released	Light Microscope	Electron Microscope	Externally Applied	Neurally Released		
Central Nervous System							
Uptake [8,59]		Fluorescence; Golgi method [51,53,60]	Reciprocal synapses [53,60]	Receptor: Dibenamine, phentolamine, & lysergic acid diethylamide block [59]		Ultrastructure of NE-mediated pathway unclear [53,60]	1
				....	Reserpine & chronic α-methyl-*m*-tyrosine reduce neural inhibition; metaraminol enhances inhibition [59]		
Uptake [12,32]		Fluorescence [1,12]	Some large, granular vesicles; predominantly small, clear, spherical vesicles in 6-hydroxydopamine-sensitive terminals [12]	Desipramine[2] potentiates depression by blocking uptake [32]		Cerebellum contains 6-10% of brain NE [1]; NE action mimicked by cAMP (which acts postsynaptically), and enhanced by phosphodiesterase inhibitors [63,64]	2
				α-Receptor: MJ-1999 blocks [32]	....		
....	....	Fluorescence [54]	Large & small granular vesicles; small clear vesicles [54]	Receptor: MJ-1999 blocks [10]	....	Reserpine stimulates antidiuretic hormone secretion [10]	3
Peripheral Nervous System							
Uptake, via Na⁺-dependent pump into presynaptic endings; second		Fluorescence; Ag stains [35]	Postsynaptic thickenings; 500- & 1000-Å dense-core vesicles [73]	Receptor: α, β, or mixed, depending on junction [33]. Reserpine depletes terminals; cocaine blocks NE		Isolated granules contain ATP, chromogranin, dopamine β-hydroxylase, & NE [9], axons have only large vesicles	4

ylimipramine.

continued

Sub-jects	Site of Synaptic Junction	Presynaptic Accumulation of NE	Neurally Evoked Release of NE	Ca²⁺ in Release	NE Evoked Ionic Conductance Changes	
					Externally Applied	Neurally Released
		[47]; synthesis by ganglion [31]	& chromogranin also released [65]			
5	Suprarenal medulla³/, as model of a sympapathetic neuron	Fluorescence [35]; chemical analysis [21]⁴/	Frequency-dependent [14]; stoichiometric amounts of ATP, chromogranin, & dopamine β-hydroxylase also released [65]⁴/	Required [21]	....	....

³/ Releases epinephrine into the blood; epinephrine acts on smooth muscle, heart, and other organs, and is inactivated by catechol *O*-methyltransferase. [5] Reserpine blocks epinephrine transfer across isolated granule membranes

Contributors: Wallace, B. G., Sargent, P. B., Mains, R. E., and Kravitz, E. A.

References

[1] Andén, N.-E., et al. 1967. Experientia 23:838.

[2] Assicot, M., and C. Bohuon. 1969. Biochem. Pharmacol. 18:1893.

[3] Axelrod, J. 1962. J. Biol. Chem. 237:1657.

[4] Axelrod, J. 1962. J. Pharmacol. Exp. Ther. 138:28.

[5] Axelrod, J. 1966. Pharmacol. Rev. 18:95.

[6] Axelrod, J. 1971. Science 173:598.

[7] Axelrod, J., and R. Tomchick. 1958. J. Biol. Chem. 233:702.

[8] Baldessarini, R. J., and I. J. Kopin. 1967. J. Pharmacol. Exp. Ther. 156:31.

[9] Banks, P., and K. B. Helle. 1971. Phil. Trans. Roy. Soc. London B261:305.

[10] Barker, J. L., et al. 1971. Science 171:208.

[11] Bisby, M. A., and M. Fillenz. 1971. J. Physiol. (London) 215:163.

[12] Bloom, F. E., et al. 1971. Brain Res. 25:501.

[13] Bradley, P. B. 1968. Int. Rev. Neurobiol. 11:1.

[14] Brown, G. L. 1965. Proc. Roy. Soc. B162:1.

[15] Carlsson, A., et al. 1962. Acta Physiol. Scand., Suppl. 196.

[16] Creveling, C. R., et al. 1962. Biochim. Biophys. Acta 64:125.

[17] Curtis, D. R., and J. M. Crawford. 1969. Annu. Rev. Pharmacol. 9:209.

[18] Dahlström, A., et al. 1965. Life Sci. 4:2071.

[19] De Potter, W. P. 1971. Phil. Trans. Roy. Soc. London B261:313.

[20] Douglas, W. W., and A. M. Poisner. 1966. J. Physiol. (London) 183:249.

[21] Douglas, W. W., and R. P. Rubin. 1963. Ibid. 167:288.

[22] Duch, D. S., et al. 1968. Biochem. Pharmacol. 17:255.

[23] Elofsson, R., et al. 1966. Z. Zellforsch. Mikrosk. Anat. 74:464.

[24] Elofsson, R., et al. 1968. Experientia 24:1159.

[25] Erwin, V. G., and L. Hellerman. 1967. J. Biol. Chem. 242:4230.

[26] Fahn, S., et al. 1969. J. Neurochem. 16:1293.

[27] Friedman, S., and S. Kaufman. 1965. J. Biol. Chem. 240:4763.

[28] Goldstein, M., et al. 1965. Ibid. 240:2066.

[29] Harada, M., et al. 1971. J. Neurochem. 18:559.

[30] Hellerman, L., and V. G. Erwin. 1968. J. Biol. Chem. 243:5234.

[31] Hildebrand, J. G., et al. 1971. J. Neurobiol. 2:231.

[32] Hoffer, B. J., et al. 1971. Brain Res. 25:523.

[33] Holman, M. E. 1970. In E. Bülbring, et al., ed. Smooth Muscle. E. Arnold, London. pp. 244-288.

[34] Iversen, L. L. 1967. The Uptake and Storage of Noradrenaline in Sympathetic Nerves. Univ. Press, Cambridge, England.

[35] Jacobowitz, D. 1970. Fed. Proc. Fed. Amer. Soc. Exp. Biol. 29:1929.

[36] Jenkinson, D. H., and I. K. M. Morton. 1967. J. Physiol. (London) 188:373.

epinephrine

Inactivation of NE		Morphology		Pharmacology of NE		Remarks	
Externally Applied	Neurally Released	Light Microscope	Electron Microscope	Externally Applied	Neurally Released		
uptake system into other sites [34]				reuptake; tyrosine analogs block synthesis & accumulation of NE.[34]		[19]; reserpine blocks NE transfer across isolated vesicle membranes [69]	
....	....	Whole suprarenal medulla is fluorescent [35]	Storage granules 3000 Å in diameter [33]	Reserpine depletes [5]; Mg^{2+} competes with Ca^{2+} for release, and Ba^{2+} produces spontaneous, rapid release [20]		Isolated granules contain ATP, chromogranin, dopamine β-hydroxylase, & NE [9]	5

[69]. 4/ Predominant catecholamine is epinephrine.

[37] Kirshner, N., and A. G. Kirshner. 1971. Phil. Trans. Roy. Soc. London B261:279.

[38] Kirshner, N., et al. 1967. Mol. Pharmacol. 3:254.

[39] Kopin, I. J., et al. 1968. J. Pharmacol. Exp. Ther. 161:271.

[40] Levin, E. Y., et al. 1960. J. Biol. Chem. 235:2080.

[41] McCaman, R. E., et al. 1965. J. Neurochem. 12:15.

[42] Millott, N. 1943. Proc. Roy. Soc. B131:362.

[43] Myhrberg, H. E. 1967. Z. Zellforsch. Mikrosk. Anat. 81:311.

[44] Nagatsu, T., et al. 1964. J. Biol. Chem. 239:2910.

[45] Nagatsu, T., et al. 1967. Biochim. Biophys. Acta 139:319.

[46] Nara, S., et al. 1966. J. Biol. Chem. 241:2774.

[47] Norberg, K. A., and B. Hamberger. 1964. Acta Physiol. Scand., Suppl. 238.

[48] Petrack, B., et al. 1968. J. Biol. Chem. 243:743.

[49] Phillis, J. W. 1970. The Pharmacology of Synapses. Pergamon Press, New York.

[50] Porter, C. C., et al. 1962. Biochem. Pharmacol. 11:1067.

[51] Powell, T. P. S., and W. M. Cowan. 1963. Nature (London) 199:1296.

[52] Prosser, C. L., and G. L. Zimmerman. 1943. Physiol. Zool. 16:77.

[53] Rall, W. 1970. In F. O. Schmitt, ed. The Neurosciences: Second Study Program. Rockefeller Univ. Press, New York. pp. 552-565.

[54] Rechardt, L. 1969. Acta Physiol. Scand., Suppl. 329.

[55] Lovenberg, W., et al. 1962. J. Biol. Chem. 237:89.

[56] Ross, S. B., and Ö. Haljasmaa. 1964. Acta Pharmacol. Toxicol. 21:205.

[57] Rude, S. 1969. J. Comp. Neurol. 136:349.

[58] Salmoiraghi, G. C., and C. N. Stefanis. 1967. Int. Rev. Neurobiol. 10:1.

[59] Salmoiraghi, G. C., et al. 1964. Amer. J. Physiol. 207:1417.

[60] Shepherd, G. M. 1970. In F. O. Schmitt, ed. The Neurosciences: Second Study Program. Rockefeller Univ. Press, New York. pp. 539-552.

[61] Schnaitman, C., et al. 1967. J. Cell Biol. 32:719.

[62] Shiman, R., et al. 1971. J. Biol. Chem. 246:1330.

[63] Siggins, G. R., et al. 1971. Brain Res. 25:535.

[64] Siggins, G. R., et al. 1971. Science 171:192.

[65] Smith, A. D. 1971. Phil. Trans. Roy. Soc. London B261:363.

[66] Smith, A. D., and H. Winkler. 1967. Biochem. J. 103:483.

[67] Smith, W. J., and N. Kirshner. 1967. Mol. Pharmacol. 3:52.

[68] Spector, S., et al. 1967. Ibid. 3:549.

[69] Stjärne, L. 1964. Acta Physiol. Scand., Suppl. 228.

[70] Tipton, K. F. 1968. Eur. J. Biochem. 4:103.

[71] Udenfriend, S. 1966. Pharmacol. Rev. 18:43.

continued

Part III. Norepinephrine

[72] Udenfriend, S., et al. 1965. Biochem. Pharmacol. 14:837.

[73] Van Orden, L. S., et al. 1966. J. Pharmacol. Exp. Ther. 154:185.

[74] Van Woert, M. H., and G. C. Cotzias. 1966. Biochem. Pharmacol. 15:275.

[75] Weiner, N., and M. Rabadjija. 1968. J. Pharmacol. Exp. Ther. 160:61.

[76] Wu, K. S. 1939. J. Exp. Biol. 16:184.

[77] Yasunobu, K. T., et al. 1968. Advan. Pharmacol. 6A:43.

[78] Youdim, M. B. H., and T. L. Sourkes. 1966. Can. J. Biochem. 44:1397.

[79] Youdim, M. B. H., et al. 1969. Nature (London) 223:626.

[80] Dahlström, A., and K. Fuxe. 1964. Acta Physiol. Scand., Suppl. 232.

continued

Part IV.   γ-Aminobutyric Acid

SYNAPTIC MODEL

Abbreviation: GABA = γ-aminobutyric acid.

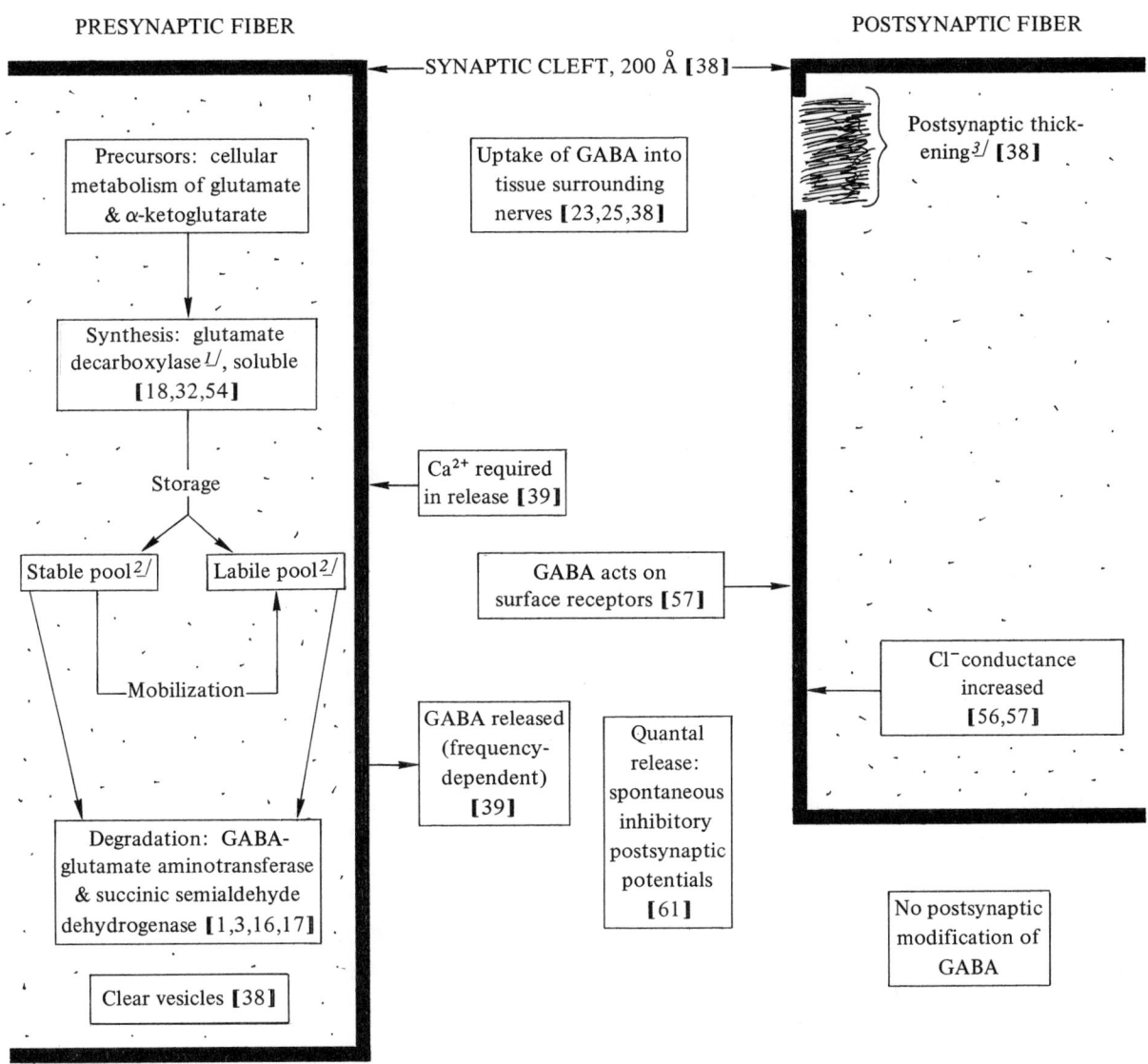

---

[1]/ Synonym: glutamic acid decarboxylase. [2]/ Control of stable and labile pools is achieved by the balance of synthesis and degradation [32]. [3]/ As "fuzz" in electron microscope.

continued

Abbreviations and Symbols: GABA = γ-aminobutyric acid; I_{50} = concentration giving 50% inhibition; K_I = dissociation constant of the enzyme-inhibitor complex; K_m = Michaelis constant, measured in moles per liter; NAD^+ = nicotinamide adenine dinucleotide, oxidized form; $NADP^+$ = nicotinamide adenine dinucleotide phosphate, oxidized form;

Glutamic acid —— Glutamate decarboxylase [4.1.1.15] ——→	GABA + Car
GABA + α-Ketoglutarate —— GABA–Glutamic acid aminotransferase ——→	Suc
Succinic semialdehyde + Pyridine nucleotide + Water —— Succinic semialde	

	Enzyme (Synonym) [Commission No.]	Tissue Source	Molecular Weight	Subcellular Localization	Cofactors + Cosubstrates	Ionic Requirements
1	Glutamate decarboxylase (Glutamic acid decarboxylase) [4.1.1.15]	Mouse brain	Gel filtration: 75,000–100,000 [54]	Concentrated in synaptosomes [48,49]	Pyridoxal 5-phosphate [48,54]	Optimum pH: 7.2 [54]
2		Mouse kidney	....	Particulate [15]	No activation by pyridoxal 5-phosphate after extensive dialysis [15]	Optimum pH: 6.1; anion activation, in order of effectiveness: Br^-, Cl^- (max 0.2 M), SO_4^{2-}, I^-, Ac^-; stimulated by aminooxyacetic acid, isethionate, pyruvate, & semicarbazide [15]
3		Lobster central nervous system	....	Soluble, cytoplasmic [32]	Pyridoxal 5-phosphate [32]	Optimum pH: 8; 75 mM K^+ required (NH_4^+, Cs^+, Rb^+ may substitute); SH required: 25 mM βME [32]
4	GABA-glutamic acid aminotransferase	Mouse brain	Sedimentation: 100,000 [62]	Mitochondrial [48]; soluble from acetone powder [62]	Pyridoxal 5-phosphate [48]	Optimum pH: 8.0; requires reduced glutathione [62]
5		Rat brain	....	....	....	Optimum pH: 8.6 [45]
6		Lobster central nervous system	....	Soluble, cytoplasmic [16]	Pyridoxal 5-phosphate [16]	Optimum pH: 8.5 [2/] [16]

[1/] Unit = μmoles substrate reacted·min^{-1}·mg protein^{-1}. [2/] Function of glutamate and GABA concentrations. [3/] Function

butyric Acid

ISTRY

SH = sulfhydryl; $s_{20,w}$ = sedimentation coefficient, corrected to water at 20°C. Figures in heavy brackets are reference numbers. Data in light brackets refer to the column heading in brackets.

bon dioxide

cinic semialdehyde + Glutamate

hyde dehydrogenase [1.2.1.16] ⟶ Succinic acid + Reduced pyridine nucleotide

Inhibitors	Apparent Michaelis Constant	Specific Activity units L/	Other Characteristics	
Anions, in order of effectiveness: I^-, NO_3^-, SO_4^{2-}, Br^-, Cl^-, Ac^-, & F^- [48, 54]; aminooxyacetic acid: 10^{-3} M [54]; light sensitivity [54]	Glutamate: 6×10^{-3} M [54]	0.67 [54]	Stabilized by pyridoxal 5-phosphate & aminoethylisothiouronium bromide; during purification, Δ optimum pH 6.7 → 7.2 and ΔK_m 3 → 6×10^{-3} M (conformational change?) [54]. Ca^{2+}-dependent binding to membranes [49].	1
β-Alanine, taurine, & succinate [15]	....	....	Activity not proportional to protein concentration; kinetics suggest two K_m values. Change in proportion of Cl^- stimulated/Cl^- inhibited during development of chick embryo brain. [15]	2
GABA: $K_I = 1.25 \times 10^{-3}$ M; glycine (aminoacetic acid): 10^{-4} M; hydroxylamine: 10^{-4} M; semicarbazide: 10^{-3} M [32]	Glutamate: 2×10^{-2} M [32]	9×10^{-3} [32]	Highly specific; β-alanine, taurine, aspartate, & succinate do not compete [32]	3
....	GABA: 1.2×10^{-2} M; α-ketoglutarate: 2.0×10^{-3} M [62]	0.5 [62]	Association-dissociation of aggregates affects activity; max activity = 0.1 mg/ml enzyme; $s_{20,w}$ = 5.0 [48,62]	4
....	GABA & α-ketoglutarate: 10^{-4} M [45]	....	....	5
Aminooxyacetic acid: 5×10^{-5} M; hydroxylamine: $I_{50} = 4 \times 10^{-4}$ M; semicarbazide: $I_{50} = 3 \times 10^{-3}$ M; α-ketoglutarate: $K_I = (0.9 - 3.4) \times 10^{-3}$ M 3/; anion: >50 mM; SH reagents [16]	At pH 7.5—GABA: 6.1×10^{-4} M; α-ketoglutarate: 1.1×10^{-4} M. At pH 8.5—GABA: 4.0×10^{-4} M; α-ketoglutarate: 2.1×10^{-4} M. [16]	....	Specific; lysine, ornithine, taurine, & α-aminobutyrate will not substitute for GABA; δ-aminovaleric acid & β-alanine will substitute. Oxalacetate & pyruvate will not substitute for α-ketoglutarate. [16]	6

of pH.

continued

	Enzyme (Synonym) [Commission No.]	Tissue Source	Molecular Weight	Subcellular Localization	Cofactors + Cosubstrates	Ionic Requirements
7	Succinic semialdehyde dehydrogenase [1.2.1.16]	Mammal brain	Gel filtration: 135,000 [46]	Mitochondrial matrix [49]	NAD⁺ (no activity with NADP⁺) [1, 45]	Optimum pH: 8.6-9.0 [1,45]; max monovalent cation activation with 50 mM NaCl or KCl; SH reagents activate [45]
8		Guinea pig kidney	....	Soluble [47]	NAD⁺ (no activity with NADP⁺) [47]	Optimum pH: 8.6-9.6; activated by SH reagents [47]
9		Lobster central nervous system	....	Soluble [17]	NAD⁺ (three times more activity than with NADP⁺) [17]	Optimum pH: 9; absolute requirement for SH [17]

[1] Unit = μmoles substrate reacted·min^{-1}·mg protein^{-1}.

Vertebrates: γ-Aminobutyric acid depresses the firing of nearly all central nervous system cells [4, 28, 29, 44] in a manner easily distinguished from that of general anesthesia [50]. Electrical or high-K⁺ stimulation of the cortex increases the rate of γ-aminobutyric acid efflux (possibly requiring Ca²⁺) [26,31,52]. Layer IV of the visual cortex, unlike layer IV in other parts of the cortex, is especially rich in γ-aminobutyric acid [20,51]. The dorsal gray column of the spinal cord contains twice as much γ-aminobutyric acid and glutamic acid decarboxylase as the ventral gray column [14].

	Subjects	Site of Synaptic Junction	Presynaptic Accumulation of GABA	Neurally Evoked Release of GABA	Ca²⁺ in Release	GABA Evoked Ionic Conductance Changes [1]	
						Externally Applied	Neurally Released
			Central Nervous System				
1	Vertebrates	Cerebellar Purkinje cell onto Deiters' nucleus cells	GABA present [34, 41]	Massive electrical stimulation of cerebellum increases efflux from 4th ventricle [35]	....	Increased conductance; same (hyperpolarized) reversal potential [36,37]	
2		? onto pyramidal cells in molecular & pyramidal layers of hippocampus	GABA & glutamic acid decarboxylase follow pattern of inhibitory inputs [53]	....	....	Spontaneous firing slowed [53]; antidromic spike invasion blocked [50]	....
			Peripheral Nervous System				
3	Invertebrates Crustaceans	Inhibitory nerve onto skeletal muscle	Inhibitory nerve cells [40]; inhibitory nerve axons [27]	Frequency-dependent [39]	Required [39]	Increased Cl⁻ [57]	

[1] Undivided column indicates similar action of externally applied and neurally released substance. [2] Data for inactivation

butyric Acid

Inhibitors	Apparent Michaelis Constant	Specific Activity units $\underline{1/}$	Other Characteristics	
m- or *p*-Hydroxybenzalde-hyde: 10^{-5} M; succinic semialdehyde: $>2 \times 10^{-5}$ M; SH reagents [1,45]	Succinic semialdehyde: 2.7×10^{-6} M [1]	....	Highly specific; two active species (?); marked temperature activation at 37°C [1,45]	7
NH_4^+: 10^{-4} M; PO_4^{2-} inac-tivates [47]	NAD^+: 5×10^{-5} M; suc-cinic semialdehyde: 7.8×10^{-5} M [47]	7×10^{-3} [47]	Highly purified preparation; specific for suc-cinic semialdehyde; abrupt activation at 27°C [47]	8
Succinic semialdehyde: $>7 \times 10^{-5}$ M; NaCl: $>5 \times 10^{-2}$ M; SH re-agents [17]	NAD^+: 1.6×10^{-4} M; suc-cinic semialdehyde: 2×10^{-5} M [17]	....	Highly specific [17]	9

OLOGY

Invertebrates: *Limulus*—γ-aminobutyric acid does not ap-pear to be the inhibitory transmitter [42]. Crustaceans—γ-aminobutyric acid (1 μg/ml) slows the heart; picrotoxin blocks. [12] In the hindgut of crayfish, picrotoxin blocks γ-aminobutyric acid depression [11]. Nematodes—in muscle tissue, γ-aminobutyric acid hyperpolarizes; Cl⁻ con-ductance increases (?). [5]

Abbreviations and Symbols: GABA = γ-aminobutyric acid; K_I = dissociation constant of the enzyme-inhibitor com-plex; ? = specific site unknown.

Inactivation of GABA Externally Applied $2/$	Morphology		Pharmacology of GABA Externally Applied & Neurally Released	Remarks	
	Light Microscope	Electron Microscope			
Central Nervous System					
Na^+-dependent up-take [24]	Golgi method [43]	Consult reference 43	Receptor: picrotoxin blocks; strychnine does not block [37]	Strong evidence of a mo-nosynaptic connection [22]	1
....	....	....	Receptor: insensitive to strych-nine; bicuculline blocks [7,37]	....	2
Peripheral Nervous System					
Na^+-dependent up-take into Schwann & endoneurial cells [23,38]	One inhibitory nerve & one excitatory nerve per fiber [59]	Postsynaptic thick-enings; flat vesi-cles [38,59]	Receptor: picrotoxin blocks (K_I = 3 μM); analogs such as β-guanido-propionic acid compete (K_I = 0.5 mM) [10,58]	GABA & glutamic acid de-carboxylase found only in inhibitory, not excita-tory, nerve cells [18]	3

of neurally released GABA are not available.

continued

	Subjects	Site of Synaptic Junction	Presynaptic Accumulation of GABA	Neurally Evoked Release of GABA	Ca^{2+} in Release	GABA Evoked Ionic Conductance Changes [1]	
						Externally Applied	Neurally Released
4		Inhibitory nerve onto peripheral sensory nerve	Synthesis by terminals [19]	....	Required [8]	Increased Cl⁻ [56]	
5	Sto-mato-pods	? onto pacemaker neurons of heart ganglion	....	....	Without Ca^{2+}, loss of inhibitory postsynaptic potential [63]	Increased Cl⁻ [63]	
6	Insects	Inhibitory neuron onto skeletal muscle	....	....	....	Increased Cl⁻ [61]	

[1] Undivided column indicates similar action of externally applied and neurally released substance. [2] Data for inactivation

Contributors: Wallace, B. G., Sargent, P. B., Mains, R. E., and Kravitz, E. A.

References

[1] Albers, R. W., and G. J. Koval. 1961. Biochim. Biophys. Acta 52:29.

[2] Atwood, H. L., et al. 1969. J. Insect Physiol. 15:529.

[3] Bessman, S. P., et al. 1953. J. Biol. Chem. 201:385.

[4] Bradley, P. B. 1968. Int. Rev. Neurobiol. 11:1.

[5] Castillo, J. del, et al. 1964. Experientia 20:141.

[6] Cohen, M. J. 1970. In F. O. Schmitt, ed. The Neurosciences: Second Study Program. Rockefeller Univ. Press, New York. pp. 798-812.

[7] Curtis, D. R., et al. 1970. Nature (London) 226:1222.

[8] Edwards, C., and S. W. Kuffler. 1959. J. Neurochem. 4:19.

[9] Faeder, I. R., and R. D. O'Brien. 1970. J. Exp. Zool. 173:203.

[10] Feltz, A. 1971. J. Physiol. (London) 216:391.

[11] Florey, E. 1956. Ibid. 156:1.

[12] Florey, E. 1957. Naturwissenschaften 15:424.

[13] Florey, E., and E. Florey. 1955. J. Gen. Physiol. 39:69.

[14] Graham, L. T., and M. H. Aprison. 1969. J. Neurochem. 16:559.

[15] Haber, B., et al. 1970. Biochem. Pharmacol. 19:1119.

[16] Hall, Z. W., and E. A. Kravitz. 1967. J. Neurochem. 14:45.

[17] Hall, Z. W., and E. A. Kravitz. 1967. Ibid. 14:55.

[18] Hall, Z. W., et al. 1970. J. Cell Biol. 46:290.

[19] Hildebrand, J. G., et al. 1971. J. Neurobiol. 2:231.

[20] Hirsch, H., and E. Robins. 1962. J. Neurochem. 9:63.

[21] Hoyle, G. 1955. Proc. Roy. Soc. B143:281.

[22] Ito, M., and M. Yoshida. 1966. Exp. Brain Res. 2:330.

[23] Iversen, L. L., and E. A. Kravitz. 1968. J. Neurochem. 15:609.

[24] Iversen, L. L., and M. J. Neal. 1968. Ibid. 15:1141.

[25] Iversen, L. L., and S. H. Snyder. 1968. Nature (London) 220:796.

[26] Iversen, L. L., et al. 1970. Brit. J. Pharmacol. 38:452P.

[27] Kravitz, E. A., and D. D. Potter. 1965. J. Neurochem. 12:323.

[28] Krnjević, K., and J. W. Phillis. 1963. J. Physiol. (London) 165:274.

[29] Krnjević, K., and S. Schwartz. 1967. Exp. Brain Res. 3:320.

[30] Mandelstam, Yu. E. 1967. In J. Salánki, ed. Neurobiology of Invertebrates. Plenum Press, New York. pp. 49-57.

butyric Acid

Inactivation of GABA Externally Applied[2/]	Morphology		Pharmacology of GABA Externally Applied & Neurally Released	Remarks	
	Light Microscope	Electron Microscope			
....	Methylene blue studies [13]	Postsynaptic thickenings; flat vesicles [59]	Receptor: picrotoxin blocks (K_I = 3 μM) [58]; analogs such as β-guanidopropionic acid compete & excite [8]	Hill coefficient for GABA action on receptor = 4 [10]	4
....	....	....	Receptor: picrotoxin blocks [63]	....	5
Uptake proportional to inhibitory innervation [33]	Claw-shaped endings [21]; procion yellow dye injection of inhibitory neurons [6]	Postsynaptic thickenings; 160- to 200-Å cleft; 350- to 500-Å vesicles [2,30]	Receptor: picrotoxin blocks [60, 61]	GABA has no effect on "fast" systems [9]	6

of neurally released GABA are not available.

[31] Mitchell, J. F., and V. Srinivasan. 1969. Nature (London) 224:663.

[32] Molinoff, P. B., and E. A. Kravitz. 1968. J. Neurochem. 15:391.

[33] Morin, W. A., and H. L. Atwood. 1969. Comp. Biochem. Physiol. 30:577.

[34] Obata, K. 1969. Experientia 25:1283.

[35] Obata, K., and K. Takeda. 1969. J. Neurochem. 16:1043.

[36] Obata, K., et al. 1967. Exp. Brain Res. 4:43.

[37] Obata, K., et al. 1970. Ibid. 11:327.

[38] Orkand, P. M., and E. A. Kravitz. 1971. J. Cell Biol. 49:75.

[39] Otsuka, M., et al. 1966. Proc. Nat. Acad. Sci. U.S. 56:1110.

[40] Otsuka, M., et al. 1967. J. Neurophysiol. 30:725.

[41] Otsuka, M., et al. 1971. J. Neurochem. 18:287.

[42] Pax, R. A., and R. C. Sanborn. 1967. Biol. Bull. 132:381.

[43] Peters, A., et al. 1970. The Fine Structure of the Nervous System. Harper and Row, New York.

[44] Phillis, J. W. 1970. The Pharmacology of Synapses. Pergamon Press, New York.

[45] Pitts, F. N., Jr., and C. Quick. 1965. J. Neurochem. 12:893.

[46] Pitts, F. N., Jr., and E. Robins. 1965. Fed. Proc. Fed. Amer. Soc. Exp. Biol. 24:350.

[47] Pitts, F. N., Jr., et al. 1965. J. Neurochem. 12:93.

[48] Roberts, E., and K. Kuriyama. 1968. Brain Res. 8:1.

[49] Salganicoff, L., and E. DeRobertis. 1965. J. Neurochem. 12:287.

[50] Salmoiraghi, G. C., and C. N. Stefanis. 1967. Int. Rev. Neurobiol. 10:1.

[51] Sisken, B., et al. 1961. J. Biol. Chem. 236:503.

[52] Srinivasan, V., et al. 1969. J. Neurochem. 16:1235.

[53] Storm-Mathisen, J., and F. Fonnum. 1971. Ibid. 18:1105.

[54] Susz, J. P., et al. 1966. Biochemistry 5:2870.

[55] Takeuchi, A., and N. Takeuchi. 1964. J. Physiol. (London) 170:296.

[56] Takeuchi, A., and N. Takeuchi. 1966. Ibid. 183:433.

[57] Takeuchi, A., and N. Takeuchi. 1967. Ibid. 191:575.

[58] Takeuchi, A., and N. Takeuchi. 1969. Ibid. 205:377.

[59] Uchizono, K. 1967. Nature (London) 214:833.

[60] Usherwood, P. N. R., and H. Grundfest. 1964. Science 143:817.

[61] Usherwood, P. N. R., and H. Grundfest. 1965. J. Neurophysiol. 28:497.

[62] Waksman, A., and E. Roberts. 1965. Biochemistry 4:2132.

[63] Watanabe, A., et al. 1968. J. Gen. Physiol. 52:908.

continued

For the biochemistry of dopamine,

PHYSI

Data are for the central nervous system of vertebrates. The action of dopamine is slow, and often inhibits more strongly than norepinephrine. Dopamine is frequently inhibitory in deep cortical cells, medulla, pons, hypothalamus, brain stem, dorsal column, and spinal interneurons. Dopamine terminals are smaller than norepinephrine terminals. [7,29]

	Site of Synaptic Junction	Presynaptic Accumulation of DA	Neurally Evoked Release of DA	DA Evoked Ionic Conductance Changes [1]		Inactivation of DA Externally Applied[2]
				Externally Applied	Neurally Released	
1	Small, intensely fluorescent cells of sympathetic ganglion onto sympathetic postganglionic cells	DA fluorescence; chemical analysis of ganglion [5]	....	Hyperpolarization; no change in conductance yet recorded [20]	....	Uptake [17]
2	Small, intensely fluorescent cells of parasympathetic ganglion onto parasympathetic postganglionic cells	Fluorescence [16]	....	Depressed rate of firing [16]	....	Uptake [16]
3	Substantia nigra onto caudate nucleus	Fluorescence [1]; biochemical analysis [33]	In-vivo stimulation of substantia nigra [30]; electrical stimulation of slices [26]	Spontaneous activity depressed [6, 9,14,25]	....
4	Thalamic centromedian nucleus onto caudate nucleus	Biochemical analysis [33]	Stimulation of centromedian nucleus [24]; electrical stimulation of slices [26]	Spontaneous & neurally evoked activity depressed [6,9,25,37]	....

[1] Undivided column indicates similar action of externally applied and neurally released substance. [2] Data for inactivation

Contributors: Wallace, B. G., Sargent, P. B., Mains, R. E., and Kravitz, E. A.

References

[1] Andén, N.-E., et al. 1966. Acta Physiol. Scand. 67: 306.
[2] Andén, N.-E., et al. 1966. Ibid. 67:313.
[3] Ascher, P. 1968. J. Physiol. (London) 198:48P.
[4] Berlind, A., et al. 1970. J. Exp. Biol. 53:669.
[5] Björklund, A., et al. 1970. Acta Physiol. Scand. 78: 334.
[6] Bloom, F. E., et al. 1965. J. Pharmacol. Exp. Ther. 150:244.
[7] Bradley, P. B. 1968. Int. Rev. Neurobiol. 11:1.
[8] Carpenter, D., et al. 1971. Int. J. Neurosci. 2:49.
[9] Connord, J. D. 1968. Science 160:899.
[10] Cooke, I. M., and M. W. Goldstone. 1970. J. Exp. Biol. 53:651.
[11] Dahl, E., et al. 1966. Z. Zellforsch. Mikrosk. Anat. 71:489.

[12] Dougan, D. F. H. 1968. Aust. J. Exp. Biol. Med. Sci. 46:P23.
[13] Elofsson, R., et al. 1966. Z. Zellforsch. Mikrosk. Anat. 74:464.
[14] Feltz, P. 1965. J. Physiol. (London) 205:8P.
[15] Feltz, P., and J. S. MacKenzie. 1969. Brain Res. 13:612.
[16] Jacobowitz, D. 1967. J. Pharmacol. Exp. Ther. 158:227.
[17] Jacobowitz, D. 1970. Fed. Proc. Fed. Amer. Soc. Exp. Biol. 29:1929.
[18] Kerkut, G. A., and N. Horn. 1968. Life Sci. 7(1): 567.
[19] Kerkut, G. A., et al. 1966. Comp. Biochem. Physiol. 18:921.

Dopamine

see Part III, Norepinephrine.

OLOGY

For information on dopamine in crustaceans, consult reference 13; in crabs, 4 and 10; in Bivalvia, 32, 34, and 38; and in snails, 3, 8, 11, 12, 18, 19, 28, 31, 32, and 35. *Abbreviation:* DA = dopamine.

Morphology		Pharmacology of DA [1]		Remarks	
Light Microscope	**Electron Microscope**	**Externally Applied**	**Neurally Released**		
Cells 8 μ in diameter [5,17, 27]; negative chromaffin reaction [36]; fluorescence [5]	Large, 900- to 1400-Å vesicles; clear vesicles in endings onto small, intensely fluorescent cells, which synapse onto ganglion cells [22,36]	Receptor: α-pharmacology [21] Reserpine depletes very slowly [27]; metaraminol not taken up [27]; norepinephrine mimics DA, but at higher concentrations [20]	....	DA-sensitive adenyl cyclase very active during ganglionic transmission [23]	1
Cells 8 μ in diameter; negative chromaffin reaction; fluorescence [16]	....	Receptor: α-pharmacology [16] Reserpine does not deplete [16]	....	....	2
Fluorescence [1]	....	Receptor: phenoxybenzamine blocks [25,37]	....	Substantia nigra projection monosynaptic [15]	3
Fluorescence [1]	....	Receptor: phenoxybenzamine blocks [25,37]		Corresponding prominent fluorescent pathway not observed [2,25,37]	4

of neurally released dopamine are not available.

[20] Kobayashi, H., and B. Libet. 1970. J. Physiol. (London) 208:353.

[21] Libet, B. 1970. Fed. Proc. Fed. Amer. Soc. Exp. Biol. 29:1945.

[22] Matus, A. I. 1970. Brain Res. 17:195.

[23] McAfee, D. A., et al. 1971. Science 171:1156.

[24] McLennan, H. 1964. J. Physiol. (London) 174: 152.

[25] McLennan, H., and D. H. York. 1967. Ibid. 189: 393.

[26] Ng, K. Y., et al. 1971. Science 172:487.

[27] Norberg, K. A., et al. 1966. Acta Physiol. Scand. 67:260.

[28] Osborne, N. N., and G. A. Cottrell. 1970. Comp. Gen. Pharmacol. 1:1.

[29] Phillis, J. W. 1970. The Pharmacology of Synapses. Pergamon Press, New York.

[30] Riddell, D., and J. C. Szerb. 1971. J. Neurochem. 18:989.

[31] Sakharov, D. A., and I. Zs.-Nagy. 1968. Acta Biol. (Budapest) 19:145.

[32] Salánki, J., ed. 1967. Neurobiology of Invertebrates. Plenum Press, New York.

[33] Snyder, S. H., et al. 1969. Brain Res. 16:469.

[34] Sweeney, D. 1963. Science 139:1051.

[35] Walker, R. J., et al. 1968. Comp. Biochem. Physiol. 24:455.

[36] Williams, T. H., and S. L. Palay. 1969. Brain Res. 15:17.

[37] York, D. H. 1967. Ibid. 5:263.

[38] Zs.-Nagy, I. 1967. Acta Biol. (Budapest) 18:1.

continued

Data are for the peripheral nervous system of invertebrates. L-Glutamate produces strong contractions of the opener muscle in *Limulus* [16] and of the hindgut in crayfish [10]. Central nervous system: Cells show specificity toward the L- form more than the D- form of glutamate; L-

	Subjects	Site of Synaptic Junction	Presynaptic Accumulation of Glutamic Acid	Neurally Evoked Release of Glutamic Acid	Ca²⁺ in Release	Glutamic Acid Evoked Ionic Conductance Changes Externally Applied & Neurally Released
1	Crusta-ceans	Excitatory axon onto skeletal muscle	Very high levels [12,18]	Not yet demon-strated [12]	Not yet demon-strated [12]	Equilibrium potential = 0-20 mV; does not involve Ca^{2+} [18]
2	Insects	Motor neuron onto skeletal muscle	Denervation [12]	Alanine, aspartate, glutamate, & glycine released [11,12,23]	Ca^{2+} increases frequency of miniature end-plate poten-tials [12,20]	Depolarization [2,21]

1/ Data for inactivation of neurally released glutamic acid are not available. 2/ Undivided column indicates similar action of

Contributors: Wallace, B. G., Sargent, P. B., Mains, R. E., and Kravitz, E. A.

References

[1] Atwood, H. L., et al. 1969. J. Insect Physiol. 15: 529.

[2] Beránek, R., and P. L. Miller. 1968. J. Exp. Biol. 49:83.

[3] Bradley, P. B. 1968. Int. Rev. Neurobiol. 11:1.

[4] Cohen, M. J. 1970. In F. O. Schmitt, ed. The Neu-rosciences: Second Study Program. Rockefeller Univ. Press, New York. pp. 798-812.

[5] Curtis, D. R., and J. M. Crawford. 1969. Annu. Rev. Pharmacol. 9:209.

[6] Faeder, I. R., et al. 1970. J. Exp. Zool. 173:187.

[7] Faeder, I. R., and M. M. Salpeter. 1970. J. Cell Biol. 46:300.

[8] Hoyle, G. 1955. Proc. Roy. Soc. B143:281.

[9] Iversen, L. L., and E. A. Kravitz. 1968. J. Neuro-chem. 15:609.

[10] Jones, H. C. 1962. J. Physiol. (London) 164:295.

[11] Kerkut, G. A., et al. 1965. Comp. Biochem. Physiol. 15:485.

[12] Kravitz, E. A., et al. 1970. Int. Meet. Neurobiol., 5th, Proc., p. 85.

Abbreviations and Symbols: DMPH₄ = 2-amino-4-hydroxy-6,7-dimethyltetrahydropteridine; K_I = dissociation constant of the enzyme-inhibitor complex; K_m = Michaelis constant, measured in moles per liter; SH = sulfhydryl; Tris = tris(hy-

tamic Acid

OLOGY

glutamate is almost universally excitatory, and is potentiated by L-aspartate. [3, 5, 13, 17] For information on the biochemistry of glutamic acid, a common cellular metabolite, *see* Part IV, γ-Aminobutyric Acid BIOCHEMISTRY.

Inactivation of Glutamic Acid Externally Applied [1]	Morphology		Pharmacology of Glutamic Acid [2]		Remarks	
	Light Microscope	Electron Microscope	Externally Applied	Neurally Released		
Na+-dependent uptake [9]	One excitatory nerve per fiber [19]	Postsynaptic thickenings; round vesicles [15,19]	Receptor: L-glutamate competes & desensitizes, and is at least 250 times more effective than D-glutamate; L-aspartate potentiates L-glutamate [12,18]		....	1
Neurilemma (Schwann cell) uptake [7]	Claw-shaped endings (?) [8]; procion yellow dye injection of motor neurons [4]	Postsynaptic thickenings; 160- to 200-Å cleft; 350- to 500-Å clear vesicles [1,14]	Receptor: L-glutamate desensitizes; D-glutamate has undetectable effects [2,22]	....	Removal of central nervous system from preparation is important [6]	2

externally applied and neurally released substance.

[13] Krnjévic, K., and J. W. Phillis. 1963. J. Physiol. (London) 165:274.

[14] Mandelstam, Yu. E. 1967. In J. Salánki, ed. Neurobiology of Invertebrates. Plenum Press, New York. pp. 49-57.

[15] Orkand, P. M., and E. A. Kravitz. 1971. J. Cell Biol. 49:75.

[16] Parnas, I., et al. 1968. Comp. Biochem. Physiol. 26: 467.

[17] Phillis, J. W. 1970. The Pharmacology of Synapses. Pergamon Press, New York.

[18] Takeuchi, A., and N. Takeuchi. 1964. J. Physiol. (London) 170:296.

[19] Uchizono, K. 1967. Nature (London) 214:833.

[20] Usherwood, P. N. R. 1963. J. Physiol. (London) 169:149.

[21] Usherwood, P. N. R., and P. Machili. 1966. Nature (London) 210:634.

[22] Usherwood, P. N. R., and P. Machili. 1968. J. Exp. Biol. 49:341.

[23] Usherwood, P. N. R., et al. 1968. Nature (London) 219:1169.

Serotonin

ISTRY

droxymethyl)aminomethane. For information on the biochemistry of DOPA decarboxylase (aromatic amino acid decarboxylase) and monoamine oxidase, *see* Part III, Norepinephrine BIOCHEMISTRY.

continued

Oxygen + Tryptophan + Reduced pteridine ———————— Tryptophan hy

5-Hydroxytryptophan ———————— DOPA decarboxyl

Serotonin + S-Adenosylmethionine ———————— Hydroxyindole O-methyl

Serotonin + Oxygen + Water ———————— Monoamine oxidase

	Enzyme	Tissue Source	Subcellular Localization	Cofactors + Cosubstrates	Ionic Requirements
1	Tryptophan hydroxylase	Cattle pineal body & rat brain stem	Soluble; concentrated in synaptosomes [30,36]	O_2 [31], reduced pteridine (tetra-hydrobiopterin) [36]	Optimum pH: 7.5 [36,40]; in crude preparation, Tris buffer gives 3 times greater activity & more alkaline optimum pH [31]. Fe^{2+} activates (due to H_2O_2 destruction?) [40].
2		Mouse mast cell	....	O_2, reduced pteridine [40]	Optimum pH: 6.8 [40]
3		Rat liver	....	O_2, reduced pteridine [40]	Optimum pH: 7.4 [40]
4	Hydroxyindole O-methyl-transferase	Vertebrate pineal body; eyes of birds, amphibians, & fishes; frog brain	Soluble [6]	S-Adenosylmethionine [5,6]	Optimum pH: 7.5-8.3; slight activation by SH [6]

1/ Unit = μmoles substrate reacted·min^{-1}·mg protein^{-1}.

PHYSI

Vertebrates: Serotonin has both excitatory and inhibitory actions in the cerebral cortex, cerebellar cortex, medulla, pons, olfactory bulb, brain stem, spinal motor neurons, and spinal interneurons [44]. Excitatory actions of serotonin are often antagonized by 2-bromolysergic acid diethylamide [10,11], and inhibitory actions of serotonin are often antagonized by lysergic acid diethylamide [10,48].

Invertebrates: Limulus—Serotonin slows heartbeat; 2-bromolysergic acid diethylamide blocks both neural and serotonin depression. [42] Crabs—Presence of serotonin in the pericardial organ is shown by fluorescence [8,14]. Leeches—Serotonin is present in Retzius cell bodies [34,

1192

Serotonin

droxylase $\longrightarrow$ 5-Hydroxytryptophan + Oxidized pteridine + Water

ase [4.1.1.26] $\longrightarrow$ 5-Hydroxytryptamine + Carbon dioxide
(serotonin)

transferase $\longrightarrow$ 5-Methoxytryptamine + S-Adenosylhomocysteine

[1.4.3.4.] $\longrightarrow$ 5-Hydroxyindoleacetic acid + Ammonia + Hydrogen
peroxide

Inhibitors	Apparent Michaelis Constant	Specific Activity unit [1/]	Other Characteristics	
p-Chloromercuribenzoate: K_I = 3×10^{-4} M [40]; p-chlorophenylalanine: $K_I = 3 \times 10^{-4}$ M [35], $(1.2\text{-}1.6) \times 10^{-5}$ M [41]; phenylalanine: $K_I = 3 \times 10^{-4}$ M [40]; 6-fluorotryptophan: $K_I = 8.6 \times 10^{-6}$ M [41]; catechols (reversed by 10^{-5} M Fe^{2+}) [40]; iron chelators: 10^{-4} M α,α'-dipyridyl & 10^{-4} M o-phenanthroline [36]	DMPH$_4$: 3×10^{-5} M [36,40]; tryptophan: 3×10^{-4} M [31,36, 40], 2×10^{-5} M [31]	9.3×10^{-5} [36]	88% of brain enzyme activity occurs in brain stem, and 12% in cerebral cortex; not detectable in cerebellum. [31] Brain stem enzyme is specific for L-tryptophan; pineal enzyme hydroxylates phenylalanine & L-tryptophan at equivalent rates. [31, 36] Tetrahydrobiopterin has lower K_m than DMPH$_4$, and gives a higher maximum velocity of reaction [36].	1
Phenylalanine: $K_I = 2 \times 10^{-4}$ M [40]	DMPH$_4$: 3×10^{-5} M; tryptophan: 4×10^{-5} M [40]	....	....	2
Phenylalanine: $K_I = 2 \times 10^{-4}$ M [40]	DMPH$_4$: 2×10^{-4} M; tryptophan: 2×10^{-3} M [40]	....	....	3
SH reagents [6]	S-Adenosylmethionine: 4.6×10^{-5} M; N-acetylserotonin: 5.4×10^{-5} M [6]	5.7×10^{-3} [6]	Methylates predominantly at position 5 [5]. N-Acetylserotonin best substrate; other 5-hydroxy-indoleamines & similar compounds methylated slowly. [6] Pineal transferase activity displays diurnal rhythm; controlled by light, with highest activity at midnight; rise blocked by puromycin. [7]	4

38,47], which sends processes to the periphery [46]. Clams—Serotonin-evoked stimulation in the rectum is blocked by methysergide [32]. Gastropods—Serotonin occurs in ganglia [12], and hyperpolarizes D and H cells; lysergic acid diethylamide, tryptamine, and serotonin block hyperpolarization. [25] In *Cryptomphallus,* serotonin stimulates CILDA neurons (cells having inhibition of long duration) [26,27]; lysergic acid diethylamide, tryptamine, and serotonin block slow excitatory postsynaptic potential and serotonin potential [27]. For a review of serotonin content in invertebrates, consult reference 55. *Abbreviation:* ? = specific site unknown.

continued

Subjects	Site of Synaptic Junction	Presynaptic Accumulation of Serotonin	Neurally Evoked Release of Serotonin	Serotonin Evoked Ionic Conductance Changes Externally Applied & Neurally Released	Inactivation of Serotonin Externally Applied [1/]
		Central Nervous System			
1 Vertebrates	Raphé nucleus in midbrain onto cerebral cortex	Fluorescence [4]; biochemical analysis [44]	Indirect evidence [2,53]	Spontaneous activity of cortical cells depressed [44,45]	Uptake [1]
2	Bulbospinal fibers onto preganglionic sympathetic neurons in lateral horn	Fluorescence [21]	From spinal cord with massive stimulation of brain stem [3]	Spontaneous activity increased [33]	....
3	? onto primary cells of lateral geniculate nucleus	Fluorescence [24]	....	Neurally evoked activity depressed [20,48,49]	....
4 Gastropods	Giant cell of metacerebral ganglion onto buccal ganglion cell	In cell body [19]	....	Buccal cell depolarized [16]	....
		Peripheral Nervous System			
5 Vertebrates	Vagal preganglionic fibers onto inhibitory portion of the myenteric plexus in the stomach	Fluorescence [11]; make from 5-hydroxytryptophan [28]	By stimulation of stomach [3/]; blocked by tetrodotoxin [11]	Increased firing rate of intrinsic ganglia in the gut [11]	Uptake [29] and/or degradation [11]
6 Mollusks	Extracardial nerve onto heart	....	Yes [13,50]	Increased frequency & amplitude of beat [39,43,50,54]	Uptake [12,52]

[1/] Data for inactivation of neurally released serotonin are not available.   [2/] Undivided column indicates similar action of ex

Contributors: Wallace, B. G., Sargent, P. B., Mains. R. E., and Kravitz, E. A.

References

[1] Aghajanian, G. K., and I. M. Asher. 1971. Science 172:1159.

[2] Aghajanian, G. K., et al. 1967. Ibid. 156:402.

[3] Andén, N.-E., et al. 1964. Life Sci. 3:473.

[4] Andén, N.-E., et al. 1966. Acta Physiol. Scand. 67:313.

[5] Axelrod, J. 1966. Pharmacol. Rev. 18:95.

[6] Axelrod, J., and H. Weissback. 1961. J. Biol. Chem. 236:211.

[7] Axelrod, J., et al. 1965. Ibid. 240:949.

[8] Berlind, A., et al. 1970. J. Exp. Biol. 53:669.

[9] Berry, C. F., and G. A. Cottrell. 1970. Z. Zellforsch. Mikrosk. Anat. 104:107.

[10] Bradley, P. B. 1968. Int. Rev. Neurobiol. 11:1.

[11] Bülbring, E., and M. D. Gershon. 1967. J. Physiol. (London) 192:823.

[12] Carpenter, D., et al. 1971. Int. J. Neurosci. 2:49.

[13] Chase, T. N., et al. 1968. Advan. Pharmacol. 6A:351.

[14] Cooke, I. M., and M. W. Goldstone. 1970. J. Exp. Biol. 53:651.

[15] Cottrell, G. A. 1970. J. Physiol. (London) 208:28P.

[16] Cottrell, G. A. 1970. Nature (London) 225:1060.

[17] Cottrell, G. A., and N. N. Osborne. 1969. Comp. Biochem. Physiol. 28:1455.

[18] Cottrell, G. A., and N. N. Osborne. 1969. Experientia, Suppl. 15:220.

[19] Cottrell, G. A., and N. N. Osborne. 1970. Nature (London) 225:470.

[20] Curtis, D. R., and R. Davis. 1962. Brit. J. Pharmacol. 18:217.

Serotonin

Morphology		Pharmacology of Serotonin [2/]		Remarks	
Light Microscope	Electron Microscope	Externally Applied	Neurally Released		
Central Nervous System					
Fluorescence [1, 4]	....	Receptor: cocaine & strychnine block [44,45]	....	....	1
Fluorescence [21]	....	....	....	Norepinephrine-depressed cells excited by serotonin [23,33]	2
Fluorescence [24]	....	....	....	Serotonin content of lateral geniculate nucleus not altered by visual deafferentiation [22]	3
....	....	....	Receptor: lysergic acid diethylamide & methysergide block [15,16]	Size of excitatory postsynaptic potential reduced after reserpine administration [15]	4
Peripheral Nervous System					
Fluorescence [11]	Autoradiography of serotonin endings; postsynaptic thickenings; dense-core vesicles [29]	Desensitization by serotonin & 5-hydroxytryptophan, and by monoamine oxidase inhibition [11]. Receptor: 2-bromolysergic acid diethylamide blocks [11]; insensitive to α, β blockade [11].	....		5
Serotonin fluorescence [17, 18]	In cephalopods, 1000-Å dense-core vesicles [9]; in *Aplysia*, 300- to 600-Å clear vesicles [52]	Receptor: 2-bromolysergic acid diethylamide & methysergide block [39,50,51]	Serotonin occurs in snail heart [12,37, 54]. Experiment with two hearts [50].		6

ternally applied and neurally released substance. [3/] Ca^{2+} required in release [11].

[21] Dahlström, A., and K. Fuxe. 1965. Acta Physiol. Scand., Suppl. 247:7.
[22] Defenu, G., et al. 1967. Exp. Neurol. 17:203.
[23] Engberg, I., and R. W. Ryall. 1966. J. Physiol. (London) 185:298.
[24] Fuxe, K. 1965. Acta Physiol. Scand., Suppl. 247: 37.
[25] Gerschenfeld, H. M. 1971. Science 171:1252.
[26] Gerschenfeld, H. M., and E. Stefani. 1966. J. Physiol. (London) 185:684.
[27] Gerschenfeld, H. M., and E. Stefani. 1968. Advan. Pharmacol. 6B:369.
[28] Gershon, M. D., and L. L. Ross. 1966. J. Physiol. (London) 186:451.
[29] Gershon, M. D., and L. L. Ross. 1966. Ibid. 186: 477.

[30] Grahame-Smith, D. G. 1968. Advan. Pharmacol. 6A:37.
[31] Green, H., and J. L. Sawyer. 1966. Anal. Biochem. 15:53.
[32] Greenberg, M. J., and T. C. Jegla. 1963. Comp. Biochem. Physiol. 9:275.
[33] Groat, W. C. de, and R. W. Ryall. 1967. Exp. Brain Res. 3(4):299.
[34] Hildebrand, J. G., et al. 1971. J. Neurobiol. 2:231.
[35] Jéquier, E., et al. 1967. Mol. Pharmacol. 3:274.
[36] Jéquier, E., et al. 1969. Biochem. Pharmacol. 18: 1071.
[37] Kerkut, G. A., and G. A. Cottrell. 1963. Comp. Biochem. Physiol. 8:53.
[38] Kerkut, G. A., et al. 1967. Ibid. 21:687.
[39] Loveland, R. E. 1963. Ibid. 9:95.

continued

Part VII. Serotonin

[40] Lovenberg, W., et al. 1968. Advan. Pharmacol. 6A: 21.

[41] McGeer, E. G., et al. 1968. Life Sci. 7(2):605.

[42] Pax, R. A., and R. C. Sanborn. 1967. Biol. Bull. 132:392.

[43] Phillis, J. W. 1966. Comp. Biochem. Physiol. 17:719.

[44] Phillis, J. W. 1970. The Pharmacology of Synapses. Pergamon Press, New York.

[45] Phillis, J. W., and D. H. York. 1967. Nature (London) 216:922.

[46] Retzius, G. 1891. Biol. Untersuch. (Stockholm) 2: 13.

[47] Rude, S., et al. 1969. J. Cell Biol. 41:832.

[48] Salmoiraghi, G. C., and C. N. Stefanis. 1967. Int. Rev. Neurobiol. 10:1.

[49] Satinsky, D. 1967. Int. J. Neuropharmacol. 6:387.

[50] S.-Rózsa, K., and L. Perenyi. 1966. Comp. Biochem. Physiol. 19:105.

[51] S.-Rózsa, K., and I. Zs.-Nagy. 1967. Ibid. 23:373.

[52] Taxi, J., and J. Gautron. 1969. J. Microsc. (Paris) 8:627.

[53] Weiss, B. L., and G. K. Aghajanian. 1971. Brain Res. 26:37.

[54] Welsh, J. H. 1957. Ann. N.Y. Acad. Sci. 66:618.

[55] Welsh, J. H., and M. Moorhead. 1960. J. Neurochem. 6:146.

Part VIII. Glycine

PHYSIOLOGY

Data are for the central nervous system of vertebrates. The concentration of glycine, a natural constituent of the cell, is four times greater in the medulla than anywhere else in the brain [2]. When interneurons are destroyed, glycine concentration decreases; γ-aminobutyric acid remains unchanged [5]. Glycine inhibits spontaneous or induced firing of almost any central nervous system cell in a manner unlike that of general anesthesia [3]. *Abbreviation:* ? = specific site unknown.

	Site of Synaptic Junction	Neurally Evoked Release of Glycine	Glycine Evoked Ionic Conductance Changes Externally Applied & Neurally Released	Inactivation of Glycine Externally Applied [1/]	Pharmacology of Glycine [2/]		Remarks
					Externally Applied	Neurally Released	
1	? onto spinal motor neurons	Increased two to six fold after stimulation with 40 mM K^+ [2]	Increased conductance with equilibrium potential near resting potential [5]	Na^+-dependent pump [2]	Receptor: strychnine blocks [1]		....
2	Distal inhibitory endings onto Mauthner's cell in goldfish	....	Increased conductance with equilibrium potential near resting potential [4]	....	Receptor: strychnine blocks slowly and only at high concentrations [4]	Receptor: strychnine blocks readily [4]	Pharmacology suggests that strychnine is not specific for glycine [4]

[1/] Data for inactivation of neurally released glycine are not available. [2/] Undivided column indicates similar action of externally applied and neurally released substance.

Contributors: Wallace, B. G., Sargent, P. B., Mains, R. E., and Kravitz, E. A.

References

[1] Curtis, D. R., et al. 1968. Exp. Brain Res. 5:235.

[2] Neal, M. J. 1971. J. Physiol. (London) 215:103.

[3] Phillis, J. W. 1970. The Pharmacology of Synapses. Pergamon Press, New York.

[4] Roper, S., et al. 1969. Nature (London) 223:1168.

[5] Werman, R., et al. 1968. J. Neurophysiol. 31:81.

142. DISTRIBUTION OF TRANSMITTER COMPOUNDS AND ENZYME SYSTEMS IN AREAS OF THE CENTRAL NERVOUS SYSTEM

To be identified as a synaptic transmitter compound, a substance must meet the following criteria: (i) the substance must be present in those neurons from which it is released; (ii) the neuron must possess the necessary enzymic mechanisms for the synthesis of the transmitter and for its release; (iii) the presence of the various precursors and intermediaries in the synthetic pathway should be demonstrable; (iv) there must be systems for the inactivation of the transmitter (these could include an enzyme system for the inactivation of the transmitter and/or a specific uptake mechanism for the reabsorption of the transmitter into the pre- or post-synaptic structures); (v) during stimulation, the substance may be detectable in extracellular fluid collected from the region of the activated synapses; (vi) when applied to the postsynaptic structure, the substance should mimic the action of the synaptically released transmitter; and (vii) pharmacological agents which interact with the synaptically released transmitter should interact with the suspected transmitter in an identical manner. For additional information, consult reference 12 in Part I.

Part I. Tissue Content: Laboratory Mammals and Pigeon

Plus/minus (±) values are standard errors.

	Tissue	Animal	Value
	Acetylcholine Content, μg/g fresh tissue [3,4,10]		
1	Cerebral cortex	Cat	2.2-4.5
2	Basal ganglia	Cat	7
3	Olfactory bulb	Cat	1.3
4	Thalamus	Cat	2-4
5	Lateral geniculate body (nucleus)	Cat	3.3
6	Hypothalamus	Cat	1.6-2.1
7	Superior colliculus	Cat	4.5
8	Cerebellar cortex	Cat	0.1-0.3
9	Pons	Cat	1.4-5.0
10	Medulla oblongata	Cat	1-2.7
	Spinal cord		
11	Dorsal horn	Dog	2.5
12	Ventral horn	Dog	1.5
13	Dorsal roots	Cat; dog	0-0.25
14	Ventral roots	Cat; dog	9-18
15	Superior cervical ganglion	Cat	18-44
16	Sciatic nerve	Cat	4-6
	Dopamine Content, μg/g tissue [1,2,8,9,11,13]		
17	Cerebral cortex	Cat	0.07
18		Dog	0.01
19	Basal ganglia	Cat	8.00
20		Dog	5.90
21	Hippocampus	Cat	0.70
22	Thalamus	Cat	0.05-0.50
23		Dog	0.05
24	Hypothalamus	Cat	0.75
25		Dog	0.26
26	Midbrain	Cat	0.53
27	Inferior & superior colliculi	Cat	1.59
28	Cerebellum	Cat	0.02
29		Dog	0.03
30	Pons	Cat	0.11
31		Dog	0.10
32	Medulla oblongata	Cat	0.08
33		Dog	0.13

	Tissue	Animal	Value
34	Spinal cord	Cat	0-0.45
	Norepinephrine (Noradrenaline) Content, μg/g tissue [1,2,8,9,13]		
35	Cerebral cortex	Cat	0.11-0.28
36		Dog	0.12
37	Basal ganglia	Cat	0.22
38		Dog	0.10
39	Hippocampus	Cat	0.17
40		Dog	0.14
41	Olfactory bulb	Dog	0.05
42	Thalamus	Cat	0.22
43		Dog	0.08
44	Lateral geniculate body (nucleus)	Dog	0.07
45	Medial geniculate body (nucleus)	Dog	0.13
46	Hypothalamus	Cat	2.05
47		Dog	1.03
48	Midbrain	Cat	0.50
49	Inferior & superior colliculi	Cat	0.40
50	Inferior colliculus	Dog	0.11
51	Superior colliculus	Dog	0.16
52	Cerebellum	Cat	0.13
53		Dog	0.06
54	Pons	Cat	0.20-0.52
55		Dog	0.20
56	Medulla oblongata	Cat	0.39
57		Dog	0.37
58	Spinal cord	Cat	0.34
59		Dog	0.18
	Acetylcholinesterase Activity, μmoles acetylthiocholine hydrolyzed·g dry wt^{-1}·hr^{-1} [5]		
	Cerebellum		
	Cortex		
60	Molecular layer	Cat	2330 ± 288
61		Guinea pig	3757 ± 285
62		Rabbit	2087 ± 216

continued

142. DISTRIBUTION OF TRANSMITTER COMPOUNDS AND ENZYME SYSTEMS IN AREAS OF THE CENTRAL NERVOUS SYSTEM

Part I. Tissue Content: Laboratory Mammals and Pigeon

	Tissue	Animal	Value		Tissue	Animal	Value
63		Rat	1000 ± 123		Cerebellum		
64		Pigeon	4293 ± 316	88	Cortex	Cat	50[2/]
65	Granular layer	Cat	3694 ± 282	89	Molecular layer	Cat	0.71 ± 0.19
66		Guinea pig	2390 ± 131	90		Guinea pig	0.62 ± 0.15
67		Rabbit	3702 ± 643	91		Rabbit	<0.1
68		Rat	1472 ± 89	92		Rat	0.98 ± 0.27
69		Pigeon	2399 ± 190	93		Pigeon	1.55 ± 0.30
70	White matter	Cat	1193 ± 211	94	Granular layer	Cat	1.23 ± 0.18
71		Guinea pig	1192 ± 69	95		Guinea pig	1.60 ± 0.22
72		Rabbit	1270 ± 126	96		Rabbit	1.52 ± 0.14
73		Rat	1620 ± 223	97		Rat	1.92 ± 0.37
74		Pigeon	775 ± 132	98		Pigeon	1.36 ± 0.19
75	Deep nuclei	Cat	1790 ± 276	99	White matter	Cat	0.44 ± 0.10
76		Guinea pig	1354 ± 189	100		Guinea pig	2.60 ± 0.08
77		Rat	1525 ± 232	101		Rabbit	1.23 ± 0.10
78		Pigeon	610 ± 58	102		Rat	2.75 ± 0.48
				103		Pigeon	1.38 ± 0.22
	Choline Acetyltransferase Activity, μg acetylcholine formed·g dry wt^{-1}·hr^{-1} [1/] [5 [1/]]			104	Deep nuclei	Cat	3.78 ± 1.08
				105		Guinea pig	10.6 ± 1.5
79	Cerebral cortex	Cat	175-270[2/]	106		Rat	6.90 ± 0.51
80	Basal ganglia	Cat	1400-2200[2/]	107		Pigeon	10.5 ± 1.5
81	Hippocampus	Dog	520[2/]		Spinal cord		
82	Olfactory bulb	Cat	440[2/]	108	Ventral horn	Goat	500[2/]
83	Thalamus	Cat	340[2/]	109	Dorsal roots	Cat; dog	0-6[2/]
84	Lateral geniculate body (nucleus)	Dog	520[2/]	110	Ventral roots	Dog	5000[2/]
				111		Goat	2500[2/]
85	Hypothalamus	Cat	50[2/]	112		Rabbit	4000[2/]
86		Dog	400[2/]	113	Superior cervical ganglion	Sheep	1600-2000[2/]
87	Superior colliculus	Cat	400[2/]	114	Sciatic nerve	Goat	300-680[2/]
				115		Rabbit	1120[2/]

[1/] Unless otherwise indicated. [2/] μg acetylcholine formed·g fresh tissue^{-1}·hr^{-1}; where necessary values for fresh tissue have been calculated on the assumption that 1 g of acetone powder is equivalent to 5 g of fresh tissue. [6,7]

Contributor: Grenell, Robert G.

References

[1] Bertler, Å., and E. Rosengren. 1959. Acta Physiol. Scand. 47:350.

[2] Carlsson, A. 1959. Pharmacol. Rev. 11:490.

[3] Deffenu, G., et al. 1967. Exp. Neurol. 17:203.

[4] Feldberg, W. 1945. Physiol. Rev. 25:596.

[5] Goldberg, A. M., and R. E. McCaman. 1967. Life Sci. 6:1493.

[6] Hebb, C. O. 1963. Handb. Exp. Pharmakol. 15:55.

[7] Hebb, C. O., and A. Silver. 1956. J. Physiol. (London) 134:718.

[8] Holzbauer, M., and M. Vogt. 1956. J. Neurochem. 1:8.

[9] Laverty, R., and D. F. Sharmon. 1965. Brit. J. Pharmacol. Chemother. 24:538.

[10] MacIntosh, F. C. 1941. J. Physiol. (London) 99:436.

[11] McGeer, P. L., et al. 1963. Arch. Neurol. (Chicago) 9:81.

[12] Phillis, J. W. 1970. The Pharmacology of Synapses. Pergamon Press, New York.

[13] Vogt, M. 1954. J. Physiol. (London) 123:451.

continued

Part II. Main Monoamine Neuron Projection Systems: Rat

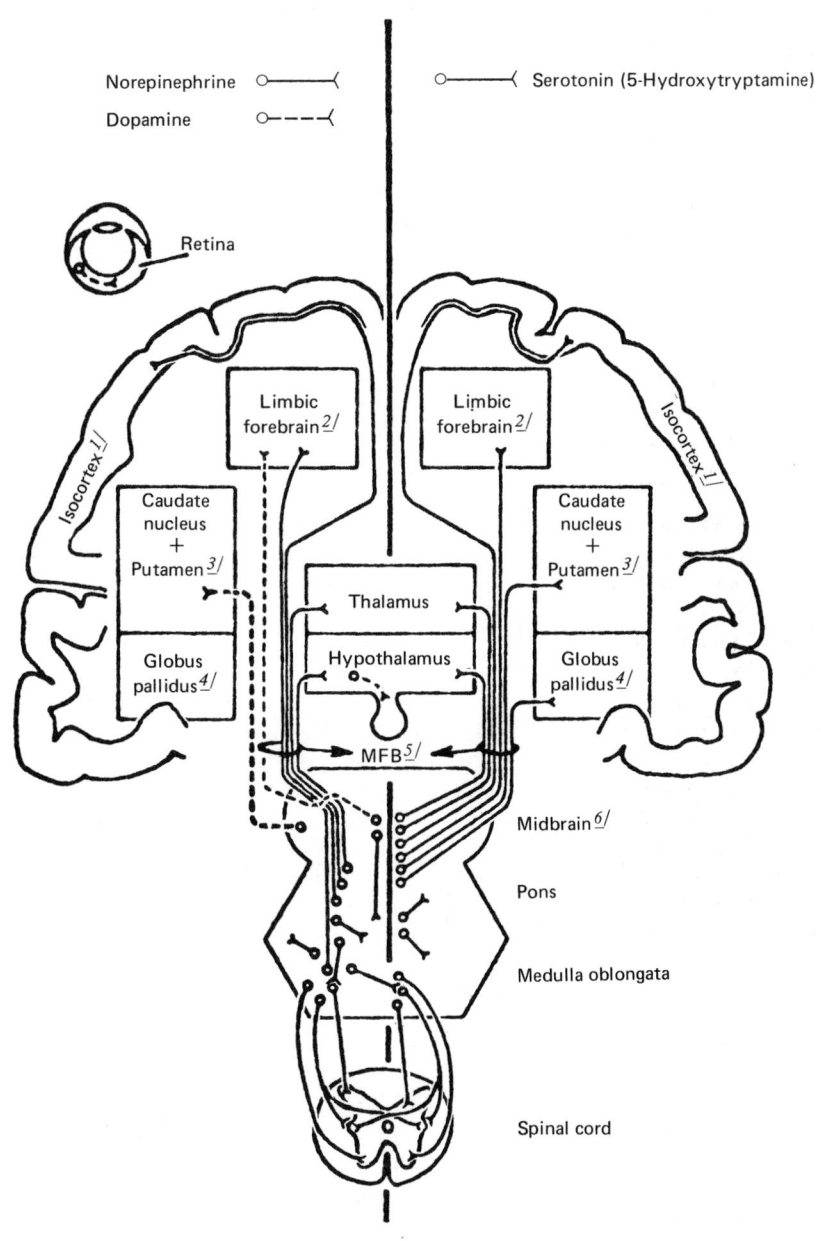

[1] Also known as the neopallium or neocortex. [2] Includes the rhinencephalon and its structures, the amygdaloid body, cingulate gyrus, fornix, and hippocampus. [3] Together, have been called the neostriatum. [4] Has been called the paleostriatum. [5] Medial forebrain bundle. [6] Also known as the mesencephalon.

Contributor: Grenell, Robert G.

Reference: Andén, N.-E., et al. 1966. Acta Physiol. Scand. 67:313.

Part I. Iontophoretically Applied Chemicals on Neurons

Chemical: GABA = γ-aminobutyric acid. **Effect:** 0 = none; − = depression; + = excitation. For additional information, consult reference 48.

	Site of Application (Synonym)	Chemical	Effect	Reference
	Spinal cord			
1	Motor neuron	Acetylcholine	0	27
2		GABA & glycine	−	24,25,72
3		Glutamate	+	24,26
4		Histamine	−	58
5		Norepinephrine	−	29,58,71
6		Serotonin	−	29,58
7	Interneuron	Acetylcholine	+;−	28,61,69
8		Dopamine	−	6
9		GABA & glycine	−	24,25,72
10		Glutamate	+	24,26
11		Histamine	−	27,58
12		Norepinephrine	+;−	6,29,69
13		Serotonin	+;−	29,58,61,69
14	Renshaw cell	Acetylcholine	+;−	19,21-23,27
15		GABA & glycine	−	24,25
16		Glutamate	+	24,26
17		Norepinephrine	+;−	4,29,61,70
18		Serotonin	−	29
19	Sympathetic pre-	Acetylcholine	0	31
20	ganglionic fibers	Norepinephrine	−	31
21		Serotonin	+	31
22	Medulla oblongata	Acetylcholine	+;−	9-11,62
23		GABA & glycine	−	10,20
24		Glutamate	+	10,20
25		Norepinephrine	+;−	9,10
26		Prostaglandins	+;−	2
27	Cochlear nucleus	Acetylcholine	+	13
28	Nucleus cuneatus	Acetylcholine	+;−	30,45,66
29	(Cuneate nucle-	Dopamine	−	30,45,66
30	us) & nucleus	GABA & glycine	−	30
31	gracilis (gracile	Glutamate	+	30,66
32	nucleus)	Histamine	−	30
33		Norepinephrine	−	30
34		Serotonin	−	30
35	Pons	Acetylcholine	+;−	10,20
36		GABA & glycine	−	10,20
37		Glutamate	+	10,20
38		Norepinephrine	+;−	9,10
39		Prostaglandins	+;−	2
40		Serotonin	+;−	10
41	Lateral vestibular	Acetylcholine	+	73
42	nucleus	GABA & glycine	−	46
43		Norepinephrine	+	73

	Site of Application (Synonym)	Chemical	Effect	Reference
	Cerebellum			
44	Cortex	Acetylcholine	+	16,39,40
45		GABA & glycine	−	34
46		Glutamate	+	34
47		Norepinephrine	−	47
48		Serotonin	−	47
49	Deep nuclei	Acetylcholine	+	12
50	Inferior colliculus	Acetylcholine	+	20
51		GABA & glycine	−	20
52		Glutamate	+	20
53	Thalamus	Acetylcholine [1]	+;−	1,41,50
54		Dopamine	−	1,49
55		Glutamate	+	1,49
56		Norepinephrine	+;−	49,50
57		Serotonin	+;−	1,49
58	Lateral geniculate	Acetylcholine	+;−	18,56,64
59	body	Dopamine	−	17,55
60		GABA & glycine	−	18
61		Glutamate	+	18,55
62		Histamine	−	17
63		Norepinephrine	+;−	17,55,64
64		Serotonin	+;−	17,55,64
65	Medial geniculate	Acetylcholine	+;−	67,68
66	body	GABA & glycine	−	67,68
67		Glutamate	+	67,68
68		Norepinephrine	+;−	67,68
69		Serotonin	+;−	67,68
70	Hypothalamus	Acetylcholine	+;−	7
71		Dopamine	+;−	7
72		Norepinephrine	+;−	7
73	Cerebral cortex	Acetylcholine	+;−	15,36,52-54
74		Dopamine	−	15,37
75		GABA & glycine	−	14,35
76		Glutamate	+	14,35
77		Histamine	+;−	37,57
78		Norepinephrine	+;−	37,51,60
79		Prostaglandins	+;−	58
80		Serotonin	+;−	15,37,59
81	Caudate nucleus	Acetylcholine	+;−	8,43
82		Dopamine	+;−	6,33,44
83		GABA & glycine	−	8
84		Glutamate	+	8
85		Norepinephrine	+;−	8

[1] Relative activities of cholinomimetic substances as excitants of thalamic neurons relative to that of acetylcholine (as +++)—carbachol (carbamyl-choline chloride): ++++; acetylcholine chloride and methacholine chloride (acetyl-β-methylcholine chloride): +++; arecoline hydrobromide, d-1-muscarine iodide, nicotine bitartrate (nicotine hydrogen tartrate), and succinylcholine chloride: ++; butyrylcholine iodide, pilocarpine hydrochloride, and propionylcholine chloride: +; and d-1-muscarone iodide: 0. [42]

continued

143. VARIOUS CHEMICALS AFFECTING THE CENTRAL NERVOUS SYSTEM: CAT

Part I. Iontophoretically Applied Chemicals on Neurons

	Site of Application (Synonym)	Chemical	Effect	Reference		Site of Application (Synonym)	Chemical	Effect	Reference
86	Globus pallidus &	Acetylcholine	+;−	74	96		Serotonin	+;−	5,32,65
87	putamen	Dopamine	+;−	74	97	Olfactory bulb	Acetylcholine	+;−	3,63
88		Glutamate	+	74	98		Norepinephrine	−	4,57
89		Norepinephrine	+;−	74	99		Serotonin	−	4,57
90		Serotonin	+;−	74	100	Piriform cortex	Acetylcholine	+;−	38
91	Hippocampus	Acetylcholine	+	5,32,65	101	(Pyriform cortex)	Dopamine	−	38
92		Dopamine	−	5	102		GABA & glycine	−	38
93		GABA & glycine	−	5,32,65	103		Glutamate	+	14,35
94		Glutamate	+	5,32,65	104		Norepinephrine	−	38
95		Norepinephrine	−	5,32,65	105		Serotonin	−	38

Contributor: Grenell, Robert G.

References

[1] Andersen, P., and D. R. Curtis. 1964. Acta Physiol. Scand. 61:85.

[2] Avanzino, G. L., et al. 1966. Brit. J. Pharmacol. Chemother. 27:157.

[3] Baumgarten, R. von, et al. 1963. Pfluegers Arch. Gesamte Physiol. Menschen Tiere 277:125.

[4] Biscoe, T. J., and D. R. Curtis. 1966. Science 151: 1230.

[5] Biscoe, T. J., and D. W. Straughan. 1966. J. Physiol. (London) 183:341.

[6] Biscoe, T. J., et al. 1966. Int. J. Neuropharmacol. 5:429.

[7] Bloom, F. E., et al. 1963. Ibid. 2:181.

[8] Bloom, F. E., et al. 1965. J. Pharmacol. Exp. Ther. 150:244.

[9] Bradley, P. B., and J. H. Wolstencroft. 1962. Nature (London) 196:840.

[10] Bradley, P. B., and J. H. Wolstencroft. 1965. Brit. Med. Bull. 21:15.

[11] Bradley, P. B., et al. 1966. J. Physiol. (London) 183: 658.

[12] Chapman, J. B., and I. McCance. 1967. Brain Res. 5:535.

[13] Comis, S. D., and I. C. Whitfield. 1966. J. Physiol. (London) 183:22.

[14] Crawford, J. M., and D. R. Curtis. 1964. Brit. J. Pharmacol. Chemother. 23:313.

[15] Crawford, J. M., and D. R. Curtis. 1966. J. Physiol. (London) 186:121.

[16] Crawford, J. M., et al. 1966. Ibid. 186:139.

[17] Curtis, D. R., and R. Davis. 1962. Brit. J. Pharmacol. Chemother. 18:217.

[18] Curtis, D. R., and R. Davis. 1963. J. Physiol. (London) 165:62.

[19] Curtis, D. R., and R. M. Eccles. 1958. Ibid. 141:435.

[20] Curtis, D. R., and K. Koizumi. 1961. J. Neurophysiol. 24:80.

[21] Curtis, D. R., and R. W. Ryall. 1966. Exp. Brain Res. (Berlin) 2:49.

[22] Curtis, D. R., and R. W. Ryall. 1966. Ibid. 2:66.

[23] Curtis, D. R., and R. W. Ryall. 1966. Ibid. 2:81.

[24] Curtis, D. R., and J. C. Watkins. 1960. J. Neurochem. 6:117.

[25] Curtis, D. R., et al. 1959. J. Physiol. (London) 146: 185.

[26] Curtis, D. R., et al. 1960. Ibid. 150:656.

[27] Curtis, D. R., et al. 1961. Ibid. 158:296.

[28] Curtis, D. R., et al. 1966. Exp. Brain Res. (Berlin) 2:97.

[29] Engberg, I., and R. W. Ryall. 1966. J. Physiol. (London) 185:298.

[30] Galindo, A., et al. 1967. Ibid. 192:359.

[31] Groat, W. C. de, and R. W. Ryall. 1967. Exp. Brain Res. (Berlin) 3:299.

[32] Herz, A., and A. C. Nacimiento. 1965. Naunyn Schmiedebergs Arch. Exp. Pathol. Pharmakol. 251: 295.

[33] Herz, A., and W. Ziegelsgänsberger. 1966. Experientia 22:839.

[34] Krnjević, K., and J. W. Phillis. 1961. J. Physiol. (London) 159:62.

[35] Krnjević, K., and J. W. Phillis. 1963. Ibid. 165:274.

[36] Krnjević, K., and J. W. Phillis. 1963. Ibid. 166:296.

[37] Krnjević, K., and J. W. Phillis. 1963. Brit. J. Pharmacol. Chemother. 20:471.

[38] Legge, K. F., et al. 1966. Ibid. 26:87.

[39] McCance, I., and J. W. Phillis. 1965. Experientia 21:108.

[40] McCance, I., and J. W. Phillis. 1968. Int. J. Neuropharmacol. 7:447.

continued

Part I. Iontophoretically Applied Chemicals on Neurons

[41] McCance, I., et al. 1968. Brit. J. Pharmacol. Chemother. 32:635.

[42] McCance, I., et al. 1968. Ibid. 32:652.

[43] McLennan, H., and D. H. York. 1966. J. Physiol. (London) 187:163.

[44] McLennan, H., and D. H. York. 1967. Ibid. 189:393.

[45] Meyer, M. 1965. Helv. Physiol. Pharmacol. Acta 23:325.

[46] Obata, K., et al. 1967. Exp. Brain Res. (Berlin) 4:43.

[47] Phillis, J. W. 1965. Experientia 21:266.

[48] Phillis, J. W. 1970. The Pharmacology of Synapses. Pergamon Press, New York. pp. 149-185.

[49] Phillis, J. W., and A. K. Tebecis. 1967. J. Physiol. (London) 192:715.

[50] Phillis, J. W., and A. K. Tebecis. 1967. Life Sci. 6:1621.

[51] Phillis, J. W., and D. H. York. 1967. Nature (London) 216:922.

[52] Phillis, J. W., and D. H. York. 1967. Brain Res. 5:517.

[53] Phillis, J. W., and D. H. York. 1968. Ibid. 10:297.

[54] Phillis, J. W., and D. H. York. 1968. Life Sci. 7:65.

[55] Phillis, J. W., et al. 1967. J. Physiol. (London) 190:563.

[56] Phillis, J. W., et al. 1967. Ibid. 192:695.

[57] Phillis, J. W., et al. 1968. Brit. J. Pharmacol. Chemother. 33:426.

[58] Phillis, J. W., et al. Unpublished. Monash Univ., Clayton, Victoria, Australia, 1970.

[59] Roberts, M. H. T., and D. W. Straughan. 1967. J. Physiol. (London) 193:269.

[60] Roberts, M. H. T., and D. W. Straughan. 1968. Naunyn Schmiedebergs Arch. Exp. Pathol. Pharmakol. 259:191.

[61] Salmoiraghi, G. C., and C. Stefanis. 1965. Arch. Ital. Biol. 103:705.

[62] Salmoiraghi, G. C., and F. A. Steiner. 1963. J. Neurophysiol. 26:581.

[63] Salmoiraghi, G. C., et al. 1964. Amer. J. Physiol. 207:1417.

[64] Satinsky, D. 1967. Int. J. Neuropharmacol. 6:387.

[65] Stefanis, C. 1964. Pharmacologist 6:171.

[66] Steiner, F. A., and M. Meyer. 1966. Experientia 22:58.

[67] Tebecis, A. K. 1967. Brain Res. 6:780.

[68] Tebecis, A. K. 1968. Aust. J. Exp. Biol. Med. Sci. 46:3P.

[69] Weight, F. F., and G. C. Salmoiraghi. 1966. J. Pharmacol. Exp. Ther. 153:420.

[70] Weight, F. F., and G. C. Salmoiraghi. 1966. Ibid. 154:391.

[71] Weight, F. F., and G. C. Salmoiraghi. 1967. Nature (London) 213:1229.

[72] Werman, R., et al. 1968. J. Neurophysiol. 31:81.

[73] Yamamoto, C. 1967. J. Pharmacol. Exp. Ther. 156:39.

[74] York, D. H. 1968. Aust. J. Exp. Biol. Med. Sci. 46:3P.

Part II. Postganglionic Responses of Sympathetic Ganglia to Drugs

Response: When recorded extracellularly between an electrode on the ganglion surface and another on the distal end of the isolated postganglionic trunk, the sequence is composed of (i) an initial negative wave, N, of about 100-ms duration; followed by (ii) a slight positive wave, P; that is terminated by (iii) a late negative wave, LN, of more than 2-s duration. The P potential can be considered an inhibitory postsynaptic potential. The LN wave represents a depolarizing, slow, excitatory, postsynaptic potential.

	Drug [Synonym]	Response	Effect
1	N-(2-Chloroethyl)dibenzamine [Dibenamine]	Synaptic transmission & N wave	None [1]
2		P wave or potential	Marked depression
3		LN wave or potential	Depression
4	Epinephrine [Adrenaline]	Synaptic transmission & N wave	Depression
5	With intra-arterially injected acetylcholine	Early discharge	No effect [1]
6		Late response	Potentiation
7	With anticholinesterases	Prolonged discharge	Depression &/or potentiation
8	Histamine	Synaptic transmission & N wave	Potentiation or depression
9	Isoproterenol [Isoprenaline]	Synaptic transmission & N wave	Potentiation

[1] Further description in reference.

continued

Part II. Postganglionic Responses of Sympathetic Ganglia to Drugs

	Drug [Synonym]	Response	Effect
10	With intra-arterially injected acetylcholine	Late response	Potentiation
11	With anticholinesterases	Prolonged discharge	Potentiation
12	"Muscarinic" acetylcholine antagonists, e.g., atropine	Synaptic transmission & N wave	No effect [1]
13		P wave or potential	Marked depression
14		LN wave or potential	Marked depression
15	With intra-arterially injected acetylcholine	Early discharge	No effect [1]
16		Late response	Marked depression
17	With anticholinesterases	Prolonged discharge	Marked depression
18	"Nicotinic" acetylcholine antagonists, e.g., hexametho-	Synaptic transmission & N wave	Marked depression
19	nium & tubocurarine	P wave or potential	Potentiation
20		LN wave or potential	Potentiation or no effect [1]
21	With intra-arterially injected acetylcholine	Early discharge	Marked depression
22		Late response	No effect [1]
23	With anticholinesterases	Prolonged discharge	Potentiation
24	Norepinephrine [Noradrenaline]	Synaptic transmission & N wave	Depression
25	With intra-arterially injected acetylcholine	Early discharge	No effect [1]
26		Late response	Depression
27	With anticholinesterases	Prolonged discharge	Depression, then potentiation
28	Serotonin [5-Hydroxytryptamine]	Synaptic transmission & N wave	Potentiation

[1] Further description in reference.

Contributor: Grenell, Robert G.

Reference: Phillis, J. W. 1970. The Pharmacology of Synapses. Pergamon Press, New York. pp. 123-148.

Part III. Various Monoamines on Thalamic Neurons

Values are percentages of neurons (with actual number of neurons in brackets) in each layer reacting in the manner specified. **Position of Neuron in Thalamus:** Position of cells was determined by acid lesioning; Superficial = cells at depths of 0-3 millimeters; Intermediate = cells at depths of 3-6 millimeters; Deep = cells at depths of 6-9 millimeters.

	Monoamine (Synonym)	Position of Neuron in Thalamus	Effect		
			Depression	Excitation	None
1	Dopamine	Superficial	86% [24]	0% [0]	14% [4]
2		Intermediate	100% [11]	0% [0]	0% [0]
3		Deep	89% [8]	0% [0]	11% [1]
4	Epinephrine (Adrenaline)	Superficial	55% [35]	9% [6]	36% [22]
5		Intermediate	57% [32]	9% [5]	34% [19]
6		Deep	61% [14]	9% [2]	30% [7]
7	Isoproterenol (Isoprenaline)	Superficial	40% [6]	7% [1]	53% [8]
8		Intermediate	61% [14]	13% [3]	26% [6]
9		Deep	36% [5]	21% [3]	43% [6]
10	Norepinephrine (Noradrenaline)	Superficial	63% [48]	3% [2]	34% [26]
11		Intermediate	35% [24]	20% [14]	45% [31]
12		Deep	37% [30]	30% [25]	33% [27]
13	Serotonin (5-Hydroxytryptamine)	Superficial	86% [24]	3% [1]	11% [3]
14		Intermediate	83% [15]	11% [2]	6% [1]
15		Deep	73% [8]	27% [3]	0% [0]

Contributor: Grenell, Robert G.

Reference: Phillis, J. W., and A. K. Tebecis. 1967. J. Physiol. (London) 192:715.

continued

Part IV. Drugs on Adrenergic Sympathetic Nerve Terminals

Diagram is a schematic representation of the sequence of events leading to norepinephrine (noradrenaline) release from a sympathetic nerve terminal. Possible sites and modes of action of various drugs in interference with adrenergic mechanisms are indicated. *Abbreviations:* DOPA = 3,4-dihydroxyphenylalanine; ATP = adenosine 5′-triphosphate.

MECHANISM OF ACTION AND INHIBITORY DRUGS

A: Circulating tyrosine, the probable precursor used by adrenergic terminals for the biosynthesis of norepinephrine, enters the adrenergic terminal possibly by a carrier-mediated process. Structurally related drugs such as α-methyl-*m*-tyrosine can take the place of tyrosine and be converted by a biosynthetic process into false adrenergic neurotransmitters.

B: The rate-limiting step in norepinephrine biosynthesis, tyrosine hydroxylase, can be inhibited by drugs such as α-methyl-*p*-tyrosine or 3-iodotyrosine.

C: False neurotransmitters can also be synthesized from drugs which are related to other intermediates in norepinephrine biosynthesis; e.g., α-methyldopa can take the place of DOPA.

D: The storage of norepinephrine in intraneuronal storage particles is prevented by drugs such as reserpine.

E: Action potentials propagate to the terminals of sympathetic postganglionic neurons. Propagation in terminal regions may be blocked by adrenergic neuron-blocking drugs such as bretylium & guanethidine. Propagation may also be blocked at the synapse between pre- & postganglionic neurons by ganglion-blocking drugs such as pempidine & chlorisondamine.

F: Many drugs can also be taken up into the adrenergic neuron by acting as substrates for the transport process. Once inside the neuron, such drugs as metaraminol, octopamine, & normetanephrine (α-methyl-noradrenaline) may displace norepinephrine from intraneuronal stores (indirect-acting sympathomimetics), and may further take the place of norepinephrine by acting as false neural transmitters.

G: The interaction of norepinephrine with α- & β-adrenergic receptors can be blocked by receptor-blocking drugs such as phenoxybenzamine, phentolamine, dichlorisoproterenol (DCI), & pronethalol. The action of norepinephrine on adrenergic receptors can be mimicked by direct-acting sympathomimetic amines such as epinephrine (adrenaline) or synephrine [1].

H: The re-uptake of norepinephrine released by nerve impulses is effected by a membrane transport process which can be inhibited by many drugs, e.g., cocaine, desipramine, & many sympathomimetic amines.

I: Free norepinephrine in the axoplasm is destroyed by monoamine oxidase situated in the intraneuronal mitochondria. This enzyme can be inhibited by a wide range of drugs such as pheniprazine, nialamide, or iproniazid.

J: The extraneuronal metabolism of norepinephrine by catechol methyltransferase can be inhibited by pyrogallol & tropolones.

ADRENERGIC TERMINAL   EFFECTOR TISSUE

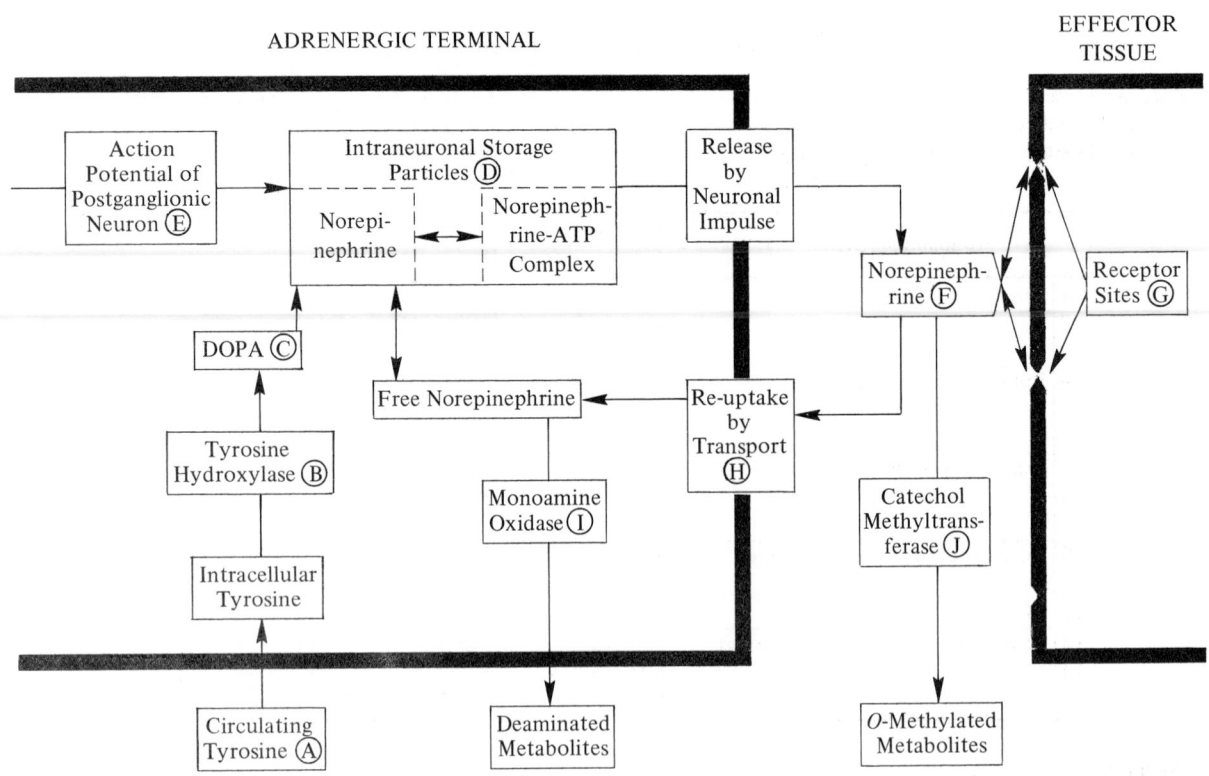

continued

143. VARIOUS CHEMICALS AFFECTING THE CENTRAL NERVOUS SYSTEM: CAT

Part IV. Drugs on Adrenergic Sympathetic Nerve Terminals

Contributor: Grenell, Robert G.

References

[1] Iversen, L. L. 1967. The Uptake and Storage of Noradrenaline in Sympathetic Nerves. Univ. Press, Cambridge.

[2] Phillis, J. W. 1970. The Pharmacology of Synapses. Pergamon Press, New York. pp. 58-59.

144. RATE OF RELEASE OF ACETYLCHOLINE FROM CEREBRAL CORTEX: CAT

Part I. From Various Areas of Unstimulated Cortex

Values are mean values of nanograms acetylcholine released per minute per square centimeter of cat cortex. Numbers in brackets are numbers of subjects.

	Condition	Area of Cortex			
		Sensorimotor	Auditory Sensory	Visual Sensory	Parietal
	Whole animal anesthetized with				
1	Allobarbital (Dial)	1.92 ± 2.14 [5]	0.65 ± 0.45 [5]	0.84 [3]	0.51 ± 0.46
2	Chloralose	0.88 ± 0.77 [7]	0.60 ± 0.42 [7]	0.37 ± 0.073 [6]	0.62 ± 0.51 [6]
3	Diethyl ether	3.04 ± 1.61 [5]	0.98 ± 0.74 [5]	0.65 ± 0.42 [5]	1.37 ± 1.1 [5]
4	Halothane	2.27 ± 2.12 [6]	0.62 ± 0.36 [6]	0.44 [3]	0.60 [2]
5	Pentobarbital sodium (Pentobarbitone sodium)	0.45 ± 0.54 [13]	0.18 ± 0.18 [9]	0.18 ± 0.17 [6]	0.34 ± 0.2 [5]
6	Isolated brain	0.45 [3]	....	....	0.20 [3]

Contributor: Grenell, Robert G.

Reference: Phillis, J. W. 1970. The Pharmacology of Synapses. Pergamon Press, New York. p. 182.

Part II. Under Various Modes of Stimulation

Values are mean percentages of increase in the rate of acetylcholine release induced by various modes of stimulation; baselines are calculated from the rates of release immediately before and after stimulation. Subjects are cats anesthetized with pentobarbital sodium (pentobarbitone sodium). Values in parentheses are ranges, estimate "c" (*see* Introduction).

	Type of Stimulation	Area of Cortex Tested				
		Left Sensorimotor	Right Sensorimotor	Auditory Sensory	Visual Sensory	Parietal
1	Left forepaw	230(25-650)	250(0-650)	320(80-650)	90(25-150)	30(5-60)
2	Left hindpaw	120(20-260)	50(0-100)	....	....	....
3	Left facial	170(30-280)	190(30-350)	130(10-250)	70(20-130)	....
4	Auditory	100(20-170)	110(0-300)	85(0-190)	80(65-100)	....
5	Visual	160(40-220)	100(10-180)	120(30-200)	110(30-200)	....
6	Reticular formation	....	180(40-320)	210(80-320)	70(30-160)	....
7	Direct cortical[1]	120(80-170)	140(100-210)	100(50-160)	30(10-60)	....

[1] Stimulation of left sensorimotor cortex within the cup; acetylcholine release determined simultaneously.

Contributor: Grenell, Robert G.

Reference: Phillis, J. W. 1970. The Pharmacology of Synapses. Pergamon Press, New York. p. 182.

Owing to the variability of tissue composition, and uncertainties in isolation or assay of the components, most of the values have been arbitrarily rounded off to two significant figures, and no analysis of variance has been attempted. Where appropriate, data have been recalculated into atomic or molecular units. **Chemical:** Phosphatidyl choline = lecithin; phosphatidyl ethanolamine = cephalin. **Concentration:** Values are for fresh weight of tissue, unless otherwise indicated.

Part I. Man

	Tissue	Age	Chemical	Concentration	Reference
1	Whole brain	Fetus, 13-14 wk	H_2O	91% by wt	32
2			Cl	72 μmoles/g	32
3			P	57 μmoles/g	32
4			K	59 μmoles/g	32
5			Na	98 μmoles/g	32
6		20-22 wk	H_2O	92% by wt	32
7			Ca	2.5 μmoles/g	32
8			Cl	73 μmoles/g	32
9			Mg	4.2 μmoles/g	32
10			P	52 μmoles/g	32
11			K	52 μmoles/g	32
12			Na	92 μmoles/g	32
13		Newborn	H_2O	90% by wt	32
14			Ca	2.4 μmoles/g	32
15			Cl	66 μmoles/g	32
16			Mg	4 μmoles/g	32
17			P	54 μmoles/g	32
18			K	58 μmoles/g	32
19			Na	81 μmoles/g	32
20		Adult	H_2O	77% by wt	12,14,26,31,32
21			Ca	2 μmoles/g	12,14,26,31,32
22			Cl	41 μmoles/g	12,14,26,31,32
23			Mg	5.7 μmoles/g	12,14,26,31,32
24			P	109 μmoles/g	12,14,26,31,32
25			K	85 μmoles/g	12,14,26,31,32
26			Na	55 μmoles/g	12,14,26,31,32
27		Unspecified	Al	0.034 μmole/g	29
28			As	0.0011 μmole/g	25
29			Ba	0.0001 μmole/g	29
30			Be	0.0002 μmole/g	27
31			Bi	0.0005 μmole/g	16
32			B	0.06 μmole/g	29
33			Cd	0.038 μmole/g	33
34			Cs	0.0002 μmole/g	33
35			Cr	0.002 μmole/g	29
36			Co	0.0001 μmole/g	33
37			Cu	0.06-0.8 μmole/g	6
38			F	0.11 μmole/g	10
39			Ga	0.0006 μmole/g	8
40			Au	0.0025 μmole/g	10
41			I	0.003 μmole/g	4
42			Pb	0.004-0.08 μmole/g	6
43			Li	0.004 μmole/g	10
44			Mn	0.02 μmole/g	28,29
45			Mo	0.002 μmole/g	29
46			Ni	0.005 μmole/g	29
47			Rb	0.17 μmole/g	33

continued

Part I. Man

	Tissue	Age	Chemical	Concentration	Reference
48			Ru	0.005 μmole/g	10
49			Se	0.024 μmole/g	9
50			Si	2.8 μmoles/g	11
51			Ag	0.0004 μmole/g	29
52			Sr	0.001 μmole/g	29
53			Tl	0.002 μmole/g	10
54			Sn	0.002 μmole/g	29
55			Ti	0.008 μmole/g	29
56			V	0.006 μmole/g	29
57			Zn	0.74 μmole/g	29
58			Zr	0.05 μmole/g	10
59	Cortex	Fetus, 4.5 mo	H_2O	900 mg/g	3
60			Phospholipid	12 mg/g	3
61			Glycolipid	2.6 mg/g	3
62			Sterol	2.3 mg/g	3
63			Protein	83 mg/g[1/]	3
64		Newborn	H_2O	900 mg/g	3
65			Phospholipid	18 mg/g	3
66			Glycolipid	4.6 mg/g	3
67			Sterol	4.7 mg/g	3
68			Protein	73 mg/g[1/]	3
69		Adult	H_2O	850 mg/g	3
70			Phospholipid	31 mg/g	3
71			Glycolipid	6.3 mg/g	3
72			Sterol	8 mg/g	3
73			Protein	105 mg/g[1/]	3
74	Cerebral	Fetus, 4 mo	Glycolipid	4.4 mg/g	3
75	white mat-	4.5 mo	H_2O	900 mg/g	3
76	ter		Phospholipid	16 mg/g	3
77			Sterol	3.7 mg/g	3
78			Protein	76 mg/g[1/]	3
79		Newborn	H_2O	900 mg/g	3
80			Phospholipid	20 mg/g	3
81			Glycolipid	4 mg/g	3
82			Sterol	6.6 mg/g	3
83			Protein	73 mg/g[1/]	3
84		Adult	H_2O	70-74% by wt; 710 mg/g	3,12,14,26,31,32
85			Ca	3.6 μmoles/g	12,14,26,31,32
86			Cl	41 μmoles/g	12,14,26,31,32
87			Mg	10.8 μmoles/g	12,14,26,31,32
88			P	127 μmoles/g	12,14,26,31,32
89			K	59-87 μmoles/g	12,14,26,31,32
90			Na	69-98 μmoles/g	12,14,26,31,32
91			Phospholipid	78 mg/g	3
92			Glycolipid	47 mg/g	3
93			Sterol	41 mg/g	3
94			Protein	124 mg/g[1/]	3
95		Unspecified	Cu	0.14-1.3 μmoles/g	6
96			Fe	5.9 μmoles/g	5
97			Mn	0.02 μmole/g	28,29
98			Phosphatidal choline	0.3 mg/g	1,19
99			Phosphatidyl choline	20 mg/g	1,19

[1/] Estimated by difference of dry weight minus total lipids.

continued

Part I. Man

	Tissue	Age	Chemical	Concentration	Reference
100			Phosphatidal & phosphatidyl choline	20 mg/g	1,19
101			Phosphatidal ethanolamine	21 mg/g	1,19
102			Phosphatidyl ethanolamine	1.3 mg/g	1,19
103			Phosphatidal & phosphatidyl ethanolamine	23 mg/g	1,19
104			Phosphatidal serine	7.8 mg/g	1,19
105			Phosphatidyl serine	2.6 mg/g	1,19
106			Phosphatidal & phosphatidyl serine	12 mg/g	1,19
107			Sphingomyelin	11-13 mg/g	1,19
108			Phosphatidyl inositols	1.4 mg/g	1,19
109			Cerebroside	31 mg/g	1,19
110			Sulfatide	7.4-8.4 mg/g	1,19
111			Cholesterol	37-43 mg/g	1,19
112			Alanine	1.6 μmoles/g	22
113			Arginine	0.38 μmole/g	22
114			Aspartic acid	1.5 μmoles/g	22
115			Cysteine	1.0 μmoles/g	22
116			Glutamic acid	7.7 μmoles/g	22
117			Glutamine	4.8 μmoles/g	22
118			Glycine	0.87 μmole/g	22
119			Histidine	0.30 μmole/g	22
120			Leucine	0.41 μmole/g	22
121			Serine	1.1 μmoles/g	22
122			Taurine	0.7 μmole/g	22
123			Threonine	0.37 μmole/g	22
124			Tyrosine	0.54 μmole/g	22
125			Valine	0.26 μmole/g	22
126			Glutathione	1.4 μmoles/g	22
127	Myelin	Unspecified	H_2O	400 mg/g[2/]	1,2,7,13,17,20, 21,23
128			Phosphatidal choline	11 mg/g	2,7,20
129			Phosphatidyl choline	0.2 mg/g	2,7,20
130			Phosphatidal & phosphatidyl choline	31 mg/g	2,7,20
131			Phosphatidal ethanolamine	15.3 mg/g	2,7,20
132			Phosphatidyl ethanolamine	0 mg/g	2,7,20
133			Phosphatidal & phosphatidyl ethanolamine	44 mg/g	2,7,20
134			Phosphatidal serine	5.3 mg/g	2,7,20
135			Phosphatidyl serine	1.9 mg/g	2,7,20
136			Phosphatidal & phosphatidyl serine	13 mg/g	2,7,20
137			Sphingomyelin	18-24 mg/g	2,7,20
138			Phosphatidyl inositols	0.6 mg/g	2,7,20
139			Cerebroside	93-96 mg/g	2,7,20
140			Sulfatide	16-20 mg/g	2,7,20
141			Sterol	107-114 mg/g	2,7,20
142			Protein	130-270 mg/g	2,7,20
143	Cerebral gray matter	Adult	H_2O	84% by wt	12,14,26,31,32
144			Ca	2.6 μmoles/g	12,14,26,31,32
145			Cl	47 μmoles/g	12,14,26,31,32
146			Mg	8.2 μmoles/g	12,14,26,31,32
147			P	71 μmoles/g	12,14,26,31,32
148			K	58-88 μmoles/g	12,14,26,31,32
149			Na	84-88 μmoles/g	12,14,26,31,32

[2/] Assumed H_2O content.

continued

Part I.   Man

	Tissue	Age	Chemical	Concentration	Reference
150		Unspecified	Cu	0.25-1.55 μmoles/g	6
151			Fe	7 μmoles/g	5
152			Mn	0.02 μmole/g	28,29
153			Phosphatidal choline	0.8 mg/g	18,19
154			Phosphatidyl choline	16 mg/g	18,19
155			Phosphatidal & phosphatidyl choline	16 mg/g	18,19
156			Phosphatidal ethanolamine	6.8 mg/g	18,19
157			Phosphatidyl ethanolamine	10 mg/g	18,19
158			Phosphatidal & phosphatidyl ethanolamine	13 mg/g	18,19
159			Phosphatidal serine	4.7 mg/g	18,19
160			Phosphatidyl serine	0.3 mg/g	18,19
161			Phosphatidal & phosphatidyl serine	5.1 mg/g	18,19
162			Sphingomyelin	3.3-4 mg/g	18,19
163			Phosphatidyl inositols	1.6 mg/g	18,19
164			Cerebroside	3.2-4.4 mg/g	18,19
165			Sulfatide	1.0-1.4 mg/g	18,19
166			Cholesterol	13 mg/g	18,19
167			Alanine	2.2 μmoles/g	22
168			Arginine	0.47 μmole/g	22
169			Aspartic acid	3.1 μmoles/g	22
170			Cysteine	2.6 μmoles/g	22
171			Glutamic acid	11 μmoles/g	22
172			Glutamine	5.4 μmoles/g	22
173			Glycine	1.3 μmoles/g	22
174			Histidine	0.43 μmole/g	22
175			Leucine	0.39 μmole/g	22
176			Serine	1.6 μmoles/g	22
177			Taurine	1.6 μmoles/g	22
178			Threonine	0.46 μmole/g	22
179			Tyrosine	0.28 μmole/g	22
180			Valine	0.39 μmole/g	22
181			Glutathione	1.2 μmoles/g	22
182	Caudate nucleus	Unspecified	Fe	10.1 μmoles/g	5
183	Globus pallidus	Unspecified	Cu	0.36-2.96 μmoles/g	6
184			Fe	2.04 μmoles/g	5
185	Putamen	Unspecified	Cu	0.55-1.89 μmoles/g	6
186	Thalamus	Unspecified	Cu	0.26-1.95 μmoles/g	6
187			Fe	7.4 μmoles/g	5
188	Cerebellar white matter	Unspecified	Cu	0.37 μmole/g	6
189	Cerebellar gray matter	Unspecified	Cu	0.30-0.52 μmole/g	6
190			Fe	4.0 μmoles/g	5
191			Mn	0.02 μmole/g	28,29
192	Brain stem	Unspecified	Cu	0.13 μmole/g	6
193	Spinal cord	0-1 yr	H_2O	82% by wt	24
194			Ca	2 μmoles/g	24
195			K	22 μmoles/g	24
196			Na	43 μmole/g	24
197		Adult	H_2O	64-74% by wt	15,24,31
198			Ca	2.5-4.5 μmoles/g	15,24,31
199			Cl	43 μmoles/g	15,24,31

continued

Part I. Man

	Tissue	Age	Chemical	Concentration	Reference
200			Mg	16 μmoles/g	15,24,31
201			P	177 μmoles/g	15,24,31
202			K	19-93 μmoles/g	15,24,31
203			Na	42-87 μmoles/g	15,24,31
204		Unspecified	Phosphatidal & phosphatidyl choline	19 mg/g	30
205			Phosphatidyl ethanolamine	26 mg/g	30
206			Phosphatidal & phosphatidyl ethanolamine	51 mg/g	30
207			Sphingomyelin	33 mg/g	30
208			Cholesterol	51 mg/g	30
209	Peripheral	Unspecified	H$_2$O	56% by wt	31
210	nerve, un-		Ca	5.4 μmoles/g	31
211	specified		P	119 μmoles/g	31
212			K	50 μmoles/g	31
213	Microsomes[3]	Unspecified	Phosphatidal & phosphatidyl choline	165 mg/g[4]	7
214			Phosphatidal & phosphatidyl ethanolamine	75 mg/g[4]	7
215			Phosphatidal & phosphatidyl serine	51 mg/g[4]	7
216			Sphingomyelin	58 mg/g[4]	7
217			Phosphatidyl inositols	32 mg/g[4]	7
218			Cerebroside	50 mg/g[4]	7
219			Sulfatide	10 mg/g[4]	7
220			Cholesterol	138 mg/g[4]	7

[3] Subcellular organelle; values recalculated, assuming total lipid content of 600 mg/g dry wt [26]. [4] Dry weight.

Contributor: Mokrasch, Lewis C.

References

[1] Adams, C. W., and A. N. Davison. 1965. In C. W. Adams, ed. Neurohistochemistry. Elsevier, Amsterdam. pp. 332-400.

[2] Autilio, L. A., et al. 1964. J. Neurochem. 11:17.

[3] Brant, G. 1949. Acta Physiol. Scand., Suppl. 63.

[4] Chilean Iodine Educational Bureau. 1952. Iodine Content of Foods; Annotated Bibliography, 1825-1851. London.

[5] Cumings, J. N. 1948. Brain 71:410.

[6] Cumings, J. N. 1959. Heavy Metals and the Brain. Blackwell, Oxford.

[7] Cuzner, M. L., et al. 1965. J. Neurochem. 12:469.

[8] Davies, F., et al. 1952. J. Physiol. (London) 118:276.

[9] Dye, W. B., et al. 1963. Anal. Chem. 35:1687.

[10] International Commission on Radiological Protection. 1964. Int. Comm. Radiol. Protect. Publ. 6.

[11] King, E. J., and T. H. Belt. 1938. Physiol. Rev. 18:329.

[12] Koch, W., and S. A. Manin. 1907. J. Physiol. (London) 36:xxxvi.

[13] Korey, S. R., et al. 1958. J. Neuropathol. Exp. Neurol. 17:430.

[14] Logan, J. E. 1961. In P. L. Altman and D. S. Dittmer, ed. Blood and Other Body Fluids. Federation of American Societies for Experimental Biology, Washington, D.C. pp. 326-328.

[15] Lowenthal, A. 1961. In J. Folch-Pi, ed. Chemical Pathology of the Nervous System. Pergamon Press, Oxford. pp. 299-306.

[16] Neufeld, A. H. 1936. Can. J. Res. B14:160.

[17] Norton, W. T., and L. A. Autilio. 1966. J. Neurochem. 13:213.

[18] Norton, W. T., et al. 1966. J. Neuropathol. Exp. Neurol. 25:582.

[19] O'Brien, J. S., and E. L. Sampson. 1965. J. Lipid Res. 6:537.

[20] O'Brien, J. S., and E. L. Sampson. 1965. Science 150:1613.

[21] O'Brien, J. S., et al. 1967. J. Neurochem. 14:357.

[22] Robinson, N., and C. B. Williams. 1965. Clin. Chim. Acta 12:311.

[23] Seminario, L. M., et al. 1964. J. Neurochem. 11:197.

[24] Siege, K., and V. Thierbach. 1961. Z. Alternsforsch. 15:46.

[25] Smoles, A. A., and B. Pate. 1952. Analyst (London) 77:196.

continued

145. CHEMICAL COMPOSITION OF NERVOUS TISSUE

Part I. Man

[26] Stewart-Wallace, A. M. 1939. Brain 62:426.
[27] Suzuki, T., and H. Hamada. 1956. Nippon Kagaku Zasshi 77:125.
[28] Tingley, A. H. 1937. J. Ment. Sci. 83:425.
[29] Tipton, I. H., and M. J. Cook. 1963. Health Phys. 9:103.

[30] Webster, F. R. 1960. Biochim. Biophys. Acta 44:109.
[31] Weil, A. 1914. Hoppe Seylers Z. Physiol. Chem. 89: 340.
[32] Widdowson, E. M., and J. W. Dickerson. 1960. Biochem. J. 77:30.
[33] Yamagata, N., et al. 1962. J. Radiat. Res. 3:4.

Part II. Vertebrates Other Than Man

	Tissue	Age	Chemical	Concentration	Reference
			Cat		
1	Nervous tissue	Newborn	Alanine	2.2 μmoles/g	18
2			β-Alanine	0.07 μmole/g	18
3			Aspartic acid	1.1 μmoles/g	18
4			Citrulline	0.04 μmole/g	18
5			L-Cystathionine	0.04 μmole/g	18
6			Glutamic acid	6.0 μmoles/g	18
7			Glutamine	3.9 μmoles/g	18
8			Glycine	2.3 μmoles/g	18
9			Histidine	0.02 μmole/g	18
10			Hydroxyproline	0.04 μmole/g	18
11			Isoleucine	0.07 μmole/g	18
12			Leucine	0.17 μmole/g	18
13			Lysine	0.07 μmole/g	18
14			Methionine	0.01 μmole/g	18
15			Ornithine	0.05 μmole/g	18
16			Phenylalanine	0.05 μmole/g	18
17			Proline	0.22 μmole/g	18
18			Serine	1.1 μmoles/g	18
19			Taurine	0.2 μmole/g	18
20			Threonine	0.58 μmole/g	18
21			Tyrosine	0.11 μmole/g	18
22			Valine	0.21 μmole/g	18
23			γ-Aminobutyric acid	1.3 μmoles/g	18
24			Ethanolamine	1.4 μmoles/g	18
25			Glutathione	0.36 μmole/g	18
26			Glycerophosphoethanolamine	0.19 μmole/g	18
27			Phosphoethanolamine	4.6 μmoles/g	18
28			Urea	9.7 μmoles/g	18
29		Adult	Alanine	0.48 μmole/g	18
30			β-Alanine	0.02 μmole/g	18
31			Aspartic acid	1.7 μmoles/g	18
32			Citrulline	0.04 μmole/g	18
33			L-Cystathionine	0.14 μmole/g	18
34			Glutamic acid	7.9 μmoles/g	18
35			Glutamine	2.8 μmoles/g	18
36			Glycine	0.78 μmole/g	18
37			Histidine	0.02 μmole/g	18
38			Hydroxyproline	0.08 μmole/g	18
39			Isoleucine	0.03 μmole/g	18

continued

Part II. Vertebrates Other Than Man

	Tissue	Age	Chemical	Concentration	Reference
40			Leucine	0.07 μmole/g	18
41			Lysine	0.08 μmole/g	18
42			Methionine	0.02 μmole/g	18
43			Ornithine	0.01 μmole/g	18
44			Phenylalanine	0.02 μmole/g	18
45			Proline	0.03 μmole/g	18
46			Serine	0.48 μmole/g	18
47			Taurine	2.3 μmoles/g	18
48			Threonine	0.17 μmole/g	18
49			Tyrosine	0.03 μmole/g	18
50			Valine	0.06 μmole/g	18
51			γ-Aminobutyric acid	1.4 μmoles/g	18
52			Ethanolamine	0.48 μmole/g	18
53			Glutathione	0.49 μmole/g	18
54			Glycerophosphoethanolamine	0.77 μmole/g	18
55			Homocarnosine	0.005 μmole/g	19
56			Phosphoethanolamine	1.2 μmoles/g	18
57			Urea	3.8 μmoles/g	18
58	Spinal cord	Unspecified	Phosphatidal & phosphatidyl choline	14 mg/g	27
59			Phosphatidal & phosphatidyl ethanolamine	45 mg/g	27
60			Phosphatidal serine	0 mg/g	27
61			Sphingomyelin	29 mg/g	27
62			Cerebroside	57 mg/g	27
63			Cholesterol	54 mg/g	27
64	Sciatic nerve	Unspecified	H_2O	68% by wt	23
65			Cl	63 μmoles/g	23
66			K	46 μmoles/g	23
67			Na	96 μmoles/g	23
			Cattle, Male Castrate[1]		
68	Whole brain	Unspecified	H_2O	78% by wt[2]	32,44
69			Phosphatidal choline	0 mg/g	35
70			Phosphatidyl choline	13 mg/g	35
71			Phosphatidal ethanolamine	9.6 mg/g	35
72			Phosphatidyl ethanolamine	5.5 mg/g	35
73			Phosphatidal serine	0 mg/g	35
74			Phosphatidyl serine	7.5 mg/g	35
75			Sphingomyelin	5.7 mg/g	35
76			Phosphatidyl inositols	1.4 mg/g	35
77			Cerebroside	16 mg/g	35
78			Sulfatide	3.5 mg/g	35
79			Cholesterol	20 mg/g	35
80	Cortex	Unspecified	H_2O	830 mg/g	9
81			Phospholipid	35 mg/g	9
82			Glycolipid	12 mg/g	9
83			Sterol	8.2 mg/g	9
84			Protein	115 mg/g[3]	9
85	White matter	Unspecified	H_2O	70% by wt[2]; 680 mg/g	9,32,44
86			Ca	4.1 μmoles/g[2]	32,44
87			Cl	50 μmoles/g[2]	32,44
88			Mg	17 μmoles/g[2]	32,44
89			P	140 μmoles/g[2]	32,44

[1] Unless otherwise indicated.   [2] Cattle.   [3] Estimated by difference of dry weight minus total lipids.

continued

Part II. Vertebrates Other Than Man

	Tissue	Age	Chemical	Concentration	Reference
90			K	64 μmoles/g[2/]	32,44
91			Na	63 μmoles/g[2/]	32,44
92			Phospholipid	87 mg/g	9
93			Phosphatidal choline	0.4 mg/g	29
94			Phosphatidyl choline	22 mg/g	29
95			Phosphatidal ethanolamine	22 mg/g	29
96			Phosphatidyl ethanolamine	4 mg/g	29
97			Phosphatidal serine	0.8 mg/g	29
98			Phosphatidyl serine	18 mg/g	29
99			Sphingomyelin	11 mg/g	29
100			Glycolipid	53 mg/g	9
101			Phosphatidyl inositols	1.5 mg/g	29
102			Cerebroside & sulfatide	4.6 mg/g	29
103			Sterol	42 mg/g	9
104			Cholesterol	41 mg/g	29
105			Protein	140 mg/g[3/]	9
106	Myelin	Unspecified	H$_2$O	400 mg/g[4/]	8,11,29
107			Phosphatidal choline	48 mg/g	8,11,29
108			Phosphatidyl choline	1.5 mg/g	8,11,29
109			Phosphatidal & phosphatidyl choline	37 mg/g	8,11,29
110			Phosphatidal ethanolamine	58 mg/g	8,11,29
111			Phosphatidyl ethanolamine	20 mg/g	8,11,29
112			Phosphatidal & phosphatidyl ethanolamine	62 mg/g	8,11,29
113			Phosphatidal serine	1.1 mg/g	8,11,29
114			Phosphatidyl serine	28 mg/g	8,11,29
115			Phosphatidal & phosphatidyl serine	35 mg/g	8,11,29
116			Sphingomyelin	31 mg/g	8,11,29
117			Phosphatidyl inositols	4.5-6.6 mg/g	8,11,29
118			Cerebroside	109 mg/g	8,11,29
119			Sulfatide	17 mg/g	8,11,29
120			Cerebroside & sulfatide	113-160 mg/g	8,11,29
121			Sterol	102-118 mg/g	8,11,29
122			Protein	117-177 mg/g	8,11,29
123	Gray matter	Unspecified	H$_2$O	82% by wt[2/]	32,44
124			Ca	3.3 μmoles/g[2/]	32,44
125			Cl	35 μmoles/g[2/]	32,44
126			Mg	9.6 μmoles/g[2/]	32,44
127			P	82 μmoles/g[2/]	32,44
128			K	87 μmoles/g[2/]	32,44
129			Na	46 μmoles/g[2/]	32,44
130	Optic nerve	Unspecified	H$_2$O	670 mg/g	9
131			Phospholipid	75 mg/g	9
132			Glycolipid	46 mg/g	9
133			Sterol	40 mg/g	9
134			Protein	170 mg/g[3/]	9
135	Spinal cord	Unspecified	H$_2$O	63-78% by wt[2/]; 640 mg/g	9,12,44
136			Ca	8 μmoles/g[2/]	12,44
137			Cl	37-44 μmoles/g[2/]	12,44
138			Mg	20 μmoles/g[2/]	12,44
139			P	167 μmoles/g[2/]	12,44
140			K	67-90 μmoles/g[2/]	12,44
141			Na	41-68 μmoles/g[2/]	12,44

[2/] Cattle. [3/] Estimated by difference of dry weight minus total lipids. [4/] Assumed H$_2$O content.

continued

Part II.   Vertebrates Other Than Man

	Tissue	Age	Chemical	Concentration	Reference
142			Phospholipid	122 mg/g	9
143			Phosphatidal & phosphatidyl choline	19 mg/g	5,9
144			Phosphatidal & phosphatidyl ethanolamine	51 mg/g	5,9
145			Sphingomyelin	30 mg/g	5,9
146			Glycolipid	71 mg/g	9
147			Phosphatidal inositols	6.3 mg/g	5,9
148			Cerebroside	60 mg/g	5,9
149			Sterol	58 mg/g	9
150			Cholesterol	48 mg/g	5,9
151			Protein	110 mg/g[3]	9
152	Spinal root, my-elin	Unspecified	H_2O	400 mg/g[4]	31
153			Phosphatidal choline	0 mg/g	31
154			Phosphatidyl choline	59 mg/g	31
155			Phosphatidal ethanolamine	55 mg/g	31
156			Phosphatidyl ethanolamine	15 mg/g	31
157			Phosphatidal serine	11 mg/g	31
158			Phosphatidyl serine	26 mg/g	31
159			Sphingomyelin	73 mg/g	31
160			Cerebroside	63 mg/g	31
161			Sulfatide	11 mg/g	31
162			Sterol	113 mg/g	31
163			Protein	145 mg/g[3]	31
164	Sciatic nerve	Unspecified	H_2O	54% by wt[2]	12
165			K	40 μmoles/g[2]	12
166			Na	93 μmoles/g[2]	12
167			Phosphatidyl choline	3.3 mg/g	38
168			Phosphatidal ethanolamine	1.2 mg/g	38
169			Phosphatidyl ethanolamine	0.5 mg/g	38
170			Phosphatidyl serine	1.1 mg/g	38
171			Sphingomyelin	1.5 mg/g	38
172			Phosphatidyl inositols	0.4 mg/g	38
173			Cerebroside & sulfatide	0.2 mg/g	38
174			Cholesterol	3.5 mg/g	38
175	Splenic nerve	Unspecified	H_2O	830 mg/g	9
176			Phospholipid	14 mg/g	9
177			Glycolipid	2.7 mg/g	9
178			Sterol	3.5 mg/g	9
179			Protein	150 mg/g[3]	9
180	Vagus nerve	Unspecified	H_2O	570 mg/g	9
181			Phospholipid	33 mg/g	9
182			Glycolipid	10 mg/g	9
183			Sterol	14 mg/g	9
184			Protein	370 mg/g[3]	9
185	Mitochondria[5]	Unspecified	Phosphatidal choline	0.3 mg/g[6]	33
186			Phosphatidyl choline	38 mg/g[6]	33
187			Phosphatidal ethanolamine	24 mg/g[6]	33
188			Phosphatidyl ethanolamine	11 mg/g[6]	33
189			Phosphatidal serine	0 mg/g[6]	33
190			Phosphatidyl serine	15 mg/g[6]	33
191			Sphingomyelin	23 mg/g[6]	33
192			Phosphatidyl inositols	3.8 mg/g[6]	33
193			Cerebroside & sulfatide	50 mg/g[6]	33
194			Cholesterol	24 mg/g[6]	33

[2] Cattle.   [3] Estimated by difference of dry weight minus total lipids.   [4] Assumed H_2O content.   [5] Subcellular organelle.   [6] Dry weight.

continued

145. CHEMICAL COMPOSITION OF NERVOUS TISSUE

Part II. Vertebrates Other Than Man

	Tissue	Age	Chemical	Concentration	Reference
			Dog		
195	Nervous tissue	Newborn	Alanine	0.76 μmole/g	18
196			β-Alanine	0.02 μmole/g	18
197			Aspartic acid	1.0 μmole/g	18
198			Citrulline	0.03 μmole/g	18
199			L-Cystathionine	0.01 μmole/g	18
200			Glutamic acid	4.3 μmoles/g	18
201			Glutamine	2.5 μmoles/g	18
202			Glycine	1.3 μmoles/g	18
203			Histidine	0.0 μmole/g	18
204			Hydroxyproline	0.14 μmole/g	18
205			Isoleucine	0.01 μmole/g	18
206			Leucine	0.05 μmole/g	18
207			Lysine	0.19 μmole/g	18
208			Ornithine	0.02 μmole/g	18
209			Phenylalanine	0.06 μmole/g	18
210			Proline	0.17 μmole/g	18
211			Serine	0.79 μmole/g	18
212			Taurine	6.8 μmoles/g	18
213			Threonine	0.86 μmole/g	18
214			Tyrosine	0.06 μmole/g	18
215			Valine	0.09 μmole/g	18
216			γ-Aminobutyric acid	0.78 μmole/g	18
217			Ethanolamine	0.18 μmole/g	18
218			Glutathione	0.57 μmole/g	18
219			Glycerophosphoethanolamine	0.28 μmole/g	18
220			Phosphoethanolamine	6.5 μmoles/g	18
221			Urea	5.4 μmoles/g	18
222		Adult	Alanine	0.99 μmole/g	18
223			β-Alanine	0.02 μmole/g	18
224			Aspartic acid	3.3 μmoles/g	18
225			L-Cystathionine	0.13 μmole/g	18
226			Glutamic acid	9.1 μmoles/g	18
227			Glutamine	4.7 μmoles/g	18
228			Glycine	0.86 μmole/g	18
229			Histidine	0.05 μmole/g	18
230			Isoleucine	0.11 μmole/g	18
231			Leucine	0.17 μmole/g	18
232			Lysine	0.23 μmole/g	18
233			Methionine	0.04 μmole/g	18
234			Ornithine	0.02 μmole/g	18
235			Phenylalanine	0.07 μmole/g	18
236			Proline	0.06 μmole/g	18
237			Serine	0.58 μmole/g	18
238			Taurine	1.3 μmoles/g	18
239			Threonine	0.29 μmole/g	18
240			Tyrosine	0.04 μmole/g	18
241			Valine	0.17 μmole/g	18
242			γ-Aminobutyric acid	3.3 μmoles/g	18
243			Ethanolamine	0.72 μmole/g	18
244			Glutathione	1.0 μmole/g	18
245			Glycerophosphoethanolamine	0.49 μmole/g	18
246			Homocarnosine	0.005 μmole/g	19

continued

Part II. Vertebrates Other Than Man

	Tissue	Age	Chemical	Concentration	Reference
247			Phosphoethanolamine	1.9 μmoles/g	18
248			Urea	3.0 μmoles/g	18
249	Whole brain	Unspecified	H_2O	76% by wt	15,42,48
250			Ca	1.1 μmoles/g	15,42,48
251			Cl	36 μmoles/g	15,42,48
252			Mg	5.7 μmoles/g	15,42,48
253			K	96 μmoles/g	15,42,48
254			Na	51 μmoles/g	15,42,48
255	White matter	Unspecified	H_2O	70% by wt	15,42,48
256			Cl	35 μmoles/g	15,42,48
257			P	122 μmoles/g	15,42,48
258			K	88 μmoles/g	15,42,48
259			Na	47 μmoles/g	15,42,48
260	Gray matter	Unspecified	H_2O	80% by wt	15,42,48
261			Cl	44 μmoles/g	15,42,48
262			P	76 μmoles/g	15,42,48
263			K	96 μmoles/g	15,42,48
264			Na	65 μmoles/g	15,42,48
265	Spinal cord	Unspecified	H_2O	67% by wt	42
266			Cl	36 μmoles/g	42
267			P	176 μmoles/g	42
268			K	71 μmoles/g	42
269			Na	59 μmoles/g	42
270	Sciatic nerve	Unspecified	H_2O	57% by wt	42
271			Cl	60 μmoles/g	42
272			P	75 μmoles/g	42
273			K	31 μmoles/g	42
274			Na	148 μmoles/g	42
			Dormouse [7/]		
275	Nervous tissue	Unspecified	Aspartic acid	2.9 μmoles/g	25
276			Glutamic acid	11 μmoles/g	25
277			Glycine	0.98 μmole/g	25
278			Histidine	0.58 μmole/g	25
279			Isoleucine	0.033 μmole/g	25
280			Leucine	0.085 μmole/g	25
281			Lysine	0.26 μmole/g	25
282			Methionine	0.14 μmole/g	25
283			Phenylalanine	0.066 μmole/g	25
284			Taurine	3.6 μmoles/g	25
285			Threonine	0.37 μmole/g	25
286			Tyrosine	0.16 μmole/g	25
287			γ-Aminobutyric acid	1.9 μmoles/g	25
			Guinea Pig		
288	Nervous tissue	Newborn	Alanine	0.04 μmole/g	4
289			β-Alanine	0.04 μmole/g	4
290			Aspartic acid	2.8 μmoles/g	4
291			L-Cystathionine	0.03 μmole/g	4
292			Glutamic acid	11 μmoles/g	4
293			Glutamine	4.9 μmoles/g	4
294			Glycine	2.6 μmoles/g	4

[7/] *Glis glis.*

continued

Part II. Vertebrates Other Than Man

	Tissue	Age	Chemical	Concentration	Reference
295			Hydroxyproline	0.05 μmole/g	4
296			Isoleucine	0.04 μmole/g	4
297			Leucine	0.08 μmole/g	4
298			Lysine	0.02 μmole/g	4
299			Methionine	0.02 μmole/g	4
300			Ornithine	0.03 μmole/g	4
301			Phenylalanine	0.03 μmole/g	4
302			Proline	0.18 μmole/g	4
303			Serine	1.8 μmoles/g	4
304			Taurine	3.4 μmoles/g	4
305			Threonine	0.31 μmole/g	4
306			Tyrosine	0.05 μmole/g	4
307			Valine	0.10 μmole/g	4
308			γ-Aminobutyric acid	1.9 μmoles/g	4
309			Ethanolamine	0.29 μmole/g	4
310			Glutathione	1.2 μmoles/g	4
311			Glycerophosphoethanolamine	0.97 μmole/g	4
312			Phosphoethanolamine	5.4 μmoles/g	4
313			Urea	3.5 μmoles/g	4
314		Adult	Alanine	0.65 μmole/g	4
315			β-Alanine	0.05 μmole/g	4
316			Aspartic acid	2.4 μmoles/g	4
317			L-Cystathionine	0.07 μmole/g	4
318			Glutamic acid	9.5 μmoles/g	4
319			Glutamine	3.9 μmoles/g	4
320			Glycine	0.98 μmole/g	4
321			Isoleucine	0.04 μmole/g	4
322			Leucine	0.07 μmole/g	4
323			Lysine	0.07 μmole/g	4
324			Methionine	0.01 μmole/g	4
325			Ornithine	0.02 μmole/g	4
326			Phenylalanine	0.04 μmole/g	4
327			Proline	0.08 μmole/g	4
328			Serine	0.68 μmole/g	4
329			Taurine	1.6 μmoles/g	4
330			Threonine	0.26 μmole/g	4
331			Tyrosine	0.07 μmole/g	4
332			Valine	0.06 μmole/g	4
333			γ-Aminobutyric acid	1.9 μmoles/g	4
334			Ethanolamine	0.32 μmole/g	4
335			Glutathione	0.93 μmole/g	4
336			Glycerophosphoethanolamine	0.66 μmole/g	4
337			Homocarnosine	0.059 μmole/g	19
338			Phosphoethanolamine	2.9 μmoles/g	4
339			Urea	7.7 μmoles/g	4
340	Whole brain	Unspecified	Phosphatidal choline	0 mg/g	14
341			Phosphatidyl choline	16 mg/g	14
342			Phosphatidal ethanolamine	8.7 mg/g	14
343			Phosphatidyl ethanolamine	6.3 mg/g	14
344			Phosphatidyl serine	6.0 mg/g	14
345			Sphingomyelin	4 mg/g	14
346			Phosphatidyl inositols	3 mg/g	14
347			Cerebroside & sulfatide	9.8 mg/g	14
348			Cholesterol	16 mg/g	14

continued

Part II. Vertebrates Other Than Man

	Tissue	Age	Chemical	Concentration	Reference
349	Spinal cord	Fetal	H_2O	91% by wt	46
350			Ca	6.0 μmoles/g	46
351			K	51 μmoles/g	46
352			Na	93 μmoles/g	46
353		Adult	H_2O	74% by wt	46
354			Ca	6 μmoles/g	46
355			K	71 μmoles/g	46
356			Na	54 μmoles/g	46
357		Unspecified	Phosphatidal & phosphatidyl choline	17 mg/g	41
358			Phosphatidal & phosphatidyl ethanolamine	42 mg/g	41
359			Sphingomyelin	21 mg/g	41
360			Cerebroside	39 mg/g	41
361			Cholesterol	48 mg/g	41
362	Mitochondria[5]	Unspecified	Phosphatidal choline	0 mg/g[6]	14
363			Phosphatidyl choline	113 mg/g[6]	14
364			Phosphatidal ethanolamine	2.7 mg/g[6]	14
365			Phosphatidyl ethanolamine	6.6 mg/g[6]	14
366			Phosphatidal serine	0 mg/g[6]	14
367			Phosphatidyl serine	17 mg/g[6]	14
368			Sphingomyelin	10 mg/g[6]	14
369			Phosphatidyl inositols	15 mg/g[6]	14
370			Cerebroside & sulfatide	9 mg/g[6]	14
371			Cholesterol	19 mg/g[6]	14
372	Nerve ending par-	Unspecified	Phosphatidal choline	0 mg/g[6]	14
373	ticles[5]		Phosphatidyl choline	14 mg/g[6]	14
374			Phosphatidal ethanolamine	5.8 mg/g[6]	14
375			Phosphatidyl ethanolamine	6.3 mg/g[6]	14
376			Phosphatidal serine	0 mg/g[6]	14
377			Phosphatidyl serine	4.6 mg/g[6]	14
378			Sphingomyelin	1.9 mg/g[6]	14
379			Phosphatidyl inositols	1.4 mg/g[6]	14
380			Cerebroside & sulfatide	5.5 mg/g[6]	14
381			Cholesterol	79 mg/g[6]	14
382	Synaptic vesi-	Unspecified	Phosphatidal & phosphatidyl choline	260 mg/g[6]	47
383	cles[8]		Phosphatidal & phosphatidyl ethanolamine	95 mg/g[6]	47
384			Phosphatidal & phosphatidyl serine	77 mg/g[6]	47
385			Sphingomyelin	70 mg/g[6]	47
386			Phosphatidyl inositols	32 mg/g[6]	47
387			Cerebroside & sulfatide	0 mg/g[6]	47
388			Cholesterol	81 mg/g[6]	47
			Hamster[9]		
389	Nervous tissue	Newborn	Alanine	0.68 μmole/g	28
390			β-Alanine	0.036 μmole/g	28
391			Arginine	0.16 μmole/g	28
392			Asparagine	0.21 μmole/g	28
393			Aspartic acid	2.01 μmoles/g	28
394			Citrulline	0.00 μmole/g	28
395			L-Cystathionine	0.063 μmole/g	28
396			Cysteine	0.054 μmole/g	28
397			Glutamic acid	8.9 μmoles/g	28
398			Glutamine	0.7 μmole/g	28

[5] Subcellular organelle. [6] Dry weight. [8] Subcellular organelle; values recalculated, assuming total lipid content of 720 mg/g dry wt [41]. [9] *Mesocricetus auratus.*

continued

	Tissue	Age	Chemical	Concentration	Reference
399			Glycine	1.2 μmoles/g	28
400			Histidine	0.63 μmole/g	28
401			Hydroxyproline	0.00 μmole/g	28
402			Isoleucine	0.09 μmole/g	28
403			Leucine	0.099 μmole/g	28
404			Lysine	0.20 μmole/g	28
405			Methionine	0.00 μmole/g	28
406			Ornithine	0.00 μmole/g	28
407			Phenylalanine	0.054 μmole/g	28
408			Proline	0.12 μmole/g	28
409			Serine	4.9 μmoles/g	28
410			Taurine	5.5 μmoles/g	28
411			Threonine	0.00 μmole/g	28
412			Tyrosine	0.072 μmole/g	28
413			Valine	0.098 μmole/g	28
414			γ-Aminobutyric acid	1.9 μmoles/g	28
415			Ammonia	0.37 μmole/g	28
416			Ethanolamine	0.11 μmole/g	28
417			Glutathione	1.1 μmoles/g	28
418			Glycerophosphoethanolamine	3.5 μmoles/g	28
419			Phosphoethanolamine	1.3 μmoles/g	28
420			Urea	11 μmoles/g	28
421		Adult	Alanine	0.63 μmole/g	28
422			β-Alanine	0.00 μmole/g	28
423			Arginine	0.08 μmole/g	28
424			Asparagine	1.06 μmoles/g	28
425			Aspartic acid	1.27 μmoles/g	28
426			Citrulline	0.00 μmole/g	28
427			L-Cystathionine	0.00 μmole/g	28
428			Cysteine	0.00 μmole/g	28
429			Glutamic acid	3.84 μmoles/g	28
430			Glutamine	0.28 μmole/g	28
431			Glycine	1.1 μmoles/g	28
432			Histidine	0.13 μmole/g	28
433			Hydroxyproline	0.00 μmole/g	28
434			Isoleucine	0.04 μmole/g	28
435			Leucine	0.067 μmole/g	28
436			Lysine	0.26 μmole/g	28
437			Methionine	0.00 μmole/g	28
438			Ornithine	0.00 μmole/g	28
439			Phenylalanine	0.02 μmole/g	28
440			Proline	0.63 μmole/g	28
441			Serine	2.7 μmoles/g	28
442			Taurine	7.3 μmoles/g	28
443			Threonine	0.00 μmole/g	28
444			Tyrosine	0.059 μmole/g	28
445			Valine	0.067 μmole/g	28
446			γ-Aminobutyric acid	0.63 μmole/g	28
447			Ammonia	0.69 μmole/g	28
448			Ethanolamine	0.059 μmole/g	28
449			Glutathione	0.83 μmole/g	28
450			Glycerophosphoethanolamine	0.078 μmole/g	28
451			Phosphoethanolamine	0.12 μmole/g	28
452			Urea	9.2 μmoles/g	28

continued

Part II. Vertebrates Other Than Man

	Tissue	Age	Chemical	Concentration	Reference
453	Astroglia	Unspecified	Alanine	0.32 μmole/g	28
454			β-Alanine	0.018 μmole/g	28
455			Arginine	0.027 μmole/g	28
456			Asparagine	0.85 μmole/g	28
457			Aspartic acid	0.13 μmole/g	28
458			Citrulline	0.00 μmole/g	28
459			L-Cystathionine	0.023 μmole/g	28
460			Cysteine	0.04 μmole/g	28
461			Glutamic acid	0.35 μmole/g	28
462			Glutamine	3.8 μmoles/g	28
463			Glycine	0.26 μmole/g	28
464			Histidine	0.027 μmole/g	28
465			Hydroxyproline	0.00 μmole/g	28
466			Isoleucine	0.45 μmole/g	28
467			Leucine	0.075 μmole/g	28
468			Lysine	0.15 μmole/g	28
469			Methionine	0.00 μmole/g	28
470			Ornithine	0.08 μmole/g	28
471			Phenylalanine	0.036 μmole/g	28
472			Proline	0.23 μmole/g	28
473			Serine	0.56 μmole/g	28
474			Taurine	0.35 μmole/g	28
475			Threonine	0.00 μmole/g	28
476			Tyrosine	0.048 μmole/g	28
477			Valine	0.032 μmole/g	28
478			γ-Aminobutyric acid	0.027 μmole/g	28
479			Ammonia	0.53 μmole/g	28
480			Ethanolamine	0.40 μmole/g	28
481			Glutathione	0.28 μmole/g	28
482			Glycerophosphoethanolamine	0.11 μmole/g	28
483			Phosphoethanolamine	0.022 μmole/g	28
484			Urea	1.4 μmoles/g	28
			Monkey		
485	Nervous tissue	Adult	Alanine	1.7 μmoles/g	39,40
486			Aspartic acid	2.7 μmoles/g	39,40
487			L-Cystathionine + cysteine	3.3 μmoles/g	39,40
488			Glutamic acid & glutamine	25 μmoles/g	39,40
489			Glycine	2.7 μmoles/g	39,40
490			Serine	1.9 μmoles/g	39,40
491			Valine	0.62 μmole/g	39,40
492			γ-Aminobutyric acid	1.9 μmoles/g	39,40
493	Sciatic nerve [10]	Unspecified	Phosphatidyl choline	5.5 mg/g	38
494			Phosphatidal ethanolamine	11.3 mg/g	38
495			Phosphatidyl ethanolamine	1.6 mg/g	38
496			Phosphatidal serine	0.6 mg/g	38
497			Phosphatidyl serine	6.8 mg/g	38
498			Sphingomyelin	12.7 mg/g	38
499			Phosphatidyl inositols	0.6 mg/g	38
500			Cerebroside & sulfatide	16.6 mg/g	38
501			Cholesterol	31 mg/g	38

[10] Rhesus monkey.

continued

Part II. Vertebrates Other Than Man

	Tissue	Age	Chemical	Concentration	Reference
			Mouse		
502	Nervous tissue	Newborn	Alanine	3.3 μmoles/g	3
503			β-Alanine	0.05 μmole/g	3
504			Arginine	0.10 μmole/g	3
505			Aspartic acid	1.9 μmoles/g	3
506			Citrulline	0.21 μmole/g	3
507			L-Cystathionine	0.05 μmole/g	3
508			Cysteine	0.007 μmole/g	3
509			Glutamic acid	4.3 μmoles/g	3
510			Glutamine	4.8 μmoles/g	3
511			Glycine	2.7 μmoles/g	3
512			Histidine	0.14 μmole/g	3
513			Hydroxyproline	0.15 μmole/g	3
514			Isoleucine	0.14 μmole/g	3
515			Leucine	0.24 μmole/g	3
516			Lysine	0.15 μmole/g	3
517			Methionine	0.02 μmole/g	3
518			Ornithine	0.05 μmole/g	3
519			Phenylalamine	0.13 μmole/g	3
520			Proline	0.49 μmole/g	3
521			Serine	1.2 μmoles/g	3
522			Taurine	16.5 μmoles/g	3
523			Threonine	0.68 μmole/g	3
524			Tyrosine	0.14 μmole/g	3
525			Valine	0.08 μmole/g	3
526			γ-Aminobutyric acid	1.4 μmoles/g	3
527			Ethanolamine	1.6 μmoles/g	3
528			Glutathione	1.4 μmoles/g	3
529			Glycerophosphoethanolamine	0.25 μmole/g	3
530			Phosphoethanolamine	4.8 μmoles/g	3
531			Urea	6.7 μmoles/g	3
532		Adult	Alanine	0.66 μmole/g	3
533			β-Alanine	0.05 μmole/g	3
534			Arginine	0.13 μmole/g	3
535			Aspartic acid	3.4 μmoles/g	3
536			L-Cystathionine	0.06 μmole/g	3
537			Cysteine	0.01 μmole/g	3
538			Glutamic acid	12 μmoles/g	3
539			Glutamine	6.6 μmoles/g	3
540			Glycine	1.6 μmoles/g	3
541			Histidine	0.05 μmole/g	3
542			Isoleucine	0.04 μmole/g	3
543			Leucine	0.08 μmole/g	3
544			Lysine	0.18 μmole/g	3
545			Methionine	0.03 μmole/g	3
546			Ornithine	0.01 μmole/g	3
547			Phenylalanine	0.04 μmole/g	3
548			Proline	0.06 μmole/g	3
549			Serine	0.78 μmole/g	3
550			Taurine	9.1 μmoles/g	3
551			Threonine	0.39 μmole/g	3
552			Tyrosine	0.05 μmole/g	3

continued

Part II. Vertebrates Other Than Man

	Tissue	Age	Chemical	Concentration	Reference
553			Valine	0.13 μmole/g	3
554			γ-Aminobutyric acid	2.5 μmoles/g	3
555			Ammonia	0.43 μmole/g	3
556			Ethanolamine	0.12 μmole/g	3
557			Glutathione	1.0 μmole/g	3
558			Glycerophosphoethanolamine	0.47 μmole/g	3
559			Phosphoethanolamine	2.0 μmoles/g	3
560			Urea	4.1 μmoles/g	3
			Rabbit		
561	Nervous tissue	Newborn	Alanine	0.94 μmole/g	2
562			β-Alanine	0.04 μmole/g	2
563			Aspartic acid	2.8 μmoles/g	2
564			L-Cystathionine	0.03 μmole/g	2
565			Glutamic acid	11.1 μmoles/g	2
566			Glutamine	4.9 μmoles/g	2
567			Glycine	2.6 μmoles/g	2
568			Hydroxyproline	0.05 μmole/g	2
569			Isoleucine	0.04 μmole/g	2
570			Leucine	0.08 μmole/g	2
571			Lysine	0.02 μmole/g	2
572			Methionine	0.02 μmole/g	2
573			Ornithine	0.03 μmole/g	2
574			Phenylalanine	0.03 μmole/g	2
575			Proline	0.18 μmole/g	2
576			Serine	1.8 μmoles/g	2
577			Taurine	3.4 μmoles/g	2
578			Threonine	0.31 μmole/g	2
579			Tyrosine	0.05 μmole/g	2
580			Valine	0.10 μmole/g	2
581			γ-Aminobutyric acid	1.9 μmoles/g	2
582			Ethanolamine	0.29 μmole/g	2
583			Glutathione	1.2 μmoles/g	2
584			Glycerophosphoethanolamine	0.97 μmole/g	2
585			Phosphoethanolamine	5.4 μmoles/g	2
586			Urea	3.5 μmoles/g	2
587		Adult	Alanine	0.65 μmole/g	2
588			β-Alanine	0.05 μmole/g	2
589			Aspartic acid	2.4 μmoles/g	2
590			L-Cystathionine	0.07 μmole/g	2
591			Glutamic acid	9.5 μmoles/g	2
592			Glutamine	3.9 μmoles/g	2
593			Glycine	0.98 μmole/g	2
594			Isoleucine	0.04 μmole/g	2
595			Leucine	0.07 μmole/g	2
596			Lysine	0.07 μmole/g	2
597			Methionine	0.10 μmole/g	2
598			Ornithine	0.02 μmole/g	2
599			Phenylalanine	0.04 μmole/g	2
600			Proline	0.08 μmole/g	2
601			Serine	0.68 μmole/g	2
602			Taurine	1.6 μmoles/g	2
603			Threonine	0.26 μmole/g	2

continued

Part II.   Vertebrates Other Than Man

	Tissue	Age	Chemical	Concentration	Reference
604			Tyrosine	0.07 μmole/g	2
605			Valine	0.06 μmole/g	2
606			γ-Aminobutyric acid	1.9 μmoles/g	2
607			Ethanolamine	0.32 μmole/g	2
608			Glutathione	0.93 μmole/g	2
609			Glycerophosphoethanolamine	0.66 μmole/g	2
610			Homocarnosine	0.11 μmole/g	19
611			Phosphoethanolamine	2.9 μmoles/g	2
612			Urea	7.7 μmoles/g	2
613	Whole brain	Unspecified	H_2O	78% by wt	6,7,17,26
614			Cl	40 μmoles/g	6,7,17,26
615			Mg	5.7 μmoles/g	6,7,17,26
616			P	106 μmoles/g	6,7,17,26
617			K	90 μmoles/g	6,7,17,26
618			Na	51 μmoles/g	6,7,17,26
619	Myelin	Unspecified	H_2O	400 mg/g[4]	11
620			Phosphatidal & phosphatidyl choline	34 mg/g	11
621			Phosphatidal & phosphatidyl ethanolamine	59 mg/g	11
622			Phosphatidal & phosphatidyl serine	26 mg/g	11
623			Sphingomyelin	32 mg/g	11
624			Phosphatidyl inositols	5.6 mg/g	11
625			Cerebroside	66 mg/g	11
626			Sulfatide	16 mg/g	11
627			Sterol	96 mg/g	11
628			Protein	230 mg/g	11
629	Cortex	Unspecified	H_2O	710 mg/g	9
630			Phospholipid	38 mg/g	9
631			Glycolipid	11 mg/g	9
632			Sterol	8.9 mg/g	9
633			Protein	230 mg/g[3]	9
634	White matter	Unspecified	H_2O	76% by wt	6,7,17,26
635			Cl	37 μmoles/g	6,7,17,26
636	Gray matter	Unspecified	H_2O	81% by wt	6,7,17,26
637			Cl	40 μmoles/g	6,7,17,26
638			K	95 μmoles/g	6,7,17,26
639			Na	52 μmoles/g	6,7,17,26
640	Spinal cord	Unspecified	H_2O	68% by wt; 630 mg/g	9,26
641			Cl	42 μmoles/g	26
642			Na	62 μmoles/g	26
643			Phospholipid	115 mg/g	9
644			Phosphatidal & phosphatidyl choline	16-17 mg/g	9,27
645			Phosphatidal & phosphatidyl ethanolamine	48-51 mg/g	9,27
646			Sphingomyelin	29-36 mg/g	9,27
647			Glycolipid	20 mg/g	9
648			Phosphatidyl inositols	6.3 mg/g	9,27
649			Cerebroside	51-54 mg/g	9,27
650			Sterol	59 mg/g	9
651			Cholesterol	48-54 mg/g	9,27
652			Protein	190 mg/g[3]	9
653	Ischial nerve	Unspecified	H_2O	650 mg/g	9
654			Phospholipid	97 mg/g	9
655			Glycolipid	19 mg/g	9

[3] Estimated by difference of dry weight minus total lipids.   [4] Assumed H_2O content.

continued

Part II. Vertebrates Other Than Man

	Tissue	Age	Chemical	Concentration	Reference
656			Sterol	41 mg/g	9
657			Protein	190 mg/g[3/]	9
658	Sciatic nerve	Unspecified	H_2O	59% by wt	24
659			Cl	50 μmoles/g	24
660			K	57 μmoles/g	24
661			Na	78 μmoles/g	24
662	Microsomes [11/]	Unspecified	Phosphatidal & phosphatidyl choline	103 mg/g[6/]	11
663			Phosphatidal & phosphatidyl ethanolamine	123 mg/g[6/]	11
664			Phosphatidal & phosphatidyl serine, & phosphatidyl inositols	68 mg/g[6/]	11
665			Sphingomyelin	52 mg/g[6/]	11
666			Cerebroside	55 mg/g[6/]	11
667			Sulfatide	5 mg/g[6/]	11
668			Cholesterol	162 mg/g[6/]	11
			Rat		
669	Whole brain	Unspecified	H_2O	78% by wt	10,20,34,49,50
670			Cl	33 μmoles/g	10,20,34,49,50
671			P	98 μmoles/g	10,20,34,49,50
672			K	89 μmoles/g	10,20,34,49,50
673			Na	50 μmoles/g	10,20,34,49,50
674			Phosphatidal choline	0.2 mg/g	45
675			Phosphatidyl choline	19 mg/g	45
676			Phosphatidal ethanolamine	10 mg/g	45
677			Phosphatidyl ethanolamine	8.3 mg/g	45
678			Phosphatidyl serine	6.5 mg/g	45
679			Sphingomyelin	2.9 mg/g	45
680			Phosphatidyl inositols	3 mg/g	45
681			Cerebroside	18 mg/g	45
682			Sulfatide	3.8 mg/g	45
683			Cholesterol	16 mg/g	45
684	Myelin	Unspecified	H_2O	400 mg/g[4/]	22,36
685			Phosphatidal choline	34 mg/g	22,36
686			Phosphatidyl choline	15 mg/g	22,36
687			Phosphatidal ethanolamine	17 mg/g	22,36
688			Phosphatidyl ethanolamine	66 mg/g	22,36
689			Phosphatidal & phosphatidyl serine	29 mg/g	22,36
690			Sphingomyelin	16 mg/g	22,36
691			Phosphatidyl inositols	8.4 mg/g	22,36
692			Cerebroside	67 mg/g	22,36
693			Sulfatide	16 mg/g	22,36
694			Sterol	78 mg/g	22,36
695			Protein	150-290 mg/g	22,36
696	Cortex	Unspecified	H_2O	800 mg/g	9
697			Phospholipid	45 mg/g	9
698			Glycolipid	12 mg/g	9
699			Sterol	9.4 mg/g	9
700			Protein	130 mg/g[3/]	9
701	White matter	Unspecified	H_2O	79% by wt	10,20,34,49,50
702	Gray matter	Unspecified	H_2O	81% by wt	10,20,34,49,50
703			Cl	30 μmoles/g	10,20,34,49,50
704			K	98 μmoles/g	10,20,34,49,50

[3/] Estimated by difference of dry weight minus total lipids.
[4/] Assumed H_2O content. [6/] Dry weight. [11/] Subcellular organelle; values recalculated, assuming total lipid content of 600 mg/g dry wt [41].

continued

Part II. Vertebrates Other Than Man

	Tissue	Age	Chemical	Concentration	Reference
705			Na	42 μmoles/g	10,20,34,49,50
706	Cerebrum	Unspecified	Alanine	0.67 μmole/g	37
707			Arginine	0.11 μmole/g	37
708			Aspartic acid	2.4 μmoles/g	37
709			L-Cystathionine	0.015 μmole/g	37
710			Cysteine	0.012 μmole/g	37
711			Glutamic acid	12 μmoles/g	37
712			Glycine	0.63 μmole/g	37
713			Histidine	0.066 μmole/g	37
714			Isoleucine	0.025 μmole/g	37
715			Leucine	0.064 μmole/g	37
716			Lysine	0.21 μmole/g	37
717			Methionine	0.037 μmole/g	37
718			Ornithine	0.02 μmole/g	37
719			Phenylalanine	0.05 μmole/g	37
720			Serine & asparagine	1.2 μmoles/g	37
721			Taurine	7.6 μmoles/g	37
722			Threonine & glutamine	2.4 μmoles/g	37
723			Tyrosine	0.073 μmole/g	37
724			Valine	0.083 μmole/g	37
725			γ-Aminobutyric acid	2.3 μmoles/g	37
726			Ammonia	0.99 μmole/g	37
727			Ethanolamine	0.11 μmole/g	37
728			Glutathione	1.1 μmoles/g	37
729			Glycerophosphoethanolamine	0.40 μmole/g	37
730			Homocarnosine	0 μmole/g	19
731			Phosphoethanolamine	2.2 μmoles/g	37
732	Pons	Unspecified	Alanine	0.56 μmole/g	37
733			Arginine	0.2 μmole/g	37
734			Aspartic acid	2.7 μmoles/g	37
735			Citrulline	0.14 μmole/g	37
736			L-Cystathionine	0.053 μmole/g	37
737			Cysteine	0.06 μmole/g	37
738			Glutamic acid	6.9 μmoles/g	37
739			Glutamine	0.77 μmole/g	37
740			Glycine	3.5 μmoles/g	37
741			Histidine	0.044 μmole/g	37
742			Isoleucine	0.034 μmole/g	37
743			Leucine	0.064 μmole/g	37
744			Lysine	0.44 μmole/g	37
745			Methionine	0.043 μmole/g	37
746			Ornithine	0.051 μmole/g	37
747			Phenylalanine	0.05 μmole/g	37
748			Serine & asparagine	0.63 μmole/g	37
749			Taurine	2.0 μmoles/g	37
750			Threonine & glutamine	1.6 μmoles/g	37
751			Tyrosine	0.059 μmole/g	37
752			Valine	0.52 μmole/g	37
753			γ-Aminobutyric acid	1.8 μmoles/g	37
754			Ammonia	0.89 μmole/g	37
755			Ethanolamine	0.18 μmole/g	37
756			Glutathione	0.62 μmole/g	37
757			Glycerophosphoethanolamine	0.60 μmole/g	37

continued

Part II. Vertebrates Other Than Man

	Tissue	Age	Chemical	Concentration	Reference
758			Homocarnosine	0.059 μmole/g	19
759			Phosphoethanolamine	0.51 μmole/g	37
760	Astrocytoma	Unspecified	Alanine	2.7 μmoles/g	28
761			β-Alanine	0.029 μmole/g	28
762			Arginine	0.18 μmole/g	28
763			Asparagine	0.44 μmole/g	28
764			Aspartic acid	0.60 μmole/g	28
765			Citrulline	0.00 μmole/g	28
766			L-Cystathionine	0.00 μmole/g	28
767			Cysteine	0.058 μmole/g	28
768			Glutamic acid	3.0 μmoles/g	28
769			Glutamine	0.72 μmole/g	28
770			Glycine	5.5 μmoles/g	28
771			Histidine	0.22 μmole/g	28
772			Hydroxyproline	0.00 μmole/g	28
773			Isoleucine	0.18 μmole/g	28
774			Leucine	0.29 μmole/g	28
775			Lysine	0.61 μmole/g	28
776			Methionine	0.00 μmole/g	28
777			Ornithine	0.00 μmole/g	28
778			Phenylalanine	0.17 μmole/g	28
779			Proline	0.85 μmole/g	28
780			Serine	2.3 μmoles/g	28
781			Taurine	1.5 μmoles/g	28
782			Threonine	0.00 μmole/g	28
783			Tyrosine	0.19 μmole/g	28
784			Valine	0.12 μmole/g	28
785			γ-Aminobutyric acid	0.14 μmole/g	28
786			Ammonia	2.8 μmoles/g	28
787			Ethanolamine	0.12 μmole/g	28
788			Glutathione	1.2 μmoles/g	28
789			Glycerophosphoethanolamine	0.55 μmole/g	28
790			Phosphoethanolamine	3.1 μmoles/g	28
791			Urea	3.6 μmoles/g	28
792	Spinal cord	Unspecified	Phosphatidal & phosphatidyl choline	21 mg/g	9,43
793			Phosphatidal & phosphatidyl ethanolamine	39 mg/g	9,43
794			Sphingomyelin	18 mg/g	9,43
795			Cerebroside	42 mg/g	9,43
796			Cholesterol	36 mg/g	9,43
797	Microsomes [11]	Unspecified	Phosphatidal & phosphatidyl choline	120 mg/g [6]	11
798			Phosphatidal & phosphatidyl ethanolamine	116 mg/g [6]	11
799			Phosphatidal & phosphatidyl serine	39 mg/g [6]	11
800			Sphingomyelin	27 mg/g [6]	11
801			Phosphatidyl inositols	14 mg/g [6]	11
802			Cerebroside	39 mg/g [6]	11
803			Sulfatide	5 mg/g [6]	11
804			Cholesterol	134 mg/g [6]	11
805	Mitochondria [5]	Unspecified	Phosphatidal & phosphatidyl choline	113 mg/g [6]	30,36
806			Phosphatidal & phosphatidyl ethanolamine	55 mg/g [6]	30,36
807			Phosphatidal & phosphatidyl serine	16 mg/g [6]	30,36
808			Sphingomyelin	25 mg/g [6]	30,36
809			Phosphatidyl inositols	9 mg/g [6]	30,36

[5] Subcellular organelle. [6] Dry Weight. [11] Subcellular organelle; values recalculated, assuming total lipid content of 600 mg/g dry wt [41].

continued

Part II.   Vertebrates Other Than Man

	Tissue	Age	Chemical	Concentration	Reference
810			Cerebroside & sulfatide	22 mg/g[6/]	30,36
811			Cholesterol	51 mg/g[6/]	30,36
812	Nerve ending par-	Unspecified	Phosphatidal & phosphatidyl choline	15 mg/g[6/]	11
813	ticles [12/]		Phosphatidal & phosphatidyl ethanolamine	8.3 mg/g[6/]	11
814			Phosphatidal & phosphatidyl serine	2.3 mg/g[6/]	11
815			Sphingomyelin	3.3 mg/g[6/]	11
816			Phosphatidyl inositols	1.1 mg/g[6/]	11
817			Cerebroside & sulfatide	3.9 mg/g[6/]	11
818			Cholesterol	97 mg/g[6/]	11
			Sheep		
819	White matter	Unspecified	Phosphatidyl choline	16 mg/g	1
820			Phosphatidal & phosphatidyl ethanolamine	21 mg/g	1
821			Phosphatidyl serine	18 mg/g	1
822			Sphingomyelin	12 mg/g	1
823			Cerebroside	41 mg/g	1
824			Cholesterol	34 mg/g	1
825	Myelin	Unspecified	H_2O	400 mg/g[4/]	11
826			Phospholipid	153 mg/g	11
827			Cerebroside & sulfatide	140 mg/g	11
828			Sterol	130 mg/g	11
829			Protein	180 mg/g	11
830	Gray matter	Unspecified	Phosphatidyl choline	12 mg/g	5
831			Phosphatidal & phosphatidyl ethanolamine	16 mg/g	5
832			Phosphatidyl serine	0.5 mg/g	5
833			Sphingomyelin	3 mg/g	5
834			Cerebroside	4.9 mg/g	5
835			Cholesterol	10 mg/g	5
836	Sciatic nerve	Unspecified	Phosphatidyl choline	4.2 mg/g	38
837			Phosphatidal ethanolamine	9.2 mg/g	38
838			Phosphatidyl ethanolamine	1.6 mg/g	38
839			Phosphatidal serine	1.1 mg/g	38
840			Phosphatidyl serine	6.2 mg/g	38
841			Sphingomyelin	10.6 mg/g	38
842			Phosphatidyl inositols	1.6 mg/g	38
843			Cerebroside & sulfatide	14.5 mg/g	38
844			Cholesterol	20 mg/g	38
			Swine		
845	Whole brain	Unspecified	H_2O	76% by wt	48
846			Cl	41 μmoles/g	48
847			Mg	6.1 μmoles/g	48
848			P	125 μmoles/g	48
849			K	76 μmoles/g	48
850			Na	61 μmoles/g	48
851			Phosphatidyl choline	18 mg/g	13
852			Phosphatidal & phosphatidyl ethanolamine	22 mg/g	13
853			Phosphatidyl serine	14 mg/g	13
854			Sphingomyelin	7.3 mg/g	13
855			Cerebroside	21 mg/g	13
856			Cholesterol	34 mg/g	13

[4/] Assumed H_2O content. [6/] Dry weight. [12/] Subcellular organelle; values recalculated, assuming total lipid content of 460 mg/g dry wt [41].

continued

Part II. Vertebrates Other Than Man

	Tissue	Age	Chemical	Concentration	Reference
			Chicken		
857	Sciatic nerve	Unspecified	Phosphatidal choline	0.5 mg/g	38
858			Phosphatidyl choline	6.0 mg/g	38
859			Phosphatidal ethanolamine	12.6 mg/g	38
860			Phosphatidyl ethanolamine	1.7 mg/g	38
861			Phosphatidyl serine	4.7 mg/g	38
862			Sphingomyelin	9.2 mg/g	38
863			Phosphatidyl inositols	0.5 mg/g	38
864			Cerebroside & sulfatide	15 mg/g	38
865			Cholesterol	20 mg/g	38
			Pigeon		
866	Whole brain, my-elin	Unspecified	H_2O	400 mg/g[4]	11
867			Phosphatidal & phosphatidyl choline	31 mg/g	11
868			Phosphatidal & phosphatidyl ethanolamine	55 mg/g	11
869			Phosphatidal & phosphatidyl serine	11 mg/g	11
870			Sphingomyelin	9.8 mg/g	11
871			Phosphatidyl inositols	3.1 mg/g	11
872			Cerebroside	77 mg/g	11
873			Sulfatide	16 mg/g	11
874			Sterol	92 mg/g	11
875			Protein	284 mg/g	11
876	Spinal cord	Unspecified	Phosphatidal & phosphatidyl choline	11 mg/g	27
877			Phosphatidal & phosphatidyl ethanolamine	26 mg/g	27
878			Sphingomyelin	18 mg/g	27
879			Cerebroside	42 mg/g	27
880			Cholesterol	39 mg/g	27
			Frog		
881	Whole brain [13]	Unspecified	H_2O	400 mg/g[4]	11
882			Phosphatidal & phosphatidyl choline	86 mg/g	11
883			Phosphatidal & phosphatidyl ethanolamine	80 mg/g	11
884			Phosphatidal & phosphatidyl serine	16 mg/g	11
885			Sphingomyelin	22 mg/g	11
886			Cerebroside	41 mg/g	11
887			Sterol	41 mg/g	11
888			Protein	234 mg/g	11
889	Sciatic nerve	Unspecified	H_2O	75% by wt	16
890			Ca	3.6 µmoles/g	16
891			Cl	32 µmoles/g	16
892			Mg	8 µmoles/g	16
893			P	33 µmoles/g	16
894			K	48 µmoles/g	16
895			Na	62 µmoles/g	16

[4] Assumed H_2O content. [13] *Rana temporaria.*

Contributor: Mokrasch, Lewis C.

continued

145. CHEMICAL COMPOSITION OF NERVOUS TISSUE

Part II. Vertebrates Other Than Man

References

[1] Adams, C. W., and A. N. Davison. 1965. In C. W. Adams, ed. Neurohistochemistry. Elsevier, Amsterdam. pp. 332-400.

[2] Agrawal, H. C., et al. 1966. Brain Res. 3:374.

[3] Agrawal, H. C., et al. 1968. J. Neurochem. 15:917.

[4] Agrawal, H. C., et al. 1969. Ibid. 15:529.

[5] Amaducci, L., et al. 1962. Ibid. 9:509.

[6] Ames, A., and F. B. Nesbitt. 1958. Ibid. 3:116.

[7] Aprison, M. H., et al. 1960. Ibid. 5:150.

[8] Autilio, L. A., et al. 1964. Ibid. 11:17.

[9] Brante, G. 1949. Acta Physiol. Scand., Suppl. 63.

[10] Crowell, C. D., et al. 1934. J. Dent. Res. 14:25.

[11] Cuzner, M. L., et al. 1965. J. Neurochem. 12:469.

[12] Davies, F., et al. 1952. J. Physiol. (London) 118:276.

[13] Dickerson, J. W., et al. 1967. Proc. Roy. Soc. B116:396.

[14] Eichberg, J., et al. 1964. Biochem. J. 92:91.

[15] Eichelberger, L., et al. 1949. Amer. J. Physiol. 156:129.

[16] Fenn, W. O., et al. 1934. Ibid. 110:74.

[17] Graves, J., and H. E. Himwich. 1955. Ibid. 180:205.

[18] Himwich, W. A., and H. C. Agrawal. 1969. Handb. Neurochem. 1:33.

[19] Kanazawa, A., and I. Sano. 1967. J. Neurochem. 14:211.

[20] Koch, M. L. 1913. J. Biol. Chem. 14:267.

[21] Koch, W., and S. A. Manin. 1907. J. Physiol. (London) 36:xxxvi.

[22] Korey, S. R., et al. 1958. J. Neuropathol. Exp. Neurol. 17:430.

[23] Krnjevic, K. 1955. J. Physiol. (London) 128:473.

[24] Lowenthal, A. 1961. In J. Folch-Pi, ed. Chemical Pathology of the Nervous System. Pergamon Press, Oxford. pp. 299-306.

[25] Mandel, P., et al. 1966. J. Neurochem. 13:533.

[26] Manery, J. F., and A. B. Hastings. 1939. J. Biol. Chem. 127:657.

[27] McColl, J. D., and R. J. Rossiter. 1952. J. Exp. Biol. 29:203.

[28] Mokrasch, L. C. 1971. Brain Res. 25:672.

[29] Norton, W. T., and L. A. Autilio. 1966. J. Neurochem. 13:213.

[30] Nussbaum, J. L., and P. Mandell. 1965. Bull. Soc. Chim. Biol. 47:365.

[31] O'Brien, J. S., et al. 1967. J. Neurochem. 14:357.

[32] Okey, R. 1945. J. Amer. Diet. Ass. 21:341.

[33] Parsons, D., and R. E. Basford. 1967. J. Neurochem. 14:283.

[34] Pappius, H. M., and K. A. C. Elliott. 1956. Can. J. Biochem. Physiol. 34:1007.

[35] Rouser, G., et al. 1963. J. Amer. Oil Chem. Soc. 40:425.

[36] Seminario, L. M., et al. 1964. J. Neurochem. 11:197.

[37] Shaw, R. K., and J. D. Heine. 1965. Ibid. 12:151.

[38] Sheltawy, A., and R. M. Dawson. 1966. Biochem. J. 100:12.

[39] Singh, S. I., and C. L. Malhotra. 1962. J. Neurochem. 9:37.

[40] Singh, S. I., and C. L. Malhotra. 1964. Ibid. 11:865.

[41] Stewart-Wallace, A. M. 1939. Brain 62:426.

[42] Tupikova, N., and R. W. Gerrard. 1937. J. Biol. Chem. 127:657.

[43] Webster, G. R. 1960. Biochim. Biophys. Acta 44:109.

[44] Weil, A. 1914. Hoppe Seylers Z. Physiol. Chem. 89:349.

[45] Wells, M. A., and J. C. Dittmer. 1967. Biochemistry 6:3169.

[46] Wender, M., and M. Hierowski. 1950. J. Neurochem. 5:105.

[47] Whittaker, V. P. 1970. Handb. Neurochem. 2:327.

[48] Widdowson, E. M., and J. W. Dickerson. 1960. Biochem. J. 77:30.

[49] Widdowson, E. M., and J. W. Dickerson. 1964. Miner. Metab. 2(A):180.

[50] Woodbury, D. M. 1958. In W. F. Windle, ed. Biology of Neuroglia. C. C. Thomas, Springfield, Ill. p. 120.

Part III. Invertebrates

Concentration: Values refer to volume of axoplasm, unless otherwise indicated.

	Class & Species	Tissue	Chemical	Concentration	Reference
			Arthropoda		
	Merostomata				
1	*Limulus* sp.	Axoplasm, peripheral nerve	K	123 μM	6
	Crustacea				
2	*Astacus* sp.	Axoplasm, stretch receptor	Cl	13 μM	1,4
3			K	180-265 μM	
4			Na	12-17 μM	
5	*Cancer* sp.	Axoplasm, peripheral nerve	K	134 μM	4

continued

Part III.   Invertebrates

	Class & Species	Tissue	Chemical	Concentration	Reference
6	*Carcinus* sp.	Axoplasm, peripheral nerve	Ca	27 μM	4,6
7			K	412-432 μM	
8			Na	30-53 μM	
9	*Homarus* sp.	Axoplasm, peripheral nerve	Ca	65 μM	4
10			Cl	107 μM	
11			K	203 μM	
12		ventral nerve	K	307 μM	
13	*Maja* sp.[1]	Axoplasm, peripheral nerve	K	132 μM	4
14	*Orconectes* sp.	Axoplasm, stretch receptor	K	180 μM	1,2
15			Na	12 μM	
16	*Procambarus* sp.	Axoplasm, stretch receptor	K	180 μM	1,2
17			Na	12 μM	
18	Lobster	Leg nerve	Phosphatidyl choline	3.3 mg/g[2]	3
19			Phosphatidal ethanolamine	1.9 mg/g[2]	
20			Phosphatidyl ethanolamine	0.7 mg/g[2]	
21			Phosphatidyl serine	0.7 mg/g[2]	
22			Sphingomyelin	1.6 mg/g[2]	
23			Phosphatidyl inositols	1.0 mg/g[2]	
24			Cholesterol	2.7 mg/g[2]	
	Insecta				
25	*Carausius* sp.	Axoplasm, nerve cord	Ca	30 μM	6
26			Cl	133 μM	
27			K	556 μM	
28			Na	86 μM	
29	*Periplaneta* sp.	Axoplasm, nerve cord	Ca	15 μM	6
30			K	225 μM	
31			Na	67 μM	
	Mollusca				
	Cephalopoda				
32	*Loligo* sp.	Axoplasm	Ca	14 μM	5,6
33			Cl	140 μM	
34			K	369 μM	
35			Na	44-65 μM	
36	*Sepioteuthis* sp.	Axoplasm, giant axon	Cl	135 μM	5
37			K	335 μM	
38			Na	52 μM	
39		neurilemma[3]	Cl	167 μM	
40			K	220 μM	
41			Na	312 μM	
	Gastropoda				
42	*Aplysia* sp.	Axoplasm	Ca	0.6 μM	6
43			Cl	46-173 μM	
44			K	232 μM	
45			Na	67 μM	

[1] Synonym: *Maia* sp.   [2] Fresh weight of nerve tissue.   [3] Synonym: Schwann cell.

Contributor: Mokrasch, Lewis C.

References

[1] Giacobini, E. 1967. Protoplasma 63:52.

[2] Giacobini, E., et al. 1967. Acta Physiol. Scand. 71: 391.

[3] Sheltawy, A., and R. M. Dawson. 1966. Biochem. J. 100:12.

[4] Treherne, J. E. 1966. The Neurochemistry of Arthropods. Univ. Press, Cambridge, England.

[5] Villegas, J., et al. 1965. J. Gen. Physiol. 49:1.

[6] Wiersma, C. A., ed. 1967. Invertebrate Nervous Systems. Univ. Chicago Press, Chicago.

146. PHYSICAL CONSTANTS OF CONTRACTILE PROTEINS

Intrinsic Sedimentation Coefficient: S = Svedberg unit (1 S = 0.1 ps). **Intrinsic Viscosity:** dl/g = deciliters/gram. **No. of Subunits:** In cases where data were controversial or uncertain, exact numbers have not been specified; subfragments with multiple cleavages not reported. **Percent α-Helix:** Averaged values determined by optical rotatory dispersion are scaled relative to paramyosin as 100. **Shape:** The width of the tails or rods are all $\sim$20 Å.

	Protein	Intrinsic Sedimentation Coefficient, S	Intrinsic Viscosity dl/g	Molecular Weight	No. of Subunits	Percent α-Helix	Shape	Reference
1	Myosin	6.4	2.1	470,000	>2	57	1400-Å tail + 2 heads 100-Å	6-8,22
	Proteolytic fragments							
2	Heavy meromyosin	7.2	0.49	340,000	...	46	400- to 500-Å tail + 2 heads	14,18,22
3	Light meromyosin	2.9	1.2	140,000	2	90	900- to 1000-Å rod	12,14,18,22
4	Single-headed heavy meromyosin	6.0	.....	215,000	...	60	400- to 500-Å tail + single head	17
5	Subfragment 1	5.8	0.06	115,000	...	33	100-Å globule	22,27
6	Subfragment 2	2.7	0.4	62,000	...	87	400- to 500-Å rod	21,22
7	Rod	3.4	2.4	220,000	2	94	1450-Å rod	2,22
	Nonproteolytic subunits							
8	Heavy chains	...	.....	200,000	...	.....	....	6,7
9	Light chains	...	.....	16,000-25,000	...	.....	....	19,25
10	G-Actin	3.3	0.04	46,000	1	26	55-Å globule	4,11,16,23,24
11	α-Actinin	6.2	0.09	190,000	2	.....	....	9
12	Troponin [1]	4.0	.....	80,000(?)	>2	35	....	5,10,13
13	Tropomyosin	2.6	0.34	65,000	2	90	400-Å rod	1,15,26
14	Paramyosin	3.1	1.9	210,000	2	100	1300-Å rod	3,20

[1] Data on native troponin incomplete.

Contributor: Lowey, Susan

References

[1] Caspar, D. L. D., et al. 1969. J. Mol. Biol. 41:87.

[2] Cohen, C., et al. 1970. Ibid. 47:605.

[3] Cohen, C., et al. 1971. Ibid. 56:223.

[4] Cohen, L. B. 1966. Arch. Biochem. Biophys. 117:289.

[5] Ebashi, S. 1971. J. Biochem. (Tokyo) 69:441.

[6] Gazith, J., et al. 1970. J. Biol. Chem. 245:15.

[7] Gershman, L. C., et al. 1969. Ibid. 244:2726.

[8] Godfrey, J. E., and W. F. Harrington. 1970. Biochemistry 9:894.

[9] Goll, D. E., et al. 1971. Biophys. Soc. Abstr. 107a.

[10] Greaser, M., and J. Gergely. 1971. J. Biol. Chem. 246:4226.

[11] Hanson, J., and J. Lowy. 1963. J. Mol. Biol. 6:46.

[12] Harrison, R. G., et al. 1971. Ibid. 59:531.

[13] Hartshorne, D. J., et al. 1969. Biochim. Biophys. Acta 175:320.

[14] Holtzer, A., et al. 1962. Symp. Fundam. Cancer Res. 15:259.

[15] Holtzer, A., et al. 1965. Biochemistry 4:2401.

[16] Johnson, P., et al. 1967. Biochem. J. 105:361.

[17] Lowey, S. 1970. Int. Congr. Biochem., 8th, Abstr., p. 28.

[18] Lowey, S., and C. Cohen. 1962. J. Mol. Biol. 4:293.

[19] Lowey, S., and D. Risby. 1971. Nature (London) 234:81.

[20] Lowey, S., et al. 1963. J. Mol. Biol. 7:234.

[21] Lowey, S., et al. 1967. Ibid. 23:287.

[22] Lowey, S., et al. 1969. Ibid. 42:1.

[23] Nagy, B. 1966. Biochim. Biophys. Acta 115:498.

[24] Rees, M. K., and M. Young. 1967. J. Biol. Chem. 242:4449.

[25] Sarkar, S., et al. 1971. Proc. Nat. Acad. Sci. U.S. 68:946.

[26] Woods, E. F. 1967. J. Biol. Chem. 242:2859.

[27] Young, D. M., et al. 1965. Ibid. 240:2428.

147. MICROSCOPIC AND ULTRAMICROSCOPIC DIMENSIONS OF MUSCLE AS RELATED TO FUNCTION: VERTEBRATES AND INVERTEBRATES

Values for A-filament lengths have been corrected for distortions occurring during tissue preparation only in *Testudo* and *Rana*; values for other species have not been corrected. In-vivo values for both length and diameter of A filaments are probably identical for all vertebrate skeletal and cardiac fibers—i.e., 1.55-1.6 μ long and 100-120 Å wide. I-filament lengths were measured from H zone to H zone across the Z line. Values for I filaments in skeletal muscle have been corrected for distortions occurring during tissue preparation; the value for *Felis* heart muscle was not corrected. I filaments in all types of muscles appear to have the same structure [8], with diameters of 50-60 Å in embedded material. Values in parentheses are ranges, estimate "c" (*see* Introduction).

	Muscle	Specification	Value	Reference
			Cavia porcellus	
1	Taenia coli	Fiber diameter	(2-4) μ	2
2			6 μ	20
3		Fiber length	0.20 mm	2
4			0.15 mm	20
5		Tension developed, whole muscle, at 37°C	1.82 ± 0.5 kg/cm²	1
6		Velocity of shortening, at 37°C	(0.32-0.4) cm/s	1
7			0.3 muscle lengths/s	1
			Felis catus	
8	Papillary	Fiber diameter	(9-10) μ	6
9	muscle of	Sarcomere length for maximum tension	2.2(2.18-2.24) μ	21
10	right ven-	A-filament length	1.5 μ	21
11	tricle	I-filament length	2.0 μ	21
12		Tension developed, whole muscle, at 29°C	1.8 ± 0.1 kg/cm²	3
13		Velocity of shortening, at 29°C	1.0 cm/s [1,2]; 1.6 cm/s [2,3]	3
14			1.5 ± 0.1 muscle lengths/s [1]; 3.0 ± 0.2 muscle lengths/s [3]	3
			Rattus norvegicus	
15	Extensor	Fiber diameter	38.6 ± 4.7 μ	4
16	digitorum	Fiber length	12.6 ± 1.1 mm	4
17	longus	I-filament length	2.15 μ	18,19
18		Tension developed, whole muscle, at 35°C	2.9 ± 0.3 kg/cm²	4
19		Velocity of shortening, at 35°C	20.8 ± 1.2 cm/s	4
20		Velocity of shortening/sarcomere, at 35°C	45.1 ± 1.9 μ/s	4
21	Soleus	Fiber diameter	39.4 ± 5.4 μ	4
22		Fiber length	14.7 ± 1.4 mm	4
23		I-filament length	2.15 μ	18,19
24		Tension developed, whole muscle, at 35°C	1.93 ± 0.24 kg/cm²	4
25		Velocity of shortening, at 35°C	10.1 ± 0.5 cm/s	4
26		Velocity of shortening/sarcomere, at 35°C	19.8 ± 1.4 μ/s	4
			Testudo graeca; T. hermanni	
27	Rectus	Fiber length	(30-40) mm	23
28	femoris	Sarcomere length for maximum tension	2.6(2.55-2.75) μ	23
29		A-filament length	1.6 μ	18
30		I-filament length	2.35 μ	18
31		Tension developed, whole muscle, at 0°C	1.9 ± 0.17 kg/cm²	23
32		Velocity of shortening, at 0°C	0.8 ± 0.33 cm/s	23
33			0.23 muscle lengths/s	23
34		Velocity of shortening/sarcomere, at 0°C	0.6 μ/s	23
			Rana temporaria	
35	Sartorius	Fiber diameter	84(19-152) μ	17
36		Fiber length	(20-25) mm	17

[1] In 2.5 m*M* Ca. [2] Value estimated from muscle lengths/s. [3] In 7.5 m*M* Ca.

continued

1232

	Muscle	Specification	Value	Reference
37		A-filament length	1.6 μ	19
38		I-filament length	1.95 μ	18
39		Tension developed, whole muscle, at 0°C	2.0 kg/cm²	10
40		at 2°C	2.5 kg/cm²	12
41		at 17°C	2.6 kg/cm²	10
42		Velocity of shortening, at 0°C	1.3 muscle lengths/s	9
43		at 2°C	4.8 cm/s	12
44		Velocity of shortening/sarcomere, at 0°C	2.6 μ/s	23
45	Semitendi-	Fiber diameter	(50-130) μ	16
46	nosus	Fiber length	(10-25) mm	16
47		Sarcomere length for maximum tension	(2.05-2.2) μ	7
48		A-filament length	1.6 μ	19
49		I-filament length	1.95 μ	18
50		Tension developed, single fiber, at 3-4°C	2.7 kg/cm²	7
51		Velocity of shortening/sarcomere, at 3-4°C	4.0 μ/s	7
52		at 4-5°C	5.9 μ/s	5

Astacus astacus (A. fluviatilis)

	Muscle	Specification	Value	Reference
53	Extensor	Fiber diameter	332(210-440) μ[4]; 458(370-560) μ[5]	24
54	carpopo-	Fiber length	14.6(12.5-17.5) mm	24
55	dite	Sarcomere length for maximum tension	10.5 μ	24
56		A-filament length	3.95 μ[6]	24
57		Tension developed, single fiber, at 18-23°C	8.2 kg/cm²	24

Mytilus edulis

	Muscle	Specification	Value	Reference
58	Anterior	Fiber diameter	(5-7) μ	22
59	byssus re-	Fiber length	$\not>$40 mm	22
60	tractor	A-filament length	30 μ	15
61		A-filament diameter	1250 Å[5]	13
62		Tension developed, whole muscle, at 20°C	10.5 kg/cm²	14
63		Velocity of shortening, at 20°C	0.6 cm/s[2]	14
64			0.25 muscle lengths/s	14

[2] Value estimated from muscle lengths/s. [4] Minimum diameter. [5] Maximum diameter. [6] A-filament diameter probably $<$200 Å, as in all other crustacean muscles studied [11].

Contributor: Page, Sally

References

[1] Åberg, A. K. G., and J. Axelsson. 1965. Acta Physiol. Scand. 64:15.

[2] Bennett, M. R., and D. C. Rogers. 1967. J. Cell Biol. 33:573.

[3] Brutsaert, D. L., et al. 1970. Circ. Res. 27:513.

[4] Close, R. 1969. J. Physiol. (London) 204:331.

[5] Edman, K. A. P., and D. W. Grieve. 1966. Ibid. 184: 21P.

[6] Fawcett, D. W., and N. S. McNutt. 1969. J. Cell Biol. 42:1.

[7] Gordon, A. M., et al. 1966. J. Physiol. (London) 184:170.

[8] Hanson, J., and J. Lowy. 1964. Proc. Roy. Soc. B160:449.

[9] Hill, A. V. 1938. Ibid. B126:136.

[10] Hill, A. V., and R. C. Woledge. 1962. J. Physiol. (London) 162:311.

[11] Hoyle, G. 1969. Annu. Rev. Physiol. 31:43.

[12] Jewell, B. R., and D. R. Wilkie. 1958. J. Physiol. (London) 143:515.

[13] Lowy, J., and J. Hanson. 1962. Physiol. Rev. 42 (Suppl. 5):34.

[14] Lowy, J., and B. M. Millman. 1963. Phil. Trans. Roy. Soc. London B246:105.

[15] Lowy, J., et al. 1964. Proc. Roy. Soc. B160:525.

[16] Lüttgau, H. C. 1963. J. Physiol. (London) 168:679.

[17] Mayeda, R. 1890. Z. Biol. 27:119.

[18] Page, S. G. 1968. J. Physiol. (London) 197:709.

[19] Page, S. G., and H. E. Huxley. 1963. J. Cell Biol. 19:369.

[20] Prosser, C. L., et al. 1960. Amer. J. Physiol. 199:545.

[21] Sonnenblick, E. H., et al. 1964. Circ. Res. 15(Suppl. 2):70.

[22] Twarog, B. M. 1967. J. Physiol. (London) 192:857.

[23] Woledge, R. C. 1968. Ibid. 197:685.

[24] Zachar, J., and D. Zacharová. 1966. Ibid. 186:596.

Part I. Nerve and Muscle: Vertebrates and Invertebrates

Length Constant: Values are measurements of the length required for an alteration in membrane potential to decline 63% $(1-1/e)$ in a preparation of infinite length. The external volume is large.

	Species	Tissue	Fiber Diameter μ	Ambient Temp °C	Length Constant mm	Time Constant ms	Membrane Resistance $\Omega\cdot cm^2$	Membrane Capacity $\mu F/cm^2$	Resistivity of Cell Interior $\Omega\cdot cm$	Reference
1	*Capra hircus*, kid	Purkinje fiber	75	37	1.9	24[1]	2000	12	105	6,7, 12
2	*Felis catus*	Tenuissimus muscle	50	37	1.2	5[1]	1430	3.5	125[2]	1
3	*Rana pipiens; R. temporaria*	Muscle (fast fibers)	80	20	1.8	21[1]	3100	6.7	200	5,8, 11
4	*Carcinus maenas*	Axon	30	16	2.5	7	7650	1	90	9
5	*Homarus americanus*	Axon	101	11	5.1	....	8260	....	80	2
6	*H. gammarus*[3]	Axon	75	15-20	2.7	2	2300	1	61	10
7	*Loligo pealei*	Axon	500	23	6.6	1	1000	1	29	3,4

[1] Frog muscle, Purkinje fibers, and probably cat muscle display 2 time constants in their membrane impedance; the shorter time constants result from additional structural complexities. Data are for the longer time constants. [2] Assumed. [3] Synonym: *H. vulgaris.*

Contributor: Freygang, Walter H., Jr.

References

[1] Boyd, I. A., and A. R. Martin. 1959. J. Physiol. (London) 147:450.
[2] Brinley, F. J., Jr. 1965. J. Neurophysiol. 28:742.
[3] Cole, K. S., and A. L. Hodgkin. 1939. J. Gen. Physiol. 22:671.
[4] Curtis, H. J., and K. S. Cole. 1938. Ibid. 21:757.
[5] Falk, G., and P. Fatt. 1964. Proc. Roy. Soc. B160:69.
[6] Fozzard, H. A. 1966. J. Physiol. (London) 182:255.
[7] Freygang, W. H., and W. Trautwein. 1970. J. Gen. Physiol. 55:524.
[8] Freygang, W. H., et al. 1967. Ibid. 50:2437.
[9] Hodgkin, A. L. 1947. J. Physiol. (London) 106:305.
[10] Hodgkin, A. L., and W. A. H. Rushton. 1946. Proc. Roy. Soc. B133:444.
[11] Schneider, M. F. 1970. J. Gen. Physiol. 56:640.
[12] Weidmann, S. 1952. J. Physiol. (London) 118:348.

Part II. Myelinated Fiber: Frog

Data are for *Rana catesbeiana, R. esculenta,* and *R. temporaria.*

	Measurement	Value		Measurement	Value
1	Fiber diameter[1]	14 μ	10	Capacity	0.6-1.5 pF
2	Myelin thickness	2 μ	11	Resting node resistance	40-80 MΩ
	Myelin sheath			Nodal membrane	
3	Capacity per unit length (c)	10-16 pF/cm	12	Area	22 μ^2[2]
4	Capacity per unit area	0.0025-0.005 $\mu F/cm^2$	13	Capacity per unit area	3-7 $\mu F/cm^2$
5	Resistance × unit area	0.1-0.16 M$\Omega\cdot cm^2$	14	Resistance × unit area	10-20 $\Omega\cdot cm^2$
6	Specific resistance	500-800 M$\Omega\cdot cm$	15	Axoplasm specific resistance	110 $\Omega\cdot cm$
7	Dielectric constant	5-10	16	Action potential	116 mV
8	Axis cylinder resistance per unit length (r)	140 MΩ/cm	17	Resting potential	71 mV
	Node(s)		18	Peak inward current density	20 mA/cm^2
9	Distance between (l)	2 mm	19	Conduction velocity	23 m/s
			20	$1/lrc$[3]	22-36 m/s

[1] Near maximal size. [2] Assumed. [3] The time for a potential to spread passively over a distance l (line 9) in a cable with resistance r (line 8) and capacity c (line 3) is proportional to l^2rc; hence, $1/lrc$ has the dimensions of a velocity.

continued

Part II. Myelinated Fiber: Frog

Contributor: Hodgkin, Alan L.

General References

[1] Huxley, A. F., and R. Stämpfli. 1949. J. Physiol. (London) 108:315.

[2] Huxley, A. F., and R. Stämpfli. 1951. Ibid. 112:476.

[3] Stämpfli, R. 1952. Ergeb. Physiol. Biol. Chem. Exp. Pharmakol. 47:70.

[4] Tasaki, I. 1955. Amer. J. Physiol. 181:639.

149. NERVE AND MUSCLE ELECTROLYTE CONCENTRATIONS, AND RESTING MEMBRANE POTENTIAL: VERTEBRATES AND INVERTEBRATES

Cell Resting Potential values are negative. Values in parentheses are ranges, estimate "c" (*see* Introduction).

	Species	Tissue	Intracellular K mM	Intracellular Na mM	Cell Resting Potential mV	Reference
1	*Homo sapiens*	Muscle	140-160	26	70-87	2,14
2	*Cavia porcellus*	Smooth muscle	114-121	3	51	3,20
3	*Rattus norvegicus*	Muscle	74-165	16	73	5,6,13
4	*Rana* sp.	Muscle	138	22	92(88-95)	1
5	*Balanus nubilis*	Muscle	168	81	67	15
6	*Carcinus maenas*	Muscle	169	52	(30-55)	9
7	*Homarus americanus*	Muscle	155	104	∿80	7
8	*Periplaneta americana*	Muscle	112	46	∿60	11,19
9		Nerve	140-182	84-103	77	17-19,21
10	*Loligo* spp.	Nerve	400	50	68(63-72)	10
11	*Sepia officinalis*	Nerve	268	32	62	12,22
12	*Anisodoris nobilis*	Nerve	235	....	(50-70)	8
13	*Helix aspersa*	Nerve	86-147	....	(33-62)	16
14	*Ascaris lumbricoides*	Muscle	99	49	33	4

Contributor: Caldwell, Peter C.

References

[1] Adrian, R. H. 1956. J. Physiol. (London) 133:631.

[2] Barnes, B. A., et al. 1957. J. Clin. Invest. 36:1239.

[3] Brading, A. F. 1971. J. Physiol. (London) 214:393.

[4] Brading, A. F., and P. C. Caldwell. 1971. Ibid. 217:605.

[5] Creese, R., et al. 1958. Ibid. 143:307.

[6] Dosekun, F. O., and D. Mendel. 1958. Ibid. 140:190.

[7] Dunham, P. B., and H. Gainer. 1968. Biochim. Biophys. Acta 150:488.

[8] Gorman, A. L. F., and M. F. Marmor. 1970. J. Physiol. (London) 210:897.

[9] Hinke, J. A. M. 1959. Nature (London) 184:1257.

[10] Hodgkin, A. L. 1958. Proc. Roy. Soc. B148:1.

[11] Hoyle, G. 1955. J. Physiol. (London) 127:90.

[12] Keynes, R. D., and P. R. Lewis. 1951. Ibid. 114:151.

[13] Liley, A. W. 1956. Ibid. 132:650.

[14] McComas, A., and R. J. Johns. 1969. In J. N. Walton, ed. Disorders of Voluntary Muscle. J. and A. Churchill, London. p. 894.

[15] McLaughlin, S. G. A., and J. A. M. Hinke. 1966. Can. J. Physiol. Pharmacol. 44:837.

[16] Moreton, R. B. 1969. J. Exp. Biol. 51:181.

[17] Narahashi, T. 1963. Advan. Insect Physiol. 1:175.

[18] Narahashi, T., and T. Yamasaki. 1960. J. Physiol. (London) 151:75.

[19] Tobias, J. M. 1948. J. Cell. Comp. Physiol. 31:125.

[20] Tomita, T. 1966. J. Physiol. (London) 183:450.

[21] Treherne, J. E. 1961. J. Exp. Biol. 38:315.

[22] Weidmann, S. 1951. J. Physiol. (London) 114:372.

Method: IP = internal perfusion; TDE = internal dialysis with isotopically labeled medium; TIE = determined from rate of loss of microinjected isotope; TSI = loading with isotope and counting vs. time or terminal countings; TSE = determined from tracer washout curves. Plus/minus (±) values are 1 standard deviation, unless otherwise indicated. Values in parentheses are ranges, estimate "c" (*see* Introduction).

Part I. Resting Ion Fluxes

Method: TDI = internal dialysis with nonradioactive medium (tracer present externally); EA = determined by tracer loading followed by counting of extruded axoplasm.

	Tissue	Source	Temp °C	Ion Concentration mmole/liter H_2O[1] Outside the Cell	Inside the Cell	Ion Flux, pmole·cm^{-2}·s^{-1} Influx Value	Method	Efflux Value	Method	Reference
				Sodium						
1	Giant axon	*Dosidicus gigas*	(9.8–14)	430	0	29.9 ± 2.5[2]	IP	....	.....	2
2		*Loligo pealei*	14(9–18)	425	82(76–89)	57 ± 25	TDI	48 ± 18	TDE	5
3			14(11–17)	429	80	43 ± 9	TDI	....	.....	6
4			16(12–17)	423	98 ± 19[3]	....	.....	78 ± 19	TIE	16
5		*Sepia officinalis*	18	486	77 ± 37[3]	32 ± 16	TSI	39 ± 11	TSE	11
6		Lobster	9(8–10)	465	46 ± 20	8.3 ± 1.6[2]	TSI	4.1 ± 1.7[2]	TSE	3
7	Muscle Diaphragm	Rat	38	145	9	10	TSI	10	TSE	9
8	Semitendinosus, single fiber	Frog	21	120	9.2 ± 1.2	3.5 ± 0.4[2]	TSI	3.7[4]	TSE	10
9	Isolated	Barnacle	13(9–16)	465	19 ± 4.1	53 ± 18	TSI	39 ± 21	TIE	4
10			22	465	19 ± 4.1	48 ± 20	TSI	....	.....	4
				Potassium						
11	Giant axon	*Loligo forbesi*	15(13–18)	10.4	351 ± 7	24	TSI	38 ± 9.6	TIE	7
12			18	10.4	....	23.7 ± 6.5	EA	....	.....	8
13		*L. pealei*	15(13–16.5)	9	304	19 ± 3.6	TDI	59 ± 11	TDE	5,14
14			17.5	10	....	29 ± 6.5	EA	....	.....	16
15		*Sepia* sp.	14(13–15)	9.7	267 ± 31	16.7 ± 7.4	TSI	58	TSE	12,13
16		Lobster	9(8–10)	10	292 ± 63	14.5 ± 3.1[2]	TSI	21.2 ± 5.5[2]	TSE	3
17	Muscle Digitorum longus	Rat	25	4.5	169 ± 4.9	4.6	TSI	4.6	TSE	18
18	Sartorius, whole	Frog	20	2.5	131	....	.....	9.6 ± 1.4	TSE	15
19			21	2.5	136 ± 5	6.2 ± 0.9	TSI	7.4 ± 1.5	TSE	17
20	Semitendinosus, single fiber	Frog	21	2.5	137 ± 2	5.4 ± 0.8[2]	TSI	8.8 ± 1.2[2]	TSE	10
21	Isolated	Barnacle	13(9–16)	10	113 ± 17	28 ± 7	TSI	60 ± 23	TIE	4
				Chloride						
22	Giant axon	*Loligo forbesi*	19	602	100	12.6 ± 3.6	TSI	6.6	TIE	7
23		Lobster	9(8–10)	533	57 ± 14	11.5 ± 3.4	TSI	9.8 ± 2.6	TSE	3
24	Sartorius muscle, whole	Frog	18(15–20)	121	(3.1–3.8)	....	.....	10	TSE	1

[1] Unless otherwise indicated. [2] Plus/minus (±) value is standard error. [3] mmole/kg axoplasm. [4] Calculated from the mean Na$^+$ concentration inside the cell and the time constant for Na$^+$ loss.

Contributor: Sjodin, Raymond A.

continued

150. ION FLUX DATA FOR NERVE AND MUSCLE: VERTEBRATES AND INVERTEBRATES

Part I. Resting Ion Fluxes

References

[1] Adrian, R. H. 1961. J. Physiol. (London) 156:623.
[2] Bezanilla, F., et al. 1970. Ibid. 207:151.
[3] Brinley, F. J., Jr. 1965. J. Neurophysiol. 28:742.
[4] Brinley, F. J., Jr. 1968. J. Gen. Physiol. 51:445.
[5] Brinley, F. J., Jr., and L. J. Mullins. 1967. Ibid. 50: 2303.
[6] Brinley, F. J., Jr., and L. J. Mullins. 1968. Ibid. 52: 181.
[7] Caldwell, P. C., and R. D. Keynes. 1970. J. Physiol. (London) 154:177.
[8] Caldwell, P. C., et al. 1960. Ibid. 152:591.
[9] Creese, R. 1968. Ibid. 197:255.
[10] Hodgkin, A. L., and P. Horowicz. 1959. Ibid. 145: 405.

[11] Hodgkin, A. L., and R. D. Keynes. 1955. Ibid. 128: 28.
[12] Keynes, R. D. 1951. Ibid. 114:119.
[13] Keynes, R. D., and P. R. Lewis. 1951. Ibid. 114: 151.
[14] Mullins, L. J., and F. J. Brinley, Jr. 1969. J. Gen. Physiol. 53:704.
[15] Mullins, L. J., and K. Noda. 1963. Ibid. 47:117.
[16] Sjodin, R. A., and L. A. Beaugé. 1968. Ibid. 51: 152s.
[17] Sjodin, R. A., and E. G. Henderson. 1964. Ibid. 47: 605.
[18] Zierler, K. L., et al. 1966. Ibid. 49:433.

Part II. Ion Fluxes During Excitatory Activity

Method: TKD = total potassium ion determination in a potassium-free solution.

| | Tissue | Source | Temp °C | Extra Ion Flux/Impulse, $pmole \cdot cm^{-2} \cdot impulse^{-1}$ | | | | | | Reference |
| | | | | Influx | | | Efflux | | | |
				Sodium	Chloride	Method	Potassium	Sodium	Method	
1	Giant axon	*Dosidicus gigas*	12	7.13	....	IP	....	....	....	1
2			15	5.23	....	IP	....	....	....	1
3			17-18	10	....	IP	....	4	IP	8
4		*Loligo forbesi*	15(13-18)	....	0.046	TSI	8.5(7.2-10.8)	....	TIE	4
5			(17-22)	5.7 ± 0.7	....	IP	....	....	....	10
6		*L. pealei*	6	....	....	....	8.8(7.0-11.6)	....	TKD	9
7			14(13-16.5)	....	....	....	2.9 ± 0.9	3.1 ± 0.6	TDE	3
8			19.5(18.5-20.3)	....	....	....	5.2 ± 0.7	....	Net K[1]	11
9							5.3 ± 0.9	....	TSE [1,2]	11
10			24	....	....	....	2.9(2.4-3.4)	....	TKD	9
11		*Sepia officinalis*	10	3.8 ± 0.4[3]	....	Net exchanges	3.6 ± 0.7	....	Net exchanges	7
12			(13-15)	10.3(5.3-15.0)	....	TSI	4.7(3.9-5.5)	6.6(4.5-9.6)	TSE	6
13		Lobster	9(8-11)	6.5 ± 0.7	−0.2 ± 0.2	TSI	5.3 ± 0.5	....	TSE	2
14	Muscle Semitendinosus	Frog	21	15.6 ± 1.8[3]	....	Net influx	9.6 ± 0.7[3]	....	Net efflux	5
15	Single fiber	Frog	21	19.6 ± 1.5[3]	....	TSI	11.4 ± 0.6[3]	....	TSE	5

[1] In a K^+-free solution. [2] Tracer is $^{42}K^+$. [3] Plus/minus (±) value is standard error.

Contributor: Sjodin, Raymond A.

References

[1] Atwater, I., et al. 1970. J. Physiol. (London) 211: 753.
[2] Brinley, F. J., Jr. 1965. J. Neurophysiol. 28:742.
[3] Brinley, F. J., Jr., and L. J. Mullins. 1967. J. Gen. Physiol. 50:2303.
[4] Caldwell, P. C., and R. D. Keynes. 1960. J. Physiol. (London) 154:177.
[5] Hodgkin, A. L., and P. Horowicz. 1959. Ibid. 145: 405.

[6] Keynes, R. D. 1951. Ibid. 114:119.
[7] Keynes, R. D., and P. R. Lewis. 1951. Ibid. 114: 151.
[8] Rojas, E., and M. Canessa-Fischer. 1968. J. Gen. Physiol. 52:240.
[9] Shanes, A. M. 1954. Amer. J. Physiol. 177:377.
[10] Shaw, T. I. 1966. J. Physiol. (London) 182:209.
[11] Sjodin, R. A., and L. J. Mullins. 1967. J. Gen. Physiol. 50:533.

All cells maintain a potential difference (membrane potential) across their membranes. Nerve cells are specialized so that this potential difference can be transiently reduced (excitatory) or increased (inhibitory) as a result of synaptic action. The time course of these transients is usually brief. For the chemical synapses considered below, the driving force for the ionic fluxes which generate synaptic potentials is the electrochemical gradient for the ions concerned. The equilibrium potential is that membrane potential at which the gradient is zero. In practice this may be difficult to determine, and instead the membrane potential at which the synaptic potential is reversed in sign (reversal potential) is determined. The direction of current injection to reverse the synaptic potential is referred to as inward or outward current, and is measured in nanoamperes. The result of synaptic action is the generation (excitation) or prevention of generation (inhibition) of action potentials. Generation requires reduction of the membrane potential to a characteristic critical level. It is assumed that the recording electrode is as close as possible to the site of generation of the potential. Data in brackets refer to the column heading in brackets. Values in parentheses are ranges, estimate "c" unless otherwise specified (see Introduction).

	Potential	Species	Site of Recording & Temp °C	Membrane Potential mV	Rise Time ms	Falling Phase Half Decay ms [Time Constant ms]	Equilibrium Potential, mV [Reversal Potential, mV or Reversal Current, nA]	Critical Membrane Potential mV	Reference
1	Endplate	*Felis catus*	Tenuissimus; 37°	−75(−90 to −60)	0.6(0.4-0.9)	0.7(0.5-1.2)	−15 mV	−46(−51 to −42)	2,5
2		*Tiliqua nigrolutea*	Scalenus muscle, en plaque innervation; (20-25°)	−71(−83 to −59)[b]	(2-5)	(3-10)	?	−60	35
3		*Chrysemys scripta elegans* [1]	Retractor capitis muscle; (15-24°)	−80(−84 to −73)	2.1(1.8-2.5)	4.6(3.3-5.9)	?	≈−40	29
4		*Testudo graeca*	Retractor capitis muscle; (15-24°)	−80(−85 to −70)	2.5(2.3-2.6)	5.7(4.1-7.0)	?	−39(−42 to −33)	29
5		*Rana temporaria*	Sartorius muscle; (16-24°)	−90(−107 to −75)	1.3(0.9-1.7)[b]	3.9(2.5-4.4)[b]	−15(−20 to −10) mV	−44(−51 to −37)	17,39
6	Excitatory, junctional	*Cavia* spp.	Vas deferens; 35°	−57(−80 to −50)	(43-160)	≈150	?	(−40 to −35)	9
7		*Rana temporaria*	Iliofibularis muscle; (15-22°)	−80(−84 to −76)[b]	(8-14)	[800]	[−7(−35 to +10) mV]	Not excitable	7,8,34,38
8		*Carassius auratus*	Muscle levator pinnae, pectoralis; (17-22°)	−75(−87 to −62)[b]	≈8	36(15-57)[b]	[−30 mV]	−60	23-25
9		*Leiurus quinquestriatus*	Closer of pedipalp; (20-28°)	−62(−72 to −57)[b]	(5-7)	(14-16)	?	−47(−57 to −37)[b]	20
10	Inhibitory, junctional	*Procambarus clarkii* [2]	Stretch receptor; (20-23°)	−75	2	(12.5-15.0)	[−70 mV [3]]	....	28
	Excitatory, postsynaptic								
11	Ganglionic	*Cavia* spp.	Sympathetic neuron; (35-37°)	(−70 to −50)	9.1(5-15)	[9.8(7-18)]	?	(−60 to −55)	3
12		*Rana pipiens*	Sympathetic neuron; (24-26°)	−65	(2.5-3.0)	[12(8-16)]	−14.3(−20 to −8) mV	−40	33
13		*Panulirus japonicus*	Large cells of cardiac ganglion; (18-20°)	−60	10	[19.5(18-24)]	−10 mV	−30	22

[1] Synonym: *Pseudemys scripta elegans*. [2] Synonym: *Cambarus clarkii*. [3] At many inhibitory junctions in crustaceans, the reversal potential is at, or close to, the membrane potential, so there may be no potential change following stimulation of an inhibitory axon.

continued

	Potential	Species	Site of Recording & Temp °C	Membrane Potential mV	Rise Time ms	Falling Phase Half Decay ms [Time Constant ms]	Equilibrium Potential, mV [Reversal Potential, mV or Reversal Current, nA]	Critical Membrane Potential mV	Reference
14	Neuronal Axosomatic & axodendritic	*Felis catus*	Spinal cord motor neuron; (37-39°)	−64(−80 to −50)[4]	(0.7-2.2)	4.9(3.5-6.0)	[3 mV][5]	−53(−59 to −46)[6,7]; −41(−51 to −25)[6,8]	11-13, 36, 37
15	Axodendritic	*Felis catus*	Trochlear motor neuron; (37-39°)	−70	2	4.5	[35(30 to 40 nA)] outward	(−58 to −55)	30
16			Purkinje cell, climbing fiber activation; (37-39°)	(−65 to −60)	1.0-1.5	6.0(5.0-7.5)	[(12 to 15) nA] outward	....	14
17		*Rana pipiens*	Spinal cord motor neuron; (15-17°)	−44(−61 to −31)[4]	3.22 (2.4-4.5)	10.3	?	−45(−67 to −40)[6,7]; −27(−36 to −22)[6,8]	16,32
18	Axosomatic	*Rana pipiens*	Spinal cord motor neuron; (15-17°)	−44(−61 to −31)[4]	2.34 (1.25-3.50)	6.5[9]	?	−48(−65 to −43)[6,7]; −31(−36 to −23)[6,8]	16,32
19	Axoaxonic	*Gasteropelecus* sp.	Mauthner cell axon[10]	−70	0.5	≈1	[≈0 mV]	−60	1
20		*Loligo pealei*	Stellate ganglion giant synapse post fiber; (8-21°)	(−65 to −60)	(1.0-1.5)	≈1.3	[45(23 to 67)[b] mV]	≈−45	6,19, 21, 26
21	Axodendritic[11]	*Aplysia* sp.	Neuron R15, excitation by L10; (21-24°)	≈−67	≈(600-800)	≈(1500-2000)	−10(−20 to 0) mV	....	4
	Inhibitory, postsynaptic neuronal								
22	Axosomatic (basket) & axodendritic (stellate)	*Felis catus*	Purkinje cell, basket & stellate cell inhibition; (37-39°)	(−65 to −60)	(12-18)	(30-40)	[≈25 nA] inward	....	15
23	Axodendritic	*Caiman sclerops*	Purkinje cell[10], stellate inhibition	≈−60	(8-15)	(25-30)	[(8 to 12) nA] inward	....	31
24	Axosomatic	*Felis catus*	Trochlear motor neuron; (37-39°)	−70	(2-3)	4.5	[(10 to 12) nA] inward	....	30
25			Spinal cord motor neuron; (35-37°)	−70(−80 to −60)[4]	(1.5-2.0)	[3.3(2.5-4.0)]	[−80(−90 to −66) mV]	....	10,13
26		*Rana pipiens*	Spinal cord motor neuron; (15-17°)	−47(−68 to −36)[4]	6.4(3.8-8.0)	[16.9(14.2-20.8)]	[−53.2(−77 to −34) mV]	....	27

[4] The membrane potential of a central nervous system neuron fluctuates with the magnitude and sign of the synaptic input, and has no characteristic value. The values given for **Critical Membrane Potential** are the depolarizations (mV) required to set up action potentials in the cell body or its axon. [5] An equilibrium or reversal potential has only rarely been determined, presumably because the site of synaptic potential generation is far from the site of recording. The great variability of rise times and falling phases in this entry has a similar basis [36]. [6] Since the critical membrane potential for the axon is lower than that for the cell body, synaptic potentials preferentially set up action potentials in the axon. [7] Value for axon. [8] Value for cell body. [9] This is the first of two time constants in the falling phase; the second is the same as that in entry 17. [10] At room temperature. [11] In these forms, the dendrites arise from the axon, not from the soma.

continued

	Potential	Species	Site of Recording & Temp °C	Membrane Potential mV	Rise Time ms	Falling Phase Half Decay ms [Time Constant ms]	Equilibrium Potential, mV [Reversal Potential, mV or Reversal Current, nA]	Critical Membrane Potential mV	Reference
27	Axoaxonic	*Astacus astacus* [12]	Giant motor fiber; ≈20°	?	3.1(1.8-4.0)	12(7.7-24)	[8.6(3 to 20) mV] [13]	....	18
28	Axodendritic [11]	*Aplysia* sp.	Neuron L1-L6, early IPSP by L10; (21-24°)	≈−45	≈800	≈(1500-2000)	[−57(−65 to −55) mV]	....	4

[11] In these forms, the dendrites arise from the axon, not from the soma. [12] Synonym: *A. fluviatilis.* [13] Mean depolarization required to reverse the potential. As the membrane potential was unknown, the absolute value of the reversal potential was also unknown.

Contributor: Hubbard, John I.

References

[1] Auerbach, A. A., and M. V. L. Bennett. 1969. J. Gen. Physiol. 53:183.

[2] Blaber, L. C. 1970. J. Pharmacol. Exp. Ther. 175:664.

[3] Blackman, J. G., and R. D. Purves. 1969. J. Physiol. (London) 203:173.

[4] Blankenship, J. E., et al. 1971. J. Neurophysiol. 34:76.

[5] Boyd, I. A., and A. R. Martin. 1956. J. Physiol. (London) 132:74.

[6] Bullock, J. H., and S. Hagiwara. 1957. J. Gen. Physiol. 40:565.

[7] Burke, W., and B. L. Ginsborg. 1956. J. Physiol. (London) 132:586.

[8] Burke, W., and B. L. Ginsborg. 1956. Ibid. 132:599.

[9] Burnstock, G., and M. E. Holman. 1961. Ibid. 155:115.

[10] Coombs, J. S., et al. 1955. Ibid. 130:326.

[11] Coombs, J. S., et al. 1955. Ibid. 130:374.

[12] Coombs, J. S., et al. 1957. Ibid. 139:232.

[13] Curtis, D. R., and J. C. Eccles. 1959. Ibid. 145:529.

[14] Eccles, J. C., et al. 1966. Ibid. 182:268.

[15] Eccles, J. C., et al. 1966. Exp. Brain Res. 1:161.

[16] Fadiga, E., and J. M. Brookhart. 1960. Amer. J. Physiol. 198:693.

[17] Fatt, P., and B. Katz. 1951. J. Physiol. (London) 115:320.

[18] Furshpan, E. J., and D. D. Potter. 1959. Ibid. 145:326.

[19] Gage, P., and J. W. Moore. 1969. Science 166:510.

[20] Gilai, A., and I. Parnas. 1970. J. Exp. Biol. 52:325.

[21] Hagiwara, S., and I. Tasaki. 1958. J. Physiol. (London) 143:114.

[22] Hagiwara, S., et al. 1959. J. Neurophysiol. 22:554.

[23] Hidaka, T., and H. Kuriyama. 1969. J. Physiol. (London) 201:61.

[24] Hidaka, T., and N. Toida. 1969. Ibid. 201:49.

[25] Hidaka, T., and N. Toida. 1969. Jap. J. Physiol. 19:130.

[26] Hodgkin, A. L., and A. F. Huxley. 1952. J. Physiol. (London) 116:424.

[27] Kubota, K., and J. M. Brookhart. 1963. Amer. J. Physiol. 204:660.

[28] Kuffler, S. W., and C. Eyzaguirre. 1955. J. Gen. Physiol. 39:155.

[29] Levine, L. 1966. J. Physiol. (London) 183:683.

[30] Llinás, R., and R. Baker. 1972. J. Neurophysiol. 35:484.

[31] Llinás, R., and C. Nicholson. 1971. Ibid. 34:532.

[32] Machne, X., et al. 1959. J. Neurophysiol. 22:484.

[33] Nishi, S., and K. Koketsu. 1960. J. Cell. Comp. Physiol. 55:15.

[34] Oomura, Y., and T. Tomita. 1960. Nature (London) 188:916.

[35] Proske, U., and P. Vaughan. 1968. J. Physiol. (London) 199:495.

[36] Rall, W. 1967. J. Neurophysiol. 30:1138.

[37] Smith, T. G., et al. 1967. Ibid. 30:1072.

[38] Stefani, E., and A. B. Steinbach. 1969. J. Physiol. (London) 203:383.

[39] Takeuchi, A., and N. Takeuchi. 1960. Ibid. 154:52.

Concentration Inside: Plasmalemma values are for cytoplasm, and tonoplast values are for vacuole. **Electrical Potential Difference:** Plasmalemma values are for cytoplasmic potential difference with respect to bathing medium, and tonoplast values are for vacuolar potential difference with respect to cytoplasm. **Ion Tracer Flux:** All ion tracer fluxes were measured in light, and only the fluxes in the direction of decreasing electrochemical potential have been reported. **Ionic Permeabilities:** Calculated according to the Goldman constant field assumption on the basis of published ion concentrations, ion fluxes, and membrane potentials. The tracer influx or efflux which opposes the active flux is taken to be entirely passive and independent; the passive influx or efflux was corrected to yield a *net passive*

influx or efflux by using the Ussing-Teorell equation. If J is the net passive flux, and if J_I and J_E are the passive tracer influx and efflux, respectively, across a membrane with a potential difference E, then $J = J_I - J_E$. $J_I/J_E = C^o/[C^i exp \cdot (zFE/RT)]$ (Ussing-Teorell equation) and $J = -P[(zFE/RT)/\langle 1-exp(zFE/RT)\rangle] \cdot [C^o - C^i exp(zFE/RT)]$ (Goldman equation) where P = permeability, C^o = outside concentration (bathing medium for plasmalemma, and cytoplasm for tonoplast), C^i = inside concentration (cytoplasm for plasmalemma, and vacuole for tonoplast), and z, F, R, T = ion charge, Faraday, gas constant, and absolute temperature, respectively. If J_E is passive, then $P = -J_E[\langle 1-exp(zFE/RT)\rangle/(zFE/RT) \cdot C^i exp(zFE/RT)]$. If J_I is passive, then $P = -J_I[1-exp(zFE/RT)/(zFEC^o/RT)]$.

	Species	Cell Membrane	Concentration mM		Electrical Potential Difference mV	Ion Tracer Flux pmole·s⁻¹·cm⁻¹		Ionic Permeability cm·s⁻¹	Reference
			Outside	Inside		Influx	Efflux		
			Potassium (K⁺)						
1	*Chaetomorpha darwinii* [1]	Plasmalemma	10	425	−72	....	200	2.7×10^{-6}	5,10
2		Tonoplast	.....	530	+77	....	200	1.2×10^{-7}	
3	*Chara corallina*	Plasmalemma	0.2	120 (?)	−140	....	0.6	2.5×10^{-7}	2,3,6,7
4		Tonoplast	.....	66	+11	170	...	1.8×10^{-6}	
5	*Enteromorpha intestinalis*	Plasmalemma	11	450	−42	....	81	5.0×10^{-7}	1
6	*Griffithsia monile*	Plasmalemma	10	400-444 [2]	−85	250 ± 10	...	7.2×10^{-6}; 4×10^{-5} [3]	8,9
7	*G. pulvinata*	Plasmalemma	10	550 [2]	−85	50-380	...	$1.4\text{-}11 \times 10^{-6}$; 3.4×10^{-5} [3]	8,9
8	*G. teges*	Plasmalemma	10	....	−80	....	...	2.5×10^{-5} [3]	8,9
9	*Hydrodictyon africanum*	Plasmalemma	0.1	93	−116	....	1.5	3.4×10^{-7}	16
10		Tonoplast	.....	40	+26	High	High	....	
11	*Nitella translucens*	Plasmalemma	0.1	119	−138	....	0.9	3.2×10^{-7}	4,14,
12		Tonoplast	.....	75	+18	100	...	1.2×10^{-6}	15,17
13	*Valonia ventricosa*	Plasmalemma	11	434 (?)	−71	....	...	....	11-13
14		Tonoplast	.....	625	+88		86	3.8×10^{-8}	
			Sodium (Na⁺)						
15	*Chaetomorpha darwinii* [1]	Plasmalemma	490	50	−72	100	...	6.7×10^{-8}	5,10
16		Tonoplast	.....	56	+77		10	5.6×10^{-8}	
17	*Chara corallina*	Plasmalemma	2.0	30 (?)	−140	0.3	...	2.7×10^{-8}	2,3,6,7
18		Tonoplast	.....	53	+11	High	High	....	
19	*Enteromorpha intestinalis*	Plasmalemma	460	260	−42	39	...	4.0×10^{-8}	1
20	*Griffithsia monile*	Plasmalemma	493.5	47-63 [2]	−85	....	...	8×10^{-8} [3]	8,9
21	*G. pulvinata*	Plasmalemma	493.5	30-90 [2]	−85	12-20	...	$0.7\text{-}1.2 \times 10^{-8}$; 7×10^{-8} [3]	8,9
22	*G. teges*	Plasmalemma	493.5	....	−80	....	...	1.3×10^{-7} [3]	8,9
23	*Hydrodictyon africanum*	Plasmalemma	1.0	51	−116	0.7	...	1.5×10^{-7}	16
24		Tonoplast	.....	17	+26	High	High	....	
25	*Nitella translucens*	Plasmalemma	1.0	14	−138	0.5	...	9.1×10^{-8}	4,14,
26		Tonoplast	.....	65	+18	High	High	....	15,17
27	*Valonia ventricosa*	Plasmalemma	485	40 (?)	−71	....	...	....	11-13
28		Tonoplast	.....	44	+88	....	3.3	2.1×10^{-8}	

[1] In positive state. [2] Average concentration of whole cell, which is mostly vacuole. [3] Calculated by authors [8,9] assuming plasmalemma resistance is due entirely to K and Na fluxes.

continued

Species	Cell Membrane	Concentration mM		Electrical Potential Difference mV	Ion Tracer Flux pmole·s⁻¹·cm⁻¹		Ionic Permeability cm·s⁻¹	Refer-ence
		Outside	Inside		Influx	Efflux		
Chloride (Cl⁻)								
29 *Chaetomorpha darwinii* [1/]	Plasmalemma	573	30	-72	200[4/]		2.0×10^{-6}	5,10
30	Tonoplast	.....	620	$+77$	200[4/]		2.1×10^{-6}	
31 *Chara corallina*	Plasmalemma	2.3	10[5/]	-140	....	0.4	7.2×10^{-9}	2,3,6,7
32	Tonoplast	.....	110	$+11$	100	...	1.9×10^{-5}	
33 *Enteromorpha intestinalis*	Plasmalemma	560	370	-42	....	54	7.2×10^{-8}	1
34 *Griffithsia monile*	Plasmalemma	573	626[2/]	-85	....	...	....	8,9
35 *G. pulvinata*	Plasmalemma	573	606-651[2/]	-85	....	4-11	....	8,9
36 *G. teges*	Plasmalemma	573	....	-80	....	...	....	8,9
37 *Hydrodictyon africanum*	Plasmalemma	1.3	58	-116	....	0.6	2.2×10^{-9}	16
38	Tonoplast	.....	38	$+26$	High	High	....	
39 *Nitella translucens*	Plasmalemma	1.3	65	-138	....	2.5[6/]	7.0×10^{-9}	4,14,
40	Tonoplast	.....	160	$+18$	180	...	2.0×10^{-6}	15,17
41 *Valonia ventricosa*	Plasmalemma	590	138 (?)	-71	....	...	....	11-13
42	Tonoplast	.....	643	$+88$	....	15	2.1×10^{-7}	

[1/] In positive state. [2/] Average concentration of whole cell, which is mostly vacuole. [4/] Cl⁻ is in flux and electrochemical equilibrium. [5/] From an electrical determination of chloride activity [2]. [6/] Influx assumed equal to efflux.

Contributor: Tyree, M. T.

References
[1] Black, D. R., and D. C. Weeks. 1972. New Phytol. 71:119.
[2] Coster, H. G. L. 1966. Aust. J. Biol. Sci. 19:545.
[3] Coster, H. G. L., and A. B. Hope. 1968. Ibid. 21:243.
[4] Dainty, J. 1969. In M. B. Wilkins, ed. The Physiology of Growth and Development. McGraw-Hill, London. pp. 455-485.
[5] Dodd, W. A., et al. 1966. Aust. J. Biol. Sci. 19:341.
[6] Findlay, G. P., and A. B. Hope. 1964. Ibid. 17:62.
[7] Findlay, G. P., et al. 1969. Biochim. Biophys. Acta 183:565.
[8] Findlay, G. P., et al. 1969. Aust. J. Biol. Sci. 22: 1163.
[9] Findlay, G. P., et al. 1970. Ibid. 23:323.
[10] Findlay, G. P., et al. 1971. Ibid. 24:731.
[11] Gutknecht, J. 1966. Biol. Bull. 130:331.
[12] Gutknecht, J. 1967. J. Gen. Physiol. 50:1821.
[13] Gutknecht, J., and J. Dainty. 1968. Oceanogr. Mar. Biol. Annu. Rev. 6:163.
[14] MacRobbie, E. A. C. 1962. J. Gen. Physiol. 45: 861.
[15] MacRobbie, E. A. C. 1964. Ibid. 47:859.
[16] Raven, J. A. 1967. Ibid. 50:1607.
[17] Spanswick, R. M., and E. J. Williams. 1964. J. Exp. Bot. 15:193.

153. ELECTRICAL CONSTANTS: PLANT CELLS

Species [Cell Type]	Cell		Concentration of Bathing Medium, mM	Resistance, kΩ·cm²		Remarks	Ref-er-ence
	Length cm	Diameter cm		Plasma-lemma	Tonoplast		
1 *Avena sativa* [Coleoptile cortical cell]	0.01-0.03	0.003	$Ca(NO_3)_2$, 1.0; $MgSO_4$, 0.25; KCl, 1.0; Na_2HPO_4, 0.048; NaH_2PO_4, 0.904	1.3		pH = 5.6-5.8; resistance increases with decreasing K⁺ or increasing Ca²⁺ concentrations	6
2	0.01	0.0035	Ca^{2+}, 0.1; K⁺, 0.1; Cl⁻, 0.3	5.1		Similar resistance reported in solution 10 times that of reported concentration. Capacity of plasmalemma + tonoplast = 2 $\mu F/cm^2$.	5

continued

	Species [Cell Type]	Cell Length cm	Cell Diameter cm	Concentration of Bathing Medium, mM	Resistance, k$\Omega \cdot$cm^2 Plasma-lemma	Tonoplast	Remarks	Ref-er-ence
3	Chaeto-morpha darwinii	....	....	CaCl$_2$, 11.5; MgCl$_2$, 25; MgSO$_4$, 25; KCl, 10; NaCl, 490; NaHCO$_3$, 2.5	0.51 [1/]; 0.75 [2/]	4.9 [1/]; 7.1 [2/]	Temp = 23°C; the more usual positive state occurs when vacuole minus outside potential = +5 mV; negative state occurs when this potential = −29 mV	4
4	Chara cor-allina	2	0.08	K$^+$, 1.9; Na$^+$, 0.1; Cl$^-$, 2.0	4		Resistance calculated from graph for small values of applied cur-rent	7,12
5				K$^+$, 0.8; Na$^+$, 1.2; Cl$^-$, 2.0	8			
6				K$^+$, 0.4; Na$^+$, 1.6; Cl$^-$, 2.0	12			
7				K$^+$, 0.1; Na$^+$, 1.9; Cl$^-$, 2.0	15			
8		1.1-1.8	0.11-0.13	Ca^{2+}, 0.5; K$^+$, 0.1; Na$^+$, 1.0; Cl$^-$, 2.1	12.1 ± 2.5	1.0 ± 0.2	Action potential of 4- to 10-s du-ration reflected in plasmalemma & tonoplast. Plasmalemma time constant = 100 ms. Capacity of plasmalemma = 1-2 μF/cm^2.	2
9	Elodea ca-nadensis [Epider-mal cells]	....	....	CaCl$_2$, 0.1; MgCl$_2$, 0.1; KCl, 0.1; NaCl, 0.5; NaOH, 0.4; Na$_2$SO$_4$, 0.05	3.1	1.0	Solution contained buffer (1.0 mM morpholinopropane sulfonic acid) at pH 7.0. Current also carried via plasmo-desmata between cells (resistance = 51 $\Omega \cdot$cm^2).	10
10	Griffith-sia spp.	0.2-0.3	0.1-0.2	CaCl$_2$, 11.5; MgCl$_2$, 25; MgSO$_4$, 25; KCl, 10; NaBr, 1; NaHCO$_3$, 2.5; NaCl, 490	0.15 [3/]; 0.19 [4/]; 0.27 [5/]	4.0 [3/]; 5.1 [4/]; 5.4 [5/]	These marine algae are prolate spheroids; quoted length is major axis, and diameter is minor axis	3
11	Nitella sp.	1.5	....	Ca^{2+}, 12.2; Mg^{2+}, 14.6; K$^+$, 16.7; Na$^+$, 29.3; Cl$^-$, 8.7; NO$_3^-$, 32.3; H$_2$PO$_4^-$, 1.1; SO$_4^{2-}$, 29.1	27 ± 12		Temp = 20°C; lower resistance from depolarized current; larger resistance from hyperpolarized current. Action potential in two spikes of 2- & 5- to 10-s duration.	1
12	N. trans-lucens	7-10	0.1	Ca^{2+}, 0.1; K$^+$, 0.1; Na$^+$, 1.0; Cl$^-$, 1.3	21.4		Space constant measured to be 2.6 cm. Capacity of plasmalemma + tonoplast = 1.0 μF/cm^2.	12
13		5-15	0.1	Ca^{2+}, 0.1; K$^+$, 0.1; Na$^+$, 1.0; Cl$^-$, 1.3	112 ± 12	6 ± 0.5	Temp = 15°C; measurements tak-en 12 hr after probe insertion. Membrane resistance reported to increase for first 5-6 hr after insertion. Resistance also pH-dependent.	8,9
14	Pelvetia fastigi-ata	....	0.0092	Ca^{2+}, 10; Mg^{2+}, 55; K$^+$, 10; Na$^+$, 483; HCO$_3^-$, 2.5; Cl$^-$, 564.5; SO$_4^{2-}$, 28; Tris HCl, 10	1.0	....	Temp = 15°C; pH = 8.0. Capacity of plasmalemma = 1 μF/cm^2. In fertilized egg, capacity remains unchanged, but membrane resistance drops to a minimum of 0.135 k$\Omega \cdot$cm^2 after 12 hr & rises to 0.26 after 27 hr.	11

[1/] Positive state. [2/] Negative state. [3/] Data for *G. monile*. [4/] Data for *G. pulvinata*. [5/] Data for *G. teges*.

Contributor: Tyree, M. T.

References

[1] Findlay, G. P. 1959. Aust. J. Biol. Sci. 12:412.

[2] Findlay, G. P., and A. B. Hope. 1964. Ibid. 17:62.

[3] Findlay, G. P., et al. 1969. Ibid. 22:1163.

[4] Findlay, G. P., et al. 1971. Ibid. 24:731.

[5] Goldsmith, M. H. M., et al. 1972. Planta 102:302.

[6] Higinbotham, N., et al. 1964. Science 143:1448.

[7] Hope, A. B., and N. A. Walker. 1961. Aust. J. Biol. Sci. 14:26.

[8] Spanswick, R. M. 1970. J. Exp. Bot. 21:617.

[9] Spanswick, R. M. 1970. J. Membr. Biol. 2:59.

[10] Spanswick, R. M. 1972. Planta 102:215.

[11] Weisenseel, M. H., and L. J. Jaffe. 1972. Develop. Biol. 27:555.

[12] Williams, E. J., et al. 1964. J. Exp. Bot. 15:1.

Complex changes in the electroencephalogram (EEG) occur with age, the most conspicuous being an increase in frequency of the alpha rhythm. Changes in occipital EEG frequencies are shown by integration of the activity recorded from the parieto-occipital areas (left parietal to left occipital plus right parietal to right occipital). Note the shift from the slower frequencies toward the adult alpha frequency and the overall voltage decrease with maturation.

Data are for 50 normal subjects, approximately five in each age group. Dots are individual subject values for the particular frequency, and the solid line is the mean.

Contributor: Bickford, Reginald G.

Reference: Corbin, L. B., and R. G. Bickford. 1955. Electroencephalogr. Clin. Neurophysiol. 7:15.

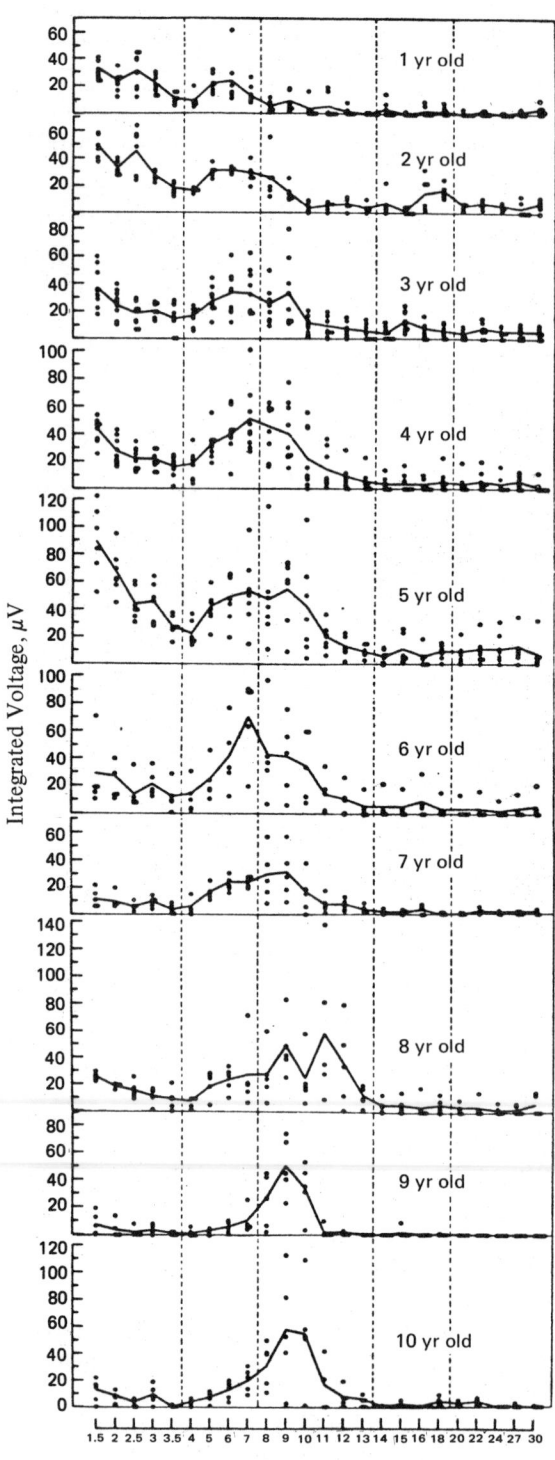

Part I. Effect of Eye Opening

The alpha rhythm has maximum amplitude in the parieto-occipital region of the scalp. Opening the eyes commonly blocks the alpha rhythm and replaces it with a low-voltage fast (so-called desynchronous) activity. Note that eye-blink artifacts appear in the upper four channels. **Lead:** T = temporal; Fp = prefrontal; F = frontal; C = central; P = parietal; O = occipital.

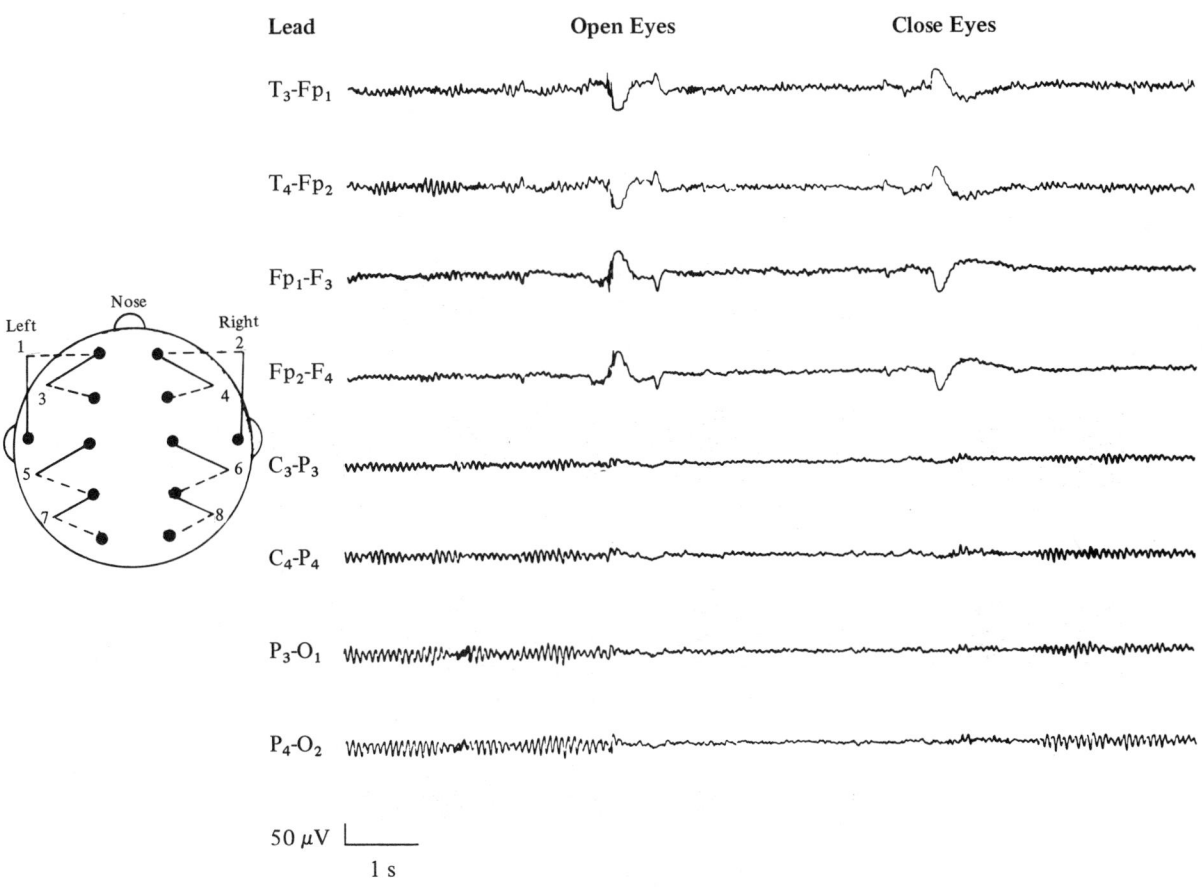

Contributor: Bickford, Reginald G.

General References
[1] Bickford, R. G., et al. 1965. A KWIC Index of EEG Literature. Elsevier, New York.
[2] Brazier, M. A. B. 1960. Electrical Activity of the Nervous System. Ed. 2. Macmillan, New York.
[3] Cooper, R., et al. 1969. EEG Technology. Butterworth, London.
[4] Kooi, K. A. 1971. Fundamentals of Electroencephalography. Harper and Row, New York.

continued

155. TYPICAL ELECTROENCEPHALOGRAM OF A NORMAL ADULT: MAN

Part II. Distribution of Alpha Rhythm

Data are for a normal, 48-year-old female. Power spectra were made on 4 seconds of electroencephalographic activity from bipolar montages in respective scalp regions. Time in 4-second increments is represented as rising vertically.

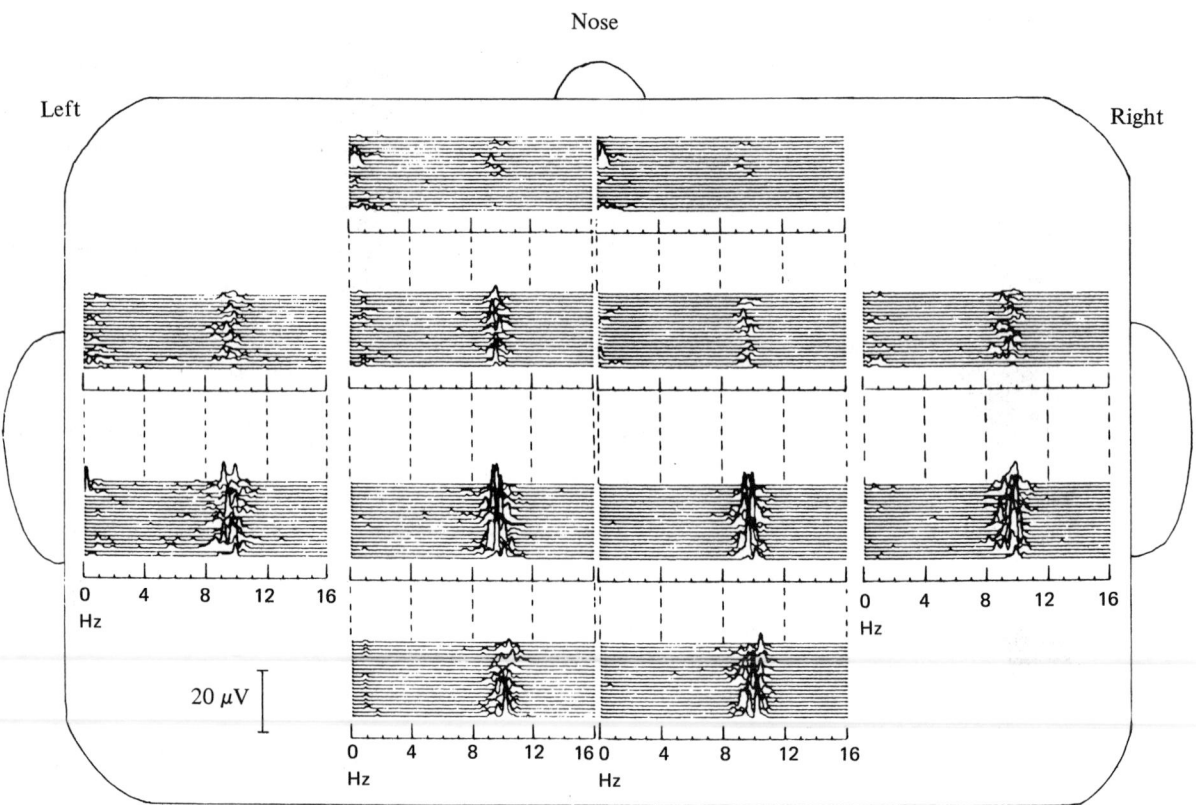

Contributor: Bickford, Reginald G.

Reference: Bickford, R. G., et al. 1972. San Diego Symp. Biomed. Eng. Proc. 11:365.

156. CHANGES IN NORMAL ELECTROENCEPHALOGRAM DURING ALERTNESS, RELAXATION, AND SLEEP: MAN

The electroencephalogram (EEG) is very sensitive to changes in alertness of the subject; drowsiness and sleep, therefore, have been classified by EEG pattern [3]. Stage 0 (or W) corresponds to the waking stage, both while relaxed and alert. Stage 1 depicts relatively low-voltage activity with mixed frequencies usually recorded during the transition from waking to sleep or after movements during sleep. Stage 2 is characterized by the occurrence of sleep spindles (12-14 Hz) and K complexes. Stage 3 is defined by slow activity (2 Hz or slower), occurring 20-50% of the time. Stage 4 is defined by the occurrence of the 2-Hz or slower activity more than 50% of the time. Stage REM (rapid eye movement) is characterized by low-voltage, mixed-frequency EEG interspersed with eye movements and a general lack of muscle tone or activity. Movement time (MT) is designated by muscle artifact or general body movement of the subject, which obscures the EEG tracing. Data are for a normal adult. Figures in brackets are reference numbers.

SLEEP STAGES

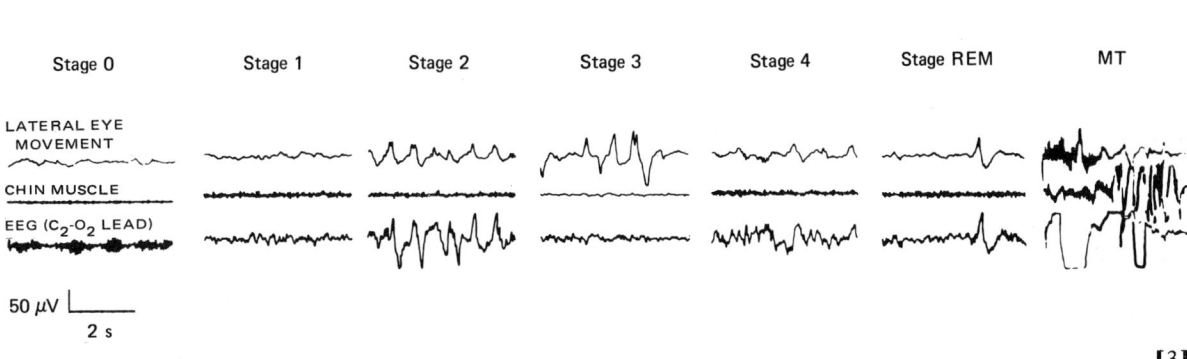

[3]

INTEGRATED EEG ACTIVITY DURING SLEEP

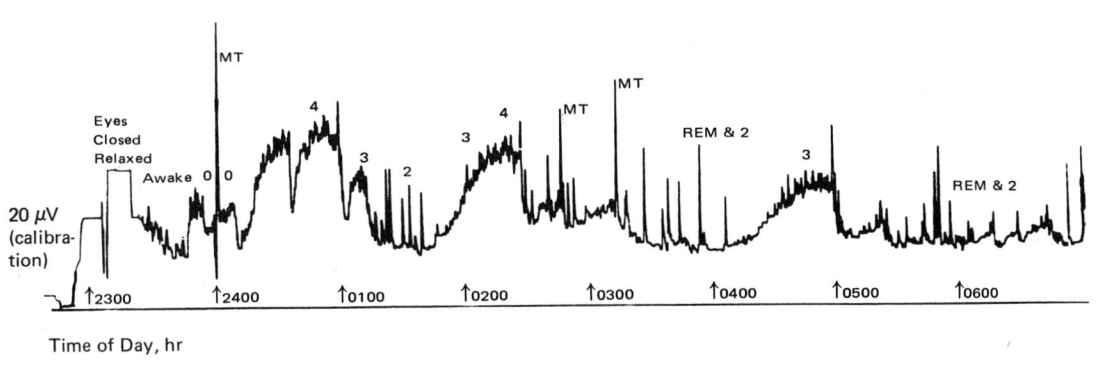

Time of Day, hr

[1,2,4]

Contributor: Bickford, Reginald G.

References

[1] Agnew, H. W. 1968. Psychophysiology 5(2):214.
[2] Maynard, D., et al. 1969. Electroencephalogr. Clin. Neurophysiol. 27:672.
[3] Rechtschaffen, A., and A. Kales, ed. 1968. Nat. Inst. Neurol. Dis. Blindness Pub. 204.
[4] Sims, J. K., et al. 1972. San Diego Symp. Biomed. Eng. Proc. 11:87.

157. CLASSIFICATION OF ELECTROENCEPHALOGRAPHIC PATTERNS: MAN

Single-channel recordings, such as those shown below, provide only minimum samples of the variety and complexity of patterns that develop and spread over the cerebral hemispheres. Multiple-channel recordings are needed to bring out the details and complexity of the electroencephalogram. Compression amplification is advantageous for revealing in a single tracing both high-voltage and low-voltage components. The instrumental characteristic of the amplification and recording system used here is shown at the right.

Figures in brackets are reference numbers.

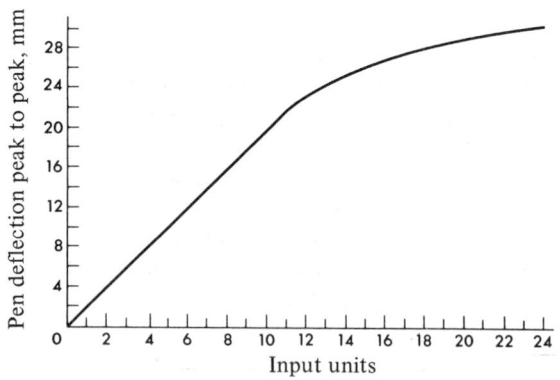

Part I.   Normal Resting, Wakeful Activity During Childhood

Occipital resting activity (eyes closed) was recorded with monopolar leads with the common reference electrode on the earlobe [3]. Activity from the anterior areas usually show less alpha activity and a greater admixture of low-voltage fast activity (beta waves).

	Age	Pattern[1]	Voltage Level & Dominant Frequency
1	3 days		Low-voltage irregular activity, no dominant frequency [1,3]
2	3 months		Low voltage, 1½-3 cycles/s [1,3]
3	5 months		Moderately high voltage, 1½-4 cycles/s [1,3]
4	6 months		High voltage, 4 cycles/s [1,3]
5	11 months		Moderate voltage, 4-5 cycles/s [1,3]
6	20 months		High voltage, 4 cycles/s, mixed with moderate voltage, 8 cycles/s alpha waves [1,3]
7	4 years		High voltage, 9 cycles/s alpha waves [1,3]

[1] 50 μV |⎽⎽⎽⎽⎽⎽⎽| = voltage × time scale for pattern.
       1 s

continued

157. CLASSIFICATION OF ELECTROENCEPHALOGRAPHIC PATTERNS: MAN

Part I. Normal Resting, Wakeful Activity During Childhood

	Age	Pattern[1/]	Voltage Level & Dominant Frequency
8	8 years		Moderate voltage, 9-10 cycles/s alpha waves [1,3]
9	12 years		High voltage, 9 cycles/s alpha waves [1,3]
10	14 years		Low voltage, 9 cycles/s alpha waves [1,3]
11	16 years		9-10 cycles/s alpha waves [1,3]
12	>16 years		9-10 cycles/s, M-shaped alpha waves[2/] [1,2]
13			9-10 cycles/s, U-shaped alpha waves[2/] [1,2]
14			10 cycles/s, spindling alpha waves [1,3]
15			Low-voltage, fast activity[3/] [1,3]

[1/] 50 μV |‾‾‾‾‾‾‾‾| = voltage × time scale for pattern. 1 s [2/] Unusual, but normal, activity. [3/] Low-voltage, fast background activity occurs in some persons with eyes closed. It also appears as a nonspecific attention response in persons with well-developed alpha waves, i.e., the alpha waves drop out with attention, leaving a background of low-voltage fast activity.

Contributors: Gibbs, Frederic A., and Gibbs, Erna L.

References
[1] Chatrian, G. E., et al. 1959. Electroencephalogr. Clin. Neurophysiol. 11:497.
[2] Gastaut, H., et al. 1952. Marseille Med. 89:296.
[3] Gibbs, F. A., and E. L. Gibbs. 1951. Atlas of Electroencephalography. Addison-Wesley, Reading, Mass. v. 1.

Part II. Normal Drowsiness and Sleep

	Depth of Sleep	Pattern[1/]	Description
1	Infantile drowsiness	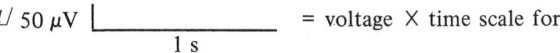	Long runs of slow activity [1]
2	Juvenile drowsiness		Paroxysmal slow activity (short runs) [1]

[1/] 50 μV |‾‾‾‾‾‾‾‾| = voltage × time scale for pattern. 1 s

continued

157. CLASSIFICATION OF ELECTROENCEPHALOGRAPHIC PATTERNS: MAN

Part II. Normal Drowsiness and Sleep

	Depth of Sleep	Pattern [1]	Description
3	Adult drowsiness		Flattening [2]
4	Sudden, brief arousal		K-Complex [3]
5	Light sleep (most marked in childhood)		Biparietal humps [1]
6	Any phase of sleep		Occipital positive sharp waves [1]
7	Deeper sleep		14 cycles/s, spindles [2]
8	Moderately deep sleep		Spindles mixed with slow activity [2]
9	Very deep sleep		2 cycles/s, slow waves [2]
10	Exceedingly deep sleep (common in childhood, rare in old age)		1 cycle/s, slow waves [2]

[1] 50 µV └──────┘ = voltage X time scale for pattern.
       1 s

Contributors: Gibbs, Frederic A., and Gibbs, Erna L.

References

[1] Gibbs, F. A., and E. L. Gibbs. 1951. Atlas of Electro-encephalography. Addison-Wesley, Reading, Mass. v. 1.

[2] Loomis, A. L., et al. 1936. J. Exp. Psychol. 19:249.

[3] Loomis, A. L., et al. 1938. J. Neurophysiol. 1:413.

continued

Part III. Nonparoxysmal Abnormalities

Electroencephalographic abnormalities are nonspecific with regard to etiology; they can be produced by anything that can injure the brain, e.g., trauma, encephalitis, vascular disease, hypoxia, hypoglycemia, metabolic disorders, toxic substances, and drugs. Light anesthesia produced by most anesthetics is associated with fast activity. Abnormal fast activity usually increases in drowsiness and light sleep.

Moderately slow activity is commonly associated with a slightly obtunded state or light stupor. Very slow activity is commonly associated with coma. A recording that appears to be flat, i.e., isoelectric, at double the usual amplification (in the absence of drugs or infection), creates a presumption of irreversible coma [3].

For additional information, consult reference 1. **Abnormality**: + signs indicate degree of severity, from mild (+) to very severe (++++).

	Abnormality	Pattern[1]	Description
1	Irritation (+) & depression		Slightly fast [1]
2	Irritation (++) & depression		Moderately fast [1]
3	Irritation (+++) & depression		Exceedingly fast [1,2]
4	Depression (+)		Slightly slow [1]
5	Depression (++)		Moderately slow [1]
6	Depression (+++)		Very slow [1]
7	Depression (++++)		Flattening & exceedingly slow [1,3]

[1] 50 μV 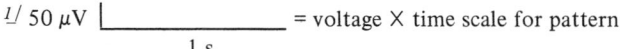 = voltage × time scale for pattern.
1 s

Contributors: Gibbs, Frederic A., and Gibbs, Erna L.

References

[1] Gibbs, F. A., and E. L. Gibbs. 1964. Atlas of Electroencephalography. Ed. 2. Addison-Wesley, Reading, Mass. v. 3.

[2] Gibbs, E. L., et al. 1950. Dis. Nerv. Syst. 11(11).

[3] Silverman, D., et al. 1970. Neurology 20:525.

continued

Part IV. Seizure Discharges

Hypsarhythmia, the 2- and 3-cycles/s spike and wave discharges, commonly appears in the awake recording and also during sleep. The grand mal type of discharge almost never appears in the awake, interseizure recording; it is seen in status epilepticus and in sleep, most often in children with petit mal variant and petit mal epilepsy. The last three "controversial" patterns most commonly appear in sleep recordings and are relatively rare in awake recordings. For additional information, consult reference 2.

	Abnormality	Pattern [1]/	Description
1	Infantile spasms		Hypsarhythmia [5]
2	Petit mal variant		2 cycles/s, spike & wave [4]
3	Petit mal		3 cycles/s, spike & wave [3]
4	Grand mal		Crescendo spiking [3]
5	Diencephalic epilepsy		14 & 6 cycles/s, positive spikes [1,8]
6			6 cycles/s, spike & wave [7,10]
7	Psychomotor variant		Mid-temporal rhythmic discharge [6,9]

[1]/ 50 μV |_____| = voltage X time scale for pattern.
         1 s

Contributors: Gibbs, Frederic A., and Gibbs, Erna L.

continued

Part IV. Seizure Discharges

References

[1] Gibbs, F. A., and E. L. Gibbs. 1951. Neurology 1: 136.

[2] Gibbs, F. A., and E. L. Gibbs. 1952. Atlas of Electro-encephalography. Addison-Wesley, Reading, Mass. v. 2.

[3] Gibbs, F. A., et al. 1935. Arch. Neurol. Psychiat. 34:1133.

[4] Gibbs, F. A., et al. 1939. Ibid. 41:1111.

[5] Gibbs, E. L., et al. 1954. Pediatrics 13:66.

[6] Gibbs, F. A., et al. 1963. Neurology 13:991.

[7] Hill, D., and G. Parr, ed. 1963. Electroencephalography. Macdonald, London.

[8] Holbrook, T. J. 1970. Clin. Electroencephalogr. 1(1):36.

[9] Lipman, I. J., and J. R. Hughes. 1969. Electroencephalogr. Clin. Neurophysiol. 27:43.

[10] Thomas, J. 1957. Neurology 7:438.

Part V. Spikes

All types of spiking are most evident in sleep, all can be focal or generalized, and all are commonly followed by one or more slow waves. Frontoparietal single spike and wave discharges are a common paroxysmal dysrhythmia [3] which is not shown below. It does not correlate with petit mal, but does correlate with convulsive disorder. Slow waves with sharp component and spike-like waves are epileptiform, but unlike the patterns below, they are not clearly epileptic. For additional information, consult reference 3.

	Occurrence	Pattern [1]	Description & Location
1	Common in premature children (associated with visual difficulties); rare in adults		Needles; usually occipital, sometimes midtemporal [2,3]
2	Often asymptomatic, but commonly associated with convulsions or psychomotor seizures; usually found in adults		Small, sharp spikes; usually diffuse or arising from multiple foci, but particularly common in the anterior temporal area [3]
3	When focal in the anterior temporal area, usually associated with psychomotor seizures		Moderate-voltage spikes; can be focal in any area [1]
4	Convulsions, either focal or generalized; occasionally found in the anterior temporal area in association with psychomotor seizures; usually seen in children		High-voltage slow spikes; can be focal in any area [3]

[1] 50 μV 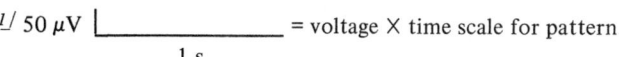 = voltage × time scale for pattern.

1 s

Contributors: Gibbs, Frederic A., and Gibbs, Erna L.

References

[1] Gibbs, E. L., et al. 1948. Arch. Neurol. Psychiat. 60:95.

[2] Gibbs, F. A., et al. 1968. Johns Hopkins Med. J. 343.

[3] Gibbs, F. A., and E. L. Gibbs. 1952. Atlas of Electro-encephalography. Addison-Wesley, Reading, Mass. v. 2.

Part I. Tabular

Electroencephalograms were recorded with bipolar leads over the telencephalon, unless otherwise indicated. The frequency band of amplification was usually ∿1-50 Hz. For additional information on electroencephalographic techniques, consult references 15, 23, 30, 35, and 60. **Experimental Condition:** Alert = alert wakefulness; Relaxed-asleep = relaxed or in the deep "slow-wave" stage of sleep; Anesthetized = surgical levels of anesthesia. **Frequency:**

Values are the number of waves which would occur in 1 second if a given wave were recurrent. Frequency measurements were made on the large "fundamental" components of the compounded electroencephalogram wave form. Values for frequency and amplitude are approximate. For additional values, consult references 5, 9, 14, 27, 37, 39, 41, 52, and 70.

	Class & Species (Synonym)	Experimental Condition	Frequency	Amplitude μV	Remarks	Reference
	Mammalia					
1	*Homo sapiens*	Alert	14-30	5-15	Theta augmented in emotional states. Alpha varies with individuals; promoted by eye closure.	38
2		Alert	10-20	10-20	Original study of dream sleep	8
3		Relaxed-asleep	1-2	50-200		
4		Anesthetized [1]	0.5-3	50-200	Injectable & inhalant anesthetics used	60
5	*Bos taurus*	Alert	9-11	25	Sleep study	57
6		Relaxed-asleep	3-4	50-150		
7	*Canis familiaris*	Alert	6-14	20-40	....	42
8		Alert	28 [2]	0-10	Subjects paralyzed with gallamine	20
9		Alert	7	20	....	49
10		Relaxed-asleep	2	50		
11		Alert	20-40	15-25	Reference electrode in frontal bone; occipital alpha	66
12		Relaxed-asleep	10-14	20-40		
13		Anesthetized [1]	5-10	15-45	Light to deep anesthesia with pentobarbital sodium; interval histograms	34
14	*Capra hircus*	Alert	20-40	20	"Activated" sleep; mastication artifact	54
15		Relaxed-asleep	2-3	50		
16		Alert	20-40	20-40	"Activated" sleep; mastication artifact. Rumen still contracts during sleep.	33
17		Relaxed-asleep	2-4	100		
18	*Cavia porcellus*	Alert	7-9	75-100	....	62
19		Alert	20-60	....	....	28
20		Relaxed-asleep	6-15	....		
21	*Condylura cristata; Scalopus*	Alert	8-20	75-100	"Activated" sleep	1
22	*aquaticus*	Relaxed-asleep	2-8	100-200		
23	*Didelphis marsupialis*	Alert	10-30	50-75	"Activated" sleep	72
24		Relaxed-asleep	4-11	100-150		
25	*D. marsupialis virginiana (D.*	Alert	6-12	20-60	Reference electrode in nasal bone & vertex	2
26	*virginiana)*	Relaxed-asleep	2-4	75-100		
27	*Equus asinus*	Alert	4-20	20-40	....	55
28		Relaxed-asleep	2-8	50-90		
29	*E. caballus*	Alert	20-30	50	Sleep study	59
30		Relaxed-asleep	2-8	100-200		
31	*Felis catus*	Alert	8-40	10-20	First full report of "activated" sleep in animals	7
32		Relaxed-asleep	2-6	30-60		
33		Alert	12-50	....	Subtle differences in frequency between alert condition & "activated" sleep	31
34		Relaxed-asleep	1-8	....		
35		Alert	20-40	50	....	22
36		Relaxed-asleep	1-3	200-500	Sleep artificially induced by electrical stimulation	

[1] Frequency and amplitude vary with depth of anesthesia and with the drug used. [2] Plus some 4- to 8-s waves.

continued

Part I. Tabular

	Class & Species (Synonym)	Experimental Condition	Frequency	Amplitude μV	Remarks	Reference
37		Anesthetized [L]	1-3	200-500	Anesthetic: alloisopropylbarbituric acid	
38		Anesthetized [L]	2-4	150	Anesthetic: thiopental sodium; plasma concentration measured	19
39	Macaca mulatta	Alert	5-20	50-75	"Activated" sleep	73
40		Relaxed-asleep	1-3	100-150		
41		Relaxed-asleep	1-3	100-200	Sleep study	74
42		Anesthetized [L]	3-6	....	....	13
43	M. nemestrina	Alert	8-20	40-75	Sleep study	50
44		Relaxed-asleep	2-3	150-350		
45	Mesocricetus auratus [3]	Alert	8-16	50-100	"Activated" sleep; cortical theta	71
46		Relaxed-asleep	2-8	100-200		
47		Alert	10-20	20-40	Anesthetic: pentobarbital sodium	6
48		Anesthetized [L]	6-12	50-100		
49	Mus musculus	Alert	20-34	50-100	....	36
50		Relaxed-asleep	3-5	200-300		
51		Alert	8-16	50-100	"Activated" sleep; cortical theta	71
52		Relaxed-asleep	2-8	100-200		
53	Oryctolagus cuniculus	Alert	16	25	....	64
54		Relaxed-asleep	4-8	50-75		
55	Ovis aries	Alert	10-15	15-30	Anesthetic: barbiturate	56
56		Anesthetized [L]	4-6	40-100		
57	Pan troglodytes (P. satyrus)	Alert	10-20	20-40	....	51
58		Alert	10-20	20-40	"Activated" sleep; K complexes during "slow-wave" sleep	12
59		Relaxed-asleep	1-2	50-100		
60	Papio sp.	Anesthetized [L]	8-12	10-20	Anesthetic: phencyclidine plus pentobarbital sodium	10
61	Rattus norvegicus	Alert	10-30	....	"Activated" sleep	67
62		Relaxed-asleep	2-10	....		
63		Alert	8-16	50-100	"Activated" sleep; cortical theta	71
64		Relaxed-asleep	2-8	100-200		
65		Alert	30-40	30	"Activated" sleep; cortical theta. Subtle differences in frequency between alert condition & "activated" sleep.	69
66		Relaxed-asleep	2-6	50-75		
67	Saimiri sp.	Alert	8-16	20-50	....	3
68	Spermophilus tridecemlineatus (C. tridecemlineatus)	Alert	8-16	50-100	"Activated" sleep; cortical theta	71
69		Relaxed-asleep	2-8	100-200		
70	Sus scrofa (S. domesticus)	Alert	7-10	80	....	53
71		Relaxed-asleep	2-3	100-200		
72		Alert	8-15	....	Sleep study	58
73		Relaxed-asleep	3-6	....		
74		Alert	7-25	15-25	....	42
75		Relaxed-asleep	2-10	50-150	....	
76		Anesthetized [L]	8-12	75-150	Anesthetic: thiopental sodium; anesthesia probably very light	
	Aves					
77	Belonopterus chilensis lampronotus	Alert	16-18	20-40	....	64
78		Relaxed-asleep	4-12	50-60		

[L] Frequency and amplitude vary with depth of anesthesia and with the drug used. [3] Flat trace during hibernation.

continued

Part I. Tabular

	Class & Species (Synonym)	Experimental Condition	Frequency	Amplitude μV	Remarks	Reference
79	*Columba livia*	Alert	8-12	....	Reference electrode over optic lobes	4
80		Relaxed-asleep[4]	2-5	80-100		
81	*Gallus gallus (G. domesticus)*	Alert	10-30	25-75	12-25 bursts of eye-motion artifact/s	44
82		Alert	10-20	50-150	....	29
83		Relaxed-asleep	2-3	100-400		
84		Alert	16-25	15-25	"Activated" sleep	32
85		Relaxed-asleep	5-9	100		
86		Alert	30-60	50	Reference electrode in comb; "activated" sleep	40
87		Relaxed-asleep	3-4	200		
88	Chick	Alert	2-6	100	....	65
89		Relaxed-asleep	1-3	100-200		
	Reptilia					
90	*Caiman crocodilus latirostris*	Alert	7-8	20	Sleep study; partial "activated" sleep	47
91	*(C. latirostris)*	Relaxed-asleep	3-7	20-40		
92	*C. sclerops*	Alert	8-15	25-50	....	25
93		Alert[4]	7-12	20-40	....	43
94		Alert	4-8	....	No "activated" sleep	48
95		Relaxed-asleep	1-2	....		
96	*Chamaeleo jacksoni; C. melleri*	Alert	10-12	30-35	Partial "activated" sleep; eye motion only	68
97		Relaxed-asleep	7-9	35-45		
98		Anesthetized[L]	4-6	60-80		
99	*Emys orbicularis (E. europea)*	Alert	40-50	10-20	....	4
100	*Iguana iguana*	Alert	30-35	50	Sleep study; no "activated" sleep	47
101		Relaxed-asleep	4-20	100		
102	*Python sebae*	Alert	20-22	50	Sleep study; no "activated" sleep	47
103		Relaxed-asleep	14-15	75		
104	*Sceloporus olivaceus*	Alert	20-25	10-25	Reference electrode lateral, in base of skull. Anesthetic: ether.	26
105		Relaxed-asleep	3-5	10-50		
106		Anesthetized[L]	0.5-1	100		
107	*Testudo marginata*	Alert	11-13	15-20	Partial "activated" sleep	21
108		Relaxed-asleep	6-8	50		
	Amphibia					
109	*Ambystoma tigrinum*	Alert	5-8, 12-14	....	Study of seizures & artifacts	45
110	*Bufo arenarum*	Alert	3-5	20-30	Reference electrode over nasal bone. In summer & spring, 8-12 fusiform responses to light/s	63
111	*Necturus maculosus*	Alert	4-8	20-40	Study of artifacts	17
112	*Notophthalmus viridescens (Triturus viridescens)*	Alert	5-7, 12-14, 18-22	....	Study of seizures & artifacts	45
113	*Rana catesbeiana*	Alert	7-9	100	....	62
114		Alert	6-15	20-30	No clear signs of sleep; rhythmic olfactory waves during motion	24
115		Relaxed-asleep	12-30	10		
	Osteichthyes					
116	*Carassius auratus*	Alert	9-14	....	Arousal responses to light; 35-40/s sometimes at end of session. Anesthetic: urethan.	61
117		Relaxed-asleep	4-8	40-70		
118		Anesthetized[L]	4-6	20		
119	*Gadus morhua (G. callarias)*	Alert	10-20	20	Arousal responses to light; alpha-like waves over midbrain; 0.3-0.8 gill artifact/s	11
120		Relaxed-asleep	8-13	25-50		

[L] Frequency and amplitude vary with depth of anesthesia and with the drug used. [4] Condition of subject not clearly identified.

continued

Part I. Tabular

	Class & Species (Synonym)	Experimental Condition	Frequency	Amplitude μV	Remarks	Reference
121	*Oncorhynchus kisutch; O. tshawytscha*	Alert	4-8	20-30	Study of olfaction in homing behavior	18
122	*Tinca tinca*	Alert	6-20	10-15	Theta sometimes; no signs of sleep	46
123	Chondrichthyes *Ginglymostoma cirratum; Negaprion brevirostris; Sphyrna tiburo*	Relaxed-asleep	4-9	30-60	Arousal responses to light in midbrain; olfactory stimuli increased frequency & amplitude in forebrain	16

Contributor: Klemm, W. R.

References

[1] Allison, T., and H. Van Twyver. 1970. Exp. Neurol. 27:564.

[2] Barratt, E. S. 1965. Electroencephalogr. Clin. Neurophysiol. 18:709.

[3] Barratt, E. S., and P. Adams. Unpublished. Univ. Texas Medical Branch, Galveston, 1971.

[4] Bremer, F., et al. 1939. J. Neurophysiol. 2:473.

[5] Bullock, T. H. 1945. Yale J. Biol. Med. 17:657.

[6] Chatfield, P. O., et al. 1951. Electroencephalogr. Clin. Neurophysiol. 3:225.

[7] Dement, W. 1958. Ibid. 10:291.

[8] Dement, W., and N. Kleitman. 1957. Ibid. 9:673.

[9] Ectors, L. 1936. Arch. Int. Physiol. 43:267.

[10] Ellington, A., et al. 1967. Baboon Med. Res. Proc. Int. Symp. 2:775.

[11] Enger, P. S. 1957. Acta Physiol. Scand. 39:55.

[12] Freemon, F. R., et al. 1969. Exp. Neurol. 25:129.

[13] Forster, F. M., and L. F. Nims. 1942. Arch. Neurol. Psychiat. 47:449.

[14] Gerard, R. W., and J. Z. Young. 1937. Proc. Roy. Soc. 122:343.

[15] Gibbs, F. A., and E. L. Gibbs. 1967. Medical Electroencephalography. Addison-Wesley, Reading, Mass.

[16] Gilbert, P. W., et al. 1964. Science 145:949.

[17] Goodman, D. A., and N. M. Weinberger. 1969. Physiol. Zool. 42:398.

[18] Hara, T. J., et al. 1965. Science 149:884.

[19] Hatch, R. C., et al. 1970. Amer. J. Vet. Res. 31:291.

[20] Herin, R. A., et al. 1968. Ibid. 29:329.

[21] Hermann, H., et al. 1964. C. R. Soc. Biol. 258:2175.

[22] Hess, R., et al. 1953. Electroencephalogr. Clin. Neurophysiol. 5:75.

[23] Hill, J. D. N., and G. Parr. 1963. Electroencephalography. Macmillan, New York.

[24] Hobson, J. A. 1967. Electroencephalogr. Clin. Neurophysiol. 22:113.

[25] Huggins, S. E., et al. 1968. Physiol. Zool. 41:371.

[26] Hunsaker, D., II, and R. W. Lansing. 1962. J. Exp. Zool. 149:21.

[27] Jouvet, D., and J. L. Valtax. 1962. C. R. Soc. Biol. 156:1411.

[28] Jouvet-Mounier, D., and L. Astic. 1966. Ibid. 160:1453.

[29] Key, B. J., and E. Marley. 1962. Electroencephalogr. Clin. Neurophysiol. 14:90.

[30] Kiloh, L. G., and J. W. Osselton. 1966. Clinical Electroencephalography. Butterworth, London.

[31] Kiyono, S., and K. Iwama. 1965. Jap. J. Physiol. 15:366.

[32] Klein, M., et al. 1964. C. R. Soc. Biol. 158:99.

[33] Klemm, W. R. 1966. Proc. Soc. Exp. Biol. Med. 121:635.

[34] Klemm, W. R. 1968. Amer. J. Vet. Res. 29:1267.

[35] Klemm, W. R. 1969. Animal Electroencephalography. Academic Press, New York.

[36] Kobayashi, T., et al. 1963. Biol. Psychiat. 5:293.

[37] Monnier, M., and H. Gangloff. 1961. Atlas for Stereotaxic Brain Research. Elsevier, New York. p. 38.

[38] Mundy-Castle, A. C. 1951. Electroencephalogr. Clin. Neurophysiol. 3:477.

[39] Muzzetto, P., et al. 1970. Clin. Vet. 93:1.

[40] Ookawa, T., and J. Gotoh. 1964. Poult. Sci. 43:1603.

[41] Pampiglione, G. 1963. Development of Cerebral Function in the Dog. Butterworth, London.

[42] Pampiglione, G. 1965. Proc. Roy. Soc. Med. 58:547.

[43] Parsons, L. C., and S. E. Huggins. 1965. Proc. Soc. Exp. Biol. Med. 119:397.

[44] Paulson, G. 1964. Electroencephalogr. Clin. Neurophysiol. 16:611.

[45] Peters, J. J., and A. R. Vonderahe. 1954. Ibid. 6:253.

[46] Peyrethon, J., and D. Dusan-Peyrethon. 1967. C. R. Soc. Biol. 161:2533.

[47] Peyrethon, J., and D. Dusan-Peyrethon. 1969. Ibid. 163:181.

[48] Rechtschaffen, A., et al. 1968. Psychophysiology 5:201.

continued

158. TYPICAL ELECTROENCEPHALOGRAMS: VERTEBRATES

Part I. Tabular

[49] Redding, R. W., and R. K. Colwell. 1964. Amer. J. Vet. Res. 25:857.

[50] Reite, M. L., et al. 1965. Arch. Neurol. (Chicago) 12:133.

[51] Reite, M. L., et al. 1966. Brain Res. 3:392.

[52] Rempel, B., and E. L. Gibbs. 1936. Science 84:334.

[53] Ruckebusch, Y. 1962. Bull. Soc. Vet. Med. Comp. 64:375.

[54] Ruckebusch, Y. 1962. C. R. Soc. Biol. 156:867.

[55] Ruckebusch, Y. 1963. Ibid. 157:840.

[56] Ruckebusch, Y. 1964. Rev. Med. Vet. 115:793.

[57] Ruckebusch, Y., and F. -R. Bell. 1970. Ann. Rech. Vet. 1:41.

[58] Ruckebusch, Y., and M. T. Morel. 1968. C. R. Soc. Biol. 162:1346.

[59] Ruckebusch, Y., et al. 1970. Ibid. 164:658.

[60] Sadove, M. S., et al. 1967. Electroencephalography for Anesthesiologists and Surgeons. J. B. Lippincott, Philadelphia.

[61] Scháde, J. P., and I. J. Weiler. 1959. J. Exp. Biol. 36:435.

[62] Schwarz, B. E., and R. G. Bickford. 1956. J. Nerv. Ment. Dis. 124:433.

[63] Segura, E. T., and A. deJuan. 1966. Electroencephalogr. Clin. Neurophysiol. 21:373.

[64] Silva, E. E., et al. 1959. Arch. Ital. Biol. 97:167.

[65] Spooner, C. E., and W. D. Winters. 1967. Int. J. Neuropharmacol. 6:109.

[66] Storm van Leeuwen, W., et al. 1968. Progr. Brain Res. 22:181.

[67] Swisher, J. E. 1962. Science 138:1110.

[68] Tauber, E. S., et al. 1966. Nature (London) 212: 1612.

[69] Timio-Iaria, C., et al. 1970. Physiol. Behav. 5:1057.

[70] Tokaji, E., and R. W. Gerard. 1939. Proc. Soc. Exp. Biol. Med. 41:653.

[71] Van Twyver, H. 1969. Physiol. Behav. 4:901.

[72] Van Twyver, H., and T. Allison. 1970. Electroencephalogr. Clin. Neurophysiol. 29:181.

[73] Weitzman, E. D. 1961. Ibid. 13:790.

[74] Weitzman, E. D., et al. 1965. Arch. Neurol. (Chicago) 12:463.

Part II. Graphic

The telencephalon of all vertebrates exhibits strikingly similar spontaneous electrical activity. This similarity plus other evidence suggests that brain waves should be considered not as a sign of the higher aspects of nervous activity, but as a reflection of some basic, primitive, common denominator of the brains of frogs and men. [3] Line numbers refer to Part I.

NORMAL WAVE FORMS

Recordings were made during the alert wakefulness stage of behavior.

MAMMALIA

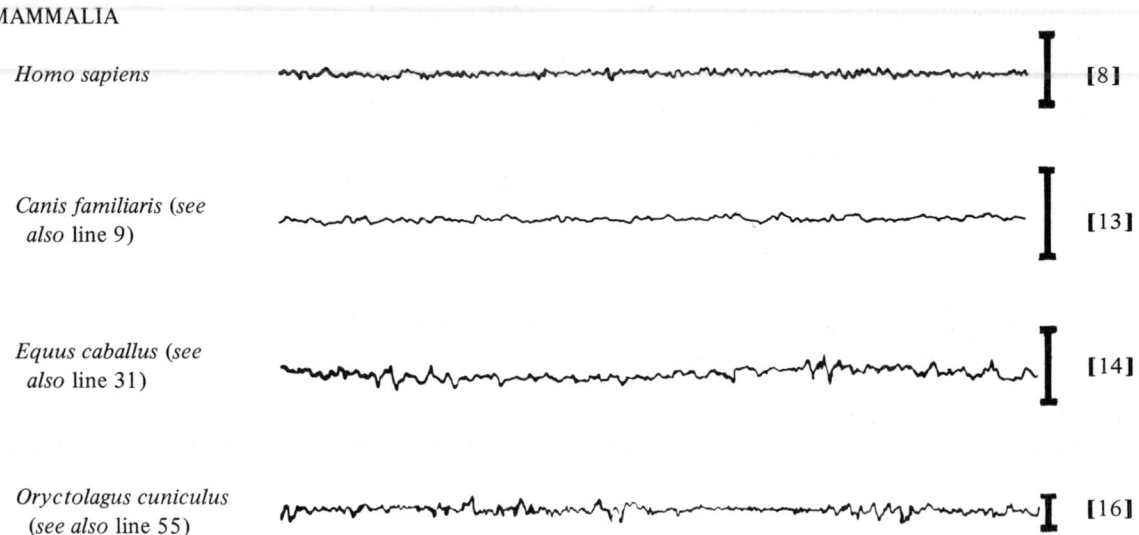

Homo sapiens [8]

Canis familiaris (*see also* line 9) [13]

Equus caballus (*see also* line 31) [14]

Oryctolagus cuniculus (*see also* line 55) [16]

continued

Part II. Graphic

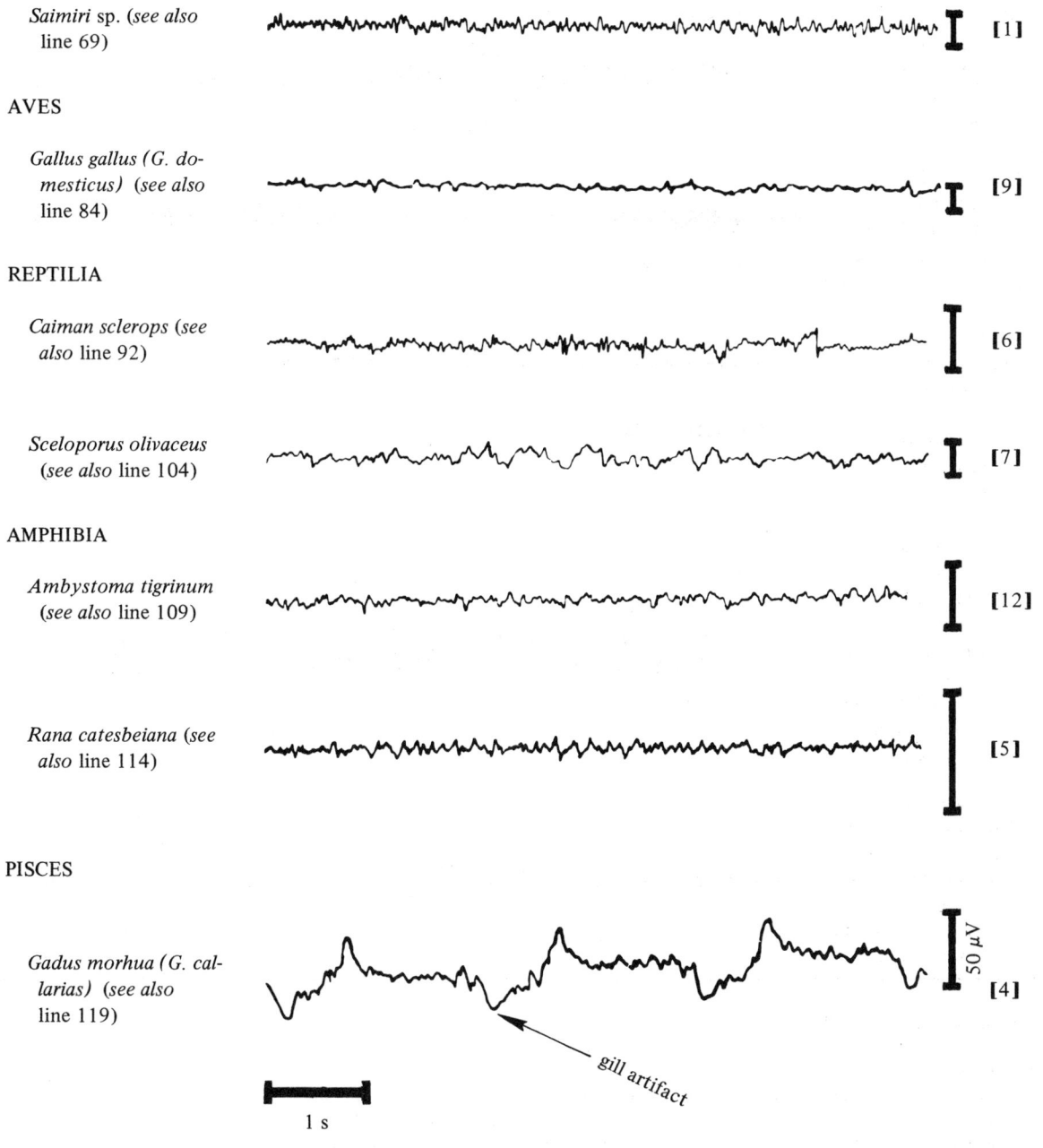

Saimiri sp. (*see also* line 69)   [1]

AVES

Gallus gallus (G. domesticus) (*see also* line 84)   [9]

REPTILIA

Caiman sclerops (*see also* line 92)   [6]

Sceloporus olivaceus (*see also* line 104)   [7]

AMPHIBIA

Ambystoma tigrinum (*see also* line 109)   [12]

Rana catesbeiana (*see also* line 114)   [5]

PISCES

Gadus morhua (G. callarias) (*see also* line 119)   50 μV   [4]

gill artifact

1 s

SPECIAL WAVE FORMS

/ / / / / / indicates waveform portion of tracing.

Homo sapiens
Alpha rhythm, 8-13/s
(Prominent in occipital regions of relaxed subjects)   [15]

continued

158. TYPICAL ELECTROENCEPHALOGRAMS: VERTEBRATES

Part II.   Graphic

K complex, slow wave followed by spindles
(Occurs during non-
activated sleep in
response to stimuli
given at arrow)

Spindles, 12-18/s
(Occur in many
mammals during
nonactivated sleep
and light anesthesia)

Isoelectric activity
(Occurs in anesthetic
overdose, cardiac
failure; if persistent,
serves as basis for
diagnosing clinical
death)

Canis familiaris
Delta waves, <4/s
(Occur in all mam-
mals during non-
activated sleep and
during anesthesia;
abnormal if present
in adults during
alert wakefulness)

Oryctolagus cuniculus
Theta rhythm, 4-7/s
(Seen in some humans
with behavior disor-
ders and in hippo-
campus of lower mam-
mals during alertness)

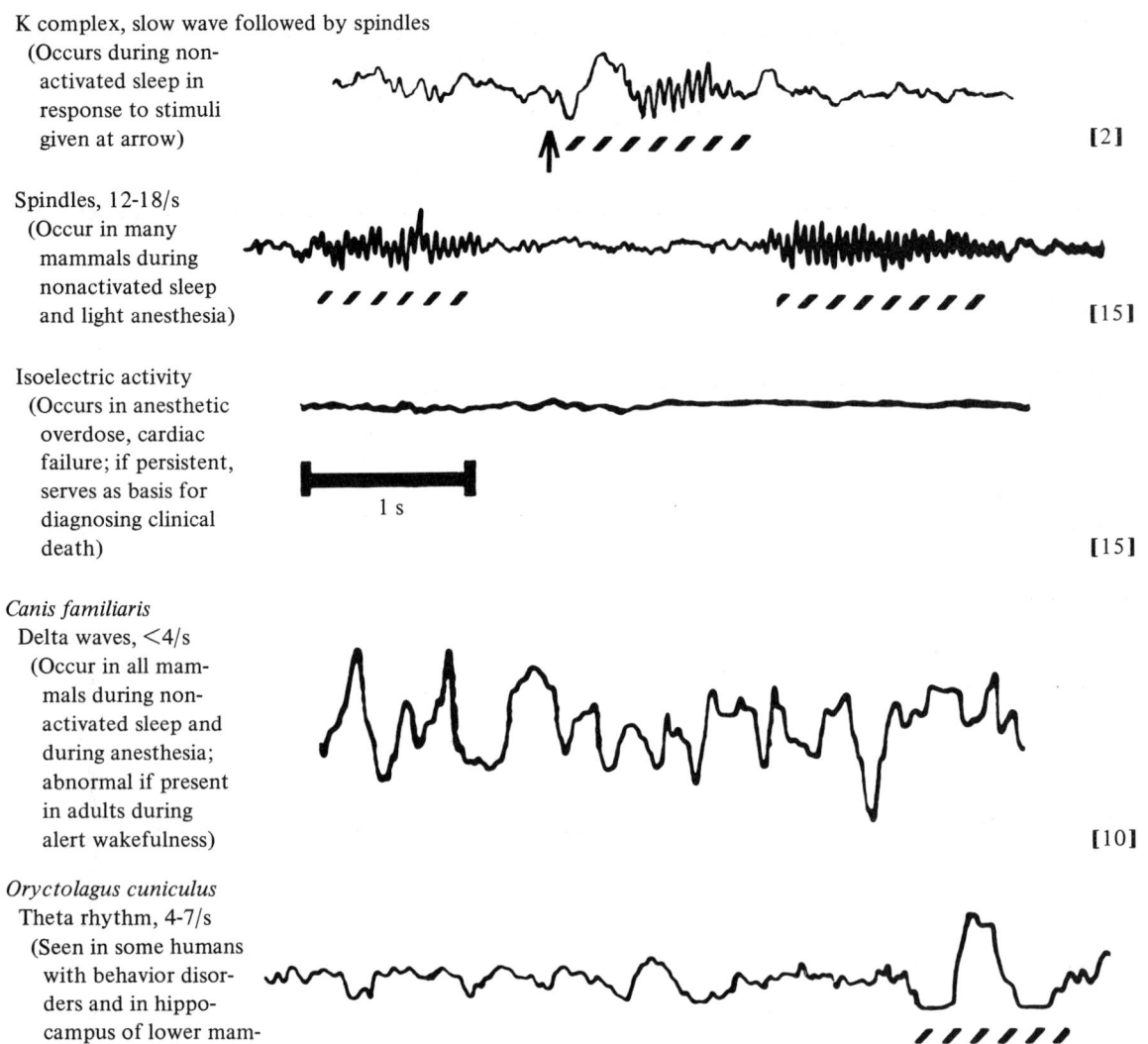

1 s

[2]

[15]

[15]

[10]

[11]

Contributor: Klemm, W. R.

References

[1] Barratt, E. S., and P. Adams. Unpublished. Univ. Texas Medical Branch, Galveston, 1971.

[2] Brazier, M. A. B. 1968. The Electrical Activity of the Nervous System. Ed. 3. Williams and Wilkins, Baltimore.

[3] Bullock, T. H. 1945. Yale J. Biol. Med. 17:657.

[4] Enger, P. S. 1957. Acta Physiol. Scand. 39:55.

[5] Hobson, J. A. 1967. Electroencephalogr. Clin. Neurophysiol. 22:113.

[6] Huggins, S. E., et al. 1968. Physiol. Zool. 41:371.

[7] Hunsaker, D., II, and R. W. Lansing. 1962. J. Exp. Zool. 149:21.

[8] Kiloh, L. G., and J. W. Osselton. 1966. Clinical Electroencephalography. Butterworth, London.

[9] Klein, M., et al. 1964. C.R. Soc. Biol. 158:99.

[10] Klemm, W. R. 1969. Animal Electroencephalography. Academic Press, New York.

[11] Monnier, M., and H. Gangloff. 1961. Atlas for Stereotaxic Brain Research. Elsevier, New York.

[12] Peter, J. J., and A. R. Vonderahe. 1954. Electroencephalogr. Clin. Neurophysiol. 6:253.

[13] Redding, R. W., and R. K. Colwell. 1964. Amer. J. Vet. Res. 25:857.

[14] Ruckebusch, Y., et al. 1970. C.R. Soc. Biol. 164:658.

[15] Sadove, M. S., et al. 1967. Electroencephalography for Anesthesiologists and Surgeons. J. B. Lippincott, Philadelphia.

[16] Silva, E. E., et al. 1959. Arch. Ital. Biol. 97:167.

	Species	Temperature Change	Nervous System Segment or Region	Function	Specific Response	Reference
1	*Homo sapiens* [1/]	Heat applied to body surface	Nerve trunk	Fiber regeneration after nerve section	Increased rate	10
2	*Bos taurus*, ♂	Heating by diathermy	Hypothalamus, anterior region	Respiration	Increased	13
3	♀	In vitro heating & in vitro cooling	Splenic nerve, isolated segment (mainly C fibers)	Action potential generation	Resting potential & afterpotential maximal at 20°C; spike height maximal at 5-10°C	19
4	*Canis familiaris*, unanesthetized	Heating with implanted electrode	Hypothalamus, preoptic area	Hydrocortisone level in blood plasma	Increased	5
5	Anesthetized	Heating & cooling with thermode, 1-2°C above & 1-2°C below body temp	Hypothalamus, temperature-sensitive neurons	Firing rate	Increased	11
6	*Capra hircus*, unanesthetized	Cooling with implanted electrode	Hypothalamus, preoptic area	Plasma-bound iodine level in blood	Increased	2
7				Temperature regulation	Shivering & peripheral vasoconstriction	1
8	*Castor canadensis*	In vitro cooling −5.0°C	Nerve trunks, isolated tissues — Caudal	Action potential	Extinction	20
9				Action potential duration	12.0 ms	
10				Absolute refractory period	10.0 ms	
11			Phrenic	Action potential duration	34.0 ms	
12				Absolute refractory period	36.0 ms	
13			Tibial	Action potential duration	15.0 ms	
14				Absolute refractory period	17.5 ms	
15		−0.7°C	Tibial	Action potential	Extinction	
16		4.5°C	Phrenic	Action potential	Extinction	
17	*Felis catus*	Localized heating or cooling	Skin thermoreceptors	Generation of nerve impulses	Changes in frequency of discharges in C fibers	12
18			Superior cervical ganglion (autonomic)	Conduction	Blockage at 10-18.5°C & 49-52°C	7
19	Unanesthetized	Cooling with implanted electrode	Hypothalamus	Temperature regulation	Cutaneous blood flow reduced; shivering initiated	17
20				Metabolic rate	Increased 11-27%	3
21	Anesthetized	Localized cooling	Spinal cord, neuronal pathways	Reflex responses & post-tetanic potentiation	Increased	15
22				Spread of discharge in reflex responses	Increased	4
23				Stretch reflex	Augmented; phasic shift in response, & earlier tension development	24

[1/] Localized cooling of brain or spinal cord regions is used for local hypothermia in surgical procedures [22].

continued

	Species	Temperature Change	Nervous System Segment or Region	Function	Specific Response	Reference
24			Motoneurons	Minimal gradient requirement for excitation	Decreased	16
25				Action potential amplitude	Decreased rise-time & fall-time	23
26				Membrane resistance	Increased	23
27				Resting membrane potential	Decreased	23
28				Postsynaptic potential	Increased	23
29			α Motoneurons	Activation	Tonic α units activated earlier than phasic	14
30			γ Motoneurons	Activation	Activated, then depressed	14
31	*F. catus, Oryctolagus cuniculus*		Vagus, splanchnic, & cervical sympathetic nerves; isolated segments	Excitation	Afferent fibers stimulated by heating, but not by cooling. Efferent fibers of similar size not stimulated by heating or cooling	6
32	*O. cuniculus*	Localized heating or cooling	Superior cervical ganglion (autonomic)	Conduction	Blockage at 19-27°C & 44-45°C	7
33	*Rattus norvegicus*	Exposure of whole body to cold	Autonomic nervous system	Norepinephrine release in myocardium	Increased	8
34				Catecholamine synthesis	Increased	9
35				Catecholamine excretion by kidney	Increased	21
36	Toad	Localized cooling	Sciatic nerve	Minimal gradient requirement for excitation	Decreased	18

Contributors: Brooks, Chandler McC., and Pinkston, J. O.

References

[1] Andersen, H. T., et al. 1962. Acta Physiol. Scand. 54:159.

[2] Andersson, B., et al. 1965. Ibid. 63:186.

[3] Betz, E., et al. 1961. Pfluegers Arch. Gesamte Physiol. Menschen Tiere 272:76.

[4] Brooks, C. McC., et al. 1955. J. Neurophysiol. 18:205.

[5] Chowers, I., et al. 1966. Amer. J. Physiol. 210:606.

[6] Euler, C. von. 1947. Acta Physiol. Scand., Suppl. 45.

[7] Eve, F. C. 1900. J. Physiol. (London) 26:119.

[8] Gordon, R., et al. 1966. J. Pharmacol. Exp. Ther. 153:440.

[9] Gutman, Y., and H. Weil-Malherbe. 1967. Brit. J. Pharmacol. 30:4.

[10] Haftek, J. 1967. Acta Med. Pol. 6:972.

[11] Hardy, J. D., et al. 1964. J. Physiol. (London) 175:242.

[12] Hensel, H., et al. 1960. Ibid. 153:113.

[13] Ingram, D. L., and G. C. Whittow. 1962. Ibid. 163:200.

[14] Klussmann, F. W., et al. 1969. Fed. Proc. Fed. Amer. Soc. Exp. Biol. 28:992.

[15] Koizumi, K., et al. 1954. Amer. J. Physiol. 179:507.

[16] Koizumi, K., et al. 1960. J. Neurophysiol. 23:421.

[17] Kundt, H. W., et al. 1957. Pfluegers Arch. Gesamte Physiol. Menschen Tiere 264:97.

[18] Lucas, K. 1907. J. Physiol. (London) 36:253.

[19] Lundberg, A. 1948. Acta Physiol. Scand., Suppl. 50.

[20] Miller, L. K. 1970. Can. J. Zool. 48:75.

[21] Nathanielsz, P. W. 1969. Diabetes 18:625.

[22] Negrin, J., Jr. 1970. Int. Surg. 54:93.

[23] Pierau, F. -K., et al. 1969. Fed. Proc. Fed. Amer. Soc. Exp. Biol. 28:1006.

[24] Ushiyama, J., and C. McC. Brooks. 1970. Ibid. 29:A391.

160. VISIBLE LIGHT AND VISION: MAN

Part I. Photometric and Radiometric Concepts

mks Unit = meter-kilogram-second unit.

	Photometric			Radiometric			Geometric Representation
	Term	Symbol	mks Unit (Synonym)	Term	Symbol	mks Unit	
1	Luminous flux	F	Lumen	Radiant flux	P	Watt	
2	Illuminance	E	Lumen/m² (lux)	Irradiance	H	Watt/m²	
3	Luminous intensity	I	Lumen/ω (candle)	Radiant intensity	J	Watt/ω	Point source Solid angle, ω
4	Luminance	B	Lumen/ω × m² (candle/m²)	Radiance	N	Watt/ω × m²	

Contributor: Nachmias, Jacob

Reference: Judd, D. B. 1951. In S. S. Stevens, ed. Handbook of Experimental Psychology. J. Wiley, New York. pp. 812-815.

Part II. Conversion Factors for Photometric Units

The total flux from a uniform point source of 1 candle is 4π lumens; the illuminance at a distance of 1 foot is 1 footcandle. An extended light source with a luminance of 1 candle/m² produces a retinal illuminance of 1 troland when the apparent area of the entrance pupil of the eye is 1 mm².

Unit (Synonym)		Value
Illuminance (Illumination), in footcandles		
1	Lumen/ft² (footcandle)	1
2	Lumen/m² (lux)	0.0929
3	Lumen/cm² (phot)	929
4	Milliphot	0.929
Luminance (Photometric Brightness), in millilamberts		
5	Microlambert	0.001

Unit (Synonym)		Value
6	Millilambert	1
7	Lambert	1000
8	Foot-lambert (equivalent footcandle)	1.076
9	Candle/in.²	487
10	Candle/ft²	3.380
11	Candle/m²	0.3142
12	Candle/cm² (stilb)	3142
13	Apostilb, international units	0.1
14	Hefner units	0.09

160. VISIBLE LIGHT AND VISION: MAN

Part II. Conversion Factors for Photometric Units

Contributor: Nachmias, Jacob

Reference: Judd, D. B. 1951. In S. S. Stevens, ed. Handbook of Experimental Psychology. J. Wiley, New York. p. 816.

Part III. Colorimetric and Photometric Specification of Stimuli

Standard conditions for the C.I.E. (Commission Internationale de l'Éclairage) system are as follows: For scotopic vision, the observer must be under 30 years of age, and light falling on the dark-adapted retina not less than 5° from the fovea. For photopic vision, field subtense = 2° (½ to 4°) with dark surround, luminance = 10^{-1} to 10^3 foot-lamberts, retinal locus = fovea and parafovea; angle of incident light and viewing angle of reflecting materials should be specified, but C.I.E. recommends 45° for the former and 90° for the latter.

Colorimetric specification: A colored stimulus may be specified by its C.I.E. tristimulus values (X, Y, Z), which represent the relative amounts of the C.I.E. primaries required by the standard observer to match that stimulus. Two stimuli of mixed wavelength are indistinguishable in color if they have the same tristimulus values. The X, Y, Z values of a stimulus producing at the cornea spectral irradiance $H(\lambda)$ are given by $X = \sum_{380}^{780} \bar{x}_\lambda \cdot H(\lambda) \cdot \Delta\lambda$, $Y = \sum_{380}^{780} \bar{y}_\lambda \cdot H(\lambda) \cdot \Delta\lambda$, $Z = \sum_{380}^{780} \bar{z}_\lambda \cdot H(\lambda) \cdot \Delta\lambda$, where $\bar{x}_\lambda$, $\bar{y}_\lambda$, and $\bar{z}_\lambda$ are the tristimulus values, sometimes called the distribution coefficients, for unit amounts of narrow-band stimuli. In practice, $H(\lambda)$

may be calculated from the spectral irradiance of the illuminating source and the spectral reflectance or transmittance of some object. For this purpose, three standard sources are defined for the C.I.E. system: illuminant A, representing a tungsten light source at 2854°K; illuminant B, representing noon sunlight; and illuminant C, approximating daylight (sunlight plus skylight). Their relative spectral irradiances are listed in columns H_A, H_B, and H_C. A colored stimulus may also be specified by its chromaticity coordinates, x and y: $x = \dfrac{X}{X+Y+Z}$ and $y = \dfrac{Y}{X+Y+Z}$.

Photometric specification: Luminous flux in lumens (F) may be obtained from $F = K_m \cdot \sum_{380}^{780} V(\lambda) \cdot P(\lambda) \cdot \Delta\lambda$, where $P(\lambda) \cdot \Delta\lambda$ is the radiant flux in watts between λ (wavelength value) and $\lambda + \Delta\lambda$, $V(\lambda)$ is the relative luminous (visual) efficiency of λ, and K_m is the maximum absolute luminous efficiency. In photopic vision, the values of $V(\lambda)$ are equivalent to $\bar{y}_\lambda$ which are listed in the **Tristimulus Values** column $\bar{y}$, and K_m = 680 lumens/watt. For scotopic vision, values of $V(\lambda)$ are listed in the **Scotopic Luminosity Function** column, and K_m = 1746. Analogous relationships apply between other photometric and radiometric quantities.

	Wave-length nm	Tristimulus Values			Relative Spectral Irradiances of Standard Sources			Scotopic Luminosity Function
		$\bar{x}$	$\bar{y}$	$\bar{z}$	H_A	H_B	H_C	
1	380	0.0014	0.0000	0.0065	9.79	22.40	33.00	0.00059
2	385	0.0022	0.0001	0.0105	10.90	26.85	39.92	0.00111
3	390	0.0042	0.0001	0.0201	12.09	31.30	47.40	0.00221
4	395	0.0076	0.0002	0.0362	13.36	36.18	55.17	0.00453
5	400	0.0143	0.0004	0.0679	14.71	41.30	63.30	0.00929
6	405	0.0232	0.0006	0.1102	16.15	46.62	71.81	0.01850
7	410	0.0435	0.0012	0.2074	17.68	52.10	80.60	0.03484
8	415	0.0776	0.0022	0.3713	19.29	57.70	89.53	0.0604
9	420	0.1344	0.0040	0.6456	21.00	63.20	98.10	0.0966
10	425	0.2148	0.0073	1.0391	22.79	68.37	105.80	0.1436
11	430	0.2839	0.0116	1.3856	24.67	73.10	112.40	0.1998
12	435	0.3285	0.0168	1.6230	26.64	77.31	117.75	0.2625
13	440	0.3483	0.0230	1.7471	28.70	80.80	121.50	0.3281
14	445	0.3481	0.0298	1.7826	30.85	83.44	123.45	0.3931
15	450	0.3362	0.0380	1.7721	33.09	85.40	124.00	0.4550
16	455	0.3187	0.0480	1.7441	35.41	86.88	123.60	0.5129
17	460	0.2908	0.0600	1.6692	37.82	88.30	123.10	0.5672
18	465	0.2511	0.0739	1.5281	40.30	90.08	123.30	0.6205
19	470	0.1954	0.0910	1.2876	42.87	92.00	123.80	0.6756

Part III. Colorimetric and Photometric Specification of Stimuli

	Wave-length	Tristimulus Values			Relative Spectral Irradiances of Standard Sources			Scotopic Luminosity
	nm	$\bar{x}$	$\bar{y}$	$\bar{z}$	H_A	H_B	H_C	Function
20	475	0.1421	0.1126	1.0419	45.52	93.75	124.09	0.7337
21	480	0.0956	0.1390	0.8130	48.25	95.20	123.90	0.7930
22	485	0.0580	0.1693	0.6162	51.04	96.23	122.92	0.8509
23	490	0.0320	0.2080	0.4652	53.91	96.50	120.70	0.9043
24	495	0.0147	0.2586	0.3533	56.85	95.71	116.90	0.9491
25	500	0.0049	0.3230	0.2720	59.86	94.20	112.10	0.9817
26	505	0.0024	0.4073	0.2123	62.93	92.37	106.98	0.9984
27	510	0.0093	0.5030	0.1582	66.06	90.70	102.30	0.9966
28	515	0.0291	0.6082	0.1117	69.25	89.65	98.81	0.9750
29	520	0.0633	0.7100	0.0782	72.50	89.50	96.90	0.9352
30	525	0.1096	0.7932	0.0573	75.79	90.43	96.78	0.8796
31	530	0.1655	0.8620	0.0422	79.13	92.20	98.00	0.8110
32	535	0.2257	0.9149	0.0298	82.52	94.46	99.94	0.7332
33	540	0.2904	0.9540	0.0203	85.95	96.90	102.10	0.6497
34	545	0.3597	0.9803	0.0134	89.41	99.16	103.95	0.5644
35	550	0.4334	0.9950	0.0087	92.91	101.00	105.20	0.4808
36	555	0.5121	1.0002	0.0057	96.44	102.20	105.67	0.4015
37	560	0.5945	0.9950	0.0039	100.00	102.80	105.30	0.3288
38	565	0.6784	0.9786	0.0027	103.58	102.92	104.11	0.2639
39	570	0.7621	0.9520	0.0021	107.18	102.60	102.30	0.2076
40	575	0.8425	0.9154	0.0018	110.80	101.90	100.15	0.1602
41	580	0.9163	0.8700	0.0017	114.44	101.00	97.80	0.1212
42	585	0.9786	0.8163	0.0014	118.08	100.07	95.43	0.0899
43	590	1.0263	0.7570	0.0011	121.73	99.20	93.20	0.0655
44	595	1.0567	0.6949	0.0010	125.39	98.44	91.22	0.0469
45	600	1.0622	0.6310	0.0008	129.04	98.00	89.70	0.03315
46	605	1.0456	0.5668	0.0006	132.70	98.08	88.83	0.02312
47	610	1.0026	0.5030	0.0003	136.34	98.50	88.40	0.01593
48	615	0.9384	0.4412	0.0002	139.99	99.06	88.19	0.01088
49	620	0.8544	0.3810	0.0002	143.62	99.70	88.10	0.00737
50	625	0.7514	0.3210	0.0001	147.23	100.36	88.06	0.00497
51	630	0.6424	0.2650	0.0000	150.83	101.00	88.00	0.003335
52	635	0.5419	0.2170	0.0000	154.42	101.56	87.86	0.002235
53	640	0.4479	0.1750	0.0000	157.98	102.20	87.80	0.001497
54	645	0.3608	0.1382	0.0000	161.51	103.05	87.99	0.001005
55	650	0.2835	0.1070	0.0000	165.03	103.90	88.20	0.000677
56	655	0.2187	0.0816	0.0000	168.51	104.59	88.20	0.000459
57	660	0.1649	0.0610	0.0000	171.96	105.00	87.90	0.0003129
58	665	0.1212	0.0446	0.0000	175.38	105.08	87.22	0.0002146
59	670	0.0874	0.0320	0.0000	178.77	104.90	86.30	0.0001480
60	675	0.0636	0.0232	0.0000	182.12	104.55	85.30	0.0001026
61	680	0.0468	0.0170	0.0000	185.43	103.90	84.00	0.0000715
62	685	0.0329	0.0119	0.0000	188.70	102.84	82.21	0.0000502
63	690	0.0227	0.0082	0.0000	191.93	101.60	80.20	0.00003533
64	695	0.0158	0.0057	0.0000	195.12	100.38	78.24	0.00002502
65	700	0.0114	0.0041	0.0000	198.26	99.10	76.30	0.00001780
66	705	0.0081	0.0029	0.0000	201.36	97.70	74.36	0.00001273
67	710	0.0058	0.0021	0.0000	204.41	96.20	72.40	0.00000914
68	715	0.0041	0.0015	0.0000	207.41	94.60	70.40	0.00000660
69	720	0.0029	0.0010	0.0000	210.36	92.90	68.30	0.00000478

Part III. Colorimetric and Photometric Specification of Stimuli

	Wave-length	Tristimulus Values			Relative Spectral Irradiances of Standard Sources			Scotopic Luminosity Function
	nm	$\bar{x}$	$\bar{y}$	$\bar{z}$	H_A	H_B	H_C	
70	725	0.0020	0.0007	0.0000	213.26	91.10	66.30	0.000003482
71	730	0.0014	0.0005	0.0000	216.12	89.40	64.40	0.000002546
72	735	0.0010	0.0004	0.0000	218.92	88.00	62.80	0.000001870
73	740	0.0007	0.0003	0.0000	221.66	86.90	61.50	0.000001379
74	745	0.0005	0.0002	0.0000	224.36	85.90	60.20	0.000001022
75	750	0.0003	0.0001	0.0000	227.00	85.20	59.20	0.000000760
76	755	0.0002	0.0001	0.0000	229.58	84.80	58.50	0.000000567
77	760	0.0002	0.0001	0.0000	232.11	84.70	58.10	0.000000425
78	765	0.0001	0.0000	0.0000	234.59	84.90	58.00	0.000000320
79	770	0.0001	0.0000	0.0000	237.01	85.40	58.20	0.000000241
80	775	0.0000	0.0000	0.0000	239.37	86.10	58.50	0.000000183
81	780	0.0000	0.0000	0.0000	241.67	87.00	59.10	0.000000139

Contributor: Nachmias, Jacob

General References

[1] Burnham, R. W., et al. 1963. Color: A Guide to Basic Facts and Concepts. J. Wiley, New York.

[2] Optical Society of America, Committee on Colorimetry. 1953. The Science of Color. T. Y. Crowell, New York.

Part IV. Visual Performance

Data are for the best, or nearly best, performance under laboratory conditions. *Abbreviations:* B = luminance (photometric brightness); E_R = peak retinal illuminance.

	Function	No. of Subjects	Experimental Conditions			Value	Reference
			Test Field	Background	Viewing		
1	Absolute threshold	22	47° angle; white light, 2400°K; 15-s duration	None	Natural pupil; no fixation	B = (0.4-2.0) × 10^{-6} candle/m² [1]; log (mean B) = 0.75 × 10^{-6} candle/m²	8
2		7	10′ arc; 510 nm light; 0.001-s duration; 20° from fixation point	None	2-mm artificial pupil	54-148 quanta at cornea [2]	4
3	Luminance increment threshold	9	2° angle; white light; 6-s duration	10° angle; subject adapted to its luminance	Natural pupil; scan of 8 possible positions	ΔB ≅ 0.01 B (370<B<0.3 candle/m²)	2
4		2	10′ arc; white light; 0.2-s duration; near center of background	30′ arc; centered 52′ arc from fixation point	Maxwellian	ΔB ≅ 0.1 B (10^6<B<30 candle/m²)	6
5	Wavelength discrimination	5	2° angle, bipartite; in fovea; indefinite exposure; B = 0.35 ± 0.3 (log candle)/m²; 2 parts of test field adjusted to equal B	None	1-mm artificial pupil	Δλ = 1-5 nm (480<λ<630 nm) Δλ = 1-2 nm (λ = 490, 590 nm)	10

[1] 50% frequency of seeing. [2] 60% frequency of seeing.

continued

Part IV. Visual Performance

	Function	No. of Sub-jects	Experimental Conditions			Value	Ref-er-ence
			Test Field	Background	Viewing		
6	Critical flicker frequency	2	19° angle; centrally fix-ated; white light; 100% square-wave modulation	35° angle; B same as test field	1.5 × 1.3-mm artificial pu-pil	cff = (12.5 log E_R + 10) cycles/s (5 < E_R < 5 × 10³ trolands) cff ≅ 58 cycles/s (E_R > 5 × 10³ trolands)	3
7		1	68° angle; centrally fix-ated; white light; 100% sinusoidal mod-ulation	None	1.5-mm artifi-cial pupil	cff = (18 log E_R + 22) cycles/s (10 < E_R < 2 × 10³ trolands) cff ≅ 85 cycles/s (E_R = 18.6 × 10³ trolands)	7
	Visual acuity						
8	Minimum visible an-gle	10	Dark line, 1° long, against sky; B ≅ 1 can-dle/m²; 1-min duration	....	Natural pupil; binocular; no fixation	0.43″ arc[3]	5
9	Minimum separable angle	1	4° angle; 1:1 grating; white light; indefinite duration; B = 282 candles/m²	30° angle	2-mm artifi-cial pupil	1.2′ arc/cycle	9
10	Minimum distinguish-able vernier	3	Vertical rods, each 40′ arc long, 107″ arc wide, separated by 20″ arc; 4-s duration; B unspecified	Entire visual field uniform-ly illuminated	Binocular; pupil un-specified; 4.6-m view-ing distance	Offset = 1.32-1.62″ arc	1
11	Minimum distinguish-able real depth dis-parity	3	Vertical rods, each 40′ arc long, 107″ arc wide, separated by 20″ arc; 4-s duration; B unspecified	Entire visual field uniform-ly illuminated	Binocular; pupil un-specified; 4.6-m view-ing distance	1.5-288″ arc	1

[3] 75% correct in 2-alternative forced choice.

Contributor: Nachmias, Jacob

References

[1] Berry, R. N. 1948. J. Exp. Psychol. 38:708.
[2] Blackwell, H. R. 1946. J. Opt. Soc. Amer. 36:624.
[3] Hecht, S., and E. L. Smith. 1936. J. Gen. Physiol. 19:979.
[4] Hecht, S., et al. 1942. Ibid. 25:819.
[5] Hecht, S., et al. 1947. J. Opt. Soc. Amer. 37:500.
[6] Heinemann, E. G. 1961. J. Exp. Psychol. 61:389.
[7] Kelly, D. H. 1961. J. Opt. Soc. Amer. 51:422.
[8] Pirene, M. H. 1952. In H. Davson, ed. The Eye. Academic Press, New York. v. 2, p. 141.
[9] Shlaer, S. 1937. J. Gen. Physiol. 21:165.
[10] Wright, W. D., and F. H. G. Pitt. 1934. Proc. Phys. Soc. London 46:459.

161. AGE AND SEX DIFFERENCES IN SCOTOPIC VISION: MAN

Values in parentheses are ranges, estimate "b" (see Introduction).

	Age, yr	Sex	Total Visual Field[1]		Absolute Threshold		Glare Recovery Time[3]	
			No. of Subjects	degrees	No. of Subjects	log intensity in $\mu\mu$lamberts[2]	No. of Subjects	seconds
1	16-19	♂	1209	174.5(166.8-182.1)	30	2.74(2.42-3.06)[4]	1189	3.7(0.9-6.5)
2		♀	746	176.4(170.8-181.9)	..	....	739	3.7(0.8-6.5)

[1] Binocular horizontal visual field [1]. [2] 1 $\mu\mu$lambert = 3.183 × 10⁻⁹ cd/m². [3] Time of dark adaptation threshold recovery after 5-s exposure to 34.44 cd/m² [2]. [4] Thresh-old luminance after 25-min dark adaptation following bleaching light; test target of 1°, 7° to right of central fixa-tion illuminated in flashes of 0.2-s duration [3].

continued

	Age, yr	Sex	Total Visual Field [1]		Absolute Threshold		Glare Recovery Time [3]	
			No. of Subjects	degrees	No. of Subjects	log intensity in $\mu\mu$lamberts [2]	No. of Subjects	seconds
3	20-29	♂	2612	174.8(167.0-182.6)	30	2.78(2.33-3.23)[4]	2563	3.9(0.8-7.0)
4		♀	1475	176.2(170.0-182.3)	103	3.22[5]	1449	3.9(1.0-6.8)
5	30-39	♂	2277	173.8(165.5-182.0)	30	2.93(2.63-3.23)[4]	2225	3.9(1.2-6.6)
6		♀	1333	175.3(168.3-182.3)	26	3.32[5]	1309	4.0(0.9-7.2)
7	40-49	♂	2082	171.7(162.4-180.9)	30	3.22(2.93-3.51)[4]	2044	4.3(1.2-7.4)
8		♀	1370	172.7(164.0-181.3)	9	3.40[5]	1345	4.3(1.2-7.4)
9	50-59	♂	1504	167.1(154.4-179.7)	30	3.55(3.02-4.08)[4]	1489	5.1(0.3-9.9)
10		♀	952	170.1(160.1-180.0)	..	....	941	4.6(0.9-8.3)
11	60-69	♂	899	160.1(144.7-175.5)	30	3.88(3.35-4.41)[4]	897	5.7(0.3-11.1)
12		♀	476	162.3(148.6-175.9)	..	....	477	5.9(0.4-11.4)
13	70-79	♂	405	151.2(133.4-168.9)	30	4.36(3.90-4.82)[4]	412	6.6(1.5-11.7)
14		♀	132	155.5(140.1-171.0)	..	....	144	6.1(1.4-10.7)
15	80+	♂	75	139.8(118.3-161.3)	30	5.01(4.35-5.67)[4]	71	7.7(1.4-14.0)
16		♀	13	138.5(116.9-160.0)	..	....	16	6.1(2.6-9.7)

[1] Binocular horizontal visual field [1]. [2] 1 $\mu\mu$lambert = 3.183×10^{-9} cd/m². [3] Time of dark adaptation threshold recovery after 5-s exposure to 34.44 cd/m² [2]. [4] Threshold luminance after 25-min dark adaptation following bleaching light; test target of 1°, 7° to right of central fixation illuminated in flashes of 0.2-s duration [3]. [5] Values are 50% frequency of seeing after 35- to 40-min dark adaptation, test conditions unspecified.

Contributor: Fisher, Kenneth D.

References

[1] Burg, A. 1968. J. Appl. Psychol. 52:10.
[2] Burg, A. 1968. Univ. Calif. Los Angeles Inst. Transp. Traffic Eng. Rep. 68-27.
[3] McFarland, R. A., et al. 1960. J. Geront. 15:149.
[4] Robertson, G. W., and J. Yudkin. 1944. J. Physiol. (London) 103:1.

162. COMPARATIVE VISUAL PIGMENT ABSORPTION AND PHOTOBEHAVIORAL ACTION SPECTRA: ANIMALS AND PLANTS

Part I. Visual Pigments: Vertebrates

Ret$_1$ and Ret$_2$: Retinal$_1$ and retinal$_2$ (retinene$_1$ and retinene$_2$), the aldehydes of Vitamin A$_1$ and Vitamin A$_2$, chemical isomers, the chromophore of the visual pigment rhodopsin.

Species (Synonym)	Absorption Max of Rhodopsin nm		Ref-er-ence		Species (Synonym)	Absorption Max of Rhodopsin nm		Ref-er-ence
	Ret$_1$	Ret$_2$				Ret$_1$	Ret$_2$	
Mammalia				8	Sciurus carolinensis pennsylvanicus (S. carolinensis leucotis)	502	....	9
1 Homo sapiens	497	....	6		Aves			
2	493	....	16					
3 Cavia porcellus	497	....	1	9 Anas platyrhynchos domesticus (A. domesticus)	502	....	2
4 Macaca mulatta	497	....	1					
5 Mus musculus domesticus	498	....	1	10 Gallus gallus	560; 500	....	17
6 Oryctolagus cuniculus	502	....	1					
7 Rattus rattus	498	....	1					

continued

162. COMPARATIVE VISUAL PIGMENT ABSORPTION AND PHOTOBEHAVIORAL ACTION SPECTRA: ANIMALS AND PLANTS

Part I. Visual Pigments: Vertebrates

Species (Synonym)	Absorption Max of Rhodopsin nm		Reference
	Ret$_1$	Ret$_2$	
11 Columba livia	544; 502	....	2,13
12 Pelecanus occidentalis	502	....	5
Reptilia			
13 Alligator mississippiensis	499	....	4,5, 18
14 Crotalus viridis helleri	500	....	5
Amphibia			
15 Necturus maculosus	.....	522	5
16 Rana pipiens	502	....	5
17 R. temporaria	502	....	10
18 Xenopus laevis	502	523	8
Osteichthyes (Pisces)			
19 Anguilla anguilla	487	523	15
20 Carassius auratus	.....	522	12

Species (Synonym)	Absorption Max of Rhodopsin nm		Reference
	Ret$_1$	Ret$_2$	
21 Conger conger	487	....	19
22 Cyprinus carpio	......	523	7
23 Dicentrarchus labrax (Morone labrax)	534	502	10
24 Gadus morhua	499	....	10
25 Melanogrammus aeglefinus (Gadus aeglefinus)	494	....	10
26 Microstomus kitt (Pleuronectes microcephalus)	510; 507	....	10
27 Platichthys flesus (Pleuronectes flesus)	501; 492	....	10
28 Salmo gairdneri	503	527	11
29 S. salar	503	527	11
30 Scomber scombrus	521; 487	....	10
Agnatha (Cyclostomata)			
31 Petromyzon marinus	497	518	3,14

Contributor: Wolken, Jerome J.

Specific References

[1] Bridges, C. D. B. 1959. Nature (London) 184:1727.
[2] Bridges, C. D. B. 1962. Vision Res. 2:125.
[3] Crescitelli, F. 1956. J. Gen. Physiol. 39:423.
[4] Crescitelli, F. 1956. Ibid. 40:217.
[5] Crescitelli, F. 1958. Proc. Annu. Biol. Colloq. Oreg. State Univ. 19:30.
[6] Crescitelli, F., and H. J. A. Dartnall. 1953. Nature (London) 172:195.
[7] Crescitelli, F., and H. J. A. Dartnall. 1954. J. Physiol. (London) 125:607.
[8] Dartnall, H. J. A. 1956. Ibid. 134:327.
[9] Dartnall, H. J. A. 1960. Nature (London) 188:475.
[10] Dartnall, H. J. A., and J. N. Lythgoe. 1965. Vision Res. 5:81.

[11] Muntz, F. W., and D. D. Beatty. 1965. Ibid. 5:1.
[12] Schwanzara, S. A. 1967. Ibid. 7:121.
[13] Sillman, A. J. 1969. Ibid. 9:1063.
[14] Wald, G. 1957. J. Gen. Physiol. 40:901.
[15] Wald, G. 1960. Comp. Biochem. 1:311.
[16] Wald, G., and P. K. Brown. 1958. Science 127:222.
[17] Wald, G., et al. 1955. J. Gen. Physiol. 38:623.
[18] Wald, G., et al. 1957. Ibid. 40:703.
[19] Walker, M. A. 1956. J. Physiol. (London) 133:56.

General Reference

[20] Lythgoe, J. N. 1972. In H. J. A. Dartnall, ed. Handbook of Sensory Physiology. J. Springer, New York. v. 7/1, pp. 567-624.

Part II. Visual Pigments: Arthropods and Mollusks

Retinal$_1$ is the chromophore of the invertebrate visual pigments. For additional information, consult General References.

Class & Species	Absorption Max nm	Reference
Arthropoda		
Merostomata		
1 Limulus polyphemus	520	8
Crustacea		
2 Callinectes sapidus	480	4

Class & Species	Absorption Max nm	Reference
3 Euphausia pacifica	462	9
4 Homarus americanus	515	12
5 Leptodora kindtii	510	13
6 Orconectes virilis	562; 510	9,12
7 Palaemonetes vulgaris	555; 496	5
8 Porcellio scaber	480	4

continued

162. COMPARATIVE VISUAL PIGMENT ABSORPTION AND PHOTOBEHAVIORAL ACTION SPECTRA: ANIMALS AND PLANTS

Part II. Visual Pigments: Arthropods and Mollusks

	Class & Species	Absorption Max nm	Reference
9	*Procambarus clarkii*	570	12
	Insecta		
10	*Apis mellifera*	440	3
11	*Blaberus giganteus*	495	14
12	*Blatta orientalis*	500	14
13	*Calliphora erythrocephala*	510; 470	10
14	*Musca domestica*	510; 437	1,11,13
15	*Periplaneta americana*	500	13,14

	Class & Species	Absorption Max nm	Reference
	Mollusca		
	Cephalopoda		
16	*Eledone moschata*	470	6
17	*Loligo pealei*	493	2
18	*Octopus ocellatus*	477	7
19	*O. vulgaris*	475	2
20	*Sepia esculenta*	486	7
21	*S. officinalis*	492	2
22	*Sepiella japonica*	500	7
23	*Todarodes pacificus*	480	7

Contributor: Wolken, Jerome J.

Specific References

[1] Bowness, J. M., and J. J. Wolken. 1959. J. Gen. Physiol. 42:779.

[2] Brown, P. K., and P. S. Brown. 1958. Nature (London) 182:1288.

[3] Goldsmith, T. H. 1958. Proc. Nat. Acad. Sci. U.S. 44:123.

[4] Goldsmith, T. H., and H. R. Fernandez. 1968. Z. Vergl. Physiol. 60:156.

[5] Goldsmith, T. H., et al. 1968. Science 161:468.

[6] Hamdorf, K., et al. 1968. Z. Vergl. Physiol. 60:375.

[7] Hara, T., and R. Hara. 1967. Nature (London) 214:572.

[8] Hubbard, R., and G. Wald. 1960. Ibid. 186:212.

[9] Kampa, E. M. 1955. Ibid. 175:996.

[10] Langer, H., and B. Thorell. 1966. Exp. Cell Res. 41:673.

[11] Marak, G. E., et al. 1970. Ophthalmol. Res. 1:65.

[12] Wald, G. 1968. J. Gen. Physiol. 51:125.

[13] Wolken, J. J. 1971. Invertebrate Photoreceptors. Academic Press, New York. p. 138.

[14] Wolken, J. J., and J. Scheer. 1963. Exp. Eye Res. 2:182.

General References

[15] Goldsmith, T. H. 1972. In H. J. A. Dartnall, ed. Handbook of Sensory Physiology. J. Springer, New York. v. 7/1, pp. 685-719.

[16] Wolken, J. J. 1971. Invertebrate Photoreceptors. Academic Press, New York.

Part III. Phototactic Spectral Response: Invertebrates and Plants

For a general discussion of phototactic spectral response in plant protists, consult reference 16. Data in brackets refer to the column heading in brackets.

	Phylum [1] & Species	Absorption Max [Secondary Absorption Max] nm	Ref- er- ence
	Invertebrates		
	Echinodermata		
1	*Asterias*	485; 405 [2]	20
2	*Diadema*	475-465	13

	Phylum [1] & Species	Absorption Max [Secondary Absorption Max] nm	Ref- er- ence
	Arthropoda		
	Crustacea [3]		
3	*Callinectes sapidus*	505	3
4	*Homarus americanus*	525	17
5	*Orconectes virilis*	565	3,17

[1] Unless otherwise indicated. [2] Without eyespot. [3] Class.

continued

Part III. Phototactic Spectral Response: Invertebrates and Plants

	Phylum[1] & Species	Absorption Max [Secondary Absorption Max] nm	Reference		Phylum & Species	Absorption Max [Secondary Absorption Max] nm	Reference
6	*Porcellio scaber*	520	3		Plants[4]		
7	*Procambarus clarkii*	570	9,17				
	Insecta[3]				Chlorophyta		
8	*Apis mellifera*	535; 440; 345	2	18	*Chlamydomonas*	504	12
9		490; 340	4	19	*Dunaliella salina*	493 [435]	6
10	*Calliphora erythrocephala*, larva	504	15	20	*Lemna trisulca*, chloroplasts	430 [485;380]	21
				21	*Mougeotia*, chloroplasts	483 [425]	7
11	*Drosophila melanogaster*	508	19	22	*Platymonas subcordiformis*	493 [435]	6
12	*Libellula*	520; 420	14	23	*Stephanoptera gracilis*	493 [435]	6
	Annelida			24	*Ulva*, gametes	485 [435]	6
13	*Arenicola*	483	12		Euglenophyta		
14	*Lumbricus*	483	12	25	*Euglena gracilis*	495 [480; 450; 425]	1
	Platyhelminthes						
15	*Dendrocoelum*	475	11	26		490 [420]	18
	Coelenterata			27	Nonphotosynthetic mutant	410 [425]	5
16	*Eudendrium*	474	10		Pyrrophyta		
	Protozoa			28	*Gonyaulax catenella*	475	6
17	*Amoeba*	515	8	29	*Peridinium*	475	6
				30	*Prorocentrum micans*	570	6

[1] Unless otherwise indicated. [3] Class. [4] Photoreceptor pigment not definitely identified.

Contributor: Wolken, Jerome J.

References

[1] Bünning, E., and G. Schneiderhöhn. 1956. Arch. Mikrobiol. 24:80.

[2] Goldsmith, T. H. 1961. In M. Rockstein, ed. Sensory Physiology. J. Wiley, New York. v. 1, pp. 357-375.

[3] Goldsmith, T. H., and H. R. Fernandez. 1968. Z. Vergl. Physiol. 60:156.

[4] Goldsmith, T. H., and P. R. Ruck. 1958. J. Gen. Physiol. 41:1171.

[5] Gössel, I. 1957. Arch. Mikrobiol. 27:288.

[6] Halldal, P. 1958. Physiol. Plant. 11:118.

[7] Haupt, W., and E. Schönbohm. 1962. Naturwissenschaften 49:42.

[8] Hitchcock, L., Jr. 1961. J. Protozool. 8:322.

[9] Kennedy, D., and M. S. Bruno. 1961. J. Gen. Physiol. 44:1089.

[10] Loeb, J., and H. Wasteneys. 1915. J. Exp. Zool. 19:23.

[11] Marriot, F. H. C. 1958. J. Physiol. (London) 143:369.

[12] Mast, S. O. 1917. J. Exp. Zool. 22:471.

[13] Millot, N., and M. Yoshida. 1957. J. Exp. Biol. 34:394.

[14] Ruck, P. R. 1965. J. Gen. Physiol. 49:289.

[15] Strange, P. H. 1961. J. Exp. Zool. 38:237.

[16] Thomas, J. B. 1965. Primary Photoprocesses in Biology. J. Wiley, New York. p. 207.

[17] Wald, G. 1968. J. Gen. Physiol. 51:125.

[18] Wolken, J. J. 1967. Euglena. Ed. 2. Appleton-Century-Crofts, New York. p. 153.

[19] Wolken, J. J. 1971. Invertebrate Photoreceptors. Academic Press, New York. p. 119.

[20] Yoshida, M., and H. Ohtsuki. 1966. Science 153:197.

[21] Zurzcki, J. 1962. Acta Soc. Bot. Pol. 31:489.

Data are for positive biological effects. Optical data on bi-refringence and dichroism are included only when directly related to orientational or electrophysiological responses to polarized light. Type of polarized light was linear. **Effects:**

e-vector = electric vector, which is the plane of polarization; λ_{max} = wavelength of maximum response. Data in brackets refer to the column heading in brackets.

Part I. Polarotaxis

Polarotaxis is a directed locomotor response to the electric vector of polarized light. This definition follows Jaffe's use of polarotropism for oriented growth responses of plants to polarized light. Such terminology also parallels the standard usage of phototropic and phototactic. Jander's term, oscillotaxis, is a synonym for polarotaxis. Basitaxis is a stereotyped orienting reflex in which the longitudinal body axis is maintained at some fixed angle or angles to a directional stimulus. For light and gravity, the angles may be at 0°, 90°, or 180°, but for *e*-vector they may include also 45° and 135°. Menotaxis is a derived variable orienting response in which a basitaxis is combined with a "command" direction to provide any momentarily required compass direction.

Polarized Light: Source—all experiments where "sky" is listed as the source were done in the field; all experiments where "filter" is listed as the source were done in the laboratory, unless otherwise indicated; **Beam Direction**—V = vertical, H = horizontal; **Wavelength**—Maximum degree of polarization of the clear blue sky occurs at approximately 460 nm (blue wavelength) measured in the sun's vertical for solar elevation angles between 0° and 75°, and at wavelengths from 625 nm to 365 nm. **Effects:** Angles, where relevant, indicate the azimuth orientation relative to the electric vector, which is the plane of polarization. *Abbreviation:* μW = microwatts or 10^{-6} watt.

Class, Order, & Species (Synonym) [Specification]	Medium	Polarized Light			Effects	Remarks	Reference
		Source	Beam Direction	Wavelength [Intensity $\mu W/cm^2$]			
Vertebrata							
Amphibia—Caudata							
1 *Ambystoma tigrinum* [Adult]	Water	Filter	V	White	Menotaxis in trained direction correspondingly altered by 90° *e*-vector rotation. Eyeless salamanders also show this response, but not when pineal-midbrain region of skull shielded with opaque screen.	....	1,55
Osteichthyes Perciformes							
2 *Pterophyllum* sp. [Adult]	Water	Filter	V	White	Marginal evidence for polarotaxis	....	61
Atheriniformes							
3 *Dermogenys pusillus* [Adult]	Water	Filter	V	White	Basitaxis near 45° & 135°	....	65
4 *Zenarchopterus buffoni* [Juvenile]	Water	Sky, water, & filter[1]	....	Blue	Menotactic orientation to sun & other cues displaced by imposed *e*-vector directions when the latter differ significantly from that of natural irradiance	....	66
5 *Z. dispar* [Juvenile]	Water	Sky, water, & filter[1]	....	Blue	Same results as for entry 4	....	66

[1] Experiments conducted in field.

continued

Part I. Polarotaxis

	Class, Order, & Species (Synonym) [Specification]	Medium	Polarized Light			Effects	Remarks	Reference
			Source	Beam Direction	Wavelength [Intensity μW/cm²]			
6						Underwater experiments show basitaxis mainly at 90°, but also at 0° to imposed *e*-vector orientation; response significantly weaker if sun's disc partially or fully covered by clouds	....	67
7		Water surface	Sky, water, & filter[1]	....	Blue	Basitaxis strong at 0° with imposed *e*-vector; this preference coincided with orientation preferred with respect to sky polarization without polarizing filter	....	19
8	Salmoniformes *Oncorhynchus nerka* [Smolt]	Water	Sky & filter[1]	....	Blue	Menotaxis to sky showed 180° reversal when sun not visible; in morning & afternoon (but not around noon), menotactic direction changed with rotation of polarizer over experimental vessel	....	31
	Arthropoda							
9	Arachnida Acarina *Arrenurus marshallae & A. megalurus* [Adult]	Water	Filter	H	White [2.75 × 10⁴]	Basitaxis at 0°, 45°, 90°, & 135°	....	39
10	Araneida *Agelena labyrinthica* [Adult]	Air	Sky	....	Blue	Menotaxis	Only anterior median eyes involved; polarotaxis & phototaxis concluded to be distinct	28,29
11	*A. similis* [Adult]	Air	Sky	....	Blue	Menotaxis	Only anterior median eyes involved	28
12	*Arctosa cinerea* [Adult]	Air	Sky	....	Blue	Menotaxis	Studies made in Arctic	44
13	*A. perita* [Adult]	Air	Sky	....	Blue	Menotaxis; regional direction learned	....	42,43
14	*A. variana* [Juvenile & adult]	Air	Sky	....	Blue	Menotaxis, innately north; local orientation learned	....	45
15	Crustacea Decapoda *Goniopsis cruentata* [Adult]	Air	Filter	V	White	Menotaxis, compensated for time of day	Artificial sun used for directional reference; polarotaxis concluded to be distinct from phototaxis	52

[1]/ Experiments conducted in field.

continued

Part I. Polarotaxis

	Class, Order, & Species (Synonym) [Specification]	Medi-um	Polarized Light			Effects	Remarks	Refer-ence
			Source	Beam Direction	Wavelength [Intensity $\mu W/cm^2$]			
16	*Ocypode ceratophthalma* [Adult & juvenile megalopa]	Air or water	Filter	V	White [304][2/]	Basitaxis at 0°, 45°, 90°, & 135°; no phototactic discrimination for different *e*-vector positions or polarized versus nonpolarized light	....	17
17	[Adult]	Air	Sky	....	Blue	Undisturbed subjects: menotaxis; disturbed subjects: basitaxis at 0°, 45°, 90°, & 135°	Polarotaxis concluded to be distinct from phototaxis	17
18						Menotaxis; directions learned	....	17
19	*Podophthalmus vigil* [Adult]	Water	Filter	V	White	Basitaxis at 0°, 45°, 90°, & 135°	Dichroic receptor molecules in rhabdom concluded to be mechanism	63
20	*Uca pugilator* [Various post-larval stages]	Air	Sky & filter[3/]	V	Blue & white	Menotaxis relative to shoreline; trained response in lab is bimodal (180°) if no other directional cues; response absent in young crabs, but becomes strong in subadult stages	Response not due to intensity pattern in surround; dorsal retina must be involved for polarotaxis	34,35
21	*U. tangeri* [Juvenile]	Air	Filter[1/]	....	Blue	Menotaxis, changeable with polarizer	....	2,3
22	Amphipoda *Hyalella azteca* [Adult]	Water	Filter	V	White [2.75 × 10⁴]	Basitaxis at 0°, 45°, 90°, & 135°	....	39
23	*Talitrus saltator* [Adult]	Air	Sky	....	Blue	Menotaxis	....	48
24	Isopoda *Tylos latreillei* [Adult]	Air	Sky	....	Blue	Menotaxis	....	46
25	Mysidacea *Mysidium gracile* [Adult]	Water	Filter	V	White [266][2/]	Basitaxis at 0° & 90°; response variable at 0°	....	4
26					White [0.1-1.79 × 10⁴][2/]	Basitaxis at 0°	Independence of polarotaxis & phototaxis shown; Autrum-von Frisch mechanism supported	39
27		Water, clear & turbid	Filter	V	White [19][2/]	Menotaxis (?); basitaxis at 90°	Synergy of polarotaxis & phototaxis shown	62

[1/] Experiments conducted in field. [2/] Calculated, by assuming that maximum luminous efficiency for white light was 0.4. [3/] Experiments conducted in both field and laboratory.

continued

Part I. Polarotaxis

Class, Order, & Species (Synonym) [Specification]	Medi- um	Polarized Light			Effects	Remarks	Refer- ence
		Source	Beam Direction	Wavelength [Intensity $\mu W/cm^2$]			
28	Water, turbid	Filter	V	White	Basitaxis at 90°	Synergy of polaro- taxis & phototaxis shown	5
29 Cyclopoida *Cyclops vernalis*	Water	Filters	V (unpo- larized) + H (hori- zontally polar- ized)	White [401][2/]	Basitaxis at 90° during beginning & end of 12 hr light in 12L:12D cycle, 0° during middle of the light period. Behavior rhythm evoked by polarization and when horizontal & vertical irradiance about equal.	Naupliar eye believed to be *e*-vector ana- lyzer	56,57
30 Calanoida *Diaptomus sho- shone*	Water	Filter	V	White [36, 401][2/]	Basitaxis at 90° with black sur- round at 401 $\mu W/cm^2$; at 0°, 45°, 90°, & 135° with black surround at 36 $\mu W/cm^2$, and with white surround at both light levels	Light contrast re- action	58
31 Cladocera *Daphnia magna* [Adult]	Water	Filter	V	White	Basitaxis at 90°; station-keeping behavior noted	....	32
32				740-320 nm, monochro- matic narrow bands (equal quanta)	Basitaxis at 90° significant from 580-360 nm (λ_{max} of 440 nm and shoulder at 520- 480 nm)	Polarotaxis different from phototaxis since λ_{max} of 540- 530 nm quoted for latter	64
33 *D. pulex* [Adult]	Water	Filter	H	White	Polarotaxis varies as $\sin 2\theta$; polaro- taxis/geotaxis ratio = 0.43. Turning tendency toward polarized light.	....	38
34			V	White	Basitaxis at 90°	....	18
35					Basitaxis at 90° found with 20 100% polarized light; 10% sometimes effective	....	64
36				White [2.75 × 10²]	Basitaxis: black sur- round or white sur- round at 0°, 45°, 90°, & 135°. Inten- sity effect noted.	Autrum-von Frisch mechanism sup- ported; results con- sistent with dichroic retinular analyzer	39
37				White [410][2/]	Basitaxis at 90° for both positive & negative photo- tactic subjects	Conflict between ex- perimental data & Fresnel mechanism hypothesized	33

2/ Calculated, by assuming that maximum luminous efficiency for white light was 0.4.

continued

Part I. Polarotaxis

	Class, Order, & Species (Synonym) [Specification]	Medium	Polarized Light			Effects	Remarks	Reference
			Source	Beam Direction	Wavelength [Intensity $\mu W/cm^2$]			
38					White [2.75 × 10⁴]	Basitaxis: black surround 90°; white surround 0°, 45°, 90°, & 135°. Light contrast reaction noted.	....	39
39	*D. schoedleri* [Adult]	Water, clear & turbid	Filter	V	White [22.8][2]	Menotaxis (?); basitaxis at 90°	Synergy of phototaxis & polarotaxis shown	62
	Insecta Hymenoptera							
40	*Andrena* sp. [Adult]	Air	Filter	V	White	Basitaxis at 45° & 135°	Subject walking	36
41	*Apis dorsata, A. florea, & A. indica* [Adult]	Air	Sky	....	Blue	Menotaxis	Subject dancing	41
42	*A. mellifera* [Adult]	Air	Sky	....	Blue	Menotaxis	Subject dancing. A 15° patch of blue sky adequate for response; zenith view not required. Retinular cell analysis mechanism supported.	20-22
43						Menotaxis	Subject flying; sun behind mountains	25
44						Menotaxis in 2 directions differing by 180°	This ambiguous activity present at dawn & dusk if zenith only visible	53
45						Menotaxis requires 20-30% polarization of sky	Calculated from data on responses to different sky areas	53
46						Menotaxis	Dancing & flying behavior. Only a direct view of blue sky & upper half of eye needed for good orientation; direct intraocular analysis is concluded.	26
47						Menotaxis	Walking & dancing behavior; directions learned. No interference with good orientation from white substrate; direct intraocular analysis is concluded.	26
48						Menotactic orientation requires ∿10% sky polarization (threshold range, 7-15%)	....	24
49					Selected bands	Menotaxis at 490-300 nm; no oriented dances at wavelengths ≮ 500 nm	....	23
50			Filter	V	White	Basitaxis at 0° & 90°	Subject walking	36
51						Basitaxis at 0°, 45°, 90°, & 135°	Quieted subjects	38

[2] Calculated, by assuming that maximum luminous efficiency for white light was 0.4.

continued

Part I. Polarotaxis

	Class, Order, & Species (Synonym) [Specification]	Medi-um	Polarized Light			Effects	Remarks	Refer-ence
			Source	Beam Direction	Wavelength [Intensity μW/cm^2]			
52	*Bombus agrorum* [Adult]	Air	Filter	V	White	Basitaxis at 0° & 90°	Subject walking	36
53						Basitaxis at 0°, 45°, 90°, & 135°	Quieted subject	38
54	*B. hypnorum* [Adult]	Air	Filter	V	White	Basitaxis at 0° & 90°	Subject walking	36
55	*B. lapidarius* [Adult]	Air	Filter	V	White	Basitaxis at 0°, 45°, 90°, & 135°	Quieted subject, walking	38
56	*B. sylvarum* [Adult]	Air	Filter	V	White	Basitaxis at 0° & 90°	Agitated subject, walking	38
57	*B. terrestris* [Adult]	Air	Filter	V	White	Basitaxis at 0° & 90°	Subject walking	36,38
58	*Camponotus lig-niperda* [Adult]	Air	Filter	V	White	Basitaxis at 0° & 90°	Subject walking	36
59	*Cataglyphis bicolor* [Adult]	Air	Sky	....	Blue	Menotaxis (trained direction) uses sky polarization 30 min before sunrise & after sunset; bimodal orientation with 180° reversal	....	68
60	*Formica fusca* [Adult]	Air	Sky & filter[1]	....	Blue	Menotaxis. Homing from blue sky dis-turbed by polarizer.	....	37
61			Filter	V	White	Menotaxis. Homing by *e*-vector learned.	....	37
62	*F. rufa* [Adult]	Air	Sky & filter[1]	V	Blue	Menotaxis. Homing from blue sky dis-turbed by polarizer.	....	37
63			Filter	V	White	Menotaxis. Homing by *e*-vector learned.	....	37
64						Basitaxis at 0°, 45°, 90°, & 135°; meno-taxis	....	36
65	*Halictus* sp. [Adult]	Air	Filter	V	White	Basitaxis at 0°, 45°, 90°, & 135°	....	36
66	*Lasius niger* [Adult]	Air	Sky & filter[1]	....	Blue	Menotaxis. Homing from blue sky dis-turbed by polarizer.	....	21,37
67			Filter	V	White	Menotaxis. Homing by *e*-vector learned; course straighter with polarized light.	....	10,37
68	*Myrmica laevi-nodis* [Adult]	Air	Sky	....	Blue	Menotaxis	....	59
69			Filter	V	White [14][2]	Menotaxis	....	59,60
70	*M. ruginodis* [Adult]	Air	Sky & filter[1]	....	Blue	Menotaxis. Homing from blue sky dis-turbed by polarizer.	....	37
71			Filter	V	White	Menotaxis. Homing by *e*-vector learned; course straighter with polarized light.	....	37,59

[1] Experiments conducted in field. [2] Calculated, by assuming that maximum luminous efficiency for white light was 0.4.

continued

Part I. Polarotaxis

	Class, Order, & Species (Synonym) [Specification]	Medi-um	Polarized Light			Effects	Remarks	Refer-ence
			Source	Beam Direction	Wavelength [Intensity $\mu W/cm^2$]			
72	*Neodiprion lecontei* [Adult]	Air	Sky & filter[1/]	....	Blue	Menotaxis	Dorsal ocelli, lateral eyes believed to be sensitive together or alone at moderate temperatures	30
73	*N. pratti banksianae* [Larva]	Air	Sky & filter[1/]	....	Blue	Menotaxis, sensitive to change in *e*-vector	Single pair of stem-mata present in larva	71
74	*N. sertifer* [Larva]	Air	Sky	....	Blue	Menotaxis	Clear zenith sky needed for a straight course	70
75	*Tapinoma erraticum* [Adult]	Air	Sky & filter[1/]	....	Blue	Menotaxis. Homing from blue sky dis-turbed by polarizer.	....	37
76			Filter	V	White	Menotaxis. Homing by *e*-vector learned.	....	37
77	*Tenthredo arcuata* [Adult]	Air	Filter	V	White	Basitaxis at 0°, 45°, 90°, & 135°	Quieted subject, walking	38
78	*Tetramorium cae-spitum* [Adult]	Air	Sky & filter[1/]	....	Blue	Menotaxis. Homing from blue sky dis-turbed by polarizer.	....	37
79			Filter	V	White	Menotaxis. Homing by *e*-vector learned.	....	37
80	*Trigona* sp. *(T. scap-totrigona)* [Adult]	Air	Filter	V	White	Basitaxis at 0° & 90°	Subject walking	36
81	*Vespula germanica (Paravespula ger-manica; Vespa germanica)* [Adult]	Air	Filter	V	White	Basitaxis at 0° & 90°	Subject walking	36
82	*Vespula vulgaris (P. vulgaris)* [Adult]	Air	Filter	V	White	Basitaxis at 0°, 45°, 90°, & 135°	Quieted subject, walking	38
	Diptera							
83	*Sarcophaga aldri-chi* [Adult]	Air	Sky	....	Blue	Menotaxis	Orientation disturbed by clouds or smoke in zenith. Either ocelli or compound eyes alone reported adequate for re-sponse.	69
	Lepidoptera							
84	*Archips cerasivora-nus* [Larva]	Air	Sky & filter[1/]	....	Blue	Menotaxis	Orientation disturbed by filter, clouds or smoke in zenith	70
85	*A. fervidanus* [Larva]	Air	Filter	H	Daylight	Polarotaxis. Orien-tation changed as *e*-vector rotated.	....	70
86	*Choristoneura fu-miferana* [Larva]	Air	Sky	....	Blue	Menotaxis	Orientation disturbed by clouds or smoke in zenith	71

[1/] Experiments conducted in field.

continued

Part I. Polarotaxis

	Class, Order, & Species (Synonym) [Specification]	Medi-um	Polarized Light			Effects	Remarks	Refer-ence
			Source	Beam Direction	Wavelength [Intensity μW/cm²]			
87	*Erannis tiliaria* [Larva]	Air	Sky	....	Blue	Menotaxis	Straight line orienta-tion stopped by clouds or smoke in zenith	70
88	*Hyphantria cunea (H. textor)* [Larva]	Air	Sky & filter[1]	....	Blue	Menotaxis	Orientation disturbed by filter, clouds or smoke in zenith	72
89	*Malacosoma amer-icanum & M. plu-viale* [Larva]	Air	Sky	....	Blue	Menotaxis	Clear zenith sky needed for orienta-tion	54
90	*M. disstria* [Larva]	Air	Sky	....	Blue	Menotaxis	Clear zenith sky needed for orienta-tion	54,71
91	Coleoptera *Geotrupes ster-corosus (G. silva-ticus)* [Adult]	Air	Sky	....	Blue	Menotaxis. Rota-tion through 90° or 180° correctable.	....	6,7
92						Menotaxis. Direc-tion changed with diurnal schedule so east maintained in A.M., west in P.M.; normal orien-tation changed when one eye covered.	....	27
93			Filter	V	White	Basitaxis at 0°, 45°, 90°, & 135°. Am-biguous response to rotation through 90°; no response to 180° rotation.	....	6,7
94					White [14][2]	Basitaxis at 0°, 45°, & 90°, preference ratios 2:3:5; no diurnal changes.	....	27
95	*Liodessus flavi-collis (Bidessus flavicollis)* [Adult]	Water	Filter	V	White [2.75×10^4]	Basitaxis at 0°, 45°, 90°, & 135°. Po-larotaxis not affected by sign change in phototaxis from positive to negative.	....	39
96	*Melolontha hippo-castani* [Adult ♀]	Air	Sky	....	Blue	Menotaxis in 2 di-rections perpen-dicular to each other	Twilight sky	14
97	*M. melolontha* [Adult ♀]	Air	Sky	....	Blue	Menotaxis in 2 di-rections perpen-dicular to each other	Twilight sky. Polaro-taxis suggested.	12,13, 15,16, 51
98	*Phaleria provin-cialis* [Adult]	Air	Sky	....	Blue	Menotaxis, modifi-able with polarizer	....	47
99	Hemiptera *Corixa punctata* [Adult]	Water	Filter	V	White [3.7-876][2]	Basitaxis at 0°, 45°, 90°, & 135°. Both positively & nega-tively phototactic animals showed the same preferential orientation to *e*-vector direction.	Ventral part of retina not involved, but dorsal (especially median dorsal) area essential for polaro-taxis	50

1/ Experiments conducted in field. 2/ Calculated, by assuming that maximum luminous efficiency for white light was 0.4.

continued

Part I. Polarotaxis

Class, Order, & Species (Synonym) [Specification]	Medium	Polarized Light			Effects	Remarks	Reference
		Source	Beam Direction	Wavelength [Intensity $\mu W/cm^2$]			
100	*Velia caprai* [Adult]	Air	Filter	V	White	Basitaxis at 0°, 30°, 60°, & 90°	....	49
101	*V. currens* [Adult]	Air & water	Sky	....	Blue	Menotaxis	....	8
Mollusca							
	Cephalopoda—Dibranchia							
102	*Euprymna morsei* [Juvenile]	Water	Filter	V	White [49.1][2/]	Basitaxis at 0°, 45°, 90°, & 135°	Polarotaxis mechanism distinct from that in phototaxis; dichroic receptor hypothesized	40
103	*Sepioteuthis lessoniana* [Larva]	Water	Filter	V	White [49.1][2/]	Basitaxis at 0°, 45°, 90°, & 135°	Polarotaxis mechanism distinct from that in phototaxis; dichroic receptor hypothesized	40
	Gastropoda—Mesogastropoda							
104	*Littorina littorea, L. neritoides, & L. saxatilis* [Adult]	Air	Filter	V	White [220][2/]	Photonegative subjects: basitaxis at 0°; photopositive subjects: basitaxis at 90°	This response present with little or no substrate reflection	11
105	*L. obtusata* [Adult]	Air & water	Filter	V	White	Basitaxis at 0°	....	9
106		Air	Filter	V	White [220][2/]	Photonegative subjects: basitaxis at 0°; photopositive subjects: basitaxis at 90°	This response present with little or no substrate reflection; no response if light incident along optic axis; response is telotaxis	11

[2/] Calculated, by assuming that maximum luminous efficiency for white light was 0.4.

Contributor: Waterman, Talbot H.

References

[1] Adler, K., and D. H. Taylor. 1971. Herpetol. Rev. 3(6):105.

[2] Altevogt, R. 1963. Naturwissenschaften 50:697.

[3] Altevogt, R., and H. von Hagen. 1964. Z. Morphol. Oekol. Tiere 53:636.

[4] Bainbridge, R., and T. H. Waterman. 1957. J. Exp. Biol. 34:342.

[5] Bainbridge, R., and T. H. Waterman. 1958. Ibid. 35:487.

[6] Birukow, G. 1953. Naturwissenschaften 40:611.

[7] Birukow, G. 1954. Z. Vergl. Physiol. 36:176.

[8] Birukow, G. 1956. Z. Tierpsychol. 13:463.

[9] Burdon-Jones, C., and G. H. Charles. 1958. Nature (London) 181:129.

[10] Carthy, J. D. 1951. Behaviour 3:275.

[11] Charles, G. H. 1961. J. Exp. Biol. 38:213.

[12] Couturier, A., and P. Robert. 1955. C. R. Acad. Sci. 240:2561.

[13] Couturier, A., and P. Robert. 1956. Ibid. 242:3121.

[14] Couturier, A., and P. Robert. 1956. Ann. Epiphyt. 3:431.

[15] Couturier, A., and P. Robert. 1958. Proc. Int. Congr. Entomol., 10th, Montreal 2:611.

[16] Couturier, A., and P. Robert. 1962. Rev. Zool. Agr. Appl. 7-9:1.

[17] Daumer, K., et al. 1963. Z. Vergl. Physiol. 47:56.

[18] Eckert, B. 1953. Cesk. Biol. 2:76.

[19] Forward, R. B., Jr., et al. 1972. Biol. Bull. 143:112.

continued

Part I. Polarotaxis

[20] Frisch, K. von. 1949. Experientia 5:142.

[21] Frisch, K. von. 1950. Ibid. 6:210.

[22] Frisch, K. von. 1951. Naturwissenschaften 38:105.

[23] Frisch, K. von. 1954. Sitzungsber. Math. Naturwiss. Kl. Bayer. Akad. Wiss. Muenchen 17:197.

[24] Frisch, K. von. 1965. Tanzsprache und Orientierung der Bienen. Springer-Verlag, Berlin. p. 411.

[25] Frisch, K. von, and M. Lindauer. 1954. Naturwissenschaften 41:245.

[26] Frisch, K. von, et al. 1960. Experientia 16:289.

[27] Geisler, M. 1961. Z. Tierpsychol. 18:389.

[28] Görner, P. 1958. Z. Vergl. Physiol. 41:111.

[29] Görner, P. 1962. Ibid. 45:307.

[30] Green, G. W. 1954. Can. Entomol. 86:371.

[31] Groot, C. 1965. Behaviour, Suppl. 14.

[32] Harris, J. E., and U. K. Wolfe. 1955. Proc. Roy. Soc. B144:329.

[33] Hazen, W. E., and E. R. Baylor. 1962. Biol. Bull. 123:243.

[34] Herrnkind, W. 1966. Amer. Zool. 6:298.

[35] Herrnkind, W. 1972. In H. E. Winn and B. L. Olla, ed. Behavior of Marine Animals. Plenum Press, New York. v. 1, p. 1.

[36] Jacobs-Jessen, U. F. 1959. Z. Vergl. Physiol. 41:597.

[37] Jander, R. 1957. Ibid. 40:162.

[38] Jander, R. 1963. Ibid. 47:381.

[39] Jander, R., and T. H. Waterman. 1960. J. Cell. Comp. Physiol. 56:137.

[40] Jander, R., et al. 1963. Z. Vergl. Physiol. 46:383.

[41] Lindauer, M. 1956. Ibid. 38:521.

[42] Papi, F. 1955. Ibid. 37:230.

[43] Papi, F. 1955. Atti Soc. Toscana Sci. Natur. Pisa Proc. Verb. Mem. B62:83.

[44] Papi, F., and J. Syrjämäki. 1963. Arch. Ital. Biol. 101:59.

[45] Papi, F., and P. Tongiorgi. 1963. Ergeb. Biol. 26:259.

[46] Pardi, L. 1954. Z. Tierpsychol. 11:175.

[47] Pardi, L. 1955. Boll. Ist. Mus. Zool. Univ. Torino 5:1.

[48] Pardi, L., and F. Papi. 1953. Z. Vergl. Physiol. 35:459.

[49] Rensing, L. 1962. Zool. Beitr., N. F. 7:447.

[50] Rensing, L., and H. Bogenschütz. 1966. Zool. Jahrb. Abt. Allg. Zool. Physiol. Tiere 72:123.

[51] Robert, P. 1963. Ergeb. Biol. 26:135.

[52] Schöne, H. 1963. Z. Vergl. Physiol. 46:496.

[53] Stockhammer, K. 1959. Ergeb. Biol. 21:23.

[54] Sullivan, C. R., and W. G. Wellington. 1953. Can. Entomol. 85:297.

[55] Taylor, D. H. 1971. Herpetol. Rev. 3(4):66.

[56] Umminger, B. L. 1968. Biol. Bull. 135:239.

[57] Umminger, B. L. 1968. Ibid. 135:252.

[58] Umminger, B. L. 1969. Crustaceana 16(2):202.

[59] Vowles, D. M. 1950. Nature (London) 165:282.

[60] Vowles, D. M. 1954. J. Exp. Biol. 31:341.

[61] Waterman, T. H. 1959. Proc. Int. Congr. Zool., 15th, London, p. 537.

[62] Waterman, T. H. 1960. Z. Vergl. Physiol. 43:149.

[63] Waterman, T. H. 1961. Proc. Int. Congr. Photobiol., 3rd, p. 214.

[64] Waterman, T. H. 1966. Funct. Organ. Compound Eye Proc. Int. Symp. 1965, p. 493.

[65] Waterman, T. H. 1972. NASA SP-262:437.

[66] Waterman, T. H., and R. B. Forward, Jr. 1970. Nature (London) 228:85.

[67] Waterman, T. H., and R. B. Forward, Jr. 1972. J. Exp. Zool. 180:33.

[68] Wehner, R., and P. Duelli. 1971. Experientia 27:1364.

[69] Wellington, W. G. 1953. Nature (London) 172:1177.

[70] Wellington, W. G. 1955. Ann. Entomol. Soc. Amer. 48:67.

[71] Wellington, W. G., et al. 1951. Can. J. Zool. 29:339.

[72] Wellington, W. G., et al. 1954. Can. Entomol. 86:529.

Part II. Electrophysiology

All subjects were adult unless otherwise indicated. **Polarized Light:** Filters were the source of polarized light; **Intensity**—μW = microwatts, or 10^{-6} watt; **Duration**—ms = milliseconds, or 10^{-3} second. **Effects:** ERG = electroretinogram; R refers to retinular cell.

Class, Order, & Species	Electrode		Polarized Light		Effects	Remarks	Reference
	Location	Type	Wavelength [Intensity μW/cm^2]	Flash Duration ms[1]			
Vertebrata							
Osteichthyes—Cypriniformes							
1 *Carassius auratus*	Extracellular	Tapered tungsten	White	500	Majority of units in optic tectum selectively sensitive to *e*-vector direction of large area stimulus to retina	....	25

[1] Unless otherwise specified.

continued

Part II. Electrophysiology

Class, Order, & Species	Electrode		Polarized Light		Effects	Remarks	Reference
	Location	Type	Wavelength [Intensity μW/cm²]	Flash Duration ms[1]			
				Arthropoda			
Arachnida—Araneida							
2 *Arctosa variana*	Extracellular	Ag-AgCl	White [1.9-38]	100, 10/s	Anterior & posterior median eyes sensitive to *e*-vector; 30% greater ERG at 90° & 270° than at 0° & 180°	Analyzer concluded to be at retinal level; strong light adaptation abolishes response to polarized light	11, 12
3 *Lycosa tarentula*	Extracellular near eye	Ag-AgCl	White [2.2-14.6]	100	ERG of posterior lateral (secondary) & posterior median (secondary) eyes sensitive to *e*-vector orientation; one max, one min in 180°, with max response to horizontal *e*-vector. Anterior lateral (secondary) & perhaps anterior median (principal) eyes lack this response.	Retinal analyzer hypothesized	10
Crustacea—Decapoda							
4 *Astacus astacus*[2]	Intracellular	Glass capillary	White	200	Max:min response sensitivity of single retinular cells averaged 6.2 for 28 cells	....	19
5 *Carcinus maenas*	Intracellular	Glass capillary	White	200	Retinular cells sensitive to *e*-vector orientation with mean max:min sensitivity ratio = 7.7 and two cell classes with max sensitivity in orthogonal directions	....	15, 19
6 *Cardisoma guanhumi*	Extracellular	Glass capillary	White [292-4744][3]	10	ERG shows selective adaptation to vertical & horizontal *e*-vectors, but not to intermediate oblique directions. Vertical & horizontal correspond to the two directions of rhabdom microvilli.	....	27
7 *Homarus americanus*	Extracellular	Glass capillary	White [292-4744][3]	....	ERG shows selective adaptation to vertical & horizontal *e*-vectors, but not to intermediate oblique directions (similar to effects for entry 6)	....	27
8 *Orconectes virilis*	Extracellular	Glass capillary	White [292-4744][3]	10	Results similar to those for entry 7	....	27
9 *Ovalipes ocellatus*	Intracellular	Glass capillary	White	200	*e*-Vector sensitivity ratios from 9:11 to 11:1 observed in a few cases	....	19
10 *Procambarus clarkii*	Intracellular	Glass capillary	740-400 nm, 18 monochromatic narrow bands (equal quanta)	400	Two classes of retinular cells present, with λ_{max} average at 594 nm & 440 nm; both sensitive to *e*-vector orientation with max or min at 0° or 90° with respect to vertical; max:min sensitivity ratios average 3.1 (range, 1.2-11.9)	....	26
Insecta Hymenoptera							
11 *Apis mellifera*	Cornea	Steel needle	White & various bands	40-100	ERG amplitude invariant with *e*-vector rotation; on-wave[4] amplitude 16-36% greater in polarized than nonpolarized light	Intraommatidial analyzer hypothesized in retinula	1

[1] Unless otherwise specified. [2] Synonym: *A. fluviatilis.* for white light was 0.4. [4] Initial negative transient component of electroretinogram.
[3] Calculated, by assuming maximum luminous efficiency

continued

Part II. Electrophysiology

| | Class, Order, & Species | Electrode | | Polarized Light | | Effects | Remarks | Reference |
		Location	Type	Wavelength [Intensity $\mu W/cm^2$]	Flash Duration ms[1]			
12	Drone	Intracellular	Double glass capillary	White & monochromatic narrow bands (equal quanta)	....	Retinular cells weakly *e*-vector sensitive (max:min = 1.288; low ratio partly accounted for by intercellular coupling). Double impalements of adjacent R's in same retinula show orthogonal response maxima and minima to *e*-vector rotation.	....	18
13	Diptera *Calliphora erythrocephala*	Extracellular	Steel needle	White & various bands	40-100	On-wave[4] of ERG invariant with changes in *e*-vector; on-wave amplitude 16-36% greater in polarized than nonpolarized light	Intraommatidial analyzer hypothesized in retinula	1
14				White [410]; or narrow bands, 606 or 505 nm	Filter rotating 3-7 cycles/s	Max receptor potential when *e*-vector is parallel to rhabdom microvilli; beam perpendicular to ommatidial axis causes receptor potential amplitude to vary with *e*-vector	Dichroic analyzer concluded to be in rhabdom	5
15		Intracellular	Glass capillary	White	220	For retinular cells 1-6 somata show some sensitivity to *e*-vector orientation, but corresponding axons do not	....	20
16				White [0.38-11,400]; or 603 or 478 nm	300	Receptor potentials vary with *e*-vector orientation; 50% decrease in intensity (max effect). 603 nm not effective.	Only some cells are sensitive	4
17				Narrow band, peak at 496 nm	200	Receptor potential amplitude varies with *e*-vector in half the cells tested. 50% decrease in intensity (max effect).	....	2
18	*C. erythrocephala,* white-eyed	Extracellular & intracellular	Glass capillary & tapered tungsten	White	....	Retinular cells 1-6 show small but detectable amplitude modulation with *e*-vector rotation; apparently 3 preferred planes for group. Second-order fibers insensitive to *e*-vector direction.	....	13
19	*C. vomitoria*	Extracellular	Steel needle	White [1410]; or narrow bands, 606 or 505 nm	Filter rotating 3-7 cycles/s	Max receptor potential when *e*-vector is parallel to rhabdom microvilli; beam perpendicular to ommatidial axis causes receptor potential amplitude to vary with *e*-vector	Dichroic analyzer concluded to be in rhabdom	5
20	*Lucilia caesar*	Extracellular or intracellular	....	White	180/min, filter rotating 0.25 cycles/s	Receptor potential showed 20% decrease; varies synchronously with *e*-vector	Rhabdomere concluded to be analyzer	8
21	*Musca domestica*	Extracellular & intracellular	Glass capillary & tapered tungsten	White & monochromatic (550-420 nm)	....	In addition to results similar to those by same authors for entry 18, higher order fibers supplied by R7-R8 are weakly *e*-vector sensitive	Presence of max and min 90° apart suggests only one of these cells is involved	13

[1] Unless otherwise specified. [4] Initial negative transient component of electroretinogram.

continued

Part II. Electrophysiology

Class, Order, & Species	Electrode		Polarized Light		Effects	Remarks	Reference
	Location	Type	Wavelength [Intensity $\mu W/cm^2$]	Flash Duration ms [1]			
22	Extracellular	Glass capillary	White & monochromatic narrow bands	30	e-Vector sensitive responses of retinular cells (R1-R6) have maxima in three planes aligned with axes of the facet mosaic. No significant e-vector sensitivity demonstrable in lamina.	....	14
23		Steel needle	White [1410]; or narrow bands, 606 or 505 nm	Filter rotating 3-7 cycles/s	Max receptor potential when e-vector is parallel to rhabdom microvilli; beam perpendicular to ommatidial axis causes receptor potential amplitude to vary with e-vector rotation in a 180° period	Dichroic analyzer concluded to be in rhabdom	5
24	Phaenicia sericata	Extracellular & intracellular	Glass capillary & tapered tungsten	White	....	Results similar to those by same authors for entry 18	....	13
	Hemiptera						
25	Gerris lacustris	Extracellular	Glass capillary & tapered steel	White & monochromatic	....	ERG weakly sensitive to e-vector orientation, two maxima (at 45° & 135° relative to dorsoventral 0°), & two minima (at 0° & 90°) in 180°. Both maxima and minima 90° apart may be distinctive, which is not true for those 180° apart; selective adaptation indicates two orthogonal e-vector sensitive channels at 0° & 90°, as in entries 6 & 35, respectively.	....	3
26	Lethocerus americanus	Intracellular	Glass capillary	White	....	Results similar to those for entry 28	....	7,24
27	L. griseus [5]	Intracellular	Glass capillary	White	....	Results similar to those for entry 28	....	7,24
28	L. insulanus	Intracellular	Glass capillary	White	....	Retinular cell receptor potentials usually vary in amplitude with e-vector direction; max:min sensitivity ratio, 3:1; two classes of units present with max response directions orthogonal. Somata & axons give same response. Not affected by adaptation.	....	7,24
29	Notonecta glauca	Cornea	....	White [228]	100[6]	ERG amplitude reduced 12-20% (max effect). Varies with e-vector rotation in a 180° period.	Gross effect concluded to be present, since ommatidia lack radial symmetry	9
	Orthoptera						
30	Locusta migratoria	Extracellular & intracellular	Glass capillary	White & monochromatic narrow bands	....	Retinular cells do not shift plane of max sensitivity or amplitude of response to e-vector direction at four wavelengths tested (672-348 nm). Most cells in lamina not e-vector sensitive, but some are weakly so.	....	17

[1] Unless otherwise specified. [5] Synonym: *Benacus griseus.* [6] A depolarizer preceded filter used as source of polarized light.

continued

Part II. Electrophysiology

| Class, Order, & Species | Electrode | | Polarized Light | | Effects | Remarks | Reference |
	Location	Type	Wavelength [Intensity μW/cm^2]	Flash Duration ms			
31	Intracellular	Glass capillary	White	....	Mean max:min e-vector sensitivity ratios for 81 retinular cells = 2.3. This is smaller than predicted on basis of observed intercellular electrical coupling.	....	18
32		Double glass capillary	White [36.5][3]	Mainly 200	Adjacent retinular cells partially coupled electrically yet somewhat sensitive to e-vector orientation with their planes of maximum sensitivity 60° apart	....	16
33	Odonata *Anax junius*[7] Intracellular	Glass capillary	White	100	Retinular cells differentially sensitive to e-vector orientation, with one max and one orthogonal min over 180°. In adults, max:min sensitivity ratios ranged from 0.1-2.5 (most units, 1.4-1.7); larval retinular cells sensitive to e-vector (best case max:min = 4.5).	....	6
34	*Libellula needhami* Intracellular	Glass capillary	White	100	Results similar to those for entry 33	....	6

Mollusca

	Cephalopoda—Dibranchia						
35	*Octopus ocellatus;* *O. vulgaris* Intracellular	Glass capillary	White	100	e-Vector rotation modulates visual cell's receptor potential with 180° period. Peak responses orthogonal, corresponding with dorsoventral & anteroposterior axes.	....	21
36	*O. vulgaris* Extracellular	Glass capillary	White	20-60 s for adapting light polarized at 0°, 45°, 90°, or 135°; flashes for test light also polarized	Adaptation at 0° reduced responses at 0° & 180°; adaptation at 90° reduced responses at 90° & 270°; adaptation at 45° or 135° had no selective effect	Orthogonal rhabdomere dichroism postulated. No regional differences found in retina.	23
37	*O. vulgaris;* *Ommastrephes sloani pacificus*[8] Extracellular	Glass capillary	White adapting light polarized at 0°, 45°, 90°, or 135°; flashes for test light polarized at intervals of 22.5° from 0-360°	20-100 s for	No differential response of ERG to 500 ms flashes polarized at 0°, 45°, 90°, or 135°. With preceding adapting light polarized at 0°: max response to test flashes at 90° & 270°, min at 0° & 180°; with adaptation at 90°: max response to test flashes at 0° & 180°, min at 90° & 270°. Adaptation at 45° & 135° had no selective effect.	Orthogonal rhabdomere dichroism postulated	22

[3] Calculated, by assuming maximum luminous efficiency for white light was 0.4. [7] Nymphs & adults. [8] Synonym: *Ommatostrephes sloani pacificus.*

Contributor: Waterman, Talbot H.

References

[1] Autrum, H., and H. Stumpf. 1950. Z. Naturforsch. 5b:116.

[2] Autrum, H., and V. von Zwehl. 1962. Z. Vergl. Physiol. 46:1.

continued

Part II. Electrophysiology

[3] Bohn, H., and U. Täuber. 1971. Ibid. 72:32.

[4] Burkhardt, D., and L. Wendler. 1960. Ibid. 43:687.

[5] Giulio, L. 1963. Ibid. 46:491.

[6] Horridge, G. A. 1969. Ibid. 62:1.

[7] Ioannides, A. C., and B. Walcott. 1971. Ibid. 71: 315.

[8] Kuwabara, M., and K. Naka. 1959. Nature (London) 184:455.

[9] Lüdtke, H. 1957. Z. Vergl. Physiol. 40:329.

[10] Magni, F. 1966. Funct. Organ. Compound Eye Proc. Int. Symp. 1965, p. 171.

[11] Magni, F., et al. 1962. Experientia 18:511.

[12] Magni, F., et al. 1965. Arch. Ital. Biol. 103:146.

[13] McCann, G. D., and D. W. Arnett. 1972. J. Gen. Physiol. 59:534.

[14] Scholes, J. 1969. Kybernetik 6:149.

[15] Shaw, S. R. 1966. Nature (London) 211:92.

[16] Shaw, S. R. 1967. Z. Vergl. Physiol. 55:183.

[17] Shaw, S. R. 1968. Symp. Zool. Soc. London 23: 135.

[18] Shaw, S. R. 1969. Vision Res. 9:999.

[19] Shaw, S. R. 1969. Ibid. 9:1031.

[20] Smola, U., and R. Gemperlein. 1972. J. Comp. Physiol. 79:363.

[21] Sugawara, K., et al. 1971. J. Fac. Sci. Hokkaido Univ., VI, 17:581.

[22] Tasaki, K., and K. Karita. 1966. Jap. J. Physiol. 16:205.

[23] Tasaki, K., and K. Karita. 1966. Nature (London) 209:934.

[24] Walcott, B. 1971. Z. Vergl. Physiol. 74:17.

[25] Waterman, T. H. 1973. Tucson Symp. Polariz. (in press).

[26] Waterman, T. H., and H. R. Fernandez. 1970. Z. Vergl. Physiol. 68:154.

[27] Waterman, T. H., and K. W. Horch. 1966. Science 154:467.

Part III. Microspectrophotometry

All animals tested were adults.

	Class, Order, & Species	Instrumentation	Effects	Reference
			Arthropoda	
	Crustacea—Decapoda			
1	*Libinia emarginata*	Double beam continuous wavelength scan (20 nm/s) from 675-400 nm, with rectangular test beam traversing one transverse layer of rhabdom microvilli	Rhodopsin λ_{max} = 493 nm measured as in *Orconectes immunis* (entry 2). Max dichroic ratio $\sim$2, as in entries 2 & 3.	1
2	*Orconectes immunis*	Double beam continuous wavelength scan (20 nm/s) from 675-400 nm, with rectangular test beam traversing one transverse layer of rhabdom microvilli	Transverse absorptance of isolated rhabdom yields rhodopsin absorption spectrum (λ_{max} = 525 nm). Dichroic ratio near 2 for microvilli in plane of observation; major absorptance parallel to axis of microvilli. Absorptance isotropic with microvilli normal to plane of observation.	9
3	*Procambarus clarkii*	Double beam continuous wavelength scan (20 nm/s) from 675-400 nm, with rectangular test beam traversing one transverse layer of rhabdom microvilli	Dichroic ratio near 2 when microvilli in plane of observation; major absorptance parallel to axis of microvilli	8

continued

Part III. Microspectrophotometry

	Class, Order, & Species	Instrumentation	Effects	Reference
	Insecta—Diptera			
4	*Calliphora erythrocephala,* mutant chalky	Double beam wavelength scanning device measured extinction from about 670-325 nm; test beam 1.0-1.5 μ in diameter	Extinction dichroic at least between 570-450 nm with maximum coinciding with an absorption maximum at 520-500 nm; dichroic ratio about 1.3. This dichroism disappears on bleaching[1].	3-6
5	*Musca domestica*	Wavelength not varied	For retinular cells 1-6, dichroic absorptance maximum parallel to axis of rhabdomere microvilli; retinular cell 7 also dichroic but max absorptance normal to axis of microvilli. Max dichroic ratio about 1.5.	2
	Mollusca			
	Cephalopoda—Dibranchia			
6	*Loligo pealei*	No details given	Dichroic ratios of 6:1 cited for rhabdoms in isolated retinal fragments; major absorption with *e*-vector parallel to axis of microvilli	7

[1] Conclude rhodopsin both receptor pigment and dichroic analyzer.

Contributor: Waterman, Talbot H.

References

[1] Hays, D., and T. H. Goldsmith. 1969. Z. Vergl. Physiol. 65:218.
[2] Kirschfeld, K. 1969. In W. Reichardt, ed. Processing of Optical Data by Organisms and by Machines. Academic Press, New York. p. 116.
[3] Langer, H. 1965. Z. Vergl. Physiol. 51:258.
[4] Langer, H. 1966. Verh. Deut. Zool. Ges. Goettingen 30:195.
[5] Langer, H., and B. Thorell. 1966. Exp. Cell Res. 41:673.
[6] Langer, H., and B. Thorell. 1966. Funct. Organ. Compound Eye Proc. Int. Symp. 1965, p. 145.
[7] Moody, M. F. 1964. Biol. Rev. 39:43.
[8] Waterman, T. H., and H. R. Fernandez. 1970. Z. Vergl. Physiol. 68:154.
[9] Waterman, T. H., et al. 1969. J. Gen. Physiol. 54:415.

Part IV. Miscellaneous

Experiments were carried out in the laboratory unless otherwise indicated. **Polarized Light: Beam Direction**—H = horizontal, V = vertical; **Intensity**—μW = microwatts or 10^{-6} watt. **Effects:** R refers to retinular cell.

	Class, Order, & Species [Specification]	Medium	Polarized Light			Effects	Reference
			Source	Beam Direction	Wavelength [Intensity μW/cm^2]		
	Vertebrata						
	Osteichthyes—Salmoniformes						
1	*Oncorhynchus nerka* [Smolt]	Water	Filter	H	White [<1.8][1]	Subject trained to discriminate between vertical and horizontal *e*-vectors. This behavior not affected by adipose eyelid removal.	3

[1] Calculated, by assuming that the maximum luminous efficiency for white light was 0.4.

continued

Part IV. Miscellaneous

Class, Order, & Species [Specification]	Medium	Polarized Light			Effects	Reference
		Source	Beam Direction	Wavelength [Intensity μW/cm²]		
Arthropoda						
Arachnida—Araneida						
2 *Aphonopelma californica*	Air	Filter	V	White	Subjects could be trained in T-maze to discriminate two orthogonal *e*-vector directions[2]	6
Crustacea Decapoda						
3 *Carcinus maenas*	Air (?)	Tungsten bulb	H	Narrow band mono-chromatic	Eye movements in response to vertical or horizontal stimulus displacement over middle spectral range (λ_{max} = 508 nm) not affected by *e*-vector direction, but with blue & red wavelengths interaction between movement direction & *e*-vector orientation present	7
4		Opto-motor drum	H	White, orange & blue	Optomotor response present when alternate stripes in drum pattern have vertical & horizontal *e*-vectors, but not when *e*-vectors are at 45° & 135°; white, orange, & blue light effective[3]	8
5 *Libinia emar-ginata*	Water	Tungsten bulb	H	White	Fine structural changes induced by light adaptation prove that in each retinula, cells 1, 4, & 5 are selectively sensitive to horizontal *e*-vector (parallel to their microvilli); cells 2, 3, 6, & 7 to vertical *e*-vector (parallel to their microvilli)	4
6 *Potamon pota-mios* [Juvenile & adults]	Air (?)	Opto-motor drum	V	White [182][1]	Overhead rotating *e*-vector induced basitaxis at 0°, 45°, 90°, & 135° (mainly 0° & 90°) in juvenile specimens (19 of 20); rarely did so in adults (1 of 10 respond at 0° & 90°)	1
7 *Uca tangeri* [Adult]	Air[4]	Filter	V	White (sunlight) [760][1]	Optomotor response to rotation of polarization plane at 6 cycles/min[5]	10
Cladocera						
8 *Daphnia magna* [Adult]	Water	2 beams at 90°	H (2)	White	Greater phototactic effect with polarized than non-polarized light	15
9 *D. pulex* [Adult]	Water	2 beams at 90°	H (2)	White	For nonpolarized light to have a phototactic effect equal to that of polarized light requires an intensity 2-3 times that of polarized; when 2 beams are non-polarized, subject chooses one having higher intensity	15
Insecta Hymenoptera						
10 *Apis mel-lifera*	Air	Opto-motor drum	H	White	Alternate polarized stripes effective in inducing opto-motor response when their *e*-vector direction is 45° & 135°[3]	9

[1] Calculated, by assuming that the maximum luminous efficiency for white light was 0.4. [2] Conclude intraocular analysis involved. [3] Angles given are *e*-vector directions relative to vertical. [4] Experiments performed in field. [5] Special function for apical ommatidia hypothesized.

continued

Part IV. Miscellaneous

Class, Order, & Species [Specification]	Medium	Polarized Light			Effects	Reference
		Source	Beam Direction	Wavelength [Intensity $\mu W/cm^2$]		
Diptera						
11 *Drosophila* sp. [Adult]	Air	2 beams	H	White	Greater phototactic effect with polarized light than with nonpolarized	15
12 *D. melanogaster,* wild type, ebony, opm 2	Air	Optomotor drum	H	White	Torques induced by attenuating stripes at 45° & 135° significantly greater than control (all stripes at 45° or 135°) for wild type & ebony, but down to 10-20% for opm 2. (These optomotor responses due to R7 & R8.)[3]	5
13 *Musca domestica*	Air	Optomotor drum	H	White & long wavelength	In white light, both 2.8° (R7, R8) and 12° (R1-R6) stripes give max discrimination with 45° or 135° *e*-vector; with long wavelengths, 2.8° stripes are like white light, but 12° response not affected by *e*-vector direction [3]	9
Orthoptera						
14 *Locusta migratoria* [5th instar]	Air	Filter	H	White [19]	Photokinetic velocity 23% greater in polarized light than in unpolarized	2
Mollusca						
Cephalopoda—Dibranchia						
15 *Octopus* sp. [Adult]	Water	Filter	H	White	Subject trained to discriminate between *e*-vectors at 0° & 90°, and 45° & 135° [6]	12, 13
16		Filters, 2 sources	H	White	Subject trained to discriminate between *e*-vectors at 0° & 90° [7]	11
17 *O. vulgaris* [Juvenile]	Water	Filter	H	White	Subject trained to discriminate between *e*-vectors at 0° & 90° [7]	14

[3] Angles given are *e*-vector directions relative to vertical. [6] Dichroic receptor mechanism hypothesized. [7] Intraocular mechanism hypothesized.

Contributor: Waterman, Talbot H.

References

[1] Bäuerlein, R. 1969. Forma Funct. 1:285.

[2] Cassier, P. 1960. Bull. Soc. Zool. Fr. 85:165.

[3] Dill, P. A. 1971. J. Fish. Res. Bd. Can. 28:1319.

[4] Eguchi, E., and T. H. Waterman. 1968. Z. Zellforsch. Mikrosk. Anat. 84:87.

[5] Heisenberg, M. 1972. J. Comp. Physiol. 80:119.

[6] Henton, W. W., and F. T. Crawford. 1966. Z. Vergl. Physiol. 52:26.

[7] Horridge, G. A. 1967. Ibid. 55:207.

[8] Kirschfeld, K. 1972. Symp. Proc. Int. Biophys. Congr., Moscow, 4th (in press).

[9] Kirschfeld, K., and W. Reichardt. 1970. Z. Naturforsch. 25b:228.

[10] Korte, R. 1965. Experientia 21:98.

[11] Moody, M. F. 1962. J. Exp. Biol. 39:21.

[12] Moody, M. F., and J. R. Parriss. 1960. Nature (London) 186:839.

[13] Moody, M. F., and J. R. Parriss. 1961. Z. Vergl. Physiol. 44:268.

[14] Rowell, C. H. F., and M. J. Wells. 1961. J. Exp. Biol. 38:827.

[15] Verkhovskaya, I. N. 1940. Byull. Mosk. Obshchest. Ispyt. Prir. Otd. Biol. 49:101.

Data are for biological effects which are either oriented by, or quantitatively dependent on, the direction of the electric vector of linearly polarized light. All responses listed have been obtained in the laboratory, using artificial light sources with polarization filters. **Polarized Light: Spectral**

Region—UV = ultraviolet; **Direction** refers to the electric vector, parallel, perpendicular, or oblique to the cell surface, the cell axis, the axis of the organ in question, or to the direction of growth. **Responses:** E = electric vector.

Part I. Movement of Cell Contents and Related Phenomena

Responses are either light-oriented displacements of chloroplasts (chloroplast phototaxis), or enhancement by light of unoriented protoplasmic streaming (photodinesis), and some cytoplasmic effects closely related to it; O = orientation of chloroplasts, C = passive displacement of chloroplasts by centrifugal forces, D = photodinesis.

	Division & Species	Medium	Polarized Light			Responses	Conclusions	Reference
			Spectral Region	Intensity	E-Vector Direction			
1	Chrysophyta *Vaucheria sessilis*	Water	Blue	High	Perpendicular to cell axis	O: Chloroplasts escape front and rear cell walls much more effectively than if E parallel to axis	Dichroic photoreceptor parallel to surface	23
2	Chlorophyta *Hormidium flaccidum*	Air, water, or oil	Red	High	Perpendicular to cell axis	O: Movement to front of cell	Dichroic chlorophyll as photoreceptor (?)	13
3			Blue	Low to medium	....	O: Movement to front or rear of cell, depending on medium and whether E parallel or perpendicular to cell axis	Dichroic photoreceptor (flavin?) parallel to surface	13
4	*Mesotaenium caldariorum*	Water-agar	Far-red	Low	....	O: Cancels response induced by red light (entry 6) independent of direction of E	....	9
5			Red	Low	Parallel to cell axis	O: Cancels response induced by E perpendicular (entry 6)	Dichroic phytochrome as photoreceptor	9
6					Perpendicular to cell axis	O: Turns from profile to face position; complete movement		
7			Red & yellow (680-540 nm)	Low	....	O: Turns from profile to face position; E perpendicular to cell axis more effective than E parallel for induction of first phase of movement	Chlorophyll, in dichroic arrangement in thylacoid membranes, assumed to be the photoreceptor, with its vector of main absorption perpendicular (red, yellow) or parallel (blue) to membranes	1
8			Blue (500-400 nm)	Low	....	O: Turns from profile to face position; E parallel to cell axis more effective than E perpendicular		
9			Blue	High	Perpendicular to cell axis	O: Turns from face to profile position	Dichroic yellow photoreceptor pigment	2
10	*Mougeotia* sp.	Water	Far-red	Low	....	O: Cancels response induced by red light with E perpendicular (entry 15)	Dichroic phytochrome parallel to cell surface in the red light (P_R) absorbing form or perpendicular to it	3,4
11						C: Reverses inhibition of red light with E perpendicular (entry 15)		17

continued

Part I. Movement of Cell Contents and Related Phenomena

	Division & Species	Medium	Polarized Light			Responses	Conclusions	Reference
			Spectral Region	Intensity	E-Vector Direction			
12				Low	Perpendicular to surface	O: With partial irradiation of cell's flank, edge of chloroplast attracted	in the far-red light absorbing form (P_{FR}); change of dichroic orientation of phytochrome molecules when converted $P_R \rightleftharpoons P_{FR}$ [4,6,8]	5,11
13				Low	Perpendicular to cell axis	O: Turns from profile to face position		3,4
14			Red	Low	Parallel to surface	O: With partial irradiation of cell's flank, edge of chloroplast pushed away		5,11
15					Parallel to cell axis	O: Cancels response induced by E perpendicular (entry 15)		3,4
16					Perpendicular to cell axis	O: Turns from profile to face position		3,4
17					cell axis	C: Drastically inhibited (fastening of chloroplast's edge to cortical cytoplasm)		17
18			Blue & near UV (470-300 nm)	Low	Parallel to cell axis	O: Turns from profile to face position as in red light (see entry 15)	Interaction of dichroic phytochrome (see entry 15) with a dichroic yellow pigment, parallel to surface & to cell axis	7
19				High	....	O: Turns from face to profile position; more effective if E parallel or perpendicular to cell axis, depending on intensity and wavelength		14-16
	Bryophyta							
20	Cirriphyllum piliferum; Mnium cuspidatum; M. medium	Water	Blue	High	Perpendicular	O: Chloroplasts gather at cell walls perpendicular to E	Dichroic photoreceptor parallel to surface	23
21	Funaria hygrometrica	Water	Blue & UV [1]	Low	Parallel	O: Chloroplasts gather at cell walls parallel to E	Dichroic photoreceptor (flavin?) parallel to surface	20
22				High	Perpendicular	O: Chloroplasts gather at cell walls perpendicular to E		
	Pteridophyta							
23	Selaginella martensii, epidermis cells	Water	Blue	Low	Parallel to cell wall	O: Chloroplast approaches cell walls parallel to E	Dichroic photoreceptor (flavin?) parallel to surface	12
24				High	Perpendicular to cell wall	O: Chloroplast approaches cell walls perpendicular to E		
	Magnoliophyta [2]							
25	Asparagus sprengeri; Polygonatum odoratum; Potamogeton crispus	....	Blue	High	Perpendicular to cell wall	O: Chloroplasts gather at cell walls perpendicular to E	Dichroic photoreceptor parallel to surface	23
26	Elodea canadensis	Water	Blue	High	....	D: Rotation of cytoplasm enhanced by light more effectively if E parallel to cell axis than if E perpendicular	Dichroic photoreceptor (flavin?) parallel to surface	18, 23
27				High	Perpendicular to cell wall	O: Chloroplasts gather at cell walls perpendicular to E		

[1] No influence of E below 400 nm. [2] Synonym: Angiospermae.

continued

Part I. Movement of Cell Contents and Related Phenomena

	Division & Species	Medium	Polarized Light			Responses	Conclusions	Ref-er-ence
			Spectral Region	Inten-sity	E-Vector Direction			
28	*Hottonia palustris; Sambucus nigra; Sedum maximum; Symphytum cordatum; Utricularia vulgaris*	....	Blue	High	Perpendicular to cell wall	O: Chloroplasts gather at cell walls perpendicular to E	Dichroic photoreceptor parallel to surface	23
29	*Lemna trisulca*	Water	Blue	Low	Parallel to cell wall	O: Chloroplasts gather at cell walls parallel to E	Same dependence on E in plasmolyzed cells. Dichroic photoreceptor parallel to surface.	10, 21-23
30				High	Perpendicular to cell wall	O: Chloroplasts gather at cell walls perpendicular to E		
31	*Vallisneria spiralis torta*	Water	Blue	High	....	D: In mesophyll cells, induction of cytoplasmic rotation by light more effective if E parallel to cell & leaf axis than if E perpendicular	Dichroic photoreceptor (flavin) parallel to surface	19
32					Parallel to cell wall	C: Facilitated by light at cell walls parallel to E		
33				High	Perpendicular to cell wall	O: Chloroplasts gather at cell walls perpendicular to E		

Contributors: Grill, Renate, and Haupt, Wolfgang.

References

[1] Dorscheid, T. 1969. Z. Pflanzenphysiol. 61:46.

[2] Gärtner, R. 1970. Ibid. 63:428.

[3] Haupt, W. 1960. Planta 55:465.

[4] Haupt, W. 1968. Z. Pflanzenphysiol. 58:331.

[5] Haupt, W. 1970. Ibid. 62:287.

[6] Haupt, W. 1970. Physiol. Veg. 8:551.

[7] Haupt, W. 1971. Z. Pflanzenphysiol. 65:248.

[8] Haupt, W., and G. Bock. 1962. Planta 59:38.

[9] Haupt, W., and R. Thiele. 1961. Ibid. 56:388.

[10] Haupt, W., and M. Weisenseel. 1967. Naturwissenschaften 54:48.

[11] Haupt, W., et al. 1969. Planta 88:183.

[12] Mayer, F. 1964. Z. Bot. 52:346.

[13] Scholz, A., and W. Haupt. 1968. Naturwissenschaften 55:186.

[14] Schönbohm, E. 1963. Z. Bot. 51:233.

[15] Schönbohm, E. 1971. Z. Pflanzenphysiol. 65:453.

[16] Schönbohm, E. 1971. Ibid. 66:20.

[17] Schönbohm, E. 1972. Ibid. 66:113.

[18] Seitz, K. 1964. Protoplasma 58:621.

[19] Seitz, K. 1967. Z. Pflanzenphysiol. 56:246.

[20] Zurzycki, J. 1967. Acta Soc. Bot. Pol. 36:143.

[21] Zurzycki, J. 1968. Ibid. 37:11.

[22] Zurzycki, J. 1969. Protoplasma 68:193.

[23] Zurzycki, J., and Z. Lelatko. 1969. Acta Soc. Bot. Pol. 38:493.

Part II. Phototropism, Polarotropism, and Induction of Polarity

Bending movements in response to direction of light (phototropism)—or in response to the polarization of the light (polarotropism)—and induction of polarity by light are due to differential or local growth under the control of light absorption. **Responses:** P = polarotropism, F = phototropism, I = induction of polarity, L = light growth response.

continued

Part II. Phototropism, Polarotropism, and Induction of Polarity

	Division & Species [Specification]	Medium	Polarized Light		Responses	Conclusions	Reference
			Spectral Region	E-Vector Direction			
1	Chlorophyta *Mougeotia* sp.	Water	Red	Perpendicular	P: Direction of cell growth perpendicular to E	Dichroic photoreceptors	17
2			Blue	Parallel	P: Direction of cell growth predominantly parallel to E		
3	Phaeophyta *Fucus furcatus; F. serratus; F. vesiculosus; Pelvetia fastigiata* [germinating zygotes]	Water	Blue & near UV	Parallel	I: Outgrowth of rhizoid parallel to E; sometimes bipolar forms induced (in unpolarized light, rhizoid at dark pole)	Dichroic photoreceptor parallel to surface	9,10, 14, 15
4	Mycota[1] *Alternaria tenuis; Cladosporium fuligineum; Trichothecium roseum*	Water-air interphase	White	....	Germination and direction of growth independent of E	....	1
5	*Botrytis cinerea* [Germinating conidia]	Water-air interphase	Blue	Parallel	I: Outgrowth of germ tube parallel to E (in unpolarized light, germ tube in a subequatorial region)	Dichroic photoreceptors perpendicular to cell surface	1,3, 13
6	[Young hyphae]	Water-air interphase	Blue	Parallel	P: Growth parallel to E (in unpolarized light, negatively phototropic)		
7	*Penicillium glaucum* [Vegetative hyphae]	Water-air interphase	Blue	Perpendicular	P: Direction of growth perpendicular to E (in unpolarized light, positively phototropic)	Dichroic photoreceptors	3
8	*Phycomyces blakesleeanus* [Stage-IV b sporangiophores]	Air	Blue & near UV	....	F & L: Light with E perpendicular approx. 15% more effective than light with E parallel	Dichroic photoreceptors oriented with some preference parallel to surface	2,11, 18
9				Oblique	F: No difference between +45° and −45°		7
10	Bryophyta *Funaria hygrometrica* [Chloronema]	Water	Red (10^{-5} to $10^{-0.5}$ erg/cm²s)	....	I: Growth direction independent of polarization (in unpolarized light, chloronema at the brightest pole)	Dichroic photoreceptor (phytochrome?) disoriented relative to surface	12
11			Red (10^0 to 10^5 erg/cm²s)	Perpendicular	P: Growth perpendicular to E (in unpolarized light, positively phototropic) I: Outgrowth perpendicular to E (in unpolarized light, chloronema at the brightest pole)	Dichroic receptors parallel to cell surface	1,12
12			Red (10^5 to 10^6 erg/cm²s)	Perpendicular to the H-vector	P: Corkscrew growth; screw axis parallel to E I: Outgrowth perpendicular to H (in unpolarized light, chloronema at the darkest pole)	Dichroic magnetically excited receptors parallel to cell surface	12

[1]/ Synonym: Fungi.

continued

Part II. Phototropism, Polarotropism, and Induction of Polarity

Division & Species [Specification]	Medium	Polarized Light		Responses	Conclusions	Ref-er-ence
		Spectral Region	E-Vector Direction			
13 *Physcomitrium turbinatum*, [Chloronema]	Water	Far-red, red, & near UV	....	F: Positive response to obliquely incident light; E parallel more effective than if E perpendicular	Dichroic phytochrome assumed to have disc-shaped absorption characteristics	16
14			Perpendicular	P: Under certain conditions of double beam irradiation, growth perpendicular to E		
15 *Sphaerocarpus donnellii*	Water	Blue & near UV	Perpendicular	P: Growth of germ tube perpendicular to E	Dichrotic photoreceptors	19, 20
Pteridophyta 16 *Dryopteris filixmas*	Water & water-air interphase	Far-red, red, blue, & near UV	Parallel	P: Growth of rhizoid parallel to E (in unpolarized light, negatively phototropic)	Dichroic phytochrome parallel (P_{660}) or perpendicular (P_{730}) to cell surface. Additional dichroic	1,4, 6, 21, 22
17			Perpendicular	P: Growth of chloronema perpendicular to E (in unpolarized light, positively phototropic)		
				photoreceptor assumed for blue light, also parallel to surface.		
18 *Equisetum arvense, E. hiemale, E. fluviatile*[2], *E. palustre,* & *E. variegatum* [germinating spores]	Water	Blue & near UV	Parallel	I: Outgrowth of rhizoid parallel to E (in unpolarized light, rhizoid at the dark pole)	Dichroic photoreceptor parallel to surface	5,8, 14, 15
19 *Mattuccia struthiopteris*[3] [Chloronema]	Water	Red	Perpendicular	P: Growth and branching perpendicular to E (in unpolarized light, positively phototropic)	Dichrotic photoreceptors	6
20 *Osmunda cinnamomea* [Germinating spores]	Water-air interphase	Blue	Parallel	I: Outgrowth of rhizoid parallel to E (in unpolarized light, rhizoid at the dark pole)	Dichroic photoreceptor parallel to surface	13

[2] Synonym: *E. limosum.* [3] Synonym: *Struthiopteris filicastrum.*

Contributors: Grill, Renate, and Haupt, Wolfgang.

References
[1] Bünning, E., and H. Etzold. 1958. Ber. Deut. Bot. Ges. 71:304.
[2] Castle, E. S. 1934. J. Gen. Physiol. 17:751.
[3] Etzold, H. 1961. Exp. Cell Res. 25:229.
[4] Etzold, H. 1965. Planta 64:254.
[5] Etzold, H., and L. Jaffe. 1963. Exp. Cell Res. 29:188.
[6] Hartmann, K. M., et al. 1965. Planta 64:363.
[7] Haupt, W., and M. Buchwald. 1967. Z. Pflanzenphysiol. 56:20.
[8] Haupt, W., and F. W. Meyer zu Bentrup. 1961. Naturwissenschaften 48:723.
[9] Jaffe, L. 1956. Science 123:1081.
[10] Jaffe, L. 1958. Exp. Cell Res. 15:282.
[11] Jaffe, L. 1960. J. Gen. Physiol. 43:897.
[12] Jaffe, L., and H. Etzold. 1965. Biophys. J. 5:715.
[13] Jaffe, L., et al. 1962. J. Cell Biol. 13:13.
[14] Meyer zu Bentrup, F. W. 1963. Planta 59:472.
[15] Meyer zu Bentrup, F. W. 1964. Ibid. 63:356.
[16] Nebel, B. J. 1969. Ibid. 87:170.
[17] Neuscheler-Wirth, H. 1970. Z. Pflanzenphysiol. 63:238.
[18] Shropshire, W., Jr. 1959. Science 130:336.
[19] Steiner, A. M. 1969. Planta 86:334.
[20] Steiner, A. M. 1969. Ibid. 86:343.
[21] Steiner, A. M. 1969. Photochem. Photobiol. 9:493.
[22] Steiner, A. M. 1969. Ibid. 9:507.

continued

Part III.   Miscellaneous

This section includes metabolic light effects and an effect on locomotion of a flagellate.

	Division & Species	Medium	Polarized Light		Responses	Conclusions	Reference
			Spectral Region	E-Vector Direction			
1	Euglenophyta *Euglena gracilis*	Water	Blue (480-450 nm)	Perpendicular to cell axis	Photophobic response (avoidance reaction when suddenly illuminated) more sensitive than if E parallel	Dichroic pigment at photoreceptor site or in shading structure	1
2			Near UV (350 nm)	Parallel to cell axis	Photophobic response more sensitive than if E perpendicular		
3	Bryophyta *Brachythecium selebrosum; Climacium dendroides*	Water	Blue	Parallel to leaf axis	Extra O_2 uptake following illumination when photosynthesis blocked by 3-(3,4-dichlorophenyl)-1,1-dimethyl-urea (DCMU); effect much more pronounced than if E perpendicular to leaf axis	....	3
4	Magnoliophyta[1] *Elodea canadensis*	Water	Blue	....	Extra O_2 uptake following illumination when photosynthesis blocked by DCMU, but only small difference between E parallel and E perpendicular, due to much less elongated shape of cells	....	3
5	*Lemna trisulca*	Water	Blue	....	Extra O_2 uptake following illumination when photosynthesis blocked by DCMU, but no influence of E direction at all, due to isodiametric shape of cells	....	3
6	*Zea mays*	Air	Far-red & red	....	Dark destruction of phytochrome in etiolated coleoptiles: After irradiation with non-saturating red light, 15-20% more destruction if E perpendicular to coleoptile axis than if E parallel. After irradiation by red light with E perpendicular followed by irradiation with far-red light with E also perpendicular, 15-20% less destruction than in same sequence with E (far-red) parallel to coleoptile axis.	Destruction depends on percentage of phytochrome which is in P_{730} form. Differences are best explained by assuming a dichroic phytochrome oriented parallel to surface.	2

[1] Synonym: Angiospermae.

Contributors: Grill, Renate, and Haupt, Wolfgang

References
[1] Diehn, B. 1969. Biochim. Biophys. Acta 177:136.
[2] Marmé, D., and E. Schäfer. 1972. Z. Pflanzenphysiol. 67:192.
[3] Zurzycki, J. 1971. Biochem. Physiol. Pflanzen 162:310.

165. AUDIBLE SOUND PRESSURE LEVELS: MAN

Part I. Thresholds of Minimum Audibility: Otologically Normal Ears

Sound Pressure Level: Point of reference is 0.0002 μbar. Values in parentheses are ranges, estimate "b" unless otherwise indicated (*see* Introduction).

	Specification	Method	No. of Subjects	Age yr	Audio-frequency cycles/s	Sound Pressure Level, dB		Reference
						♂	♀	
	Air Coupling							
	Minimum audible field							
1	Monaural	Sound pressure levels	10♂, 4♀	......	100 [1]	32.0(21.8-42.2)		11
2		developed in free	[14		200	19.0(5.8-32.2)		
3		field at center of	ears]		500	8.0(−7.8 to +23.8)		
4		head, the head re-			1000	4.0(−13.4 to +21.4)		
5		moved from field			1500	−2.0(−20.2 to +16.6)		
6					2000	−5.0(−21.0 to +11.0)		
7					3000	−7.0(−19.6 to +5.6)		
8					4000	−8.0(−23.2 to +7.2)		
9					6000	3.0(−10.6 to +16.6)		
10					8000	8.0(−5.0 to +21.0)		
11					10,000	11.0(−12.0 to +34.0)		
12					15,000	21.0(−15.4 to +57.4)		
13	Binaural	Sound pressure levels	13♂, 2♀	......	60 [1]	44.0(39.0-49.9)		11
14		developed in free	[30		100 [1]	32.0(21.4-42.6)		
15		field at center of	ears]		200	19.0(3.4-34.6)		
16		head, the head re-			500	8.0(−4.4 to +20.4)		
17		moved from field			1000	4.0(−8.0 to +16.0)		
18					1500	0(−12.0 to +12.0)		
19					2000	−2.0(−11.0 to +7.0)		
20					3000	−4.0(−14.0 to +6.0)		
21					4000	−6.0(−18.0 to +6.0)		
22					6000	3.0(−9.4 to +15.4)		
23					8000	10.0(−1.0 to +21.0)		
24					10,000	14.0(−1.0 to +29.0)		
25					15,000	27.0(4.0-50.0)		
	Minimum audible pressure							
26	Children	Western Electric 705A	∿2800♂♀	5-14	250	30.2(14.6-45.8)		3
27		earphone in National			500	18.2(2.4-34.0)		
28		Bureau of Standards 9A coupler			1000	12.3(−3.5 to +28.1)		
29	White	Western Electric 705A	∿950♂,	5-14	250	31.2(16.6-45.8)	31.1(14.9-47.3)	3
30		earphone in National	∿950♀		500	18.5(3.7-33.3)	18.0(2.2-33.8)	
31		Bureau of Standards			1000	12.4(−3.0 to +27.8)	11.7(−4.3 to +27.7)	
32		9A coupler			2000	12.5(−4.1 to +29.1)	12.3(−4.1 to +28.7)	
33					4000	2.8(−16.8 to +22.4)	11.3(−5.7 to +28.3)	
34					6000	18.0(−4.8 to +40.3)	14.5(−3.5 to +32.5)	
35					8000	19.8(−3.4 to +43.0)	17.9(−2.7 to +38.5)	
36	Nonwhite	Western Electric 705A	∿460♂,	5-14	250	30.7(15.9-45.9)	32.0(15.0-49.0)	3
37		earphone in National	∿480♀		500	18.0(2.8-33.2)	18.0(0.2-35.8)	
38		Bureau of Standards			1000	12.5(−1.7 to +26.7)	12.6(−4.6 to +29.8)	
39		9A coupler			2000	14.2(−2.2 to +30.6)	14.0(−4.0 to +32.0)	
40					4000	12.4(−4.8 to +29.6)	12.8(−7.0 to +32.6)	
41					6000	15.1(−4.5 to +34.7)	16.6(−3.8 to +37.0)	
42					8000	17.5(−4.1 to +39.1)	16.8(−4.8 to +38.4)	

[1] Thresholds at 125 cycles/s and below are probably masked, not true, thresholds, due to physiological noise generated under the earphone [10].

continued

Part I. Thresholds of Minimum Audibility: Otologically Normal Ears

	Specification	Method	No. of Subjects	Age yr	Audio-frequency cycles/s	Sound Pressure Level, dB ♂	Sound Pressure Level, dB ♀	Reference
43	Young adults[2/]	Telephone receiver & thermophone, the latter calibrated with a 1-cm³ coupler and a condenser transmitter	41♂♀ [82 ears]	......	50[1/]	64.0		4
44					125[1/]	43.0		
45					250	30.0		
46					500	17.0		
47					1000	10.0		
48					1500	7.0		
49					2000	9.0		
50					3000	7.0		
51					4000	7.0		
52		Telephone receiver & loudspeaker; sound pressure level measured in ear canal with probe tube	11♂♀ [22 ears]	......	50[1/]	74.0		11
53					125[1/]	53.0		
54					250	27.0		
55					500	11.0		
56					1000	5.0		
57					1500	6.0		
58					2000	8.0		
59					3000	7.0		
60					4000	7.0		
61					6000	15.0		
62					8000	20.0		
63					10,000	22.0		
64					15,000	41.0		
65		Standard Telephones & Cables, Ltd. 4026A earphone in British Standard 2042 artificial ear	45♂, 54♀	18-25	125[1/]	45.5(31.9-59.1)		2
66					250	28.0(13.4-42.6)		
67					500	12.5(−0.5 to +25.5)		
68					1000	5.5(−5.9 to +16.9)		
69					1500	8.5(−3.7 to +20.7)		
70					2000	10.5(−1.7 to +22.7)		
71					3000	7.0(−4.8 to +18.8)		
72					4000	9.5(−4.3 to +23.3)		
73					6000	10.5(−7.7 to +28.7)		
74					8000	9.0(−8.4 to +26.4)		
75					10,000	17.0(−1.0 to +35.0)		
76					12,000	20.5(1.3-39.7)		
77					15,000	39.0(17.6-60.4)		
78	Adults	TRACOR high-frequency audiometer	100♂	17-23	8000	−3(−12.4 to +6.4)	....	5
79					9000	4.7(−4.9 to +14.3)	....	
80					10,000	2.2(−7.0 to +11.2)	....	
81					11,000	0.0(−9.9 to +9.9)	....	
82					12,000	2.0(−9.1 to +13.1)	....	
83					13,000	13.1(−1.6 to +27.8)	....	
84					14,000	18.1(2.8-33.4)	....	
85					15,000	24.5(7.4-41.6)	....	
86					16,000	29.0(11.4-46.6)	....	
87					18,000	72.5(57.0-88.0)	....	

[1/] Thresholds at 125 cycles/s and below are probably masked, not true, thresholds, due to physiological noise generated under the earphone [10]. [2/] Probable error of 20-30% in determination of pressure variation.

continued

Part I. Thresholds of Minimum Audibility: Otologically Normal Ears

	Specification	Method	No. of Subjects	Age yr	Audio-frequency cycles/s	Sound Pressure Level, dB		Ref-er-ence
						♂	♀	
88	Adults[3/]	Telephonics TDH-39 earphone in NBS 9A coupler (For information on American audiometer "Zero," consult reference 7)	6672♂♀	18-79	500	23.6	23.6	7
89					1000	14.2	13.7	
90					2000	20.0	18.0	
91					3000	27.6	18.6	
92					4000	35.3	19.3	
93					6000	45.5	33.0	
94	College-town residents (Screened sample of individuals minimally exposed to high-intensity noise)	Permoflux PDR-8 earphone in National Bureau of Standards 9A coupler	36♂, 42♀	18-24	250	26.8(22.2-31.4)	24.3(20.0-28.6)	1
95					500	11.1(5.3-16.9)	9.6(4.6-14.6)	
96					1000	5.9(1.5-10.3)	4.8(0.3-9.3)	
97					1500	7.2(2.0-12.4)	5.2(0.3-10.1)	
98					2000	7.7(1.5-13.9)	4.7(0.1-9.3)	
99					3000	12.8(5.6-20.0)	7.7(1.9-13.5)	
100					4000	14.0(5.4-22.6)	10.1(4.7-15.6)	
101					6000	30.0(16.9-43.1)	21.1(14.8-27.4)	
102					8000	27.6(13.6-41.6)	19.4(11.0-27.8)	
103			62♂, 146♀	26-32	250	27.2(21.0-33.4)	25.6(18.5-32.7)	
104					500	13.3(7.3-19.3)	13.1(2.3-23.9)	
105					1000	6.7(1.8-11.6)	7.1(−0.1 to +15.2)	
106					1500	7.8(−0.6 to +16.2)	8.0(0.4-15.6)	
107					2000	8.8(−0.7 to +18.3)	7.8(−0.2 to +15.8)	
108					3000	13.6(−1.2 to +28.4)	7.7(0.0-15.4)	
109					4000	21.5(1.8-40.8)	8.3(1.7-14.9)	
110					6000	38.8(19.8-57.8)	21.7(11.3-32.1)	
111					8000	33.6(16.7-50.5)	23.9(11.8-36.0)	
112			64♂, 158♀	34-40	250	24.5(17.8-31.2)	26.1(19.9-32.3)	
113					500	12.1(5.4-18.8)	11.9(5.8-18.0)	
114					1000	7.8(3.7-11.9)	7.3(1.8-12.8)	
115					1500	8.2(0.4-16.0)	7.8(0.6-15.0)	
116					2000	10.0(−0.2 to +20.2)	8.8(1.6-16.0)	
117					3000	16.3(2.4-30.2)	8.9(1.1-16.7)	
118					4000	23.7(6.2-41.2)	13.2(4.2-22.2)	
119					6000	38.9(20.5-57.3)	26.4(15.0-37.8)	
120					8000	36.8(17.4-56.2)	26.7(16.2-37.2)	
121			50♂, 82♀	43-49	250	31.0(11.6-50.4)	25.3(14.5-36.1)	
122					500	14.0(−4.5 to +32.5)	10.6(−0.2 to +21.4)	
123					1000	12.5(−4.6 to +29.6)	8.0(−2.6 to +18.6)	
124					1500	18.6(3.2-34.0)	13.2(4.5-21.9)	
125					2000	19.1(3.6-34.6)	13.8(3.5-24.1)	
126					3000	28.6(11.5-45.7)	15.0(1.8-28.2)	
127					4000	34.6(19.9-49.3)	19.6(7.5-31.7)	
128					6000	48.2(29.8-66.6)	35.9(22.3-49.5)	
129					8000	40.1(21.4-59.8)	30.9(15.0-46.8)	
130			132♂, 172♀	51-57	250	28.3(20.4-36.2)	31.7(21.1-42.3)	
131					500	15.8(7.9-23.7)	18.8(8.6-29.0)	
132					1000	12.4(3.4-21.4)	14.4(3.6-25.2)	
133					1500	14.9(3.5-26.3)	15.4(4.7-26.1)	
134					2000	18.7(6.1-31.3)	18.2(7.8-28.6)	
135					3000	28.4(11.6-45.2)	20.1(9.1-31.1)	
136					4000	35.3(15.7-54.9)	24.7(11.0-38.4)	

[3/] Distributions are skewed, so that standard errors are not informative.

continued

Part I. Thresholds of Minimum Audibility: Otologically Normal Ears

	Specification	Method	No. of Subjects	Age yr	Audio-frequency cycles/s	Sound Pressure Level, dB		Reference
						♂	♀	
137					6000	47.7(26.7-68.7)	38.0(24.4-51.6)	
138					8000	46.6(25.9-67.3)	37.7(21.2-54.2)	
139			149♂, 154♀	59-65	250	34.3(24.9-44.7)	36.5(25.7-47.3)	
140					500	18.6(9.9-27.3)	20.4(8.0-32.8)	
141					1000	15.5(5.0-26.0)	14.5(1.3-27.7)	
142					1500	20.1(7.2-33.0)	17.2(4.2-30.2)	
143					2000	25.9(9.6-42.2)	20.0(6.3-33.7)	
144					3000	39.8(18.9-60.7)	23.7(10.0-37.4)	
145					4000	48.8(28.0-69.6)	27.6(12.9-42.3)	
146					6000	61.3(38.7-83.9)	45.2(27.4-63.0)	
147					8000	55.3(32.7-77.9)	43.7(23.2-64.2)	
148	Low-noise environment (Mabaan tribe, Sudan)	High-frequency loudspeaker coupled to eardrum with circumaural cushion	117♂♀	10-19	12,000	20.1		9
149					14,000	24.9		
150					16,000	40.4		
151					18,000	64.5		
152					20,000	No response		
153			119♂♀	20-29	12,000	28.5		
154					14,000	31.9		
155					16,000	52.1		
156					18,000	82.0		
157			107♂♀	30-39	12,000	29.9		
158					14,000	37.5		
159					16,000	68.1		
160			108♂♀	40-49	12,000	37.5		
161					14,000	54.0		
162					16,000	84.6		
163			108♂♀	50-59	12,000	53.6		
164					14,000	67.5		
165			102♂♀	60-69	12,000	61.7		
166					14,000	81.2		
167			101♂♀	70-79	12,000	80.5		
168					14,000	91.8		
169	City dwellers (New York, Düsseldorf, Cairo)	High-frequency loudspeaker coupled to eardrum with circumaural cushion	338♂♀	10-19	12,000	22.6		9
170					14,000	26.1		
171					16,000	46.4		
172					18,000	74.6		
173					20,000	>91.2		
174			372♂♀	20-29	12,000	29.2		
175					14,000	36.6		
176					16,000	57.8		
177					18,000	89.4		
178			325♂♀	30-39	12,000	41.4		
179					14,000	57.5		
180					16,000	86.6		
181			320♂♀	40-49	12,000	59.1		
182					14,000	85.5		
183					16,000	No response		
184			328♂♀	50-59	12,000	84.4		
185					14,000	No response		

continued

Part I. Thresholds of Minimum Audibility: Otologically Normal Ears

	Specification	Method	No. of Subjects	Age yr	Audio-frequency cycles/s	Sound Pressure Level, dB ♂	Sound Pressure Level, dB ♀	Reference
186			324♂♀	60-69	12,000	No response		
187					14,000	No response		
188			336♂♀	70-79	12,000	No response		
189					14,000	No response		
190	Low-noise occupations (Individuals with no history of military or acoustic trauma)	Telephonics TDH earphone in National Bureau of Standards 9A coupler	35♂	20-29	500	14.8(7.2-22.4)	....	8
191					1000	7.9(0.1-15.7)	....	
192					2000	12.7(1.1-24.3)	....	
193					4000	10.6(−8.2 to +29.4)	....	
194			41♂	30-39	500	15.5(2.7-28.3)	....	
195					1000	10.2(−6.6 to +27.0)	....	
196					2000	15.6(−1.0 to +32.2)	....	
197					4000	15.3(−5.7 to +36.3)	....	
198			26♂	40-49	500	14.0(6.6-21.4)	....	
199					1000	9.1(−0.9 to +19.1)	....	
200					2000	15.7(−2.9 to +34.3)	....	
201					4000	20.8(−0.8 to +42.4)	....	
202			22♂	50-59	500	17.2(4.8-29.6)	....	
203					1000	18.6(5.2-32.0)	....	
204					2000	16.5(−2.7 to +35.7)	....	
205					4000	27.1(−8.5 to +62.7)	....	
			Water Coupling					
206	Adults	Self-contained underwater breathing apparatus; wet suit 0.47 cm thick, face mask, hood pushed back	4♂	20-22	250	62(57-65)c	....	6
207					500	68(65-71)c	....	
208					1000	68(70-82)c	....	
209					1500	80(72-93)c	....	
210					2000	78(68-85)c	....	
211					3000	76(70-83)c	....	
212					4000	70(65-76)c	....	
213					6000	70(64-75)c	....	
214		Dry suit, face mask, no hood; "Zero" or better on audiometer; listening at depth of 9.14 m in freshwater, 4.57 m from source	4♂	Young	250	60	....	
215					500	52	....	
216					1000	53	....	
217					2000	53	....	
218					4000	55	....	

Contributor: Harris, J. Donald

References

[1] Corso, J. 1963. Arch. Otolaryngol. 77:385.
[2] Dadson, R. S., and J. H. King. 1952. J. Laryngol. Otol. 66:366.
[3] Eagles, E. L., et al. 1963. Laryngoscope (Suppl.).
[4] Fletcher, H., and R. L. Wegel. 1922. Phys. Rev. 19: 553.
[5] Harris, J. D., and C. K. Myers. 1971. J. Acoust. Soc. Amer. 49:600.
[6] Montague, W. F., and J. F. Strickland. 1961. Ibid. 33:1376.
[7] National Center for Health Statistics. 1965. U.S. Pub. Health Serv. Publ. Vital Health Statist. Ser. 1000-11-11.
[8] Nixon, J. C., et al. 1962. J. Laryngol. Otol. 76:288.
[9] Rosen, S., et al. 1964. Arch. Otolaryngol. 79:18.
[10] Rudmose, W. 1962. Int. Congr. Acoust., 4th, Copenhagen (H52).
[11] Sivian, L. J., and S. D. White. 1933. J. Acoust. Soc. Amer. 4:288.

continued

Part II. Thresholds of Minimum Audibility: Audiometric Systems

Sound Pressure Level: Point of reference is 0.0002 μbar. For information on American audiometer "Zero," consult Part I, reference 7.

	Audiometer "Zero"	Method	Audio-frequency cycles/s	Sound Pressure Level dB	Ref-er-ence		Audiometer "Zero"	Method	Audio-frequency cycles/s	Sound Pressure Level dB	Ref-er-ence
1	British	Standard Tele-	125 L/	47	3	36			2000	9	
2		phones & Cables,	250	28		37			3000	10.5	
3		Ltd., 4026A ear-	500	11.5		38			4000	11.5	
4		phone in British	1000	5.5		39			6000	18.5	
5		Standard 2042	1500	6.5		40			8000	9.5	
6		artificial ear	2000	9.		41	Proposed,	Western Electric	125 L/	45.5	1
7			3000	8		42	Interna-	705A earphone in	250	24.5	
8			4000	9.5		43	tional	National Bureau	500	11.0	
9			6000	8		44	Standards	of Standards 9A	1000	6.5	
10			8000	10		45	Organiza-	coupler	1500	6.5	
11	French	PTT earphone in	125 L/	44.4	3	46	tion, 1964		2000	8.5	
12		CNET coupler	250	27.5		47			3000	7.5	
13			500	11.5		48			4000	9.0	
14			1000	5.5		49			6000	8.0	
15			1500	4.5		50			8000	9.5	
16			2000	4.5		51		Permoflux PDR	125 L/	43.4	1
17			3000	6		52		earphone in Na-	250	23.7	
18			4000	8		53		tional Bureau of	500	10.7	
19			6000	17		54		Standards 9A	1000	6.3	
20			8000	14.5		55		coupler	1500	7.4	
21	German	Beyer DT48 ear-	125 L/	47.5	3	56			2000	8.6	
22		phone in National	250	28.5		57			3000	8.6	
23		Bureau of Stan-	500	14.5		58			4000	8.6	
24		dards 9A coupler	1000	8		59			6000	12.7	
25			1500	7.5		60			8000	16.9	
26			2000	8		61		Telephonics TDH	125 L/	42.8	1
27			3000	6		62		earphone in Na-	250	24.4	
28			4000	5.5		63		tional Bureau of	500	10.3	
29			6000	8		64		Standards 9A	1000	6.3	
30			8000	14.5		65		coupler	1500	8.2	
31	Russian	TD earphone in	125 L/	55	3	66			2000	9.5	
32		UY-3 coupler	250	33		67			3000	8.8	
33			500	14.5		68			4000	8.2	
34			1000	8.5		69			6000	11.9	
35			1500	8.5		70			8000	15.4	

L/ Thresholds at 125 cycles/s and below are probably masked, not true, thresholds, due to the physiological noise generated under the earphone [2].

Contributor: Harris, J. Donald

References

[1] Davis, H., and F. W. Kranz. 1964. J. Speech Hear. Res. 7:7.

[2] Rudmose, W. 1962. Int. Congr. Acoust., 4th, Copenhagen (H52).

[3] Weissler, P. G. 1968. J. Acoust. Soc. Amer. 44:264.

continued

165. AUDIBLE SOUND PRESSURE LEVELS: MAN

Part III. Thresholds of Maximum Audibility

Sound Pressure Level: Point of reference is 0.0002 μbar. Values in parentheses are ranges, estimate "a" (*see* Introduction).

	Threshold	No. of Subjects	Specification	Audio-frequency cycles/s	Sound Pressure Level dB	Reference
1	Annoyance for bands of noise	162	Individuals exposed to high noise level in jobs; at least 5 judged each band of noise (absolute judgments)	150-394	91(84-98)	2
2				670-1000	92(80-104)	
3				1420-1900	94(86-102)	
4				2450-3120	91(81-101)	
5				4000-5100	95(85-105)	
6				6600-9000	92(80-104)	
7			Individuals not exposed to high noise level in jobs; at least 10 judged each band of noise (absolute judgments)	150-394	78(68-88)	
8				670-1000	78(71-85)	
9				1420-1900	78(66-90)	
10				2450-3120	79(72-86)	
11				4000-5100	79(69-89)	
12				6600-9000	73(61-85)	
13	Discomfort for pure tones	4♂, 5♀ [16 ears]	Normal-hearing adults, 16-42 yr; judgments recorded after initial exposure (first value) and after 5½ experimental sessions (second value)	250	113.0; 124.5	1
14				500	109.5; 129.0	
15				1000	111.0; 122.0	
16				1400	109.0; 121.0	
17				2000	108.0; 119.0	
18				2800	111.0; 117.0	
19				4000	105.0; 120.0	
20				5600	113.0; 128.0	
21	"Tickle" for pure tones	4♂, 5♀ [16 ears]	Normal-hearing adults, 16-42 yr; judgments recorded after initial exposure (first value) and after 5½ experimental sessions (second value)	250	136.0; 144.0	1
22				500	134.0; >145.0	
23				1000	129.0; >145.0	
24				1400	128.0; 144.0	
25				2000	128.0; >142.0	
26				2800	130.5; 144.5	
27				4000	140.0; >140.0	
28				5600	>140.0; 139.0	
29	Pain for pure tones	4♂, 5♀ [16 ears]	Normal-hearing adults, 16-42 yr; judgments recorded after initial exposure (first value) and after 5½ experimental sessions (second value)	250	142.5; 144.0	1
30				500	138.5; >145.0	
31				1000	139.0; >145.0	
32				1400	139.5; 144.0	
33				2000	137.5; >145.0	
34				2800	144.0; >145.0	
35				4000	>140.0; >140.0	
36				5600	>140.0; >139.0	

Contributor: Harris, J. Donald

References

[1] Silverman, S. R., et al. 1947. Ann. Otol. Rhinol. Laryngol. 56:658.

[2] Spieth, W. 1956. J. Acoust. Soc. Amer. 28:872.

Diagram summarizes the effects of pulsed ultrasonic radiation (on 0.5 second, off 1.0 second) on the height of action potentials and conduction velocities in different fiber groups of sensory cutaneous nerves. Higher frequency or higher dosage will produce greater temperature rises. However, the intensity and cycle of the ultrasound and attenuation coefficients, and the specific heat, heat diffusion properties, and mass of the medium surrounding the nerve, also determine actual temperatures attained at the nerve. Changes in the action potential or conduction velocity are related to the actual temperature rather than to the ultrasonic parameters. Curves for the action potential and conduction velocity of Aδ fibers lie between those of Aβ and C fibers, but have been omitted for clarity. Comparable results have been obtained for unitary spike potentials of the median and lateral giant fibers of *Lumbricus terrestris.*

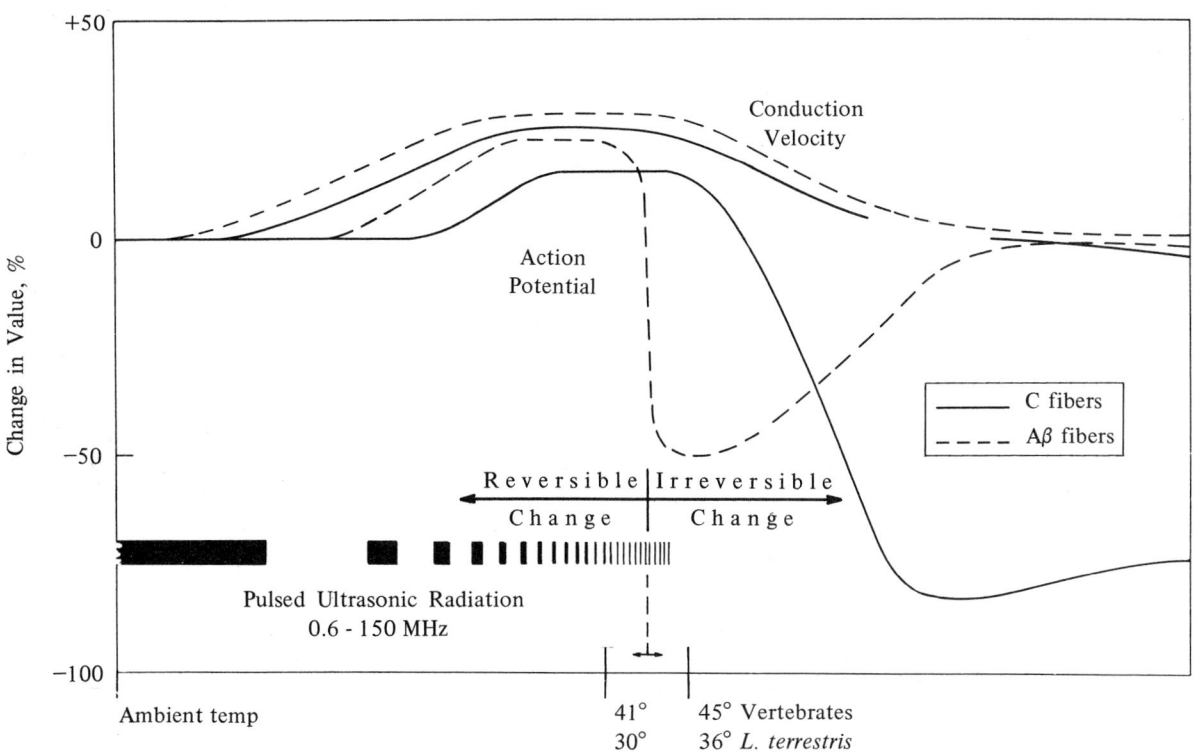

Surface Temperature of Nerves, °C

Contributor: Lele, P. P.

General References: [1] Lele, P. P. 1963. Exp. Neurol. 8:47. [2] Singer, R., and P. P. Lele. Unpublished. Massachusetts Institute of Technology, Cambridge, 1972.

167. RESPONSE OF THE NERVOUS SYSTEM TO INCREASED AND DECREASED OXYGEN LEVEL: POIKILOTHERMIC VERTEBRATES AND INVERTEBRATES

Changes in respiratory rhythm and/or onset of convulsions are the reflections of effects on the nervous system of exposure to lower or higher oxygen levels than are normally encountered in the environment. In many cases these effects may be mediated through specialized receptors of the nervous system. Usually the responses are closely related to temperature, carbon dioxide pressure, pH, and/or other variables. **Effects:** Po_2 = oxygen pressure. For additional information, consult references 1, 20, 22, and 23.

	Class & Animal (Synonym)	Effects	
		Oxygen Pressure Lower than Normal	Oxygen Pressure Higher than Normal
	Chordata		
	Reptilia		
1	*Chrysemys scripta elegans (Chelopus elegans, Pseudemys scripta elegans)*	....	At 1 atm & 37.5°C, no convulsions, although respiratory pattern changes [14]
2	*Clemmys guttata (Chelopus guttatus)*	....	At 1 atm & 37.5°C, no convulsions, although respiratory pattern changes [14]
3	*Lacerta*	Pulmonary ventilation constant down to 10% O_2, then decreased [25]	....
4	Alligator [1]	....	At ~2-5 atm, no convulsions for 1.5 hr[2] [34]
5	Alligator; snake; turtle	Generally, low Po_2 results in increased ventilatory movements (indicative of CNS activity) [28]	....
	Amphibia		
6	*Ambystoma mexicanum (Siredon mexicanum)*	....	At 7.5 atm, convulsions within 15 hr [35]
7	*Triturus vulgaris*	....	At 9.0 atm, convulsions within 10 hr [35]
8	Frog	Reportedly survives for 24 hr in absence of O_2 [26,28]; generally, low Po_2 results in increased ventilatory movements (indicative of CNS activity) [28]	At 1 atm for 40-52 da, no obvious CNS effects [3], but succumbs due, apparently, to other effects [33]; at 1-50 atm, conduction velocity & spike amplitude decrease, rheobase increases [11]; at 3-4 atm, convulsions for 22-25 hr (anesthetic depresses convulsions, but frog dies) [2]; at 3.5 atm, survives for 65 hr [9]; at 7.0 atm, survives for 7.5 hr, but evidence of "strychnine-type" convulsions [35]
	Dipnoi		
9	*Neoceratodus*	Hypoxic water yields strong stimulus for both branchial & aerial respiration [18]	....
10	*Protopterus aethiopicus*	Little or no ventilatory response to deoxygenated water. [17] Young (65 g) survive for 36 hr submerged in air-equilibrated water; hypoxia increases lung breathing more than gill breathing. [16]	Hyperoxia decreases both gill & lung breathing [16]
	Osteichthyes (Pisces)		
11	*Carassius auratus*	Little effect on respiratory frequency with changes in Po_2 [10]	Relatively high Po_2 (6.2 ml/liter) decreases opercular movements for "considerable periods" [10]
12	*Lepisosteus*	Little or no significant effect of Po_2 on breathing interval [29]	....
13	*Leuciscus*	Surfaces to gulp air (generally dependent on Po_2 of water for ventilatory activity) [12,37]	....

[1] Single subject. [2] Guinea pig and dog died during simultaneous exposure to same conditions [34].

continued

167. RESPONSE OF THE NERVOUS SYSTEM TO INCREASED AND DECREASED OXYGEN LEVEL: POIKILOTHERMIC VERTEBRATES AND INVERTEBRATES

	Class & Animal (Synonym)	Effects	
		Oxygen Pressure Lower than Normal	Oxygen Pressure Higher than Normal
14	*Tinca*	Increases frequency & amplitude of breathing [30]; reportedly survives indefinitely in water with only 0.3 ml O_2/liter [28]. As Po_2 decreases, respiratory neuron activity increases [39].	....
15	Eel	....	At 2.0 atm, survives for 72 hr; at 4.1 atm, convulsions after 22 hr; at 5 atm, dies in <20 hr; at 6.5 atm, dies in >27 hr; at 15 atm, dies [2]
16	Loach	....	Respiratory movements very slow after 4-5 hr [35]
17	Pike	....	Similar to effect on loach (entry 16); at 25 atm, some survive 5-15 hr with no convulsions [35]
18	Chondrichthyes *Squalus*	Bradycardia; initial slowing, late acceleration of breathing [31]	....
19	Agnatha Cyclostomata[3]/	....	At 7 atm, survives 31.5 hr; swimming movements affected after 12 hr [35]
		Echinodermata	
20	Holothuroidea *Holothuria*	Pumping rate increases as Po_2 decreases; at 60-70% of air saturation, pumping ceases [24]	....
		Arthropoda	
21	Arachnida	....	At 11 atm, quiescent, no convulsions after 5 hr [2]
22	Crustacea	Most species show increased scaphognathite & pleopod movement [20,28]	....
23	*Callianassa californiensis*	Hyperventilation with low Po_2 [13]	....
24	*Ligia*	No specific movement response to low Po_2 [15]	Slow pleopod movement as Po_2 increases [15]
25	*Procambarus*	Slight increase in ventilation [21]	....
26	Insecta	Low Po_2 may excite, depress, or have no effect [28]	....
27	*Drosophila*	....	Maintenance of balance was first function permanently lost; maximal lethal effects after 2 hr at 10 atm [36]
28	*Musca*	O_2 concentration of <2% induces spiracle opening [5]	....
29	*Sphinx ligustri*	....	At 7.5 atm for 9-12.5 hr, apparent destruction of anterior "higher" centers and death of larva [35]
30	*Stilpnotia salicis*	....	At 7.5 atm for 9-12.5 hr, apparent destruction of anterior "higher" centers and death of larva [35]
31	Ant; bee; butterfly; cockroach; dragonfly; flea; fly; termite	....	At 11 atm for 0.5 hr, fly relatively sensitive to Po_2, others quiescent with no evidence of convulsions after 5 hr [2]

[3]/ Subclass.

continued

167. RESPONSE OF THE NERVOUS SYSTEM TO INCREASED AND DECREASED OXYGEN LEVEL: POIKILOTHERMIC VERTEBRATES AND INVERTEBRATES

	Class & Animal (Synonym)	Effects	
		Oxygen Pressure Lower than Normal	Oxygen Pressure Higher than Normal
32	Chilopoda; Diplopoda	....	At 11 atm, quiescent; no convulsions after 5 hr [2]
		Annelida	
33	Hirudinea *Haemopis*	....	At 7.5 atm for 9 hr, evidence of damage to anterior ganglia [35]
34	Oligochaeta *Limnodrilus*	Paralyzed with $Po_2 < 1.5$ mm Hg (this asphyxia is reversible) [4,27]	....
35	*Lumbricus*	Migrated toward higher Po_2 [22]	At 7.5 atm for 9 hr, damage to anterior ganglia [35]
36	*Tubifex*	Maintains self anaerobically for 60 hr [32]; tail undulates toward higher Po_2 [28]	....
		Mollusca	
37	Cephalopoda	Respiratory movements probably stimulated by low Po_2 [28,32,38]	....
38	*Aplysia; Helix*	Neuron membrane depolarization with decrease in intracellular Po_2 [6,7]; greatest effect when Na^+ injected into neuron [19]	Hyperpolarization with increased Po_2 [6,7]
39	Freshwater pulmonates	Mean interval breathing periods 3 times longer in water with 6.4 ml O_2/liter than in water with 1.7 ml O_2/liter [8]. Breathing rhythm appears to be a function of O_2 needs [4].	....

Contributor: Smith, Charles W.

References

[1] Bean, J. W. 1945. Physiol. Rev. 25:1.

[2] Bert, P. 1943. Barometric Pressure. College Book, Columbus, O. pp. 709-779, 1009-1035.

[3] Benet, L., and M. Bochet. 1963. J. Physiol. (Paris) 55:405.

[4] Bullard, R. W. 1964. Handb. Physiol., Sect. 4, Adapt. Environ., p. 683.

[5] Case, J. F. 1956. Physiol. Zool. 29:163.

[6] Chalazonitis, N. 1969. J. Physiol. (London) 202:2P.

[7] Chalazonitis, N., et al. 1965. C. R. Soc. Biol. 159(12):2451.

[8] Cheatum, E. P. 1934. Trans. Amer. Microsc. Soc. 53:348.

[9] Cleveland, L. R. 1925. Biol. Bull. 48:455.

[10] Crozier, W. J., and T. B. Stier. 1925. J. Gen. Physiol. 7:699.

[11] Cymerman, A., and S. F. Gottlieb. 1970. Aerosp. Med. 41:36.

[12] Dolk, H. E., and N. Postma. 1927. J. Gen. Physiol. 5:417.

[13] Farley, R. D., and J. F. Case. 1968. Biol. Bull. 134:261.

[14] Faulkner, J. M., and C. A. L. Binger. 1927. J. Exp. Med. 45:865.

[15] Fox, H. M., and M. L. Johnson. 1934. J. Exp. Biol. 11:1.

[16] Jesse, M. J., et al. 1967. Resp. Physiol. 3:267.

[17] Johansen, K., and C. Lenfant. 1968. J. Exp. Biol. 49:453.

[18] Johansen, K., et al. 1967. Comp. Biochem. Physiol. 20:835.

[19] Kerkut, G. A., and B. York. 1969. Ibid. 28:1125.

[20] Kinne, O. 1964. Handb. Physiol., Sect. 4, Adapt. Environ., p. 669.

[21] Larimer, J. L. 1961. Physiol. Zool. 34:158.

[22] Ledebur, J. F. 1939. Ergeb. Biol. 16:173.

[23] Ledebur, J. F. 1939. Ibid. 16:262.

[24] Newell, R. C., and W. A. M. Courtney. 1965. J. Exp. Biol. 42:45.

[25] Nielsen, B. 1962. Ibid. 39:107.

[26] Pflüger, E. 1875. Pfluegers Arch. Gesamte Physiol. Menschen Tiere 10:251.

[27] Prosser, C. L. 1955. Biol. Rev. Cambridge Phil. Soc. 30:229.

continued

[28] Prosser, C. L., and F. A. Brown, Jr. 1961. Comparative Animal Physiology. Ed. 2. W. B. Saunders, Philadelphia. pp. 153-237.

[29] Rahn, H., et al. 1971. Resp. Physiol. 11:285.

[30] Randall, D. J., and G. Shelton. 1963. Comp. Biochem. Physiol. 9:229.

[31] Satchell, G. H. 1961. J. Exp. Biol. 38:531.

[32] Scheer, B. T. 1948. Comparative Physiology. J. Wiley, New York. pp. 179, 239.

[33] Smith, C. W. Unpublished. Ohio State Univ., Dep. Physiology, Columbus, 1972.

[34] Thompson, G. E. 1889. Med. Rec. 36:1.

[35] Viono-Yasenetskii, A. V. 1960. Clear. House Fed. Sci. Tech. Inform. (Springfield) TT-60-51068.

[36] Williams, C. M., and H. K. Beecher. 1944. Amer. J. Physiol. 140:566.

[37] Winterstein, H. 1908. Pfluegers Arch. Gesamte Physiol. Menschen Tiere 125:73.

[38] Winterstein, H. 1925. Z. Vergl. Physiol. 2:315.

[39] Young, S. 1969. J. Physiol. (London) 200:85P.

APPENDIXES

Appendix I.   SCIENTIFIC NAMES AND CORRESPONDING COMMON NAMES

Common names are not available for some organisms in this book. In such cases, the name of the general category to which the organism has been assigned, modified by its geographic distribution (if known), has been included to assist the user unacquainted with the scientific name. Protozoa and nonvascular plants have not been included.

Part I.   Animals

Scientific Name	Common Name
Abacion magnum	Millipede
Abacomorphus asperulus	Ground beetle
Abax ater	Ground beetle
A. ovalis	Ground beetle
A. parallelus	Ground beetle
Abbreviata caucasica	Parasitic nematode worm
Acanthocephala declivis	Leaf-footed bug
A. femorata	Leaf-footed bug
A. granulosa	Leaf-footed bug
Acanthocoris sordidus	Bed bug
Acanthomyops claviger	Smaller yellow ant
Acanthophis antarcticus	Death adder
Acanthoscelides obtectus	Bean weevil
Acanthurus glaucopareius	Surgeonfish
A. triostegus	Indopacific convict fish
Acilius sulcatus	Predacious diving beetle
Acinopus sp.	Ground beetle
Acipenser sp.	Sturgeon
Acris crepitans	Cricket frog
Actinia equina	Sea anemone
Actinopyga agassizi	Sea cucumber
Adamsia palliata	Cloak anemone
Aedes aegypti	Yellow-fever mosquito
A. albopictus	Mosquito
A. communis	Mosquito
A. hexodontus	Mosquito
A. leucocelaenus	Mosquito
A. polynesiensis	Mosquito
A. triseriatus	Mosquito
Aelia fieberi	Stink bug
Aeronautes saxatilis	White-throated swift
Aetobatus narinari	Spotted eagle ray
Aetomylaeus nichofii	Eagle ray
Agabus bipustulatus	Predacious diving beetle
A. sturmi	Predacious diving beetle
Agelaius phoeniceus	Red-winged blackbird
A. tricolor	Tricolored blackbird
Agelena labyrinthica	Funnel-web spider
A. similis	Funnel-web spider
Agkistrodon contortrix	Copperhead, or dryland moccasin
A. halys	Old World pit viper, or mamushi
A. piscivorus	Cottonmouth, or water moccasin
Agonum assimilis	Ground beetle
A. dorsalis	Ground beetle
A. marginatum	Ground beetle
A. moestum	Ground beetle
A. sexpunctatum	Ground beetle
A. viduum	Ground beetle
Agriopocoris frogatti	Bed bug
Agriotes obscurus	Click beetle

Scientific Name	Common Name
Alabama argillacea	Cotton leafworm
Albula vulpes	Bonefish
Alligator mississippiensis	American alligator
Alopex lagopus	Arctic fox
Alphitobius diaperinus	Lesser mealworm
Altica sp.	Flea beetle
Aluterus scriptus	Scrawled filefish
Amara familiaris	Ground beetle
A. similata	Ground beetle
Amauris lobengula	Milkweed butterfly
Amblyomma americanum	Lone-star tick
Amblytelus curtus	Ground beetle
Ambystoma maculatum	Spotted salamander
A. mexicanum	Axolotl
A. tigrinum	Tiger salamander
Amitermes herbertensis	Desert termite
A. laurensis	Desert termite
A. vitiosus	Desert termite
Ammospermophilus leucurus	White-tailed antelope squirrel
Amorbus alternatus	Bed bug
A. rhombifer	Bed bug
A. rubiginosus	Bed bug
Amphibolips confluenta	Gall wasp
Amphibolurus barbatus	Australian bearded lizard
A. ornatus	Ornate lizard
Amphiuma means	Amphiuma
A. tridactylum	Three-toed amphiuma
Anabrus simplex	Mormon cricket
Anagasta kuehniella	Mediterranean flour moth
Anas acuta	Common pintail duck
A. platyrhynchos	Mallard duck
A. platyrhynchos domesticus	Pekin or domestic duck
Anasa tristis	Squash bug
Anastatus bifasciatus	Chalcid wasp
Anax imperator	European dragonfly
A. junius	Green darner
Ancylostoma braziliense	Creeping-eruption hookworm
A. caninum	Dog hookworm
A. duodenale	Old World hookworm
A. tubaeforme	Hookworm
Andrena sp.	Bee
Anemonia sulcata	Sea anemone
Anguina tritici	Wheat nematode
Anguis fragilis	Slowworm
Anisodactylus binotatus	Ground beetle
Anisodoris nobilis	Dorid sea slug
Anisomorpha buprestoides	Walking stick
Anisotarsus sp.	Beetle
Anniella pulchra	California legless lizard
Anolis carolinensis	Green anole lizard
A. homolechis	Cuban anole lizard

continued

1309

Part I. Animals

Scientific Name	Common Name	Scientific Name	Common Name
Anopheles funestus	Mosquito	*A. colombica*	Leaf-cutting ant
A. gambiae	Mosquito	*A. laevigata*	Leaf-cutting ant
A. pseudopunctipennis	Mosquito	*A. robusta*	Leaf-cutting ant
A. quadrimaculatus	Common malaria mosquito	*A. sexdens*	Leaf-cutting ant
Antheraea pernyi	Giant silkworm moth	*A. texana*	Texas leaf-cutting ant
Anthia venator	Domino beetle	*Attagenus megatoma*	Skin beetle, or dermestid
Anthonomus grandis	Boll weevil	*Aulacosternum nigrorubrum*	Bed bug
Anthrenus verbasci	Varied carpet beetle	*Aurelia aurita*	Moon jellyfish
Antrozous pallidus	Pallid bat	*Azteca* sp.	Ant
Apamea monoglypha	Cutworm		
Apanteles glomeratus	Little braconid		
Aphaenogaster longiceps	Ant	*Badister bipustulatus*	Ground beetle
Aphelenchoides besseyi	Summer-dwarf nematode of strawberry	*Balanus nubilis*	Giant barnacle
		Balistoides conspicillum	Spotted triggerfish
A. fragariae	Spring-dwarf nematode of strawberry	*Baronia brevicornis*	Swallowtail butterfly
		Battus philenor	Pipevine swallowtail butterfly
Aphelinus semiflavus	Chalcid fly		
Apheloria corrugata	Flat-backed millipede	*B. polydamus*	Swallowtail butterfly
Aphis nerii	Aphid	*Bdellonyssus sylviarum*	Northern fowl mite
A. pomi	Apple aphid	*Belonolaimus longicaudatus*	Sting nematode
Aphodius sp.	Dung-feeding scarab beetle	*Belonopterus chilensis lampronotus*	Southern lapwing
Aphonopelma californica	Trap-door spider		
Apis cerana indica	Honey bee	*Bembidion andreae*	Ground beetle
A. dorsata	Southeast Asian honey bee	*B. lampros*	Ground beetle
A. florea	Southeast Asian honey bee	*B. quadriguttatum*	Ground beetle
A. indica	Southeast Asian honey bee	*Beroe cucumis*	Comb jelly
A. mellifera	Honey bee	*Bertiella studeri*	Traemioid tapeworm
Aplysia californica	California sea hare	*Biprorulus bibax*	Stink bug
Apodemus flavicollis	European wood mouse	*Bitis arietans*	Puff adder
Aprion virescens	Blue-gray snapper	*B. gabonica*	Gabon viper
Apus apus	Swift	*Blaberus giganteus*	Cockroach
Arbacia punctulata	Common Atlantic sea urchin	*Blaniulus guttulatus*	Juliform millipede
Archips cerasivoranus	Uglynest caterpillar	*Blaps gigas*	Darkling beetle
A. fervidanus	Oak webworm	*B. lethifera*	Darkling beetle
Archiulus sabulosus	Juliform millipede	*B. mortisaga*	Darkling beetle
Arctia caja	Tiger moth	*B. mucronata*	Darkling beetle
A. villica	Tiger moth	*B. requienii*	Darkling beetle
Arctosa cinerea	Wolf spider	*B. sulcata*	Darkling beetle
A. perita	Wolf spider	*B. wiedemanni*	Darkling beetle
A. variana	Wolf spider	*Blatta orientalis*	Oriental cockroach
Arenicola sp.	Lugworm	*Blattella germanica*	German cockroach
Argas persicus	Fowl tick	*Blissus leucopterus*	Chinch bug
Argyroploce leucotreta	False codling moth	*Bohadschia argus*	Sea cucumber
Argyrotaenia velutinana	Red-banded leafroller	*Bombus agrorum*	European bumble bee
Armigeres sp.	Mosquito	*B. hypnorum*	European bumble bee
Aromia moschata	Long-horned beetle	*B. lapidarius*	European bumble bee
Arothron hispidus	Puffer	*B. sylvarum*	Bumble bee
A. meleagris	White-spotted puffer	*B. terrestris*	European bumble bee
A. nigropunctatus	Black-spotted puffer	*Bombyx mori*	Silkworm moth
Arrenurus marshallae	Water mite	*Bonasa umbellus*	Ruffed grouse
A. megalurus	Water mite	*Boophilus annulatus*	Cattle tick
Artemia salina	Brine shrimp	*Bos indicus*	Brahman cattle
Arthropterus sp.	Ground beetle	*B. taurus*	Cattle
Asaphidion flavipes	Ground beetle	*Bothrops alternatus*	Urutú, or víbora de la cruz
Ascaridia galli	Large roundworm of chicken	*B. atrox*	Fer-de-lance, or barba amarilla
Ascaris lumbricoides	Large roundworm		
A. suum	Large roundworm of swine	*B. jararaca*	Jararaca
Aspiculuris tetraptera	Pinworm	*B. nummifer*	Jumping viper, or mano de piedra
Astacus astacus	Crayfish		
Asterias forbesi	Starfish	*Brachinus crepitans*	Bombardier beetle
Asthenosoma varium	Leather sea urchin	*B. explodens*	Bombardier beetle
Atta bisphaerica	Leaf-cutting ant	*B. sclopeta*	Bombardier beetle
A. capiguara	Leaf-cutting ant		

continued

Part I. Animals

Scientific Name	Common Name	Scientific Name	Common Name
Brachyiulus ulineatus	Juliform millipede	*Cambala hubrichti*	Juliform millipede
Bracon brevicornis	Braconid wasp	*Cambarus virilis*	Cave crayfish
B. juglandis	Braconid wasp	*Camnula pellucida*	Clear-winged grasshopper
Branchiostoma caribaeum	Lancelet	*Camponotus ligniperda*	European carpenter ant
Brochymena quadripustu-	Stink bug	*C. maculatus*	Carpenter ant
lata		*Cancer* sp.	Crab
Broscus cephalotes	Ground beetle	*Canis familiaris*	Dog
Bruchus pisorum	Pea weevil	*Canthigaster margaritatus*	Sharp-nosed puffer
Brugia malayi	Malayan filarial worm	*C. rivulatus*	Sharp-nosed puffer
Bryotropha similis	Moth	*Capillaria hepatica*	Capillary nematode worm
Bufo alvarius	Colorado river toad	*Capra hircus*	Goat
B. americanus	American toad	*Carabus auratus*	Ground beetle
B. arenarum	Argentine toad	*C. granulatus*	Ground beetle
B. asper	Asian rough toad	*C. problematicus*	Ground beetle
B. blombergi	Blomberg's toad	*Caranx hippos*	Jack
B. bufo	Common European toad	*Carassius auratus*	Goldfish
B. bufo bufo	Common European toad	*Carausius morosus*	Stick insect
B. bufo formosus	Formosan toad	*Carcharhinus* sp.	Requiem shark
B. bufo gargarizans	Cantor's toad	*Carcinus maenas*	Green crab
B. granulosus fernandeze	Gallardo's toad	*Cardisoma guanhumi*	West Indies land crab
B. marinus	Marine toad	*Carduelis carduelis*	European goldfinch
B. mauritanicus	North African toad	*C. chloris*	Greenfinch
B. melanostictus	Southeast Asian black-spined	*C. spinus*	European siskin
	toad	*Carenum bonelli*	Ground beetle
B. peltocephalus	Cuban toad	*C. interruptum*	Ground beetle
B. quercicus	Oak toad	*C. tinctillatum*	Ground beetle
B. regularis	African leopard toad	*Carpocapsa pomonella*	Codling moth
B. spinulosus	South American spiny toad	*Carpocoris purpureipennis*	Stink bug
B. valliceps	Gulf Coast toad	*Carpodacus mexicanus*	House finch
B. viridis viridis	Green toad	*Cassiopea frondosa*	Upside down jellyfish
B. woodhousei fowleri	Fowler's toad	*Castelnaudia superba*	Ground beetle
Bungarus caeruleus	Indian krait	*Castor canadensis*	North American beaver
Bunostomum trigonoceph-	Sheep hookworm	*Cataglyphis bicolor*	Ant
alum		*Catocala* sp.	Underwing moth
Byrsotria fumigata	West Indies cockroach	*Causus rhombeatus*	Common night adder
		Cavernularia obesa	Sea pen
		Cavia porcellus	Guinea pig
Cadra cautella	Almond moth	*Centropogon australis*	Waspfish
Caiman sclerops	Spectacled caiman	*Cephalopholis argus*	Blue-spotted argus
Calathus fuscipes	Ground beetle	*Cephus cinctus*	Wheat stem sawfly
C. melanocephalus	Ground beetle	*C. pygmaeus*	European wheat stem sawfly
Calendra maidis	Maize billbug	*Ceratitis capitata*	Mediterranean fruit fly
Callianassa californiensis	Burrowing shrimp	*Ceratocoris cephalicus*	European stink bug
Callinectes sapidus	Blue crab	*Ceratophyllus fasciatus*	European rat flea
Calliphora erythrocephala	Bluebottle blow fly	*Cercaertus nanus*	Dormouse possum
C. vomitoria	Blow fly	*Cerura vinula*	Prominent moth
Callistus lunatus	Ground beetle	*Cervus nippon*	Japanese deer, or sika
Callorhinus sp.	Fur seal	*Chalcodermus aeneus*	Cowpea curculio
Calloselasma rhodostoma	Malayan pit viper	*Chamaeleo jacksoni*	Jackson's chameleon
Calosoma affini	Caterpillar hunter beetle	*C. melleri*	Meller's chameleon
C. alternans sayi	Caterpillar hunter beetle	*Chaoborus* sp.	Phantom midge
C. externum	Caterpillar hunter beetle	*Chauliognathus lecontei*	Soldier beetle
C. macrum	Caterpillar hunter beetle	*Chelydra serpentina*	Common snapping turtle
C. marginalis	Caterpillar hunter beetle	*Cherokia georgiana*	Flat-backed millipede
C. oceanicum	Caterpillar hunter beetle	*Cheyletiella parasitivorax*	Rabbit mite
C. parvicollis	Caterpillar hunter beetle	*Chicobolus spinigerus*	Millipede
C. perigrinater	Caterpillar hunter beetle	*Chilomycterus spinosus*	Spiny boxfish
C. prominens	Caterpillar hunter beetle	*Chimaera* sp.	Rabbit-fish
C. schayeri	Caterpillar hunter beetle	*Chionaspis furfura*	Scurfy scale
C. scrutator	Caterpillar hunter beetle	*Chirodiscoides caviae*	Fur mite of guinea pig
C. sycophanta	Caterpillar hunter beetle	*Chironex fleckeri*	Sea wasp
Calypte anna	Anna's hummingbird	*Chiropsalmus quadrigatus*	Sea wasp

continued

Scientific Name	Common Name
Chlaenius australis	Ground beetle
C. bipunctatus	Ground beetle
C. chrysocephalus	Ground beetle
C. cordicollis	Ground beetle
C. festivus	Ground beetle
C. tristus	Ground beetle
C. vestitus	Ground beetle
Chloeia flava	Bristle worm
Chloris chloris	Greenfinch
Chordeiles minor	Nighthawk
Choridactylus multibarbis	Scorpionfish
Choristoneura fumiferana	Spruce budworm
C. rosaceana	Spruce budworm
Chrysaora hysoscella	Sea nettle
Chrysemys picta	Painted turtle
C. picta belli	Western painted turtle
C. scripta elegans	Red-eared turtle
Chrysobothris femorata	Flatheaded apple tree borer
Chrysomya bezziana	Blow fly
C. chloropyga	Blow fly
Chrysops sp.	Deer fly
Cimex lectularius	Common bed bug
Circulifer tenellus	Beet leafhopper
Cladius isomerus	Bristly rose-slug
Clemmys guttata	Spotted turtle
Clethrionomys glareolus	Bank vole
C. rufocanus	Gray-sided vole
C. rutilus	Ruddy vole
Clivina basalis	Ground beetle
C. fossor	Ground beetle
Clonorchis sinensis	Chinese liver fluke
Clupanodon thrissa	Shad
Clupea harengus	Herring
Cnemidophorus tigris	Western whiptail
Coccus hesperidum	Brown soft scale
Cochliomyia hominivorax	Screwworm fly
C. macellaria	Secondary screwworm fly
Colinus virginianus	Bobwhite quail
Colius striatus	Mouse bird
Colomesus psittacus	Freshwater puffer
Coluber constrictor	Black snake
Columba livia	Street pigeon
C. oenas	Pigeon
C. palumbus	Wood pigeon
Colymbetes fuscus	Predacious diving beetle
Condylura cristata	Star-nosed mole
Conger conger	European conger eel
Conotrachelus nenuphar	Plum curculio
Conus aulicus	Court cone
C. californicus	California cone
C. catus	Cat cone
C. geographus	Geographer cone
C. imperialis	Imperial cone
C. lividus	Bluish cone
C. magus	Indopacific oak cone
C. marmoreus	Marbled cone
C. quercinus	Indopacific cone
C. spurius	Alphabet cone
C. striatus	Striated cone
C. textile	Woven cone
C. tulipa	Tulip cone
Copelatus ruficollis	Predacious diving beetle

Scientific Name	Common Name
Coregonus clupeaformis	Lake whitefish
Corixa dentipes	Water boatman
C. punctata	Water boatman
Corvus brachyrhynchos	American common crow
C. cryptoleucus	White-necked raven
Corythuca arcuata	Oak lace bug
Costelytra zealandica	Beetle
Cotinis sp.	June beetle
Coturnix coturnix	European quail
Craspedophorus sp.	Ground beetle
Crassostrea virginica	Virginia oyster
Cratoferonia phylarchus	Ground beetle
Cratogaster melas	Ground beetle
Crematogaster africana	Ant
C. peringueyi	Ant
Cricetus cricetus	European hamster
Criconemoides sp.	Ring nematode
Crinia signifera	South Australian leptodactylid frog
Crotalus adamanteus	Eastern diamondback rattlesnake
C. atrox	Western diamondback rattlesnake
C. cerastes	Sidewinder
C. durissus durissus	Tropical rattlesnake, or cascabel
C. durissus terrificus	Tropical rattlesnake, or cascabel
C. scutulatus	Mohave rattlesnake
C. viridis	Prairie rattlesnake
Crotaphytus collaris	Collared lizard
Cryptolestes ferrugineus	Rusty grain beetle
C. pusillus	Flat grain beetle
Cryptomphallus sp.	Argentine land snail
Ctenocephalides canis	Dog flea
C. felis	Cat flea
Cuclotogaster heterographus	Chicken head louse
Cucumaria echinata	Sea cucumber
Culex pipiens	Northern house mosquito
C. pipiens quinquefasciatus	Southern house mosquito
C. tarsalis	Mosquito
Culicoides grahamii	Punkie
C. impunctatus	Punkie
Cyanea capillata	Giant jellyfish
Cybister confusus	Predacious diving beetle
C. lateralimarginalis	Predacious diving beetle
C. limbatus	Predacious diving beetle
C. tripunctatus	Predacious diving beetle
Cychrus rostratus	Ground beetle
Cyclops serulatus	Cyclopoid copepod
C. vernalis	Cyclopoid copepod
C. viridis	Cyclopoid copepod
Cycnia tenera	Tiger moth
Cylas formicarius elegantulus	Sweet potato weevil
Cylindroiulus teutonicus	Juliform millipede
Cyprinus carpio	Carp
Dactylometra quinquecirrha	Pink-fringed jellyfish
Danaus chrysippus	Milkweed butterfly
D. gilippus berenice	Milkweed butterfly
D. gilippus strigosus	Milkweed butterfly

continued

Scientific Name	Common Name	Scientific Name	Common Name
D. plexippus	Monarch butterfly	*Dipsosaurus dorsalis*	Desert iguana
Daphnia magna	Waterflea	*Dipylidium caninum*	Double-pored dog tapeworm
D. pulex	Waterflea	*Dirofilaria immitis*	Dog heart worm
D. schoedleri	Waterflea	*Dispholidus typus*	Boomslang
Dasyatis brevis	Stingray	*Dissosteira longipennis*	High plains grasshopper
D. dipterura	Diamond stingray	*Ditylenchus destructor*	Potato-rot nematode
D. guttata	Stingray	*D. dipsaci*	Bulb and stem nematode
D. longus	Stingray	*Dolichoderus clarki*	Ant
D. pastinaca	Stingray	*D. dentata*	Ant
D. sabina	Roughtail stingray	*D. scabridus*	Ant
D. violacea	Violet stingray	*Dolichodorus heterocepha-lus*	Awl nematode
Dasypus novemcinctus	Nine-banded armadillo		
Demansia nuchalis affinis	Dugite snake	*Dolichonyx oryzivorus*	Bobolink
D. nuchalis nuchalis	Western brown snake	*Dolycoris baccarum*	Stink bug
D. textilis	Eastern or common brown snake	*Doratogonus annulipes*	Millipede
		Dorymyrmex bicolor	Ant
Demodex canis	Dog follicle mite	*D. pyramicus*	Pyramid ant
Dendroaspis angusticeps	Eastern green mamba	*D. pyramicus flavopectus*	Pyramid ant
D. polylepis	Black mamba	*Dosidicus gigas*	Squid
Dendrobates auratus	Golden arrow-poison frog	*Dracunculus medinensis*	Guinea worm
Dendrocoelum sp.	Triclad	*Drepanotermes rubriceps*	Termite
Dendroctonus brevicomis	Western pine beetle	*Drosophila deflexa*	Fruit fly
D. frontalis	Southern pine beetle	*D. funebris*	Fruit fly
D. ponderosae	Mountain pine beetle	*D. melanogaster*	Fruit fly
D. valens	Red turpentine beetle	*D. pseudoobscura*	Fruit fly
Denisonia superba	Australian copperhead	*D. repleta*	Fruit fly
Dermacentor andersoni	Rocky Mountain wood tick	*D. subobscura*	Fruit fly
D. variabilis	American dog tick	*Drymaplaneta communis*	Cockroach
Dermanyssus gallinae	Chicken mite	*D. semivitta*	Cockroach
Dermatobia hominis	Human bot fly	*D. shelfordi*	Cockroach
Dermestes maculatus	Hide beetle	*Drypta dentata*	Ground beetle
Dermogenys pusillus	Wrestling halfbeak	*Dugesia gonocephala*	Freshwater planarian
Desmognathus fuscus	Dusky salamander	*Dysdercus intermedius*	Red bug
Desmosteria scripta	Cockroach	*Dytiscus latissimus*	Predacious diving beetle
Diabrotica duodecimpunc-tata	Twelve-spotted cucumber beetle	*D. marginalis*	Diving beetle
D. undecimpunctata	Western spotted cucumber beetle	*Echidnophaga gallinacea*	Sticktight flea
		Echinococcus granulosus	Hydatid tapeworm
D. vittata	Striped cucumber beetle	*E. multilocularis*	Multilocular hydatid tape-worm
Diachromus germanus	Ground beetle		
Diacrisia mendica	Tiger moth	*Echinothrix calamaris*	Sea urchin
Diadema setosum	Black sea urchin	*E. diadema*	Sea urchin
Diaperis boleti	Darkling beetle	*Echis carinatus*	Sawscaled or carpet viper
D. maculata	Darkling beetle	*E. coloratus*	Arabian sawscaled or Burton's viper
Diaphoromenus edwardsi	Ground beetle		
Diaptomus shoshone	Freshwater copepod	*Egernia stokesii*	Stokes' skink
Dicaelus dilatatus	Ground beetle	*Eisenia* sp.	Earthworm
D. purpuratus	Ground beetle	*Elaphrus ripareus*	Ground beetle
D. splendidus	Ground beetle	*Electrophorus electricus*	Electric eel
Dicentrarchus labrax	European bass	*Eledone aldrovandi*	Octopus
Dichirotrichus obsoletus	Ground beetle	*E. moschata*	Mediterranean musk octopus
Dicrochile brevicollis	Ground beetle	*Eleodes hispilabris*	Darkling beetle
D. goryi	Ground beetle	*E. longicollis*	Darkling beetle
Dicrostonyx groenlandicus	Collared lemming	*Empoasca fabae*	Potato leafhopper
Dictyocaulus filaria	Thread lungworm of sheep	*Emys orbicularis*	European pond turtle
D. viviparus	Lungworm of cattle	*Endria inimica*	Painted leafhopper
Didelphis marsupialis	Large American opossum	*Engraulis japonicus*	Anchovy
D. marsupialis virginiana	Common opossum	*Enhydrina schistosa*	Beaked or Malayan sea snake
Diodon holocanthus	Balloonfish	*Enterobius vermicularis*	Pinworm
Diphyllobothrium latum	Fish tapeworm	*Ephestia elutella*	Tobacco moth
Diploptera punctata	Pacific beetle cockroach	*Epibolus insidiator*	Indopacific long-jawed wrasse
Diprion hercyniae	European spruce sawfly		

continued

Part I.   Animals

Scientific Name	Common Name
Epicauta adspersa	Blister beetle
E. gorhami	Blister beetle
E. pestifera	Margined blister beetle
E. pilma	Blister beetle
E. ruficeps	Blister beetle
E. vittata	Striped blister beetle
E. waterhousei	Blister beetle
Epilachna varivestis	Mexican bean beetle
Epinephelus fuscoguttatus	Mottled grouper
Epitrix cucumeris	Potato flea beetle
E. hirtipennis	Tobacco flea beetle
Eptesicus fuscus	Big brown bat
Equus asinus	Donkey
E. caballus	Horse
Erannis tiliaria	Linden looper
Erinaceus europaeus	European hedgehog
E. europaeus roumanicus	Roumanian hedgehog
Eriophyes pyri	Pear leaf blister mite
Eriosoma lanigerum	Woolly apple aphid
Erithacus rubecula	European robin
Esox lucius	Northern pike
Eudalia macleayi	Ground beetle
Eudendrium sp.	Hydroid
Eumeces obsoletus	Great Plains skink
Eunice aphroditois	Bristle worm
E. schemacephala	West Indian palolo worm
Euphausia pacifica	Euphausiacean krill
Euprymna morsei	Western Pacific squid
Eurostopodus guttatus	Spotted nightjar
Eurycotis biolleyi	Cockroach
E. decipiens	Cockroach
E. floridana	Florida cockroach
Eurydema ornatum	European ornate vegetable bug
Eurygaster integriceps	Stink bug
Eurylychnus blagravei	Ground beetle
E. olliffi	Ground beetle
Eurythoe brasiliensis	Bristle worm
Euschistus servus	Brown stink bug
Eutamias umbrinus	Uinta chipmunk
Euthynnus pelamis	Skipjack tuna
Eutrombicula alfreddugesi	Chigger
Euzosteria nobilis	Cockroach
Fasciola hepatica	Liver fluke
Fasciolopsis buski	Large intestinal fluke
Felis catus	Cat
Fimbria fimbria	Sea hare
Floridobolus penneri	Millipede
Forelius foetidus	Ant
Forficula auricularia	European earwig
Formica cinerea	Ant
F. fusca	Silky ant
F. polyctena	Ant
F. rufa	Red ant
F. sanguinea	Ant
Fringilla coelebs	Chaffinch
F. montifringilla	Bramble finch
Fugu basilevskianus	Puffer
F. chrysops	Puffer
F. niphobles	Puffer

Scientific Name	Common Name
F. ocellatus	Puffer
F. pardalis	Puffer
F. pseudommus	Puffer
F. rubripes	Puffer
F. stictonotus	Puffer
F. vermicularis	Puffer
F. xanthopterus	Puffer
Fundulus heteroclitus	Mummichog
Gadus morhua	Atlantic cod
Galba sp.	Freshwater snail
Galleria mellonella	Greater wax moth
Gallus gallus	Chicken
Galumna sp.	Grass mite
Gasteropelecus sp.	Hatchetfish
Gasterophilus haemorrhoidalis	Horse bot fly
G. intestinalis	Horse bot fly
G. nasalis	Horse bot fly
Gasterosteus aculeatus	Three-spined stickleback
Gastrodiscoides hominis	Amphistome fluke
Gehyra variegata	Gecko
Geotrupes stercorosus	Common European dung beetle
Gerrhonotus multicarinatus	Southern alligator lizard
Gerris lacustris	Water strider
Ginglymostoma cirratum	Nurse shark
Glaucomys volans	Southern flying squirrel
Gliricola porcelli	Slender guinea pig louse
Glis glis	Fat dormouse
Glomeris marginata	European pill millipede
Glossina morsitans	Tsetse fly
Glycera dibranchiata	Blood worm
G. ovigera	Blood worm
Gnaptor spinimanus	Darkling beetle
Gnathodentex aureolineatus	Snapper
Gomphodesmus pavani	Flat-backed millipede
Goniobasis sp.	Freshwater snail
Goniopsis cruentata	Mangrove crab
Gopherus agassizii	Desert tortoise
Graphidium strigosum	Nematode of rabbit
Graphium marcellus	Zebra swallowtail butterfly
G. sarpedon choredon	Swallowtail butterfly
Graphoderus cinereus	Predacious diving beetle
Grapholitha molesta	Oriental fruit moth
Graphosoma rubrolineatum	Stink bug
Gryllus campestris	Field cricket
Gymnothorax flavimarginatus	Moray eel
G. javanicus	Moray eel
G. meleagris	White-spotted moray eel
G. pictus	Speckled moray eel
G. undulatus	Brown moray eel
Gymnura marmorata	Butterfly stingray
Gyropus ovalis	Oval guinea pig louse
Haemagogus spegazzinii	Mosquito
Haematobia irritans	Horn fly
Haematopinus eurysternus	Shortnosed cattle louse
H. suis	Hog louse

continued

Part I. Animals

Scientific Name	Common Name	Scientific Name	Common Name
Haemodipsus ventricosus	Rabbit louse	*Hybomitra* sp.	Horse fly
Haemonchus contortus	Twisted stomach worm	*Hydrachna cruenta*	Water mite
H. placei	Large stomach worm of cattle	*Hydroporus pallustris*	Predacious diving beetle
		Hygia opaca	Bed bug
Haemopis sp.	Leech	*Hyla arborea*	European green tree frog
Haliclona viridis	Green sponge	*H. caerulea*	White's tree frog
Halictus sp.	Bee	*H. crucifer*	Spring peeper
Harmolita grandis	Wheat strawworm	*H. regilla*	Pacific tree frog
H. tritici	Wheat jointworm	*Hylemya antiqua*	Onion maggot
Harpalus atratus	Ground beetle	*Hylocichla guttata*	Hermit thrush
H. azurus	Ground beetle	*Hylotrupes bajulus*	Oldhouse borer
H. caliginosus	Ground beetle	*Hylurgopinus rufipes*	Native elm bark beetle
H. dimidiatus	Ground beetle	*Hymenolepis carioca*	Thread tapeworm
H. distinguendus	Ground beetle	*H. diminuta*	Rat tapeworm
H. griseus	Ground beetle	*H. nana*	Dwarf tapeworm
H. luteicornis	Ground beetle	*Hyocephalus* sp.	Hyocephalid bug
H. pubescens	Ground beetle	*Hyphantria cunea*	Fall webworm
H. tardus	Ground beetle	*Hypoderma bovis*	Cattle grub
Harpogoxenus sublaevis	Ant	*H. lineatum*	Common cattle grub
Helicotylenchus sp.	Spiral nematode		
Heliothis virescens	Tobacco budworm		
H. zea	Corn earworm or bollworm	*Ictalurus punctatus*	Channel catfish
Helix aspera	Dented garden snail, or petit-gris	*Ilyanassa obsoleta*	Common mud snail
		Ilybius fenestratus	Predacious diving beetle
Helluo costatus	Ground beetle	*I. fuliginosus*	Predacious diving beetle
Helluomorphoides ferrugi-neus	Ground beetle	*Ilyocoris cimicoides*	Creeping water bug
		Inimicus didactylus	Devil scorpionfish
H. latitarsus	Ground beetle	*I. japonicus*	Devil scorpionfish
Heloderma suspectum	Gila monster	*Ips confusus*	California five-spined ips
Helops aeneus	Darkling beetle	*Iridomyrmex conifer*	Ant
H. quisquilus	Darkling beetle	*I. detectus*	Ant
Hemachatus haemachatus	Ringhals spitting cobra	*I. humilis*	Argentine ant
Hermodice carunculata	Bristle worm	*I. itinerans*	Ant
Hesperiphona vespertina	Evening grosbeak	*I. myrmecodiae*	Ant
Heterakis gallinarum	Cecal worm of poultry	*I. nitidus*	Ant
Heterodera glycines	Soybean cyst nematode	*I. pruinosus*	Ant
H. rostochiensis	Golden nematode of potato	*I. pruinosus analis*	Ant
H. schactii	Sugar-beet nematode	*I. rufoniger*	Ant
Heterohyrax brucei	Bruce's yellow-spotted dassie	*Isia isabella*	Banded woollybear
Heteropachyloidellus ro-bustus	Daddy-longlegs	*Ixodes ricinus scapularis*	Blacklegged tick
Himasthla quissetensis	Fluke		
Hippodamia convergens	Convergent lady beetle	*Julus heserus*	Juliform millipede
Hirschmanniella oryzae	Rice root nematode	*J. terrestris*	Juliform millipede
Holomelina aurantiaca	Moth	*Junco hyemalis*	Slate-colored junco
H. ferruginosa	Moth	*J. hyemalis hyemalis*	Slate-colored junco
H. fragilis	Moth	*J. oreganus montanus*	Oregon junco
H. immaculata	Moth	*J. oreganus oreganus*	Oregon junco
H. laeta	Moth	*J. oreganus pinosus*	Oregon junco
H. lamae	Moth	*J. oreganus shufeldti*	Oregon junco
H. nigicans	Moth	*J. oreganus thurberi*	Oregon junco
H. rubicundaria	Moth		
Holothuria atra	Sea cucumber		
H. impatiens	Sea cucumber	*Laccopterum foveigerum*	Ground beetle
H. poli	Sea cucumber	*Lacerta agilis*	European fence lizard
Homarus americanus	American lobster	*L. sicula*	Mediterranean lizard
H. gammarus	European lobster	*Lachesis muta*	Pit viper or bushmaster
Homo sapiens	Man	*Lactophrys bicaudalis*	Spotted trunkfish
Hoplolaimus galeatus	Lance nematode	*L. trigonus*	Trunkfish
Horistonotus sp.	Wireworm	*Lactoria cornutus*	Boxfish
Hyalella azteca	Amphipod	*L. fornasini*	Boxfish
Hyalomma sp.	Tick		

continued

Part I. Animals

Scientific Name	Common Name
Lagocephalus laevigatus inermis	Smooth puffer
L. lunaris	Smooth puffer
L. sceleratus	Smooth puffer
Lagopus lagopus	Willow ptarmigan
Lampetia equestris	Narcissus bulb fly
Lampropeltis getulus	King snake
Lampyris sp.	Lightning bug
Laothoe populi	Sphinx moth
Larus canus	Common gull
Lasioderma serricorne	Cigarette beetle
Lasius alienus	Holarctic field ant
L. alienus americanus	Cornfield ant
L. fuliginosus	Holarctic field ant
L. niger	Holarctic field ant
L. spathepus	Holarctic field ant
L. umbratus	Holarctic field ant
Laspeyresia pomonella	Codling moth
Laticauda semifasciata	Broadbanded blue sea snake, or erabu
Latimeria chalumnae	Coelacanth
Lebia chlorocephala	Ground beetle
Leistus ferrugineus	Ground beetle
Leiurus quinquestriatus	Old World desert scorpion
Lemmus trimucronatus	Brown lemming
Lepas fascicularis	Goose barnacle
Lepisma saccharina	Silverfish
Lepisosteus sp.	Gar
Lepomis macrochirus purpurescens	Bluegill
Leptinotarsa decemlineata	Colorado potato beetle
Leptocottus armatus	Pacific staghorn sculpin
Leptodactylus pentadactylus	Tropical American bullfrog
Leptodora kindtii	Water flea, or cladoceran
Leptoglossus clypeatus	Leaf-footed bug
Leptopsylla segnis	Mouse flea
Leptoterna dolabrata	Plant bug
Lepus alleni	Antelope jackrabbit
L. americanus	American hare, or snowshoe rabbit
L. brachyurus	Japanese hare
L. californicus	Black-tailed jackrabbit
L. timidus	Blue hare
Lestes sponsa	European damselfly
Lestrimelitta limao	Bee
Lethocerus americanus	Giant water bug
L. griseus	Giant water bug
L. insulanus	Giant water bug
Lethrinus miniatus	Snapper-like fish
Leucania conigera	Old World noctuid moth
L. impura	Old World noctuid moth
L. pallens	Old World noctuid moth
Leuciscus sp.	Chub
Leucophaea maderae	Madeira cockroach
Leucosticte atrata	Black rosy finch
L. tephrocotis littoralis	Gray-crowned rosy finch
L. tephrocotis tephrocotis	Gray-crowned rosy finch
Libellula needhami	Dragonfly
Libinia emarginata	Spider crab
Libyaspis anglolensis	Bed bug
Licinus nitidior	Ground beetle
Ligia baudiniana	Rock slater

Scientific Name	Common Name
Limnodrilus sp.	Aquatic oligochaete worm
Limnodynastes tasmaniensis	Marbled frog
Limonius californicus	Sugar-beet wireworm
Limulus polyphemus	Horseshoe or king crab
Linguatula serrata	Tongue worm
Linognathus vituli	Long-nosed cattle louse
Liodessus flavicollis	American water beetle
Liometopum microcephalum	Western American ant
Liposcelis divinatorius	Book louse
Liriomyza pusilla	Serpentine leaf miner
Littorina neritoides	European periwinkle
L. obtusata	Northern yellow periwinkle
L. saxatilis	Rough periwinkle
Loa loa	African filarial worm
Locusta migratoria	Migratory locust
L. migratoria gallica	French migratory locust
Loligo forbesi	Northern Atlantic squid
L. pealei	Common American or Atlantic squid
Longidorus sp.	Nematode
Lophius sp.	Goosefish
Loricera pilicornis	Ground beetle
Loxandrus longiformis	Ground beetle
Loxodactylus carinulatus	Ground beetle
Lucilia caesar	Blow fly
Lumbricus terrestris	Earthworm
Lutjanus bohar	Twinspot snapper
L. gibbus	Red snapper
L. monostigmus	Snapper
L. nematophorus	Chinaman fish or snapper
L. vaigiensis	Red snapper
Lycorea ceres ceres	Butterfly
Lycosa tarentula	Tarantula, or wolf spider
Lydella thompsoni	Tachinid fly
Lygus lineolaris	Tarnished plant bug
Lymnaea stagnalis	Freshwater snail
Lytocarpus nuttingi	Feather hydroid
Macaca mulatta	Rhesus monkey
M. nemestrina	Pig-tailed macaque
Macracanthorhynchus hirudinaceus	Thorny-headed worm of swine
Macroscytus sp.	Burrower bug
Macrosiphum avenae	English grain aphid
Macrosteles divisus	Six-spotted leafhopper
Magicicada septendecim	Periodical cicada
Maja sp.	Spider crab
Malacosoma americanum	Eastern tent caterpillar
M. disstria	Forest tent caterpillar
M. pluviale	Western tent caterpillar
Manduca quinquemaculata	Tomato hornworm
Manica rubida	Ant
Mansonia sp.	Mosquito
Maoriblatta novaeseelandiae	Cockroach
Marmosa mexicana isthmica	Murine opossum
M. microtarsus	Murine opossum
Marmota marmota	Eurasian marmot
M. monax	Woodchuck
Martes americana	American marten
M. zibellina	Sable

continued

Part I. Animals

Scientific Name	Common Name	Scientific Name	Common Name
Masticophis flagellum flagellum	Eastern coachwhip snake	*Mya arenaria*	Soft-shell clam
Mastigoproctus giganteus	American whip scorpion	*Mycteroperca venenosa*	Yellowfin grouper
Mastotermes darwiniensis	Termite	*Myliobatis californica*	Bat ray
Mauremys caspica leprosa	European pond turtle	*Myobia musculi*	Fur mite of mouse
Mayetiola destructor	Hessian fly	*Myocoptes musculinus*	Myocoptic mange mite
Megachile latimanus	Leaf-cutting bee	*Myotis daubentonii*	Daubenton's bat
Megazosteria patula	Cockroach	*M. lucifugus*	Little brown bat
Melanogrammus aeglefinus	Haddock	*M. myotis*	Common brown bat
Melanoplus bivittatus	Two-striped grasshopper	*M. sodalis*	Brown bat
M. differentialis	Differential grasshopper	*M. yumanensis*	Brown bat
M. femur-rubrum	Red-legged grasshopper	*Myrmecia gulosa*	Australian ant
M. sanguinipes	Migratory grasshopper	*Myrmecina graminicola*	European ant
Melanotus sp.	Wireworm	*Myrmica brevinodis brevinodis*	Ant
Melasoma populi	Leaf beetle	*M. laevinodis*	Ant
Meleagris gallopavo	Turkey	*M. lobicornis fracticornis*	Ant
Melittobia chalybii	Parasitic wasp	*M. punctiventris punctiventris*	Ant
Meloidogyne sp.	Root-knot nematode		
Melolontha hippocastani	European melolontha	*M. rubra*	Ant
M. melolontha	European scarab beetle	*M. ruginodis*	Ant
Melophagus ovinus	Sheep ked	*M. sabuleti americana*	Ant
Menacanthus stramineus	Chicken body louse	*M. sabuleti sabuleti*	Ant
Menida scotti	Stink bug	*M. scabrinodis*	Ant
Menopon gallinae	Shaft louse of chicken	*Myrmicaria natalensis*	Ant
Mercenaria mercenaria	Northern quahog	*Mysidium gracile*	Opossum-shrimp
Merodon equestris	Narcissus bulb fly	*Mystropomus regularis*	Ground beetle
Mesocricetus auratus	Golden or Syrian hamster	*Mytilus edulis*	Edible mussel
Metastrongylus apri	Swine lungworm	*Myxine* sp.	Hagfish
Methana convexa	Cockroach	*Myzus persicae*	Green peach aphid
Microcentrum rhombifolium	Broad-winged katydid		
Microdipodops pallidus	Pale kangaroo mouse	*Naja haje*	Egyptian cobra
Micropterus salmoides	Largemouth bass	*N. naja*	Indian or common cobra
Microstomus kitt	Lemon sole	*N. naja atra*	Chinese cobra
Microtus agrestis	Short-tailed vole	*N. nigricollis*	Blacknecked spitting cobra, or m'fesi
M. arvalis	Common vole		
M. montanus	Montane vole	*Nannaria* sp.	Flat-backed millipede
M. nivalis wagneri	Snow vole	*Nanophyetus salmincola*	Salmon-poisoning fluke
M. oeconomus	Tundra vole	*Narceus annularis*	Juliform millipede
Micrurus corallinus	South American coral snake	*N. gordanus*	Juliform millipede
M. fulvius	Eastern coral or harlequin snake	*Nasonia vitripennis*	Chalcid wasp
		Nasutitermes exitiosus	Nasutiform termite
M. nigrocinctus	Central American coral snake	*N. graveolus*	Nasutiform termite
		N. longipennis	Nasutiform termite
Mictis caja	Bed bug	*N. magnus*	Nasutiform termite
M. profana	Bed bug	*N. triodiae*	Nasutiform termite
Millepora alcicornis	Stinging coral	*N. walkeri*	Nasutiform termite
Minous monodactylus	Scorpionfish	*Natrix sipedon*	North American water snake
Mola mola	Ocean sunfish	*Nebria livida*	Ground beetle
Molops elatus	Ground beetle	*Necator americanus*	Hookworm
Molothrus ater	Brown-headed cowbird	*Nectophrynoides* sp.	Frog
M. ater obscurus	Brown-headed cowbird	*Necturus maculosus*	Mud puppy
Moniliformis dubius	Thorny-headed worm of rat	*Negaprion brevirostris*	Lemon shark
Morisisa planta tingitiana	Darkling beetle	*Neoceratodus* sp.	Lungfish
Murgantia histrionica	Harlequin bug	*Neodiprion lecontei*	Conifer sawfly
Mus musculus	House mouse	*N. pratti banksianae*	Jack-pine sawfly
M. musculus domesticus	Common house mouse	*N. sertifer*	European pine sawfly
Musca domestica	House fly	*Neofibularia nolitangere*	Brown sponge
Muscardinus avellanarius	Common dormouse	*Neotoma fuscipes*	Dusky-footed wood rat
Musgraveia sulciventris	Stink bug	*N. lepida*	Desert wood rat
Mustela furo	European ferret	*Nezara antennata*	Stink bug
M. rixosa	Least weasel	*N. viridula*	Southern green stink bug
M. vison	Mink		
Mustelus sp.	Dogfish		

continued

Appendix I. SCIENTIFIC NAMES AND CORRESPONDING COMMON NAMES

Part I. Animals

Scientific Name	Common Name	Scientific Name	Common Name
N. viridula var. *smaragdula*	Stink bug	*Orthoporus conifer*	Juliform millipede
Nochtia nochti	Stomach worm	*O. flavior*	Juliform millipede
Nosopsyllus fasciatus	Northern rat flea	*O. punctilliger*	Juliform millipede
Notechis scutatus	Tiger snake	*Oryctolagus cuniculus*	European rabbit
Notesthes robusta	Bullrout	*Oryzaephilus surinamensis*	Saw-toothed grain beetle
Notiophilus biguttatus	Ground beetle	*Oscinella frit*	Frit fly
Notoedres cati	Notoedric mange mite	*Osmerus mordax*	Rainbow smelt
N. muris	Ear mange mite of rat	*Ostertagia circumcincta*	Medium or brown stomach worm of sheep
Notonecta glauca	Backswimmer		
Notonomus angustibasis	Ground beetle	*O. ostertagi*	Medium stomach worm of sheep
N. crenulatus	Ground beetle		
N. miles	Ground beetle	*Ostracion lentiginosus*	Blue trunkfish
N. muelleri	Ground beetle	*Ostrinia nubilalis*	European corn borer
N. opulentus	Ground beetle	*Otobius megnini*	Spinose ear tick
N. rainbowi	Ground beetle	*Otodectes cyanotis*	Ear mite
N. scotti	Ground beetle	*Ovalipes ocellatus*	Lady crab
N. triplogenioides	Ground beetle	*Ovis aries*	Sheep
N. variicollis	Ground beetle	*Oxidus gracilus*	Flat-backed millipede
Notophthalmus viridescens	Common newt	*Oxyuranus scutellatus*	Taipan
Nyctalus noctula	Noctule bat		
		Pachybolus laminatus	Millipede
Octopus apollyon	Octopus	*Pachycolpura manca*	Bed bug
O. bimaculatus	Octopus	*Pachydesmus crassicutus*	Flat-backed millipede
O. fitchi	Octopus	*Paederus fuscipes*	Rove beetle
O. flindersi	Octopus	*Pagellus erythrinus*	Porgy
O. macropus	Giant Pacific octopus	*Pagrus pagrus*	Porgy
O. ocellatus	Western Pacific octopus	*Palaemonetes vulgaris*	Common prawn
O. vulgaris	Common octopus	*Paleacrita vernata*	Spring cankerworm
Ocypode ceratophthalma	Ghost crab	*Palomena viridissima*	Stink bug
Odacantha melanura	Ground beetle	*Paltothyreus tarsatus*	Ant
Odocoileus virginianus	White-tailed deer	*Pamborus alternans*	Ground beetle
Oebalus pugnax	Rice stink bug	*P. guerini*	Ground beetle
Oecoptoma thorica	Carrion beetle	*P. pradieri*	Ground beetle
Oesophagostomum aculeatum	Nodular worm	*P. viridis*	Ground beetle
		Pan troglodytes	Chimpanzee
O. apiostomum	Monkey nodular worm	*Panagaeus bipistulatus*	Ground beetle
O. bifurcum	Nodular worm	*Panonychus ulmi*	European red mite
O. columbianum	Sheep nodular worm	*Panstrongylus* sp.	Conenosed bug
Oestrus ovis	Sheep bot fly	*Panulirus japonicus*	Japanese spiny lobster
Ommastrephes sloani pacificus	Pacific common squid	*Papilio aegeus aegeus*	Swallowtail butterfly
		P. anactus	Swallowtail butterfly
Omophron limbatum	Round sand beetle	*P. antimachus*	Swallowtail butterfly
Onchocerca volvulus	Blinding filarial worm	*P. aristolochiae*	Swallowtail butterfly
Oncopeltus fasciatus	Large milkweed bug	*P. cresphontes*	Orange-dog swallowtail butterfly
Oncorhynchus keta	Chum salmon		
O. kisutch	Coho salmon	*P. demodocus*	Swallowtail butterfly
O. nerka	Sockeye salmon	*P. demoleus sthenelus*	Swallowtail butterfly
O. tshawytscha	Chinook salmon	*P. glaucus*	Swallowtail butterfly
Opatroides punctulatus	Darkling beetle	*P. machaon*	Swallowtail butterfly
Ophioderma brevispinum	Brittle starfish	*P. palamedes*	Swallowtail butterfly
Ophiophagus hannah	King cobra, or hamadryad	*P. polyxenes*	Black swallowtail butterfly
Orchestoidea benedicti	Pacific coast sand hopper	*P. troilus*	Spice-bush swallowtail butterfly
O. californiana	Pacific coast sand hopper		
O. corniculata	Pacific coast sand hopper	*Papio* sp.	Baboon
Orconectes immunis	Freshwater crayfish	*Paragonimus westermanii*	Human lung fluke
O. pellucidus	Cave crayfish	*Parascaris equorum*	Roundworm of horse
O. rusticus	Freshwater crayfish	*Paraspidodera uncinata*	Intestinal worm
O. virilis	Freshwater crayfish	*Paratylenchus* sp.	Plant root nematode
Oribatula sp.	Beetle mite	*Parupeneus chryserydros*	Surmullet
Oriulus delus	Juliform millipede	*Parus major*	Great tit
Ornithonyssus bacoti	Tropical rat mite	*Pasimachus duplicatus*	Ground beetle
Ornithoptera priamus	Swallowtail butterfly	*Passalurus ambiguus*	Pinworm
Ornithorhynchus anatinus	Platypus	*Passer domesticus*	House sparrow

continued

Part I. Animals

Scientific Name	Common Name	Scientific Name	Common Name
P. montanus montanus	European tree sparrow	*P. coolgardiensis*	Cockroach
P. montanus saturatus	European tree sparrow	*P. jungi*	Cockroach
Passerella iliaca	Fox sparrow	*P. morosa*	Cockroach
Patinapta ooplax	Sea cucumber	*P. nitidella*	Cockroach
Pecten sp.	Scallop	*P. occidentalis*	Cockroach
Pectinophora gossypiella	Pink bollworm	*P. ruficeps*	Cockroach
Pedicinus eurygaster	Monkey louse	*P. scabra*	Cockroach
Pediculus humanus	Head and body louse	*P. scabrella*	Cockroach
Pelagia noctiluca	Jellyfish	*P. soror*	Cockroach
Pelamis platurus	Yellow-bellied or pelagic sea snake	*P. stradbrokensis*	Cockroach
		Plecotus auritus	Long-eared bat
Pelecanus occidentalis	Brown pelican	*Plectropomus oligacanthus*	Grouper
Pelmatosilpha coriacea	Cockroach	*Plethodon cinereus*	Red-backed salamander
Pennaria tiarella	Stinging hydroid	*P. glutinosus*	Slimy salamander
Perca flavescens	Yellow perch	*Plinachtus bicoloripes*	Bed bug
Peridroma saucia	Variegated cutworm	*Plodia interpunctella*	Indian meal moth
Periplaneta americana	American cockroach	*Plutella maculipennis*	Diamondback moth
Perognathus hispidus	Plains or pale pocket mouse	*Podophthalmus vigil*	Long-eyed swimming crab
P. intermedius	Rock pocket mouse	*Poecilometis strigatus*	Stink bug
P. longimembris	Little pocket mouse	*Poekilocerus bufonius*	Short-horned grasshopper
Peromyscus leucopus	White-footed mouse	*P. pictus*	Short-horned grasshopper
P. maniculatus	Deer mouse	*Pogonomyrmex badius*	Florida harvester ant
Petromyzon marinus	Sea lamprey	*P. barbatus*	Texas harvester ant
Phaenicia sericata	Blow or greenbottle fly	*P. californicus*	California harvester ant
Phalacrocorax auritus	Double-crested cormorant	*P. desertorum*	Harvester ant
Phalaenoptilus nuttallii	Poor-will	*P. occidentalis*	Western harvester ant
Phaleria provincialis	European darkling beetle	*P. rugosus*	Harvester ant
Phascolarctos cinereus	Koala	*Polia nebulosa*	Moth caterpillar
Phasianus colchicus	Ring-necked pheasant	*P. persicariae*	Moth caterpillar
Pheucticus ludovicianus	Rose-breasted grosbeak	*Polydesmus collaris collaris*	Flat-backed millipede
Philaenus spumarius	Meadow spittlebug	*P. vicinus*	Flat-backed millipede
Philophloeus australis	Ground beetle	*P. virginiensis*	Flat-backed millipede
Philoscaphus tuberculatus	Ground beetle	*Polyodon* sp.	Paddlefish
Phlebotomus papatasii	Sand fly	*Polyphylla* sp.	June beetle
Phlogophora meticulosa	Moth	*Polyplax serrata*	Common mouse louse
Phoca sp.	Seal	*P. spinulosa*	Spined rat louse
Phoenicurus phoenicurus	European redstart	*Polypterus* sp.	Bichir
Phormia regina	Black blow fly	*Polystichus connexus*	Ground beetle
Phormosoma bursarium	Sea urchin	*Polyzosteria cuprea*	Cockroach
Phospuga atrata	Carrion beetle	*P. limbata*	Cockroach
Phoxinus phoxinus	Minnow	*P. mitchelli*	Cockroach
Phragmatobia fuliginosa	Tiger moth	*P. oculata*	Cockroach
Phrynosoma cornutum	Horned lizard	*P. pulchra*	Cockroach
Phyllophaga sp.	June beetle	*P. viridissima*	Cockroach
Phyllotreta sp.	Flea beetle	*Popillia japonica*	Japanese beetle
Phylloxera vitifoliae	Grape plant louse	*Porcellio laevis*	Sow bug
Phylodecta vittelinae	Leaf beetle	*P. scaber*	Scabby sow bug, or wood louse
Phymateus baccatus	Short-horned grasshopper		
P. viridipes	Short-horned grasshopper	*Porites clavaria*	Stony coral
Physalia physalis	Portuguese man-of-war	*Porthetria dispar*	Gypsy moth
Physaloptera tumefaciens	Intestinal worm	*Potamon potamios*	Freshwater crab
Pieris rapae	Imported cabbageworm	*Potamotrygon falkneri*	Freshwater stingray
Piezodorus teretipes	Stink bug	*P. hystrix*	Freshwater stingray
Pimelia confusa	Darkling beetle	*P. labradori*	Freshwater stingray
Pipilo erythrophthalmus	Rufous-sided towhee	*P. motoro*	Freshwater stingray
Pipistrellus pipistrellus	European brown bat	*Praon palitans*	Braconid wasp
Pituophis catenifer	Bull snake	*Pratylenchus mahogani*	Meadow nematode
Placopecten magellanicus	Giant scallop	*P. scribneri*	Scribner's meadow nematode
Plagiodera sp.	Leaf beetle	*Prionychus ater*	Comb-clawed beetle
Planococcus citri	Citrus mealybug	*Procambarus clarkii*	Louisiana red crayfish
Platichthys flesus	Flounder	*Procris geryon*	Leaf skeletonizer moth
Platyzosteria armata	Cockroach	*Procyon cancrivorus*	Crab-eating raccoon
P. castenea	Cockroach	*P. lotor*	Raccoon

continued

1319

Part I. Animals

Scientific Name	Common Name	Scientific Name	Common Name
Promecoderus sp.	Ground beetle	*Rhantus exsoletus*	Predacious diving beetle
Prosopognus harpaloides	Ground beetle	*Rhinocricus insulatus*	Flat-backed millipede
Protopterus aethiopicus	African lungfish	*Rhinolophus ferrumequi-num*	Greater horseshoe bat
Prunella modularis	Hedge sparrow		
Psallus seriatus	Cotton fleahopper	*Rhipicephalus sanguineus*	Brown dog tick
Psammechinus microtuber-culatus	Sea urchin	*Rhizoglyphus echinopus*	Bulb mite
		Rhizophysa eysenharti	Jellyfish or siphonophore
Pseudacris triseriata	Chorus frog	*Rhizostoma pulmo*	Jellyfish, or cabbage bleb
Pseudaletia unipuncta	Armyworm	*Rhodnius* sp.	Assassin bug
Pseudechis australis	King brown or mulga snake	*Rhombomys* sp.	Gerbil
P. papuanus	Papuan black snake	*Rhopalosiphum padi*	Aphid
P. porphyriacus	Red-bellied black snake	*Rhytisternus laevilaterus*	Ground beetle
Pseudoceneus iridescens	Ground beetle	*Riptortus clavatus*	Bed bug
Pseudopleuronectes ameri-canus	Winter flounder	*Romalea microptera*	Short-horned walking stick
		Rotylenchulus reniformis	Reniform nematode
Pseudopolydesmus serratus	Flat-backed millipede	*Rotylenchus* sp.	Spiral nematode
Pseudosmittia arenaria	European midge	*Rypticus saponaceus*	Greater soapfish
Pseudosuccinea sp.	Freshwater snail		
Psila rosae	Carrot rust fly		
Psorergates simplex	Follicle mite of mouse	*Sagartia elegans*	Sea anemone
Psorophora confinnis	Rice-field mosquito	*Sagitta elegans*	Arrowworm
Psoroptes cuniculi	Ear canker mite of rabbit	*Saimiri* sp.	Monkey
P. equi ovis	Sheep scab mite	*Saissetia oleae*	Black scale insect
Psylla pyricola	Pear psylla	*Salamandra salamandra*	Common fire salamander
Pternistria bispina	Bed bug	*Salmo clarki clarki*	Cutthroat trout
Pterois antennata	Lionfish	*S. gairdneri*	Rainbow trout
P. radiata	Lionfish	*S. salar*	Atlantic salmon
P. volitans	Turkeyfish or lionfish	*S. trutta*	Brown trout
Pterophyllum sp.	Angelfish	*Salpa maxima*	North Atlantic tunicate
Pterostichus cupreus	Ground beetle	*Salvelinus fontinalis*	Brook trout
P. macer	Ground beetle	*Samia cynthia*	Cynthia moth
P. melas	Ground beetle	*Sanninoidea exitiosa*	Peachtree borer
P. metallicus	Ground beetle	*Saperda candida*	Roundheaded apple tree borer
P. niger	Ground beetle		
P. vulgaris	Ground beetle	*Sarcophaga aldrichi*	Flesh fly
Ptinus ocellus	Spider beetle	*Sarcoptes scabiei*	Itch mite
Pulex irritans	Human flea	*Sarsia tubulosa*	Stinging hydroid
Pygopus baileyi	Scaly-foot legless lizard	*Sarticus cyaneocinctus*	Ground beetle
Pyrrhula pyrrhula	Bullfinch	*Scalopus aquaticus*	Eastern American mole
Python sebae	Common African python	*Scaphinotus andrewsi gerari*	Ground beetle
		S. andrewsi montana	Ground beetle
		S. viduus	Ground beetle
Quadraspidiotus perniciosus	San Jose scale insect	*S. webbi*	Ground beetle
		Scaphiopus hammondi	Western spadefoot toad
		Scaptocoris divergens	Burrower bug
Radfordia affinis	Fur mite of mouse	*Scardafella inca*	Inca dove
R. ensifera	Fur mite of rat	*Scarus coeruleus*	Blue parrotfish
Radopholus similis	Burrowing nematode	*S. microrhinos*	Parrotfish
Raillietina cesticillus	Broad-headed tapeworm	*Scaurus uncinus*	Darkling beetle
Raja ocellata	Winter or big skate	*Sceloporus graciosus*	Sagebrush lizard
Ramosia tipuliformis	Currant borer	*S. occidentalis*	Pacific fence lizard, or igua-na
Rana catesbeiana	Bullfrog		
R. esculenta	European edible frog	*S. occidentalis biseriatus*	Great basin fence lizard
R. pipiens	Leopard frog	*S. olivaceus*	Texas spiny lizard
R. temporaria	Common European frog	*Schistocerca gregaria*	Desert locust
Rattus norvegicus	Norway rat	*Schistosoma haematobium*	Human blood fluke
R. rattus	Black rat	*S. japonicum*	Oriental blood fluke
Reduvius personatus	Assassin bug or masked hunter	*S. mansoni*	Manson's blood fluke
		Schizura leptinoides	Prominent moth
Reticulitermes flavipes	Eastern subterranean termite	*Sciurus carolinensis penn-sylvanicus*	Common gray squirrel
Rhabdoscelus obscurus	New Guinea sugarcane weevil		
Rhadinaphelenchus co-cophilus	Coconut palm nematode	*S. vulgaris*	European red squirrel
Rhagoletis pomonella	Apple maggot	*Scolytus multistriatus*	Smaller European elm bark beetle

continued

1320

Appendix I. SCIENTIFIC NAMES AND CORRESPONDING COMMON NAMES

Part I. Animals

Scientific Name	Common Name	Scientific Name	Common Name
Scomber scombrus	Atlantic mackerel	*Stenaptinus catoirei*	Ground beetle
Scorpaena guttata	California scorpionfish	*S. verticalis*	Ground beetle
S. plumieri	Sculpin	*Steneotarsonemus pallidus*	Cyclamen mite
S. porcus	Small-scaled scorpionfish	*Stenocoris apicalis*	Bed bug
S. scrofa	Orange scorpionfish	*Stenolophus mixtus*	Ground beetle
Scorpaenopsis diabolis	Scorpionfish	*Stenus bipunctatus*	Rove beetle
Scotinophora lurida	Stink bug	*Sternotherus odoratus*	Musk turtle
Scutellonema sp.	Spiral nematode	*Stichopus variegatus*	Sea cucumber
Scutigerella immaculata	Garden symphylan	*Stilpnotia salicis*	Satin moth
Scyliorhinus sp.	Shark	*Stomoxys calcitrans*	Stable fly
Selasphorus sasin	Allen's hummingbird	*Streptopelia turtur*	Turtledove
Sepia esculenta	Western Pacific cuttlefish	*Streptopharagus pigmenta-*	Parasitic nematode worm
S. officinalis	Cuttlefish	*tus*	
Sepiella japonica	Western Pacific cuttlefish	*Strongylocentrotus purpu-*	Western purple sea urchin
Sepioteuthis lessoniana	Palk Bay squid	*ratus*	
Serinus canarius	Canary	*Strongyloides papillosus*	Sheep intestinal threadworm
Setifer setosus	Hedgehog tenrec	*S. stercoralis*	Human intestinal thread-
Setonix brachyurus	Scrub wallaby		worm
Siagonyx blackburni	Ground beetle	*Strongylus equinus*	Double-toothed strongyle
Sicista betulina	Birch mouse	*S. vulgaris*	Single-toothed strongyle
Sigara falleni	Water boatman	*Struthio camelus*	African ostrich
Silpha obscura	Carrion beetle	*Sturnella magna*	Eastern meadowlark
Simulium arcticum	Black fly	*Sturnus vulgaris*	Common starling
S. occidentale	Black fly	*Sula dactylatra*	Blue-faced booby gannet
S. ornatum	Black fly	*Sus scrofa*	Swine
S. slossonae	Black fly	*Sylvilagus transitionalis*	New England cottontail
S. vittalum	Black fly	*Synanceja horrida*	Stonefish
Sistrurus catenatus	Ground rattlesnake or east-	*S. verrucosa*	Stonefish
	ern massasauga	*Syphacia muris*	Pinworm
Sitophilus granarius	Granary weevil	*S. obvelata*	Cecal worm of rodents
S. oryza	Rice weevil		
Sminthurus viridis	Lucerneflea		
Spermophilus citellus	Souslik	*Tabanus* sp.	Horse fly
S. columbianus	Columbian ground squirrel	*Tachyglossus aculeatus*	Spiny anteater
S. franklini	Franklin ground squirrel	*Tadarida brasiliensis*	Free-tail bat
S. lateralis	Golden-mantled ground	*Taenia pisiformis*	Serrated dog tapeworm
	squirrel	*T. saginata*	Beef tapeworm
S. leucurus	White-tailed antelope squirrel	*T. solium*	Pork tapeworm or pork
S. mohavensis	Mohave ground squirrel		bladder worm
S. richardsonii	Richardson's ground squirrel	*T. taeniaeformis*	Cyclophyllidean tapeworm
S. tereticaudus	Roundtail ground squirrel	*Talitrus saltator*	Sand hopper
S. tridecemlineatus	Thirteen-lined ground	*Tamias striatus*	Eastern chipmunk
	squirrel	*Tapinoma erraticum*	Odorous ant
S. undulatus	Arctic ground squirrel	*T. sessile*	Odorous ant
Sphallomorpha colymbe-	Ground beetle	*T. simrothi karavaievi*	Odorous house ant
toides		*Taricha torosa*	California newt
Sphenodon sp.	Tuatara	*Tedania ignis*	Fire sponge
Sphinx ligustri	Privet hawkmoth	*T. toxicalis*	Reddish-brown sponge
Sphodrosomus saisseti	Ground beetle	*Tenebrio molitor*	Yellow mealworm
Sphoeroides annulatus	Bullseye puffer	*T. obscurus*	Dark mealworm
S. maculatus	Northern puffer	*Tenebroides mauritanicus*	Cadelle
S. spengleri	Bandtail puffer	*Tenrec ecaudatus*	Tenrec
Sphyraena barracuda	Great barracuda	*Tenthredo arcuata*	Sawfly
Sphyrna tiburo	Atlantic bonnet shark	*Ternidens deminutus*	Intestinal worm
Spilopsyllus cuniculi	European rabbit flea	*Terrapene carolina*	Eastern box turtle
Spirostreptidus castaneus	Millipede	*Tessaratoma aethiops*	Stink bug
Spirostreptus virgator	Millipede	*Testudo graecea*	European tortoise
Spissistilus festinus	Three-cornered alfalfa bug	*T. hermanni*	Hermann's tortoise
Spisula solidissima	Atlantic surf clam	*T. marginata*	Tortoise
Spizella arborea	Tree sparrow	*Tetragonurus cuvieri*	Smalleye squaretail
Spodoptera eridania	Southern armyworm	*Tetramorium caespitum*	Pavement ant
S. frugiperda	Fall armyworm	*Tetranychus urticae*	Two-spotted spider mite
Squalus acanthias	Spiny dogfish	*Tetraodon lineatus*	Puffer

continued

Scientific Name	Common Name	Scientific Name	Common Name
Thalarctos maritimus	Polar bear	*Tylenchulus semipenetrans*	Citrus nematode
Thamnophis radix	Plains garter snake	*Tyloderma fragariae*	Strawberry crown borer
Thermobia domestica	Firebrat	*Tylos latreillei*	European beach isopod
Thrips tabaci	Onion thrips	*Tyria jacobaeae*	Tiger moth
Thunnus sp.	Tuna		
Thyridopteryx ephemerae-formis	Bagworm	*Uca maracoani*	Fiddler crab
Tiliqua nigrolutea	Australian blotched blue-tongue skink	*U. pugilator*	Atlantic fiddler crab
		U. pugnax	Fiddler crab
Tinca tinca	Golden tench	*U. rapax*	Fiddler crab
Tiphia vernalis	Japanese beetle parasite wasp	*U. tangeri*	Fiddler crab
Todarodes pacificus	Western Pacific cuttlefish	*Uma notata*	Colorado desert fringe-toed lizard
Torpedo marmorata	Marbled electric ray		
Torquigener hamiltoni	Puffer	*Upeneus arge*	Surmullet
Toxascaris leonina	Large nematode of carnivores	*Uroblaniulus canadensis*	Juliform millipede
Toxocara canis	Large roundworm of carnivores	*Urolophus armatus*	Round stingray
		U. halleri	Round stingray
Toxopneustes pileolus	Sea urchin	*Ursus americanus*	Black bear
Trachinus araneus	Weever	*Uta stansburiana*	Side-blotched lizard, or ground uta
T. draco	Greater weever		
T. radiatus	Weever	*U. stansburiana hesperis*	California side-blotched lizard
T. vipera	Lesser weever		
Trialeurodes vaporariorum	Greenhouse whitefly	*Utetheisa bella*	Tiger moth
Triatoma sanguisuga	Bloodsucking conenose, or assassin bug		
Tribolium castaneum	Red flour beetle	*Velia caprai*	Water strider
T. confusum	Confused flour beetle	*V. currens*	Water strider
Trichinella spiralis	Trichina worm	*Veromessor andrei*	Black harvester ant
Trichodectes canis	Dog-biting louse	*Vespula germanica*	Wasp
Trichodorus christiei	Stubby-root nematode	*V. vulgaris*	Hornet
Trichoecius romboutsi	Mouse mite	*Vipera berus*	Common European viper or adder
Trichogramma cacoeciae	Egg-parasitic fly, or trichogrammatid		
T. minutum	Egg-parasitic fly, or trichogrammatid	*V. berus bosniensis*	Balkan cross adder
		V. russellii	Russell's viper, or daboia
Trichoplusia ni	Cabbage looper	*V. xanthina palestinae*	Palestinian or Near East viper
Trichosternus nudipes	Ground beetle	*Vulpes fulva*	Red fox
Trichostrongylus axei	Minute stomach worm		
T. colubriformis	Black scour worm	*Wohlfahrtia vigil*	Flesh fly
Trichuris suis	Whipworm of swine	*Wuchereria bancrofti*	Bancroft's filarial worm
T. trichiura	Human whipworm		
T. vulpis	Dog whipworm		
Trigona postica	Bee	*Xantusia vigilis*	Yucca night lizard
T. tubiba	Bee	*Xenopsylla cheopis*	Oriental rat flea
Trigonoiulus lumbricinus	Juliform millipede	*Xenopus laevis*	Clawed toad
Trimeresurus flavoviridis	Okinawan pit viper, or habu	*Xiphinema* sp.	Dagger nematode
T. mucrosquamatus	Formosan pit viper, or habu		
Trionyx spinifer	Spiny softshell turtle	*Zapus hudsonius*	Meadow jumping mouse
Trioxys utilis	Braconid wasp	*Zenaidura macroura*	Mourning dove
Trioza tripunctata	Blackberry plant louse or psyllid	*Z. macroura carolinensis*	Mourning dove
		Zenarchopterus buffoni	Indo-Australian halfbeak
Tripneustes gratilla	Sea urchin	*Z. dispar*	Indo-Australian halfbeak
Triturus vulgaris	Smooth newt	*Zonioploca bicolor*	Cockroach
Troglodytes aedon	Northern house wren	*Z. pallida*	Cockroach
Trogoderma granarium	Khapra or carpet beetle	*Zonotrichia albicollis*	White-throated sparrow
T. inclusum	Carpet beetle	*Z. atricapilla*	Golden-crowned sparrow
Troides helena	Swallowtail butterfly	*Z. leucophrys gambelii*	White-crowned sparrow
Trombicula akamushi	Redbug	*Z. leucophrys leucophrys*	White-crowned sparrow
Tropidechis carinatus	Rough-scaled snake	*Z. leucophrys nuttalli*	White-crowned sparrow
Tubifex sp.	Sludge worm	*Z. leucophrys pugetensis*	White-crowned sparrow
Tumulitermes pastinator	Termite	*Z. querula*	Harris' sparrow
Tylenchorhynchus sp.	Stunt nematode		

continued

Part I.   Animals

Scientific Name	Common Name
Zoogonus rubellus	Intestinal fluke of fish
Zophobas rugipes	Darkling beetle
Zosterops japonica japonica	Japanese white-eye, or spectacle bird

Scientific Name	Common Name
Zygaena filipendulae	Leaf skeletonizer moth
Z. lonicerae	Leaf skeletonizer moth
Z. trifolii	Leaf skeletonizer moth

Part II.   Plants

Scientific Name	Common Name
Abies alba	Silver fir
A. grandis	Grand fir
A. guatemalensis	Fir
Abrus praecatorius	Rosary pea
Abutilon theophrasti	Piemarker abutilon
Acacia dealbata	Silvergreen-wattle acacia
A. farnesiana	Sweet acacia
Acer campestre	Hedge maple
A. platanoides	Norway maple
A. pseudoplatanus	Plane-tree or sycamore maple
A. rubrum	Red maple
Aconitum napellus	Aconite monkshood
Adiantum capillus-veneris	Southern maidenhair fern
Aesculus hippocastanum	Common horse chestnut
Agave americana	Century plant agave
Agropyron cristatum	Crested wheatgrass
A. desertorum	Desert wheatgrass
A. intermedium	Intermediate wheatgrass
A. repens	Quack grass
Agrostemma githago	Common corn cockle
Agrostis alba	Redtop
A. palustris	Creeping bentgrass
Allium cepa	Garden onion
Alnus glutinosa	European alder
A. incana	Speckled alder
A. rubra	Red alder
Althaea rosea	Hollyhock
Amaranthus retroflexus	Green amaranth
Ambrosia trifida	Giant ragweed
Anagallis arvensis	Scarlet pimpernel
Ananas comosus	Pineapple
Andropogon scoparius	Little bluestem
Anthriscus cerefolium	Salad chervil
Antiaris toxicaria	Upas tree
Antirrhinum majus	Common snapdragon
Apium graveolens	Wild celery
A. graveolens dulce	Garden celery
Arabidopsis thaliana	Mouse-ear cress
Arachis hypogaea	Peanut
Arbutus unedo	Strawberry madrone
Asclepias curassavica	Milkweed
Asparagus acutifolius	Asparagus
A. officinalis	Garden asparagus
A. sprengeri	Sprenger asparagus
Asplenium glandulosum	Spleenwort
Astragalus sp.	Locoweed, or poison vetch
Athyrium filix-femina	Lady fern
A. spinulosa	Fern
Atriplex sp.	Saltbush

Scientific Name	Common Name
Atropa belladonna	Belladonna
Avena byzantina	Red oat
A. fatua	Wild oat
A. sativa	Common oat
Avicennia marina	Black mangrove
Batis maritima	Maritime saltwort
Bauhinia monandra	Butterfly bauhinia
Begonia decandra	Begonia
Beta saccharifera	Sugar beet
B. vulgaris	Common beet
B. vulgaris var. *cicla*	Swiss chard
Betula alba	Old World canoe-birch
B. lenta	Sweet birch
B. papyrifera	Paper birch
B. tauschii	Birch
Billbergia elegans	Airbrom
Blighia sapida	Akee
Bougainvillea glabra	Lesser brougainvillea
Bouteloua gracilis	Blue grama
Brassica campestris	Bird rape
B. fruticulosa	Mediterranean brassica
B. hirta	White mustard
B. juncea	India mustard
B. kaber	Charlock
B. kaber var. *pinnatifida*	Charlock
B. napus var. *napobrassica*	Rutabaga
B. nigra	Black mustard
B. oleracea	Wild cabbage
B. oleracea var. *botrytis*	Cauliflower
B. oleracea var. *capitata*	Cabbage
B. oleracea var. *gemmifera*	Brussels sprouts
B. oleracea var. *gongylodes*	Kohlrabi
B. oleracea var. *italica*	Broccoli
B. pekinensis	Chinese cabbage
B. rapa	Turnip
Bromus inermis	Hungarian bromegrass
Bryophyllum crenatum	Kalanchoe
B. daigremontianum	Kalanchoe
Camellia japonica	Common camellia
C. sinensis	Common tea
Campanula longestyla	Bellflower
C. medium	Canterbury bells
Cannabis sativa	Hemp
Capsicum annuum	Common pepper
C. frutescens	Bush red pepper
Carthamus tinctorius	Safflower

continued

Scientific Name	Common Name	Scientific Name	Common Name
Carya illinoensis	Pecan	*Cuscuta* sp.	Dodder
C. ovata	Shagbark hickory	*Cyclamen* sp.	Cyclamen
Cassia obtusifolia	Sickle senna	*Cynodon dactylon*	Bermuda grass
Castanea dentata	American chestnut	*Cyperus rotundus*	Nut grass sedge
C. sativa	European chestnut		
Catalpa sp.	Catalpa		
Cedrus atlantica	Atlas cedar	*Dactylis glomerata*	Orchard grass
C. deodara	Deodar cedar	*Dahlia* sp.	Dahlia
C. libani	Cedar of Lebanon	*Datura metel*	Hindu datura
Celtis sp.	Hackberry	*D. stramonium*	Jimsonweed
Centaurea cyanus	Cornflower	*Daucus carota*	Carrot
C. nigra	Knapweed	*Delphinium ajacis*	Rocket larkspur
Cephalanthus sp.	Buttonbush	*D. consolida*	Forking larkspur
Ceratonia siliqua	Carob	*Desmodium* sp.	Tick clover
Cestrum nocturnum	Night-blooming cestrum	*Dianthus barbatus*	Sweet william
Chamaecyparis thyoides	Southern white cedar	*D. carthusianorum*	Carthusian pink
Chamaerops humilis	Mediterranean palm	*D. caryophyllus*	Clove pink
Chenopodium amaranti-color	Goosefoot	*D. purpurea*	Pink
		Dicranopteris pectinata	Leptosporangiate fern
C. botrys	Jerusalem-oak goosefoot	*Digitalis lanata*	Grecian foxglove
C. capitatum	Blight goosefoot	*D. purpurea*	Common foxglove
C. quinoa	Quinoa	*Dioscorea hispida*	Wild yam
C. rubrum	Red goosefoot	*Diospyros virginiana*	Common persimmon
Chondrilla juncea	Rush skeleton weed	*Dryopteris filix-mas*	Male fern
Chrysanthemum maximum	Pyrenees chrysanthemum		
C. morifolium	Florist's chrysanthemum		
Cicer arietinum	Gram chickpea	*Echinochloa crusgalli*	Barnyard grass
Cichorium endivia	Endive	*Elodea callitrichoides*	South American waterweed
C. intybus	Common chicory	*E. canadensis*	Canada waterweed
Cicuta sp.	Water hemlock	*Equisetum arvense*	Common horsetail
Cirsium arvense	Canada thistle	*E. fluviatile*	Water horsetail
Citrullus colocynthis	Colocynth	*E. hiemale*	Scouring rush
C. vulgaris	Watermelon	*E. palustre*	Meadow horsetail
Citrus aurantium	Sour orange	*E. variegatum*	Variegated horsetail
C. hassaku	Citrus	*Eremochloa ophiuroides*	Centipede grass
C. jambhiri	Rough lemon	*Erica carnea*	Spring heath
C. limetioides	Sweet lime	*E. tetralix*	Cross-leaf heath
C. limon	Lemon	*Erythroxylon coca*	Coca tree
C. nobilis var. *unshiu*	Satsuma orange	*Eucalyptus globulus*	Tasmanian blue eucalyptus or gum
C. sinensis	Sweet orange		
Clivia sp.	Kafir lily	*Euphorbia lathyris*	Caper spurge, or mole plant
Cocos nucifera	Coconut	*E. resinifera*	Gum euphorbia
Coffea arabica	Arabian coffee	*Euterpe globosa*	Euterpe palm
Colchicum autumnale	Autumn-crocus		
Coleus blumei	Common coleus		
Commelina communis	Common dayflower	*Fagopyrum sagittatum*	Buckwheat
Conium maculatum	Poison hemlock	*Fagus grandifolia*	American beech
Conocarpus erecta	Button mangrove	*F. sylvatica*	European beech
Coreopsis grandiflora	Bigflower coreopsis	*Festuca elatior*	Meadow fescue
Cornus florida	Flowering dogwood	*F. elatior arundinacea*	Alta or reed fescue
Corylus avellana	European filbert	*F. glauca*	Blue fescue
C. cornuta	Beaked filbert	*F. rubra*	Red fescue
Cosmos sulphureus	Yellow cosmos	*Ficus carica*	Fig
Crataegus sp.	Hawthorn	*Fragaria ananassa*	Pine strawberry
Crotalaria sp.	Crotalaria	*F. chiloensis*	Chiloe strawberry
Croton tiglium	Purging croton	*F. virginiana*	Virginia strawberry
Cucumis melo	Muskmelon	*Fraxinus excelsior*	European ash
C. sativus	Cucumber		
Cucurbita maxima	Winter squash		
C. pepo	Pumpkin	*Gardenia jasminoides*	Cape jasmine
Cupressus lusitanica	Mexican cypress	*Geum urbanum*	Siberian geum
C. sempervirens	Italian cypress	*Gilia* sp.	Gilia

continued

Part II. Plants

Scientific Name	Common Name	Scientific Name	Common Name
Ginkgo biloba	Ginkgo	*Lathyrus cicera*	Flat-pod pea
Gladiolus sp.	Gladiolus	*L. clymenum*	Peavine
Glaux maritima	Sea milkwort	*L. montanus*	Mountain peavine
Gleditsia triacanthos	Common honey locust	*L. odoratus*	Sweet pea
Gloxinia grandiflora	Gloxinia	*L. sativus*	Grass pea
Glycine max	Soybean	*Laurus nobilis*	Grecian laurel
Gomphrena globosa	Common globe amaranth	*Lemna perpusilla*	Minute duckweed
Gossypium arboreum	Asiatic tree cotton	*L. trisulca*	Star duckweed
G. barbadense	Sea Island cotton	*Lens culinaris*	Common lentil
G. hirsutum	Upland cotton	*Leontodon* sp.	Hawkbit
Guarea trichilioides	American muskwood	*Lepidium densiflorum*	Prairie pepperweed
		L. virginicum	Virginia pepperweed
		Lespedeza sp.	Lespedeza
Haemanthus katharinae	Katharine blood lily	*Leucaena leucocephala*	Lead tree
Hedysarum carnosum	Sweet vetch	*Ligustrum* sp.	Privet
Helianthus annuus	Common sunflower	*Lilium longiflorum*	Easter lily
Helleborus niger	Christmas rose	*Linnaea borealis*	Twinflower
Hibiscus esculentus	Okra	*Linum austriacum*	Austrian flax
Hippomane mancinella	Manchineel	*L. usitatissimum*	Common flax
Hordeum vulgare	Barley	*Liquidambar styraciflua*	American sweetgum
Hottonia palustris	European featherfoil	*Liriodendron tulipifera*	Yellow poplar, or tulip tree
Humulus lupulus	Common hop	*Lolium multiflorum*	Italian ryegrass
Hydrangea macrophylla	Bigleaf hydrangea	*L. perenne*	Perennial ryegrass
Hyoscyamus niger	Black henbane	*L. temulentum*	Darnel ryegrass
		Lotus corniculatus	Bird's-foot trefoil
		Lunaria annua	Dollar plant
Ilex aquifolium	English holly	*Lupinus angustifolius*	Tree lupine
I. opaca	American holly	*L. luteus*	European yellow lupine
Impatiens balsamina	Garden balsam	*Lycopersicon esculentum*	Common tomato
I. parviflora	Small-flowered snapweed		
I. sultani	Sultan snapweed		
Ipomoea batatas	Sweet potato	*Magnolia grandiflora*	Southern magnolia
I. hederacea	Ivy-leaf morning glory	*Malus pumila*	Apple
I. nil	White-edge morning glory	*Malva alcea* var. *fastigiata*	Hollyhock mallow
Iresine sp.	Bloodleaf	*M. neglecta*	Mallow
Iris chamaeiris	Crimean iris	*M. parviflora*	Little mallow
		Mangifera indica	Common mango
		Manihot esculenta	Cassava
Jatropha curcas	Barbados nut	*Matthiola incana*	Common stock
J. multifida	Coral plant	*Mattuccia struthiopteris*	Ostrich fern
Juglans nigra	Eastern black walnut	*Medicago sativa*	Alfalfa
J. regia	Persian walnut	*M. tribuloides*	Barrel medic
Juniperus scopularum	Western red cedar	*Melia azedarach*	Chinaberry
J. virginiana	Eastern red cedar	*Melilotus alba*	White sweet clover
		M. officinalis	Yellow sweet clover
		Mentha cardiaca	Little-leaf mint
Kalanchoe blossfeldiana	Kalanchoe	*M. piperita*	Peppermint
K. daigremontiana	Kalanchoe	*M. spicata*	Spearmint
K. fedtschenkoi	Fedtschenko kalanchoe	*Metopium toxiferum*	Florida poisonwood tree
K. pinnata	Air plant kalanchoe	*Musa acuminata*	Banana
Kalmia angustifolia	Lambkill	*M. sapientum*	Common banana
K. latifolia	Mountain laurel	*Myosotis alpestris*	Alpine forget-me-not
Lactuca sativa	Lettuce	*Narcissus* sp.	Narcissus
L. sativa var. *longifolia*	Romaine lettuce	*Nephrolepis biserrata*	Purplestalk sword fern
Laguncularia racemosa	False mangrove	*N. rivularis*	Sword fern
Lamium amplexicaule	Henbit dead nettle	*Nerium oleander*	Common oleander
Larix dahurica	Larch	*Nicandra physaloides*	Apple of Peru
L. decidua	European larch	*Nicotiana alata*	Winged tobacco
L. leptolepis	Japanese larch	*N. clevelandii*	Tobacco
L. occidentalis	Western larch	*N. debneyi*	Tobacco

continued

Scientific Name	Common Name	Scientific Name	Common Name
N. glutinosa	Tobacco	*P. strobus*	Eastern white pine
N. megalosiphon	Tobacco	*P. sylvestris*	Scotch pine
N. rustica	Aztec tobacco	*P. taeda*	Loblolly pine
N. rustica var. *brasilia*	Aztec tobacco	*P. virginiana*	Virginia pine
N. sylvestris	Tobacco	*Piper nigrum*	Black pepper
N. tabacum	Common tobacco	*Piptadeniastrum africanum*	African piptadenia
Nyssa sylvatica	Black tupelo	*Pisum sativum*	Garden pea
		P. sativum arvense	Field pea
		Platanus acerifolia	London plane tree
Ocimum basilicum	Sweet basil	*P. occidentalis*	American sycamore
Oenothera biennis	Common evening primrose	*P. orientalis*	Oriental plane tree
O. parviflora	Small-flower evening prim-rose	*P. racemosa*	California sycamore
		P. wrightii	Arizona sycamore
Olea europaea	Common olive	*Poa annua*	Annual bluegrass
O. europaea sativa	Olive	*P. nipponica*	Bluegrass
Oleandra articulata	Fern	*P. pratensis*	Kentucky bluegrass
Ononis sp.	Ononis	*Polygonatum odoratum*	Solomon seal
Opuntia sp.	Prickly pear	*Polygonum lapathifolium*	Curl-top lady's thumb
Oryza sativa	Rice	*Polypodium australe*	Polypody fern
Osmunda cinnamomea	Cinnamon fern	*P. vulgare*	Common polypody fern
Ostrya virginiana	American hop hornbeam	*Polystichum lobatum*	Fern
Oxalis acetosella	Wood sorrel oxalis	*Poncirus trifoliata*	Trifoliate orange
		Populus deltoides	Eastern poplar
		P. grandidentata	Large-toothed aspen
Panicum virgatum	Switch grass	*P. nigra*	Black poplar
Papaver somniferum	Opium poppy	*P. tremuloides*	Quaking aspen
Parthenium argentatum	Guayule parthenium	*Potamogeton crispus*	Curly pondweed
Paspalum dilatatum	Dallis grass	*Prosopis glandulosa* var. *glandulosa*	Honey mesquite
P. notatum	Pensacola bahia grass		
Pastinaca sativa	Parsnip	*P. juliflora*	Mesquite
Pelargonium hortorum	Fish pelargonium	*Prunus amygdalus*	Almond
Pennisetum glaucum	Pearl millet	*P. armeniaca*	Apricot
Perilla frutescens var. *crispa*	Purple common perilla	*P. avium*	Mazzard cherry
Persea americana	American avocado	*P. cerasus*	Sour cherry
Petroselinum crispum	Parsley	*P. domestica*	Garden plum
Petunia hybrida	Common petunia	*P. laurocerasus*	Common laurel cherry
Phalaris canariensis	Canary grass	*P. mahaleb*	Mahaleb cherry
P. tuberosa	Bulb canary grass	*P. padus*	Bird cherry
Phaseolus limensis	Lima bean	*P. persica*	Peach
P. lunatus	Sieva bean	*P. serotina*	Black cherry
P. vulgaris	Kidney bean	*P. virginiana*	Common chokecherry
Phleum pratense	Timothy	*Pseudotsuga menziesii*	Douglas fir
Phlox paniculata	Summer phlox	*P. taxifolia*	Common Douglas fir
Phoenix canariensis	Canary date	*Psidium* sp.	Guava
P. dactylifera	Date	*Pteridium aquilinum*	Bracken fern
Phytolacca americana	Pokeberry	*Pyrus communis*	Pear
Picea abies	Norway spruce		
P. glehnii	Sakhalin spruce		
P. mariana	Black spruce	*Quercus alba*	White oak
Pinus banksiana	Jack pine	*Q. coccinea*	Scarlet oak
P. cembra	Swiss stone pine	*Q. falcata*	Southern red oak
P. cembroides	Mexican piñon pine	*Q. ilex*	Holly oak
P. contorta	Shore pine	*Q. palustris*	Pin oak
P. echinata	Shortleaf pine	*Q. robur*	English oak
P. halepensis	Aleppo pine	*Q. suber*	Cork oak
P. monticola	Western white pine	*Q. velutina*	Black oak
P. mugo	Swiss mountain pine		
P. palustris	Longleaf pine		
P. parviflora	Japanese white pine	*Raphanus sativus*	Garden radish
P. pinea	Italian stone pine	*Rheum rhaponticum*	Garden rhubarb
P. ponderosa	Ponderosa pine	*Rhododendron brachycarpum*	Fujiyama rhododendron
P. resinosa	Red pine		

continued

Appendix I.   SCIENTIFIC NAMES AND CORRESPONDING COMMON NAMES

Part II.   Plants

Scientific Name	Common Name
R. ferrugineum	Rock rhododendron
R. hirsutum	Garland rhododendron
R. obtusum var. *amoenum*	Amoena azalea
Ribes aureum	Golden currant
R. nigrum	European black currant
Ricinus communis	Castor bean
Robinia pseudoacacia	Black locust
Rosa multiflora	Japanese rose
R. pendulina	Drop hip rose
Rubus idaeus	Red raspberry
Rudbeckia newmanii	Coneflower
Rumex acetosa	Garden sorrel
R. conglomeratus	Clustered dock
R. patientia	Patience dock
Ruscus aculeatus	Butcher's-broom
Saccharum officinarum	Sugarcane
Salicornia europaea	Marshfire glasswort
S. virginica	Woody glasswort
Salix koriyanagi	Willow
S. viminalis	Basket willow
Salvia horminum	Joseph's sage
Sambucus nigra	European elder
Secale cereale	Rye
Sedum acre	Goldmoss stonecrop
S. maximum	Great stonecrop
S. reflexum	Jenny stonecrop
Senecio sp.	Groundsel
Sequoia sempervirens	Redwood
Sesbania exaltata	Hemp sesbania
Setaria italica	Foxtail millet
Silene armeria	Sweet william silene
Silphium trifoliatum	Rosinweed
Solanum melongena	Eggplant
S. nodiflorum	Nightshade
S. pseudocapsicum	Jerusalem cherry
S. tuberosum	Potato
Solidago altissima	Tall goldenrod
Sorbus aucuparia	European mountain ash
Sorghum bicolor	Sorghum
S. halepense	Johnson grass
S. vulgare sudanense	Sudan grass
Spartina patens	Marsh hay cordgrass
Spinacia oleracea	Spinach
Stellaria media	Common chickweed
Strophanthus kombe	Strophanthus
S. sarmentosus	Arrow poison strophanthus
Strychnos castelnaei	Amazon poison nut
S. nuxvomica	Nux vomica poison nut
S. toxifera	Curare poison nut
Symphytum cordatum	Comfrey
Tagetes patula	French marigold
Taraxacum officinale	Dandelion
Taxodium distichum	Common bald cypress
Taxus baccata	English yew
T. cuspidata	Japanese yew
Thelypteris deltoidea	Wood fern

Scientific Name	Common Name
Thevetia peruviana	Yellow oleander
Thuja occidentalis	Eastern arborvitae
T. plicata	Giant arborvitae
Tilia sp.	Linden
Toxicodendron radicans	Poison ivy
T. vernix	Poison sumac
Trachycarpus fortunei	Fortune's windmill palm
Tradescantia virginiana	Virginia spiderwort
Trifolium hybridum	Alsike clover
T. incarnatum	Crimson clover
T. pratense	Red clover
T. repens	White clover
T. subterraneum	Subterranean clover
Triglochin maritima	Shore podgrass
Triodia flava	Purpletop
Triticum aestivum	Wheat
T. durum	Durum wheat
Tropaeolum majus	Common nasturtium
Tsuga canadensis	Eastern hemlock
T. heterophylla	Western hemlock
Tulipa gesneriana	Common tulip
Ulmus americana	American elm
U. parvifolia	Chinese elm
Utricularia vulgaris	Common bladderwort
Vaccinium corymbosum	Highbush blueberry
Vallisneria spiralis	Spiral wild celery
V. spiralis torta	Spiral wild celery
Veratrum viride	Green hellebore
Verbesina encelioides	Golden crownbeard
Veronica beccabunga	Beccabunga speedwell
V. tournefortii	Tournefort speedwell
Viburnum burkwoodii	Burkwood viburnum
V. tinus	Laurestinus viburnum
Vicia faba	Broad bean
V. faba minor	Small horsebean
Vigna sinensis	Common cowpea
Viola hirta	Hairy violet
V. odorata	Sweet violet
Vitis vinifera	European grape
Weigela florida var. *varigata*	Old-fashioned weigela
Wittrockia superba	Brazilian wittrockia
Xanthium orientale	Oriental cocklebur
X. strumarium	Cocklebur
Xanthosoma sagittifolium	Yautia malanga
Yucca flaccida	Weakleaf yucca
Zea mays	Corn
Zigadenus sp.	Death camas
Zinnia sp.	Zinnia
Zoysia japonica	Japanese lawn grass

Common names are not available for some organisms in this book. In such cases, the name of the general category to which the organism has been assigned, modified by its geographic distribution (if known), has been included to assist the user unacquainted with the scientific name. Protozoa and nonvascular plants have not been included.

Part I. Animals

Common Name	Scientific Name
Adder, Balkan cross	*Vipera berus bosniensis*
common night	*Causus rhombeatus*
death	*Acanthophis antarcticus*
puff	*Bitis arietans*
Alfalfa bug, three-cornered	*Spissistilus festinus*
Alligator, American	*Alligator mississippiensis*
Amphipod	*Hyalella azteca*
Amphiuma	*Amphiuma means*
three-toed	*A. tridactylum*
Anchovy	*Engraulis japonicus*
Anemone, cloak	*Adamsia palliata*
Angelfish	*Pterophyllum* sp.
Ant	*Aphaenogaster longiceps*
	Azteca sp.
	Cataglyphis bicolor
	Crematogaster africana
	C. peringueyi
	Dolichoderus clarki
	D. dentata
	D. scabridus
	Dorymyrmex bicolor
	Forelius foetidus
	Formica cinerea
	F. polyctena
	F. sanguinea
	Harpogoxenus sublaevis
	Iridomyrmex conifer
	I. detectus
	I. itinerans
	I. myrmecodiae
	I. nitidus
	I. pruinosus
	I. pruinosus analis
	I. rufoniger
	Manica rubida
	Myrmica brevinodis brevinodis
	M. laevinodis
	M. lobicornis fracticornis
	M. punctiventris punctiventris
	M. rubra
	M. ruginodis
	M. sabuleti americana
	M. sabuleti sabuleti
	M. scabrinodis
	Myrmicaria natalensis
	Paltothyreus tarsatus
Argentine	*Iridomyrmex humilis*
Australian	*Myrmecia gulosa*
black harvester	*Veromessor andrei*
California harvester	*Pogonomyrmex californicus*
carpenter	*Camponotus maculatus*
cornfield	*Lasius alienus americanus*
European	*Myrmecina graminicola*
European carpenter	*Camponotus ligniperda*
Florida harvester	*Pogonomyrmex badius*
harvester	*P. desertorum*
	P. rugosus

Common Name	Scientific Name
Holarctic field	*Lasius alienus*
	L. fuliginosus
	L. niger
	L. spathepus
	L. umbratus
leaf-cutting	*Atta bisphaerica*
	A. capiguara
	A. colombica
	A. laevigata
	A. robusta
	A. sexdens
odorous	*Tapinoma eraticum*
	T. simrothi karavaievi
odorous house	*T. sessile*
pavement	*Tetramorium caespitum*
pyramid	*Dorymyrmex pyramicus*
	D. pyramicus flavopectus
red	*Formica rufa*
silky	*F. fusca*
smaller yellow	*Acanthomyops claviger*
Texas harvester	*Pogonomyrmex barbatus*
Texas leaf-cutting	*Atta texana*
western American	*Liometopum microcephalum*
western harvester	*Pogonomyrmex occidentalis*
Anteater, spiny	*Tachyglossus aculeatus*
Aphid	*Aphis nerii*
	Rhopalosiphum padi
apple	*Aphis pomi*
English grain	*Macrosiphum avenae*
green peach	*Myzus persicae*
woolly apple	*Eriosoma lanigerum*
Argus, blue-spotted	*Cephalopholis argus*
Armadillo, nine-banded	*Dasypus novemcinctus*
Armyworm	*Pseudaletia unipuncta*
fall	*Spodoptera frugiperda*
southern	*S. eridania*
Arrowworm	*Sagitta elegans*
Assassin bug	*Rhodnius* sp.
or masked hunter	*Reduvius personatus*
Axolotl	*Ambystoma mexicanum*
Baboon	*Papio* sp.
Backswimmer	*Notonecta glauca*
Bagworm	*Thyridopteryx ephemeraeformis*
Balloonfish	*Diodon holocanthus*
Barnacle, giant	*Balanus nubilis*
goose	*Lepas fascicularis*
Barracuda, great	*Sphyraena barracuda*
Bass, European	*Dicentrarchus labrax*
largemouth	*Micropterus salmoides*
Bat, big brown	*Eptesicus fuscus*
brown	*Myotis sodalis*
	M. yumanensis
common brown	*M. myotis*
Daubenton's	*M. daubentonii*
European brown	*Pipistrellus pipistrellus*
free-tail	*Tadarida brasiliensis*

continued

Common Name	Scientific Name	Common Name	Scientific Name
greater horseshoe	*Rhinolophus ferrumequinum*		*C. perigrinater*
little brown	*Myotis lucifugus*		*C. prominens*
long-eared	*Plecotus auritus*		*C. schayeri*
noctule	*Nyctalus noctula*		*C. scrutator*
pallid	*Antrozous pallidus*		*C. sycophanta*
Bear, black	*Ursus americanus*	cigarette	*Lasioderma serricorne*
polar	*Thalarctos maritimus*	click	*Agriotes obscurus*
Beaver, North American	*Castor canadensis*	Colorado potato	*Leptinotarsa decemlineata*
Bed bug	*Acanthocoris sordidus*	comb-clawed	*Prionychus ater*
	Agriopocoris frogatti	common European dung	*Geotrupes stercorosus*
	Amorbus alternatus	confused flour	*Tribolium confusum*
	A. rhombifer	convergent lady	*Hippodamia convergens*
	A. rubiginosus	darkling	*Blaps gigas*
	Aulacosternum nigrorubrum		*B. lethifera*
	Hygia opaca		*B. mortisaga*
	Libyaspis anglolensis		*B. mucronata*
	Mictis caja		*B. requienii*
	M. profana		*B. sulcata*
	Pachycolpura manca		*B. wiedemanni*
	Plinachtus bicoloripes		*Diaperis boleti*
	Pternistria bispina		*D. maculata*
	Riptortus clavatus		*Eleodes hispilabris*
	Stenocoris apicalis		*E. longicollis*
common	*Cimex lectularius*		*Gnaptor spinimanus*
Bee	*Andrena* sp.		*Helops aeneus*
	Halictus sp.		*H. quisquilus*
	Lestrimelitta limao		*Morisisa planta tingitiana*
	Trigona postica		*Opatroides punctulatus*
	T. tubiba		*Pimelia confusa*
bumble	*Bombus sylvarum*		*Scaurus uncinus*
European bumble	*B. agrorum*		*Zophobas rugipes*
	B. hypnorum	diving	*Dytiscus marginalis*
	B. lapidarius	domino	*Anthia venator*
	B. terrestris	dung-feeding scarab	*Aphodius* sp.
honey	*Apis cerana indica*	European darkling	*Phaleria provincialis*
	A. mellifera	European scarab	*Melolontha melolontha*
leaf-cutting	*Megachile latimanus*	flat grain	*Cryptolestes pusillus*
Southeast Asian honey	*Apis dorsata*	flea	*Altica* sp.
	A. florea	ground	*Abacomorphus asperulus*
	A. indica		*Abax ater*
Beetle	*Anisotarsus* sp.		*A. ovalis*
	Costelytra zealandica		*A. parallelus*
American water	*Liodessus flavicollis*		*Acinopus* sp.
blister	*Epicauta adspersa*		*Agonum assimilis*
	E. gorhami		*A. dorsalis*
	E. pilma		*A. marginatum*
	E. ruficeps		*A. moestum*
	E. waterhousei		*A. sexpunctatum*
bombardier	*Brachinus crepitans*		*A. viduum*
	B. explodens		*Amara familiaris*
	B. sclopeta		*A. similata*
carpet	*Trogoderma inclusum*		*Amblytelus curtus*
carrion	*Oecoptoma thorica*		*Anisodactylus binotatus*
	Phospuga atrata		*Arthropterus* sp.
	Silpha obscura		*Asaphidion flavipes*
caterpillar hunter	*Calosoma affini*		*Badister bipustulatus*
	C. alternans sayi		*Bembidion andreae*
	C. externum		*B. lampros*
	C. macrum		*B. quadriguttatum*
	C. marginalis		*Broscus cephalotes*
	C. oceanicum		*Calathus fuscipes*
	C. parvicollis		*C. melanocephalus*

continued

Common Name	Scientific Name	Common Name	Scientific Name
	Callistus lunatus		*N. rainbowi*
	Carabus auratus		*N. scotti*
	C. granulatus		*N. triplogenioides*
	C. problematicus		*N. variicollis*
	Carenum bonelli		*Odacantha melanura*
	C. interruptum		*Pamborus alternans*
	C. tinctillatum		*P. guerini*
	Castelnaudia superba		*P. pradieri*
	Chlaenius australis		*P. viridis*
	C. bipunctatus		*Panagaeus bipistulatus*
	C. chrysocephalus		*Pasimachus duplicatus*
	C. cordicollis		*Philophloeus australis*
	C. festivus		*Philoscaphus tuberculatus*
	C. tristus		*Polystichus connexus*
	C. vestitus		*Promecoderus* sp.
	Clivina basalis		*Prosopognus harpaloides*
	C. fossor		*Pseudoceneus iridescens*
	Craspedophorus sp.		*Pterostichus cupreus*
	Cratoferonia phylarchus		*P. macer*
	Cratogaster melas		*P. melas*
	Cychrus rostratus		*P. metallicus*
	Diachromus germanus		*P. niger*
	Diaphoromenus edwardsi		*P. vulgaris*
	Dicaelus dilatatus		*Rhytisternus laevilaterus*
	D. purpuratus		*Sarticus cyaneocinctus*
	D. splendidus		*Scaphinotus andrewsi gerari*
	Dichirotrichus obsoletus		*S. andrewsi montana*
	Dicrochile brevicollis		*S. viduus*
	D. goryi		*S. webbi*
	Drypta dentata		*Siagonyx blackburni*
	Elaphrus ripareus		*Sphallomorpha colym-*
	Eudalia macleayi		*betoides*
	Eurylychnus blagravei		*Sphodrosomus saisseti*
	E. olliffi		*Stenaptinus catoirei*
	Harpalus atratus		*S. verticalis*
	H. azurus		*Stenolophus mixtus*
	H. caliginosus		*Trichosternus nudipes*
	H. dimidiatus	hide	*Dermestes maculatus*
	H. distinguendus	Japanese	*Popillia japonica*
	H. griseus	June	*Cotinis* sp.
	H. luteicornis		*Phyllophaga* sp.
	H. pubescens		*Polyphylla* sp.
	H. tardus	khapra or carpet	*Trogoderma granarium*
	Helluo costatus	leaf	*Melasoma populi*
	Helluomorphoides ferrugi-		*Phylodecta vittelinae*
	neus		*Plagiodera* sp.
	H. latitarsus	long-horned	*Aromia moschata*
	Laccopterum foveigerum	margined blister	*Epicauta pestifera*
	Lebia chlorocephala	Mexican bean	*Epilachna varivestis*
	Leistus ferrugineus	mountain pine	*Dendroctonus ponderosae*
	Licinus nitidior	native elm bark	*Hylurgopinus rufipes*
	Loricera pilicornis	potato flea	*Epitrix cucumeris*
	Loxandrus longiformis	predacious diving	*Acilius sulcatus*
	Loxodactylus carinulatus		*Agabus bipustulatus*
	Molops elatus		*A. sturmi*
	Mystropomus regularis		*Colymbetes fuscus*
	Nebria livida		*Copelatus ruficollis*
	Notiophilus biguttatus		*Cybister confusus*
	Notonomus angustibasis		*C. lateralimarginalis*
	N. crenulatus		*C. limbatus*
	N. miles		*C. tripunctatus*
	N. muelleri		*Dytiscus latissimus*
	N. opulentus		*Graphoderus cinereus*

continued

Appendix II. COMMON NAMES AND CORRESPONDING SCIENTIFIC NAMES

Part I. Animals

Common Name	Scientific Name
	Hydroporus pallustris
	Ilybius fenestratus
	I. fuliginosus
	Rhantus exsoletus
red flower	*Tribolium castaneum*
red turpentine	*Dendroctonus valens*
round sand	*Omophron limbatum*
rove	*Paederus fuscipes*
	Stenus bipunctatus
rusty grain	*Cryptolestes ferrugineus*
saw-toothed grain	*Oryzaephilus surinamensis*
skin, or dermestid	*Attagenus megatoma*
smaller European elm bark	*Scolytus multistriatus*
soldier	*Chauliognathus lecontei*
southern pine	*Dendroctonus frontalis*
spider	*Ptinus ocellus*
striped blister	*Epicauta vittata*
striped cucumber	*Diabrotica vittata*
tobacco flea	*Epitrix hirtipennis*
twelve-spotted cucumber	*Diabrotica undecimpunc-tata*
varied carpet	*Anthrenus verbasci*
western pine	*Dendroctonus brevicomis*
western spotted cucumber	*Diabrotica undecimpunc-tata*
Bichir	*Polypterus* sp.
Billbug, maize	*Calendra maidis*
Blackbird, red-winged	*Agelaius phoeniceus*
tricolored	*A. tricolor*
Blood worm	*Glycera dibranchiata*
	G. ovigera
Bluegill	*Lepomis macrochirus pur-purescens*
Bobolink	*Dolichonyx oryzivorus*
Bollworm, pink	*Pectinophora gossypiella*
Bonefish	*Albula vulpes*
Boomslang	*Dispholidus typus*
Borer, currant	*Ramosia tipuliformis*
European corn	*Ostrinia nubilalis*
flatheaded apple tree	*Chrysobothris femorata*
oldhouse	*Hylotrupes bajulus*
peachtree	*Sanninoidea exitiosa*
roundheaded apple tree	*Saperda candida*
strawberry crown	*Tyloderma fragariae*
Boxfish	*Lactoria cornutus*
	L. fornasini
spiny	*Chilomycterus spinosus*
Braconid, little	*Apanteles glomeratus*
Brine shrimp	*Artemia salina*
Bristle worm	*Chloeia flava*
	Eunice aphroditois
	Eurythoe brasiliensis
	Hermodice carunculata
Budworm, spruce	*Choristoneura fumiferana*
	C. rosaceana
tobacco	*Heliothis virescens*
Bulltrout	*Notesthes robusta*
Burrower bug	*Macroscytus* sp.
	Scaptocoris divergens
Butterfly	*Lycorea ceres ceres*
black swallowtail	*Papilio polyxenes*
milkweed	*Amauris lobengula*
	Danaus chrysippus

Common Name	Scientific Name
	D. gilippus berenice
	D. gilippus strigosus
monarch	*D. plexippus*
orange-dog swallowtail	*Papilio cresphontes*
pipevine swallowtail	*Battus philenor*
spice-bush swallowtail	*Papilio troilus*
swallowtail	*Baronia brevicornis*
	Battus polydamus
	Graphium sarpedon choredon
	Ornithoptera priamus
	Papilio aegeus aegeus
	P. anactus
	P. antimachus
	P. aristolochiae
	P. demodocus
	P. demoleus sthenelus
	P. machaon
	P. palamedes
	Troides helena
tiger swallowtail	*Papilio glaucus*
zebra swallowtail	*Graphium marcellus*
Cabbageworm, imported	*Pieris rapae*
Cadelle	*Tenebroides mauritanicus*
Caiman, spectacled	*Caiman sclerops*
Canary	*Serinus canarius*
Cankerworm, spring	*Paleacrita vernata*
Carp	*Cyprinus carpio*
Cat	*Felis catus*
Caterpillar, eastern tent	*Malacosoma americanum*
forest tent	*M. disstria*
uglynest	*Archips cerasivoranus*
western tent	*Malacosoma pluviale*
Catfish, channel	*Ictalurus punctatus*
Cattle	*Bos taurus*
Brahman	*B. indicus*
Cecal worm of poultry	*Heterakis gallinarum*
Cecal worm of rodents	*Syphacia obvelata*
Chameleon, Jackson's	*Chamaeleo jacksoni*
Meller's	*C. melleri*
Chicken	*Gallus gallus*
Chigger	*Eutrombicula alfreddugesi*
Chimpanzee	*Pan troglodytes*
Chinaman fish or snapper	*Lutjanus nematophorus*
Chinch bug	*Blissus leucopterus*
Chipmunk, eastern	*Tamias striatus*
uinta	*Eutamias umbrinus*
Chub	*Leuciscus* sp.
Cicada, periodical	*Magicicada septendecim*
Clam, Atlantic surf	*Spisula solidissima*
soft-shell	*Mya arenaria*
Cobra, blacknecked spitting, or m'fesi	*Naja nigricollis*
Chinese	*N. naja atra*
Egyptian	*N. haje*
Indian or common	*N. naja*
king, or hamadryad	*Ophiophagus hannah*
ringhals spitting	*Hemachatus haemachatus*
Cockroach	*Blaberus giganteus*
	Desmosteria scripta
	Drymaplaneta communis
	D. semivitta
	D. shelfordi
	Eurycotis biolleyi

continued

Common Name	Scientific Name	Common Name	Scientific Name
	E. decipiens	Cormorant, double-crested	*Phalacrocorax auritus*
	Euzosteria nobilis	Cottonmouth or water moc-	*Agkistrodon piscivorus*
	Maoriblatta novaeseeland-	casin	
	iae	Cottontail, New England	*Sylvilagus transitionalis*
	Megazosteria patula	Cowbird, brown-headed	*Molothrus ater*
	Methana convexa		*M. ater obscurus*
	Pelmatosilpha coriacea	Crab	*Cancer* sp.
	Platyzosteria armata	Atlantic fiddler	*Uca pugilator*
	P. castenea	blue	*Callinectes sapidus*
	P. coolgardiensis	fiddler	*Uca maracoani*
	P. jungi		*U. pugnax*
	P. morosa		*U. rapax*
	P. nitidella		*U. tangeri*
	P. occidentalis	freshwater	*Potamon potamios*
	P. ruficeps	ghost	*Ocypode ceratophthalma*
	P. scabra	green	*Carcinus maenas*
	P. scabrella	horseshoe or king	*Limulus polyphemus*
	P. soror	lady	*Ovalipes ocellatus*
	P. stradbrokensis	long-eyed swimming	*Podophthalmus vigil*
	Polyzosteria cuprea	mangrove	*Goniopsis cruentata*
	P. limbata	spider	*Libinia emarginata*
	P. mitchelli		*Maja* sp.
	P. oculata	West Indies land	*Cardisoma guanhumi*
	P. pulchra	Crayfish	*Astacus astacus*
	P. viridissima	cave	*Cambarus virilis*
	Zonioploca bicolor		*Orconectes pellucidus*
	Z. pallida	freshwater	*O. immunis*
American	*Periplaneta americana*		*O. rusticus*
Florida	*Eurycotis floridana*		*O. virilis*
German	*Blattella germanica*	Louisiana red	*Procambarus clarkii*
Madeira	*Leucophaea maderae*	Cricket, field	*Gryllus campestris*
Oriental	*Blatta orientalis*	Mormon	*Anabrus simplex*
Pacific beetle	*Diploptera punctata*	Crow, American common	*Corvus brachyrhynchos*
West Indies	*Byrsotria fumigata*	Curculio, cowpea	*Chalcodermus aeneus*
Cod, Atlantic	*Gadus morhua*	plum	*Conotrachelus nenuphar*
Coelacanth	*Latimeria chalumnae*	Cuttlefish	*Sepia officinalis*
Comb jelly	*Beroe cucumis*	Western Pacific	*S. esculenta*
Cone, alphabet	*Conus spurius*		*Sepiella japonica*
bluish	*C. lividus*		*Todarodes pacificus*
California	*C. californicus*	Cutworm	*Apamea monoglypha*
cat	*C. catus*	variegated	*Peridroma saucia*
court	*C. aulicus*		
geographer	*C. geographus*		
imperial	*C. imperialis*	Daddy-longlegs	*Heteropachyloidellus robus-*
Indopacific	*C. quercinus*		*tus*
Indopacific oak	*C. magus*	Damselfly, European	*Lestes sponsa*
marbled	*C. marmoreus*	Darner, green	*Anax junius*
striated	*C. striatus*	Dassie, Bruce's yellow-	*Heterohyrax brucei*
tulip	*C. tulipa*	spotted	
woven	*C. textile*	Deer, Japanese, or sika	*Cervus nippon*
Conenose, bloodsucking, or	*Triatoma sanguisuga*	white-tailed	*Odocoileus virginianus*
assassin bug		Dog	*Canis familiaris*
Conenosed bug	*Panstrongylus* sp.	Dogfish	*Mustelus* sp.
Convict fish, Indopacific	*Acanthurus triostegus*	spiny	*Squalus acanthias*
Copepod, cyclopoid	*Cyclops serulatus*	Donkey	*Equus asinus*
	C. vernalis	Dormouse, common	*Muscardinus avellanarius*
	C. viridis	fat	*Glis glis*
freshwater	*Diaptomus shoshone*	Dove, Inca	*Scardafella inca*
Copperhead, Australian	*Denisonia superba*	mourning	*Zenaidura macroura*
or dryland	*Agkistrodon contortrix*		*Z. macroura carolinensis*
mocassin		Dragonfly	*Libellula needhami*
Coral, stinging	*Millepora alcicornis*	European	*Anax imperator*
stony	*Porites clavaria*	Duck, common pintail	*Anas acuta*

continued

Common Name	Scientific Name	Common Name	Scientific Name
mallard	*A. platyrhynchos*	Oriental blood	*S. japonicum*
Pekin or domestic	*A. platyrhynchos domesticus*	salmon-poisoning	*Nanophyetus salmincola*
		Fly, black	*Simulium arcticum*
			S. occidentale
Earthworm	*Eisenia* sp.		*S. ornatum*
	Lumbricus terrestris		*S. slossonae*
Earwig, European	*Forficula auricularia*		*S. vittalum*
Earworm or bollworm,	*Heliothis zea*	black blow	*Phormia regina*
corn		blow	*Calliphora vomitoria*
Eel, brown moray	*Gymnothorax undulatus*		*Chrysomya bezziana*
electric	*Electrophorus electricus*		*C. chloropyga*
European conger	*Conger conger*		*Lucilia caesar*
moray	*Gymnothorax flavimargina-*	blow or bluebottle	*Calliphora erythrocephala*
	tus	blow or greenbottle	*Phaenicia sericata*
	G. javanicus	carrot rust	*Psila rosae*
speckled moray	*G. pictus*	chalcid	*Aphelinus semiflavus*
white-spotted moray	*G. meleagris*	deer	*Chrysops* sp.
		egg-parasitic, or tri-	*Trichogramma cacoeciae*
		chogrammatid	
Fer-de-lance	*Bothrops atrox*		*T. minutum*
Ferret, European	*Mustela furo*	flesh	*Sarcophaga aldrichi*
Filarial worm, African	*Loa loa*		*Wohlfahrtia vigil*
Bancroft's	*Wuchereria bancrofti*	frit	*Oscinella frit*
blinding	*Onchocerca volvulus*	fruit	*Drosophila deflexa*
Malayan	*Brugia malayi*		*D. funebris*
Filefish, scrawled	*Aluterus scriptus*		*D. melanogaster*
Finch, black rosy	*Leucosticte atrata*		*D. pseudoobscura*
brambling	*Fringilla montifringilla*		*D. repleta*
bull-	*Pyrrhula pyrrhula*		*D. subobscura*
chaf-	*Fringilla coelebs*	greenhouse white-	*Trialeurodes vaporariorum*
European gold-	*Carduelis carduelis*	Hessian	*Mayetiola destructor*
gray-crowned rosy	*Leucosticte tephrocotis lit-*	horn	*Haematobia irritans*
	toralis	horse	*Hybomitra* sp.
	L. tephrocotis tephrocotis		*Tabanus* sp.
green-	*Carduelis chloris*	horse bot	*Gasterophilus haemorrhoi-*
	Chloris chloris		*dalis*
house	*Carpodacus mexicanus*		*G. intestinalis*
Firebrat	*Thermobia domestica*		*G. nasalis*
Flea, cat	*Ctenocephalides felis*	house	*Musca domestica*
dog	*C. canis*	human bot	*Dermatobia hominis*
European rabbit	*Spilopsyllus cuniculi*	Mediterranean fruit	*Ceratitis capitata*
European rat	*Ceratophyllus fasciatus*	narcissus bulb	*Lampetia equestris*
human	*Pulex irritans*		*Merodon equestris*
mouse	*Leptopsylla segnis*	sand	*Phlebotomus papatasii*
northern rat	*Nosopsyllus fasciatus*	screwworm	*Cochliomyia hominivorax*
Oriental rat	*Xenopsylla cheopis*	secondary screwworm	*C. macellaria*
sticktight	*Echidnophaga gallinacea*	sheep bot	*Oestrus ovis*
water, or cladoceran	*Leptodora kindtii*	stable	*Stomoxys calcitrans*
Fleahopper, cotton	*Psallus seriatus*	tachinid	*Lydella thompsoni*
Flounder	*Platichthys flesus*	tsetse	*Glossina morsitans*
winter	*Pseudopleuronectes ameri-*	Fox, arctic	*Alopex lagopus*
	canus	red	*Vulpes fulva*
Fluke	*Himasthla quissetensis*	Frog	*Nectophrynoides* sp.
amphistome	*Gastrodiscoides hominis*	bull-	*Rana catesbeiana*
Chinese liver	*Clonorchis sinensis*	chorus	*Pseudacris triseriata*
human blood	*Schistosoma haematobium*	common European	*Rana temporaria*
human lung	*Paragonimus westermanii*	cricket	*Acris crepitans*
intestinal, of fish	*Zoogonus rubellus*	European edible	*Rana esculenta*
large intestinal	*Fasciolopsis buski*	European green tree	*Hyla arborea*
liver	*Fasciola hepatica*	golden arrow-poison	*Dendrobates auratus*
Manson's blood	*Schistosoma mansoni*	leopard	*Rana pipiens*

continued

Part I. Animals

Common Name	Scientific Name
marbled	*Limnodynastes tasmaniensis*
Pacific tree	*Hyla regilla*
South Australian lepto-dactylid	*Crinia signifera*
tropical American bull-	*Leptodactylus pentadacty-lus*
White's tree	*Hyla caerulea*
Gannet, blue-faced booby	*Sula dactylatra*
Gar	*Lepisosteus* sp.
Gecko	*Gehyra variegata*
Gerbil	*Rhombomys* sp.
Gila monster	*Heloderma suspectum*
Goat	*Capra hircus*
Goldfish	*Carassius auratus*
Goosefish	*Lophius* sp.
Grasshopper, clear-winged	*Camnula pellucida*
differential	*Melanoplus differentialis*
high plains	*Dissosteira longipennis*
migratory	*Melanoplus sanguinipes*
red-legged	*M. femur-rubrum*
short-horned	*Phymateus baccatus*
	P. viridipes
	Poekilocerus bufonius
	P. pictus
two-striped	*Melanoplus bivittatus*
Grosbeak, evening	*Hesperiphona vespertina*
rose-breasted	*Pheucticus ludovicianus*
Grouper	*Plectropomus oligacanthus*
mottled	*Epinephelus fuscoguttatus*
yellowfin	*Mycteroperca venenosa*
Grouse, ruffed	*Bonasa umbellus*
Grub, cattle	*Hypoderma bovis*
common cattle	*H. lineatum*
Guinea pig	*Cavia porcellus*
Guinea worm	*Dracunculus medinensis*
Gull, common	*Larus canus*
Haddock	*Melanogrammus aeglefinus*
Hagfish	*Myxine* sp.
Halfbeak, Indo-Australian	*Zenarchopterus buffoni*
	Z. dispar
wrestling	*Dermogenys pusillus*
Hamster, European	*Cricetus cricetus*
golden or Syrian	*Mesocricetus auratus*
Hare, American, or snow-shoe rabbit	*Lepus americanus*
blue	*L. timidus*
Japanese	*L. brachyurus*
Harlequin bug	*Murgantia histrionica*
Hatchetfish	*Gasteropelecus* sp.
Hawkmoth, privet	*Sphinx ligustri*
Heart worm, dog	*Dirofilaria immitis*
Hedgehog, European	*Erinaceus europaeus*
Roumanian	*E. europaeus roumanicus*
Herring	*Clupea harengus*
Hookworm	*Ancylostoma tubaeforme*
	Necator americanus
creeping-eruption	*Ancylostoma braziliense*
dog	*A. caninum*

Common Name	Scientific Name
Old World	*A. duodenale*
sheep	*Bunostomum trigonocepha-lum*
Hornet	*Vespula vulgaris*
Hornworm, tomato	*Manduca quinquemaculata*
Horse	*Equus caballus*
Hummingbird, Allen's	*Selasphorus sasin*
Anna's	*Calypte anna*
Hydroid	*Eudendrium* sp.
feather	*Lytocarpus nuttingi*
stinging	*Pennaria tiarella*
	Sarsia tubulosa
Hyocephalid bug	*Hyocephalus* sp.
Iguana, desert	*Dipsosaurus dorsalis*
Intestinal worm	*Paraspidodera uncinata*
	Physaloptera tumefaciens
	Ternidens deminutus
Ips, California five-spined	*Ips confusus*
Isopod, European beach	*Tylos latreillei*
Jack	*Caranx hippos*
Jararaca	*Bothrops jararaca*
Jellyfish	*Pelagia noctiluca*
giant	*Cyanea capillata*
moon	*Aurelia aurita*
pink-fringed	*Dactylometra quinquecirrha*
or cabbage bleb	*Rhizostoma pulmo*
or siphonophore	*Rhizophysa eysenharti*
upside down	*Cassiopea frondosa*
Jointworm, wheat	*Harmolita tritici*
Junco, Oregon	*Junco oreganus montanus*
	J. oreganus oreganus
	J. oreganus pinosus
	J. oreganus shufeldti
	J. oreganus thurberi
slate-colored	*J. hyemalis*
	J. hyemalis hyemalis
Katydid, broad-winged	*Microcentrum rhombifolium*
Koala	*Phascolarctos cinereus*
Krait, Indian	*Bungarus caeruleus*
Krill, euphausiacean	*Euphausia pacifica*
Lace bug, oak	*Corythuca arcuata*
Lamprey, sea	*Petromyzon marinus*
Lancelet	*Branchiostoma caribaeum*
Lapwing, southern	*Belonopterus chilensis lam-pronotus*
Leaf-footed bug	*Acanthocephala declivis*
	A. femorata
	A. granulosa
	Leptoglossus clypeatus
Leafhopper, beet	*Circulifer tenellus*
painted	*Endria inimica*
potato	*Empoasca fabae*
six-spotted	*Macrosteles divisus*
Leaf miner, serpentine	*Liriomyza pusilla*

continued

Common Name	Scientific Name
Leafroller, red-banded	*Argyrotaenia velutinana*
Leafworm, cotton	*Alabama argillacea*
Leech	*Haemopis* sp.
Lemming, brown	*Lemmus trimucronatus*
collared	*Dicrostonyx groenlandicus*
Lightning bug	*Lampyris* sp.
Lionfish	*Pterois antennata*
	P. radiata
Lizard, Australian bearded	*Amphibolurus barbatus*
California legless	*Anniella pulchra*
California side-blotched	*Uta stansburiana hesperis*
collared	*Crotaphytus collaris*
Colorado desert fringe-toed	*Uma notata*
Cuban anole	*Anolis homolechis*
European fence	*Lacerta agilis*
great basin fence	*Sceloporus occidentalis biseriatus*
green anole	*Anolis carolinensis*
horned	*Phrynosoma cornutum*
Mediterranean	*Lacerta sicula*
ornate	*Amphibolurus ornatus*
Pacific fence, or iguana	*Sceloporus occidentalis*
sagebrush	*S. graciosus*
scaly-foot legless	*Pygopus baileyi*
side-blotched, or ground uta	*Uta stansburiana*
southern alligator	*Gerrhonotus multicarinatus*
Texas spiny	*Sceloporus olivaceus*
yucca night	*Xantusia vigilis*
Lobster, American	*Homarus americanus*
European	*H. gammarus*
Japanese spiny	*Panulirus japonicus*
Locust, desert	*Schistocerca gregaria*
French migratory	*Locusta migratoria gallica*
migratory	*L. migratoria*
Looper, cabbage	*Trichoplusia ni*
linden	*Errannis tiliaria*
Louse, blackberry plant or psyllid	*Trioza tripunctata*
book	*Liposcelis divinatorius*
chicken body	*Menacanthus stramineus*
chicken head	*Cuclotogaster heterographus*
common mouse	*Polyplax serrata*
dog-biting	*Trichodectes canis*
grape plant	*Phylloxera vitifoliae*
head and body	*Pediculus humanus*
hog	*Haematopinus suis*
long-nosed cattle	*Linognathus vituli*
monkey	*Pedicinus eurygaster*
oval guinea pig	*Gyropus ovalis*
rabbit	*Haemodipsus ventricosus*
shortnosed cattle	*Haematopinus eurysternus*
slender guinea pig	*Gliricola porcelli*
spined rat	*Polyplax spinulosa*
Louse of chicken, shaft	*Menopon gallinae*
Lucerneflea	*Sminthurus viridis*
Lugworm	*Arenicola* sp.
Lungfish	*Neoceratodus* sp.
African	*Protopterus aethiopicus*

Common Name	Scientific Name
Lungworm, swine	*Metastrongylus apri*
Lungworm of cattle	*Dictyocaulus viviparus*
Lungworm of sheep, thread	*D. filaria*
Macaque, pig-tailed	*Macaca nemestrina*
Mackerel, Atlantic	*Scomber scombrus*
Maggot, apple	*Rhagoletis pomonella*
onion	*Hylemya antiqua*
Mamba, black	*Dendroaspis polylepis*
eastern green	*D. angusticeps*
Man	*Homo sapiens*
Marmot, Eurasian	*Marmota marmota*
Marten, American	*Martes americana*
Meadowlark, eastern	*Sturnella magna*
Mealworm, dark	*Tenebrio obscurus*
lesser	*Alphitobius diaperinus*
yellow	*Tenebrio molitor*
Mealybug, citrus	*Planococcus citri*
Melolontha, European	*Melolontha hippocastani*
Midge, European	*Pseudosmittia arenaria*
phantom	*Chaoborus* sp.
Milkweed bug, large	*Oncopeltus fasciatus*
Millipede	*Abacion magnum*
	Chicobolus spinigerus
	Doratogonus annulipes
	Floridobolus penneri
	Pachybolus laminatus
	Spirostreptidus castaneus
	Spirostreptus virgator
European pill	*Glomeris marginata*
flat-backed	*Apheloria corrugata*
	Cherokia georgiana
	Gomphodesmus pavani
	Nannaria sp.
	Oxidus gracilus
	Pachydesmus crassicutus
	Polydesmus collaris collaris
	P. vicinus
	P. virginiensis
	Pseudopolydesmus serratus
	Rhinocricus insulatus
juliform	*Archiulus sabulosus*
	Blaniulus guttulatus
	Brachyiulus ulineatus
	Cambala hubrichti
	Cylindroiulus teutonicus
	Julus heserus
	J. terrestris
	Narceus annularis
	N. gordanus
	Oriulus delus
	Orthoporus conifer
	O. flavior
	O. punctilliger
	Trigonoiulus lumbricinus
	Uroblaniulus canadensis
Mink	*Mustela vison*
Minnow	*Phoxinus phoxinus*
Mite, beetle	*Oribatula* sp.
bulb	*Rhizoglyphus echinopus*
chicken	*Dermanyssus gallinae*

continued

1335

Common Name	Scientific Name	Common Name	Scientific Name
cyclamen	*Steneotarsonemus pallidus*	cynthia	*Samia cynthia*
dog follicle	*Demodex canis*	diamondback	*Plutella maculipennis*
ear	*Otodectes cyanotis*	false codling	*Argyroploce leucotreta*
European red	*Panonychus ulmi*	flour Mediterranean	*Anagasta kuehniella*
grass	*Galumna* sp.	giant silkworm	*Antheraea pernyi*
itch	*Sarcoptes scabiei*	greater wax	*Galleria mellonella*
mouse	*Trichoecius romboutsi*	gypsy	*Porthetria dispar*
myocoptic mange	*Myocoptes musculinus*	Indian meal	*Plodia interpunctella*
northern fowl	*Bdellonyssus sylviarum*	leaf skeletonizer	*Procris geryon*
notoedric mange	*Notoedres cati*		*Zygaena filipendulae*
pear leaf blister	*Eriophyes pyri*		*Z. lonicerae*
rabbit	*Cheyletiella parasitivorax*		*Z. trifolii*
sheep scab	*Psoroptes equi ovis*	Old World noctuid	*Leucania conigera*
tropical rat	*Ornithonyssus bacoti*		*L. impura*
two-spotted spider	*Tetranychus urticae*		*L. pallens*
water	*Arrenurus marshallae*	oriental fruit	*Grapholitha molesta*
	A. megalurus	prominent	*Cerura vinula*
	Hydrachna cruenta		*Schizura leptinoides*
Mite of guinea pig, fur	*Chirodiscoides caviae*	satin	*Stilpnotia salicis*
Mite of mouse, follicle	*Psorergates simplex*	silkworm	*Bombyx mori*
fur	*Myobia musculi*	sphinx	*Laothoe populi*
	Radfordia affinis	tiger	*Arctia caja*
Mite of rabbit, ear canker	*Psoroptes cuniculi*		*A. villica*
Mite of rat, ear mange	*Notoedres muris*		*Cycnia tenera*
fur	*Radfordia ensifera*		*Diacrisia medica*
Mole, eastern American	*Scalopus aquaticus*		*Phragmatobia fuliginosa*
star-nosed	*Condylura cristata*		*Tyria jacobaeae*
Monkey	*Saimiri* sp.		*Utetheisa bella*
rhesus	*Macaca mulatta*	tobacco	*Ephestia elutella*
Mosquito	*Aedes albopictus*	underwing	*Catocala* sp.
	A. communis	Mouse, birch	*Sicista betulina*
	A. hexodontus	common house	*Mus musculus domesticus*
	A. leucocelaenus	deer	*Peromyscus maniculatus*
	A. polynesiensis	European wood	*Apodemus flavicollis*
	A. triseriatus	house	*Mus musculus*
	Anopheles funestus	little pocket	*Perognathus longimembris*
	A. gambiae	meadow jumping	*Zapus hudsonius*
	A. pseudopunctipennis	pale kangaroo	*Microdipodops pallidus*
	Armigeres sp.	plains or pale pocket	*Perognathus hispidus*
	Culex tarsalis	rock pocket	*P. intermedius*
	Haemagogus spegazzinii	white-footed	*Peromyscus leucopus*
	Mansonia sp.	Mouse bird	*Colius striatus*
common malaria	*Anopheles quadrimaculatus*	Mud puppy	*Necturus maculosus*
northern house	*Culex pipiens*	Mummichog	*Fundulus heteroclitus*
rice-field	*Psorophora confinnis*	Mussel, edible	*Mytilus edulis*
southern house	*Culex pipiens quinquefasciatus*		
yellow-fever	*Aedes aegypti*	Nematode	*Longidorus* sp.
Moth	*Bryotropha similis*	awl	*Dolichodorus heterocephalus*
	Holomelina aurantiaca	bulb and stem	*Ditylenchus dipsaci*
	H. ferruginosa	burrowing	*Radopholus similis*
	H. fragilis	citrus	*Tylenchulus semipenetrans*
	H. immaculata	coconut palm	*Rhadinaphelenchus cocophilus*
	H. laeta		
	H. lamae	dagger	*Xiphinema* sp.
	H. nigricans	lance	*Hoplolaimus galeatus*
	H. rubicundaria	meadow	*Pratylenchus mahogani*
	Phlogophora meticulosa	plant root	*Paratylenchus* sp.
almond	*Cadra cautella*	potato-rot	*Ditylenchus destructor*
caterpillar	*Polia nebulosa*	reniform	*Rotylenchulus reniformis*
	P. persicariae	rice root	*Hirschmanniella oryzae*
codling	*Carpocapsa pomonella*	ring	*Criconemoides* sp.
	Laspeyresia pomonella		

Part I. Animals

Common Name	Scientific Name
root-knot	*Meloidogyne* sp.
Scribner's meadow	*Pratylenchus scribneri*
soybean cyst	*Heterodera glycines*
spiral	*Helicotylenchus* sp.
	Rotylenchus sp.
	Scutellonema sp.
sting	*Belonolaimus longicaudatus*
stubby-root	*Trichodorus christiei*
stunt	*Tylenchorhynchus* sp.
sugar-beet	*Heterodera schactii*
wheat	*Anguina tritici*
Nematode of carnivores, large	*Toxascaris leonina*
Nematode of potato, golden	*Heterodera rostochiensis*
Nematode of rabbit	*Graphidium strigosum*
Nematode of strawberry, spring-dwarf	*Aphelenchoides fragariae*
Nematode of strawberry, summer-dwarf	*A. besseyi*
Nematode worm, capillary	*Capillaria hepatica*
parasitic	*Abbreviata caucasica*
	Streptopharagus pigmentatus
Newt, California	*Taricha torosa*
common	*Notophthalmus viridescens*
smooth	*Triturus vulgaris*
Nighthawk	*Chordeiles minor*
Nightjar, spotted	*Eurostopodus guttatus*
Nodular worm	*Oesophagostomum aculeatum*
	O. bifurcum
monkey	*O. apiostomum*
sheep	*O. columbianum*
Octopus	*Eledone aldrovandi*
	Octopus apollyon
	O. bimaculatus
	O. fitchi
	O. flindersi
	O. vulgaris
common	*O. macropus*
giant Pacific	
Mediterranean musk	*Eledone moschata*
western Pacific	*Octopus ocellatus*
Oligochaete worm, aquatic	*Limnodrilus* sp.
Opossum, common	*Didelphis marsupialis virginiana*
large American	*D. marsupialis*
murine	*Marmosa mexicana isthmica*
	M. microtarsus
Opossum-shrimp	*Mysidium gracile*
Ostrich, African	*Struthio camelus*
Oyster, Virginia	*Crassostrea virginica*
Paddlefish	*Polyodon* sp.
Palolo worm, West Indian	*Eunice schemacephala*
Parrotfish	*Scarus microrhinos*
blue	*S. coeruleus*
Peeper, spring	*Hyla crucifer*
Pelican, brown	*Pelecanus occidentalis*
Perch, yellow	*Perca flavescens*
Periwinkle, European	*Littorina neritoides*

Common Name	Scientific Name
northern yellow	*L. obtusata*
rough	*L. saxatilis*
Pheasant, ring-necked	*Phasianus colchicus*
Pigeon	*Columba oenas*
street	*C. livia*
wood	*C. palumbus*
Pike, northern	*Esox lucius*
Pinworm	*Aspiculuris tetraptera*
	Enterobius vermicularis
	Passalurus ambiguus
	Syphacia muris
Pit viper, or bushmaster	*Lachesis muta*
Formosan, or habu	*Trimeresurus mucrosquamatus*
Malayan	*Calloselasma rhodostoma*
Okinawan, or habu	*Trimeresurus flavoviridis*
Old World, or mamushi	*Agkistrodon halys*
Planarian, freshwater	*Dugesia gonocephala*
Plant bug	*Leptoterna dolabrata*
tarnished	*Lygus lineolaris*
Platypus	*Ornithorhynchus anatinus*
Poor-will	*Phalaenoptilus nuttallii*
Porgy	*Pagellus erythrinus*
	Pagrus pagrus
Portuguese man-of-war	*Physalia physalis*
Possum, dormouse	*Cercaertus nanus*
Prawn, common	*Palaemonetes vulgaris*
Psylla, pear	*Psylla pyricola*
Ptarmigan, willow	*Lagopus lagopus*
Puffer	*Arothron hispidus*
	Fugu basilevskianus
	F. chrysops
	F. niphobles
	F. ocellatus
	F. pardalis
	F. pseudommus
	F. rubripes
	F. stictonotus
	F. vermicularis
	F. xanthopterus
	Tetraodon lineatus
	Torquigener hamiltoni
bandtail	*Sphoeroides spengleri*
black-spotted	*Arothron nigropunctatus*
bullseye	*Sphoeroides annulatus*
freshwater	*Colomesus psittacus*
northern	*Sphoeroides maculatus*
sharp-nosed	*Canthigaster margaritatus*
	C. rivulatus
smooth	*Lagocephalus laevigatus inermis*
	L. lunaris
	L. sceleratus
white-spotted	*Arothron meleagris*
Punkie	*Culicoides grahamii*
	C. impunctatus
Python, common African	*Python sebae*
Quahog, northern	*Mercenaria mercenaria*
Quail, bobwhite	*Colinus virginianus*
European	*Coturnix coturnix*

continued

Common Name	Scientific Name
Rabbit, antelope jack-	*Lepus alleni*
black-tailed jack-	*L. californicus*
European	*Oryctolagus cuniculus*
Rabbit-fish	*Chimaera* sp.
Raccoon	*Procyon lotor*
crab-eating	*P. cancrivorus*
Rat, black	*Rattus rattus*
desert wood	*Neotoma lepida*
dusky-footed wood	*N. fuscipes*
Norway	*Rattus norvegicus*
Raven, white-necked	*Corvus cryptoleucus*
Ray, bat	*Myliobatis californica*
eagle	*Aetomylaeus nichofii*
marbled electric	*Torpedo marmorata*
spotted eagle	*Aetobatus narinari*
Red bug	*Dysdercus intermedius*
	Trombicula akamushi
Redstart, European	*Phoenicurus phoenicurus*
Robin, European	*Erithacus rubecula*
Roundworm, large	*Ascaris lumbricoides*
Roundworm of carnivores, large	*Toxocara canis*
Roundworm of chicken, large	*Ascaridia galli*
Roundworm of horse	*Parascaris equorum*
Roundworm of swine, large	*Ascaris suum*
Sable	*Martes zibellina*
Salamander, common fire	*Salamandra salamandra*
dusky	*Desmognathus fuscus*
red-backed	*Plethodon cinereus*
slimy	*P. glutinosus*
spotted	*Ambystoma maculatum*
tiger	*A. tigrinum*
Salmon, Atlantic	*Salmo salar*
chinook	*Oncorhynchus tshawytscha*
chum	*O. keta*
coho	*O. kisutch*
sockeye	*O. nerka*
Sand hopper	*Talitrus saltator*
Pacific coast	*Orchestoidea benedicti*
	O. californiana
	O. corniculata
Sawfly	*Tenthredo arcuata*
conifer	*Neodiprion lecontei*
European pine	*N. sertifer*
European spruce	*Diprion hercyniae*
European wheat stem	*Cephus pygmaeus*
jack-pine	*Neodiprion pratti banksianae*
wheat stem	*Cephus cinctus*
Scale, brown soft	*Coccus hesperidum*
scurfy	*Chionaspis furfura*
Scale insect, black	*Saissetia oleae*
San Jose	*Quadraspidiotus perniciosus*
Scallop	*Pecten* sp.
giant	*Placopecten magellanicus*
Scorpion, American whip	*Mastigoproctus giganteus*
Old World desert	*Leiurus quinquestriatus*
Scorpionfish	*Choridactylus multibarbis*
	Minous monodactylus
	Scorpaenopsis diabolis

Common Name	Scientific Name
California	*Scorpaena guttata*
devil	*Inimicus didactylus*
	I. japonicus
orange	*Scorpaena scrofa*
small-scaled	*S. porcus*
Scour worm, black	*Trichostrongylus colubriformis*
Sculpin	*Scorpaena plumieri*
Pacific staghorn	*Leptocottus armatus*
Sea anemone	*Actinia equina*
	Anemonia sulcata
	Sagartia elegans
Sea cucumber	*Actinopyga agassizi*
	Bohadschia argus
	Cucumaria echinata
	Holothuria atra
	H. impatiens
	H. poli
	Patinapta ooplax
	Stichopus variegatus
Sea hare	*Fimbria fimbria*
California	*Aplysia californica*
Sea nettle	*Chrysaora hysoscella*
Sea pen	*Cavernularia obesa*
Sea slug, dorid	*Anisodoris nobilis*
Sea urchin	*Echinothrix calamaris*
	E. diadema
	Phormosoma bursarium
	Psammechinus microtuberculatus
	Toxopneustes pileolus
	Tripneustes gratilla
black	*Diadema setosum*
common Atlantic	*Arbacia punctulata*
leather	*Asthenosoma varium*
western purple	*Strongylocentrotus purpuratus*
Sea wasp	*Chironex fleckeri*
	Chiropsalmus quadrigatus
Seal	*Phoca* sp.
fur	*Callorhinus* sp.
Shad	*Clupanodon thrissa*
Shark	*Scyliorhinus* sp.
Atlantic bonnet	*Sphyrna tiburo*
lemon	*Negaprion brevirostris*
nurse	*Ginglymostoma cirratum*
requiem	*Carcharhinus* sp.
Sheep	*Ovis aries*
Sheep ked	*Melophagus ovinus*
Shrimp, burrowing	*Callianassa californiensis*
Sidewinder	*Crotalus cerastes*
Silverfish	*Lepisma saccharina*
Siskin, European	*Carduelis spinus*
Skate, winter or big	*Raja oscellata*
Skink, Australian blotched bluetongue	*Tiliqua nigrolutea*
Great Plains	*Eumeces obsoletus*
Stokes'	*Egernia stokesii*
Slater, rock	*Ligia baudiniana*
Slowworm	*Anguis fragilis*
Sludge worm	*Tubifex* sp.
Slug, bristly rose-	*Cladius isomerus*
Smelt, rainbow	*Osmerus mordax*

continued

Part I. Animals

Common Name	Scientific Name	Common Name	Scientific Name
Snail, Argentine land	*Cryptomphallus* sp.	golden-crowned	*Zonotrichia atricapilla*
common mud	*Ilyanassa obsoleta*	Harris'	*Z. querula*
dented garden, or	*Helix aspersa*	hedge	*Prunella modularis*
petit-gris		house	*Passer domesticus*
freshwater	*Galba* sp.	tree	*Spizella arborea*
	Goniobasis sp.	white-crowned	*Z. leucophrys gambelii*
	Lymnaea stagnalis		*Z. leucophrys leucophrys*
	Pseudosuccinea sp.		*Z. leucophrys nuttalli*
Snake, beaked or Malayan	*Enhydrina schistosa*		*Z. leucophrys pugetensis*
sea		white-throated	*Z. albicollis*
black	*Coluber constrictor*	Spider, funnel-web	*Agelena labyrinthica*
broadbanded blue	*Laticauda semifasciata*		*A. similis*
sea, or erabu		trap-door	*Aphonopelma californica*
bull	*Pituophis catenifer*	wolf	*Arctosa cinerea*
Central American	*Micrurus nigrocinctus*		*A. perita*
coral			*A. variana*
dugite	*Demansia nuchalis affinis*	Spittlebug, meadow	*Philaenus spumarius*
eastern coachwhip	*Masticophis flagellum flagellum*	Sponge, brown	*Neofibularia nolitangere*
		fire	*Tedania ignis*
eastern coral or har-	*Micrurus fulvius*	green	*Haliclona viridis*
lequin		reddish-brown	*Tedania toxicalis*
eastern diamond-	*Crotalus adamanteus*	Squaretail, smalleye	*Tetragonurus cuvieri*
back rattle-		Squash bug	*Anasa tristis*
eastern or common	*Demansia textilis*	Squid	*Dosidicus gigas*
brown		common American or	*Loligo pealei*
ground rattle-, or	*Sistrurus catenatus*	Atlantic	
eastern massasauga		northern Atlantic	*L. forbesi*
king	*Lampropeltis getulus*	Pacific common	*Ommastrephes sloani pacificus*
king brown or mulga	*Pseudechis australis*		
Mohave rattle-	*Crotalus scutulatus*	Palk Bay	*Sepioteuthis lessoniana*
North American	*Natrix sipedon*	western Pacific	*Euprymna morsei*
water		Squirrel, Arctic ground	*Spermophilus undulatus*
Papuan black	*Pseudechis papuanus*	Columbian ground	*S. columbianus*
plains garter	*Thamnophis radix*	common gray	*Sciurus carolinensis pennsylvanicus*
prairie rattle-	*Crotalus viridis*		
red-bellied or Aus-	*Pseudechis porphyriacus*	European red	*S. vulgaris*
tralian black		Franklin ground	*Spermophilus franklini*
rough-scaled	*Tropidechis carinatus*	golden-mantled	*S. lateralis*
South American	*Micrurus corallinus*	ground	
coral		Mohave ground	*S. mohavensis*
tiger	*Notechis scutatus*	Richardson's	*S. richardsonii*
tropical rattle-, or	*Crotalus durissus durissus*	ground	
cascabel		roundtail ground	*S. tereticaudus*
	C. durissus terrificus	southern flying	*Glaucomys volans*
western brown	*Demansia nuchalis nuchalis*	thirteen-lined	*Spermophilus tridecemlineatus*
western diamond-	*Crotalus atrox*	ground	
back rattle-		white-tailed ante-	*Ammospermophilus leucurus*
yellow-bellied or	*Pelamis platurus*	lope	
pelagic sea			*Spermophilus leucurus*
Snapper	*Gnathodentex aureolineatus*	Starfish	*Asterias forbesi*
	Lutjanus monostigmus	brittle	*Ophioderma brevispinum*
blue-gray	*Aprion virescens*	Starling, common	*Sturnus vulgaris*
red	*Lutjanus gibbus*	Stick insect	*Carausius morosus*
	L. vaigiensis	Stickleback, three-spined	*Gasterosteus aculeatus*
twinspot	*L. bohar*	Stingray	*Dasyatis brevis*
Snapper-like fish	*Lethrinus miniatus*		*D. guttata*
Soapfish, greater	*Rypticus saponaceus*		*D. longus*
Sole, lemon	*Microstomus kitt*		*D. pastinaca*
Souslik	*Spermophilus citellus*	butterfly	*Gymnura marmorata*
Sow bug	*Porcellio laevis*	diamond	*Dasyatis dipterura*
scabby, or wood louse	*P. scaber*	freshwater	*Potamotrygon falkneri*
Sparrow, European tree	*Passer montanus montanus*		*P. hystrix*
	P. montanus saturatus		*P. labradori*
fox	*Passerella iliaca*		

continued

Common Name	Scientific Name	Common Name	Scientific Name
	P. motoro	rat	*Hymenolepis diminuta*
roughtail	*Dasyatis sabina*	serrated dog	*Taenia pisiformis*
round	*Urolophus armatus*	thread	*Hymenolepis carioca*
	U. halleri	traemioid	*Bertiella studeri*
violet	*Dasyatis violacea*	Tarantula, or wolf spider	*Lycosa tarentula*
Stink bug	*Aelia fieberi*	Tench, golden	*Tinca tinca*
	Biprorulus bibax	Tenrec	*Tenrec ecaudatus*
	Brochymena quadripustulata	hedgehog	*Setifer setosus*
	Carpocoris purpureipennis	Termite	*Drepanotermes rubriceps*
	Dolycoris baccarum		*Mastotermes darwiniensis*
	Eurygaster integriceps		*Tumulitermes pastinator*
	Graphosoma rubrolineatum	desert	*Amitermes herbertensis*
	Menida scotti		*A. laurensis*
	Musgraveia sulciventris		*A. vitiosus*
	Nezara antennata	eastern subterranean	*Reticulitermes flavipes*
	N. viridula var. *smaragdula*	nasutiform	*Nasutitermes exitiosus*
	Palomena viridissima		*N. graveolus*
	Piezodorus teretipes		*N. longipennis*
	Poecilometis strigatus		*N. magnus*
	Scotinophora lurida		*N. triodiae*
	Tessaratoma aethiops		*N. walkeri*
brown	*Euschistus servus*	Thorny-headed worm of rat	*Moniliformis dubius*
European	*Ceratocoris cephalicus*	Thorny-headed worm of	*Macracanthorhynchus hiru-*
rice	*Oebalus pugnax*	swine	*dinaceus*
southern green	*Nezara viridula*	Threadworm, human intes-	*Strongyloides stercoralis*
Stomach worm	*Nochtia nochti*	tinal	
minute	*Trichostrongylus axei*	sheep intestinal	*S. papillosus*
twisted	*Haemonchus contortus*	Thrips, onion	*Thrips tabaci*
Stomach worm of cattle,	*H. placei*	Thrush, hermit	*Hylocichla guttata*
large		Tick	*Hyalomma* sp.
Stomach worm of sheep,	*Ostertagia ostertagi*	American dog	*Dermacentor variabilis*
medium		blacklegged	*Ixodes ricinus scapularis*
Stomach worm of sheep,	*O. circumcincta*	brown dog	*Rhipicephalus sanguineus*
medium or brown		cattle	*Boophilus annulatus*
Stonefish	*Synanceja horrida*	fowl	*Argas persicus*
	S. verrucosa	lone-star	*Amblyomma americanum*
Strawworm, wheat	*Harmolita grandis*	Rocky Mountain wood	*Dermacentor andersoni*
Strongyle, double-toothed	*Strongylus equinus*	spinose ear	*Otobius megnini*
single-toothed	*S. vulgaris*	Tit, great	*Parus major*
Sturgeon	*Acipenser* sp.	Toad, African leopard	*Bufo regularis*
Sunfish, ocean	*Mola mola*	American	*B. americanus*
Surgeonfish	*Acanthurus glaucopareius*	Argentine	*B. arenarum*
Surmullet	*Parupeneus chryserydros*	Asian rough	*B. asper*
	Upeneus arge	Blomberg's	*B. blombergi*
Swift	*Apus apus*	Cantor's	*B. bufo gargarizans*
white-throated	*Aeronautes saxatilis*	clawed	*Xenopus laevis*
Swine	*Sus scrofa*	Colorado river	*Bufo alvarius*
Symphylan, garden	*Scutigerella immaculata*	common European	*B. bufo*
			B. bufo bufo
		Cuban	*B. peltocephalus*
Taipan	*Oxyuranus scutellatus*	Formosan	*B. bufo formosus*
Tapeworm, beef	*Taenia saginata*	Fowler's	*B. woodhousei fowleri*
broad-headed	*Raillietina cesticillus*	Gallardo's	*B. granulosus fernandeze*
cyclophyllidean	*Taenia taeniaeformis*	green	*B. viridis viridis*
double-pored	*Dipylidium caninum*	Gulf Coast	*B. valliceps*
dog		marine	*B. marinus*
dwarf	*Hymenolepis nana*	North African	*B. mauritanicus*
fish	*Diphyllobothrium latum*	oak	*B. quercicus*
hydatid	*Echinococcus granulosus*	South American spiny	*B. spinulosus*
multilocular	*E. multilocularis*	southeast Asian	*B. melanostictus*
hydatid		black-spined	
pork or pork	*Taenia solium*	western spadefoot	*Scaphiopus hammondi*
bladder worm		Tongue worm	*Linguatula serrata*

continued

Part I. Animals

Common Name	Scientific Name
Tortoise	*Testudo marginata*
desert	*Gopherus agassizii*
European	*Testudo graecea*
Hermann's	*T. hermanni*
Towhee, rufous-sided	*Pipilo erythrophthalmus*
Trichina worm	*Trichinella spiralis*
Triclad	*Dendrocoelum* sp.
Triggerfish, spotted	*Balistoides conspicillum*
Trout, brook	*Salvelinus fontinalis*
brown	*Salmo trutta*
cutthroat	*S. clarki clarki*
rainbow	*S. gairdneri*
Trunkfish	*Lactophrys trigonus*
blue	*Ostracion lentiginosus*
spotted	*Lactophrys bicaudalis*
Tuatara	*Sphenodon* sp.
Tuna	*Thunnus* sp.
skipjack	*Euthynnus pelamis*
Tunicate, North Atlantic	*Salpa maxima*
Turkey	*Meleagris gallopavo*
Turkeyfish or lionfish	*Pterois volitans*
Turtle, common snapping	*Chelydra serpentina*
eastern box	*Terrapene carolina*
European pond	*Emys orbicularis*
	Mauremys caspica leprosa
musk	*Sternotherus odoratus*
painted	*Chrysemys picta*
red-eared	*C. scripta elegans*
spiny softshell	*Tironyx spinifer*
spotted	*Clemmys guttata*
western painted	*Chrysemys picta belli*
Turtledove	*Streptopelia turtur*
Urutú, or víbora de la cruz	*Bothrops alternatus*
Vegetable bug, European	*Eurydema ornatum*
ornate	
Viper, Arabian or Burton's	*Echis colaratus*
common European,	*Viper berus*
or common Euro-	
pean adder	
Gabon	*Bitis gabonica*
jumping	*Bothrops nummifer*
Palestinian or Near	*Vipera xanthina palestinae*
East	
Russell's	*V. russellii* or *daboia*
sawscaled or carpet	*Echis carinatus*
Vole, bank	*Clethrionomys glareolus*
common	*Microtus arvalis*
gray-sided	*Clethrionomys rufocanus*
montane	*Microtus montanus*
ruddy	*Clethrionomys rutilus*
short-tailed	*Microtus agrestis*
snow	*M. nivalis wagneri*
tundra	*M. oeconomus*

Common Name	Scientific Name
Walking stick	*Anisomorpha buprestoides*
short-horned	*Romalea microptera*
Wallaby, scrub	*Setonix brachyurus*
Wasp	*Vespula germanica*
braconid	*Bracon brevicornis*
	B. juglandis
	Praon palitans
	Trioxys utilis
chalcid	*Anastatus bifasciatus*
	Nasonia vitripennis
gall	*Amphibolips confluenta*
Japanese beetle parasite	*Tiphia vernalis*
parasitic	*Melittobia chalybii*
Waspfish	*Centropogon australis*
Water boatman	*Corixa dentipes*
	C. punctata
	Sigara falleni
Water bug, creeping	*Ilyocoris cimicoides*
giant	*Lethocerus americanus*
	L. griseus
	L. insulanus
Waterflea	*Daphnia magna*
	D. pulex
	D. schoedleri
Water strider	*Gerris lacustris*
	Velia caprai
	V. currens
Weasel, least	*Mustela rixosa*
Webworm, fall	*Hyphantria cunea*
oak	*Archips fervidanus*
Weever	*Trachinus araneus*
	T. radiatus
greater	*T. draco*
lesser	*T. vipera*
Weevil, bean	*Acanthoscelides obtectus*
boll	*Anthonomus grandis*
granary	*Sitophilus granarius*
New Guinea sugar-	*Rhabdoscelus obscurus*
cane	
pea	*Bruchus pisorum*
rice	*Sitophilus oryza*
sweet potato	*Cylas formicarius elegantulus*
Whiptail, western	*Cnemidophorus tigris*
Whipworm, dog	*Trichuris vulpis*
human	*T. trichiura*
Whipworm of swine	*T. suis*
White-eye, Japanese, or	*Zosterops japonica japonica*
spectacle bird	
Whitefish, lake	*Coregonus clupeaformis*
Wireworm	*Horistonotus* sp.
	Melanotus sp.
sugar-beet	*Limonius californicus*
Woodchuck	*Marmota monax*
Woollybear, banded	*Isia isabella*
Wrasse, Indopacific long-	*Epibolus insidiator*
jawed	
Wren, northern house	*Troglodytes aedon*

Part II. Plants

Common Name	Scientific Name
Abutilon, piemarker	*Abutilon theophrasti*

Common Name	Scientific Name
Acacia, silvergreen-wattle	*Acacia dealbata*

continued

Appendix II. COMMON NAMES AND CORRESPONDING SCIENTIFIC NAMES

Part II. Plants

Common Name	Scientific Name	Common Name	Scientific Name
sweet	*A. farnesiana*	Butcher's-broom	*Ruscus aculeatus*
Agave, century plant	*Agave americana*	Buttonbush	*Cephalanthus* sp.
Airbrom	*Billbergia elegans*		
Akee	*Blighia sapida*		
Alder, European	*Alnus glutinosa*	Cabbage	*Brassica oleracea* var. *capitata*
red	*A. rubra*	Chinese	*B. pekinensis*
speckled	*A. incana*	wild	*B. oleracea*
Alfalfa	*Medicago sativa*	Camas, death	*Zigadenus* sp.
Almond	*Prunus amygdalus*	Camellia, common	*Camellia japonica*
Amaranth, common globe	*Gomphrena globosa*	Canterbury bells	*Campanula medium*
green	*Amaranthus retroflexus*	Carob	*Ceratonia siliqua*
Apple	*Malus pumila*	Carrot	*Daucus carota*
Apple of Peru	*Nicandra physaloides*	Cassava	*Manihot esculenta*
Apricot	*Prunus armeniaca*	Catalpa	*Catalpa* sp.
Arborvitae, eastern	*Thuja occidentalis*	Cauliflower	*Brassica oleracea* var. *botrytis*
giant	*T. plicata*	Cedar, atlas	*Cedrus atlantica*
Ash, European	*Fraxinus excelsior*	deodar	*C. deodara*
Asparagus	*Asparagus acutifolius*	eastern red	*Juniperus virginiana*
garden	*A. officinalis*	western red	*J. scopularum*
Sprenger	*A. sprengeri*	Cedar of Lebanon	*Cedrus libani*
Aspen, large-toothed	*Populus grandidentata*	Celery, garden	*Apium graveolens dulce*
quaking	*P. tremuloides*	wild	*A. graveolens*
Avocado, American	*Persea americana*	Cestrum, night-blooming	*Cestrum nocturnum*
Azalea, amoena	*Rhododendron obtusum* var. *amoenum*	Chard, Swiss	*Beta vulgaris* var. *cicla*
		Charlock	*Brassica kaber*
			B. kaber var. *pinnatifida*
		Cherry, bird	*Prunus padus*
Balsam, garden	*Impatiens balsamina*	black	*P. serotina*
Banana	*Musa acuminata*	common choke-	*P. virginiana*
common	*M. sapientum*	common laurel	*P. laurocerasus*
Barbados nut	*Jatropha curcas*	Jerusalem	*Solanum pseudocapsicum*
Barley	*Hordeum vulgare*	mahaleb	*Prunus mahaleb*
Basil, sweet	*Ocimum basilicum*	mazzard	*P. avium*
Bauhinia, butterfly	*Bauhinia monandra*	sour	*P. cerasus*
Bean, broad	*Vicia faba*	Chervil, salad	*Anthriscus cerefolium*
castor	*Ricinus communis*	Chestnut, American	*Castanea dentata*
kidney	*Phaseolus vulgaris*	European	*C. sativa*
lima	*P. limensis*	Chickpea, gram	*Cicer arietinum*
sieva	*P. lunatus*	Chickweed, common	*Stellaria media*
small horse-	*Vicia faba minor*	Chicory, common	*Cichorium intybus*
Beech, American	*Fagus grandifolia*	Chinaberry	*Melia azedarach*
European	*F. sylvatica*	Christmas rose	*Helleborus niger*
Beet, common	*Beta vulgaris*	Chrysanthemum, florist's	*Chrysanthemum morifolium*
sugar	*B. saccharifera*	Pyrenees	*C. maximum*
Begonia	*Begonia decandra*	Citrus	*Citrus hassaku*
Belladonna	*Atropa belladonna*	Clover, alsike	*Trifolium hybridum*
Bellflower	*Campanula longestyla*	crimson	*T. incarnatum*
Birch	*Betula tauschii*	red	*T. pratense*
Old World canoe-	*B. alva*	subterranean	*T. subterraneum*
paper	*B. papyrifera*	white	*T. repens*
sweet	*B. lenta*	white sweet	*Melilotus alba*
Bladderwort, common	*Utricularia vulgaris*	yellow sweet	*M. officinalis*
Bloodleaf	*Iresine* sp.	Coca tree	*Erythroxylon coca*
Blood lily, Katharine	*Haemanthus katharinae*	Cocklebur	*Xanthium strumarium*
Blueberry, highbush	*Vaccinium corymbosum*	oriental	*X. orientale*
Bluestem, little	*Andropogon scoparius*	Coconut	*Cocos nucifera*
Bougainvillea, lesser	*Bougainvillea glabra*	Coffee, Arabian	*Coffea arabica*
Brassica, Mediterranean	*Brassica fruticulosa*	Coleus, common	*Coleus blumei*
Broccoli	*B. oleracea* var. *italica*	Colocynth	*Citrullus colocynthis*
Brussels sprouts	*B. oleracea* var. *gemmifera*	Comfrey	*Symphytum cordatum*
Buckwheat	*Fagopyrum sagittatum*	Coneflower	*Rudbeckia newmanii*
		Coral plant	*Jatropha multifida*

continued

Common Name	Scientific Name	Common Name	Scientific Name
Coreopsis, bigflower	*Coreopsis grandiflora*	ostrich	*Mattuccia struthiopteris*
Corn	*Zea mays*	polypody	*Polypodium australe*
cockle, common	*Agrostemma githago*	purplestalk sword	*Nephrolepis biserrata*
Cornflower	*Centaurea cyanus*	southern maidenhair	*Adiantum capillus-veneris*
Cosmos, yellow	*Cosmos sulphureus*	sword	*Nephrolepis rivularis*
Cotton, Asiatic tree	*Gossypium arboreum*	wood	*Thelypteris deltoidea*
Sea Island	*G. barbadense*	Fescue, alta or reed	*Festuca elatior arundinacea*
upland	*G. hirsutum*	blue	*F. glauca*
Cowpea, common	*Vigna sinensis*	meadow	*F. elatior*
Cress, mouse-ear	*Arabidopsis thaliana*	red	*F. rubra*
Crocus, autumn	*Colchicum autumnale*	Fig	*Ficus carica*
Crotalaria	*Crotalaria* sp.	Filbert, beaked	*Corylus cornuta*
Croton, purging	*Croton tiglium*	European	*C. avellana*
Crownbeard, golden	*Verbesina encelioides*	Fir	*Abies guatemalensis*
Cucumber	*Cucumis sativus*	common Douglas	*Pseudotsuga taxifolia*
Currant, European black	*Ribes nigrum*	Douglas	*P. menziesii*
golden	*R. aureum*	grand	*Abies grandis*
Cyclamen	*Cyclamen* sp.	silver	*A. alba*
Cypress, common bald	*Taxodium distichum*	Flax, Austrian	*Linum austriacum*
Italian	*Cupressus sempervirens*	common	*L. usitatissimum*
Mexican	*C. lusitanica*	Forget-me-not, alpine	*Myosostis alpestris*
		Foxglove, common	*Digitalis purpurea*
Dahlia	*Dahlia* sp.	Grecian	*D. lanata*
Dandelion	*Taraxacum officinale*		
Date	*Phoenix dactylifera*	Geum, Siberian	*Geum urbanum*
canary	*P. canariensis*	Gilia	*Gilia* sp.
Datura, Hindu	*Datura metel*	Ginkgo	*Ginkgo biloba*
Dayflower, common	*Commelina communis*	Gladiolus	*Gladiolus* sp.
Dead nettle, henbit	*Lamium amplexicaule*	Glasswort, marshfire	*Salicornia europaea*
Dock, clustered	*Rumex conglomeratus*	woody	*S. virginica*
patience	*R. patientia*	Gloxinia	*Gloxinia grandiflora*
Dodder	*Cuscuta* sp.	Goldenrod, tall	*Solidago altissima*
Dogwood, flowering	*Cornus florida*	Goosefoot	*Chenopodium amaranticolor*
Dollar plant	*Lunaria annua*	blight	*C. capitatum*
Duckweed, minute	*Lemna perpusilla*	Jerusalem-oak	*C. botrys*
star	*L. trisulca*	red	*C. rubrum*
		Grama, blue	*Bouteloua gracilis*
		Grape, European	*Vitis vinifera*
Eggplant	*Solanum melongena*	Grass, annual blue-	*Poa annua*
Elder, European	*Sambucus nigra*	barnyard	*Echinochloa crusgalli*
Elm, American	*Ulmus americana*	Bermuda	*Cynodon dactylon*
Chinese	*U. parvifolia*	blue-	*Poa nipponica*
Endive	*Cichorium endivia*	bulb canary	*Phalaris tuberosa*
Eucalyptus or gum, Tas-	*Eucalyptus globulus*	canary	*P. canariensis*
manian blue		centipede	*Eremochloa ophiuroides*
Euphorbia, gum	*Euphorbia resinifera*	creeping bent-	*Agrostis palustris*
Evening primrose, common	*Oenothera biennis*	crested wheat-	*Agropyron cristatum*
small-	*O. parviflora*	dallis	*Paspalum dilatatum*
flower		darnel rye-	*Lolium temulentum*
		desert wheat-	*Agropyron desertorum*
		Hungarian brome-	*Bromus inermis*
Featherfoil, European	*Hottonia palustris*	intermediate wheat-	*Agropyron intermedium*
Fern	*Athyrium spinulosa*	Italian rye-	*Lolium multiflorum*
	Oleandra articulata	Japanese lawn	*Zoysia japonica*
	Polystichum lobatum	Johnson	*Sorghum halepense*
bracken	*Pteridium aquilinum*	Kentucky blue-	*Poa pratensis*
cinnamon	*Osmunda cinnamomea*	marsh hay cord-	*Spartina patens*
common polypody	*Polypodium vulgare*	orchard	*Dactylis glomerata*
lady	*Athyrium filix-femina*	Pensacola bahia	*Paspalum notatum*
leptosporangiate	*Dicranopteris pectinata*	perennial rye-	*Lolium perenne*
male	*Dryopteris filix-mas*	quack	*Agropyron repens*

continued

Common Name	Scientific Name	Common Name	Scientific Name
shore pod-	*Triglochin maritima*	mountain	*Kalmia latifolia*
Sudan	*Sorghum vulgare sudanense*	Lead tree	*Leucaena leucocephala*
switch	*Panicum virgatum*	Lemon	*Citrus limon*
Groundsel	*Senecio* sp.	rough	*C. jambhiri*
Guava	*Psidium* sp.	Lentil, common	*Lens culinaris*
		Lespedeza	*Lespedeza* sp.
		Lettuce	*Lactuca sativa*
Hackberry	*Celtis* sp.	romaine	*L. sativa* var. *longifolia*
Hawkbit	*Leontodon* sp.	Lily, Easter	*Lilium longiflorum*
Hawthorn	*Crataegus* sp.	Lime, sweet	*Citrus limettioides*
Heath, cross-leaf	*Erica tetralix*	Linden	*Tilia* sp.
spring	*E. carnea*	Locoweed, or poison vetch	*Astragalus* sp.
Hellebore, green	*Veratrum viride*	Locust, black	*Robinia pseudoacacia*
Hemlock, eastern	*Tsuga canadensis*	common honey	*Gleditsia triacanthos*
poison	*Conium maculatum*	Lupine, European yellow	*Lupinus luteus*
water	*Cicuta* sp.	tree	*L. angustifolius*
western	*Tsuga heterophylla*		
Hemp	*Cannabis sativa*		
Henbane, black	*Hyoscyamus niger*	Madrone, strawberry	*Arbutus unedo*
Hickory, shagbark	*Carya ovata*	Magnolia, southern	*Magnolia grandiflora*
Holly, American	*Ilex opaca*	Malanga, yautia	*Xanthosoma sagittifolium*
English	*I. aquifolium*	Mallow	*Malva neglecta*
Hollyhock	*Althaea rosea*	hollyhock	*M. alcea* var. *fastigiata*
Hop, common	*Humulus lupulus*	little	*M. parviflora*
Hornbeam, American hop	*Ostrya virginiana*	Manchineel	*Hippomane mancinella*
Horse chestnut, common	*Aesculus hippocastanum*	Mango, common	*Mangifera indica*
Horsetail, common	*Equisetum arvense*	Mangrove, black	*Avicennia marina*
meadow	*E. palustre*	button	*Conocarpus erecta*
variegated	*E. variegatum*	false	*Laguncularia racemosa*
water	*E. fluviatile*	Maple, hedge	*Acer campestre*
Hydrangea, bigleaf	*Hydrangea macrophylla*	Norway	*A. platanoides*
		plane-tree or syca-more	*A. pseudoplatanus*
Iris, Crimean	*Iris chamaeiris*	red	*A. rubrum*
Ivy, poison	*Toxicodendron radicans*	Marigold, French	*Tagetes patula*
		Medic, barrel	*Medicago tribuloides*
		Mesquite	*Prosopis juliflora*
Jasmine, cape	*Gardenia jasminoides*	honey	*P. glandulosa* var. *glandulosa*
Jimsonweed	*Datura stramonium*	Milkweed	*Asclepias curassavica*
		Milkwort, sea	*Glaux maritima*
		Millet, foxtail	*Setaria italica*
Kafir lily	*Clivia* sp.	pearl	*Pennisetum glaucum*
Kalanchoe	*Bryophyllum crenatum*	Mint, little-leaf	*Mentha cardiaca*
	B. daigremontianum	Monkshood, aconite	*Aconitum napellus*
	Kalanchoe blossfeldiana	Morning glory, ivy-leaf	*Ipomoea hederacea*
	K. daigremontiana	white-edge	*I. nil*
air plant	*K. pinnata*	Mountain ash, European	*Sorbus aucuparia*
fedtschenko	*K. fedtschenkoi*	Muskmelon	*Cucumis melo*
Knapweed	*Centaurea nigra*	Muskwood, American	*Guarea trichilioides*
Kohlrabi	*Brassica oleracea* var. *gongylodes*	Mustard, black	*Brassica nigra*
		India	*B. juncea*
		white	*B. hirta*
Lady's thumb, curl-top	*Polygonum lapathifolium*		
Lambkill	*Kalmia angustifolia*	Narcissus	*Narcissus* sp.
Larch	*Larix dahurica*	Nasturtium, common	*Tropaeolum majus*
European	*L. decidua*	Nightshade	*Solanum nodiflorum.*
Japanese	*L. leptolepis*		
western	*L. occidentalis*		
Larkspur, forking	*Delphinium consolida*	Oak, black	*Quercus velutina*
rocket	*D. ajacis*	cork	*Q. suber*
Laurel, Grecian	*Laurus nobilis*	English	*Q. robur*

continued

1344

Common Name	Scientific Name	Common Name	Scientific Name
holly	*Q. ilex*	Scotch	*P. sylvestris*
pin	*Q. palustris*	shore	*P. contorta*
scarlet	*Q. coccinea*	shortleaf	*P. echinata*
southern red	*Q. falcata*	Swiss mountain	*P. mugo*
white	*Q. alba*	Swiss stone	*P. cembra*
Oat, common	*Avena sativa*	Virginia	*P. virginiana*
red	*A. byzantina*	western white	*P. monticola*
wild	*A. fatua*	Pineapple	*Ananas comosus*
Okra	*Hibiscus esculentus*	Pink	*Dianthus purpurea*
Oleander, common	*Nerium oleander*	carthusian	*D. carthusianorum*
yellow	*Thevetia peruviana*	clove	*D. caryophyllus*
Olive	*Olea europaea sativa*	Piptadenia, African	*Piptadeniastrum africanum*
common	*Olea europaea*	Plane tree, London	*Platanus acerifolia*
Onion, garden	*Allium cepa*	oriental	*P. orientalis*
Ononis	*Ononis* sp.	Plum, garden	*Prunus domestica*
Orange, Satsuma	*Citrus nobilis* var. *unshiu*	Poison nut, Amazon	*Strychnos castelnaei*
sour	*C. aurantium*	curare	*S. toxifera*
sweet	*C. sinensis*	nux vomica	*S. nuxvomica*
trifoliate	*Poncirus trifoliata*	Poisonwood tree, Florida	*Metopium toxiferum*
Oxalis, wood sorrel	*Oxalis acetosella*	Pokeberry	*Phytolacca americana*
		Pondweed, curly	*Potamogeton crispus*
		Poplar, black	*Populus nigra*
Palm, Euterpe	*Euterpe globosa*	eastern	*P. deltoides*
Fortune's windmill	*Trachycarpus fortunei*	yellow, or tulip tree	*Liriodendron tulipifera*
Mediterranean	*Chamaerops humilis*	Poppy, opium	*Papaver somniferum*
Parsley	*Petroselinum crispum*	Potato	*Solanum tuberosum*
Parsnip	*Pastinaca sativa*	Prickly pear	*Opuntia* sp.
Parthenium, guayule	*Parthenium argentatum*	Privet	*Ligustrum* sp.
Pea, field	*Pisum sativum arvense*	Pumpkin	*Cucurbita pepo*
flat-pod	*Lathyrus cicera*	Purpletop	*Triodia flava*
garden	*Pisum sativum*		
grass	*Lathyrus sativus*		
rosary	*Abrus praecatorius*	Quinoa	*Chenopodium quinoa*
sweet	*Lathyrus odoratus*		
Peach	*Prunus persica*		
Peanut	*Arachis hypogaea*	Radish, garden	*Raphanus sativus*
Pear	*Pyrus communis*	Ragweed, giant	*Ambrosia trifida*
Peavine	*Lathyrus clymenum*	Rape, bird	*Brassica campestris*
mountain	*L. montanus*	Raspberry, red	*Rubus idaeus*
Pecan	*Carya illinoensis*	Redtop	*Agrostis alba*
Pelargonium, fish	*Pelargonium hortorum*	Redwood	*Sequoia sempervirens*
Pepper, black	*Piper nigrum*	Rhododendron, Fujiyama	*Rhododendron brachycarpum*
bush red	*Capsicum frutescens*	garland	*R. hirsutum*
common	*C. annuum*	rock	*R. ferrugineum*
Peppermint	*Mentha piperita*	Rhubarb, garden	*Rheum rhaponticum*
Pepperweed, prairie	*Lepidium densiflorum*	Rice	*Oryza sativa*
Virginia	*L. virginicum*	Rose, drop hip	*Rosa pendulina*
Perilla, purple common	*Perilla frutescens* var. *crispa*	Japanese	*R. multiflora*
Persimmon, common	*Diospyros virginiana*	Rosinweed	*Silphium trifoliatum*
Petunia, common	*Petunia hybrida*	Rutabaga	*Brassica napus* var. *napobrassica*
Phlox, summer	*Phlox paniculata*		
Pimpernel, scarlet	*Anagallis arvensis*	Rye	*Secale cereale*
Pine, aleppo	*Pinus halepensis*		
eastern white	*P. strobus*		
Italian stone	*P. pinea*	Safflower	*Carthamus tinctorius*
jack	*P. banksiana*	Sage, Joseph's	*Salvia horminum*
Japanese white	*P. parviflora*	Saltbush	*Atriplex* sp.
loblolly	*P. taeda*	Saltwort, maritime	*Batis maritima*
longleaf	*P. palustris*	Scouring rush	*Equisetum hiemale*
Mexican piñon	*P. cembroides*	Sedge, nut grass	*Cyperus rotundus*
ponderosa	*P. ponderosa*	Senna, sickle	*Cassia obtusifolia*
red	*P. resinosa*		

continued

Common Name	Scientific Name	Common Name	Scientific Name
Sesbania, hemp	*Sesbania exaltata*		*N. debneyi*
Silene, sweet william	*Silene armeria*		*N. glutinosa*
Skeleton weed, rush	*Chondrilla juncea*		*N. megalosiphon*
Snapdragon, common	*Antirrhinum majus*		*N. sylvestris*
Snapweed, small-flowered	*Impatiens parviflora*	Aztec	*N. rustica*
sultan	*I. sultani*		*N. rustica* var. *brasilia*
Solomon seal	*Polygonatum odoratum*	common	*N. tabacum*
Sorghum	*Sorghum bicolor*	winged	*N. alata*
Sorrel, garden	*Rumex acetosa*	Tomato, common	*Lycopersicon esculentum*
Soybean	*Glycine max*	Trefoil, bird's-foot	*Lotus corniculatus*
Spearmint	*Mentha spicata*	Tulip, common	*Tulipa gesneriana*
Speedwell, beccabunga	*Veronica beccabunga*	Tupelo, black	*Nyssa sylvatica*
tournefort	*V. tournefortii*	Turnip	*Brassica rapa*
Spiderwort, Virginia	*Tradescantia virginiana*	Twinflower	*Linnaea borealis*
Spinach	*Spinacia oleracea*		
Spleenwort	*Asplenium glandulosum*		
Spruce, black	*Picea mariana*	Upas tree	*Antiaris toxicaria*
Norway	*P. abies*		
Sakhalin	*P. glehnii*		
Spurge, caper, or mole	*Euphorbia lathyris*	Vetch, sweet	*Hedysarum carnosum*
plant		Viburnum, burkwood	*Viburnum burkwoodii*
Squash, winter	*Cucurbita maxima*	laurestinus	*V. tinus*
Stock, common	*Matthiola incana*	Violet, hairy	*Viola hirta*
Stonecrop, goldmoss	*Sedum acre*	sweet	*V. odorata*
great	*S. maximum*		
Jenny	*S. reflexum*		
Strawberry, chiloe	*Fragaria chiloensis*	Walnut, eastern black	*Juglans nigra*
pine	*F. ananassa*	Persian	*J. regia*
Virginia	*F. virginiana*	Watermelon	*Citrullus vulgaris*
Strophanthus	*Strophanthus kombe*	Waterweed, Canada	*Elodea canadensis*
arrow poison	*S. sarmentosus*	South American	*E. callitrichoides*
Sugarcane	*Saccharum officinarum*	Weigela, old-fashioned	*Weigela florida* var. *varigata*
Sumac, poison	*Toxicodendron vernix*	Wheat	*Triticum aestivum*
Sunflower, common	*Helianthus annuus*	durum	*T. durum*
Sweetgum, American	*Liquidambar styraciflua*	White cedar, southern	*Chamaecyparis thyoides*
Sweet potato	*Ipomoea batatas*	Wild celery, spiral	*Vallisneria spiralis*
Sweet william	*Dianthus barbatus*		*V. spiralis torta*
Sycamore, American	*Platanus occidentalis*	Willow	*Salix koriyanagi*
Arizona	*P. wrightii*	basket	*S. viminalis*
California	*P. racemosa*	Wittrockia, Brazilian	*Wittrockia superba*
Tea, common	*Camellia sinensis*	Yam, wild	*Dioscorea hispida*
Thistle, Canada	*Cirsium arvense*	Yew, English	*Taxus baccata*
Tick clover	*Desmodium* sp.	Japanese	*T. cuspidata*
Timothy	*Phleum pratense*	Yucca, weakleaf	*Yucca flaccida*
Tobacco	*Nicotiana clevelandii*	Zinnia	*Zinnia* sp.

INDEX

To facilitate identification, the index includes the taxonomic order for animals and the family for plants, unless otherwise specified. As a further aid, the index lists the animals and plants as they are presented in the tables. Entries for a particular organism may therefore be found under the common name, under the scientific name, or under both. Where information is available under both, cross-references (and Appendixes I and II if more than one genus is applicable to a common name) make the data easily accessible. In some tables, only the formula for a chemical compound appears. When this occurs, the formula is listed in the index, with a cross-reference to the chemical name.

* indicates diagram or graph
fn indicates footnote material
hn indicates headnote material

Abacion magnum, CHORDEUMIDA, 690
Abacomorphus asperulus, COLEOPTERA, 682
Abax ater, COLEOPTERA, 682
A. ovalis, 682
A. parallelus, 682
Abbreviata caucasica, SPIRURIDA, 1055
Abbreviations, xix
Abdominal aorta; abdominal aortic plexus, 1146, 1151*
Abdominal distress
       organophosphates, 750
       parasites, 1056-1057, 1078
       plant toxins, 736-738, 740-742, 744
       reptile venoms, 705, 707, 709-712, 717
Abdominal ganglion, 611, 1164
Abdominal viscera, 1077, 1152, 1152 fn
Abdominal wall, 1043, 1077
Abducens nerve, 1128*, 1130-1132
Abies spp., Pinaceae, 889, 1113
A. alba, 813, 885
A. grandis, 813
A. guatemalensis, 813
*A. mariana (*see *Picea mariana)*
Abomasum, 1078-1080
A-bomb: radiation effects, 934, 937
Abortion
       hormones, 635, 639, 655
       parasites, 1081, 1093-1098
       plant toxins, 736, 741, 744-745
Abric acid, 735
L-Abrine, 735
Abrus praecatorius, Fabaceae, 735
Abscesses, 1077, 1097
Abscission, leaf, 1009-1012, 1014
Absidia corymbifera, Mucoraceae, 1122
A. ramosa, 1122
Abutilon theophrasti, Malvaceae, 767
Ac (*see* Actinium)
Acacia spp., Fabaceae, 889
A. dealbata, 815
A. farnesiana, 773
Acanthocephala declivis, HEMIPTERA, 686
A. femorata, 686
A. granulosa, 686
Acanthocoris sordidus, HEMIPTERA, 686
Acanthomyops claviger, HYMENOPTERA, 681, 694
Acanthophis antarcticus, SQUAMATA, 711
Acanthoscelides obtectus, COLEOPTERA, 696, 822-823, 1065
Acanthurus glaucophareius, PERCIFORMES, 726

A. triostegus, 726
Acarians, 762 (*see also* specific genus)
Accessory nerve, 1128*
Acclimation: temperature tolerance, 785-787, 794-798
Acer spp., Aceraceae
       parasites, 1099, 1113-1114
       soil pH, 889
       sulfur dioxide, 1008
A. campestre, 816
A. platanoides, 1008
A. pseudoplatanus, 1008
A. rubrum, 892
Acetabularia mediterranea, Dasycladaceae, 663
A. wettsteinii, 663
Acetaldehyde, 686-688
Acetamidophenoxyacetic acids, 668
Acetamidophenylacetic acids, 669
Acetate
       antimetabolites, 676
       Carboxin, 761
       hormones, 645, 652
Acetic acid
       arthropod secretion, 680, 683, 686
       bacteria: thermal death, 842
       synaptic transmission, 1161 hn
Acetone
       air pollution, 943
       hormones, 632, 634, 636, 648, 656
16β-Acetoxy-14,15β-epoxyperiplogenin, 724
3-Acetoxypalmitic acid, 726
Acetylandromedol, 739
Acetylcholine
       arthropod secretion, 693
       autonomic nervous system, 1150
       central nervous system, 1197, 1200
       circadian rhythm: brain tissue, 1041
       fish toxin, 727
       release: cerebral cortex, 1205
       sympathetic ganglia, 1202-1203
       synaptic transmission, 1153 hn, 1154, 1159*, 1160-1167
Acetylcholine chloride, 1200 fn
Acetylcholinesterase
       air pollution, 980
       central nervous system, 1197-1198
       reptile venoms, 698 hn, 699-719
       synaptic transmission, 1154, 1159*, 1161-1165, 1167
Acetyl coenzyme A, 1159*, 1160-1161

† Class

Birch, 1067 (see also *Betula*)
Birds (*see also* Aves; specific genus)
    hormones, 636-638, 642
        parasites, arthropod, 1059-1061, 1063
            bacterial, 1093-1098
            fungal, 1121-1123
            helminth, 1078-1080
            protozoan, 1076, 1082
            rickettsial, 1092-1093
    plant toxins, 739, 742
    synaptic transmitters, 1192
    viral diseases, 1082-1083
Bishydroxycoumarin, 672, 677 hn, 678
Bismuth [Bi], 930, 1206
Bitis arietans, SQUAMATA, 706
B. gabonica, 707
*B. lachesis (*see *B. arietans)*
Bittner virus, 1082
Bivalvia, 610, 1135, 1189 hn (*see also* specific genus)
Bk (*see* Berkelium)
Blaberus giganteus, ORTHOPTERA, 1270
Blackberry, 1109 fn
Blackbird, 660 (see also *Agelaius*)
Black canker, 1117
Black chaff disease, 1100
Blackfly pests, 1063 fn
Black-hull disease, 1102
Black knot, plant, 1108-1109, 1117
Blackleg, 1094, 1100, 1102
Black locust mosaic virus, 858
Black mold, 1101, 1105, 1109
Black piedra, 1121
Black pit, 1099
Black pox, 1106
Black rot, 1102-1107, 1109-1111
Black rust, 1099
Black sap, 1105
Black smut, 1105
Black spot canker, 1106, 1109
Black stem diseases, 1102, 1107, 1111, 1118
Bladder, urinary, 1077, 1146-1149, 1151*
Blakeslea trispora, Choanephoraceae, 666 fn
Blaniulus guttulatus, JULIDA, 690
Blaps gigas, COLEOPTERA, 686
B. lethifera, 686
B. mortisaga, 686
B. mucronata, 686
B. requienii, 686
B. sulcata, 686
B. wiedemanni, 686
Blast disease, 1099, 1101, 1107
Blastomyces dermatitidis, Moniliaceae, 847, 1123
Blastomycosis, 1123
Blastospore(s), 1120, 1123
Blatta orientalis, ORTHOPTERA, 1270
Blattella germanica, ORTHOPTERA, 823
Blebs, 703, 715, 717 (*see also* Blisters)
Bleeding, 700-702, 707-709, 717-719 (*see also* Hemorrhage)
Bleeding canker, 1114-1115, 1119
Blepharisma lateritia, HETEROTRICHIDA, 807
Blepharoptosis, 704

Blighia sapida, Sapindaceae, 736
Blight, plant, 1099-1100, 1102-1103, 1108-1110
Blindness, 703-704, 737-740, 749
Blissus leucopterus, HEMIPTERA, 1065
Blister canker, 1106
Blister rust, 1116
Blisters, 699-700, 706-707, 709-710 (*see also* Blebs)
Blister scab, 1099
Blister spot, 1099
Blood (*see also* Erythrocytes; Plasma; Serum)
    air pollutants: complex, 992
            gases & vapors, 968, 970-971, 973,
              977-978, 980, 982, 985-986
            particulates, 949-950, 959, 965
    anticoagulants, 676-679
    bacteria: thermal death, 835
    circadian rhythms, 1039-1040
    hormones, 631, 633, 635-640, 643, 645, 647-654,
      656-657
    parasites, fungal, 1123
            helminth, 1074 fn, 1075, 1077
            protozoan, 830-831, 1077, 1079, 1081-
            1082
    radiation, 934, 936-938
    reptile venoms, 700, 702, 704, 707-709, 711, 714,
      717-719
    synaptic transmitters, 1162 hn, 1178 fn
    viral diseases, 1083
Blood flow, cutaneous, 1261
Blood pressure (*see also* Hypertension; Hypotension)
    afferent dorsal root fibers, 1149
    hormones, 638-639, 647, 649, 651, 656
    reptile venoms, 718
    synaptic transmitters, 1154-1155
    toxins, marine animal, 730-731
        plant, 736, 738, 742, 744
Blood sugar, 608-609, 783 hn
Blood urea nitrogen [BUN], 959, 978, 1040
Blood vessels (*see also* specific vessel)
    autonomic nerve impulses, 1152-1153, 1152 fn-
      1153 fn
    autonomic nervous system, 1151*
    central nervous system nuclei, 1131
    endocrine tissues, 613, 615
    helminth parasites, 1077
    hormones, 649-651
    radiation, 939
Blood volume, 733
Blossom blight, 1100, 1103, 1108, 1111
Blossom pests, 1067
Blotch scurf, 1110
Blueberry, 850-851, 1068 (see also *Vaccinium*)
Blue mold, 1106-1107, 1111-1112
Blue tongue virus, 855, 1082
BMR (*see* Basal metabolic rate)
Body cooling, 781 hn (*see also* Evaporative cooling)
Body fat
    hormones, 637, 645, 647, 649, 655
    photoperiodic control, 1048, 1050-1051
    radiation, 923
Body heat storage, 781 hn
Body insulation, 781*-783*

Branchiostoma caribaeum, CEPHALOCHORDATA‡, 803
B. lanceolatus (see *B. caribaeum*)
Brassicaceae, 1067, 1090 (*see also* specific genus)
Brassica sp., Brassicaceae, 773, 1088
B. campestris, 893, 897, 1000
B. fruticulosa, 818
B. hirta
        amitrole, 765
        flowering, 893, 897
        osmotic potential, 918
B. juncea, 897
B. kaber, 1000
B. kaber var. *pinnatifida,* 765
B. napobrassica (see *B. napus* var. *napobrassica*)
B. napus, 817, 1000
B. napus var. *napobrassica,* 889, 897 (*see also* Rutabaga)
B. nigra, 889, 897
B. oleracea
        fungal parasites, 1102
        herbicides, 768, 770, 774
        sulfur dioxide, 1000
        temperature tolerance, 821
B. oleracea var. *botrytis* (*see also* Cauliflower)
        bromacil, 766
        flowering, 897
        soil pH, 889
        temperature tolerance, 821
B. oleracea var. *capitata* (*see also* Cabbage)
        anthocyanin synthesis, 891
        flowering, 897
        soil pH, 889
        temperature tolerance, 821
B. oleracea var. *gemmifera,* 889
B. oleracea gongylodes, 897
B. oleracea var. *italica,* 889 (*see also* Broccoli)
B. pekinensis, 897 (*see also* Chinese cabbage)
B. rapa (*see also* Turnip)
        anthocyanin synthesis, 891
        bromacil, 766
        flowering, 897
        pollen dispersion, 885
        soil pH, 889
        sulfur dioxide, 1001
B. rapa esculenta (see *B. rapa*)
Breathing (*see also* Respiration entries)
        oxygen levels, 1304-1306
        plant toxins, 736-737, 740-742
        reptile venoms, 717
Breeding: photoperiodic control, 1047 hn, 1048
Bremia lactucae, Peronosporaceae, 1105
Bretylium, 1204
Bristle worm, 731 (*see also* specific genus)
Broad bean viruses, 1088-1089
Broccoli, 1071 (see also *Brassica oleracea* var. *italica*)
Brochymena quadripustulata, HEMIPTERA, 687
"Brockman's" bodies, 618-619
Broken-neck syndrome, 704
Bromacil, 766
Brome mosaic virus, 1089
Bromide(s) [Br⁻], 747, 1182-1183
Bromine [Br], 924

5-Bromodeoxyuridine, 675
2-Bromolysergic acid diethylamide, 1156, 1192 hn, 1195
Bromomethane (see Methyl bromide)
Bromophenoxyacetic acids, 668-670
Bromophenylacetic acids, 669
(4-Bromophenyl)butyric acid, 670
5-Bromouracil, 675
Bromoviruses, 1089
Bromoxynil, 766
Bromus sp., Poaceae, 885
B. inermis, 898
Bronchi; bronchial tissue
        air pollutants: complex, 991, 995
                gases & vapors, 967-968, 971-972,
                    974-977, 980, 985-986, 988
                particulates, 949-957, 961-962,
                    964-965
        autonomic nerve impulses, 1152
        central nervous system nuclei, 1131
        parasites, fungal, 1123
                helminth, 1077
                nematode, 1078-1079
Bronchiolitis, 974, 979, 981
Bronchiolitis obliterans, 951, 968
Bronchitis
        air pollutants: complex, 992-993
                gases & vapors, 968, 974, 980-981
                particulates, 950, 961
        parasites, bacterial, 1095-1096
                nematode, 1079
Bronchoconstriction, 750
Bronchodilator, 748
Bronchogenic carcinoma, 953, 960
Bronchopneumonia
        air pollutants, 965-968, 974, 981, 991, 994
        bacterial parasites, 1094-1095
Broscus cephalotes, COLEOPTERA, 682
Brown butt rot, 1113-1114, 1116-1117, 1119
Brown felt snow mold, 1113, 1116-1119
Brown heartrot, 1116-1117
Brown leaf spot, 1103, 1106, 1112
Brown patch, 1101, 1103-1104, 1108
Brown pocket rot, 1118
Brown root rot, 1071
Brown rot, 1099, 1100, 1103, 1106, 1108-1109
Brown spot, 1104, 1107, 1116-1117
Brown stem rot, 1104
Brucella, Brucellaceae, 753-754
B. abortus, 834, 1094
B. canis, 1094
B. melitensis, 834, 1094
B. suis, 834, 1094
Brucellosis, 1094
Bruchus pisorum, COLEOPTERA, 876
Brucine, 742
Brugia malayi, SPIRURIDA, 1074-1075
Bryophyllum crenatum, Crassulaceae, 893
B. daigremontianum, 893 (see also *Kalanchoe daigremontiana*)
Bryotropha similis, LEPIDOPTERA, 696
Bryum sp., Bryaceae, 812

---

‡ Subphylum

glucagon, 645
iontophoretically applied chemicals, 1200
synaptic transmitters: acetylcholine, 1197, 1205
                  choline acetyltransferase,
                    1198
                  dopamine, 1197
                  norepinephrine, 1176 hn,
                    1197
                  serotonin, 1192 hn, 1193-
                    1194
Cerebral edema, 749
Cerebral ganglia, 610-611, 1135, 1137
Cerebral gray matter, 1208-1209
Cerebral hemorrhage, 701
Cerebral white matter, 1207-1208
Cerebroside: nervous tissue
    man, 1208-1209
    other vertebrates, 1212-1214, 1217-1218, 1220,
        1223-1224, 1226-1228
Cerebrum, 1138, 1225 (*see also* Cerebral entries)
Cerium [Ce], 928, 938
Cerura vinula, LEPIDOPTERA, 681
Cervical ganglion, 1197-1198, 1262
Cervical lymph nodes, 1095
Cervical sympathetic nerve, 1262
Cervicitis, 1095
Cervus nippon, ARTIODACTYLA, 1047
Cesium [Cs], 927, 1182, 1206
Cestodes, 611, 1064, 1134 (*see also* specific genus)
Cestrum nocturnum, Solanaceae, 894, 1029, 1033
Cetyl acetate, 690, 696
Cf (*see* Californium)
Chaetomium sp., Chaetomiaceae, 850
C. olivaceum, 850
C. thermophile var. *coprophile,* 850
Chaetomorpha cannabina, Cladophoraceae, 809
C. darwinii, 1241-1243
C. linum, 809
Chagas' disease, 1063, 1081
Chalaropsis thielavioides, Dematiaceae, 1119
Chalcodermus aeneus, COLEOPTERA, 876
Chamaecyparis thyoides, Pinaceae, 889
Chamaeleo jacksoni, SQUAMATA, 1256
C. melleri, 1256
Chamaerops humilis, Arecaceae, 815
Chaoborus spp., DIPTERA, 609, 1019
Chara corallina, Characeae, 1241-1243
Charcoal rot, 1104, 1107, 1110, 1112
Charlock, 996 fn
Chauliognathus lecontei, COLEOPTERA, 682
Chayote, 1072
Cheek pouch temperature, 784 hn, 790
Chelating agents, 748, 751
Chelonia, 613, 615, 617, 619, 621 (*see also* specific genus)
Chelopus elegans (see *Chrysemys scripta elegans*)
Chelopus guttatus (see *Clemmys guttata*)
Chelydra serpentina, CHELONIA, 796
Chemicals (*see* specific chemical)
Chemoreceptors, 1147 fn
Chemotaxis, 661-665
Chemotropism, 662, 665-667
Chenopodium, Chenopodiaceae, 1092

C. album, 774
C. amaranticolor, 858, 860-861
C. botrys, 1031
C. capitatum, 860
C. quinoa, 857-858
C. rubrum
    circadian rhythms, 1034, 1037-1038
    flowering, 891, 894
Cherokia georgiana, POLYDESMIDA, 690
Cherry, 820 fn, 1065, 1089
Chestnut blight, 883,1118
Cheyletiella parasitivorax, ACARINA, 1053
Chick, 671, 756, 758
Chick embryo, 744, 1183
Chicken (see also *Gallus*)
    parasites, 1060-1062, 1078-1081, 1095-1097
    prolactin-releasing hormone, 660
    rubratoxin B, 745
    sciatic nerve, 1228
    temperature, 863, 865
    viral diseases, 1082
Chicobolus spinigerus, SPIROBOLIDA, 690
Chilomycterus spinosus, TETRAODONTIFORMES, 728
Chilopoda, 609, 1136, 1306 (*see also* specific genus)
Chimaera, CHIMAERIFORMES, 622
Chimpanzee, 1074 (see also *Pan*)
Chinchilla, 1120
Chinese cabbage, 1088 (see also *Brassica pekinensis*)
Chin muscle movement, 1247*
Chionaspis furfura, HOMOPTERA, 876
Chirodiscoides caviae, ACARINA, 1053
Chironex fleckeri, CUBOMEDUSAE, 732
Chiropsalmus quadrigatus, CUBOMEDUSAE, 732
Chlaenius australis, COLEOPTERA, 683
C. bipunctatus, 683
C. chrysocephalus, 683
C. cordicollis, 683
C. festivus, 683
C. tristus, 683
C. vestitus, 683
Chlamydia psittaci, Chlamydiaceae, 1092
C. trachomatis, 1092
Chlamydomonas, Chlamydomonadaceae, 1271
C. eugametos, 662
C. moewusii rotunda, 662
C. nivalis, 808
C. paupera, 662
C. reinhardi, 662, 1022
C. suboogamum, 662
Chlamydospore(s), 852, 1120-1121, 1123
Chloeia flava, POLYCHAETA†, 731
Chloralose, 1205
Chloramben, 767
Chloramphenicol, 680, 753
Chlordane, 756
Chlorella, Chlorellaceae, 767
C. pyrenoidosa, 765, 771, 808
C. sorokiniana, 808
C. vulgaris, 764, 770, 775
Chloride [Cl⁻]
    flux: nerve & muscle, 1236-1237
    hormones, 639, 653, 655

---

† Class

1365

C. *yellowstonensis*, 808
Chrysanthemum spp., Asteraceae, 1070, 1099
C. *maximum*, 894
C. *morifolium*
       flowering, 892, 894, 898
       osmotic potential, 917
       soil pH, 889
       viruses, 858, 1086
Chrysaora hysoscella, SEMAEOSTOMAE, 732
Chrysemys, CHELONIA, 617
C. *belli* (see C. *picta belli*)
C. *marginata belli* (see C. *picta belli*)
C. *picta belli*, 796
C. *scripta*, 796
C. *scripta elegans*, 796, 1238, 1304
Chrysobothris femorata, COLEOPTERA, 1065
Chrysomya bezziana, DIPTERA, 1061 fn
C. *chloropyga*, 1063 fn
Chrysomyxa arctostaphyli, Melampsoraceae, 1116
Chrysophyceae, 808 (*see also* specific genus)
Chrysops spp., DIPTERA, 1061, 1074
Chrysotile, 950
Chymotrypsin
       hormones, 632, 638, 642
       synaptic transmitters, 1161
Ciboria acerina, Sclerotiniaceae, 1113
C. *betulae*, 1114
Ciborinia bifrons, Sclerotiniaceae, 1117
Cicer arietinum, Fabaceae, 898
Cichorium sp., Asteraceae, 1001
C. *endivia*, 821, 898, 1001
C. *intybus*, 898
Cicuta spp., Apiaceae, 737
Cicutoxin, 737
Ciguatera poisoning, 726-727
Ciguatoxin, 726
Ciliary ganglion, 1151 fn
Ciliary muscle, 1151*, 1152
Cimex lectularius, HEMIPTERA, 686, 823, 1061
1,8-Cineole, 685
Cingulate cortex, 1129
Cingulate gyrus, 1199 fn
Cinobufagin, 724
Cinobufaginol, 724
Cinobufotalin, 724
Cinobufotoxin, 724
Circadian rhythms
       invertebrates, 1019-1023
       plants, 1024-1038
       vertebrates: continuous light & darkness, 1015-1018
          hormones, 633, 639
          physiological variables, 1039-1047
Circinella sp., Mucoraceae, 851
Circulation
       hormones, 633, 641, 649
       reptile venoms, 706, 711, 713, 717
       toxins, fish, 727
          plant, 742, 744
          toad, 723 hn
Circulifer tenellus, HOMOPTERA, 876, 880
Circumoral nerve ring, 607
Cirrhosis, 1055, 1080

Cirriphyllum piliferum, Brachytheciaceae, 1291
Cirsium arvense, Asteraceae, 764-765
Citellus (see *Spermophilus*)
Citral, 681, 694
Citrate, 653, 763, 769
Citrate-cleavage enzyme (*see* ATP citrate lyase)
Citric acid, 775, 838, 842
Citronellal, 681, 694-695
Citronellol, 681, 694
Citrulline, 1211, 1215, 1218-1221, 1225-1226
Citrullus colocynthis, Cucurbitaceae, 817
C. *vulgaris*, 819, 885, 889
Citrus (see also *Citrus hassaku*)
       arthropod pests, 1065, 1067
       viruses, 882, 1091
Citrus spp., Rutaceae, 1070-1072, 1099, 1103
C. *aurantium*, 902, 1009
C. *hassaku*, 1009
C. *jambhiri*, 903
C. *limettioides*, 903
C. *limon*, 819, 889
C. *nobilis* var. *unshiu*, 1009
C. *sinensis*, 819, 889, 916 (see also Orange rust)
Cl (*see* Chlorine)
Cl⁻ (*see* Chloride)
Cladionia foliaceae, Cladoniaceae, 811
C. *pocillum*, 811
C. *pyxidata*, 811
C. *rangiferina*, 811
C. *rangiformis*, 811
Cladius isomerus, HYMENOPTERA, 1065
Cladophora fracta, Cladophoraceae, 808
C. *graminea*, 916
C. *hamosa*, 809
C. *prolifera*, 810
C. *rupestris*, 809
C. *spinulosa*, 809
C. *suhriana*, 663
C. *trichotoma*, 809
Cladosporium sp., Dematiaceae, 745, 851
C. *carpophilum*, 1108-1109
C. *carrionii*, 1124
C. *cucumerinum*, 1103
C. *effusum*, 1103, 1114
C. *fuligineum*, 1293
C. *fulvum*, 848, 1105
C. *herbarum*, 848
C. *werneckii*, 1120
Cladostephus spongiosus, Cladostephaceae, 664
Clam, 1166, 1193 hn
Claviceps purpurea, Clavicipitaceae, 744, 1104, 1110-1111
Cleft (*see* Synaptic cleft)
Clemmys guttata, CHELONIA, 1304
Clethrionomys glareolus, RODENTIA, 1016
C. *rufocanus*, 1016
C. *rutilus*, 1016
Climacium dendroides, Climaciaceae, 1295
Clinker rot, 1114
Clitocybe tabescens, Agaricaceae, 1103, 1106, 1109
Clivia sp., Amaryllidaceae, 917
Clivina basalis, COLEOPTERA, 683
C. *fossor*, 683
Cloacal temperature, 792, 794-798

† Class

1376

† Class

_____
† Class

---

† Class

† Class

† Class

1397

O. capillare, 808
Oenothera biennis, Onagraceae, 889, 894, 900
O. parviflora, 900
Oesophagostomum aculeatum, STRONGYLIDA, 1055
O. apiostomum, 1055
O. bifurcum, 1055
O. columbianum, 828, 1079
Oestrus ovis, DIPTERA, 1063
Oidium spp., Moniliaceae, 1107
Okra, 840, 1067, 1070-1072 (see also *Hibiscus*)
Olea europaea, Oleaceae, 815, 819, 1107
O. europaea sativa, 815, 820
Oleandra articulata, Polypodiaceae, 813
Oleandrigenin, 724
Olefins, 944
Oleic acid, 983
Olfactory bulb, 1128*, 1128, 1162 hn, 1192 hn, 1197-1198, 1201
Olfactory cortex, 1128
Olfactory tract, 1128*, 1176
Oligochaeta, 610, 1135 (*see also* specific genus)
Olive, 1072
Olive (nuclear), 1129-1130, 1132, 1137, 1139
Olivocerebellar tract, 1137
Olivo-cochlear bundle, 1130
Olpidium brassicae, Olpidiaceae, 1111
Ommastrephes sloani pacificus, DIBRANCHIA, 1285
Ommochromes, 607
Omophron limbatum, COLEOPTERA, 684
Omphalanthus filiformis, Lejeuneaceae, 811
Onchocerca volvulus, SPIRURIDA, 1074-1075
Oncopeltus fasciatus, HEMIPTERA, 825
Oncorhynchus, SALMONIFORMES, 1080
O. keta, 801
O. kisutch, 1257
O. nerka, 1273, 1287
O. tshawytscha, 801, 1257
Oncornaviruses, 853 hn, 854
Onion, 883-884, 1064, 1066, 1068, 1070 (see also *Allium*)
Ononis, Fabaceae, 1088
Onychophorans, 610, 1166-1167
Oocyst, 830-831, 1076, 1081 fn
Oogonia, 663, 665-666
Oospora pustulans, Moniliaceae, 1110
Oospore, 849
Opatroides punctulatus, COLEOPTERA, 686
Ophiobolus graminis, Pseudosphaeriaceae, 1101, 1111
O. sativus, 848
Ophioderma brevispinum, OPHIURIDA, 803
Ophiophagus hannah, SQUAMATA, 717
Ophonus azurus (see *Harpalus azurus*)
Ophthalmia, 713, 1092, 1096
Ophthalmoplegia, 704, 713, 715, 718
Opossum, 1081 (*see also* specific genus)
Optic glands: Cephalopoda, 610
Optic nerve, 1128*, 1129, 1162 hn, 1213 (*see also* Eye)
Optomotor response: polarized light, 1288
Opuntia sp., Cactaceae, 817
Orange rust disease, 1108-1109
Orchestoidea benedicti, AMPHIPODA, 1019
O. californiana, 1019
O. corniculata, 1019
Orchid, 1085-1086
Orconectes sp., DECAPODA, 1230

O. immunis, 1286
O. pellucidus, 1019
O. rusticus, 804
O. virilis, 1269-1270, 1282
Organic acids: air pollution, 945-948
Organic gases: air pollution, 943
Organophosphates, 750, 1154, 1161
Oribatula, ACARINA, 1080
Oriulus delus, JULIDA, 690
Ornithine
       nervous tissue: composition, 1211-1212, 1215, 1217, 1219-1222, 1225-1226
       synaptic transmission, 1183
Ornithine-ketoacid transaminase, 1042
Ornithonyssus bacoti, ACARINA, 1053
Ornithoptera priamus, LEPIDOPTERA, 693
Ornithorhynchus anatinus, MONOTREMATA, 789
Orotic acid, 675
Oroya fever, 1092
Orthomyxoviruses, 854
Orthoporus conifer, SPIROSTREPTIDA, 689
O. flavior, 689
O. punctilliger, 689
Orthoptera, 758-759 (*see also* specific genus)
Oryctolagus cuniculus, LAGOMORPHA (*see also* Rabbit)
      anticoagulants, 677-679
      electroencephalogram, 1255, 1258*, 1260*
      nerves, 1262
      temperature, 786, 789, 1262
      visual pigment absorption, 1268
Oryzaephilus surinamensis, COLEOPTERA, 825, 1067
Oryza sativa, Poaceae (*see also* Rice)
      flowering, 895, 900
      herbicides, 769, 772, 774-775
      parasites, 1099, 1107
      pollen dispersion, 886
      soil pH, 890
Os (*see* Osmium)
Oscillatoria amphibia, Oscillatoriaceae, 808
O. filiformis, 808
O. formosa, 808
O. geminata, 808
O. okeni, 808
O. proboscidea, 808
O. tenuis tergestina, 808
Oscinella frit, DIPTERA, 1067
Osmerus mordax, SALMONIFORMES, 801
Osmium [Os], 929
Osmoregulation, 610
Osmotic potential: plant tissues, 909 hn, 909-919
Osmotic pressure: blood plasma, 639
Osmunda cinnamomea, Osmundaceae, 1294
Osteitis, 934-935
Osteogenic malignancies, 935, 937-939
Osteomyelitis, 935, 1097
Osteoporosis, 635, 643, 651, 653
Ostertagia circumcincta, STRONGYLIDA, 829, 1079
O. ostertagi, 1079
Ostracion lentiginosus, TETRAODONTIFORMES, 726
Ostrich, 863 (see also *Struthio*)
Ostrinia nubilalis, LEPIDOPTERA
      diapause, 874
      dispersion, 878, 880
      lethal temperature, 825

■ Order

1404

Plasma (*see also* Blood)
    air pollutants, 970, 973, 986
    constituents: circadian rhythms, 1040
    hormones, 630, 632, 634-636, 638, 640-643, 645-
      647, 649, 651-652, 655
Plasma-bound iodine level, 1261
Plasmalemma, 772, 1241-1243
Plasmin, 698 hn, 699-710, 712-717
Plasminogen activators, 698 hn, 699-700, 707-708
Plasmodiophora brassicae, Plasmodiophoraceae, 1102
Plasmodium cathemerium, EUCOCCIDA, 831
P. falciparum, 1076-1077
P. malariae, 1076-1077
P. vivax, 831, 1076-1077
Plasmopara halstedii, Peronosporaceae, 1104
P. viticola, 1111
Platanus spp., Platanaceae, 890
P. acerifolia, 1010
P. occidentalis, 1010
P. orientalis, 815, 1010
P. racemosa, 1010
P. wrightii, 1010
Platichthys flesus, PLEURONECTIFORMES, 1269
Platinum [Pt], 929-930
Platyhelminthes, 1056, 1074-1077 (*see also* specific genus)
Platymonas subcordiformis, Chlamydomonadaceae, 1271
*Platynus (*see *Agonum)*
Platyzosteria sp., ORTHOPTERA, 689
P. armata, 689
P. castenea, 689
P. coolgardiensis, 689
P. jungi, 689
P. morosa, 689
P. nitidella, 689
*P. novaeslandiae (*see *Maoriblatta novaeslandiae)*
P. occidentalis, 689
P. ruficeps, 689
P. scabra, 689
P. scabrella, 689
P. soror, 689
P. stradbrokensis, 689
Plecotus auritus, CHIROPTERA, 871
Plectropomus oligacanthus, PERCIFORMES, 727
Plenodomus destruens, Sphaeropsidaceae, 1105
Pleospora betae, Pseudosphaeriaceae, 1102
Plethodon cinereus, CAUDATA, 798
P. glutinosus, 798
Pleural abnormalities, 950-951
Pleural ganglia, 611
Pleurochaete squarrosa, Pottiaceae, 812
*Pleuronectes flexus (*see *Platichthys flesus)*
*P. microcephalus (*see *Microstomus kitt)*
Pleurotus ostreatus, Agaricaceae, 1114-1116, 1118-1119
P. ulmarius, 1119
Plinachtus bicoloripes, HEMIPTERA, 687
Plodia interpunctella, LEPIDOPTERA, 696, 825, 1067
Ploioderma lethale, Hypodermataceae, 1117
Plum, 820 fn, 1065, 1087, 1108 (see also *Prunus domes-
tica)*
Plutella maculipennis, LEPIDOPTERA, 1067
Plutonium [Pu], 932, 937, 939
Pm (*see* Promethium)
PMA (*see* Phenylmercuric acetate)
Pneumococci, 753

Pneumonia
    air pollutants, 949, 985, 992-993
    parasites, 1078-1079, 1081, 1095-1098
Pneumonitis, 949, 954-955, 957-959, 962, 972, 974, 978, 994
Po (*see* Polonium)
Po$_2$ (*see* Oxygen pressure)
PO$_4$ (*see* Phosphate)
Poa annua, Poaceae, 998
P. nipponica, 998
P. pratensis
    flowering, 895, 900
    fungal parasites, 1108
    soil pH, 890
    sulfur dioxide, 998
Pod diseases, 1100, 1102, 1104
Podophthalmus vigil, DECAPODA, 1274
Podosphaera leucotricha, Erysiphaceae, 1106, 1109
P. oxyacanthae, 1108
Poecilometis strigatus, HEMIPTERA, 688
*Poecilus cupreus (*see *Pterostichus cupreus)*
Poekilocerus bufonius, ORTHOPTERA, 693
P. pictus, 693
Pogonomyrmex badius, HYMENOPTERA, 695
P. barbatus, 695
P. californicus, 695
P. desertorum, 695
P. occidentalis, 695
P. rugosus, 695
Polarized light, 1290-1295
Polarotaxis, 1272-1280
Polarotropism, 1292 hn, 1293-1294
Polia nebulosa, LEPIDOPTERA, 696
P. persicariae, 696
Poliovirus, 855, 1083
Pollen, 741, 885-888
Polonium [Po], 930
Polychaeta, 610, 1135 (*see also* specific genus)
Polydesmus collaris collaris, POLYDESMIDA, 690
P. vicinus, 690
P. virginiensis, 690
Polyethylene glycol, 907, 911-913, 915
Polygonatum odoratum, Liliaceae, 1291
Polygonum lapathifolium, Polygonaceae, 767
Polymyxin, 753-754
Polyodon, ACIPENSERIFORMES, 618
Polyoma, 854, 1082
Polyphenols, 731
Polyphylla spp., COLEOPTERA, 1067 fn
Polyplacophora, 1134
Polyplax serrata, ANOPLURA, 1054
P. spinulosa, 1054
Polypodium australe, Polypodiaceae, 813
*P. serratum (*see *P. australe)*
P. vulgare, 813, 813 fn
Polyporus spp., Polyporaceae, 1106
P. anceps, 1116
P. balsameus, 1113
P. berkeleyi, 1117-1118
P. dryadeus, 1114, 1118-1119
P. dryophilus, 1118
P. glomeratus, 1114-1115, 1118
P. hispidus, 1118
P. lucidus, 1115, 1118-1119

circadian rhythm: plants, 1025, 1027
    endocrine organs, 607, 610-611
    photoperiodic control, 1048, 1049 hn
    radiation, 936
Reptiles (see also specific genus)
    endocrine tissues, 613, 615, 617, 619, 621, 623, 625
    hormones, 637-638, 642
    temperature tolerance, 794-796
    venoms, 698-719
    viral diseases, 1083
Reserpine
    adrenergic sympathetic nerve terminals, 1204
    synaptic transmitters, 1155-1156, 1177, 1178 fn, 1179, 1189, 1195
    toxicity, 750
Resibufogenin, 724-725
Respiration
    afferent dorsal root fibers, 1147
    air pollutants: mammals, 965, 969-970, 973, 975-979, 984, 992
        plants, 1002, 1004
    herbicides, 765, 768, 771-772, 774-776
    hormones, 635, 641, 653
    oxygen levels, 1304-1306
    reptile venoms, 703-704, 710-719
    temperature: animals, 784, 788, 870-872, 1261
        plants, 810 hn
    toxins, animal, 726, 728, 730-733
        neuro-, 747-751
        plant, 735-742, 744
Respiratory diseases, 854, 935, 1082-1083
Respiratory evaporative cooling, 865
Respiratory tract; respiratory tissue
    air pollutants, 949-951, 978, 980
    parasites, 1059, 1094-1098
Resting potentials, 1196, 1234-1235, 1241-1242, 1261-1262
Reticular formation, 1128-1132, 1162, 1164, 1205
Reticulitermes flavipes, ISOPTERA, 1068
Reticulocerebellar tract, 1138
Reticulocyte(s), 1040
Reticuloendothelium, 1077, 1081
Retina; retinal tissue
    polarized light, 1276, 1282-1285, 1287-1289
    protozoan parasite, 1077
    standard visual conditions, 1263 hn-1264 hn, 1266 hn
    synaptic transmitters, 1162 hn, 1199*
Retinal chromatophorotropin, 607
Retractor capitis muscle, 1238
Retzius cell bodies, 1192 hn
Reversal potentials, 1164, 1184
Rh (see Rhodium)
Rhabdocline pseudotsugae, Stictidaceae, 1117
Rhabdocnemis obscura (see Rhabdoscelus obscurus)
Rhabdom microvilli, 1282-1283, 1286-1287
Rhabdonema adriaticum, Fragillariaceae, 662
Rhabdoscelus obscurus, COLEOPTERA, 879
Rhabdoviruses, 855, 1083
Rhadinaphelenchus cocophilus, TYLENCHIDA, 1071
Rhagoletis pomonella, DIPTERA, 879, 1068
Rhantus exsoletus, COLEOPTERA, 685
Rhenium [Re], 929

Rheobase, 1304
Rheumatic fever, 1098
Rheumatoid arthritis, 748
Rheum sp., Polygonaceae, 1006
R. rhaponticum, 821, 1100
Rhinencephalon, 1199 fn
Rhinesomus bicaudalis (see Lactophrys bicaudalis)
Rhinitis, 985, 1094
Rhinocricus insulatus, SPIROBOLIDA, 690
Rhinolophus ferrumequinum, CHIROPTERA, 1017
Rhinosporidium seeberi, Synchytriaceae, 1125
Rhinovirus, 855, 1083
Rhipicephalus, ACARINA, 1081
R. sanguineus, 1060
Rhizina undulata, Pezizaceae, 1113, 1115-1116
Rhizoctonia sp., Mycelia Sterilia■, 1102-1103, 1106, 1114
R. carotae, 1103
R. crocorum, 1119
R. oryzae, 1107
R. solani
    fungicides, 761, 763
    plant parasite, 1101 hn, 1101-1108, 1110, 1116-1117, 1119
Rhizoglyphus echinopus, ACARINA, 1064
Rhizogonium spiniforme, Rhizogoniaceae, 811
Rhizophysa eysenharti, SIPHONOPHORA, 732
Rhizopus spp., Mucoraceae, 761, 1103
R. arrhizus, 1102, 1122
R. chinensis, 851
R. equinus, 847
R. nigricans, 666, 1106, 1109
R. oryzae, 851, 1102, 1122
R. stolonifer, 849, 851, 1102, 1105, 1108
Rhizostoma pulmo, RHIZOSTOMAE, 732
Rhodium [Rh], 926
Rhodnius, HEMIPTERA, 1081
Rhododendron sp., Ericaceae, 894
R. brachycarpum, 815
R. fauriai (see R. brachycarpum)
R. ferrugineum, 814
R. hirsutum, 814
R. obtusum var. amoenum, 890
Rhodopsin, 1268-1269, 1286, 1287 fn
Rhodospirillum rubrum, Anthiorhodaceae, 770
Rhodotorula mucilaginosa, Cryptococcaceae, 851
Rhodymenia pacifica, Rhodymeniaceae, 809
R. pertusa, 809
Rhoecocoris sulciventris (see Musgraveia sulciventris)
Rhombomys, RODENTIA, 1081
Rhopalosiphum padi, HOMOPTERA, 858 fn
Rhynchocephalia, 613, 619 (see also Sphenodon)
Rhynchocoela, 611
Rhynchosporium secalis, Moniliaceae, 1105
Rhytisma acerinum, Phacidiaceae, 1114
R. punctatum, 1114
Rhytisternus laevilaterus, COLEOPTERA, 684
Ribes aureum, Saxifragaceae, 1100
R. nigrum, 821
Ribgrass mosaic virus, 1085
Riboflavin, 671
Ribonucleases
    pichloram, 773
    reptile venoms, 698-704, 706, 708-709, 712, 715-719

■ Order

Ribonucleic acid [RNA]
 air pollutants, 972-973, 982
 antibiotics, 752 hn, 754
 antimetabolite action, 675
 circadian rhythm, 1045-1047
 herbicides, 764 hn, 764, 766-771, 773-774, 776
 hormones, 631 hn, 635, 637, 641, 643
 viruses, 1082-1084, 1088
Ribonucleic acid, messenger [mRNA]
 herbicides, 766-767
 hormones, 631 hn, 645, 647 fn, 653
Ribonucleic acid nucleotidyltransferase, 754, 769
Ribonucleic acid, ribosomal [rRNA], 767
Ribonucleic acid, transfer [tRNA], 753, 755, 762, 767, 774
Ribosome(s)
 biocides, 753-755, 762, 765, 769
 viral diseases, 1083
Rice, 744-745, 1069, 1071 (see also *Oryza*)
Rice viruses, 1091
Ricinus communis, Euphorbiaceae, 741, 890, 900, 1006
Rickettsia, Rickettsiaceae, 753
R. akari, 1093
R. australis, 1093
R. conori, 1093
R. prowazeki, 833, 1093
R. quintana, 833
R. rickettsi, 833, 1093
R. sibirica, 1093
R. tsutsugamushi, 1093
R. typhi, 1054, 1093
Rickettsiae, 753-755, 833, 1092-1093
Ricolesia bovis, Chlamydiaceae, 1093
R. conjunctivae, 1093
Rinderpest, 854, 1083
Ring rot, plant, 1099-1100
Ring spot, plant, 1085-1086, 1089, 1091, 1102, 1105
Riptortus clavatus, HEMIPTERA, 687
Rn (see Radon)
RNA (see Ribonucleic acid)
Robinia spp., Fabaceae, 890
R. pseudoacacia, 815
Rochalimaea quintana, Rickettsiaceae, 1093
Rocky Mountain spotted fever, 1059, 1093
Rodents (see also specific genus)
 parasites, arthropod, 1059, 1062-1064
  bacterial, 1095, 1098
  fungal, 1123
  helminth, 1074, 1080
  protozoan, 1076, 1081
  rickettsial, 1093
 viral diseases, 1083
Romalea microptera, ORTHOPTERA, 688
Ronnel, 759
Root; root tissue
 arthropod pests, 1065-1068
 fungal diseases, 1101 hn, 1101-1109, 1111-1112, 1114, 1116-1117, 1119
 herbicides, 765-772, 774-776
 nematode parasites, 1069 hn, 1070-1072
 osmotic potential, 909-910, 912-915, 918-919
 sulfur dioxide, 1000, 1006, 1014 fn
 temperature, 820-821, 898, 908 fn
 toxins, 735-742
Rosa, Rosaceae, 764, 890

R. multiflora, 1100
R. pendulina, 816
Rose, 1065, 1072, 1091
Rosellinia herpotrichioides, Xylariaceae, 1117
Rot, plant, 1070-1071, 1099, 1102, 1104, 1112
Rotylenchulus reniformis, TYLENCHIDA, 1072
Rotylenchus spp., TYLENCHIDA, 1070 fn
Rous sarcoma virus, 854, 1082
rRNA (see Ribonucleic acid, ribosomal)
Ru (see Ruthenium)
Rubella, 1083
Rubidium [Rb], 924, 1182, 1206
Rubrocerebellar tract, 1138
Rubrospinal tract, 1130, 1132
Rubus spp., Rosaceae, 821, 890, 1109
R. idaeus, 821 (see also Raspberry)
Rudbeckia newmanii, Asteraceae, 892
Ruffini end organs, 1143
Rumen, 1254
Rumex acetosa, Polygonaceae, 890
R. conglomeratus, 772
R. patientia, 917
Ruminants, 745, 1078, 1080-1082 (see also specific animal)
Ruscus aculeatus, Liliaceae, 818
Rust, plant, 1099, 1102, 1104-1112
Rutabaga, 1071 (see also *Brassica napus* var. *napobrassica*)
Ruthenium [Ru], 926, 1207
Rye, 883, 1065, 1067, 1069, 1091 (see also *Secale*)
Ryegrass, 745, 1007 fn
Rypticus saponaceus, PERCIFORMES, 726

S (see Sulfur)
Saccharomyces anomalous (see *Hansenula anomala*)
S. cerevisiae, Saccharomycetaceae
 herbicides, 764, 770, 772
 sex hormones, 666-667
 thermal death, 852
S. ellipsoideus, 852
Saccharum officinarum, Poaceae, 890, 1109
Sacral nerves, 1150
Sagartia elegans, ACTINIARIA, 733
Sagitta elegans, CHAETOGNATHA††, 804
Saimiri sp., PRIMATES 1255, 1259* (see also Monkey)
Saissetia oleae, HOMOPTERA, 879
Salamandra, CAUDATA, 617, 619, 621
S. maculosa (see *S. salamandra*)
S. salamandra, 798
Salicornia europaea, Chenopodiaceae, 916
S. pacifica (see *S. virginica*)
S. virginica, 916
Salicylates, 680
Salientia, 613, 619, 621 (see also specific genus)
Saliva; salivation
 evaporative cooling, 865
 reptile venoms, 713-715, 718
 toxins, animal, 723 hn, 728, 731
  neuro-, 750
  plant, 736-737, 739, 742
Salivary glands, 640, 647, 1152
Salix spp., Salicaceae, 890, 1011, 1100, 1118
S. koriyanagi, 816

†† Phylum

■ Order

---

† Class

Tetrodotoxin, 751, 1194
Texas root rot, 1102-1103, 1114, 1118-1119
Th (*see* Thorium)
Thalamus; thalamic tissue, 1127*
      chemical response, 1200, 1203
      composition, 1209
      nuclei, 1128-1132
      synaptic transmitters, 1197-1198, 1199*
Thalarctos maritimus, CARNIVORA, 871 (*see also* Bear)
Thallium [Tl], 751, 930, 1207
Thamnophis radix, SQUAMATA, 794
T. sirtalis, 794
Thea sinensis (see *Camellia sinensis*)
Theileria annulata, EUCOCCIDA, 1081
Thelypteris deltoidea, Polypodiaceae, 813
Thermal death (*see* Temperature: tolerance)
Thermal inactivation (*see* Temperature: virus inactivation)
Thermobia domestica, THYSANURA, 1068
Thermomyces sp., Dematiaceae, 852
Thermoneutrality zone: homoiothermic animals, 863 hn,
  863
Thermoreceptors, 1141 hn
Thermosphere: characteristics, 941, 943 hn
Theta rhythm, 1254-1255, 1257, 1260*
Thevetia peruviana, Apocynaceae, 742
Thiamine, 671
Thielaviopsis basicola, Dematiaceae, 1102, 1104, 1107
T. paradoxa, 852
Thigh tissue, 1132, 1142 fn
Thiopental sodium, 1255
Thirst
      reptile venoms, 701, 703, 708, 715
      toxins, 733, 737-738, 740-741, 744
Thoracic aorta, 1146
Thoracic ganglion, 608
Thorax, 1131-1132
Thorium [Th], 931
Thorium oxide, 939
Thraustotheca spp., Saprolegniaceae, 665-666
Thread blight, plant, 1103
Threonine
      antimetabolite action, 673
      hormones, 634 fn-635 fn, 640, 642 fn-644 fn,
        646 fn, 656 fn
      nervous tissue, 1208-1209, 1211-1212, 1215-1217,
        1219-1222, 1225-1226
Thrips tabaci, THYSANOPTERA, 1068
Throat
      temperature: birds, 793
      toxins, 714, 723 hn, 726, 735-739, 741-742
Thrombin time, 677
Thromboangiitis, 750
Thrombocid, 678 fn
Thrombocytopenia, 701, 708, 940
Thrombophlebitis, 707, 1098
Thrombosis; thrombi, 701, 706, 709, 712, 717-719
Thuidium urceolatum, Thuidiaceae, 812
Thuja occidentalis, Cupressaceae, 889
T. plicata, 1118-1119
Thulium [Tm], 929
Thunnus, PERCIFORMES, 616
Thylacoid membrane, plants, 1290
Thymus, 617, 619, 642, 653, 658
Thyridopteryx ephemeraeformis, LEPIDOPTERA, 1068

Thyrocalcitonin, 618-619, 642-643
Thyroglobulin, 640-641
Thyroid; thyroid tissue
      acetylsalicylic acid, 747
      comparative anatomy, 616-619
      hormones, 633, 636-637, 640-643
      mimosine, 740
      radiation, 921, 927, 929 fn, 930, 934
Thyrotropic hormone, 630-631, 636-637, 641
Thyrotropin, 660, 1042
Thyrotropin-releasing hormone, 630-631, 637, 660
Thyroxine, 630, 636, 640-642, 645
D-Thyroxine sodium salt, 680
Thyroxinogenesis, 607
Ti (*see* Titanium)
Tick, 607, 1083, 1084, 1092-1094 (*see also* specific
  genus)
Tick-borne viruses, 856, 1060, 1092
Tidal volume: air pollutant inhalation, 969-970, 973, 975,
  977-979, 981-982, 985, 992
Tilia spp., Tiliaceae, 890
Tiliqua nigrolutea, SQUAMATA, 1238
Tilletia brevifaciens, Tilletiaceae, 1111
T. caries, 1111
T. foetida, 1111
T. horrida, 1107
T. tritici, 884
Timothy, 1065 (see also *Phleum*)
Tin [Sn], 926, 1207
Tinca, CYPRINIFORMES, 1305
T. tinca, 1257
Tinea barbae, 1120-1121
Tinea capitis, 1120-1121
Tinea corporis, 1120-1121
Tinea cruris, 1120-1121
Tinea imbricata, 1121
Tinea nigra, 1120
Tinea pedis, 1120-1121
Tinea unguium, 1120-1121
Tinea versicolor, 1120
TiO_2 (*see* Titanium dioxide)
Tiphia vernalis, HYMENOPTERA, 879
Tissues (*see* specific organ)
Titanium [Ti], 1207
Titanium dioxide, 963
Tl (*see* Thallium)
Tm (*see* Thulium)
Toad, 623, 723-725, 1262 (*see also* specific genus)
Tobacco (see also *Nicotiana*)
      arthropod pests, 1066-1067
      fungal disease, 884
      nematode parasite, 1071-1072
      viruses, 860, 861 fn, 1085, 1087-1089, 1091 fn,
        1092
Tobacco amblyopia, 750
Tobamoviruses, 1085
Tobraviruses, 1085
α-Tocopherol, 672, 983 fn (*see also* Vitamin E)
Todarodes pacificus, DIBRANCHIA, 1270
Toe, 1143 fn
Togaviruses, 856, 1083
Tolbutamide, 645, 680
Tomato (see also *Lycopersicon*)
      arthropod pests, 1066-1068

---

† Class